Environmental Engineering

Principles and Practice

Environmental Engineering

Principles and Practice

Richard O. Mines, Jr.
School of Engineering, Mercer University, Macon, Georgia, USA

WILEY Blackwell

This edition first published 2014 ©2014 by John Wiley & Sons, Ltd

Registered office: John Wiley & Sons, Ltd, The Atrium, Southern Gate, Chichester, West Sussex, PO19 8SQ, UK

Editorial offices: 9600 Garsington Road, Oxford, OX4 2DQ, UK
The Atrium, Southern Gate, Chichester, West Sussex, PO19 8SQ, UK
111 River Street, Hoboken, NJ 07030-5774, USA

For details of our global editorial offices, for customer services and for information about how to apply for permission to reuse the copyright material in this book please see our website at www.wiley.com/wiley-blackwell.

Library of Congress Cataloging-in-Publication Data

Mines, Richard O.
Environmental engineering : principles and practice / Richard O. Mines, Jr.
pages cm
Includes index.
Summary: "The content is accessible to sophomore engineering students with content appropriate for upper classmen and working engineers" – Provided by publisher.
ISBN 978-1-118-80145-1 (hardback)
1. Environmental engineering – Textbooks. I. Title.
TA170.M55 2014
628 – dc23

2013047872

A catalogue record for this book is available from the British Library.

Wiley also publishes its books in a variety of electronic formats. Some content that appears in print may not be available in electronic books.

Cover illustration courtesy of Angela Mines Plunkett

Set in 9/11pt Minion by Laserwords Private Limited, Chennai, India

1 2014

This book is dedicated with sincere appreciation and gratitude
to my family; my wife, Beth Pehle Mines, and my sons;
R. Andrew Mines and Daniel C. Mines.

Contents

Contributing Authors

André J. Butler
Associate Professor and Chair
Environmental Engineering Department
Mercer University
Macon, GA, USA

John C. Little, Ph.D., P.E.
Charles Edward Via, Jr. Professor
Department of Civil and Environmental Engineering
Virginia Tech
Blacksburg, VA, USA

Zhe Liu, Ph.D.
Department of Civil and Environmental Engineering
Virginia Tech
Blacksburg, VA, USA

Philip T. McCreanor, Ph.D.
Associate Professor
Mercer University School of Engineering
Macon, GA, USA

John T. Novak, Ph.D.
Professor of Civil and Environmental Engineering
Virginia Tech
Blacksburg, VA, USA

Paige J. Novak, Professor
Resident Fellow of the Institute on the Environment
Department of Civil Engineering
University of Minnesota
Minneapolis, MN, USA

Arthur B. Nunn, III, QEP
President
The Air Compliance Group, LLC
Roanoke, VA, USA

Peter Vikesland, Ph.D.
Professor of Civil and Environmental Engineering
Virginia Tech
Blacksburg, VA, USA

Preface

Intended Audience

Environmental Engineering: Principles and Practice is written for an advanced undergraduate course or a first-semester graduate course in environmental engineering. The target audience is students pursuing the civil and/or environmental engineering curriculum. However, the text may be used by students in chemical and mechanical engineering where several environmental concepts are of interest; especially those on water and wastewater treatment, air pollution, and sustainability. The text is a good resource for practicing engineers, since it covers the major environmental topics and provides numerous, step-by-step examples to facilitate learning and problem solving.

Goals and Motivation

The text was written to provide a clear and concise understanding of the major topic areas facing environmental professionals. For each topic, the theoretical principles are introduced, followed by numerous examples illustrating the proper process design approach. This approach is thought by the author to be an effective educational method. It was desired that this text be more practical, methodical, and functional for students – providing knowledge and background, as well as opportunities for application, through problems and examples that would facilitate understanding. Another motivation for writing the text evolved from the frustration of using texts that students do not read because of their complexity and emphasis on the theoretical aspects, while providing insufficient detail in the example problems for students to follow.

Unique Features

Several items unique to our text include:

1. **Problem Solving Scheme**: Chapter 1 provides a reliable problem solving scheme for students along with introducing them to statistical analysis. Proper analysis and communication of data are incorporated into the chapter. From our experience, many upper-level engineering students are still struggling with these concepts. Also, graduate students often find this type of material most beneficial. We are not aware of another environmental engineering textbook that introduces these critical tools to students.
2. **Example Problems with both US and SI Units**: The text is enriched by a multitude of example problems that incorporate both SI and the more customary English unit systems. We feel that many texts fall short in both these areas by not providing students with examples that help explain difficult technical concepts, and by only focusing on one system of units only. More importantly, our text provides step-by-step procedures for solving the difficult concepts, where other texts skip several steps before arriving at the final answer.
3. **Water and Wastewater Design Chapters**: Substantial material is presented on the selection and design of unit operations and processes used to treat drinking water and domestic wastewater. Numerous step-by-step examples lead the novice or the seasoned practitioner through the process.
4. **Sustainability**: A complete chapter has been devoted to sustainability, which is currently a hot topic. Special emphasis is placed on the causes and consequences of global climate change. Ecological Footprint is defined, and alternative energies, including solar power, wind energy, and geothermal energy, are presented. Carbon sequestration is explained, along with the capture, compression, transport, and storage of carbon dioxide. An example illustrating the Ecological Footprint and Life Cycle Assessment procedure is provided.
5. **Public Health Chapter**: Many environmental engineering text books omit public health topics. We provide an expansive view of this topic in a chapter entitled "Environmental Health". Toxicological pathways relevant to chemical exposures are described, in addition to pertinent organ and cellular systems involved in the ingestion, inhalation, dermal absorption, and placental transfer of toxic substances. Utilization of potency factors and chronic daily intake values for lifetime risk, and LOAEL, NOAEL, and RfD values are discussed. The epidemiology triangle, to describe the dissemination of an infectious disease, is discussed along with the thirteen factors responsible for disease dissemination that affect emergence and reemergence.

About the Book

Salient points of each chapter are presented below:

1. **Introduction to Environmental Engineering and Problem Solving**: Chapter 1 provides the historical context of environmental engineering. A six-step problem-solving method is introduced. Analyzing experimental error and calculating error estimates is presented. Most importantly, statistical analyses using linear-regression, correlation coefficient, coefficient of determination, and application of the Student's t-test and one-way Analysis of Variance is presented.

2 **Essential Chemistry Concepts**: A review of essential chemical concepts, along with thermodynamic principles and the kinetics of reactions, is discussed. Numerous examples are provided so that the novice and practitioner can apply the appropriate chemistry to various environmental applications.

3 **Water and Wastewater Characteristics**: Conventional physical, chemical, and biological characteristics of water and wastewater are described. A brief presentation of the analytical procedures for measuring traditional water quality parameters is presented. Numerous tables comparing the characteristics of ground water, surface water, wastewater, and stormwater are provided.

4 **Essential Biology Concepts**: Chapter 4 provides the background material necessary for understanding the role of organisms, and especially how microorganisms are used for treating wastewater, sludges, contaminated ground waters, and soils. A brief discussion of prokaryotic and eukaryotic cells is presented, followed by a presentation of the relationship between energy and synthesis reactions involving heterotrophic and autotrophic organisms. Ecological topics are addressed with particular emphasis on the carbon, nitrogen, phosphorus, and sulfur cycles, and their role in biology. The Streeter-Phelps dissolved oxygen model, along with limnological terms such as eutrophication, stratification, and turnover, are discussed.

5 **Environmental Systems: Modeling and Reactor Design**: This chapter serves as the foundation for modeling and designing reactor systems for treating water and wastewater. The chapter begins with a discussion of material balances and the application of the Law of Conservation of Matter. This is followed by how one determines the rate and order of reaction. Particular emphasis is placed on the design of batch, complete-mix, and plug flow reactors containing reactive and non-reactive contaminants. The chapter ends with the application of the Law of Conservation of Energy, which is essential in performing energy balances.

6 **Design of Water Treatment Systems**: Chapter 6 provides detailed information on the objectives of water treatment and proper selection of appropriate unit operations and processes necessary to meet primary and secondary drinking water standards. Numerous examples are presented for designing conventional and advanced water treatment systems, along with systems for handling water treatment residuals. A discussion of various disinfectants and what to consider when selecting the appropriate disinfectant is presented.

7 **Design of Wastewater Treatment Systems**: Analogous to the water treatment systems chapter, Chapter 7 deals with the design of conventional and advanced systems used for treating municipal wastewater. This chapter focuses on the use of biological wastewater systems, such as trickling filters, rotating biological contactors (RBCs), activated sludge processes, sequencing batch reactors (SBRs), and single-sludge biological nutrient removal (BNR) processes for removing nitrogen and phosphorus from wastewater. Sludge volume and mass relationships are discussed, along with various sludge thickening, dewatering, and stabilizing processes. The characteristics and use of indicator organisms as surrogate parameters for pathogens is presented, along with a guide to selecting the appropriate chemicals for disinfecting wastewater effluent.

8 **Municipal Solid Waste Management**: Chapter 8 identifies and describes the primary regulations for solid waste disposal in the United States. Solid waste generation and disposal trends along with the composition of generated, disposed, and recycled solid waste streams in the US are described. Examples showing how to calculate the composition of a generated waste, the moisture content of MSW, the energy content of a waste, and how to size a landfill are presented. The chapter concludes with the procedure for determining the volume and rate of landfill gas production.

9 **Air Pollution**: The types, sources, and effects of air pollutants, including local and global impacts is presented in Chapter 9. Meteorological fundamentals and impact on the evaluation of air pollutant emissions and basis for atmospheric dispersion modeling are discussed. Various examples illustrating the basic design and function of various types of air pollution control technologies for gaseous and particulate air pollutants are provided.

10 **Environmental Sustainability**: Chapter 10 provides an overview on sustainability. The causes and consequences of global climate change are presented, along with the detrimental effects of rapid human development. Estimating ecological footprints and evaluation of alternative forms of renewable energy are addressed. Carbon sequestration is explained. An example of using Ecological Footprint and Life Cycle Assessment method is presented.

11 **Environmental Public Health**: Toxicological pathways relevant to chemical exposures are described, in addition to pertinent organ and cellular systems involved in the ingestion, inhalation, dermal absorption, and placental transfer of toxic substances. Disease, epidemiology, and toxicology are defined. Acute and chronic exposures are compared and contrasted. Utilization of potency factors and chronic daily intake values for lifetime risk and LOAEL, NOAEL, and RfD values are discussed. The epidemiology triangle to describe the dissemination of an infectious disease is discussed, along with the thirteen factors responsible for disease dissemination that affect emergence and reemergence.

12 **Hazardous Waste Management**: Chapter 12 presents an overview of what constitutes a hazardous compound or hazardous waste. Common groups of hazardous wastes are presented, in addition to important physical and chemical characteristics of a contaminant that determine its fate, treatment options available, and the likely risk and pathways for exposure. Henry's law, the octanol-water partition coefficient, and Darcy's law are discussed, as well

as how they relate to hazardous waste management. Various remediation and treatment methods, along with what considerations are relevant for deciding on an appropriate treatment option. Advantages and disadvantages of *in situ* and *ex situ* treatment are delineated.

Course Suggestions

We will provide guidance, describing how a faculty member might use the text for an upper-level undergraduate course and as a first-semester graduate course.

Instructor Resources

The following resources will be available to instructors on the book website:

- **Solutions Manual**– Complete solutions for end of chapter problems will be provided to instructors who require the text for their course. It will be provided on the book website as PDF files.
- **Image Gallery**– Images from the text in electronic form, suitable for use in lecture slides. This includes all figures and all tables from the book.

Acknowledgements

The preparation of a book can take on a life of its own. It consumes every wakening moment until the entire project is completed. There are few gratifying experiences in life that can compete with publishing a book!

I would like to acknowledge the following individuals and their associations that have had a profound and positive impact on my life and career: Donald K. Jamison (VMI), Cal Sawyer (VMI), Clinton Parker (UVA), Roy Burke III (UVA), Joseph Sherrard (VPI), John Novak (VPI), William Knocke (VPI), Clifford Randall (VPI), Mel Anderson (USF), Glen Palen (CH_2M Hill), and Douglas Smith (Black & Veatch). These persons encouraged, mentored, and inspired me along this long and arduous journey.

Sincere appreciation is extended to my contributing authors: Arthur Nunn, Peter Vikesland, John Little, Zhe Liu, Paige Novak, John Novak, Philip McCreanor, and André J. Butler. Without their assistance and perseverance, this book would not have been possible.

We would like to thank our families, colleagues, students, Wiley folks, reviewers, contributors, accuracy reviewers, and instructor/student resource authors for making this project come to fruition.

About the Cover Artist

Angela Mines Plunkett, sister of Richard Mines, is an Associate Faculty Member at Saddleback College in Mission Viejo, California. Angela has a Bachelor of Arts from Mary Baldwin College in Staunton, Virginia; a Master of Education from the University of Virginia, Charlottesville; a Bachelor of Fine Arts from California State University Long Beach; and a Master of Fine Arts from California State University Long Beach.

About the Author

Richard O. Mines, Jr. is a Professor of Environmental Engineering and Director of the MSE/MS Programs in the School of Engineering at Mercer University in Macon, Georgia. He graduated from the Virginia Military Institute with a BS in Civil Engineering in 1975, from the University of Virginia with a ME in Civil Engineering in 1977, and from Virginia Polytechnic Institute and State University with a PhD in Civil Engineering in 1983.

Dr. Mines has had over 26 years of teaching experience at the Virginia Military Institute, University of South Florida, and Mercer University. He has over seven years of experience with the following engineering consultants: Hodges, Harbin, Newberry, and Tribble; CH_2M Hill; BLACK & VEATCH; and William Matotan & Associates. During his consulting years, he served as an Adjunct Professor at Santa Fe Community College in Gainesville, Florida, and at the University of South Florida, in Tampa.

Dr. Mines is a registered Professional Engineer in Florida (Retired), New Mexico, and Virginia (Inactive); a licensed private pilot; and a certified scuba diver. He has authored/co-authored over 100 technical and educational papers. Dr. Mines is the senior author of *Introduction to Environmental Engineering*, published by Prentice-Hall. He has taught and conducted research on: oxygen uptake rate and oxygen transfer in wastewater treatment systems; aerator testing; suspended growth biological systems using the following processes: Modified Ludzack-Ettinger process, Virginia Initiative Process, and sequencing batch reactor for accomplishing nutrient removal; modeling of bionutrient removal systems; and engineering education.

During his career, Dr. Mines has been a member of the following professional societies: American Society of Civil Engineers (ASCE), American Water Works Association (AWWA), Water Environment Federation (WEF), and American Society for Engineering Education (ASEE).

His awards and honors include election as a Fellow in ASCE (2007) and a Fellow in the Environmental and Water Resources Institute (2013). He received the Tony Tilmans Section Service Award for the ASEE-SE section in 2011. Dr. Mines was a Mercer on Mission Fellow to Malawi, Africa during summer 2010, and Mercer University Commons Fellow in 2008. He received the following awards from the American Society for Engineering Education: 2003 – 04 Outstanding ASEE Zone II Campus Representative Award; 2003 – 04 Outstanding ASEE Campus Representative Award for Southeastern Section; 2002 – 03 Outstanding ASEE Zone II Campus Representative Award; and 2002 – 03 Outstanding ASEE Campus Representative Award for Southeastern Section; and 2001 – 02 Outstanding ASEE Campus Representative Award for Southeastern Section. Dr. Mines was selected a Baylor University Faith in Vocations Fellow in 2002 and a 2001 National Teaching Effectiveness Institute Fellow of ASEE.

Dr. Mines is a member of Chi Epsilon Honor Society (1992) and the Order of the Engineer (1992). He received teaching awards from the American Society of Civil Engineers Student Chapter at the University of South Florida as Professor of the Year in 1994 – 95 and 1996 – 1997. Dr. Mines was selected by the faculty at the University of South Florida for a Teaching Incentive Program (TIP) Award in 1995 – 96, 1996 – 1997, and 1997 – 1998. For his many years of service to the Florida Water and Environment Association, he received the FWEA Service Award in 1998. During the summer 1997, he was a Faculty Research Program Fellow, Air Force Office of Scientific Research at Eglin Air Force Base, Ft. Walton Beach, Florida.

Not only is Dr. Mines a scholar, but a true athlete. He was elected into the Bath County, Virginia Athletic Hall of Fame in January 2007 having letter in football, basketball, and baseball while in high school. While attending the Virginia Military Institute, he began running and lettered in cross-country, indoor track, and outdoor track. He was selected Captain of the VMI Cross-country team, Co-captain of the indoor track team, and Tri-captain of the outdoor track team. Over the last 40 years, Dr. Mines has logged approximately 70,000 miles and successfully completed 56 marathons, five of which were at Boston.

He and his family reside in Macon, Georgia where they are active members of Martha Bowman Memorial United Methodist Church.

About the Companion Website

This book is accompanied by a companion website:

www.wiley.com/go/mines/environmentalengineeringprinciples

The website includes:

- Complete solutions for end of chapter problems
- Powerpoints of all figures from the book for downloading
- PDFs of all tables from the book for downloading

Chapter 1

Introduction to environmental engineering and problem solving

André J. Butler

Learning Objectives

After reading this chapter, you should be able to:

- appreciate the historical context of environmental engineering;
- discuss significant national and international environmental concerns;
- outline prominent environmental statutes;
- apply a six-step problem-solving method to engineering problems;
- effectively communicate analysis results in tables and figures;
- analyze experimental data;
- estimate experimental error;
- apply the concepts of variance and standard deviation to describe the uncertainty in experimental data;
- calculate error estimates associated with experimental data;
- use linear-regression analysis to describe the strength of relationship between two variables;
- quantify the magnitude of the relationship between two variables using the sample correlation coefficient and the coefficient of determination;
- apply the Student's t-test and the one-way Analysis of Variance (ANOVA) to determine if sample groups are statistically different.

1.1 History of environmental engineering

The civilizations of our ancient history were deeply and innately connected to the realm of religion and spirituality. Profoundly influenced by the natural world, there is no religion left untouched by our spiritual connection to the environment. Since the beginning of time, humans have sought comfort and healing in the earth. It is where we find sustenance, protection and beauty – but we struggle to live in harmony with it.

Our reverence for the world around us is exemplified by the prophets of our oldest religions and their desire to seek and renew their connection to the Divine through retreating into nature. Moses, Jesus, Buddha, and Muhammad all independently retreated to nature for renewal. Similarly, while living in the forest, the Hindu sages wrote the Vedic scriptures that fueled today's democratic pluralistic Indian civilization. Finally, three to five thousand years ago, Taoism and Confucianism encouraged their followers to mimic the patterns of nature, while Aristotle taught that one could understand life through imitating nature.

Further evidence of our storied relationship with the environment is in our sacred scriptures. For instance, the Old Testament of the Bible teaches that God expects humans to be the stewards of nature, and that we can learn from it:

> *The land is mine and you are but aliens and my tenants. Throughout the country that you hold as a possession, you must provide for the redemption of the land*
>
> *(Lev. 25:23–24).*

Environmental Engineering: Principles and Practice, First Edition. Richard O. Mines, Jr.

But ask the animals, and they will teach you, or the birds in the sky, and they will tell you, or speak to the earth, and it will teach you, or let the fish in the sea inform you

(*Job 12: 7–8*).

Islam's Holy book, the Qur'an, stresses balance and proportion and challenges the rights of man to change this balance and destroy Allah's creation:

And produced therein all kinds of things in due balance

(*Surah 15, verse 19*).

Verily, all things have We created in proportion and measure

(*Surah 54, verse 49*).

Buddhist scripture suggests how humans should treat the natural world, and the result of living selfishly in light of communion with it:

Rajah Koravya had a king banyan tree called Steadfast, and the shade of its widespread branches was cool and lovely. Its shelter broadened to twelve leagues ... None guarded its fruit, and none hurt another for its fruit. Now there came a man who ate his fill of fruit, broke down a branch, and went his way. Thought the spirit dwelling in that tree, "How amazing, how astonishing it is, that a man should be so evil as to break off a branch of the tree, after eating his fill. Suppose the tree were to bear no more fruit." And the tree bore no more fruit.

(*Anguttara Nikaya iii.368*).

Confucianism also stresses care for the environment, encouraging its followers to pursue a better way of living, not just for the earth but for the rest of their community:

If you do not allow nets with too fine a mesh to be used in large ponds, then there will be more fish and turtles than they can eat; if hatchets and axes are permitted in the forests on the hills only in the proper seasons, then there will be more timber than they can use ... This is the first step along the kingly way.

(*Mencius I.A.3*)

As might be expected, the bulk of this book is focused on:

1 the basic science and math principles required to describe and analyze the environment, and
2 the design and application of traditional unit operations and processes that are commonly used by environmental engineers to protect human health and the environment.

Before we learn to design the future, however, let us recall and learn from our past. We must recognize that modern civilizations were not the first to encounter and search for solutions to environmental problems or to embrace environmental activism. We start this discussion by providing a brief overview of environmental history in an effort to add perspective to our current state, in the hope that one day the relationship between humans and the environment will no longer prove to be one of continued turmoil, but one of unity and wholeness.

The backbone of this historical discussion has been built from the work of Mark Neuzil and William Kovarik and their book entitled *Mass Media and Environmental Conflict*. If this brief snapshot of environmental history peaks your interest, you are encouraged to begin further exploration through this excellent resource. An overview of the environmental history timeline from the book can be accessed online at http://www.runet.edu/~wkovarik/envhist/.

From the beginning of human history, humans have aimed to rise above and separate themselves from their animal counterparts. No longer content living at the mercy of nature, we sought to harness the ground in order to provide dependable nourishment for ourselves and our family. This new-found agricultural consistency soon blossomed into communities and cities, each with their own language, culture and religion. However, as these groups grew, it quickly became obvious that humankind had a responsibility to reconcile our lifestyle within a healthy, thriving earth; otherwise, we could face a lack of clean air, water, and other natural resources necessary for life. We needed to find ways to engineer the environment so the earth could continue to support our need to create, control and expand.

The following sections of this chapter outline humankind's pursuit of a better earth throughout written history, up to our present day regulations and standards.

1.1.1 Early civilization (the rise of humankind to ancient Rome)

Early civilization experienced many scientific and social developments that continue to fascinate modern-day engineers and experts – creations like Stonehenge and the Great Pyramid, the development of agriculture, written language, and mathematics. These advancements marked the rise of culture and structured society. Humans were no longer simply coexisting with the earth – rather, we were reaching out to conquer and control our environment.

1.1.1.1 Solid Waste Management

One of the earliest issues surrounding mankind and the environment is garbage and its disposal. When humans lived as nomadic hunter-gatherers, they could leave their solid waste wherever was most convenient for them, often buried. Since the hunter-gatherer lifestyle prompted continual movement, rodents, insects, and natural processes had time and opportunity to degrade the waste and return it to the soil. As we abandoned the nomadic lifestyle in favor of permanent settlement, however, the disposal practices followed for generations before were no longer appropriate. Leaving waste from food and fuel outside the home would invite vectors to thrive and breed among the urban developments, encouraging disease and destruction of food stored by the community.

1.1.1.2 Sustainability and Public Health

As civilization developed, resource conservation also became a growing need, with a focus on encouraging sustainable lifestyles for the people inhabiting certain areas. If they were careless, whole communities could lose access to resources staple to their development, health, and economy. As early as 6000 BCE, deforestation was blamed for the rise and fall of communities in southern Israel and, in 2700 BCE, it factored into the demise of Sumeria (now southern Iraq). Large-scale commercial timbering continued with the Phoenicians (2600 BCE), the Minoans (1450 BCE) and within the ancient city-state of Troy (1200 BCE).

1.1.1.3 Water Management

Clean, fresh water for drinking, bathing, and agriculture is vital for the support of a healthy, growing community. While most areas depended on surface water for their needs (a great example would be those who settled between the Tigris and Euphrates Rivers in the area known as the Fertile Crescent), the Chinese were using wells as deep as 1,500 ft to obtain drinking water for centuries before the modern era, and possibly used alum to clarify their water (Symons, 2001).

There is evidence that the Romans built dams, sometimes as high as 15 m, to create drinking water reservoirs. Ancient Rome is well known for its elaborate network of sewers and aqueducts, which, although built c. 500 BCE, provided a new standard of public health that was unmatched until the mid-18th century. Roman ruins show the use of open channels to carry human excreta from public baths and palaces. Disinfection of surface water was also exhibited in ancient times by the Egyptians, as they were using some form of chlorine for disinfection as early as 3000 BCE.

1.1.1.4 Air Quality

Air pollution from natural sources, such as volcanic eruption and forest fires caused by lightning, occurred long before communal living. However, it is important to note that when humans harnessed fire, anthropogenic activity became a major contributor to indoor and outdoor air pollution. Air pollution was complicated further as humans began to pursue metalworking, smelting lead, silver and copper to make coins, jewelry and tools.

Mineral deposits are typically harvested as sulfides (Cu_2S, PbS, ZnS). The smelting process produces primary ore, sulfur dioxide and other pollutants. For example, the smelting process for lead begins by pulverizing and heating such that:

$$2PbS + 3O_2 \rightarrow 2PbO + 2SO_2$$

The ore is further reduced by heating with carbon:

$$PbO + C \rightarrow Pb + CO$$

The end products, metallic lead and carbon monoxide, are toxic. Additionally, air pollution resulted from burning wood, tanning, and from decaying trash.

Rome's poor air quality led Emperor Nero's tutor and self-proclaimed Socratic philosopher, Seneca, to state in 61 CE:

> *"As soon as I escaped from the oppressive atmosphere of the city, and from that awful odor of reeking kitchens which, when in use, pour forth a ruinous mess of steam and soot, I perceived at once that my health was mending."*

1.1.2 The Middle Ages (500–1500)

The Middle Ages gave rise to such great things as algebra, the printing press, the clock, and the Magna Carta (an article that heavily influenced the United States Constitution and the Bill of Rights). Despite these contributions, this time period is one of the more somber for western culture. The Middle Ages were marked by a dramatic increase in urban population, due to feudalism and, thus, an influx of poor people who owned no land. Subsequently, any problems with waste, water, and air pollution rose exponentially and, coupled with several famines and a widespread series of pestilences, it became obvious that drastic change was required in order to provide for the health and welfare of these communities.

1.1.2.1 Solid Waste Management

The heavy influx of people into cities posed a tremendous solid waste problem. There was little room to dispose of garbage, which was either burned, buried, or left in the streets – practices that encouraged poor air quality and served as a breeding ground for pests and vermin, bringing with them filth and disease. In light of the Black Death – a pandemic that destroyed 30–60% of the European population during that time – it was time to start rethinking what practices should be permitted for the sake of public and occupational health. By 1366, the city of Paris required butchers to dispose of animal carcasses outside the city instead of in the streets.

1.1.2.2 Water Management

Along with the fall of Rome came the fall of support for its extensive system that provided water to the public. Communities had to revert to wells and surface water, which were often contaminated due to overuse and overpopulation. Much of this contamination resulted from human excreta, which were not disposed of in a sanitary fashion. In fact, a common practice of the 13th and 14th centuries was to empty chamber pots from upstairs windows into the street gutters below.

Sewage covered the streets of Europe during the Middle Ages, often contaminating water supplies. Improvements began in 1589, when Sir John Harrington invented the water closet in England and published a book entitled *The Metamorphosis of Ajax*, in which he provided a complete description. While Parliament had passed an act in 1388 forbidding the throwing of filth and garbage into ditches, rivers, and waterways, there was still a lack of proper sewerage, and his invention was largely ignored. It was not until the 1690s that Paris became the first European city to build an extensive sewer system and such tools as the water closet became more practical in everyday life.

1.1.2.3 Air Quality

Burning wood, coal and garbage was common practice on the streets and in homes during this time, which exacerbated the poor air quality. From 1560 to 1600, England was rapidly becoming industrialized, and wood shortages led to an increased burning of sea-coal. Sea-coal has a high sulfur content and, when burned, produces air pollutants such as sulfur dioxide, carbon monoxide, carbon dioxide, nitrogen oxides, soot, and particulate matter. In 1661, John Evelyn proposed a solution to London's air pollution through his pamphlet *Fumifugium, or the Inconvenience of the Aer and Smoake of London Dissipated.* Evelyn wrote several notable passages, including:

> *The immoderate use of, and indulgence to, sea-coale in the city of London exposes it to one of the fowlest inconveniences and reproaches that can possibly befall so noble and otherwise incomparable City ... Whilst they are belching it forth their sooty jaws, the City of London resembles the face rather of Mount Aetna, the Court of Vulcan ... or the suburbs of Hell [rather] than an assembly of rational creatures.*

Later, in 1684, he wrote that the smoke was so severe that "*hardly could one see across the street, and this filling lungs with its gross particles exceedingly obstructed the breast, so as one would scarce breathe.*"

The situation was more hopeful when James I succeeded Queen Elizabeth and demanded that harder, cleaner-burning coal from Scotland be used within his household, but it would be decades before air pollution was addressed as a public health issue.

1.1.2.4 Public Health

While modern solid waste management practices were unavailable, legislation like the aforementioned restrictions on dumping waste marked a greater government involvement in the health of its citizens. Also, during this time period, the growing development of industry began to muster some concerns about the health impacts of common occupational practices. In 1473, Ulrich Ellenbog wrote the first pamphlet on occupational disease and injuries among goldsmiths. Ellenbog's pamphlet was followed by a similar treatise in 1556 by Agricola (Georgius Bauer), which further outlined the techniques and occupational hazards of assaying, mining, and smelting a variety of metals.

1.1.3 The Age of Enlightenment (1650–1800)

The Age of Enlightenment gave rise to a new era of scientific, philosophical, and humanitarian study. During this time, the United States achieved independence from Britain, Leibniz and Newton invented calculus, we lent our ear to the musical stylings of Beethoven, Mozart, and Bach and investigated the writings of John Locke, Immanuel Kant, and Voltaire. The newfound pursuit of reason yielded many familiar names into the world of science, like Fahrenheit, Celsius, Bernoulli, Kepler, and Venturi, to name a few. This quest for knowledge also marked the beginning of better public health, as humanitarians felt it was necessary to spread the recent discoveries in science and medicine to provide good health and hygiene to everyone.

1.1.3.1 Solid Waste Management

During this time period, populations within urban environments continued to blossom, further encouraging illness and disease despite past regulations and development. Public health laws were inconsistent from state to state and city to city, so people continued to live in the manner that they felt best. These practices varied immensely, depending on the values and lifestyles of the citizens in the region. For example, while Benjamin Franklin led a group in Philadelphia in an attempt to regulate waste disposal and water pollution, Jonathan Swift noted the despicable contents within London's gutters, "sweeping from butchers' stalls, dung, guts, and blood ... "

1.1.3.2 Sustainability

As the world was developing new scientific and technological discoveries, many people were beginning to use this knowledge to enhance the health and sustainability of our environment. By 1762, Jared Eliot was promoting soil conservation in his *Essays on Field Husbandry in New England.* History even recalls men and women of the Bishnoi faith protesting the demolition of a grove of khejri trees at the hands of an Indian maharaja, even forfeiting their lives to stop the destruction. The villagers who died that day in 1730 knew of the immense value of the khejri tree, which they used for medicine, firewood, and food, among many other beneficial uses. The maharaja halted the demolition, but not before 363 people lost their lives. Conservation was also alive and well in the Western world, evidenced by Benjamin Franklin's plea to France and Germany in 1784, urging them to switch from wood to coal in an attempt to save their forests.

1.1.3.3 Water Management

The Age of Enlightenment provided a few great advances in the realm of water quality. Filtration became a more viable treatment technique, with Frenchman Joseph Army creating a filter out of a perforated box filled with sponges in 1746 (*The Quest for Pure Water*, 1981). In 1791, an upflow sand filter with a downflow washing system was invented by James Peacock. In 1744, a limited distribution public water supply was started in New York City. In 1772, Providence, RI also began distributing a public water supply within the city. Public water became even more readily available as London installed the first modern municipal sewers in 1800.

1.1.3.4 Air Quality

By the 1700s, sea coal importation to London had grown dramatically, and similar developments occurred in cities worldwide as business prepared for the Industrial Revolution. In 1804, Presley Neville wrote regarding Pittsburgh: "the general dissatisfaction which prevails and the frequent complaints which are exhibited, in consequence of the Coal Smoke from many buildings in the Borough, particularly from smithies and blacksmith shops ... " The smoke affected the "comfort, health and ... peace and harmony" of the new city. At this time, the best remedy was to build higher chimneys.

1.1.3.5 Public Health

The push for intellectual advancement in the 18th century helped to provide insight into the causes and effects of illness in humans. In 1723, the lead in alcohol stills was shown to cause serious abdominal pains. In 1775, English scientist Percival Potts noted that chimney sweeps had an unusually high incidence of cancer. Further attempts at monitoring occupational health were reflected as Sir Thomas Percival formed the Manchester Board of Health to supervise textile mills and recommended hours and working conditions.

1.1.4 The Industrial Revolution (1760–1890)

The Industrial Revolution, while not a revolution in the sense of an *immediate* overhaul of the traditional way of doing things, did indeed, quietly and consistently, transform and shape the way business and the world look today. Major developments included mechanizing the textile industry, creation of the steam engine, and an overhaul and increase in the field of metallurgy. The 19th century introduced many great names in the fields of science and technology; Thomas Edison, Charles Darwin, Alexander Graham Bell, and Nikola Tesla provided us with many great ideas and developments that we still follow today. Indeed, it could even be argued that the field of environmental engineering began during this time, marked by the initiation of the American Society of Civil Engineers, as well as the contributions by people like John Snow (the father of modern epidemiology) and Henri-Philibert-Gaspard Darcy.

1.1.4.1 Air Quality

One of the greatest environmental issues resulting from the Industrial Revolution was the immense burden placed on the atmosphere due to a dramatic increase in the amount of goods available for human consumption, which implied a proportional increase in the burning of coal and other fuels. William Wordsworth wrote about the Industrial Revolution in 1770, commenting that it was an "outrage done to nature", and he was angered that the common people were no longer "breathing fresh air." In 1859, Svante August Arrhenius predicted global warming from fossil fuel induced carbon dioxide build-up. In 1880, a temperature inversion led to an air pollution episode in London, killing 700.

Drastic measures had to be taken in order to prevent industry from harming the public. In 1881, Chicago became the first American city to create a local smoke ordinance regulating smoke discharge, but it would be decades before federal mandates would even begin to address air quality concerns adequately.

1.1.4.2 Sustainability

Despite the marked increase in resource consumption, many voices arose in protest during the Industrial Revolution, seeking to preserve the world in light of the seemingly mindless destruction of thousands of square miles of natural resources. People like Thoreau and Emerson pushed the public, encouraging a more sustainable way of life. As a result, the idea of National Parks was proposed by George Caitlin in 1832, and a bill was passed in 1864 to protect trees in the Yosemite Valley. Developments were even made in sustainable power creation, as the first hydroelectric plant began operation in 1882 in Appleton, Wisconsin.

1.1.4.3 Water Management

Sanitation and water supply systems developed rapidly during the 19th century, as the delivery of pure drinking water became important in fighting typhoid and cholera epidemics. During this time, sewers for transporting human waste and storm runoff were constructed in Brooklyn, Chicago, and Jersey City (Tarr, 1985). By 1905, all US towns with a population of over 4,000 had city sewers. In that same year, Henri-Philibert-Gaspard Darcy published his observations about the field of public water supply. He also proposed Darcy's Law, that describes flow through porous medium, as well as the design of a relatively modern rapid sand filter. Also of note during this time period is the fact that the city of Lenox, Massachusetts began using combined surface and subsurface irrigation to treat sewage and the use of sedimentation for water treatment became common practice.

In 1841, Thomas Henry patented the process to remove hardness from water through the addition of lime resulting in the precipitation of calcium carbonate (Merrill, 1962). Two years later, coagulation before filtration was proposed by James Simpson to treat river water in England (*The Quest for Pure Water*, 1981). The forerunner of the modern filter – a self-cleaning slow sand filter using upward flow for washing – made its debut during the early 19th century in Greenock, Scotland. In 1885, British scientists discovered the benefits of filtering drinking water through sand filters to reduce bacteria. Soon after this development, Hiram Mills designed intermittent sand filters in response to typhoid fever epidemics in 1890 and 1891, which were used to treat water from the Merrimack River, and removed 98 percent of the bacteria from the polluted river water (MSBH, 1894). In the early 1900s, sand filters were being used in several US cities and copper sulfate, chlorine, and ozone were known as treatments to kill typhoid and cholera bacteria. The first continuous chlorination system was put into place in 1908 in Jersey City.

Data from Tarr *et al.* (1980) were used to develop Table 1.1, which shows the enormous growth of water treatment and distribution and wastewater treatment from 1880 through

Table 1.1 Growth of water distribution, wastewater sewerages and treatment in the USA from 1880 through 1900.

Year	Urban population	Total US population	Population in cities treatment sewage	Population served by sewerage systems	Population supplied with treated drinking water
1880	14,129,735	50,155,783	5,000	9,500,000	30,000
1890	22,106,265	62,947,710	100,000	16,100,000	310,000
1900	30,159,921	75,994,575	1,000,000	24,500,000	1,860,000

Source: Data from Tarr *et al.*, 1980.

Table 1.2 Miles of sewers and death rates for selected US cities in 1880, 1890, and 1900.

	1880		1890		1900	
City	**Miles of sewers**	**Deaths per 100,000 people**	**Miles of sewers**	**Deaths per 100,000 people**	**Miles of sewers**	**Deaths per 100,000 people**
Atlanta		66.8	24	72		74.6
Chicago	337	33.9	525	72.2	1,453	21.1
Louisville, KY	41	65.7	52	75.7	97	64.0
Nashville		133.7	24	64		49.5
Newark, NJ	47	52.7	87	99.5	180	21.1
Philadelphia	200	58.7	376	73.6	887	37.2
Pittsburgh	22.5	134.9	87	127.4	275	144.3
Richmond		61.3	35	61		104.6
Rochester, NY		23.4	138	39.6		17.2
Salt Lake City			5			39.2
San Francisco	126	53.4	193	55.5	307	30.3
Spokane			3			45.9
Toledo		35.9	61	36	156	41.0
Trenton		17.1	4	16		32.7
Washington, DC	169	53.4	266	86.8	405	79.7

Source: Data from Tarr *et al.*, 1980.

1900. Table 1.2 also uses data from Tarr *et al.* to show the relationship between sewerage disposal practices and death rates from typhoid fever. Death rates from typhoid fever did not drop as drastically as expected from the installation of sewer systems but, rather, death rates declined as cities began to use chlorination in treating drinking water.

1.1.4.4 Public Health

In 1851, the first international health conference was held in Paris to address the arrival and spread of cholera and other pestilent diseases. Shortly after the conference, in 1854, John Snow conducted the first epidemiological study of a waterborne disease. By conducting interviews with local residents, he traced the source of a cholera epidemic to a contaminated water pump on Broad Street in London's Soho district, providing insight that water pollution carried disease. Between 1860 and 1880, Louis Pasteur's theory of disease revolutionized the theories associated with public health and, in 1882, tuberculosis and cholera were isolated by German physician Robert Koch.

1.1.5 The Progressive Era (1890–1920)

The Progressive Era marked the beginning of the period known as the Machine Age. Disgusted by the developments and concessions made by the government and big business during the Industrial Revolution, middle-class Americans began to speak up, demanding reformation in all areas of life – politically, socially, economically, and morally. Contemporary environmental engineering principles began to take root during this time, which served as further evidence of progressive ideologies.

1.1.5.1 Sustainability

One of the greatest examples of the need for conservation during this period was the wild buffalo (bison) population, which dropped from 30 million to less than 40 animals in less than a century. Unfortunately, the buffalo were not the only species endangered due to our overindulgent lifestyle. It soon became obvious to most that government intervention was required in order to keep American wildlife alive.

In 1891, the Forest Reserve Act was passed, and 17.5 million acres were set aside. In 1902, Congress established the Bureau of Reclamation and George Washington Carver wrote *How to Build Up Worn Out Soils*. Also during this time, Charles Van Hise wrote *The Conservation of Natural Resources* (1909), urging business to be careful with its use of oil, metals, lumber, and land. We can also see, during this period, the use of the law to protect and preserve the environment. For example, the state of Missouri sued the state of Illinois for polluting the Mississippi River, and the state of Georgia sued the Tennessee Copper Co. in an effort to limit the amount of noxious fumes they would be permitted to discharge.

1.1.5.2 Water Management

In 1893, the first trickling filter for wastewater treatment was developed by Joseph Corbett in Salford, England. Biological treatment of wastewater was being developed aggressively during the mid-1890s by engineers in England and at the Lawrence Experiment Station in Massachusetts. As a result of sewage cleanup in the Thames River, some fish species returned in 1895, providing hope for the redemption of wasteful human living.

1.1.5.3 Air Quality

While the Progressives pushed for a better life, industry continued to have a huge impact on the lives of the people living in the cities where manufacturing occurred. In 1892, 1,000 Londoners died in a smog-related episode and, in 1909, winter-time atmospheric inversions caused smoke accumulations, killing over 1,000 more citizens. Not only could air pollution harm humans in the short term, evidence was also presented during this time that emphasized the danger of air pollution to future generations. In 1896, the Swedish chemist Svante August Arrhenius projected the effect of carbon dioxide in the atmosphere. In his article, published in the *Philosophical Magazine and Journal of Science*, entitled "On the influence of Carbonic Acid in the Air upon the Temperature of the Ground", he predicted a global temperature increase of 8–9° F for a doubling of carbon dioxide in the atmosphere (available online at http://www.globalwarmingart.com/images/1/18/Arrhenius.pdf).

1.1.5.4 Public Health

The last serious European cholera outbreak occurred in Hamburg, Germany in 1892, and killed 8,600 people. This event, while destructive, paled in comparison to other cholera pandemics in past history, which often claimed anywhere between 50,000–250,000 lives within the course of a few years. This improvement could be attributed to the development of sewage and water treatment systems. In 1904, lead poisoning in children was first linked to lead-based paints, prompting France, Belgium, and Austria to ban white-lead interior paint. The League of Nations – an international organization formed following World War I, with the goals of preserving the peace through collective action and promoting international cooperation in economic and social affairs – would go on to ban white-lead based paint in 1922.

1.1.6 The Great Depression and World War II (1920–1945)

The second half of the American Machine Age was wrought with a whirlwind of change and turmoil. The fall of the global economy and, subsequently, the American stock market, had many devastating effects on its citizens. Soon afterwards, the United States plunged into World War II, seeking to promote peace to all the world, but ultimately dropping atomic bombs on the Japanese cities of Hiroshima and Nagasaki in August of 1945. With so much upheaval and a focus on wartime efforts taking precedence over personal and social needs, it is no surprise that environmental efforts were put on the back-burner during this time in history.

1.1.6.1 Sustainability

By 1941, over 25,000 solar water heaters were being used primarily in Florida and California. The Soil Conservation Society was formed in 1944 by Hugh Bennett.

1.1.6.2 Air Quality

The first large-scale air pollution study was conducted in Salt Lake City in 1926. In 1930, A three-day inversion in the Meuse River Valley in Belgium caused an air pollution episode that killed 63 persons and made 6,000 ill. The term "greenhouse effect" was coined by Glen Thomas Trewartha in his book entitled *An Introduction to Weather and Climate*. St Louis suffered from a smog episode in 1939, where it was reported that the smog was so thick that lanterns were needed during daylight hours for a week.

1.1.6.3 Public Health

In 1922, the League of Nations banned white-lead interior paint, but the US did not adopt the measure. In 1923, leaded gasoline first went on sale in Dayton, Ohio. The Oil Pollution Act was passed in 1924. By 1932, the *British Medical Journal* had reported that leaded gasoline was dangerous to human health.

1.1.7 Post-war era (1945 to present)

After a relatively quick transition between the tension of war and a time of peace, the years after World War II provided a growth in the economy, resulting in a more affluent lifestyle for many Americans. Many of today's modern comforts were created in this time – vacuum cleaners, toasters, televisions, and central heat and air. Society was on the upswing; we raced to the Moon and enjoyed discovering the full benefits of technology.

Prior to 1945, public health or nuisance laws were legislated, controlled, and poorly coordinated and enforced by state and local governments. Congress only acted when a public health problem was particularly visible or obvious. In subsequent years, however, the federal government began to take a more dedicated interest in public and environmental health, passing some of the first legislation influential to the field of environmental engineering – the Federal Water Pollution Control Act (passed over President Nixon's veto), Coastal Zone Management Act, Ocean Dumping Act, Marine Mammal Protection Act, the Federal Insecticide, Fungide, Rodenticide Act (FIFRA), and the Toxic Substance Control Act (TSCA), to name a few.

1.1.7.1 Solid Waste Management

In 1976, Congress passed the Resource Conservation and Recovery Act (RCRA) to regulate hazardous waste and municipal solid waste. In 1978, Lois Gibbs initiated a community battle against local, state, and federal governments, claiming that the land she and her neighbors lived on in Niagara Falls, NY, was causing health problems for their children. Their neighborhood – eventually dubbed Love Canal – was built on top of a leaking chemical waste dump previously owned by Hooker Chemicals and Plastics Corp. Her efforts led to the cleanup of the Love Canal and was instrumental to the creation of The Comprehensive Environmental Response, Compensation and Liability Act (CERCLA), also known as Superfund, created by the EPA in 1980 to cleanup abandoned toxic waste sites.

1.1.7.2 Sustainability

Following the War era, a greater emphasis was placed on the conservation of our natural resources. The books, *Silent Spring* and *The Silent World*, by Rachel Carson and Jacques Cousteau, respectively, encouraged greater concern for the environment. Carson's book, in particular, is often credited as launching the modern environmental movement. In addition to RCRA and CERCLA, Congress also passed the Federal Land Management Act and the Whale Conservation and Protective Study Act in 1976. Two years later, they also approved the National Energy Act, the Endangered American Wilderness Act, and the Antarctic Conservation Act. The United States was not the only nation seeking to preserve our natural resources, however. In 1991, an international treaty was signed that prohibited mining, protected wildlife, and limited pollution in Antarctica.

1.1.7.3 Water Management

Congress first addressed water pollution in 1948 by enacting the Federal Water Pollution Control Act, in order to "enhance the quality and value of our water resources and to establish a national policy for the prevention, control and abatement of water pollution." Congress further granted the federal government a greater responsibility for the public's drinking water through passing the Safe Drinking Water Act in 1972. The Safe Drinking Water Act was amended to set standards for an additional 83 contaminants and ban the use of lead pipes and solder in new drinking water systems.

1.1.7.4 Air Quality

Several air pollution episodes occurred in both London and in the US during the post-war era. Most notable would be the 1948 episode in Donora, PA, where 20 people died and more than 7,000 were hospitalized due to severe air pollution. Other notable episodes are summarized in Table 1.3. In response, the Air Pollution Control Act was passed by Congress in 1955. In 1963, the Clean Air Act was passed, and the Environmental Protection Agency (EPA) was created in 1970, which were significant milestones in our efforts to reduce smog and air pollution by promulgating federal standards.

In 1950, Dr. Arie Haagen-Smit discovered that smog was caused by reactions between hydrocarbons and nitrogen oxides. F. Sherwood Rowland and Mario J. Molina reported that chloroflorocarbons (CFCs) are instrumental in depleting the stratospheric ozone layer in 1974. In 1985, Joe Farman discovered the hole in the ozone layer over Antarctica. The Montreal Protocol was approved by 24 countries in 1987, and provided an international agreement to phase out the use of ozone-depleting chemicals. In 1988, DuPont announced an end to CFC production. In December of 1997, the Kyoto Protocol was adopted by the USA and 121 other nations in an effort to reduce carbon dioxide emission. In 2001, the US opted out of participation in the Kyoto Protocol.

1.1.7.5 Public Health

A recollection of recent history regarding environmental and public health concerns would be remiss without mentioning the incidents at Three Mile Island, Bhopal, and Chernobyl – events that have colored public perception of technology and government standards for occupational health. In 1979, the Three Mile Island nuclear power plant partially melted down after losing coolant. While no lives were lost and there were no (confirmed) illnesses due to the radiation emitted, it is considered the most significant accident in the history of the American nuclear power industry. In the 1984 Bhopal disaster, a Union Carbide fertilizer plant located in Bhopal, India leaked methyl isocyanate, resulting in over 2,000 deaths and over 100,000 injuries. As a result of this incident, the Emergency Planning and Community Right to Know Act was passed in 1986, requiring manufacturers to report releases of toxic chemicals. Also in 1986, the Chernobyl nuclear reactor exploded, causing 31 deaths immediately. It is estimated that over 4,200 deaths occurred as a result of this disaster, with anywhere between 10,000 – 50,000 cases of cancer occurring as a result of radiation.

1.1.8 Present day

By the advent of the 21st century, Environmental Engineering had become a well-defined discipline, with over 50 ABET, Inc. accredited degree programs within the US. Environmental engineering remains inherently multidisciplinary. As a result, we tend frequently to work in teams having members from the other engineering disciplines and also from the applied and social sciences. As our work naturally interfaces between the environment (natural systems) and engineering systems, it follows that environmental engineers work hand-in-hand with biologists, environmental toxicologists, and chemists to unravel multifaceted environmental problems in an effort to secure solutions.

The environmental engineer must also be aware of the social implications resulting from historical and current environmental regulatory practices. In many instances, the environmental engineer must work in consultation with social psychologists, lawyers, and politicians. For example, consider the implications of siting a nuclear waste repository, an incinerator, a wind farm, or an airport. The Not In My Back Yard (NIMBY) acronym describes the community resistance and backlash often associated with many newly proposed neighborhood projects, even when it is recognized that the project may be good for the masses. Regretfully, literature on environmental racism indicates a disproportionate balance between pollution exposure and environmental benefits for people of color.

The environmental engineer may work on a broad range of application-oriented projects that are often dependent on the targeted media (air, soil, surface water, etc). Other projects require a more macroscopic approach. For example, a sustainability or risk assessment task may naturally encompass work conducted in a variety of specific media. The work of an environmental engineer may be focused on local, regional, national, or even global environmental problems.

As an example, the environmental engineering consultant may work directly with a local textile industry to study the impact of perfluorooctanoic acid (PFOA, a surfactant) on the surrounding surface and ground waters. After quantifying the environmental impact related to the emissions, the environmental engineer may then develop, design, and test a mitigation system. This type of project would encompass a broad spectrum of talented people, ranging from other engineering disciplines, scientists, community leaders, and state and federal regulators. Equally fascinating, an environmental

Table 1.3 Summary of air pollution episodes.

Air pollution episode	Date	Cause	Consequences	Additional reading
London, UK	1892	SO_2 and particulate matter from coal combustion/ temperature inversion.	1,000 deaths from smog incident.	Schwartz, J., Marcus, A. (1990). Mortality and air pollution in London: a time series analysis. *Am J Epidemiol* **131**, 185–94.
Meuse Valley, Belgium	December 1930	High concentrations of SO_2, H_2SO_4, fluorinated compounds, particulates from factories/temperature inversion.	Death of 63 people, lethal damage to cattle, birds, and vegetation, thousands of persons sick.	Nemery, B., Hoet, P.H.M., Nemmar, A. (2001). The Meuse Valley fog of 1930: an air pollution disaster. *The Lancet* **357**(9257), 704–708. http://whqlibdoc.who.int/monograph/WHO_MONO_46_(p159).pdf
St. Louis, USA	1939	Coal combustion/ temperature inversion.	Thick smog and a reduction of visibility	
Los Angeles, USA	1943 – ?			http://whqlibdoc.who.int/monograph/WHO_MONO_46_(p159).pdf
Donora, USA	October 1948	Uncontrolled industrial emissions, /temperature inversion and fog.	42.7% of the population suffered from the incident, which killed 20 people and many animals.	Ciocco, A., Thompson, D.J. (1961). A Follow Up on Donora Ten Years After: Methodology and Findings. *Am J Public Health Nations Health* **51**(2), 155–164. http://whqlibdoc.who.int/monograph/WHO_MONO_46_(p159).pdf
Pozarica, Mexico	1950	Industrial accident releasing hydrogen sulfide/temperature inversion.	22 persons dead and many animals, hundreds of people suffered from acute topic poisoning.	http://whqlibdoc.who.int/monograph/WHO_MONO_46_(p159).pdf
London, UK	December 1952	SO_2 and particulate matter from coal combustion/ subsidence temperature inversion.	~4,000 deaths	Bell, M.L., Davis, D.L. (2001). Reassessment of the Lethal London Fog of 1952: Novel Indicators of Acute and Chronic Consequences of Acute Exposure to Air Pollution, *Environmental Health Perspectives* **109**(3). http://whqlibdoc.who.int/monograph/WHO_MONO_46_(p159).pdf
London, UK	1962			Schwartz, J., Marcus, A. (1990). Mortality and air pollution in London: a time series analysis. *Am J Epidemiol* **131**, 185–94.
New York, USA	1966	Temperature inversion.	Resulted in the deaths of 168 persons.	

engineer may be focused globally on technology development to capture and sequester carbon emissions from a coal-fired power plant that may be transported thousands of miles to another nation.

1.2 Significant national and international environmental concerns

In a world replete with great wealth and resources, it is hard to imagine that the majority of the Earth's population suffers from a lack of even the basics essential to good health. It is estimated that nearly 20% of the world's population does not have access to potable water, 40% lacks adequate sanitation, and adequate housing is not available to 20%. Considering these staggering statistics, the role of environmental engineers in securing our future cannot be overstated. Now consider that during the next two decades, the world's population is projected to increase by two billion, with 95% of the growth occurring in underdeveloped countries. This growth will create unparalleled demands for energy, food, water, transportation, waste disposal, health care, environmental cleanup, and infrastructure.

The American Society of Civil Engineers (ASCE) periodically evaluates and grades the infrastructure quality in the US. In 2013, America's overall infrastructure was awarded a grade of D+, with a total required investment of $3.6 trillion by 2020. The infrastructure reviewed and its associated grades include aviation (D), bridges (C+), dams (D), drinking water (D), energy (D+), hazardous waste (D), inland waterways (D−), public parks and recreation (C−), rail (C+), roads (D), schools (D), solid waste (B), transit (D), and wastewater (D).

1.2.1 Water quality

Since the passage of the Clean Water Act in 1972, the US government has invested $72 billion to construct publicly owned sewage treatment works (POTWs). Unfortunately, the physical condition of many of the systems is poor, due to inadequate investment in treatment plant maintenance, updated equipment, and capital improvements. Many of the 16,000 treatment plants have reached the end of their design lives and are plagued with chronic overflow problems during heavy snowmelt and rainstorms that result in raw sewage being discharged into surface waters. The EPA estimates that, to meet increasing demands, $390 billion should be spent over the next 20 years to upgrade and replace existing systems and to build new systems. The case for federal investment is compelling to ensure public and environment health.

There are approximately 54,000 drinking water systems in the US. The investment shortfall to replace aging facilities and to comply with federal water regulations is estimated as $11 billion. This estimate assumes that there will be no increased demand for safe drinking water over the next 20 years, which is likely erroneous. In 2001, the EPA estimated that funding for POTW infrastructure needs over the next 20 years is $151 billion. The Congressional Budget Office (CBO) concluded in 2003 that the current funding for the nation's drinking water infrastructure was inadequate.

1.2.2 Solid waste management

In 1980, President Jimmy Carter signed into law the Comprehensive Environmental Response, Compensation, and Liability Act of 1980 (CERLA, or "Superfund"). The resulting program allowed for the federal government to manage and clean up the nation's uncontrolled hazardous waste sites. Over the following 25 years, the EPA funded the development of innovative cleanup technologies, completed cleanup construction programs at 966 sites and has been working actively at an additional 422 sites. Also during this time period, the EPA enforcement program secured approximately $24 billion dollars for cleanup activities from parties responsible for the contamination. Despite the progress, one in four Americans lives within four miles of a potentially hazardous Superfund site (www.epa.gov/superfund), and the quantity of hazardous waste generated in 2007 was reported as 46,693,284 tons.

In 2007, the average American generated 4.6 pounds per day of municipal solid waste, resulting in about 254 million tons of trash produced during that year. Of the trash generated, 85 million tons (33.4%) of the material was either recycled or composted. On average, Americans recycle or compost 1.5 pounds of our individual waste generation per person per day (US EPA, 2008).

According to the count conducted by the EPA, there were 7,683 municipal solid waste landfills in the United States in 1986. In 1991, stringent new regulations were imposed on landfill design and operation by the EPA. As a result, a decline in landfills was noted by the agency and, in 2001, ten years after regulations were imposed, only 1,858 landfills were operational – a drop of 78%. Although capacity remains nearly constant, the decline in facilities that accept solid waste has contributed to the increase in shipping municipal solid waste across state lines for disposal. The decline has also prompted experimentation with bioreactor landfills designed to increase the rate of waste degradation and stabilization.

1.3 Prominent federal environmental statues – an overview

Historically, population growth and technological advancement are known to have been two key anthropogenic causative agents of environmental concerns. The uncertainties of cumulative risk and of cause and effect, and of the potential for catastrophic effects, are additional important drivers for contemporary environmental concerns. Modern efforts geared toward technical solutions of environmental problems have had beneficial effects. Our initial efforts were focused on solving some of the easier problems to identify; for example, burning rivers (Stradling and Stradling, 2008), Love Canal, and strip mining. Engineers and scientists continue to uncover new problems and to seek solutions that meet our national health standard goals.

Due to widespread agreement on the importance of good stewardship, the environment has been labeled a consensual issue. Critics of environmental policy argue that although some environmental goals have been realized, it has come at too high a cost, and also that it has ignored some significant issues. As environmental policy matures, discussion revolves around a more holistic approach, on long-term environmental threats, and on cost, as well as the benefits of the environmental regulations being considered.

Federal environmental laws are intended to protect human health and the environment. Laws originate when a group of citizens, a business, or a government agency contacts a congressperson with a highlighted concern. This concern may lead to a Bill, which is then passed by Congress and signed into law by the President. The Environmental Protection Agency (EPA) is then tasked with developing the associated regulations and the technical and operational procedures required to implement the law.

Regulation development begins through the formation of a workgroup that identifies what data are needed, the anticipated costs, and the required expertise. The workgroup then identifies appropriate options through the evaluation of current environmental practices and technologies and incentives. If the workgroup decides that a regulation is needed, a proposal is drafted and published in the Federal Register for public comment. After review of the public comments, a final regulation is issued. Figure 1.1 provides an overview of the environmental regulatory process, while Figure 1.2 shows a timeline of many environmental laws and executive orders. A selection of major environmental laws that protect the environment and public health are discussed.

1.3.1 Regulations of chemical manufacturing

1.3.1.1 The Toxic Substances Control Act (TSCA)

In an effort to respond to public concern over the use and disposal practices of several dangerous chemicals (including vinyl chloride, mercury, lead, PCBs, propellants, and fluorocarbons), congress enacted the Toxic Substance Control Act (TSCA; pronounced TAHS-ka) in 1976. The Act granted the EPA authority to create a regulatory structure to evaluate and control the risks associated with chemicals, and to set standards that are applicable during any period of a chemical's life cycle.

The Act provides an outline of intent for TSCA policies that include:

1. Data adequate to describe an effect on health and the environment of a chemical or chemical mixture must be developed. The manufacturer and other users are charged with providing these data.
2. Chemical substances and mixtures that present an unreasonable risk of injury to health or the environment should be adequately regulated.
3. Regulatory authority of chemical substances and mixtures should not create unnecessary economic barriers to technological innovation.

TSCA Section 4 requires rules and testing for manufacturers, importers, and processors of new or existing chemical substances and mixtures regarding their effect on human health and the environment. Section 5 allows TSCA to regulate new chemicals prior to their manufacture, import, processing, or distribution, and regulates existing chemicals substances for significant new uses.

An inventory of chemical substances has been developed under Section 5. If a chemical is not already on the inventory, and has not already obtained exclusion by TSCA, a premanufacture notice (PMN) must be submitted to the EPA prior to manufacture or import. The PMN must identify the compound and provide health and environmental effects. If the data are inadequate to evaluate the chemical's effect, restrictions may be imposed until additional information on its health and environmental effects can be assessed. Section 6 of TSCA allows for regulating the manufacture or distribution of a chemical that EPA has determined to pose an unreasonable risk to human health or the environment. Chemicals currently regulated under Section 6 include asbestos, chlorofluorocarbons, polychlorinated biphenyls (PCBs), and lead (Bergeson, 2000).

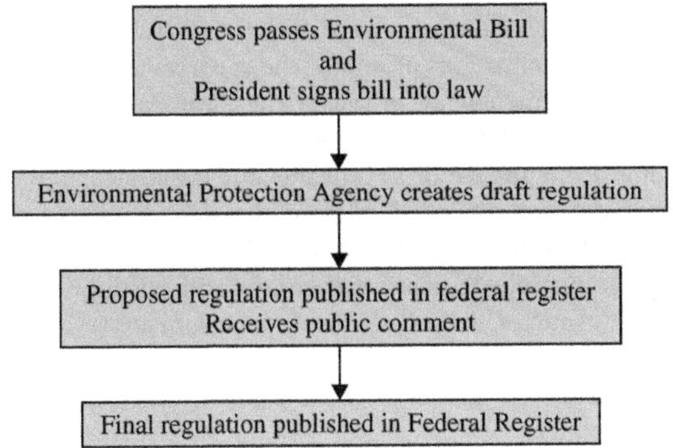

Figure 1.1 The regulatory process.

1.3.1.2 The Federal Insecticide, Fungicide, and Rodenticide Act (FIFRA)

In 1910, the first pesticide law was passed. It focused on protecting consumers from ineffective products and deceptive labeling. When The Federal Insecticide, Fungicide, and Rodenticide Act (FIFRA; pronounced fif'-rah) was enacted in 1947, its impetus was to establish registration and labeling procedures for pesticides with the US Department of Agriculture. Pesticide efficacy, not use, was the primary focus of the law. FIFRA was rewritten in 1972, and the new legislation provided EPA the authority to promulgate a regulatory framework for targeted chemicals (this provision was woefully absent from the original law). Since that time, the law has been amended numerous times by the Federal Environmental Pesticide Control Act (FEPCA) and the Food Quality Protection Act (FQPA).

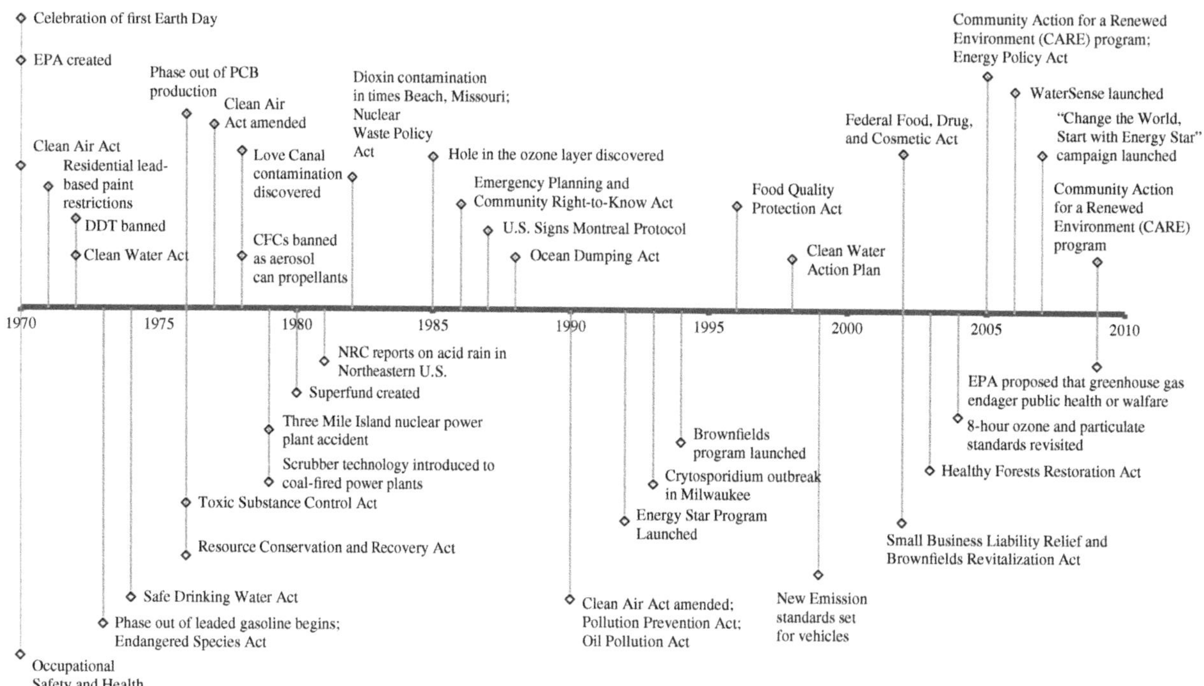

Figure 1.2 Laws and Executive Orders that Influenced Environmental Progress. *Source:* www.epa.gov.

The FIFRA now provides a framework for the regulation, sale, distribution, and use of pesticides in the US. FIFRA authorizes the EPA to ban or suspend a registered product if new information shows that continued use would pose unreasonable risk, enforce compliance against banned and unregistered products, and register pesticides for specified uses. Additional key requirements imposed by FIFRA include that each manufacturer must register each pesticide and its label with EPA before it can be manufactured, transported, or sold for commercial use. Manufacturers intending to produce pesticides are required to submit a registration application, a proposed label that can be easily understood, a statement of all claims, directions for use, the chemical formula, and testing protocol that substantiates the manufacturer's claims. The distribution of an unregistered or improperly labeled pesticide is prohibited through this regulation.

1.3.1.3 The Occupational Safety and Health Act (OSH) Act

The Occupational Safety and Health Act of 1970 was enacted to impose workplace safety regulations. Authority was given to create the Occupational Safety and Health Administration (OSHA), which was tasked to ensure that employers provide their workers a work environment free from safety and health hazards. Standards were imposed on exposure to toxic chemicals, excessive noise levels, mechanical dangers, heat or cold stress, or unsanitary conditions. In some instances, standards imposed are very detailed about the way a specific job must be done to avoid an injury, while a general duty clause is provided as a catch-all in those situations for which no standard exists.

The Act was controversial because it removed much of the workplace regulatory authority from individual states. In addition, the law authorized federal compliance officers from the US Department of Labor to make surprise inspections of workplaces. It also resulted in recommendations to upgrade worker protection, provide higher disability benefits, and unlimited medical care with rehabilitation. An OSHA research institute called The National Institute for Occupational Safety and Health (NIOSH) was also created by the Act. The primary responsibility of NIOSH is to establish workplace health and safety standards. The US Department of Labor is tasked to oversee OSHA actions. Compliance to OSH A standards requires detailed record keeping. Incidents must generally be reported and the agency may impose fines if safe and healthful working conditions are not maintained.

1.3.2 Regulation of discharges to air, water, and soil

1.3.2.1 Clean Air Act (CAA)

Smog advisories and concerns of diminished visibility caused by particulate emissions prompted the congress to promulgate a major environmental law in 1963 – The Clean Air Act (CAA). The Act gives and defines the EPA responsibility for protecting and improving the nation's air quality and the stratospheric ozone layer. As a result, federal programs were developed that established air quality goals and set end-of-pipe pollution control technology requirements on new and existing stationary sources. Since it was first enacted, major revisions to the Act have been passed by Congress in 1970, 1977 and 1990. The Clean Air Act is detailed in the United States Code as Title 42, Chapter 85.

Provisions of the CAA allow the EPA to establish national ambient air quality standards (NAAQS) intended to protect public health and welfare. Through the Act, the EPA has established a variety of emissions limitations for air pollutants for both stationary and mobile sources. For example, the EPA has promulgated primary and secondary NAAQS for a set of criteria pollutants. The primary standards are intended to protect public health while secondary standards are in place to protect against environmental and property damage. Currently, there are six criteria pollutants: nitrogen oxides, sulfur oxides, particulate matter, ozone, carbon monoxide, and lead.

Compliance to NAAQS is federally mandated, yet states are tasked with developing strategies for achieving and maintaining compliance. Regions within states that meet the NAAQS are deemed *attainment areas*, while those exceeding the limitations of one or more of the criteria pollutants are deemed *non-attainment* areas. States are required to prepare a State Implementation Plan (SIP) that describes how each region will achieve or maintain the NAAQS. Federal highway funds are often withheld until an appropriate SIP is prepared.

New Source Performance Standards (NSPS) have also been developed as a result of the CAA. The NSPS regulations ensure that federally-mandated, pollution-control technology is used on new sources or modified stationary sources that result in an increase in emissions, regardless of their location. NSPS have been established for over 70 point source categories, including the air emissions from wastewater treatment plants, petroleum refineries, boilers, and landfills.

National Emission Standards for Hazardous Air Pollutants (NESHAPS) have been developed for 188 hazardous air pollutants (HAPs) emitted from a variety of industries. HAPs include tetrachloroethylene, a compound commonly used in the dry-cleaning industry, toluene (a constituent in gasoline), and heavy metals such as mercury and chromium. Major sources of these pollutants are subject to federally-mandated emission control standards based on Maximum Achievable Control Technologies (MACT).

The CAA is a complex law. In addition to those items already discussed, it also contains programs that focus on motor vehicle emissions, acid rain precursors, Title V permitting, stratospheric ozone layer, and noise pollution.

1.3.2.2 Clean Water Act (CWA)

The principal law governing pollution of the nation's surface waters and adjoining shorelines is the Federal Water Pollution Control Act, or Clean Water Act (CWA). Originally enacted in 1948, it was totally revised by amendments in 1972 that gave the act its current form. The objective of the CWA was to create ambitious programs "to restore and maintain the chemical, physical, and biological integrity of the Nation's waters" (§ 101(a)). Goals concerning the health of US waters were established in support of this objective that include a complete elimination of pollutants discharged into navigable waters, the ban of noxious pollutants in toxic amounts, and the control of both point and nonpoint source pollution (§ 101(a)). Congress further revised the law in 1977, 1981, and 1987.

Through federally mandated standards, states are responsible for implementing and enforcing day-to-day activities. This Act has focused on regulatory requirements for industrial and municipal dischargers. Historically, emphasis was placed on controlling discharges of conventional pollutants such as suspended solids and biochemical oxygen demand. More recently, emphasis has been placed on controlling the release of toxic substances into navigable waters. As a result of the permitting requirements promulgated by the Act, the release of point source pollutants into our nation's waters has been significantly reduced. Permits stipulate that the water released from industrial or municipal point sources must meet both EPA and state-established criteria for maintaining water quality sufficient for designated public uses. In general, waters must be fit for fishing and swimming as a minimum requirement. Permitting requirements are influenced by the technological feasibility of abatement and the economic demands placed on the emitting industry.

The permitting requirements placed on industries are thought to have controlled point source discharges from pipes and outfalls reasonably well. Unfortunately nonpoint source pollution derived from stormwater runoff from agricultural lands, forests, construction sites, and urban areas continue to be a major source of water pollution. Title VI of the Act provides financial assistance for the construction of new municipal sewage treatment plants. The law has civil, criminal, and administrative enforcement provisions.

1.3.2.3 Resource Conservation and Recovery Act (RCRA)

The Resource Conservation and Recovery Act (RCRA; pronounced "rick-rah" or "wreck-rah") was enacted in 1976 to protect human health and the environment from the potential hazards associated with waste disposal. As a result of the Act, a regulatory structure for the classification, transportation, storage, and disposal of solid and hazardous wastes was developed. RCRA provides for two main categories of solid waste management, which are commonly differentiated through the names 'subtitle C', which deals with hazardous waste, and 'subtitle D', which handles municipal solid waste.

RCRA subtitle C provides for the identification and listing of hazardous wastes and establishes a "cradle to grave" tracking system. RCRA hazardous wastes are defined as either a listed waste or by waste characteristic. *Listed wastes* include F (non-specific source wastes from common manufacturing and industrial processes), K (source-specific wastes from 17 different industries), P (acutely hazardous chemicals), and U (acutely toxic) wastes that are specifically listed in RCRA.

As defined by RCRA, the four characteristics of hazardous waste include toxicity, reactivity, ignitability, and corrosivity (TRIC). The Act requires a permitting system for hazardous waste generators and transporters, as well as Treatment, Storage, and Disposal Facilities (TSDFs). Hazardous wastes are tracked from the point of generation to the point of disposal and generators are required to be registered and obtain a generator identification number. A six-part manifest is used to track hazardous wastes. The manifest is originated by the generator and is transferred to the transporters and then to the TSDF. Copies of the manifest are provided to the Department of Toxic Substances Control by both the generator and the TSDF. The generator is also required to provide a biennial report regarding the generation of hazardous waste.

RCRA Subtitle D provides guidance for state solid waste programs. This title instituted a ban on creating new open dumps and developed federal criteria for Municipal Solid Waste Landfills (MSWLFs). Open dumping was prohibited except at sanitary landfills, and the EPA must provide minimum regulatory standards for MSWLFs. The Hazardous and Solid

Waste Amendments of 1984 (HSWA) provide regulations for landfills, which require landfill lining, leachate collection systems, and landfill closure plans that ensure it will not fail. It also banned land disposal of hazardous waste for which alternatives are available. Measures for corrective action were also identified in the HSWA. Subtitle I was also added in 1984, and this focused on underground storage tanks (UST). The number of leaking underground storage tanks (LUST) is estimated to be 1.4 million. New regulations stipulate design, installation, operation requirements, and closure for USTs.

1.3.3 Cleanup, disclosure, and pollution prevention

1.3.3.1 The Comprehensive Environmental Response, Compensation, and Liability Act (CERCLA)

The Comprehensive Environmental Response, Compensation, and Liability Act (CERCLA) is commonly referred to as Superfund. The Act was passed in 1980 in response to the growing awareness of our improper disposal practices of hazardous waste materials, and is highlighted by the Love Canal incident that first prompted Congress to act. Initially, the Act was designed to be a five-year program. Its goals included identifying and establishing prohibitions and requirements for closed and abandoned hazardous waste sites, site remedy, and providing for strict, joint, and several liability of parties responsible for the release of hazardous waste, resulting in the responsible parties paying for the cleanup action. Potentially Responsible Parties (PRPs) include the current facility owner or operator, the owner or operator when the site was contaminated, transporters of waste to the facility, and generators of wastes. PRPs may include individuals, corporate officers, corporations, and landlords. PRPs may be pursued by the government and other PRPs, and may be subject to citizen suits. In addition, the Act established a trust fund to support cleanup actions when the responsible party could not be identified.

The Superfund Reauthorization and Amendments (SARA) Act was established in 1986 to increase the total fund allotment to $8.5 billion and to create more stringent cleanup requirements. In summary, the Acts allowed the EPA to clean up, order abatement, and to recover the costs of cleanup.

Two types of response actions were authorized by the law: short-term removal and long-term remedial response actions. Short-term *removal* actions were established to address immediate threats triggered by a release of a hazardous substance into the environment. For instance, removal actions would address a tanker truck accident or a spill from a damaged drum. These actions are designed to address emergencies and their scope is limited in cost (less than $2 million) and duration (less than 12 months). A CERCLA *remedial* action is designed to provide a permanent solution for contaminated sites that pose no immediate threat to human health and the environment; these tend to be longer-term and more costly than short-term removal actions, and are only conducted at sites listed on EPA's National Priorities List (NPL).

The Hazardous Ranking System (HRS) is used to determine if a site qualifies as a Superfund site under federal jurisdiction. The HRS system assigns a numerical value to three variables that are relative to the risk associated with the conditions at the site. The three variable categories are:

1. The probability that a site has released or will release hazardous substances into the environment;
2. The characteristics of the waste;
3. The population and environment affected by a release.

Four pathways of release are considered in the HRS determination:

1. Migration of groundwater affecting drinking water;
2. Migration of surface water affecting potential drinking water sources and sensitive environments;
3. Exposure from soil by resident populations;
4. Population affected by air migration.

Scores for each pathway are calculated and used to determine the overall score. An electronic scoring tool that can be used to determine the HRS, called Quickscore, is available through the EPA website. Figure 1.3 provides an overview of the steps involved in a remedial action.

Additional provisions of CERCLA include the National Oil and Hazardous Substances Pollution Contingency Plan, commonly called the National Contingency Plan (NCP). The NCP was developed in response to the 1967 *Torrey Canyon* oil tanker spill that occurred off the coast of England. The provision provided the first plan developed within the US to ensure a coordinated approach to responding to both oil spills and hazardous substance releases. The initial plan provided for a system of accident reporting, spill containment, and cleanup. The NCP also established National and Regional Response Teams. The EPA has recommended revisions to the NCP based on the Deepwater Horizon oil spill in April 2010.

1.3.3.2 The Emergency Planning and Community Right to Know Act (EPCRA – part of SARA)

The Emergency Planning and Community Right to Know Act (EPCRA) was established as Title III of SARA in 1986 (42 USC 11001–11050). The creation of EPCRA was motivated by the 1984 disaster at the Union Carbide pesticide production facility in Bhopal, India. The incident resulted in the loss of thousands of lives, and countless additional related injuries were caused by

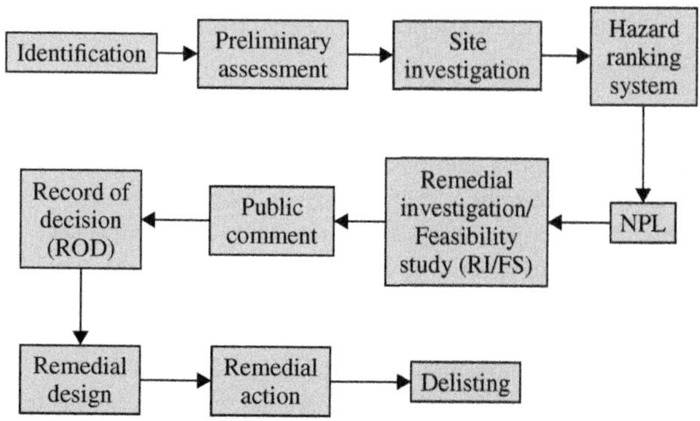

Figure 1.3 Overview of CERCLA Remedial Action Process. *Source*: United States Environmental Protection Agency.

a sudden, unexpected release of over 20 tons of methyl isocyanate – a toxic gas. Subtitle A of EPCRA provided the EPA authority to require local government officials and businesses to develop and implement response procedures in the event of a release of a harmful compound, an explosion, or a fire. Local businesses are required to inform responsible government officials of specific hazardous chemicals used or stored and to alert them quickly in the event of an accident. The release of a hazardous substance must also be reported to the National Response Center governed by CERCLA.

Reporting requirements for facilities using or handling hazardous materials are established under Subtitle B of EPCRA. Industries must report annually a summary of their hazardous chemical inventory to local emergency planning committees (LEPCs) and their local fire departments. The report must include an estimate of:

1. the maximum amount of targeted chemicals present during the preceding year;
2. average material present daily; and
3. the general location of the chemicals within the facility.

Each facility using, manufacturing, or processing hazardous chemicals must report annually to the EPA the amount of chemicals either transported offsite or released to the environment (air, land, or water). The EPA is mandated to develop the Toxic Release Inventory (TRI) that summarizes the release of toxic chemicals by reporting facilities. The TRI is annually distributed nationwide by the EPA.

1.3.3.3 Pollution Prevention Act (PPA) of 1990

The Pollution Prevention Act (PPA) was promulgated in 1990 to address the millions of tons of pollution produced annually in the United States (42 USC 13101). The Act mandated the EPA to establish an Office of Pollution Prevention and also offered a unique way to begin considering processes and the resulting pollution. It was recognized that the traditional end-of-pipe technologies developed and implemented as a result of current regulations successfully reduced targeted emissions and allowed environmental sectors damaged from historical emissions to begin the process of repair. In contrast, the PPA encourages a reduction of pollutant generation from the source and is focused on specific industries instead of specific pollutants.

Source reduction or pollution prevention (P2) is defined as any practice which:

1. reduces the amount of waste entering the environment prior to being recycled, treated, or disposed; and
2. reduces the hazards to public health and the environment that are associated with the release of pollutants.

In an effort to promote source reduction, the EPA is required to develop, test, and disseminate source reduction procedures for broad spectrum applications, to encourage research, and to broadly disseminate information regarding source reduction through a clearing house. The PPA requires industries to include in their EPCRA reports information regarding the facility's efforts to incorporate source reduction and recycling into their activities. For example, facilities must report the P2 practices implemented, techniques used to identify source reduction opportunities, and the quantity of toxic substances recycled on- or off-site.

1.4 An approach to problem solving: a six-step method

The goal of this section is to present a consistent approach to problem solving. This method can be used to solve typical engineering problems and should be used when solving homework problems. The problem solving method consists of six steps. As you proceed through the steps, remember that your work should be presented in a legible, easy-to-follow, logical fashion.

Step 1. Problem identification Clearly define the problem in language you can understand. For homework problems, this will require reading, thinking, and understanding the system described. Identify what is to be determined (for example, find pressure (Pa) and the volumetric flowrate (m^3/s) at the pipe exit).

Step 2. Accumulate relevant data and sketch system Gather all pertinent information, verify accuracy, and take care to include compatible units. A clear, simple sketch of the system with appropriate labels is almost always useful.

Step 3. Select appropriate theory and document all relevant equations Write appropriate theory or principle in equation form, if applicable. Identify each symbol used, indicate units, and specify any constraints.

Step 4. Make necessary simplifications and assumptions For many engineering problems, perfect solutions are impossible to achieve. Simplifications are often necessary and any missing data may force you to make certain assumptions. Note that any simplifications and assumptions must be made with care in order to stay within acceptable bounds of accuracy.

An empirical equation is observation-based and requires no relationship with theory. In situations where this type of equation is used, the units may not cancel properly, so therefore it is essential that the appropriate units are substituted for each parameter. The Hazen-Williams formula is an empirical equation in which the proper units must be substituted for each parameter. One form of the Hazen-Williams equation is used to calculate the volumetric flow rate of water in a pipe. Using English units, the Hazen-Williams equation is expressed as follows:

$$Q = 0.279\, C_{HW} D^{2.63} S^{0.54} \qquad (1.1)$$

where:

Q = volumetric flow rate, millions of gallons per day (MGD)
C_{HW} = Hazen-Williams roughness coefficient depending on pipe material, dimensionless

D = pipe diameter, ft
S = slope of energy grade line, feet per foot of length (ft/ft).

The units do not cancel properly. The correct answer for flow rate can only be determined by substituting the pipe diameter with units of feet and slope in terms of feet per foot to yield the appropriate flow rate in million gallons per day.

Step 5. Solve problem Use the identified theory and data to solve the problem. Attach proper units to all numerical values and double-check that all units cancel properly and that signs have been considered. Common solution techniques include mathematical models, graphical techniques, and trial and error. It is particularly important in an academic setting to show all the steps used to obtain a solution, as the audience of your work is assigning a grade in relation to your understanding of the theory. When using software such as *Excel* or *Mathcad* to solve problems, be sure to show your work. Example hand calculations should be provided, and a well-documented table or figure should summarize your results.

Step 6. Validate solution and communicate results Double-check that your results make physical sense. If your answer does not make sense, you may need to revisit the theory. Consider the units and conversions used when solving the problem. Make it a habit to check your results; verification might be accomplished by:

1 making simplifying assumptions and estimating the answer;
2 comparing your solution to literature results;
3 repeating the calculation;
4 working the problem backwards; or
5 asking a more seasoned engineer if they concur with your theory and the assumptions used to solve the problem.

Finally, a clearly communicated result may simply refer to underlining or boxing your answer(s), or you may choose to present your result visually, using a figure or table. Section 1.5.2 provides guidelines on how to clearly present solutions to engineering problems using tables and figures.

Although it would be comfortable if all problem solutions were obtained linearly by following steps 1 through 6 in sequence, this is not always the case. Learning often occurs during both problem solving and as we evaluate solutions, causing the progression of problem solving to proceed in an iterative fashion, as illustrated in Figure 1.4.

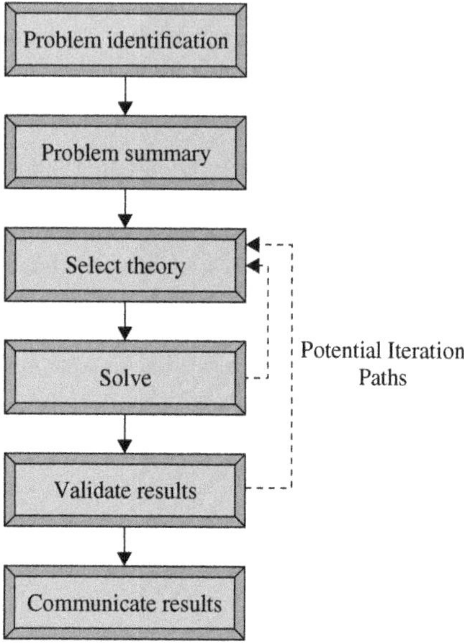

Figure 1.4 Six-step problem solving approach showing potential iteration paths.

Example 1.1 Use the six-step method to solve the following problem

As energy demands continue to rise, coupled with concerns of carbon dioxide emissions contributing to global climate change, clean and efficient (and when possible, renewable) energy resources are in demand. For example, coal combustion processes release approximately 1 kg of CO_2 per kilowatt-hour of power generated, while nuclear, hydroelectric, and wind power produce approximately 0.004 kg CO_2 per kilowatt-hour (Andrews & Jelley, 2007).

Hydroelectric power is often used during peak demand periods. Figures 1.5a and 1.5b show schematics outlining how power is produced at a hydroelectric dam and at pumped-storage plants. In both instances, water (and

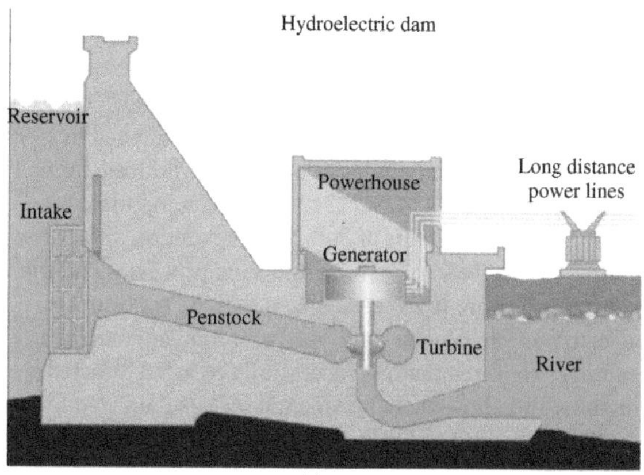

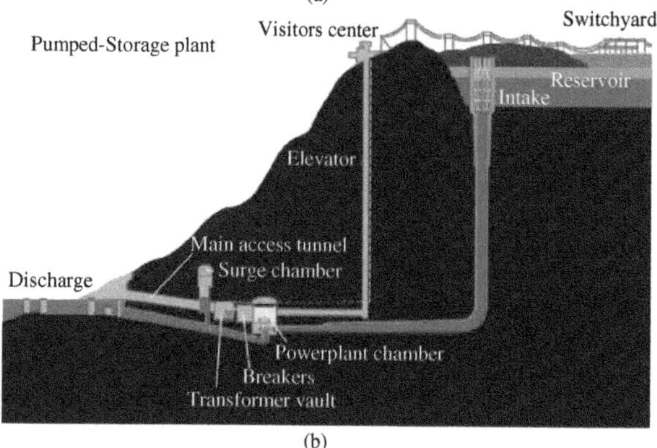

Figure 1.5 (a) Schematic of a hydroelectric dam. (b) Schematic of a pumped-storage hydroelectric plant. *Source*: Tennessee Valley Authority.

therefore potential energy) is held in a reservoir and, when released, the water spins turbine blades which rotate a shaft inside an electromagnetic coil, generating electricity. The Tennessee Valley Authority (TVA) operates the Raccoon Mountain pump-storage facility in Chattanooga, Tennessee. During low demand periods such as nights and weekends, energy is stored by pumping water from the lower to the upper reservoir.

The reservoir at the top of the mountain has 528 acres of water surface. The dam at Raccoon Mountain's upper reservoir is 230 feet high and 8,500 feet long. The water drops 990 feet to the turbine and then flows into the lower reservoir. When the upper reservoir is full, the pumped-storage plant can provide 22 hours of continuous power generation by using half of the water stored in the reservoir.

Assuming a generation efficiency of 80%, estimate the power output from the facility.

Identify and summarize the problem:

Use a schematic of the pump-storage facility to help identify and summarize the problem.

Determine:

the Power Output (MW) for the generating facility.

Known values:

- Water drops 990 ft
- Upper dam height = 230 ft
- Upper dam length = 8500 ft
- Working capacity = 50% of the reservoir volume
- Surface area of upper reservoir = 528 ac
- Generation efficiency = η = 80%
- Time of operation = t = 22 hours
- Water density = $\rho = 1\frac{\text{g}}{\text{mL}} = 1000\frac{\text{kg}}{\text{m}^3}$
- Gravitation acceleration constant = g = 9.81 m/s^2

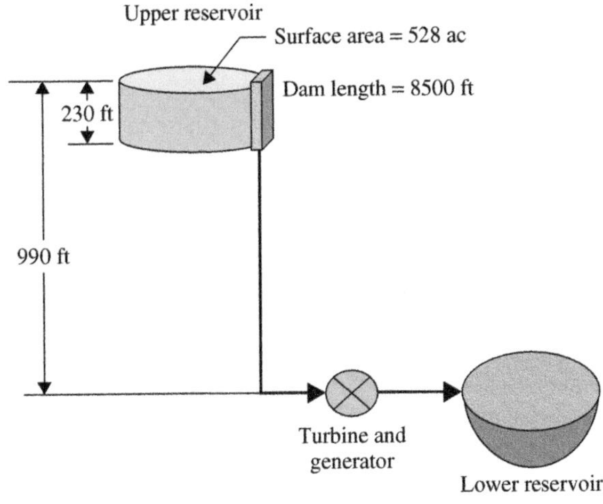

Assumptions:

- Banks of the reservoir are vertical (i.e., not sloped)
- Constant head of 990 ft

Recall that:

$$\text{MW} = 1000\ \text{KW} = 10^6\ \text{W}$$

$$\text{W} = 1\frac{\text{J}}{\text{s}}; \qquad \text{J} = \frac{\text{kg}\cdot\text{m}^2}{\text{s}^2}$$

Theory and Relevant Equations:

Power output from hydroelectric generation depends on the potential energy of the stored water. The gravitational potential energy (*PE*) is based on a material's mass (m) and height (h) from some reference point. In this example, we are concerned with the vertical distance between the turbine and the upper reservoir water surface. This height is often referred to as the *head*. The PE can be calculated as follows:

$$PE = mgh$$

where g is the gravitational constant. Power generation (P) depends on the period of time (t) over which the water is discharged through the known head such that:

$$P = \frac{PE}{t} = \frac{mgh}{t}$$

The mass of the water can be expressed as a volume (V), knowing the fluid density (ρ).

$$m = \rho V$$

Power can now be expressed in terms of mass flow ($\dot{m} = m/t$) or volumetric flow ($Q = V/t$) such that:

$$P = \dot{m}gh = \rho gQh$$

The typical efficiencies (η) for hydroelectric power facilities transforming potential energy into electricity range between 80% and 85%. Our problem stipulated that the efficiency of conversion is 80%. Our final equation describing power generation for the pump-storage facility is as follows:

$$P = \eta\dot{m}gh = \eta\rho gQh$$

Solve:

Begin by estimating the available working volume of water stored in the upper reservoir. It is assumed that the water available for power generation is one half of that stored.

$$V_{water} = \frac{1}{2} \times \text{surface area of reservoir} \times \text{depth of reservoir}$$

$$V_{water} = \frac{1}{2}(528 \text{ ac}) \left(\frac{43,560 \text{ ft}^2}{1 \text{ ac}}\right)(230 \text{ ft}) = 2.64 \times 10^9 \text{ ft}^3$$

The volumetric flow rate of the water can be determined: $Q = \frac{V}{t}$

$$Q = \frac{V_{water}}{t} = \left(\frac{2.64 \times 10^9 \text{ ft}^3}{22 \text{ h}}\right)\left(\frac{1 \text{ h}}{60 \text{ min}}\right)\left(\frac{1 \text{ min}}{60 \text{ s}}\right) \times \left(\frac{7.48 \text{ gal}}{\text{ft}^3}\right)\left(\frac{3.785 \text{ L}}{\text{gal}}\right)\left(\frac{1 \text{ m}^3}{1000 \text{ L}}\right)$$

$$Q = 944 \frac{\text{m}^3}{\text{s}}$$

Estimate the power generation:

$$P = \eta \rho g Q h$$

$$P = (0.80)\left(1000 \frac{\text{kg}}{\text{m}^3}\right)\left(9.81 \frac{\text{m}}{\text{s}^2}\right) \times \left(944 \frac{\text{m}^3}{\text{s}}\right)(990 \text{ ft})\left(\frac{1 \text{ m}}{3.28 \text{ ft}}\right)$$

$$P = 2.2 \times 10^9 \frac{\text{kg m}^2}{\text{s}^3}\left(\frac{1 \text{ J}}{\text{kg} \cdot \text{m}^2/\text{s}}\right) \times \left(\frac{1 \text{ W}}{1 \text{ J/s}}\right) = 2.2 \times 10^9 \text{ W}$$

Communicate Final Answer:

$$P = 2.2 \times 10^9 \text{ W}\left(\frac{1 \text{ MW}}{10^6 \text{ W}}\right) = 2200 \text{ MW}$$

Validate:

According to online resources, the generating capacity of Raccoon Mountain is about 1,600 megawatts of electricity (www.tva.gov). Our answer corresponds reasonably well with the TVA published generating value, but perhaps we overestimated the volume of water available in the upper reservoir or the efficiency of the turbine and generator system. We may also need to reevaluate our assumption that the reservoir banks were vertical, and that it was appropriate to model this system with a constant head of 990 ft.

When considering various sources of power, you will find it convenient to know that a large coal-fired or nuclear power plant has an output of about 1,000 MW. By comparison, the Raccoon Mountain facility output is substantial.

1.5 Data collection, analysis, interpretation, and communication

The importance of understanding data collection, followed by proper analysis and clear communication of results, cannot be overstated. Analysis for engineering design and problem solving involves identifying the motivation for the posed problem and defining the question. Mathematical modeling is often used to predict results. The engineer is responsible for analyzing the consequences of the model results and the communication of sound, supported decisions.

Each analysis should consider the reason and appropriate technical questions, stated assumptions with relevant mathematical model, the solution or simulation that predicts the result, a presentation of the results using both a written explanation and appropriate graphs and tables, a discussion of the meaning, limitations, and impact of the results, and provide a conclusion/decision that describes the final implications of the analysis.

The communication of the work accomplished is critical. The assumptions and model used to analyze the problem must be concisely presented. Briefly explaining supporting equations adds clarity to your work. Explaining the steps used in problem solving adds credibility to your work and value to your written deliverable. The following section provides guidelines for presenting, interpreting, and communicating data and analysis solutions.

1.5.1 Analysis of experimental data

There are seventeen known species of penguins. This is obviously an exact measurement or numerical statement. Regrettably, measurements obtained from laboratory experiments or from field studies do not lend themselves to an exact nature and therefore present some degree of uncertainty. In an attempt to find an exact quantity or measurement, we compensate by taking multiple measurements with the same or varying instruments. It is common to report a range of values that are believed to include the true measurement. Proper units are required in Equation (1.2).

$$\text{measurement} = (\text{measured value} \pm \text{uncertainty}) \tag{1.2}$$

There are two types of errors – *random* and *systematic*. Random error is the scatter observed from repeated measurements and is, by definition, unpredictable and bidirectional, such that it changes from one measurement to the next. In this context, error does not imply a mistake. Random error can be caused by unsteady-state experimental conditions, inherent randomness or fluctuations in the measuring instrument, unobserved external influences on the measurement, or by operator error. Operator error is easily understood. Consider two students conducting a series of titrations; each student would invariably visually interpolate the incremented pipette scale differently, resulting in a random error.

Systematic errors occur to the same degree each time the measurement is made – they are unidirectional. Calibration error, or the offset of the measurement device, is a very common form of systematic error. Systematic errors may also result from a change in environmental conditions, such as temperature or pressure. For example, suppose a scale had been improperly tared. Each subsequent measure would be off by the inconsistent magnitude of the tare. If many measurements are made, they will all be off by the systematic error. Such errors are difficult to detect and very difficult to estimate in hindsight. However, if detected early, systematic errors are relatively easy to manage. Multiple measurements are used to average out or reduce random error, but this strategy will not affect systematic error. In this way, systematic error is a form of statistical bias.

It is also important to understand the concepts of accuracy and precision. Accuracy relates to how close the measurement is to the correct or true value. Inaccuracy is the amount of measurement error; the deviation of the measurement from the true value. Precision relates to the repeatability of the measurement or to how close independent measurements of the same quantity are to each other (note that reproducibility does not relate to or indicate the true value). Precision relates the reliability or reproducibility of the results. Measurements taken that are characterized by small random errors provide high precision results. Measurements obtained with a systematic error are characterized with low accuracy. Measurements obtained with a small systematic error and small random error are quite precise. Unfortunately, measurements obtained with a relatively large systematic error and a small random error are precise, yet inaccurate.

Error associated with experimental observations should always be considered when decisions are made based on experimental results. Those reviewing your conclusion from your experimental work should have the opportunity to consider experimental error when deciding to agree or disagree with your conclusions. When no error estimate is presented, the implication is that the data are perfect – and this is rarely, if ever, the case. How we deal with and communicate experimental error is dependent on how often a parameter is measured. Figure 1.6 shows that if a field or laboratory parameter is measured only once, then the error is estimated and reported as a relative uncertainty. If multiple measurements were made of the parameter, statistics can be used to interpret and present data. Finally, if your desired result is a function of more than one measured parameter, error should be calculated by propagation techniques.

1.5.1.1 Standard error estimate – for a single measurement

After collection of data, it is important to properly characterize and report results. This section describes common techniques that engineers use to present results using a single number called the central tendency, as well as the variation or distribution in the data.

The measurement of a quantity should include both the measured value and its uncertainty, and this is commonly

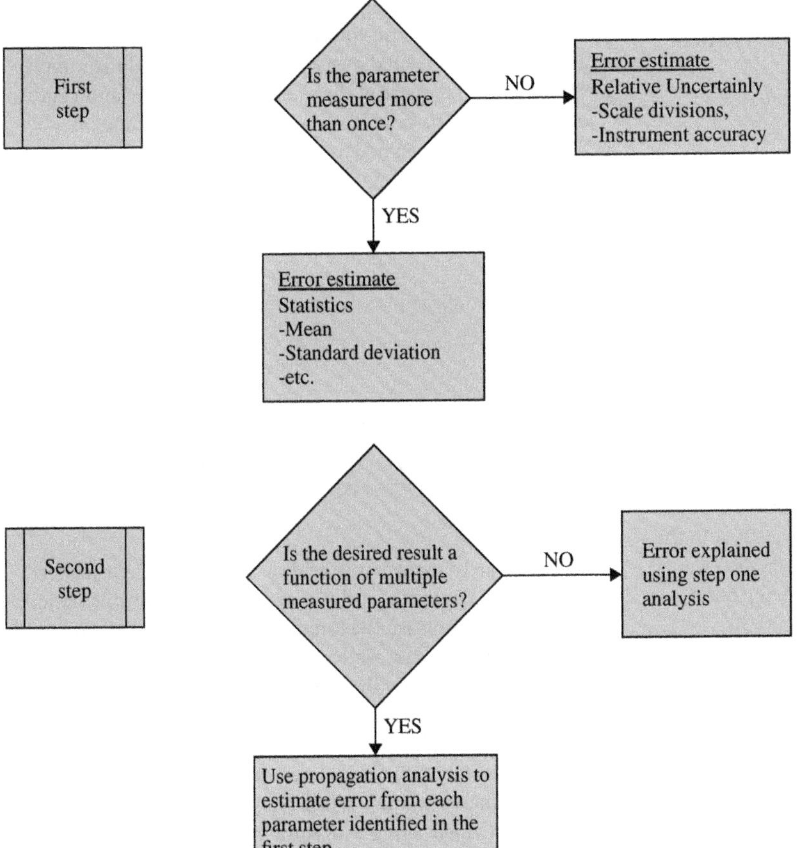

Figure 1.6 Process for error determination.

presented in the form $x = x_{measured} \pm \Delta x$. Here, $x_{measured}$ represents our best estimate of the true value and Δx is an estimate of the relative uncertainty. For example, the pH of a solution may be presented as pH = 7.0 ± 0.2 (Note the awareness of proper use of significant figures). The fractional or relative uncertainty provides information regarding the precision of the measurement.

$$\text{Relative Uncertainty} = \Delta x = \left|\frac{\text{uncertainty}}{x_{measured}}\right| \quad (1.3)$$

Consider an example: many pipettes used for performing titrations are graduated in 1 mL increments. The uncertainty of any measurement instrument is about half of its smallest scale division. So, for our example, the uncertainty associated with reading the pipette is 0.5 mL. If the total volume of titrant used was 20 mL, then the precision of this measurement is reported as a relative uncertainty of:

$$\Delta x = \left|\frac{\text{uncertainty}}{x_{measured}}\right| = \left|\frac{0.5\text{ mL}}{20\text{mL}}\right| = 0.025 = 2.5\%$$

As shown above, the relative uncertainty can also be represented as a percentage. The relative error is often used to quantify accuracy.

$$\text{Relative Error} = \frac{(x_{measured} - \text{expected value})}{\text{expected value}} \quad (1.4)$$

If titration results were expected to yield 18.5 mL of titrant used, experimental accuracy is reported as:

$$\text{Relative Error} = \frac{(x_{measured} - \text{expected value})}{\text{expected value}}$$
$$= \frac{(20 - 18.5)\text{ mL}}{18.5\text{ mL}} = 0.081 = 8.1\%$$

1.5.1.2 Estimating and Reporting Uncertainty – Repeated Measurements

Before we proceed further, we must introduce two important concepts in basic statistical analysis: population and sample. A *population* is a very large set of N observations. For example, if you were interested in determining the average January temperature in Macon, Georgia, then you would need to know the 31 daily January temperatures from the current year, the corresponding values from last year, etc,. all the way back to the beginning of time. If we had access to all those data, then we could calculate a true average (or mean) January temperature, as well as the associated true uncertainty (or spread). By convention, the true mean value of a population is denoted by the Greek symbol μ (mu), and the true spread (or standard deviation) is likewise denoted by the Greek symbol σ (sigma). These two values (μ and σ) are known as population *parameters*, and can be regarded as fixed, numerical characteristics of the population.

In most cases, the population (e.g., January temperatures, heights of Brazilian teenagers, IQs of private school dropouts, etc.) is so large and unwieldy that the parameters cannot be determined. We get around this by taking samples. A *sample* is a subset of the population that contains $n \ll N$ observations. The arithmetic mean of a sample is denoted as $\bar{x}$, and the sample standard deviation is denoted as s_x. If well-chosen, a sample allows us to make appropriate inferences about a population. The values $\bar{x}$ and s_x are known as sample *statistics*.

Returning to our discussion of instrument variability, we understand that, due to inherent random errors in our measurement technique, repeated measured values of x will vary. If the laboratory or field test system allows for repeated measurements of a value, it is recommended to take multiple measurements and average the results in an effort to obtain the best estimate of the true population parameter. The precision of the measurement is improved with increased measurement repetitions. What this means is, as n approaches N, $\bar{x}$ will approach μ.

The average (or sample mean) value of measurements of x can be determined as:

$$\bar{x} = \frac{1}{n}\sum_{i}^{n} x_i \quad (1.5)$$

where:
$\bar{x}$ = sample average value of x
x_i = the value of the ith measurement of x, and
n = number of measurements of x taken.

As mentioned previously, the sample standard deviation (s_x) provides an estimate of the uncertainty associated with your best estimate (the average) of x. In other words, it provides an indication of how close your measured values are to the true average value. A related measure of spread (called the sample variance) is the square of the sample standard deviation and is denoted as s_x^2. The sample variance and sample standard deviation of x are defined as:

$$s_x^2 = \text{variance} = \frac{\sum_{i}^{n}(x_i - \bar{x})^2}{n-1} \quad (1.6)$$

and

$$s_x = \text{standard deviation} = \sqrt{\frac{\sum_{i}^{n}(x_i - \bar{x})^2}{n-1}} \quad (1.7)$$

Fortunately, many hand-held calculators and spreadsheet software packages can easily compute this tedious calculation. Note that the units for the sample variance are the square of the units for the x_i values, while the units for the sample standard deviation are precisely the same as those for the x_i values. For this reason, s_x is somewhat easier to interpret and is used more often in engineering discussions.

Repeated laboratory measurements of a given property will produce different values. When reporting the measurement, the mean is provided, along with some type of uncertainty estimate. Frequently, in science and engineering, the distribution of

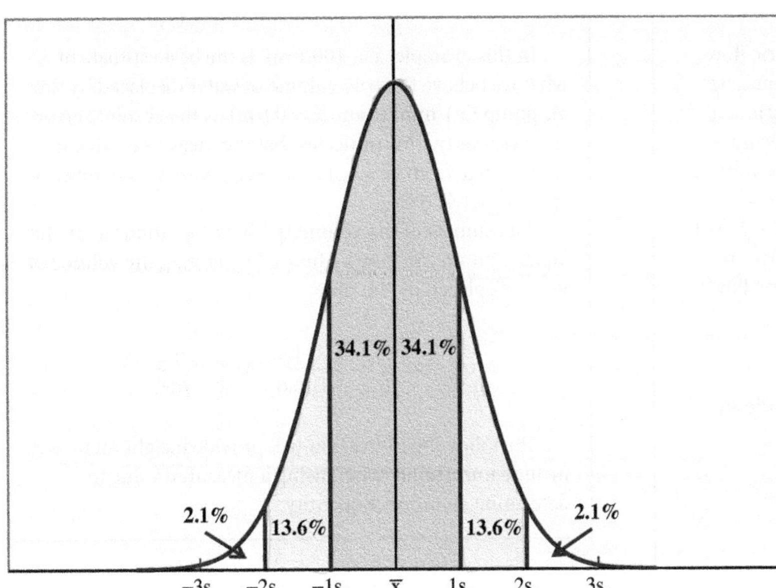

Figure 1.7 A normal or Gaussian distribution has approximately 95% of its distribution within ± 2 standard deviations of the mean.

measured values around the mean is characterized as a Gaussian distribution (common synonyms include normal distribution and bell-shaped curve). For a set of measurements, the Gaussian distribution is mathematically given by:

$$P(x) = \frac{1}{\sigma\sqrt{2\pi}} e^{\frac{-(x_i - \bar{x})^2}{2s_x^2}}$$

As shown in Figure 1.7, the resulting curve is peaked at and is symmetric about $\bar{x}$. Using this theory, we often talk about a measurement being a certain number of standard deviations from the population mean. For this Gaussian distribution, approximately 68.2% of the measured values lie within one standard deviation (1σ) from the mean; 95.4% lie with two standard deviations (2σ), and 99.7% lie within three standard deviations (3σ).

The standard deviation is related to the error associated with each measurement. However, we may be more interested in the error associated with our best estimate or the average of $\bar{x}$. If the errors in the measurements of x are random and independent, and are not influenced by systematic errors, then the error in our best estimate of x is less than s. In this instance, the standard error of the mean (also commonly referred to as the standard deviation of the mean) is a function of the standard deviation of the individual measurements and the total number of measurements.

$$s_m = \frac{s_x}{\sqrt{n}} \tag{1.8}$$

where:

s_m = is the standard deviation of the mean

n = is the number of individual results averaged to obtain s_x.

The value s_m is called the standard error, and is $\sqrt{n}$ times smaller than the standard deviation s_x. In essence, if we are interested in how a given sample x_i value is distributed about a true population mean (μ), then we assess the variability with s_x. If, however, we want to know how a particular sample mean $\bar{x}$ is distributed about μ, then we assess the variability with s_m, which is *smaller* than s_x. This result is a statement of the Central Limit Theorem.

Example 1.2 Standard Deviation

In an attempt to determine the gas flow rate produced from an aquarium pump inexpensively, students devised a simple laboratory scheme that included a water-filled bucket, a 2 L graduated cylinder and a stop watch. As shown in Figure 1.8, the air from the pump was collected into an inverted, water-filled graduated cylinder whose opening was held under the water surface in a bucket.

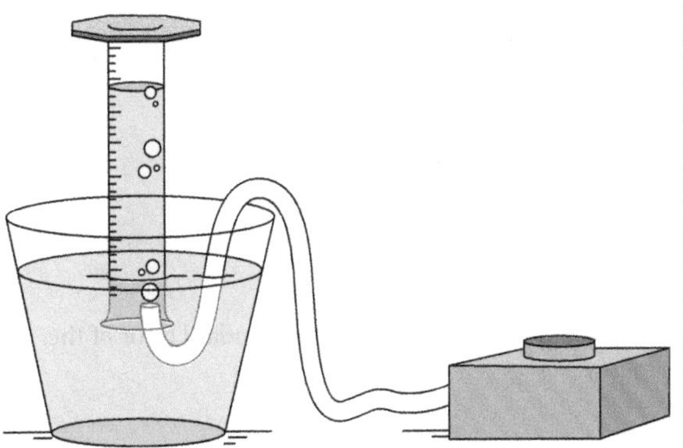

Figure 1.8 Schematic for Example 1.2.

Assuming that the head created by the column of water negligibly affects the pump performance, the rate at which the air displaces the water from the cylinder provides sufficient information to determine the volumetric flow rate (obviously, a more accurate measurement could be obtained with a gas flowmeter). The simple experiment was repeated ten times in an effort to increase accuracy; each trial was allowed to run for one minute. The volume (mL) of water displaced per run was 98, 103, 105, 97, 99, 102, 99, 100, 97, and 101. Determine the average value of water displaced in one minute, the standard deviation, the standard error of the mean, and the air volumetric flow rate provided by the pump.

Solution

Use Equation (1.5) to find the mean or average volume of water displaced in a minute.

$$\bar{x} = \frac{1}{n}\sum_{i}^{n} x_i$$

$$\bar{x} = \frac{98+103+105+97+99+102+99+100+97+101}{10}$$

$$\bar{x} = 100.1\,\text{mL}$$

Prepare a table to assist in the calculation of the standard deviation. Referring to Equation (1.6), the following table was made in an *Excel* spreadsheet to assist with data manipulation.

Displaced fluid, mL	$(x_i - \bar{x})^2$
98	4.41
103	8.41
105	24.01
97	9.61
99	1.21
102	3.61
99	1.21
100	0.01
97	9.61
101	0.81
$\bar{x} = 100.1$ mL	$\sum (x_i - \bar{x})^2 = 62.9$

Now the standard deviation is easily obtained as follows:

$$s_x = \sqrt{\frac{\sum_{i}^{n} (x_i - \bar{x})^2}{n-1}} = \sqrt{\frac{62.9}{9}} = 2.6\,\text{mL}$$

Use Equation (1.8) to find the standard error of the mean.

$$s_m = \frac{s}{\sqrt{n}} = \frac{2.6\,\text{mL}}{\sqrt{10}} = 0.8\,\text{mL}$$

The laboratory result would be displayed in the form:

$$\text{water displaced (mL)} = \bar{x} \pm \Delta x = (100.1 \pm 0.8)\text{mL}$$

In this example, $\bar{x} = 100.1$ mL is the best estimate of what we believe the true volume of water displaced by the air pump in 1 minute and $\bar{x} = 0.8$ mL is the absolute error or uncertainty. This indicates that the measured value lies in the range from $(\bar{x} - s_m)$ to $(\bar{x} + s_m)$ with a probability of approximately 68%.

An estimate of the volumetric flow, Q, produced by the aquatic pump can now be found by dividing the volume of water displaced by the time.

$$Q = \frac{V}{t} = \frac{100.1\,\text{mL}}{1\,\text{min}} = 100\,\frac{\text{mL}}{\text{min}}$$

The following discussion will provide insight on how to include uncertainty when using a measured value to determine an indirect quantity.

1.5.1.3 Calculation Error Estimates – Propagation of Independent Errors

Laboratory measurements are often used to find the quantity of interest (the indirect measurement). However, as you now know, laboratory measurements are not exact and have errors associated with them. Using the estimated uncertainties of the measured quantities, this section provides guidance for determining the uncertainty of the calculated result. This is often referred to as the propagation of direct measurement errors.

In the discussion, y will represent the quantity to be determined (indirect quantity), which depends on several variables named $x_1, x_2, \ldots, x_i$. We can determine the error associated with y if we measure $x_1, x_2, \ldots$ with errors of Δx_1, $\Delta x_2, \ldots$ The Δx values are dependent on circumstances and are either standard deviations, standard errors, or the absolute uncertainty (measurement error).

The general equation for the propagation of error is given by (Andraos, 1996):

$$\Delta y = \pm\sqrt{\sum_{i=1}^{n}\left(\frac{\partial y}{\partial x_i}\right)^2 (\Delta x_i)^2} \tag{1.9}$$

The equation allows for the determination of uncertainty in the function Δy if the error or uncertainty (Δx_i) is known for each variable (x_i) in the function.

For example, consider the addition or subtraction of two mutually independent quantities such as the function $y = x_1 + x_2$. Allow the uncertainty in x_1 and x_2 to be Δx_1 and Δx_2. Consider Equation (1.9); the partial derivative with respect to each variable gives: $\frac{\partial y}{\partial x_1} = 1$ and $\frac{\partial y}{\partial x_2} = 1$. Substituting these results into Equation (1.9) gives the uncertainty in y as:

$$\Delta y = \pm\sqrt{(1)^2(\Delta x_1)^2 + (1)^2(\Delta x_2)^2} = \pm\sqrt{(\Delta x_1)^2 + (\Delta x_2)^2} \tag{1.10}$$

The result is the same when $y = x_1 - x_2$.

Now consider the function $y = x_1 \times x_2$ (the result is the same for $y = x_1/x_2$). Again make the assumption that the error

Table 1.4 Summary of error propagation techniques.

Function	Uncertainty
Addition and subtraction $y = x_1 + x_2 - x_3$	$\Delta y = \pm\sqrt{(\Delta x_1)^2 + (\Delta x_2)^2 + (\Delta x_3)^2}$
Function with a constant $y = ax_1$	$\Delta y^2 = a^2 \Delta x_1^2$
Product or quotient $y = x_1 \times x_2 / x_3$	$\frac{\Delta y}{y} = \pm\sqrt{\left(\frac{\Delta x_1}{x_1}\right)^2 + \left(\frac{\Delta x_2}{x_2}\right)^2 + \left(\frac{\Delta x_3}{x_3}\right)^2}$
Exponential function (assuming no uncertainty in a) $y = x_i^a$	$\frac{\Delta y}{y} = a\left(\frac{\Delta x_i}{x_i}\right)$
Logarithmic function $y = ae^{\pm bx_1}$ $y = a\ln(\pm bx_1)$	$\frac{\Delta y}{y} = ab\Delta x_1$ $\Delta y = a\frac{\Delta x_1}{x_1}$

Δy = the uncertainty in the result of calculating the function y
y = result of the calculation
x_i = real variables obtained experimentally that are used to calculate y
Δx_i = the uncertainty in numbers (x_i) used for calculation of y
a, b = constants having no uncertainty
Source: Data from Andraos, J. 1996.

associated with x_1 and x_2 is Δx_1 and Δx_2. Taking the partial derivative with respect to each variable gives: $\frac{\partial y}{\partial x_1} = x_2$ and $\frac{\partial y}{\partial x_2} = x_1$. Substitution into Equation (1.9) gives: $\Delta y = \pm\sqrt{(x_2)^2(\Delta x_1)^2 + (x_1)^2(\Delta x_2)^2}$. Dividing this result by $y = x_1 \times x_2$ provides results in the form that is most commonly presented:

$$\frac{\Delta y}{y} = \pm\sqrt{\left(\frac{\Delta x_1}{x_1}\right)^2 + \left(\frac{\Delta x_2}{x_2}\right)^2} \quad (1.11)$$

A summary of techniques used to determine how error propagates through an experimental procedure is provided in Table 1.4.

Example 1.3 Error Propagation

Consider the experiment outlined in Example 1.2. The uncertainty of the experimental time was assumed to be 1 second (1/60 min). Use the rules for error propagation to determine the uncertainty in Q, where $Q = \frac{V}{t}$ with $V = (100.1 \pm 0.8)\text{mL}$ and $t = \left(1 \pm \frac{1}{60}\right)$ min.

Solution

Use Equation (1.11) to solve for the uncertainty in Q.

$$\frac{\Delta y}{y} = \pm\sqrt{\left(\frac{\Delta x_1}{x_1}\right)^2 + \left(\frac{\Delta x_2}{x_2}\right)^2} = \frac{\Delta Q}{Q}$$

$$= \pm\sqrt{\left(\frac{\Delta V}{V}\right)^2 + \left(\frac{\Delta t}{t}\right)^2}$$

$$\Delta Q = \pm Q \times \sqrt{\left(\frac{\Delta V}{V}\right)^2 + \left(\frac{\Delta t}{t}\right)^2}$$

$$= \pm\left(100.1\ \frac{\text{mL}}{\text{min}}\right) \times \sqrt{\left(\frac{0.8}{100.1}\right)^2 + \left(\frac{1/60}{1}\right)^2}$$

$$\Delta Q = 1.9\ \text{ml/min}$$

The volumetric flow rate, Q, obtained from these experimental results should be expressed as:

$$Q = \frac{(100.1 \pm 1.9)\text{mL}}{\text{min}}$$

Example 1.4 Error Propagation

A wide variety of manufacturers, including the textile, food, paper and cosmetic industries, utilize azo dyes in their production processes. As a result, color-laden industrial wastewater is frequently discharged into the environment and produces an obnoxious appearance that is not favorably perceived by regulators or the public. In general, azo dyes are not considered hazardous, but are often considered recalcitrant in traditional wastewater treatment processes. In the United States, color is federally regulated by a secondary maximum contaminant level under the Drinking Water Act (40 CFR 143.2), while some states mandate an additional level of compliance.

As a result of regulatory and public relation problems, industries discharging color have recently begun focusing on novel decolorization processes such as ozonation, catalytic oxidation, adsorption, and hybrid technologies that combine ozonation with biological treatment schemes. Acid yellow 17 dye ($C_{16}H_{12}Cl_2N_4O_7S_2 \cdot 2Na$; MW = 551.29) is a common additive found in ordinary household products such as shampoo, bubble bath, shower gel, liquid soap, multipurpose cleanser, dishwashing liquid and alcohol-based perfumes. As regulations associated with dyestuff are tightened, associated industries are faced with finding economically viable water treatment solutions.

During laboratory experiments with dye-laden wastewater, the concentration of dye required routine measurement. A calibration curve that related absorbance to concentration was used. The dye concentration was measured, using a device called a spectrophotometer, at a wavelength of 400 nm, and a standard curve was prepared, showing the absorbance versus concentration of acid yellow 17 dye. From the calibration curve, the concentration of the dye in the wastewater sample was found to be 250 ± 5 mg/L. The solution was prepared by dissolving 0.0625 g of the dye into 250.0 ± 0.5 mL of water. Find the uncertainty of the dye weight that was actually added to make the stock solution.

Solution

Use Equation (1.10) to find the uncertainty of this weight:

$$\frac{\Delta y}{y} = \pm\sqrt{\left(\frac{\Delta x_1}{x_1}\right)^2 + \left(\frac{\Delta x_2}{x_2}\right)^2} = \frac{\Delta W}{0.00625}$$

$$= \pm\sqrt{\left(\frac{5}{250}\right)^2 + \left(\frac{0.5}{250}\right)^2}$$

$$\Delta W = \pm 0.0001 \text{ g}$$

The weight of the unknown is 0.0063 ± 0.0001 g

One final comment regarding uncertainty calculations: the calculated uncertainty should not be stated with too much precision. For example, if titration results provide an answer of 14.28 mL with an uncertainty of 0.1 mL, then the result should be presented as 14.3 ± 0.1 mL. Similarly, if the uncertainty is 1 mL, then the result should be presented as 14 ± 1 mL. Note that the answer and the uncertainty have the same order of magnitude and are presented with the same decimal position. When more than one experimental variable is used to find the result, the accuracy of the answer is limited by the least accurate measurement. With this in mind, an overview of "rules" for working with significant figures is shown in Table 1.5.

Table 1.5 Overview of significant figures for experimental measurements.

Guidelines for assigning significant figures	Example
Rule 1: All nonzero digits are significant.	103°C has three significant figures. 0.67 has two significant figures.
Rule 2: All digits are significant between and including the first non-zero digit from the left, through the last digit. Zeros are significant except when they are used to locate the decimal point.	0.0025 has two significant figures. 0.55 has two significant figures. 2.17 has three significant figures. 0.0030 has two significant figures.
Rule 3: Zeros between nonzero digits are significant.	1.05 has three significant figures. 1005 has four significant figures.
Rule 4: For numbers that end with a zero to the left of the decimal point, zeros may or may not be significant. Use scientific notation to clearly express the number of significant figures.	1400 as written here could have 2, 3, or 4 significant figures. 1.4×10^2 clearly shows two significant figures. 1.40×10^2 shows three significant figures.

1.5.2 Using tables and figures to analyze and communicate results

Tables and figures can be used as powerful tools to analyze and to communicate the physical context of your results. When using these visual communication tools, the honest documentation of your results is crucial. Avoid the dangers associated with forcing your data to fit your anticipated results. Report them as simply and honestly as possible. Both tables and figures should be located on the page they are called and discussed, or on the following page of text, and both should be numbered consecutively. Graphs and figures are labeled and captioned at the bottom. Conversely, the caption and label for tables is always placed at the top. Although reported data within a figure or a table provides an excellent visual summary, it does not substitute for the complete discussion of your results; showing results in a table or a figure only is inadequate. Discussion of your findings as presented in the figure is still necessary.

Table 1.6 provides an example of the type of data that are easily communicated in tabular form. The table provides a list of polycyclic aromatic hydrocarbons (PAHs) having environmental interest in relation to their molecular weight,

Table 1.6 Characteristics of polycyclic aromatic hydrocarbons with environmental significance.

Hydrocarbon	Molecular weight (g/mol)	Solubility (mg/L)	Soil-water partition coefficient
Benzene	78.11	1780	97
Toluene	92.1	500	242
o-xylene	106.17	170	363
Ethyl benzene	106.17	150	622
Naphthalene	128.16	31.7	1300
Acenophthene	154.21	3.93	2580
Acenaphthylene	152.2	3.93	3814
Fluorene	166.2	1.98	5835
Fluoranthene	202	0.275	19000
Phenanthrene	178.23	1.29	23000
Anthracene	178.23	0.073	26000
Pyrene	202.26	0.135	63000
Benzo(a)anthracene	228	0.014	125719
Benzo(a)pyrene	252.3	0.0038	282185
Chrysene	228.2	0.006	420108
Benzo(b)fluoranthe	252	0.0012	1148497
Benzo(g,h,i)perylene	276	0.00026	1488389
Dibenz(a,h)anthracene	278.35	0.00249	1668800
Benzo(k)fluoranthene	252	0.00055	2020971

Source: Public Domain: Fetter, C.W., Contaminant Hydrology, 2nd edition, Prentice Hall, Upper Saddle River, NJ, 1999.

solubility, and soil-water partition coefficient. Polycyclic aromatic hydrocarbons are comprised of fused benzene rings, are found in gasoline, are prevalent in the environment as a result of incomplete combustion of fossil fuels, and are characteristic of the hydrocarbons found in coal tar. Important factors that may influence the environmental fate of a PAH compound include the compound's aqueous solubility, its vapor pressure, and its soil-water partition coefficient. It is clear from the tabulated data that, as the molecular weight of PAH increases, the solubility decreases while the soil-water partition coefficient increases.

Within a saturated soil system, the soil-water partition coefficient quantifies the ratio of the organic compound absorbed or adsorbed in the solid soil phase to the concentration of the dissolved organic in the aqueous phase. In general, as the compound molecular weight increases, the more likely it is to absorb or adsorb to the solid phase, thus characterized by an increasing partition coefficient.

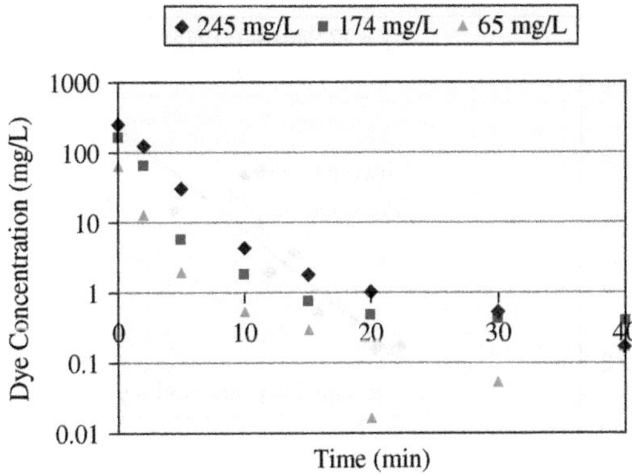

Figure 1.9 Example figure illustration that shows how dye concentration is changing as ozonation time increases.

Observe that Table 1.6 is properly labeled with a caption at the top of the table and that columns are labeled logically and legibly. Note that compounds are listed in order of increasing molecular weight. This helps the reader note relationships between molecular weight, solubility, and the soil-water partition coefficient.

Engineers often use *x-y* graphs or scatterplots to communicate the strength of relationship between two variables. Pie and bar charts may also be useful given the nature of the data. When making a chart, maximize the effectiveness by incorporating as much related data as possible on a given figure without compromising clarity. For example, data in Figure 1.9 describe how acid yellow 17 dye concentration changes with respect to treatment time with ozone. Results are presented from three different experiments on this one figure to allow for comparison. Displaying data in this matter also allows for a visual representation of the relationship between variables.

A review of Figure 1.9 provides some guidelines for effectively plotting experimental results. Note that the figure is clearly labeled at the bottom of the caption. The plot is descriptive, yet simple, and the *x*-and *y*-axes are clearly labeled and have associated numerical values with units. Since multiple data sets are presented, a legend is provided. Notice that different data markers are used for each data set.

1.5.3 Linear regression and the correlation coefficient

Scatterplots (*x-y* graphs) are useful to view the form, direction, and strength of relationship between two variables. Often, the relationship is linear. The correlation coefficient can be used to quantify both the direction and strength of the observed linear relationship, and it serves as a "goodness of fit" measure. The equation for a linear regression line has the form:

$$y = mx + b \tag{1.12}$$

where:

y = the dependent variable (the variable to be predicted)
x = independent variable

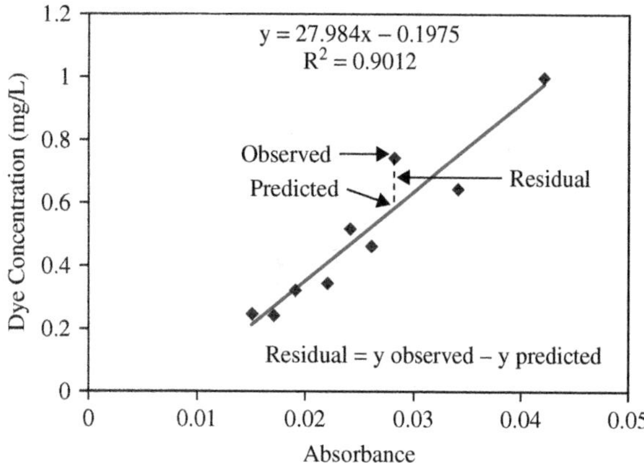

Figure 1.10 The least-squares regression line of dye concentration on absorbance is the line that minimizes the sum of squared errors.

m = the slope of the linear regression line
b = the intercept, the value of y when $x = 0$.

The least-squares method is commonly used for fitting a regression line. Using this method, the best fit line for the data is determined by minimizing the sum of the squares of the vertical deviations from each data point to the line (i.e., minimizing the sum of the residuals). Figure 1.10 shows a visual representation of the residuals as the dotted line between the observed data points and the corresponding predicted values. The vertical distance is zero when a point lies directly on a line. As the deviations are squared prior to the summation, the positive and negative deviations do not negate or cancel out their values.

Using this method, the slope and the y-intercept for the best fit regression line can be determined as follows:

$$m = \frac{\sum_{i}^{n}(x_i - \overline{x})(y_i - \overline{y})}{\sum_{i}^{n}(x_i - \overline{x})^2} \tag{1.13}$$

$$m = r\frac{s_y}{s_x} \tag{1.14}$$

$$b = \overline{y} - m \times \overline{x} \tag{1.15}$$

where:

b = the constant in the regression equation that represents the y-intercept when the independent variable, x, is equal to zero
x_i = the value of x of observation i
$\overline{x}$ = the mean of x
$\overline{y}$ = the mean of y
m = slope of the regression line that explains how y changes for a unit change in x
r = the correlation between x and y (Pearson's r)
s_y and s_x = the standard deviation of y and x, respectively.

Once m and b are known, they can be substituted into Equation (1.12) to describe the linear regression line. To make a prediction for an unmeasured x, just substitute in to the developed equation and solve for y. From Equation (1.15), note that the least-squares regression line always passes through the point $(\overline{x}, \overline{y})$.

The sample correlation coefficient, r, quantifies the direction and the magnitude of the relationship between two variables. Correlation coefficient values range from −1 to 1. A positive correlation indicates that as one variable increases, the other variable also increases. A negative correlation suggests that as one variable increases, the other decreases. The greater the absolute value of the coefficient, the stronger the linear relationship. Note that if the correlation coefficient was zero, this only indicates that there is no linear relationship, but does not indicate that there is no relationship between the variables. The strongest linear relationship is indicated from values of −1 and 1. In both instances, all of the observed data fall directly on the regression line. From Equation (1.14), note that the change of 1 standard deviation in x corresponds to a change of r standard deviations in y. The sample correlation coefficient can be calculated as follows:

$$r = \frac{1}{n-1}\sum_{i=1}^{n}\left(\frac{x_i - \overline{x}}{s_x}\right)\left(\frac{y_i - \overline{y}}{s_y}\right) \tag{1.16}$$

Computer or calculator software packages often provide the coefficient of determination (R^2) as an output from regression analysis. The R^2 value is the square of the correlation coefficient and represents the fraction of the variation in the values of y that is explained by the least squares regression of y on x; this value is a measure of how successfully the regression explains the observed response. The coefficient ranges from 0 to 1. An R^2 of zero indicates that the dependent variable cannot be predicted from the independent variable, while an R^2 of 1 means the dependent variable can be predicted without error from the independent variable. The coefficient of determination for a linear regression model with one independent variable is:

$$R^2 = \frac{1}{n-1}\sum_{i=1}^{n}\left\{\left(\frac{x_i - \overline{x}}{s_x}\right)\left(\frac{y_i - \overline{y}}{s_y}\right)\right\}^2 = r^2 \tag{1.17}$$

Fortunately, there exists a variety of software programs and calculators that make these regression calculations very simple. In the example below, we show how to determine m, b, and R^2 by hand and by using an *Excel* spreadsheet.

Example 1.5 Least-squares Linear Regression

Many wastewater treatment plants use microbiological processes to degrade the organic contaminants commonly associated with municipal wastewater. Two measurements often used by municipalities and regulators to characterize the organic content in the wastewater include the biochemical oxygen demand (BOD) and the chemical oxygen demand (COD) parameters. The BOD_5 measurement quantifies how much oxygen the aerobic microbes within the treatment system require to oxidize the degradable organics in the water during a five-day reaction period. The COD measures the amount of oxygen required to oxidize chemically all of the organic material in the wastewater. A more detailed description of these two important parameters will be covered in Chapter 3. At this juncture, you should observe the correlations between these parameters in this example problem].

Given the data in Table 1.7 to describe the relationship between the influent COD and BOD_5 for the wastewater treatment plant; use the data to calculate means, standard deviations for BOD_5 and COD, correlation r, coefficient of determination R^2, slope m, intercept b, and the equation of the least-squares line for this case.

Table 1.7 Monthly BOD_5 and COD averages from a wastewater treatment plant located in Atlanta, GA.

Month	BOD_5 average (mg/L)	COD average (mg/L)
January	228.2	674.0
February	198.7	448.0
March	200.2	583.0
April	179.0	575.0
May	190.8	486.0
June	175.6	420.0
July	176.8	459.0
August	177.4	382.0
September	175.0	382.0
October	182.3	391.0
November	219.2	596.0
December	239.9	608.0

Solution

Use Equation (1.5) to find the mean for both the BOD and COD data. For the BOD data:

$$\overline{BOD_5} = \frac{1}{n}\sum_i^n BOD_{5,i}$$

$$\overline{BOD_5} = \frac{\left(\begin{array}{l}228.2 + 198.7 + 200.2 + 179.0 + 190.8 \\ +175.6 + 176.8 + 177.4 + 175.0 \\ +182.3 + 219.2 + 239.9\end{array}\right)}{12}$$

$$\overline{BOD_5} = 195.3\,\text{mg/L}$$

In similar fashion, the COD mean was found to be $\overline{COD} = 500.3\,\text{mg/L}$.

Referring to Equation (1.7), prepare a calculation table to assist in the determination of the standard deviation of COD.

COD, mg/L	$(COD_i - \overline{COD})^2$
674	30171.7
448	2735.3
583	6839.3
575	5580.1
486	204.5
420	6448.1
459	1705.7
382	13994.9
382	13994.9
391	11946.5
596	9158.5
608	11599.3
$\overline{COD} = 500.3$ mg/L	$\Sigma(COD_i - \overline{COD})^2 = 114378.7$

Now the standard deviation is easily obtained as follows:

$$s_{COD} = \sqrt{\frac{\sum_i^n (COD_i - \overline{COD})^2}{n-1}}$$

$$= \sqrt{\frac{114378.7}{11}} = 102\,\text{mg/L}$$

Similarly, the standard deviation for BOD was determined as $s_{BOD} = 22.6$ mg/L.

Use Equation (1.16) and an associated calculation table to find the sample correlation coefficient, r.

Month	COD (mg/L)	BOD_5 (mg/L)	$\left(\frac{COD_i - \overline{COD}}{s_{COD}}\right)$	$\left(\frac{BOD_{5,i} - \overline{BOD_5}}{s_{BOD_5}}\right)$	$\left(\frac{COD_i - \overline{COD}}{s_{COD}}\right)\left(\frac{BOD_{5,i} - \overline{BOD_5}}{s_{BOD_5}}\right)$
Jan	674	228.2	1.70	1.462	2.49
Feb	448	198.7	−0.513	0.151	−0.077
Mar	583	200.2	0.811	0.218	0.177
Apr	575	179.0	0.732	−0.724	−0.531
May	486	190.8	−0.140	−0.2	0.028
Jun	420	175.6	−0.787	−0.876	0.689
Jul	459	176.8	−0.405	−0.822	0.333
Aug	382	177.4	−1.160	−0.796	0.923
Sep	382	175.0	−1.160	−0.902	1.046
Oct	391	182.3	−1.072	−0.578	0.619
Nov	596	219.2	0.938	1.062	0.997
Dec	608	239.9	1.056	1.982	2.093
average	500.3	195.3			
std dev	102	22.5			Sum = 8.79

$$r = \frac{1}{n-1}\sum_{i=1}^n \left(\frac{COD_i - \overline{COD}}{s_{COD}}\right)\left(\frac{BOD_{5,i} - \overline{BOD_5}}{s_{BOD_5}}\right)$$

$$= \frac{8.79}{11} = 0.80$$

Calculating the coefficient of determination is now very simple if we recognize that $R^2 = r^2$ as shown in Equation (1.17):

$$R^2 = \left\{\frac{1}{n-1}\sum_{i=1}^{n}\left(\frac{x_i-\bar{x}}{s_x}\right)\left(\frac{y_i-\bar{y}}{s_y}\right)\right\}^2 = r^2$$

$$R^2 = 0.64$$

This result indicates that 64% of the variance in BOD_5 is predictable from the COD.

Complete this example by plotting the data as shown in Figure 1.11 and by using Equations (1.12) through (1.14) to characterize the least-squares regression line for the BOD_5 and COD data.

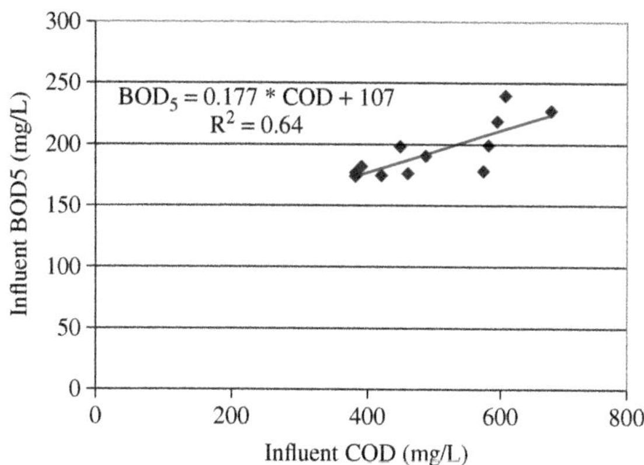

Figure 1.11 Scatterplot of monthly influent COD and BOD5 averages for an Atlanta, GA wastewater treatment plant. A least-squared regression analysis was performed to find the characteristic linear relationship between the parameters.

Using Equation (1.14), the slope of the regression line is determined to be 0.177.

$$m = r\frac{s_y}{s_x} = 0.80\frac{22.6}{102} = 0.177$$

Now use Equation (1.15) to determine the y-intercept of the linear regression line.

$$b = \bar{y} - m \times \bar{x} = \overline{BOD_5} - m \times \overline{COD}$$

$$b = 195.3 - (0.177 \times 500.3) = 107$$

These results lead to the least-squares linear regression line that describes the COD versus BOD_5 relationship for this particular wastewater influent. This is shown in the form of $y = mx + b$ as follows:

$$BOD_5 = 0.177 \times COD + 107$$

Interestingly, literature suggests an approximate ratio for COD and BOD_5 of settled, municipal wastewater (Mara & Horan, 2003):

$$\frac{COD}{BOD_5} \cong 2$$

This indicates that there are about twice as many non-biodegradable organics in the wastewater compared to those that are susceptible to microbial degradation. With a COD of 500 mg/L, our regression line predicts $BOD_5 = 0.177 \times 500 + 107 = 195.5$ mg/L. So our predicted $\frac{COD}{BOD_5}$ ratio is:

$$\frac{COD}{BOD_5} = \frac{500}{195.5} = 2.6,$$

showing that our regression result reasonably supports this well-established trend.

With our example problem solution in mind, we provide some final thoughts associated with linear regression:

- The goal of linear regression is to find an equation (slope and *y*-intercept) that best fits the measured experimental values.
- Least-squares regression analysis is the most common method used to fit a line to scatterplot data.
- The equation developed can be used to predict the value of *y* for any value of *x* by substituting *x* into the equation. For our case, the expected influent BOD_5 can be estimated by substituting a measured COD value into the regression equation.
- Extrapolating beyond the range of *x* values used to develop the regression equation to predict *y* values is risky. In Example 1.5, the lowest observed average influent COD was 382 mg/L. In this instance, it is inadvisable to extrapolate extensively beyond this value.
- The intercept, *b*, has no statistical value unless *x* can actually take values near zero.
- The goal of correlation analysis is to see whether the variability in *x* and *y* values is predictive, and to measure the strength of any relationship between the variables. The results of correlation are often expressed as an *r*-value (correlation coefficient) or an R^2 value (coefficient of determination).
- The concepts of regression and correlation are directly connected.

1.5.4 Interpreting linear, power, and exponential equations

Many simple nonlinear functions can be algebraically manipulated into linear form. For example, the decolorization of dye in a batch system by ozonation may be well modeled using kinetics. According to first-order kinetic theory, the dye concentration at any time during the ozonation process can be determined from the following equation:

$$C_t = C_0 e^{-kt} \tag{1.18}$$

where:
C_t = dye concentration at time, t, mg/L
C_0 = dye concentration at $t = 0$, mg/L
t = reaction time, min
k = first-order kinetic coefficient, min^{-1}.

If, during experimentation, the time of ozonation and the dye concentration is periodically recorded, the first-order rate constant that characterizes the system can be easily estimated by recognizing that the defining equation can be linearized. The linear result, as shown in Equation (1.19), can be obtained from Equation (1.18) as a result of algebraic manipulation.

$$\ln(C_t) = \ln(C_0) - kt$$
$$y = b + mx \tag{1.19}$$

where:
y = dependent variable
b = y-axis intercept
m = slope
x = independent variable.

Equation (1.19) can also be written as a base-10 logarithm. The conversion between a base-10 logarithm and the natural logarithm is readily identified by the ratio:

$$\frac{\log 10}{\ln 10} = \frac{1}{2.303} \tag{1.20}$$

To convert a natural logarithm to a base-10 logarithm, divide by the conversion factor 2.303. Rewriting Equation (1.19) using base-10 logarithms gives:

$$2.303 \times \log(C_t) = 2.203 \times \log(C_0) - k\,t \tag{1.21}$$

There are many additional systems encountered by environmental engineers that are described using nonlinear models that can be easily manipulated into a linear format. For example, the Monod model (Monod, 1949), used to predict the state of a continuous flow bioreactor or chemostat, is described using two variables: the specific growth rate of the microbes (μ) and the growth-limiting substrate (S).

$$\mu = \frac{\mu_{max} S}{K_s + S} \tag{1.22}$$

where:
μ = specific growth rate, h^{-1}
μ_{max} = maximum specific growth rate, h^{-1}
K_s = half-saturation constant, g/L
S = limiting substrate concentration, g/L.

When Equation (1.22) is linearized, μ_{max} and K_s can be estimated from experimental results. Consider the simple algebraic manipulation of Equation (1.22):

$$\mu = \frac{\mu_{max} S}{K_s + S}$$

Cross-multiply and obtain:

$$\frac{1}{\mu} = \frac{K_s}{\mu_{max} S} + \frac{S}{\mu_{max} S}$$

With further algebraic simplification:

$$\frac{1}{\mu} = \frac{K_s}{\mu_{max}}\left(\frac{1}{S}\right) + \frac{1}{\mu_{max}}$$
$$y = mx + b \tag{1.23}$$

Figure 1.12 shows that by using a double reciprocal plot of the specific growth rate (dependent variable) with the limiting substrate concentration (independent variable), both μ_{max} and K_s can be determined using linear regression analysis. With the determination of system constants, the dependence of growth rate on the limiting substrate concentration is described.

Other examples of nonlinear models that lend to simple linear manipulation include Deacon's Power Law, which relates wind velocity to altitude, the Arrhenius equation that correlates a rate of a reaction and its temperature, the Langmuir and Freundlich isotherms that describe adsorption processes, and the Michaelis-Menten model that is used to estimate enzyme kinetics. Although these physical systems are not described linearly, model parameters can be determined from the linearization of experimental data. Table 1.8 provides a generalized summary of several common functions that are routinely evaluated in their linear form.

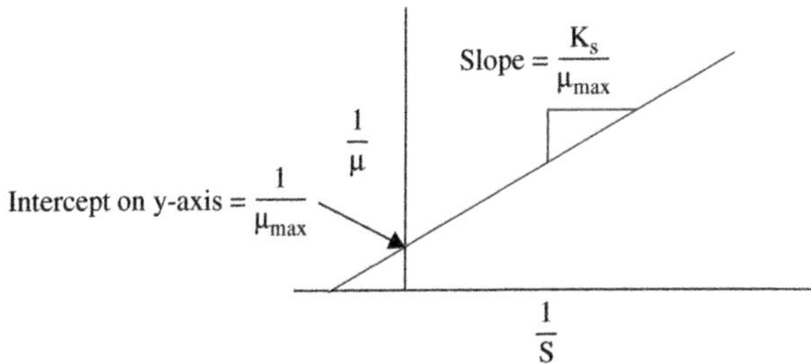

Figure 1.12 Double reciprocal plot used to find the biokinetic coefficients for the Monod model of microbial growth kinetics.

Table 1.8 Generic linearized functions having the form $y = mx + b$.

$F(z) \Rightarrow$ linear function	y	m	x	b
$az^c \Rightarrow \ln(f(z)) = c\ln(z) + \ln(a)$	$\ln[f(z)]$	c	$\ln(z)$	$\ln(a)$
$\frac{az}{1+cz} \Rightarrow \frac{1}{f(z)} = a\left(\frac{1}{z}\right) + \frac{c}{a}$	$\frac{1}{f(z)}$	a	$\frac{1}{z}$	$\frac{c}{a}$
$\frac{az}{c+z} \Rightarrow \frac{1}{f(z)} = \frac{c}{a}\left(\frac{1}{z}\right) + \frac{1}{a}$	$\frac{1}{f(z)}$	$\frac{c}{a}$	$\frac{1}{z}$	$\frac{1}{a}$

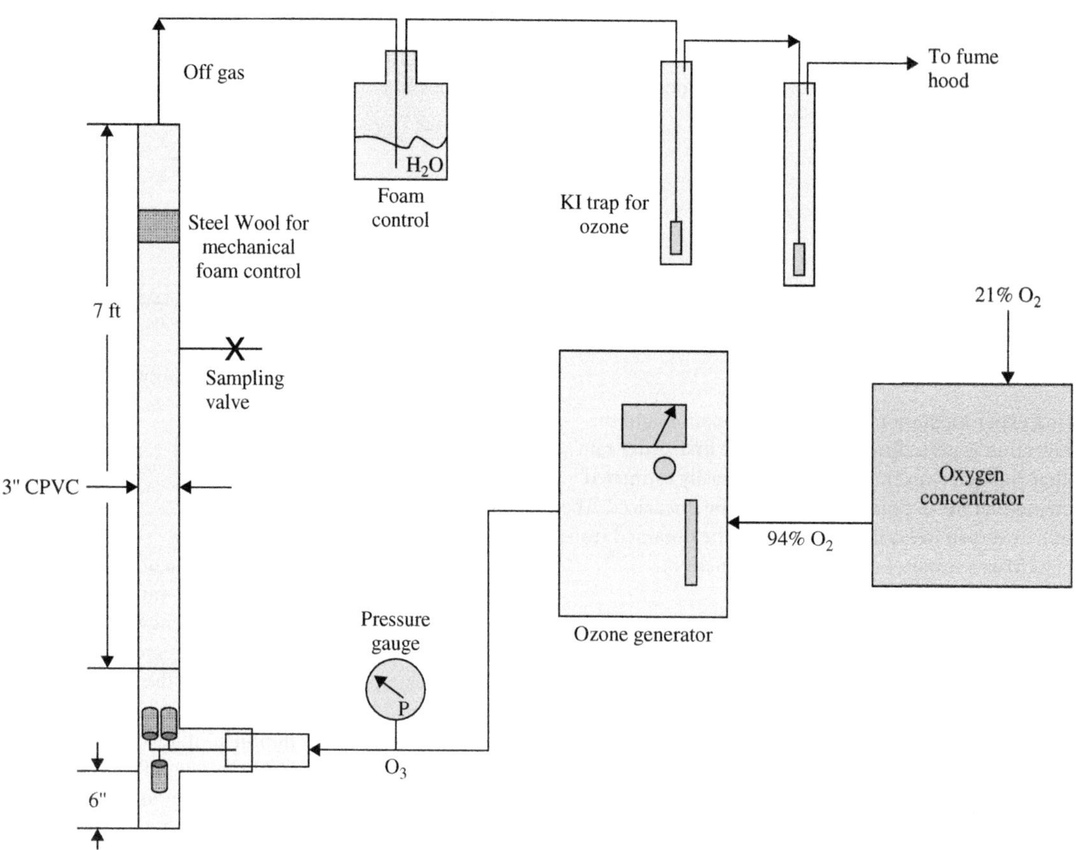

Figure 1.13 Schematic of experimental apparatus used to collect data for Example 1.6.

Example 1.6 Linear Manipulation of Non-linear Models

A semi-batch bubble column was used to test the affect of ozone on a synthetically prepared wastewater that contained acid yellow 17 dye. The carrier gas flow rate was maintained at a constant flow, resulting in uniform ozone introduction to the system. The gas feed was sparged through three stainless steel filters that entered the side of the reactor near the bottom of the column. Figure 1.13 provides a schematic of the laboratory test equipment. Color concentration was measured spectrophotometrically using a Spectronic 20D+ at a wavelength of 400 nm.

Experimental results showing the removal of color from the wastewater are shown in the table below.

Experimental Results

Ozonation time, min	Dye concentration, mg/L
0.5	247.2
2	200.3
5	90.0
10	24.5
15	12.2
20	6.7
30	0.5
40	0.1

Using the data provided and the equation shown below (assume the data are well modeled using first order kinetics), find the first-order rate constant, *k*, which characterizes the decolorization of the wastewater, and estimate the initial dye concentration.

$$C_t = C_0 e^{-kt}$$

where:
C_t = is the dye concentration at some time, t
C_0 = represents the initial dye concentration
k = is the first-order rate constant.

Solution

Algebraically manipulate the first-order rate equation into a linear format by taking the natural log of both sides.

$$C_t = C_0 e^{-kt}$$

$$\ln(C_t) = \ln(C_0) - kt$$

$$y = b + (-mx)$$

Use a spreadsheet to quickly derive this calculation table.

Ozonation time, min	Dye concentration (C_t), mg/L	ln (C_t)
0.5	247.2	5.51
2	200.3	5.30
5	90.0	4.50
10	24.5	3.20
15	12.2	2.50
20	6.7	1.90
30	0.5	−0.69
40	0.1	−2.30

Plot time versus the natural log of the dye concentration, as shown in Figure 1.14. Use linear regression analysis to determine the slope and y-intercept of the line that best fits the data. The resulting regression line is $y = -0.20\,x + 5.54$, with an R^2 value of 0.994.

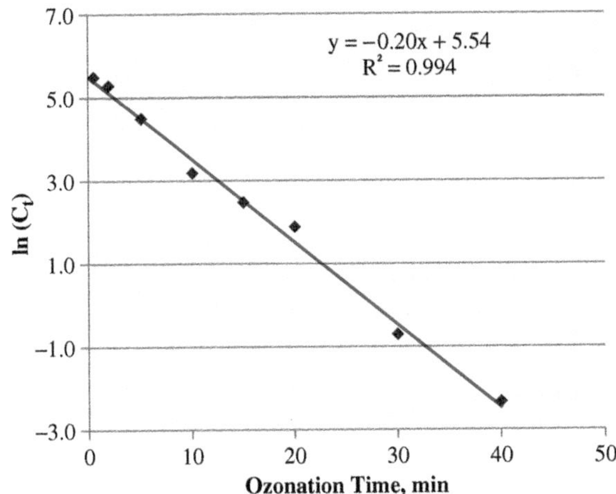

Figure 1.14 Manipulate the first-order rate equation and use experimental results to determine the rate coefficient, k.

Interpreting these results gives:

$$\ln(C_t) = -0.20t + 5.54$$

Algebraically manipulate this result back into the original format shown for first order kinetics.

$$\ln(C_t) = -0.20\,t + 5.54$$
$$e^{\ln(C_t)} = e^{-0.20t} + e^{5.54}$$
$$C_t = 255\,e^{-0.20\,t}$$

From this, we know that the first-order rate constant, k is:

$$k = 0.20 \text{ min}^{-1}$$

and the estimated concentration of dye at time zero is $C_0 =$ 255 mg/L.

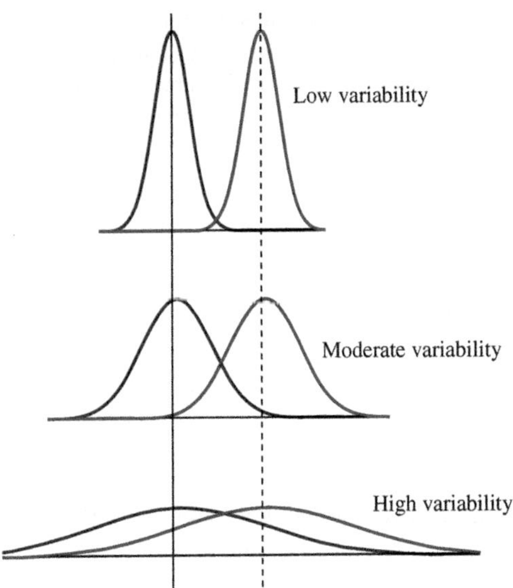

Figure 1.15 Three groups of Gaussian or bell-shaped curves are presented. The difference between the means for the grouped data is the horizontal distance between the vertical lines. In each case, the means are exactly the same. Note that for the case characterized with low variability or spread, it is easy to conclude that the two groups of means are different.

1.5.5 Student's t-test

The Student's two-tailed t-test[1] is used to compare the means of two sample groups by determining if the two groups are statistically different from each other. The t-test assumes that the measurements within the two groups are normally distributed and that the variances in the two groups are equal.

Figure 1.15 shows three very different scenarios, but the difference between the means is the same for all situations depicted. The top example shows two sets of data characterized by low variability. The second set of data indicates moderate variability within each group, while the bottom case is an example of high variability.

From visual inspection of the overlapping regions within the three examples, it is reasonable to conclude that average scores of the low-variability case appear to be most distinct. In this case, minimal overlap occurs between the data. In the high variability case, however; the difference between the means appears less evident because the two normal-distribution curves exhibit appreciable overlap. With this physical representation in mind, let us explore how the t-test statistic is used to compare the means of two samples.

To conduct the t-test analysis, and most other statistical tests of this type, do the following four steps:

1 **Establish the hypotheses.** Two hypotheses are typically identified; the null (H_0) and alternative (H_1) hypotheses.

[1] "Student" was a pen name used by William Gosset in the early 1900s. During this time, Gosset was employed by the Guinness brewery; they did not allow their scientists to publish proprietary information and did not want their competitors to know that they were using statistics to facilitate quality control in the brewing of beer. In 1908, Student (William Sealy Gosset) published The Probable Error of a Mean in *Biometrika* Vol 6, No. 1.

The null hypothesis states that the difference between the control (or accepted) and treatment samples is zero (i.e., the means of the two samples are the same). The alternative hypothesis states that the difference between the control and treatment sample means is not zero.

2 **Set the significance or α level.** The alpha level represents the significance of the results. For example, for an alpha level of 0.05, the probability that the results are due to chance is only five in a hundred. The lower the significance level, the stronger the evidence.

3 **Calculate the t-statistic.** The knowledge required to calculate the t-statistic includes four items:
   a. The average of the control sample or other known value (expected value)
   b. The average of the treatment sample, $\bar{x}$, (Equation (1.5))
   c. The standard deviation of the averages, s_x (Equation (1.7))
   d. The number of observations, n.

4 **Compare the calculated t value, t-statistic to the tabulated t value, t-critical.** Find t-critical (t_{crit}) by using Table 1.9, which provides tabulated t-values as a function of significance level (α) and the number of degrees of freedom ($df = n-1$). The α value measures the probability of exceeding t_{crit}. Compare t-statistic (t_{calc}) to the appropriate t_{crit}. If $t_{calc} > t_{crit}$, reject the null hypothesis in favor of the alternative hypothesis. If $t_{crit} > t_{calc}$, we fail to reject the null hypothesis and are forced to accept that the means of the two samples are indistinguishable. Referring to Table 1.9 for $\alpha = 0.05$ (a 95% confidence limit), a one-sided test and $df = 6$, we find $t_{crit} = 1.94$.

The t-test can be used to compare a sample mean to an accepted value such as a population mean or a legislated value produced by a governing body such as the EPA; or it can be used to compare the means of two sample sets. When comparing one sample with a population mean or a known value, Equation (1.25) is used. When comparing two samples for significance of difference, Equation (1.26) is used. Notice that in both formulae, the numerator is the difference between the means and the denominator is the standard error of the difference in the means.

Recall from Equation (1.8), that the standard error is a function of the standard deviation of the data. In Equations (1.25) and (1.26), the t-test statistic, t_{calc}, increases as the difference between the means increases. As the sample standard deviation decreases or the sample size increases, the standard error of the difference of the means (the denominator) gets smaller. As a result, t_{calc} gets larger as the means get farther apart, the variances get smaller, and/or the sample sizes increase.

When comparing one sample with a sample mean for significance of difference:

$$t_{calc} = \frac{\bar{x}_t - \bar{x}_c}{s_m} = \frac{\bar{x}_t - \bar{x}_c}{s_t/\sqrt{n_t}} \quad (1.25)$$

When comparing two samples for significance of difference:

$$t_{calc} = \frac{|\bar{x}_t - \bar{x}_c|}{s_{(\bar{x}_t - \bar{x}_c)}} = \frac{|\bar{x}_t - \bar{x}_c|}{\frac{s_t}{\sqrt{n_t}} + \frac{s_c}{\sqrt{n_c}}} \quad (1.26)$$

Table 1.9 Distribution of Critical t values, t_c, shown with significance level α and degrees of freedom, df.

	Significance level (α)			
One-sided	0.05	0.025	0.005	0.0005
Two-sided	0.1	0.05	0.01	0.001
Degrees of freedom, df				
1	6.31	12.71	63.66	636.62
2	2.92	4.3	9.93	31.6
3	2.35	3.18	5.84	12.92
4	2.13	2.78	4.6	8.61
5	2.02	2.57	4.03	6.87
6	1.94	2.45	3.71	5.96
7	1.89	2.37	3.5	5.41
8	1.86	2.31	3.36	5.04
9	1.83	2.26	3.25	4.78
10	1.81	2.23	3.17	4.59
11	1.8	2.2	3.11	4.44
12	1.78	2.18	3.06	4.32
13	1.77	2.16	3.01	4.22
14	1.76	2.14	2.98	4.14
15	1.75	2.13	2.95	4.07
16	1.75	2.12	2.92	4.02
17	1.74	2.11	2.9	3.97
18	1.73	2.1	2.88	3.92
19	1.73	2.09	2.86	3.88
20	1.72	2.09	2.85	3.85
21	1.72	2.08	2.83	3.82
22	1.72	2.07	2.82	3.79
23	1.71	2.07	2.82	3.77
24	1.71	2.06	2.8	3.75
25	1.71	2.06	2.79	3.73
26	1.71	2.06	2.78	3.71
27	1.7	2.05	2.77	3.69
28	1.7	2.05	2.76	3.67
29	1.7	2.05	2.76	3.66
30	1.7	2.04	2.75	3.65
40	1.68	2.02	2.7	3.55

Table 1.9 *(continued)*

	Significance level (α)			
60	1.67	2	2.66	3.46
120	1.66	1.98	2.62	3.37
infinity	1.65	1.96	2.58	3.29

Source: The National Institute of Standards and Technology, US Department of Commerce.

where:
x_t = mean for the treatment data
x_c = mean for the control group or an accepted value
s_t = standard deviation of the treatment sample
s_c = standard deviation of the control group
n_t = number of data points in the treatment group, and
n_t = number of data point in the control group.

Example 1.7 One-sided t-test

Monthly lead concentrations measured in a lake are presented in the table below. Use the one-sample *t*-test to compare the average lead concentration in the lake with the EPA guideline for lead in drinking water. For drinking water, the EPA guidelines state that lead concentrations should be ≤ 15 μg/L (www.epa.gov).

Monthly averages for lake lead concentrations, μg/L
17 12 18 19 20 15 8 23 19 22 17 15 11 12 15 22 20

Solution

Step 1. Begin the solution by stating the null hypothesis (H_0) and the alternative hypothesis (H_1). The null hypothesis states that there is no difference between the means. The alternative hypothesis is adopted if the null hypothesis is shown to be untrue. A one-sided hypothesis, as shown below, claims that a parameter is either larger or smaller than the value given by the null hypothesis.

Null hypothesis	H_0: Lake lead concentration = EPA guideline
	H_0: Lake lead concentration = 15 μg/L
Alternative hypothesis	H_1: Lake lead concentration > EPA guideline
	H_1: Lake lead concentration > 15 μg/L

Step 2. Set the significance or α level. For this example, we assume an alpha level of 0.05. $\alpha = 0.05$.

Step 3. Calculate the appropriate value of *t*. For this case, use the one-tailed hypothesis. Use Equation (1.25) to determine t_{calc}.

$$t_{calc} = \frac{\bar{x}_t - \bar{x}_c}{s_m} = \frac{\bar{x}_t - \bar{x}_c}{s_t/\sqrt{n_t}}$$

Use Equation (1.5) to find the mean lead concentration.

$$\bar{x}_t = \frac{1}{n}\sum_i^n x_i = 16.8\frac{\mu g}{L}$$

Prepare a calculation table to assist in the computation of the standard deviation. Referring to Equation (1.6), the following table was made in an *Excel* spreadsheet to assist with data manipulation.

Lead concentration, μg/L	$(x_i - \bar{x})^2$
17	0.06
12	22.70
18	1.53
19	5.00
20	10.47
15	3.11
8	76.82
23	38.88
19	5.00
22	27.41
17	0.06
15	3.11
11	33.23
12	22.70
15	3.11
22	27.41
20	10.47
$\bar{x}_t = 16.8\frac{\mu g}{L}$	$\sum (x_i - \bar{x})^2 = 291.1$

Find the standard deviation:

$$s_x = \sqrt{\frac{\sum_i^n (x_i - \bar{x})^2}{n-1}} = \sqrt{\frac{291.1}{17-1}} = 4.3\frac{\mu g}{L}$$

The standard error of the mean is found by using Equation (1.7).

$$s_m = \frac{s_x}{\sqrt{n_t}} = \frac{4.3\,\mu g/L}{\sqrt{17}} = 1.04\frac{\mu g}{L}$$

Calculate *t*-statistic:

$$t_{calc} = \frac{\bar{x}_t - \bar{x}_c}{s_m} = \frac{\bar{x}_t - \bar{x}_c}{s_t/\sqrt{n_t}} = \frac{16.8 - 15}{1.04} = 1.73$$

Step 4. Evaluate t_{crit}.

Use Table 1.9 to determine the t_{crit} value for the specified α. Select the column with the desired probability. For this example, $\alpha = 0.05$ for a one-sided *t*-test. Select the appropriate row for degrees of freedom, *df*. For a one-sided *t*-test, the degree of freedom is $df = n - 1$. For this example, $df = n - 1 = 17 - 1 = 16$.

Step 5. Compare the value found in the table (t_{crit})with the value calculated (t_{calc}) to reject or fail to reject the null hypothesis. If $t_{calc} > t_{crit}$, then the null hypothesis is rejected and the alternate hypothesis is accepted. From Table 1.9, a level of significance of 0.05 and a one-sided t-test with 16 degrees of freedom gives $\alpha_{0.05,\,16} = 1.75$. For our example, $t_{calc} = 1.73 < t_{crit} = 1.75$, so we fail to reject the null hypothesis and we conclude that our sample mean is not statistically larger than the accepted limit of 15 µg/L.

Example 1.8 Two-sided t-test

A pulp and paper mill wastewater treatment plant (WWTP) consists of primary settling for solids removal, anaerobic treatment, followed by an aerated basin and a settling pond. Wastewater from the paper mill is characterized by low volume (2 MGD) and high organic loading. During typical operation of the WWTP, the anaerobic system removes approximately 70% of the COD load, with the balance being treated by the aeration basin.

During operation, the paper mill inadvertently discharged toxic materials to the WWTP, which killed the anaerobic system. While the anaerobic system was off-line, an on-site biofermentation process was used to increase and maintain sufficient microbes in the aerobic basin to prevent the WWTP COD effluent from increasing, and to maintain NPDES compliance (http://www.waste-water .com). Refer to Figure 1.16 for WWTP process diagrams that describe the flow before and after failure of the anaerobic system.

Two sets of effluent COD data from the WWTP are provided below. The first column provides COD concentrations from the plant during proper operation of the anaerobic treatment system. This scenario is called the "control." The second column provides effluent COD data for the period of time after the anaerobic system had been killed and taken off-line. During this time period, the biofermentation process was used to supplement the aerobic treatment, and these data are referred to as the "treatment." Compare the means of the two sample groups and determine if the effluent COD concentrations are statistically different from each other.

Control effluent COD, mg/L	Treatment effluent COD, mg/L
354	404
307	406
255	345
266	394
356	399
380	343
401	368
319	307
334	335
407	441
315	557
354	542
335	465
289	470
834	
638	
440	
569	

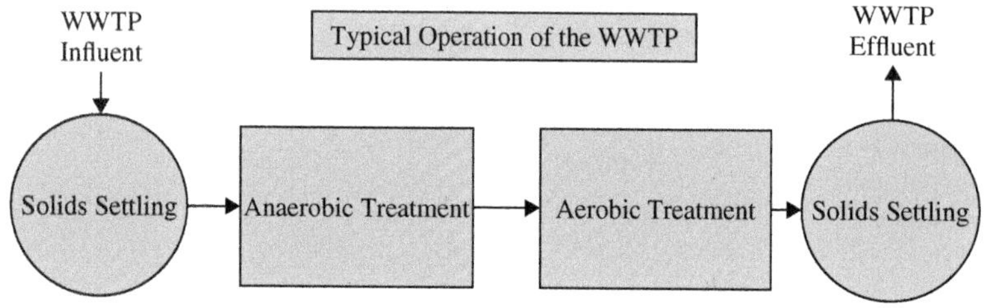

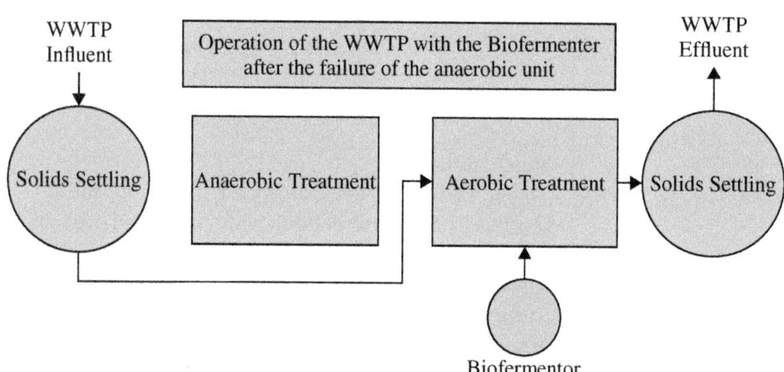

Figure 1.16 Schematic for Example 1.8.

Solution

Step 1. Begin the solution by stating the null hypothesis (H_0) and the alternative hypothesis (H_1). The null hypothesis states that there is no difference between the means. The alternative hypothesis is adopted if the null hypothesis is shown to be untrue. When the two-tailed t-statistic is used, the alternate hypothesis is simply the opposite of the null hypothesis.

Null hypothesis: H_0: Effluent COD_c = Effluent COD_t
Alternative hypothesis: H_1: Effluent $COD_c \neq$ Effluent COD_t

Step 2. Set the significance level, $\alpha = 0.5$.

Step 3. Calculate the appropriate value of t_{calc}. For this case, use the two-tailed hypothesis and Equation (1.26) to determine the t value. An *Excel* spreadsheet table is used to assist in the calculations.

	Control effluent COD, mg/L	Treatment effluent COD, mg/L	*Excel* function
	354	404	
	307	406	
	255	345	
	266	394	
	356	399	
	380	343	
	401	368	
	319	307	
	334	335	
	407	441	
	315	557	
	354	542	
	335	465	
	289	470	
	834		
	638		
	440		
	569		
Sample number, n	$n_c = 18$	$n_t = 14$	COUNT
Sample average, $\bar{x}$	$\bar{x}_c = 397.4$	$\bar{x}_t = 412.6$	AVERAGE
Standard deviation, s	$s_c = 146.3$	$s_t = 75.3$	STDEV
t_{crit}	$t_{crit} = 2.04$		TINV

Use the *Excel* COUNT function to determine the sample number, n, for both the treatment and control data. To enter a formula in *Excel*, begin by using the "=" sign in the desired cell. After the = sign, any formula or function can be entered. When using the COUNT function, *Excel* expects a range of cells to be entered. It is easy to insert a column by highlighting the desired cells. It is also easy to highlight a single cell for use in an equation. Or, a single cell or a range of cells may be used, by calling a single cell, (e.g., B4) or a range such as C4:C30.

Use the *Excel* function AVERAGE to calculate both sample means. The STDEV function was used to determine the sample standard deviations of both data sets. Determine t-statistic using Equation (1.26).

$$t_{calc} = \frac{|\bar{x}_t - \bar{x}_c|}{s_{(\bar{x}_t - \bar{x}_c)}} = \frac{|\bar{x}_t - \bar{x}_c|}{\frac{s_t}{\sqrt{n_t}} + \frac{s_c}{\sqrt{n_c}}} = \frac{|412.6 - 397.4|}{\frac{75.3}{\sqrt{14}} + \frac{146.3}{\sqrt{18}}}$$
$$= 0.278$$

Use Table 1.9 or the Excel function TINV to find t-critical for n–2 (df = 14 + 18 – 2 = 30) degrees of freedom. Since $t_{calc} = 0.278 < t_{crit} = 2.04$, then we fail to reject the null hypothesis. We conclude that when the anaerobic system went off-line and treatment was supplemented using the biofermentor process, the WWTP effluent COD did not significantly change and the effluent COD_c = effluent COD_t.

1.5.6 One-way analysis of variance (ANOVA)

The One-way Analysis of Variance (ANOVA) is a commonly used statistical method employed to detect differences of three or more means of grouped data. Like the Student t-test, the ANOVA method is used to test the equality of the means rather than to predict new values of a dependent variable (regression analysis is often used for this type of analysis). There are various kinds of ANOVA that match with different experimental designs, but only the one-way ANOVA is described here. Although the ANOVA equations are not mathematically challenging, the calculations should be routinely done using spreadsheet or statistical software. The procedure for conducting an ANOVA is presented below, followed by an example problem.

Consider that the groundwater at a nearby Department of Defense facility is contaminated with the chlorinated solvent, trichloroethylene (TCE). Your boss has requested that you determine whether the groundwater aqueous-phase TCE concentration is the same at a variety of different sampling wells. Assume that there are S sampling wells and that each well has D_i data points or TCE concentrations each. Allow d_{ij} to represent the sample for the ith well and the jth data point. Prior to beginning the statistical analysis, ensure that three assumptions for the one-way ANOVA are met:

1 The sample populations are normally distributed,
2 The samples are independent, and
3 The variances of the populations are equal.

Step 1. State the null hypothesis, H_o, which indicates that all population means are equal.

Step 2. Determine the grand mean of all the samples or concentrations, $\overline{X}_{GM}$. The grand mean is calculated by summing all the data values and dividing by the total sample size.

$$\overline{X}_{GM} = \frac{\sum_{i=1}^{S}\sum_{j=1}^{D_i} d_{ij}}{N} \quad (1.27)$$

where:
D_i = is the number of data points for the ith well
S = is the number of treatments, (wells, in this case)
N = is the total number of observations.

$$N = \sum_{i=1}^{S} D_i \quad (1.28)$$

Step 3. Now find the sum of the squares between treatment means and the grand mean.

$$SS_b = SS_{wells} = \sum_{i=1}^{S} D_i(\overline{D}_i - \overline{X}_{GM})^2 \quad (1.29)$$

where $\overline{D}_i$ is the mean of the ith group:

$$\overline{D}_i = \frac{1}{D_i}\sum_{j=1}^{D_i} D_{ij}$$

The corresponding degrees of freedom $df_b = S - 1$.
Step 4. Compute the total sum of the squares, SS_t.

$$SS_t = \sum_{i=1}^{S}\sum_{j=1}^{D_i}(d_{ij} - \overline{X}_{GM})^2 \quad (1.30)$$

with corresponding degrees of freedom: $df_t = N - 1$.
Step 5. Determine the sum of squares for within each treatment, SS_w.

$$SS_w = \sum_{i=1}^{S}\sum_{j=1}^{D_i}(d_{ij} - \overline{D}_i)^2 \quad (1.31)$$

with corresponding degrees of freedom: $df_w = N - S$.

The SS_w is a measure of the variability within the treatments. By definition, the total degrees of freedom df_t and the total sum of squares SS_t is:

$$df_t = df_b + df_w$$

$$SS_t = SS_b + SS_w$$

Step 6. Test the null hypothesis of equal means between all treatments by determining the F, or Fisher, statistic.

$$F = \frac{MSS_b}{MSS_w} \quad (1.32)$$

In this equation, MSS_b is the between treatments mean square variance, and MSS_w is the within treatment mean square variance, such that:

$$MSS_b = \frac{SS_b}{df_b}, \text{ and } MSS_w = \frac{SS_w}{df_w}.$$

When $F \gg 1$, this suggests that there are differences between class (well) means, which provides us with sufficient statistical evidence to reject the null hypothesis, H_o. To test the hypothesis of equal means statistically, compare the calculated F value to the tabulated F_{crit} with $S-1$ and $N-S$ degrees of freedom. If $F > F_{crit}$, reject the hypothesis of equal means. However, if $F_{crit} > F$, then conclude that there is no significant difference between mean concentrations at the S wells. It is customary to summarize ANOVA data in tabular form, as shown in Table 1.10.

Example 1.9 One-way ANOVA

Dissolved oxygen (DO) was measured in lake water using three common techniques: the Azide-Winkler titration method, a DO probe and meter, and with a field kit. Use one-way analysis of variance to determine if there is sufficient evidence that at least one technique provides significantly different concentrations from at least one other method. Each method was used ten times and data from all three analysis techniques are provided in Table 1.11.

Table 1.10 Typical ANOVA table summary.

Variation Source	Sum of squares, SS	Degrees of freedom, df	Mean squares, MSS	F
Between treatments	$SS_b = \sum_{i=1}^{S} D_i(\overline{D}_i - \overline{X}_{GM})^2$	$df_b = S - 1$	$MSS_b = \frac{SS_b}{df_b}$	$F = \frac{MSS_b}{MSS_w}$
Error (within treatments)	$SS_w = \sum_{i=1}^{S}\sum_{j=1}^{D_i}(d_{ij} - \overline{D}_i)^2$	$df_w = N - S$	$MSS_w = \frac{SS_w}{df_w}$	
Total	$SS_t = \sum_{i=1}^{S}\sum_{j=1}^{D_i}(d_{ij} - \overline{X}_{GM})^2$	$df_t = N - 1$		

Table 1.11 Dissolved oxygen values of lake surface water measured using three common techniques (measurements were taken during the month of August).

Dissolved oxygen concentration (mg/L)		
Azide-Winkler Method	**DO probe and meter**	**Field kit**
7.9	7.9	6.7
7.8	7.8	7.5
7.9	7.8	8.0
7.7	7.7	7.8
7.9	7.8	8.2
6.6	7.8	7.2
7.9	7.8	7.1
7.8	7.6	6.7
7.8	7.3	6.9
7.6	7.6	6.5

Solution

Step 1. Begin by stating the null and alternative hypotheses.

The null hypothesis can be stated as:

H_0: There is no significant difference among the dissolved oxygen concentrations obtained by these three analytical techniques.

If desired, an alternative hypothesis can be stated:

H_1: At least one of these measurement techniques provides results significantly different from at least one other.

Step 2. Use Equation (1.27) to determine the grand mean of all the samples. Use the following table to assist in this calculation.

Azide-Winkler method	DO probe and meter	Field kit
7.9	7.9	6.7
7.8	7.8	7.5
7.9	7.8	8.0
7.7	7.7	7.8
7.9	7.8	8.2
6.6	7.8	7.2
7.9	7.8	7.1
7.8	7.6	6.7
7.8	7.3	6.9
7.6	7.6	6.5
Sum = 76.9	Sum = 77.1	Sum = 72.6
$D_i = 10$	$D_i = 10$	$D_i = 10$
$\overline{D}_i = 7.69$	$\overline{D}_i = 7.71$	$\overline{D}_i = 7.26$

$$\overline{X}_{GM} = \frac{\sum_{i=1}^{S}\sum_{j=1}^{D_i} d_{ij}}{N} = \frac{(76.9 + 77.1 + 72.6)}{(10 + 10 + 10)} = 7.55$$

Step 3. Find the sum of the squares between treatments and the grand mean. Use Equation (1.29).

$$SS_b = \sum_{i=1}^{S} D_i(\overline{D}_i - \overline{X}_{GM})^2 = 10(7.69 - 7.55)^2$$
$$+ 10(7.71 - 7.55)^2 + 10(7.26 - 7.55)^2$$
$$SS_b = 1.293$$

The degrees of freedom between the treatments is $df_b = S - 1 = 3 - 1 = 2$.

Step 4. Find the total sum of the squares. Use Equation (1.30).

$$SS_t = \sum_{i=1}^{S}\sum_{j=1}^{D_i}(d_{ij} - \overline{X}_{GM})^2 = (7.9 - 7.55)^2 + (7.8 - 7.55)^2$$
$$+ (7.9 - 7.55)^2 + \ldots + (6.5 - 7.55)^2$$
$$SS_t = 6.115$$

The total degrees of freedom is $df_t = N - 1 = 30 - 1 = 29$.

Step 5. Find the within treatment sum of squares. Knowing the relationship $SS_t = SS_b + SS_w$, the SS_w can be found by:

$$SS_w = SS_t - SS_b = 6.115 - 1.293 = 4.822$$

The degrees of freedom with the classes is $df_w = N - S = 30 - 3 = 27$.

Step 6. Test the null hypothesis by determining the F statistic.

$$F = \frac{MSS_b}{MSS_w} = \frac{SS_b/df_b}{SS_w/df_w} = \frac{1.293/2}{4.882/27} = 3.58$$

Compare to the tabulated F_{crit} with $S - 1 = 2$ and $N - S = 27$ degrees of freedom. Referring to Table 1.12, $F_{crit} = 3.354$.

Since the value calculated for the F statistic exceeds $F_{crit} = 3.354$, the null hypothesis can be rejected at the 0.05 level of significance. From this, we can conclude that at least one of the three dissolved oxygen measurement techniques differs from the others.

The one-way ANOVA results are summarized in tabular form as shown below:

Variation Source	Sum of squares, SS	Degrees of freedom, *df*	Mean squares, MSS	F
Between treatments	1.293	2	0.647	F = 3.58
Error (within treatments)	4.822	27	0.181	
Total	6.115	29		

ANOVA analysis is relatively simple to conduct in Microsoft *Excel*. Begin by entering the data, such that each group is in a separate column. Label each column appropriately, as shown below.

Click on Data > Data analysis > Anova: single factor. Input the desired range of data, including the labels, by highlighting within the spreadsheet. If the data analysis option is not available in the Data tab, use the *Excel* option feature to add-in the Analysis Toolpak.

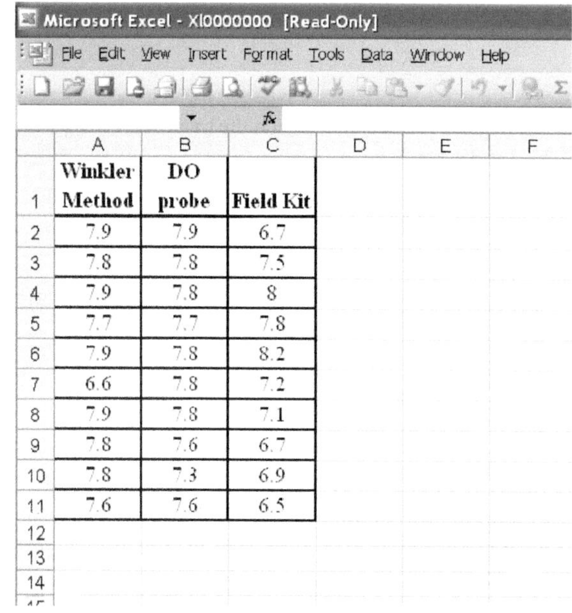

	A	B	C
1	Winkler Method	DO probe	Field Kit
2	7.9	7.9	6.7
3	7.8	7.8	7.5
4	7.9	7.8	8
5	7.7	7.7	7.8
6	7.9	7.8	8.2
7	6.6	7.8	7.2
8	7.9	7.8	7.1
9	7.8	7.6	6.7
10	7.8	7.3	6.9
11	7.6	7.6	6.5

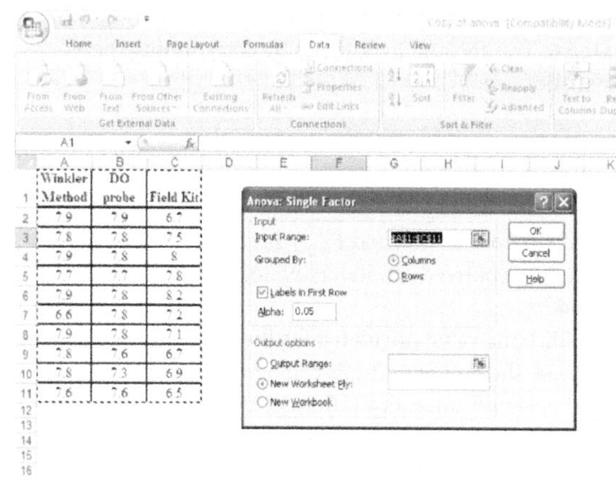

Click OK after selecting the data. Excel provides the following tabulated results.

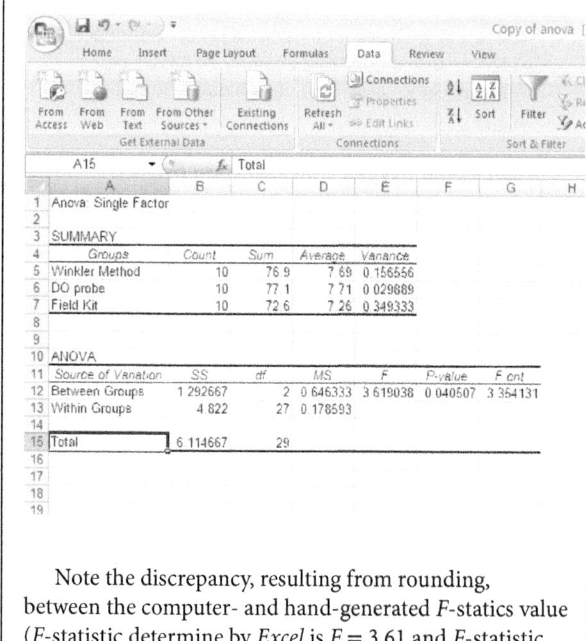

Anova: Single Factor

SUMMARY

Groups	Count	Sum	Average	Variance
Winkler Method	10	76.9	7.69	0.156556
DO probe	10	77.1	7.71	0.029889
Field Kit	10	72.6	7.26	0.349333

ANOVA

Source of Variation	SS	df	MS	F	P-value	F crit
Between Groups	1.292667	2	0.646333	3.619038	0.040507	3.354131
Within Groups	4.822	27	0.178593			
Total	6.114667	29				

Note the discrepancy, resulting from rounding, between the computer- and hand-generated *F*-statics value (*F*-statistic determine by *Excel* is $F = 3.61$ and *F*-statistic from hand calculations was 3.58).

Summary

- The study of environmental engineering aptly began with the perspective of how man has historically interacted with the environment.
- Modern day efforts to protect the environment via legislation were introduced.
- The Environmental Protection Agency was founded in 1970 and, since that time, numerous regulations have been promulgated to protect human health and the environment.
- Methodically solving engineering problems is important. A six-step method was introduced in this chapter that provides a logical scheme for the user. A summary of the six-step scheme follows:
  - **Step 1.** Problem identification. Clearly define the problem.
  - **Step 2.** Collect all relevant data and make an appropriate sketch.
  - **Step 3.** Select applicable theory and document all relevant equations.
  - **Step 4.** Make appropriate simplifications and assumptions.
  - **Step 5.** Solve the problem.
  - **Step 6.** Verify solution and clearly communicate results.
- The importance of understanding data collection, followed by proper analysis and clear communication of results was discussed.

Table 1.12 Critical values of the F distribution for df_b degrees of freedom and df_w degrees of freedom with a significance level of 5%.

$df_w \backslash df_b$	1	2	3	4	5	6	7	8	9	10	12	15	20	24	30	40	60	120	inf
1	161.4	199.5	215.7	224.6	230.2	234.0	236.8	238.9	240.5	241.9	243.9	245.9	248.0	249.1	250.1	251.1	252.2	253.3	254.3
2	18.51	19.00	19.16	19.25	19.30	19.33	19.35	19.37	19.38	19.40	19.41	19.43	19.45	19.45	19.46	19.47	19.48	19.49	19.50
3	10.13	9.55	9.28	9.12	9.01	8.94	8.89	8.85	8.81	8.79	8.74	8.70	8.66	8.64	8.62	8.59	8.57	8.55	8.53
4	7.71	6.94	6.59	6.39	6.26	6.16	6.09	6.04	6.00	5.96	5.91	5.86	5.80	5.77	5.75	5.72	5.69	5.66	5.63
5	6.61	5.79	5.41	5.19	5.05	4.95	4.88	4.82	4.77	4.74	4.68	4.62	4.56	4.53	4.50	4.46	4.43	4.40	4.36
6	5.99	5.14	4.76	4.53	4.39	4.28	4.21	4.15	4.10	4.06	4.00	3.94	3.87	3.84	3.81	3.77	3.74	3.70	3.67
7	5.59	4.74	4.35	4.12	3.97	3.87	3.79	3.73	3.68	3.64	3.57	3.51	3.44	3.41	3.38	3.34	3.30	3.27	3.23
8	5.32	4.46	4.07	3.84	3.69	3.58	3.50	3.44	3.39	3.35	3.28	3.22	3.15	3.12	3.08	3.04	3.01	2.97	2.93
9	5.12	4.26	3.86	3.63	3.48	3.37	3.29	3.23	3.18	3.14	3.07	3.01	2.94	2.90	2.86	2.83	2.79	2.75	2.71
10	4.96	4.10	3.71	3.48	3.33	3.22	3.14	3.07	3.02	2.98	2.91	2.85	2.77	2.74	2.70	2.66	2.62	2.58	2.54
11	4.84	3.98	3.59	3.36	3.20	3.09	3.01	2.95	2.90	2.85	2.79	2.72	2.65	2.61	2.57	2.53	2.49	2.45	2.40
12	4.75	3.89	3.49	3.26	3.11	3.00	2.91	2.85	2.80	2.75	2.69	2.62	2.54	2.51	2.47	2.43	2.38	2.34	2.30
13	4.67	3.81	3.41	3.18	3.03	2.92	2.83	2.77	2.71	2.67	2.60	2.53	2.46	2.42	2.38	2.34	2.30	2.25	2.21
14	4.60	3.74	3.34	3.11	2.96	2.85	2.76	2.70	2.65	2.60	2.53	2.46	2.39	2.35	2.31	2.27	2.22	2.18	2.13
15	4.54	3.68	3.29	3.06	2.90	2.79	2.71	2.64	2.59	2.54	2.48	2.40	2.33	2.29	2.25	2.20	2.16	2.11	2.07
16	4.49	3.63	3.24	3.01	2.85	2.74	2.66	2.59	2.54	2.49	2.42	2.35	2.28	2.24	2.19	2.15	2.11	2.06	2.01
17	4.45	3.59	3.20	2.96	2.81	2.70	2.61	2.55	2.49	2.45	2.38	2.31	2.23	2.19	2.15	2.10	2.06	2.01	1.96
18	4.41	3.55	3.16	2.93	2.77	2.66	2.58	2.51	2.46	2.41	2.34	2.27	2.19	2.15	2.11	2.06	2.02	1.97	1.92
19	4.38	3.52	3.13	2.90	2.74	2.63	2.54	2.48	2.42	2.38	2.31	2.23	2.16	2.11	2.07	2.03	1.98	1.93	1.88
20	4.35	3.49	3.10	2.87	2.71	2.60	2.51	2.45	2.39	2.35	2.28	2.20	2.12	2.08	2.04	1.99	1.95	1.90	1.84
21	4.32	3.47	3.07	2.84	2.68	2.57	2.49	2.42	2.37	2.32	2.25	2.18	2.10	2.05	2.01	1.96	1.92	1.87	1.81
22	4.30	3.44	3.05	2.82	2.66	2.55	2.46	2.40	2.34	2.30	2.23	2.15	2.07	2.03	1.98	1.94	1.89	1.84	1.78
23	4.28	3.42	3.03	2.80	2.64	2.53	2.44	2.37	2.32	2.27	2.20	2.13	2.05	2.01	1.96	1.91	1.86	1.81	1.76
24	4.26	3.40	3.01	2.78	2.62	2.51	2.42	2.36	2.30	2.25	2.18	2.11	2.03	1.98	1.94	1.89	1.84	1.79	1.73
25	4.24	3.39	2.99	2.76	2.60	2.49	2.40	2.34	2.28	2.24	2.16	2.09	2.01	1.96	1.92	1.87	1.82	1.77	1.71
26	4.23	3.37	2.98	2.74	2.59	2.47	2.39	2.32	2.27	2.22	2.15	2.07	1.99	1.95	1.90	1.85	1.80	1.75	1.69
27	4.21	3.35	2.96	2.73	2.57	2.46	2.37	2.31	2.25	2.20	2.13	2.06	1.97	1.93	1.88	1.84	1.79	1.73	1.67
28	4.20	3.34	2.95	2.71	2.56	2.45	2.36	2.29	2.24	2.19	2.12	2.04	1.96	1.91	1.87	1.82	1.77	1.71	1.65
29	4.18	3.33	2.93	2.70	2.55	2.43	2.35	2.28	2.22	2.18	2.10	2.03	1.94	1.90	1.85	1.81	1.75	1.70	1.64
30	4.17	3.32	2.92	2.69	2.53	2.42	2.33	2.27	2.21	2.16	2.09	2.01	1.93	1.89	1.84	1.79	1.74	1.68	1.62
40	4.08	3.23	2.84	2.61	2.45	2.34	2.25	2.18	2.12	2.08	2.00	1.92	1.84	1.79	1.74	1.69	1.64	1.58	1.51
60	4.00	3.15	2.76	2.53	2.37	2.25	2.17	2.10	2.04	1.99	1.92	1.84	1.75	1.70	1.65	1.59	1.53	1.47	1.39
120	3.92	3.07	2.68	2.45	2.29	2.17	2.09	2.02	1.96	1.91	1.83	1.75	1.66	1.61	1.55	1.50	1.43	1.35	1.25
inf	3.84	3.00	2.60	2.37	2.21	2.10	2.01	1.94	1.88	1.83	1.75	1.67	1.57	1.52	1.46	1.39	1.32	1.22	1.00

Source: The National Institute of Standards and Technology, US Department of Commerce http://www.itl.nist.gov/div898/handbook/; LWL).

- When considering experimental data, there are two types of errors – *random* and *systematic.*
  - Random error is the scatter observed from repeated measurements and is, by definition, unpredictable and bidirectional, such that it changes from one measurement to the next.
  - Systematic errors occur to the same degree each time the measurement is made – they are unidirectional.
- When evaluating experimental data, the average or mean value of measurements of x, is often reported as $\bar{x}$.
- The measure of the spread or dispersion of the measured x_i values is often provided by the variance, s^2, or the standard deviation, s.
- Least-squares regression analysis is used to determine the best fit line for two related variables. If the relationship is linear, the equation for a linear regression line has the form $y = mx + b$. In the equation, $y =$ is the dependent variable (the variable to be predicted), $x =$ independent variable, m = the slope of the linear regression line, and $b =$ the intercept (the value of y when $x = 0$).
- The R^2 value represents the fraction of the variation in the values of y that is explained by the least squares regression of y on x; this value is a measure of how successfully the regression explains the observed response.
  - The coefficient ranges from 0 to 1. An R^2 of zero indicates that the dependent variable cannot be predicted from the independent variable while an R^2 of 1 means the dependent variable can be predicted without error from the independent variable.
- The Student's t-test is used to compare the difference between two data sets, or between one set and a "true" value. When comparing one sample with a sample mean (or "true" value) for significance of difference:

$$t_{calc} = \frac{\bar{x}_t - \bar{x}_c}{s_m} = \frac{\bar{x}_t - \bar{x}_c}{s_t/\sqrt{n_t}}$$

  - The Student's two-tailed t-test is used to compare the means of two sample groups by determining if the two groups are statistically different from each other. When comparing two samples for significance of difference:

$$t_{calc} = \frac{|\bar{x}_t - \bar{x}_c|}{s_{(\bar{x}_t - \bar{x}_c)}} = \frac{|\bar{x}_t - \bar{x}_c|}{\frac{s_t}{\sqrt{n_t}} + \frac{s_c}{\sqrt{n_c}}}$$

  - The One-way Analysis of Variance (ANOVA) is similar to a t-test but it is used to detect differences among multiple data sets simultaneously. The ANOVA uses the F-distribution instead of the t-distribution.
  - The assumption is made that all of the data sets have equal variances.

Key words

ABET
age of enlightenment
alternative hypothesis
analysis of variance
ANOVA
ASCE
average
CERCLA
CERLA
Clean Air Act
Clean Water Act
correlation coefficient
early civilization
EPCRA
exponential equations
F statistic
F-critical
FIFRA
great depression
industrial revolution
linear equations
linear regression
mean
middle ages
non-linear models
null hypothesis
one-sided t-test
OSHA
Pollution Prevention Act
post-war era
power equations
present day
progressive era
propagation of error
random errors
RCRA
relative uncertainty
SARA
six-step method
standard deviation
standard error
Student's t-test
systematic errors
TINV function
TSCA
t-statistic
two-sided t-test
uncertainty
variance
World War II

References

American Society of Civil Engineers (ASCE) (2005). *Infrastructure Report Card 2005.* Available online www.asce.org/reportcard/2005.

Anderson, W.C. (2002). *A History of Environmental Engineering in the United States, Environmental and Water Resources History.*

Andraos, J. (1996). On the Propagation of Statistical Errors for a Function of Several Variables. *J. Chem. Educ* **73**, 150–154.

Andrews, J., Jelley, N. (2007). *Principles, technologies, and impacts.* Oxford University Press, Oxford, UK.

Bergeson, L.L., Campbell, L.M., Rothenberg, L. (2000). TSCA and the Future of Chemical Regulation. *EPA Administrative Law Reporter* **15**(4). Accessed online at http://www.lawbc.com/other_pdfs/tsca.pdf, 14 January, 2009.

Mara, D, Horan, N.J. (2003). *The handbook of water and wastewater microbiology.* Academic, San Diego, CA.

Fetter, C.W. (1999) *Contaminant Hydrology*, 2nd edition, Prentice Hall, Upper Saddle River, NJ.

Hong, S., Candelong, J-P., Paterson, C.C., Boutron, C.F. (1994). Greenland Ice Evidence of Hemispheric Lead Pollution Two Millennia Ago by Greek and Roman Civilizations. *Science* **265**(5180), 1841–1843.

Massachusetts State Board of Health (MSBH) (1894). Twenty-fifth annual report of the state board of health.

Melosi, M.V. (1981) *Garbage in the Cities: Refuse, Reform and the Environment, 1880–1980.* Texas A&M University Press, College Station, Texas.

Merrill, R.L. (1962). *Hoover's Water Supply and Treatment, Ninth Edition, Bulletin 211.* National Lime Association, Washington, DC.

Monod, J. (1949). The Growth of Bacterial Cultures. *Annual Review of Microbiology* **3**, 371–394.

Neuzil, M., Kovarik, W. (1996). *Mass Media and Environmental Conflict: America's Green Crusades.* Ed. Astrid Virding, SAGE Publications, Thousand Oaks, CA.

Seneca, L.A. *Moral Epistles.* Translated by Richard M. Gummere. The Loeb Classical Library. Cambridge, Mass.: Harvard UP, 1917–25. 3 vols.: Volume III.

Stradling, D., Stradling, R. (2008). Perceptions of the Burning River: Deindustrialization and Cleveland's Cuyahoga River. *Environmental History.* http://www.historycooperative.org/journals/eh/13.3/stradling.html (14 Jan. 2009).

Symons, G.E. (2001). The Origins of Environmental Engineering: Prologue to the 20th Century. *Journal of the New England Water Works Association* **115**(4), 253–287.

Tarr, J.A. (1985). Sewerage and the Development of the Networked City of the United States, 1850-1930. In Tarr, J.A. & DuPuy, G. (eds). *Technology and the Rise of the Networked City in Europe and America*, pp. 159–185. Temple University Press, Philadelphia, PA.

Tarr, J.A. *et al.* (1980). Chapter 3, The Development and Impact of Urban Wastewater Technology: Changing Concepts of Water Quality Control 1850–1930. In Melosi, M.V. (ed), *Pollution and Reform in American Cities*, 1870–1930, University of Texas Press, Austin, TX and London, UK.

American Water Works Association (1981). *The Quest for Pure Water*, 2nd ed. Denver, CO.

US EPA (2008a). *Municipal Solid Waste in the United States: 2007 Facts and Figures.* United States Environmental Protection Agency, Office of Solid Waste, (5306P) EPA530-R-08-010, November 2008 www.epa.gov.

US EPA (2008b) *The National Biennial RCRA Hazardous Waste Report (Based On 2007 Data).* United State Environmental Protection Agency, Solid Waste and Emergency Response (5305P) EPA530-R-08-012, November 2008.

Problems

1 Determine the efficiency of energy utilization for a pump. Assume the following efficiencies in the energy conversions:

- Crude oil to fuel oil is 90%.
- Fuel to electricity is 40%.
- Electricity transmission and distribution is 90%.
- Conversion of electrical energy into mechanical energy of the fluid being pumped is 40%.

Hint: the overall efficiency for the primary energy source is the product of all the individual conversion efficiencies.

2 Replacing automobiles that have internal combustion engines with electric-powered vehicles is considered by some as the best solution to urban smog and tropospheric ozone. Write a short paper (1–2 pages double-spaced) on the likely effects of this transition on industrial production of fuels. Assume that the amount of energy required per mile traveled is roughly the same for each kind of vehicle. Consider the environmental impacts of using different kinds of fuel for the electricity generation to satisfy the demand from electric vehicles. This analysis does not need to include the loss of power over the lines/grid.

3 Visit the Office of Solid Waste and Emergency Response on the EPA website (www.epa.gov) and determine how close you live to a designated Superfund site.

4 You have been tasked with determining the volume of a rectangular box.

a. Using Equation (1.8) as a starting place, derive an equation that describes the minimum uncertainty expected in the box volume, based solely on the uncertainties in the measured dimensions.

b. Assume the box has dimensions of x, y, and z. A meter stick was used to obtain the following measurements: $x = (10 \pm 1)$ cm; $y = (20 \pm 1)$ cm; and $z = (5 \pm 1)$ cm. Find the uncertainty in the volume of the box.

c. A caliper was used to measure the dimensions of a box having dimensions of x, y, and z. The following measurements were obtained: $x = (10 \pm 0.01)$ cm; $y = (20 \pm 0.01)$ cm; and $z = (5 \pm 0.01)$ cm. Find the uncertainty in the volume of the box.

5 The Manning Equation is an empirical relationship that relates open-channel flow velocity with channel characteristics.

$$V = \frac{1}{n}\left(\frac{A}{P}\right)^{\frac{2}{3}} S^{1/2}$$

where:
V = mean velocity, m/s
n = Manning's roughness coefficient, dimensionless
S = slope of the energy grade line (m/m)
A = cross-sectional area of the channel, m^2
P = wetted perimeter of the flow channel, m.

Assume that the Manning surface roughness, the area, and the wetted perimeter are uncertain. Using the standard rules for error propagation, derive an equation of error for velocity.

6 Review the data shown in Table 1.6 that describes several chemical properties of polycyclic aromatic hydrocarbons.

a. Use the tabulated data to calculate means, standard deviations for molecular weight and solubility, correlation r, coefficient of determination R^2, slope m, intercept b, and the equation of the least-squares line for this case. Hint: consider plotting molecular weight on the abscissa and the logarithm of solubility on the ordinate.

b. Use an internet search engine to define the soil-water partition coefficient. Use the tabulated data to calculate means, standard deviations for solubility and the soil-water partition coefficient, correlation r, coefficient of determination R^2, slope m, intercept b, and the equation of the least-squares line for this case. Hint: consider plotting the logarithm of solubility on the abscissa and the logarithm of the soil-water partition coefficient on the ordinate.

c. Use the tabulated data to calculate means, standard deviations for molecular weight and the soil-water partition coefficient, correlation *r*, coefficient of determination R^2, slope *m*, intercept *b*, and the equation of the least-squares line for this case. Hint: consider plotting molecular weight on the abscissa and the logarithm of soil-water partition coefficient on the ordinate. It is recommended that a spreadsheet software such as MS *Excel* be used to graph the requested relationships and to perform all calculations.

d. Use the TINV function in *Excel* to generate a table of critical *t* values for 95% confidence level.

Chapter 2

Essential chemistry concepts

Richard O. Mines, Jr.

Learning Objectives

After reading this chapter, you should be able to:

- understand the commonly used systems of units and the difference between fundamental and derived units;
- explain the difference between dimension and unit;
- balance chemical equations and apply stoichiometry to environmental applications.
- define what is meant by oxidation and reduction, and develop half-reactions for oxidation-reduction reactions;
- learn the fundamental concepts of thermodynamics and apply them to chemical equilibria.
- evaluate acid-base systems for determining pH and all chemical species involved at equilibrium;
- comprehend the importance of the carbonate system and be able to calculate the alkalinity of a water sample;
- calculate the solubility product and determine if a solution is undersaturated, saturated, or supersaturated with respect to specific chemical species;
- use and apply the gas phase laws, with special emphasis on Henry's Law;
- explain the difference between the three major groups of hydrocarbons and explain the major functional groups of organic compounds and their importance in environmental engineering.

2.1 Introduction

Environmental engineers and scientists must have a strong understanding of chemical concepts. Chemicals such as alum, lime, soda ash, and sodium hydroxide are routinely added to hard water in order to precipitate calcium and magnesium ions. Ferrous sulfate, ferric chloride, and lime may be added to wastewater for removing orthophosphate. Chlorine and ozone are often used for oxidizing organic compounds and as primary disinfectants in various environmental applications.

The gas laws are important, since oxygen or nitrogen may be added to a reactor to enhance biological oxidation or to strip out unwanted gases. The chemistry and solubility of gases such as carbon dioxide, oxygen, and ammonia in water affect the rate at which they can be transferred into or out of solution and the saturation concentration of each. The formation of smog and other photochemical species relates to atmospheric chemistry. As engineers, we are not only concerned about the final or equilibrium concentrations that can be achieved during a chemical reaction but, more importantly, the rate of reaction (i.e., the kinetics of the reaction that will govern the design of most remediation systems, since contact time or detention time is often the primary design parameter). These concepts, along with a brief introduction to organic chemistry, will be presented in this chapter. Reaction kinetics is discussed in Chapter 5.

2.2 Dimensions, units, and conversions

Engineers work with measurements of physical quantities on a daily basis. Specifically, physical quantities such as length, time, temperature, pressure, velocity, and weight are some of the

Environmental Engineering: Principles and Practice, First Edition. Richard O. Mines, Jr.

Table 2.1 Four commonly used systems of units.

System	Length	Mass	Time	Temperature	Force	Energy
Absolute system (CGS)	centimeter	gram	second	K, °C	dyne	joule or calorie
British gravitational system (BG)	foot	slug	second	°R, °F	pound force, lb_f	BTU ft·lb
English engineering system (EE)	foot	pound mass, lb_m	second	°R, °F	pound force, lb_f	BTU
International system (SI)	meter	kilogram	second	K, °C	newton	watt

most widely used measurements. These types of measurements are used to describe an object or system. Not only is it important to specify the magnitude or numerical value of the physical quantity, it is essential that the dimension being expressed be accompanied by the appropriate set of units. The units provide the increments necessary to quantify the dimension; for example, the length of an object is a dimension. The units used to measure the length of the object could be expressed in meters, miles, feet, or yards.

The fundamental dimensions are generally considered to be length (L), mass (m), time (t), and temperature (T). All other dimensions (i.e., derived dimensions) are developed or derived from the fundamental dimensions. Examples of some commonly derived dimensions include velocity, (L/t), volume (L^3), density (m/L^3), and pressure (F/L^2); where force is denoted by F. Table 2.1 is a list of fundamental dimensions, derived dimensions, and their associated units for four commonly used systems. Prefixes that are used to describe multiples and fractions of SI units are presented in Table 2.2. A list of conversion factors for energy, length, mass, power, pressure, temperature, and volume is provided in Table 2.3.

Table 2.2 Prefixes for SI units.

Prefix	Symbol	Factor by which unit is multiplied
tera	T	10^{12}
giga	G	10^9
mega	M	10^6
kilo	k	10^3
hecto	h	10^2
deka	da	10
deci	d	10^{-1}
centi	c	10^{-2}
milli	m	10^{-3}
micro	μ	10^{-6}
nano	n	10^{-9}
pico	p	10^{-12}
femto	f	10^{-15}
atto	a	10^{-18}

Table 2.3 Conversion factors.

Units	To convert from	To	Multiply by
Energy	kWh	BTU	3412.8
Energy	kWh	calorie, cal	8.6057×10^5
Energy	kWh	hp·hr	1.341
Energy	calories, cal	joules, J	4.184
Energy	BTU	joules, J	1055
Length	meters, m	feet, ft	3.281
Length	yards, yd	feet, ft	3.0
Length	miles	feet, ft	5280
Length	feet, ft	inches, in	12
Length	inches, in	centimeters, cm	2.54
Length	Angstrom, Å	meters, m	10^{-10}
Length	miles	kilometers, km	1.609
Mass	pounds	grams	453.6
Mass	kilograms, kg	pounds	2.2
Mass	pounds	ounces	16
Mass	short tons	pounds	2000
Mass	metric tons	kilograms, kg	1000
Power	kilowatts, kW	joules/s	1000
Power	kilowatts, kW	hp	1.34
Power	BTU/s	kilowatts, kW	1.055
Power	kilowatts, kW	$(ft \cdot lb_f)/s$	737.56
Pressure	atmospheres, atm	mm Hg	760
Pressure	atmospheres, atm	psi	14.7
Pressure	atmospheres, atm	kPa	101.37
Pressure	atmospheres, atm	bar	1.013
Pressure	atmospheres, atm	N/m^2	1.01325×10^5
Pressure	psi	ft of water @ 4°C	2.307
Temperature	°F	°C	5/9(°F − 32)

Table 2.3 (*continued*)

Units	To convert from	To	Multiply by
Temperature	°R	°F	°F + 459.67
Temperature	°C	K	°C + 273.15
Volume	ft^3	gal	7.48
Volume	ft^3	L	28.32
Volume	gal	L	3.785
Volume	m^3	L	1000
Volume	m^3	yd^3	1.308
Volume	m^3	ft^3	35.31
Volume	cm^3	ml	1

2.2.1 Density

The density of a substance is defined as mass per unit volume. It is a derived parameter and may be expressed mathematically as follows:

$$\rho = \frac{m}{V} \qquad (2.1)$$

where:
ρ = density of substance, $lb_m/ft^3(kg/m^3)$
m = mass of substance, $lb_m(kg)$
V = volume, $ft^3(m^3)$

In the SI system, the density of water is approximately $1{,}000\ kg/m^3$ or $1\ g/cm^3$ at a temperature of 4°C, whereas in the English Engineering system, water has a density of $62.4\ lb_m/ft^3$ at 39.2°F. The density can vary widely between various fluids but, for liquids, pressure and temperature variations have only a small effect. For gases, however, both pressure and temperature have a significant effect on density. This will be discussed in more detail in the section on gas laws. Table 2.4 is a list of densities for various substances.

Table 2.4 Densities of common substances at 20°C.

Substance	Physical state	Density, g/cm^3
Aluminum	Solid	[a]2.64
Benzene	Liquid	[b]0.88
Crude oil	Liquid	[b]0.86
Ethanol	Liquid	[b]0.79
Gasoline	Liquid	[b]0.68
Hydrogen	Gas at 1 atmosphere & 0°C	[a]9.0×10^{-3}
Iron	Solid	[c]7.87
Jet fuel (JP-4)	Liquid	[b]0.77
Magnesium	Solid	[c]1.74
Mercury	Liquid	[b]13.55
Oxygen	Gas at 1 atmosphere & 0°C	[a]1.4×10^{-3}
Silver	Solid	[c]10.5
Sodium chloride	Solid	[c]2.16
Water	Liquid	1.000

[a] Density values from Eide *et al.* (1979). *Engineering Fundamentals and Problem Solving*, pp. 425–426.
[b] Density values from Vennard and Street (1982). p. 11.
[c] Density values from Zumdahl (2000). *Introductory Chemistry: A Foundation.* 4th Edition, p. 47.

Example 2.1 The effect of impurities on the density of water

The density of a pure solution of sodium chloride is 2.165 g/ml and the density of water at 0°C is 0.9998 g/ml. Determine the density of a 10% by weight solution of sodium chloride if the total mass of the solution is 1,000 g.

Solution

The mass of sodium chloride is calculated as follows:

$$m_{NaCl} = 0.10 \times 1000\,g = 100.0\,g$$

The volume of sodium chloride in the solution is calculated by dividing the mass of sodium chloride by the density, as shown below.

$$V_{NaCl} = \frac{100.0\,g}{2.165\,g/ml} = 46.19\,ml$$

The mass of water in the solutions is calculated as follows:

$$m_{H_2O} = 1000\,g - 100\,g = 900\,g$$

The volume of water in the solution is calculated by dividing the mass of water by the density of water as shown below.

$$V_{H_2O} = \frac{900.0\,g}{0.9998\,g/ml} = 900.2\,ml$$

The density of the 10% by weight sodium chloride solution is determined by dividing the total mass of sodium chloride and water by the total volume of the solution.

$$\rho_{NaCl\ Solution} = \frac{100.0\,g\ NaCl + 900.0\,g\ H_2O}{46.19\,ml\ NaCl + 900.2\,ml\ H_2O} = \boxed{1.057\,\frac{g}{ml}}$$

Table 2.5 Specific gravities of several compounds at 20°C.

Compound	Specific gravity
Acetone	0.785 @ 25°C
Alcohol-ethyl	[a]0.79
Benzene	[a]0.88
Butyric acid	0.953 @ 25°C
Crude Oil	[a]0.86
Chloroform	1.479 @ 25°C
Formic acid	1.220
Gasoline	[a]0.68
Glycerine	[a]1.26
Jet fuel (JP-4)	[a]0.77 @ 15.6°C
Mercury	[a]13.57
Methyl iodide	2.28
Naphthalene	1.025
Steel, cold drawn	[b]7.83
Steel, machine	[b]7.80
Toluene	0.86
Uranium	[b]18.7

Specific gravities from *CRC Handbook of Chemistry and Physics* (2010), 91st Edition, pp. 3-4, 3-74, 3-262, 3-268, 3-306, 3-382, 3-440, 3-486, 3-492
[a] Specific gravities from Vennard and Street (1982) p. 11.
[b] Specific gravities from Eide *et al.* (1979). *Engineering Fundamentals and Problem Solving*, pp. 425–426.

2.2.2 Specific gravity

The specific gravity of a substance is another important concept. Mathematically, it is defined as follows:

$$\text{Specific Gravity (S.G.)} = \frac{\rho_{\text{substance}}}{\rho_{\text{water at } 4°\text{C}}} \tag{2.2}$$

The density of water at 4°C serves as the standard against which the density of solids and liquids are referenced, while gases are sometimes referenced to air at standard conditions. Specific gravity is dimensionless. Table 2.5 presents the specific gravities of several compounds.

Example 2.2 Calculating specific gravity and density

The density of titanium is approximately 281 lb/ft^3.

Determine:
a) Specific gravity of titanium
b) Mass in kg of 10 cm^3 of titanium

Solution part a

The specific gravity of a substance in defined by Equation (2.2).

$$\text{Specific Gravity (S.G.)} = \frac{\rho_{\text{substance}}}{\rho_{\text{water at } 4°\text{C}}} = \frac{281\ \frac{\text{lb}}{\text{ft}^3}}{62.4\ \frac{\text{lb}}{\text{ft}^3}} = \boxed{4.50}$$

Solution part b

The density of a substance is defined as the mass per unit volume.

$$\rho = \frac{m}{V\!\!\!\!-}$$

Therefore, the mass is determined by multiplying the density by the volume of material, i.e., rearranging Equation (2.1). The density of water is approximately equal to 1,000 kg/m^3.

$$m = \rho \times V\!\!\!\!- = 4.50 \times \left(\frac{1000\text{ kg}}{\text{m}^3}\right)(10\text{ cm}^3)\left(\frac{\text{m}}{100\text{ cm}}\right)^3$$

$$m = \boxed{0.045\text{ kg}}$$

2.2.3 Concentrations

There are various ways to express the concentration of a substance in a solution: mass or weight per unit volume, mass or weight percentage, volume percent, moles per liter, mole fraction, and equivalents per liter. Each of these will be described briefly. Concentration is a derived dimension.

2.2.3.1 Mass or Weight per Unit Volume

The most widely used means of expressing the concentration of a substance is to present it gravimetrically on a **mass or weight per unit volume** basis. Concentration is normally expressed in units of mg/L, which is equivalent to parts per million (ppm) when the specific gravity of the solution is equal to 1. For dilute aqueous solutions, it is not unusual to have concentrations expressed in micrograms per liter (μg/L) or parts per billion (ppb). The concentration of substance *A* in a mixture of *A* and *B* is:

$$C_A\left(\frac{\text{mg}}{\text{L}}\right) = \frac{m_A}{V\!\!\!\!-_A + V\!\!\!\!-_B} \tag{2.3}$$

where:
C_A = concentration of substance *A*, mg/L
m_A = mass of substance *A*, mg
$V\!\!\!\!-_A$ = volume of substance *A*, L
$V\!\!\!\!-_B$ = volume of substance *B*, L.

2.2.3.2 Mass or Weight Percentage

Concentrations are also expressed on a **mass or weight percentage** basis. For instance, the mass percent of substance *A* in solution is mathematically expressed as follows:

$$\text{mass}_A \text{ or weight}_A(\%) = \left(\frac{m_A}{m_A + m_B}\right) \times 100 \quad (2.4)$$

where m_B = mass of substance B, mg, g, lb_m

2.2.3.3 *Volume Percent*

The **volume percent** concentration is analogous to the mass percent concentration as presented in Equation (2.4); however, the mass of each substance is replaced with the volume of each substance, as shown in Equation (2.5).

$$\text{volume}_A(\%) = \left(\frac{V_A}{V_A + V_B}\right) \times 100 \quad (2.5)$$

2.2.3.4 *Moles per Liter*

The **molar concentration** (M) of a solution represents the number of moles of a substance per liter. By definition, a mole of an element or compound is equal to the atomic mass of an element or the molecular mass of a compound. For example, one mole of hydrochloric acid, HCl, consists of 36.6 g of HCl (atomic mass of H = 1 and Cl = 35.5).

To prepare a 1 M solution of HCl, 36.6 g are added to a 1 L volumetric flask and filled to the 1 L mark with either distilled or deionized water. A mole (mol) is also defined by Avogadro's number (6.02×10^{23}) such that there are 6.02×10^{23} molecules of an element or compound per mole. A **molal solution** consists of dissolving 1 gram molecular weight of the solute in 1 liter of water, resulting in a final volume slightly greater than 1 liter. Figure 2.1 is a representation of a 1 molar and 1 molal solution of hydrochloric acid.

The molar concentration is expressed as follows:

$$C_A(M) = \frac{n_A}{V} \quad (2.6)$$

where:
n_A = moles of element or compound A
V = total volume of solution, L.

2.2.3.5 *Mole Fraction*

Another way of expressing concentration is in regard to **mole fraction**. Mole fraction is calculated by dividing the number of moles of an element or compound by the total number of moles of the solution; this is expressed mathematically as Equation (2.7).

$$x_A = \frac{n_A}{n_A + n_B + \dots n_n} = \frac{n_A}{\text{total moles of solution}} \quad (2.7)$$

where:
x_A = mole fraction of element or compound A
n_A = moles of element or compound A
n_B = moles of element or compound B
n_n = moles of element or compound n.

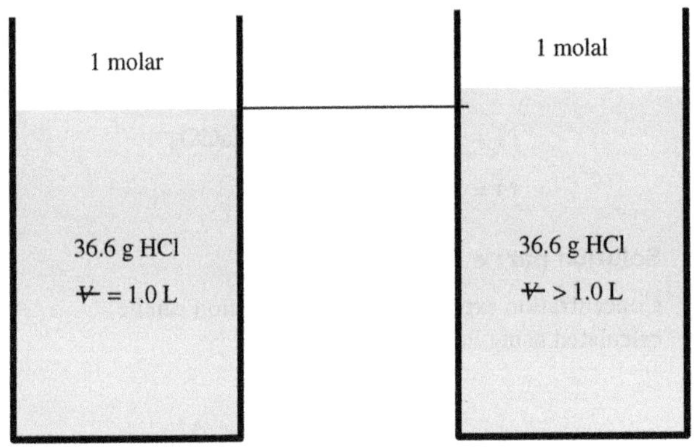

Figure 2.1 Comparison of 1 molar and 1 molal solution of HCl.

2.2.3.6 *Normality*

The **normality (N)** of a solution is the number of equivalents per liter of solution and is calculated from Equation (2.8).

$$C_A(N) = \frac{\text{Equivalents}_A}{V} = \frac{Eq_A}{V} \quad (2.8)$$

where:
$\text{Equivalents}_A = \frac{m_A}{EW_A}$
V = total volume of solution, L
m_A = mass of substance A, g.

The equivalent weight (EW) of an element or radical is calculated using Equations (2.9) and (2.10). Equation (2.9) is used for elements and Equation (2.10) for compounds and radicals; z represents the assumed valence or charge of the ion. In acid-base reactions, the equivalent weight is related to the number of protons or hydroxyl radicals that react. For oxidation-reduction reactions, the equivalent weight is based on the number of moles of electrons that are transferred.

$$EW_A = \frac{AW}{z} \quad (2.9)$$

$$EW_A = \frac{MW}{z} \quad (2.10)$$

where:
AW = atomic weight of ion, g
MW = molecular weight of radical, g.

The relationship between normality and molarity is given in Equation (2.11).

$$N = z \times M \quad (2.11)$$

Although normality is typically defined as equivalents per liter, in many environmental engineering applications it is expressed in milli-equivalents per liter (meq/L), since we often work with dilute solutions. Table 2.6 lists some of the common radicals encountered in environmental chemistry.

Example 2.3 Calculating solution concentration

A graduate student adds 2.5 g of calcium carbonate ($CaCO_3$) to a 0.5 L volumetric flask. Enough distilled water

Table 2.6 Common radicals encountered in environmental chemistry.

Radical	Chemical formula	AW or MW, g/mol	EW, g/eq
Aluminum	Al^{3+}	27	9
Ammonium	NH_4^+	18	18
Calcium	Ca^{2+}	40	20
Bicarbonate	HCO_3^-	61	61
Carbonate	CO_3^{2-}	60	30
Cyanide	CN^-	26	26
Chloride	Cl^-	35.3	35.5
Chromate	CrO_4^{2-}	116	58
Dichromate	CrO_7^{2-}	216	108
Fluoride	F^-	19	19
Hydrogen ion	H^+	1	1
Hydrogen sulfate	HSO_4^-	97	97
Hydroxyl	OH^-	17	17
Hypochlorite	OCl^-	51.5	51.5
Magnesium	Mg^{2+}	24.3	12.2
Nitrate	NO_3^-	62	62
Nitrite	NO_2^-	46	46
Orthophosphate	PO_4^{3-}	95	31.7
Permanganate	MnO_4^-	119	119
Sodium	Na^+	23	23
Sulfate	SO_4^{2-}	96	48
Sulfide	S^{2-}	32	16

is added to reach the 0.5 L mark on the volumetric. Determine the concentration of calcium carbonate in solution expressing your answer in the units specified below.

a) mg/L
b) mass %
c) molarity
d) normality
e) mole fraction

Solution part a

MW of $CaCO_3 = 40 + 12 + 3(16) = 100$ g/mole
From Equation (2.3):

$$\text{concentration}\left(\frac{\text{mg}}{\text{L}}\right) = \frac{2.5\text{ g}}{0.5\text{ L}} \times \left(\frac{1000\text{ mg}}{\text{g}}\right) = \boxed{\frac{5000\text{ mg}}{\text{L}}}$$

Solution part b

Assume that the density of the solution is 1 g/ml. From Equation (2.4):

$$\text{mass}_A \text{ or weight}_A(\%) = \left(\frac{m_A}{m_A + m_B}\right) \times 100$$

$$m_a + m_b = 0.5\text{ L} \times \left(\frac{1000\text{ ml}}{\text{L}}\right) \times \left(\frac{1\text{ g}}{\text{ml}}\right) = 500\text{ g}$$

$$\text{mass}_{CaCO_3}(\%) = \left(\frac{2.5\text{ g}}{500\text{ g}}\right) \times 100 = \boxed{0.5\ \%}$$

Solution part c

The molarity of a solution is the number of moles of solute per liter of solution. Equation (2.6):

$$C_A\,(\text{M}) = \frac{n_A}{V}$$

$$C_{CaCO_3}\,(\text{M}) = \frac{(2.5\text{ g }CaCO_3)\left(\dfrac{\text{mole }CaCO_3}{100\text{ g }CaCO_3}\right)}{0.5\text{ L solution}} = \boxed{0.05\text{ M}}$$

Solution part d

The normality of a solution is the number of equivalents of solute per liter of solution expressed by Equation (2.8):

$$C_A(\text{N}) = \frac{\text{Equivalents}_A}{V} = \frac{\text{Eq}_A}{V}$$

$$\text{EW}_A = \frac{\text{MW}}{z}$$

$$\text{EW}_{CaCO_3} = \frac{100\text{ g }CaCO_3}{2\text{ eq}} = \frac{50\text{ g }CaCO_3}{\text{eq}}$$

$$C_{CaCO_3}(\text{N}) = \frac{2.5\text{ g }CaCO_3\left(\dfrac{\text{eq}}{50\text{ g }CaCO_3}\right)}{0.5\text{ L}} = \boxed{\frac{0.1\text{ eq}}{\text{L}} = 0.1\text{ N}}$$

Alternatively, normality can be calculated using Equation (2.11):

$$\text{N} = z \times \text{M} \qquad z = 2 \text{ for } CaCO_3$$

$$\text{N} = 2 \times 0.05\text{ M} = \boxed{0.1\text{ N}}$$

Solution part e

Concentration expressed as a mole fraction can be calculated using Equation (2.7).

$$x_A = \frac{n_A}{n_A + n_B + \ldots\, n_n} = \frac{n_A}{\text{total moles of solution}}$$

$$n_{CaCO_3} = 2.5\,g\,CaCO_3 \times \left(\frac{mole\ CaCO_3}{100\,g\,CaCO_3}\right) = 0.025\ moles$$

$$n_{H_2O} = 500\,g\,H_2O \times \left(\frac{mole\ H_2O}{18\,g\,H_2O}\right) = 27.8\ moles$$

$$x_{CaCO_3} = \frac{0.025\ moles}{0.025\ moles + 27.8\ moles} = \boxed{9.0 \times 10^{-4}}$$

2.2.4 Mass and volumetric flow rates

In environmental engineering, the flow rate may be expressed either on a mass (gravimetric) or a volumetric basis. Recall that the quantity of mass can be calculated by rearranging Equation (2.1) as presented below:

$$m = \rho \times \forall \tag{2.12}$$

The mass-based flow rate represents the quantity of mass flowing through a pipe or system per unit of time. Equation (2.12) can be rewritten as Equation (2.13) to indicate the mass flow rate and as shown, it is not independent from the volumetric flow rate.

$$\dot{m} = \rho \times Q \tag{2.13}$$

where:
$\dot{m}$ = mass rate of flow, slugs/s, lb_m/s (kg/s)
ρ = mass density, slugs/ft^3, lb_m/ft^3 (kg/m^3)
Q = volumetric flow rate, ft^3/s (m^3/s).

The mass flow rate of some component A in the volumetric flow can be determined as follows:

$$\dot{m}_A = C_A \times Q \tag{2.14}$$

where:
$\dot{m}_A$ = mass flow rate of component A, lb_m/s (kg/s)
C_A = concentration of component A, lb_m/ft^3(mg/L).

Example 2.4 Calculating mass flow rate (SI) units

A mountain stream flowing at a rate of 10 m^3/s contains 10 mg/L of suspended solids. Determine the mass flow rate of the suspended solids in units of kg/min.

Solution

From Equation (2.14), the mass flow rate of some component can be calculated as follows:

$$\dot{m}_A = C_A \times Q$$

$$\dot{m}_{SS} = C_{SS} \times Q$$

$$\dot{m}_{SS} = \frac{10\,mg\,SS}{L} \times \left(\frac{10\,m^3}{s}\right) \times \left(\frac{kg}{10^6\,mg}\right) \times \left(\frac{1000\,L}{m^3}\right) \times \left(\frac{60\,s}{min}\right)$$

$$\dot{m}_{SS} = \boxed{\frac{6.0\,kg}{min}}$$

Example 2.5 Calculating mass flow rate (EE) units

Advanced wastewater treatment (AWT) plants typically must meet an effluent permit requirement of 5 mg/L or less for five-day biochemical oxygen demand (BOD_5). The Howard F. Current AWT in Tampa, Florida has a design flow capacity of approximately 90 million gallons of day (MGD). Determine the mass flow rate of the BOD_5 discharged into Tampa Bay in units of lb_m/d.

Solution

From Equation (2.14), the mass flow rate of some component can be calculated as follows:

$$\dot{m}_A = C_A \times Q$$

$$\dot{m}_{BOD_5} = C_{BOD_5} \times Q$$

$$\dot{m}_{BOD_5} = \frac{5\,mg\,BOD_5}{L} \times \frac{90\,MG}{d} \times \left[\left(\frac{g}{1000\,mg}\right) \times \left(\frac{lb_m}{454\,g}\right) \times \left(\frac{10^6 gal}{MG}\right) \times \left(\frac{3.785\,L}{gal}\right)\right]$$

$$\dot{m}_{BOD_5} = \boxed{3.75 \times 10^3 \frac{lb_m}{d}}$$

Alternatively, the mass flow rate in lb_m/d can be calculated using the following equation, when the concentration is expressed in mg/L and the volumetric flow rate in million gallons per day (MGD), by using the conversion factor $\frac{8.34\,lb_m}{\frac{mg}{L} \cdot MG}$.

$$\dot{m}_A = C_A\left(\frac{mg}{L}\right) \times Q\left(\frac{MG}{d}\right) \times \frac{8.34\,lb_m}{\frac{mg}{L} \cdot MG}$$

$$\dot{m}_{BOD_5} = \left(\frac{5\,mg}{L}\right) \times \left(\frac{90\,MG}{d}\right) \times \frac{8.34\,lb_m}{\frac{mg}{L} \cdot MG}$$

$$\dot{m}_{BOD_5} = \boxed{3.75 \times 10^3 \frac{lb_m}{d}}$$

2.2.5 Detention time

A fundamental concept in environmental engineering is detention time. Traditionally, this is defined as the average unit of time that a fluid particle remains in a system (e.g., reactor, pipe, or tank). Alternatively, it is defined as the time required filling a tank or container. Detention time is also known as retention time or residence time. Mathematically, detention time is determined by dividing the volume of the tank or container by the volumetric flow rate (Q) as shown below.

$$\tau = \frac{\forall}{Q} \tag{2.15}$$

where:
τ = detention, retention, or residence time, min, h, d
$\forall$ = volume of tank or container, ft^3, m^3, gal, L
Q = volumetric flow rate, ft^3/s (m^3/s).

Example 2.6 Calculating detention time

Two, rectangular sedimentation basins operating in parallel process a total flow of 0.20 m^3/s of coagulated water. Each basin has the following dimensions: 4.5 m by 30 m by 7.5 m. Determine the detention time of each basin in units of hours.

Solution

First, draw a schematic of the basins showing how the flow is distributed.

$Q = 0.10\ m^3/s$ → Sedimentation Basin →

$Q = 0.10\ m^3/s$ → Sedimentation Basin →

Both basins and the total volumetric flow rate may be used for calculating the detention time. The detention time is calculated using Equation (2.15) as follows:

$$\tau = \frac{\forall}{Q} = \frac{[4.5\ \text{m} \times 30\ \text{m} \times 7.5\ \text{m}] \times 2}{\frac{0.20\ \text{m}^3}{\text{s}} \times \frac{60\ \text{s}}{\text{min}} \times \frac{60\ \text{min}}{\text{h}}} = \boxed{2.8\ \text{h}}$$

Alternatively, we can calculate the detention based on the flow rate entering only one of the basins. The answer should be the same as above.

$$\tau = \frac{\forall}{Q} = \frac{[4.5\ \text{m} \times 30\ \text{m} \times 7.5\ \text{m}]}{\frac{0.10\ \text{m}^3}{\text{s}} \times \frac{60\ \text{s}}{\text{min}} \times \frac{60\ \text{min}}{\text{h}}} = \boxed{2.8\ \text{h}}$$

2.3 Balancing reactions

A chemical equation expresses what happens during a chemical reaction or chemical change. Instead of trying to explain in words what happens during a chemical reaction, chemical elements, compounds, and symbols are used to convey this in a clear and concise manner using a chemical equation. Chemical equations must be verified by experimentation in the laboratory.

Environmental engineers use chemical equations for estimating the quantities of chemicals that are added to water, wastewater, sludge, contaminated gas streams, etc. For example, these equations can also be used to estimate the quantity of precipitate that is generated during softening of hard water.

Before using any chemical equation, one must examine it closely to determine if the equation is properly balanced. An equation is balanced when the law of conservation of mass is satisfied; i.e., the mass of all elements on both sides of the equation are the same and the net charge of all compounds is zero. It is important to remember that atoms are neither created nor destroyed during a chemical reaction.

A properly balanced equation represents the stoichiometry of the equation; that is, it shows the relationship between the number moles of each reactant and the number of moles of the various products that are produced. Table 2.7 shows some of the symbols commonly used for writing chemical equations, and the major steps used in writing and balancing chemical equations are presented in Table 2.8. Examples 2.7 and 2.8 illustrate how to balance a chemical equation.

Table 2.7 Symbols commonly used for writing chemical equations.

Symbol	Meaning
(aq)	Aqueous solution, substance dissolved in water, $Mg^{2+}_{(aq)}$
(g)	Gaseous state, $CO_{2(g)}$
(*l*)	Liquid
(s)	Solid or precipitate, $CaCO_{3(s)}$
↑	Gas, when written immediately after a substance
↓	Solid or precipitate, written immediately after a substance
↔	Reversible reaction
→	Irreversible reaction – proceeds in the direction of the arrow.

Table 2.8 Major steps used in writing and balancing chemical equations.

Step	Explanation
1	Write the correct formula for all reactants and products based on valence, oxidation number, or experimental data. Show their appropriate states.
2	Write the unbalanced equation, showing reactants to the left of the arrow and products to the right.
3	Balance the equation by inspection, starting with the most complex compound. Assume that only one mole of this compound is involved in the reaction. Balance all elements of this compound one at a time.
4	It is generally advantageous to balance all elements other than hydrogen and oxygen first; then balance hydrogen, and then finally oxygen.
5	Check all coefficients used in the equation to see if the same number of atoms exists on both sides of the arrow. An atom may be present in an element, compound, or ion.
6	Ensure that the coefficients used are the smallest by removing fractions and simplifying.

Example 2.7 Balancing chemical equation 1

Balance the following chemical equation:

$$CaCl_2 + Na_2CO_3 \rightarrow CaCO_3 + NaCl$$

Solution

Select soda ash, i.e., sodium carbonate (Na_2CO_3) as the most complex compound and balance all elements associated with it. Assume there is one mole of soda ash. This results in two atoms of sodium on the left side of the equation and only one on the right. Therefore, a number 2 must be placed before sodium chloride (NaCl) to balance the sodium atoms on the left side of the equation. There is one carbonate ion (CO_3^{2-}) on the left side of the equation and one carbonate ion on the right side, incorporated into calcium carbonate ($CaCO_3$). Therefore, the sodium atoms and carbonate ions are balanced.

$$CaCl_2 + Na_2CO_3 \rightarrow CaCO_3 + 2\,NaCl$$

Check the number of calcium atoms (Ca) on each side of the equation. There is one calcium atom in calcium chloride ($CaCl_2$) on the left side of the equation and one calcium atom associated with calcium carbonate ($CaCO_3$) on the right side. So the equation is balanced with respect to calcium.

Check the number of chloride ions (Cl^-) on each side of the equation. There are two chloride ions associated with calcium chloride ($CaCl_2$) on the left side of the equation and two chloride ions associated with the two moles of sodium chloride (NaCl) on the right side of the equation. Therefore, chloride is balanced.

As a final check, count the number of atoms of each element on the left and right sides of the equation to ensure that it is balanced.

Element	Left side	Right side
Ca	1	1
Cl	2	2
Na	2	2
C	1	1
O	3	3

Example 2.8 Balancing chemical equation 2

Balance the following chemical equation involving the oxidation of liquid ethanol with gaseous oxygen to form carbon dioxide and water vapor.

$$C_2H_5OH_{(l)} + O_{2(g)} \rightarrow CO_{2(g)} + H_2O_{(g)}$$

Solution

Select ethanol (C_2H_5OH) as the most complex compound and balance all elements associated with it. Assume there is one mole of ethanol. There are two atoms of carbon on the left side, so a 2 must be placed before carbon dioxide to yield two carbon atoms on the right side. Next consider hydrogen; there are six atoms on the left side, so a 3 is placed before water (H_2O), yielding six atoms of hydrogen:

$$C_2H_5OH_{(l)} + O_{2(g)} \rightarrow 2\,CO_{2(g)} + 3\,H_2O_{(g)}$$

Finally, the oxygen atoms may be balanced. On the right side of the equation there are four atoms of oxygen associated with the two moles of carbon dioxide and three atoms of oxygen associated with the three moles of water, resulting in a total of seven atoms of oxygen. To balance the oxygen atoms on the left side of the equation, there is one atom associated with ethanol (C_2H_5OH), so we must place a 3 before the gaseous oxygen ($O_{2(g)}$) on the left side of the equation:

$$C_2H_5OH_{(l)} + 3\,O_{2(g)} \rightarrow 2\,CO_{2(g)} + 3\,H_2O_{(g)}$$

As a final check, count the number of atoms of each element on the left and right sides of the equation to ensure that the equation is balanced.

Element	Left side	Right side
C	2	2
H	6	6
O	7	7

2.4 Oxidation-reduction reactions

In oxidation-reduction or **redox** reactions, atoms or ions undergo a change in their oxidation number and this is always accompanied by a transfer in electrons. Recall that **oxidation** is defined as the loss of electrons or a gain in valence. The opposite of oxidation is **reduction**, which is defined as a gain in electrons or a loss in valence, i.e., a loss in oxidation number. If one element releases electrons, then another element must be present to accept the electrons.

Ionic compounds are formed when a metal reacts with a nonmetal. The formation of sodium chloride results from the reaction of sodium metal with chlorine gas, as shown below.

$$2\,Na_{(s)} + Cl_{2(g)} \rightarrow 2\,NaCl_{(s)}$$

This is a redox reaction in which an electron is transferred from each sodium atom to a chlorine atom forming sodium chloride. In the formation of NaCl, sodium loses an electron and is present as the sodium ion (Na^+), having a net charge of +1; whereas chlorine accepts an electron and is present as the chloride ion (Cl^-), having a net charge of −1. This is illustrated in Figure 2.2.

With these concepts in mind, an **oxidizing agent** is any substance that can add electrons. Alternatively, any substance that can give up electrons is called a **reducing agent**. Table 2.9 lists several oxidizing and reducing agents. Elements that have an intermediate oxidation state may serve either as an oxidizing or reducing agent.

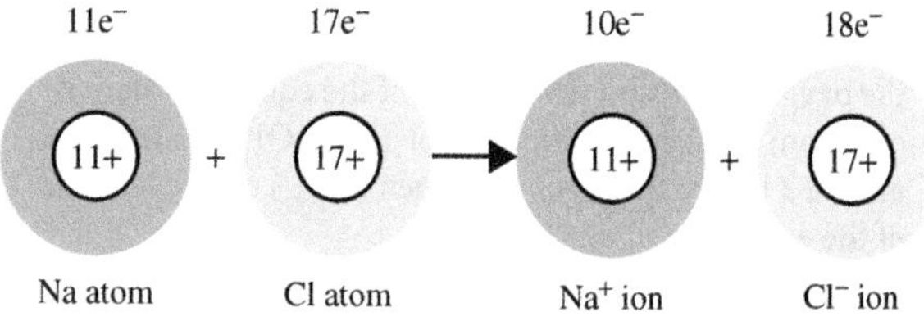

Figure 2.2 Illustration of electron transfer during formation of sodium chloride.

Table 2.9 Common oxidizing and reducing agents.

Oxidizing agents	Reducing agents
O(0)	H(0)
Cl(0)	Fe(0)
Fe(III)	Mg(0)
Cr(IV)	Fe(II)
Mn(IV)	Cr(II)
Mn(VII)	Mn(IV)
N(V)	N(III)
N(III)	Cl(−I)
S(0)	S(0)
S(IV)	S(−II)
S(VI)	S(IV)

Follow the steps presented below for balancing complex oxidation-reduction equations.

1. Check to see whether any chemical species is being oxidized or reduced.
2. Write and balance half-reactions.
3. Fill in the remainder of the chemical equation.
4. Check mass balance on both sides of the equation.

Examples 2.9 and 2.10 illustrate typical oxidation-reduction reactions that occur in water chemistry.

Example 2.9 Oxidation-reduction reaction I

Balance the reaction below:

$$KClO_3 \rightarrow KCl + O_2$$

Solution

1. Assign oxidation numbers to all elements in the reaction as follows:

$$\overset{1+}{K}\overset{5+}{Cl}\overset{3(2-)}{O_3} \rightarrow \overset{1+}{K}\overset{1-}{Cl} + \overset{0}{O_2}$$

2. Separate the basic equation into two, half-reactions: one should illustrate the oxidation step and the other the reduction.

$$\text{Oxidation:}\quad O^{2-} \rightarrow O^0 + 2\,e^-$$

$$\text{Reduction:}\quad Cl^{5+} + 6\,e^- \rightarrow Cl^-$$

3 Balance all the atoms in the half-reactions. Multiply the oxidation equation by 3 and then add the equations together.

$$3\,O^{2-} \rightarrow 3\,O^0 + 6\,e^-$$
$$\underline{Cl^{5+} + 6\,e^- \rightarrow Cl^-}$$
$$3\,O^{2-} + Cl^{5+} \rightarrow 3\,O^0 + Cl^-$$

4 Fill in the remainder of the chemical reaction.

$$KClO_3 \rightarrow KCl + \frac{3}{2}\,O_2$$

Multiply both sides of the equation to remove the fraction.

$$2\,KClO_3 \rightarrow 2\,KCl + 3\,O_2$$

5 Check mass balance on both sides of the equation.

Element	Left side	Right side
K	2	2
Cl	1	1
O	6	6

The chemical equation is balanced!

Example 2.10 Oxidation-reduction reaction 2

Balance the reaction below.

$$Zn^0 + CuSO_4 \rightarrow ZnSO_4 + Cu^0$$

Solution

1 Assign oxidation numbers to all elements in the reaction as follows.

$$Zn^0 + \overset{2+}{Cu}\,\overset{6+}{S}\,\overset{4(2-)}{O_4} \rightarrow \overset{2+}{Zn}\,\overset{6+}{S}\,\overset{4(2-)}{O_4} + Cu^0$$

2 Separate the basic equation into two, half-reactions; one should illustrate the oxidation step and the other the reduction.

$$\text{Oxidation:}\quad Zn^0 \rightarrow Zn^{2+} + 2\,e^-$$
$$\text{Reduction:}\quad Cu^{2+} + 2\,e^- \rightarrow Cu^0$$

3 Balance all the atoms in the half-reactions and add the equations together.

$$Zn^0 \rightarrow Zn^{2+} + 2\,e^-$$
$$\underline{Cu^{2+} + 2\,e^- \rightarrow Cu^0}$$
$$Zn^0 + Cu^{2+} \rightarrow Zn^{2+} + Cu^0$$

4 Fill in the remainder of the chemical reaction.

$$Zn + CuSO_4 \rightarrow ZnSO_4 + Cu$$

5 Check mass balance on both sides of the equation.

Element	Left side	Right side
Zn	1	1
Cu	1	1
S	1	1
O	4	4

The chemical equation was balanced as originally written!

2.5 Thermodynamic equilibrium

Thermodynamics is the branch of science dealing with changes in energy that accompanies physical and chemical processes. Thermodynamics allows environmental engineers and chemists to determine whether or not a chemical reaction will or will not occur. However, it will not allow us to predict the rate of reaction. The thermodynamic laws allow us to predict equilibrium of a reaction. Other importance aspects of thermodynamics in environmental engineering relate to biological energetics, sludge drying and incineration, and the cooling of thermal wastes. Some definitions that are useful in working with thermodynamics are presented in Table 2.10.

2.5.1 First law of thermodynamics

The first law of thermodynamics states that energy may neither be created nor destroyed, only transferred or changed from one form to another. In equation form, the first law of thermodynamics is presented below. The change in internal energy of the system is related to the flow of heat into or out of the system and the work done by the system.

$$\Delta E = q - w \qquad (2.16)$$

Table 2.10 Definitions useful in working with thermodynamics.

Closed system: mass may not enter or leave the system.

Energy: the capacity to do work.

Equilibrium state: a state in which the macroscopic properties of the system, i.e., temperature, density, chemical composition, are well defined and do not change with time.

Open system: both mass and energy may enter or exit the system.

Standard state: elements in the standard state give a reference point from which to measure change.

Surroundings: the remainder of the physical world that surrounds the system being examined.

System: the portion of the physical world that is being examined thermodynamically.

where:
ΔE = change in internal energy of the system
q = heat added to the system
w = work done by the system.

If heat is added to the system, q has a positive value; if the system gives off heat, q is negative. Likewise, if work is done on the system by the surroundings, w has a negative value. However, w is positive when the system does work on the surroundings. In chemical systems, work involves expansion work and is measured in terms of force times distance. In a closed system, this is represented by a pressure times the change in volume. Mathematically, this is expressed as Equation (2.17):

$$\int_0^w dW = \int_{V_1}^{V_2} P dV \quad (2.17)$$

$$w = \int_{V_1}^{V_2} P dV \quad (2.18)$$

Substituting Equation (2.18) into Equation (2.16) results in the equation below:

$$\Delta E = q - \int_{V_1}^{V_2} P dV \quad (2.19)$$

If the reaction takes place in a closed system, volume is constant and Equation (2.19) reduces to:

$$\Delta E = q - 0 \quad \text{at constant volume} \quad (2.20)$$

$$\Delta E = q_V \quad (2.21)$$

When ΔE has a negative value, it means that the reaction is exothermic or heat is given off. A positive value for ΔE indicates an endothermic reaction and that heat is absorbed.Most chemical reactions are not performed or conducted at a constant volume, but at a constant pressure of 1 atm. For constant pressure, it is necessary to define a new function of state, enthalpy (H), as presented below.

$$H = E + PV \quad (2.22)$$

where:
H = enthalpy of the system
H = total energy content of an element or compound
E = internal energy of the system
P = pressure on the system
V = volume of the system.

A change in enthalpy is expressed as:

$$\Delta H = \Delta E + \Delta(PV) \quad (2.23)$$

$$\Delta H = q - w + \Delta(PV) \quad (2.24)$$

For changes that occur at constant pressure, recall that work is defined as follows:

$$w = P\Delta V \quad (2.25)$$

From calculus:

$$\Delta(PV) = P\,\Delta V + V\Delta P \quad (2.26)$$

$$V\Delta P = 0 \quad \text{at constant pressure} \quad (2.27)$$

$$\Delta(PV) = P\,\Delta V + 0 \quad (2.28)$$

$$\Delta(PV) = P\,\Delta V \quad (2.29)$$

Substituting Equations (2.25) and (2.29) into Equation (2.24) results in:

$$\Delta H = q - w + \Delta(PV) \quad (2.24)$$

$$\Delta H = q - P\,\Delta V + P\,\Delta V \quad (2.30)$$

The change in enthalpy is equal to the heat absorbed when the reaction is carried out at constant pressure:

$$\Delta H = q_P \quad (2.31)$$

A negative ΔH value indicates an exothermic reaction, whereas a positive value means the reaction is endothermic. For liquids and solids, $\Delta H \cong \Delta E$, since there is very little change in volume.For gases, the following equations are applicable:

$$PV = nRT \quad (2.32)$$

where:
P = pressure, atm
V = volume, L
n = number of moles of gas
R = universal gas constant, $\frac{0.08206 \text{ atm} \cdot \text{L}}{\text{mol} \cdot \text{K}}$
T = temperature of gas, K.

$$\Delta(P\mathcal{V}) = nRT \quad (2.33)$$

$$\Delta H = \Delta E + \Delta(P\mathcal{V}) \quad (2.23)$$

$$\Delta H = \Delta E + \Delta n RT \quad (2.34)$$

Where Δn is the change in the number of moles of gas due to the chemical reaction at a fixed temperature.

Example 2.11 Calculating changes in enthalpy and energy

When 1 mole of ice melts at 0°C at a constant pressure of 1 atm, 1440 cal of heat are absorbed by the system. The molar volumes of ice and water are 0.0196 and 0.0180 liters, respectively. Calculate the change in enthalpy, ΔH, and change in energy, ΔE.

Solution

From Equation (2.31):

$$\Delta H = q_p = 1440\ \text{cal}$$

$$\Delta(P\mathcal{V}) = P\Delta\mathcal{V} + \mathcal{V}\Delta P = P\Delta\mathcal{V} + 0$$

$$\Delta(P\mathcal{V}) = P\Delta\mathcal{V} = P(\mathcal{V}_2 - \mathcal{V}_1)$$

$$\Delta(P\mathcal{V}) = 1\ \text{atm}(0.0180\ \text{L} - 0.0196\ \text{L})$$

$$= -1.6 \times 10^{-3}\text{L} \cdot \text{atm}$$

$$\Delta(P\mathcal{V}) = -1.6 \times 10^{-3}\text{L} \cdot \text{atm}\left(\frac{\text{mol} \cdot \text{K}}{0.08206\ \text{atm} \cdot \text{L}}\right)$$

$$\times \left(\frac{1.98\ \text{cal}}{\text{K} \cdot \text{mol}}\right) = -0.039\ \text{cal}$$

$$\Delta H = \Delta E + \Delta(P\mathcal{V})$$

$$\Delta H = 1440\ \text{cal} - 0.039\ \text{cal}$$

Therefore, $\Delta H \cong \Delta E = \boxed{1440\ \text{cal}}$

Example 2.12 Calculating changes in energy

For the decomposition reaction of magnesium carbonate given by the reaction below:

$$MgCO_{3(s)} \rightarrow MgO_{(s)} + CO_{2(g)}$$

The change in enthalpy, $\Delta H = 26,000$ cal at 900 K and 1 atm of pressure. If the molar volume of $MgCO_3$ is 0.028 L and that of MgO is 0.011 L, determine the change in energy, ΔE.

Solution

$$\Delta(P\mathcal{V}) = \Delta(P\mathcal{V})_{(s)} + \Delta(P\mathcal{V})_{(g)}$$

$$\Delta(P\mathcal{V})_{(s)} = 1\ \text{atm}(0.011\ \text{L} - 0.028\ \text{L}) = -0.017\ \text{L} \cdot \text{atm}$$

$$\Delta(P\mathcal{V})_{(s)} = -0.017\ \text{L} \cdot \text{atm}\left(\frac{\text{mol} \cdot \text{K}}{0.08206\ \text{atm} \cdot \text{L}}\right)$$

$$\times \left(\frac{1.98\ \text{cal}}{\text{K} \cdot \text{mol}}\right) = -0.41\ \text{cal}$$

$$\Delta(P\mathcal{V})_{(g)} = nRT$$

$$\Delta(P\mathcal{V})_{(g)} = (1\ \text{mole})\left(\frac{1.98\ \text{cal}}{\text{K} \cdot \text{mol}}\right)(900\ \text{K}) = 1780\ \text{cal}$$

$$\Delta(P\mathcal{V}) = -0.41\ \text{cal} + 1780\ \text{cal} = 1780\ \text{cal}$$

$$\Delta H = \Delta E + \Delta(P\mathcal{V})$$

$$\Delta E = \Delta H - \Delta(P\mathcal{V}) = 26,000\ \text{cal} - 1780\ \text{cal}$$

$$= \boxed{24,200\ \text{cal}}$$

The total enthalpy of a system is difficult to measure, so scientists and engineers are generally interested in a change in enthalpy. A system has been developed so that the change in enthalpy, or the heat of formation of a compound or element, can be determined at standard conditions of 25°C and 1 atm pressure. At standard conditions, compounds whose standard state is gas, liquid, or crystal or solid are assigned enthalpies equal to zero. In this text, the standard enthalpy of formation will be represented as ΔH_f°. The change in standard enthalpy for a chemical reaction is given below:

$$\Delta H^\circ = \sum \Delta H^\circ_{f\,(\text{products})} - \sum \Delta H^\circ_{f\,(\text{reactants})} \quad (2.35)$$

where:

ΔH° = change in standard enthalpy for a reaction, kcal/mole, kJ/mole

$\sum \Delta H^\circ_{f\,(\text{products})}$ = sum of the change in standard enthalpy for products, kcal/mole, kJ/mole

$\sum \Delta H^\circ_{f\,(\text{reactants})}$ = sum of the change in standard enthalpy for reactants, kcal/mole, kJ/mole.

Table 2.11 lists standard enthalpy values for various compounds.

Example 2.13 Calculating ΔH° for the oxidation of glucose

For standard conditions, calculate the change in enthalpy or heat required for oxidizing glucose to carbon dioxide and water vapor as shown in the equation given below.

Table 2.11 Standard enthalpies of formation for various compounds at 25°C.

Species	State	ΔH_f°, kcal/mol	Species	State	ΔH_f°, kcal/mol
Ca^{2+}	*aq*	−130	H_2O	*g*	−57.798
$CaCO_3$	c	−288.4	HS^-	*aq*	−4.22
CaF_2	c	−290.3[a]	H_2S	*g*	−4.815
$Ca(OH)_2$	c	−235.8[a]	H_2S	*aq*	−9.4
$CaSO_4 \cdot 2H_2O$	c	−483.0[a]	H_2SO_4	*l*	−193.9[a]
CH_4	*g*	−17.889	Mg^{2+}	*aq*	−53.3
CH_3CH_3	*g*	−20.23[a]	Na^+	*aq*	−57.3[a]
CH_3COOH	*aq*	−116.7[a]	NH_3	*g*	−11.04
CH_3COO^-	*aq*	−116.84	NH_3	*aq*	−19.32
CH_3CH_2OH	*l*	−66.37[a]	NH_4^+	*aq*	−31.74
CO_2	*g*	−94.05	NO_2^-	*aq*	−25.4[a]
CO_2	*aq*	−98.69	NO_3^-	*aq*	−49.372
CO_3^{2-}	*aq*	−161.63	OH^-	*aq*	−54.957
F^-	*aq*	−78.66[a]	S^{2-}	*aq*	10.0[a]
HCO_3^-	*aq*	−165.18	SO_4^{2-}	*aq*	−216.90
H_2CO_3*	*aq*	−167.0	Zn^{2+}	*aq*	−36.4[a]
H_2O	*l*	−68.3174	ZnS	c	−48.5[a]

Enthalpy values from Snoeyink & Jenkins (1980) p. 64.
[a] Enthalpy values from Sawyer *et al.* (1994) p. 53.
1 cal = 4.184Joules; *aq*-aqueous; *l*-liquid; *c*-crystal; *g*-gas

The standard enthalpy of formation for glucose, $C_6H_{12}O_{6(s)}$, is −304.3 kcal/mol.

$$C_6H_{12}O_{6(s)} + 6\,O_{2(g)} \rightarrow 6\,CO_{2(g)} + 6\,H_2O_{(g)}$$

Solution

From Equation (2.35) and using ΔH_f° values from Table 2.11, calculate ΔH°.

$$\Delta H^\circ = \sum \Delta H^\circ_{f\,(\text{products})} - \sum \Delta H^\circ_{f\,(\text{reactants})}$$

$$\sum \Delta H^\circ_{f\,(\text{products})} = 6 \times \left(-94.05\,\frac{\text{kcal}}{\text{mol}}\right) + 6 \times \left(-57.8\,\frac{\text{kcal}}{\text{mol}}\right)$$

$$\sum \Delta H^\circ_{f\,(\text{products})} = -911.1\,\frac{\text{kcal}}{\text{mol}}$$

$$\sum \Delta H^\circ_{f\,(\text{reactants})} = 1 \times \left(-304.3\,\frac{\text{kcal}}{\text{mol}}\right) + 6 \times \left(0\,\frac{\text{kcal}}{\text{mol}}\right)$$

$$\sum \Delta H^\circ_{f\,(\text{reactants})} = -\,304.3\,\frac{\text{kcal}}{\text{mol}}$$

$$\Delta H^\circ = -911.1\,\frac{\text{kcal}}{\text{mol}} - \left(-304.3\,\frac{\text{kcal}}{\text{mol}}\right) = \boxed{-606.8\,\frac{\text{kcal}}{\text{mol}}}$$

2.5.2 Second law of thermodynamics

The second law of thermodynamics states that all systems tend to approach a state of equilibrium. Entropy is the concept developed by physical chemists used to judge whether or not a chemical process is at equilibrium or proceeding to equilibrium. Work can only be obtained from systems that have not reached equilibrium. In a reversible process, the entropy of the universe is constant. In an irreversible process, the entropy of the universe increases. An increase in entropy relates to a decrease in the orderliness of a system, while a decrease in entropy causes an increase in the orderliness of the system. In equation form, the change in entropy is defined as follows:

$$\Delta S = \int \frac{dq_{rev}}{T} \tag{2.36}$$

where:
ΔS = change in entropy
q_{rev} = heat absorbed by the system during a reversible chemical reaction
T = absolute temperature of the system.

The entropy of a substance at 0°K is zero. For an isothermal process, Equation (2.36) simplifies to:

$$q_{rev} = T\Delta S \quad (2.37)$$

Changes in entropy at standard conditions (25°C) can be calculated using entropy values for the products and reactants in chemical reactions. The change in entropy at standard conditions is calculated according to Equation (2.38).

$$\Delta S^\circ = \sum S^\circ(\text{products}) - \sum S^\circ(\text{reactants}) \quad (2.38)$$

where:

ΔS° = change in standard entropy for the reaction, J/K

$\sum S^\circ(\text{products})$ = sum of the standard entropies of products, $\frac{\text{J}}{\text{K} \cdot \text{mol}}$

$\sum S^\circ(\text{reactants})$ = sum of the standard entropies of reactants, $\frac{\text{J}}{\text{K} \cdot \text{mol}}$.

Table 2.12 presents standard molar entropy values for various compounds.

Example 2.14 Calculating changes in entropy at standard conditions

For the following chemical reaction, show how to calculate the change in entropy at standard conditions.

$$aA + bB \rightleftharpoons lL + mM$$

Solution

From Equation (2.38):

$$\Delta S^\circ = \sum S^\circ(\text{products}) - \sum S^\circ(\text{reactants})$$

$$\Delta S^\circ = lS^\circ_L + mS^\circ_M - aS^\circ_A + bS^\circ_B$$

Table 2.12 Standard molar entropy values for various compounds at 25°C.

Compound or element	Entropy, S°, $\frac{\text{J}}{\text{K·mol}}$	Compound or element	Entropy, S°, $\frac{\text{J}}{\text{K·mol}}$
$C_{(C)}$(graphite)	5.74 ± 0.10	$H_2O_{(l)}$	69.95 ± 0.03
$Na_{(c)}$	51.30 ± 0.20	$H_2O_{(g)}$	188.835 ± 0.010
$P_{(c)}$(white)	41.09 ± 0.25	$NH_{3(g)}$	192.77 ± 0.05
$S_{(c)}$(rhombic)	32.054 ± 0.050	$H_2O_{2(l)}$	109.6
$Ag_{(c)}$	42.55 ± 0.20	$CH_3OH_{(l)}$	239.865
$He_{(g)}$	126.153 ± 0.002	$CHCl_{3(l)}$	201.7
$Ne_{(g)}$	146.328 ± 0.003	$CH_3Cl_{(g)}$	234.6
$Ar_{(g)}$	154.846 ± 0.003	$CO_{(g)}$	197.660 ± 0.004
$Kr_{(g)}$	164.085 ± 0.003	$CO_{2(g)}$	213.785 ± 0.010
$Xe_{(g)}$	169.685 ± 0.003	$NO_{(g)}$	210.745
$H_{2(g)}$	130.680 ± 0.003	$NO_{2(g)}$	240.166
$N_{2(g)}$	191.609 ± 0.004	$N_2O_{4(g)}$	304.4
$O_{2(g)}$	205.152 ± 0.005	$SO_{2(g)}$	248.223 ± 0.050
$F_{2(g)}$	202.791 ± 0.005	$CH_{4(g)}$	186.369
$Cl_{2(g)}$	223.081 ± 0.010	$C_2H_{2(g)}$	200.927
$Br_{2(g)}$	245.468 ± 0.005	$C_2H_{4(g)}$	219.316
$I_{2(g)}$	260.687 ± 0.005	$C_2H_6O_{(g)}$	281.622

Entropy values from *CRC Handbook of Chemistry and Physics* (2010), 91st Edition, pp. 5-1, 5-2, 5-3, 5-13, 5-16, 5-19, 5-20, 5-46, 5-47, 5-49, 5-61.
1 cal = 4.184 Joules.
aq-aqueous *l*-liquid *c*-crystal *g*-gas

Example 2.15 Calculating changes in entropy at standard conditions

Determine the change in entropy at standard conditions for the following chemical reaction:

$$CO_{(g)} + 3\,H_{2(g)} \rightarrow CH_{4(g)} + H_2O_{(g)}$$

Solution

From Equation (2.38):

$$\Delta S^\circ = \sum S^\circ(\text{products}) - \sum S^\circ(\text{reactants})$$

Obtain entropy values from Table 2.12:

$$\sum S^\circ(\text{products}) = 1 \times 186.369 \frac{\text{J}}{\text{K} \cdot \text{mol}} + 1 \times 188.835 \frac{\text{J}}{\text{K} \cdot \text{mol}}$$

$$\sum S^\circ(\text{products}) = 375.204 \frac{\text{J}}{\text{K}}$$

$$\sum S^\circ(\text{reactants}) = 1 \times 197.66 \frac{\text{J}}{\text{K} \cdot \text{mol}} + 3 \times 130.68 \frac{\text{J}}{\text{K} \cdot \text{mol}}$$

$$\sum S^\circ(\text{reactants}) = 589.7 \frac{\text{J}}{\text{K}}$$

$$\Delta S^\circ = 375.204 \frac{\text{J}}{\text{K}} - 589.7 \frac{\text{J}}{\text{K}} = \boxed{-214.496}$$

2.5.3 Gibbs free energy, ΔG

Free energy is that component of the total energy of a system which can do work under isothermal conditions. For irreversible processes, ΔS increases; however, the change in free energy ΔG decreases. In equation form, free energy is defined as follows:

$$G = H - TS \tag{2.39}$$

where:
G = Gibbs free energy
H = enthalpy of system
T = absolute temperature of system
S = entropy of system.

Equation (2.39) may be expressed in differential form as:

$$dG = dH - T\,dS - S\,dT \tag{2.40}$$

At constant temperature, $dT = 0$ and at constant pressure, $dH = dq$, Equation (2.40) reduces to:

$$dG = dq - T\,dS \tag{2.41}$$

$$\Delta G = \Delta q - T\,\Delta S \quad \text{or} \quad \Delta G = \Delta H - T\,\Delta S \tag{2.42}$$

Free energy calculations allow us to predict whether or not a chemical reaction will occur. When $\Delta G = 0$, the reaction is at equilibrium. If $\Delta G > 0$ (i.e., a positive value), the reaction will occur spontaneously or proceed in the forward direction as written. If $\Delta G < 0$ (i.e., a negative value), the reaction will not occur spontaneously in the forward direction, but will proceed in the reverse.

The change in energy at standard conditions can be calculated using Equation (2.43). Standard conditions mean 1 atm pressure and a temperature of 25°C.

$$\Delta G^\circ = \sum \Delta G_f^\circ(\text{products}) - \sum \Delta G_f^\circ(\text{reactants}) \tag{2.43}$$

where:
ΔG° = change in free energy of formation of reaction, kcal/mol, kJ/mol
$\sum \Delta G_f^\circ(\text{products})$ = sum of the changes in the free energy of products, kcal/mol, kJ/mol
$\sum \Delta G_f^\circ(\text{reactants})$ = sum of the changes in the free energy of reactants, kcal/mol, kJ/mol.

Table 2.13 presents standard free energy of formation values for various compounds at 25°C.

Example 2.16 Calculating changes in free energy at standard conditions

Calculate the change in free energy for the combustion of ethane at 25°C.

$$C_2H_{6(g)} + \frac{7}{2}\,O_{2(g)} \rightarrow 2\,CO_{2(g)} + 3\,H_2O_{(l)}$$

Solution

From Equation (2.43):

$$\Delta G^\circ = \sum \Delta G_f^\circ(\text{products}) - \sum \Delta G_f^\circ(\text{reactants})$$

$$\Delta G^\circ = 2\Delta G_f^\circ(CO_2)_{(g)} + 3\Delta G_f^\circ(H_2O)_{(l)} - \Delta G_f^\circ(C_2H_6)_{(g)} - \frac{7}{2}\Delta G_f^\circ(O_2)_{(g)}$$

Obtain free energy values for the various elements and compounds from Table 2.13.

$$\Delta G^\circ = 2 \times \left(-94.26 \frac{\text{kcal}}{\text{mole}}\right) + 3 \times \left(-56.69 \frac{\text{kcal}}{\text{mole}}\right) - 1 \times \left(-7.86 \frac{\text{kcal}}{\text{mole}}\right) - \frac{7}{2} \times (0)$$

$$\Delta G^\circ = \boxed{-350.73 \frac{\text{kcal}}{\text{mole}}}$$

Table 2.13 Standard free energy of formation for various compounds at 25°C.

Species	State	ΔG_f°, kcal/mol	Species	State	ΔG_f°, kcal/mol
Ca^{2+}	*aq*	−132.18	H_2O	*l*	−56.69
$CaCO_3$	c	−269.78	H_2O	*g*	−54.6357
CaF_2	c	−277.7[a]	HS^-	*aq*	−3.01
$Ca(OH)_2$	c	−214.33[a]	H_2S	*g*	−7.892
$CaSO_4 \cdot 2H_2O$	c	−429.2[a]	H_2S	*aq*	−6.54
CH_4	*g*	−12.140	Mg^{2+}	*aq*	−108.99
CH_3CH_3	*g*	−7.86[a]	Na^+	*aq*	−62.6[a]
CH_3COOH	*aq*	−95.5[a]	NH_3	*g*	−3.976
CH_3COO^-	*aq*	−89	NH_3	*aq*	−6.37
CH_3CH_2OH	*l*	−41.8[a]	NH_4^+	*aq*	−19.00
$C_6H_{12}O_6$	*aq*	−217.0[a]	NO_2^-	*aq*	−8.25[a]
CO_2	*g*	−94.26	NO_3^-	*aq*	−26.43
CO_2	*aq*	−92.31	OH^-	*aq*	−37.595
CO_3^{2-}	*aq*	−126.22	S^{2-}	*aq*	20.0[a]
F^-	*aq*	−66.1[a]	SO_4^{2-}	*aq*	−177.34
HCO_3^-	*aq*	−140.31	Zn^{2+}	*aq*	−35.2[a]
H_2CO_3*	*aq*	−149.0	ZnS	c	−47.4[a]

Free energy values from Snoeyink and Jenkins (1980), p. 64.
[a] Free energy values from Sawyer et *al.* (1994), p. 53.
1 cal = 4.184 Joules.
aq-aqueous *l*-liquid *c*-crystal *g*-gas

2.5.4 Extent of reaction, equilibrium constant (K)

The equilibrium constant (K) can be determined from free energy data. Given the following reaction:

$$aA + bB \leftrightarrow cC + dD \tag{2.44}$$

where:
a = moles of substance A
b = moles of substance B
c = moles of substance C
d = moles of substance D.

The equilibrium constant represents the ratio of the concentration of products to reactants at equilibrium, i.e., the concentration of the reactants and products have reached equilibrium and do not change with time. Mathematically, the equilibrium constant is defined by the following equation:

$$K = \frac{[C]^c[D]^d}{[A]^a[B]^b} \tag{2.45}$$

where:
K = equilibrium constant
A = concentration of substance A, moles/L
B = concentration of substance B, moles/L
C = concentration of substance C, moles/L
D = concentration of substance D, moles/L.

Free energy is related to the equilibrium constant by the following equation:

$$\Delta G = \Delta G^\circ + RT\ln(K) \tag{2.46}$$

where:
R = universal gas constant and can be expressed in various ways
$R = 0.08206\ (L \cdot atm)/mol$
$R = 8.3144\ J/(K \cdot mol)$
T = absolute temperature, K (273.15 + °C).

At equilibrium when $\Delta G = 0$, Equation (2.46) reduces to the following:

$$\Delta G^\circ = -RT\ln(K) \tag{2.47}$$

2.5.4.1 Temperature Dependency of Equilibrium Constant

The equilibrium constant is dependent upon temperature. A differential equation showing the relationship to temperature is shown below:

$$\frac{d\ln(K)}{dT} = \frac{\Delta H^\circ}{RT^2} \quad (2.48)$$

Over a limited temperature range, the change in enthalpy can be assumed to not be a function of temperature. Therefore, integration of Equation (2.48) yields:

$$\int d\ln(K) = \frac{\Delta H^\circ}{R}\int_{T_1}^{T_2}\frac{1}{T^2}dT \quad (2.49)$$

$$\int d\ln(K) = \frac{\Delta H^\circ}{R} \times -\frac{1}{T}\int_{T_1}^{T_2}\frac{1}{T}dT \quad (2.50)$$

$$\ln\frac{K_1}{K_2} = \frac{\Delta H^\circ}{R}\left(\frac{1}{T_2} - \frac{1}{T_1}\right) \quad (2.51)$$

$$\ln\frac{K_1}{K_2} = -\frac{\Delta H^\circ}{R}\left(\frac{T_1 - T_2}{T_1 T_2}\right) \quad (2.52)$$

Example 2.17 Correcting the equilibrium constant for temperature

Calculate the equilibrium constant (K_1) at 10°C for carbonic acid assuming that $K_1 = 4.35 \times 10^{-7}$ at 25°C. ΔH_f° for $H^+_{(aq)} = 0$.

$$H_2CO_{3(aq)} \leftrightarrow H^+_{(aq)} + HCO^-_{3(aq)}$$

Solution

First calculate ΔH° From Equation (2.35) using the enthalpy values from Table 2.11:

$$\Delta H^\circ = \sum \Delta H^\circ_{f\,(products)} - \sum \Delta H^\circ_{f\,(reactants)}$$

$$\Delta H^\circ = 1 \times (0) + 1 \times \left(-165.18\,\frac{kcal}{mol}\right)$$

$$-1 \times \left(-167.0\,\frac{kcal}{mol}\right) = 1.82\ kcal = 1820\ cal$$

$$\ln\frac{K_2}{K_1} = -\frac{\Delta H^\circ}{R}\left(\frac{T_1 - T_2}{T_1 T_2}\right)$$

$$\ln\frac{K_{283}}{K_{298}} = -\frac{1820\ cal}{1.98\ \frac{cal}{K\cdot mol}}\left(\frac{298\ K - 283\ K}{298\ K \times 283\ K}\right)$$

$$\ln\frac{K_{283}}{K_{298}} = -0.163$$

$$\frac{K_{283}}{K_{298}} = e^{-0.163} = 0.85$$

$$K_{283} = 0.85K_{298} = 0.85(4.35 \times 10^{-7}) = \boxed{3.7 \times 10^{-7}}$$

2.6 Acid-base chemistry

Chemists often use the "p scale" for expressing concentrations and equilibrium constants. This is based on common logarithms (base 10) and is a convenient way of expressing numbers with small magnitudes. In the "p scale" system, if N represents a number, then:

$$pN = -\log N \quad (2.53)$$

Or the value of N can be calculated if the value of pN is known.

$$N = 10^{-pN} \quad (2.54)$$

Rather than expressing the hydrogen ion concentration in moles/L, Sorensen in 1909 proposed the concept of pH (Sawyer and McCarty, 1978). The pH scale is usually represented by a scale ranging from 0 to 14, with a neutral pH indicated by a value of 7. Solutions are acidic when the pH is below 7 and basic or alkaline when the pH is higher than 7. The pH of water at 25°C is equal to 7, therefore, it is neutral. Mathematically, pH is defined as the negative of the logarithm of the hydrogen ion concentration.

$$pH = -\log[H^+] \quad (2.55)$$

where $[H^+]$ = hydrogen ion concentration, moles/L.

The hydrogen ion concentration, as monitored by pH, is an important parameter in environmental engineering. Rates of biological and chemical reactions are impacted by pH and, outside a specific range, they may not proceed. Water and wastewater treatment processes generally function best within a pH range of 6 to 8.5. Chemical coagulants and polymers each have an optimum pH where they function best. Fish and most aquatic life do not thrive at pH values lower than 5.

The "p scale" may be used to express other constituents in a solution. In equilibrium chemistry, the concentration of hydroxide may be expressed in terms of pOH, analogous to the pH concept. Mathematically, pOH is defined as the negative of the logarithm of the hydroxide ion concentration, as shown below.

$$pOH = -\log[OH^-] \quad (2.56)$$

where $[OH^-]$ = hydroxide ion concentration, moles/L.

Table 2.14 lists several household items and their estimated pH value.

Acid-base chemistry is best described by the Bronsted-Lowry model. According to this model, an acid is a proton (H^+) donor, and a base is a proton acceptor. This can best be illustrated with the following reaction:

$$HA_{(aq)} + H_2O \leftrightarrow H_3O^+_{(aq)} + A^-_{(aq)} \quad (2.57)$$

Table 2.14 Estimate of pH for several common items.

Item	pH
1 M HCl	0
Lemonade	2.2–3.0
Raw apples	2.2–3.0
Milk	6.4–7.6
Pure water	7.0
Drinking water	6.5–8.5
Ammonia	11–12
1 M NaOH	14

Based on EPA website on acid rain at: http://www.epa.gov/acidrain/measure/ph.html
pH of 1 M HCl and 1 M NaOH were calculated.

where:

HA = represents some generic acid dissolved in water

H_2O = represents a base in this reaction

H_3O^+ = hydronium ion, typically shown only as a hydrogen ion, $H^+.H_3O^+$ represents a conjugate acid in this reaction

A^- = represents the conjugate base formed in this reaction.

The conjugate acid forms when the proton is transferred from the acid to the base; while the conjugate base results from the loss of the proton from the acid. In Equation (2.57), the acid-base pairs are: HA (acid) and A^- (base), and H_2O (base) and H_3O^+ (acid). A better example of the conjugate acid-base paring is shown in Equation (2.58).

$$HCl_{(aq)} + H_2O_{(l)} \leftrightarrow H_3O^+_{(aq)} + Cl^-_{(aq)} \tag{2.58}$$

In the above reaction, hydrochloric acid loses a proton to form chloride, its conjugate base, while water (base) accepts a proton forming the hydronium ion, its conjugate acid.

Amphoteric compounds are those that can react either as an acid or a base. The following amphoteric compounds are encountered in environmental systems: H_2O, HCO_3^-, HSO_4^-, and NH_3. Water can ionize according to the following reaction:

$$H_2O_{(l)} + H_2O_{(l)} \leftrightarrow H_3O^+_{(aq)} + OH^-_{(aq)} \tag{2.59}$$

Water acts as both an acid and a base. As mentioned previously, the hydronium ion and hydrogen ion are used interchangeably. The hydrogen ion does not actually exist in aqueous systems, but combines with a molecule of water to form the hydronium ion (H_3O^+). Equation (2.59) is typically abbreviated as shown below:

$$H_2O_{(l)} \leftrightarrow H^+_{(aq)} + OH^-_{(aq)} \tag{2.60}$$

The equilibrium constant (K) at 25°C for water is 1.8×10^{-16}.

$$K = \frac{[H^+][OH^-]}{[H_2O]} = 1.8 \times 10^{-16} \tag{2.61}$$

The concentration of water in dilute solutions is relatively constant; in 1 L of water, there are 55.5 moles, as shown in Equation (2.62).

$$\frac{1000 \text{ g } H_2O/L}{18 \text{ g } H_2O/\text{mole}} = 55.5 \frac{\text{moles}}{L} \tag{2.62}$$

$$K = \frac{[H^+][OH^-]}{[H_2O]} = 1.8 \times 10^{-16} \tag{2.63}$$

$$[H^+][OH^-] = (1.8 \times 10^{-16}) \times [H_2O] = K_w \tag{2.64}$$

$$[H^+][OH^-] = (1.8 \times 10^{-16}) \times \left(55.6 \frac{\text{moles}}{L}\right) = K_w = 1.0 \times 10^{-14} \tag{2.65}$$

$$[H^+][OH^-] = K_w = 1.0 \times 10^{-14} \tag{2.66}$$

The dissociation constant (K_w) for water at 25°C is 1×10^{-14} or 10^{-14}. Using the "p scale", Equation (2.66) can be expressed as:

$$pH + pOH = pK_w \tag{2.67}$$

$$pH + pOH = 14 \tag{2.68}$$

Example 2.18 Calculate the pH of pure water at 25°C

The dissociation constant (K_w) at 25°C for water is 10^{-14}. Calculate the pH of pure water.

Solution

The charge balance for a solution containing pure water is given below:

$$[H^+] = [OH^-]$$

Substituting into Equation (2.66) yields a hydrogen ion concentration of 1.0×10^{-7}moles/L.

$$[H^+][OH^-] = K_w = 1.0 \times 10^{-14}$$

$$[H^+][H^+] = K_w = 1.0 \times 10^{-14}$$

$$[H^+]^2 = K_w = 1.0 \times 10^{-14}$$

$$[H^+] = \sqrt{K_w} = \sqrt{1.0 \times 10^{-14}} = 1.0 \times 10^{-7}$$

The pH is then determined by substitution into Equation (2.55):

$$pH = -\log [H^+]$$

$$pH = -\log (1.0 \times 10^{-7}) = \boxed{7.0}$$

Example 2.19 Calculate the pH, pOH, and $[OH^-]$ in a solution

Determine the pH, pOH, and $[OH^-]$ in a solution in which $[H^+] = 6.00 \times 10^{-4}$.

Solution

From Equation (2.55):

$$pH = -\log[H^+]$$

$$pH = -\log(6.00 \times 10^{-4}) = \boxed{3.22}$$

From Equation (2.68):

$$pH + pOH = 14$$

$$pOH = 14 - pH = 14 - 3.22 = \boxed{10.78}$$

From Equation (2.56):

$$pOH = -\log[OH^-]$$

$$[OH^-] = 10^{-pOH} = 10^{-10.78} = \boxed{1.66 \times 10^{-11}}$$

Alternatively, the hydroxide concentration can be calculated by rearranging Equation (2.66):

$$[H^+][OH^-] = K_w = 1.0 \times 10^{-14}$$

$$[OH^-] = \frac{1.0 \times 10^{-14}}{[H^+]} = \frac{1.0 \times 10^{-14}}{6.0 \times 10^{-4}} = \boxed{1.66 \times 10^{-11}}$$

Example 2.20 Calculate the $[H^+]$ in a solution

Determine the $[H^+]$ concentration in a solution with a pH of 9.6.

Solution

Rearrange Equation (2.55) to solve for $[H^+]$.

$$pH = -\log[H^+]$$

$$[H^+] = 10^{-pH} = 10^{-9.6} = \boxed{2.5 \times 10^{-10}}$$

2.6.1 Strong acid and base solutions

Water has the ability to cause many substances to split apart into charged species. This process is known as dissociation or ionization. Strong electrolytes dissociate completely in water. Almost all salts are completely dissociated in water. Examples of salts that dissociate readily include KBr, NaCl, $CaCl_2$, $Ca(NO_3)_2$, $Fe(ClO_4)_3$, and $Cd(BrO_3)_2$.

Example 2.21 Dissociation of sodium chloride in water

Calculate the molar concentration of the sodium and chloride ions when 2 moles of sodium chloride are added to distilled water to form 1 liter of solution.

Solution

Assume complete dissociation of the sodium chloride.

$$NaCl \rightarrow Na^+ + Cl^-$$

Since stoichiometry indicates that 1 mole of sodium chloride yields 1 mole of sodium ion and 1 mole of chloride ion, the resulting solution will contain the following:

$$[NaCl] = \boxed{0} \qquad [Na^+] = \boxed{2.0\ M} \qquad [Cl^-] = \boxed{2.0\ M}$$

The dissociation of strong acids and bases is almost complete. Few acids are completely dissociated or ionized in water. The dissociation constant of an acid or base is used to quantify its strength. The dissociation constant for an acid is represented by K_a and for a base by K_b. Strong acids have large dissociation constants or ΔG values that are very negative.

Strong acids generally have a $pK_a < 2$, where $pK_a = -\log[K_a]$ (Mihelcic, 1999, page 88). Examples of strong acids ($K_a \approx \infty$) include: HCl, HNO_3, H_2SO_4, HBr, HI, and $HClO_4$.

Acids are proton donors. When an acid dissociates in water, a hydrogen ion is released forming a conjugate acid and a conjugate base as shown in Equation (2.69).

$$HA_{(aq)} + H_2O \leftrightarrow H_3O^+_{(aq)} + A^-_{(aq)} \tag{2.69}$$

where:
$H_3O^+{}_{(aq)}$ = hydronium ion or conjugate acid
$A^-{}_{(aq)}$ = conjugate base.

In most environmental engineering texts, Equation (2.69) is simplified so that water is omitted from the reaction and the hydronium ion is replaced with the hydrogen ion. Equation (2.70) is the general reaction for the dissociation of a generic acid.

$$HA \leftrightarrow H^+ + A^- \tag{2.70}$$

where:
HA = generic acid
H^+ = hydrogen ion
A^- = conjugate base ion.

Equation (2.71) is used to calculate the acid dissociation constant, K_a.

$$K_a = \frac{[H^+][A^-]}{[HA]} \qquad (2.71)$$

where:
$[HA]$ = concentration of generic acid, moles/L
$[H^+]$ = hydrogen ion concentration, moles/L
$[A^-]$ = concentration of conjugate base, moles/L
K_a = dissociation constant for a generic acid.

Example 2.22 Dissociation of electrolytes in water

Calculate the molar concentration of the ions in a solution prepared by diluting 0.10 moles of $Ca(NO_3)_2$, 0.15 moles of HCl, and 0.30 moles of $CaCl_2$ to 1 liter with distilled water.

Solution

Assume complete dissociation of the hydrochloric acid since it is a strong acid and complete dissociation of calcium nitrate and calcium chloride since they are strong salts.

$$HCl \rightarrow H^+ + Cl^-$$

The hydrogen ion concentration is equal to 0.15 moles/L, since the hydrochloric acid is completely dissociated. This reaction also yields 0.15 moles/L of chloride ion.

$$CaCl_2 \rightarrow Ca^{2+} + 2\,Cl^-$$

The dissociation of calcium chloride yields 0.30 moles/L of calcium ion and 0.6 moles/L of chloride ion (2 × 0.30 moles/L).

$$Ca(NO_3)_2 \rightarrow Ca^{2+} + 2\,NO_3^-$$

The dissociation of calcium nitrate yields 0.10 moles/L of calcium ion and 0.20 moles/L of nitrate ion (2 × 0.10 moles/L). The final concentrations of ionic species in solution are presented below:

$$[H^+] = \boxed{0.15\text{ M}}$$

$$[Ca^{2+}] = 0.10 + 0.30 = \boxed{0.40\text{ M}}$$

$$[Cl^-] = 0.15 + 2 \times 0.30 = \boxed{0.75\text{ M}}$$

$$[NO_3^-] = 2 \times 0.10 = \boxed{0.20\text{ M}}$$

Also show that the solution is electrically neutral.

$$[H^+] + 2[Ca^+] = [NO_3^-] + [Cl^-]$$

$$0.15 + 2(0.40) = 0.20 + 0.75$$

$$0.95 = 0.95$$

Bases are proton acceptors. When a base dissociates in water, a hydroxide ion is released and a conjugate acid is formed as shown in Equation (2.72). BOH represents a generic base.

$$BOH \leftrightarrow B^+ + OH^- \qquad (2.72)$$

where:
BOH = generic base
B^+ = conjugate acid
OH^- = hydroxide ion.

Equation (2.73) is used to calculate base dissociation constant, K_b.

$$K_b = \frac{[B^+][OH^-]}{[BOH]} \qquad (2.73)$$

where:
$[BOH]$ = concentration of generic base, moles/L
$[B^+]$ = conjugate acid concentration, moles/L
$[OH^-]$ = hydroxide ion concentration, moles/L
K_b = dissociation constant for a generic base.

Strong bases that dissociate readily ($K_b \approx \infty$) include NaOH, KOH, LiOH, and $Ba(OH)_2$. Table 2.15 shows the equilibrium or dissociation constants for several acids and bases used in environmental applications.

2.6.2 Weak acids and bases

A weak electrolyte is a compound which does not completely dissociate in water. Most acids and almost all organic acids are weak electrolytes. Examples of weak acids include H_2S, H_2CO_3, HF, H_3PO_4, HCN, and CH_3COOH. Weak bases also result in incomplete dissociation in water. Examples of weak bases include NH_3, N_2H_4 (hydrazine), and organic bases such as $C_6N_5H_2$ (aniline) and $C_2N_5NH_2$ (ethylhyamine). Complexes or coordination compounds, which are species of a central metal ion and one or more associated groups are weak electrolytes. Examples of these include: $Ag(NH_3)_2^+$, $Fe(CNS)^{2+}$, $Hg(CN)_9^+$, and $Fe(CN)_6^{3+}$. Water is a weak electrolyte, since it ionizes only slightly into hydrogen and hydroxide ions.

When solving weak acid-base equilibrium problems, follow the steps below:

- Write all chemical reactions that occur.
- Write all equilibrium constant expressions.
- Write the electroneutrality equation.

Table 2.15 Equilibrium (dissociation) constants for several acids and bases.

Acid	Name	pK_a	Base	Name	pK_b
$HClO_4$	Perchloric	−7	ClO_4^-	Perchlorate ion	21
HCl	Hydrochloric	~−3	Cl^-	Chloride ion	17
H_2SO_4	Sulfuric	~−3	HSO_4^-	Bisulfate ion	17
HNO_3	Nitric	0	NO_3^-	Nitrate ion	14
H_3O^+	Hydronium ion	0	H_2O	Water	14
HSO_4^-	Bisulfate ion	2.0	SO_4^{2-}	Sulfate ion	12
H_3PO_4	Phosphoric	2.1	$H_2PO_4^-$	Dihydrogen phosphate	11.9
$[Fe_3(H_2O_6)]^{3+}$	Ferric ion	2.2	$[Fe_3(H_2O)_5(OH)]^{2+}$	Hydroxo iron (III) complex	11.8
CH_3COOH	Acetic	4.7	CH_3COO^-	Acetate ion	9.3
$[Al_3(H_2O)_6]^{3+}$	Aluminum ion	4.9	$[Al(H_2O)_5(OH)]^{2+}$	Hydroxo aluminum (III) complex	9.1
$H_2CO_3^*$	Carbon dioxide and carbonic acid	6.3	HCO_3^-	Bicarbonate	7.7
H_2S	Hydrogen sulfide	7.1	HS^-	Bisulfide	6.9
$H_2PO_4^-$	Dihydrogen phosphate	7.2	HPO_4^{2-}	Monohydrogen phosphate	6.8
$HOCl$	Hypchlorous	7.5	OCl^-	Hypochorite ion	6.4
HCN	Hydrocyanic	9.3	CN^-	Cyanide ion	4.7
H_3CO_3	Boric	9.3	$B(OH)_4^-$	Borate ion	4.7
NH_4^+	Ammonium ion	9.3	NH_3	Ammonia	4.7
H_4SiO_4	O-Silicic acid	9.5	H_3SiO_4	Trihydrogen silicate ion	4.5
HCO_3^-	Bicarbonate	10.3	CO_3^{2-}	Carbonate	3.7
H_2O_2	[1]Hydrogen peroxide	11.7	HO_2^-	[1]Hydrogen peroxide anion	2.3
HPO_4^{2-}	Monohydrogen phosphate	12.3	PO_4^{3-}	Phosphate	1.7
HS^-	Bisulfide ion	14	S^{2-}	Sulfide	0
H_2O	Water	15.74[a]	OH^-	Hydroxide	−1.74[a]
NH_3	Ammonia	~23	NH_2^-	Amide ion	−9
OH^-	Hydroxide	~24	O^{2-}	Oxide	−10

pK_a and pK_b values from Snoeyink & Jenkins (1980). *Water Chemistry*, p. 90 and 91.

[a] pK_a and pK_b values from Stumm & Morgan (1996). *Aquatic Chemistry: Chemical Equilibria and Rate in Natural Waters*, 3rd Edition, p. 96.

- Write the material balance equation.
- Solve algebraically.

2.6.2.1 Electroneutrality

All solutions must be electrically neutral; therefore, a charge balance must exist. The positive ions (cations) must equal to the negative ions (anions) for electroneutrality. A complete water analysis should produce a solution that is electrically neutral. The electroneutrality or charge balance equation is useful in solving acid-base equilibria problems. Two examples are presented below to illustrate how to determine the charge balance in a solution.

Example 2.23 Calculating the charge balance for acetic acid

Determine the charge balance for a solution of acetic acid (HAc).

Solution

Write all chemical reactions that occur:

$$HAc \leftrightarrow H^+ + Ac^-$$

$$H_2O \leftrightarrow H^+ + OH^-$$

The following chemical species are present at equilibrium: HAc, Ac^-, H_2O, H^+, and OH^-

The charge balance is shown below:

$$[H^+] = [OH^-] + [Ac^-]$$

Example 2.24 Calculating the charge balance for sodium carbonate

Determine the charge balance for a solution of sodium carbonate (Na_2CO_3).

Solution

Write all chemical reactions that occur:

$$Na_2CO_3 \rightarrow 2\,Na^+ + CO_3^{2-}$$
$$H_2CO_3^* \leftrightarrow H^+ + HCO_3^-$$
$$HCO_3^- \leftrightarrow H^+ + CO_3^{2-}$$
$$H_2O \leftrightarrow H^+ + OH^-$$

The following chemical species are present at equilibrium: Na^+, $H_2CO_3^*$, HCO_3^-, CO_3^{2-}, H_2O, H^+, and OH^-

The charge balance is shown below:

$$[H^+] + [Na^+] = [OH^-] + [HCO_3^-] + 2[CO_3^{2-}]$$

2.6.2.2 Proton Condition Equation

When working with acid-base equilibria problems, it is necessary to introduce the concept of the proton condition. The proton condition or equation is developed by performing a mass balance on protons. To start with, however, a reference level or zero level must be established. The zero level represents the chemical species present which are added to a solution or that are present initially. As the solution moves towards equilibrium, some chemical species gain protons and others lose protons. The proton condition equation is then developed by equating all chemical species that have gained a proton to those that have lost a proton relative to the zero or reference level. Two examples are presented below to illustrate how to develop the proton equation. The chemical species present at the reference level do not appear in the proton equation.

Example 2.25 Calculating the proton condition equation for acetic acid

Determine the proton condition equation for a solution of acetic acid (HAc).

Solution

Write all chemical reactions that occur:

$$HAc \leftrightarrow H^+ + Ac^-$$
$$H_2O \leftrightarrow H^+ + OH^-$$

The following chemical species are present at equilibrium: HAc, Ac^-, H_2O, H^+, and OH^-

The following chemical species are initially present in the solution (reference or zero level): HAc and H_2O.

Chemical species that gain a proton include H^+:

$$H_2O + H^+ \leftrightarrow H_3O^+ \quad \text{(hydronium ion)}$$

which is generally represented as a hydrogen ion or H^+.

Chemical species that lose a proton include Ac^- and OH^-:

$$HAc \leftrightarrow H^+ + Ac^-$$
$$H_2O \leftrightarrow H^+ + OH^-$$

The proton condition equation is shown below:

$$[H^+] = [OH^-] + [Ac^-]$$

For acetic acid, the proton condition is the same as the electroneutrality equation.

Example 2.26 Calculating the proton condition equation for sodium bicarbonate

Determine the proton condition equation for a solution of sodium bicarbonate ($NaHCO_3$).

Solution

Write all chemical reactions that occur:

$$NaHCO_3 \leftrightarrow Na^+ + HCO_3^-$$
$$H_2CO_3^* \leftrightarrow H^+ + HCO_3^-$$
$$HCO_3^- \leftrightarrow H^+ + CO_3^{2-}$$
$$H_2O \leftrightarrow H^+ + OH^-$$

The following chemical species are present at equilibrium: Na^+, $H_2CO_3^*$, HCO_3^-, CO_3^{2-}, H^+, and OH^-

The following chemical species are initially present in the solution (reference or zero level): HCO_3^- and H_2O.

Sodium bicarbonate is a strong base and will completely dissociate into Na^+ and HCO_3^-, so $NaHCO_3$ is not shown as one of the species at the reference level.

Chemical species that gain a proton include: H^+ and $H_2CO_3^*$.

$$H_2O + H^+ \leftrightarrow H_3O^+ \quad \text{(hydronium ion)}$$

$$HCO_3^- \leftrightarrow H^+ + H_2CO_3^*$$

Chemical species that lose a proton include CO_3^{2-} and OH^-

$$H_2O \leftrightarrow H^+ + OH^-$$

$$HCO_3^- \leftrightarrow H^+ + CO_3^{2-}$$

The proton condition equation is shown below:

$$[H^+] + [H_2CO_3^*] = [OH^-] + [CO_3^{2-}]$$

Example 2.27 Calculating the pH in a solution of weak acid

Calculate the hydrogen ion concentration and pH in a solution prepared by diluting 0.10 mole of HCN to 1.0 L with distilled water. The dissociation constant for HCN is 7.2×10^{10}.

Solution

Write all chemical reactions that occur:

$$HCN \leftrightarrow H^+ + CN^-$$

$$H_2O \leftrightarrow H^+ + OH^-$$

Write all equilibrium constant expressions:

$$\frac{[H^+][CN^-]}{[HCN]} = 7.2 \times 10^{-10}$$

$$[H^+][OH^-] = K_w = 1.0 \times 10^{-14}$$

Write the electroneutrality equation:

$$[H^+] = [CN^-] + [OH^-]$$

Write the material balance equation:

$$[HCN] + [CN^-] = 0.10\,M$$

Solve algebraically.

Since the solution is acidic, assume that $[H^+] >> [OH^-]$, therefore, $[H^+] \cong [CN^-]$. Substituting this into the material balance equation yields:

$$[HCN] + [H^+] = 0.10\,M$$

Because K_a is very small, $[HCN] >> [H^+]$, therefore, $[HCN] = 0.10$ M.

Substituting into the equilibrium expression yields the hydrogen ion concentration.

$$\frac{[H^+][CN^-]}{[HCN]} = 7.2 \times 10^{-10}$$

$$\frac{[H^+][H^+]}{0.10\,M} = 7.2 \times 10^{-10}$$

$$[H^+]^2 = 7.2 \times 10^{-10} \times (0.10\,M)$$

$$[H^+] = \sqrt{7.2 \times 10^{-11}} = 8.5 \times 10^{-6}$$

The pH of the solution is calculated from Equation (2.55).

$$pH = -\log [H^+]$$

$$pH = -\log (8.5 \times 10^{-6}) = \boxed{5.1}$$

Check major assumptions. First compare the hydrogen and hydroxide concentrations.

$$[H^+][OH^-] = K_w = 1.0 \times 10^{-14}$$

$$[OH^-] = \frac{1.0 \times 10^{-14}}{[H^+]} = \frac{1.0 \times 10^{-14}}{8.5 \times 10^{-6}} = 1.2 \times 10^{-9}$$

Therefore, $[H^+] >> [OH^-]$, checks.

Next, compare [HCN] and $[H^+]$ concentrations.

$$[HCN] = 0.10\,M \qquad [H^+] = 8.5 \times 10^{-6}$$

Therefore,

$$[HCN] >> [H^+], \text{checks.}$$

Finally, check the electroneutrality equation.

$$[H^+] = [CN^-] + [OH^-]$$

$$\frac{[H^+][CN^-]}{[HCN]} = 7.2 \times 10^{-10}$$

$$[CN^-] = \frac{7.2 \times 10^{-10}\,[HCN]}{[H^+]}$$

$$= \frac{7.2 \times 10^{-10}(0.10\,M)}{8.5 \times 10^{-6}} = 8.47 \times 10^{-6}$$

$$8.5 \times 10^{-6} = 8.47 \times 10^{-6} + 1.2 \times 10^{-9}$$

$$8.5 \times 10^{-6} = 8.5 \times 10^{-6}, \text{checks.}$$

Example 2.28 Calculating the pH in a solution of weak base

Calculate the hydrogen ion concentration, pH, and hydroxide concentration in a solution prepared by diluting 0.10 mole of NaCN to 1.0 L with distilled water. The dissociation constant for HCN is 7.2×10^{10}.

Solution

Write all chemical reactions that occur.

$$NaCN \leftrightarrow NA^+ + CN^-$$

$$CN^- + H_2O \leftrightarrow HCN + OH^-$$

$$H_2O \leftrightarrow H^+ + OH^-$$

Write all equilibrium constant expressions:

$$\frac{[HCN][OH^-]}{[CN^-]} = K_b = 1.4 \times 10^{-5}$$

$$[H^+][OH^-] = K_w = 1.0 \times 10^{-14}$$

Write the electroneutrality equation.

$$[Na^+] + [H^+] = [CN^-] + [OH^-]$$

Write the material balance equation.

$$[Na^+] = 0.10\,M$$

$$[HCN] + [CN^-] = 0.10\,M$$

Solve algebraically.
Since the solution is basic, assume that $[OH^-] >> [H^+]$. Substituting this into the electroneutrality equation yields:

$$[Na^+] + [H^+] = [CN^-] + [OH^-]$$

$$0.10 = [CN^-] + [OH^-]$$

Equate the above equation with the material balance equation for CN as follows:

$$[HCN] + [CN^-] = 0.10 = [CN^-] + [OH^-]$$

$$[HCN] = [OH^-]$$

Because K_b is very small, assume that $[CN^-] >> [OH^-]$, therefore, $[CN^-] = 0.10$ M.

$$\frac{[HCN][OH^-]}{[CN^-]} = K_b = 1.4 \times 10^{-5}$$

$$\frac{[OH^-][OH^-]}{0.10\,M} = K_b = 1.4 \times 10^{-5}$$

$$\frac{[OH^-]^2}{0.10\,M} = K_b = 1.4 \times 10^{-5}$$

$$[OH^-]^2 = (0.10\,M)1.4 \times 10^{-5}$$

$$[OH^-] = \sqrt{1.4 \times 10^{-6}} = \boxed{1.18 \times 10^{-3}\,M}$$

$$[H^+][OH^-] = K_w = 1.0 \times 10^{-14}$$

$$[H^+] = \frac{1.0 \times 10^{-14}}{1.18 \times 10^{-3}} = \boxed{8.45 \times 10^{-12}\,M}$$

The pH of the solution is calculated from Equation (2.55).

$$pH = -\log[H^+]$$

$$pH = -\log(8.45 \times 10^{-12}) = \boxed{11.1}$$

Check major assumptions. First compare the hydrogen and hydroxide concentrations:

$$[OH^-] >> [H^+]$$

$$1.18 \times 10^{-3} >> 8.45 \times 10^{-12}, \text{checks.}$$

Next, compare $[CN^-]$ and $[OH^-]$ concentrations.

$$[CN^-] >> [OH^-]$$

$$0.10\,M >> 1.18 \times 10^{-3}, \text{checks.}$$

Finally, check the electroneutrality equation.

$$0.10 + 8.45 \times 10^{-12} = 0.10 + 1.18 \times 10^{-3}$$

$$\frac{(0.10 + 8.45 \times 10^{-12}) \times 100}{(0.10 + 1.18 \times 10^{-3})}$$

$$= 98.2\% \text{ or } 1.2\% \text{ error, close enough}$$

Example 2.29 Calculating the pH in a solution of weak polyprotic acid

Calculate the $[H^+]$, $[HS^-]$, $[S^{2-}]$, and pH in a 0.10 M H_2S solution. The dissociation constants for H_2S and HS^- are 1.1×10^{-7} and 1.0×10^{-14}, respectively.

Solution

Write all chemical reactions that occur:

$$H_2S \leftrightarrow H^+ + HS^-$$
$$HS^- \leftrightarrow H^+ + S^{2-}$$
$$H_2O \leftrightarrow H^+ + OH^-$$

Write all equilibrium constant expressions:

$$\frac{[H^+][HS^-]}{[H_2S]} = 1.1 \times 10^{-7} = K_1$$
$$\frac{[H^+][S^{2-}]}{[HS^-]} = 1.0 \times 10^{-14} = K_2$$
$$[H^+][OH^-] = K_w = 1.0 \times 10^{-14}$$

Write the electroneutrality equation:

$$[H^+] = [HS^-] + 2[S^{2-}] + \quad [OH^-]$$

Write the material balance equation:

$$[H_2S] + [HS^-] + [S^{2-}] = 0.10\ M$$

Solve algebraically.

Since the solution is acidic, assume that $[H^+] >> [OH^-]$. Substituting this into the electroneutrality equation yields:

$$[H^+] = [HS^-] + 2[S^{2-}]$$

Because K_2 is very small, $[HS^-] >> [S^{2-}]$. The above equation therefore becomes:

$$[H^+] = [HS^-]$$

Since K_1 is also very small, $[H_2S] >> [HS^-]$; then the material balance equation becomes:

$$[H_2S] = 0.10\ M$$

Making the substitution into the first equilibrium constant expression results in the following:

$$\frac{[H^+][H^+]}{0.10\ M} = 1.1 \times 10^{-7}$$
$$[H^+]^2 = (0.10\ M)1.1 \times 10^{-7}$$
$$[H^+] = \sqrt{1.1 \times 10^{-8}} = \boxed{1.05 \times 10^{-4}\ M}$$
$$pH = -\log[H^+]$$
$$pH = -\log(1.05 \times 10^{-4}) = \boxed{3.98}$$
$$[HS^-] = \boxed{1.05 \times 10^{-4}\ M}$$

From the K_2 equilibrium constant expression, solve for $[S^{2-}]$ as follows:

$$\frac{(1.05 \times 10^{-4})[S^{2-}]}{(1.05 \times 10^{-4})} = 1.0 \times 10^{-14}$$
$$[S^{2-}] = \boxed{1.0 \times 10^{-14}\ M}$$
$$[H^+][OH^-] = K_w = 1.0 \times 10^{-14}$$
$$[OH^-] = \frac{1.0 \times 10^{-14}}{1.05 \times 10^{-4}} = \boxed{9.52 \times 10^{-11}\ M}$$

Check major assumptions. First compare the hydrogen and hydroxide concentrations.

$$[H^+] >> [OH^-]$$
$$1.05 \times 10^{-4} >> 9.52 \times 10^{-11}, \text{checks.}$$

Next, compare $[HS^-]$ and $[S^{2-}]$ concentrations.

$$[HS^-] >> [S^{2-}]$$
$$1.05 \times 10^{-4} >> 1.0 \times 10^{-14}, \text{checks.}$$

Compare $[H_2S]$ and $[HS^-]$ concentrations.

$$[H_2S] >> [HS^-] \qquad 0.10 >> 1.05 \times 10^{-4}, \text{checks.}$$

Finally, check the electroneutrality equation.

$$[H^+] = [HS^-] + 2[S^{2-}] + \quad [OH^-]$$
$$1.05 \times 10^{-4} = 1.05 \times 10^{-4} + 2(1.0 \times 10^{-14}) + 9.52 \times 10^{-11}$$
$$1.05 \times 10^{-4} = 1.05 \times 10^{-4}, \text{checks!}$$

2.6.3 pC-pH diagrams

Graphical techniques that make use of pC-pH diagrams are useful for solving more difficult acid-base equilibrium problems. The main advantage offered by this technique is that it shows which chemical species are negligible and which are not. The procedure involves selecting a "master variable" and then drawing curves to show how the concentrations of the various species change as we change the master variable. In acid-base problems, the master variable is $[H^+]$ or pH. On the plot, the log concentration, log $[C]$, of the various chemical species is plotted versus pH. These diagrams may be constructed by hand or plotted using spreadsheet software. The major steps in constructing a pC-pH diagram, as suggested by

Sawyer *et al.* (1994), page 123, are listed below. This procedure is used for manually constructing the diagram:

1 Draw a horizontal line at the log [C].
2 Locate the system point which occurs at pH = pK_A and draw two 45° lines sloping to the left and right through the system point.
3 Locate a point 0.3 logarithmic units below the system point and connect the horizontal and 45° lines with curves passing through this point.
4 Lines for the log $[H^+]$ and log $[OH^-]$ are also drawn at 45° lines. They intersect at a pH of 7 and where log [C] equals −7.
5 Changing the concentration of the solution shifts the diagram up or down.

A pC-pH diagram is going to be constructed to illustrate this graphical technique in the following example. A spreadsheet was used in drawing the diagram and therefore, the five-step procedure presented above is modified such that equations can be written and used within the spreadsheet.

Example 2.30 Construction of pC-pH diagram for a weak acid

Given a 0.1 M solution of an acetic acid represented by HAc and $K_A = 1.8 \times 10^{-5}$, develop a pC-pH diagram.

Solution

The chemical species that are present in the solution include: [HAc], $[H^+]$, $[OH^-]$, and $[Ac^-]$. A plot will be made of the log of the concentration of each species as a function of pH.

Step 1

A plot of log $[H^+]$ versus pH is easily accomplished by using the following equation. It produces a straight line on the pC-pH diagram.

$$pH = -\log [H^+]$$

$$\log [H^+] = -pH \qquad (1)$$

For pH = 1, log $[H^+]$ = −pH = −1. The table below shows log $[H^+]$ versus pH.

log $[H^+]$	pH
−1	1
−3	3
−5	5
−7	7
−9	9
−11	11
−13	13

Step 2

A plot of log $[OH^-]$ versus pH is made using the following equation. It produces a straight line on the pC-pH diagram.

$$[H^+][OH^-] = 1 \times 10^{-14}$$

$$[OH^-] = \frac{1 \times 10^{-14}}{[H^+]}$$

$$\log [OH^-] = -14 - \log [H^+]$$

$$\log [OH^-] = -14 + pH$$

$$\log [OH^-] = pH - pK_w \qquad (2)$$

For pH = 2, log $[OH^-]$ = −12. The table below shows log $[OH^-]$ versus pH

log $[OH^-]$	pH
−12	2
−10	4
−8	6
−6	8
−4	10
−2	12

Step 3

To plot log [HAc] and log $[Ac^-]$, we must make use of the equilibrium expression and the conservation of mass equation as listed below. Develop an equation for log [HA].

$$\frac{[H^+][Ac^-]}{[HAc]} = K_A = 1 \times 10^{-5} \qquad [Ac^-] = \frac{K_A[HAc]}{[H^+]}$$

$$[HAc] + [Ac^-] = C_A = 0.1 \text{ M}$$

Solving for [HAc]:

$$[HAc] + \frac{K_A[HAc]}{[H^+]} = C_A$$

$$[HAc]\left(1 + \frac{K_A}{[H^+]}\right) = C_A$$

$$[HAc] = \frac{C_A}{\left(1 + \frac{K_A}{[H^+]}\right)} = \frac{[H^+]}{[H^+]}\frac{C_A}{\left(1 + \frac{K_A}{[H^+]}\right)}$$

$$[HAc] = \frac{C_A[H^+]}{([H^+] + K_A)} \qquad (3)$$

Develop an equation for log $[Ac^-]$.

$$\frac{[H^+][Ac^-]}{[HAc]} = K_A = 1 \times 10^{-5} \qquad [HAc] = \frac{[Ac^-][H^+]}{K_A}$$

$$[HAc] + [Ac^-] = C_A = 0.1 \text{ M}$$

$$\frac{[Ac^-][H^+]}{K_A} + [Ac^-] = C_A$$

$$[Ac^-]\left(\frac{[H^+]}{K_A} + 1\right) = C_A$$

$$[Ac^-]\left(\frac{[H^+] + K_A}{K_A}\right) = C_A$$

$$[Ac^-] = \frac{C_A}{\left(\frac{[H^+] + K_A}{K_A}\right)} = \frac{K_A C_A}{[H^+] + K_A}$$

$$[Ac^-] = \frac{K_A C_A}{[H^+] + K_A} \qquad (4)$$

The table given below shows the pH and the log concentrations of the various species used in constructing the pC-pH diagram (Figure 2.3) for a 0.1 M solution of acetic acid.

pH	log [H⁺] Equation (1)	log [OH⁻] Equation (2)	log [HAc] Equation (3)	log [Ac⁻] Equation (4)
0	0	−14	−1.00	−5.74
1	−1	−13	−1.00	−4.74
2	−2	−12	−1.00	−3.75
3	−3	−11	−1.01	−2.75
4	−4	−10	−1.07	−1.82
5	−5	−9	−1.45	−1.19
6	−6	−8	−2.28	−1.02
7	−7	−7	−3.26	−1.00
8	−8	−6	−4.26	−1.00
9	−9	−5	−5.26	−1.00
10	−10	−4	−6.26	−1.00
11	−11	−3	−7.26	−1.00
12	−12	−2	−8.26	−1.00
13	−13	−1	−9.26	−1.00
14	−14	0	−10.26	−1.00

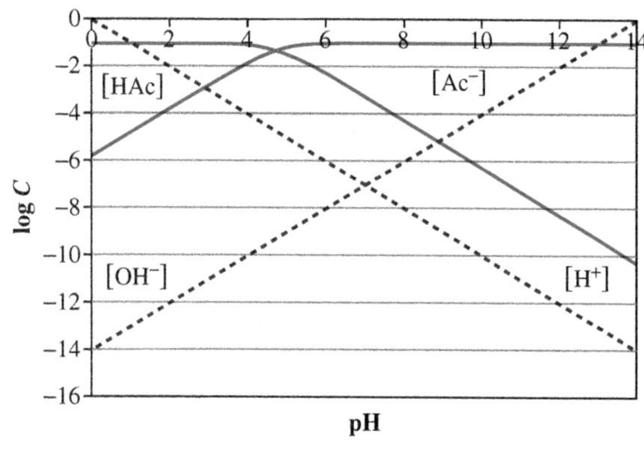

Figure 2.3 pC-pH diagram for 0.1 M acetic acid solution.

Example 2.31 Estimating equilibrium pH from pC-pH diagram

Estimate the equilibrium pH of a 25°C solution of 0.1 M solution of an acetic acid. The pC-pH diagram for a 0.1 M acetic acid solution was developed in Example 2.30 and is presented below as Figure 2.4 for this problem.

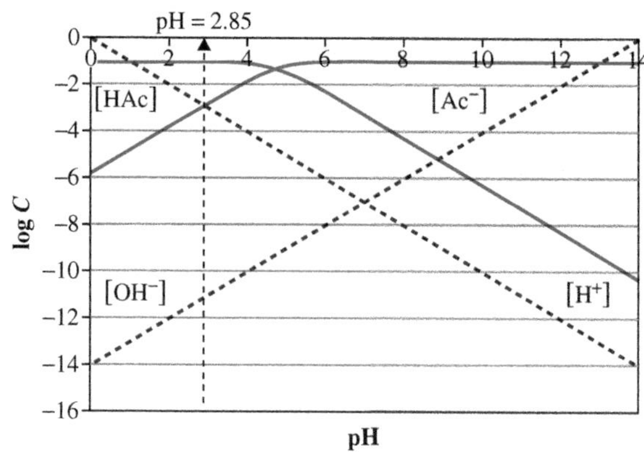

Figure 2.4 pC-pH diagram for 0.1 M acetic acid solution.

Solution

A proton equation is developed by looking at all species that either gain or lose a hydrogen ion or proton.

$$[H^+] = [Ac^-] + [OH^-]$$

The solution to the problem will be found on the pC-pH diagram where the proton condition is satisfied. First, examine the intersection of the hydrogen ion and hydroxide ion lines (See Figure 2.4). At this point, $[H^+] = [OH^-] = 10^{-7}$ and $[Ac^-] = 0.1$, therefore this is not the correct solution.

Next, examine the intersection of the $[H^+]$ and $[Ac^-]$ lines. At this location, the pH is 2.85, $\log [H^+] = \log [Ac^-] = -2.85$, and $\log [OH^-] = -11.15$. This satisfies the proton equation:

$$[H^+] = [Ac^-] + [OH^-]$$

$$10^{-2.85} = 10^{-2.85} + 10^{-11.15}$$

$$1.41 \times 10^{-3} = 1.41 \times 10^{-3}$$

The pH of the solution is approximately 2.85.

Example 2.32 Estimating equilibrium pH from pC-pH diagram

Estimate the equilibrium pH of a 25°C solution of 0.1 M solution of sodium acetate as represented by NaAc. The pC-pH diagram developed for a 0.1 M acetic acid solution and presented as Figure 2.5 can be used in arriving at the solution.

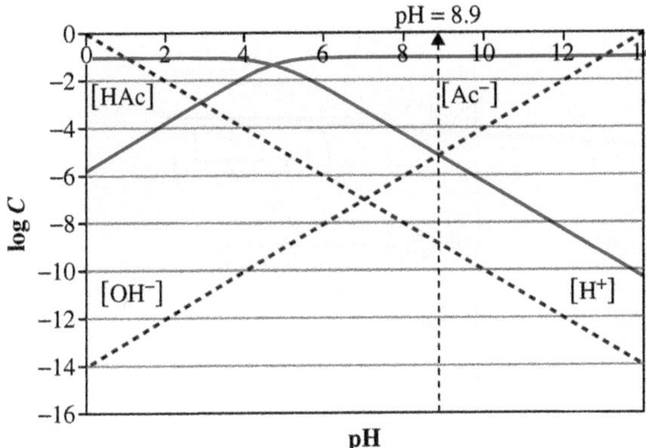

Figure 2.5 pC-pH diagram for 0.1 M acetic acid solution.

Solution

NaAc is assumed to completely dissociate into cations and anions as shown below:

$$NaAc \rightarrow Na^+ + Ac^-$$

Water dissociates into hydrogen and hydroxide ions.

$$H_2O \leftrightarrow H^+ + OH^-$$

Some of the hydrogen ions react with the anion, Ac^-, to form HAc.

$$H^+ + Ac^- \leftrightarrow HAc$$

The proton condition equation is based on those species that either gain or lose a proton or hydrogen ion.

$$[HAc] + [H^+] = [OH^-]$$

The solution to the problem is obtained by locating the point on the pC-pH diagram where the proton condition is satisfied (See Figure 2.5). Try the intersection of the $[H^+]$ and $[OH^-]$ lines intersection which is at a pH = 7.0. The proton condition will not be solved since the concentration of $[HAc] = 10^{-3.26} = 5.5 \times 10^{-4}$ M is much greater than the concentration of $[H^+]$ and $[OH^-] = 10^{-7}$M. Try the intersection of the [HAc] and $[OH^-]$ lines which occurs at a pH of 8.9. Solve the proton condition as shown below.

$$[HAc] + [H^+] = [OH^-]$$

$$10^{-5.1} + 10^{-8.9} = 10^{-5.1}$$

$$7.9 \times 10^6 = 7.9 \times 10^6 \text{ Checks!}$$

The pH of the sodium acetate solution is approximately 8.9.

Example 2.33 Construction of pC-pH diagram for a weak base

Given a 0.01 M solution of ammonia represented by NH_3 and $pK_B = 4.74$, develop a pC-pH diagram.

Solution

The chemical species that are present in the solution include: $[NH_4^+]$, $[H^+]$, $[OH^-]$, and $[NH_3]$. A plot will be made of the log of the concentration of each species as a function of pH (See Figure 2.6).

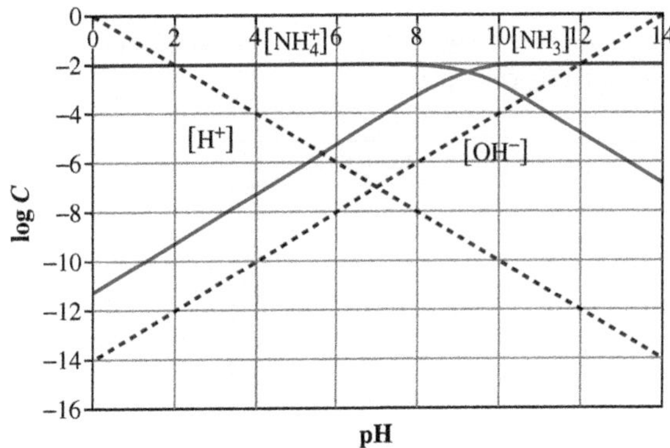

Figure 2.6 pC-pH diagram for 0.01M ammonia solution.

Step 1

A plot of log $[H^+]$ versus pH is easily accomplished by using the following equation. It produces a straight line on the pC-pH diagram.

$$pH = -\log [H^+]$$

$$\log [H^+] = -pH \qquad (1)$$

For pH = 1, log [H^+] = −pH = −1. The table below shows log [H^+] versus pH.

log [H^+]	pH
−1	1
−3	3
−5	5
−7	7
−9	9
−11	11
−13	13

Step 2

A plot of log [OH^-] versus pH is made using the following equation. It produces a straight line on the pC-pH diagram.

$$[H^+][OH^-] = 1 \times 10^{-14}$$

$$[OH^-] = \frac{1 \times 10^{-14}}{[H^+]}$$

$$\log [OH^-] = -14 - \log [H^+]$$

$$\log [OH^-] = -14 + pH$$

$$\log [OH^-] = pH - pK_w \quad (2)$$

For pH = 2, log [OH^-] = −12. The table below shows log [OH^-] versus pH.

log [OH^-]	pH
−12	2
−10	4
−8	6
−6	8
−4	10
−2	12

Step 3

To plot log[NH_4^+] and log [NH_3], we must make use of the equilibrium expression and the conservation of mass equation as listed below. Develop an equation for log[NH_4^+].

$$\frac{[NH_4^+][OH^-]}{[NH_3]} = K_B = 10^{-4.74} = 1.8 \times 10^{-5}$$

$$[NH_3] = \frac{[NH_4^+][OH^-]}{K_B}$$

$$[NH_4^+] + [NH_3] = C_A = 0.01\ M$$

Solving for [NH_4^+]:

$$[NH_4^+] + \frac{[NH_4^+][OH^-]}{K_B} = C_A$$

$$[NH_4^+]\left(1 + \frac{[OH^-]}{K_B}\right) = C_A$$

$$[NH_4^+] = \frac{C_A}{\left(1 + \frac{[OH^-]}{K_B}\right)}$$

$$[NH_4^+] = \frac{K_B C_A}{K_B\left(1 + \frac{[OH^-]}{K_B}\right)}$$

$$[NH_4^+] = \frac{K_B C_A}{(K_B + [OH^-])}$$

$$[H^+][OH^-] = K_w = 10^{-14}$$

$$[OH^-] = \frac{K_w}{[H^+]}$$

$$[NH_4^+] = \frac{K_B C_A}{\left(K_B + \frac{K_w}{[H^+]}\right)} \quad (3)$$

Develop an equation for log [NH_3].

$$\frac{[NH_4^+][OH^-]}{[NH_3]} = K_B = 10^{-4.74} = 1.8 \times 10^{-5}$$

$$[NH_4^+] = \frac{K_B[NH_3]}{[OH^-]}$$

$$[NH_4^+] + [NH_3] = C_A = 0.01\ M$$

$$\frac{K_B[NH_3]}{[OH^-]} + [NH_3] = C_A$$

$$[NH_3]\left(\frac{K_B}{[OH^-]} + 1\right) = C_A$$

$$[NH_3] = \frac{C_A}{\left(\frac{K_B}{[OH^-]} + 1\right)}$$

$$[OH^-] = \frac{K_w}{[H^+]}$$

$$[NH_3] = \frac{C_A}{\left(\frac{K_B[H^+]}{K_w} + 1\right)} \quad (4)$$

The table given below shows the pH and the log concentrations of the various species used in constructing the pC-pH diagram (Figure 2.6) for a 0.01 M solution of ammonia.

pH	log [H^+] Equation (1)	log [OH^-] Equation (2)	log [NH_4^+] Equation (3)	log [NH_3] Equation (4)
0	0	−14	−2.00	−11.26
1	−1	−13	−2.00	−10.26
2	−2	−12	−2.00	−9.26
3	−3	−11	−2.00	−8.26
4	−4	−10	−2.00	−7.26
5	−5	−9	−2.00	−6.26
6	−6	−8	−2.00	−5.26
7	−7	−7	−2.00	−4.26
8	−8	−6	−2.02	−3.28
9	−9	−5	−2.19	−2.45
10	−10	−4	−2.82	−2.07
11	−11	−3	−3.75	−2.01
12	−12	−2	−4.75	−2.00
13	−13	−1	−5.74	−2.00
14	−14	0	−6.74	−2.00

2.6.3.1 Polyprotic Acids

Graphical procedures may also be used for solving acid-base problems involving polyprotic acids and bases. An example will be presented to show the concepts that are involved. For further information on pC-pH diagrams, the reader should consult the following references:

- Stumm, W., Morgan, J.J. (1996). *Aquatic Chemistry: Chemical Equilibria and Rates in Natural Waters.* John Wiley and Sons, Inc.
- Snoeyink, V.L., Jenkins, D. (1980). *Water Chemistry.* John Wiley and Sons, Inc.
- Sawyer, C.N., McCarty, P.L., Parkin, G.F. (1994). *Chemistry for Environmental Engineering*, by, McGraw-Hill.

Example 2.34 Construction of pC-pH diagram for a polyprotic acid

Given a 0.1 M solution of a diprotic acid represented by H_2X and $K_1 = 10^{-3}$ and $K_2 = 10^{-7}$ develop a pC-pH diagram and determine the pH of the solution.

Solution

The chemical species that are present in the solution include [H_2X], [HX^-], [X^{2-}], [H^+], [OH^-] and [H_2O]. A plot will be made of the log of the concentration of each species as a function of pH.

$$H_2X \leftrightarrow H^+ + HX^- \qquad K_1 = 10^{-3}$$

$$HX^- \leftrightarrow H^+ + X^{2-} \qquad K_1 = 10^{-7}$$

Step 1

A plot of log [H^+] versus pH is easily accomplished by using the following equation. It produces a straight line on the pC-pH diagram.

$$pH = -\log[H^+]$$

$$\log[H^+] = -pH \qquad (1)$$

Step 2

A plot of log [OH^-] versus pH is made using the following equation. It produces a straight line on the pC-pH diagram.

$$[H^+][OH^-] = 1 \times 10^{-14}$$

$$[OH^-] = \frac{1 \times 10^{-14}}{[H^+]}$$

$$\log[OH^-] = -14 - \log[H^+]$$

$$\log[OH^-] = -14 + pH$$

$$\log[OH^-] = pH - pK_w \qquad (2)$$

Step 3

To plot log [H_2X], log [HX^-] and log [X^{2-}], we must make use of the equilibrium expression and the conservation of mass equation as listed below. Develop an equation for log [H_2X].

$$\frac{[H^+][HX^-]}{[H_2X]} = K_1 = 1 \times 10^{-3}$$

$$[HX^-] = \frac{K_1[H_2X]}{[H^+]}$$

$$\frac{[H^+][X^{2-}]}{[HX^-]} = K_2 = 1 \times 10^{-7}$$

$$[X^{2-}] = \frac{K_2[HX^-]}{[H^+]}$$

$$[X^{2-}] = \frac{K_1K_2[H_2X]}{[H^+]^2}$$

$$[H_2X] + [HX^-] + [X^{2-}] = C_X = 0.1\ M$$

Solving for [H_2X]:

$$[H_2X] + \frac{K_1[H_2X]}{[H^+]} + \frac{K_1K_2[H_2X]}{[H^+]^2} = C_X$$

$$[H_2X]\left(1 + \frac{K_1}{[H^+]} + \frac{K_1K_2}{[H^+]^2}\right) = C_X$$

$$[H_2X] = \frac{C_X}{\left(1 + \frac{K_1}{[H^+]} + \frac{K_1K_2}{[H^+]^2}\right)} = \alpha_0 C_X \qquad (3)$$

Develop an equation for $[HX^-]$.

$$[H_2X] + [HX^-] + [X^{2-}] = C_X = 0.1\ M$$

$$\frac{[H^+][HX^-]}{[H_2X]} = K_1 = 1 \times 10^{-3}$$

$$[H_2X] = \frac{[H^+][HX^-]}{K_1}$$

$$[X^{2-}] = \frac{K_2[HX^-]}{[H^+]}$$

$$\frac{[H^+][HX^-]}{K_1} + [HX^-] + \frac{K_2[HX^-]}{[H^+]} = C_X$$

$$[HX^-]\left(\frac{[H^+]}{K_1} + 1 + \frac{K_2}{[H^+]}\right) = C_X$$

$$[HX^-] = \frac{C_X}{\left(\frac{[H^+]}{K_1} + 1 + \frac{K_2}{[H^+]}\right)} = \alpha_2 C_X \qquad (4)$$

Develop an equation for $[X^{2-}]$.

$$[H_2X] + [HX^-] + [X^{2-}] = C_X = 0.1\ M$$

$$[H_2X] = \frac{[H^+][HX^-]}{K_1}$$

$$[X^{2-}] = \frac{K_2[HX^-]}{[H^+]}$$

$$[HX^-] = \frac{[H^+][X^{2-}]}{K_2}$$

$$[H_2X] = \frac{[H^+]^2[X^{2-}]}{K_1K_2}$$

$$\frac{[H^+]^2[X^{2-}]}{K_1K_2} + \frac{[H^+][X^{2-}]}{K_2} + [X^{2-}] = C_X$$

$$[X^{2-}]\left(\frac{[H^+]^2}{K_1K_2} + \frac{[H^+]}{K_2} + 1\right) = C_X$$

$$[X^{2-}] = \frac{C_X}{\left(\frac{[H^+]^2}{K_1K_2} + \frac{[H^+]}{K_2} + 1\right)} = \alpha_2 C_X \qquad (5)$$

The table given below shows the pH and the log concentrations of the various species used in constructing the pC-pH diagram (Figure 2.7) for a 0.1 M solution of H_2X.

pH	log $[H^+]$ Equation (1)	log $[OH^-]$ Equation (2)	log $[H_2X]$ Equation (3)	log $[HX^-]$ Equation (4)	log $[X^{2-}]$ Equation (5)
0	0	−14	−1.00	−4.00	−11.00
1	−1	−13	−1.00	−3.00	−9.00
2	−2	−12	−1.04	−2.04	−7.04
3	−3	−11	−1.30	−1.30	−5.30
4	−4	−10	−2.04	−1.04	−4.04
5	−5	−9	−3.00	−1.01	−3.01
6	−6	−8	−4.00	−1.04	−2.04
7	−7	−7	−5.00	−1.30	−1.30
8	−8	−6	−6.00	−2.04	−1.04
9	−9	−5	−7.00	−3.00	−1.00
10	−10	−4	−8.00	−4.00	−1.00
11	−11	−3	−9.00	−5.00	−1.00
12	−12	−2	−10.00	−6.00	−1.00
13	−13	−1	−11.00	−7.00	−1.00
14	−14	0	−12.00	−8.00	−1.00

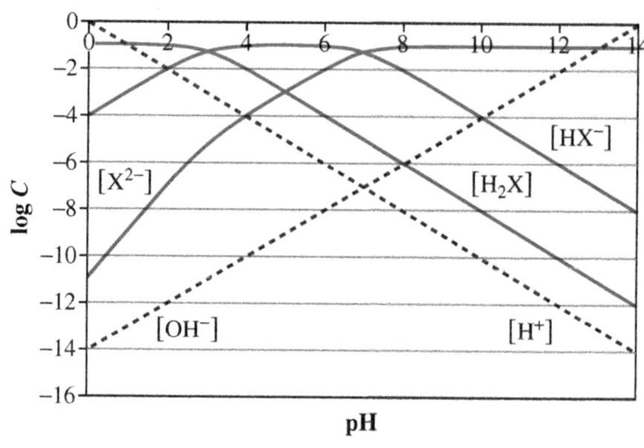

Figure 2.7 pC-pH diagram for 0.01M H_2X solution.

2.6.4 Buffers

As previously mentioned, strong acids and bases are considered to dissociate completely or ionize in dilute solutions, while weak acids and bases are only partially ionized. When an acid is added to pure water, the hydrogen ion concentration will increase. Similarly, the hydroxide ion concentration increases when a base is added to pure water.

Buffers are substances in a solution that resist a change in pH when either an acid or base is added to the solution. Buffered solutions are very important in water chemistry, where we desire to maintain a specific pH. When coagulants are added to raw water to remove turbidity or to precipitate a specific ion, a narrow pH range is necessary for this to occur under optimal conditions. During biological nitrification in activated sludge systems, both nitrous and nitric acid are produced, which can lower the pH of the wastewater, thereby inhibiting the process.

Example 2.35 Calculating the pH of an unbuffered solution

Calculate the change in pH when 0.01 mol of HCl is added to 1.0 L of pure water. Assume that the pH of the pure water is 7.0.

Solution

Hydrochloric acid is a strong acid and will completely dissociate or ionize in the water according to the following reaction.

$$\underset{0.01M}{HCl} \rightarrow \underset{0.01M}{H^+} + \underset{0.01M}{Cl^-}$$

The final pH of the solution will be 2.0. Adding this strong acid causes the pH to change by 5 units.

$$pH = -\log[H^+] = -\log(0.01) = \boxed{2.0}$$

Example 2.36 Calculating the pH of an unbuffered solution

Calculate the final pH when 0.005 mol of NaOH is added to 5.0 L of pure water. Assume that the pH of the pure water is 7.0.

Solution

Sodium hydroxide is a strong base and will completely dissociate or ionize in the water according to the following reaction. The initial concentration of NaOH before dissociation is 0.005M/5 L = 0.001 M.

$$\underset{0.001M}{NaOH} \rightarrow \underset{0.001M}{Na^+} + \underset{0.001M}{OH^-}$$

The final pH of the solution will be 11.0. Adding this strong base raises the pH from 7 to 11.

$$[OH^-] = 0.001M$$

$$[H^+][OH^-] = 10^{-14}$$

$$[H^+] = \frac{10^{-14}}{[OH^-]} = \frac{10^{-14}}{0.001} = 1.0 \times 10^{-11} M$$

$$pH = -\log[H^+] = -\log(1.0 \times 10^{-11}) = \boxed{11.0}$$

A solution is buffered by the presence of a weak acid and its salt (conjugate base) or alternatively, by a weak base and its salt (conjugate acid). Therefore, weak acids and bases and their salts are very effective as buffers within ±1 pH unit of their pK value (Sawyer and McCarty (1994), page 143). Assume that a weak acid (HA) will be used as a buffer and its dissociation reaction is shown below.

$$HA \leftrightarrow H^+ + A^- \tag{2.74}$$

$$K_A = \frac{[H^+][A^-]}{[HA]} \tag{2.75}$$

where:

K_A = equilibrium, ionization, or dissociation constant of weak acid

$[H^+]$ = hydrogen ion concentration, moles/L

$[A^-]$ = conjugate base concentration, moles/L

$[HA]$ = concentration of weak acid, moles/L.

Rearranging Equation (2.75) results in:

$$\frac{1}{[H^+]} = \frac{1}{K_A}\frac{[A^-]}{[HA]} \tag{2.76}$$

$$pH = pK_A + \log\frac{[A^-]}{[HA]} \tag{2.77}$$

Equation (2.77) shows that the final pH of a buffered solution made from a weak acid is dependent upon the ratio of the conjugate base or salt to the acid concentration. The larger the concentration of weak acid, the greater will be the buffering capacity of the solution. In biological systems, phosphoric acid is often used, since its second pK_A value is 7.2. The following example illustrates buffering capacity.

Example 2.37 Buffering capacity of phosphate buffer

A buffer solution consisting of 100 mg/L of monobasic potassium salt (KH_2PO_4) and 200 mg/L of dibasic potassium salt (K_2HPO_4) will be used to maintain a neutral pH during the biochemical oxygen demand (BOD) test.

Determine:
a) initial pH of the buffer solution
b) pH after adding 20 mg/L of HCl
c) pH after adding 20 mg/L of NaOH

Solution

$$KH_2PO_4 \rightarrow K^+ + H_2PO_4^-$$

$$K_2HPO_4 \rightarrow 2K^+ + HPO_4^{2-}$$

$$H_2PO_4^- \leftrightarrow H^+ + HPO_4^{2-} \qquad pK_A = 7.2$$

$$pH = pK_A + \log\frac{[A^-]}{[HA]}$$

$$pH = 7.2 + \log\frac{[HPO_4^{2-}]}{[H_2PO_4^-]}$$

$$\text{MW of } H_2PO_4 = 2(1) + 31 + 4(16) = 97$$
$$\text{MW of } KH_2PO_4 = 39 + 2(1) + 31 + 4(16) = 136$$
$$\text{MW of } K_2HPO_4 = 2(39) + 1 + 31 + 4(16) = 174$$

$$[KH_2PO_4] = \frac{100\text{ mg}}{L}\left(\frac{1\text{ g}}{1000\text{ mg}}\right)\left(\frac{\text{mole}}{136\text{ g}}\right)$$
$$= 7.35 \times 10^{-4}\text{ M}$$

$$\underset{7.35\times10^{-4}\text{ M}}{KH_2PO_4} \rightarrow K^+ + \underset{7.35\times10^{-4}\text{ M}}{H_2PO_4^-}$$

$$[KH_2PO_4] = [H_2PO_4^-] = 7.35 \times 10^{-4}\text{ M}$$

$$[K_2HPO_4] = \frac{200\text{ mg}}{L}\left(\frac{1\text{ g}}{1000\text{ mg}}\right)\left(\frac{\text{mole}}{174\text{ g}}\right)$$
$$= 1.15 \times 10^{-3}\text{ M}$$

$$K_2HPO_4 \rightarrow 2\,K^+ + HPO_4^{2-}$$
$$[K_2HPO_4] = [HPO_4^{2-}] = 1.15 \times 10^{-3}\text{ M}$$

$$pH = 7.2 + \log\frac{[HPO_4^{2-}]}{[H_2PO_4]}$$
$$pH = 7.2 + \log\frac{(1.15 \times 10^{-3})}{(7.35 \times 10^{-4})} = \boxed{7.4}$$

Solution part b

$$\text{MW of HCl} = 1 + 35.5 = 36.6$$
$$[HCl] = \frac{20\text{ mg}}{L}\left(\frac{1\text{ g}}{1000\text{ mg}}\right)\left(\frac{\text{mole}}{36.6\text{ g}}\right)$$
$$= 5.46 \times 10^{-4}\text{ M}$$
$$HCl \rightarrow H^+ + Cl^-$$
$$[HCl] = [H^+] = 5.46 \times 10^{-4}\text{ M}$$

The pH of an unbuffered system by adding 20 mg/L of HCl would be 3.3.

$$pH = -\log[H^+] = -\log(5.46 \times 10^{-4}) = 3.3$$
$$H^+ + HPO_4^{2-} \rightarrow H_2PO_4^-$$

	$[H_2PO_4^-]$	$[HPO_4^{2-}]$
Initially	7.35×10^{-4} M	1.15×10^{-3} M
HCl addition	$+5.46 \times 10^{-4}$ M	-5.46×10^{-4} M
	1.28×10^{-3} M	6.04×10^{-4} M

$$pH = 7.2 + \log\frac{[HPO_4^{2-}]}{[H_2PO_4^-]}$$
$$pH = 7.2 + \log\frac{(6.04 \times 10^{-4})}{(1.28 \times 10^{-3})} = \boxed{6.87}$$

Solution part c

$$\text{MW of NaOH} = 23 + 16 + 1 = 40$$
$$[NaOH] = \frac{20\text{ mg}}{L}\left(\frac{1\text{ g}}{1000\text{ mg}}\right)\left(\frac{\text{mole}}{40\text{ g}}\right)$$
$$= 5.00 \times 10^{-4}\text{ M}$$

$$\underset{0.0005M}{NaOH} \rightarrow \underset{0.0005M}{Na^+} + \underset{0.0005M}{OH^-}$$
$$\underset{0.0005M}{H^+} \rightarrow \underset{0.0005M}{OH^-} + \underset{0.0005M}{H_2O}$$
$$\underset{-0.0005M}{H_2PO_4^-} \rightarrow \underset{-0.0005M}{H^+} + \underset{+0.0005M}{HPO_4^{2-}}$$

	$[H_2PO_4^-]$	$[HPO_4^{2-}]$
Initially	7.35×10^{-4} M	1.15×10^{-3} M
NaOH addition	-5.00×10^{-4} M	$+5.00 \times 10^{-4}$ M
	2.35×10^{-4} M	1.65×10^{-3} M

$$pH = 7.2 + \log\frac{[HPO_4^{2-}]}{[H_2PO_4^-]}$$
$$pH = 7.2 + \log\frac{(1.65 \times 10^{-3})}{(2.35 \times 10^{-4})} = \boxed{8.05}$$

2.6.5 Carbonate system and alkalinity

The carbonate systems is the most important acid-base system in water chemistry, since it serves as the basis of providing natural buffering capacity in water. Alkalinity is defined as the buffering capacity of a system to resist a change in pH with the addition of an acid. Alternatively, the buffering capacity of a system to resist a change in pH with the addition of a base is called acidity. Traditionally, alkalinity is determined analytically by titrating a sample of water with 0.02 *N* H_2SO_4 to a pH of approximately 4.5. In natural water systems, the total alkalinity of a system is primarily a function of the carbonate system and the hydroxide and hydrogen ions.

There are several chemical species that make up the carbonate system which are presented below:

- Gaseous carbon dioxide, $CO_{2(g)}$
- Aqueous dissolved carbon dioxide, $CO_{2(aq)}$
- Calcium carbonate solids, $CaCO_{3(s)}$

- Bicarbonate, HCO_3^-
- Carbonate, CO_3^{2-}

Carbon dioxide is an important participant in the carbonate system. It is released into the atmosphere from fossil fuel combustion, volcanoes, biological respiration, and also from supersaturated waters. Autotrophic organisms use CO_2 in synthesizing biomass. Atmospheric CO_2 dissolves in water, with the ocean being a major reservoir of aqueous CO_2 in equilibrium with carbonic acid.

The concentration of carbonates and bicarbonates in surface and ground waters is highly dependent upon the geological formations and bedrock to which the water is exposed. Regions consisting of dolomite ($CaMg(CO_3)_2$) or calcite ($CaCO_3$) bedrock will contain high concentrations of carbonate species, including calcium (Ca^{2+}) and magnesium (Mg^{2+}) ions, whereas regions containing igneous rocks will have significantly low levels of these and will exhibit low alkalinity with minimal buffering capacity.

The chemical reactions and equilibrium expressions for the major species involved in the carbonate system are presented next.

$$CO_{2(g)} \leftrightarrow CO_{2(aq)} \tag{2.78}$$

$$K_H = \frac{[CO_{2(aq)}]}{[CO_{2(g)}]} = 10^{-1.47} \frac{\text{mol}}{\text{L} \cdot \text{atm}} \text{ at } 25°\text{C} \tag{2.79}$$

Where $K_H = H =$ Henry's law constant.

$$CO_{2(g)} + H_2O \leftrightarrow H_2CO_3 \tag{2.80}$$

$$K_M = \frac{[H_2CO_3]}{[CO_{2(aq)}]} = 10^{-2.8} \text{ at } 25°\text{C} \tag{2.81}$$

From the equation above, carbonic acid concentration is only 0.16% of the aqueous carbon dioxide concentration. Therefore:

$$[CO_{2(aq)}] >> [H_2CO_3] \text{ and } [H_2CO_3^*] \cong [CO_{2(aq)}].$$

$$H_2CO_3 \leftrightarrow H^+ + HCO_3^- \tag{2.82}$$

$$K_1' = \frac{[H^+][HCO_3^-]}{[H_2CO_3]} = 10^{-3.5} \text{ at } 25°\text{C} \tag{2.83}$$

Analytically, it is impossible to differentiate between the true carbonic acid concentration, $[H_2CO_3]$, and the dissolved or aqueous carbon dioxide concentration, $[CO_{2(aq)}]$; therefore, a new term called $[H_2CO_3^*]$ is used to denote the sum of these two species:

$$[H_2CO_3^*] = [H_2CO_3] + [CO_{2(aq)}] \tag{2.84}$$

$$H_2CO_3^* \leftrightarrow H^+ + HCO_3^- \tag{2.85}$$

$$K_1 = \frac{[H^+][HCO_3^-]}{[H_2CO_3^*]} = 10^{-6.35} \text{ at } 25°\text{C} \tag{2.86}$$

$$H_2CO_3^- \leftrightarrow H^+ + HCO_3^{2-} \tag{2.87}$$

$$K_2 = \frac{[H^+][CO_3^{2-}]}{[HCO_3^-]} = 10^{-10.33} \text{ at } 25°\text{C} \tag{2.88}$$

In a closed system, i.e., not open to the atmosphere or the rate of dissolution from atmospheric carbon dioxide into water is negligible, the total concentration of carbonate species (C_T) is constant and is calculated as follows:

$$C_T = [H_2CO_3^*] + [HCO_3^-] + [CO_3^{2-}] \tag{2.89}$$

Equations for calculating $[H_2CO_3^*]$, $[HCO_3^-]$, and $[CO_3^{2-}]$ are derived as follows.

Equations (2.86) and (2.88) are substituted into Equation (2.89) and rearranged to develop an equation for $[H_2CO_3^*]$.

$$C_T = [H_2CO_3^*] + [HCO_3^-] + [CO_3^{2-}] \tag{2.89}$$

$$C_T = [H_2CO_3^*] + \frac{K_1[H_2CO_3^*]}{[H^+]} + \frac{K_2[HCO_3^-]}{[H^+]} \tag{2.90}$$

$$C_T = [H_2CO_3^*] + \frac{K_1[H_2CO_3^*]}{[H^+]} + \frac{K_1K_2[H_2CO_3^*]}{[H^+]\ [H^+]} \tag{2.91}$$

$$C_T = [H_2CO_3^*]\left[1 + \frac{K_1}{[H^+]} + \frac{K_1K_2}{[H^+]^2}\right] \tag{2.92}$$

$$[H_2CO_3^*] = \frac{C_T}{1 + \frac{K_1}{[H^+]} + \frac{K_1K_2}{[H^+]^2}} \tag{2.93}$$

An equation for the bicarbonate ion concentration is derived as follows.

$$C_T = [H_2CO_3^*] + [HCO_3^-] + [CO_3^{2-}] \tag{2.89}$$

$$C_T = \frac{[H^+][HCO_3^-]}{K_1} + [HCO_3^-] + \frac{K_2[HCO_3^-]}{[H^+]} \tag{2.94}$$

$$C_T = [HCO_3^-]\left[\frac{[H^+]}{K_1} + 1 + \frac{K_2}{[H^+]}\right] \tag{2.95}$$

$$[HCO_3^-] = \frac{C_T}{\frac{[H^+]}{K_1} + 1 + \frac{K_2}{[H^+]}} \tag{2.96}$$

An equation for the carbonate ion concentration is derived as follows.

$$C_T = [H_2CO_3^*] + [HCO_3^-] + [CO_3^{2-}] \tag{2.89}$$

$$C_T = \frac{[H^+][HCO_3^-]}{K_1} + \frac{[H^+][CO_3^{2-}]}{K_2} + [CO_3^{2-}] \tag{2.97}$$

$$C_T = \frac{[H^+][H^+][CO_3^{2-}]}{K_1\,K_2} + \frac{[H^+][CO_3^{2-}]}{K_2} + [CO_3^{2-}] \tag{2.98}$$

$$C_T = [CO_3^{2-}]\left[\frac{[H^+][H^+]}{K_1 K_2} + \frac{[H^+]}{K_2} + 1\right] \quad (2.99)$$

$$[CO_3^{2-}] = \frac{C_T}{\frac{[H^+]^2}{K_1 K_2} + \frac{[H^+]}{K_2} + 1} \quad (2.100)$$

Example 2.38 Calculating carbonate species in closed system

The pH of a water sample is 4.66 and the $[HCO_3^-] = 0.001$ M for a closed system.

Determine:
a) $[H_2CO_3^*]$
b) $[CO_3^{2-}]$
c) C_T

Solution part a

From Equation (2.86):

$$K_1 = \frac{[H^+][HCO_3^-]}{[H_2CO_3^*]} = 10^{-6.35} \text{ at } 25°C$$

$$[H_2CO_3^*] = \frac{[H^+][HCO_3^-]}{K_1} = \frac{(10^{-4.66})(0.001)}{10^{-6.35}}$$

$$= \boxed{0.049\text{ M}}$$

Solution part b

From Equation (2.88):

$$K_2 = \frac{[H^+][CO_3^{2-}]}{[HCO_3^-]} = 10^{-10.33} \text{ at } 25°C$$

$$[CO_3^{2-}] = \frac{K_2[HCO_3^-]}{[H^+]} = \frac{(10^{-10.33})(0.001)}{10^{-4.66}}$$

$$= \boxed{2.14 \times 10^{-9}\text{ M}}$$

Solution part c

From Equation (2.89):

$$C_T = [H_2CO_3^*] + [HCO_3^-] + [CO_3^{2-}]$$

$$C_T = 0.049\text{ M} + 0.001\text{ M} + 2.14 \times 10^{-9}\text{ M}$$

$$C_T = \boxed{0.05\text{ M}}$$

Example 2.39 Calculating carbonate species in open system

The pH of a water sample is 8.0 in an open system. The temperature is 20°C, the Henry's constant for carbon dioxide is $10^{-1.41}\ \frac{\text{mol}}{\text{L·atm}}$, and pK_1 and pK_2 for the dissociation of carbonic acid at 20°C is 6.38 and 10.38, respectively (Snoeyink & Jenkins (1980), page 157). The partial pressure of carbon dioxide is $10^{-3.5}$atm.

Determine:
a) $[H_2CO_3^*]$
b) $[HCO_3^-]$
c) $[CO_3^{2-}]$
d) C_T

Solution part a

Assume that:

$$[H_2CO_3^*] = [CO_{2(aq)}].$$

$$[CO_{2(aq)}] = K_H P_{CO_2} = 10^{-1.41}\frac{\text{mol}}{\text{L}\cdot\text{atm}}(10^{-3.5}\text{ atm})$$

$$[CO_{2(aq)}] = 1.23 \times 10^{-5}\text{ M} = [H_2CO_3^*]$$

$$[H_2CO_3^*] = \boxed{1.23 \times 10^{-5}\text{ M}}$$

Solution part b

$$K_1 = \frac{[H^+][HCO_3^-]}{[H_2CO_3^*]} = 10^{-6.38} \text{ at } 20°C$$

$$[HCO_3^-] = \frac{[H_2CO_3^*]10^{-6.38}}{[H^+]} = \frac{(1.23 \times 10^{-5})10^{-6.38}}{10^{-8.0}}$$

$$[HCO_3^-] = \boxed{5.13 \times 10^{-4}\text{ M}}$$

Solution part c

$$K_2 = \frac{[H^+][CO_3^{2-}]}{[HCO_3^-]} = 10^{-10.38} \text{ at } 20°C$$

$$[CO_3^{2-}] = \frac{K_2[HCO_3^-]}{[H^+]} = \frac{10^{-10.38}(5.13 \times 10^{-4})}{10^{-8.0}}$$

$$= \boxed{2.14 \times 10^{-6}\text{M}}$$

Solution part d

From Equation (2.89):

$$C_T = [H_2CO_3^*] + [HCO_3^-] + [CO_3^{2-}]$$

$$C_T = 1.23 \times 10^{-5}\text{M} + 5.13 \times 10^{-4}\text{M} + 2.14 \times 10^{-6}\text{M}$$

$$C_T = \boxed{5.27 \times 10^{-4}\text{M}}$$

2.6.5.1 Alkalinity

The equation for alkalinity can be derived using the chemical equilibrium equations for the carbonate species reacting with a strong base as represented by BOH. A strong base will completely dissociate in water according to the following reaction.

$$BOH \rightarrow B^+ + OH^- \quad (2.101)$$

The following chemical species are initially present in the solution prior to dissociating: BOH, H_2CO_3, and H_2O. After dissociation, the following chemical species exist: H^+, OH^-, B^+, HCO_3^-, and CO_3^{2-}. For electroneutrality, the following equation can be written:

$$[B^+] + [H^+] = [OH^-] + [HCO_3^-] + 2[CO_3^{2-}] \quad (2.102)$$

Rearranging and solving Equation (2.102) for $[B^+]$ results in:

$$[B^+] = 2[CO_3^{2-}] + [OH^-] + [HCO_3^-] - [H^+] \quad (2.103)$$

Since $[B^+]$ represents the concentration of strong base added, we can replace it with [ALK] to represent the alkalinity. Equation (2.103) can be written in terms of equivalents per liter rather than moles/L as shown below.

$$[ALK]_e = [CO_3^{2-}]_e + [OH^-]_e + [HCO_3^-]_e - [H^+]_e \quad (2.104)$$

Where $[\ ]_e$ = concentration in eq/L.

Recall that equivalents react with equivalents. To calculate the concentration in equivalents per liter (eq/L), divide the molar concentration by the equivalent weight of the compound. Alkalinity is normally expressed in units of mg/L as $CaCO_3$.

Example 2.40 Calculating concentrations in equivalents per liter

The pH of a water sample is 6.5. Express the concentration of the hydrogen and hydroxide ions in terms of equivalents per liter and in terms of mg/L as $CaCO_3$.

Solution part a

$$pH = -\log[H^+]$$

$$[H^+] = 10^{-pH} = 10^{-6.5} = 3.2 \times 10^{-7} M$$

The equivalent weight of the hydrogen ion $= \frac{MW}{z} = \frac{1\ g\ H^+}{1\ eq}$

$$[H^+]_e = 3.2 \times 10^{-7} M \left(\frac{1\ g\ H^+}{1\ mole\ H^+}\right)\left(\frac{1\ eq}{1\ g\ H^+}\right)$$

$$= \boxed{3.2 \times 10^{-7} \frac{eq}{L}}$$

The equivalent weight of the hydroxide ion $= \frac{MW}{z} = \frac{17\ g\ OH^-}{1\ eq}$

$$[H^+][OH^-] = 10^{-14}$$

$$[OH^-] = \frac{10^{-14}}{[H^+]} = \frac{10^{-14}}{3.2 \times 10^{-7} M} = 3.2 \times 10^{-8}$$

$$[OH^-]_e = 3.2 \times 10^{-8} M \left(\frac{17\ g\ OH^-}{1\ mole\ OH^-}\right)\left(\frac{1\ eq}{17\ g\ OH^-}\right)$$

$$= \boxed{3.2 \times 10^{-8} \frac{eq}{L}}$$

Solution part b

The equivalent weight of the $CaCO_3$:

$$\frac{MW}{z} = \frac{100\ g\ CaCO_3}{2\ eq} = \frac{50\ g\ CaCO_3}{eq} \text{ or } \frac{50\ mg\ CaCO_3}{meq}$$

$$H^+ = 3.2 \times 10^{-7} \frac{eq}{L}\left(\frac{50\ g\ CaCO_3}{eq}\right)\left(\frac{1000\ mg}{1\ g}\right)$$

$$= \boxed{1.6 \times 10^{-2} \frac{mg}{L}}$$

$$OH^- = 3.2 \times 10^{-8} \frac{eq}{L}\left(\frac{50\ g\ CaCO_3}{eq}\right)\left(\frac{1000\ mg}{1\ g}\right)$$

$$= \boxed{1.6 \times 10^{-3} \frac{mg}{L}}$$

2.6.5.2 Calculating Alkalinity Species from pH and Alkalinity Measurements

In the laboratory, alkalinity measurements are made by titrating a sample of water with an acid to an endpoint of approximately 4.5. Both the phenolphthalein and total alkalinities are determined from titrations and hydroxide, carbonate, and bicarbonate alkalinities are then calculated. In order to determine the various alkalinity forms, this method assumes that hydroxide and bicarbonate alkalinity cannot exist together in the same sample. This is an incorrect assumption but, for all practical purposes, it does not introduce significant error.

Standard Methods (1998) recommends using 0.02 *N* H_2SO_4, since 1 ml of titrant will neutralize 1 mg of alkalinity as $CaCO_3$. During the titration, the hydrogen ions react with the alkalinity in the water according to the following equations:

$$H^+ + OH^- \leftrightarrow H_2O \quad (2.105)$$

$$H^+ + CO_3^{2-} \leftrightarrow HCO_3^- \quad (2.106)$$

$$H^+ + HCO_3^- \leftrightarrow H_2CO_3 \quad (2.107)$$

Figure 2.8 shows a generalized titration curve for an alkalinity determination. There are two inflection points that occur at pH values of approximately 8.3 and 4.5. The conversion of all the hydroxide to water is complete at pH 8.3, Equation (2.105). Also, at pH 8.3, the conversion of the carbonate to bicarbonate is essentially complete, Equation (2.106); however, an equivalent amount of acid must be added to neutralize the HCO_3^-, i.e., conversion of HCO_3^- to H_2CO_3 according to Equation (2.107). Therefore, the neutralization of carbonate is only 50% complete.

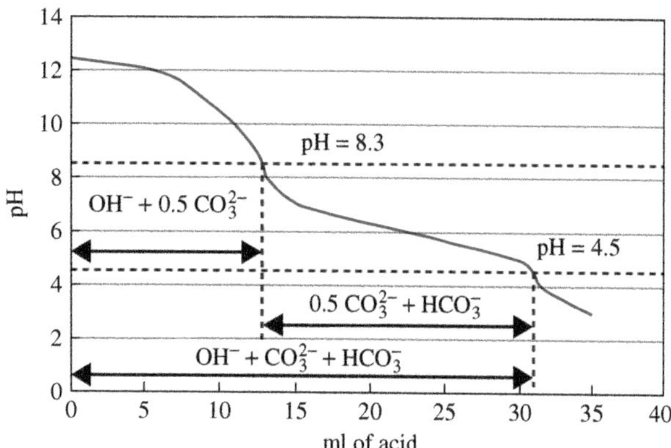

Figure 2.8 Alkalinity titration curve.

At pH 4.5, the neutralization of all alkalinity species is complete. The quantity of standard acid required to titrate a water sample to pH 4.5 is used to calculate the total alkalinity of the water. Equation (2.108) is used to calculate alkalinity from a titration.

$$\text{Alkalinity}\left(\frac{\text{mg}}{\text{L}}\text{ as CaCO}_3\right)=\frac{(\text{ml of }0.02N\text{ H}_2\text{SO}_4)\times 1000}{\text{ml of sample}} \tag{2.108}$$

The following generalizations can be made concerning the various forms of alkalinity when following this procedure, i.e., determining the various forms of alkalinity based on phenolphthalein and total alkalinities along with pH measurements. In these statements, P is equal to the quantity of acid required to titrate to pH 8.3, and M represents the total acid required in the titration to reach pH 4.5. The phenolphthalein alkalinity is calculated based on the amount of acid required to reach pH 8.3, whereas the total alkalinity is based on the total amount of acid required to reach pH 4.5.

- If $P = M$, all alkalinity is in the hydroxide form.
- If $P = M/2$, all alkalinity is in the carbonate form.
- If $P = 0$, all alkalinity is in the bicarbonate form and pH < 8.3.
- If $P < M/2$, the predominant alkalinity species are carbonate and bicarbonate.
- If $P > M/2$, the predominant alkalinity species are hydroxide and carbonate.

Example 2.41 illustrates how to determine the various forms of alkalinity based on titration measurements, pH, and the generalizations given above, assuming that both hydroxide and bicarbonate cannot exist together.

Example 2.41 Alkalinity calculated from titration measurements

A 100 ml water sample is titrated with 0.02 N H_2SO_4. The initial pH is 11.0, and 10.0 ml of acid is required to reach pH 8.3. An additional 5.5 ml of acid is required to reach pH 4.5. Determine the phenolphthalein and total alkalinities, hydroxide, carbonate, and bicarbonate alkalinities expressed in mg/L as $CaCO_3$.

Solution part a

From Equation (2.108):

$$\text{Alkalinity}\left(\frac{\text{mg}}{\text{L}}\text{ as CaCO}_3\right)=\frac{(\text{ml of }0.02N\text{ H}_2\text{SO}_4)\times 1000}{\text{ml of sample}}$$

Recall that the phenolphthalein alkalinity is based on the quantity of acid required to reach a pH of 8.3 during the titration. If the pH of the sample is below 8.3, there is no phenolphthalein alkalinity. The phenolphthalein alkalinity is calculated below.

$$\text{Phenolphthalein Alkalinity}\left(\frac{\text{mg}}{\text{L}}\text{ as CaCO}_3\right)=\frac{(10.0\text{ ml})\times 1000}{100\text{ ml}}=\boxed{100\frac{\text{mg}}{\text{L}}\text{ as CaCO}_3}$$

The total alkalinity is based on the total quantity of acid required to reach a pH of 4.5 during the titration. The total alkalinity is calculated below.

$$\text{Total Alkalinity}\left(\frac{\text{mg}}{\text{L}}\text{ as CaCO}_3\right)=\frac{(10.0\text{ ml}+5.5\text{ ml})\times 1000}{100\text{ ml}}=\boxed{155\frac{\text{mg}}{\text{L}}\text{ as CaCO}_3}$$

Solution part b

In this problem, $P = 10.0$ ml and $M = 15.5$ ml. Therefore, $P > M/2$ (10.0 ml > 15.5 ml/2), so both carbonate and hydroxide are the predominant alkalinity species that are present.

$$[H^+][OH^-]=10^{-14}$$

$$[OH^-]=\frac{10^{-14}}{[H^+]}=\frac{10^{-14}}{10^{-11}}=1.0\times10^{-3}$$

$$OH^-=1.0\times10^{-3}\text{ M}\left(\frac{17\text{ g OH}^-}{1\text{ mole OH}^-}\right)\left(\frac{1\text{ eq}}{17\text{ g OH}^-}\right)\times\left(\frac{50\text{ g CaCO}_3}{\text{eq}}\right)\left(\frac{1000\text{ mg}}{1\text{ g}}\right)$$

$$OH^-=\boxed{50\frac{\text{mg}}{\text{L}}\text{ as CaCO}_3}$$

Solution part c

Determine the quantity of acid required for the neutralization of hydroxide by rearranging Equation (2.108).

$$50\ \frac{mg}{L}\ \text{as}\ CaCO_3 = \frac{(X\ ml)\times 1000}{100\ ml}$$

$$X\ \text{ml of acid} = \frac{50\ \frac{mg}{L}\ \text{as}\ CaCO_3 \times 100\ ml}{1000} = 5$$

The quantity of acid required to neutralize 50% of the carbonate alkalinity is equal to:

$$\text{ml of acid for carbonate neutralization} = 2\times(10.0\ ml - 5\ ml) = 10\ ml$$

From Equation (2.108), the carbonate alkalinity is calculated as follows:

$$\text{Carbonate Alkalinity}\left(\frac{mg}{L}\ \text{as}\ CaCO_3\right) = \frac{(10\ ml)\times 1000}{100\ ml} = \boxed{100\ \frac{mg}{L}\ \text{as}\ CaCO_3}$$

Solution part d

Finally, the quantity of acid required to neutralize the bicarbonate is calculated as the difference between the additional acid added to reach 4.5 and the quantity of acid required to neutralize 50% of the carbonate alkalinity.

$$\text{ml of acid for bicarbonate neutralization} = (5.5\ ml - 5.0\ ml) = 0.5\ ml$$

From Equation (2.108), the bicarbonate alkalinity is calculated as follows:

$$\text{Bicarbonate Alkalinity}\left(\frac{mg}{L}\ \text{as}\ CaCO_3\right) = \frac{(0.5\ ml)\times 1000}{100\ ml} = \boxed{5\frac{mg}{L}\ \text{as}\ CaCO_3}$$

Check to see if all alkalinity species add up to the total alkalinity.

$$\text{Total Alkalinity}\left(\frac{mg}{L}\ \text{as}\ CaCO_3\right) = OH^-\ \text{Alkalinity} + CO_3^{2-}\ \text{Alkalinity} + HCO_3^-\ \text{Alkalinity}$$

$$\text{Total Alkalinity}\left(\frac{mg}{L}\ \text{as}\ CaCO_3\right) = 50\ \frac{mg}{L} + 100\frac{mg}{L} + 5\ \frac{mg}{L} = \boxed{155\ \frac{mg}{L}}\ \text{Agrees!}$$

2.6.5.3 Calculating Alkalinity Species from pH, Alkalinity, and Equilibrium Equations

Alkalinity measurements based on titration and pH measurements and using equilibrium equations allows a more accurate determination of the various alkalinity species.

The following equations may be used for a temperature of 25°C. Hydroxide alkalinity is calculated from Equation (2.109).

$$OH^-\ \text{Alkalinity}\left(\frac{mg}{L}\ \text{as}\ CaCO_3\right) = 50{,}000\times 10^{(pH-pK_w)} \quad (2.109)$$

where:
OH^- Alkalinity = hydroxide alkalinity, mg/L as $CaCO_3$
pK_w = log of the dissociation constant for water at 25°C, 14.

Carbonate alkalinity is calculated from Equation (2.110).

$$CO_3^{2-}\ \text{Alkalinity}\left(\frac{mg}{L}\ \text{as}\ CaCO_3\right) = \frac{50{,}000[(ALK)/50{,}000 + [H^+] - (K_w/[H^+])]}{1 + ([H^+]/2K_2)} \quad (2.110)$$

where:
CO_3^{2-} Alkalinity = carbonate alkalinity, mg/L as $CaCO_3$
ALK = total alkalinity of water sample, mg/L as $CaCO_3$
K_w = 1×10^{-14} at 25°C
K_2 = $10^{-10.33}$ at 25°C.

Bicarbonate alkalinity is calculated from Equation (2.111):

$$HCO_3^-\ \text{Alkalinity}\left(\frac{mg}{L}\ \text{as}\ CaCO_3\right) = \frac{50{,}000[(ALK)/50{,}000 + [H^+] - (K_w/[H^+])]}{1 + (2K_2/[H^+])} \quad (2.111)$$

where HCO_3^- Alkalinity = bicarbonate alkalinity, mg/L as $CaCO_3$.

Example 2.42 Alkalinity calculated from pH, titration, and equilibrium equations

Using the data given in Example 2.41, determine total alkalinity, hydroxide, carbonate, and bicarbonate alkalinities expressed in mg/L as $CaCO_3$ based on equilibrium equations.

Solution part a

The total alkalinity is based on the total quantity of acid required to reach a pH of 4.5 during the titration, Equation (2.108).

$$\text{Total Alkalinity}\left(\frac{mg}{L}\ \text{as}\ CaCO_3\right) = \frac{(10.0\ ml + 5.5\ ml)\times 1000}{100\ ml} = \boxed{155\ \frac{mg}{L}\ \text{as}\ CaCO_3}$$

Solution part b

Hydroxide alkalinity is calculated from Equation (2.109).

$$OH^-\ \text{Alkalinity}\left(\frac{mg}{L}\ \text{as}\ CaCO_3\right) = 50{,}000\times 10^{(pH-pK_w)}$$

$$\text{OH}^- \text{ Alkalinity}\left(\frac{\text{mg}}{\text{L}} \text{ as CaCO}_3\right) = 50,000 \times 10^{(11-14)}$$

$$= \boxed{50 \frac{\text{mg}}{\text{L}} \text{ as CaCO}_3}$$

Solution part c

Carbonate alkalinity is calculated from Equation (2.110).

$$\text{CO}_3^{2-} \text{ Alkalinity}\left(\frac{\text{mg}}{\text{L}} \text{ as CaCO}_3\right)$$

$$= \frac{50,000[(\text{ALK})/50,000 + [\text{H}^+] - (K_w/[\text{H}^+])]}{1 + ([\text{H}^+]/2K_2)}$$

$$\text{CO}_3^{2-} \text{ Alkalinity}\left(\frac{\text{mg}}{\text{L}} \text{ as CaCO}_3\right)$$

$$= \frac{50,000\left[\left(155\frac{\text{mg}}{\text{L}}\right)/50,000 + 10^{-11} - (10^{-14}/10^{-11})\right]}{1 + (10^{-11}/2 \times 10^{-10.33})}$$

$$\text{CO}_3^{2-} \text{ Alkalinity}\left(\frac{\text{mg}}{\text{L}} \text{ as CaCO}_3\right)$$

$$= \boxed{95 \frac{\text{mg}}{\text{L}} \text{ as CaCO}_3}$$

Solution part d

Bicarbonate alkalinity is calculated from Equation (2.111).

$$\text{HCO}_3^- \text{ Alkalinity}\left(\frac{\text{mg}}{\text{L}} \text{ as CaCO}_3\right)$$

$$= \frac{50,000[(\text{ALK})/50,000 + [\text{H}^+] - (K_w/[\text{H}^+])]}{1 + (2K_2/[\text{H}^+])}$$

$$\text{HCO}_3^- \text{ Alkalinity}\left(\frac{\text{mg}}{\text{L}} \text{ as CaCO}_3\right)$$

$$= \frac{50,000\left[\left(155\frac{\text{mg}}{\text{L}}\right)/50,000 + 10^{-11} - (10^{-14}/10^{-11})\right]}{1 + (2 \times 10^{-10.33}/10^{-11})}$$

$$\text{HCO}_3^- \text{ Alkalinity}\left(\frac{\text{mg}}{\text{L}} \text{ as CaCO}_3\right)$$

$$= \boxed{10 \frac{\text{mg}}{\text{L}} \text{ as CaCO}_3}$$

Check to see if all alkalinity species add up to equal the total alkalinity obtained from the titration.

$$\text{Total Alkalinity}\left(\frac{\text{mg}}{\text{L}} \text{ as CaCO}_3\right)$$

$$= \text{OH}^- \text{ Alkalinity} + \text{CO}_3^{2-} \text{ Alkalinity} + \text{HCO}_3^- \text{ Alkalinity}$$

$$\text{Total Alkalinity}\left(\frac{\text{mg}}{\text{L}} \text{ as CaCO}_3\right)$$

$$= 50 \frac{\text{mg}}{\text{L}} + 95\frac{\text{mg}}{\text{L}} + 10 \frac{\text{mg}}{\text{L}} = \boxed{155 \frac{\text{mg}}{\text{L}} \text{ as CaCO}_3}$$

The values calculated from the equilibrium expressions are in close agreement with the values determined from the approximations, as presented in Example 2.41.

2.7 Solubility (solubility product)

So far, we have dealt with aqueous solutions in which the chemical species are highly soluble. In this section, our focus will be on liquid-solid species that are partially soluble or insoluble. All solids, no matter how seemingly insoluble, *are* soluble to some degree. When a solid is placed in water, the ions at the surface of the solid will migrate into the water. This is called dissolution. Simultaneously, ions in the solution will be redeposited on the surface of the solid; this is known as precipitation. Equilibrium will be reached between the crystals of the compound in the solid state and its ions in solution.

In general, the solubility of most compounds increases with increasing temperature. Snoeyink & Jenkins (1980, page 251) indicate that the solubilities of $CaCO_3$, $Ca_3(PO_4)_2$, $CaSO_4$, and $FePO_4$ do not increase as temperature increases.
Equation (2.112) shows the general equation of a solid compound dissolving in pure water to form its constituent ions.

$$\text{A}_Z\text{B}_{Y(S)} \underset{\text{Precipitation}}{\overset{\text{Dissolution}}{\rightleftharpoons}} Z\,\text{A}^{Y+} + Y\,\text{B}^{Z-} \tag{2.112}$$

The equilibrium expression is written as follows:

$$K = \frac{[\text{A}^{Y+}]^Z[\text{B}^{Z-}]^Y}{[(\text{A}_Z\text{B}_Y)_{(S)}]} \tag{2.113}$$

As described by Sawyer & McCarty (1994, page 37), at equilibrium or saturation, the surface area of the solid is the only portion that is in equilibrium with the ions in solution. Therefore, the concentration of solid as represented by $[(A_ZB_Y)_{(S)}]$ in the denominator of Equation (2.113) can be considered a constant (K_s) in equilibrium solubility problems. Equation (2.114) is rewritten to show the development of the solubility-product constant, K_{sp}.

$$[\text{A}^{Y+}]^Z[\text{B}^{Z+}]^Y = K \times K_s = K_{sp} \tag{2.114}$$

$$K = \frac{[\text{A}^{Y+}]^Z[\text{B}^{Z+}]^Y}{[(\text{A}_Z\text{B}_Y)_{(S)}]} = \frac{[\text{A}^{Y+}]^Z[\text{B}^{Z+}]^Y}{K_s} \tag{2.115}$$

When $[A^{Y+}]^Z[B^{Z+}]^Y = K_{sp}$, the solution is saturated or at equilibrium.

When $[A^{Y+}]^Z[B^{Z+}]^Y < K_{sp}$, the solution is under-saturated and no solids species are present.

When $[A^{Y+}]^Z[B^{Z+}]^Y > K_{sp}$, the solution is super-saturated and solid species are being formed.

The solubility-product constants for several solids of significance in environmental engineering are presented in Table 2.16. Partially soluble salts have small K_{sp} values, while soluble salts have relatively large K_{sp} values. Comparing the K_{sp} values of sparingly soluble solids does not predict relative solubility if the number of ions produced by dissolution is different. Solids that dissolve into two or three ions will generally have higher solubilities. Several examples will be presented to illustrate the concepts presented in this section.

Table 2.16 Typical Solubility-Product Constants at 25°C.

Compound	Equation	pK_{sp}
Calcium sulfate	$CaSO_{4(s)} \rightleftharpoons Ca^{2+} + SO_4^{2-}$	4.59[a]
Magnesium carbonate	$MgCO_{3(s)} \rightleftharpoons Mg^{2+} + CO_3^{2-}$	5.0[a]
Calcium hydroxide	$Ca(OH)_{2(s)} \rightleftharpoons Ca^{2+} + 2\,OH^-$	5.3
Calcium hydrogen phosphate	$CaHPO_{4(s)} \rightleftharpoons Ca^{2+} + 2\,HPO_4^{2-}$	6.5[b]
Calcium carbonate (calcite)	$CaCO_{3(s)} \rightleftharpoons Ca^{2+} + CO_3^{2-}$	8.34[a]
Silver chloride	$AgCl_{(s)} \rightleftharpoons Ag^+ + Cl^-$	10[a]
Barium sulfate	$BaSO_{4(s)} \rightleftharpoons Ba^{2+} + SO_4^{2-}$	10[a]
Calcium fluoride	$CaF_{2(s)} \rightleftharpoons Ca^{2+} + 2\,F^-$	10.3[a]
Zinc carbonate	$ZnCO_{3(s)} \rightleftharpoons Zn^{2+} + CO_3^{2-}$	10.9[c]
Magnesium hydroxide	$Mg(OH)_{2(s)} \rightleftharpoons Mg^{2+} + 2\,OH^-$	11.0[b]
Manganous hydroxide	$Mn(OH)_{2(s)} \rightleftharpoons Mn^{2+} + 2\,OH^-$	12.8[a]
Lead carbonate	$PbCO_{3(s)} \rightleftharpoons Pb^{2+} + CO_3^{2-}$	13.5[c]
Ferrous hydroxide	$Fe(OH)_{2(s)} \rightleftharpoons Fe^{2+} + 2\,OH^-$	14.5[a]
Nickel hydroxide	$Ni(OH)_{2(s)} \rightleftharpoons Ni^{2+} + 2\,OH^-$	15.7[b]
Zinc hydroxide	$Zn(OH)_{2(s)} \rightleftharpoons Zn^{2+} + 2\,OH^-$	17.1[a]
Copper hydroxide	$Cu(OH)_{2(s)} \rightleftharpoons Cu^{2+} + 2\,OH^-$	19.3[a]
Calcium phosphate	$Ca_3(PO_4)_{2(s)} \rightleftharpoons Ca^{2+} + 2\,PO_4^{3-}$	27[b]
Chromium hydroxide	$Cr(OH)_{3(s)} \rightleftharpoons Cr^{3+} + 3\,OH^-$	30.2[c]
Aluminum hydroxide	$Al(OH)_{3(s)} \rightleftharpoons Al^{3+} + 3\,OH^-$	33[a]
Manganic hydroxide	$Mn(OH)_{3(s)} \rightleftharpoons Mn^{3+} + 3\,OH^-$	36[b]
Ferric hydroxide	$Fe(OH)_{3(s)} \rightleftharpoons Fe^{3+} + 3\,OH^-$	37.2[b]

[a] pK_{sp} values from Snoeyink & Jenkins (1980). *Water Chemistry*, p. 249.
[b] pK_{sp} values Sawyer et *al.* (1994). *Chemistry for Environmental Engineering*, p. 38.
[c] pK_{sp} values Benefield & Morgan (1990). *Chemical Precipitation*, in *Water Quality and Treatment: A Handbook of Community Water Supplies*, pp. 10.4–10.5.

Example 2.43 Calculating the solubility of solids

Calculate the solubility of silver chromate (Ag_2CrO_4), given a K_{sp} value of 1.9×10^{-12}.

Solution

$$Ag_2CrO_{4(s)} \leftrightarrow 2Ag^+_{(aq)} + CrO^{2-}_{4(aq)}$$

$$X \text{ moles } Ag_2CrO_4 = 2X \text{ moles } Ag^+ + X \text{ moles } CrO_4^{2-}$$

$$[Ag^+] = 2[CrO_4^{2-}]$$

$$K_{sp} = 1.9 \times 10^{-12} = [Ag^+]^2\,[CrO_4^{2-}]^1$$

$$K_{sp} = 1.9 \times 10^{-12} = (2[CrO_4^{2-}])^2\,[CrO_4^{2-}]^1 = 4[CrO_4^{2-}]^3$$

$$[CrO_4^{2-}] = \sqrt[3]{\frac{1.9 \times 10^{-12}}{4}} = \boxed{7.8 \times 10^{-5}\ M}$$

$$[Ag^+] = 2[CrO_4^{2-}] = (2)(7.8 \times 10^{-5}) = \boxed{1.6 \times 10^{-4}\ M}$$

The solubility of Ag_2CrO_4 is also equal to 7.8×10^{-5}M.

Example 2.44 Using solubility to calculate K_{sp}

The solubility of barium sulfate ($BaSO_4$) is approximately 1.1×10^{-5}M at 20°C. Calculate the solubility-product constant, K_{sp}.

Solution

$$BaSO_{4(s)} \leftrightarrow Ba^{2+}_{(aq)} + SO^{2-}_{4(aq)}$$

$$X \text{ moles } BaSO_{4(s)} = X \text{ moles } Ba^{2+} + X \text{ moles } SO_4^{2-}$$

$$X = 1.1 \times 10^{-5}M$$

$$K_{sp} = [Ba^{2+}]\,[SO_4^{2-}] = (X)(X) = X^2$$

$$K_{sp} = (1.1 \times 10^{-5})^2 = \boxed{1.2 \times 10^{-10}}$$

Example 2.45 Calculating pH from solubility and K_{sp}

The dissolution of copper hydroxide [$Cu(OH)_{2(s)}$] produces a copper ion concentration of 0.002 M at $T = 25$°C. If $K_{sp} = 2 \times 10^{-19}$, calculate the pH.

Solution

$$Cu(OH)_{2(s)} \leftrightarrow Cu^{2+}_{(aq)} + 2OH^-_{(aq)}$$

$$K_{sp} = 2.0 \times 10^{-19} = [Cu^{2+}]\,[OH^-]^2$$

$$[Cu^{2+}] = 0.002\ M$$

$$[OH^-]^2 = \frac{2.0 \times 10^{-19}}{[Cu^{2+}]} = \frac{2.0 \times 10^{-19}}{0.002\ M} = 1.0 \times 10^{-16}$$

$$[OH^-] = 1.0 \times 10^{-8}$$

$$[H^+][OH^-] = 1.0 \times 10^{-14}$$

$$[H^+] = \frac{1.0 \times 10^{-14}}{[OH^-]} = \frac{1.0 \times 10^{-14}}{1.0 \times 10^{-8}} = 1.0 \times 10^{-6}$$

$$pH = -\log[H^+] = -\log(1.0 \times 10^{-6}) = \boxed{6.0}$$

2.8 Gas phase laws

2.8.1 Boyle's law

Robert Boyle (1627–1691) conducted experiments with a J-tube to determine the relationship between the pressure of trapped gas and the volume of the gas. In his original experiments, Boyle measured the volume of gas (in^3) as a function of pressure (measured in inches of mercury). He varied the quantity of mercury added to the J-tube, so that the pressure exerted on the trapped gas would increase. Equation (2.116) shows the relationship that Boyle derived from his experiments.

$$P \times V = k_B \tag{2.116}$$

where:
P = pressure of gas at constant temperature, in of Hg
V = volume of gas at constant temperature, in^3
k_b = Boyle's law constant, in $Hg \cdot in^3$.

Equation (2.116) is only valid at a specific temperature (i.e., temperature remains constant), and for a given amount of gas. Boyle's data indicated that the volume of the gas decreased as the pressure of the gas increased. Stating this another way, volume and pressure are inversely proportional; when one of the parameters increases, the other decreases. Boyle's law can be written to show how the pressure and volume at two different measurements are related.

$$P_1 \times V_1 = k_B = P_2 \times V_2 \tag{2.117}$$

Rearranging Equation (2.117) yields the following equation that is used to calculate the new volume of the gas, knowing the original volume and pressure of the gas.

$$V_2 = \frac{P_1}{P_2} \times V_1 \tag{2.118}$$

where:
V_1 = volume of gas for a given amount of gas at P_1, in^3
V_2 = volume of gas for a given amount of gas at P_2, in^3
P = pressure of gas, in Hg.

Other units may be used in Boyle's law, as long as a consistent set is used.

Example 2.46 Using Boyle's law to calculate pressure

Given the following sets of pressure-volume data, calculate the missing quantity, assuming that the mass of the gas and temperature remain constant.

$$V = 19.3\ \text{L at } 102.1\ \text{kPa}; \quad V = 10.0\ \text{L at } P = ?\ \text{kPa}$$

Solution

Rearranging Equation (2.118):

$$P_2 = P_1 \times \frac{V_1}{V_2}$$

$$P_2 = 102.1\ \text{kPa} \times \frac{19.3\ \text{L}}{10.0\ \text{L}} = \boxed{197\ \text{kPa}}$$

2.8.2 Charles's law

The French physicist Jacques Charles (1746–1823) showed that the volume of a given amount of gas at constant pressure increases as the temperature of the gases increases. Mathematically, **Charles's law** is stated as follows:

$$V = cT \tag{2.119}$$

where:
V = volume of gas for a given amount of gas at constant pressure, L
c = Charles's proportionality constant, dimensionless
T = temperature of gas, K.

Charles's law may be arranged to show the relationship between the volume and temperature of a given amount of gas at two different measurements. The equation is given below.

$$\frac{V_1}{T_1} = c = \frac{V_2}{T_2} \tag{2.120}$$

where:
V_1 = volume of gas for a given amount of gas at T_1, L
V_2 = volume of gas for a given amount of gas at T_2, L
T = temperature of gas, K.

Example 2.47 Using Charles's law to calculate volume

Given the following sets of temperature-volume data, calculate the missing quantity, assuming that the mass of the gas and temperature remain constant.

$$V = 2.5\ \text{L at } T = 15°\text{C}; \quad V = ?\ \text{L at } T = 40°\text{C}$$

Solution

First, the temperature data must be converted to kelvins as follows:

$$T_1 = 15°\text{C} + 273 = 288\ \text{K} \qquad T_2 = 40°\text{C} + 273 = 313\ \text{K}$$

Rearranging Equation (2.120).

$$\frac{V_1}{T_1} = c = \frac{V_2}{T_2}$$

$$V_2 = \frac{T_2}{T_1} \times V_1 \qquad V_2 = \frac{313\text{ K}}{288\text{ K}} \times 2.5\text{ L} = \boxed{2.7\text{ L}}$$

2.8.3 Avogadro's law

The relationship between the volume of a gas and the number of molecules (6.02×10^{23}) present in the gas at constant pressure and temperature was first postulated by Amadeo Avogadro in 1811. Mathematically, this is expressed by the following equation:

$$V = a\,n \tag{2.121}$$

where:
V = volume of gas at constant pressure and temperature, L
a = Avogadro's proportionality constant, dimensionless
n = number of moles of gas, mol.

This equation means that the volume of a gas at constant temperature and pressure is directly proportional to the number of moles of gas. Equation (2.121) can be arranged to solve for either the number of moles of gas or the volume as follows:

$$\frac{V_1}{n_1} = a = \frac{V_2}{n_2} \tag{2.122}$$

where:
V_1 = volume of gas for a given amount of gas at n_1, L
V_2 = volume of gas for a given amount of gas at n_2, L.

Example 2.48 Using Avogadro's law to calculate volume

If 0.25 moles of argon gas occupies a volume of 0.65L at a specified pressure and temperature, what volume would 0.40 moles of argon gas occupy under the same pressure and temperature?

Solution

From Equation (2.122):

$$\frac{V_1}{n_1} = a = \frac{V_2}{n_2}$$

$$V_2 = \frac{n_2}{n_1} V_1 = \frac{0.40\text{ mol}}{0.25\text{ mol}} \times 0.65\text{ L} = \boxed{1.04\text{ L}}$$

2.8.4 Ideal gas law

Avogadro's, Boyle's, and Charles's laws show how the volume of a gas depends on the number of moles of gas, the pressure, and the temperature. Each of these laws is based on experimental data. The familiar **Ideal Gas law** is derived by combining these three laws as follows:

$$V = n\,a = c\,T = \frac{k_B}{P} \tag{2.123}$$

$$V = \frac{n\,T(a\,c\,k_B)}{P} \tag{2.124}$$

$$V = \frac{n\,T(R)}{P} \tag{2.125}$$

$$P V = n\,R\,T \tag{2.126}$$

where:
P = absolute pressure, atm,
V = volume occupied by the gas, L
n = moles of gas,
R = universal gas law constant, $\frac{\mathbf{0.08206}\text{ atm}\cdot\text{L}}{\text{mol}\cdot\text{K}}$
R = universal gas law constant, $\frac{1.986\text{ cal}}{\text{mol}\cdot\text{K}}$,
R = universal gas law constant, $\frac{1.986\text{ BTU}}{\text{lb mol}\cdot°\text{R}}$
T = temperature, K (273.15 + °C).

To solve Equation (2.126), only three of the four parameters are required. According to Zumdahl (2000), the Ideal Gas law is applicable to most gases when the pressure is approximately 1 atm or lower and the temperature is 0°C or higher. Other equations of state must be used at high pressures and lower temperatures.

Example 2.49 Using the Ideal Gas law

Assume we have 0.25 moles of ammonia gas at 25°C with a volume of 3.5 L at a pressure of 1.75 atm. The gas is then compressed to a volume of 1.5 L at 25°C. Use Boyle's law and the Ideal Gas law to solve for pressure.

Solution part a

From Equation (2.117):

$$P_1 \times V_1 = k_B = P_2 \times V_2$$

$$1.75\text{ atm} \times 3.5\text{ L} = P_2 \times 1.5\text{ L}$$

$$P_2 = \boxed{4.1\text{ atm}}$$

Solution part b

The Ideal Gas law can be used for solving almost any type of problem involving gases. Boyle's law is the easiest method, but the Ideal Gas law is more applicable to a wide array of cases. When using it, all variables that do not change should be shown on one side of the equation, with the remaining variables on the other side. For this problem, the number of moles of gas, the universal gas constant, and temperature remain the same, so they should be placed on one side of the equation.

From Equation (2.126):

$$P \forall = nRT$$

$$P_1 \forall_1 = nRT = P_2 \forall_2$$

$$P_2 = \frac{nRT}{\forall_2} = \frac{0.25\,\text{mol}\left(\frac{0.08206\,\text{atm}\cdot\text{L}}{\text{mol}\cdot\text{K}}\right)(273.15 + 25\,\text{K})}{1.5\,\text{L}}$$

$$= \boxed{4.1\,\text{atm}}$$

2.8.5 Dalton's law

John Dalton was one of the first scientists to study mixtures of gases. Based on his studies, Dalton stated that, for a mixture of gases in a container, the total pressure exerted is the sum of the partial pressure of each of the gases that are present. **Dalton's law** is presented mathematically as follows:

$$P_T = P_1 + P_2 + P_3 + \ldots + P_i \qquad (2.127)$$

where:

P_i = partial pressure of gaseous component i, atm

P_1, P_2, P_3 = partial pressure of gaseous component 1, 2, and 3, atm

P_T = total pressure of the system, atm.

The partial pressure of a gas is the pressure that the gas would exert if it were the only gas present in the container, i.e., it is proportional to the percentage by volume of that gas in the mixture. Mathematically, the partial pressure of component i is equal to the mole fraction of component i times the total pressure:

$$P_i = y_i P_T \qquad (2.128)$$

where:

y_i = mole fraction of component i in the gas phase

$y_i = \frac{n_i}{n} = \frac{P_i}{P_T}$

n_i = the number of moles of gaseous component i

n = the total number of moles of gas in the system.

Example 2.50 Calculating partial pressure and using Dalton's law

A mixture of gases at 760 mm Hg pressure contains 60% nitrogen, 18.0% oxygen, and 22% carbon dioxide by volume. Determine the partial pressure of each of the gaseous components and the total pressure of the system.

Solution

From Equation (2.128):

$$P_i = y_i P_T$$

$$P_{N_2} = 0.60 \times (760\,\text{mm Hg}) = \boxed{456\,\text{mm Hg}}$$

$$P_{O_2} = 0.18 \times (760\,\text{mm Hg}) = \boxed{137\,\text{mm Hg}}$$

$$P_{CO_2} = 0.22 \times (760\,\text{mm Hg}) = \boxed{167\,\text{mm Hg}}$$

From Equation (2.127):

$$P_T = P_1 + P_2 + P_3 + \ldots + P_i$$

$$P_T = 456\,\text{mm Hg} + 137\,\text{mm Hg} + 167\,\text{mm Hg}$$

$$= \boxed{760\,\text{mm Hg}}$$

Example 2.51 Calculating partial pressure and using Dalton's law

If 4.0 g of oxygen gas and 4.0 g of helium gas are placed in a 5.0 L container at 65°C, determine the partial pressure of each gas and the total pressure in the container.

Solution

First, it is necessary to determine the moles of each gas. The molecular weight of oxygen is 32.0 g per mole.

$$n_{O_2} = \frac{4.0\,\text{g}\ O_2}{32\,\text{g/mol}} = 0.13\,\text{mol}$$

The molecular weight of helium is 4.0 g per mole.

$$n_{He} = \frac{4.0\,\text{g He}}{4.0\,\text{g/mol}} = 1.0\,\text{mol}$$

From Equation (2.126):

$$P\forall = nRT \qquad P = \frac{nRT}{\forall}$$

$$P_{O_2} = \frac{(0.13\,\text{mol})\left(\frac{0.08206\,\text{atm}\cdot\text{L}}{\text{mol}\cdot\text{K}}\right)(273 + 65\,\text{K})}{5.0\,\text{L}}$$

$$= \boxed{0.72\,\text{atm}}$$

$$P_{He} = \frac{(1.0\,\text{mol})\left(\frac{0.08206\,\text{atm}\cdot\text{L}}{\text{mol}\cdot\text{K}}\right)(273 + 65\,\text{K})}{5.0\,\text{L}}$$

$$= \boxed{5.5\,\text{atm}}$$

The total pressure is determined from Equation (2.127) as follows:

$$P_T = P_1 + P_2 + P_3 + \dots + P_i$$

$$P_T = 0.72 \text{ atm} + 5.5 \text{ atm} = \boxed{6.2 \text{ atm}}$$

2.8.6 Raoult's law

When a liquid is in contact with air, equilibrium conditions exist when the rate of molecules leaving the liquid as vapor equals the rate of molecules from the air dissolving into the liquid. Molecules leave the liquid as vapor through evaporation. The vapor pressure of the liquid is the pressure that is exerted by the vapor on the liquid once equilibrium has been achieved. The vapor pressure of a substance is defined as the partial pressure of the gaseous phase of the substance in equilibrium with the liquid phase of the substance at a specified temperature. This relationship was discovered by Francois Raoult in 1886 and is called **Raoult's law**.

Raoult's law relates the partial pressure of the gaseous component present above a liquid to the vapor pressure of the pure component and the mole fraction of the pure component in the liquid phase. Mathematically, Raoult's law is expressed as:

$$P_A = P_A^* x_A \tag{2.129}$$

where:
P_A^* = vapor pressure of component A, atm
P_A = partial pressure of component A, atm
x_A = mole fraction of component A in the liquid phase.

Several organic compounds are very volatile, i.e., they have a high vapor pressure. Volatilization is often used interchangeably with evaporation in environmental work; it is the transfer of a compound from the liquid phase to the gaseous phase. Table 2.17 lists the vapor pressures of several chemicals as a function of temperature.

Table 2.17 Vapor pressure of several chemical components at specified temperature.

Chemical	Formula	Vapor pressure, kPa	T, °C
Acetone	C_3H_6O	100	55.7
Ammonia	NH_3	100	−33.6
Benzene	C_6H_6	100	79.7
Chloroform	$CHCl_3$	100	60.8
Toluene	C_7H_8	100	110.1
Tetrachloroethylene	C_2Cl_4	100	120.7
Trichloroethylene	C_2HCl_3	100	86.8
2,4,6-Trichlorophenol	$C_6H_3Cl_3O$	100	245.7
Water	H_2O	100	99.6

Vapor pressure and temperature values from *CRC Handbook of Chemistry and Physics*, (2010), 91st Edition, pp. 6-90, 6-91, 6-93, 6-94, 6-96, 6-101, 6-105.

2.8.7 Henry's law

Henry's law is a special case of Raoult's law that is applied to dilute solutions, used for calculating the solubility of a gas in a liquid or to describe equilibrium between the gas and liquid phases. There are several forms of Henry's law, depending on the units associated with the Henry's constant. Henry's law can be expressed to show the relationship between the liquid mole fraction concentration (x_i) of a chemical that is in equilibrium to the gas phase concentration. Equation (2.130) shows this relationship.

$$P_A = H(\text{atm})\, x_A \tag{2.130}$$

where:
P_A = partial pressure of component A, atm
$H(\text{atm})$ = Henry's constant expressed in atm
x_A = mole fraction of component A in the liquid phase.

For pure solutions, $P_A^* = H(\text{atm})$. Henry's law also can be expressed to show the relationship between the aqueous or liquid phase concentration of a substance in moles/L with the equilibrium gaseous phase concentration. This form of Henry's law is presented as Equation (2.131).

$$P_A = H\left(\frac{\text{L} \cdot \text{atm}}{\text{mol}}\right) [\text{A}] \tag{2.131}$$

where:
P_A = partial pressure of component A, atm
$H\left(\frac{\text{L·atm}}{\text{mol}}\right)$ = Henry's constant expressed in $\frac{\text{L·atm}}{\text{mol}}$
[A] = concentration of component A in the liquid phase, moles/L.

A third way of expressing Henry's law is to use the unitless Henry's constant (H_u), as shown in Equation (2.132):

$$C_g = H_u\, C_l \tag{2.132}$$

where:
C_g = concentration of component A in the gas phase, µg/m³, mg/L
H_u = Henry's constant, unitless
C_l = concentration of component A in the liquid phase, µg/m³, mg/L.

The following equations show the relationships between the various Henry's constants:

$$H\left(\frac{\text{L} \cdot \text{atm}}{\text{mol}}\right) = \frac{H(\text{atm})}{\dfrac{55.6 \text{ mol } H_2O}{L_{H_2O}}} \tag{2.133}$$

$$H_u = \frac{H(\text{atm})}{4.56 \times T} \quad (2.134)$$

Where temperature, T, is expressed in kelvin.

A detailed discussion and comparison of the various forms of Henry's law and their associated Henry's constants is found in the following references: Metcalf & Eddy (2003), pp. 66–69, and MWH (2005), pp. 1167–1174.

Henry's constant is a function of temperature. The following modified form of the van't Hoff-Arrhenius equation may be used for estimating Henry's constant at different temperatures (Metcalf & Eddy (2003), page 68).

$$\log H = \frac{-A}{T} + B \quad (2.135)$$

where:

H = Henry's constant at temperature T, atm
A, B = Empirical coefficients for chemical or gas
T = Absolute temperature, $K = (273.15 + °C)$.

Table 2.18 presents Henry's constant for several gases of environmental concern, along with the empirical coefficients necessary for making temperature corrections.

Example 2.52 Saturation concentration of carbon dioxide in water

Determine the saturation or equilibrium concentration of carbon dioxide in water in contact with dry air at 1 atm and 20°C, assuming that the carbon dioxide concentration in the atmosphere is 0.03% by volume.

Solution

From Table 2.17, $H = 1.49 \times 10^3$ atm for carbon dioxide at 20°C.

From Equation (2.128):

$$P_i = y_i P_T$$

$$P_{CO_2} = 1 \text{ atm} \times \frac{0.03}{100} = 3.00 \times 10^{-4} \text{ atm}$$

Now, solve for the aqueous phase mole fraction by rearranging Equation (2.130):

$$P_A = H(\text{atm})\, x_A$$

$$x_{CO_2} = \frac{P_{CO_2}}{H(\text{atm})} = \frac{3.00 \times 10^{-4}\text{atm}}{1.49 \times 10^3\text{atm}} = 2.01 \times 10^{-7}$$

$$x_{CO_2} = 2.01 \times 10^{-7} = \frac{n_{CO_2}}{n_{CO_2} + n_{H_2O}}$$

In 1.0 L of water, there are 1000 g of water and approximately 55.6 moles of water.

$$n_{H_2O} = \frac{1000 \text{ g } H_2O}{18 \text{ g } H_2O/\text{mole}} = 55.6 \text{ moles}$$

$$x_{CO_2} = 2.01 \times 10^{-7} = \frac{n_{CO_2}}{n_{CO_2} + 55.6 \text{ moles}}$$

Table 2.18 Henry's constant at 20°C (293.15 K) for various gases.

Gas	ΔH°_{dis}, $\frac{kcal}{kmole} \times 10^3$	K_C	T, K	H_u, unitless	H, atm	H, $\frac{L\cdot atm}{mol}$
Ammonia	8.63	1526	293.15	0.00056	0.75	0.013
Benzene	8.47	357,678	293.15	0.17	231	4.2
Carbon dioxide	4.77	4013	293.15	1.1	1490	27
Carbon tetrachloride	9.32	8,580,096	293.15	0.97	1290	23
Chlorine	4.01	420	293.15	0.43	575	10
Chlorine dioxide	6.75	4300	293.15	0.040	53	0.96
Chloroform	9.21	940,789	293.15	0.13	171	3.1
Hydrogen sulfide	4.26	567	293.15	0.38	505	9.1
Methane	3.55	12,402	293.15	28.0	37387	672
Oxygen	3.34	9627	293.15	31.1	41619	749
Ozone	5.80	38,848	293.15	3.97	5311	95.5
Sulfur dioxide	5.53	358	293.15	0.027	36	0.65
Tetrachloroethylene	9.88	17,926,362	293.15	0.77	1031	19
Trichloroethylene	7.85	290,732	293.15	0.41	545	9.8

Henry's constants calculated using enthalpy change and K_C values from MWH (2005), *Water Treatment: Principles and Design*, p. 1173.

Assume that $n_{H_2O} >> n_{CO_2}$, therefore $x_{CO_2} =$

$$2.01 \times 10^{-7} \cong \frac{n_{CO_2}}{55.6 \text{ moles}}$$

$$n_{CO_2} = (55.6 \text{ moles})2.01 \times 10^{-7} \cong 1.12 \times 10^{-5} \text{moles}$$

The concentration of carbon dioxide in water is calculated below. The molecular weight of carbon dioxide is 44 g per mole.

$$CO_{2(20°C)} = 1.12 \times 10^{-5} \frac{\text{moles } CO_2}{L} \times \left(\frac{44 \text{ g } CO_2}{\text{mole } CO_2}\right) \times \left(\frac{1000 \text{ mg}}{\text{g}}\right)$$

$$CO_{2(20°C)} = 0.49 \frac{\text{mg } CO_2}{L}$$

2.9 Organic chemistry overview

Organic chemistry is the study of carbon-containing compounds and their properties. Originally, it was believed that organic compounds only could be formed by plants and animals. This was known as the vital-force theory, and German chemist Friedrich Wohler (1800–1882) disproved this theory when he was able to produce urea, a component of urine, by heating ammonium cyanate according to the following reaction.

$$NH_4OCN \xrightarrow{Heat} N_2H_4CO \qquad (2.136)$$

All organic compounds contain carbon in combination with one or more elements. Hydrocarbons consist only of carbon and hydrogen. Many organic compounds contain carbon, hydrogen, and oxygen; yet others contain nitrogen, phosphorus, and sulfur in addition to carbon. Organic compounds are important since they:

- form the basis for all life;
- are used in the production of pesticides, herbicides, insecticides, polymers, antibiotics, hormones, and alcohols;
- cause deleterious effects on the environment, since many of the synthesized organics contain halogens and metals which may be toxic and or carcinogenic.

According to Sawyer & McCarty (1994, p. 189), there are seven major differences between organic and inorganic compounds:

- Organics are usually combustible.
- In general, organics have lower boiling and melting points.
- Organics are usually less soluble in water than are inorganics.
- Several organic compounds may exist for a given formula; this is known as isomerism.
- Chemical reactions involving organic compounds are often quite slow, since these reactions are molecular rather than ionic.
- Organic compounds have high molecular weights, often greater than 1,000.
- Most organic compounds can serve as an energy and carbon source for bacteria and other microorganisms.

There are three major types of organic compounds: aliphatic, aromatic and heterocyclic:

- Aliphatic organic compounds (open-chain structures) contain carbon-carbon bonds with functional groups linked to a straight or branched chain.
- Aromatic organics consist of six-member carbon ring structures that contain double bonds rather than single-covalent bonds.
- Finally, the heterocyclic organic compounds are those that have a ring structure in which at least one element is other than carbon.

2.9.1 Introduction to hydrocarbons (alkanes, alkenes, and alkynes)

Hydrocarbons consist only of hydrogen and carbon, and they may be saturated or unsaturated. A saturated hydrocarbon means that adjacent carbon atoms are joined together by a single covalent bond and all other bonds are made with hydrogen. Recall that a covalent bond is one in which a pair of electrons is shared between atoms. Figure 2.9 shows the structural formula of ethane.

Unsaturated hydrocarbons are those that have at least two carbon atoms that are joined together by more than a single covalent bond; the remaining bonds are formed with hydrogen, i.e., they contain double and triple bonds. Figure 2.10 shows the structural formula of ethylene.

2.9.1.1 Alkanes

The alkanes are hydrocarbons in which all carbon-carbon bonds are single bonds (saturated compounds). They are represented by the following formula: C_nH_{2n+2}. Alkanes are also called the paraffins or aliphatic hydrocarbons (open-chain structures, i.e., straight or branched chains). Table 2.19 shows

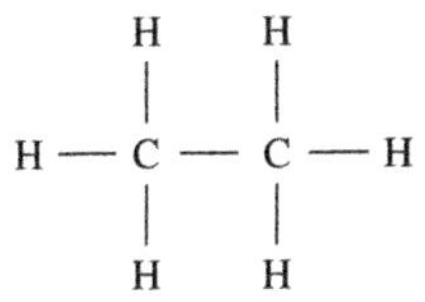

```
    H   H
    |   |
H — C — C — H
    |   |
    H   H
```

Figure 2.9 Structural formula of the saturated hydrocarbon ethane.

Table 2.19 Formulas of the first ten straight-chain alkanes or paraffins.

Name	Formula	Structure	BP, °C	MP, °C
Methane	CH_4	CH_4	−161.5	−183
Ethane	C_2H_6	CH_3CH_3	−88.3	−172
Propane	C_3H_8	$CH_3CH_2CH_3$	−42.2	−187.1
Butane	C_4H_{10}	$CH_3CH_2CH_2CH_3$	−0.6	−135
Pentane	C_5H_{12}	$CH_3CH_2CH_2CH_2CH_3$	36.2	−130
Hexane	C_6H_{14}	$CH_3CH_2CH_2CH_2CH_2CH_3$	69.0	−94.3
Heptane	C_7H_{16}	$CH_3CH_2CH_2CH_2CH_2CH_2CH_3$	98.5	−90.5
Octane	C_8H_{18}	$CH_3CH_2CH_2CH_2CH_2CH_2CH_2CH_3$	125.8	−56.5
Nonane	C_9H_{20}	$CH_3CH_2CH_2CH_2CH_2CH_2CH_2CH_2CH_3$	150.7	−53.7
Decane	$C_{10}H_{22}$	$CH_3CH_2CH_2CH_2CH_2CH_2CH_2CH_2CH_2CH_3$	174	−30

Boiling point (BP) and melting point (MP) values from Sawyer *et al.* (1994) *Chemistry for Environmental Engineering*, p. 195.

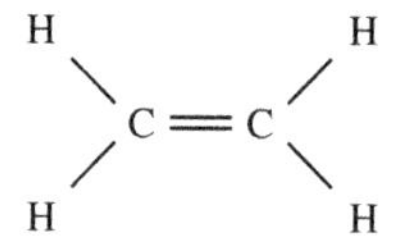

$$\begin{array}{ccc} H & & H \\ & C=C & \\ H & & H \end{array}$$

Figure 2.10 Structural formula of the unsaturated hydrocarbon ethylene.

the molecular formula, boiling point (BP), melting point (MP), and the nomenclature used for naming the compound as used by the International Union of Pure and Applied Chemists (IUPAC). The suffix, *–ane*, is used in naming hydrocarbons associated with the alkanes.

According to Sawyer & McCarty (1994, p. 193), the alkanes are colorless, practically odorless, and quite insoluble in water for those with five or more carbon atoms. They dissolve readily in most organic solvents. At room temperature, alkane members through C_5 are gases, those from C_5 to C_{17} are liquids, and those above C_{17} are solids (Sawyer & McCarty, 1994, p. 193).

Alkanes are widely used as fuels. At relatively high temperatures, alkanes may be combusted to carbon dioxide and water. Equation (2.137) shows the combustion of propane.

$$C_3H_{8(g)} + 5\ O_{2(g)} \rightarrow 3\ CO_{2(g)} + 4\ H_2O_{(g)} \quad (2.137)$$

Alkanes may also undergo substitution reactions, in which one or more of the hydrogen atoms are replaced by a different atom. An example of a substitution reaction is shown below, in which methane reacts with chlorine in the presence of ultraviolet light (hv) to produce chloromethane. These types of reactions are extremely important in the treatment of drinking water. When chlorine is used as the primary disinfectant in potable water treatment, it can react with organic compounds in the water to form trihalomethanes and haloacetic acid compounds that are potentially carcinogenic. The US Environmental Protection Agency has established maximum contaminant levels (MCL) for these types of chlorinated species.

$$CH_4 + \ Cl_2 \xrightarrow{hv} CH_3Cl + HCl \quad (2.138)$$

Besides combustion and substitution reactions, alkanes can also undergo dehydrogenation reactions, in which hydrogen atoms are removed from a saturated hydrocarbon to form an unsaturated hydrocarbon. An example of this is the dehydrogenation of ethane to ethylene, according to the following reaction. A temperature of 500°C and chromium (III) oxide catalyst are required.

$$CH_3CH_3 \underset{500^{\circ}C}{\overset{Cr_2O_3}{\rightarrow}} CH_2{=}\ CH_2 + H_2 \quad (2.139)$$

2.9.1.2 Alkenes

Alkenes are hydrocarbons that contain at least one double bond and are represented by the following formula: C_nH_{2n}. They are also called olefins, and their names all end with the suffix *–ene*. Table 2.20 lists the formula, boiling point, melting point, and IUPAC name of selected alkenes.

2.9.1.3 Alkynes

Alkynes are hydrocarbons that contain at least one triple bond and are represented by the following formula: C_nH_n. They are also called olefins, and their names all end with the suffix, *–yne*. Table 2.21 lists the formula, boiling point, melting point, and IUPAC name of selected alkynes.

2.9.1.4 Halogenated Hydrocarbons

Halogenated hydrocarbons, such as those involving the substitution of a hydrogen atom with a chlorine molecule to alkanes and alkenes, include several commercial products used as solvents, refrigerants, lubricants, hydraulic fluids, aerosol propellants, polymers, and pesticides (Baum, 1978,

Table 2.20 Formulas of selected alkenes.

Name	Formula	Structure	BP, °C	MP, °C
Ethene	C_2H_4	C_2H_4	−103.9**	−169.4**
Propene	C_3H_6	$CH_2 = CHCH_3$	−47*	−185*
1-Butene	C_4H_8	$CH_2 = CHCH_2CH_3$	−6.5*	−130*
2-Methylpropene	C_4H_8	$(CH_3)_2C = CH_2$	−6.9*	−141*
1-Pentene	C_5H_{10}	$CH_2 = CH(CH_2)_2CH_3$	30*	−138*
1-Hexene	C_6H_{12}	$CH_2 = CH(CH_2)_3CH_3$	64.1**	−98.5**
2-Methyl-2-pentene	C_6H_{12}	$CH_3CH_2CH = C(CH_3)_2$	67–69***	−135***
1-Heptene	C_7H_{14}	$CH_2 = CH(CH_2)_4CH_3$	95**	−120**
2,4-Dimethyl-2-pentene	C_7H_{14}	$(CH_3)_2CHC(CH_3) = CHCH_3$	83****	−128****

*Boiling point (BP) and melting point (MP) values Baum (1978) *Introduction to Organic and Biological Chemistry*, p. 37.
**Boiling point (BP) and melting point (MP) values Sawyer et *al.* (1994) *Chemistry for Environmental Engineering*, p. 197.
***Boiling point (BP) and melting point (MP) values from:
http://www.sigmaaldrich.com/catalog/product/aldrich/m67303?lang=en®ion=US
****Boiling point (BP) and melting point (MP) values from:
http://www.chemsynthesis.com/base/chemical-structure-15925.html

Table 2.21 Formulas of selected alkynes.

Name	Formula	BP, °C	MP, °C
Ethyne	H–C≡C–H	−84.7 sp	−80.8 (triple point)
Propyne	CH_3 – C≡C–H	−23.2	−102.7
1-Butyne	CH_3CH_2–C≡C–H	8.08	−125.7
2-Butyne	CH_3–C≡C–CH_3	26.9	−32.2
1-Pentyne	$CH_3CH_2CH_2$–C≡C–H	40.1	−90
2-Pentyne	CH_3CH_2–C≡C–CH_3	56.1	−109.3

Boiling point (BP) and melting point (MP) values from *CRC Handbook of Chemistry and Physics*, (2010), 91st Edition, pp. 3-6, 3-84, 3-418, 3-446.
sp -sublimation point.

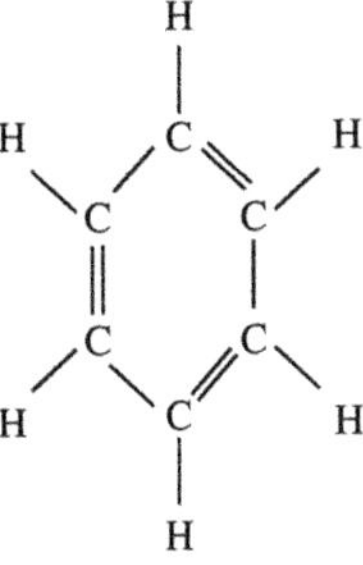

Figure 2.11 Lewis structure for benzene ring.

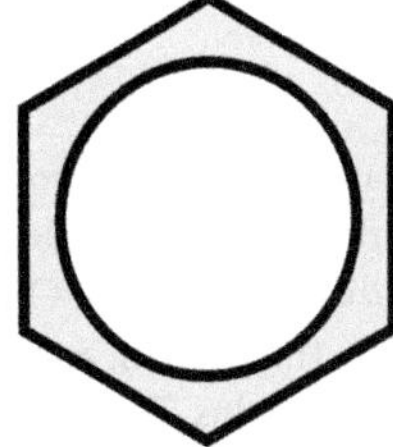

Figure 2.12 Simplified structure of benzene.

pages 77–85). Many of the compounds are known or suspected carcinogens, and regulatory cleanup requirements often approach detectable limits. Some of the more pervasive chlorinated aliphatic compounds that are found in ground water and at hazardous waste sites are listed in Table 2.22.

2.9.2 Aromatic hydrocarbons

An aromatic hydrocarbon, or arene or aryl hydrocarbon, is a hydrocarbon that contains alternating single and double bonds between carbons. The aromatic organic carbons all contain a six-membered ring of carbon atoms called the benzene ring. Benzene has the formula C_6H_6 and is the simplest of the aromatic hydrocarbons. Figure 2.11 is the Lewis structure for benzene showing the double bonds between alternate carbon atoms in the ring. Figure 2.12 is the simplified formula for benzene often used by chemists.

Some of the aromatic hydrocarbons have pleasant odors: cinnamon, vanillin, and wintergreen are examples. Table 2.23 shows some of the important aromatic hydrocarbons commonly encountered in environmental engineering.

Aromatic compounds are important to industry and living systems. Chemicals and polymers such as styrene, phenol,

Table 2.22 Halogenated alkanes and alkenes of environmental importance.

Name	Formula	BP, °C	MP, °C
Carbon tetrachloride (CT)	http://en.wikipedia.org/wiki/Carbon_tetrachloride	76.8	−22.62
Chloroform (CF)	http://en.wikipedia.org/wiki/File:Chloroform_displayed.svg	61.17	−63.41
Dichloroethylene (DCE)	http://en.wikipedia.org/wiki/File:Cis-1,2-dichloroethene.png	60.1	−80
Tetrachloroethylene (PCE)	http://en.wikipedia.org/wiki/File:Tetrachloroethylene.svg	121.3	−22.3
Trichloroethylene (TCE)	http://en.wikipedia.org/wiki/File:Trichloroethene.svg	87.21	−84.7
Vinyl chloride (VS)	http://en.wikipedia.org/wiki/File:Vinyl-chloride-2D.png	−13.8	−153.84

Boiling point (BP) and melting point (MP) values from *CRC Handbook of Chemistry and Physics*, (2010), 91st Edition, pp. 3-100, 3-154, 3-470, 3-492.

aniline, polyester, and nylon are made from aromatic compounds. Arene compounds are produced during oil refining and from the distillation of tar. Other aromatics such as histidine, phenylalanine, tryptophan, and tyrosine, serve as basic building blocks of proteins. The genetic code of all organisms, DNA and RNA, consists of adenine, thymine, cytosine, guanine, and uracil, which are aromatic purines or pyrimidines.

Benzene, toluene, ethyl benzene, and xylene collectively referred to as BTEX is commonly found at contaminated sites. They are associated with petroleum products and often enter the environment from leaking underground storage tanks (LUSTs).

2.9.2.1 Polycyclic Aromatic Hydrocarbons

Polycyclic aromatic hydrocarbons (PAHs) or polynuclear aromatic hydrocarbons are molecules containing two or more simple aromatic rings fused together by sharing two neighboring carbon atoms. PAHs are formed from the incomplete combustion of fossil fuels and some are known to be carcinogenic, mutagenic, and teratogenic. PAHs are lipophilic and therefore mix more readily with oil than water. They generally have high molecular weights, low water solubility, and low volatility, i.e., they do not evaporate easily. Table 2.24 shows some of the more commonly found polycyclic aromatic hydrocarbons.

2.9.2.2 Polychorinated Biphenyls (PCBs)

Polychlorinated biphenyls (PCBs) make up a family of man-made chlorinated hydrocarbons. Their chemical formula is represented by $C_{12}H_{10-n}Cl_n$, where *n* is the number of chlorine atoms. They are made from biphenyl compounds that contain from one to five chlorine atoms bonded to each of the aromatic rings instead of hydrogen atoms. There are approximately 210 different PCB compounds, with commercial mixtures varying from 40 to 60% chlorine by weight. The EPA

Table 2.23 Names and characteristics of several aromatic hydrocarbons of importance.

Compound name	Structure	Molecular weight	Henry's law constant, $\frac{atm \cdot m^3}{mol}$	T, °C
Aniline	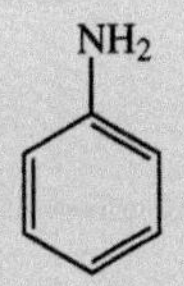 NH_2 http://en.wikipedia.org/wiki/File:Aniline.svg	89.094	1.38×10^{-1}	25
Benzene	 http://en.wikipedia.org/wiki/File:Benzene_circle.svg	[a]78.12	[a]5.69×10^{-3}	25
Benzoic acid	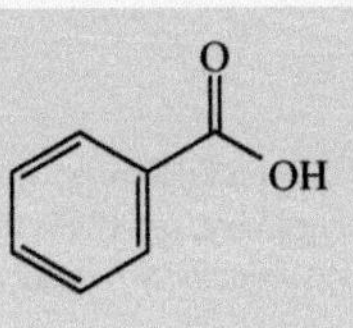 O OH http://en.wikipedia.org/wiki/File:Benzoic_acid.svg	[b]122.1	[b]1.82×10^{-8}	20
Biphenyl	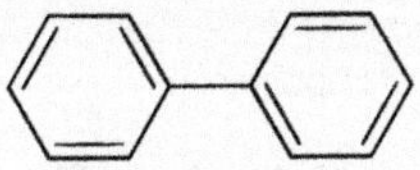 http://en.wikipedia.org/wiki/File:Bifenyl.svg	154.21	2.76×10^{-4}	25
Ethylbenzene	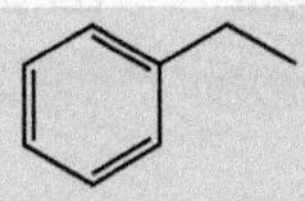 http://en.wikipedia.org/wiki/File:Ethylbenzene-2D-skeletal.png	106.165	8.32×10^{-3}	25
Nitrobenzene	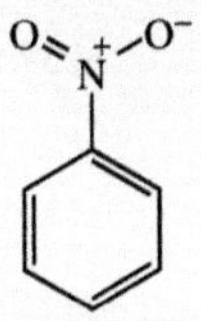 $O{=}N^{+}{-}O^{-}$ http://en.wikipedia.org/wiki/File:Nitrobenzol.svg	[b]123.1	[b]1.3×10^{-5}	25
Phenol	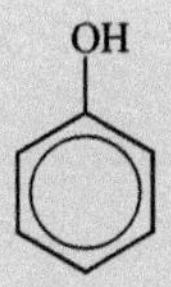 OH http://en.wikipedia.org/wiki/File:Phenol-2D-skeletal.png	[a]94.12	[a]1.32×10^{-6}	25
2-Phenylhexane	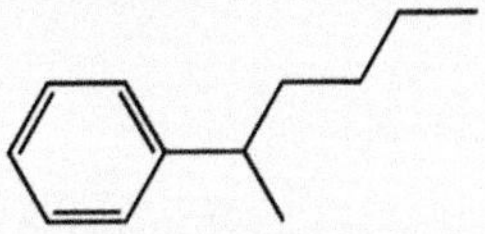 http://en.wikipedia.org/wiki/File:2-phenyl-hexane.png	162.27	......	......
Toluene	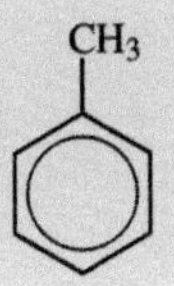 CH_3 http://en.wikipedia.org/wiki/File:Toluene.svg	92.139	6.51×10^{-3}	25

(continued overleaf)

Table 2.23 (*continued*)

Compound name	Structure	Molecular weight	Henry's law constant, $\frac{atm \cdot m^3}{mol}$	T, °C
m-Xylene	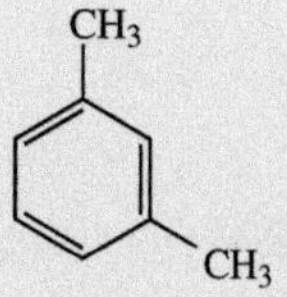 CH_3 CH_3 http://en.wikipedia.org/wiki/File:M-Xylene.png	106.165	7.20×10^{-3}	25
p-Xylene	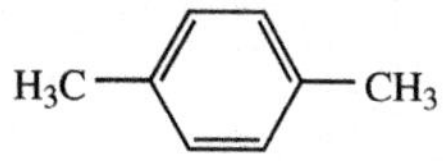 H_3C CH_3 http://en.wikipedia.org/wiki/File:P-Xylene.svg	106.165	6.81×10^{-3}	25

Molecular weight (MW) and Henry's constant values from *CRC Handbook of Chemistry and Physics* (2010), 91st Edition, pp. 8-90, 8-101, 8-118, 8-120.
[a]Molecular weight (MW) and Henry's constant values from LaGrega *et al.* (1994) *Hazardous Waste Management*, pp. 1040, 1048.
[b]Molecular weight (MW) and Henry's constant values from EPA (1990), *CERCLA Site Discharges to POTWs Treatability manual, EPA 540/2-90-007, Office of Water*, Table 8-1.

Table 2.24 Names and characteristics of several polycyclic aromatic hydrocarbons of importance.

Compound name	Structure	Molecular weight	Henry's law constant, $\frac{atm \cdot m^3}{mol}$	T, °C
Anthracene	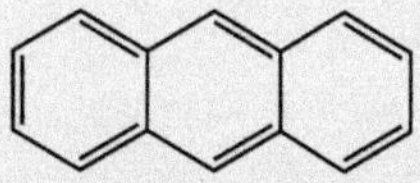 http://en.wikipedia.org/wiki/File:Anthracene-2D-Skelctal.png	178.2229	3.91×10^{-5}	25
Benzo[a]pyrene	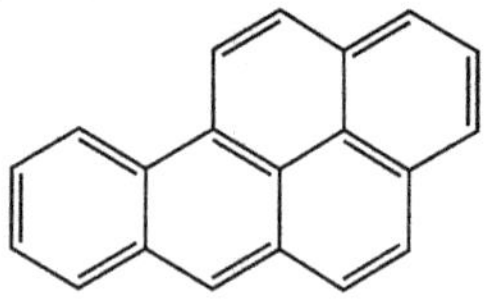 http://en.wikipedia.org/wiki/File:Benzo-a-pyrene.svg	252.309	4.59×10^{-7}	25
Naphthalene	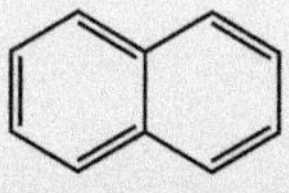 http://en.wikipedia.org/wiki/File:Naphthalene-2D-Skeletal.svg	128.171	4.24×10^{-4}	25
Phenanthrene	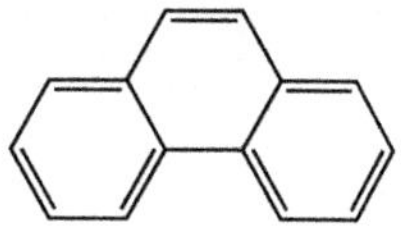 http://en.wikipedia.org/wiki/File:Phenanthrene.svg	178.229	3.20×10^{-5}	25
Pyrene	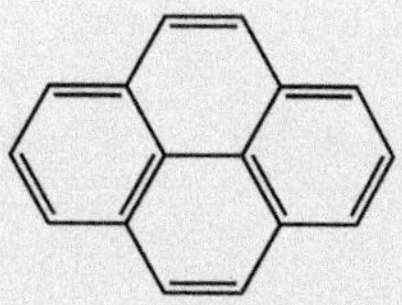 http://en.wikipedia.org/wiki/File:Pyrene.svg	202.250	9.08×10^{-6}	25

Molecular weight (MW) and Henry's constant values from *CRC Handbook of Chemistry and Physics*, (2010), 91st Edition, pp. 8-88, 8-89, 8-111, 8-114, 8-115.
http://en.wikipedia.org/wiki/Aromatic_hydrocarbons.
http://www.epa.gov/epawaste/hazard/tsd/pcbs/pubs/congeners.htm.

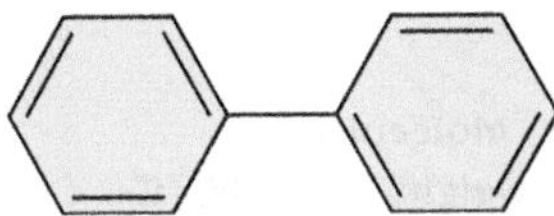

Figure 2.13 Structure of biphenyl molecule.
http://en.wikipedia.org/wiki/Biphenyl

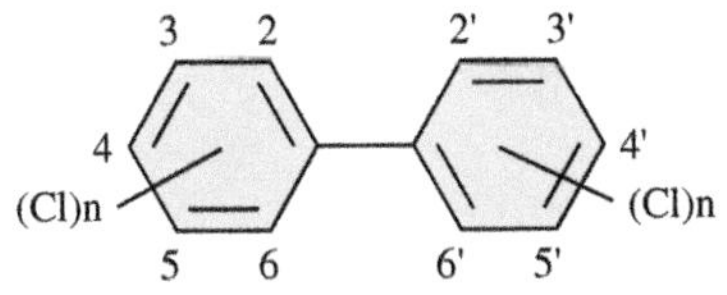

Figure 2.14 Chemical structure of PCBs.
http://en.wikipedia.org/wiki/Polychlorinated_biphenyl

banned production of PCBs in 1979. Figures 2.13 and 2.14 show the structure of biphenyl and general structure of a chlorinated biphenyl, respectively.

PCBs were manufactured domestically from 1929 until their manufacture was banned in 1979. They have a range of toxicity and vary in consistency from thin, light-colored liquids to yellow or black waxy solids. Due to their non-flammability, chemical stability, high boiling point, and electrical insulating properties, PCBs were used in hundreds of industrial and commercial applications including electrical, heat transfer, and hydraulic equipment. They were also used as plasticizers in paints, plastics, and rubber products; in pigments, dyes, and carbonless copy paper; and many other industrial applications.

In the US, PCBs were manufactured as a mixture of various PCB congeners or isomers. The most common series was called Aroclors and a numbering system was established. The first two digits represent the number of carbon atoms in the phenyl rings; for PCBs, this is 12. The second number refers to the percentage of chlorine by mass in the mixture. For example, Aroclor 1210 means that the mixture contains approximately 10% chlorine by mass, whereas, for Aroclor 1268, the mixture contains approximately 68% chlorine by mass.

2.9.3 Heterocyclic organic compounds

Heterocyclic organic compounds have at least one other element in their ring structure other than carbon. These compounds may be either aliphatic or aromatic in nature. Most of the heterocyclic organic compounds in environmental engineering relate to biological processes, i.e., synthesis and other biochemical reactions. Some five-membered ring aliphatic heterocyclic compounds include furaldehyde, pyrrole, and phyrolidine. Some six-membered aromatic heterocyclic compounds are: purine, pyrimidine, adenine, guanine, cytosine, uracil, and thymine.

These compounds are important in the synthesis of proteins and deoxyribonucleic acid (DNA). Chapter 4 provides more details on protein synthesis and DNA. Table 2.25 shows the structure of some of the heterocyclic aliphatic and aromatic compounds.

2.9.4 Functional groups of environmental importance

As previously mentioned, organic compounds generally contain hydrogen, oxygen, nitrogen, sulfur, and other elements in addition to carbon. The vast array of organic compounds that exist are actually hydrocarbon derivatives that contain additional atoms or groups of atoms called functional groups. These functional groups cause the chemical and physical properties of the hydrocarbon derivative to change significantly. Alcohols are characterized by the presence of the hydroxide group (–OH). When alcohols react with a strong acid, they act as bases, i.e., accept protons. The simplest aromatic alcohol is phenol which has a hydroxide group attached to the benzene ring. In general, phenols are more acidic than alcohols and water. Aldehydes and ketones contain the carbonyl group (C = O). Compounds containing the –COOH group are weak acids (carboxylic acids). Esters are derived from carboxylic acids and have the general structural formula as follows:

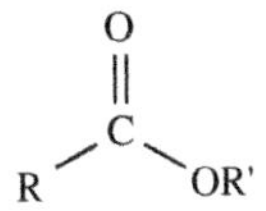

$$R-C(=O)-OR'$$

where:
R = represents an alkyl group
R' = represents an aryl group.

Amines are considered derivatives of ammonia with one, two, or three hydrogen atoms replaced by an alkyl group. The amines resemble ammonia and act as weak bases. Amides may be produced from ammonia and carboxylic acids. Some common functional groups and examples of each are presented in Table 2.26.

Summary

- Environmental engineers must have a firm foundation in chemical concepts, because these are applied to various treatment systems when dealing with water, wastewater, residuals, and air pollution problems.
- A dimension is a physical quantity expressing length, time, temperature, pressure, velocity, and weight. It must be accompanied by the appropriate set of units to quantify the dimension.
  - Fundamental dimensions include: length (L), mass (m), time (t), and temperature (T).
  - Derived dimensions are those developed or derived from fundamental dimensions. Examples include velocity (L/t), volume (L^3), density (m/L^3), and pressure (F/L^2).
  - Density (ρ) and specific gravity (S.G.) are two important derived dimensions that are widely used in chemical systems.
- The concentration of a substance in a solution can be expressed in a number of ways.

Table 2.25 Heterocyclic organic compounds of significance in environmental engineering.

Compound name	Structure	*Molecular weight	*BP, °C	*MP, °C
Furaldehyde	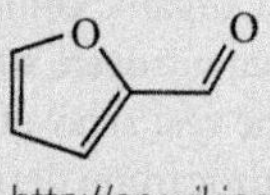 http://en.wikipedia.org/wiki/File:Furfural_structure.png	96.085	161.7	−38.1
Pyrrole	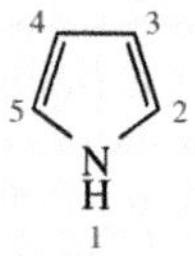 http://en.wikipedia.org/wiki/File:Pyrrole-2D-numbered.svg	67.090	129.79	−23.39
Pyrrolidine	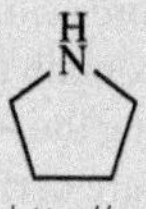 http://en.wikipedia.org/wiki/File:Pyrrolidine.png	71.121	86.56	−57.79
Pyridine	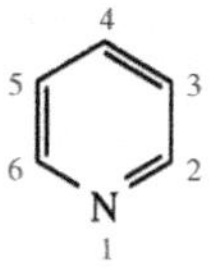 http://en.wikipedia.org/wiki/File:Pyridine_numbers.svg	79.101	115.23	−41.70
Purine	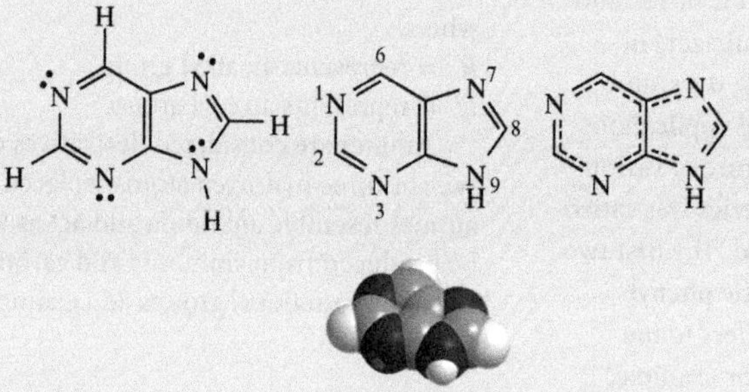 http://en.wikipedia.org/wiki/File:Purine_chemical_structure.png	120.113	....	216.5
Pyrimidine	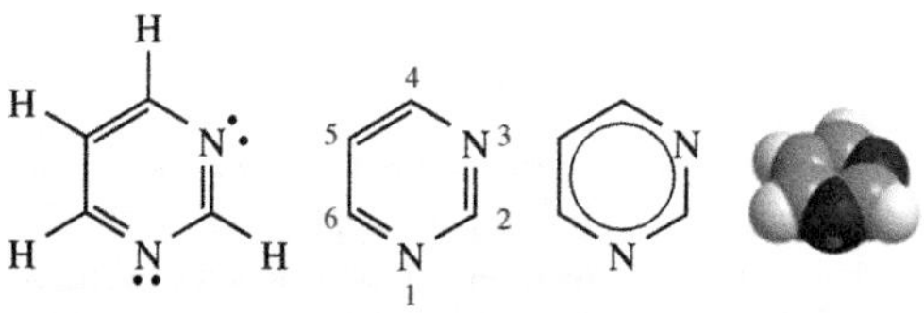 http://en.wikipedia.org/wiki/File:Pyrimidine_chemical_structure.png	80.088	123.8	22
Adenine	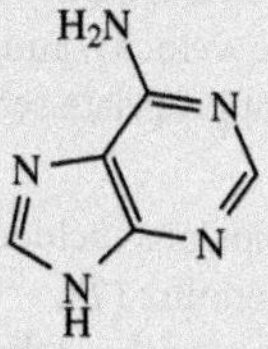 http://en.wikipedia.org/wiki/File:Adenine_chemical_structure.png	135.128	sub 220	360 dec
Guanine	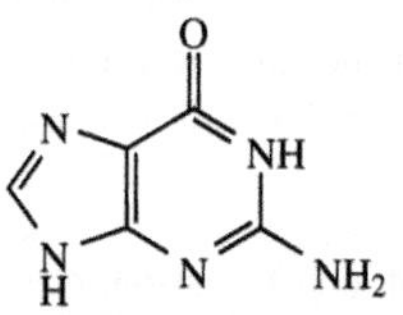 http://en.wikipedia.org/wiki/File:Guanin.svg	151.127	sublimes	360 dec

Table 2.25 *(continued)*

Compound name	Structure	*Molecular weight	*BP, °C	*MP, °C
Cytosine	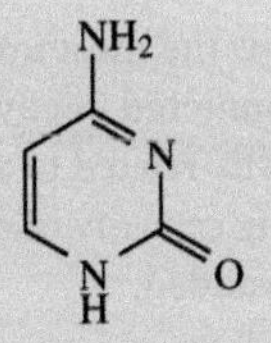 http://en.wikipedia.org/wiki/File:Cytosine_chemical_structure.png	111.102	....	322 dec
Uracil	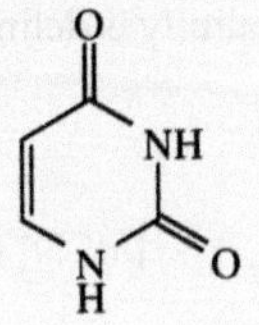 http://en.wikipedia.org/wiki/File:Uracil.svg	112.087	---------	338
Thymine	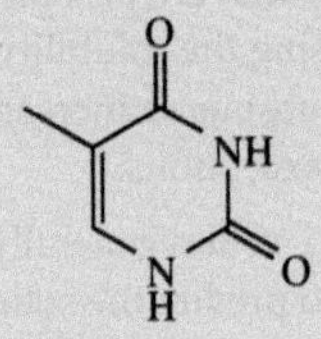 http://en.wikipedia.org/wiki/File:Thymine_skeletal.svg	126.11	----------	316

*Molecular weight (MW), boiling point (BP), and melting point (MP) values from *CRC Handbook of Chemistry and Physics* (2010), 91st Edition, pp. 3-8,3-134, 3-266, 3-270, 3-448, 3-452, 3-486, 3-516.
deg – decomposition observed at stated temperature.
sub – solid has a significant sublimation at ambient temperature.

Table 2.26 Common functional groups and examples encountered in environmental engineering.

Class	Functional group	General formula	Example
Alcohol	–OH	R–OH	Methanol: CH_3OH
Aldehyde	$-C^{=O}_{-H}$	$R-C^{=O}_{-H}$	Acetaldehyde: C_2H_4O
Alkyl halide	–X(Br, Cl, F, I)	R–X	Chloroform: $CHCl_3$
Amide	$-C^{=O}_{-NH_2}$	$R-C^{=O}_{-NH_2}$	Urea: CH_4N_2O
Amine	$-NH_2$	$R-C-NH_2$	Methylamine: CH_5N
Carboxylic acid	$-C^{=O}_{-OH}$	$R-C^{=O}_{-OH}$	Acetic acid: CH_3COOH
Ester	$-C^{=O}_{-O-}$	$R-C^{=O}_{-O-}$	Methylethanoate: $C_3H_6O_2$
Ether	–O–	R–O–R′	Dimethylether C_2H_6O
Ketone	$-C^{=O}_{-}$	$R-C^{=O}_{-}$	Acetone: C_3H_6O
Mercaptan	–SH	R–SH	Methylmercaptan: CH_3SH
Phenol	–OH	R–OH	Phenol: C_6H_6O
Sulfonic acid	$-SO_3H$	$R-SO_3H$	Ethanesulfonic acid: $C_2H_6O_3S$

R and R′ represents any functional group or hydrogen.

- Typically, concentration is given in terms of moles/liter (M) or milligrams per liter (mg/L).
- Concentration is sometimes expressed in terms of mass/volume, mass/mass as a percentage, volume/volume as a percentage, or mole fraction.
- The normality (N) of a solution represents the number of equivalents of substance per liter of solution.

- In environmental engineering, the flow rate may be given either on a mass or volumetric basis.
  - The mass flow rate is equal to the density of the solution multiplied by the volumetric flow rate (Q).
  - The mass flow rate of some component in the flow is determined by multiplying the concentration of the component by the volumetric flow rate.
- Detention time (τ) is a fundamental concept in environmental engineering and it represents the average unit of time that a fluid particle remains in the system. Detention time is calculated by dividing the volume of the tank or container by the volumetric flow rate.
- Engineers must make sure that all chemical equations are properly balanced before using to calculate chemical dosages and quantities.
  - The law of conservation of mass must be satisfied, i.e., the mass of elements on the left side of the equation must equal the mass of elements on the right side.
  - The solution must be electrically neutral, i.e., a charge balance should exist.
- In oxidation-reduction or redox reactions, electrons are transferred and the oxidation number of a chemical species changes.
  - Oxidation refers to a loss of electrons or a gain in valence.
  - Reduction means a gain in electrons or a loss in valence.
- Thermodynamics allows engineers and chemists to predict whether or not a chemical reaction will proceed as written. It does not allow one to determine the rate at which a reaction takes place.
  - The following thermodynamic parameters are used in thermodynamic equilibria: change in enthalpy (ΔH), change in entropy (ΔS); and change in free energy (ΔG).
- The equilibrium constant (K) can be determined from free energy data. It can be used to determine the concentration of various chemical species at equilibrium.
  - Free energy is related to K in the following equation:

$$\Delta G = \Delta G^{\circ} + RT\ln(K)$$

  - Temperature affects the equilibrium rate constant. The van't Hoff equation is used to make temperature corrections:

$$\ln\frac{K_1}{K_2} = -\frac{\Delta H^{\circ}}{R}\left(\frac{T_1 - T_2}{T_1 T_2}\right)$$

- Acids are proton donors and bases accept protons.
- Acid-base chemistry involves estimating the concentration of various chemical entities and pH when either an acid or base is added to a solution.
  - The pH of a solution can be solved analytically or graphically by drawing a pC-pH diagram.
- The carbonate system is the most important natural buffering system encountered in environmental engineering. It is directly related to alkalinity, which is defined as the buffering capacity of a water to resist a change in pH when an acid is added.
  - Alkalinity is measured by titrating a sample of water with 0.02 N H_2SO_4 to a pH of approximately 4.5.
  - Mathematically, alkalinity is defined by the following equation:

$$[ALK]_e = [CO_3^{2-}]_e + [HCO_3^-]_e + [OH^-]_e - [H^+]_e$$

- Liquid-solid species that are partially soluble or insoluble can be explained using the solubility product (K_{sp}). The solubility of most substances increases with temperature; however, there are exceptions:
  - Salts with low K_{sp} values generally have low solubilities.
  - It is impossible to predict the solubilities of various compounds solely based on K_{sp} values since solids that dissolve into two or three ions will generally have higher solubilities.
- Proper application of gas phase laws is essential for designing gas transfer systems, gas strippers, determining saturation concentration of dissolved gases in aqueous solutions, and understanding the relationship between pressure and volume of a gas and between temperature and volume of a gas.
  - The ideal gas law is used to solve for pressure, volume, moles, or temperature, given three of these four parameters.
  - Dalton's law of partial pressure states that the total pressure exerted by a mixture of gases is the sum of the partial pressures of each gas present.
  - Raoult's law relates the partial pressure of the gaseous component present above a liquid to the vapor pressure of the pure component and the mole fraction of the pure component in the liquid phase.
  - Henry's law is used for calculating the solubility of a gas in a liquid, or to describe the equilibrium condition between the gas and liquid phases.
  - There are several forms of Henry's law, so special attention should be placed on the dimensions on the Henry's law constant H, in order to apply the appropriate equation expressing Henry's law.
- Organic chemistry is the study of carbon-containing compounds and their properties. Organic compounds are important because: they form the basis of all life; they are used in the production of pesticides, herbicides, insecticides, polymers, antibiotics, hormones, and alcohols; and they cause detrimental affects to the environment, since many of them are toxic and/or carcinogenic.

- Hydrocarbons are organic compounds consisting only of hydrogen and carbon.
- A saturated hydrocarbon is one in which adjacent carbon atoms are joined together by a covalent bond and all of the bonds are made with hydrogen.
- Unsaturated hydrocarbons have a least two carbon atoms, joined together by a double or triple bond.
- Alkanes are aliphatic hydrocarbons represented by the formula C_nH_{2n+2}.
- Alkenes are hydrocarbons that have one double bond and are represented by the formula C_nH_{2n}.
- Alkynes are hydrocarbons with at least one triple bond and are represented by the formula C_nH_n.
- Aromatic organic carbon compounds all contain a six-membered ring of carbon atoms, called the benzene ring. Aromatic compounds of significance in environmental engineering include benzene, toluene, ethylene benzene, and xylene.
- Polycyclic aromatic hydrocarbon (PAHs) contain two or more aromatic rings, fused together by sharing two adjacent carbon atoms.
- Polychlorinated biphenyls (PCBs) are chlorinated hydrocarbons made up of biphenyl compounds, containing from one to five chlorine atoms bonded to each aromatic ring instead of hydrogen atoms.
- Heterocyclic organic compound are those that contain at least one element other than carbon in their ring structure. Purine, pyrimidine, adenine, guanine, cytosine, uracil, and thymine are six-membered aromatic heterocyclic compounds involved in the synthesis of proteins and DNA.
- Functional groups are atoms or groups of atoms attached to organic compounds which change the chemical and physical properties of the hydrocarbon derivatives.

Key Words

acidity
acids
aliphatic HC
alkalinity
alkanes
alkenes
alkynes
aromatic HC
Avogadro's law
balancing equations
bases
benzene
Boyle's law
BTEX
buffers
carbonate system
Charles's law
concentration
conjugate acid
conjugate base
Dalton's law
density
derived dimensions
detention time
dimension
dissociation
electroneutrality
enthalpy
entropy
equilibrium chemistry
equilibrium constant
equivalent weight
ethyl benzene
equivalents
free energy
functional groups
fundamental dimensions
halogenated HC
HC
heat
Henry's constant
Henry's law
heterocyclic organics
hydrocarbon
Ideal gas law
ionization
mass balance
mass flow rate
molal solution
mole fraction
normal solution
organic chemistry
oxidation
oxidizing agent
PAHs
partial pressure
PCBs
pC-pH diagram
pH
pOH
radicals
Raoult's law
redox
reducing agent
reduction
saturated HC
solubility
solubility product
specific growth
toluene
units
universal gas constant
unsaturated HC
volumetric flow rate
work

References

Baum, S.J. (1978). *Introduction to Organic and Biological Chemistry*, p. 47, 77–85. Macmillan Publishing Co., Inc., New York, NY

Davis, M.L., Masten, S.J. (2009). *Principles of Environmental Engineering and Science*, p. 64. McGraw-Hill, New York, NY.

Eide, A., Jenison, R.D., Mashaw, L.H., Northup, L.L. (1979). *Engineering Fundamentals and Problem Solving*, pp. 425-426. McGraw Hill, New York, NY.

LaGrega, M.D., Buckingham, P.L., Evans, J.C. (1994). *Hazardous Waste Management*, pp. 1040–1048. McGraw-Hill, New York, NY.

Metcalf and Eddy (2003). *Wastewater Engineering: Treatment and Reuse*, p. 67, 68, pp. 66-69. McGraw-Hill, New York, NY.

Mihelcic, J.R. (1999). *Fundamentals of Environmental Engineering*, p. 88, 89. John Wiley & Sons, Hoboken, NJ.

MWH (2005). *Water Treatment: Principles and Design*, pp. 1167–1174. John Wiley & Sons, Inc., Hoboken, NJ.

Mines, R.O., Lackey, L.W. (2009). *Introduction to Environmental Engineering*, p. 43, 46. Prentice Hall, Upper Saddle River, NJ.

Sawyer, C.N., McCarty, P.L. (1978). *Chemistry for Environmental Engineering*, p. 16, 32, 95–96, 100, 467. McGraw-Hill, New York, NY.

Sawyer, C.N., McCarty, P.L., Parkin, G.F. (1994). *Chemistry for Environmental Engineering*, p. 123, 143. McGraw-Hill, New York, NY.

Snoeyink, V.L., Jenkins, D. (1980). *Water Chemistry*, p. 157, 251. John Wiley and Sons, Inc., New York, NY.

Stumm, W., Morgan, J.J. (1996). *Aquatic Chemistry: Chemical Equilibria and Rate in Natural Waters*, 3rd Edition, p. 96. John Wiley and Sons, New York, NY.

APHA (1998). *Standard Methods for the Examination of Water and Wastewater*. American Public Health Association, Washington, DC.

Zumdahl, S.S. (2000). *Introductory Chemistry: A Foundation*, p. 47, 632. Houghton Mifflin Company, Boston, MA.

Problems

1 Air weighs 8 pounds per 100 ft^3. What is the density in (a) grams per ft^3 and (b) grams/L?

2 A block of wood weighs 4 pounds and has the following dimensions: 10 in by 6.0 in by 2.0 in. Calculate the density in (a) lb/ft^3 and (b) g/cm^3.

3 The specific gravity of mercury is 13.6. Calculate the density in (a) g/cm^3 and (b) lb/ft^3.

4 A bottle used for determining specific gravity weighs 200 g when empty, 400 g when filled with water, and

334 g when filled with gasoline. Determine (a) the capacity of the bottle in ml and (b) the specific gravity of the gasoline.

5 Balance the following equations:
a. $K_{(s)} + H_2O_{(l)} \rightarrow H_{2(g)} + KOH_{(aq)}$
b. $FeS_{(s)} + HCl_{(l)} \rightarrow FeCl_{2(s)} + H_2S_{(g)}$
c. $NH_{3(g)} + O_{2(g)} \rightarrow NO_{(g)} + H_2O_{(g)}$
d. $SiO_{2(s)} + HF_{(aq)} \rightarrow SiF_{4(g)} + H_2O_{(l)}$

6 Balance the following equations:
a. $FeCl_{3(aq)} + KOH_{(aq)} \rightarrow Fe(OH)_{3(s)} + KCl_{(aq)}$
b. $Cl_{2(g)} + KI_{(aq)} \rightarrow I_{2(s)} + KCl_{(aq)}$
c. $CaF_{2(s)} + H_2SO_{4(l)} \rightarrow CaSO_{4(s)} + HF_{(g)}$
d. $MnO_{2(s)} + NaCl_{(s)} + H_2SO_{4(l)} \rightarrow MnSO_{4(g)} + H_2O_{(l)} + Cl_{2(g)} + Na_2SO_{4(s)}$

7 Balance the following equations:
a. $ZnSO_4 + NaCl \rightarrow ZnCl_2 + Na_2SO_4$
b. $KMnO_4 + H_2C_2O_4 \rightarrow K_2CO_3 + MnO_2 + H_2O + CO_2$
c. $KMnO_4 + FeSO_4 + H_2SO_4 \rightarrow Fe_2(SO_4)_3 + K_2SO_4 + MnSO_4 + H_2O$
d. $KH(IO_3)_2 + KI + H_2SO_4 \rightarrow I_2 + K_2SO_4 + H_2O$

8 Determine the pH of a 0.07 M HCl solution. $pK_A = -3.0$.

9 Determine the pH of a $2.0 \times 10^{-3} M$ $Ba(OH)_2$ solution.

10 Determine the pH of a $3.0 \times 10^{-4} M$ H_2SO_4 solution.

11 Calculate the $[H^+]$, $[HS^-]$, $[S^{2-}]$, and pH in a 0.01 M H_2S solution. The dissociation constants for H_2S and $[HS^-]$ are $10^{-7.1}$ and $10^{-6.9}$, respectively.

12 Given the following sets of pressure/volume data, calculate the missing quantity assuming that the mass of the gas and temperature remain constant.
a. $V = 42$ in^3 at $P = 85.5$ in Hg; $V =?$ in^3 at $P = 75.9$ in Hg
b. $V = 1.04$ L at 759 mm Hg; $V = 2.24$ L at $P =?$ mm Hg
c. $V = 53$ L at 785 mm Hg; $V =?$ L at $P = 690$ mm Hg
d. $V = 4.0$ mL at 140 atm; $V = 10.0$ mL at $P =?$ mm Hg
e. $V = 2.7$ L at 101 kPa; $V = 3.0$ L at $P =?$ mm Hg
f. $V = 1.0$ L at 760 mm Hg; $V = 50.0$ mL at $P =?$ atm

13 Given the following sets of temperature/volume data, calculate the missing quantity assuming that the mass of the gas and pressure remain constant.
a. $V = 45$ mL at $T = 26.5°C$; $V =?$ mL at $T = 56.5°C$
b. $V = 1.10$ L at $T = 20°C$; $V =?$ L at $T = 45°C$
c. $V = 4.5$ L at $T = 25°C$; $V =?$ L at $T = -270°C$
d. $V = 45$ mL at $T = 298$ K; $V =?$ mL at $T = 5$ K
e. $V = 44.5$ mL at $T = 298$ K; $V =?$ mL at $T = 0°C$
f. $V = 2.8$ L at $T = -50°C$; $V = 5.0$ L at $T = ?°C$

14 If 2.0 g of helium gas occupies a volume of 12.0 L at 25°C, what volume will 6.5 g of helium gas occupy under the same pressure and temperature?

15 If 3.5 moles of argon gas occupies a volume of 100 L, what volume will 14.5 mol of argon gas occupy under the same pressure and temperature?

16 If 46.5 g of oxygen gas occupies a volume of 100 L at a specific pressure and temperature, what volume will 6.0 g of oxygen gas occupy under the same pressure and temperature?

17 Given the following values for three of the four gas variables, determine the unknown quantity.
a. $P = 23$ atm at $V = 145$ mL; $n = 0.45$ mol at $T =?$ K
b. $P =?$ atm at $V = 1.5$ mL; $n = 0.00015$ mol at $T = 293$ K
c. $P = 755$ mm Hg at $V =?$ mL; $n = 0.45$ mol at $T = 135°C$
d. $P = 1.5$ mm Hg at $V =?$ L; $n = 0.75$ mol at $T = 155°C$
e. $P = 1.05$ atm at $V = 22.5$ mL; $n = 0.0045$ mol at $T =?$ K
f. $P =?$ atm at $V = 1.75$ mL; $n = 0.00015$ mol at $T = 185$ K

18 A gaseous mixture contains 7.00 g of nitrogen gas, 4.50 g of oxygen gas, and 3.00 g of helium gas. What volume (L) does the mixture occupy at 30°C and 1.1 atm?

19 A mixture of helium and oxygen gas is contained in a 5.0 L pressurized vessel. Fifteen liters of oxygen gas at 25°C and 1.0 atm and 45 liters of helium gas at 25°C and 1.0 atm was pumped into the vessel. Determine the partial pressure of each gas and the total pressure in the pressurized vessel at 25°C.

20 A 50.0 L tank contains 5.5 kg of nitrogen gas and 4.5 kg of oxygen gas. What is the pressure in the tank at 23°C?

21 A tank contains a mixture of 3.0 mol of nitrogen gas, 2.0 mol of oxygen gas, and 2.0 mol of carbon dioxide gas at 25°C and a total pressure of 12 atm. Determine the partial pressure (in torr) of each gas in the mixture. 1 torr = 1 mm Hg.

22 A covered pure oxygen activated sludge process is operated at 2.0 atm of pressure. If the composition of the gas above the liquid is 80% oxygen, 5% carbon dioxide, and 15% nitrogen, determine the equilibrium concentration (mg/L) of oxygen in the liquid phase at 20°C.

23 Calculate the saturation or equilibrium concentration (mg/L) of oxygen, nitrogen, and carbon dioxide in water at 0, 20, and 40°C, assuming that their percentage by volume in the atmosphere is 21%, 79%, and 0.03%, respectively at 1 atm.

24 Compare the saturation concentrations (mg/L) of oxygen at 20°C at elevations of 0, 1,600 m, 2,200 m,

and 4,300 m. Assume that oxygen is 21% by volume in the atmosphere. At sea level, the atmospheric pressure is 760 mm Hg = 1 atm = 101 kPa. The change in pressure as a function of elevation can be calculated from the following equation:

$$P_b = P_a \exp\left[\frac{-gM\left(z_b - z_a\right)}{RT}\right]$$

Metcalf and Eddy, 2003, page 1738

P_b = atmospheric pressure at elevation z_b

(meters above sea level) in kPa,

P_a = atmospheric pressure at elevationz_a

(meters above sea level) in kPa,

$g = 9.81\ \text{m/s}^2$,

$R = 8314\frac{\text{N}\cdot\text{m}}{\text{kg}\cdot\text{mol}\cdot\text{K}} = 8314\frac{\text{kg}\cdot\text{m}^2}{\text{s}^2\cdot\text{kg}\cdot\text{mol}\cdot\text{K}}$,

T = temperature, K(273.15 + °C), and

M = molecular weight of air, (28.97kg/kg · mole).

25 Determine the solubility-product constant given the solubility of the following compounds:
a. 6.0×10^{-5} *M* of $Mg_3(PO_4)_2$
b. 6.0×10^{-9} *M* of FeS
c. 2.0×10^{-7} *M* of $Zn_3(PO_4)_2$
d. 7.0×10^{-3} *M* of CuF_2

26 Calculate the pH of a 0.1 *M* H_2X solution using a pC-pH diagram. $pK_1 = 3.0$ and $pK_2 = 7.0$.

27 Calculate the pH of a 0.1 *M* NaHX solution using a pC-pH diagram. $pK_1 = 3.0$ and $pK_2 = 7.0$.

28 Calculate the pH of 0.1 *M* Na_2X using a pC-pH diagram. $pK_1 = 3.0$ and $pK_2 = 7.0$.

29 Calculate the pH of a solution prepared by diluting 0.1 mole of $(NH_4)_2X$ to 1 liter with distilled water using a pC-pH diagram. $pK_A = 9.3$.

30 Calculate the pH of a solution prepared by diluting 0.1 mole of NaA to 1 liter with distilled water using a pC-pH diagram. $pK_A = 4.7$.

31 Calculate the pH of a 0.01 *M* acetic acid solution at 25°C using a pC-pH diagram. $pK_A = 4.7$

32 Calculate the pH of a 0.01 *M* potassium acetate solution at 25°C using a pC-pH diagram. $pK_A = 4.7$.

33 Calculate the pH of a 0.01 *M* ammonium chloride $(NH_4)Cl$ solution at 25°C using a pC-pH diagram.

34 Calculate the pH of a 0.01 *M* sodium bicarbonate solution at 25°C using a pC-pH diagram.

35 Calculate the pH of a 0.01 *M* potassium bicarbonate solution at 25°C using a pC-pH diagram.

36 Calculate the pH of a 0.01 *M* hydrogen sulfide solution using a pC-pH diagram.

37 Calculate the pH of solutions containing 200 mg/L of each of the following weak acids or salts of weak acids:
a. Acetic acid
b. Hypochlorous acid
c. Ammonia
d. Hydrocyanic acid

38 A $10^{-2.5}$ *M* solution of ammonia is prepared. Calculate the equilibrium concentration for each chemical species in the solution.

39 Calculate the pH of a $10^{-3}M$ HCl solution. $K = 10^3$.

40 Calculate the pH of a $10^{-7}M$ HCl solution. $K = 10^3$.

41 Explain the difference between an aliphatic and aromatic organic compound.

42 Give the functional group associated with each of the following organic compounds: alcohol, aldehyde, phenol, ketone, ester, ether, amine, amide, mercaptan, halide, and sulfonic acid.

43 Discuss the characteristics and relevance of PAHs and PCBs after conducting a "Google" search on the internet.

44 Determine the solubility constant for carbon dioxide gas in water at 25°C from free energy calculations. $CO_{2(g)} \leftrightarrow CO_{2(aq)}$.

45 Determine the first ionization constant (K_1) for carbonic acid at 25° from free energy calculations. $H_2CO_{3(aq)} \leftrightarrow H^+_{(aq)} + HCO^-_{3(aq)}$

46 Determine the second ionization constant (K_2) for carbonic acid at 25°C from free energy calculations. $HCO^-_{3(aq)} \leftrightarrow H^+_{(aq)} + CO^{2-}_{3(aq)}$

47 Acetate (CH_3COO^-) can be biologically oxidized either by aerobic or anaerobic conditions. Using the following stoichiometric equation, determine the standard free energy of formation for each reaction. Would a larger population of microorganisms be supported under aerobic or anaerobic conditions? Explain your answer. ΔG°_f for $O_{2(g)} = 0$.

Aerobic:

$$CH_3COO^-_{(aq)} + 2O_{2(g)} \leftrightarrow HCO^-_{3(aq)} + H_2O_{(l)} + CO_{2(g)}$$

Anaerobic:

$$CH_3COO^-_{(aq)} + H_2O_{(l)} \leftrightarrow HCO^-_{3(aq)} + CH_{4(g)}$$

48 Calculate the enthalpy change for the formation of water vapor from the combination of hydrogen gas and oxygen gas as shown in the following equation.

$$H_{2(g)} + \frac{1}{2}O_{2(g)} \leftrightarrow H_2O_{(g)}$$

49 Calculate the enthalpy change for the formation of water from production of water vapor given the following equations. The combination of hydrogen gas and oxygen gas as shown in the following equation.

$$\text{Equation (1)}: \; H_{2(g)} + \frac{1}{2}O_{2(g)} \leftrightarrow H_2O_{(g)}$$

$$\text{Equation (2)}: \; H_2O_{(g)} \leftrightarrow H_2O_{(l)}$$

50 Calculate the quantity of heat that is liberated during the slaking of 1 kg of quicklime (CaO) according to the following reaction.

$$CaO_{(s)} + H_2O_{(l)} \leftrightarrow Ca(OH)_{2(s)}$$

51 Calculate the enthalpy change for the formation of carbon dioxide from the oxidation of graphite for the following reaction.

$$C_{(graphite)} + O_{2(g)} \rightarrow CO_{2(g)}$$

52 Calculate the enthalpy change for the formation of hydrogen iodide gas for the following reaction.

$$H_{2(g)} + I_{2(crystal)} \rightarrow 2HI_{(g)}$$

Chapter 3

Water and wastewater characteristics

Richard O. Mines, Jr.

Learning Objectives

After reading this chapter, you should be able to:

- list and describe the three water quality parameter categories;
- list and explain the importance of five physical waster quality parameters;
- list and explain the importance of five chemical waster quality parameters;
- list and explain the importance of five biological waster quality parameters;
- identify and calculate the various species of alkalinity, given the initial pH, titration data, and/or equilibrium equations and constants;
- calculate the theoretical oxygen demand of a chemical compound;
- draw a schematic of the BOD test procedure and calculate the BOD of a "seeded" or "unseeded" sample of water or wastewater;
- discuss the various types of solids analyses and calculate the concentration of each, given appropriate data;
- defend the use of indicator organisms as surrogate parameters for pathogens;
- explain and calculate the most probable number (MPN) for a water or wastewater sample;
- assess the quality of a water as "good" or "poor", based on the water quality parameters presented in this chapter.

3.1 Overview

The quality of water determines its usefulness. Take, for instance, a river: will it serve as a drinking water source or an irrigation source? Perhaps it will serve as a recreational resource; however, it may be the ultimate disposal location for industrial or municipal wastewater. Normally, one may picture a clear, fast-moving, pristine creek in the Rocky Mountains as an excellent source of drinking water. However, is it really safe to drink just because it appears to be clean and refreshing? In reality, the water probably contains some contaminants such as the dangerous disease-causing microorganisms *Giardia lamblia* and/or *Cryptosporidium parvum*. These microscopic organisms can cause the inexperienced backpacker severe abdominal distress and a memorable, yet unpleasant experience.

The conventional physical, chemical, and biological characteristics of water and wastewater will be discussed in this chapter. Where appropriate, a brief presentation of the analytical procedures used for measuring some of the traditional water quality parameters will be presented. Tables showing water quality characteristics for surface and groundwater will be presented for comparison with stormwater and municipal wastewater. Proper characterization of water quality is of paramount importance if sustainable, water and wastewater treatment facilities are to be designed by environmental engineers.

3.2 Water quality parameters

The constituents found in water may be classified as physical, chemical, or biological. Table 3.1 presents the common parameters used in water quality assessment. *Standard Methods*

Environmental Engineering: Principles and Practice, First Edition. Richard O. Mines, Jr.

Table 3.1 Physical, chemical, and biological characteristics of water.

Physical	Inorganic chemical	Organic chemical	Biological
Color	Ammonia	BOD_5	Bacteria
Odor	Nitrite	COD	Helminths
Solids	Nitrate	TOC	Protozoa
Temperature	Organic nitrogen	Specific organic compounds	Viruses
Absorbance and transmittance	Total Kjeldhal nitrogen		
Turbidity	Total phosphorus		
	Inorganic phosphorus		
	Organic phosphorus		
	Metals		
	Alkalinity		
	pH		
	Dissolved oxygen		

for the Examination of Water and Wastewater (2012) is the authoritative handbook consulted in the United States for conducting water quality analyses.

Physical parameters primarily relate to our senses and the aesthetics of the water. Many of the chemical parameters (organic and inorganic) have a detrimental affect on health. Finally, the microbiological quality of water is related to its biological characteristics. In subsequent sections of this chapter, the measurement and significance of the most important parameters encountered in water quality management will be presented. First, the analyses used to determine the organic content of a sample will be presented.

3.3 Lumped parameter organic quantification

A non-specific analytic procedure is used to measure the organic content of a water sample. Typical analyses performed on water and wastewater samples include biochemical oxygen demand (BOD), chemical oxygen demand (COD), and total organic carbon (TOC). To highlight the difference in each of these parameters, a brief discussion on theoretical oxygen demand (ThOD) is presented first.

3.3.1 Theoretical oxygen demand (ThOD)

The theoretical oxygen demand of an organic substance relates to the quantity of oxygen required to oxidize the organic matter to carbon dioxide and water. The ThOD can be calculated only if the formula of the organic compound is known. However, this is not normally the case in most environmental engineering applications. Example 3.1 illustrates how to calculate the ThOD of a 500 mg/L glucose solution.

Example 3.1 Calculation of Theoretical Oxygen Demand of a hydrocarbon

Calculate the ThOD of a 500 mg/L glucose ($C_6H_{12}O_6$) solution using the balanced stoichiometric equation for the oxidation of glucose to carbon dioxide and water, as given below.

$$C_6H_{12}O_6 + 6\ O_2 \rightarrow 6\ CO_2 + 6\ H_2O$$

First, check to make sure that the chemical reaction is properly balanced. The number of moles of carbon, hydrogen, and oxygen must be the same on both the left side of the reaction (reactants) and the right side of the reaction (products). Here, the chemical equation is properly balanced.

Second, determine the molecular weight of glucose and oxygen as follows:

The molecular weight of glucose is: $6 \times 12 + 12 \times 1 + 6 \times 16 = 180$ g/mole.

The molecular weight of oxygen is: $2 \times 16 = 32$ g/mole.

Finally, use the stoichiometric ratio observed in the balanced oxidation reaction to calculate the ThOD concentration in mg/L.

$$\text{ThOD} = \frac{500\ \text{mg}\ C_6H_6O_6}{\text{L}}\left(\frac{1\ \text{g}\ C_6H_{12}O_6}{1000\ \text{mg}\ C_6H_{12}O_6}\right)$$
$$\times\left(\frac{1\ \text{mol}\ C_6H_{12}O_6}{180\ \text{g}\ C_6H_{12}O_6}\right)\left(\frac{6\ \text{mol}\ O_2}{1\ \text{mol}\ C_6H_{12}O_6}\right)$$
$$\times\left(\frac{32\ \text{g}\ O_2}{1\ \text{mol}\ O_2}\right)\left(\frac{1000\ \text{mg}}{1\ \text{g}}\right)$$
$$\text{ThOD} = \boxed{533\ \text{mg/L}}$$

To calculate the ThOD of an organic compound containing nitrogen, assume that the organic nitrogen is oxidized to nitrate (NO_3^-) and that the organic matter (carbon) is oxidized to carbon dioxide and water, as previously presented. Example 3.2 shows how to calculate the ThOD of a 300 mg/L solution containing $C_5H_7O_2N$.

Example 3.2 Calculation of ThOD of an organic containing nitrogen

Calculate the ThOD of a 300 mg/L solution of $C_5H_7O_2N$.

Begin by developing a balanced stoichiometric equation for the oxidation of the organic compound to carbon dioxide, water, and nitrate. The number of moles of carbon, hydrogen, nitrogen, and oxygen must be the same on the left and right side of the reaction. The chemical equation is properly balanced as follows:

$$C_5H_7O_2N + 7.25\ O_2 \rightarrow 5\ CO_2 + 3.5\ H_2O + NO_3^-$$

Second, determine the molecular weights of $C_5H_7O_2N$ and oxygen as follows:

The molecular weight of $C_5H_7O_2N$ is: $5 \times 12 + 7 \times 1 + 2 \times 16 + 1 \times 14 = 113$ g/mole.

The molecular weight of oxygen is: $2 \times 16 = 32$ g/mole.

Third, calculate the ThOD concentration in mg/L by setting up a ratio based on the atomic and molecular weights calculated in step 2.

$$\text{ThOD} = \frac{300\ \text{mg}\ C_5H_7O_2N}{\text{L}}\left(\frac{1\ \text{g}\ C_5H_7O_2N}{1000\ \text{mg}\ C_5H_7O_2N}\right)$$
$$\times\left(\frac{1\ \text{mol}\ C_5H_7O_2N}{113\ \text{g}\ C_5H_7O_2N}\right)\left(\frac{7.25\ \text{mol}\ O_2}{1\ \text{mol}\ C_5H_7O_2N}\right)$$
$$\times\left(\frac{32\ \text{g}\ O_2}{1\ \text{mol}\ O_2}\right)\left(\frac{1000\ \text{mg}}{1\ \text{g}}\right)$$
$$\text{ThOD} = \boxed{616\ \text{mg/L}}$$

3.3.2 Biochemical oxygen demand (BOD)

The most widely used method for estimating the biodegradable organic concentration of a water sample is to conduct a five-day BOD test. The numerical value obtained from the BOD test represents the quantity of oxygen required by bacteria to oxidize and stabilize the biodegradable organic material in the sample. The end products of oxidation are carbon dioxide, water, and nitrate, unless an inhibitor is added to prevent nitrification. *Standard Methods for the Examination of Water and Wastewater* (1998) is the authoritative handbook used in environmental engineering, and this provides the accepted and approved procedures for performing water quality analyses, including BOD.

Normally, the BOD test is based on a five-day period. According to Horan (1991), the United Kingdom Royal Commission recommended in 1912 that an incubation period of five days be employed for the BOD test, since smaller experimental error is introduced than that incurred at longer incubation periods. If the BOD test is conducted longer than five days, a secondary reaction involving nitrifying bacteria will occur, resulting in the oxidation of ammonia to nitrate. This is known as the nitrogenous oxygen demand (NOD).

In the United States, most regulatory agencies have established effluent standards from wastewater treatment plants (WWTPs) in terms of carbonaceous biochemical oxygen demand (CBOD). The CBOD is the five-day BOD test performed with inhibitors added to prevent the nitrification reaction. Therefore, it represents the quantity of oxygen required just to oxidize the organic matter in the sample.

3.3.2.1 Carbonaceous BOD Derivation

Mathematically, a first-order removal reaction is used to model the BOD reaction. The organic carbon or substrate for the bacteria is assumed to be removed at an exponential rate, according to the reaction shown in Equation (3.1), where C represents the concentration of organic matter. In environmental engineering, it is customary to use L, rather than C, where L, which represents the BOD remaining (See Equation (3.2)).

$$\frac{dC}{dt} = -kC \tag{3.1}$$

$$\frac{dL}{dt} = -kL \tag{3.2}$$

Rearranging and integrating Equation (3.2) yields the following:

$$\frac{dL}{L} = -kt \tag{3.3}$$

$$\int_{L_o}^{L_t} \frac{dL}{L} = -kt \tag{3.4}$$

$$\ln L_t - \ln L_o = -kt \tag{3.5}$$

$$L_t = L_o\, e^{-kt} \tag{3.6}$$

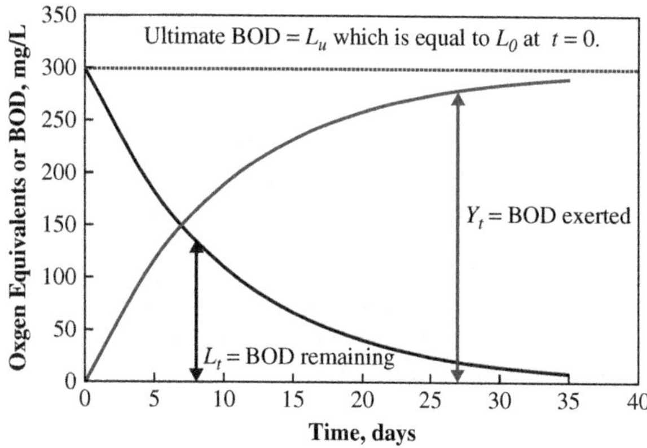

Figure 3.1 Relationships between BOD remaining, BOD exerted, and Ultimate BOD for Carbonaceous Biochemical Oxygen Demand.

where:

L_o = oxygen equivalent of organics or BOD remaining at time = 0, mg/L

L_t = oxygen equivalent of organics or BOD remaining at time t, mg/L

k = base "e" BOD reaction rate constant, days^{-1}

t = time, days.

Figure 3.1 shows the relationship between BOD remaining (L) and the BOD exerted (Y). In engineering applications, the BOD exerted is the parameter of interest. The BOD exerted at time equal to "t" is defined as Y_t. At time equal to zero, the **ultimate BOD, L_u**, is equal to L_o.

Equation (3.6) may be modified as follows for calculating the BOD exerted.

$$L_t = L_u e^{-kt} \tag{3.7}$$

$$L_u = Y_t + L_t \tag{3.8}$$

$$L_u - Y_t = L_u e^{-kt} \tag{3.9}$$

$$Y_t = L_u (1 - e^{-kt}) \tag{3.10}$$

Equation (3.10) is commonly expressed as Equation (3.11) to make it easier to remember and use.

$$\boxed{BOD_t = BOD_u (1 - e^{-kt})} \tag{3.11}$$

where:

BOD_t = BOD exerted at any time t, mg/L

BOD_u = ultimate BOD of the sample, mg/L.

Example 3.3 Calculation of BOD_5 and BOD_7

Calculate the five-day and seven-day BOD for a sample of treated wastewater assuming a BOD rate constant, $k = 0.10$ days^{-1} and an ultimate BOD of 350 mg/L.

First, substitute a time of 5 days into Equation (3.11) to calculate the five-day BOD.

$$BOD_t = BOD_u \left(1 - e^{-kt}\right)$$

$$BOD_5 = 350 \frac{mg}{L}\left(1 - e^{-0.10\,days^{-1} \times 5\,days}\right) = \boxed{138\,mg/L}$$

Next, substitute a time of 7 days into Equation (3.11) to calculate the 7-day BOD.

$$BOD_7 = 350 \frac{mg}{L}\left(1 - e^{-0.10\,days^{-1} \times 7\,days}\right) = \boxed{176\,mg/L}$$

Finally, calculate the percent of ultimate BOD exerted at 5 days.

$$\%\ BOD_u\ Exerted = \frac{138}{350} \times 100 = \boxed{39.4\%}$$

For domestic wastewater with a negligible industrial waste component, the five-day BOD typically ranges from 66–68% of the BOD_u.

3.3.2.2 Laboratory Procedure for BOD Determination

This section outlines the procedures used for determining BOD. Detailed procedures are found in *Standard Methods* (1998). The analysis involves incubating a series of diluted wastewater samples and blanks in the dark at a temperature of 20°C for five days. Typically, 300 ml glass bottles are used in BOD determinations. The test is conducted in the dark to preclude the growth of algae and the associated effects of oxygen production via photosynthesis that would yield inaccurate results. Approximately 4.0 mg/L of dissolved oxygen (DO) should be consumed during the five-day incubation period. A minimum of 1.0 mg/L of DO should remain after the incubation period. The fraction of wastewater to be added to the BOD bottle is calculated by dividing the anticipated BOD of the sample by 4.0 mg/L of DO consumed.

Dilution water used in BOD analyses consists of distilled water, to which the following chemicals are added: KH_2PO_4, K_2HPO_4, Na_2HPO_4, NH_4Cl, $MgSO_4$, $CaCl_2$, and $FeCl_3$. These chemicals provide the nutrients necessary for bacterial growth during the BOD test. Since microorganisms are used in the test, it is considered a bioassay procedure. The dilution water is also saturated with oxygen, so that there is sufficient DO during the incubation period. Water or wastewater samples that do not contain bacteria must be "seeded" with an inoculum of bacteria. Either primary clarifier effluent or non-chlorinated secondary effluent from WWTPs is typically used as the inoculum.

Figures 3.2 and 3.3 are diagrams illustrating the BOD procedure using the dilution method. Dilution water and sample are mixed together in a 1 L graduated cylinder or calibrated container before transferring into 300 ml BOD bottles.

Alternatively, dilution water and sample may be added directly to the BOD bottle; this is known as the direct method procedure. To achieve the required dilution, V_S ml of sample and V_{DW} ml of dilution water are mixed.

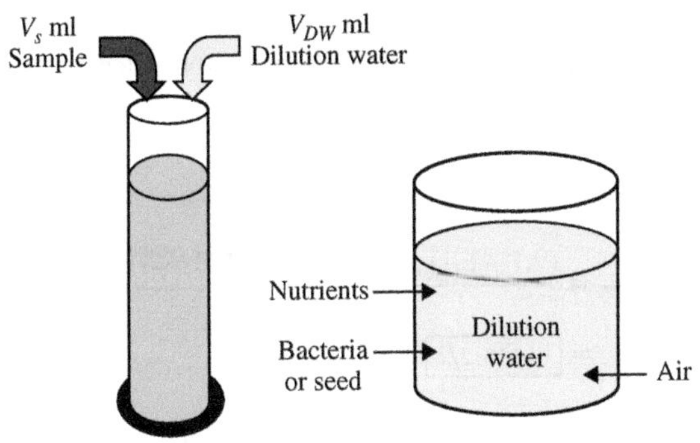

Figure 3.2 Diagram of dilution method BOD procedure. *Source:* Reprinted by permission of Pearson Education, Inc., Upper Saddle River, NJ.

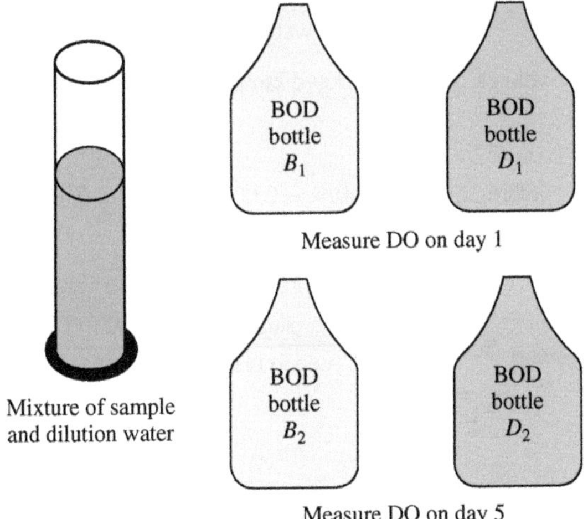

Figure 3.3 Diagram of BOD procedure using "seeded" dilution water. *Source:* Reprinted by permission of Pearson Education, Inc., Upper Saddle River, NJ.

Normally, duplicate or triplicate BOD bottles are used for each dilution made. If three 300 ml BOD bottles were to be filled and a 5% dilution was desired, 45 ml of sample would be mixed with 855 ml of dilution water. In Figure 3.3, B_1 and B_2 are BOD bottles filled only with dilution water, whereas, D_1 and D_2 are BOD bottles containing the mixture of sample and dilution water. The DO concentration in the diluted sample is measured either using a calibrated DO meter and probe or by titration using the Azide Modification of the Winkler method. Equation (3.12) is used for calculating the BOD of an "unseeded" water or wastewater sample.

$$\text{BOD}_t = \frac{(D_1 - D_2)}{P} \quad (3.12)$$

where:

BOD_t = biochemical oxygen demand of the sample at time "t", mg/L

D_1 = DO concentration of diluted sample immediately after preparation, mg/L

D_2 = DO concentration of diluted sample after "t" days of incubation, mg/L

t = incubation time, normally five days

P = decimal fraction of sample used

$$P = \frac{\text{volume of sample}}{\text{volume of sample plus dilution water}}.$$

Example 3.4 Calculation of "Unseeded" BOD

A five-day BOD test is performed on an "unseeded" primary effluent wastewater sample. 10 ml of primary effluent are added to each 300 ml BOD bottle, to which dilution water is added. A total of four BOD bottles are used in this particular test. The average DO concentration of the diluted wastewater samples at the beginning and end of the BOD test is 9.2 mg/L and 4.3 mg/L, respectively. Calculate the BOD_5 of the primary effluent.

First, calculate the decimal fraction of sample used in the BOD analysis as follows.

$$P = \frac{\text{volume of sample}}{\text{volume of sample plus dilution water}} = \frac{\text{10 ml of primary effluent}}{\text{300 ml total volume}} = 0.0333$$

Next, substitute the DO values along with the value for P into Equation (3.12).

$$\text{BOD}_5 = \frac{(D_1 - D_2)}{P} = \frac{(9.2 - 4.3)\ \text{mg/L}}{0.0333} = \boxed{147\ \text{mg/L}}$$

Example 3.5 Calculation of volume of sample to be used in BOD test

A five-day BOD test is to be performed on a sample of lake water. Estimate the volume of sample that should be used in the BOD test if the anticipated BOD of the lake water is 35 mg/L.

Make the assumption that at least 4.0 mg/L of DO will be consumed during the test.

First, calculate the fraction of sample that must be used.

$$\text{fraction of sample} = \frac{4.0\ \text{mg/L}}{35\ \text{mg/L}} = 0.114$$

Next, calculate the volume of lake water sample to be added to a 300 ml BOD bottle.

$$\text{volume of sample} = 0.114 \times 300\ \text{ml} = \boxed{34.2\ \text{ml}}$$

Therefore, add 34 ml of lake water to a BOD bottle and dilute with 266 ml of dilution water.

The BOD procedure using the dilution method, in which the dilution water must be "seeded", is presented in Figure 3.3. In this case, BOD bottles called "blanks" are filled with the "seeded" dilution water, so that the BOD value can be corrected for the seed material. Equation (3.12) must be modified to account for the oxygen demand associated with blanks (B_1 and B_2). Equation (3.13) is used to calculate the BOD for "seeded" samples.

$$BOD_t = \frac{(D_1 - D_2) - (B_1 - B_2)f}{P} \quad (3.13)$$

where:

B_1 = DO concentration of seeded dilution water immediately after preparation, mg/L

B_2 = DO concentration of seeded dilution water after "t" days of incubation, mg/L

f = ratio of the seed in the diluted sample to the seed in control or blank

$$f = \frac{(\%\text{ seed in diluted sample})}{(\%\text{ seed in control or blank})} = \frac{(\text{volume of seed in diluted sample})}{(\text{volume of seed in control or blank})}$$

Example 3.6 Calculation of "Seeded" BOD

A five day BOD test is performed on a secondary, dechlorinated wastewater effluent sample. 20 ml of the secondary, dechlorinated effluent is added to 300 ml BOD bottles, to which dilution water is added. The average DO concentration of the diluted wastewater samples at the beginning and end of the BOD test is 9.1 mg/L and 6.1 mg/L, respectively. The average DO concentration in the blanks containing the "seeded" dilution water at the beginning and end of the test is 9.2 mg/L and 7.2 mg/L, respectively. 20 liters of dilution water were prepared, containing 40 ml of "seed" material. Calculate the BOD_5 of the secondary, chlorinated effluent.

First, calculate the decimal fraction of sample used in the BOD analysis as follows.

$$P = \frac{\text{volume of sample}}{\text{volume of sample plus dilution water}} = \frac{20\text{ ml of secondary effluent}}{300\text{ ml total volume}} = \boxed{0.00667}$$

Second, calculate the "f" value.

$$f = \frac{(\%\text{ seed in diluted sample})}{(\%\text{ seed in control or blank})} = \frac{(280\text{ ml}/300\text{ ml})}{(300\text{ ml}/300\text{ ml})} = \boxed{0.933}$$

Finally, substitute into Equation (3.13) to calculate the BOD.

$$BOD = \frac{(D_1 - D_2) - (B_1 - B_2)f}{P}$$

$$BOD_5 = \frac{(9.1 - 6.1)\text{ mg/L} - (9.2 - 7.2)\text{ mg/L} \times 0.933}{0.0667} = \boxed{17.0\text{ mg/L}}$$

Alternatively, "f" could have been calculated as follows:

$$\%\text{ seed in dilution water} = \frac{40\text{ ml of seed}}{20\text{ L of dilution water} \times \frac{1000\text{ ml}}{\text{L}}} = 0.002$$

$$\text{volume of seed in diluted sample} = 0.002 \times 280\text{ ml} = 0.56\text{ ml}$$

$$\text{volume of seed in blank} = 0.002 \times 300\text{ ml} = 0.6\text{ ml}$$

$$f = \frac{(\text{volume of seed in diluted sample})}{(\text{volume of seed in control or blank})} = \frac{(0.56\text{ ml seed in diluted sample}/300\text{ ml})}{(0.6\text{ ml seed in blank}/300\text{ ml})} = \boxed{0.933}$$

3.3.2.3 Graphical Determination of BOD Constants

Quite often it is necessary to determine the ultimate BOD concentration and **BOD rate constant, *k***, for specific environmental engineering applications. For instance, the ultimate BOD value for a particular wastewater is necessary for estimating oxygen requirements and sizing the aeration systems for biological treatment facilities. In modeling river systems, both the BOD_u and k are required for estimating the DO concentration as a function of travel time down the stream.

To determine k and ultimate BOD, a series of BOD measurements are conducted as a function of time, rather than just the traditional five-day incubation period. Once the data are collected, there are numerous methods such as the least-squares, Thomas (1950), and Fujimoto (1961) that can be used to estimate k and BOD_u. The least-squares and Fujimoto methods are presented in Metcalf & Eddy (2003). The Thomas method is summarized in Benefield & Randall (1980) and Davis & Cornwall (2008).

The Fujimoto method involves making an arithmetic plot of BOD_{t+1} versus BOD_t. On the same graph, a line with a slope of 1 is plotted. The BOD_u can be read from the graph at the intersection of these two lines. The rate constant, k, can then be estimated by substituting into Equation (3.11). An example illustrating the Fujimoto method is presented below.

Example 3.7 Determination of BOD_u and *k*

Using the Fujimoto method and the data given below, determine the ultimate BOD concentration and the BOD reaction rate constant, *k*, of the wastewater.

t (days)	0	1	2	3	4	5	6	7	8	9
BOD_t (mg/L)	0	20	32	40	42	46	48	49	50	50

First, prepare an arithmetic plot of BOD_{t+1} versus BOD_t. Draw the line of best fit through the data and then draw a line with a slope of 1. The BOD value at the intersection of these two lines is $\mathbf{BOD_u}$, which is approximately 50 mg/L. See the plot shown below.

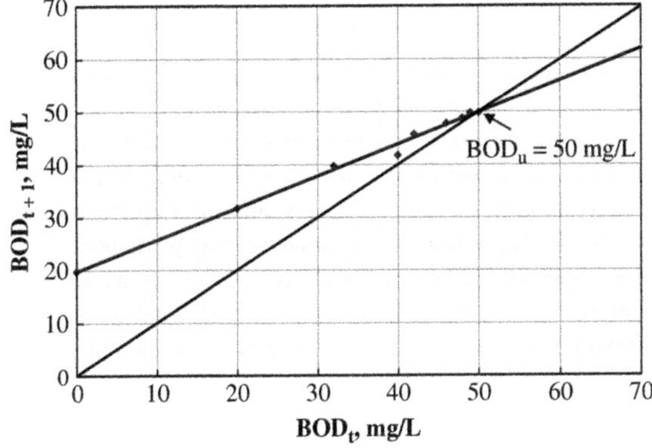

Next, estimate the BOD reaction rate constant, *k*, by substituting into Equation (3.11). At time equals 5 days, the BOD is 46 mg/L.

$$BOD_t = BOD_u(1 - e^{-kt})$$

$$46\,\text{mg/L} = 50\,\text{mg/L}(1 - e^{-k\times 5\,\text{days}})$$

$$\frac{46}{50} = (1 - e^{-k\times 5\,\text{days}})$$

$$0.92 - 1.00 = -e^{-5k}$$

$$-0.08 = -e^{-5k}$$

$$\ln(0.08) = -5k$$

$$k = \frac{\ln(0.08)}{-5} = \boxed{0.505\,\text{d}^{-1}}$$

Although the calculations are not presented, averaging individual *k* values for days 1 through 7 resulted in a BOD reaction rate constant of $0.510\,\text{d}^{-1}$. Use of the average value is preferred.

3.3.3 Nitrogenous oxygen demand (NOD)

Nitrification is usually modeled as a sequential, two-step process involving the genera *Nitrosomonas* and *Nitrobacter*. These are autotrophic bacteria that use inorganic carbon (CO_3^{2-} and HCO_3^-) as their source of carbon. Nitrification typically occurs between 5–8 days, when ammonia nitrogen has been introduced into a water body (at 20°C), exerting an additional oxygen demand due to the oxidation of ammonia to nitrate. The following three reactions show the nitrification process that may occur during the BOD test if the test is conducted longer than five days. Equation (3.16) represents the overall nitrification reaction.

These reactions ignore the quantity of ammonia that is used in synthesizing nitrifier biomass; since the nitrifiers make up only a small fraction of the total biomass, the amount of ammonia incorporated into cellular components is negligible. Nitrification is an important biological process that is used widely in biological wastewater treatment processes as a first step in removing nitrogen from wastewater. It will be discussed more thoroughly in the chapter on wastewater treatment.

$$2NH_4^+ + 3O_2 \xrightarrow{\text{Nitrosomonas}} 2NO_2^- + 4H^+ + 2H_2O \quad (3.14)$$

$$2NO_2^- + O_2 \xrightarrow{\text{Nitrobacter}} 2NO_3^- \quad (3.15)$$

$$NH_4^+ + 2O_2 \xrightarrow{\text{Nitrifiers}} NO_3^- + 2H^+ + H_2O \quad (3.16)$$

The nitrogenous oxygen demand (NOD) can be estimated using Equation (3.16). For each mole of ammonium nitrogen, two moles of oxygen are required for complete oxidation of the ammonium to nitrate. Alternatively, on a weight basis, $(2\times 32)/14 = 4.57$ grams of oxygen are required per gram of ammonium nitrogen oxidized during the nitrification process.

Figure 3.4 presents BOD versus time curves showing both the carbonaceous and nitrogenous oxygen demands that may be observed. It should also be mentioned that nitrification will only occur if sufficient oxygen and alkalinity are present to support nitrifier growth.

3.3.4 Chemical oxygen demand (COD)

The chemical oxygen demand (COD) test measures the oxygen equivalents of organic matter oxidized to carbon dioxide and water under acid conditions. The procedure involves refluxing (vaporizing and condensing) a sample of water or wastewater containing organic matter with sulfuric acid and excess standardized potassium dichromate. During the reflux period, organic matter that is chemically oxidized reduces a stoichiometric equivalent quantity of dichromate. The quantity of dichromate remaining in solution after refluxing is measured by titration with ferrous ammonium sulfate. Equation (3.17), as

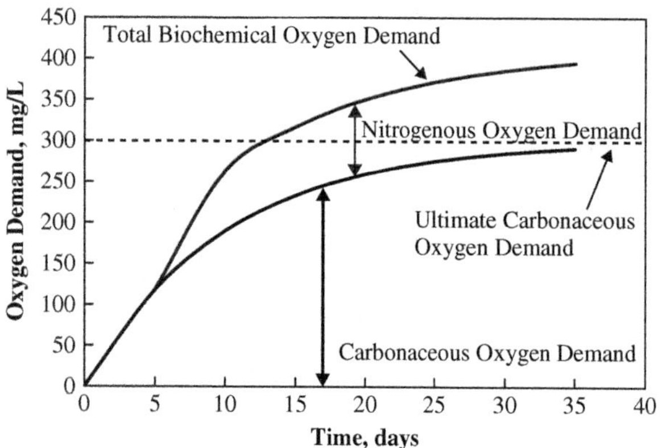

Figure 3.4 Carbonaceous and nitrogenous oxygen demand curves.

presented by Sawyer *et al.* (1994), illustrates the oxidation of organic matter and organic nitrogen.

$$C_nH_aO_bN_c + d\,Cr_2O_7^{2-} + (8d + c)H^+ \rightarrow n\,CO_2 + \frac{a + 8d - 3c}{2}H_2O + c\,NH_4^+ + 2d\,Cr^{3+} \qquad (3.17)$$

where $d = \frac{2n}{3} + \frac{a}{6} - \frac{b}{3} - \frac{c}{2}$.

An advantage of the COD test over the traditional BOD test is that it can be performed in three hours (two hours for refluxing and one hour for cooling), compared with five days. A disadvantage is that COD is incapable of distinguishing between biodegradable organics and those that are not. COD is calculated using Equation (3.18).

$$\mathrm{COD}\left(\frac{\mathrm{mg}}{\mathrm{L}}\right) = \frac{8000(B - A)[\text{Normality of FAS}]}{V} \qquad (3.18)$$

where:
FAS = ferrous ammonium sulfate titrant, $Fe(NH_4^+)_2(SO_4)_2$
B = amount of FAS added to the blank, ml
A = amount of FAS added to the sample, ml
V = volume of sample used, ml.

3.3.4.1 Other Comments on BOD and COD

COD testing is becoming more popular, since it can be determined rapidly and the results are reproducible. It is possible to develop a relationship between BOD and COD, but only for a given type of wastewater. For domestic wastewaters, the five-day BOD is approximately two-thirds of the ultimate BOD of the wastewater. This rule of thumb cannot, and should not, be applied to industrial wastewaters.

Theoretically, the ultimate BOD and COD of a sample should be essentially the same. However, remember that COD measures total oxidizable organics and, therefore, will generally have a larger magnitude than BOD, which only measures the biodegradable fraction. Certain samples may indicate a high COD value, whereas the BOD may be low or close to zero. This may be attributed to toxic substances in the sample that kill the bacteria necessary to exert an oxygen demand, or the sample may contain complex, non-biodegradable organics that cannot be readily oxidized, i.e., organochlorine pesticides.

Both BOD and COD can be subdivided into particulate and soluble fractions. This is generally accomplished by filtering the sample through a 0.45 μm fiberglass filter. When a BOD or COD analysis is performed on the filtrate, it is termed a **soluble BOD** (SBOD) or **soluble COD** (SCOD). A **total BOD** (TBOD) or **total COD** (TCOD) is a BOD or COD analysis performed on the total sample that includes the particulate or solid matter that would otherwise be removed upon filtration.

In biological treatment studies performed for bionutrient removal applications, the COD particulate and soluble fractions can be fractionated into other categories such as: readily biodegradable SCOD; slowly biodegradable colloidal and particulate COD; non-biodegradable SCOD; and non-biodegradable colloidal and particulate COD. Characterization and fractionation of wastewater can be found in Metcalf & Eddy (2003) and Henze *et al.* (1987).

3.3.5 Total organic carbon (TOC)

The organic carbon in a water or wastewater sample can be measured conveniently and directly with a total organic carbon (TOC) analyzer; performing a TOC analysis requires only 10–15 minutes or less. A TOC analysis may be divided into three stages: acidification; oxidation; and detection and quantification. TOC methods may use heat, catalysts, oxygen, ultraviolet irradiation, chemical oxidants, or a combination of these for the conversion of organic carbon to CO_2.

Typically, aqueous samples containing organic matter are injected into a furnace, where the water is evaporated and the organic carbon is catalytically oxidized to CO_2. A carrier gas, usually oxygen, carries the CO_2 through an infrared analyzer, which measures and records its concentration. Results from samples processed in this manner actually yield total carbon (TC). To obtain the TOC concentration, the sample must first be acidified to convert inorganic carbon to carbon dioxide and purged with a pure gas before injection into the furnace. Water samples that have been filtered through a 0.45 μm fiberglass filter, acidified, and purged before injecting into the TOC analyzer, will yield dissolved total organic carbon (DTOC). It is also possible to relate TOC to both BOD and COD.

3.4 Physical parameters

Some important physical parameters used in water quality management include: color, odor, solids, temperature, transmittance, and turbidity.

3.4.1 Color

The **true color** of a sample of water or wastewater is primarily attributed to dissolved and colloidal substances. **Apparent**

color is due to suspended substances that can be removed by settling and filtration. Color associated with dissolved and colloidal substances, however, can only be removed by adding a chemical that causes the substance to be precipitated or coagulated. Powerful oxidizing agents, such as chlorine and ozone, can be used to oxidize color compounds into innocuous end-products.

Most surface waters contain organic or vegetable extracts from the decomposition of leaves, pine needles, and lignin. In some cases, the water may appear dark brown – even black, if the water originated from a swamp. Not only is colored water aesthetically displeasing to consumers, but many organic compounds, such as humic and fulvic acids, will be converted to trihalomethanes (THMs) during chlorination of water. Trihalomethanes are suspected carcinogens, and the United States Environmental Protection Agency (EPA) has established a maximum contaminant level (MCL) of 0.08 mg/L for total trihalomethanes (TTHMs).

When brought to the surface, groundwater that contains iron may exhibit a reddish-brown color due to oxidation of reduced iron (Fe^{2+}) to ferric iron (Fe^{3+}). Manganese in its reduced form (Mn^{2+}) will also be oxidized to Mn^{3+} when groundwater is pumped from a well to the surface.

Domestic wastewater generally appears to be dark brown to black. Many industrial wastewaters, such as textile, and pulp and paper mill wastes, will be highly colored. Conventional wastewater treatment methods will not remove the contaminants causing color in these types of wastewater. Advanced treatment using activated carbon adsorption or advanced oxidation methods must be used.

Typically, color is determined by visual comparison with a known standard solution of potassium chloroplatinate or calibrated color disks. The color produced by 1 mg/L of platinum (as K_2PtCl_6) serves as the standard unit of color. A stock solution of potassium dichloroplatinate, containing 500 mg/L of platinum, is made. From this stock solution, color standards ranging from 5–70 color units are prepared by diluting the stock with distilled water. The color value of water is pH dependent, so the pH should be measured and noted when reporting color.

An alternative method for measuring color of a filtered or centrifuged sample is to measure the transmittance at various wavelengths. The procedures for these and other methods for color determination are found in *Standard Methods* (1998).

3.4.2 Taste and odor

Both organic and inorganic contaminants may cause taste and odor problems in water. Unfortunately, domestic and industrial wastewaters usually have odors ranging from humus-like to that of rotten eggs. In this discussion, we will focus on taste and odor concerns in drinking water.

Taste and odor compounds are difficult to quantify because they are subjective parameters. Consumers are displeased with water that has an odor or an unpleasant taste; they consider the water to be contaminated. However, something that tastes or smells good to one individual might not be appealing to another. Several inorganic or mineral compounds will produce a taste in water without creating an odor. Normally, substances that produce an odor in water will also impart a taste as well. Many organic compounds tend to produce taste and odor problems in drinking water. Certain species of algae (blue-green) secrete organic compounds causing tastes and odors in water.

Quantitative tests have been established that rely on the human senses of smell and taste. The threshold odor test and the flavor threshold test use a panel of testers to quantify the smell or taste of a sample of water. Several dilutions of the sample are made, so that the total volume in the glass beaker is 200 ml. In the threshold odor test, the sample is diluted with distilled water until the tester can just detect an odor. The threshold odor number (TON) is the greatest dilution of the sample with odor-free, distilled water which yields a perceptible odor. TON is calculated as follows:

$$\text{TON} = \frac{A+B}{A} \tag{3.19}$$

where:
TON = threshold odor number, unitless
A = ml of sample
B = ml of odor-free, distilled water used.

Example 3.8 Determination of threshold odor number

Calculate the threshold odor number (TON) for a 25 ml sample that is diluted to 200 ml before no odor is detected.

Substitute the above values into Equation (3.19):

$$\text{TON} = \frac{A+B}{A} = \frac{25\text{ ml sample} + 175\text{ ml distilled water}}{25\text{ ml sample}}$$

$$= \boxed{8}$$

The higher the TON value, the more odiferous the sample.

The flavor threshold test is performed similar to that for odor, the difference being that a panel of testers will actually taste and characterize the flavor of a sample of water. Dilutions of the sample are prepared so that a flavor cannot be detected. Equation (3.20) is used to calculate the flavor threshold number (FTN).

$$\text{FTN} = \frac{A+B}{A} \tag{3.20}$$

where:
FTN = flavor threshold number, unit less.
A = ml of sample, and
B = ml of odor-free, distilled water used.

Example 3.9 Flavor threshold number

If the flavor threshold number (FTN) of a water sample is 100, estimate the volume of sample diluted to 200 ml so that no taste is observed.

First, rearrange Equation (3.20) to solve for the volume *A* as follows:

$$\text{FTN} = \frac{A+B}{A}$$

Note that $A + B = 200$ ml
$\text{FTN} \times A = 200$ ml

$$A = \frac{200\text{ ml}}{\text{FTN}} = \frac{200\text{ ml}}{100} = \boxed{2\text{ ml}}$$

3.4.3 Solids

Environmental engineers and scientists are concerned about the measurement of solid matter in water, wastewater, and sludge. The **solids content** of a sample of water or wastewater can be divided into several subcategories. Figure 3.5 illustrates the relationships between the various solids categories.

By definition, "**total solids**" refers to matter that remains after evaporation and drying at a temperature of 103°C to 105°C. Materials that have a high vapor pressure will be lost during the analysis. To perform a total solids analysis, a sample is placed into a clean aluminum pan or a porcelain dish that has been tared. Next, the pan or dish is placed on a water bath to evaporate the water, and then transferred to a drying oven at a temperature between 103 – 105°C until a constant weight is obtained. Alternatively, the pan or dish can be transferred directly to a drying oven to accomplish evaporation and drying.

The formulae for calculating total and all additional types of solids involve determining the weight difference before and after drying in an oven divided by the sample volume used. The **total solids (TS)** concentration is determined by subtracting the weight of the empty tare from the weight of the dried solids plus tare and dividing by the sample volume, as shown in Equation (3.21).

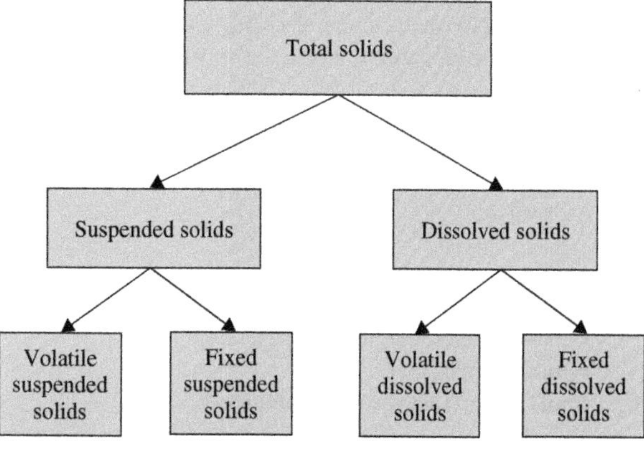

Figure 3.5 Relationships between the various solids categories.

$$\text{Total Solids (TS)} = \frac{W_{TS+Tare} - W_{Tare}}{V_S} \qquad (3.21)$$

where:
TS = total solids or residue remaining after evaporation and drying, mg/L or g/L
$W_{TS + Tare}$ = weight of total solids plus tare remaining after evaporation and drying at 103°C to 105°C, mg or g
W_{Tare} = weight of empty tare or dish, mg or g
V_S = volume of sample used, L.

The **total suspended solids (TSS)** or filterable solids concentration in a water sample is determined by passing a known volume of the sample through a clean, tared fiberglass filter pad. Pore sizes of filters range from 0.45 μm to 2.0 μm, with the 0.45 μm filter typically being used. Once the sample has been filtered, the dirty filter is placed back into the weighed aluminum pan and placed in the oven to be dried at a temperature of 103 – 105°C. Upon drying, the weight of TSS may be determined by subtracting the difference in weights of clean filter pad and tare, from the weight of the dried solids, filter, and tare. Equation (3.22) is used for calculating the total TSS concentration.

$$\text{Total Suspended Solids (TSS)} = \frac{W_{TSS+Tare+Filter} - W_{Tare+Filter}}{V_S} \qquad (3.22)$$

where:
TSS = total suspended solids or residue remaining after evaporation and drying, mg/L or g/L
$W_{TSS + Tare + Filter}$ = weight of total suspended solids or residue remaining after evaporation and drying at 103°C to 105°C plus weight of tare and filter, mg or g
$W_{Tare + Filter}$ = weight of clean, empty tare plus clean filter, mg or g
V_S = volume of sample used, L.

The dissolved solids or non-filterable solids are those that pass through the filter. The **total dissolved solids concentration (TDS)** is determined by placing a sample of the filtrate into a clean porcelain dish or aluminum pan that is then dried in an oven at 103°C to 105°C. Equation (3.23) is used for calculating the TDS concentration of a sample.

$$\text{Total Dissolved Solids (TDS)} = \frac{W_{TDS+Tare} - W_{Tare}}{V_S} \qquad (3.23)$$

where:
TDS = total dissolved or non-filterable solids remaining after evaporation and drying, mg/L or g/L
$W_{TDS + Tare}$ = weight of total dissolved solids remaining after evaporation and drying at 103°C to 105°C plus weight of tare, mg or g
W_{Tare} = weight of clean empty tare, mg or g
V_S = volume of sample used, L.

Alternatively, the TDS concentration may be calculated by subtracting the TSS concentration from the TS concentration.

Each of the solids categories above has an organic (volatile) and inorganic (mineral or fixed) fraction that can be measured. A volatile solids determination is made to obtain a measurement of the organic content present in a sample. This is accomplished by combusting the organic matter to CO_2 and water, under controlled temperature conditions, so that decomposition and volatilization of inorganic compounds does not occur.

The standard procedure in environmental applications is to conduct volatile solids analysis at a temperature of 550°C. The residue that remains after burning or combustion is referred to as fixed solids, while the organic matter that was combusted and/or volatilized is called volatile solids. There are volatile and fixed fractions for total, suspended, and dissolved solids. The following equations show how the volatile and fixed fractions relate to each of the three categories (TS, TSS, and TDS).

$$\mathrm{TS} = \mathrm{TVS} + \mathrm{TFS} \tag{3.24}$$

where:
TVS = total volatile solids concentration, mg/L or g/L
TFS = total fixed solids concentration, mg/L or g/L.

The **total volatile solids (TVS)** concentration can be calculated as follows:

$$\text{Total Volatile Solids (TVS)} = \frac{W_{TS+Tare} - W_{TFS+Tare}}{V_S} \tag{3.25}$$

where:
$W_{TS+Tare}$ = weight of total solids plus tare, mg or g
$W_{TFS+Tare}$ = weight of total fixed solids plus tare after ignition at 550°C, m or g.

Calculation of the **volatile suspended solids (VSS)** can be performed using Equation (3.26):

$$\mathrm{VSS} = \frac{W_{VSS+Filter+Tare} - W_{FSS+Filter+Tare}}{V_S} \tag{3.26}$$

where:
$W_{VSS+Filter+Tare}$ = weight of volatile suspended solids, filter, and tare, mg or g
$W_{FSS+Filter+Tare}$ = weight of fixed suspended solids and filter plus tare after ignition at 550°C, mg or g.

Summation of the volatile suspended solids and **fixed suspended solids (FSS)** yields the total suspended solids as shown in Equation (3.27).

$$\mathrm{TSS} = \mathrm{VSS} + \mathrm{FSS} \tag{3.27}$$

where:
VSS = volatile suspended solids concentration, mg/L or g/L, and
FSS = fixed suspended solids concentration, mg/L or g/L.

Summation of the **volatile dissolved solids (VDS)** and **fixed dissolved solids (FDS)** yields the total dissolved solids, as shown in the following equation.

$$\mathrm{TDS} = \mathrm{VDS} + \mathrm{FDS} \tag{3.28}$$

where:
VDS = volatile dissolved solids concentration, mg/L or g/L
FDS = fixed dissolved solids concentration, mg/L or g/L.

Example 3.10 Calculating various types of solids concentrations

The following test results were obtained on an influent sample to an industrial wastewater treatment plant. All solid analyses were performed using a sample volume of 50 ml. Determine the concentration of:

a) total solids
b) total volatile solids
c) total suspended solids
d) volatile suspended solids
e) total dissolved solids.

Tare weight of evaporating dish =	54.6423 g
Weight of evaporating dish plus residue after evaporation @ 105°C =	54.7148 g
Weight of evaporating dish plus residue after ignition @ 550°C =	54.6818 g
Tare weight of Whatman glass fiber filter =	1.5434 g
Weight of Whatman glass fiber filter and residue after drying @ 105°C =	1.5625 g
Weight of Whatman glass fiber filter and residue after ignition @ 550°C =	1.5531 g

First, calculate the total solids concentration using Equation (3.21).

$$\mathrm{TS} = \frac{W_{TS+Tare} - W_{Tare}}{V_S} = \frac{(54.7148\,\mathrm{g} - 54.6423\,\mathrm{g})}{50\,\mathrm{ml}} \times \left(\frac{1000\,\mathrm{mg}}{\mathrm{g}}\right)\left(\frac{1000\,\mathrm{ml}}{\mathrm{L}}\right) = \boxed{1450\,\frac{\mathrm{mg}}{\mathrm{L}}}$$

Next, calculate the total volatile solids concentration using Equation (3.25).

$$\mathrm{TVS} = \frac{W_{TS+Tare} - W_{TFS+Tare}}{V_S}$$

$$\mathrm{TVS} = \frac{W_{TS+Tare} - W_{TFS+Tare}}{V_S} = \frac{(54.7148\,\mathrm{g} - 54.6818\,\mathrm{g})}{50\,\mathrm{ml}} \times \left(\frac{1000\,\mathrm{mg}}{\mathrm{g}}\right)\left(\frac{1000\,\mathrm{ml}}{\mathrm{L}}\right) = \boxed{660\,\frac{\mathrm{mg}}{\mathrm{L}}}$$

Now, using Equation (3.22) calculate the total suspended solids (TSS) concentration.

$$\text{TSS} = \frac{W_{TSS+Tare+Filter} - W_{Tare+Filter}}{V_S}$$
$$= \frac{(1.5625\text{ g} - 1.5434\text{ g})}{50\text{ ml}}\left(\frac{1000\text{ mg}}{\text{g}}\right)\left(\frac{1000\text{ ml}}{\text{L}}\right)$$
$$= \boxed{382\ \frac{\text{mg}}{\text{L}}}$$

Next, calculate the volatile suspended solids (VSS) concentration using Equation (3.26).

$$\text{VSS} = \frac{W_{VSS+Filter+Tare} - W_{FSS+Filter+Tare}}{V_S}$$
$$= \frac{(1.5625\text{ g} - 1.5531\text{ g})}{50\text{ ml}}\left(\frac{1000\text{ mg}}{\text{g}}\right)\left(\frac{1000\text{ ml}}{\text{L}}\right)$$
$$= \boxed{188\ \frac{\text{mg}}{\text{L}}}$$

The total dissolved solids (TDS) concentration is determined by subtraction as follows:

$$\text{TDS} = \text{TS} - \text{TSS} = 1450 - 382 = \boxed{1068\ \frac{\text{mg}}{\text{L}}}$$

3.4.4 Temperature

Although sometimes overlooked, the measurement of temperature is a critical parameter that affects biological and chemical reactions. The solubility of gases and chemical species are temperature dependent, and discharging heated water into ecosystems can have a detrimental impact on aquatic life. The solubility of dissolved oxygen (DO) increases as temperature decreases. The saturation concentration of DO at 0°C, 10°C, 20°C and 30°C is 14.6 mg/L, 11.3 mg/L, 9.09 mg/L, and 7.56 mg/L, respectively (*Standard Methods*, 1998).

A rule of thumb used in estimating biological and chemical rates of reaction is that for every 10°C increase in temperature, the rate of reaction doubles. The Arrhenius-van't Hoff equation is used to derive a **temperature correction coefficient**, usually referred to as theta (θ), that is often used to correlate reaction rate with temperature. The derivation of θ is as follows:

The reaction rate for many chemical and biochemical reactions increases rapidly (exponentially) with an increase in temperature as shown by the Arrhenius relationship, Equation (3.29).

$$k = Ae^{-(E_a/RT)} \tag{3.29}$$

where:
k = reaction rate constant at temperature T
A = a constant that is independent of temperature for a specific reaction
E_a = activation energy, cal/mol
R = ideal gas constant, 1.98 cal/mol·K
T = reaction temperature, K

If the reaction rate constant is known at a given temperature, the Arrhenius relationship can be used to predict the rate constant at another temperature. For example, consider the Arrhenius equation for temperatures identified as T_1 and T_2:

$$\text{For temperature } T_1 : k_1 = Ae^{-(E_a/RT_1)} \tag{3.30}$$

$$\text{For temperature } T_2 : k_2 = Ae^{-(E_a/RT_2)} \tag{3.31}$$

Dividing Equation (3.31) by Equation (3.30) yields:

$$\frac{k_2}{k_1} = \frac{Ae^{-(E_a/RT_2)}}{Ae^{-(E_a/RT_1)}} \tag{3.32}$$

Taking the natural log of both sides of Equation (3.32) and rearranging yields:

$$\ln\left(\frac{k_2}{k_1}\right) = \frac{-E_a}{RT_2} + \frac{E_a}{RT_1} \tag{3.33}$$

Simplifying Equation (3.33) produces:

$$\ln\left(\frac{k_2}{k_1}\right) = \frac{E_a}{R}\frac{(T_2 - T_1)}{T_2 T_1} \tag{3.34}$$

For the temperature ranges encountered in most situations in environmental engineering, the term $\frac{E_a}{R\,T_2\,T_1}$ may be considered a constant (C) and Equation (3.34) can be rearranged as follows:

$$\ln\left(\frac{k_2}{k_1}\right) = C(T_2 - T_1) \tag{3.35}$$

Taking the antilog of both sides of Equation (3.35) results in the following equation:

$$\frac{k_2}{k_1} = e^{C(T_2 - T_1)} \tag{3.36}$$

Replacing e^C with the temperature correction coefficient θ results in the following equation that is commonly used for correcting biochemical and chemical reactions for temperature variations.

$$\frac{k_2}{k_1} = \theta^{(T_2 - T_1)} \tag{3.37}$$

where:
θ = temperature correction coefficient for a specific application, dimensionless
k_2 = reaction rate constant at temperature T_2
k_1 = reaction rate constant at temperature T_1.

Absolute temperature values must be used in other forms of the Arrhenius equation, but the use of Celsius is acceptable in Equation (3.37), since only a difference in temperature is involved. For making temperature corrections to BOD data, a θ value of 1.135 is used for the temperature range of 4 to 20°C and 1.056 for the temperature range of 20 to 30°C (Metcalf & Eddy, 2003), whereas a value of 1.02 is routinely used to correct oxygen transfer data for mechanical aeration systems (Reynolds & Richards, 1996).

Example 3.11 Temperature effect on BOD rate constant

A θ value of 1.047 is typically used for making temperature corrections to the biochemical oxygen demand (BOD) rate constant, *k*. If the BOD rate constant at 20°C is 0.16 d^{-1}, determine the value of the BOD rate constant, *k*, for a temperature of 25°C.

Solve Equation (3.37) for k_2. Make appropriate substitutions and solve as follows.

$$k_2 = k_1\ \theta^{(T_2-T_1)}$$

$$k_2 = 0.16\,d^{-1}(1.047)^{(25°C-20°C)} = \boxed{0.20\,d^{-1}}$$

The value of the BOD rate constant *k* at a temperature of 25°C is equal to 0.20 d^{-1}. The rate of reaction for many systems is strongly dependent on temperature. In this example, increasing the temperature by 5°C increased the reaction rate by 25 percent.

Example 3.12 Estimate the Temperature Correction Coefficient

Estimate the temperature correction coefficient, θ, for a mechanical aeration system if the oxygen transfer coefficient at 20°C and 30°C is 2.54 hr^{-1} and 1.98 hr^{-1}, respectively.

Substitute into Equation (3.37) and rearrange to solve for θ as follows.

$$k_2 = k_1\ \theta^{(T_2-T_1)}$$

$$\frac{k_2}{k_1} = \theta^{(T_2-T_1)}$$

$$\frac{1.98\,hr^{-1}}{2.54\,hr^{-1}} = \theta^{(30°C-20°C)}$$

$$0.7795 = \theta^{(10°C)}$$

Raising both sides of the equation by the reciprocal of the power yields the value of θ.

$$0.7795^{(1/10)} = \theta^{10\times 1/10}$$

$$\theta = \boxed{0.975}$$

3.4.5 Absorbance and transmittance

The **absorbance** of a solution is a measure of the amount of light that is absorbed by the constituents in the solution at a given wavelength, λ. A wavelength of 254 nanometers (nm) is typically used for measuring absorbance in aqueous solutions. A colorimeter or spectrophotometer is an instrument which makes possible a quantitative measure of light passing through a solution.

The quantity of monochromatic light transmitted by a solution is related to the solute concentration and is frequently described by the Beer-Lambert law:

$$\log\left(\frac{I_o}{I}\right) = A = a\ b\ c \qquad (3.38)$$

where:

A = absorbance

I = intensity of light (at a specified λ) after passing through a solution of known thickness containing constituents of interest, mW/cm^2

I_0 = intensity of light (at specified λ) after passing through a blank solution, mW/cm^2

a = absorptivity, a constant for a given solution and a given wavelength

b = path length, cm

c = concentration of light absorbing solute, g/L.

The **transmittance** of a solution is defined as the fractional relationship of the intensity of the transmitted light through a solution and pure solvent.

$$T = \frac{I}{I_0} \times 100 \qquad (3.39)$$

where T = Transmittance of solution, %.

The optical density of the solution is directly proportional to concentration and can be expressed as:

$$-\log T = \log\frac{1}{T} = \text{optical density (O.D.)} \qquad (3.40)$$

Most spectrophotometers are capable of providing data on O.D. measurement (absorbance), which makes use of a logarithmic scale, and a %T displayed using an arithmetic scale. If quantifying color using the Beer-Lambert law, the logarithmic or absorbance-scale would be most beneficial. Absorbance values for various types of water (MWH, 2005) are presented in Table 3.2.

3.4.6 Turbidity

Turbidity is a measure of the extent to which light is either scattered or absorbed by colloidal and suspended solids in water (Peavy *et al.*, 1985). Turbidity is caused by suspended and colloidal material in the water. Clay, silt, finely divided organic and inorganic matter, plankton, and microorganisms are

Table 3.2 Absorbance values for various types of water ($\lambda = 254$ nm).

Type of water	Absorbance value
Natural water	0.125–0.025
Treated and filtered water	0.08–0.05
Microfiltered water	0.04–0.025
Reverse osmosis water	0.025–0.01

Based on MWH (2005). *Water Treatment: Principles and Design*, p. 45.

Table 3.3 Types and size range of typical particles.

Type of particle	Size
Suspended solids	1 to 10^3 μm
Settleable solids	10^1 to $>10^3$ μm
[a] Bacteria	0.1 to 10 μm
[a] Viruses	0.01 to 0.1 μm
Colloidal	10^{-3} to 1 μm
Dissolved solids	10^{-5} to 10^{-3} μm

Based on Tchobanoglous and Schroeder (1985). *Water Quality: Characteristics, Modeling, Modification*, p. 58.
[a] Based on Davis (2010). *Water and Wastewater Engineering: Design Principles and Practice*, p. 3–2.

examples of matter causing turbidity. Colloidal particles will not settle out of solution by the force of gravity. They are very small particles with low mass and large surface area.

Turbidity is easily measured in the laboratory; turbidimeters are photometers that measure the intensity of scattered light. Units for turbidity are expressed in **Nephelometric turbidity units (NTUs)**. Table 3.3 presents the types and size range of typical particles found in water and wastewater.

3.5 Inorganic chemical parameters

Important inorganic chemical parameters include pH and alkalinity as well as the nitrogen, phosphorus, metal, and gas content of the water or wastewater. A description of each parameter is presented below.

3.5.1 pH

The pH of water is a function of the hydrogen ion, H^+, concentration. Mathematically, pH is defined as the negative logarithm of the molar concentration of H^+.

$$pH = -\log[H^+] \tag{3.41}$$

The pH scale is usually represented from 0 to 14. At 25°C, a pH value of 7 is considered neutral. Solutions with pH values below 7 are called "acidic" and their hydrogen ion concentration exceeds the concentration of hydroxide ion, OH^-. Basic solutions have pH values higher than 7 and their OH^- concentration exceeds their H^+ concentration. A pH between 6.5 and 8.5 is generally considered favorable for drinking water and wastewater effluent discharges.

The importance of pH in environmental engineering cannot be overstated. All biological processes for treating water, wastewater, and sludge must be maintained within a specified range to ensure survival of the microorganisms involved. Typically, this pH range is 6.5–8.5. Both biological and chemical reactions have an optimum pH value. Operating systems at suboptimal pH leads to inefficiency and increased costs.

Disinfection of certain pathogens can be accomplished by varying the pH. Maintaining a pH above 10 is very effective at killing pathogens; lime stabilization of sludge is an example of this. The relationship between pH, alkalinity, and acidity and other important concepts are addressed in subsequent sections of this chapter.

Measurement of pH is accomplished by measuring the hydrogen ion activity (concentration) using a glass electrode and reference electrode. The procedures for measuring pH are presented in Standard Methods (1998).

3.5.2 Overview of alkalinity and acidity

The ability of a water system to resist a change in pH when an acid or base is added is an important, yet difficult, concept to comprehend. **Alkalinity** refers to the buffering capacity of water to resist a change in pH when an acid is added. Conversely, **acidity** is defined as water's ability to resist a change in pH when a base or alkaline substance is added. Both of these terms are important in water chemistry, especially in the design of water treatment systems, where chemicals such as aluminum sulfate ($Al_2(SO_4)_3 \cdot 18H_2O$), ferrous sulfate ($FeSO_4 \cdot 7H_2O$), or ferric chloride ($FeCl_3$) are added to water to coagulate organic compounds and colloidal particles that may include bacteria and color containing compounds.

During biological treatment of wastewater, the addition of chemicals to increase the alkalinity concentration is often necessary, so that the process of nitrification – the conversion of ammonia to nitrate – is not inhibited. Typically, alkaline reagents such as lime, in the form of calcium oxide (CaO) or sodium hydroxide (NaOH), may be added to increase the alkalinity. To better understand the chemistry behind alkalinity and acidity, a brief review of the carbonate buffering system and acid-base chemistry is required.

3.5.2.1 Acid-Base chemistry

As previously discussed in Chapter 2, in which several key chemistry concepts were reviewed, an acid and base are best defined by the Bronsted-Lowry theory. An acid is a substance that can give up or donate a hydrogen ion, H^+ (or proton, as it is frequently called) while a base is a substance that can take up or accept a hydrogen ion. A generalized chemical reaction illustrating the concept of an acid-base conjugate base pair is presented below:

$$HA + B^- \leftrightarrow HB + A^- \quad (3.42)$$

where:
HA = is an acid capable of donating a proton
B^- = is a base capable of accepting a proton
HB = is the new acid formed during the reaction
A^- = is a new base formed during the reaction.

The hydrogen ion donor, HA, and the new base that is formed, A^-, are called an acid-conjugate base pair. Similarly, HB and B^- also represent an acid-conjugate base pair.

Water can either serve as an acid or a base, depending on the chemical reaction. Substances that can function as both an acid and base are **amphoteric**. Recall the dissociation of water, as shown in the following reaction.

$$H_2O + H_2O \leftrightarrow H_3O^+ + OH^- \quad (3.43)$$

In aqueous solutions, the hydrogen ion or proton is more correctly written as H_3O^+ (hydrated proton or hydronium ion), since it is bounded rather strongly to a water molecule. According to Snoeyink & Jenkins (1980), there is strong evidence that there are three additional water molecules attached to the hydronium ion, such that the symbolic representation would be $H_9O_4^+$ or $H^+ \cdot 4H_2O$. Although not shown in Equation (3.43), the hydroxide ion, OH^- (or hydroxyl ion) is bonded or hydrated with water molecules so that the chemical formula is more accurately portrayed as $H_7O_4^-$or $OH^- \cdot 3H_2O$. In most texts however, the proton is simply denoted as H^+, while the hydroxyl ion is represented as OH^-; and the dissociation or auto-ionization of water is typically shown as Equation (3.44).

$$H_2O \leftrightarrow H^+ + OH^- \quad (3.44)$$

The equilibrium ionization constant for the dissociation of water can be expressed as:

$$(K_a)_{eq} = \frac{[H^+][OH^-]}{[H_2O]} = 1.8 \times 10^{-16}\,\text{mol/L} \quad (3.45)$$

At 25°C, $(K_a)_{eq}$ for the ionization of water has been found to be 1.8×10^{-16} moles/L (Butler, 1973; Benefield *et al.*, 1982). Knowing that the density of water at 25°C is essentially 1.0 g/ml, the molar concentration of water $[H_2O]$ can be determined as:

$$\left(\frac{1.0\,\text{g}\,H_2O}{\text{ml}}\right)\left(\frac{1000\,\text{ml}}{\text{L}}\right)\left(\frac{1\,\text{mol}\,H_2O}{18.0\,\text{g}\,H_2O}\right) = 55.6\,\text{mol/L} \quad (3.46)$$

In dilute solutions, the concentration of water $[H_2O]$ is virtually constant. With this knowledge, it is possible to determine the ion product of water, K_w.

$$K_w = (K_a)_{eq}\,[H_2O] \quad (3.47)$$

$$K_w = (1.8 \times 10^{-16}\,\text{mol/L})(55.6\,\text{mol/L}) = 1.0 \times 10^{-14} \quad (3.48)$$

Substitution of Equation (3.45) into Equation (3.47) shows that the dissociation of water in dilute solutions at 25°C can be expressed as:

$$\boxed{K_w = [H^+][OH^-]} \quad (3.49)$$

where:
K_w = the ion product of water or water constant
$\boldsymbol{K_w}$ = $\mathbf{10^{-14}}$ at 25°C
$[H^+]$ = hydrogen ion or proton concentration, moles/L
$[OH^-]$ = hydroxyl ion concentration, moles/L.

Table 3.4 shows that the water constant, K_w, is temperature dependent.

3.5.2.2 Carbonate System

In water chemistry applications, the **carbonate system** is of paramount importance and it also serves as the basis for the concepts of alkalinity and acidity. A more detailed discussion of the carbonate system may be found in Snoeyink & Jenkins (1980), Stumm & Morgan (1966), Henry & Heinke (1996) and Benefield *et al.* (1982). There are several chemical species that make up the carbonate system which control the pH of most natural waters. Listed below are the major species.

- Gaseous carbon dioxide, $CO_{2\,(g)}$
- Aqueous or dissolved carbon dioxide, $CO_{2(aq)}$
- Carbonic acid, H_2CO_3
- $H_2CO_3^* = H_2CO_{3(aq)} + CO_{2(aq)}$
- Bicarbonate, HCO_3^-
- Carbonate, CO_3^{2-}
- Calcium carbonate solid, $CaCO_{3\,(s)}$

Table 3.4 Temperature effects on K_w.

T, 0°C	K_w, moles/L
0	1.17×10^{-15}
10	2.95×10^{-15}
20	6.76×10^{-15}
[a]25	1.0×10^{-14}
30	1.48×10^{-14}
50	5.50×10^{-14}

Based on Stumm & Morgan (1996) *Aquatic Chemistry: Chemical Equilibria and Rates in Natural Waters*, p.57.
[a] The water ionization constant is typically quoted as 10^{-14} at 25°C.

Diffusion of gaseous carbon dioxide from the atmosphere into water, along with the release of CO_2 from biological processes from microorganisms and aquatic life, impact the concentration of aqueous CO_2 in water. The equilibrium concentration of aqueous CO_2 in water resulting from diffusion from the atmosphere can be determined using Henry's Law.

$$CO_{2(g)} \leftrightarrow CO_{2(aq)} \quad (3.50)$$

$$[(CO_2)_{(aq)}] = K_H P_g \quad (3.51)$$

where:
$[(CO_2)_{(aq)}]$ = concentration of aqueous CO_2, moles/L
K_H = Henry's law coefficient for CO_2
K_H = $10^{-1.47}$ moles/L·atm at 25°C
P_g = partial pressure of CO_2 in air, atm
P_g = typical pressure of CO_2 in air is $10^{-3.5}$ atm.

The dissolved or aqueous carbon dioxide forms aqueous carbonic acid according to the following reaction:

$$CO_{2(aq)} + H_2O \leftrightarrow H_2CO_{3(aq)} \quad (3.52)$$

Rearranging Equation (3.52) and noting that the concentration of water $[H_2O]$ is equal to 55.6 mol/L, the dissociation constant, K_m is solved as follows:

$$K_m = \frac{[H_2CO_{3(aq)}]}{[CO_{2(aq)}]} \quad (3.53)$$

where:
K_m = equilibrium dissociation constant for aqueous CO_2 in water, moles/L
$K_m = 10^{-2.8}$ moles/L at 25°C.

Since K_m is much less than "1", note that, in solution at 25°C, the concentration of $CO_{2(aq)}$ is much greater than the concentration of $H_2CO_{3(aq)}$. Analytically, the concentration of aqueous carbonic acid is difficult to distinguish from aqueous carbon dioxide, $CO_{2(aq)}$. Therefore, the convention is to use the hypothetical species called $H_2CO_3^*$, which is the sum of $[CO_{2(aq)}]$ and $[H_2CO_{3(aq)}]$.

$$[H_2CO_3^*] = [CO_{2(aq)}] + [H_2CO_{3(aq)}] \quad (3.54)$$

Equilibria for the remaining carbonate species can be shown as follows. The dissociation of aqueous carbonic acid is represented by the following reaction.

$$H_2CO_{2(aq)} \leftrightarrow H^+ + HCO_3^- \quad (3.55)$$

$$K_1' = \frac{[H^+][HCO_3^-]}{[H_2CO_{3(aq)}]} = 10^{-3.5} \text{ at } 25°C \quad (3.56)$$

where:
K_1' = equilibrium constant for the dissociation of aqueous carbonic acid
$K_1' = 10^{-3.5}$ at 25°C.

The dissociation of $H_2CO_3^*$ occurs as follows.

$$H_2CO_3^* \leftrightarrow H^+ + HCO_3^- \quad (3.57)$$

$$K_{a,1} = \frac{[H^+][HCO_3^-]}{[H_2CO_3^*]} \quad (3.58)$$

where $K_{a,1}$ equilibrium constant for the dissociation of $H_2CO_3^*$.

The equilibrium constant for the dissociation of $H_2CO_3^*$ can be derived by substituting Equation (3.54) into Equation (3.58) and multiplying the numerator and denominator of the equation by $[H_2CO_{3(aq)}]$. Equations (3.53) and (3.56) must also be used in the derivation to yield Equation (3.59).

$$K_{a,1} = \frac{[H^+][HCO_3^-]}{[H_2CO_{3(aq)}]}$$

$$= \frac{[H^+][HCO_3^-]}{[H_2CO_{3(aq)}] + [CO_{2(aq)}]} \times \frac{[H_2CO_{3(aq)}]}{[H_2CO_{3(aq)}]}$$

$$K_{a,1} = \frac{[H^+][HCO_3^-]/[H_2CO_{3(aq)}]}{([H_2CO_{3(aq)}] + [CO_{2(aq)}])/[H_2CO_{3(aq)}]}$$

$$K_{a,1} = \frac{[H^+][HCO_3^-]/[H_2CO_{3(aq)}]}{(1 + [CO_{2(aq)}]/[H_2CO_{3(aq)}])} = \frac{K_1'}{1 + \frac{1}{K_m}} \quad (3.59)$$

Notice that $1/K_m$ is much greater than 1 ($1/K_m = 1/10^{-2.8} = 631$). As a result, the 1 in the denominator of the previous equation is typically neglected, and an estimate for $K_{a,1}$ can be determined.

$$K_{a,1} = \frac{K_1'}{1 + \frac{1}{K_m}} = \frac{K_1'}{\frac{1}{K_m}} = K_1' \times K_m = 10^{-3.5} \times 10^{-2.8} = 10^{-6.3} \quad (3.60)$$

Comparing this result for $K_{a,1}$ with the equilibrium constant for $H_2CO_{3(aq)}$ ($K_1' = 10^{-3.5}$) shows that H_2CO_3 is a much stronger acid than $H_2CO_3^*$ (by a factor of 631 at 25°C and 1 bar).

The reaction for the dissociation of bicarbonate into a proton and carbonate ion occurs as follows.

$$HCO_3^- \leftrightarrow H^+ + CO_3^{2-} \quad (3.61)$$

$$K_{a,2} = \frac{[H^+][CO_3^{2-}]}{[HCO_3^-]} \quad (3.62)$$

where:
$K_{a,2}$ = equilibrium constant for dissociation of bicarbonate into carbonate
$K_{a,2} = 10^{-10.3}$ at 25°C.

The hydroxyl ion (OH^-) is produced when carbonate ions react with water.

$$CO_3^{2-} + H_2O \leftrightarrow HCO_3^- + OH^- \quad (3.63)$$

The pH dependence on the concentration of carbonate species in solution is evident from the equilibrium and dissociation equations [Equations (3.56), (3.58), (3.62)], because of the presence of H^+ in these equations. For example, when significant algal blooms occur, an increase in pH to values from 9 to 10 may occur, due to the consumption of bicarbonate shifting Equation (3.63) from left to right.

For simplicity, the fraction of carbonate species in solution is often expressed as an **α value**, as shown in Equations (3.64 to 3.66). In the equations, α_0 correlates to $[H_2CO_3^*]$, α_1 relates to $[HCO_3^-]$, and α_2 relates to $[CO_3^{2-}]$, all mole fractional solution concentrations.

$$\alpha_0 = \frac{[H_2CO_3^*]}{[H_2CO_3^*] + [HCO_3^-] + [CO_3^{2-}]} = \frac{[H_2CO_3^*]}{C_T} \quad (3.64)$$

$$\alpha_1 = \frac{[HCO_3^-]}{[H_2CO_3^*] + [HCO_3^-] + [CO_3^{2-}]} = \frac{[HCO_3^-]}{C_T} \quad (3.65)$$

$$\alpha_2 = \frac{[CO_3^{2-}]}{[H_2CO_3^*] + [HCO_3^-] + [CO_3^{2-}]} = \frac{[CO_3^{2-}]}{C_T} \quad (3.66)$$

$$C_T = [H_2CO_3^*] + [HCO_3^-] + [CO_3^{2-}] \quad (3.67)$$

If a solid carbonate species, such as calcium carbonate, $CaCO_{3(s)}$, is present, then the following equation and solubility product are applicable:

$$CaCO_{3(s)} \leftrightarrow Ca^{2+} + CO_3^{2-} \quad (3.68)$$

$$[Ca^{2+}][CO_3^{2-}] = K_{sp} = 10^{-8.34} \quad (3.69)$$

where:

$CaCO_{3(s)}$ = is the solid carbonate species
$[Ca^{2+}]$ = concentration of calcium ions in solution, moles/L
$[CO_3^{2-}]$ = concentration of carbonate ions in solution, moles/L
K_{sp} = solubility product constant for $CaCO_{3(s)} = 10^{-8.34}$ at 25°C.

In dealing with the carbonate buffering system, there are four major systems that may be encountered in water chemistry:

1. An open system with no solid or precipitate present.
2. A closed system with no solid or precipitate present.
3. An open system with a solid or precipitate present.
4. A closed system with a solid or precipitate present.

An open system refers to one in which carbon dioxide is freely able to diffuse into and out of the water, i.e., it is exposed to the atmosphere. In a closed system, carbon dioxide cannot diffuse into water. A detailed discussion of typical open/closed systems encountered during water and wastewater treatment may be found in Stumm & Morgan (1996), Henry & Heinke (1996), and Snoeyink & Jenkins (1980).

3.5.3 Alkalinity

Alkalinity is water's capacity to neutralize strong acids. In natural water systems, the major chemicals attributed to alkalinity include bicarbonate, carbonate, and hydroxide ions. Other chemical constituents contributing to alkalinity which are present at low concentrations include silicates ($HSiO_3^-$), borates ($H_2BO_3^-$), phosphates (HPO_4^{2-}, $H_2PO_4^-$), hydrogen sulfide (HS^-), and ammonia (NH_3).

Constituents causing alkalinity primarily result from dissolution of mineral substances by water, due to weathering and passage through the soil and other geologic formations. Phosphates from detergents in wastewater discharges or from fertilizers and insecticides from agricultural stormwater runoff are the main contributors of alkalinity in surface water. Transfer of carbon dioxide from the atmosphere along with microbial and aquatic life respiration releases CO_2 into water. The decomposition of organic matter by microorganisms also releases hydrogen sulfide and ammonia into water. Mathematically, alkalinity is defined as follows:

$$\text{Alkalinity}(eq/L) = [HCO_3^-] + 2[CO_3^{2-}] + [OH^-] - [H^+] \quad (3.70)$$

where [ ] denotes the molar concentration of the various species.

Analytically, alkalinity is determined by titrating a sample of water with a standard strong acid solution, such as 0.02 *N* H_2SO_4, to an approximate endpoint of 4.5 to 4.8. If a 0.02 *N* diprotic acid is used during titration, 1 ml of titrant corresponds to 1 mg of alkalinity as $CaCO_3$. During the titration, the following series of reactions are assumed to reach completion at their associated pH endpoints. The pH values listed in Equations (3.71 to 3.73) are valid if the alkalinity in the system is derived primarily from the carbonate system.

$$OH^- + H^+ \rightarrow H_2O; \quad pH_{CO_3^{2-}} \approx 10.3 \quad (3.71)$$

$$CO_3^{2-} + H^+ \rightarrow HCO_3^-; \quad pH_{HCO_3^-} \approx 8.3 \quad (3.72)$$

$$HCO_3^- + H^+ \rightarrow H_2CO_3^*; \quad pH_{H_2CO_3^*} \approx 4.5 \quad (3.73)$$

At the onset of titration, add phenolphthalein indicator to the sample. If no pink occurs, the sample pH is less than 8.3 and all the alkalinity in the sample is in the form of bicarbonate. If the initial pH of the sample is greater than approximately 10.3, titrant is added to complete the reaction represented by Equation (3.71). At this endpoint, essentially all the alkalinity can be represented as CO_3^{2-}, caustic alkalinity. Because the $pH_{CO_3^{2-}}$ endpoint is solution-dependent, caustic alkalinity is not accurately determined by titration. Caustic alkalinity can be determined algebraically, however, if the sample's pH is known.

As titration continues, phenolphthalein indicates the acidification of carbonate to bicarbonate in solution and continues until Equation (3.72) has gone to completion. The solution will change from pink to colorless, as a result of the phenolphthalein indicator, to signify a solution pH of approximately 8.3. At this endpoint, phenolphthalein alkalinity (PA) is indicated and the volume of titrant should be noted. Should the original pH of a sample be below 8.3, the sample contains no PA. The hydroxyl and carbonate ions are essentially the primary species that contribute to phenolphthalein alkalinity.

From titration results, the carbonate (pH ~8.3) and bicarbonate (pH ~4.5) endpoints can be identified by using the inflection point titration (IPT) or the Gran function plot (Gran) method (Rounds, 2006; Gran, 1952). The IPT method is adequate for most waters. When using the IPT method, titrate cautiously on both sides of the expected equivalence points. For example, if the initial sample pH is greater than 8.1, titrate carefully until reaching a pH of approximately 8.0. Use small increments of acid and record both titrant volume and resulting solution pH. At this point, titration can be conducted rapidly, using relatively large acid additions, until a pH of approximately 5.5 is reached. Carefully collect titration data from pH = 5.5 to a sample pH = 3.5 to 4.0.

With an increase in organic acid concentration with respect to carbonate species, or with a decrease in alkalinity, the IPT method may be difficult to use. In situations where the expected total carbonate alkalinity is low (< approximately 4 meq/L (20 mg/L as $CaCO_3$)), in sea water, when the conductivity is less than 100 μS/cm, or organic compounds are appreciable, the Gran method should be used. If carbonate speciation is desirable, titrate carefully throughout the entire pH range. A good rule of thumb is to collect data for every 0.2 to 0.3 pH units. Additional titration methodology is available in the USGS *National Field Manual for the Collection of Water-Quality Data* (Rounds, 2006).

Alkalinity is customarily expressed in terms of mg/L as $CaCO_3$. Equation (3.74) is used to calculate the phenolphthalein alkalinity (PA) based on titration of a water sample with standard 0.02 *N* sulfuric acid.

$$\text{P.A.}\left(\frac{\text{mg CaCO}_3}{\text{L}}\right) = \frac{(\text{ml } 0.02\,N\ \text{H}_2\text{SO}_4 \text{ to reach pH 8.3}) \times 1000}{\text{ml sample}} \quad (3.74)$$

Titration to the methyl-orange endpoint of pH ~4.5, identified by a change in solution color from yellow to red, is used to indicate completion of the acidification reaction that converts bicarbonate into aqueous carbon dioxide or $H_2CO_3^*$, as shown in Equation (3.73). Bromocresol-green is also commonly used to indicate this endpoint, with a change in solution color from yellow to red as the reaction reaches completion.

If the initial pH of the solution is less than 8.3, the sample contains only Methyl Orange Alkalinity (M.O.A.). If standard 0.02 *N* sulfuric acid is used for the titration, the M.O.A. can be easily determined as shown in Equation (3.75).

$$\text{M.O}\cdot\text{A}\cdot\left(\frac{\text{mg CaCO}_3}{\text{L}}\right) = \frac{\left(\begin{array}{c}\text{ml } 0.02N\ \text{H}_2\text{SO}_4 \text{ to lower}\\ \text{solutution pH from 8.3 to 4.5}\end{array}\right) \times 1000}{\text{ml sample}} \quad (3.75)$$

Titration to the pH = 4.5 endpoint defines total alkalinity. The actual pH endpoint for the titration depends upon the total alkalinity in the sample and the types of constituents. *Standard Methods* (1998) provides the specific endpoint pH values for various concentrations of total alkalinity; remember to use either the IPT or the Gran method to determine the endpoints for titration more specifically.

Equation (3.76) is used to calculate the **Total Alkalinity (ALK)**:

$$\text{ALK}\left(\frac{\text{mg CaCO}_3}{\text{L}}\right) = \frac{(\text{Total ml } 0.02\,N\ \text{H}_2\text{SO}_4 \text{ to reach pH 4.5}) \times 1000}{\text{ml sample}} \quad (3.76)$$

The methyl orange alkalinity will equal the total alkalinity when the initial pH of the sample is below 8.3 and there is no phenolphthalein alkalinity. An example is presented below to illustrate how alkalinity can be calculated from titration data.

Example 3.13 Calculation of alkalinity from titration data

A 100 ml sample of water from a mountain spring was titrated with 0.02 *N* sulfuric acid. 10 ml of standard sulfuric acid were added to the sample to reach a pH of 8.3. An additional 25 ml of standard sulfuric acid were added to reach a pH of 4.5. Calculate the phenolphthalein, methyl orange, and total alkalinities in mg/L as $CaCO_3$ for the water sample.

First, substitute the ml of titrant used to reach the phenolphthalein endpoint into Equation (3.74).

$$\text{P.A.}\left(\frac{\text{mg CaCO}_3}{\text{L}}\right) = \frac{\left(\begin{array}{c}\text{ml } 0.02\ N\,\text{H}_2\text{SO}_4\\ \text{to reach pH 8.3}\end{array}\right) \times 1000}{\text{ml sample}}$$

$$\text{P.A.}\left(\frac{\text{mg CaCO}_3}{\text{L}}\right) = \frac{(10\text{ ml}) \times 1000}{100\text{ ml}} = \boxed{100}$$

Next, the methyl orange alkalinity may be calculated by using Equation (3.75) to use the milliliters of standard acid consumed during the titration from pH 8.3 to 4.5.

$$\text{M.O.A.}\left(\frac{\text{mg CaCO}_3}{\text{L}}\right) = \frac{\left(\begin{array}{c}\text{ml 0.02 N H}_2\text{SO}_4\\ \text{to reach pH 4.5}\end{array}\right) \times 1000}{\text{ml sample}}$$

$$\text{M.O.A.}\left(\frac{\text{mg CaCO}_3}{\text{L}}\right) = \frac{(25\text{ ml}) \times 1000}{100\text{ ml}} = \boxed{250}$$

Finally, the total alkalinity is determined by summing the phenolphthalein and methyl orange alkalinities.

$$\text{Total Alkalinity} = 100 + 250 = \boxed{350\frac{\text{mg}}{\text{L}}\text{ as CaCO}_3}$$

Alternatively, the total alkalinity could be calculated by substituting into Equation (3.76):

$$\text{ALK}\left(\frac{\text{mg CaCO}_3}{\text{L}}\right) = \frac{\left(\begin{array}{c}\text{total ml 0.02 NH}_2\text{SO}_4\\ \text{to reach pH 4.5}\end{array}\right) \times 1000}{\text{ml sample}}$$

$$\text{ALK}\left(\frac{\text{mg CaCO}_3}{\text{L}}\right) = \frac{(10 + 25\text{ ml}) \times 1000}{100\text{ ml}} = \boxed{350}$$

3.5.3.1 *Speciation of Alkalinity–Hydroxide, Carbonate, and Bicarbonate*

The results obtained from a titration analysis where the total and phenolphthalein alkalinities are determined allow for the classification of three forms of alkalinity that are often related to many types of water and wastewater systems. As suggested by Equations (3.71)–(3.73), total alkalinity is the sum of the caustic (or hydroxide), carbonate, and bicarbonate alkalinities. The sample equations in Table 3.5 are derived from a mass balance on the system, and these can be used to transform titration results into alkalinity concentrations that are meaningful. These tabulated equations are generalized such that the alkalinity titration can be conducted, and results quantified, using any standardized strong acid solution.

If the sample alkalinity is derived predominantly from carbonate species, and by recognizing a set of additional assumptions detailed below, concentration of specific carbonate species can be easily approximated from laboratory data.

1. Alkalinity is present and the concentration of H^+ ions is insignificant.
2. At the phenolphthalein endpoint (pH = 8.3), half of the $[CO_3^{2-}]$ has been neutralized, i.e., Equation (3.72) has gone to completion.
3. The hydroxide ion and bicarbonate ions are incompatible in a system and therefore do not exist simultaneously.
4. When PA exists and is less than the ALK, carbonate alkalinity is present.
5. When PA exists and is more than one-half the ALK, hydroxide alkalinity is present.
6. When PA exists and is less than one-half the ALK, bicarbonate alkalinity is present.
7. When no PA exists, all the alkalinity is assumed to be bicarbonate alkalinity.

Table 3.5 Alkalinity definition and relationships used to transform titration results into meaningful alkalinity concentrations.

Alkalinity definition	Governing equation (eq/L)[a]
Caustic or hydroxide alkalinity = $[OH^-]$	$[OH^-] = \frac{K_w}{[H^+]} = 10^{(pH-pK_w)}$
Phenolphthalein alkalinity (P.A.) = $\alpha_2 C_T + [OH^-]$ (equivalence of acid required to bring the sample to a pH = 8.3)	$\text{P.A.} = \frac{V_p \times N}{V}$
Carbonate alkalinity = $2\alpha_2 C_T$ (alkalinity contribution from CO_3^{2-})	$CO_3^{2-}\text{ alkalinity} = 2\left(\frac{V_p \times N}{V} - \frac{K_w}{[H^+]}\right)$
Bicarbonate alkalinity = $\alpha_1 C_T$ (= total – carbonate alkalinities)	$HCO_3^-\text{ alkalinity} = \frac{(V_{end} - 2V_p)N}{V} + \frac{K_w}{[H^+]}$
Total alkalinity = ALK = $(2\alpha_2 + \alpha_1)C_T + [OH^-] - [H^+]$ (If sample pH < 10.5, Total ALK ~ carbonate + bicarbonate alkalinities)	$\text{ALK} = \frac{V_{end} \times N}{V}$

Key:
[a] To report as mg/L $CaCO_3$, multiply results measured in eq/L by 50,000.
V_p = volume of titrant (ml) used to reach the phenolphthalein endpoint (pH = 8.3).
V_{end} = volume of titrant (ml) used to reach the methyl orange endpoint (pH = 4.3).
V = initial volume of sample (ml).
N = normality of strong acid titrant.

Table 3.6 From titration results, five conditions exist that dictate the approximate concentration of alkalinity species.

Approximate concentration of alkalinity species

Results	**Alkalinity species present**	**Species (moles/L)**[a]
$V_p = 0$ (P.A. = 0; all the alkalinity is HCO_3^-)	HCO_3^-	$[HCO_3^-] = V_{mo}(N/V)$
$V_{mo} = 0$ (All the alkalinity is OH^-)	OH^-	$[OH^-] = V_p(N/V)$
$V_{mo} = V_p$ (P.A. = $\frac{1}{2}$ ALK; all the alkalinity is carbonate)	CO_3^{2-}	$[CO_3^{2-}] = V_p(N/V)$
$V_p < V_{mo}$ (P.A. is less than $\frac{1}{2}$ALK)	CO_3^{2-} HCO_3^-	$[CO_3^{2-}] = V_p(N/V)$ $[HCO_3^-] = (V_{mo} - V_p)(N/V)$
$V_{mo} < V_p$ (P.A. is greater than $\frac{1}{2}$ ALK)	OH^- CO_3^{2-}	$[OH^-] = (V_p - V_{mo})(N/V)$ $[CO_3^{2-}] = V_{mo}(N/V)$

Key:
[a] To report as mg/l $CaCO_3$, convert from moles/L to eq/L and multiply by 50,000.
V_p = volume of titrant (ml) used to reach the phenolphthalein endpoint (pH = 8.3).
V_{mo} = volume of titrant (ml) used to reach the methyl orange endpoint (pH = 4.3) from the phenolphthalein endpoint (pH = 8.3).
V = sample volume (ml).
N = normality of titrant.
Based on Snoeyink and Jenkins (1980). *Water Chemistry*, p. 176.

8 When titration has proceeded to the carbonic acid equivalence point (pH = 4.5), $H_2CO_3^*$ is the predominant species in solution.

If the above conditions suitably describe the water system, and if the phenolphthalein and total alkalinity titration endpoints are known, the equations presented in Table 3.6 can be used to approximate the hydroxide, carbonate, and bicarbonate alkalinities. Five conditional possibilities are summarized in the table.

Example 3.14 Alkalinity determination from laboratory data

To determine the alkalinity of a river water sample, a titration method was used in the laboratory. The titrant used was 0.02 N H_2SO_4. The initial sample pH = 9.8 and a 50 ml sample volume was used in the titration. The phenolphthalein endpoint (pH = 8.3) was reached after 3.2 ml of acid were added. An additional 6.3 ml of acid were required to reach a pH = 4.5, corresponding to the bromocresol green endpoint. Use these laboratory data:

a) to identify prevalent alkalinity species present in the sample and to estimate their concentrations; and
b) to determine total alkalinity.

Present the results in eq/L and mg/L $CaCO_3$.

Solution part a

From titration results, we know that $V_p = 3.2$ ml and $V_{mo} = 6.3$ ml. From Table 3.6, we know that, when $V_p < V_{mo}$, the primary alkalinity species are CO_3^{2-} and HCO_3^-. Determining the approximate concentration of each species gives:

$$[CO_3^{2-}] = V_p\left(\frac{N}{V}\right) = \left(3.2\,\text{ml}\left(\frac{0.02\,\text{eq/L}}{50\,\text{ml}}\right)\right)$$
$$\times 0.0013\frac{\text{mol}\,CO_3^{2-}}{\text{L}}$$

Converting units into eq/L and mg/L as $CaCO_3$ is shown as:

$$0.0013\frac{\text{mol}\,CO_3^{2-}}{\text{L}} \times \frac{2\,\text{eq}}{\text{mol}\,CO_3^{2-}} = 0.0026\frac{\text{eq}}{\text{L}} \text{ or } 2.6\frac{\text{meq}}{\text{L}}$$

$$2.6\frac{\text{meq}}{\text{L}} \times \frac{50\,\text{mg}\,CaCO_3}{\text{meq}} = \boxed{130\frac{\text{mg}}{\text{L}} \text{ as } CaCO_3}$$

Similarly, for the bicarbonate species:

$$[HCO_3^-] = (V_{mo} - V_p)\left(\frac{N}{V}\right) = (6.3 - 3.2)\text{ml}\left(\frac{0.02\,\text{eq/L}}{50\,\text{ml}}\right)$$
$$= 0.0012\frac{\text{mol}\,HCO_3^-}{\text{L}}$$

Converting to mg/L as $CaCO_3$ gives:

$$0.0012\frac{\text{mol } HCO_3^-}{\text{L}} \times \frac{1 \text{ eq}}{1 \text{ mol } HCO_3^-} \times \frac{50 \text{ g } CaCO_3}{\text{eq}} \times \frac{1000 \text{ mg}}{\text{g}} = \boxed{60\frac{\text{mg}}{\text{L}} \text{ as } CaCO_3}$$

Solution part b

The total alkalinity is equal to the equivalence of titrant required to reach a pH = 4.5.

$$\text{Total Alkalinity} = (V_{mo} + V_p)\left(\frac{N}{V}\right) = (6.3 + 3.2)\text{ml} \times \left(\frac{0.02 \text{ eq/L}}{50 \text{ ml}}\right) = 0.0038\frac{\text{eq}}{\text{L}}$$

Converting gives:

$$0.0038\frac{\text{eq}}{\text{L}} \times \frac{50 \text{ g } CaCO_3}{\text{eq}} \times \frac{1000 \text{ mg}}{\text{g}} = \boxed{190\frac{\text{mg}}{\text{L}} \text{ as } CaCO_3}$$

Notice that the total alkalinity is equal to the sum of the carbonate and bicarbonate alkalinities.

3.5.3.2 Advanced Speciation of Hydroxide, Carbonate, and Bicarbonate Alkalinity

The various forms of alkalinity may also be determined using the carbonate equilibrium expressions plus total alkalinity and pH measurements. Figure 3.6 shows how the various forms of alkalinity are affected by pH. This figure was developed for a total alkalinity of 100 mg/L as $CaCO_3$ at 25°C. The hydroxide alkalinity is calculated from an initial pH measurement and using the dissociation constant (K_w) for water.

Rearranging and substituting K_w and the hydrogen ion concentration into Equation (3.49), while applying appropriate conversions, yields Equation (3.77):

$$[H^+][OH^-] = K_w$$

$$[OH^-] = \frac{K_w}{[H^+]} = \frac{K_w}{10^{-pH}} = K_w 10^{pH} = 10^{-pK_w} 10^{pH} = 10^{(pH-pK_w)}$$

$$\text{Hydroxide Alkalinity} = 10^{(pH-pK_w)} \times \frac{17 \text{ g } OH^-}{\text{L}} \times \frac{1 \text{ eq}}{17 \text{ g } OH^-} \times \frac{50 \text{ g } CaCO_3}{\text{eq}} \times \frac{1000 \text{ mg}}{\text{g}}$$

$$\text{Hydroxide Alkalinity} = 50{,}000 \times 10^{(pH-pK_w)} \quad (3.77)$$

where hydroxide alkalinity is expressed in mg/L as $CaCO_3$.

The carbonate alkalinity can be calculated by simultaneous solution of Equations (3.49), (3.62), and (3.70), along with a

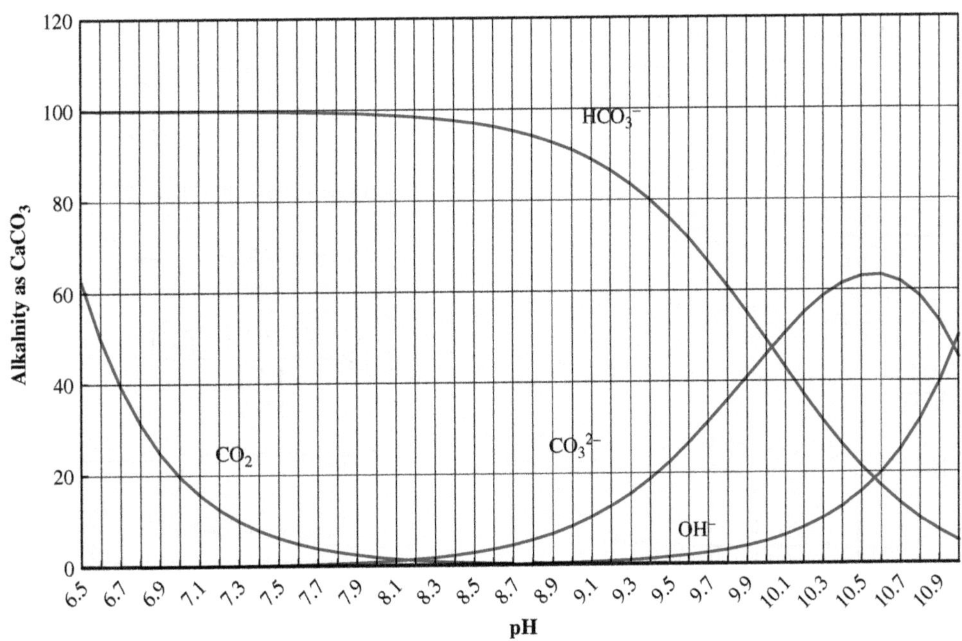

Figure 3.6 Relationship between alkalinity species and pH.

determination of hydrogen ion concentration from a pH measurement, and knowing the total alkalinity.

$$ALK(eq/L) = [HCO_3^-] + 2[CO_3^{2-}] + [OH^-] - [H^+]$$

$$\frac{[H^+][CO_3^{2-}]}{[HCO_3^-]} = K_{a,2} \text{ or } [HCO_3^-] = \frac{[H^+][CO_3^{2-}]}{K_{a,2}}$$

$$[OH^-] = \frac{K_w}{[H^+]}$$

$$ALK = \frac{[H^+][CO_3^{2-}]}{K_{a,2}} + 2[CO_3^{2-}] + \frac{K_w}{[H^+]} - [H^+]$$

$$ALK = [CO_3^{2-}]\left[\frac{[H^+]}{K_{a,2}} + 2\right] + \frac{K_w}{[H^+]} - [H^+]$$

$$[CO_3^{2-}] = \frac{ALK + [H^+] - \frac{K_w}{[H^+]}}{\frac{[H^+]}{K_{a,2}} + 2} = \frac{ALK + [H^+] - \frac{K_w}{[H^+]}}{\frac{[H^+]}{2K_{a,2}} + 1}$$

$$\text{Carbonate Alkalinity}\left(\frac{\text{mg CaCO}_3}{\text{L}}\right) = (50{,}000) \times \frac{ALK + [H^+] - \frac{K_w}{[H^+]}}{\frac{[H^+]}{2K_{a,2}} + 1} \quad (3.78)$$

where:
ALK = Total alkalinity in equivalents per liter
$K_{a,2} = 10^{-10.3}$ at 25°C

Carbonate alkalinity expressed in mg/L as $CaCO_3$.

An equation for calculating the bicarbonate alkalinity may be derived by substituting Equation (3.62) into Equation (3.70), along with the relationship between hydrogen ion concentration and pH, if the total alkalinity is known.

$$ALK(eq/L) = [HCO_3^-] + 2[CO_3^{2-}] + [OH^-] - [H^+]$$

$$K_{a,2} = \frac{[H^+][CO_3^{2-}]}{[HCO_3^-]}$$

$$[CO_3^{2-}] = \frac{[HCO_3^-]K_{a,2}}{[H^+]}$$

$$[OH^-] = \frac{K_w}{[H^+]}$$

$$ALK = [HCO_3^-] + \frac{2K_{a,2}[HCO_3^-]}{[H^+]} + \frac{K_w}{[H^+]} - [H^+]$$

$$ALK = [HCO_3^-]\left[1 + \frac{2K_{a,2}}{[H^+]}\right] + \frac{K_w}{[H^+]} - [H^+]$$

$$[HCO_3^-] = \frac{ALK + [H^+] - \frac{K_w}{[H^+]}}{1 + \frac{2K_{a,2}}{[H^+]}} = \frac{ALK + [H^+] - \frac{K_w}{[H^+]}}{1 + \frac{2K_{a,2}}{[H^+]}}$$

$$\text{Bicarbonate Alkalinity}\left(\frac{\text{mg CaCO}_3}{\text{L}}\right) = (50{,}000) \times \frac{ALK + [H^+] - \frac{K_w}{[H^+]}}{1 + \frac{2K_{a,2}}{[H^+]}} \quad (3.79)$$

Example 3.15 Calculating the various forms of alkalinity using pH and alkalinity measurements with equilibrium chemistry

The total alkalinity of a softened water is 100 mg/L as $CaCO_3$. The initial pH of the sample is 9.5 and the temperature is 25°C. Determine the hydroxide, carbonate, and bicarbonate alkalinities of the softened water in terms of mg/L as $CaCO_3$.

The hydroxide alkalinity is calculated from the pH measurement and substituting into Equation (3.77).

$$\text{Hydroxide Alkalinity} = 50{,}000 \times 10^{(pH - pK_w)}$$

$$\text{Hydroxide Alkalinity} = 50{,}000 \times 10^{(9.5-14.0)}$$

$$= \boxed{1.58 \frac{\text{mg}}{\text{L}} \text{ as CaCO}_3}$$

The carbonate alkalinity is calculated from Equation (3.78). First, the total alkalinity must be converted to equivalents per liter (eq/L) as follows:

$$100 \frac{\text{mg CaCO}_3}{\text{L}} \times \frac{1 \text{ g}}{1000 \text{ mg}} \times \frac{1 \text{ eq}}{50 \text{ g CaCO}_3} = 2.00 \times 10^{-3} \frac{\text{eq}}{\text{L}}$$

$$pH = -\text{Log}[H^+][H^+] = 10^{-9.5} \qquad K_{a,2} = 10^{-10.3}$$

$$\text{Carbonate Alkalinity}\left(\frac{\text{mg CaCO}_3}{\text{L}}\right) = (50{,}000) \times \frac{ALK + [H^+] - \frac{K_w}{[H^+]}}{\frac{[H^+]}{2K_{a,2}} + 1}$$

$$\text{Carbonate Alkalinity}\left(\frac{\text{mg CaCO}_3}{\text{L}}\right) = (50{,}000) \times \frac{2.00 \times 10^{-3} + 10^{-9.5} - \frac{10^{-14}}{10^{-9.5}}}{\frac{10^{-9.5}}{2 \times 10^{-10.3}} + 1}$$

$$\text{Carbonate Alkalinity}\left(\frac{\text{mg CaCO}_3}{\text{L}}\right) = \boxed{23.7}$$

The bicarbonate alkalinity is calculated from Equation (3.79).

$$\text{Bicarbonate Alkalinity}\left(\frac{\text{mg CaCO}_3}{\text{L}}\right) = (50{,}000)\frac{\text{ALK} + [\text{H}^+] - \dfrac{K_w}{[\text{H}^+]}}{1 + \dfrac{2K_{a,2}}{[\text{H}^+]}}$$

$$\text{Bicarbonate Alkalinity}\left(\frac{\text{mg CaCO}_3}{\text{L}}\right) = (50{,}000)\frac{2.00 \times 10^{-3} + 10^{-9.5} - \dfrac{10^{-14}}{10^{-9.5}}}{\dfrac{2 \times 10^{-10.3}}{10^{-9.5}} + 1}$$

$$\text{Bicarbonate Alkalinity}\left(\frac{\text{mg CaCO}_3}{\text{L}}\right) = \boxed{74.7}$$

Finally, perform a check of the calculations to see if the hydroxide, carbonate, and bicarbonate alkalinities add up to 100 mg/L as $CaCO_3$, the total alkalinity of the water.

$$\text{Total Alkalinity} = 100\frac{\text{mg CaCO}_3}{\text{L}} = \text{OH}^- + \text{CO}_3^{2-} + \text{HCO}_3^-$$

$$\text{Total Alkalinity} = 100\frac{\text{mg CaCO}_3}{\text{L}} \cong 1.58 + 23.7 + 74.7$$

$$= \boxed{99.98\frac{\text{mg}}{\text{L}} \text{ as CaCO}_3} \text{ Checks}$$

3.5.4 Nitrogen

Nitrogen is an essential nutrient for all living organisms. The most common forms of nitrogen are as follows: ammonia (NH_3, valence of N^{3-}), ammonium ($NH_4{}^+$, valence of N^{3-}), nitrogen gas (N_2, valence of N^0), nitrite ion (NO_2^-valence of N^{3+}), and nitrate ion (NO_3^-, valence of N^{5+}). Most bacteria prefer to assimilate ammonium nitrogen and algae typically use nitrate. Some bacteria (*cyanobacteria*) are capable of fixing nitrogen gas from the atmosphere to form ammonium salts.

When present in excessive concentrations in surface waters, nitrogen contributes to abundant algal growth or cultural eutrophication. Nitrogen in the form of nitrate, found in fertilizers, enters the water supply through agricultural and urban runoff. High concentrations of ammonia and nitrate are often found in receiving waters, due to industrial and domestic wastewater treatment discharges. The discharge of ammonia to surface waters can stimulate autotrophic bacteria to oxidize ammonia to nitrate, a process called nitrification, which can lead to anaerobic conditions.

Three processes that have been successfully used for removing ammonia from wastewater are: biological nitrification followed by biological denitrification; ammonia stripping; and ion exchange. Details for designing these processes can be found in Metcalf & Eddy (2003), Reynolds & Richards (1996), EPA (1993), and WEF(1998).

Total **Kjeldahl nitrogen (TKN)** is the sum of the ammonia and organic nitrogen, and can be expressed mathematically as:

$$\text{TKN} = \text{NH}_3 - \text{N} + \text{Organic} - \text{N} \quad (3.80)$$

Organic compounds that contain nitrogen are called organic nitrogen, and they typically have a negative 3 valence (N^{3-}). Proteins, amino acids, and urea (NH_2CONH_2) are forms of organic nitrogen. During degradation of organic nitrogen through deamination reactions, ammonia is released.

The two oxidized forms of nitrogen of primary concern are nitrite and nitrate. Nitrification is routinely modeled as a two-step biological process by which *Nitrosomonas* oxidizes ammonia to nitrite and *Nitrobacter* oxidizes nitrite to nitrate. It is common to see the **total oxidized forms** of **nitrogen** represented as (**NO_x**), which is the sum of NO_2^- and NO_3^-. The primary drinking water standard for nitrate is 10 mg/L, as N because it is linked to infant methemoglobinemia (blue baby syndrome).

The **total nitrogen** (**TN**) concentration in a sample consists of the TKN and oxidized nitrogen as shown below.

$$\text{Total nitrogen (TN)} = \text{TKN} + \text{NO}_X \quad (3.81)$$

The ammonium ion exists in equilibrium with ammonia according to the following equilibrium expression. At pH values normally encountered in environmental engineering, the ammonium ion is the predominant species whereas, ammonia predominates at pH values greater than 10.

$$\text{NH}_4^+ \leftrightarrow \text{NH}_3 + \text{H}^+ \quad (3.82)$$

$$\frac{[\text{NH}_3]\,[\text{H}^+]}{[\text{NH}_4^+]} = K_a = 10^{-9.25} \text{ at } 25°\text{C} \quad (3.83)$$

$$\text{NH}_3(\%) = \frac{[\text{NH}_3]\ 100}{[\text{NH}_3] + [\text{NH}_4^+]} = \frac{100}{1 + [\text{NH}_4^+]/[\text{NH}_3]} \quad (3.84)$$

If the pH of an aqueous system at equilibrium is known, the percentage of the nitrogen in the form of ammonia can be determined.

$$\text{NH}_3\,(\%) = \frac{100}{1 + [\text{H}^+]/K_a} = \frac{100}{1 + [\text{H}^+]/10^{-9.25}} \quad (3.85)$$

$$[\text{NH}_3 - \text{N}] = [\text{NH}_4^+] + [\text{NH}_3] \quad (3.86)$$

where $[NH_3 - N]$ = concentration of ammonium and ammonia which is obtained from performing an ammonia nitrogen analysis according to Standard Methods (1998).

Ammonia nitrogen is measured by a colorimetric procedure using Nessler's Reagent (potassium mercuric iodine, K_2HgI_4). Nessler's Reagent combines with NH_3 to form a yellowish-brown colloid. The procedures for performing the various types of nitrogen analyses are found in Section 4500-*N* of *Standard Methods* (1998).

3.5.5 Phosphorus

Phosphorus is a nutrient required by all living organisms. It, too, is linked to cultural eutrophication of surface waters if present in excessive amounts. Phosphorus enters the water supply from phosphorus-based detergents, corrosion inhibitors added to drinking water systems, and domestic and industrial wastewater discharges. The **total phosphorus** concentration (**TP**) is the sum of the organic-phosphorus (**organic-P**) and inorganic- phosphorus (**inorganic-P**).

$$\text{Total Phosphorus (TP)} = \text{Inorganic-P} + \text{Organic-P} \quad (3.87)$$

Organic-P is typically only a minor consideration, and it is found in proteins and amino acids. Inorganic-P species include orthophosphates, polyphosphates, and metaphosphates. Polyphosphates and metaphosphates are also known as condensed phosphates. Condensed phosphates and organic phosphorus must be converted to orthophosphate before they can be measured.

3.5.5.1 Orthophosphate

Principal orthophosphates include: trisodium phosphate (Na_3PO_4), disodium phosphate (Na_2HPO_4), monosodium phosphate (NaH_2PO_4), and diammonium phosphate $(NH_4)_2HPO_4$. Orthophosphates are used by microorganisms and can be precipitated out of solution during wastewater treatment by adding chemicals such as lime or alum. Enhanced biological phosphorus removal (EBPR) processes may also be used for the removal of orthophosphate from wastewater. Design of chemical and biological processes for the removal of phosphorus from wastewater are addressed in Metcalf & Eddy (2003), WEF (1998), and EPA (1987).

Orthophosphate is typically measured by one of three colorimetric methods. The phosphate ion (PO_4^{3-}) combines with ammonium molybdate under acidic conditions to form molybdophosphate complex according to Equation (3.89). The blue-colored sol produced by the addition of stannous chloride is proportional to the amount of phosphate present, as illustrated qualitatively in Equation (3.90).

$$PO_4^{3-} + 12(NH_4)_2MoO_4 + 24H^+ \rightarrow (NH_4)_3PO_4 \cdot 12MoO_3 + 21NH_4^+ + 12H_2O \quad (3.89)$$

$$(NH_4)_3PO_4 \cdot 12MoO_3 + Sn^{2+} \rightarrow (\text{Molybdenum blue}) + Sn^{4+} \quad (3.90)$$

3.5.5.2 Condensed Phosphate

Major condensed polyphosphates include: sodium hexametaphosphate [$Na_3(PO_4)_6$], sodium tripolyphosphate [$Na_5P_3O_{10}$]; and tetrasodium pyrophosphate [$Na_4P_2O_7$]. Condensed phosphates are converted to orthophosphates by boiling samples that have been acidified with H_2SO_4 for 90 minutes. The orthophosphate formed from the condensed phosphates is measured in the presence of the original orthophosphate concentration in the sample by one of the colorimetric procedures, yielding the total inorganic phosphate concentration. The concentration of condensed phosphate is then determined by difference using Equation (3.91).

$$\text{Condensed Phosphate} = \text{Total Inorganic Phosphate} - \text{Orthophosphate} \quad (3.91)$$

3.5.5.3 Organic Phosphate

Prior to quantifying, organic phosphorus must be destroyed by wet oxidation or digestion using one of the following chemicals: perchloric acid, nitric-sulfuric acid, or persulfate. As a result of this transformation, all forms of phosphorus are measured in an organic determination (Total Phosphorus). After digestion, a colorimetric analysis is performed to quantify the total orthophosphate. The organic-P fraction of the total concentration is then determined by difference using Equation (3.91).

$$\text{Organic Phosphorus} = \text{Total Phosphorus} - \text{Total Inorganic Phosphorus} \quad (3.92)$$

The procedures for performing the various types of phosphorus analyses are found in Section 4500-P of *Standard Methods* (1998).

3.5.6 Metals

Both the type of metal and concentration are important in water quality management. Some metals are essential for microbial growth. According to Crites & Tchobanoglous (1998), the following metals are required for biological growth: calcium, chromium, cobalt, copper, iron, lead, magnesium, manganese, molybdenum, nickel, potassium, sodium, tungsten, vanadium, and zinc. Unfortunately, some of these metals can be toxic at high concentrations, and several metals are classified as priority pollutants, so it is necessary to monitor and control the concentration of these in water, wastewater, and sludge. Table 3.7 lists some of the metals classified as priority pollutants along with their potential health affects.

Standard Methods (1998) lists the following procedures for performing metals analyses: atomic absorption spectrometry (including flame, electrothermal, hydride, and cold vapor techniques); flame photometry; inductively coupled plasma emission spectrometry; inductively coupled plasma mass spectrometry; and anodic stripping voltammetry. Some of the typical metal species encountered in water quality management are briefly described below.

3.5.6.1 Calcium (Ca)

Calcium is generally present in water as the free ion Ca^{2+}. Common mineral forms of calcium include: calcite or aragonite ($CaCO_3$), gypsum ($CaSO_4 \cdot H_2O$), anhydrite ($CaSO_4$) and fluorite (CaF_2). Calcium ions are a significant source of hardness in water.

Table 3.7 Typical metals classified as priority pollutants and potential health effects.

Metal	Health effects
Arsenic (As)	Skin damage or problems with circulatory systems, nay increase risk of cancer
Cadmium (Cd)	Kidney damage
Chromium (Cr)	Allergic dermatitis
Cooper (Cu)	Gastrointestinal distress
Lead (Pb)	Kidney problems, birth defects, high blood pressure
Mercury (Hg)	Kidney damage
Selenium (Se)	Hair of fingernail loss, circulatory problems, numbness in fingers or toes

Source: Based on http://water.epa.gov/drink/contaminants/index.cfm#List and http://water.epa.gov/scitech/methods/cwa/pollutants.cfm Accessed April 23, 2012. United States Environmental Protection Agency.

3.5.6.2 Iron (Fe)

Iron is found in rocks, soils, and waters in a variety of forms and oxidation states. Mineral forms of iron include: hematite (Fe_2O_3), ferric hydroxide [$Fe(OH)_3$], pyrite and marcasite (Fe_2S), siderite ($FeCO_3$), and magnetite (Fe_3O_4).

3.5.6.3 Magnesium (Mg)

Magnesium is generally present in water as the free ion Mg^{2+}. Magnesium sulfate ($MgSO_4$) and magnesium chloride ($MgCl_2$) are also found in solution. Magnesium is not as prevalent in water as calcium, but it, too, causes hardness in water.

3.5.6.4 Manganese (Mn)

Manganese is found in rocks and soils primarily in the form of manganese oxides and hydroxides. The predominate form is the divalent metallic cation, Mn^{2+}.

3.5.6.5 Potassium (K)

The concentration of potassium in natural waters is much lower than sodium. Some common industrial potassium salts include $KHCO_3$, potassium chlorate ($KClO_3$), potassium ferricyanide ($K_3\,Fe\,(CN)_6$), potassium thiocyanate (KSCN), potassium fluoride (KF), and potassium permanganate ($KMnO_4$).

3.5.6.6 Sodium (Na)

Sodium compounds comprise about 3% of earth's crust and are commonly found in water as the free ion Na^+. The addition of the following chemicals increases the sodium content in water: NaOCl, NaOH, Na_2CO_3, and sodium silicate. Several complexes and ion pairs containing sodium found in natural waters include sodium carbonate, sodium chloride, sodium bicarbonate and sodium sulfate.

3.5.7 Gases

Some gases that may be dissolved in water and wastewater include ammonia (NH_3), carbon dioxide (CO_2), hydrogen sulfide (H_2S), methane (CH_4), nitrogen (N_2), and oxygen (O_2). The concentration of a gas in solution is affected by:

1 temperature;
2 concentration of other impurities, salinity, suspended solids;
3 solubility of the gas;
4 partial pressure of the gas.

Gas bubbles interfere with sedimentation and filtration, and may interfere with water quality measurements. The significance of each of the above gases will be discussed briefly. This section will conclude with a review of Henry's Law and how it is used to calculate the solubility of a gas in water.

3.5.7.1 Ammonia (NH_3)

Ammonia is primarily released during the decomposition of organic matter by bacteria and other microorganisms. The degradation of proteins and amino acids by deamination reactions releases ammonia. Un-ionized ammonia nitrogen (NH_3) is toxic to fish at a concentration of 0.02 mg/L. The release of wastewater containing ammonia into surface water can result in anaerobic conditions due to oxygen consumption by nitrifying bacteria. In aqueous environments, ammonia is in equilibrium with the ammonium ion as shown in Equation (3.82).

3.5.7.2 Carbon Dioxide (CO_2)

Carbon dioxide, the primary form of carbon found in the atmosphere, is a result of respiration from humans and animals, and from the burning of fossil fuels. Aqueous carbon dioxide occurs from the diffusion of CO_2 from the atmosphere into water and from the oxidation of organic compounds by microorganisms in aquatic environments. Carbon dioxide is an important component of the carbonate buffering system. Typically, air stripping is used for removing aqueous CO_2 from water if the concentration is greater than or equal to 10 mg/L. Lime, soda ash, or caustic (NaOH) may be added to water or wastewater to neutralize aqueous carbon dioxide.

3.5.7.3 Fluorene (F_2)

Fluorene is the most active element known, and it is not used in its elemental form in environmental engineering applications (Sawyer *et al.*, 1994). It is a pale, yellowish-brown, poisonous gas. Fluorides are compounds that combine fluorene with other positively charged ions, primarily metals. In nature, fluoride exists as fluorite (CaF_2) and apalite ($(Ca_5FCPO_4)_3$).

Since 1945, fluoride has been added to drinking water in the United States to prevent tooth decay. Typical forms added to water include sodium fluoride, sodium silicofluoride, hydrofluorosilicic acid, or ammonium silicofluoride. The normal dose is around 1 mg/L. If the concentration is too high, >1 mg/L, mottling of teeth occurs.

3.5.7.4 Hydrogen Sulfide (H_2S)

Hydrogen sulfide is often formed from the decomposition of organic matter that contains sulfur compounds, or from the reduction of sulfates to sulfide under anaerobic conditions. It is a colorless, flammable gas, with a rotten egg odor. Death can occur at H_2S concentrations of 300 ppm by volume in air.

Aeration or gas stripping may be used to remove H_2S, although oxidants such as chlorine and potassium permanganate are frequently used for removing sulfides from drinking water. Hydrogen sulfide in wastewater leads to the corrosion of hand railings, screens, and other metal components at the headworks of wastewater treatment facilities.

Hydrogen sulfide is a precursor to the formation of sulfuric acid, which can be accomplished by oxidizing bacteria (*Thiobacillus*) ubiquitous in nature. Chlorine, oxygen, hydrogen peroxide, and metal salts have been used for controlling sulfides in wastewater collections systems (EPA, 1985).

3.5.7.5 Methane (CH_4)

Methane is a colorless, odorless, combustible gas which forms primarily from the decomposition of organic matter under anaerobic conditions. Anaerobic digestion of wastewater sludge is a biological process used widely in the wastewater industry. Methane is a by-product of the process that can be collected, scrubbed, and then burned to generate electricity or to produce heat for maintaining the appropriate temperature in the digesters, which operates at approximately 98°F (37°C). In lakes and swamps, methane can be released into the water column and air because of anaerobic processes that occur within sediments.

3.5.7.6 Nitrogen (N_2)

Nitrogen is the most prevalent gas found in the atmosphere and its concentration is approximately 78% by volume. It may be found in water as dissolved or aqueous N_2 or N_2O. In the laboratory, nitrogen gas is sometimes used to strip other gases from water, such as dissolved oxygen, a process known as deaeration. Nitrogen is a nutrient required by all living organisms for growth and synthesis of biochemical compounds. Humans consume proteins and amino acids to acquire nitrogen. Certain bacteria and cyanobacteria (blue-green algae) are capable of using nitrogen gas as their source of nitrogen used in biochemical reactions. This process is called nitrogen fixation wherein, N_2 is reduced to ammonium, and then the ammonium is converted to an organic nitrogen form to be used in synthesis and biochemical pathways.

3.5.7.7 Oxygen (O_2)

Oxygen in one form or another is required by all living organisms to sustain their metabolic processes. Monitoring of the dissolved oxygen (DO) concentration in surface water is one means of assessing water quality to ensure that aerobic conditions exist. When organic material is discharged into a water body (wastewater discharges), microorganisms in the water will oxidize organic matter and ammonia, consuming large quantities of oxygen. If the rate at which oxygen is transferred from the atmosphere to the surface water is less than the rate of oxygen consumed by microbes, benthic organisms, and other aquatic life, anaerobic conditions will develop.

During summer months, when water temperatures increase and flow rates in streams decrease, anaerobic conditions can easily occur. Regulatory agencies often establish stringent effluent requirements for municipal and industrial wastewater discharges during summer months, based on the ten-year, seven-day low flow period. Most aquatic organisms require an aerobic environment if they are to flourish. Game fish such as trout require 5 mg/L or more of DO in order to survive.

Anaerobic conditions lead to the production of hydrogen sulfide, methane, mercaptans, and other disagreeable odors. As noted earlier, DO is the key parameter that is measured during the BOD test.

The solubility of oxygen or saturation concentration in water is a function of temperature and pressure. At sea level (1 atm of pressure) and 20°C, the solubility of oxygen in fresh water is approximately 9.09 mg/L. At a temperature of 10°C, the DO saturation increases to 11.29 mg/L, whereas, at 30°C, the solubility of oxygen decreases to 7.56 mg/L. DO saturation increases with an increase in pressure or decreases with an increase in elevation. An increase in the concentration of dissolved solids and salts results in a decrease in the dissolved oxygen concentration. The level of DO found in surface waters determines whether aerobic or anaerobic conditions will exist.

Aerobic microorganisms, primarily bacteria, require DO to oxidize organic carbon in the activated sludge and aerobic digestion processes. Environmental engineers are called upon to design efficient aeration systems to provide the necessary oxygen required during these biological treatment processes. A minimum of 2.0 mg/L of DO is normally required for stabilization of the BOD in wastewater, and to ensure that nitrification is uninhibited in the activated sludge process.

Dissolved oxygen concentration is a contributing factor in the corrosion of iron and steel. Corrosion can be a significant problem in water distribution systems and in steam boilers. Chemical and physical means can be used to remove DO from water to make it less corrosive.

DO is measured either using a calibrated membrane probe and DO meter or by an iodometric method such as the Azide Modification of the Winkler method. The iodometric method is the most reliable and precise. Membrane electrodes are widely used in the field. The procedures for measuring DO are presented in Standard Methods (1998).

3.5.7.8 Henry's Law

In waters exposed to the atmosphere (open system), the equilibrium or saturation concentration of a dissolved gas is a function of the partial pressure of the gas in the atmosphere and the specific gas. This relationship is quantified by **Henry's law** and is expressed mathematically by Equation (3.93):

$$p_g = \frac{H}{P_T} x_g \tag{3.93}$$

where:

p_g = mole fraction of gas in air, mole gas/mole of air

H = Absorption coefficient or Henry's law constant, $\frac{\text{atm (mole gas/mole air)}}{\text{(mole gas/mole water)}}$

x_g = mole fraction of gas in water, mole gas/mole water

$$x_g = \frac{\text{mole gas in the water}(n_g)}{\text{mole gas in the water}(n_g) + \text{mole water}(n_w)}$$

P_T = total pressure, usually 1.0 atm.

The most common way of expressing Henry's law in the literature is presented as Equation (3.94). In this form, the concentration of the gas is expressed as the partial pressure of the gas (P_g).

$$P_g = Hx_g \tag{3.94}$$

where P_g = partial pressure of gas in air, atm.

Detailed discussions on the use of Henry's law and the various units given for Henry's constant are found in Metcalf & Eddy (2003) and Davis & Cornwell (2008).

Example 3.16 Calculation of solubility of oxygen in water

Estimate the solubility of oxygen in water, assuming that the atmosphere contains 21% oxygen by volume, the pressure is 1 atmosphere, and the Henry's Law constant is 4.11×10^4 atm at 20°C.

First, determine the partial pressure of oxygen in the atmosphere. The partial pressure is calculated by multiplying the total pressure by the fraction that the gas occupies by volume.

$$P_g = \left(\frac{21\%}{100\%}\right) \times (1.0\,\text{atm}) = 0.21\,\text{atm}$$

The mole fraction of the gas in water is determined by substituting into Equation (3.94).

$$P_g = Hx_g$$

$$0.21\,\text{atm} = (4.11 \times 10^4\,\text{atm})x_g$$

$$x_g = 5.11 \times 10^{-6} = \frac{\text{mol } O_2}{\text{mol } O_2 + \text{mol } H_2O}$$

The number of moles of water in 1 liter of water is calculated as follows:

$$\frac{1000\,\text{g water}}{(18\,\text{g water/mole})} = 55.6\,\text{mol water}$$

$$5.11 \times 10^{-6} = \frac{\text{mol } O_2}{\text{mol } O_2 + \text{mol } H_2O} = \frac{\text{mol } O_2}{\text{mol } O_2 + 55.6}$$

Since 55.6 ≫ moles of O_2, the number of moles of O_2 in the denominator on the right side of the above equation can be neglected. Therefore, the number of moles of oxygen are determined to be 2.84×10^{-4} moles/L.

$$5.11 \times 10^{-6} = \frac{\text{mol } O_2}{55.6}$$

$$\text{moles of } O_2 = 2.84 \times 10^{-4}\,\text{moles/L}$$

Finally, the concentration of dissolved oxygen is determined by multiplying the moles of oxygen per liter by the atomic weight of oxygen (32 g/mole).

$$\text{Concentration of } O_2 = 2.84 \times 10^{-4}\,\frac{\text{moles}}{\text{L}} \times \left(\frac{32\,\text{g } O_2}{\text{mole}}\right) \times \left(\frac{1000\,\text{mg}}{\text{g}}\right) = \boxed{9.09\,\frac{\text{mg}}{\text{L}}}$$

The DO concentration listed in *Standard Methods* (1998) for a temperature of 20°C and at sea level is 9.092 mg/L, which agrees favorably with our calculated value of 9.09 mg/L.

3.6 Biological and microbiological characteristics

All natural bodies of water support biological ecosystems that include specific organisms which are harmful or pathogenic to humans and animals. A variety of pathogens exist in wastewater so, therefore, it must be treated to a high level before being reused. Many human diseases are considered waterborne, since they are primarily transmitted by the consumption of water that has been contaminated with fecal matter from infected people or animals. In some instances, mere contact with pathogen contaminated water through the eyes, skin, and ears may infect a person. Water treatment plants are designed around a two to three phase barrier approach so that each unit operation or process used to treat drinking water will reduce pathogens.

Under the United States Environmental Protection Agency's Surface Water Treatment Rule, surface water or ground water under the direct influence of surface water must be disinfected and filtered during the treatment process. Also, the National Primary Drinking Water Standards require the following: 99% (2 Log reduction) of *Cryptosporidium*, 99.9 % (3 Log reduction) of *Giardia lamblia*, 99.99% (4 Log reduction) enteric viruses, and not more than 5% of the water samples collected for total coliform bacteria can be positive.

The four primary pathogenic groups of organisms of concern during water and wastewater treatment include bacteria, helminths, protozoa, and viruses. Each of these pathogens will be discussed and the tests used to indicate the presence of pathogens will be presented. An excellent overview of emerging pathogens in drinking water is provided by Lindquist, in Chapter 16 of *Control of Microorganisms in Drinking Water*, Edited by Srinivasa Lingireddy (2002).

3.6.1 Bacteria

Bacteria are the pathogenic group with the highest population of microorganisms encountered in water and wastewater treatment. Bacteria (singular: bacterium) are microscopic, single-cell organisms that use soluble food. A bacterium

Table 3.8 Waterborne diseases from bacteria.

Bacterium	Associated disease
Salmonella typhosa	Typhoid fever
Salmonella paratyphi	Paratyphoid fever
Salmonella schottinulleri	Paratyphoid fever
Salmonella hirschfeldi C.	Paratyphoid fever
Shigella flexneri	Bacillary dysentery
Shigella dysenteriae	Bacillary dysentery
Shigella sonnei	Bacillary dysentery
Shigella paradysinteriae	Bacillary dysentery
Vibrio comma	Cholera
Vibrio cholerae	Cholera
Pasteurella tularensis	Tularemia
Brucella melitensis	Brucellosis
Burkholderia pseudomomallei	Melioidosis
Leptospira icterohaemorrhagiae	Leptospirosis
Enteropathogenic *Escherichia coli*	Gastroenteritis

Source: EPA *Guidance Manual, Alternative Disinfectants and Oxidants* (1999), p. 2–3. United States Environmental Protection Agency.

normally replicates by binary fission, wherein the cell divides to form two identical "daughter" cells. Bacterial cells are spherical (cocci), spiral, or rod-shaped. Typical dimensions range in length from 1 µm to 50 µm and diameter from 0.3 µm to 4.0 µm (Metcalf & Eddy, 2003). Table 3.8 lists typical waterborne diseases caused by bacteria.

3.6.2 Viruses

Viruses are generally smaller in size (require an electron microscope for viewing) and simpler in structure and composition than other microorganisms. They primarily consist of protein and nucleic acid. According to Viessman *et al.* (2009), enteric viruses range in size from 20 nm to 100 nm. Brock (1979) defines a virus as a genetic element that contains either deoxyribonucleic acid (DNA) or ribonucleic acid (RNA), which can alternate between an extracellular infectious state and intracellular state. Viruses are submicroscopic particles that are metabolically inert, require no enzymes, use no nutrients, and produce no energy. Virus particles are called virions and they can only replicate inside a host (intracellular state). There are over one hundred enteric viruses that are excreted by humans that can infect and cause disease or cause some hereditable changes in cells. Table 3.9 lists typical waterborne diseases caused by human enteric viruses.

3.6.3 Protozoa

Protozoa are single-cell eukaryotic microorganisms that do not have a cell wall. Protozoa generally range in size from 10 µm to 50 µm and their food may consist of alga cells, bacteria, and fungi. They may be found in aqueous environments and in the soil. Several species are parasitic and live in host organisms.

These parasites exist in two forms: as trophozites and cysts or oocysts. While in the trophozite form, they are in a proliferative stage in which they are actively feeding and growing within the host. The cyst or oocyst form is a dormant stage which allows protozoa to survive harsh environmental conditions outside their host. As the trophozites pass through the intestinal systems of humans and warm-blooded animals, they transform morphologically to the cyst form.

Most protozoa reproduce by binary fission; however, some species reproduce sexually or asexually. The most common protozoan diseases are related to gastrointestinal problems such as diarrhea and dysentery. *Entamoeba histolytica*, *Giardia lamblia*, and *Cryptosporidium parvum* are the most prevalent and have the greatest impact on children, the elderly, and persons who have compromised immune systems. Table 3.10 lists waterborne diseases caused by protozoans.

3.6.4 Helminths

Parasitic worms or helminths are classified as cestodes, nematodes, or trematodes. Asano *et al.* (2007) state that, worldwide, helminths are one of the major causative agents of human disease, collectively with an estimated 4.5 billion illnesses per year. Helminths and helminth ova (eggs) are found in untreated wastewater. *Ascarsis lumbricoides* (an intestinal roundworm) causes the infectious disease ascariasis that results in abdominal pain, vomiting, and the discharge of live worms in vomit and feces. *Schistomsoma mansoni* causes schistosomiasis which is a debilitating infection of the liver and bladder. Table 3.11 list some of the most common parasitic worms associated with waterborne diseases and those found in feces.

3.6.5 Indicator organisms

In most instances, only a few pathogenic organisms are found in water, wastewater, and/or polluted water, making it difficult to isolate and identify them. Quite often, large water samples must be collected, filtered and concentrated, and expensive tests performed for their identification and enumeration. Enumeration and identification of viruses in water may require a 1,000 gallon sample, whereas, only 1 – 2 gallons of wastewater may be required (Metcalf & Eddy, 2003). Most water quality professionals are not trained to work with pathogenic organisms. Therefore, in lieu of trying to isolate, identify, and enumerate specific types of pathogens, a surrogate or **indicator organism** is used, whose presence indicates the possibility of harmful or pathogenic organisms being present. The ideal indicator organism should have the following characteristics (Metcalf & Eddy, 2003; Lin, 2002):

Table 3.9 Waterborne diseases from human enteric viruses.

Group	Subgroup	Associated disease
Enterovirus	Poliovirus	Muscular paralysis, aseptic meningitis, febrile episode
Enterovirus	Echovirus	Muscular paralysis, aseptic meningitis, exanthema, respiratory disease, diarrhea, epidemic myalgia, hepatitis, pericarditis and myocarditis
Enterovirus	Coxsackie virus	Herpangina, febrile episode with sores in mouth
Enterovirus	A	Acute lymphatic pharyngitis, muscular paralysis, aseptic meningitis, hand-foot-mouth disease, infantile diarrhea, hepatitis, pericarditis and myocarditis
Enterovirus	B	Pleurodynia, muscular paralysis, aseptic meningitis, meningoencephalitis, pericarditis, endocarditis, myocarditis, respiratory disease, hepatitis or rash, spontaneous abortion, insulin-dependent diabetes, congenital heart anomalies
Reovirus	–	Not well known
Adenovirus		Respiratory diseases, acute conjunctivitis, acute appendicitis, intussusception, subacute thyroiditis, sarcoma in hamsters
Hepatitis		Infectious hepatitis, serum hepatitis, Down's syndrome

Based on *EPA Guidance Manual, Alternative Disinfectants and Oxidants* (1999), pp. 2–4, 2–5.

Table 3.10 Waterborne diseases from protozoans.

Protozoan	Associated disease
Acanthamoeba castellani	Amoebic meningoencephalitis
Cryptosporidium muris	Cryptosporidiosis
Cryptosporidium parvum	Cryptosporidiosis
Entamoeba histolytica	Amoebic dysentery
Giardia lamblia	Giardiasis (gastroenteritis)

Source: *EPA Guidance Manual, Alternative Disinfectants and Oxidants* (1999), p. 2–3. United States Environmental Protection Agency.

1 Must be present when pathogen is present and absent when it is not.

2 The number of indicator organisms present should be equal to or exceed the number of target pathogenic organisms.

3 The indicator organism should have similar survival characteristics in the environment as the target pathogens.

4 The indicator organism must be easily and inexpensively quantified.

5 The indicator organism should be in the intestinal tract of warm-blooded animals.

6 The testing procedure used should not pose a threat to laboratory workers.

Total coliform (TC) bacteria and fecal coliform (FC) bacteria have been widely used to indicate the presence of pathogenic organisms. The intestinal tract of humans contains a large population of rod-shape coliform bacteria. Viessman *et al.* (2009) indicate that humans excrete approximately 50 million coliform bacteria per gram of fecal matter daily.

Since pathogens are found in excrement from infected individuals, coliform bacteria should be an excellent indicator organism. *Standard Methods* (1998) defines the coliform group as anaerobic, gram-negative, non-spore forming, rod-shaped bacteria that ferment lactose broth within 48 hours at 35°C when the multiple tube fermentation test is used for detection.

Table 3.11 Waterborne diseases from helminths.

Helminth	Associated disease
Ancylostoma duodenale	Hookworm
Ascaris lumbricoides	Ascariasis
Hymenolepis nana (dwarf tapeworm)	Hymenolepiasis
Nector americanus (hookworm)	Hookwork
Schistosoma mansoni	Schistosomiasis
Strongyloides stercoralis (threadworm)	Strongyloidiasis
Taenia saginata (beef tapeworm)	Taeniasis
Trichuris trichiura (whipworm)	Trichuriasis

Based on *EPA Guidance Manual, Alternative Disinfectants and Oxidants* (1999), p. 2 – 6.

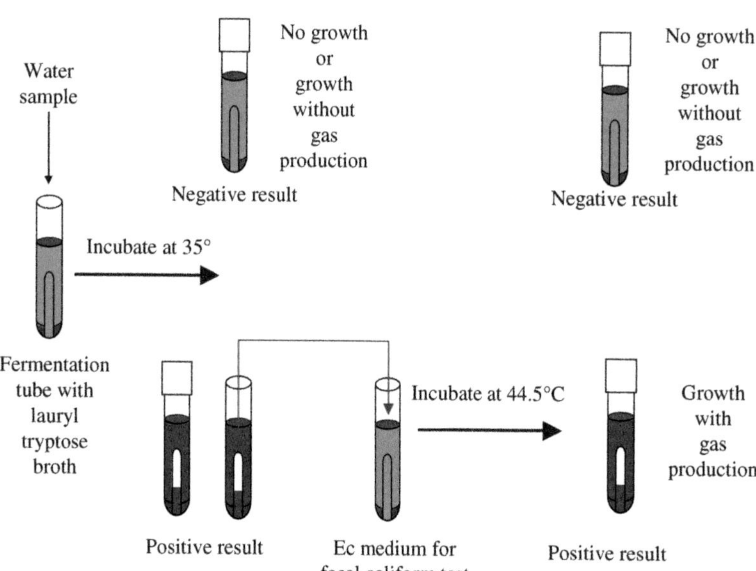

Figure 3.7 Presumptive and confirmed coliform testing procedure.

Other microorganisms that have been identified as potential indicator organisms include *Bacteroides, Klebsiella, Escherichia coli, Fecal streptococci, Enterococci, Clostridium perfringens, Pseudomonas aeruginosa*, and *Aeromonas hyrophila* (Metcalf & Eddy, 2003).

3.6.5.1 Enumeration and Indication of Bacterial Indicator Organisms

Enumeration and identification of indicator organisms, which are bacteria, may be accomplished by direct microscopic counts, pour plates, spread plates, the multiple-tube fermentation method, or the membrane filter (MF) technique.

Direct Microscopic Counts Counting bacterial cells may be accomplished by direct measurement with a Petroff-Hauser counting chamber and microscope or electronic particle counter. In either method, it is difficult to differentiate between living and non-living cells.

Pour and Spread Plates The pour and spread plate method involves adding a diluted sample of water or wastewater to a Petri dish containing agar culture media-specific to the type of organism that is being cultured, identified, and enumerated. In the pour plate method, the serially diluted sample is added to a Petri dish, then warm liquid agar is added. The sample and media are mixed and then placed in an incubator under specified conditions of time and temperature (*Standard Methods*, 1998).

After incubation, the number of individual colonies that have grown are counted and expressed in **Colony Forming Units (CFUs)** per unit volume of sample used, typically, CFU/ml. The spread plate procedure is almost identical, with the exception that the diluted water sample is placed into a sterilized Petri dish already containing agar. A sterilized glass rod or wire is used to spread the sample over the agar media before incubating. After incubation, colonies are counted and converted to CFU/ml. If the sample is inappropriately diluted, however, too many colonies to count accurately may be grown, so the test must be repeated at a higher dilution so that individual colonies are distinguishable.

Multiple-Tube Fermentation Technique A popular method of detecting coliform bacteria is the multiple-tube fermentation technique. This procedure involves inoculating a series of test tubes filled with lauryl tryptose broth with varying amounts of sample. The standard test is carried out through the presumptive, confirmed, and completed phases according to Section 9221 of *Standard Methods* (1998). Figure 3.7 shows a diagram of the presumptive and confirmed phases for the basic total coliform test.

Normally, 15 test tubes are used, with sample volumes of 10 ml, 1.0 ml, and 0.1 ml being added to the test tubes. Coliform bacteria ferment lactose broth under anaerobic conditions, resulting in gas production (CO_2), and bacterial growth as indicated by a cloudy broth. Statistical tables in *Standard Methods* (1998) or the Thomas Equation (Thomas, 1942), Equation (3.95), can be used for calculating the **most probable number (MPN)** per 100 ml of sample. The MPN is a statistical estimate of the number of coliforms that are potentially present, since a number of other microbial species can also ferment lactose.

$$\frac{\text{MPN}}{100\text{ ml}} = \frac{\text{Number of Positive Tubes} \times 100}{[(\text{ml sample neg. tubes}) \times (\text{ml sample all tubes})]^{0.5}} \tag{3.95}$$

Example 3.17 Calculating the most probable number

Calculate the MPN in a set of 15 test tubes with the following number of positive tubes: 3, 1, and 1 for sample sizes of 10.0 ml, 1.0 ml, and 0.1 ml, respectively.

First, determine the volume of sample used in all the test tubes.

$$\text{ml of sample in all tubes} = 5(10.0) + 5(1.0) + 5(0.1)$$
$$= 55.5$$

Next, determine the volume of sample used in the test tubes that were negative.

$$\text{ml of sample in neg. tubes} = 2(10.00) + 4(1.0) + 4(0.1)$$
$$= 24.4$$

Finally, estimate the Most Probable Number (MPN) per 100 ml of sample by substituting into the Thomas Equation (3.95).

$$\frac{\text{MPN}}{100\text{ ml}} = \frac{\text{Number of Positive Tubes} \times 100}{\begin{bmatrix}(\text{ml sample neg. tubes})\\ \times(\text{ml sample all tubes})\end{bmatrix}^{0.5}}$$

$$\frac{\text{MPN}}{100\text{ ml}} = \frac{(7 \times 100)}{[(24.4\text{ ml}) \times (55.5\text{ ml})]^{0.5}} = \boxed{19}$$

From Table 9221.IV of Standard Methods (1998), the MPN value per 100 ml of sample is given as 14, which is based on the Poisson distribution equation. The Thomas equation is an approximation of the Poisson distribution equation.

The primary drinking water standards require that no more than 5.0% of water samples test positive for total coliform (TC) in a given month. For water systems that collect fewer than 40 routine samples per month, no more than one sample can test positive for TC. Every sample that tests positive for TC must also be analyzed for either fecal coliforms or *Escherichia coli*. If two consecutive water samples test positive for TC and one is also positive for *E. coli* or fecal coliform, the water system has an acute maximum contaminant level (MCL) violation. The maximum contaminant level goal (MCLG) for total coliform including fecal coliform and *E. coli* is zero.

Figure 3.8 Membrane filter testing apparatus.

Membrane Filter Technique In the membrane filter (MF) technique, a known volume of water or wastewater is passed through a sterilized filter pad (0.45 μm pores) that retains the indictor organisms. The filter pad is then placed right-side-up (grid side up) in a Petri dish containing Endo-type media containing lactose. *Standard Methods* (1998), Section 9222, describes the MF technique procedures to be followed for either agar-based media or liquid medium. The Petri dish is inverted and incubated at 35°C for 24 hours. Bacteria that produce a red colony with a golden, metallic sheen are considered members of the coliform group. Only membrane filters containing 20% to 80% coliform colonies, and no more than 200 colonies of all types, should be counted. The coliform density is then estimated, using the following equation, and expressed in terms of colony forming units (CFUs) per 100 ml of sample. A photograph of a MF apparatus and associated filter pad is shown in Figure 3.8.

$$\frac{\text{Total coliforms}}{100\text{ ml}} = \frac{\text{colonies counted} \times 100}{\text{ml sample filtered}} \quad (3.96)$$

If no colonies appear on the membrane after incubation, the results should be reported as less than 1 coliform per 100 ml.

3.7 Sample water quality data

This section presents typical water quality data for illustrative and comparative purposes. Water quality data from the Hillsborough River is presented in Table 3.12. The Hillsborough River Water Treatment Plant withdraws water from the river to supply residents in Tampa and some residents in Hillsborough County, Florida with potable water.

Water characteristics from Sulfur Springs in Tampa, Florida are presented in Table 3.13. The notable differences between this surface water and groundwater source include higher color, total organic carbon (TOC), and turbidity for the river as compared with the spring, versus higher concentrations of total dissolved solids (TDS), specific conductivity, and sulfate for the groundwater as compared with the river.

Table 3.14 compares the characteristics of several types of wastewater to rainfall. In general, higher concentrations of each constituent are observed from stormwater runoff to combined wastewater to municipal wastewater, with the exception of total suspended solids. Combined wastewater consists of stormwater runoff and municipal wastewater. Untreated wastewater or raw sewage from municipalities has similar characteristics, unless there is a significant industrial waste component.

Table 3.15 shows the untreated influent and treated effluent characteristics associated with the Rocky Creek Wastewater Treatment Plant in Macon, Georgia.

The effluent characteristics from the Howard F. Curren Advanced Wastewater Treatment Plant in Tampa, Florida are presented in Table 3.16. This reclaimed wastewater is of high-quality and has the potential to be used as a source of drinking water.

Table 3.12 Hillsborough River, Florida, water quality characteristics, January 2, 2000.

Parameter	Unit	Value
BOD_5	mg/L	3.0
Chloride	mg/L	18.3
Color	PCU	34
Dissolved oxygen	mg/L	10.0
Kjeldahl nitrogen	mg/L	0.4
Ammonia nitrogen	mg/L	<0.1
Nitrite nitrogen	mg/L	<0.005
Nitrate nitrogen	mg/L	0.19
Total nitrogen	mg/L	0.59
Ortho phosphate	mg/L	<0.05
Total phosphate	mg/L	0.2
pH	pH Units	8.11
Specific conductivity	μmhos/cm	443
Sulfate	mg/L	47.6
Temperature	°C	16.0
Total Dissolved Solids (TDS)	mg/L	288
Total Suspended Solids (TSS)	mg/L	7
Total Organic Carbon (TOC)	mg/L	4
Turbidity	NTU	2.8

Table 3.13 Sulfur Spring, Florida, water quality characteristics, January 2, 2000.

Parameter	Unit	Value
BOD_5	mg/L	<2
Chloride	mg/L	1390
Color	PCU	11
Dissolved oxygen	mg/L	0.4
Kjeldahl nitrogen	mg/L	0.2
Ammonia nitrogen	mg/L	0.1
Nitrite nitrogen	mg/L	0.021
Nitrate nitrogen	mg/L	0.37
Total nitrogen	mg/L	0.69
Ortho phosphate	mg/L	0.17
Total phosphate	mg/L	0.2
pH	pH units	7.20
Specific conductivity	μmhos/cm	5270
Sulfate	mg/L	383
Temperature	°C	24.4
Total Dissolved Solids (TDS)	mg/L	3175
Total Suspended Solids (TSS)	mg/L	<1
Total Organic Carbon (TOC)	mg/L	2.2
Turbidity	NTU	0.7

Table 3.14 Characteristics of various types of wastewater.

Parameter	Unit	[a]Raw sanitary	[b]Landfill leachate	[c]Raw domestic	[d]Urban runoff
BOD	mg/L	200	10,040	190	7.8–10.0
COD	mg/L	–	7,500	430	70–73
Fecal coliform bacteria	MPN/100 ml	–	–	$10^4 - 10^6$	–
Total nitrogen as N	mg/L	35		40	–
TKN	mg/L	–	350	–	0.965–1.9
NH_3	mg/L	–	–	25	–
Nitrate	mg/L	–	–	0	–
Nitrite + Nitrate	mg/L	–	–	–	0.54–0.736
Total phosphorus	mg/L	10	–	7	0.121–0.383
Soluble phosphorus	mg/L	7	–	5	0.026–0.143
Total solids	mg/L	800	–	720	67–101
Suspended solids	mg/L	240	900	210	–
Volatile suspended solids	mg/L	180	–	160	–

[a] Based on Hammer (1986) *Water and Wastewater Technology*, p. 324.
[b] Based on Horan (1990) *Biological Wastewater Treatment Systems: Theory and Operation*, p. 29.
[c] Based on Metcalf & Eddy (2003) *Wastewater Engineering: Treatment and Reuse*, p. 186.
[d] EPA (1983) *Results of the Nationwide Urban Runoff Program*, p. 6–31.

Table 3.15 2003 calendar year average influent and effluent wastewater characteristics to Rocky Creek Wastewater Treatment Plant, Macon, Georgia.

Parameter	Unit	Influent value	Effluent value
BOD_5	mg/L	130.33	4.46
TSS	mg/L	200.83	9.42
Dissolved oxygen	mg/L	N.M.	7.30
Ammonia nitrogen	mg/L	N.M.	0.16
pH (maximum)	pH units	N.M.	7.62
pH (minimum)	pH units	N.M.	6.66
Fecal coliform	MPN/100 ml	N.M.	11.64

N.M. – Not measured

Table 3.16 Effluent characteristics from the Howard F. Curren Advanced Wastewater Treatment Plant in Tampa, Florida.

Year	BOD (mg/L)	TSS (mg/L)	TN (mg/L)	NO_x (mg/L)
1980	4.5	3.0	2.7	0.80
1981	2.0	2.2	2.5	1.05
1982	4.8	2.7	2.8	1.23
1983	4.6	3.9	2.9	0.90
1984	2.6	2.5	2.4	0.86
1985	2.2	2.6	2.6	0.83
1986	2.2	2.4	2.6	1.10
1987	3.5	2.7	3.0	1.14
1988	4.2	3.3	3.5	1.66
1989	2.6	1.5	2.9	1.12
1990	2.6	1.9	2.6	0.83
1991	2.1	1.6	2.3	0.74
1992	2.5	1.9	2.6	0.83
1993	2.6	1.8	2.5	0.84
Average	3.1	2.4	2.7	1.00

Reprinted with permission from *Biological and Chemical Systems for Nutrient Removal*, 1998.

Summary

- Physical, chemical, and biological parameters are used for assessing water quality and for monitoring and operating water and wastewater treatment facilities.
- Aggregate or lumped sum parameters such as biochemical oxygen demand (BOD), chemical oxygen demand (COD), and total organic carbon (TOC) do not measure a specific type of contaminant. However, they provide an indirect means of measuring the quantity of oxygen required to oxidize organic compounds.
- Physical water quality parameters include color, taste and odor, solids, temperature, absorbance and transmittance, and turbidity.
  - Solids analyses are routinely performed on water, wastewater, and sludge.
- Alkalinity and the carbonate buffering system are important in chemical reactions involved in both water and wastewater treatment to prevent significant changes in pH when an acid or base is added.
- The predominant forms of nitrogen encountered in environmental systems include nitrogen gas (N_2), ammonium (NH_4^+), ammonia (NH_3), total Kjeldahl nitrogen (TKN), organic nitrogen (Organic-N), nitrite (NO_2^-), and nitrate (NO_3^-).
  - All forms of life require nitrogen for survival.
  - Microbes prefer ammonium as their source of nitrogen, whereas algae prefer nitrates.
  - The discharge of wastewaters containing (ammonium/ammonia) may lead to oxygen depletion in receiving streams, due to the process of nitrification, which is the oxidation of ammonium to nitrate.
  - The discharge of wastewaters containing nitrate into surface waters can lead to cultural eutrophication, i.e., excessive growth of algae.
- Major forms of phosphorus are orthophosphates (compounds containing PO_4^{3-}), condensed phosphorus and organic phosphorus.
  - Phosphorus is an essential nutrient for all life forms.
  - Microbes use orthophosphate as their source of phosphorus for growth and synthesis.
  - Chemicals such as lime and alum may be added in order to precipitate orthophosphates from solution.
  - The discharge of wastewater containing phosphorus into receiving streams may lead to cultural eutrophication.
- Metals that are required for biological growth include calcium, iron, magnesium, manganese, potassium, and sodium.
  - At high concentrations, these pose potential health risks due to their toxic effects.
- Seven gases that are important in environmental engineering include ammonia, carbon dioxide, fluorene, hydrogen sulfide, methane, nitrogen, and oxygen.
  - Henry's law is used to calculate the solubility or concentration of a gas in water.
- Bacteria, viruses, protozoa, and helminths are the major categories of pathogens encountered in water quality management.
  - Indicator organisms, such as total and fecal coliform bacteria, are monitored since their presence indicates the likelihood of pathogenic organisms being present, too.

- The multiple tube fermentation and membrane filter techniques are two widely used methods for identifying the presence of indicator organisms.
- Actual water quality data from surface and groundwater sources along with actual effluent from a secondary and advanced wastewater treatment plant are presented for comparison with typical characteristics of untreated rainfall, stormwater, combined wastewater, and municipal wastewater.

Key Words

α value
absorbance
acidity
alkalinity
ammonia (NH_3)
ammonium (NH_4^+)
apparent color
bacteria
biochemcial oxygen demand (BOD)
biological characteristics
BOD rate constant, k
calcium
carbon dioxide
carbonate system
chemical characteristics
chemical oxygen demand (COD)
colony forming unit (CFUs)
condensed polyphosphates
fixed suspended solids (FSS)
fluoride
helminths
Henry's law
hydrogen sulfide
indicator organism
inorganic phosphorus
iron
magnesium
manganese
membrane filter (MF) technique
metal
methane
methyl orange alkalinity
most probable number (MPN)
multiple tube fermentation technique
Nephelometric turbidity units (NTUs)
nitrate (NO_3^-)
Nitrification
nitrite (NO_2^-)
Nitrobacter
nitrogen
nitrogenous oxygen demand (NOD)
Nitrosomonas
organic phosphorus
orthophosphate
oxygen
pH
phenolphthalein alkalinity
phosphorus
physical characteristics
potassium
protozoa
sodium
solids content
soluble BOD (SBOD)
soluble COD (SCOD)
Standard Methods
taste and odor compounds
temperature correction coefficient (θ)
theoretical oxygen demand (ThOD)
total alkalinity
total BOD (TBOD)
total COD (TCOD)
total dissolved solids (TDS)
Total Kjeldhal nitrogen
total nitrogen(TN)
total organic carbon (TOC)
total oxidized nitrogen (NO_X)
total phosphorus (TP)
total solids (TS)
total suspended solids (TSS)
total volatile solids (TVS)
transmittance
trihalomethanes (THMs)
true color
turbidity
ultimate BOD, L_u
viruses
volatile suspended solids (VSS)
water constant (K_w)

References

APHA, AWWA, WEF (1998). *Standard Methods for the Examination of Water and Wastewater*. American Public Health Association, 1015 Fifteenth Street, NW, Washington, DC 20005, pp. 2-1 to 2-6, 3-1, 5-2 to 5-6, 4-86 to 5-91, 4-103 to 4-112, 4-129 to 4-136, 9-37 to 9-38, 9-47, and 9-52.

APHA, AWWA, WEF (2012) *Standard Methods for the Examination of Water and Wastewater*, 22nd Edition, American Public Health Association, 1015 Fifteenth Street, NW, Washington, DC 20005.

Asano, T. Burton, F.L., Leverenz, H.L., Tsuchihashi, R., Tchobanoglous, G. (2007). *Water Reuse: Issues, Technologies, and Application*. McGraw Hill, New York, NY, p. 89.

Benefield, L.D., Randall, C.W. (1980). *Biological Process Design for Wastewater Treatment*. Prentice Hall, Englewood Cliffs, NJ, pp. 68-71.

Benefield, L.D., Judkins, J. F., Weand, B.L. (1982). *Process Chemistry for Water and Wastewater Treatment*. Prentice Hall, Englewood Cliffs, NJ, pp. 42, 83-105.

Brock, T.D. (1979). *Biology of Microorganisms*, 3rd Edition. Prentice Hall, Englewood Cliffs, NJ, p. 309.

Butler, J.N. (1973). *Solubility and pH Calculations*. Addison-Wesley Publishing Company, Inc., Reading, MA, p. 32.

Crites, R., Tchobanoglous, G. (1998). *Small and Decentralized Wastewater Management Systems*. McGraw-Hill, Boston, p. 55.

Davis, Mackenzie L. (2010). *Water and Wastewater Engineering: Design Principles and Practice*. McGraw Hill, New York, NY, pp. 3-2.

Davis, M.L., Cornwell, D.A. (1998). *Introduction to Environmental Engineering*, 3rd Edition. WCB/McGraw Hill, New York, NY, pp. 210-212.

Davis, M.L.. Cornwell, D.A. (2008). *Introduction to Environmental Engineering*, 4th Edition. McGraw-Hill, New York, NY, pp. 365-367.

Environmental Protection Agency (1987)., EPA/625/1-87/001, *Design Manual Phosphorus Removal*. Center for Environmental Research Information, Cincinnati, OH, pp. 15-80.

Environmental Protection Agency (1993)., EPA/625/R-93/010, *Manual Nitrogen Control*. Center for Environmental Research Information, Cincinnati, OH, pp.129-309.

EPA (1999). *Alternative Disinfectants and Oxidants Guidance Manual*. EPA 815-R-99-014, Office of Water, pp. 2-3 to 2-6.

EPA (1985). *Design Manual: Odor and Corrosion Control in Sanitary Sewerage Systems and Treatment Plants*. EPA/625/1-85/018, Center for Environmental Research Information, Cincinnati, OH, pp. 35-66.

Environmental Protection Agency EPA 841 – B-05-005, (2005). *Handbook for Developing Watershed Plans to Restore and Protect Our Waters*.

Fujimoto, Y. (1961). Graphical Use of First Stage BOD Equation. *Journal Water Pollution Control Federation* **36**(1). 69.

Gran, G. (1952). Determination of the equivalence point in potentiometric titrations. *Analyst* **77**(920). 661-671.

Henry, J.G. and Heinke, G.W. (1996). *Environmental Science and Engineering*, Prentice Hall, Upper Saddle River, NJ, **161**, 162-169.

Henze, M., Grady, C.P.L., Gujer, W., Marais, G.V.R., Matsuo, T. (1987). A General Model for Single-Sludge Wastewater Treatment Systems. *Water Research*. **21**(5). 505-515.

Horan, N.J. (1991). *Biological Wastewater Treatment Systems Theory and Operation*. John Wiley, Inc., Chichester, West Sussex, England, p. 6.

Lin, S.D. (2002). The Indicator Concept and its Application in Water Supply. In: Lingireddy, S. (ed.). *Control of Microorganisms in*

Drinking Water, ASCE, 1801 Alexander Bell Drive, Reston, VA, pp. 51–79.
Lindquist, H.D.A. (2002). Emerging Pathogens of Concern in Drinking Water. In: Lingireddy, S. (ed.). *Control of Microorganisms in Drinking Water*, ASCE, 1801 Alexander Bell Drive, Reston, VA, pp. 273–290.
Metcalf & Eddy (2003). *Wastewater Treatment and Reuse.* McGraw Hill, New York, NY, pp. 66–69, 79, 86, 88–91, 108, 115–116, 191, 500–514, 616–623, 666–676, 749–798, 799–815, 1178–1180, 1189.
MWH (2005). *Water Treatment: Principles and Design.* John Wiley & Sons, Inc., Hoboken, NJ., p. 45.
Peavy, H.S., Rowe, D.R., Tchobanoglous, G. (1985). *Environmental Engineering*. McGraw-Hill, NY, p. 17.
Reynolds, T.D., Richards, P.A. (1996). *Unit Operations and Processes in Environmental Engineering*. PWS Publishing, Boston, MA, pp. 327–347, 384–390, and 504.
Rounds, S.A. (2006). Alkalinity and acid neutralizing capacity (version 3.0).: U.S. Geological Survey Techniques of Water-Resources Investigations, book 9, chap. A6., July, accessed December 19, 2008 from http://pubs.water.usgs.gov/twri9A6/.
Sawyer, C.N., McCarty, P.L., Parkin, G.F. (1994). *Chemistry for Environmental Engineering*. McGraw Hill, New York, NY, pp. 547, 586
Snoeyink, V.L., Jenkins, D. (1980). *Water Chemistry*. John Wiley & Sons, Inc., New York, NY, pp. 156–191, p. 87.
Stumm, W., Morgan, J.J. (1996). *Aquatic Chemistry: Chemical Equilibria and Rates in Natural Waters.* John Wiley & Sons, Inc., New York, NY, pp. 148–179.
Tchobanoglous, G., Schroeder, E.D. (1985). *Water Quality: Characteristics, Modeling, Modification*. Addison-Wesley Publishing Company, Reading, MA, p. 58.
Thomas, H.A. (1950). Graphical Determination of BOD Curve Constants. *Water and Sewage Works* 123–124.
Thomas, H.A. (1942). Bacterial Densities from Fermentation Tube Tests. *Journal American Water Works Association* **34**,(4). 572.
Viessman, W., Hammer, M.J., Perez, E.M., Chadik, P.A. (2009). *Water Supply and Pollution Control*, 3rd Edition. Pearson/Prentice Hall, Upper Saddle River, NJ, pp. 257, 258, and 262.
Water Environment Federation (1998). *Biological and Chemical Systems for Nutrient Removal.* Alexandria, VA, pp. 39–56, 101–189, and 193–258.

Problems

1 Water quality is classified by physical, chemical, biological or microbiological, and radiological parameters. Go to the United States Environmental Protection Agency's website at www.epa.gov under the Office of Drinking Water. List four water quality parameters in each of the four categories (physical, chemical, microbiological, and radiological), along with the current maximum contaminant level (MCL), concentration, or treatment technique (TT). Select from both primary and secondary standards.

2 The following data were determined in the laboratory by incubation at 20°C.

Time (days)	0	1	2	3	4	5	6	7
BOD (mg/L)	0	72	120	155	182	202	220	237

Determine the BOD reaction rate constant, k, and ultimate BOD, BOD_u.

3 Solids analysis is one of the most widely used parameters for assessing water quality. Use the following data for calculating total solids (TS), volatile solids (VS), dissolved solids (DS), total suspended solids (TSS), and total volatile suspended solids (TVSS). A sample volume of 150 ml was used in performing all solids analyses.
Tare mass of evaporating dish = 24.3520 g
Mass of evaporating dish plus residue after evaporation @ 105°C = 24.3970 g
Mass of evaporating dish plus residue after ignition @ 550°C = 24.3850 g
Mass of Whatman filter and tare = 1.5103 g
Mass of Whatman filter and tare after drying @ 105°C = 1.5439 g
Residue on Whatman filter and tare after ignition @ 550°C = 1.5199 g

4 Use the following data for calculating total solids (TS). volatile solids (VS), dissolved solids (DS), total suspended solids (TSS), and total volatile suspended solids (TVSS). A sample volume of 200 ml was used in performing all solids analyses.
Tare mass of evaporating dish = 25.334 g
Mass of evaporating dish plus residue after evaporation @ 105°C = 25.439 g
Mass of evaporating dish plus residue after ignition @ 550°C = 25.385 g
Mass of Whatman filter and tare = 1.5103 g
Mass of Whatman filter and tare after drying @ 105°C = 1.5439 g
Residue on Whatman filter and tare after ignition @ 550°C = 1.5199 g

5 A solids analysis is to be performed on a wastewater sample. The abbreviated procedure is outlined as follows:

a. A Gooch crucible and filter pad are dried @105°C to a constant mass of 25.439 g.
b. 200 ml of a well-mixed sample of the wastewater is passed through the filter pad.
c. The crucible, filter pad, and solids collected on the pad are dried at 105°C to a constant mass of 25.645 g.
d. 100 ml of the filtrate that passes through the filter pad in Step (b) above is placed in an evaporation dish that had been pre-weighed at 275.410 g.
e. The sample in Step (d) is evaporated to dryness at 105°C and the dish and residue are weighed at 276.227 g.
f. Both the crucible from Step (c). and the evaporation dish from Step (e) are placed in a muffle furnace at 550°C for an hour. After cooling in a dessicator, the mass of the crucible is 25.501 g and the mass of the dish is 275.944 g.

Determine the following: suspended solids (mg/L), dissolved solids (mg/L), total solids (mg/L), organic or volatile fraction of the suspended solids (mg/L), and the organic or volatile fraction of the dissolved solids (mg/L).

6 An ammonia nitrogen analysis performed on a wastewater sample yielded 30 mg/L as nitrogen. If the pH of the sample is 8.5, determine the ammonium nitrogen concentration (mg/L) in the sample assuming a temperature of 25°C.

7 A 100 ml sample of water is titrated with 0.02*N* H_2SO_4. The initial pH is 9.8 and 6.5 ml of acid are required to reach the pH 8.3 endpoint. An additional 10.1 ml of acid are required to reach the pH 4.5 endpoint. Determine the following: total alkalinity, hydroxide alkalinity, carbonate alkalinity, and bicarbonate alkalinity, in terms of mg/L as $CaCO_3$.

8 A 200 ml sample of water with an initial pH of 7.8 is titrated with 0.02*N* H_2SO_4. 16 ml of acid are added to reach the endpoint pH of 4.5. Determine the species of alkalinity present and the concentration in mg/L as $CaCO_3$

9 Calculate the theoretical oxygen demand of a solution containing 550 mg/L of glutamic acid ($C_5H_9NO_4$).

10 Calculate the theoretical oxygen demand (mg/L). of a solution containing 450 mg of glucose ($C_6H_{12}O_6$). in 2 L of distilled water.

11 The result of a seven-day BOD test performed on a sample of water from an oligotrophic lake was 10 mg/L. The base "e" BOD rate constant determined from previous studies was estimated to be 0.10 d^{-1}. Determine the ultimate BOD and five-day BOD of the sample taken from the lake.

12 A five-day BOD test is performed on an industrial wastewater sample that contains no bacteria; therefore, a "seeded" BOD test is run. 10 ml of "seed" are added to 20 L of dilution water. 30 ml of industrial wastewater are added to a 300 ml BOD bottle and the remaining volume consists of "seeded" dilution water. The average dissolved oxygen concentration of the diluted wastewater samples and blanks (seeded dilution water) on the first day of the test is 7.5 mg/L and 9.0 mg/L, respectively. After incubating separate BOD bottles at 20°C for five days, the average DO concentration of the diluted wastewater BOD bottles and seeded dilution water BOD bottles is 3.1 and 8.5 mg/L, respectively. Use Equation (3.13) for calculating the five-day BOD of the industrial wastewater.

13 A multiple tube fermentation test was performed on a sample of water obtained from the Ocmulgee River in Macon, Georgia during the summer. A set of 15 test tubes with sample sizes of 10, 1.0, and 0.1 ml were used in the analysis resulting in the following number of positive tubes: 5-5-1. Estimate the most probable number (MPN) per 100 ml of sample using Equation (3.95). Compare your answer to the value presented in *Standard* Methods (1998).

14 A multiple tube fermentation test was performed on a sample of water obtained from the Occoquan Reservoir near Manassas, Virginia during the winter. A set of 15 test tubes with sample sizes of 1.0, 0.1, and 0.01 ml were used in the analysis, resulting in the following number of positive tubes: 1-1-0. Estimate the most probable number (MPN) per 100 ml of sample using Equation (3.95). Compare your answer to the value presented in *Standard Methods* (1998).

15 Discuss in your own words what is meant by an indicator organism. Which group of organisms are used as indicator organisms and why? List three groups of pathogenic organisms that may be found in water and wastewater.

Chapter 4

Essential biology concepts

Richard O. Mines, Jr.

Learning Objectives

After reading this chapter, you should be able to:

- describe how various organisms and specifically, how microorganisms are used in the treatment of wastewaters, sludges, contaminated ground waters, and soils;
- differentiate between prokaryotic and eukaryotic cells;
- understand the relationship between energy and synthesis reactions involving heterotrophic and autotrophic organisms;
- how environmental factors such as pH, DO, moisture, and temperature affect microbial growth rate;
- be familiar with the biochemical principles involved in the production of carbohydrates, proteins, lipids, and nucleic acids in organisms;
- understand the role of enzymes and Michaelis-Menten enzyme kinetics;
- define ecology and primary productivity;
- draw and discuss the carbon, nitrogen, phosphorus, and sulfur cycles, and their role in biology;
- discuss how population growth has impacted the environment in regard to pollution, carrying capacity, and ecological footprint;
- explain how total maximum daily loads and the discharge of organic wastes affects water quality in rivers;
- use the Streeter-Phelps dissolved oxygen model for predicting the lowest DO concentration in a river;
- communicate the understanding of eutrophication, turnover, stratification, and other limnological terms.

4.1 Introduction to microbiological unit processes

Microorganisms are used in numerous applications in environmental engineering for treating water, wastewater, and sludges. Biotechnology involves the use of microorganisms and biological systems to produce biodiesel, alcohols, other fermentation products such as methane, antibiotics, and other pharmaceuticals, and systems to degrade specific organic contaminants and wastes. A basic understanding of the types of cells and how organisms obtain energy and synthesize biomass is critical for the proper design and operation of biological processes.

Organisms are typically classified according to the Linnaean system, a descending hierarchical organization of organisms based on the structure of their bodies, starting with the least specific to the most specific. Table 4.1 shows the Linnaean system that culminates with the genus and species.

When referenced in the literature, the genus and species names for given organisms are italicized or underlined, with the genus capitalized. For instance, *Clostridium botulinum* is a species of bacteria that causes botulism; a severe type of food poisoning. *Clostridium* is the genus name and *botulinum* is the species name. Several species of clostridia ferment sugar, amino acids, and cellulose.

Current taxonomic systems include phylogeny, which classifies organisms based on their genetic characteristics and evolutionary history. To the taxonomist, structural details

Environmental Engineering: Principles and Practice, First Edition. Richard O. Mines, Jr.

Table 4.1 Linnaean Classification System of Organisms.

Domain
Kingdom
Phylum
Class
Order
Family
Genus
Species

(morphology) and physiology are the two most important characteristics used in classifying microorganisms (Gaudy & Gaudy, 1988, p. 343).

4.2 Cell basics

The **cell** is the fundamental unit of all living organisms. A single cell is an entity that is separated from other cells by a cell wall or membrane. Microbial cells are distinct from animal and plant cells, because they are capable of carrying out all life processes (growth, respiration, and reproduction) independently from other cells. There are two types of microbial cells: prokaryotic and eukaryotic. Organisms that contain cells that do not have a nucleus are called **prokaryotes** (Figure 4.1). Their genetic material, deoxyribonucleic acid (DNA), is contained in a single plasmid. Prokaryotic cells contain few internal structures. Bacteria and cyanobacteria (Blue-green algae) are eukaryotic microbes.

Eukaryotes contain cells with a true nucleus; this means their DNA is enclosed in a membrane structure. Eukaryotic cells are more complex (Figure 4.2). Table 4.2 lists some of the major differences between these two types of cells.

4.3 Energy and synthesis (carbon and energy transformations)

Living organisms must have the capacity to use and transform energy. Microorganisms consume substrate, use nutrients, and carry out oxidation-reduction reactions for growth and cellular

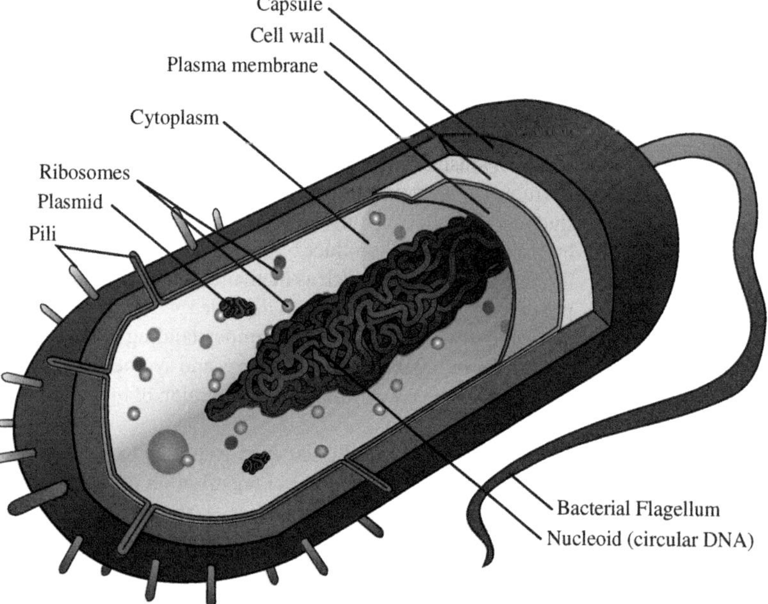

Figure 4.1 Prokaryotic cell. *Source:* http://en.wikipedia.org/wiki/File:Average_prokaryote_cell-_en.svg. Reproduced by kind permission of LadyofHats (Mariana Ruiz Villarreal).

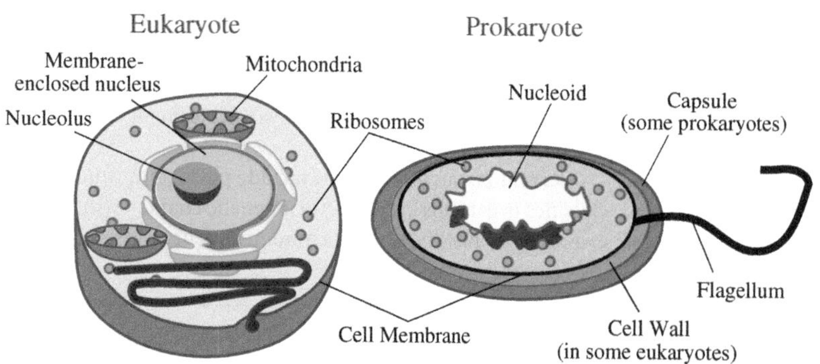

Figure 4.2 Eukaryotic cell. *Source:* http://en.wikipedia.org/wiki/File:Celltypes.svg. Science Primer (National Center for Biotechnology Information). Vectorized by Mortadelo 2005.

Table 4.2 Comparison of prokaryotic and eukaryotic cells.

Characteristic	Prokaryotic	Eukaryotic
Cell wall	Composed of peptidoglycan	Absent of peptidoglycan
DNA	Nucleoid or plasmids	Chromosomal DNA
Energy generation	Part of cytoplasmic membrane	Mitochondria
Nucleus	Absent	Present
Photosynthetic pigments	Chloroplasts absent	Chloroplasts present
Phylogenic group	Archea, bacteria, cyanobacteria (blue-green algae)	Single cell: algae, fungi, and protozoan. Multicellular: animals, fungi, and plants.
Size range	1–2 μm by 1–4 μm	>5 μm

Source: Henze *et al.* (2008) pp. 10–12; Metcalf and Eddy (2003) page 106; Pelczar *et al.* (1977) pp. 7–8; Rittman and McCarty (2001) pp. 10–12.

maintenance. **Metabolism** is the term that encompasses all the reactions that cells use to process substrate (food materials) for obtaining energy, and for producing the compounds for building new cellular components. Degradative reactions that produce energy and building blocks for cellular growth are called **catabolic** reactions. **Anabolic** reactions are biosynthetic reactions that produce compounds necessary for growth and cell maintenance.

Coupled oxidation/reduction reactions are used to produce energy for growth, mobility, and producing adenosine triphosphate (ATP). The quantity of energy produced by coupled reactions depends on the chemical properties of the electron donor and electron acceptor. Table 4.3 shows the free energy of release ($\Delta G°$) at standard conditions and a temperature of 25°C for the oxidation of glucose ($C_6H_{12}O_6$) under various conditions.

A significant increase in the quantity of free energy released is realized when glucose is oxidized by aerobic respiration, $\left(-120\frac{kJ}{e^{-}eq}\right)$ versus fermentation $\left(-10\frac{kJ}{e^{-}eq}\right)$. Anaerobic respiration results in less free energy released, approximately $\left(-17\frac{kJ}{e^{-}eq}\right)$ when CO_2 is used as the electron acceptor, versus $\left(-20\frac{kJ}{e^{-}eq}\right)$ when SO_4^{2-} serves as the electron acceptor. Fermentation, aerobic respiration, and anaerobic respiration are discussed in Section 4.3.1.

Microbes obtain carbon for cell growth from either organic carbon or carbon dioxide (inorganic carbon). Organisms that use organic carbon (relatively complex, reduced organic compounds) for synthesizing biomass are called **heterotrophs**, while those that use inorganic carbon (carbon dioxide) for producing cell mass are called **autotrophs**. Heterotrophs require less energy than autotrophs, since the incorporation of inorganic carbon into cellular compounds is a reductive process requiring more energy.

Energy needed for cell synthesis and cell maintenance activities is primarily provided by oxidizing chemical compounds or by capturing light. Microorganisms are classified as **phototrophs** if they use light as their energy source and as **chemotrophs** if they use oxidation-reduction reactions for supplying their energy. Chemotrophs may be further classified

Table 4.3 Comparison of free energy yields for oxidation of glucose by various means.

Oxidation/reduction reaction	Process simulated	Free energy released, $\Delta G°\left(\frac{kJ}{e^{-}eq}\right)$
$\frac{1}{24}C_6H_{12}O_6 + \frac{1}{4}O_2 = \frac{1}{4}CO_2 + \frac{1}{4}H_2O$	Fermentation and aerobic respiration, O_2 electron acceptor	−120.10
$\frac{1}{24}C_6H_{12}O_6 + \frac{1}{5}NO_3^- + \frac{1}{5}H^+ = \frac{1}{4}CO_2 + \frac{1}{10}N_2 + \frac{7}{20}H_2O$	Fermentation and anaerobic respiration, NO_2^- electron acceptor	−113.63
$\frac{1}{24}C_6H_{12}O_6 + \frac{1}{8}SO_4^{2-} + \frac{3}{16}H^+ = \frac{1}{16}H_2S + \frac{1}{16}HS^- + \frac{1}{4}CO_2 + \frac{1}{4}H_2O$	Fermentation and anaerobic respiration, SO_4^{2-} electron acceptor	−20.69
$\frac{1}{24}C_6H_{12}O_6 = \frac{1}{8}CO_2 + \frac{1}{8}CH_4$	Fermentation and anaerobic respiration, CO_2 electron acceptor	−17.85
$\frac{1}{24}C_6H_{12}O_6 = \frac{1}{12}CO_2 + \frac{1}{12}CH_3CH_2OH$	Fermentation to ethanol	−10.17

Adapted from Metcalf and Eddy (2003), pp. 572–573.

Table 4.4 Microbial reactions and their nutritional classification.

Microbial reaction	Nutritional classification
$5CO_2 + 3H_2O + NH_3 \xrightarrow{light} C_5H_7O_2N + 5O_2 + H_2O$	Autotrophic, photosynthetic
$C_5H_7O_2N + 5O_2 \rightarrow 5CO_2 + 2H_2O + NH_3$	Cellular respiration, aerobic
$C_6H_{12}O_6 + 6O_2 \rightarrow 6CO_2 + 6H_2O$	Heterotrophic, aerobic (aerobic oxidation)
$C_6H_{12}O_6 \rightarrow 2C_2H_6O + 2CO_2$	Heterotrophic, anaerobic (ethanol fermentation)
$5C_6H_{12}O_6 + 24NO_3^- + 24H^+ \rightarrow 30CO_2 + 12N_2 + 42H_2O$	Heterotrophic, anoxic (denitrification)
$2NH_3 + 4O_2 \rightarrow 2HNO_3 + 2H_2O$	Autotrophic, aerobic (nitrification)
$C_6H_{12}O_6 \rightarrow 3CO_2 + 3CH_4$	Heterotrophic, anaerobic (methanogenesis)

Adapted from Benefield and Randall (1980) page 26; Rittman and McCarty (2001) page 133.

depending on the type of chemical compound oxidized, i.e., the electron donor. For instance, a **chemoorganotroph** is an organism that oxidizes complex organic compounds such as glucose ($C_6H_{12}O_6$) as their electron donor, whereas chemoautotrophs oxidize inorganic compounds such as ammonia, hydrogen sulfide, and nitrite. Table 4.4 lists some typical microbial reactions and their nutritional classification.

4.3.1 Heterotrophs and energy capture

Heterotrophs use the same organic material as their source of carbon and energy. The oxidative reactions carried out by the heterotrophs are actually dehydrogenations, since they involve the loss of hydrogen atoms. Remember, oxidation means the loss of electrons. When a hydrogen ion is removed from a compound, an electron is lost, too.

Energy released during catabolic reactions is recovered in two ways: substrate-level phosphorylation and oxidative phosphorylation. **Phosphorylation** is the term used for describing the formation of the high energy carrier adenosine triphosphate (ATP) from adenosine diphosphate (ADP) and inorganic phosphate (P_i). Energy is conserved in high-energy phosphate bonds, each yielding a free energy of approximately 8.3 kcal per mole. In substrate-level phosphorylation, a portion of the energy released during the oxidation or dehydrogenation of the organic compound is used to drive the endergonic phosphorylation reaction of the conversion of ADP to ATP according to the following reaction. An **endergonic** reaction requires energy, whereas, an **exergonic** reaction generates energy.

$$\text{1, 3-diphospho-glyceric acid} + \text{ADP} \rightarrow \text{3-phospho-glyceric acid} + \text{ATP} \quad (4.1)$$

Substrate-level phosphorylation is not the major mechanism for producing ATP in aerobic organisms.

In **oxidative phosphorylation**, electrons that are produced during the oxidation of the electron donor are passed through an electron-transport system to a terminal electron acceptor, typically oxygen. The coenzyme nicotinamide-adenine dinucleotide (NAD) is involved in electron transfer during catabolic reactions. NAD^+ represents the oxidized form of NAD, while $NADH + H^+$ represents the reduced form.

Another coenzyme, nicotinamide-adenine dinucleotide phosphate (NADP), is another electron carrier involved in anabolic or biosynthetic reactions. $NADP^+$ is the oxidized form of NADP and NADPH is the reduced form. FAD/$FADH_2$ also serves as an electron carrier. FAD stands for flavin adenine dinucleotide; its oxidized form is represented by FAD and its reduced form is $FADH_2$. When electrons enter the electron-transport system or chain at the level of NAD, three moles of ATP per pair of electrons will be produced. However, if the electrons enter at the level of FAD, only two moles of ATP per pair of electrons will be created.

In eukaryotic organisms, the electron-transport system is located in the mitochondria. Figure 4.3 shows the electron-transport system used by most bacteria. It uses a series of electron carriers that capture the energy released during the oxidation of the electron donor, which drives the endergonic phosphorylation reaction of the conversion of ADP to ATP.

There are three major pathways used by heterotrophic microorganisms for converting organic substrate into cellular components and for energy. These are: fermentation, aerobic respiration, and anaerobic respiration.

4.3.1.1 Fermentation

Fermentation is one method by which a heterotrophic microbe can obtain energy, and it occurs in the absence of external electron acceptors. In fermentation, the organic substrate serves both as the electron donor and electron acceptor. The organic compound is degraded by a series of enzyme mediated reactions, with substrate-level phosphorylation used to capture a portion of the energy in the form of ATP. The overall

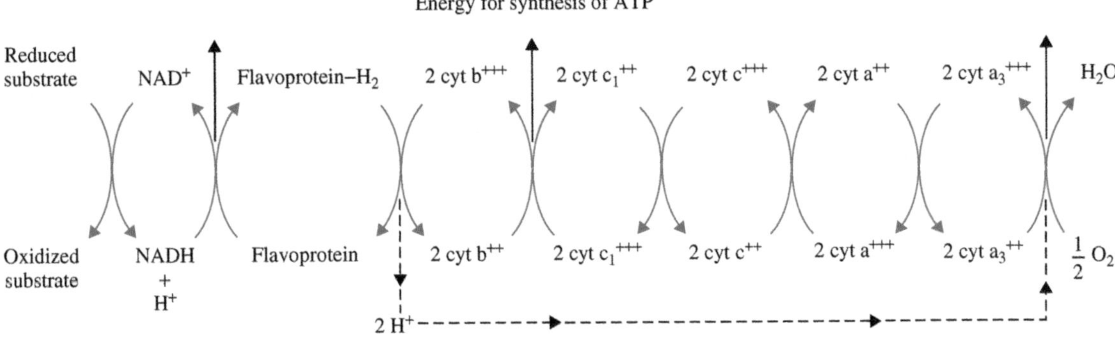

Figure 4.3 Electron transport system common to most bacteria. *Source*: Benefield & Randall (1980), page 38.

fermentation of glucose can be expressed by the following equation:

$$\text{Glucose} + 2\text{ADP} \rightarrow 2\,\text{Ethanol} + 2\,CO_2 + 2\text{ATP} \qquad (4.2)$$

Figure 4.4 shows the fermentation of glucose under anaerobic conditions with no external electron acceptors. The series of enzyme-mediated steps, beginning with glucose and ending with pyruvate, is called "glycolysis", or the Embden-Meyerhof-Parnas (EMP) pathway.

The complete oxidation of glucose to water and carbon dioxide yields approximately 686 kcal of free energy (Gaudy & Gaudy, 1988, page 320). Assuming that each mole of ATP conserves about 8.3 kcal of free energy, only 2.4% of the energy is stored during glycolysis ($2 \times 8.3\,\text{kcal}/686\,\text{kcal} \times 100 = 2.4\%$).

4.3.1.2 Aerobic Respiration

In aerobic respiration, the organic carbon source is first oxidized or dehydrogenized by passing through the EMP pathway or glycolysis to form pyruvate, i.e., fermentation pathway. When oxygen is present, pyruvate is then converted to acetyl-coenzyme A, as illustrated in the following reaction:

$$\text{Pyruvate} + \text{CoASH} + \text{NAD} \rightarrow \text{Acetyl CoA} + \text{NADH} + CO_2 + H^+ \qquad (4.3)$$

This is an irreversible reaction that requires the participation of six cofactors in order to proceed: coenzyme A, thiamine pyrophosphate (TPP), lipoic acid, NAD, FAD, and magnesium. Acetyl CoA then enters the **tricarboxylic acid** (TCA) cycle, where it undergoes oxidation to CO_2 and H_2O. Figure 4.5 shows the overall TCA cycle.

A total of 38 moles of ATP can be produced by oxidizing glucose to carbon dioxide and water by fermentation and the TCA cycle. Table 4.5 is an accounting of the ATPs produced. Detailed discussions of this and the TCA cycle can be found in the following references: Baum (1978), Brock (1979), and Gaudy & Gaudy (1988). Once acetyl CoA enters the TCA cycle and when oxygen is present, complete oxidation of the organic compound to carbon dioxide and water is possible. Oxygen serves as the ultimate electron acceptor.

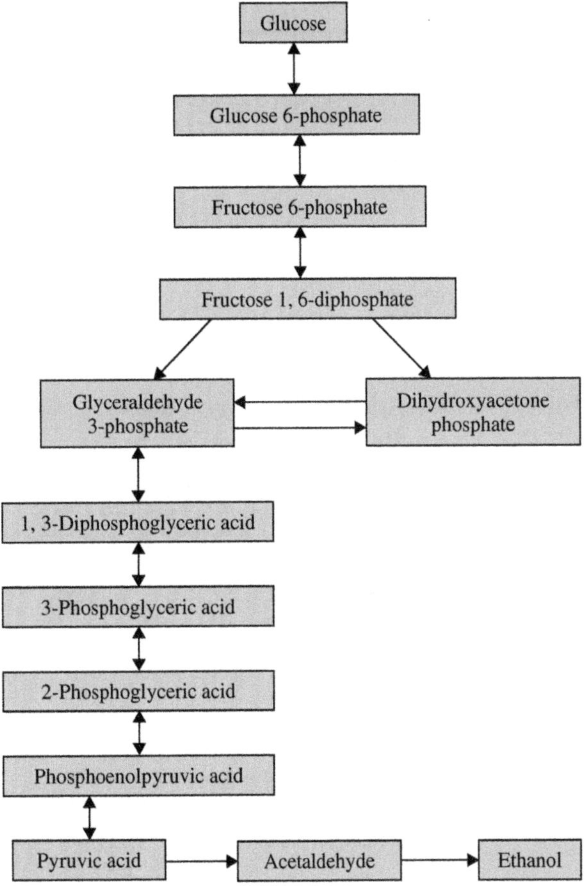

Figure 4.4 Glucose fermentation by yeast under anaerobic conditions. The steps from glucose to pyruvate is known as "glycolysis" or as the Embden-Meyerhoff-Parnas pathway. *Source*: Adapted from Brock, Thomas D. (1979). Biology of Microorganisms, Prentice Hall, Englewood Cliffs, NJ, pp. 99.

4.3.1.3 Anaerobic Respiration

Certain microorganisms can use other inorganic compounds, rather than oxygen, as the external electron acceptor. Typical inorganic electron acceptors include CO_2, SO_4^{2-}, NO_2^-, and NO_3^-. Lower ATP production from the TCA will occur because

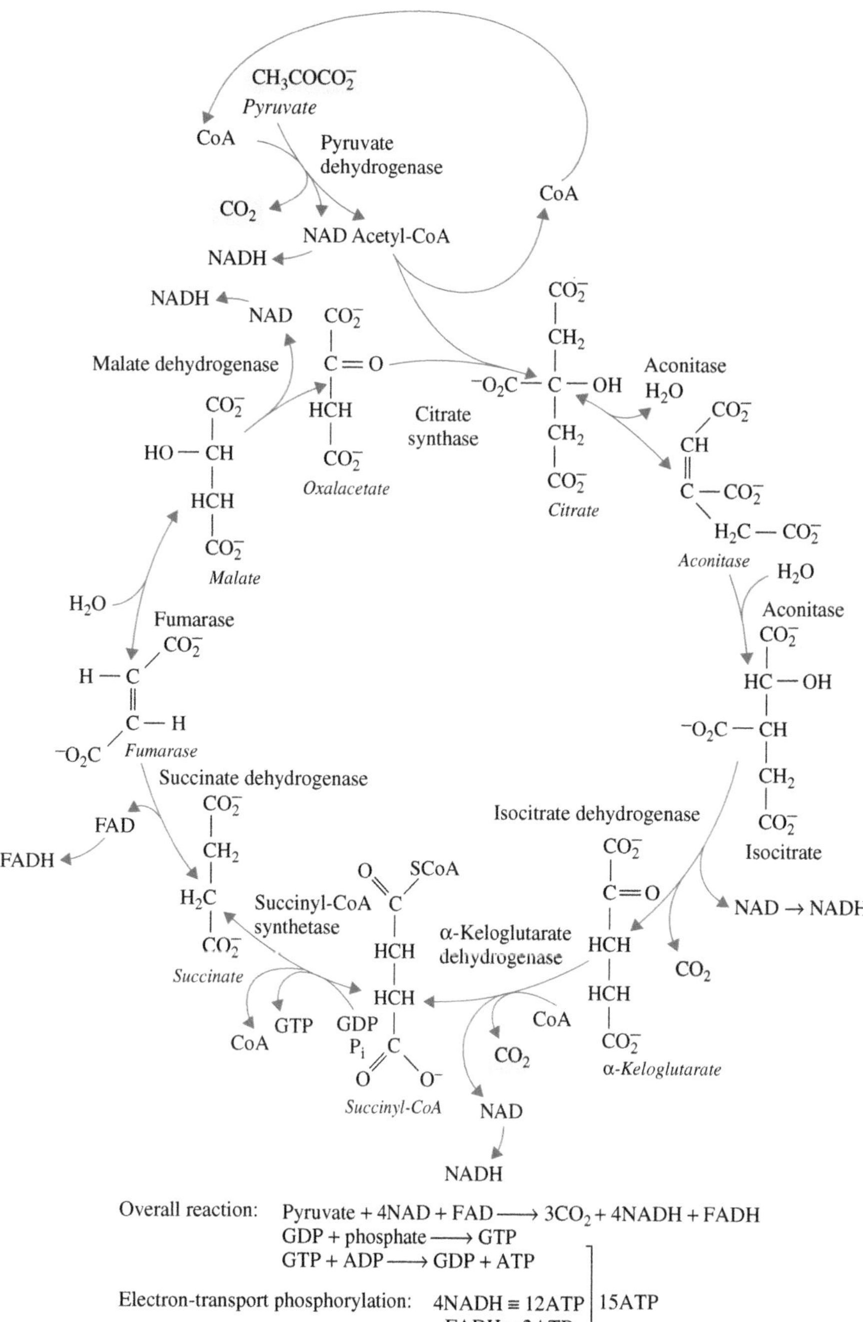

Figure 4.5 Tricarboxylic acid cycle. *Source*: Adapted from Brock, Thomas D. (1979). Biology of Microorganisms, Prentice Hall, Englewood Cliffs, NJ, pp. 99, 111, 158, 751.

the redox potential between the electron donor and these alternative electron acceptors is not as great as can be achieved using oxygen. Oxidation with these alternative electron acceptors is called anaerobic respiration. Figure 4.6 is a schematic of anaerobic respiration, in which alternative electron acceptors capture electrons that are passed along the electron transport chain. The reduced organic compound is denoted as $\mathbf{DH_2}$ and the oxidized form as **D**.

4.3.2 Autotrophs

Organisms that use inorganic carbon (carbon dioxide or bicarbonate) as their carbon source for synthesizing biomass and cellular material are called autotrophs. They derive energy from oxidizing inorganic compounds (chemoautotrophic), or from sunlight (photoautotrophic). The aerobic nitrifying bacteria, collectively called nitrifiers, consisting of the genera *Nitrosomonas* and *Nitrobacter*, are examples of two chemoautotrophic organisms. The oxidation of ammonium to nitrate (neglecting cell synthesis) is considered a two-step, sequential reaction and is presented below:

$$NH_4^+ + 1.5O_2 \xrightarrow{Nitrosomonas} NO_2^- + 2H^+ + H_2O \qquad (4.4)$$

$$\Delta G^0 = -274.74\,\frac{\text{kJ}}{\text{mol}}$$

Table 4.5 ATP yield from fermentation and aerobic respiration.

Pathway or process	Mechanism for ATP production	Total ATPs produced
Fermentation to pyruvate	2 NADH produced	$2 \times \frac{3\,ATPs}{NADH} = 6\,ATPs$
	Substrate-level phosphorylation	2 ATPs
		Subtotal = 8 ATPs
Aerobic respiration (TCA) cycle	4 NADH produced	$4 \times \frac{3\,ATPs}{NADH} = 12\,ATPs$
	1 $FADH_2$ produced	$1 \times \frac{2\,ATPs}{NADH} = 2\,ATPs$
	GTP + ADP → GDP + ATP	1 ATP
		Subtotal = 15 ATPs pyruvate
	2 moles of pyruvate enter TCA	$2 \times \frac{15\,ATPs}{pyruvate} = 30\,ATPs$
		Total = 38 ATPs

Adapted from Brock (1979), page 111.

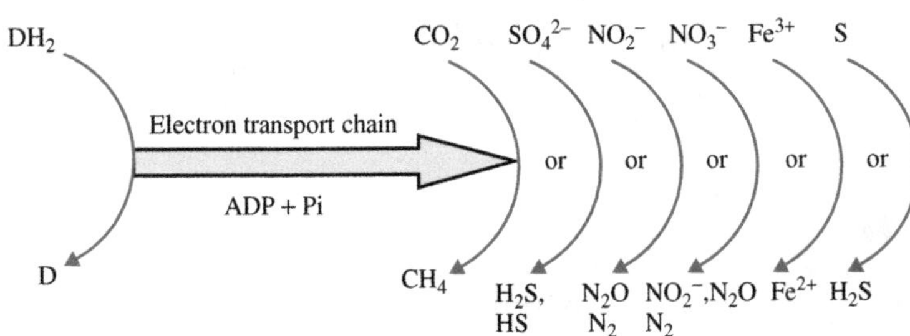

Figure 4.6 Electron transport chain for anaerobic respiration. *Source:* Benefield and Randall, page 39 also reference Wilkinson, J.F., introduction to Microbiology, Halsted Press, a division of John Wiley, New York, 1975.

$$NO_2^- + 0.5\,O_2 \xrightarrow{Nitrobacter} NO_3^- \quad \Delta G^0 = -74.14\,\frac{kJ}{mol} \tag{4.5}$$

$$NH_4^+ + 2O_2 \xrightarrow{Nitrifiers} NO_3^- + 2H^+ + H_2O \quad \Delta G^0 = -348.88\,\frac{kJ}{mol} \tag{4.6}$$

As shown in Equation (4.4), there is more energy available from the oxidation of ammonium than from the oxidation of nitrite. This could lead to a build-up of nitrite until the growth of the nitrite oxidizers (*Nitrobacter* species) have a chance to catch up. The overall nitrification reaction, including synthesis of cellular material, can be modeled (Crites & Tchobanoglous, 1998, page 437) by Equation (4.7). Due to rounding of the coefficients, the equation does not balance exactly; however, according to Crites & Tchobanoglous, the error introduced is negligible.

$$NH_4^+ + 1.731\,O_2 + 1.962\,HCO_3^- \rightarrow 0.038\,C_5H_7O_2N + 0.962\,NO_3^- + 1.769\,H_2CO_3 + 1.077\,H_2O \tag{4.7}$$

Almost all autotrophic organisms use the **Calvin Cycle** for synthesizing organic matter from carbon dioxide. Figure 4.7 is a schematic of the Calvin Cycle.

The overall equation for the synthesis of glucose from CO_2 is shown in the following equation (Rittman & McCarty, 2001, page 78):

$$6CO_2 + 18ATP + 12NADPH + 12H^+ \rightarrow C_6H_{12}O_6 + 18ADP^+ + 12NADP^+ + 18P_i + 6H_2O \tag{4.8}$$

Rittman & McCarty (2001) indicate that approximately 3,204 kJ of energy is required for producing one mole of glucose.

4.3.3 Energy

Organisms acquire their energy from light or by oxidizing inorganic or organic compounds. The ultimate source of energy for all life on earth is the sun. Photosynthesis is the process by which green plants and various types of microorganisms use light energy for synthesizing organic compounds from water and carbon dioxide. Green plants, cyanobacteria, algae, and other genera of bacteria can use sunlight. These organisms use energy with a wavelength between 400 and 700 nm, which is used to synthesize organic compounds and release oxygen from carbon dioxide and water. The overall simplified equation for

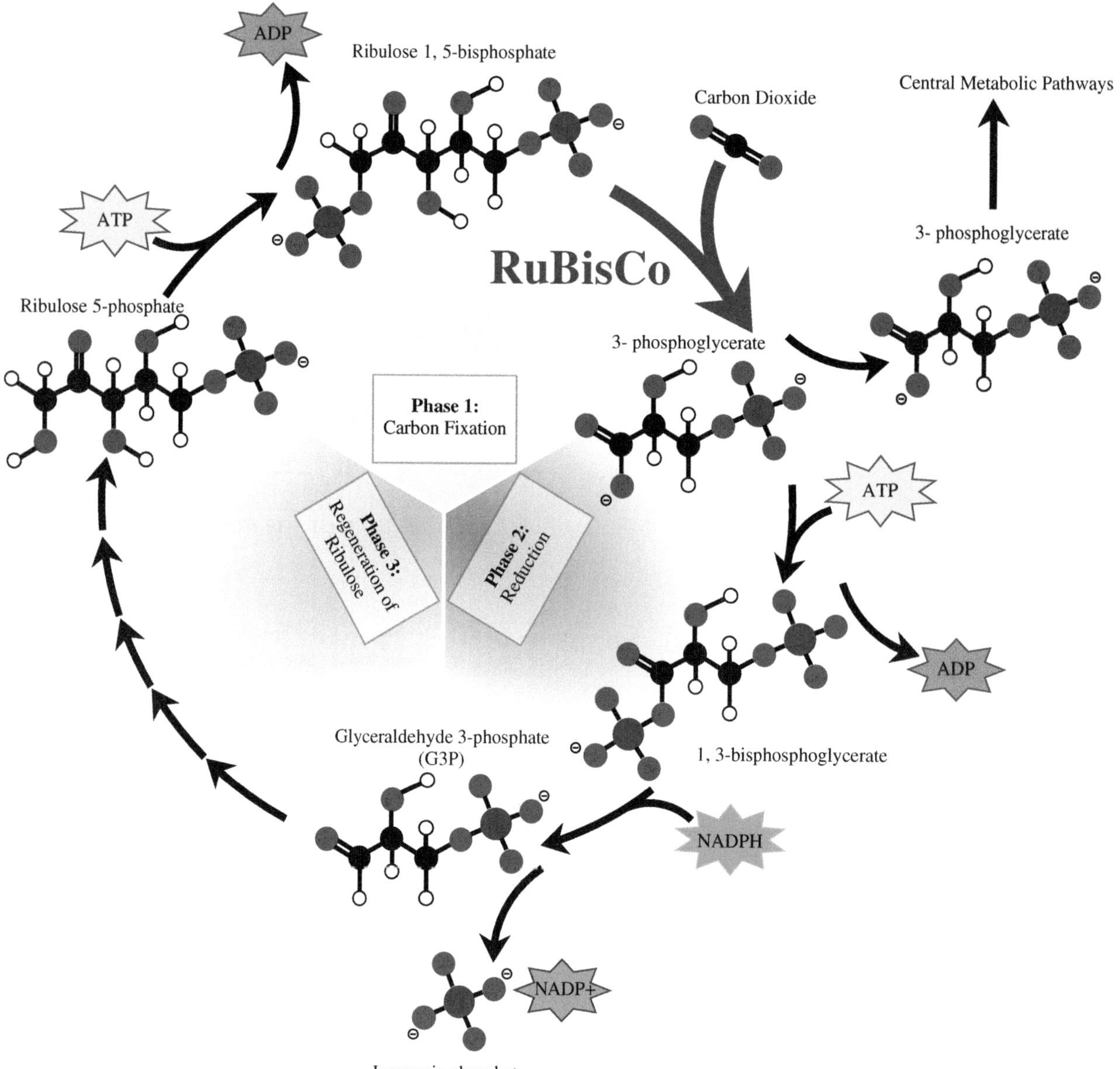

Figure 4.7 The Calvin Cycle. *Source*: http://en.wikipedia.org/wiki/File:Calvin-cycle4.svg. Mike Jones (Adenosine).

photosynthesis for organisms containing chlorophyll is presented below:

$$6\,CO_2 + 6\,H_2O + 2800\text{ kJ energy from sun} \xrightarrow{\text{chlorophyll}} C_6H_{12}O_6 + 6\,O_2 \quad (4.9)$$

Some bacteria that use sunlight as their energy source do not produce oxygen as a by-product of photosynthesis. When sunlight is not available, or when the organism is unable to carry out photosynthesis (e.g., during the night), Equation (4.9) proceeds in the reverse direction, so that energy can be produced by aerobic respiration.

4.3.3.1 Chemosynthesis

Another biochemical process similar to photosynthesis is chemosynthesis. In chemosynthesis, chemoautotrophic organisms convert carbon dioxide and/or methane into organic matter in the absence of sunlight. Rather than using sunlight as a source of energy, these organisms oxidize inorganic compounds for energy. Examples of chemoautotrophic organisms include sulfur oxidizing bacteria, iron oxidizing bacteria, and the nitrifying bacteria. Equation (4.10) shows a chemosynthetic reaction carried out by purple sulfur bacteria, in which glucose and sulfuric acid are produced from carbon dioxide.

$$6CO_2 + 6H_2O + 3H_2S \leftrightarrow C_6H_{12}O_6 + 3H_2SO_4 \quad (4.10)$$

4.3.4 Growth regulating environmental factors (temperature, pH, moisture, oxygen)

Several environmental factors impact the growth of microorganisms. The engineer must provide the proper environment in order to optimize and ensure proper treatment in biological systems. Temperature, pH, moisture, and oxygen are the principal environmental parameters that must be controlled.

4.3.4.1 Temperature Effects

Temperature affects both biological and chemical reactions. As a rule of thumb, a ten degree increase in temperature typically results in doubling the rate of a biological or chemical reaction. The growth rate of a microorganism will increase with an increase in temperature until the maximum rate of growth is achieved. Figure 4.8 shows the relationship between microbial growth rate and temperature.

The temperature at which maximum growth occurs is called the optimum temperature. Once this is achieved, the growth rate will decrease at a significant rate within a small increase in temperature. The decrease in the growth rate is attributed to the denaturation of proteins, amino acids, and enzymes, ultimately leading to the death of the cell.

Microbes can also be classified according to the temperature range in which they function best. Table 4.6 shows the normal and optimum temperature ranges for growth.

4.3.4.2 pH

Most microorganisms function best at a pH near neutrality, from 6.5 to 7.5 although there are certain types of microbes that can operate at extreme pH values – those above 9 and those below 4. Figure 4.9 shows the relationship between microbial growth rate and pH.

Some general comments concerning pH and microbial growth as noted by Gaudy & Gaudy (1988) are listed in Table 4.7.

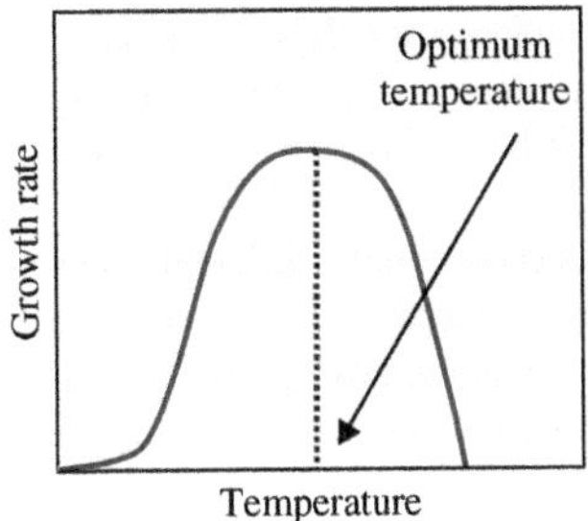

Figure 4.8 Effect of temperature on microbial growth rate.

Table 4.6 Temperature classification of microorganisms.

Classification	Temperature range, °C
Psychrophiles	10 to 30* −7 to 30** −5 to 20***
Mesophiles	20 to 50 25 to 45** 8 to 45***
Thermophiles	35 to 75 45 to 60** 40 to 70***
Hyperthermophiles	65 to 110***

Source: *Metcalf & Eddy (2003), page 559; **Pelczar et *al.* (1977), pp. 111–112; ***Rittman and McCarty (2001), page 16.

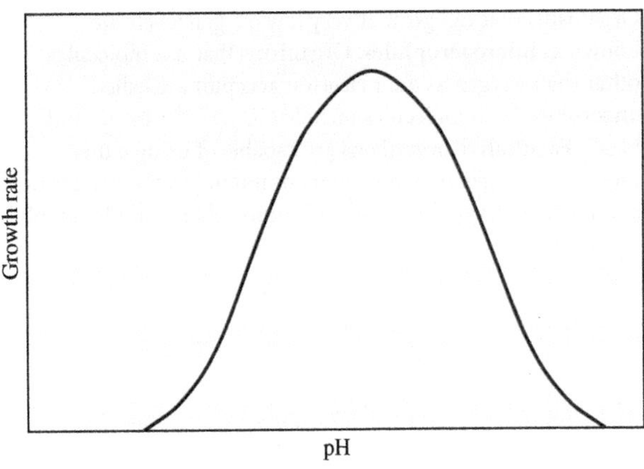

Figure 4.9 Microbial growth rate versus pH.

Table 4.7 General comments concerning the effect of pH on various microorganisms.

Classification	pH range	Optimal pH
*Bacteria	Minimum at 4 and maximum at 9.5	6.5 to 7.5
**Fungi	Prefer acid environment and have minimum between 1 to 3	5
***Protozoa	5–8	7

Adapted from *Metcalf and Eddy (2003), page 559; **Gaudy and Gaudy (1988), page 183; ***Pelczar et *al.* (1977), page 358.

4.3.4.3 Moisture Content

Water is a major component of microbial protoplasm, and an adequate supply must be maintained in order for microbes to grow. Water also is the transport medium that allows nutrients, organics, inorganics, and dissolved oxygen into microbial cells, while carrying away waste products. In aquatic ecosystems and biological wastewater treatment systems, this does not pose a problem. In bioremediation treatment processes, however, the moisture of the soil must be in the range of 50–75% of the soil's field capacity (US EPA, 1985) for microbes to function at an optimum level.

4.3.4.4 Oxygen

Some microorganisms require oxygen in order to survive. Organisms that use molecular oxygen (O_2) as their electron acceptor are called **aerobes**, whereas organisms that cannot grow in the presence of oxygen are called obligate anaerobes.

Organisms that can grow at very low oxygen levels are known as **microaerophiles**. Organisms that use molecules other than oxygen as their electron acceptor are called **anaerobes**. Such molecules include CO_2, SO_4^{2-}, NO_2^-, and NO_3^-. **Facultative organisms** are capable of using either molecular oxygen or some other compound as their electron acceptor; however, their growth is more efficient under aerobic conditions.

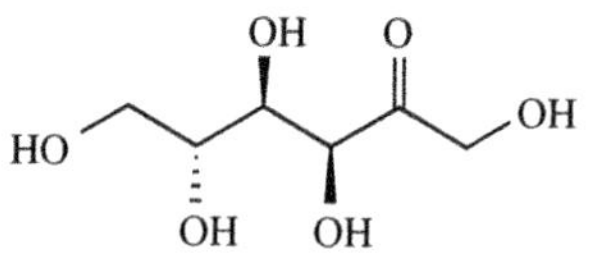

4.3.5 Biochemistry overview

All living cells are essentially composed of the same types of compounds: carbohydrates, lipids, proteins, and nucleic acids (RNA and DNA). **Anabolism** is a synthesis process in which energy is primarily provided from the high-energy phosphate bond of ATP. Biochemical reactions are used for synthesizing cellular material. Microorganisms use carbohydrates for energy and for synthesizing cell tissue. Some bacteria store carbohydrates as polysaccharides, within the cell or on the outside, in the form of a slime layer or capsule. As discussed previously, simple sugars such as glucose can be metabolized for energy by the glycolysis pathway, resulting in the production of pyruvic acid.

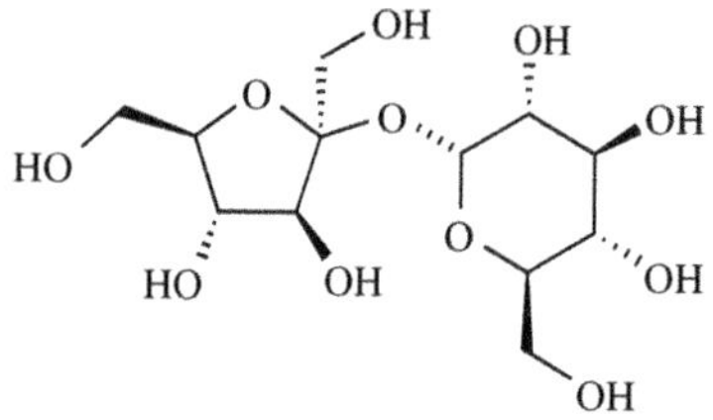

4.3.5.1 *Carbohydrates*

Carbohydrates are compounds that contain carbon, hydrogen, and oxygen. Their molecular formula is represented as $C_nH_{2n}O_n$. They are generally grouped into three categories:

1 simple sugars or monosaccharides;

2 complex sugars or disaccharides;

3 polysaccharides.

Carbohydrates serve as building blocks for other organic compounds or can be metabolized for providing energy for synthesis or for cell maintenance. Simple sugars or monosaccharides contain large numbers of hydroxyl groups (OH). They all contain a carbonyl group ($-C=O$) in the form of an aldehyde or a keto group. Glucose is an important example of a hexose sugar containing six carbons and an aldehyde group (See Figure 4.10).

Fructose is also a hexose sugar containing six carbons, but it has a keto group rather than an aldehyde group (See Figure 4.11).

When two simple sugars are joined together by a glycosidic bond, they form a disaccharide. There are three sugars with a general formula of $C_{12}H_{22}O_{11}$: sucrose, maltose, and lactose. Sucrose is an important disaccharide and is commonly known as "table sugar." The structure of sucrose is presented in Figure 4.12.

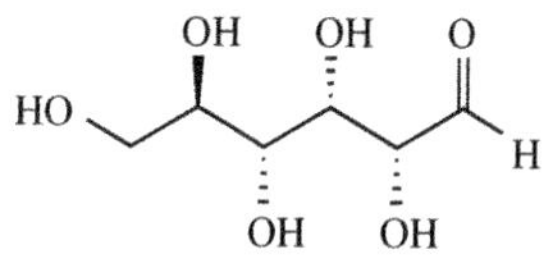

Figure 4.10 Structure of glucose. *Source:* http://en.wikipedia.org/wiki/File:Glucose_chain_structure.svg

Figure 4.11 Structure of fructose. *Source:* http://en.wikipedia.org/wiki/Fructose

Figure 4.12 Structure of sucrose. *Source:* http://en.wikipedia.org/wiki/File:Sucrose_CASCC.png

Polysaccharides are large macromolecules consisting of either monosaccharides or derived sugars linked by glycosidic bonds. There are numerous kinds of polysaccharides, but the most abundant are those derived from glucose. Starch, glycogen, and cellulose are the most important polysaccharides. The generic formula for starch is $(C_6H_{10}O_5)_x$, where x is the number of monomers per molecule and ranges from 100 to 1,000. When starch is hydrolyzed, the monosaccharide glucose is formed.

Glycogen is a class of polysaccharides synthesized by animals and some microbes as a source of stored energy and carbon (Gaudy & Gaudy, 1988, page 97). Cellulose is the most abundant carbohydrate and source of organic carbon on earth. Cellulose does not degrade easily. A few species of microorganisms can produce cellulose, some species can metabolize cellulose anaerobically, while a fewer number can metabolize it aerobically.

4.3.5.2 *Lipids*

Organic compounds that are soluble in organic solvents such as ether, benzene, or acetone, and sparingly insoluble in water, are called 'lipids' (Gaudy & Gaudy, 1988, page 74). Simple lipids are fats and oils or waxes. Each of these is an ester; esters being organic compounds that are formed by the reaction of an acid and alcohol. Fats and oils are esters of the trihydroxy alcohols, glycerol (See Figure 4.13). Waxes, in turn, are esters of long-chain monohydroxyalcohols and alcohols. The main functions of lipids are to store energy and to serve as a structural component of cell membranes.

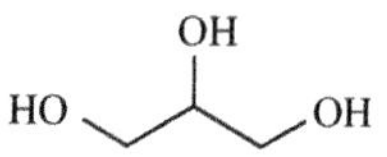

Figure 4.13 Structure of glycerol. *Source:* http://en.wikipedia.org/wiki/File:Glycerin_Skelett.svg

4.3.5.3 Proteins

Proteins are complex molecules consisting of carbon, hydrogen, oxygen, and nitrogen. A few proteins also contain phosphorus and sulfur. Amino acids are the building blocks of proteins; there are approximately 20 amino acids that are found in proteins. Most plants and bacteria are capable of synthesizing amino acids which, in turn, are used for constructing proteins. Higher animals cannot synthesize certain amino acids and, therefore, to obtain them they must consume plant and animal protein as a part of their diet. Free amino acids behave like acids and also like bases, since they contain an amino group.

Amino acids are joined together by peptide links. When two molecules of amino acids are linked, the molecule formed is known as a peptide. If more than three molecules of amino acids are joined, it is called a polypeptide. The general formula of an amino acid is shown below in Figure 4.14.

Table 4.8 lists some of the most common amino acids and their structural formula.

4.3.5.4 Nucleic Acids

Nucleic acids are macromolecular polymers that contain the genetic information of all living organisms. This genetic code provides the directions that govern the growth of organisms and the synthesis of proteins in organisms. There are two types of nucleic acids, deoxyribonucleic acid (DNA) and ribonucleic acid (RNA). Nucleic acids are comprised of a five-carbon sugar, either ribose or 2-deoxyribose; phosphate, and a nitrogen base (purine or pyrimidine). The nitrogen bases include adenine (A), cytosine (C), guanine (G), thymine (T), and uracil (U). Table 4.9 lists the five nitrogen bases, nucleic acids formed, and the structure of each nitrogen base.

Normally, DNA is a double-stranded nucleic acid with two nucleic acid chains linked by hydrogen bonding between nitrogen bases. Some viruses are known to have only a single-stranded DNA. The nitrogen bases found in DNA include A, C, G, and T. Adenine pairs with thymine and cytosine with guanine. Figure 4.15 shows a single strand of DNA.

RNA is normally a single-stranded nucleic acid; however, some viruses have been known to have double-stranded RNA. The nitrogen bases found in RNA include adenine, cytosine, guanine, and uracil; they can pair in various combinations. An RNA monomer is presented in Figure 4.16 alongside DNA for comparison. In-depth discussions of the role of nucleic acids on growth and protein synthesis may be found in the following references: Rittman & McCarty (2001), Baum (1978), and Metcalf & Eddy (2003).

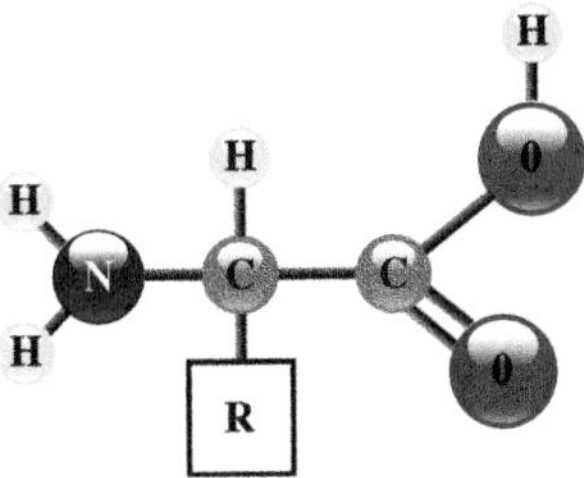

Figure 4.14 General structure of an alpha amino acid. *Source:* http://en.wikipedia.org/wiki/File:AminoAcidball.svg

Table 4.8 Common amino acids.

Name	Structure
Aspartic acid	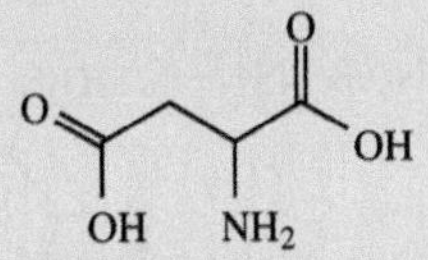 http://en.wikipedia.org/wiki/Aspartic_acid
Glycine	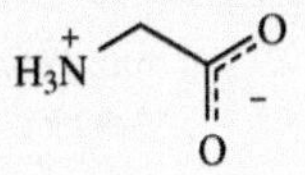 http://en.wikipedia.org/wiki/Glycine
Lysine	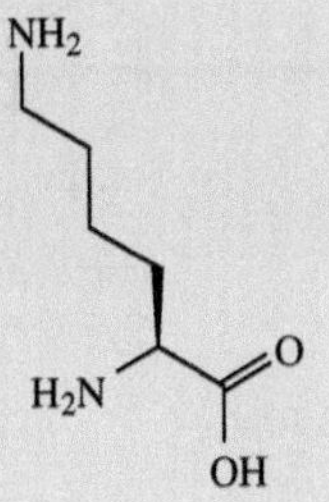 http://en.wikipedia.org/wiki/Lysine
Tyrosine	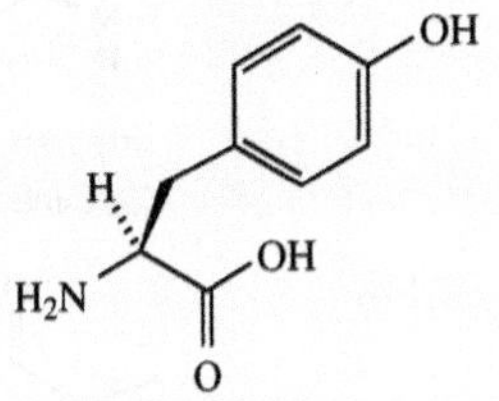 http://en.wikipedia.org/wiki/Tyrosine

4.4 Michaelis-Menten enzyme kinetics

Enzymes are complex organic catalysts (protein molecules) produced by living cells. Recall that the definition of a catalyst is a substance that increases the rate of a reaction and is completely recovered or recycled. Enzymes that function within the cell are known as "intracellular", while those that are secreted outside of the cell are called "extracellular". Enzymes are specific and will catalyze only certain kinds of reactions; they will only act on one specific substrate. The enzyme only combines with the substrate momentarily for a few hundredths of a second. During the combination, a chemical reaction occurs, forming a new compound, and the enzyme is released to combine with another substrate molecule.

Overall rates of biological reactions are dependent on the enzyme concentration and activity. The Michaelis-Menten equation describes the enzyme kinetics for a single reaction and single substrate. The same form of this equation has been successfully used to model the kinetics observed in biological

Table 4.9 Nucleic acid bases and structure.

Nitrogen base name	Nucleic acid	Nitrogen base structure
Adenine	DNA & RNA	http://en.wikipedia.org/wiki/Adenine
Cytosine	DNA & RNA	http://en.wikipedia.org/wiki/Cytosine
Guanine	DNA & RNA	http://en.wikipedia.org/wiki/Guanine
Thymine	DNA	http://en.wikipedia.org/wiki/Thymine
Uracil	RNA	http://en.wikipedia.org/wiki/Uracil

wastewater treatment plants and in the bioremediation of hazardous wastes. In these instances, a heterogeneous culture of microorganisms and diverse array of substrates are found. The derivation of the Michaelis-Menten equation is presented below.

Equation (4.11) shows the reversible reaction relating the concentration of free enzyme [E] and substrate concentration [S] to the concentration of the enzyme-substrate complex [ES] where K_1 is the rate constant for the forward reaction.

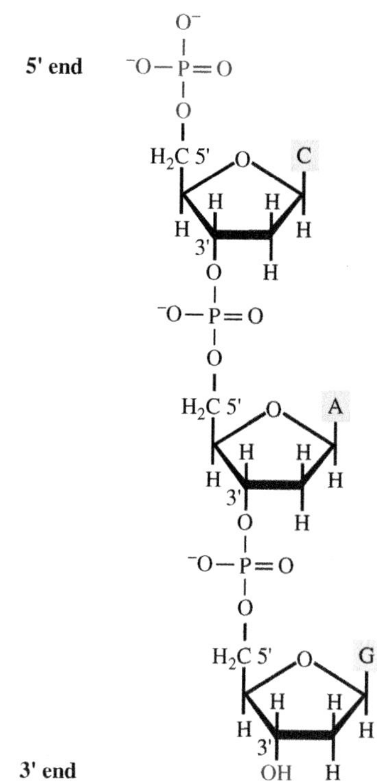

Figure 4.15 Single strand of DNA containing cytosine (C), adenine (A), and guanine (G). *Source*: Molecular Cell Biology. 4th edition. Lodish H, Berk A, Zipursky SL, *et al.* New York: W. H. Freeman; 2000.

The concentration of [ES] then decreases due to the reversible reaction with a rate constant of K_2.

$$E + S \underset{K_2}{\overset{K_1}{\rightleftharpoons}} ES \qquad (4.11)$$

where:

[E] = concentration of free enzyme, mass/volume
[S] = substrate concentration, mass/volume
[ES] = concentration of enzyme-substrate complex, mass/volume.

The concentration of ES is also being removed due to the production of product, P, as shown below. The rate constant for this reaction is K_3.

$$ES \xrightarrow{K_3} E + P \qquad (4.12)$$

Where [P] is the concentration of product expressed as mass/volume.

The overall change in the concentration of the enzyme-substrate complex [ES] with time is given by the following equation:

$$\frac{d[ES]}{dt} = K_1[E][S] - K_2[ES] - K_3[ES] \qquad (4.13)$$

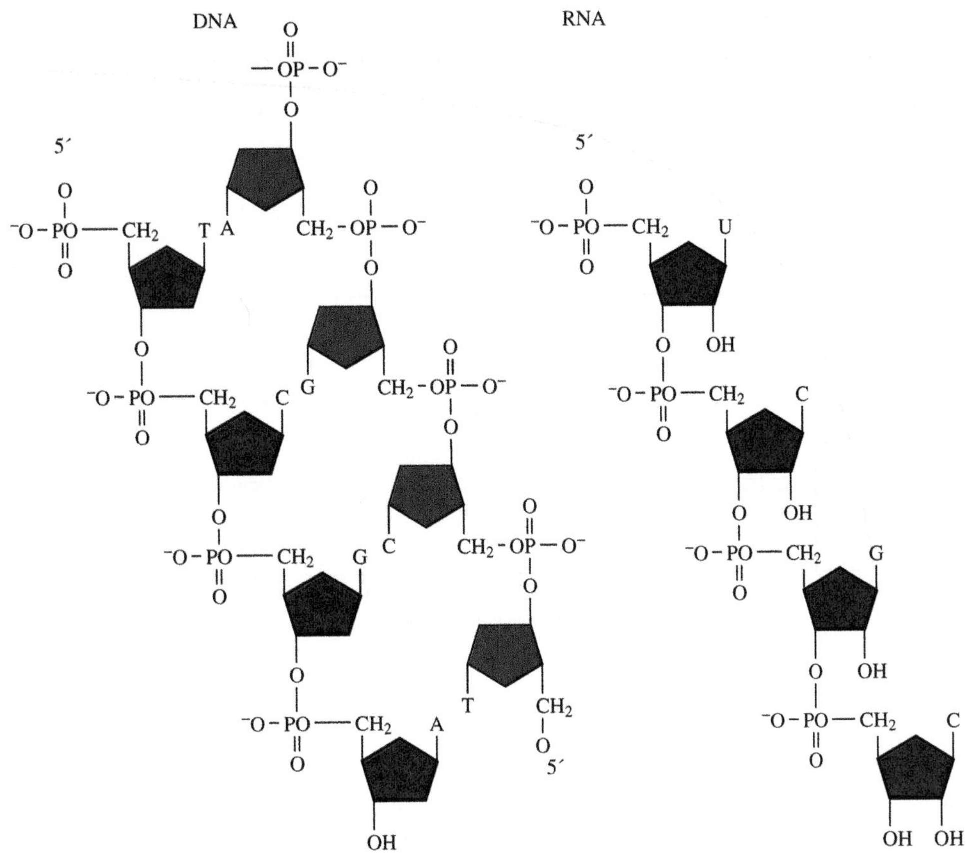

Figure 4.16 Structure of DNA and RNA. *Source:* http://www.biology.arizona.edu/biochemistry/problem_sets/large_molecules/06t.html

At steady-state, the concentration of all reactants and products remains the same; therefore, the change in the concentration [ES] with time can be set equal to zero and Equation (4.15) can be rearranged as Equation (4.16):

$$\frac{d[ES]}{dt} = 0 \qquad (4.14)$$

$$K_1[E][S] = K_2[ES] + K_3[ES] \qquad (4.15)$$

$$[ES] = \frac{K_1[E][S]}{K_2 + K_3} \qquad (4.16)$$

The Michaelis-Menten constant, K_M, is defined as follows:

$$K_M = \frac{K_2 + K_3}{K_1} \qquad (4.17)$$

Where K_M = Michaelis-Menten or saturation constant, mass/volume.

K_M is also equal to the substrate concentration when the rate of reaction is equal to one-half the maximum rate of reaction. See Figure 4.17 for this relationship.

Substituting Equation (4.17) into Equation (4.16) yields the following:

$$[ES] = \frac{[E][S]}{K_M} \qquad (4.18)$$

The total concentration of enzyme present at any time is given below.

$$[E]_T = [E] + [ES] \qquad (4.19)$$

Where $[E]_T$ is the total concentration of enzyme in the system at any given time expressed as, mass/volume.

Or, the free enzyme concentration at any time is determined by solving Equation (4.19) for [E].

$$[E] = [E]_T - [ES] \qquad (4.20)$$

An equation can be developed for rate of reaction, r, as follows:

$$r = \frac{dP}{dt} = -\frac{dS}{dt} = K_3[ES] \qquad (4.21)$$

where:

r = overall rate of the reaction, mass/(volume · time)

$\frac{dP}{dt}$ = change in product concentration with time, mass/(volume · time)

$\frac{dS}{dt}$ = change in substrate concentration with time, mass/(volume · time).

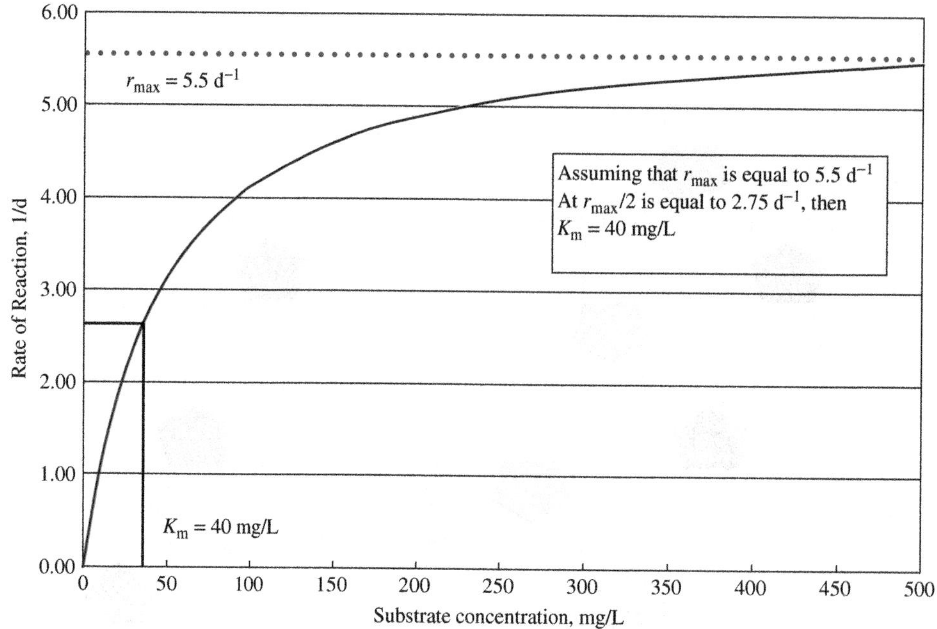

Figure 4.17 Rate of reaction versus substrate concentration.

The overall rate of the formation of product is given as Equation (4.26); it was developed by first substituting Equation (4.20) into Equation (4.18) as follows:

$$[ES] = \frac{[E][S]}{K_M} \quad (4.18)$$

$$[ES] = \frac{([E]_T - [ES])[S]}{K_M} = \frac{[E]_T[S] - [ES][S]}{K_M} \quad (4.22)$$

$$[ES]K_M = [E]_T[S] - [ES][S] \quad (4.23)$$

$$[ES](K_M + [S]) = [E]_T[S] \quad (4.24)$$

$$[ES] = \frac{[E]_T[S]}{K_M + [S]} \quad (4.25)$$

Then Equation (4.25) is substituted into Equation (4.21), yielding Equation (4.26).

$$r = \frac{dP}{dt} = -\frac{dS}{dt} = K_3[ES] \quad (4.21)$$

$$r = \frac{K_3[E]_T[S]}{K_M + [S]} \quad (4.26)$$

The maximum rate of product formation, r_{max}, occurs when the enzyme-substrate complex concentration [ES] is equal to the total enzyme concentration, $[E]_T$. This is expressed mathematically as follows, and occurs only at high substrate concentration wherein no free enzyme exists – it is all bound:

$$r_{max} = K_3[E]_T \quad (4.27)$$

where r_{max} is the maximum rate of reaction, mass/(volume · time).

Utilizing Equation (4.27), Equation (4.26) can be written as follows:

$$r = \frac{r_{max}[S]}{K_M + [S]} \quad (4.28)$$

4.4.1 Limiting conditions

There are two limiting conditions for the Michaelis-Menten equation. At high substrate concentrations, $[S] \gg K_M$, and K_M may be ignored in the denominator of Equation (4.28). This reduces the Equation to a zero-order reaction with respect to the substrate concentration, i.e., the rate of reaction is equal to the maximum rate of reaction and it is independent of the substrate concentration.

$$r = \frac{r_{max}[S]}{0 + [S]} = r_{max} \quad (4.29)$$

When the substrate concentration is very low, $K_M \gg [S]$, [S] can be ignored in the denominator of Equation (4.28). Equation (4.28) then reduces to a first-order reaction with respect to the substrate concentration, as shown by Equation (4.30). Figure 4.17 shows, graphically, the relationship between the rate of reaction, r, and the substrate concentration S.

$$r = \frac{r_{max}[S]}{K_M + 0} = \frac{r_{max}[S]}{K_M} \quad (4.30)$$

4.5 Introduction to ecology

Ecology is the branch of biology that deals with the interrelationships among plants and animals (biota) and their interactions with the physical and chemical environment (abiotic or non-living environment). Animals, microorganisms, and plants are the biotic (living components) of ecological systems or ecosystems. In ecosystems, the term **population** represents a group of organisms belonging to one species. A number of populations that live within an ecosystem is called a **community**.

Whittaker (1975, page 2) defined an ecosystem as "a community and its environment treated together as a functional system of complementary relationships, and transfer and circulation of energy and matter." Ecosystems vary in size, ranging from very small to large areas that occupy much of the Earth's surface. For instance, a small ecosystem could be represented by a compost pile in one's own back yard, or a small pond. Large ecological systems include rain forests, alpine and arctic tundra, temperate grasslands, deserts, continental shelf, and the open ocean. Figure 4.18 is a photo of a freshwater pond ecosystem in Alaska.

Ecosystems may be artificial instead of natural. Constructed wetlands are being designed and built to provide tertiary treatment of municipal wastewater, and for treating stormwater runoff from urban and agricultural areas. Growing crops on agricultural land is another form of artificial ecosystem. Best management practices to minimize the effects of erosion and the release of nutrients (nitrogen and phosphorus) from these sites is critical for maintaining high water quality in receiving streams.

The cycling of nutrients and energy in ecosystems is critical to their sustainability, as well as for Earth's continual survival. Energy flows up through ecological systems through food chains. Energy is used, and some is lost in the form of waste heat as it moves through ecosystems. Ultimately, solar radiation provides the main energy source for autotrophic organisms here on Earth, while chemosynthesis provides a much smaller portion of energy to bacteria and other organisms. Nutrients are continually used and recycled, as plants and animals grow and die within ecosystems. These nutrient cycles will be discussed more thoroughly in later sections of this chapter.

Figure 4.18 Freshwater pond ecosystem in Alaska.

Human activity can have adverse effects on ecosystems. All ecosystems naturally change over time, because of changes in climate, e.g., precipitation, temperature, and nutrients. Volcanic activity, forest fires, and earth quakes can affect ecosystems, destroying certain species. Although it is not always recognized by the general public, humans can have a detrimental impact on the environment through various venues. For instance, the production of food from agricultural operations results in the release of nutrients from fertilizers, pesticides, and animal wastes being released into the environment. Carbon dioxide and other greenhouse gases are released into the atmosphere from tractors and other farming machinery. Dust and erosion from tilling operations can lead to other detrimental effects.

The construction of hydroelectric facilities can have a negative effect on river ecosystems; not only do they affect the migration of certain fish populations, but they can affect the number of fish populations due to changes in water temperature and dissolved oxygen concentration. Another important factor that receives less consideration during the construction and operation of hydroelectric facilities is the loss of habitat and displacement of animals and other organisms.

Harvesting trees for the pulp and paper industry and mining coal from strip mines are two other activities that can have negative consequences on ecosystems if they are not properly managed. For all of these examples, best management practices must be employed to minimize adverse effects and provide sustainable solutions that benefit human beings while protecting our ecosystems.

4.5.1 Food chains

The sequence of steps or trophic levels that are involved in transferring energy in ecosystems may be represented by food chains and food webs. In the classical food chain, plants (primary producers) are consumed or eaten by herbivores (primary consumers) which, in turn, are consumed by carnivores (secondary consumers). Figure 4.19 represents the classical food chain.

The primary producers (plants and algae) make up the first trophic level (level of nourishment) in the classical food chain. Plants and algae are autotrophic organisms, since they obtain their carbon from inorganic carbon sources such as carbon dioxide and bicarbonate. They are also called phototrophs if they contain chlorophyll for the conversion of sunlight into chemical energy.

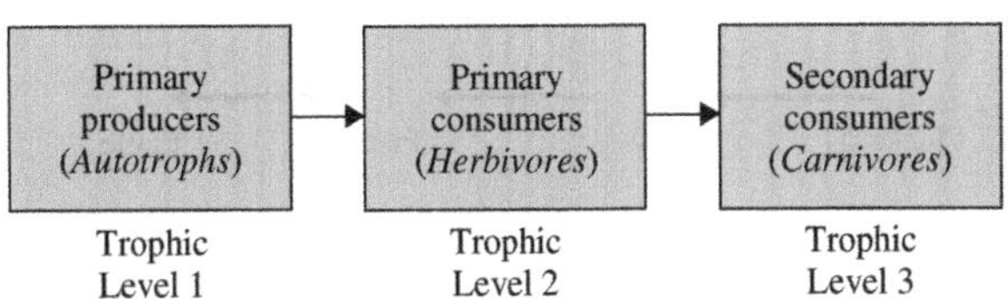

Figure 4.19 Representation of the classical food chain.

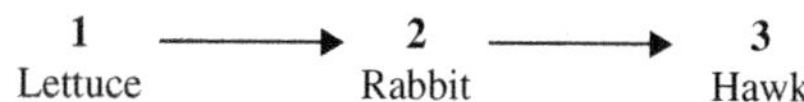

Figure 4.20 Simplified food chain for a terrestrial ecosystem.

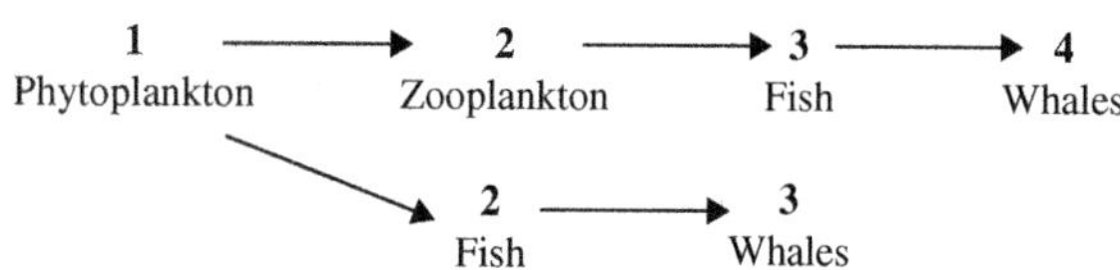

Figure 4.21 Simplified, parallel, food chain for a marine ecosystem.

As we move up to the next trophic level, primary consumers (herbivores) eat the primary producers. Herbivores or primary consumers are chemoheterotrophic organisms. Incidentally, higher levels of energy are expended as we move up through the food chain. This primarily results due to the expenditure of more energy by the organism at a higher trophic level. Herbivores, for example, must roam around looking for sources of plants and vegetation to consume.

At the third trophic level, carnivores or secondary consumers eat the herbivores. A general rule of thumb is that only 10% of the energy that is consumed will be converted to biomass (Davis & Masten, 2009, page 186). Nutrients are recycled in ecosystems by the decomposers, which consist primarily of various species of bacteria and fungi (chemoheterotrophs) that feed upon dead and decaying plant and animal remains.

Figure 4.20 shows a very simplified food chain for a terrestrial ecosystem. For this terrestrial environment example, a rabbit (primary consumer) eats lettuce (primary producer), which in turn is eaten by a red-tail hawk (secondary consumer). A simplified, parallel, food chain is presented in Figure 4.21. In the oceans, phytoplankton are the primary producers; they are consumed by zooplankton and fish at the second trophic level. At trophic level 3, fish consume zooplankton and whales consume the fish. At trophic level 4, whales consume fish.

Example 4.1 Draw a simplified food chain

Draw a simplified food chain diagram with a minimum of four trophic levels or organisms in a fresh-water pond feeding on plankton.

Solution

1 Plankton → 1 Frog → 1 Fish → 1 Eagle

Figure E4.1

A **food web** differs from a food chain in that it is much more complex and it shows the actual pattern of food consumption in an ecosystem. Figure 4.22 is a representation of a food web, showing the major pathways which food (energy) moves through the ecosystem. In this diagram, vegetation and plankton serve as the primary producers, whereas forage fish such as Chub, Alewife, Smelt, and Sculpin are the primary consumers. Waterfowl, cormorant, herring gull, snapping turtle, lake trout and salmon, and eagle are secondary consumers. Humans are at the highest trophic level; in other words, they are at the top of the food chain.

Example 4.2 Energy balance and efficiency

A 500 kg horse consumes 5 kg of feed containing 75% alfalfa meal and 25% oats each day. The energy content of the feed meal is 4.5 kcal/g. Approximately 35% of the total energy ingested is excreted as undigested material or lost to methane gas production. Of the 65% of the total energy that is digested, 95% is metabolized or lost through heat production. The remaining 5% of this energy is converted to body tissue (data adapted from Pagan & Hintz (1986)).

Determine:

a) Energy converted into body tissue each day (kcal/d).
b) Percentage of energy consumed that is converted into body tissue (%).

Solution part a

The energy content of the feed is calculated as follows:

$$\frac{5\text{ kg feed}}{\text{d}}\left(\frac{1000\text{ g}}{\text{kg}}\right)\left(\frac{4.5\text{ kcal}}{\text{g}}\right) = 2.25\times10^4\ \frac{\text{kcal}}{\text{d}}$$

Draw a schematic to show the energy balance.

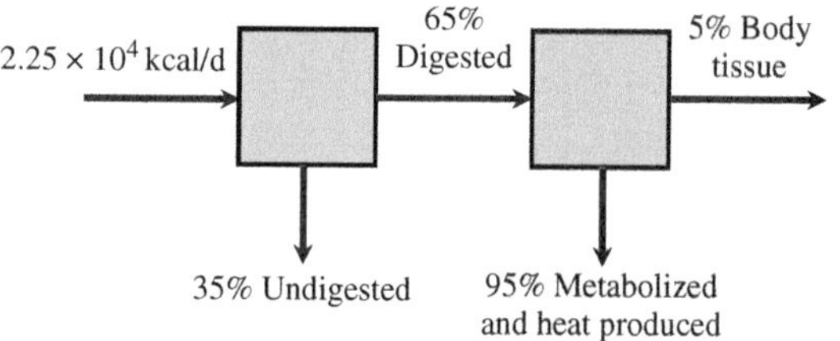

Figure E4.2

The quantity of energy digested is calculate as:

$$2.25\times10^4\ \frac{\text{kcal}}{\text{d}}\times0.65 = 1.46\times10^4\ \frac{\text{kcal}}{\text{d}}$$

The quantity of energy that is converted into body tissue is calculated below.

$$1.46\times10^4\ \frac{\text{kcal}}{\text{d}}\times0.05 = \boxed{730\ \frac{\text{kcal}}{\text{d}}}$$

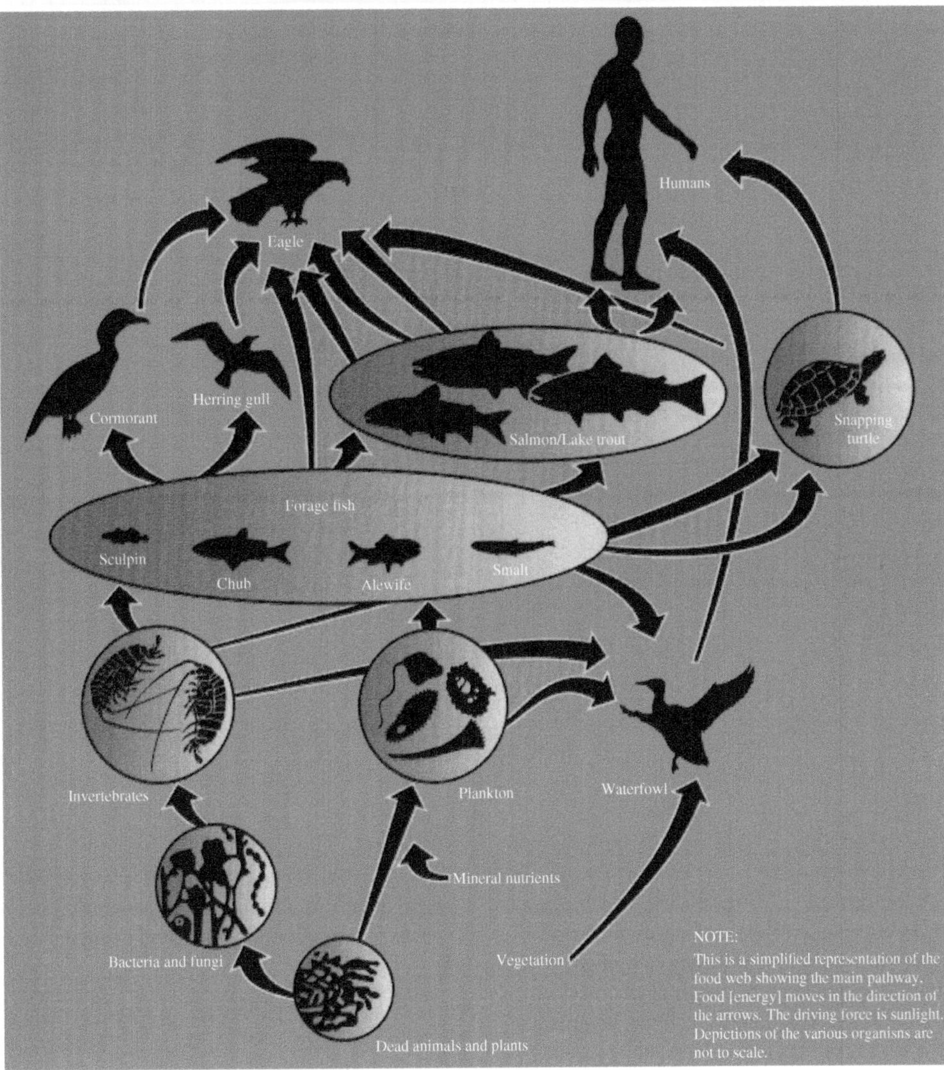

Figure 4.22 Example of a food web. *Source*: http://www.epa.gov/glnpo/atlas/images/big05.gif. United States Environmental Protection Agency.

Solution part b

The percentage of energy consumed that is converted into body tissue is calculated below.

$$\frac{730\,\frac{\text{kcal}}{\text{d}}}{2.25\times10^{4}\,\frac{\text{kcal}}{\text{d}}}(100)=\boxed{3.2\%}$$

The transfer of energy through a food chain is inefficient, and much of the energy is converted to heat energy which is lost to the environment. Typically, the energy available for use at the next trophic level is only 5–20% of the input energy. The actual energy that is available varies with different species. Figure 4.23 is a hypothetical energy pyramid showing how the quantity of energy and the number of organisms decreases as one travels up the food chain. In this pyramid, the energy available for each trophic level is assumed to be 10% of the input energy.

Research conducted by Lindeman (1942) on Cedar Bog Lake in Minnesota, as presented by Henry & Heinke (1996, page 311), shows the flow of energy expressed in kJ/(m^2Ayr) through the food chain (Figure 4.24). The percentage of the energy available for the herbivores from consuming the autotrophs is approximately 13.5% (630/4660 × 100 = 13.5). The carnivores have only 19.8% (125/630 × 100) of the input energy from the herbivore production.

In general, less energy is lost when there are fewer trophic levels to contend with in a food chain. Or in other words, a shorter food chain results in increased efficiency in the transfer of energy, since less energy is wasted (e.g., lost to heat, respiration, and decomposition).

4.6 Primary productivity

The rate at which plants and photosynthetic organisms produce organic carbon compounds from atmospheric or aquatic carbon dioxide is defined as **primary productivity**. To a much

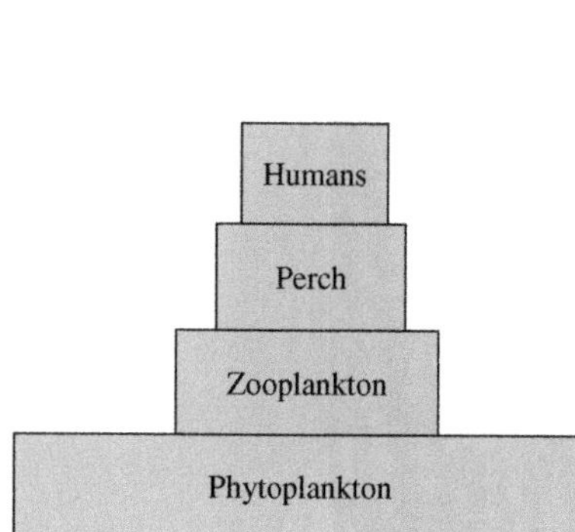

	Usable energy (kcal)	Number of organisms
Tertiary consumers	10	1
Secondary consumers	100	100
Primary consumers	1000	10,000
Primary producers	10,000	1,000,000

Figure 4.23 Hypothetical energy pyramid for a typical food chain. *Source*: From Miller (1975). *Living in the Environment*. Brooks/Cole, a part of Cengage Learning, Inc. Reproduced by permission: www.cengage.com /permissions.

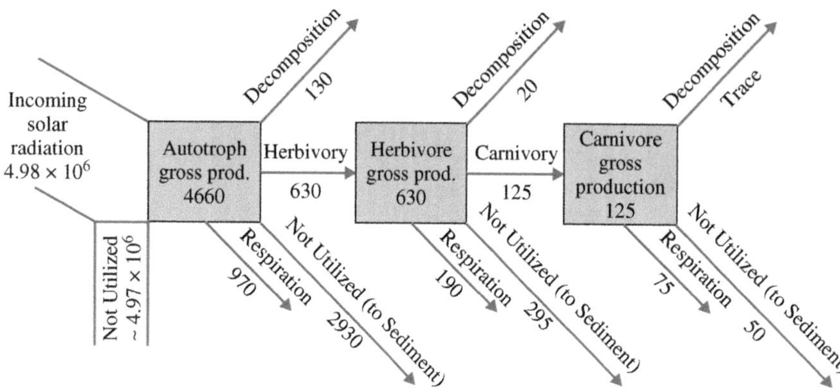

Figure 4.24 Fate of energy through food chain in Cedar Bog Lake, Minnesota. *Source*: Adapted from Henry and Heinke (1996). Environmental Science and Engineering, Prentice Hall, Englewood Cliffs, NJ, page 311.

less extent, chemosynthesis is a viable pathway in which some organisms, especially bacteria, convert inorganic compounds into organic compounds without the presence of sunlight. Primary productivity involves the production of chemical energy in the form of organic compounds by living organisms, primarily using the energy of sunlight to drive the reaction.

A simplified chemical reaction illustrating photosynthesis is presented below. Plants consume carbon dioxide and water in the presence of sunlight to produce organic compounds (carbohydrates, CH_2O) and oxygen. Simple carbohydrates and sugars produced from photosynthesis are then, in turn, used for synthesizing more complex molecules, including amino acids, proteins, lipids, and nucleic acids.

$$CO_2 + H_2O + \text{sunlight} \rightarrow CH_2O + O_2 \quad (4.31)$$

An example of the production of carbohydrate by chemosynthesis is shown below. Organisms that obtain all their carbon from carbon dioxide are called chemoautotrophs.

$$CO_2 + O_2 + 4H_2S \rightarrow CH_2O + 4S + 3H_2O \quad (4.32)$$

Primary productivity may be expressed as gross primary production (GPP) or net primary production (NPP). GPP is the rate at which an ecosystem produces organic compounds in the form of biomass in a given length of time. Net primary production accounts for the loss in energy that cannot be captured and stored as biomass since it is used by an organism for cellular respiration and maintenance functions (R). Net primary production is calculated using the following equation:

$$\text{NPP} = \text{GPP} - \text{R} \quad (4.33)$$

The primary productivity of a community in an ecosystem is expressed as the amount of biomass produced per unit of area and time. Primary productivity is normally expressed in units of dry organic matter, kg C/(m^2Ayr), or in units of energy, joules/(m^2Ayr). Biomass is defined as the mass of organisms per unit of area, and is typically expressed in units of dry organic matter, tons/ha or g/m^2. It may also be expressed in units of energy, joules/m^2. In some instances, biomass may be expressed in units of mass per unit volume. Figure 4.25 shows the global oceanic and terrestrial photoautotroph abundance. This is an approximate indicator of primary production.

Standing crop is another important parameter, and is a measure of the quantity of biomass at a single point in time for a given area. It is expressed in units of calories or grams per m^2. Dividing standing crop by the production rate yields the turnover rate as expressed in the following equation:

$$\text{Turnover Rate(yr)} = \frac{\text{Standing Crop(kg/m}^2)}{\text{Production(kg/(m}^2 \cdot \text{yr))}} \quad (4.34)$$

Turnover rate is expressed in units of time and represents the residence time.

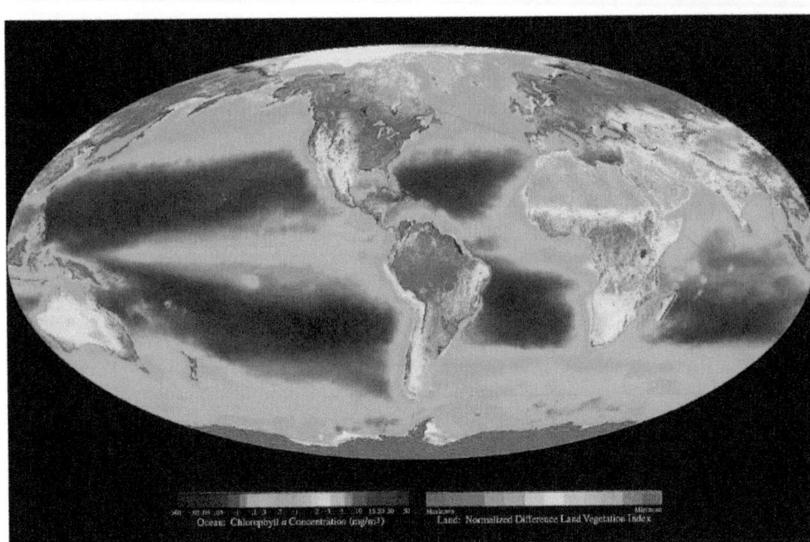

Figure 4.25 Global oceanic and terrestrial photoautotroph abundance, from September 1997 to August 2000. *Source*: http://upload.wikimedia.org/wikipedia/commons/4/44/Seawifs_global_biosphere.jpg. Provided by the SeaWiFS Project, NASA/Goddard Space Flight Center and ORBIMAGE.

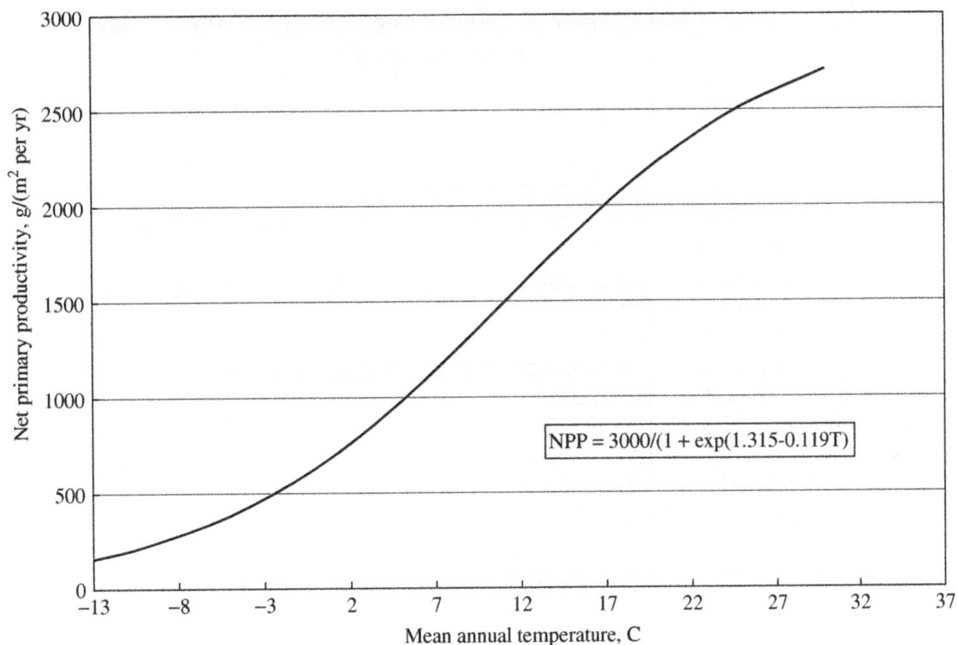

Figure 4.26 Relationship between terrestrial NPP and mean annual temperature. *Source*: Adapted from Whittaker (1975), page 204.

4.6.1 Distribution on earth

Productivity in the ecosystems of the world varies tremendously. The climate and availability of nutrients in a given area control the rate of primary production. Primary productivity is greatest in the oceans followed by terrestrial production. Figures 4.26 and 4.27 illustrate the effects of temperature and precipitation on primary productivity.

Figure 4.26 shows the relationship between terrestrial NPP and mean annual temperature represented by the line of best fit as given in Equation (4.35).

$$\mathrm{NPP} = \frac{3000}{(1 + \exp^{(1.315-0.119 \times T)})} \tag{4.35}$$

where:

NPP = net primary productivity, above and below ground, $\frac{\mathrm{g}}{(\mathrm{m}^2 \cdot \mathrm{yr})}$

T = mean annual temperature, °C.

Figure 4.27 shows the relationship between terrestrial NPP and mean annual precipitation represented by the line of best fit as given in Equation (4.36).

$$\mathrm{NPP} = 3000(1 - \exp^{-0.000644 \times P}) \tag{4.36}$$

where P = mean annual precipitation, mm.

Equations (4.35) and (4.36) are based on the work conducted by Lieth (1973) as presented in Whittaker (1975). Essentially, areas that are warm and receive adequate

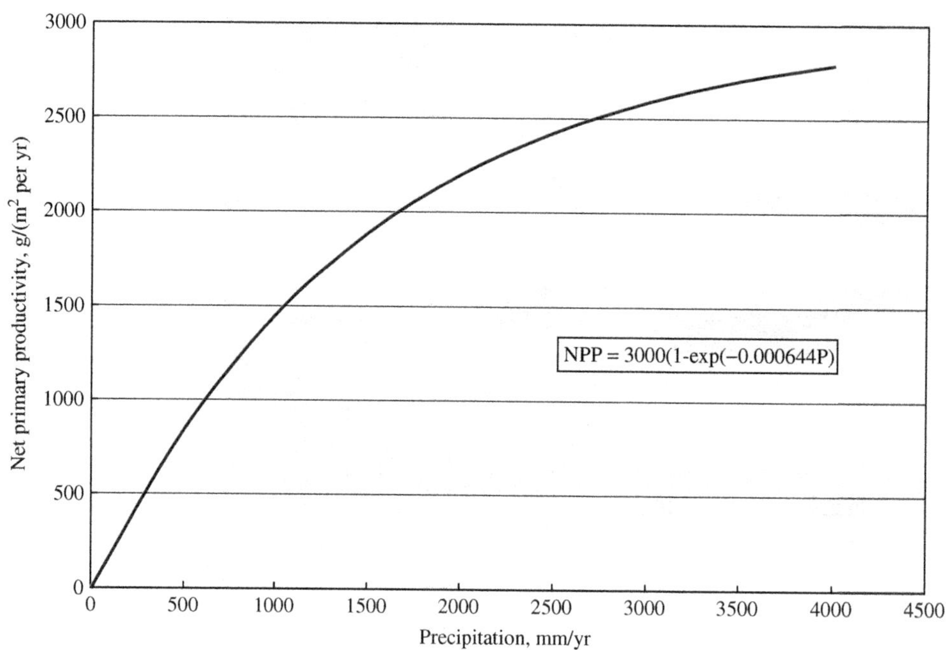

Figure 4.27 Relationship between terrestrial NPP and mean annual precipitation. *Source*: Adapted from Whittaker (1975), page 202.

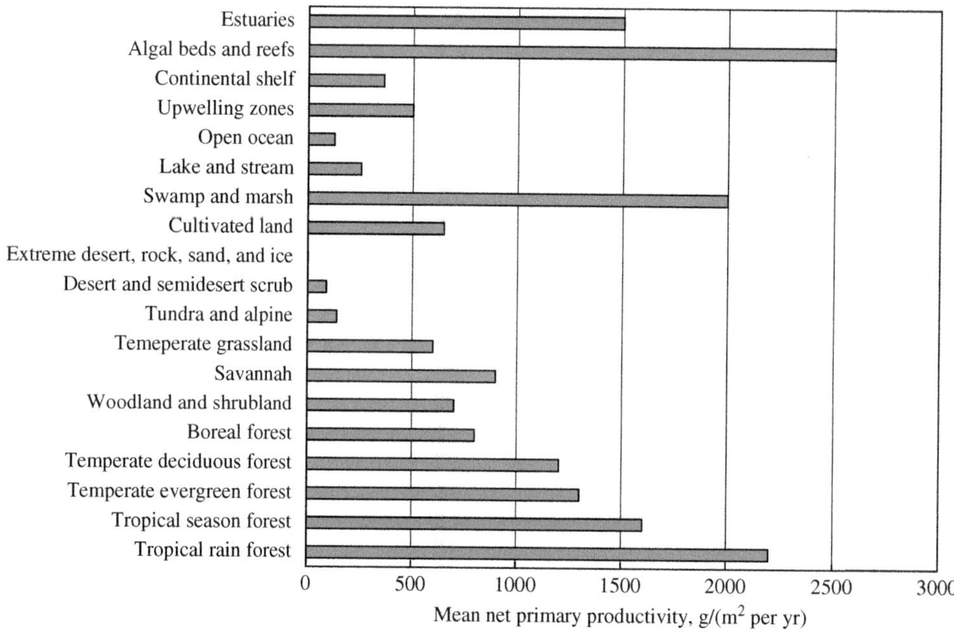

Figure 4.28 Mean net primary productivity for various aquatic and terrestrial ecosystems. *Source*: Adapted from Whittaker (1975), page 224.

precipitation are more productive than those that do not, e.g., deserts and polar regions.

Figure 4.28 shows the mean NPP, g/($m^2 \cdot$ yr), for various aquatic and terrestrial ecosystems.

On the land, the most productive ecosystems are swamps and marshes, tropical rain forests, and temperate rain forests (see Figure 4.28). Almost all primary production is performed by vascular plants (those having special tissue for conducting water). As mentioned earlier, climate and nutrient availability affect the growth of plants. Plants and trees need water for growth and for photosynthesis. The amount of sunlight and precipitation for a given region also impact productivity. Plant growth is curtailed in areas where extreme temperatures and where limited nutrients and water exist.

The mean World NPP, Gt/yr, is plotted as a function of the type of ecosystem in Figure 4.29. As shown in Figures 4.28 and

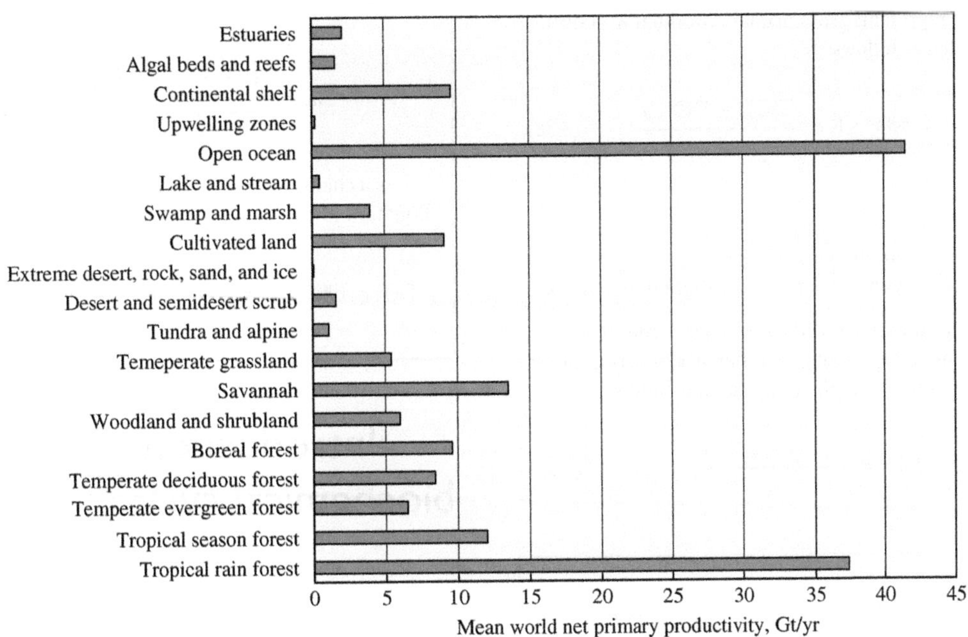

Figure 4.29 Mean world net primary productivity for various terrestrial and aquatic ecosystems. *Source*: Adapted from Whittaker (1975), page 224.

4.29, the tundra and desert are areas that are not nearly as productive as other areas, such as temperate and tropical rain forests.

Unlike terrestrial ecosystems, phytoplankton and other autotrophic organisms are responsible for primary production in the oceans. Green algae, brown algae and red algae, along with vascular plants such as sea grasses, are the major contributors to primary productivity. Light and nutrients must be available to the phytoplankton and autotrophic organisms, so that primary productivity is not limited. The photic or sunlight zone of the ocean is the depth to which there is sufficient light available for photosynthesis to occur.

Mixing of the vertical water column occurs due to wind energy at the surface of the ocean. Mixing is important by providing nutrients such as nitrogen and phosphorus for growth of the phytoplankton. However, the deeper the layer of mixing, the lower the amount of light that is available to the phytoplankton for photosynthesis. As organisms die and settle to the bottom of the oceans, they decay and bacteria consume organics, releasing nutrients that can re-enter the water column. Stratification at various locations in the ocean may occur during the summer or winter, resulting in decreased productivity due to restricted mixing of the water column. Stratification can restrict the recycling of nutrients and inorganic compounds, since mixing is limited.

4.6.2 Measurement of Productivity

In aquatic systems, there are four major *in vitro* methods used for measuring primary productivity (Bender *et al.*, 1986).

1. ^{14}C assimilation into organic matter.
2. DO light-dark bottle production and consumption.
3. CO_2 light-dark bottle consumption and production.
4. ^{18}O tracer method.

Littler (1973) compared pH, DO, and ^{14}C data with productivity data. He found that the pH and DO methods were more useful and reliable than the ^{14}C methods.

Limnologists and oceanographers typically measure the dissolved oxygen (DO) concentration variation in samples of water contained in sealed bottles. DO is usually measured by titration using the Winkler method or with a DO probe and meter. The procedure involves collecting several samples of water in identical bottles. For illustrative purposes, three samples of water are collected from a lake or water body. For the first sample, the DO concentration is measured immediately and the value recorded. The remaining samples are incubated for some fixed time period, one in light and the other in the dark. After incubation, the DO of each of these samples is measured and the values recorded. The DO measurement for the bottle incubated in the dark is used for determining the respiration rate (R). The respiration rate is determined from the following equation:

$$R = \frac{(DO_{IB} - DO_{DB})}{t} \tag{4.37}$$

where:
R = respiration rate, mg/(LAh)
DO_{IB} = dissolved oxygen concentration in initial bottle, mg/L
DO_{DB} = dissolved oxygen concentration in dark bottle after incubation, mg/L
t = incubation time, h.

In the bottle that is exposed to light, both photosynthesis and respiration occur. The difference in DO between the light and dark bottles represents gross primary production (GPP), which is calculated as follows:

$$GPP = NPP + R = \frac{(DO_{LB} - DO_{DB})}{t} \tag{4.38}$$

where:
GPP = gross primary production, mg/(LAh)
NPP = net primary production, mg/(LAh)
DO_{LB} = dissolved oxygen concentration in light bottle, mg/L.

Net primary production (NPP) is estimated from the following equation. It represents the difference between the DO in the light bottle and DO in the initial sample bottle.

$$NPP = \frac{(DO_{LB} - DO_{IB})}{t} \tag{4.39}$$

Example 4.3 Calculating primary production in a water sample

Water samples were collected from Peak Creek in Virginia during late spring. Determine the respiration rate (R), gross primary production (GPP), and net primary production (NPP) in Peak Creek, given the following data. The DO concentrations of the water for the initial sample, light bottle after incubation, and dark bottle after incubation are 7 mg/L, 12 mg/L, and 4 mg/L, respectively. The incubation period is 2 hours.

Solution

The respiration rate is calculated by subtracting the DO concentration in the dark bottle from the DO concentration in the initial bottle and multiplying by the incubation time, Equation (4.37).

$$R = \frac{(DO_{IB} - DO_{DB})}{t}$$

$$R = \frac{\left(\frac{7\,\text{mg}}{\text{L}} - \frac{4\,\text{mg}}{\text{L}}\right)}{2\,\text{h}} = \boxed{\frac{1.5\,\text{mg}}{\text{L}\cdot\text{h}}}$$

Gross primary production is determined using Equation (4.38):

$$GPP = NPP + R = \frac{(DO_{LB} - DO_{DB})}{t}$$

$$GPP = NPP + R = \frac{\left(\frac{12\,\text{mg}}{\text{L}} - \frac{4\,\text{mg}}{\text{L}}\right)}{2\,\text{h}} = \boxed{\frac{4\,\text{mg}}{\text{L}\cdot\text{h}}}$$

Net primary production is determined from Equation (4.39):

$$NPP = \frac{(DO_{LB} - DO_{IB})}{t}$$

$$NPP = \frac{\left(\frac{12\,\text{mg}}{\text{L}} - \frac{7\,\text{mg}}{\text{L}}\right)}{2\,\text{h}} = \boxed{\frac{2.5\,\text{mg}}{\text{L}\cdot\text{h}}}$$

As a check, recall that GPP = NPP + R, Equation (4.38).

$$GPP = NPP + R = \frac{2.5\,\text{mg}}{\text{L}\cdot\text{h}} + \frac{1.5\,\text{mg}}{\text{L}\cdot\text{h}} = \boxed{\frac{4\,\text{mg}}{\text{L}\cdot\text{h}}}.\ \text{checks!}$$

4.7 Introduction to biochemical cycles

4.7.1 Carbon cycle

Carbon is the building block of all life on earth. It is found in the atmosphere, in living organisms, in fossil fuels, rocks, soil, and in the oceans. The major reservoirs of carbon include the oceans, geological formations, and plants. Table 4.10 shows the quantity of carbon in gigatons.

Carbon dioxide is the ultimate source of carbon in organic matter. Organic matter is primarily derived from CO_2, and CO_2 enters the food chain from photosynthesis. Green plants, photosynthetic algae and autotrophic bacteria use carbon dioxide from the atmosphere to produce organic carbon and other organic molecules. A simplified equation for photosynthesis was presented as Equation (4.9).

$$6CO_2 + 6H_2O + 2800\,\text{kJ energy from sun} \xrightarrow{\text{chlorophyll}} C_6H_{12}O_6 + 6O_2 \tag{4.9}$$

Carbon dioxide is returned to the atmosphere by human and animal respiration, by oxidation of organic matter by decomposers consisting primarily of bacteria and fungi, by forest fires, combustion of fossil and other fuels, and by diffusion from the oceans. Respiration can be illustrated by the reversal of Equation (4.9). Carbon compounds such as glucose are oxidized by animals and other organisms, resulting in the release of CO_2, H_2O, and energy in the form of heat.

Table 4.10 Major reservoirs of carbon.

Source	Quantity, gigatons (Gt)	%
Atmosphere	720	1.6
Fossil fuels	4,130	9.1
Oceans	38,400	84.9
Terrestrial biosphere	2,000	4.4
Total	45,250	100.0

Source: Falkowski et al. (2000), page 29.

As stated previously, oceans are a major reservoir or sink for carbon. Carbon dioxide gas from the atmosphere diffuses into the oceans, resulting in dissolved carbon dioxide, $CO_{2(aq)}$. Dissolved CO_2 gas is in equilibrium with the carbonate (CO_3^{2-}) and bicarbonate (HCO_3^-) ions, which make up the carbonate buffering system. Phytoplankton and zooplankton (which are primary producers) use the inorganic carbon forms to produce organic carbon. When plankton die and settle to the bottom of the ocean, this represents a carbon sink. Cycling of this carbon occurs due to mixing and upwelling of water from the bottom of the oceans.

Bacteria, fungi, and other decomposers are responsible for oxidizing plant and animal matter under both anaerobic and aerobic conditions, resulting in the release of nutrients and CO_2. In water, the CO_2 becomes dissolved carbon dioxide gas and, in terrestrial environments, the CO_2 is released to the atmosphere.

The release of CO_2 from anthropogenic sources such as the combustion of fossil fuels (i.e., coal, oil, peat, and natural gas) has resulted in an increase in the concentration of CO_2 in the atmosphere from 280 parts per million by volume (ppmv), pre-industrial (circa 1850), to 380 ppmv in 2006 (NOAA, 2007). This is partially offset by the diffusion of CO_2 into the oceans but, as the concentration of CO_2 increases in the oceans, the capacity of the oceans to take up more CO_2 decreases. The loss of rain forests and other changes in land use globally have also contributed to an increase in the CO_2 concentration in the atmosphere. Recall that photosynthesis involves the utilization of carbon dioxide with the production of organic carbon and release of oxygen as a by-product.

Many scientists believe that there is a direct relationship between the increase in the CO_2 concentration in the atmosphere and the increase in global temperature. Major detrimental effects include the loss of glaciers, a rise in ocean levels, and the extinction of various animals, plants, and other species. This is a complex phenomenon, and one that will be explored in more detail in the section on Global Warming in Chapter 9.

Figure 4.30 shows the global carbon cycle, i.e., the flow of carbon from one reservoir to another and the estimated quantity stored in each reservoir in terms of gigatonnes. One gigatonne (Gt) of carbon is equal to 10^9 tonnes or 10^{12} kg. According to Figure 4.30, 750 Gt of carbon is in the atmosphere and 2,190 GtC is sequestered in terrestrial vegetation, soils, and detritus. The carbon stored in the oceans is slightly less than 39,973 GtC. This quantity is approximately 53 times the quantity of carbon found in the atmosphere.

4.7.2 Nitrogen

Nitrogen is an essential nutrient, since it is found in all forms of life. Based on the elemental composition of a bacterium, as represented by $C_5H_7O_2N$, nitrogen accounts for approximately 12% (14 × 100/113) of the composition. Nitrogen is an essential element in proteins and in amino acids, which are the building blocks of proteins. For example, the amino acid gylcine (CH_2NCH_2COOH) has a molecular weight of 87 grams per mole and nitrogen represents 16% (14 × 100/87) of its composition by weight. All proteins contain carbon, nitrogen, hydrogen, and oxygen. Some also contain phosphorus and sulfur. Typically, nitrogen comprises 15 to 18%, averaging 16% of most proteins (Sawyer *et al.*, 1994, page 250).

Nitrogen exists in several forms in the biosphere. The movement and transformation of nitrogen compounds in the biosphere is called the nitrogen cycle. Figure 4.31 is a generalized characterization of the nitrogen cycle, and several

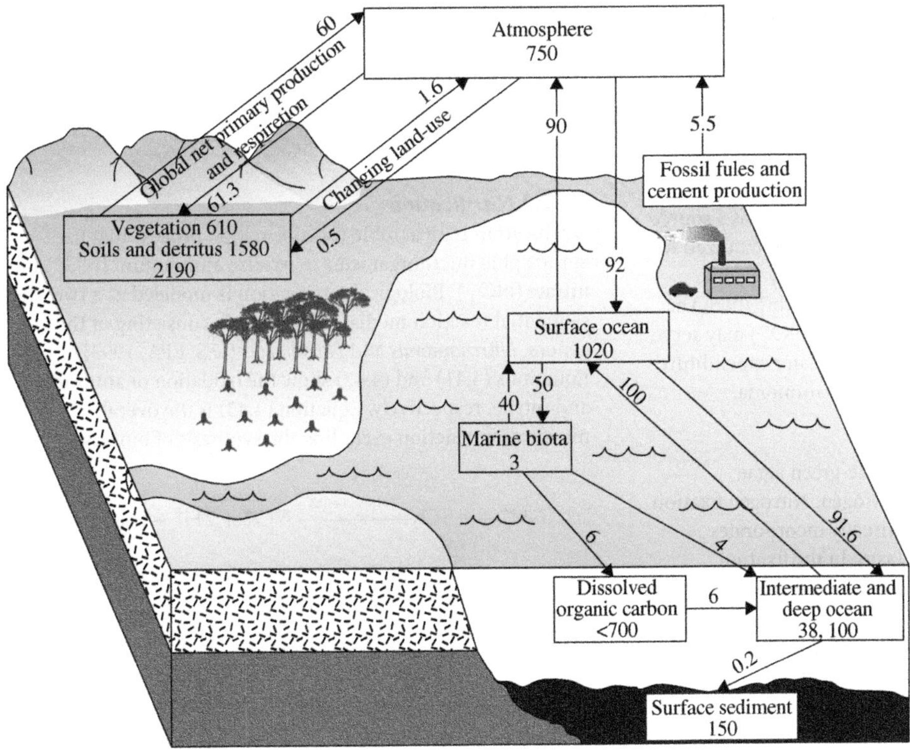

Figure 4.30 The global carbon cycle, showing carbon reservoirs (GTc) and flux (GtC/yr), based on annual averages over the period 1980 to 1989. *Source*: Intergovernmental Panel on Climate Change (IPCC), 1995, page 77. Reproduced by permission of Cambridge University Press.

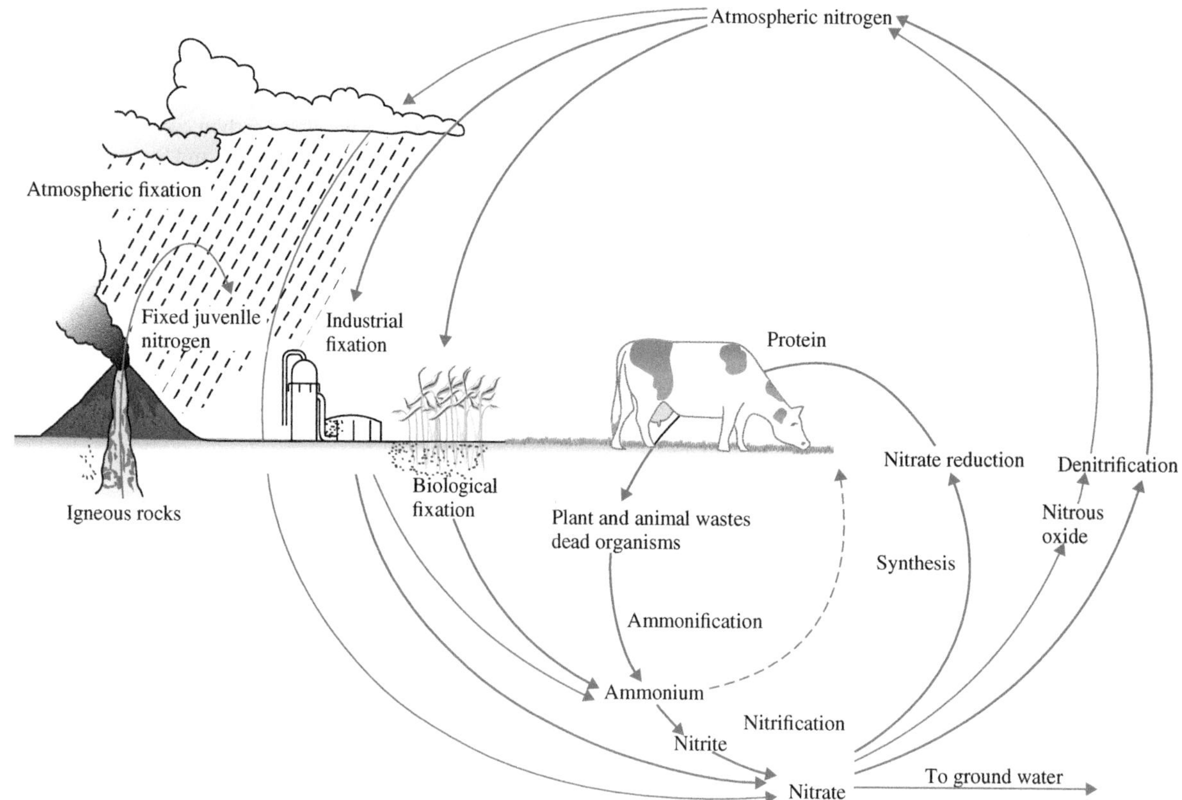

Figure 4.31 Generalized characterization of the nitrogen cycle. *Source*: U.S. Environmental Protection Agency, 1993, page 5. Environmental Protection Agency.

mechanisms are involved in the transformation of nitrogen as it passes through the cycle. Some of the most important ones include fixation, nitrification, denitrification, and ammonification.

The major reservoir of nitrogen on Earth is the atmosphere, which is the major source of nitrogen. The composition of nitrogen (N_2) in dry air is 78.03% by volume and 75.47% by weight (Metcalf & Eddy, 2003, page 1737). Other sources of nitrogen are nitrogenous compounds of plant and animal origin, potassium nitrate, sodium nitrate in mineral deposits, and nitrogenous compounds found in animal excrement.

Plants cannot directly use diatomic nitrogen (N_2) as a source of nitrogen. It must be first reduced to ammonia or oxidized to nitrate before it can be taken up and assimilated by plants. Microorganisms prefer ammonium/ammonia (NH_4^+/NH_3) as their source of nitrogen. Alternatively, nitrate (NO_3^-) may serve as their nitrogen source, but this requires a greater expenditure of energy, since it must then be converted to ammonia.

4.7.2.1 Nitrogen Fixation

Several species of bacteria, fungi, and blue-green algae (cyanobacteria) are capable of fixing nitrogen. Nitrogen fixation is the process by which an organism directly incorporates atmospheric nitrogen into its protoplasm. In the fixation process, N_2 is first reduced to ammonium before it is converted to an organic form.

There are two types of nitrogen-fixing microorganisms: those that form a symbiotic relationship with plants and are associated with the roots of legumes such as peas, soybeans, and clover, and those that are called "free-living" nitrogen fixers.

Nitrogen fixation requires energy, and the following equation illustrates the overall reaction for nitrogen fixation for *Clostridium* (Brock, 1979, page 158).

$$6H^+ + 6e^- + N_2 + (18-24)ATP \rightarrow 2NH_3 + (18-24)ADP + (18-24)P_i \tag{4.40}$$

where P_i = inorganic phosphate.

4.7.2.2 Nitrification

Nitrification is an aerobic transformation process that uses autotrophic microorganisms to oxidize ammonium (NH_4^+) into nitrate (NO_3^-). Biological nitrification is modeled as a two-step sequential reaction mediated by bacteria consisting of the genera, *Nitrosomonas* and *Nitrobacter* (U.S. EPA, 1993). Equations (4.41) and (4.42) show the oxidation of ammonium and nitrite, respectively. Equation (4.43) is the overall nitrification reaction excluding the synthesis of biomass.

$$NH_4^+ + 1.5O_2 \xrightarrow{\text{Nitrosomonas}} NO_2^- + 2H^+ + H_2O \tag{4.41}$$

$$NO_2^- + 0.5O_2 \xrightarrow{\text{Nitrobacter}} NO_3^- \tag{4.42}$$

$$NH_4^+ + 2O_2 \xrightarrow{\text{Nitrifiers}} NO_3^- + 2H^+ + H_2O \tag{4.43}$$

Equation (4.43) indicates that 4.57 g of O_2 are required per g of ammonium nitrogen ($2 \times 32/14 = 4.57$) oxidized to nitrate.

Using the equivalent weight of $CaCO_3$ as 50 g per equivalent, 7.14 grams of alkalinity expressed as $CaCO_3$ are required per gram of ammonium nitrogen oxidized $((2 \times 1/14) \times 50 = 7.14)$. If an insufficient concentration of alkalinity is present, nitrification will be inhibited, since the nitrifiers require inorganic carbon as their carbon source for synthesizing biomass.

Example 4.4 Nitrification stoichiometric coefficients

Calculate the stoichiometric coefficients for oxygen consumption and alkalinity consumption during nitrification using Equation (4.43).

Solution

For oxygen consumption:

$$\left(\frac{2\text{ moles } O_2}{1\text{ mole } NH_4^+ - N}\right)\left(\frac{32\text{ g } O_2}{1\text{ mole } O_2}\right)\left(\frac{1\text{ mole } NH_4^+ - N}{14\text{ g}N}\right) = 4.57\,\frac{\text{g } O_2 \text{ consumed}}{\text{g } NH_4^+ - N \text{ oxidized}}$$

For alkalinity consumption:

$$\left(\frac{2\text{ moles } H^+}{1\text{ mole } NH_4^+ - N}\right)\left(\frac{1\text{ g } H^+}{1\text{ mole } H^+}\right)\left(\frac{1\text{ mole } NH_4^+ - N}{14\text{ g}N}\right) \times \left(\frac{1\text{ eq}}{1\text{ g}H^+}\right)\left(\frac{50\text{ g Alkalinity as } CaCO_3}{\text{eq}}\right) = \boxed{7.14\,\frac{\text{g Alkalinity as } CaCO_3 \text{ consumed}}{\text{g } NH_4^+ - N \text{ oxidized}}}$$

4.7.2.3 *Denitrification*

Denitrification is a biologically mediated, anoxic process that involves the reduction of nitrate (NO_3^-) into nitrogen gas (N_2). It is a widely used process in advanced wastewater treatment for the removal of nitrogen from wastewater. A carbon source is required since the denitrifiers are heterotrophic organisms. Biological dissimilatory denitrification is typically modeled as a two-step sequential reaction as follows when synthesis of biomass is excluded and methanol (CH_3OH) is used as the carbon source.

$$6NO_3^- + 2CH_3OH \rightarrow 6NO_2^- + 2CO_2 + 4H_2O \quad (4.44)$$

$$6NO_2^- + 3CH_3OH \rightarrow 3N_2 + 3CO_2 + 3H_2O + 6OH^- \quad (4.45)$$

The overall denitrification reaction is summarized in Equation (4.46).

$$6NO_3^- + 5CH_3OH \xrightarrow{\text{Denitrifiers}} 3N_2 + 5CO_2 + 7H_2O + 6OH^- \quad (4.46)$$

A significant point about denitrification is that a portion of the alkalinity that is destroyed or consumed during the nitrification process is restored. Based on the stoichiometric equation (4.46) for the overall denitrification process, 3.57 g of alkalinity as $CaCO_3$ are produced per g of nitrate nitrogen reduced (see Example 4.5).

Example 4.5 Denitrification stoichiometric coefficients

Calculate the stoichiometric coefficient for alkalinity production during denitrification using Equation (4.46).

Solution

$$\left(\frac{6\text{ moles } OH^-}{6\text{ moles } NO_3^- - N}\right)\left(\frac{17\text{ g } OH^-}{1\text{ mole } OH^-}\right)\left(\frac{1\text{ mole } NO_3^- - N}{14\text{ g}N}\right) \times \left(\frac{1\text{ eq}}{17\text{ g}OH^-}\right)\left(\frac{50\text{ g Alkalinity as } CaCO_3}{\text{eq}}\right) = \boxed{3.57\,\frac{\text{g Alkalinity as } CaCO_3 \text{ produced}}{\text{g } NO_3^- - N \text{ reduced}}}$$

4.7.2.4 *Ammonification*

Ammonification is the conversion of organic nitrogen to ammonium/ammonia nitrogen NH_4^+/NH_3. When plants and animals die, proteins are first hydrolyzed by hydrolytic enzymes produced by bacteria. Specific types of bacteria are capable of removing the amino group (NH_2) from amino acids, either under aerobic or anaerobic conditions, resulting in the release of ammonia nitrogen and making it available for recycling in the biosphere.

Figure 4.32 is a representation of the nitrogen cycle in surface water. Nitrogen (NH_4^+/NH_3, organic – N, and NO_3^-) primarily enters the water column from wastewater treatment plant (WWTP) discharges and stormwater runoff. Some nitrogen will enter surface water through precipitation and dustfall.

Nitrogen fixation, in which certain species of bacteria and blue-green algae convert nitrogen gas into NH_4^+/NH_3 and NO_3^-, is another mechanism involved in the nitrogen cycle. In the water column itself, nitrification and denitrification may take place depending on the dissolved oxygen levels. Nitrification typically requires a DO concentration of approximately 2.0 mg/L, whereas denitrification requires extremely low DO levels, approaching 0 mg/L. Ammonification will occur in both the water column and sediments.

Figure 4.33 is a representation of the nitrogen cycle in soil and ground water. Nitrogen enters the soil from the application

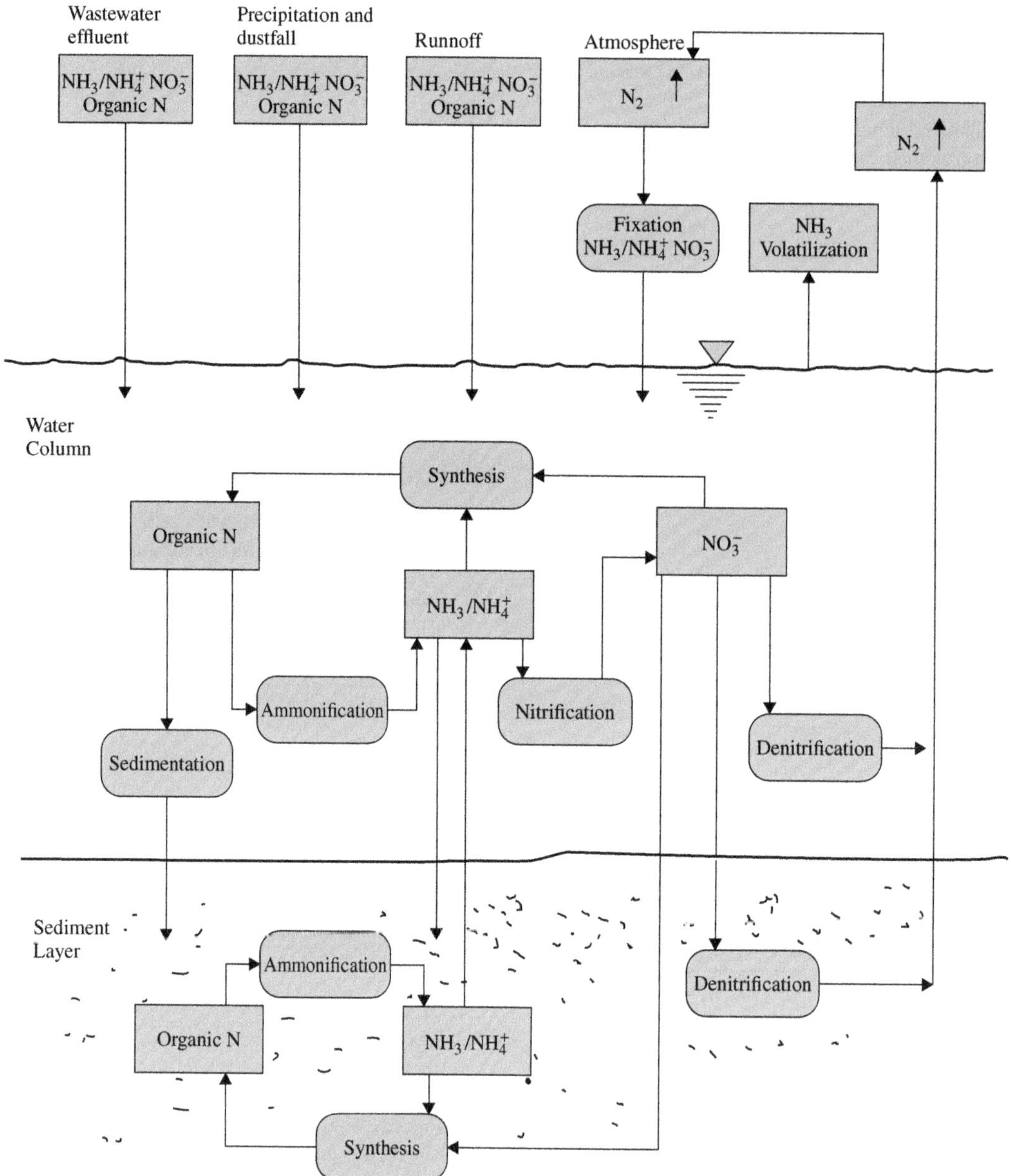

Figure 4.32 The nitrogen cycle in surface water. *Source:* U.S. Environmental Protection Agency, 1993, page 7.

of fertilizer, wastewater effluent, precipitation, dustfall, plant and animal remains, and from fixation. Normally, more than 90% of the nitrogen present in the soil is in the organic form (EPA, 1993, page 6). Nitrate levels in soil are usually low, since nitrate is removed by synthesis reactions, leached by water percolating through the soil, or denitrified under anoxic conditions.

Most state regulatory agencies have established a ground water standard of 10 mg/l $NO_3^- - N$, since synthesis and denitrification cannot remove all nitrates from the soil. This can lead to elevated nitrate levels in ground water at agricultural sites where nitrogen loadings may exceed the nitrogen uptake rate of the crop.

The global nitrogen cycle for 1970 is presented in Figure 4.34. Flux rates are expressed in teragrams (Tg) per year. A Tg is equal to 10^{12} grams. Salient points from Figure 4.34 are that nitrogen fixation in aquatic systems ranged from 20 – 120 Tg/yr, the net transport of nitrogen compounds into the oceans through runoff was estimated at 13 – 24 Tg/yr, and approximately 38 Tg of organic nitrogen are permanently trapped in the sediment.

Table 4.11 shows the quantity of nitrogen found in various reservoirs on a global basis. The percentages of nitrogen found in terrestrial, oceanic, and atmospheric reservoirs are 98, 0.012, and 2.0 %, respectively.

4.7.3 Phosphorus

Phosphorus is also an essential element required by all organisms for growth. Nucleic acids, phospholipids, and several phosphorylated compounds contain phosphorus. As discussed

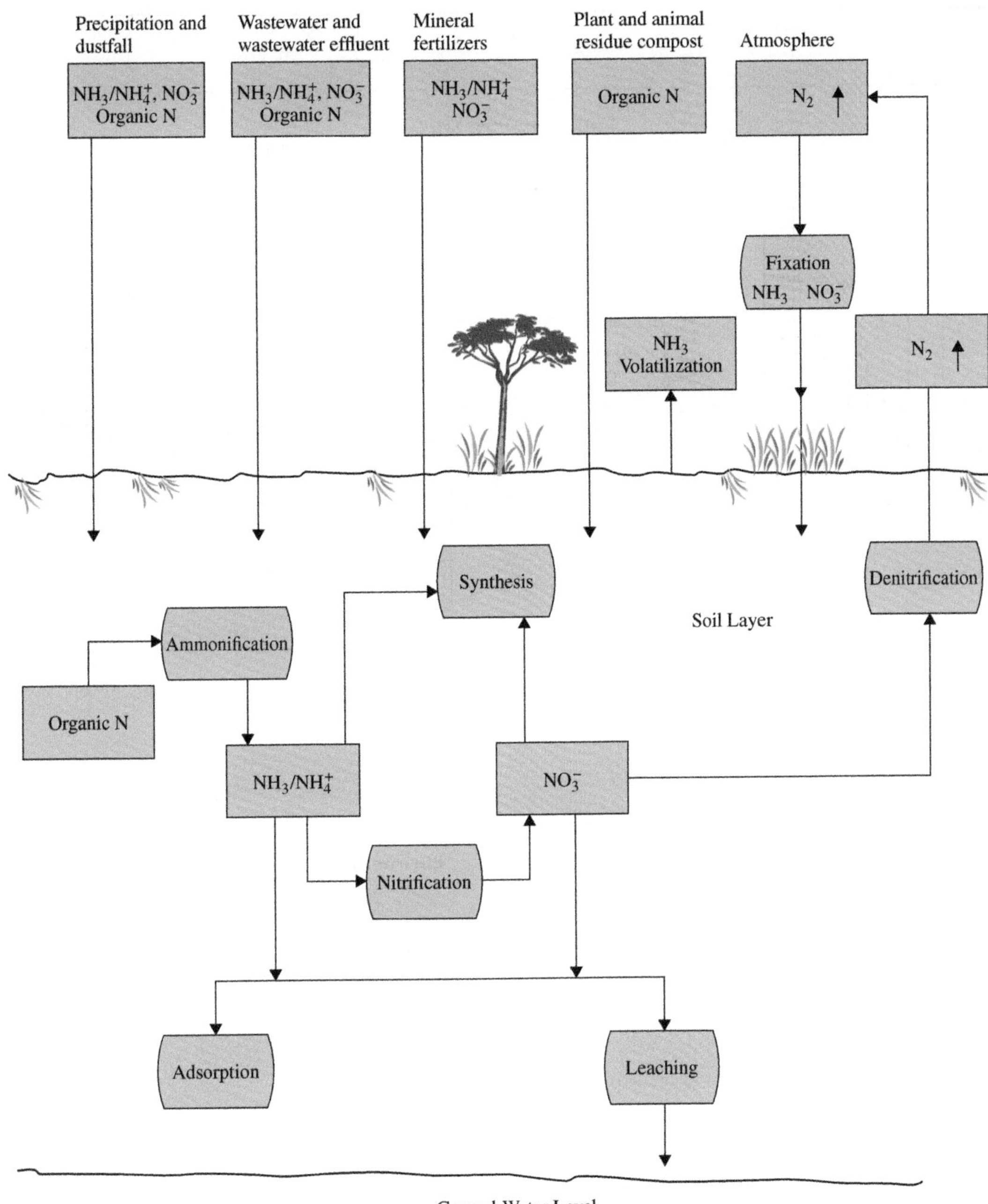

Figure 4.33 The nitrogen cycle in soil and ground water. *Source*: U.S. Environmental Protection Agency, 1993, page 8.

previously, energy is stored in high-energy phosphate bonds during the synthesis of ATP. Substrate-level and electron-transport phosphorylation are two processes by which ATP is synthesized during oxidation-reduction reactions. The cycling of phosphorus in the biosphere is important, since it tends to be the limiting nutrient that must be controlled to prevent and reduce the excessive growth of algae and aquatic plants in lakes and rivers.

Microorganisms, plants, and algae require phosphorus in the dissolved or soluble form as an inorganic orthophosphate species (HPO_4^{2-}, PO_4^{3-}, etc.). Once taken up by the organism, it is converted into organic phosphorus. Using the following chemical formula for a typical microorganism ($C_{60}H_{87}O_{23}N_{12}P$), phosphorus accounts for approximately 2.3% of the biomass by weight ($31 \times 100/1374$).

In nature, phosphorus is primarily found in soils and rocks as calcium phosphate [$Ca(PO_4)_2$] and as hydroxyapatite [$Ca_5(PO_4)_3(OH)$]. Since these compounds are only slightly soluble, the concentration of phosphorus found in natural waters may be as low as 1 ppb (Hutchinson, 1996, page 318). Higher concentrations of phosphorus exist in polluted waters due to human activity. Agricultural stormwater runoff contains phosphorus from the application of fertilizers, which consist of nitrogen, phosphorus, and potassium, along with phosphorus that is found in animal feces. The discharge of municipal and industrial wastewater also contributes a phosphorus load to

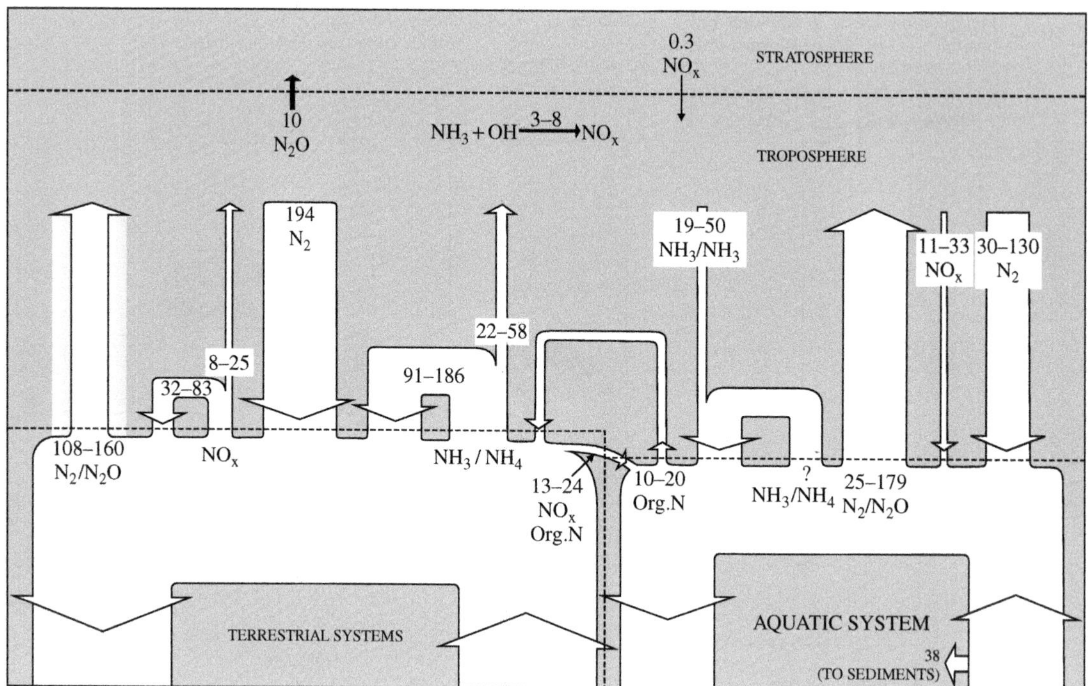

Figure 4.34 Global nitrogen cycle. *Source*: Soderlund and Svensson, 1976, page 27. Reproduced with permission from John Wiley & Sons Ltd.

Table 4.11 Global nitrogen reservoirs.

Environment	TgN
TERRESTRIAL	
Plant biomass	1.1×10^4 to 1.4×10^4
Animal biomass	2.00×10^2
Litter	1.9×10^3 to 3.3×10^3
Soil:	
organic matter	3.0×10^5
insoluble inorganic	1.6×10^4
soluble inorganic	N.A.
microorganisms	5.0×10^2
Rocks	1.9×10^{11}
Sediments	4.0×10^8
Coal deposits	1.2×10^5
Subtotal:	1.904×10^{11}
OCEANIC	
Plant biomass	3.0×10^2
Animal biomass	1.7×10^2
Dead organic matter:	
dissolved	5.3×10^5
particulate	3.0×10^3 to 2.4×10^4
N_2 (dissolved)	2.2×10^7
N_2O	2.0×10^2
NO_3^-	5.7×10^5
NO_2^-	5.0×10^2
NH_4^+	7.0×10^3
Subtotal:	2.31×10^7
ATMOSPHERIC	
N_2	3.9×10^9
N_2O	1.3×10^3
NH_3	0.9
NH_4^+	1.8
NO_X	1 to 4
NO_3^-	0.5
Org-N	1.0
Subtotal:	3.92×10^9
Total	1.943×10^{11}

Table 4.11 (*continued*)

N.A.: not available.
Source: Adapted from Soderlund and Svensson, 1976, page 30. Reproduced by permission of John Wiley & Sons Ltd.

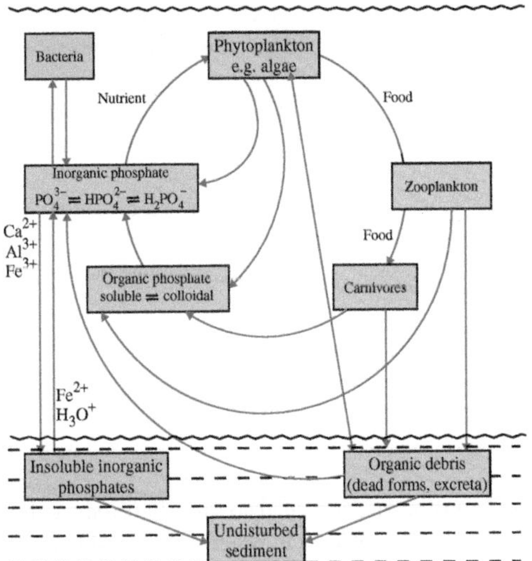

Figure 4.35 The water-based phosphate cycle. *Source*: WEF (1998), page 18.

receiving waters. Phosphorus excreted by humans has been estimated at 0.2 to 1.0 kg phosphorus/(capita · yr), and the amount of phosphorus in wastewater attributed to detergents has been estimated at 0.3 kg phosphorus/(capita · yr) (WEF, 1998, page 17).

Figure 4.35 shows a simplified diagram of the phosphorus cycle in the water column. The most significant attribute of the phosphorus cycle compared to the carbon and nitrogen cycles is that no gaseous phosphorus compounds exist. Therefore, since there is no volatile component, phosphates are only found in the soil and aquatic environments. In lakes and rivers, much of the phosphorus is bound to the sediments. Therefore, sediment in water bodies acts as a sink for organic and inorganic phosphorus. Anaerobic activity in the sediments releases orthophosphates back into the water column. Soil erosion and stormwater runoff containing high levels of fertilizer has a significant impact on the phosphorus cycle.

In the northern hemisphere, phosphorus is returned to the water column during lake mixing that occurs during spring and fall turnover. As noted in Figure 4.35, algae, phytoplankton, and zooplankton are active participants in the cycling of phosphorus in ecosystems. Limnological aspects will be discussed in a subsequent section in this chapter.

Orthophosphate can be removed from industrial and municipal wastewater by adding aluminum and ferric salts for precipitation as metal phosphates or as phosphate hydroxides. Lime may also be used to precipitate phosphorus as hydroxyapatite ($Ca_5(PO_4)_3(OH)$). Phosphorus-accumulating organisms (PAOs) and bacteria such as *Acinetobacter* can be used for removing orthophosphates from wastewater in enhanced biological phosphorus removal (EBPR) systems.

The natural pre-anthropogenic global phosphorus cycle is presented in Figure 4.36. Phosphorus reservoirs are shown in units of teragrams (10^{12}) of phosphorus (TgP) and fluxes in TgP per year. According to Filippelli (2002, page 393), nearly 98% of the 122,600 TgP in the soil/biota system is held in soils in a variety of forms, and the flux rate of soluble phosphorus to the oceans is estimated at 2–3 TgP/year for pre-human or pre-anthropogenic times.

Figure 4.37 shows the global phosphorus cycle and reservoirs associated with human activities. Fillippelli (2002, page 395) estimates that the flux of soluble phosphorus now entering the oceans has doubled, from 2–3 TgP/yr to 4–6 TgP/yr, because of human activities (e.g., use of commercial fertilizers, deforestation and soil erosion, discharges from wastewater treatment plants, and other wastes). This has led to cultural eutrophication, which will be discussed in the limnological section.

4.7.4 Sulfur

The cycling of sulfur through the biosphere is an important consideration in environmental systems. Prior to the Industrial Revolution, the effect of sulfur on the environment was quite

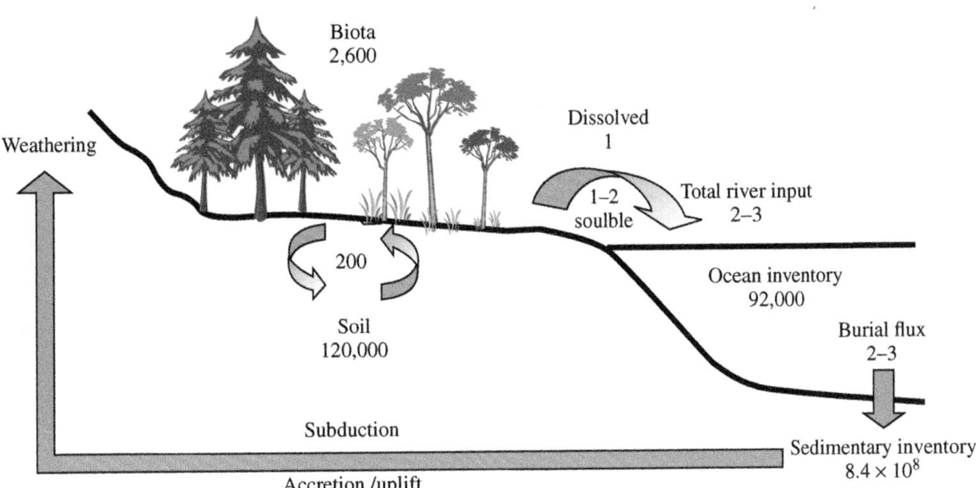

Figure 4.36 The natural pre-anthropogenic global phosphorus cycle. *Source*: Filippelli, 2002, page 392. Reproduced by permission of the Mineralogical Society of America.

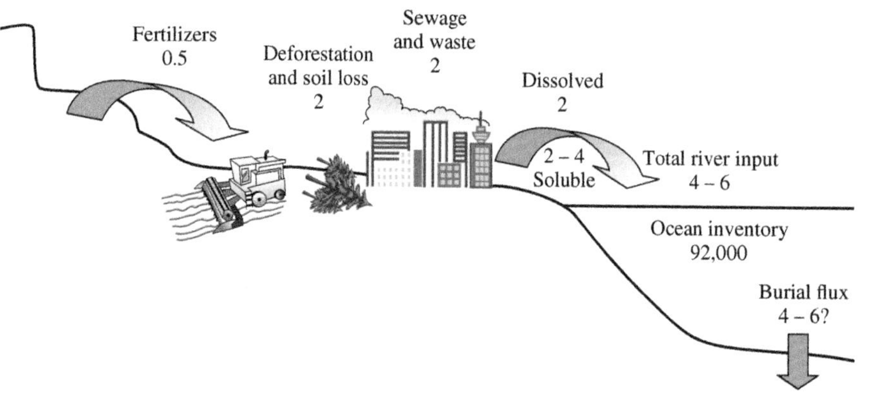

Figure 4.37 The modern phosphorus cycle, showing reservoirs (in Tg P) and fluxes (denoted by arrows, in Tg P/yr) associated with human activities. *Source:* Filippelli, Gabriel M. (2002), page 398. Reproduced by permission of the Mineralogical Society of America.

minimal. Human activities have increased the release of sulfur into the environment from the combustion of fossil fuels, mining activities, and metals processing. The three predominant forms of sulfur found in nature include sulfates (SO_4^{2-}), sulfides (S^{2-}), and organic forms. Sulfur is a component of some amino acids and proteins. It is used in the synthesis processes of plants, animals, and microorganisms.

The sulfate ion is one of the major anions found in water supplies and the oceans. The Secondary Drinking Water Standard for sulfate is 250 mg/L, since it can produce a cathartic effect on humans at higher concentrations. Environmental scientists and engineers are concerned about sulfates, since they can cause taste and odor problems in water supplies and create odor and corrosion problems in sewerage systems and at the headworks of wastewater treatment facilities. The release of sulfur oxides into the atmosphere creates air pollution problems related to malodorous gases and acid rain.

The major reservoir for sulfur in the biosphere is the oceans which primarily contain sulfate and, to a much lesser extent, living and dead organic matter containing sulfur. Sulfur is found in sediments and rocks as gypsum ($CaSO_4$) and pyrite (FeS_2). Figure 4.38 shows the global transport of sulfur.

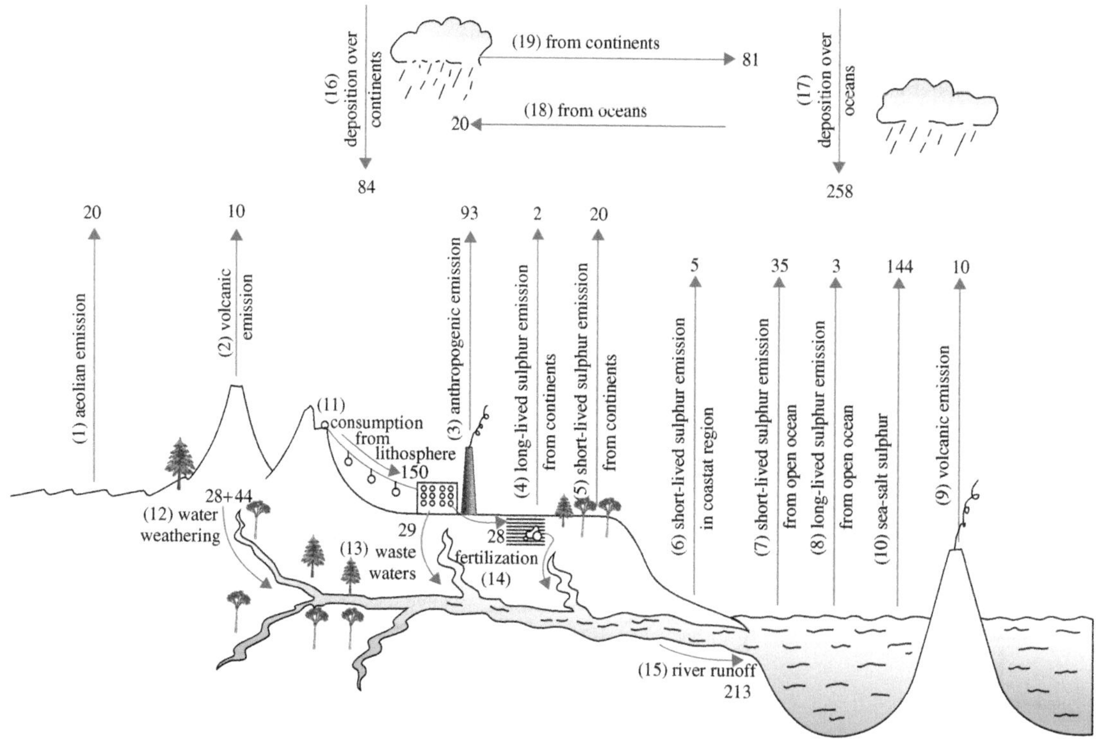

Figure 4.38 Global sulfur cycle. *Source:* Brimblecombe *et al.*, 1989, page 82. Reproduced with permission of John Wiley & Sons Ltd.

Table 4.12 Scheme of the global sulfur cycle during the mid-1980s.

Number	Route	Sulfur compounds	Flux, TgS/yr
1	Aeolian emission	SO_4^{2-}	20
2	Volcanic emission into the continental atmosphere	SO_2, H_2S, SO_4^{2-}	10
3	Anthropogenic emission into the atmosphere	SO_2, H_2S, SO_4^{2-}	93
4	Emission of long-lived sulfur compounds into continental atmosphere	COS, CS_2	2
5	Emission of short-lived sulfur compounds into continental atmosphere	H_2S, DMS, etc.	20
6	Emission of short-lived sulfur compounds into the atmosphere from coastal regions of the ocean	H_2S, DMS	5
7	Emission of short-lived sulfur compounds into the atmosphere from the open ocean	DMS, etc.	35
8	Emission of long-lived sulfur compounds into oceanic atmosphere	COS, CS_2	3
9	Volcanic emission into the oceanic atmosphere	SO_2, H_2S, SO_4^{2-}	10
10	Emission of sea salt aerosol sulfur from the ocean	SO_4^{2-}	144
11	Anthropogenic output from the lithosphere	SO_4^{2-}, S^{2-}	150
12	Weathering and water erosion	SO_4^{2-}	72
13	Wastewaters	SO_4^{2-}	29
14	Mineral fertilizers	SO_4^{2-}	28
15	River runoff into the ocean	SO_4^{2-}	213
16	Scavenging from the atmosphere on the continental surface	SO_2, SO_4^{2-}	84
17	Scavenging from the atmosphere on the oceanic surface	SO_2, SO_4^{2-}	258
18	Transport from the oceanic atmosphere into the continental atmosphere	SO_2, SO_4^{2-}	20
19	Transport from the continental atmosphere into the oceanic atmosphere	SO_2, SO_4^{2-}	81

COS = carbonyl sulfide; CS_2 = carbon disulfide; DMS = dimethyl sulfide.
Source: Brimblecombe, *et al.*, 1989, page 83. Reproduced by permission of John Wiley & Sons Ltd.

Table 4.12 is a listing of all the routes, sulfur compounds, and flux rates associated with Figure 4.38.

4.8 Population dynamics

Population dynamics focuses on how populations (organisms, animals, and people) change with time, and the models used for estimating/predicting population growth. For instance, zoologists and forest rangers may be interested in estimating the carrying capacity of the number of elephants that could be supported in the Liwonde National Park in Malawi, Africa (Figure 4.39). Environmental engineers are concerned about designing systems that use heterogeneous cultures of microorganisms for treating industrial and municipal wastewater or the clean-up of toxic chemicals that have contaminated groundwater and soil. Worldwide human population growth continues to be a main concern and is it

Figure 4.39 Elephants in Liwonde National Park, Malawi, Africa, June, 2010.

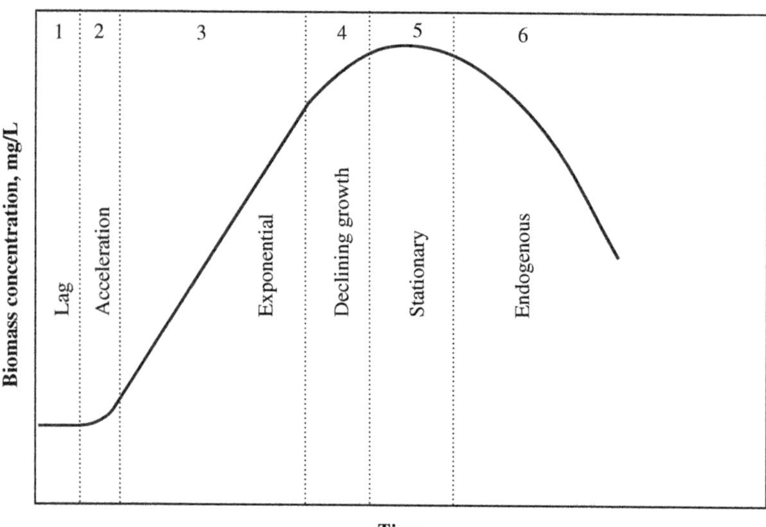

Figure 4.40 Generalized bacterial growth curve for a batch culture.

sustainable? We will primarily look at population models for microbial systems with a follow-up discussion on human population and growth rate.

4.8.1 Bacterial population growth

4.8.1.1 Pure Culture Growth in Batch Systems

Growth of individual cells is defined as an increase in its size and weight which usually precedes cell division. In some instances, cell division may not occur, but the cell's size and weight increase. Population growth involves the increase in the number of cells resulting from cell division. The growth of a population of cells typically occurs at an exponential rate. This means that a cell divides to form two daughter cells which, in turn, divide to produce two new cells. The rate of exponential growth may be expressed in terms of a generation time, t_g or doubling time, t_d, (e.g., the time it takes the cell to divide and form two daughter cells). The generation time for *Escherichia coli* ranges from 15–20 minutes (Henry & Heinke, 1996, page 262).

If a single species of viable bacterial cells or organisms is introduced into a batch system that contains the appropriate substrate, nutrients and environmental conditions, a generalized growth curve as shown in Figure 4.40 will result. Microbiologists often make similar plots based on the number of organisms per unit volume versus time or plots of the biomass measured in terms of optical density versus time. Environmental scientists and engineers typically measure the concentration of biomass in dry mass of solids per unit volume expressed either as total suspended solids (TSS) or volatile suspended solids (VSS). Figure 4.40 is a semi-log plot showing the six distinct phases of growth.

Depending on the microbial culture, environmental conditions, and environmental constraints, one or more of the phases may not occur. Each of the phases will be described briefly. Jacques Monod (1949), the famous French microbiologist, was one of the first to present research on bacterial cultures.

1 **Lag phase:** When a microbial population is inoculated into a fresh medium, some period of time passes before growth begins. This is known as the lag phase, and it may be a brief or extended period of time. There is a null growth rate during the lag phase. Cells taken from another culture must acclimate and adapt to their new environment. If the medium is different from the old medium, the microbes will have to produce appropriate enzymes to metabolize the new medium resulting in a lag in growth. The growth phase from which the inoculum is taken will affect the length of the lag phase. If cells are taken from a culture growing in the stationary phase, a longer lag period will be observed. Cells taken from cells growing at an exponential rate generally do not exhibit a lag phase when transferred to the new medium.

2 **Acceleration phase:** This is normally a brief phase in which the generation time is shorter and the growth rate increases.

3 **Exponential phase:** The maximum rate of growth occurs in this phase and is constant. Substrate and nutrients are in excess, so growth in uninhibited. Cells grow and divide according to the generation or doubling time. Maximum substrate utilization is observed.

4 **Declining growth phase:** The growth rate decreases due to gradual depletion of substrate and the accumulation of toxic waste products.

5 **Stationary phase:** In a batch or closed system, a population of organisms cannot grow at an exponential rate indefinitely. Some limiting nutrient, carbon, energy, nitrogen, phosphorus, or other growth factor will cause the population growth rate to cease. In this phase, there is a null growth rate. The substrate and nutrients have been depleted from the medium.

6 **Endogenous phase:** In this phase, the cells may remain alive and rely on endogenous metabolism, but most often they die. Cell lysis typically accompanies cell death, which leads to a decrease in the number of cells as well as in cell mass. As this phase proceeds, the death rate increases at an exponential rate.

Bacteria multiply by binary fission, and their generation or doubling rate may be as little as 15–20 minutes; this is the time it takes a bacterium cell to divide into two daughter cells. The process is then repeated. The maximum growth of bacteria occurs during the exponential growth phase. Exponential or logarithmic growth can be expressed by the following equation:

$$N_t = N_0 2^{k \times t} \tag{4.47}$$

where:
N_0 = number of organisms initially
N_t = number of organisms at a future time, t
K = growth rate constant, number of doublings per unit time
t_d = doubling time = t_g= generation time = $1/k$, time
t = time period.

Rearranging and taking the natural logarithm of both sides of Equation (4.47) yields.

$$\ln(N_t) - \ln(N_0) = k\ln(2)t \tag{4.48}$$

A plot of $\ln(N_t)$ versus time produces a straight line if exponential growth is occurring. The slope of the line of best fit is equal to $k \times \ln(2)$ and the y-intercept is $\ln(N_0)$. Example 4.6 illustrates how one determines the growth rate constant and doubling time using data collected from a batch system with a pure culture of microorganisms.

Example 4.6 Determining growth rate constant from a batch system

A pure culture of microorganisms was inoculated into an enclosed vessel containing a special medium. One thousand microbes were added to the batch system initially. The number of microbes was measured hourly and the data are presented in the table below.

Determine:

a) The growth rate constant, k, with units of h^{-1}
b) The doubling time, t_d, hours.

Time, t (h)	X (numbers)
0	1.00E+03
1	8.00E+03
2	6.40E+04
3	5.10E+05
4	4.10E+06
5	3.30E+07
6	2.60E+08
7	2.10E+09
8	1.70E+10
9	1.30E+11
10	1.10E+12

Solution part a

Determine the natural logarithm of X and then make a plot of $\ln(X)$ versus time. The slope of the line of best fit is equal to the growth rate constant k times $\ln(2)$.

Time, t (h)	X (numbers)	$\ln(X)$
0	1.00E+03	6.9
1	8.00E+03	9.0
2	6.40E+04	11.1
3	5.10E+05	13.1
4	4.10E+06	15.2
5	3.30E+07	17.3
6	2.60E+08	19.4
7	2.10E+09	21.5
8	1.70E+10	23.6
9	1.30E+11	25.6
10	1.10E+12	27.7

The slope of the line of best fit from the plot is 2.0798.

$$\text{Slope} = k \times \ln(2) = 2.0798$$

$$k = \frac{2.0798}{\ln(2)} = \boxed{3.00\ \text{hr}^{-1}}$$

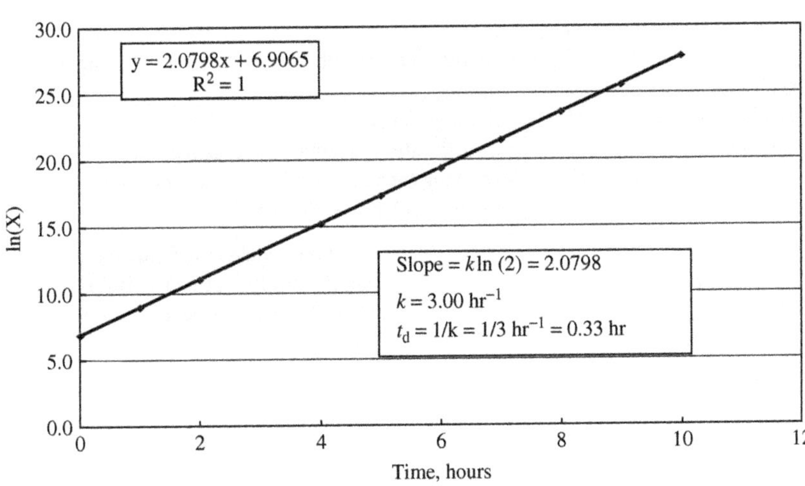

Figure E4.3

Solution part b

The doubling time or generation time is equal to the reciprocal of k.

$$t_d = \frac{1}{k} = \frac{1}{3.00\ \text{hr}^{-1}} = \boxed{0.33\ \text{hr}}$$

4.8.1.2 Exponential Growth

Environmental engineers and scientists usually quantify the population of microorganisms in terms of the quantity of biomass expressed in terms of mg/L of TSS or VSS, rather than the number of organisms. The exponential growth of a population of organisms can be expressed by the following equations:

$$\left(\frac{dX}{dt}\right)_G = \mu X \tag{4.49}$$

where:

$\left(\frac{dX}{dt}\right)_G$ = organism growth rate, mass/(unit volume · time)

μ = specific growth rate of organism, time^{-1}

X = organism or biomass concentration, mass/unit volume.

Rearranging and integrating Equation (4.49) from X_0 at $t = 0$ to X_t at some future time, t, results in the logarithmic form of the equation.

$$\ln(X_t) - \ln(X_0) = \mu t \tag{4.50}$$

Taking the antilog of both sides of Equation (4.50) yields the more familiar, exponential form of the equation.

$$X_t = X_0 e^{\mu t} \tag{4.51}$$

where:

X_t = biomass concentration at time, t, mg/L

X_0 = initial biomass concentration at time equal to zero, mg/L

μ = specific growth rate of the organism, h^{-1} or d^{-1}

t = time, h or d.

An equation relating the time required for the biomass concentration to double, t_d, is derived below. This assumes that the microbial population is growing at an exponential rate. In Equation (4.51), t_d is substituted for t, and $X_t = 2 \times X_0$, resulting in the following equation.

$$2 \times X_0 = X_0 e^{\mu t_d} \tag{4.52}$$

$$\frac{2 \times X_0}{X_0} = e^{\mu t_d} \tag{4.53}$$

$$\ln\left(\frac{2 \times X_0}{X_0}\right) = \ln(e^{\mu t_d}) \tag{4.54}$$

$$\ln(2) = \mu t_d \tag{4.55}$$

$$\mu = \frac{\ln(2)}{t_d} = \frac{0.693}{t_d} \tag{4.56}$$

where t_d = doubling time, h or d.

Example 4.7 illustrates exponential growth of different microorganisms. Each of the microbes presented have a different specific growth rate, which gives each a unique growth curve.

Example 4.7 Plotting biomass concentrations for various specific growth rates

Make a plot of the biomass concentration in mg/L versus time in days for three different microbial species with specific growth rates of 0.05 d^{-1}, 0.10 d^{-1}, and 0.15 d^{-1}, respectively. Vary the time allocated for growth from 0–20 days in increments of one day. Assume that the microbes have excess substrate and nutrients, and are experiencing exponential growth. The initial biomass concentration for each organism is 5 mg/L.

Solution

See Figure E4.4 on page 173.

4.8.1.3 Logistic Growth Model

In natural ecosystems, the logistic growth model provides a more realistic approach of how populations and communities of organisms grow. Exponential growth in nature seldom occurs, since nutrients, substrate, predators, and other environmental factors, in addition to space limitations, will limit the organism growth rate until some ultimate capacity is reached. This carrying capacity, K, is defined as the size of the population or community of organisms that can be sustained by the ecosystem. The logistic growth model equation is presented as follows:

$$\left(\frac{dN}{dt}\right) = kN\left(1 - \frac{N}{K}\right) \tag{4.57}$$

where:

$\left(\frac{dN}{dt}\right)$ = change in the number or concentration of organisms in time

N = number of organisms

k = growth rate constant of the organism, time^{-1}

K = carrying capacity, number or concentration of organisms for a given area.

Integration of Equation (4.57) yields the following equation that is used for calculating the number or concentration of organisms at a specific time. A consistent set of units must be used in Equation (4.58) for N, K, k, and t.

$$N_t = \frac{K}{1 + \left[\frac{K - N_o}{N_o}\right] e^{-kt}} \tag{4.58}$$

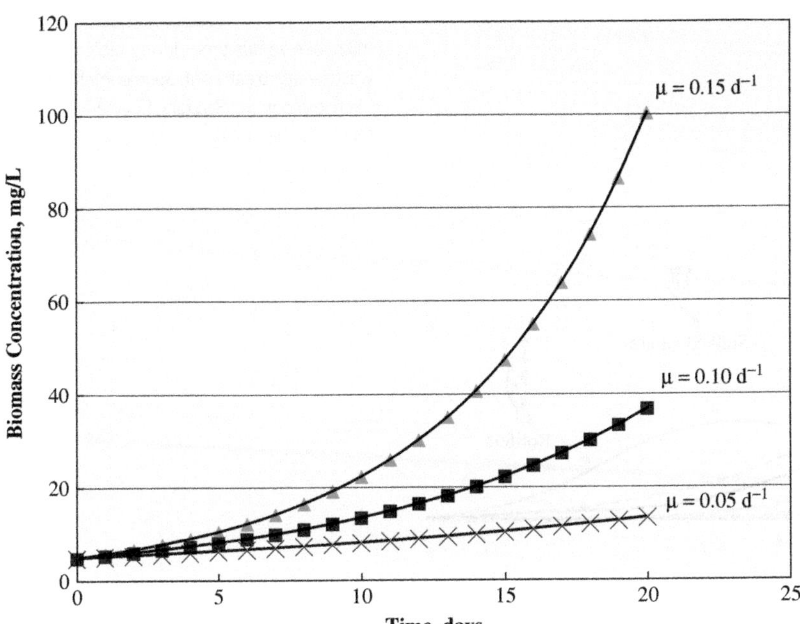

Figure E4.4

where:
N_0 = initial number of organisms at time zero
N_t = number of organisms at time, t
K = carrying capacity = number of organisms at capacity
R = specific growth rate of the organisms, t^{-1}
t = time, h or d.

Example 4.8 Comparison of exponential and logistic growth models

Compare the exponential growth and logistic growth models for a specific microbial species given the following data: $\mu = 0.70\ d^{-1}$, $X_0 = 10$ mg/L, $K = 2{,}000$ mg/L, and t will be varied from 0 to 20 days in increments of 1 day.

Solution

See Figure E4.5.

From this plot, one can see there is a significant difference in the biomass concentration predicted at day 20 between the two models. The models begin to diverge around six days. In nature, the logistic model is more appropriate. Predator-prey relationships and competition, both within and between species, will affect populations. These and other factors are addressed by Whittaker (1975).

4.8.1.4 Growth in Mixed Cultures

Growth in mixed cultures is similar to that presented in the preceding discussions by a single population of organisms. In natural and engineered ecosystems, heterogeneous cultures of organisms coexist. Predator-prey relationships also exist, and

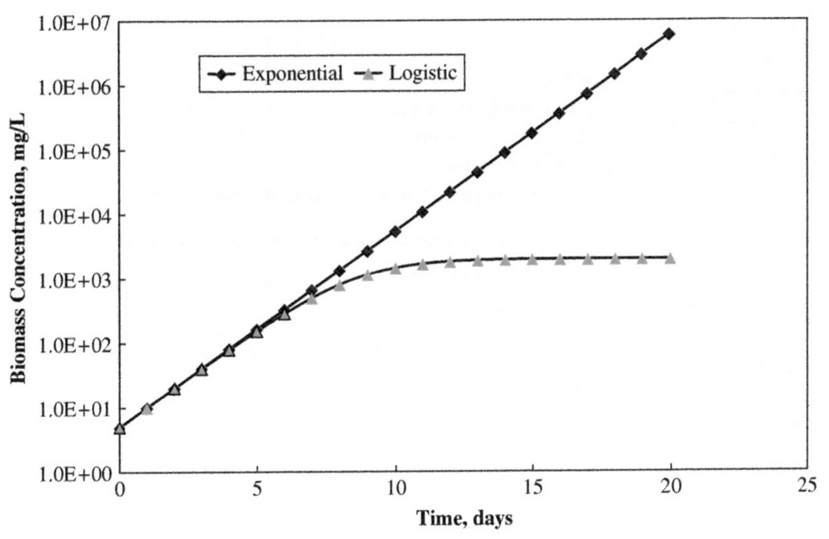

Figure E4.5

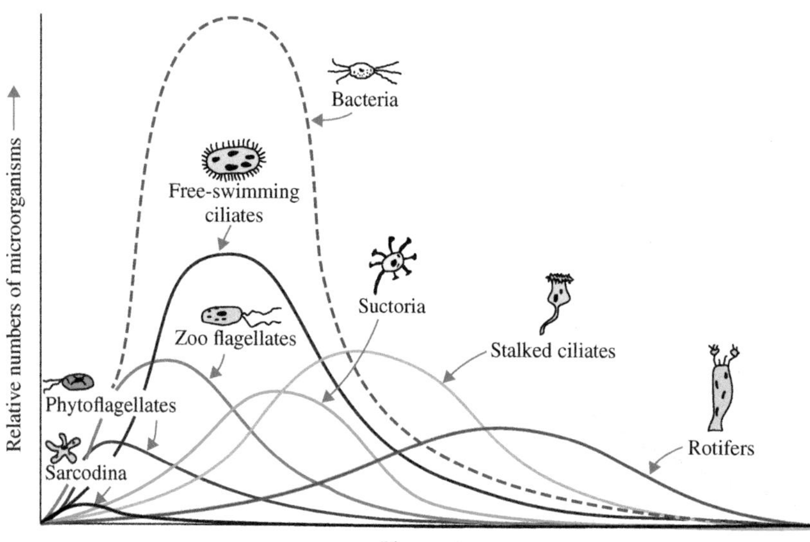

Figure 4.41 Relative numbers of microbes versus time during biological wastewater treatment. *Source:* McKinney's Microbiology for Sanitary Engineers, McGraw-Hill, 1962.

there is competition among various species for substrate, nutrients, and water. Figure 4.41 shows the relative numbers of various organisms as a function of time that would be expected in a biological treatment process. As shown in the figure, bacteria are the predominant microbe present, and the concentration of the various microbial species varies with the residence time in the system.

Modeling of continuous flow, mixed culture systems, such as activated sludge, is presented in Chapter 7. The following equation, which is developed in Chapter 7 and includes a Monod type function, is the most widely used for modeling growth rate in biological wastewater treatment systems. In these systems, a specific substrate or nutrient will limit growth rate rather than the carrying capacity of the ecosystem.

$$\frac{1}{\theta_c} = Y\frac{kS_e}{K_s + S_e} - k_d \tag{4.59}$$

where:
θ_c = mean cell residence time, d
Y = yield coefficient, g VSS/g BOD_5
k = maximum specific substrate utilization rate, d^{-1}
K_s = half-saturation coefficient, substrate concentration at which half the maximum specific substrate utilization rate occurs, mg BOD_5/L
S_e = effluent soluble substrate concentration, mg BOD_5/L
k_d = endogenous decay coefficient, d^{-1}.

4.8.2 Human Population Dynamics

Estimating and predicting populations is a critical aspect in the determination of design flow rates for water supply reservoirs, wastewater treatment facilities, and water treatment facilities. It is also imperative to predict future populations, so that natural resources and pollution management plans may be developed for a given area.

4.8.2.1 Historical Population Growth and Population Growth Models

Some of the models that have been used for predicting human population growth are presented below. Environmental engineers require population data, which serve as the basis for developing design flows for water and wastewater treatment facilities, the capacity of water supply reservoirs, and for estimating quantities of solid wastes that must be handled, recycled, and eventually disposed of. There are several methods used for making population projections. Three of the most widely used mathematical projections are based on the logistic or S-shaped logistic curve (Figure 4.42). As seen in this figure, population growth may occur at an exponential rate, an arithmetic rate, or at a decreasing rate of increase when the carrying capacity or saturation population is being approached.

Exponential Growth Exponential growth can be modeled as a first-order or exponential equation as shown below:

$$\frac{dP}{dt} = k_g P \tag{4.60}$$

where:
dP = change in population
dt = change in time
P = population
k_g = geometric growth rate constant, persons/yr.

Integrating Equation (4.60) between time (t_1) and some future time (t_2) yields:

$$\int_{P_1}^{P_2} \frac{1}{P} dP = k_g \int_{t_1}^{t_2} dt \tag{4.61}$$

$$\ln P_2 - \ln P_1 = k_g(t_2 - t_1) \tag{4.62}$$

$$P_2 = P_1 e^{k_g(t_2 - t_1)} \tag{4.63}$$

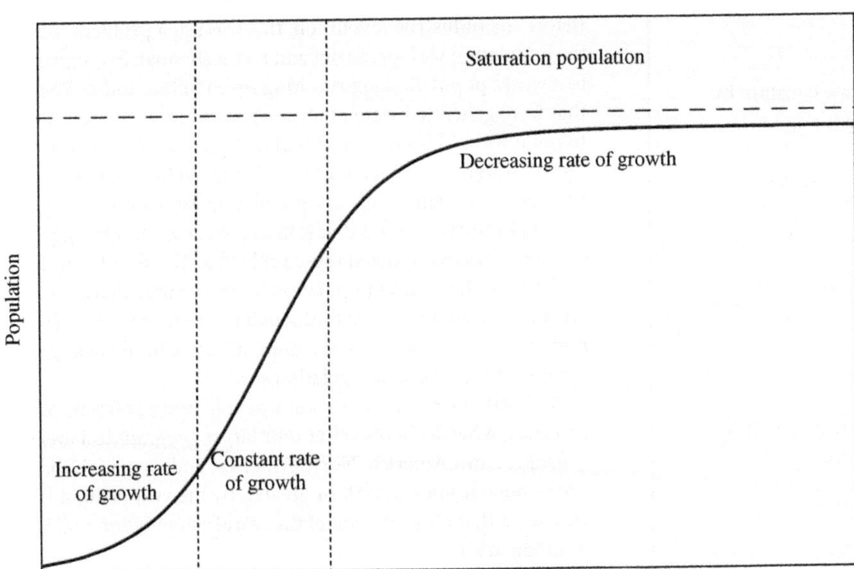

Figure 4.42 Generalized S-shape population growth curve.

Arithmetic Growth In the middle of the logistic growth curve, a uniform rate of growth (k_u) can occur. Arithmetic growth is modeled using Equation (4.64).

$$\frac{dP}{dt} = k_u \quad (4.64)$$

Integrating Equation (4.64) between time (t_1) and some future time (t_2) yields:

$$\int_{P_1}^{P_2} dP = k_u \int_{t_1}^{t_2} dt \quad (4.65)$$

$$P_2 - P_1 = k_u(t_2 - t_1) \quad (4.66)$$

$$P_2 = P_1 + k_u(t_2 - t_1) \quad (4.67)$$

Decreasing Rate of Increase Growth As the population approaches the carrying capacity or saturation population of a given area, the decreasing rate of increase model is used for population forecasting. The rate of growth varies and is denoted by r_d. Mathematically, the decreasing rate of increase growth model is expressed as follows:

$$\frac{dP}{dt} = k_d(Z - P) \quad (4.68)$$

where Z = saturation population.

Integrating Equation (4.68) yields the following equation:

$$\int_{P_1}^{P_2} \frac{dP}{(Z - P)} = k_d \int_0^t tdt \quad (4.69)$$

$$-\ln\left(\frac{Z - P_2}{Z - P_1}\right) = k_d(t_2 - t_1) \quad (4.70)$$

Rearranging the above equation results in:

$$Z - P_2 = (Z - P_1)e^{-k_d(t_2 - t_1)} \quad (4.71)$$

Next, subtracting both sides of Equation (4.71) from ($Z - P_1$) yields:

$$(Z - P_1) - (Z - P_2) = (Z - P_1) - (Z - P_1)e^{-k_d(t_2 - t_1)} \quad (4.72)$$

Rearranging Equation (4.72) results in Equation (4.74), which is useful for predicting future populations in the short-term:

$$Z - P_1 - Z + P_2 = (Z - P_1)(1 - e^{-k_d(t_2 - t_1)}) \quad (4.73)$$

$$P_2 = P_1 + (Z - P_1)(1 - e^{-k_d(t_2 - t_1)}) \quad (4.74)$$

$$Z = \frac{2P_0P_1P_2 - P_1^2(P_0 + P_2)}{P_0P_2 - P_1^2} \quad (4.75)$$

where:

Z = saturation population

P_0, P_1, P_2 = population at t_0, t_1, t_2.

Example 4.9 Predicting future populations by various models

Estimate the 2010 population using three population models available for short-term forecasting: arithmetic, geometric, and declining.

Census date	Population
1980	10,000
1990	15,000
2000	18,000

Solution part a

First determine the arithmetic growth rate constant by rearranging Equation (4.67) and solving for k_u.

$$k_u = \frac{P_2 - P_1}{t_2 - t_1} = \frac{18{,}000 - 15{,}000}{2000 - 1990} = 300\ \frac{\text{people}}{\text{yr}}$$

$$P_t = P_2 + k_u(t_t - t_2)$$

$$P_{(2010)} = 18{,}000 + 300\ \frac{\text{people}}{\text{yr}}(2010 - 2000) = \boxed{21{,}000}$$

Solution part b

First determine the geometric growth rate constant by rearranging Equation(4.62) and solving for k_g.

$$k_g = \frac{\ln(P_2) - \ln(P_1)}{t_2 - t_1}$$

$$k_g = \frac{\ln(18{,}000) - \ln(10{,}000)}{2000 - 1980} = \frac{9.798 - 9.210}{20}$$

$$= 0.029\ \text{yr}^{-1}$$

$$P_{2010} = (18{,}000)\,e^{(0.029\ \text{yr}^{-1}\cdot(2010-2000))} = \boxed{24{,}056}$$

Solution part c

First, determine the saturation population from Equation (4.75):

$$Z = \frac{2P_0P_1P_2 - P_1^2(P_0 + P_2)}{P_0P_2 - P_1^2}$$

$$= \frac{2(10000)(15000)(18000) - (15000)^2(10000 + 18000)}{(10000)(18000) - (15000)^2}$$

$$Z = 20{,}000$$

Next, determine the decreasing rate of growth rate constant by rearranging Equation (4.70) and solving for k_d:

$$k_d = \frac{-\ln\left(\frac{Z - P_2}{Z - P_1}\right)}{(t_2 - t_1)} = \frac{-\ln\left(\frac{20{,}000 - 18{,}000}{20{,}000 - 15{,}000}\right)}{(2000 - 1990)}$$

$$= 0.0916\ \text{yr}^{-1}$$

$$P_t = P_2 + (Z - P_2)(1 - e^{-k_d\cdot(t_t - t_2)})$$

$$P_{2010} = 18000 + (20000 - 18000)(1 - e^{-0.0916\cdot 10\ \text{yr}})$$

$$= \boxed{19{,}200}$$

4.8.3 World population growth

World population growth has had a significant impact on the utilization and exploitation of Earth's natural resources. Before the industrial revolution, this was not a problem, since the population was spread out and relatively small in comparison to a world population approaching seven billion today. Wastes that were generated were easily assimilated (e.g., the solution to pollution is dilution). As population increases, however, the air, water, and soil cannot handle the environmental problems associated with this population explosion.

The industrialization that is taking place in developing countries has exacerbated these problems. The shift from the majority of the world's population living in rural areas to urban areas has created new environmental problems resulting from a higher density of people and a more industrialized society. Figure 4.43 shows this population shift.

In 2010, the rural population was still around 60% in Africa and Asia, whereas in the other four larger geographical areas (Europe, Latin America, North America, and Oceania), the urban population was 70% or greater. By the year 2050, it is estimated that 60% or more of the world's population will live in urban areas.

As the quality of life improves, citizens tend to waste more. In the past, we have not managed our natural resources in a sustainable way. The depletion of our fossil fuel reserves and the relatively, uncontrolled emission of greenhouse gases are two classic examples illustrating non-sustainable development.

The historical growth of the world population is presented in Figure 4.44. A J-shaped curve representing exponential growth is observed.

The world's population from 1950 through 2050 is presented in Figure 4.45. In the 40-year period between 1959 and 1999, the world population doubled from 3 billion to 6 billion people. The Census Bureau's latest projections suggest that the world population will grow more slowly in the 21st century. By 2044, the world population is projected to grow to 9 billion, which is an increase of 50 percent in a 45-year span.

Human population growth is often characterized as exponential growth according to Equation (4.76).

$$\frac{dP}{dt} = rP \tag{4.76}$$

where:
dP = change in population
dt = change in time
P = population
r = growth rate constant.

Integrating Equation (4.76) between time (t_1) and some future time (t_2) yields:

$$\int_{P_1}^{P_2} \frac{1}{P}dP = k_g \int_{t_1}^{t_2} dt \tag{4.77}$$

$$\ln P_2 - \ln P_1 = k_g(t_2 - t_1) \tag{4.78}$$

$$P_2 = P_1\,e^{k_g(t_2 - t_1)} \tag{4.79}$$

The growth rate constant is typically expressed as a percent increase per year or as the increase in the number of people per 1,000 people for a given population in a year. The growth rate of a population is dependent upon four major components: birth rate, death rate, immigration rate, and emigration rate.

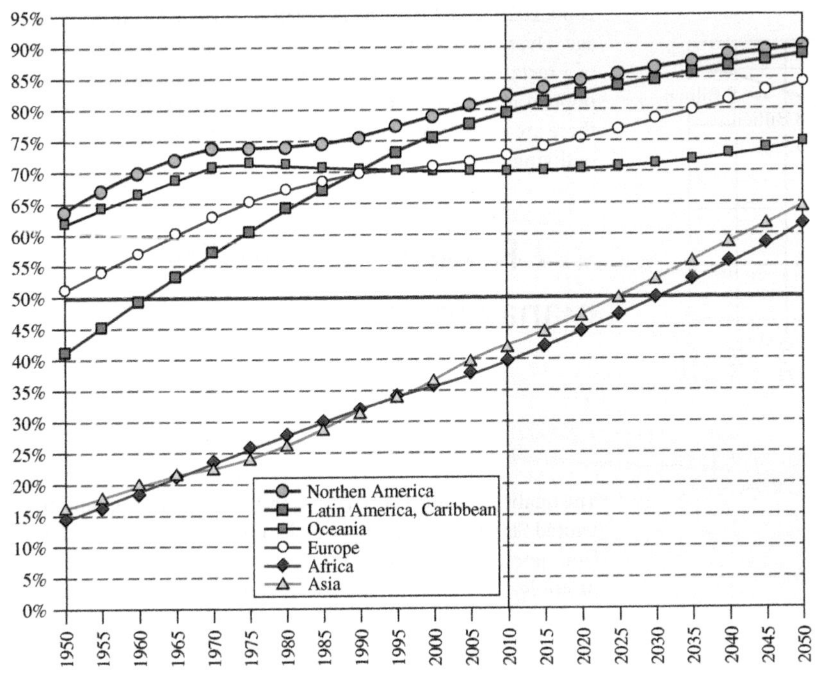

Figure 4.43 World urban population by major geographical area (% of total population). *Source:* http://esa.un.org/unpd/wup/Fig_1.htm Accessed January 31, 2011. United Nations, Department of Economic and Social Affairs, Population Division: *World Urbanization Prospects, the 2009 Revision.* New York, 2010. Reproduced with permission.

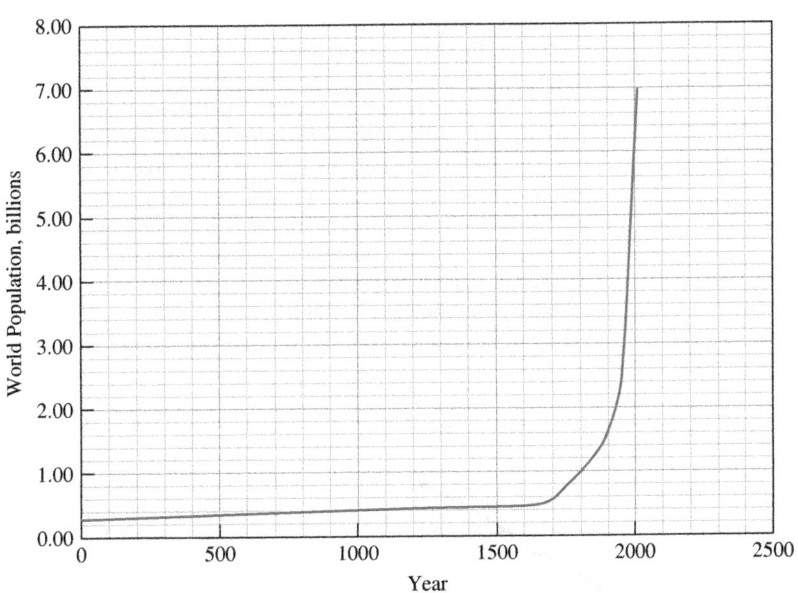

Figure 4.44 World population growth through history. *Source:* http://www.prb.org/Articles/2002/HowManyPeopleHaveEverLivedonEarth.aspx

Growth rate can be defined by the following equation:

$$r = b - d + i - e \tag{4.80}$$

where:
b = birth rate
d = death rate
i = immigration rate
e = emigration rate.

All of the above terms are expressed as % increase per year or increase # of persons/(1,000 people per year).

Figure 4.46 indicates that the world population growth rate has been declining since the early 1960s. In the year 2010, the world population growth rate was approximately 1.2% per year.

The natural increase in population is defined as the difference between excess births and the number of deaths. Net migration is the difference between immigration and emigration rates for a given area. Gender ratio, fertility rate, age structure, and cultural factors also affect population growth. The total fertility rate represents the total number of children that a woman is expected to bear during her lifetime. The average worldwide net reproductive rate from 2005 through 2010 is 1.2 (http://esa.un.org/unpd/wpp2008/tab-sorting_fertility.htm). Figure 4.47 shows the predicted world population using different fertility rates.

Population pyramids are often used to show how the age and gender of various populations vary over time. Figures 4.48, 4.49, and 4.50 are population pyramids for the United States in

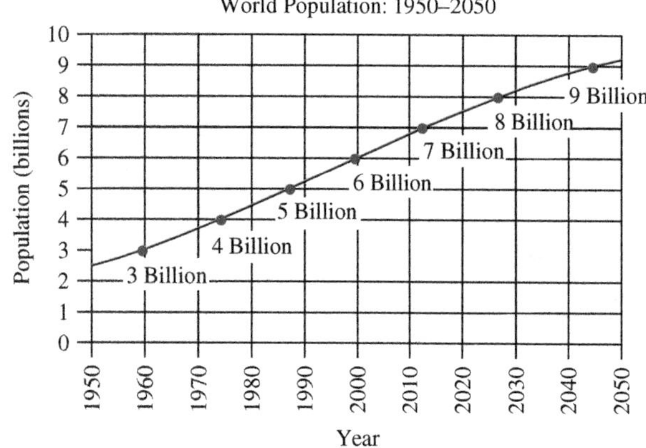

Figure 4.45 World population from 1950 through 2050. *Source*: http://www.census.gov/ipc/www/idb/worldpopgraph.php Accessed January 30, 2011. US Census Bureau.

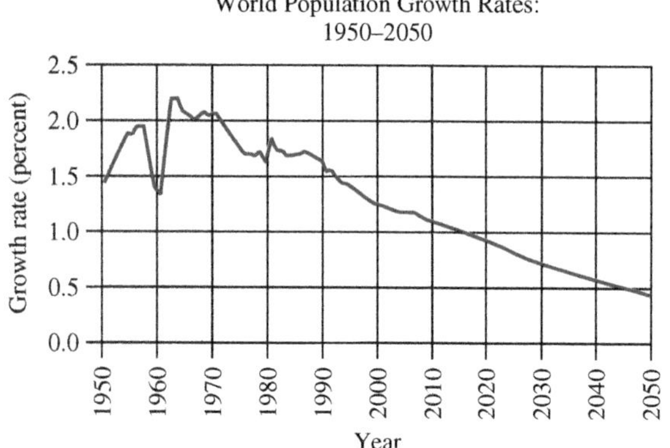

Figure 4.46 The world population growth rate from 1950–2050. *Source*: http://www.census.gov/ipc/www/idb/worldgrgraph.php Accessed January 30, 2011. US Census Bureau.

2000, 2025, and 2050. The population of the US continues to age. Over time, the pyramid changes from a "pyramid" to more of a rectangular shape. This represents a slow-growing population. The top of the pyramids (the older ages) indicate there are decreasing numbers of people and that females are outliving males.

4.9 River water quality management

4.9.1 Total maximum daily load (TMDL)

The quality of some lakes, rivers, and streams throughout the United States has been impaired by the discharge of pollutants from point and non-point sources of pollution. Since the promulgation of the federal Clean Water Act (1972), the water quality of receiving waters has greatly improved by the regulation of point sources of pollution. This has been accomplished by permits issued by the US EPA or state agencies under the National Pollution Discharge Elimination System (NPDES). These permits are required for discharging municipal and industrial effluent to receiving waters. The concentration and quantity of various pollutants that may be discharged are established in the permit; this is known as the waste load allocation (WLA) from a point source. To better manage our water resources, a holistic approach is now being practiced by regulatory agencies, wherein an entire watershed is considered when establishing the pollutant load that can be allowed in a given stream or lake.

Non-point sources of pollution, such as those resulting from runoff associated with rainfall from agricultural areas and

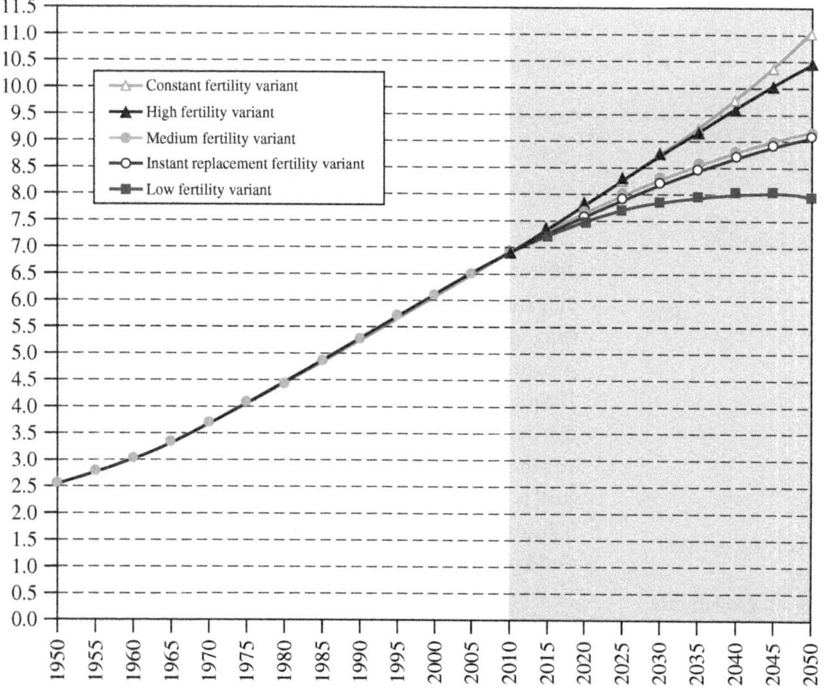

Figure 4.47 Population of the world (in billions) 1950–2050 according to different projections in fertility rate. *Source*: http://esa.un.org/unpd/wpp/Analytical-Figures/htm/fig_1.htm Accessed February 27, 2012.

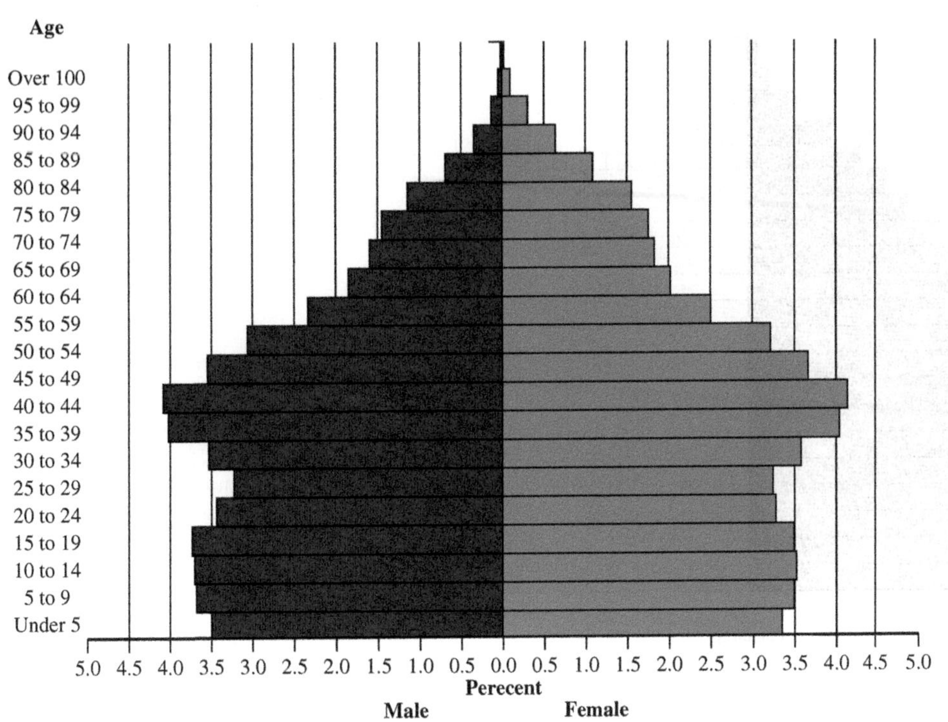

Figure 4.48 Population pyramid for the United States in 2000. *Source*: http://www.census.gov/population/www/projections/np_p2.gif Accessed January 31, 2011. US Census Bureau.

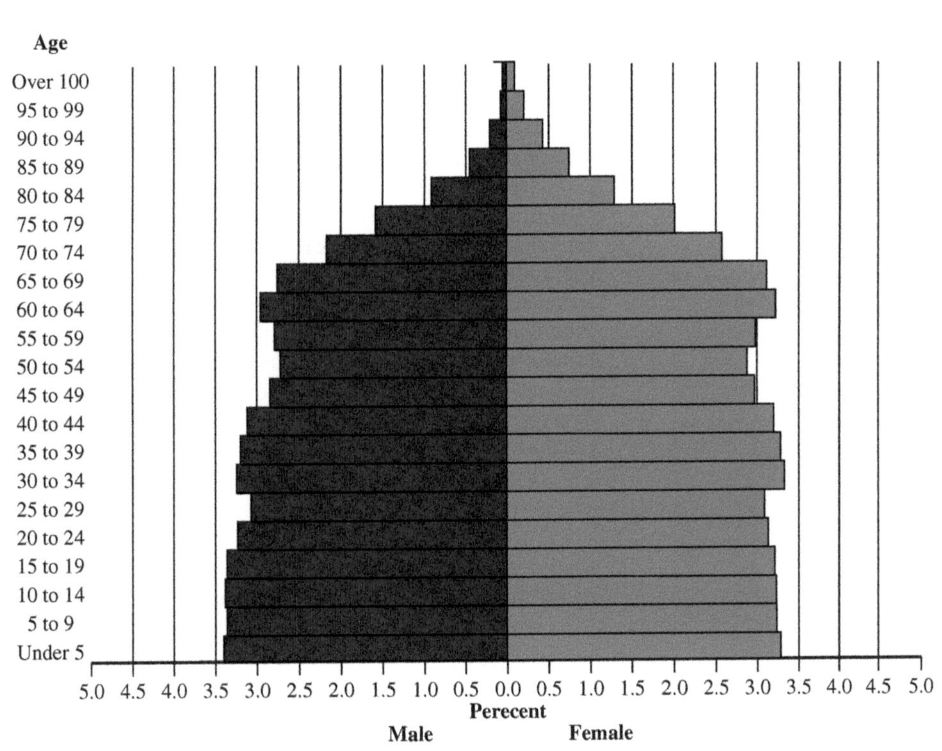

Figure 4.49 Population pyramid for the United States in 2025. *Source*: http://www.census.gov/population/www/projections/np_p3.gif Accessed January 31, 2011. US Census Bureau.

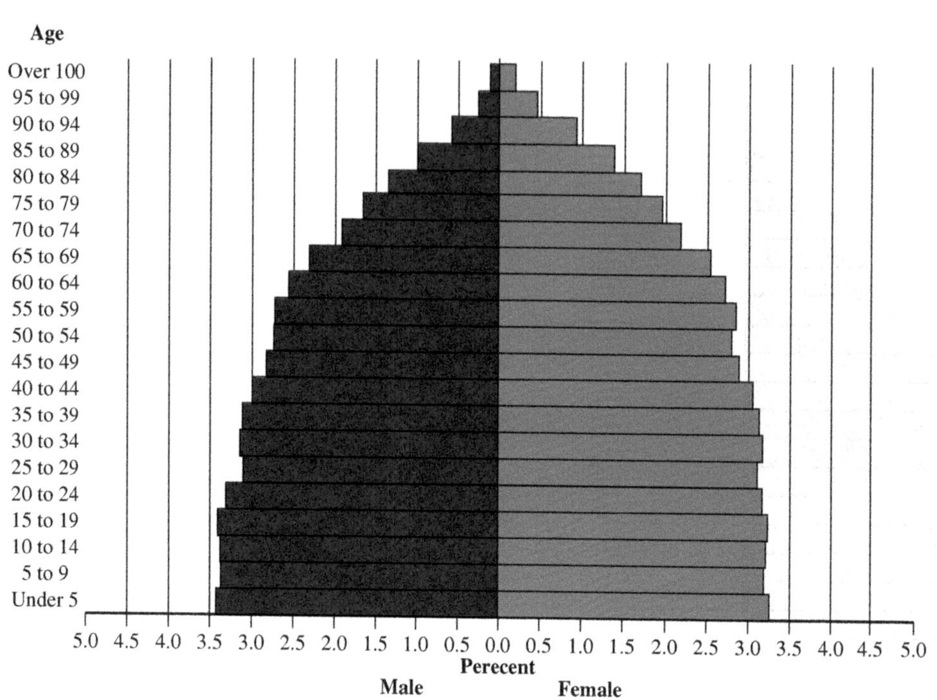

Figure 4.50 Population pyramid for the United States in 2050. *Source*: http://www.census.gov/population/www/projections/np_p4.gif Accessed January 31, 2011. US Census Bureau.

urban areas, contribute significant quantities of pollutants into receiving waters. These diffuse sources are sometimes difficult to identify and quantify. In some cases, even when the effluent standards from a wastewater treatment facility are met, water quality or stream standards are still not achieved.

Under section 303(d) of the Clean Water Act, states, territories, and authorized Indian tribes are required to develop lists of impaired waters. The law requires that these jurisdictions establish priority rankings for these impaired waters and to develop total maximum daily loads (TMDLs). The TMDL rule was promulgated in 2000 (US EPA, 2000). A TMDL is the maximum amount of a pollutant that a water body can receive and still safely meet water quality standards. It represents the sum of the waste-load allocation from individual wastewater treatment plants (point sources of pollution), waste-load allocation from non-point sources (agricultural and stormwater runoff), natural background levels in the water body, and a safety margin (US EPA, 2011). The calculation of a TMDL is determined as follows: w

$$TMDL = WLA + LA + MOS \quad (4.81)$$

where:
$TMDL$ = total maximum daily load
WLA = waste load allocation (point sources)
LA = load allocation (non-point sources)
MOS = margin of safety.

4.9.2 Dissolved oxygen depletion in streams

Dissolved oxygen (DO) concentration is an important water quality parameter, because fish and other aquatic organisms require oxygen for survival. In streams, a minimum of 2 mg/L of DO is required for higher life forms (Peavy *et al.*, 1985, page 83). A DO concentration of 4 mg/L or more is necessary to support game fish. The discharge of degradable organic wastes into surface waters results in oxygen depletion, due primarily to bacterial metabolism. Oxygen utilization is modeled by the biochemical oxygen demand (BOD) test, as previously discussed in Chapter 3. Oxygen is primarily transferred into a stream by dissolution from the atmosphere (reaeration) and, to a much lesser extent, by algal photosynthesis. Overall, algal production of oxygen does not exceed the quantity of oxygen used by algae during the night.

Streams have natural assimilative abilities to oxidize organic and nitrogen compounds discharged into rivers by indigenous heterotrophic and autotrophic bacteria. When oxygen consumption exceeds the reaeration rate, the water body's assimilative capacity is exceeded. This leads to unfavorable conditions for aquatic life, often resulting in "fish kills." The DO concentration will decrease with time and distance downstream from the point of discharge. This process is known as deoxygenation, and it is modeled by a first-order removal reaction as shown below:

$$r_D = k_D L \quad (0.5)$$

where:
r_D = rate of deoxygenation, which is equal to the rate at which oxygen is being removed from a stream, mg/(L · d)
k_D = deoxygenation rate coefficient (base e), which is equal to the BOD rate constant (k), d^{-1}
L = ultimate biochemical oxygen demand (BOD), mg/L.

The ultimate BOD represents the total quantity of oxygen consumed by bacteria at 20°C for the oxidation of organic compounds to carbon dioxide and water.

Table 4.13 Solubility of oxygen in water exposed to water-saturated air at atmospheric pressure.

Temperature (°C)	Oxygen solubility (mg/L)	Temperature (°C)	Oxygen solubility (mg/L)	Temperature (°C)	Oxygen solubility (mg/L)
0	14.62	11	11.03	22	8.74
1	14.22	12	10.78	23	8.58
2	13.83	13	10.54	24	8.42
3	13.46	14	10.31	25	8.26
4	13.11	15	10.08	26	8.11
5	12.77	16	9.87	27	7.97
6	12.45	17	9.66	28	7.83
7	12.14	18	9.47	29	7.69
8	11.84	19	9.28	30	7.56
9	11.56	20	9.09	31	7.43
10	11.29	21	8.92	32	7.30

Developed using on-line program from the USGS: http://water.usgs.gov/software/DOTABLES/

4.9.2.1 Reaeration

The equilibrium concentration of dissolved oxygen in water is temperature dependent and can be calculated from Henry's law. The solubility of oxygen increases as temperature decreases, as seen in Table 4.13. The actual DO concentration may be less than or greater than the equilibrium concentration.

In most streams, the actual DO concentration is less because of microbial activity which uses the DO for degrading the organics and nitrogenous compounds that have been discharged into it. The difference between the equilibrium dissolved concentration and actual DO concentration is called the deficit, as shown in Equation (4.83).

$$D = C_s - C \tag{4.83}$$

where:
D = dissolved oxygen deficit, mg/L
C_s = equilibrium dissolved oxygen concentration, mg/L
C = actual dissolved oxygen concentration, mg/L.

Reaeration is the process by which oxygen is transferred by diffusion across the water surface exposed to the atmosphere. Water traveling over rocks and through rapids entrains more oxygen than does a slower moving river. The rate of reaeration is modeled as a first-order removal reaction as shown below:

$$r_R = -k_R D \tag{4.84}$$

where:
r_R = rate of reaeration, equal to the rate at which oxygen is transferred into the stream or water body, mg/(L · d)
k_R = reaeration rate coefficient (base e), d^{-1}
D = dissolved oxygen deficit, mg/L.

A negative sign precedes the reaeration coefficient, because it increases the supply of oxygen to the river while reducing the oxygen deficit. This will become more apparent when examining Equation (4.87). The reaeration rate coefficient is stream-specific, i.e., turbulence of stream, stream velocity, and temperature. O'Connor and Dobbins (1958) developed the following equation for estimating k_R.

$$k_R = \frac{(D_L \times U)^{0.5}}{H^{1.5}} \tag{4.85}$$

where:
D_L = oxygen diffusivity coefficient in water = 8.1×10^{-5} ft²/h at 20°C
U = stream flow velocity, ft/h
H = depth of flow, ft.

The Arrhenius-van't Hoff model is used for correcting the deoxygenation and reaeration rate coefficients for temperature variations:

$$k_{T°C} = k_{20°C}(\theta)^{T°C-20°C} \tag{4.86}$$

where:
$k_{T°C}$ = rate coefficient at temperature, T, in °C, d^{-1}
$k_{20°C}$ = rate coefficient at 20°C, d^{-1}
θ = temperature correction coefficient, dimensionless
T = actual temperature of the receiving waters, °C.

A temperature correction coefficient (θ) value of 1.047 was recommended by O'Connor & Dobbins (1958) for correcting the reaeration rate coefficient. Peavy *et al.* (1985, page 86) recommend a value of 1.016 for θ. Most recently, a theta value of 1.024 has been used for correcting the reaeration coefficient for temperature (Davis & Masten, 2009, page 363). Reaeration

Table 4.14 Reaeration Coefficients.

Water body	k_R at 20°C, d^{-1}
Small ponds and backwaters	0.10–0.23
Sluggish streams and large lakes	0.23–0.35
Large streams at low velocity	0.35–0.46
Large streams at normal velocity	0.46–0.69
Swift streams	0.69–1.15
Rapids and waterfalls	>1.15

Reaeration coeffients, k_R, were calculated using self-purification constants, f, and a deoxygenation constant k of 0.23 d^{-1} from Fair, Geyer, and Okun (1968) page 33–26.

rate coefficients for various types of water bodies are presented in Table 4.14. Metcalf & Eddy (1991, page 76) recommend temperature correction coefficients of 1.056 in the temperature range between 20 and 30°C and 1.135 in the temperature range between 4 and 20°C.

4.9.2.2 Streeter-Phelps DO Sag Model

The concept of "dissolved oxygen sag" was first described by Streeter and Phelps (1925). They provided what is now considered the classical approach for modeling the impact of organic wastes discharged into receiving streams. The basic model assumes that there is no dispersion of the organic pollutant as it moves downstream, and that there is complete mixing of the pollutant with depth and along each cross-section of the river. The oxygen deficit within a stream is a function of both oxygen consumption and reaeration. Reaeration causes the oxygen deficit to decrease, while oxygen consumption increases the deficit. The overall rate of change in the oxygen deficit depends on the deoxygenation and reaeration reactions. The Streeter-Phelps model describes the simultaneous transfer and uptake of oxygen in a river by the following differential equation:

$$\frac{dD}{dt} = k_D L - k_R D \tag{4.87}$$

where $\frac{dD}{dt}$ = change in DO deficit (D) with time, mg/(L · d).

A plot of the DO concentration versus time shows the characteristic dissolved oxygen sag curve (Figure 4.51). The time at which the deoxygenation rate equals the reaeration rate results in the greatest oxygen deficit and the lowest DO concentration in the stream, and is known as the critical time, t_c.

All other terms have been previously defined. The ultimate BOD (L) as a function of time "t" is modeled as a first-order reaction as shown below. This equation was derived in Chapter 3. The value of the BOD rate coefficient, k, is used for k_D:

$$L = L_o e^{-kt} = L_o e^{-k_D t} \tag{4.88}$$

where:
L_0 = ultimate BOD concentration after the stream and wastewater discharge have mixed, mg/L

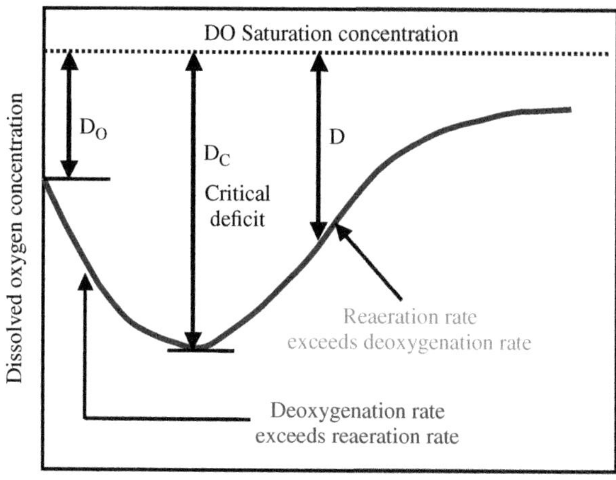

Figure 4.51 Dissolved oxygen sag curve.

K = BOD rate constant, d^{-1}
t = time of travel of wastewater discharge downstream, d
e = base "e", 2.71828.

The ultimate BOD concentration (L_0) is calculated using Equation (4.89).

$$L_0 = \frac{Q_S(L_S) + Q_{ww}(L_{ww})}{Q_S + Q_{ww}} \tag{4.89}$$

where:
Q_S = flow rate in stream, volume/time
Q_{ww} = flow rate of wastewater discharged into stream, volume/time
L_S = ultimate BOD concentration of stream prior to wastewater discharge, mg/L
L_{ww} = ultimate BOD concentration of wastewater, mg/L.

Substituting Equation (4.88) into Equation (4.87) results in Equation (4.90):

$$\frac{dD}{dt} = k_D L_0 e^{-k_D t} - k_R D \tag{4.90}$$

Equation (4.90) may be integrated using the boundary conditions at $t = 0$, $D = D_0$, and $L = L_0$, and at $t = t$, $D = D_t$, and $L = L_t$, resulting in the general form of the Streeter-Phelps Equation (4.91), used for estimating the DO deficit at downstream locations from the point of discharge.

$$D_t = \frac{k_D L_0}{k_R - k_D}[e^{(-k_D t)} - e^{(-k_R t)}] + D_0 e^{(-k_R t)} \tag{4.91}$$

where:
D_t = DO deficit at any time t, downstream of the discharge point, mg/L
t = time of travel from discharge to any point downstream, d
D_0 = DO deficit at the point of discharge, mg/L.

To calculate the oxygen deficit at the point of discharge, the DO concentration at the point of discharge must be calculated using Equation (4.92):

$$DO_0 = \frac{Q_S(DO_S) + Q_{ww}(DO_{ww})}{Q_S + Q_{ww}} \quad (4.92)$$

where:
DO_S = DO concentration upstream of the wastewater discharge, mg/L
DO_{ww} = DO concentration of wastewater, mg/L.

The DO deficit (D_0) at the point of discharge can then be calculated using Equation (4.93).

$$D_0 = C_s - DO_0 \quad (4.93)$$

The maximum dissolved oxygen deficit will occur where the reaeration rate equals the deoxygenation rate. This point is known as the critical point and the time required to reach it can be determined by differentiating Equation (4.91) and setting the derivative equal to zero. Solving for time gives Equation (4.94).

$$t_c = \frac{1}{k_R - k_D}\ln\left[\frac{k_R}{k_D}\left(1 - \frac{D_0\,(k_R - k_D)}{k_D\,L_0}\right)\right] \quad (4.94)$$

where t_c time of travel to the critical deficit point in stream, d.

The distance traveled (x) for a river flowing at a constant velocity (u) can be determined by multiplying the stream velocity by the travel time (t). Example 4.10 illustrates how to use the Streeter-Phelps equations for locating the critical deficit point in a stream.

Example 4.10 Calculating critical DO concentration in a stream

A municipal WWTP discharges 0.5 m^3/s of secondary effluent, at a temperature of 25°C, that contains 30 mg/L of ultimate BOD. To meet regulatory requirements, the effluent is aerated to achieve 5.0 mg/L of dissolved oxygen prior to discharge. The stream flow is 3.0 m^3/s and the temperature upstream of the discharge point is 15°C. The background ultimate BOD in the stream is 15.0 mg/L. The reaeration (k_R) and deoxygenation (k_D) rate coefficients are 0.40 d^{-1} and 0.20 d^{-1}, respectively, at 20°C. In-stream standards require a minimum of 7.0 mg/L of DO at all times. Determine the following:

a) temperature of the combined wastewater and stream;
b) dissolved oxygen concentration of the mixture of wastewater and stream;
c) DO deficit of the mixture of wastewater and stream;
d) ultimate BOD concentration of mixture of wastewater and stream;
e) critical time to reach point of minimum DO concentration, and
f) minimum DO concentration in stream.

Solution part a

The temperature of the mixture of the stream and wastewater is determined as follows:

$$T_0 = \frac{Q_S(T_S) + Q_{ww}(T_{ww})}{Q_S + Q_{ww}}$$

$$= \frac{3.0\,\text{m}^3/\text{s}(15°\text{C}) + 0.5\,\text{m}^3/\text{s}(25°\text{C})}{3.0\,\text{m}^3/\text{s} + 0.5\,\text{m}^3/\text{s}} = \boxed{16.4°\text{C}}$$

Solution part b

Determine the DO concentration of the combined water and wastewater at the point of discharge. Assume that the stream is saturated with oxygen. Therefore, at $T = 15°$C, the DO saturation is 10.08 mg/L from Table 4.13.

$$DO_0 = \frac{Q_S(DO_S) + Q_{ww}(DO_{ww})}{Q_S + Q_{ww}}$$

$$= \frac{3.0\,\text{m}^3/\text{s}(10.08\,\text{mg/L}) + 0.5\,\text{m}^3/\text{s}(5.0\,\text{mg/L})}{3.0\,\text{m}^3/\text{s} + 0.5\,\text{m}^3/\text{s}}$$

$$DO_0 = \boxed{9.35\,\text{mg/L}}$$

Solution part c

Calculate the initial oxygen deficit (D_0) at the point of discharge. First, estimate the DO saturation value (C_S) at a temperature of 16.4°C by interpolation of the values in Table 4.13. The DO saturation concentration of water at 16.4°C (the temperature of the mixture of water and wastewater), which is 9.79 mg/L.

$$D_0 = (C_S)_0 - DO_0 = 9.79\,\text{mg/L} - 9.35\,\text{mg/L}$$

$$= \boxed{0.44\,\text{mg/L}}$$

Solution part d

Determine the ultimate BOD concentration (L_0) of the mixture of water and wastewater at the point of discharge.

$$L_0 = \frac{Q_S(L_S) + Q_{ww}(L_{ww})}{Q_S + Q_{ww}}$$

$$= \frac{3.0\,\text{m}^3/\text{s}(15.0\,\text{mg/L}) + 0.5\,\text{m}^3/\text{s}(30\,\text{mg/L})}{3.0\,\text{m}^3/\text{s} + 0.5\,\text{m}^3/\text{s}}$$

$$L_0 = \boxed{17.1\,\text{mg/L}}$$

Solution part e

Next, correct the reaeration and deoxygenation rate coefficients for the temperature of 16.4°C. Use theta values (θ) of 1.024 and 1.135 for correcting k_R and k_D, respectively.

$$k_{(T°\text{C})} = k_{(20°\text{C})}(\theta)^{(T°\text{C}-20°\text{C})}$$

$$k_{R(16.4°\text{C})} = k_{R(20°\text{C})}(1.024)^{(16.4°\text{C}-20°\text{C})}$$

$$= 0.40\,\text{d}^{-1}(1.024)^{(22°\text{C}-20°\text{C})} = 0.37\,\text{d}^{-1}$$

$$k_{D(16.4°C)} = k_{D(16.4°C)}(1.135)^{(16.4°C-20°C)}$$
$$= 0.20\,d^{-1}(1.135)^{(16.4°C-20°C)} = 0.13\,d^{-1}$$

Solution part f

Determine the time of travel (t_c) to reach the point of minimum DO in the stream.

$$t_c = \frac{1}{k_R - k_D}\ln\left[\frac{k_R}{k_D}\left(1 - \frac{D_0\,(k_R - k_D)}{k_D\,L_0}\right)\right]$$

$$t_c = \frac{1}{0.37\,d^{-1} - 0.13\,d^{-1}}$$
$$\ln\left[\frac{0.37\,d^{-1}}{0.13\,d^{-1}}\left(1 - \frac{0.44\,\frac{mg}{L}\,(0.37 - 0.13\,d^{-1})}{0.13\,d^{-1}(17.1\,mg/L)}\right)\right]$$

$$t_c = \boxed{4.2\,d}$$

Solution part g

The value of the minimum deficit is found by substituting the critical time (t_c) into Equation (4.91).

$$D_c = \frac{k_D\,L_0}{k_R - k_D}[e^{(-k_D\,t_c)} - e^{(-k_R\,t_c)}] + D_0 e^{(-k_R\,t_c)}$$

$$D_c = \frac{0.13\,d^{-1}(17.1\,mg/L)}{0.37\,d^{-1} - 0.13\,d^{-1}}$$
$$\times [e^{(-0.13\,d^{-1} \times 4.2\,d)} - e^{(-0.37\,d^{-1} \times 4.2\,d)}]$$
$$+ 0.44\,\frac{mg}{L}\,e^{(-0.37\,d^{-1} \times 4.2\,d)}$$

$$D_c = 3.50\,mg/L$$

The minimum DO concentration in the stream may be calculated by substituting into Equation (4.83) and rearranging it:

$$D_c = C_S - DO_c$$
$$DO_c = C_S - D_c = 9.79\,mg/L - 3.50\,mg/L = \boxed{6.29\,mg/L}$$

The in-stream standard of 7.0 mg/L will be violated. To remedy this, the WWTP must achieve a higher level of BOD removal or reduce the volume of wastewater discharged into the stream.

4.10 Limnology

4.10.1 Introduction

Limnology is the study of the biological, chemical, and physical characteristics of fresh water, e.g., lakes and rivers. Primary productivity in lakes is attributed to plankton. Plankton, in turn, are subdivided into phytoplankton, which are plant species such as algae, while zooplankton are animal species, such as crustacean, rotifers, and protozoa. Large aquatic plants (macrophytes), that may be either free-floating or attached to the bottom (emergent), also contribute to primary productivity. Water quality data for various lakes are accessible by the internet at: http://waterdata.usgs.gov/nwis. Typical data that are collected include conductivity, dissolved oxygen (DO) concentration, pH, and temperature as a function of depth.

It is important for environmental engineers and scientists to understand limnological concepts, since lakes and rivers are used for water supplies, recreational areas, and sources of food (fish). In the United States, approximately 80% of all the water used by municipal water treatment systems comes from surface supplies (USGS, 2005). Proper management of land use in the drainage basin or watershed of the lake or river is critical for maintaining a high quality source of water. Precipitation that falls in the watershed provides the primary source of water for lakes and rivers and contributes to ground water flow. The stormwater that runs off the watershed is directly related to land use in the drainage basin.

Where there is sufficient vegetation and ground cover, the quality of runoff will be higher than from an area where trees and vegetation do not exist. This is a significant problem in many developing countries in Africa, since trees are cut down by the natives for heating, cooking, and making charcoal. As a result, many drainage basins have been deforested and denuded, causing extreme erosion and leading to poor water quality and low productivity in freshwater systems. Deforestation has not only affected water quality in these areas, but contributes to the devastation and loss of animal habitat. Lightning strikes that cause forest fires can result in poor water quality during storm events until vegetation and ground cover have been reestablished. In areas where there is heavy agricultural activities, runoff from the land may contain high concentrations of nutrients such as nitrogen, phosphorus, and potassium from fertilizers, soil particles from erosion during tilling and planting operations, and biological pathogens that may be found in animal excrement if it is used as a fertilizer.

Urbanization has affected both the volume and quantity of stormwater runoff, since there are more impervious areas, i.e., pavements, highways, sidewalk, roofs, and parking lots that hinder the percolation of rainfall into the ground. Trash, litter, oil, soil particles, and fertilizers are all found in stormwater runoff in cities. Eventually, stormwater runoff from rural and urban areas reaches lakes and rivers, thereby impacting the water quality.

4.10.2 Light transmission in lakes

Colloidal and suspended solid particles that make their way into lakes and rivers can affect the clarity of the water by interfering with the transmission of light and the transfer of oxygen into the water. The depth to which light penetrates into a lake or water body will have a tremendous impact on primary productivity and the aquatic life. In general, the light intensity will vary from 100% at the surface to less than 1% as depth increases. Light is absorbed by the water, by substances dissolved in the water such as organic matter, and by particles suspended in the water.

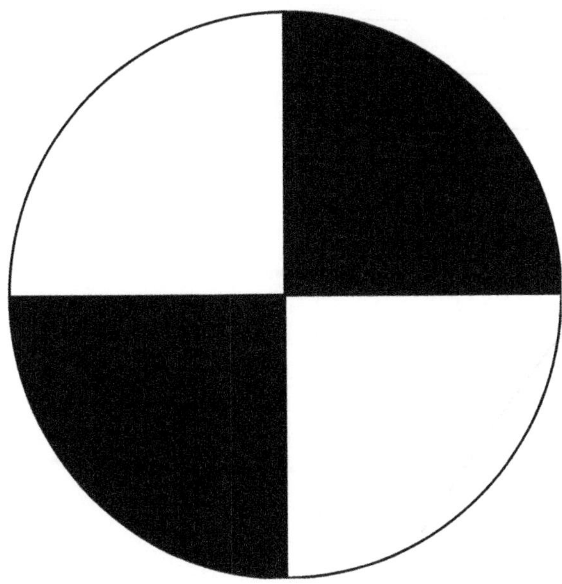

Figure 4.52 Secchi disk. *Source*: http://en.wikipedia.org/wiki/File:Secchi_disk_pattern.svg. Courtesy of Mysid (Oona Räisänen).

The depth to which algal growth occurs is called the eutrophic zone. In shallow lakes, this zone may extend down to the bottom but, in most cases, only about 50% of the mass of water within a lake is available for primary productivity (Henry & Heinke, 1996, page 324). Fish, bacteria, fungi, and zooplankton, reside below the eutrophic zone for short time periods, if not permanently. In the bottom sediments, or benthos, benthic invertebrates and anaerobic organisms predominate. Bacteria and fungi use the organic matter from dead algal cells, thereby releasing nitrogen and phosphorus back into the water column to be recycled and reused by phytoplankton. In freshwaters, benthic micro-invertebrates that may be found include crayfish, clams, snails, and aquatic worms.

The Secchi Disk, a circular disk painted with a black and white design, as shown in Figure 4.52, is used for measuring the clarity or transparency of water. The disk is attached to a pole or line, then lowered into the water until the pattern on the disk is no longer visible; this depth is reported as the Secchi depth. The clarity of the water is thus correlated to the Secchi depth. For example, if Secchi disk readings were taken from two different lakes with values of 6 meters and 20 meters, the water clarity in the lake with a Secchi disk reading of 20 meters would be much better than the clarity of the water in the lake with a reading of 6 meters.

4.10.3 Nutrient management

Over time, any freshwater body naturally ages and becomes enriched with nutrients. Eventually, it fills up with detritus and debris from dead animal and plant matter, and also from sediments that are carried in from the watershed (e.g., a peat bog forms). This aging process normally takes thousands of years to complete and is known as natural eutrophication. Stimulation of algal growth is primarily due to the nutrients, nitrogen and phosphorus, that are released from sediments in rivers and lakes, and from the decay of animal and plant matter.

Cultural eutrophication is the term used when natural eutrophication is accelerated due to human activities. Point sources of pollution from municipal and industrial wastewater plant discharges have had created adverse affects on receiving waters. In the past, organic compounds, as measured by BOD and COD, in addition to suspended solids, were the primary culprits. More recently, ammonia, nitrate, and phosphate, and chlorine residual have been major parameters that were targeted for removal to enhance water quality. Most recently, however, regulators have been concerned with removing microconsituents that involve very low concentrations of pharmaceuticals, beauty and health care products, human and animal hormones, and other endocrine-disrupting chemicals that are not adequately treated by conventional wastewater treatment methods.

The level of nutrients in a lake affects primary productivity and the types of fish and aquatic species that can be supported. Plankton are the small, freely-suspended organisms that live in the water column of fresh waters, seas, and the oceans. Plankton are divided into two groups: phytoplankton, which represent the plant species – primarily algae and other photosynthetic organisms; and zooplankton, the animal species, consisting of crustaceans, rotifers, and protozoa that feed on phytoplankton. As discussed earlier in the chapter, productivity in aquatic ecosystems is primarily attributed to plankton. Some fish and whales feed off plankton in marine environments.

Lakes may be classified according to their nutrient levels, temperature profile, clarity, or biological productivity. Oligotrophic lakes contain low levels of nutrients, low biological productivity (algae and macrophytes), high clarity, and abundant oxygen. A lake with high nutrient levels, high productivity, poor clarity, and low oxygen levels in the hypolimnion is called a eutropohic lake. Algal blooms are observed frequently during the summer in this type of lake, and cold-water fish are absent. Lakes that contain moderate levels of nutrients, biological productivity, and oxygen are referred to as mesotrophic.

To reduce the adverse affects of cultural eutrophication on water bodies, nutrient release into the environment, especially into rivers and lakes, must be controlled. Nutrient removal processes are being implemented routinely at industrial and municipal wastewater treatment facilities to meet stringent limits for nitrogen and phosphorus for discharges into receiving waters to minimize algal stimulation. Stormwater runoff from agricultural and urban areas, at a minimum, should employ best management practices to reduce nutrients and sediments from entering aquatic ecosystems. These diffuse sources of pollution are the main contributors to nutrient enrichment of receiving waters.

Research suggests that the most important nutrient to control is phosphorus, since there are several species of bacteria that are capable of fixing nitrogen from the atmosphere. Lieberg's Law of the Minimal states that the growth rate and amount of biomass that can be produced is dependent on the nutrient that will limit growth. Phosphorus is the growth-limiting nutrient in most aquatic ecosystems, and every effort should be made to reduce the quantity of phosphorus released into the environment by human activities. According to Mihelcic (1999, page 296), it is generally accepted that the

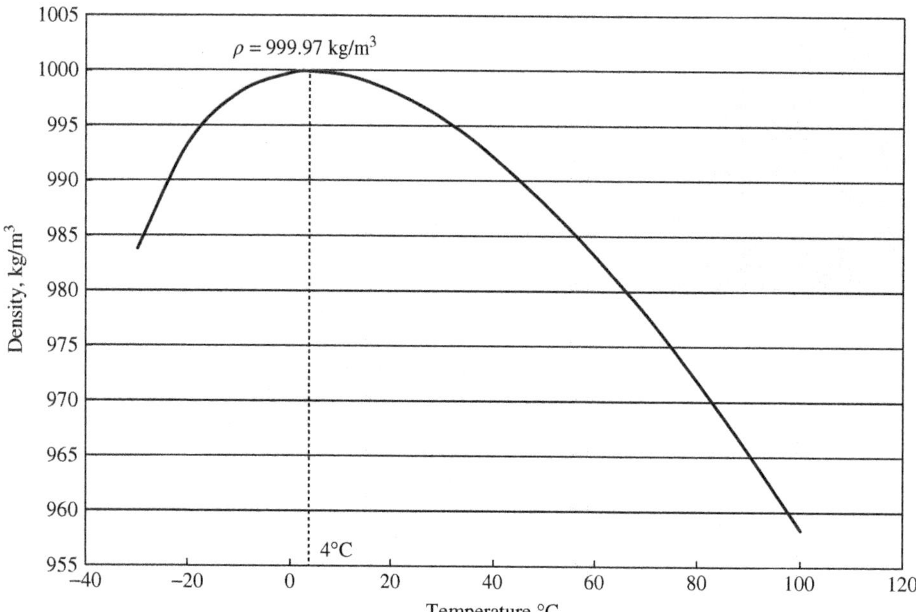

Figure 4.53 Density of water as a function of temperature.

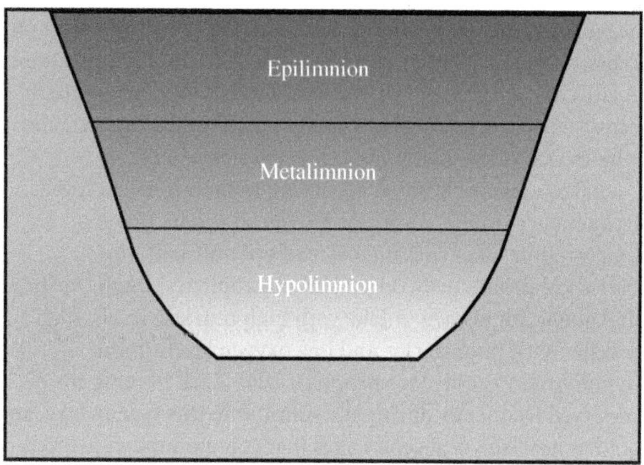

Figure 4.54 Lake stratification.

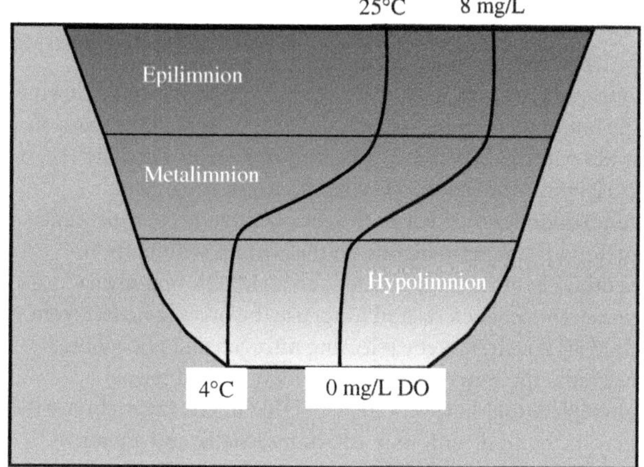

Figure 4.55 Temperature and DO profile during summer stratification.

boundaries between oligotrophy and mesotrophy, and between mesotrophy and eutrophy, are based on total phosphorus concentrations of 0.01 and 0.02 mg/L as P, respectively.

4.10.4 Stratification and turnover in lakes

Changes in temperature cause changes in the density of water in lakes, resulting in stratification or formation of layers. Climate and the changing of the seasons result in temperature changes. The density of water, as shown in Figure 4.53, increases from 983.8 kg/m^3 at −30°C to 999.9 kg/m^3 at 4°C (39°F), before decreasing to 958.4 kg/m^3 at 100°C. That is why a lake may freeze over during the winter, allowing ice to float over a layer of water that is warmer and denser. This allows fish and other aquatic life to survive beneath frozen lakes.

The sun provides the energy for heating the water in the lake, resulting in thermal stratification or formation of three distinct layers (Figure 4.54).

At and near the surface, where the temperature is highest, a layer called the **epilimnion** forms. It is well-mixed, and primary productivity is greatest in this layer, with high dissolved oxygen (DO) levels due to photosynthesis. Beneath the epilimnion is the **metalimnion**, which is a transition layer with a significant decline in temperature with depth. The significant drop in temperature ($\geq$ 1.0°C per meter of depth) that is observed in this layer is known as the **thermocline**. The **hypolimnion** is the layer at the bottom of the lake and below the metalimnion. It is generally well-mixed and DO levels tend to remain low, resulting in anaerobic conditions in the sediments. Water density is the highest at the bottom of the lake, since the temperature there is lowest, with exception of ice formation at the surface during the winter.

In temperate latitudes, lakes undergo stratification during the summer and winter. Figure 4.55 shows the DO and temperature profile in a lake during summer stratification.

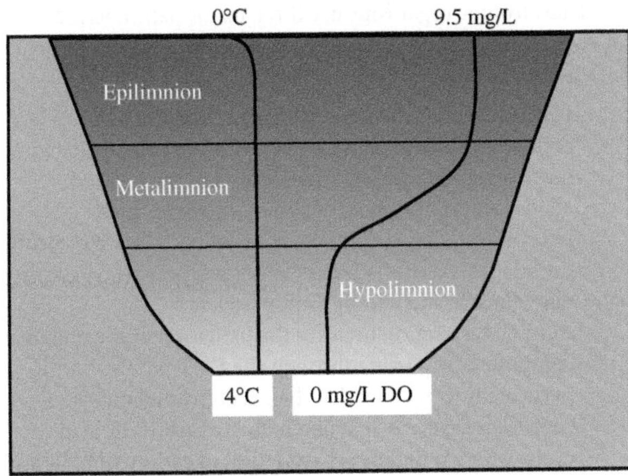

Figure 4.56 Temperature and DO profile during winter stratification.

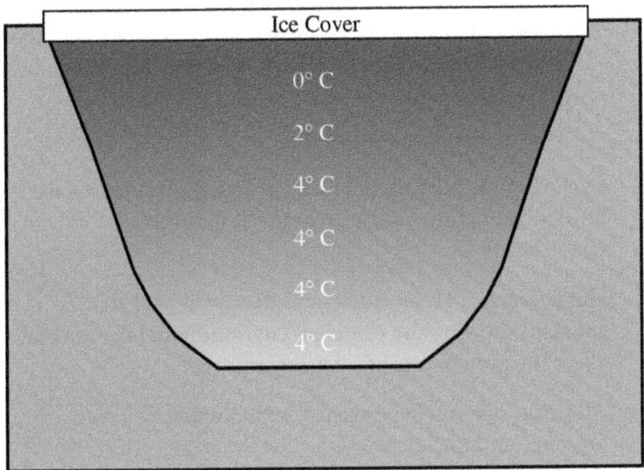

Figure 4.57 Temperature profile during winter stratification.

The DO concentration will vary, from saturation at the surface to zero at the bottom of the lake. The temperature profile decreases from approximately 25°C at the surface to 4°C at the bottom. The thermocline in the metalimnion layer results in a temperature gradient of ≥ 1.0°C per meter of depth.

During winter stratification, the DO profile is similar to that observed in the summer, decreasing from approximately 9.5 mg/L to 0 mg/L at the bottom (Figure 4.56). The major difference is that the temperature gradient slightly increases from the surface, from 0°C to 4°C with increasing depth. Recall that ice is less dense than liquid water, causing it to float on top of denser water that is typically around 4°C.

During spring and fall, complete-mixing of the lake occurs, which causes nutrients and sediments to re-enter the water column. This is knows as **turnover**, resulting in uniform levels of DO, nutrients, and temperature throughout the water column. Figures 4.57 and 4.58 show the temperature profiles that can be expected during winter stratification and spring turnover.

Algal blooms often follow spring turnovers, since nitrogen and phosphorus have been re-introduced into the water column. It is not unusual for a lake to appear as "pea green soup" due to algal growth. Depending on the type of algae

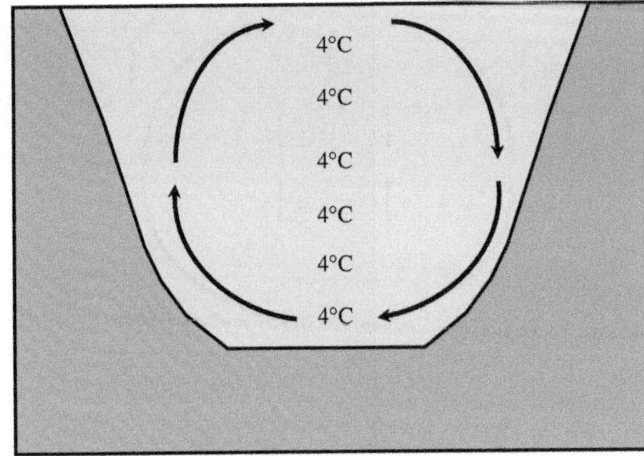

Figure 4.58 Mixing profile during spring turnover.

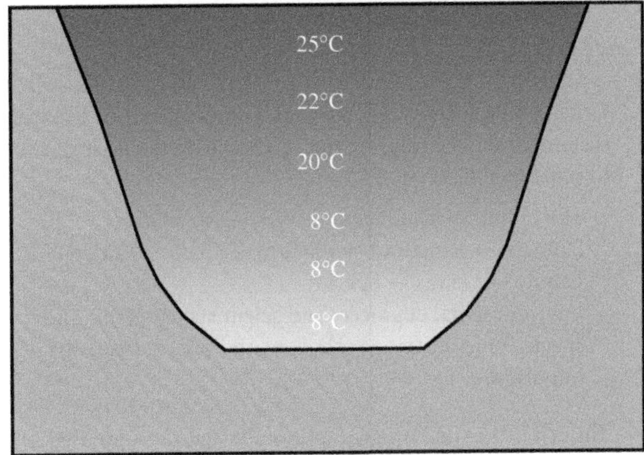

Figure 4.59 Temperature profile during summer stratification.

present, the color of the lake may take on a red or brownish color.

As winter transitions into spring, energy from the sun caused the ice to melt and the water temperature in the epilimnion to rise above 0°C. Recall that the density of water increases from 0°C to 4°C. The resulting increase in temperature in the epilimnion causes this denser water to sink towards the bottom of the lake, creating currents that bring colder water to the surface. Eventually, the entire contents of the lake become mixed. With the onset of summer, thermal stratification again occurs, resulting in the temperature profile shown in Figure 4.59.

As the seasons change, the cycle is repeated. During the fall, the temperature of the air decreases, causing the temperature in the epilimnion to decrease. This results in water with a higher density than the water beneath it. Ultimately, mixing currents are created, as this denser water settles to the bottom of the lake (Figure 4.60).

Once again, complete mixing of the lake occurs (fall turnover), allowing organics, nutrients, and sediments to re-enter the water column. Color and turbidity changes accompany these turnover events. As the air temperature continues to drop as winter approaches, the lake cools and ice may form on the surface. Regardless, the lake will undergo thermal stratification as depicted in Figure 4.57, with a temperature profile varying from 0°C or below at the surface to 4°C at the bottom.

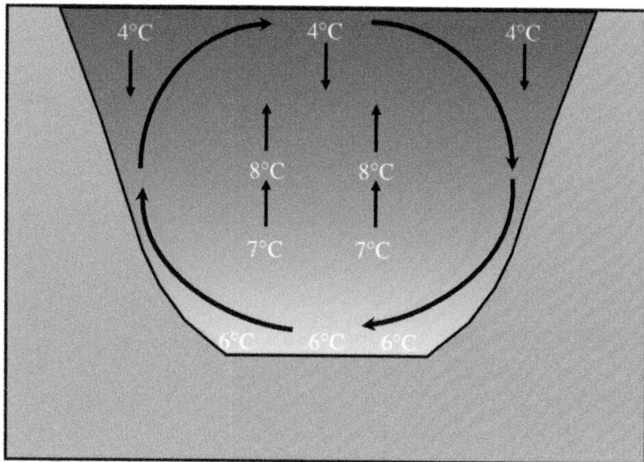

Figure 4.60 Mixing profile during fall turnover.

Summary

- Organisms are typically classified according to the Linnaean system, a taxonomic grouping starting with the least specific to most specific.
  - The Linnaean classification system is as follows: Domain → Kingdom → Phylum → Class → Order → Family → Genus → Species.
  - Environmental engineers and scientists use genus and species in identifying pathogens and microorganisms of importance.
- The cell is the fundamental unit of life and there are two types of microbial cells:
  - Prokaryotic cells do not have a nucleus and their DNA is contained in a single plasmid.
  - Eukaryotic cells have a true nucleus and are more complex than prokaryotic cells.
- Living organisms consume substrate, utilize nutrients, and carry out oxidation-reduction reactions for growth and maintenance functions.
- Microorganisms are further classified according to their carbon and energy source:
  - Microbes that use organic carbon as their source of carbon for synthesis and for oxidation for energy are called heterotrophs.
  - Autotrophic microbes use inorganic carbon as their source of carbon for synthesis of biomass and typically oxidize inorganic compounds for energy.
  - Microbes that use light as their source of energy are called phototrophs.
- Heterotrophs use the same organic material as their source of carbon and energy. Energy that is released during catabolic reactions is recovered by substrate-level phosphorylation and oxidative phosphorylation.
  - In substrate-level phosphorylation, energy released during the oxidation or dehydrogenation of organic compounds is used to convert ADP to ATP.
  - In oxidative phosphorylation, electrons that are produced during oxidation of the electron donor are then passed through an electron-transport system to a terminal electron acceptor, typically oxygen.
- Heterotrophic organisms use three major pathways for converting organic compounds into energy and synthesis reactions:
  - Fermentation is the process by which an organic compound is degraded by a series of enzyme mediated reactions and the energy is captured in the form of adenosine triphosphate (ATP) by substrate-level phosphorylation. This pathway is called "Glycolyis" or the Embden-Meyerhof-Parnas (EMP) pathway when starting with glucose and ending with pyruvate.
  - Aerobic respiration involves the oxidation of the organic compound by passing through the EMP pathway until pyruvate is formed, which then is converted into acetyl-coenzyme A that enters the tricarboxylic acid cycle, where it undergoes oxidation to carbon dioxide and water. Oxygen serves as the ultimate electron acceptor.
  - Anaerobic respiration is the process by which certain microorganisms use inorganic compounds such as carbon dioxide, sulfate, nitrite, and nitrate as the external electron acceptor rather than oxygen. These inorganic compounds accept electrons that are passed through the electron transport chain. Lower ATP production will be realized than what can be achieved during aerobic respiration.
- Organisms acquire their energy from light (phototrophs) or by oxidizing inorganic or organic compounds (chemotrophs).
- Temperature, pH, moisture, and dissolved oxygen concentration are the principal environmental factors that affect the growth rate of microorganisms.
- All living cells are composed of carbohydrates, lipids, proteins, and nucleic acids.
  - Carbohydrates are compounds that contain carbon, hydrogen, and oxygen; they are used as energy sources or in synthesis of cellular components.
  - Lipids are organic compounds that are soluble in organic solvents and sparingly soluble in water. The main functions of lipids are to store energy and serve as a structural component of cell membranes.
  - Proteins are complex molecules consisting of carbon, hydrogen, oxygen, and nitrogen. Amino acids are the building blocks of proteins.
  - Nucleic acids are macromolecular polymers that contain the genetic information of all living organisms. There are two types of nucleic acids: deoxyribonucleic acid (DNA) and ribonucleic acid (RNA).
- The Michaelis-Menten equation is used for modeling enzyme kinetics for a single reaction and single substrate.
- Ecology is the branch of biology that deals with the interrelationships among plants and animals (biota) and their interactions with the physical and chemical environment (abiotic or non-living environment).
- The sequence of steps or trophic levels involved in transferring energy in ecosystems may be represented by food chains and food webs.
- Primary productivity is the rate at which plants and photosynthetic organisms produce organic carbon compounds from atmospheric or aquatic carbon dioxide.

- Primary productivity may be defined on a gross primary productivity (GPP) or net primary productivity (NPP) basis.
- Terrestrial net primary productivity generally increases with an increase in mean annual temperature or an increase in mean annual precipitation.
- The mean world net primary productivity is the greatest for the oceans and is approximately 40 Gt/yr.

- Major biochemical cycles important in environmental engineering include: carbon, nitrogen, phosphorus, and sulfur.
  - The major reservoir for carbon on Earth is found in terrestrial systems and represents approximately 86% of the total carbon available.
  - The movement and transformation of nitrogen compounds in the biosphere is called the nitrogen cycle. Some of the most important transformation mechanisms include fixation, nitrification, denitrification, and ammonification.
  - Phosphorus is also an essential element required by all organisms for growth. Energy is stored in high-energy phosphate bonds during the synthesis of ATP.
  - Environmental scientists and engineers are concerned about the sulfur cycle since sulfates cause taste and odor problems in water supplies and create odor and corrosion problems in sewerage systems, and the release of sulfur oxides into the atmosphere creates air pollution problems related to malodorous gases and acid rain.
- Population dynamics focuses on how populations (organisms, animals, and people) change with time and the models used for estimating/predicting population growth.
- The six phases of growth involving pure cultures of microorganisms in batch systems are: lag, acceleration, exponential, declining, stationary, and endogenous.
- Most microorganisms, especially bacteria, multiply by binary fission, forming two new daughter cells every 15–20 minutes.
- The logistic growth model is used for estimating populations in ecosystems that are limited by some carrying capacity, e.g., land area.
- Human population projections are often modeled using arithmetic, exponential, and increasing rate of decrease mathematical models.
  - Engineers use human population projections for developing design flows for water and wastewater treatment facilities, the capacity of water supply reservoirs, and for estimating quantities of solid wastes that must be handled, recycled, and eventually disposed.
- Various factors such as birth rate, death rate, immigration rate, and emigration rate affect the overall growth rate.
- The world population is approaching seven billion people, and this impacts our ecosystems and the utilization of our natural resources in numerous ways. Earth's resources have been exploited and, as the world's populations have shifted from rural to urban areas, higher levels of pollution are now being generated and ecosystems are being compromised.
  - Population pyramids are drawn to show how various characteristics such as gender and age change with time.
- Water quality management is now based on a holistic viewpoint, in which an entire watershed is evaluated.
  - Total maximum daily loads (TMDLs) are being established for water bodies in the United States to improve and protect water quality.
  - A TMDL is the maximum amount of a pollutant that a water body can receive and still safely meet water quality standards.
- The Streeter-Phelps dissolved oxygen sag model serves as the basis for modeling the discharge of organic wastes into lakes and receiving streams.
  - This model is used to determine the distance downstream from the point of discharge to the location where the lowest dissolved oxygen concentration will occur.
  - Lakes and streams require high levels of dissolved oxygen (>5 mg/L) if game fish and other aquatic life are going to flourish.
- Limnology is the study of the biological, chemical, and physical characteristics of fresh water.
  - Lakes may be classified according to their nutrient levels, temperature profile, clarity, or biological productivity.
  - In the northern hemisphere, most deep lakes turnover in the fall and spring, causing nutrients and sediments to be reintroduced into the water column.
  - In these same lakes, there is stratification, or layers form, at different temperatures during the summer and winter.
  - Due to the unique characteristic of water having its maximum density at 4°C, lakes freeze from the surface down. This allows fish and other aquatic organisms to survive under a layer of ice during the winter.

Key Words

ADP
aerobes
aerobic respiration
amino acids
ammonification
anaerobes
anaerobic respiration
ATP
autotroph
binary fission
Calvin cycle
carbohydrates
carbon cycle
carrying capacity
chemosynthesis
chemotroph
citric acid cycle
community
cultural eutrophication
decomposers
denitrification
deoxygenation
ecology
electron transport chain
Embden-Meyerhof-Parnas
enzyme kinetics
epilimnion
eukaryotic
eutrophication
exponential growth
facultative organisms
FAD
$FADH_2$
fats
fermentation
food chain
food web
free energy
glycolysis
growth pyramids
heterotroph
hypolimnion
limnology
Linnaean
lipids
logistic growth
mesophiles
mesotrophic
metalimnion
Michaelis-Menten
NAD
NADP
net primary productivity
nitrification
nitrogen cycle
nitrogen fixation
nucleic acids
oils
oligotrophic
optimum pH
optimum temperature
phosphorus cycle
photosynthesis
phototroph

phytoplankton
population
population dynamics
primary consumers
primary producers
primary productivity
prokaryotic
proteins
psychrophiles
reaeration
Secchi disk
secondary consumers
standing crop
stratification
Streeter-Phelps
sulfur cycle
TCA
thermocline
thermophiles
TMDL
trophic level
turnover
turnover rate
zooplankton

References

Baum, Stuart J. (1978). *Introduction to Organic and Biological Chemistry*, MacMillan Publishing Co., Inc., New York, NY.

Bender, M., Grande, K., Johnson, K., Marra, J., Williams, P.J. LeB, Sieburth, J., Pilson, M., Langdon, C., Hitchcock, G., Orchardo, J., Hunt, C., Donaghay, P., Heinemann, K. (1986). A Comparison of four methods for determining planktonic community production, *Limnol Oceanorgr* **32**(5), 1085–1098.

Benefield, L.D. and Randall, C.W. (1980). *Biological Process Design for Wastewater Treatments*, pp. 26, 38, 39. Prentice Hall, Englewood Cliffs, NJ.

Brimblecombe, P., Hammer, C., Rodhe, H., Ryaboshapko, A., Boutron, C.F. (1989). *Evolution of the Global Biogeochemical Sulphur Cycle* Chapter 5, Human Influence on the Sulphur Cycle, pp. 82–84. SCOPE. John Wiley and Sons, Inc.

Brock, Thomas D. (1979). *Biology of Microorganisms*, pp. 99, 111, 158, 751. Prentice Hall, Englewood Cliffs, NJ.

Crites, R., Tchobanoglous, G. (1998). *Small and Decentralized Wastewater Management Systems*, p. 437. McGraw-Hill, New York, NY.

Davis, M.L., Masten, S.J. (2009). *Principles of Environmental Engineering and Science*, pp. 186, 363. McGraw-Hill, New York, NY.

Fair, Gordon M., Geyer, John C. and Okun, Daniel A. (1968). *Water and Wastewater Engineering, Volume 2. Water Purification and Wastewater Treatment and Disposal*, John Wiley & Sons, Inc., New York, pp. 33–26.

Falkowski, P., Scholes, R. J., Boyle, E., Canadell, J., Canfield, D., Elser, J., Gruber, N., Hibbard, K., Hogberg, P., Linder, S., Mackenzie, F. T., Moore III, B., Pedersen, T., Rosenthal, Y., Seitzinger, S., Smetacek, V., and Stefen, W. (2000). The Global carbon cycle: a test of our knowledge of Earth as a system, *Science*, **290**(5490), p. 291.

Filippelli, G.M. (2002). *Global Phosphorus Cycle*, pp. 392, 393, 395, 398. DOI: 10.2138/rmg.2002.48.10.

Gaudy, A.F., Gaudy, E.T. (1988). *Microbiology for Environmental Scientists and Engineers*, pp. 74, 97, 102–105, 183, 320, 343. McGraw-Hill, New York, NY.

Henry, J.G., Heinke, G.W. (1996). *Environmental Science and Engineering*, 2nd Edition, pp. 22, 262, 311, 318, 324. Prentice Hall, Upper Saddle River, New Jersey.

Henze, M., van Loosdrecht, M.C.M., Ekama, G.A., and Brdjanovic, D. (2008). *Biological Wastewater Treatment: Principles, Modelling and Design*, pp. 10–12, IWA Publishing, London.

Hutchinson, T.C. (1996). Ecology. In Henry, J.G., Heinke, G.W. (eds.), *Environmental Science and Engineering*, pp. 311, 318. Prentice Hall, Upper Saddle River, NJ.

Intergovernmental Panel on Climate Change (1995). Climate Change 1995, *The Science of Climate Change*, page 77.

Lieth, H. (1973). Primary Production: Terrestrial Ecosystems. *Human Ecology* **1**, 303–332.

Lindeman, R.L. (1942). The Trophic-Dynamic Aspect of Ecology. *Ecology* **23**, 399–418.

Littler, M.M. (1973). The Productivity of Hawaiian Fringing – Reef *Crustose Corallinaceae* and an Experimental Evaluation of Production Methodology. *Limnology and Oceanography*, **18**(6), 936–952.

Lodish, H., Berk, A., Zipursky, S.L., Matsudaira, P., Baltimore, D., and Darnell, J. (2000). *Molecular Cell Biology*, 4th Edition, W.H. Freeman, New York.

Metcalf and Eddy (1991). *Wastewater Engineering: Treatment, Disposal, and Reuse*, 3rd Edition, pp. 76, 369. McGraw-Hill, New York, NY.

Metcalf and Eddy (2003). *Wastewater Engineering: Treatment and Reuse*, 4th Edition, pp. 106, 369, 559, 572–573, 1737. McGraw-Hill, New York, NY.

Mihelcic, J.R. (1999). *Fundamentals of Environmental Engineering*, p. 296. John Wiley and Sons, Inc., Hoboken, NJ.

Miller, G.T. (1975). *Living in the Environment Concepts, Problems, and Alternatives*, p. 63. Wadsworth Publishing Company, Inc., Belmont, California.

Monod, J. (1949). The Growth of Bacterial Cultures. *Annual Review of Microbiology* **III**, 371–394.

National Oceanic and Atmosphere Administration (2007). Carbon Dioxide, Methane Rise Sharply *in 2007*. http://www.noaanews.noaa.gov/stories2008/20080423_methane.html; accessed on February 12, 2011.

O'Connor, D.J., Dobbins, W.E. (1958). Mechanism of Reaeration in Natural Streams, *American Society of Civil Engineers Transactions* **153**, 641.

Pagan, J.D., Hintz, H.F. (1986). Equine Energetics I. Relationship between Body Weight and Energy in Horses. *J. Anim Sci.* **63**, 815–821.

Peavy, H.S., Rowe, D.R., Tchobanoglous, G. (1985). *Environmental Engineering*, pp. 83, 86. McGraw-Hill, New York, NY.

Pelczar, Michael J., Reid, Roger D., and Chan, E.C.S. (1977). *Microbiology*, McGraw-Hill, New York, pp. 111–112, 358.

Rittman, B.E., McCarty, P.L. (2001). *Environmental Biotechnology: Principles and Applications*, pp. 10–12, 16, 133. McGraw-Hill, New York, NY.

Sawyer, C.N., McCarty, P.L., Parkin, G.F. (1994). *Chemistry for Environmental Engineering*, p. 250. McGraw-Hill, New York, NY.

Soderlund, R., Svenson, B.H. (1976). The Global Nitrogen Cycle, in *Nitrogen, Phosphorus, and Sulphur Global Cycles*, SCOPE Report 7, *Ecol. Bull.* **22**, 23–73.

Streeter, H.W., Phelps, E.B. (1925). A Study of the Pollution and Natural Purification of the Ohio River. *US Public Health Service Bulletin No.* **146**.

US EPA (1985). *EPA Guide for Identifying Cleanup Alternatives at Hazardous Wastes Sites and Spills: Biological Treatment.* US Environmental Protection Agency, EPA 600/3–83/063, Washington, DC.

US EPA (1993). *Manual Nitrogen Control*, pp. 5–8. EPA/625/R-93/010, Washington, DC.

US EPA (2000). *Total Maximum Daily Load (TMDL)*, US Environmental Protection Agency, EPA 841-F-00-009, Washington, DC.

US Environmental Protection Agency (2011). *Impaired Waters and Total Maximum Daily Loads.* http://water.epa.gov/lawsregs/lawsguidance/cwa/tmdl/index.cfm. Accessed February 12, 2011.

US Geological Survey (2005). *Estimated Water Use in the United States, 2005.* http://pubs.usgs.gov/fs/2009/3098/. Accessed February 11, 2011.

WEF (1998). *Biological and Chemical Systems for Nutrient Removal*, pp. 17–18. Alexandria, VA.

Whittaker, R.H. (1975). *Communities and Ecosystems*, pp. 2, 202, 204, 224. MacMillan Publishing Co., Inc., New York, NY.

Problems

1 Discuss some of the major differences between prokaryotic and eukaryotic cells.

2 Define the following terms:
a. Heterotroph
b. Autotroph
c. Phototroph
d. Chemotroph
e. Chemosynthesis
f. Trophic level
g. Food chain
h. Nitrification
i. Denitrification
j. Eutrophication
k. Ecology

3 Define phosphorylation and explain what is meant by substrate-level phosphorylation versus oxidative phosphorylation.

4 Prepare a brief paragraph discussing how pH, temperature, oxygen, and moisture affect microbial growth rate.

5 Perform a "Google" search on nucleic acids. Discuss the major differences between DNA and RNA.

6 The Michealis-Menten equation describes the enzyme kinetics for a single reaction and single substrate. It is also used for modeling the kinetics observed during biological wastewater treatment. Given the following data collected in the laboratory, determine the values for r_{max} and K_M by rearranging Equation (4.28) and plotting $1/r$ versus $1/S$.

Substrate concentration (mg/L)	Rate of reaction, h^{-1}
4.0	0.28
5.0	0.33
6.6	0.40
10.0	0.50
20.0	0.66

7 A pure culture of *Nitrosomonas* was grown at 20°C under batch conditions. The following experimental data were collected.

Time (h)	*Nitrosomonas* Concentration, X_t (mg/L)
2.0	102
5.0	105
10.0	117
20.0	123
25.0	130
30.0	137

Rearranging of Equation (4.50) results in the following equation which can be used for determining the specific growth rate of the culture.

$$\ln(X_t) = \ln(X_0) + \mu t$$

A plot of $\ln(X_t)$ versus time (t) will yield a straight line with the slope equal to μ and the y-intercept equal to $\ln(X_0)$ or $X_0 = e^{Y\text{-Intercept}}$. Determine the specific growth rate μ.

8 Draw a simplified food chain with three trophic levels for a fresh-water pond.

9 Draw a simplified food chain with three trophic levels for an ecosystem in a tropical rain forest.

10 How does climate affect primary productivity? Give examples of ecosystems (both aquatic and terrestrial) that yield the highest net primary productivity levels in Gt/y.

11 Water samples were collected from Mountain Lake near Blacksburg, Virginia during early summer. Determine the respiration rate, gross primary production, and net primary production in the lake given the following data: the dissolved oxygen concentrations of the water for the initial sample, light bottle after incubation, and dark bottle incubation are 9 mg/L, 15 mg/L, and 5 mg/L, respectively; the incubation period was one hour.

12 List, by rank (most to least), the six major global reservoirs of carbon in terms of GtC.

13 Draw a sketch of the generalized nitrogen cycle, showing nitrogen fixation, nitrification, ammonification, and denitrification.

14 Nitrification is an aerobic process in which ammonium/ammonia nitrogen is transformed into nitrate. The genera, *Nitrosomonas* and *Nitrobacter* are the two bacterial organisms that mediate the process. Equations (4.41), (4.42), and (4.43) show the reactions when synthesis of biomass is neglected. Metcalf & Eddy (2003) show the following overall nitrification reaction including synthesis of biomass. Due to rounding of the coefficients, the equation does not balance exactly, but this error is negligible:

$$NH_4^+ + 1.863\,O_2 + 0.098\,CO_2 \rightarrow 0.0196\,C_5H_7O_2N + 0.98\,NO_3^- + 0.941\,H_2O + 1.98\,H^+$$

a. Calculate the quantity (grams) of alkalinity as $CaCO_3$ consumed per gram of ammonium nitrogen (NH_4^+-N) oxidized?
b. How does this compare with the alkalinity consumption from Equation (4.43)?
c. Calculate the quantity (grams) of oxygen required per gram of –N oxidized.
d. How does this compare with the oxygen consumption from Equation (4.43)?

15 The following stoichiometric equation shows the overall denitrification reaction when the biodegradable

organic matter in the incoming wastewater represented as $C_{10}H_{19}O_3N_1$ is used rather than adding an external source such as methanol or acetate.

$$C_{10}H_{19}O_3N_1 + 10NO_3^- \rightarrow 5N_2\uparrow + 10CO_2 + 3H_2O + NH_3 + 10\,OH^-$$

a. Calculate the grams of alkalinity as $CaCO_3$ produced per gram of nitrate nitrogen utilized.
b. Determine the grams of organic matter required per gram of $NO_3^- - N$ converted to nitrogen gas.

16 The following stoichiometric equation shows the overall denitrification reaction when acetate serves as the organic carbon source for the denitrifiers rather than methanol.

$$5CH_3COOH + 8NO_3^- \rightarrow 4N_2\uparrow + 10CO_2 + 6H_2O + 8\,OH^-$$

a. Calculate the grams of alkalinity as $CaCO_3$ produced per gram of nitrate nitrogen utilized.
b. Determine the grams of organic matter required per gram of $NO_3^- - N$ converted to nitrogen gas.

17 Based on the global nitrogen reservoir data presented in Table 4.11, calculate the % terrestrial, oceanic, and atmospheric nitrogen, based on the total quantity of 1.943×10^{11} TgN.

18 A municipal WWTP discharges 20 million gallons per day of secondary effluent containing 30 mg/L of ultimate BOD at 27°C with 2.0 mg/L of dissolved oxygen. The stream flow is 150 cubic feet per second (cfs) at a velocity of 1.5 feet per second (fps) and an average depth of 5 ft. The temperature of the stream before the wastewater enters the river is 19°C. The stream is 90% saturated with oxygen and has an ultimate BOD of 8.0 mg/L. The reaeration (k_R) and deoxygenation (k_D) rate coefficients are 0.50 d^{-1} and 0.19 d^{-1}, respectively, at 20°C. Determine the following:

a. wastewater flow rate in cfs;
b. temperature of the combined wastewater and stream;
c. dissolved oxygen concentration of the mixture of wastewater and stream;
d. DO deficit of the mixture of wastewater and stream;
e. ultimate BOD concentration of mixture of wastewater and stream;
f. time of travel to the point of minimum DO concentration;
g. minimum DO concentration in stream;
h. plot the DO deficit versus time, starting at 0 and going to 3.0 days in increments of 0.1 days.

19 A municipal WWTP discharges 0.75 m^3/s of secondary effluent at a temperature of 28°C that contains 32 mg/L of ultimate BOD. To meet regulatory requirements, the effluent is aerated to achieve 6.0 mg/L of dissolved oxygen prior to discharge. The stream flow is 3.5 m^3/s and the temperature upstream of the discharge point is 18°C. The background ultimate BOD in the stream is 8.0 mg/L. The reaeration (k_R) and deoxygenation (k_D) rate coefficients are 0.35 d^{-1} and 0.22 d^{-1}, respectively, at 20°C. In-stream standards require a minimum of 5.0 mg/L of DO at all times. Determine the following:

a. temperature of the combined wastewater and stream;
b. dissolved oxygen concentration of the mixture of wastewater and stream;
c. DO deficit of the mixture of wastewater and stream;
d. ultimate BOD concentration of mixture of wastewater and stream;
e. critical time to reach point of minimum DO concentration;
f. minimum DO concentration in stream.

20 A municipal WWTP discharges 22.5 million gallons per day of secondary effluent containing 30 mg/L of ultimate BOD at 25°C with 2.0 mg/L of dissolved oxygen. The stream flow is 160 cubic feet per second (cfs) at a velocity of 1.5 feet per second (fps) and an average depth of 5 ft. The temperature of the stream before the wastewater enters the river is 20°C. The stream is 85% saturated with oxygen and has an ultimate BOD of 5.0 mg/L. The reaeration (k_R) and deoxygenation (k_D) rate coefficients are 0.35 d^{-1} and 0.20 d^{-1}, respectively at 20°C. Determine the following:

a. wastewater flow rate in cfs;
b. temperature of the combined wastewater and stream;
c. dissolved oxygen concentration of the mixture of wastewater and stream;
d. DO deficit of the mixture of wastewater and stream;
e. ultimate BOD concentration of mixture of wastewater and stream;
f. time of travel to the point of minimum DO concentration;
g. minimum DO concentration in stream;
h. plot the DO deficit versus time starting at 0 and going to 3.0 days in increments of 0.1 days.

21 A municipal WWTP discharges 0.5 m^3/s of secondary effluent at a temperature of 22°C that contains 30 mg/L of ultimate BOD. To meet regulatory requirements, the effluent is aerated to achieve 5.0 mg/L of dissolved oxygen prior to discharge. The stream flow is 3.0 m^3/s and the temperature upstream of the discharge point is 18°C. The background ultimate BOD in the stream is 10.0 mg/L. The reaeration (k_R) and deoxygenation (k_D) rate coefficients are 0.40 d^{-1} and 0.20 d^{-1}, respectively at 20°C. In-stream standards require a minimum of 6.0 mg/L of DO at all times. Determine the following:

a. temperature of the combined wastewater and stream;
b. dissolved oxygen concentration of the mixture of wastewater and stream;
c. DO deficit of the mixture of wastewater and stream;
d. ultimate BOD concentration of mixture of wastewater and stream;
e. critical time to reach point of minimum DO concentration;
f. minimum DO concentration in stream.

22 Use the following data from Norris Lake, Tennessee, to plot temperature profile. Estimate the magnitude of the thermocline.

Depth, m	Temperature, °C
0.0	29.1
0.9	29.1
2.1	29.1
3.0	29.1
4.0	29.1
4.9	29.1
6.1	29.1
7.0	28.6
7.9	27.2
9.1	25.7
10.1	24.5
11.0	23.7
11.9	22.3
13.1	21.2
14.0	20.0
14.9	19.0
15.8	18.0
17.1	17.2
18.0	16.4
18.9	15.9
20.1	15.1
21.0	14.5
21.9	14.0
22.9	13.6
24.1	13.4
25.0	13.1
25.9	12.9
27.1	12.7
28.0	12.5
29.0	12.2
29.9	12.0

23 Use the following world population data and make an appropriate plot, using a spreadsheet, to determine the world population growth rate as a percentage. How does the rate that you determined compare to the growth rate shown in Figure 4.46?

Year	World population (billions)
1959	3.0
1974	4.0
1987	5.0
1999	6.0
2011	7.0

24 Estimate the 2020 population of a city, using three population models available for short term forecasting: arithmetic, exponential, and declining.

Census date	Population
1990	25,000
2000	39,000
2010	55,000

25 Estimate the 2020 population of a city using three population models available for short term forecasting: arithmetic, exponential, and declining.

Census date	Population
1990	40,000
2000	55,000
2010	60,000

Chapter 5

Environmental systems: modeling and reactor design

Richard O. Mines, Jr.

Learning Objectives

After reading this chapter, you should be able to:

- apply the Law of Conservation of Matter in performing material balances on processes or systems;
- determine the rate of reaction and order of reaction;
- describe complete-mix, plug, and dispersed-plug flow regimes;
- design complete-mix batch reactors, complete-mix flow reactors, and plug flow reactors containing reactive and non-reactive contaminants;
- apply the Law of Conservation of Energy in performing energy balances on natural and engineered systems.

5.1 Introduction

Environmental engineers use the mass or materials balance concept to model environmental systems, something similar or analogous to balancing a checkbook. The mass of materials entering a control volume must be equal to the mass of materials leaving the control volume, plus any mass that is accumulated within the control volume. Two important laws are used in conjunction with performing material balances:

- The **Law of Conservation of Matter** states that matter cannot be created nor destroyed. The exception is in the case of nuclear reactions, where a portion of matter is changed into energy.
- Equally important is the **Law of Conservation of Energy**, which states that energy cannot be created or destroyed. However, the *form* of energy can change.

As energy conservation continues to be a priority world-wide, it is essential that environmental engineers be able to perform energy balances around all types of systems. The ability to perform energy balances on an ecosystem or coal-fired power plant is a skill that all engineers should possess. The kinetics (how fast a reaction occurs) of biological and chemical reactions that occur within a system will influence the efficiency and sizing of engineered systems. For instance, in designing an aeration basin to treat municipal wastewater, knowledge of the oxygen uptake and substrate utilization rates are required to size the aeration system properly and to determine the dimensions of the reactor. When chlorine is added to drinking water, it is essential to know the required contact time to ensure efficient pathogen inactivation, i.e., the kinetics of chlorine disinfection. The kinetics of reactions in natural systems is important, too. An example of this involves the transfer of oxygen from the atmosphere to water, known as the process of reaeration in river systems.

5.2 Material balances

Material or mass balances are used for modeling the mass rate of flow of materials entering and exiting a control volume ($C\not{V}$). The control volume is the specific region of space on which the mass balance is performed. Examples of control volumes include reactors, lakes, landfills, rivers, tanks, aquifers, and the

Environmental Engineering: Principles and Practice, First Edition. Richard O. Mines, Jr.

air basin above a city. Normally, a dashed line is placed around the control volume to indicate a boundary on which the mass balance is being performed.

There are three potential fates for a substance entering the control volume. The substance on which the materials balance is being performed may be mass, energy, momentum, or some other property. A substance may accumulate within the control volume; some of it may leave the control volume without undergoing any type of change, or some of it may be transformed into a different compound, e.g., a gas, a precipitate, or synthesized into biomass. Note there are inputs and outputs to the control volume, as well as the possibility of accumulating mass within it.

Equation (5.1) is a qualitative equation that expresses what happens in a materials balance when no reaction is occurring, i.e., a non-reactive or conservative process. This word "equation" is typically used by environmental engineers when starting a materials balance.

$$[\text{accumulation}] = [\text{inputs}] - [\text{outputs}] \qquad (5.1)$$

The accumulation term is similar to the balance in a checkbook. When the flow and concentration in each stream entering and exiting the control volume remain constant, steady-state conditions exist and the accumulation term is equal to zero. During transient or nonsteady-state conditions, the accumulation term will increase or decrease. Inputs (deposits) represent all flows and materials entering the system, whereas outputs (withdrawals) represent all flows and materials exiting the system.

Chemical and mechanical engineers often use another form of the equation in which each term is expressed on a mass per unit time ($\dot{m}$) basis. The conservation of mass principle for a control volume or system undergoing a process without a reaction can be expressed as follows:

$$\begin{bmatrix}\text{net change}\\ \text{in mass within}\\ C\!\!\!/V\end{bmatrix} = \begin{bmatrix}\text{total mass}\\ \text{entering}\\ C\!\!\!/V\end{bmatrix} - \begin{bmatrix}\text{total mass}\\ \text{exiting}\\ C\!\!\!/V\end{bmatrix} \qquad (5.2)$$

Or:

$$\frac{dm_{C\!\!\!/V}}{dt} = \dot{m}_i - \dot{m}_e \qquad (5.3)$$

where:

$\frac{dm_{C\!\!\!/V}}{dt}$ = net change of mass contained within the control volume, $\frac{\text{mass}}{\text{time}}$

$\dot{m}_i$ = mass flow rate across inlet, $\frac{\text{mass}}{\text{time}}$

$\dot{m}_e$ = mass flow rate across outlet, $\frac{\text{mass}}{\text{time}}$

Figure 5.1 illustrates the conservation of mass principle for a two-inlet and one-exit steady-state flow process.

Remember that at steady-state, the mass within the $C\!\!\!/V$ remains constant, therefore $\frac{dm_{C\!\!\!/V}}{dt}$ or the accumulation term is set equal to zero. The total mass flow rate exiting the $C\!\!\!/V$ is equal to the sum of the mass flow rates entering the $C\!\!\!/V$.

A detailed presentation of the six-step method discussed in Chapter 1 will be used in solving the first three examples in

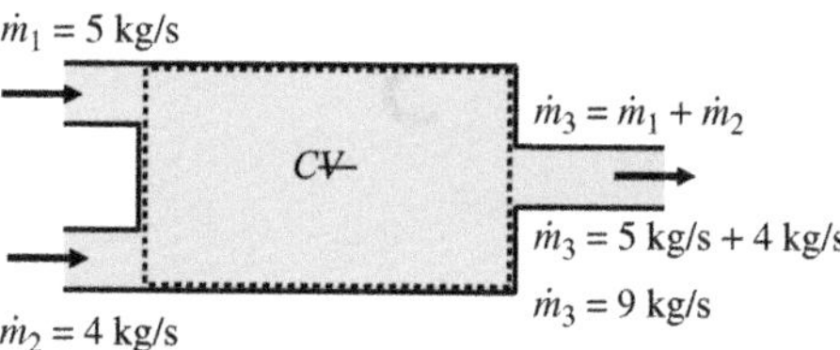

Figure 5.1 Conversation of mass for a two-inlet and one-exit steady-state flow system.

Chapter 5. In subsequent examples, however, an abbreviated problem-solving approach will be followed. Example 5.1 shows a simple mass balance on a swimming pool that has a leak in the bottom. This is an example of a "conservative" or non-reactive substance, since no biochemical or chemical reaction occurs in the control volume (swimming pool).

Example 5.1 Mass balance on flows

The owner of a swimming pool desires to fill it up with three garden hoses. The dimensions of the pool are 2-meters deep, 10-meters long, and 5-meters wide. Garden hose #1 is flowing at 1,000 cm^3/min, garden hose #2 is flowing at 15,000 cm^3/min, and garden hose #3 is flowing at 2,000 cm^3/min. To the owner's dismay, a small crack in the pool results in a leak that flows from the bottom of the pool at a rate of 750 cm^3/min. Determine how many hours it will take to fill the pool if all the hoses are turned on and the leak begins at the same time.

Solution

Identify and summarize the problem:
Draw a diagram showing all flows entering and exiting the pool.

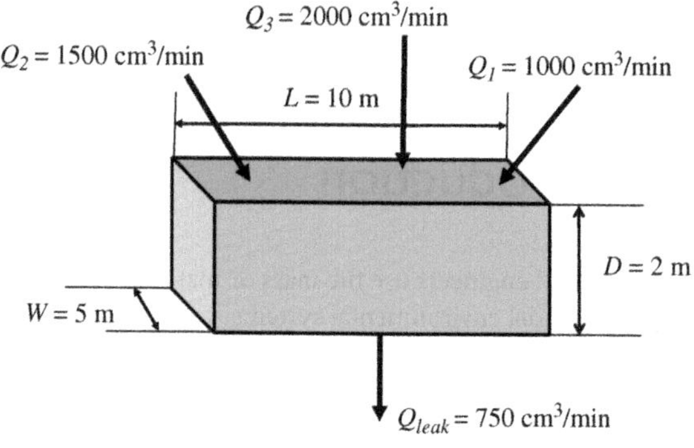

Figure E5.1

Determine: how many hours it will take to fill the pool up if all the hoses are turned on and the leak begins at the same time.

Known values:

- Pool length = 10 m
- Pool depth = 2 m
- Pool width = 5 m
- Flow in garden hose #1 = 1000 cm^3/min
- Flow in garden hose #2 = 1500 cm^3/min
- Flow in garden hose #3 = 2000 cm^3/min
- Leak in pool bottom results in a flow = 750 cm^3/min

Assumptions:

- No leaks develop in the garden hoses
- No other leaks develop in the pool
- Density of water (ρ) = 1 g/cm^3

Theory and relevant equations:
The volume of the pool is equal to length times width times depth. Calculate the volume of the pool using the following equation:

$$\forall = L \times W \times D$$

The volumetric flow rate, Q, is equal to the volume of water divided by time. This is expressed mathematically as follows:

$$Q = \frac{\forall}{t}$$

The mass flow rate ($\dot{m}$) is equal to density times flow, or expressed mathematically as follows:

$$\dot{m} = \rho \times Q$$

Recall that a material balances may be qualitatively expressed as Equation (5.1) when no reaction is occurring within the system.

$$[\text{accumulation}] = [\text{inputs}] - [\text{outputs}]$$

Solve:
Sufficient information and theory is available for solving the problem. First, calculate the volume of the pool in cubic meters and convert to cubic centimeters.

$$\forall = L \times W \times D = 10\,\text{m} \times 5\,\text{m} \times 2\,\text{m} = 100\,\text{m}^3$$

$$\forall = 100\,\text{m}^3 \times \left(\frac{100^3\,\text{cm}^3}{1\,\text{m}^3}\right) = 1.00 \times 10^8\,\text{cm}^3$$

Calculate the mass of water in the pool when it is full. Recall that mass is equal to density times the volume. This value represents the accumulation term in the mass balance Equation (5.1):

$$m_{accum} = \rho \times \forall$$

$$m_{accum} = \frac{1\,\text{g}}{\text{cm}^3}(1.00 \times 10^8\,\text{cm}^3) = 1.00 \times 10^8\,\text{g}$$

The mass of water flowing into and out of the pool is equal to the volumetric flow rate multiplied by the density of the water.

$$\dot{m}_1 = \rho \times Q_1 = \frac{1\,\text{g}}{\text{cm}^3} \times \frac{1000\,\text{cm}^3}{\text{min}} = \frac{1000\,\text{g}}{\text{min}}$$

$$\dot{m}_2 = \rho \times Q_2 = \frac{1\,\text{g}}{\text{cm}^3} \times \frac{1500\,\text{cm}^3}{\text{min}} = \frac{1500\,\text{g}}{\text{min}}$$

$$\dot{m}_3 = \rho \times Q_3 = \frac{1\,\text{g}}{\text{cm}^3} \times \frac{2000\,\text{cm}^3}{\text{min}} = \frac{2000\,\text{g}}{\text{min}}$$

$$\dot{m}_{leak} = \rho \times Q_{leak} = \frac{1\,\text{g}}{\text{cm}^3} \times \frac{750\,\text{cm}^3}{\text{min}} = \frac{750\,\text{g}}{\text{min}}$$

Next, substitute the mass flow rates of water calculated above into Equation (5.1), realizing that the inputs and outputs must be multiplied by time, *t*, so that the units on both sides of the equation are in terms of mass of water.

$$[\text{accumulation}] = [\text{inputs}] - [\text{outputs}]$$

$$m_{accum} = (\dot{m}_1 \times t) + (\dot{m}_2 \times t) + (\dot{m}_3 \times t) - (\dot{m}_{leak} \times t)$$

$$1.00 \times 10^8\,\text{g} = \left(1000\,\frac{\text{g}}{\text{min}} \times t\right) + \left(1500\,\frac{\text{g}}{\text{min}} \times t\right) + \left(2000\,\frac{\text{g}}{\text{min}} \times t\right) - \left(750\,\frac{\text{g}}{\text{min}} \times t\right)$$

$$1.00 \times 10^8\,\text{cm}^3 = 3750\,\frac{\text{g}}{\text{min}} \times t$$

$$t = \frac{1.00 \times 10^8\,\text{g}}{3750\,\text{g/min}} = 2.67 \times 10^4\,\text{min}$$

Communicate final answer:

$$\boxed{\text{time} = 2.67 \times 10^4\,\text{min} \times \left(\frac{1\,\text{h}}{60\,\text{min}}\right)\left(\frac{1\,\text{d}}{24\,\text{h}}\right) = 18.5\,\text{d}}$$

Check or validate answer:
Using the time of 2.67×10^4 minutes, calculate the volume of water that should accumulate in the pool by multiplying the time by the flow into minus the flow out of the pool. It should equal to the volume of the pool which is 1.00×10^8 cm^3. See calculations below.

$$\forall = (Q_1 \times t) + (Q_2 \times t) + (Q_3 \times t) - (Q_{leak} \times t)$$

$$\forall = (Q_1 + Q_2 + Q_3 - Q_{leak}) \times t$$

$$1.00 \times 10^8\ \text{cm}^3 = \left(1000\ \frac{\text{cm}^3}{\text{min}} + 1500\ \frac{\text{cm}^3}{\text{min}} + 2000\ \frac{\text{cm}^3}{\text{min}} - 750\ \frac{\text{cm}^3}{\text{min}}\right) \times 2.67 \times 10^4\ \text{min}$$

$$1.00 \times 10\ \text{cm}^3 = 1.00 \times 10^8\ \text{cm}^3$$

The time of 2.67×10^4 minutes is correct!

It is not unusual for reactions to be occurring within the *CV*. In some systems, a destructive type of reaction will occur, resulting in the removal or reduction of a certain chemical or contaminant. Often, the reaction will result in the production or synthesis of a compound, so the concentration of the compound leaving the reactor will be greater than the concentration entering it. Sometimes, the reactant is transformed to a new product. The biological mediated reaction carried out by the genus *Nitrosomonas* involves the transformation (oxidation) of ammonium (NH_4^+) to nitrite (NO_2^-). In most environmental engineering applications, mass balances are performed on all flow and material streams entering and exiting the control volume. Energy balances are usually performed separately and in some instances may not be required. Equation (5.4) is a modification of Equation (5.1) to account for reactions that occur within the process.

$$[\text{accumulation}] = [\text{inputs}] - [\text{outputs}] + [\text{reaction}] \quad (5.4)$$

Depending on the type of system or process, a chemical or biochemical reaction may take place, resulting in the production of or the removal of a substance. The positive sign in front of the reaction term may actually be negative if the reaction involves the removal or destruction of a constituent. If the reaction results in the formation of a product, then it remains positive. In biological treatment systems, microorganisms are grown, so the reaction term would be positive. Simultaneously, substrate, in the form of biochemical oxygen demand (BOD), is being removed from these systems. Therefore, a negative sign precedes the reaction term when material balances are performed on substrate in biological processes.

5.2.1 Steady-state modeling

In several situations, steady-state or equilibrium conditions will be assumed when performing the mass balance. For steady-state to exist, all flow rates, concentrations, pressures, temperatures, etc. entering and exiting the control volume must remain constant. Although steady-state conditions are seldom achieved for extended periods of time, it simplifies the mathematically modeling of the system and represents a good approximation of the system. Process designs are normally based on the steady-state assumption. The accumulation term in Equations (5.1), (5.2), (5.3), and (5.4) can then be set equal to zero. Assuming steady-state conditions for a conservative substance, i.e., nonreactive, Equation (5.1) can be simplified as follows:

$$[\text{inputs}] = [\text{outputs}] \quad (5.5)$$

This type of analysis is often used to determine the concentration of a conservative pollutant when two or more flow streams combine. The following equation is used for such and Example 5.2 illustrates the use of Equation (5.6):

$$C_3 = \frac{Q_1C_1 + Q_1C_2}{Q_1 + Q_2} = \frac{Q_1C_1 + Q_2C_2}{Q_3} \quad (5.6)$$

where:

C_3 = concentration of conservative substance in combined streams, mass/volume

C_1 = concentration of conservative substance in stream #1, mass/volume

C_2 = concentration of conservative substance in stream #2, mass/volume

Q_1 = flow rate in stream #1, volume/time

Q_2 = flow rate in stream #3, volume/time

Q_3 = flow rate of combined streams #1 and #2, volume/time.

In the following example, the effluent from a domestic wastewater treatment plant (WWTP) is discharged into a river. The concentration of suspended solids is to be determined at the point of discharge.

Example 5.2 Material balance on suspended solids

Determine the concentration of suspended solids (SS) in the river at the point of discharge from the WWTP. Five million gallons of wastewater containing 20 mg/L of SS are discharged into the river each day. The SS concentration in the river upstream of the discharge is 8 mg/L, and the flow rate in the river upstream from the point of discharge is 10 million gallons per day (10 MGD).

Solution

Identify and summarize the problem:
Draw a diagram showing the river and the discharge from the WWTP along with the flow rates and suspended solids concentration values in each.

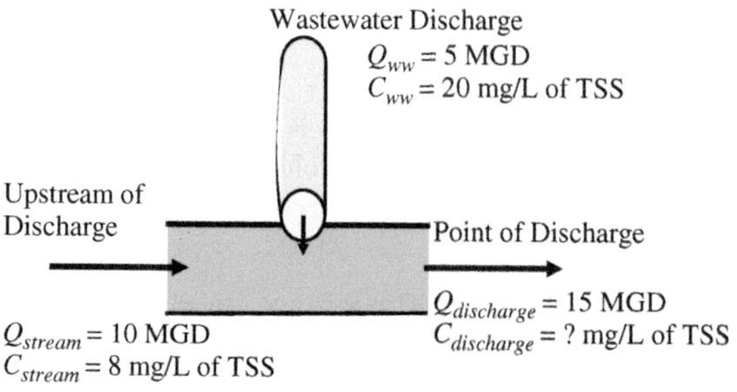

Figure E5.2

Determine: the suspended solids concentration in the river at the point of discharge.

Known values:

- Flow rate in river upstream of discharge = 10 MGD
- Flow rate in WWTP discharge = 5 MGD
- Concentration of SS in river upstream of discharge = 8 mg/L
- Concentration of SS in WWTP effluent = 20 mg/L

Assumptions:

- Suspended solids in the effluent do not immediately degrade due to biological activity so there are no reactions occurring within the control volume.
- Steady-state conditions prevail, meaning that the flow rates and concentration of suspended solids in each of the streams is not changing with time.

Theory and relevant equations:
Recall, that the materials balance equation may be qualitatively expressed as Equation (5.1).

$$[\text{accumulation}] = [\text{inputs}] - [\text{outputs}]$$

Since steady-conditions exist and the suspended solids concentrations are assumed to be a conservative substance, Equation (5.1) reduces to Equation (5.5).

$$[\text{inputs}] = [\text{outputs}]$$

Solve:
Solve the problem by rearranging Equation (5.6) and substituting appropriate values.

$$Q_1C_1 + Q_2C_2 = Q_3C_3$$

$$10\,\text{MGD} \times \frac{8\text{ mg SS}}{\text{L}} + 5\,\text{MGD} \times \frac{20\text{ mg SS}}{\text{L}} = (10\,\text{MGD} + 5\,\text{MGD})C_3$$

Communicate final answer:

$$C_3 = \frac{(10\,\text{MGD}) \times (8\text{ mg/L}) + (5\,\text{MGD}) \times (20\text{ mg/L})}{(10\,\text{MGD} + 5\,\text{MGD})}$$

$$= \boxed{12\,\frac{\text{mg SS}}{\text{L}}}$$

A box is often used to model air quality problems in a simplistic, idealistic way. Example 5.3 illustrates the mass balance concept for a conservative pollutant (particulate matter) wherein steady-state conditions are assumed and a reaction that generates particulate matter is occurring.

Example 5.3 Material balance on particulate matter

Estimate the rate of particulate matter generated from a forest fire. Particulate matter is a form of air pollution that can be retained in the lungs, causing permanent lung damage in addition to aggravating asthma and emphysema. The fire is burning 200 hectares of forest, resulting in a particulate concentration of 9,500 μg/m^3 in the air above the ground. The particulates rise to a height of 700 meters above the ground and the wind velocity is constant at 0.5 meters per second.

Solution

Identify and summarize the problem:
Draw a diagram showing the concentration of particulates entering and exiting the box above the forest fire area.

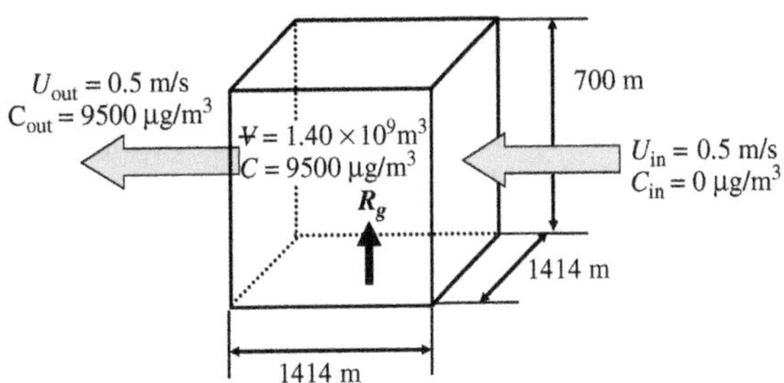

Figure E5.3

Determine: the rate of particulate generation from the forest fire (kg/s).

Known values:

- Wind speed into the box above the area (U_{in}) = 0.5 m/s
- Wind speed out of the box above the area (U_{out}) = 0.5 m/s
- Concentration of particulates upwind of the forest fire = 0 μg/m^3
- Concentration of particulates downwind of the forest fire = 9500 μg/m^3
- Height of plume rise = 700 meters
- Length and width of forest = 1,414 meters

Assumptions:

- The volume of air above the area is completely and uniformly mixed. Therefore, the concentration of particulate matter within the box and immediately downwind of the box is the same. Complete-mix reactors will be discussed in more detail later in this chapter.
- Steady-state conditions prevail, meaning that the air speed and concentration of particulate matter in the air upwind, downwind, and within the box, remains the same.

Theory and relevant equations:
The volume of the box is determined by multiplying the length by width by height.

$$\forall = L \times W \times H$$

Recall that the materials balance equation may be qualitatively expressed as Equation (5.4).

$$[\text{accumulation}] = [\text{inputs}] - [\text{outputs}] + [\text{reaction}]$$

Since steady-conditions exist and the accumulation term is zero, the reaction term or in this case – the rate of particulate generation – will be equal to the difference between the input and output streams.

$$0 = [\text{inputs}] - [\text{outputs}] + [\text{reaction}]$$

The continuity equation states that volumetric flow rate, Q, is calculated by multiplying the velocity by the cross-sectional area using the following equation.

$$Q = A \times U$$

The mass flow rate of any contaminant is estimated by multiplying the volumetric flow rate by the concentration of the contaminant.

$$\dot{m} = Q \times C$$

Solve:
First, calculate the volume of the air above the area denoted as the forest fire.

$$\forall = L \times W \times H = 1414\,\text{m} \times 1414\,\text{m} \times 700\,\text{m}$$
$$= 1.40 \times 10^9\,\text{m}^3$$

Next, calculate the air flow rate as follows:

$$A = L \times W = 1414\,\text{m} \times 700\,\text{m} = 9.90 \times 10^5\,\text{m}^2$$
$$Q = A \times V = 9.0 \times 10^5\,\text{m}^2 \times 0.5\,\text{m/s} = 4.95 \times 10^5\,\text{m}^3/\text{s}$$

Determine the mass of particulates entering and exiting the box.

$$\dot{m}_{in} = Q_{in} \times C_{in} = 4.95 \times 10^5\,\text{m}^3/\text{s} \times 0\,\mu\text{g/m}^3 = 0$$
$$\dot{m}_{out} = Q_{out} \times C_{out} = 4.95 \times 10^5\,\text{m}^3/\text{s} \times 9500\,\mu\text{g/m}^3$$
$$= 4.70 \times 10^9\,\mu\text{g/s}$$

Solve for the rate of particulate generation due to the forest fire by substituting the mass flow rates into the materials balance equation as shown below.

$$0 = [\text{inputs}] - [\text{outputs}] + [\text{reaction}]$$
$$[\text{reaction}] = [\text{outputs}] - [\text{inputs}] = 4.70 \times 10^9\,\mu\text{g/s} + 0$$

Communicate final answer:

Rate of particulate generation

$$= 4.70 \times 10^9\,\frac{\mu\text{g}}{\text{s}} \times \frac{1\,\text{g}}{10^6\,\mu\text{g}} \times \frac{1\,\text{kg}}{1000\,\text{g}} = \boxed{4.7\,\frac{\text{kg}}{\text{s}}}$$

5.2.2 Dynamic or transient modeling

Thus far we have primarily focused on steady-state modeling of environmental systems. More often than not, however, nonsteady-state or transient conditions prevail. Although zero, first, and second order reactions will be discussed in significant detail in future sections, it is imperative to illustrate a nonsteady-state application to show how it is handled. Let us examine a continuous flow reactor with a first-order decay reaction occurring within as shown in Figure 5.2.

We will assume that biochemical oxygen demand (BOD) is being removed in a complete-mix flow reactor (CMFR) at a first-order rate, where $r = -kC$. When performing a mass balance, begin with our qualitative equation (Equation (5.4)), as shown below.

$$[\text{accumulation}] = [\text{inputs}] - [\text{outputs}] + [\text{reaction}] \quad (5.4)$$

$$\left(\frac{dC}{dt}\right)\forall = QC_0 - QC + r\forall \quad (5.7)$$

We will need to solve for the steady-state solution in order to solve the differential equation that results during transient or nonsteady-state conditions. Recall that the accumulation term is equal to zero when all parameters remain constant with respect to time, i.e., steady-state conditions, and the effluent concentration will be C_∞ at time equal to infinity.

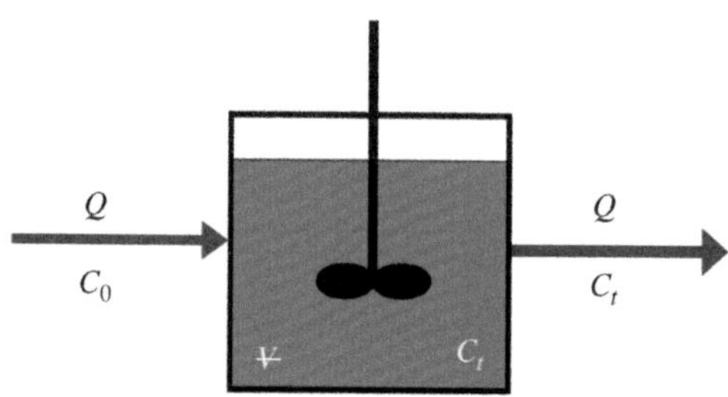

Figure 5.2 Nonsteady-state CMFR with first-order removal reaction.

$$0 = QC_0 - QC_\infty - kC_\infty V \quad (5.8)$$

$$0 = QC_0 - C_\infty(Q + kV) \quad (5.9)$$

$$C_\infty = \frac{QC_0}{(Q + kV)} \quad (5.10)$$

To solve the differential equation as presented as Equation (5.7), it must be rearranged in the following manner in order to arrive at the solution:

$$\left(\frac{dC}{dt}\right) = \frac{QC_0 - QC - kCV}{V} = \frac{(-kV - Q)C}{V} + \frac{QC_0}{V} \quad (5.11)$$

$$\left(\frac{dC}{dt}\right) = -\left(k + \frac{Q}{V}\right)C + \frac{QC_0}{V}\frac{\left(k + \frac{Q}{V}\right)}{\left(k + \frac{Q}{V}\right)} \quad (5.12)$$

$$\left(\frac{dC}{dt}\right) = -\left(k + \frac{Q}{V}\right) \times \left[C - \frac{QC_0}{V}\frac{1}{\left(k + \frac{Q}{V}\right)}\right] \quad (5.13)$$

$$\left(\frac{dC}{dt}\right) = -\left(k + \frac{Q}{V}\right) \times \left[C - \frac{QC_0}{(kV + Q)}\right] \quad (5.14)$$

Next, replace $\frac{QC_0}{(kV+Q)}$ in Equation (5.14) with Equation (5.10).

$$\left(\frac{dC}{dt}\right) = -\left(k + \frac{Q}{V}\right) \times [C - C_\infty] \quad (5.15)$$

One way to solve this differential equation is to make a change in variables as follows:

$$y = C - C_\infty \quad (5.16)$$

Differentiating y with respect to time results in Equation (5.17).

$$\frac{dy}{dt} = \frac{dC}{dt} \quad (5.17)$$

Substituting Equations (5.16) and (5.17) into Equation (5.15) yields Equation (5.18).

$$\left(\frac{dy}{dt}\right) = -\left(k + \frac{Q}{V}\right) \times y \quad (5.18)$$

Rearranging Equation (5.18) for integration results in the following equation, with limits of $y = y_0$ at time is equal to zero, and $y = y_t$ at time is equal to "t":

$$\int_{y_0}^{y_t} \frac{1}{y} dy = -\left(k + \frac{Q}{V}\right) \int_0^t dt \quad (5.19)$$

Integrating Equation (5.19) yields:

$$\ln y_t - \ln y_0 = -\left(k + \frac{Q}{V}\right) t \quad (5.20)$$

From Equation (5.16), $y_t = C_t - C_\infty$, therefore, $y_0 = C_0 - C_\infty$. Substituting these into Equation (5.20) yields the following equation:

$$\ln\left(\frac{C_t - C_\infty}{C_0 - C_\infty}\right) = -\left(k + \frac{Q}{V}\right) t \quad (5.21)$$

Taking the antilog of both sides of Equation (5.21) results in Equation (5.22).

$$\frac{(C_t - C_\infty)}{(C_0 - C_\infty)} = \exp\left[-\left(k + \frac{Q}{V}\right) t\right] \quad (5.22)$$

Equation (5.22) can be rearranged and solved for C_t as follows:

$$C_t = C_\infty + (C_0 - C_\infty)\exp\left[-\left(k + \frac{Q}{V}\right) t\right] \quad (5.23)$$

At time is equal to zero, the exponential portion of Equation (5.23) is equal to 1 and, therefore, $C_t = C_0$. Alternatively, at time is equal to infinity, the exponential function is equal to zero, and the concentration of C_t then becomes equal to C_∞.

Example 5.4 Nonsteady-state lagoon with non-conservative pollutant

A wastewater lagoon with a volume of 1,500 ft^3 receives industrial wastewater containing a non-conservative pollutant at a concentration of 20 mg/L at a flow rate of 100 ft^3/day. The first-order removal rate coefficient k for the pollutant is 0.20 d^{-1}. Determine the following:

a) The concentration of the pollutant leaving the lagoon, assuming that completely-mixed conditions exist in the lagoon and that steady-state conditions exist.
b) The concentration of the pollutant leaving the lagoon seven days after the influent concentration is suddenly increased to 120 mg/L.

Solution

Part a

Start with Equation (5.4) to derive a first-order removal equation assuming that steady-state conditions exist.

[accumulation] = [inputs] − [outputs] + [reaction] (5.4)

$$\left(\frac{dC}{dt}\right)V = QC_0 - QC - kCV \quad (5.24)$$

$$0 = QC_0 - QC - kCV \quad (5.25)$$

$$QC_0 - QC = kCV \quad (5.26)$$

$$QC_0 = C(kV + Q) \quad (5.27)$$

$$C = \frac{QC_0}{k\forall + Q} = \frac{\frac{100\ \text{ft}^3}{\text{d}}\left(\frac{20\ \text{mg}}{\text{L}}\right)}{\left(0.20\ \text{d}^{-1} \times 1500\ \text{ft}^3 + \frac{100\ \text{ft}^3}{\text{d}}\right)} \quad (5.28)$$

$$\boxed{C = 5.0\frac{\text{mg}}{\text{L}}}$$

Part b

Calculate the effluent pollutant at steady-state or infinity when the influent pollutant concentration increases to 120 mg/L by substituting into Equation (5.10).

$$C_\infty = \frac{QC_0}{(Q + k\forall)} = \frac{\frac{100\ \text{ft}^3}{\text{d}}\left(\frac{120\ \text{mg}}{\text{L}}\right)}{\left(\frac{100\ \text{ft}^3}{\text{d}} + 0.20\ \text{d}^{-1} \times 1500\ \text{ft}^3\right)} \quad (5.10)$$

$$C_\infty = \frac{30.0\ \text{mg}}{\text{L}}$$

Finally, calculate the final effluent pollutant concentration from the lagoon seven days after the influent pollutant concentration is increased to 120 mg/L by substituting into Equation (5.23).

The influent pollutant concentration is equal to 5.0 mg/L from Part a, since the lagoon is assumed to have reached steady-state before the influent pollutant concentration is increased to 120 mg/L.

$$C_t = C_\infty + (C_0 - C_\infty)\exp\left[-\left(k + \frac{Q}{\forall}\right)t\right] \quad (5.23)$$

$$C_7 = \frac{30\ \text{mg}}{\text{L}} + \left(\frac{5\ \text{mg}}{\text{L}} - \frac{30\ \text{mg}}{\text{L}}\right) \times \exp\left[-\left(0.2\ \text{d}^{-1} + \frac{100\ \text{ft}^3/\text{d}}{1500\ \text{ft}^3}\right)7\ \text{d}\right]$$

$$\boxed{C = 26.1\ \frac{\text{mg}}{\text{L}}}$$

5.3 Reaction kinetics

The study of the rates of reactions and the mechanisms that cause them is referred to as kinetics. Chemical kinetics deals with how fast a reaction occurs. Chemical reactions involving oxidation/reduction and acid-base neutralizations generally occur rapidly. However, biological reactions mediated by microorganisms that degrade organic compounds found in wastewater or contaminated groundwater generally do not reach equilibrium quickly. Thermodynamic principles are used to determine which direction a process or reaction will proceed of its own accord (e.g., will it occur spontaneously, requiring no additional energy to proceed?). These principles will not tell us how fast the process will reach equilibrium. Environmental engineers are concerned about how fast chemical and biological reactions occur in nature and in engineered systems.

5.3.1 Rates of reaction

The **rate of a reaction** (r) is the rate of formation or disappearance of a chemical compound or species. Homogeneous reactions are those that occur within a single phase, i.e., liquid, gas, or solid. Examples of homogeneous reactions include biochemical oxygen demand (BOD) by bacteria during biological treatment of wastewater, and the chemical reactions between aqueous chlorine and pathogens during disinfection of drinking water. Heterogeneous reactions occur between two phases, such as the air-water interface or solid-water interface. Examples of heterogeneous reactions are the transfer of oxygen into water or wastewater to increase the dissolved oxygen (DO) concentration, and the adsorption of organic contaminants onto activated carbon.

The rate of a reaction is affected by concentration, pressure, and temperature. In most applications in environmental engineering, the pressure and temperature are assumed to remain constant; therefore, the concentration of chemical species will primarily control the reaction rate. Temperature effects cannot be overlooked when designing biological treatment and oxygen transfer systems. Engineers must ensure that these systems will operate properly under both low and high temperature conditions.

The units for the rate of reaction (r) are normally expressed in moles per unit of volume per unit time, $\frac{\text{moles}}{(\text{unit volume}\cdot\text{time})}$, or mass per unit of volume per unit time, $\frac{\text{mass}}{(\text{unit volume}\cdot\text{time})}$. The rate equation can be written in several ways. The chemical reaction rate, r, can be expressed as the time rate of change of any of the reactants or products. For example, given the simple stoichiometric reaction as shown in Equation (5.29); one mole of species A reacts with one mole of species B to form one mole of product, C.

$$\text{A} + \text{B} \rightarrow \text{C} \quad (5.29)$$

The rate of appearance of product equals the disappearance of the reactants. Mathematically, this is expressed as follows:

$$r = \frac{d[\text{C}]}{dt} = -\frac{d[\text{A}]}{dt} = -\frac{d[\text{B}]}{dt} \quad (5.30)$$

where:
r = rate of reaction, moles/(L· time)
[ ] = concentration, moles/L or M.

When the stoichiometric coefficients are not all unity, as in Equation (5.29), it is customary to divide the change in concentration or derivative by the stoichiometric coefficient.

$$2\text{A} \rightarrow 3\text{C} \quad (5.31)$$

The rate of reaction for Equation (5.31) can be expressed as follows:

$$r = \frac{1}{3}\frac{d[C]}{dt} = -\frac{1}{2}\frac{d[A]}{dt} \quad (5.32)$$

5.3.2 Rate law and order of reaction

The **rate law** (written in the form of an equation) expresses the mathematical relationship between the rate of the reaction and the concentration of species involved in the reaction. Determination of the rate law is important to understanding the kinetics of a reaction. The rate law for any reaction must be determined by experiment, and cannot be inferred from the reaction stoichiometry. However, for elementary reactions, it is possible to determine the reaction order from the stoichiometric equation. The reader is referred to other texts to learn more about determining rate laws (Levenspiel, 1999; Snoeyink & Jenkins, 1980; Eisenberg & Crothers, 1979).

Equation (5.33) represents a stoichiometric equation for an irreversible reaction, where a, b, and c represent the stoichiometric coefficients or the number of moles of species A, B, and C, respectively.

$$aA + bB \rightarrow cC \quad (5.33)$$

The theoretical rate equation for Equation (5.33) can be expressed in the following manner.

$$r = -k[A]^{\alpha}[B]^{\beta} = k[C]^{\gamma} \quad (5.34)$$

Alpha (α), beta (β), and gamma (γ) represent the order of the reaction with regard to each individual chemical species. The overall order of the reaction presented in Equation (5.34) based on the reactants A and B is $\alpha + \beta$; or the reaction is said to be α-order with respect to A and β-order with respect to B. Alternatively, the overall reaction order based on product formation is γ and the reaction is said to be γ-order with respect to species C. Fractional reaction orders are possible too; often an integer value is assumed or used when writing the rate law.

Example 5.5 Determination of reaction order

For the rate law equation given below, determine the order of the reaction with respect to each chemical species and give the overall order of the reaction.

$$r = \frac{d[A]}{dt} = k[A][B]^2$$

Solution

The reaction is considered to be first-order with respect to A and second-order with respect to B. The overall order of the reaction is third.

Example 5.6 Determination of rate equation

(Ladon, 2001, http://pages.towson.edu/ladon/kinetics.html; Accessed March 5, 2012)

As noted earlier, the rate law must be determined experimentally. Consider the following overall chemical reaction:

$$2\,NO_{(g)} + O_{2(g)} \rightleftharpoons 2\,NO_2(g)$$

The proposed rate law for this reaction is as follows:

$$r = k[NO]^2[O_2]$$

Three experimental runs were performed to confirm the proposed rate law presented above. The table below show the data collected during each run.

Run Number	Rate of reaction, M/s	Initial [NO] concentration, M	Initial [O_2] concentration, M
1	1.2×10^{-8}	0.10	0.10
2	2.4×10^{-8}	0.10	0.20
3	1.08×10^{-7}	0.30	0.10

Solution

To develop the rate law, pick the first two runs and determine how the rate of reaction is affected by doubling the concentration of oxygen since the concentration of nitrous oxide [NO] is held constant. The rate of reaction has doubled by doubling the concentration of [O_2]. Showing this mathematically as follows:

$$2 = 2^x$$

Therefore, the value of x must be equal to 1 and the order of the reaction with respect to oxygen is 1.

Now select Runs 1 and 3, in which the concentration of oxygen is held constant at 0.10 M: and the rate of reaction increases by a factor of 9 $\left(\frac{1.08\times10^{-7}}{1.2\times10^{-8}} = 9\right)$. The rate of reaction increases nine-fold, while the concentration of nitrous oxide is tripled. This is expressed mathematically as follows:

$$9 = 3^x$$

Therefore, the value of x must be 2; this means that the order of reaction with respect to [NO] is 2. The proposed rate law is correct.

Determine the value of the reaction rate constant, k, using data from one of the runs. Let us use the data from Run #3:

$$r = k[NO]^2[O_2]$$

$$k = \frac{r}{[NO]^2[O_2]} = \frac{1.08 \times 10^{-7}\ \text{M/s}}{(0.30\ \text{M})^2(0.10\ \text{M})}$$

$$= 1.2 \times 10^{-5}\ \text{M}^{-2}\text{s}^{-1}$$

The relationship among rate of reaction (r), concentration of reactant (*conc*), and reaction order (n), can be simply expressed as Equation (5.35).

$$r = (Conc)^n \tag{5.35}$$

Taking the natural logarithm of both sides of the equation yields the following:

$$\ln(r) = n \ln(Conc) \tag{5.36}$$

Equation (5.35) can be applied to experimental data by taking the natural log (ln) of the instantaneous rate of change of the reactant concentration as a function of time, plotted against the natural log (ln) of the reactant concentration. Plotting equation (5.36) yields a straight line with slope equal to the reaction order.

Figure 5.3 shows a ln-ln plot for determining the order of a reaction. Note that there is a difference in the increments used on the X and Y scales. A zero-order reaction results in a horizontal line and the rate of reaction is independent of the concentration of reactant. For a first-order reaction, the rate of reaction is directly proportional to the reactant concentration raised to the first power whereas, for second-order reactions, the rate is proportional to the square of the reactant concentration. Fractional orders may exist, too; however, for developing rate equations in environmental applications, integers values are typically assumed or used.

5.3.3 Reactions

Reactions that proceed according to simple, irreversible reactions are typically zero-, first-, and second-, and perhaps third-order with respect to one or more of their reactants. Table 5.1 presents the general form of the equation that can be derived for zero-, first-, and second-order removal and production reactions. These equations only apply to unimolecular, irreversible reactions that can be described as a single reactant, A, being converted to a single product, P, according to the following equation: $A \rightarrow P$. The reaction rate constant is denoted as "k" and the reaction time as "t". "C" is the concentration of a given species. A "removal" or "consumption" reaction denotes the removal or conversion of a given species during the reaction or process. Similarly, the increase in concentration of a given species would signify a "production" or "generation" reaction.

The equations in Table 5.1 may be used to determine the required reaction time for a specified degree of conversion within a reactor or process, or they may be used to predict the effluent from a process, if the reaction time and reaction rate

Table 5.1 General form of zero-, first-, and second-order reactions.

Order	Removal	Production
Zero	$C_t = C_0 - kt$	$C_t = C_0 + kt$
First	$C_t = C_0e^{-kt}$	$C_t = C_0e^{kt}$
Second	$\frac{1}{C_t} = \frac{1}{C_0} + kt$	$\frac{1}{C_t} = \frac{1}{C_0} - kt$

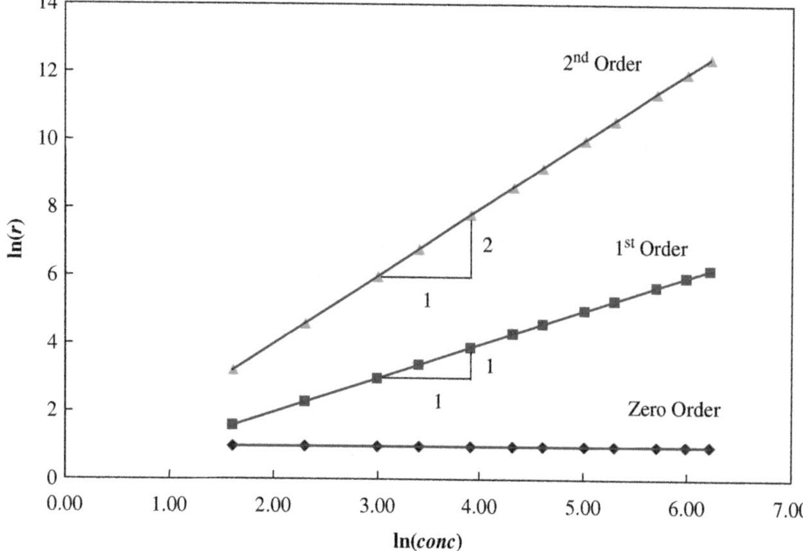

Figure 5.3 Determination of reaction order by ln-ln plot.

constant are known. The following sections illustrate the derivation of some of these equations.

5.3.4 Zero-order reactions

A zero-order reaction is one that proceeds at a rate that is independent of the concentration of any reactant. For example, consider the irreversible conversion of a single reactant (A) to a single product (P), according to Equation (5.37):

$$A \rightarrow P \tag{5.37}$$

Assuming a zero-order **removal** reaction, the rate law equation is written as follows:

$$r = -\frac{dC}{dt} = kC^0 = k \tag{5.38}$$

where:

$\frac{dC}{dt}$ = rate of change in the concentration of A with time or rate of disappearance of A, $\frac{\text{mass}}{(\text{volume}\cdot\text{time})}$

k = reaction rate constant, $\frac{\text{mass}}{(\text{volume}\cdot\text{time})}$.

Equation (5.38) is a differential equation and can be expressed in integral form as Equation (5.39) with limits from C_0, the initial concentration of A at time equal to 0 to C_t, the concentration of A at any time, t, as shown in Equation (5.39).

$$\int_{C_0}^{C_t} dC = -k\int_0^t dt \tag{5.39}$$

The integrated form of Equation (5.39) is presented below:

$$C_t - C_0 = -kt \tag{5.40}$$

Environmental engineers apply Equation (5.40) in one of two ways. Knowing the rate constant, k, and the initial and final concentrations to a specific process, it is possible to determine the required reaction time, t, as shown in Equation (5.41).

$$t = \frac{C_0 - C_t}{k} \tag{5.41}$$

Alternatively, Equation (5.40) can be used as a predictive or modeling tool by rearranging and solving for the final concentration for a given reaction time and given reaction rate constant as shown in Equation (5.42):

$$C_t = C_0 - kt \tag{5.42}$$

A zero-order **removal** reaction will plot as a straight line on arithmetic paper with a negative slope (Figure 5.4). The value of the reaction rate constant, k, is equal to the slope of the line. The value of the Y-intercept for this plot is equal to the initial concentration of reactant, C_0. Similarly, a zero-order **production** reaction will plot as a straight line on arithmetic paper with a positive slope.

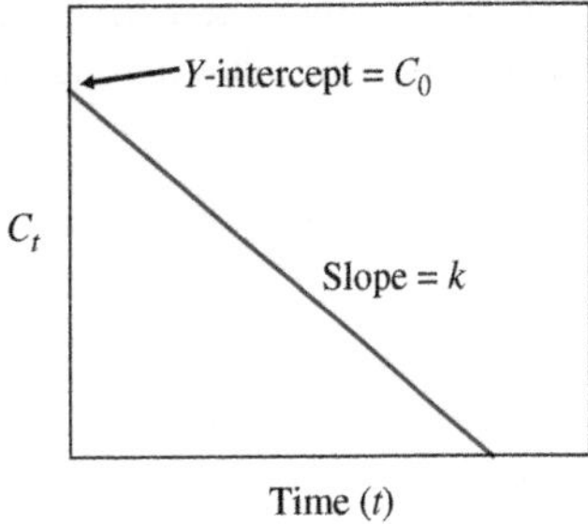

Figure 5.4 Plot of zero-order removal reaction.

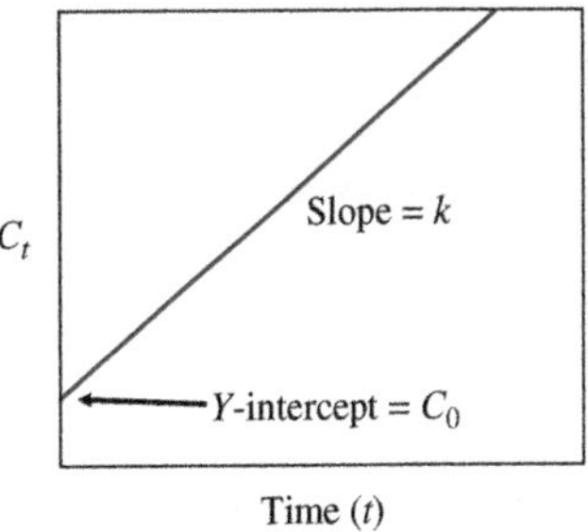

Figure 5.5 Plot of zero-order production reaction.

The value of k is equal to the slope of the line and the Y-intercept is equal to C_0 (Figure 5.5).

5.3.5 First-order reactions

Reactions that proceed at a rate that is proportional to the concentration of one of the reactants raised to the first power are called first-order. Since the concentration of the reactant changes with time, an arithmetic plot of product concentration, C_t, for a first-order **removal** reaction will be decreasing exponentially, as shown in Figure 5.6.

Consider the irreversible conversion of a single reactant (A) to a single product (P), as shown by Equation (5.43).

$$A \rightarrow P \tag{5.43}$$

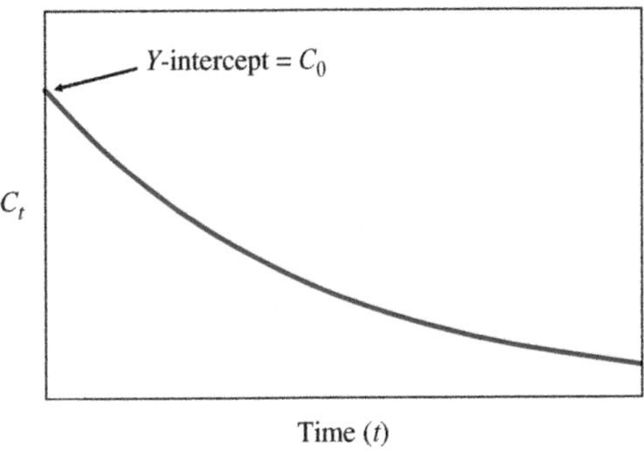

Figure 5.6 Arithmetic plot of first-order removal reaction.

Writing the rate law equation, assuming a first-order **removal** reaction, results in the following equation:

$$r = -\frac{dC}{dt} = kC^1 \tag{5.44}$$

The integral form of Equation (5.44), with the same integration limits as used for the zero-order derivation, where C_0 = initial concentration of A at time equal to zero and C_t = concentration of A at any time, t, results in Equation (5.45):

$$\int_{C_0}^{C_t} \frac{1}{C} dC = -k \int_0^t dt \tag{5.45}$$

The integrated form of Equation (5.45) is presented below.

$$\ln(C_t) - \ln(C_0) = -kt \tag{5.46}$$

Taking the antilog of both sides of Equation (5.46) results in the familiar form of a first-order or exponential equation, Equation (5.47).

$$C_t = C_0 e^{-kt} \tag{5.47}$$

where:
C_t = concentration of A at time t, mg/L
C_0 = concentration of A at time zero, mg/L
K = reaction rate constant, time^{-1}.

A plot of the natural logarithm of concentration versus time for a first-order **removal** reaction, as shown in Figure 5.7, will produce a straight line with a negative slope. The value of the reaction rate constant, k, is equal to the slope of line and the Y-intercept is equal to the natural logarithm of the initial concentration, $\ln[C_0]$.

Figure 5.8 is an arithmetic plot of a first-order **production** reaction showing how the concentration of the product increases at an exponential rate.

Figure 5.9 shows a plot of the natural log of concentration versus time for a first-order **production** reaction that results in a straight line with positive slope. The slope of the line is equal to k and the Y-intercept is equal to $\ln[C_0]$.

For radioactive substances, the decay or decomposition rate of the substance is defined in terms of the half-life.

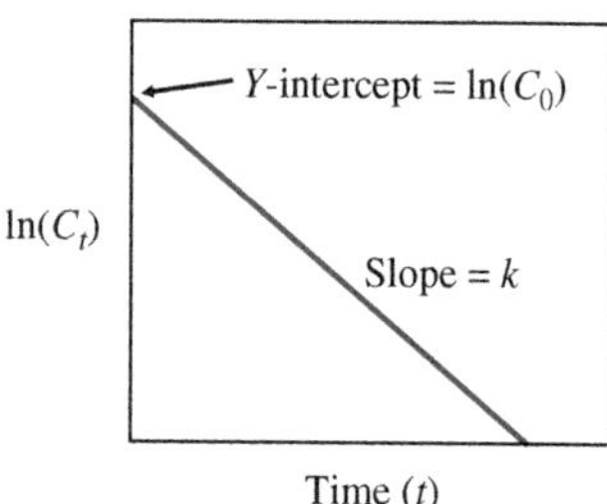

Figure 5.7 Semilog plot of first-order removal reaction.

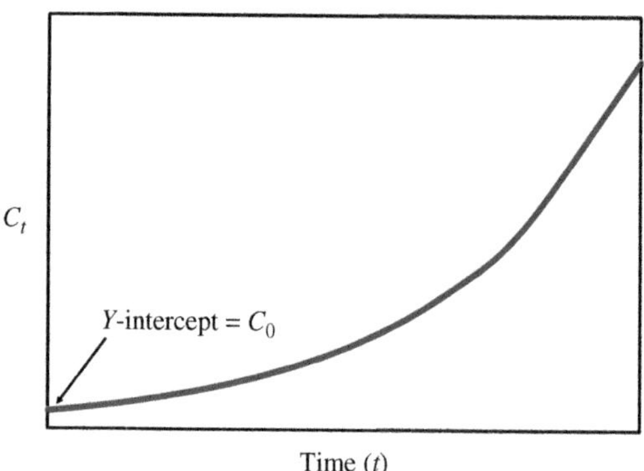

Figure 5.8 Arithmetic plot of first-order production reaction.

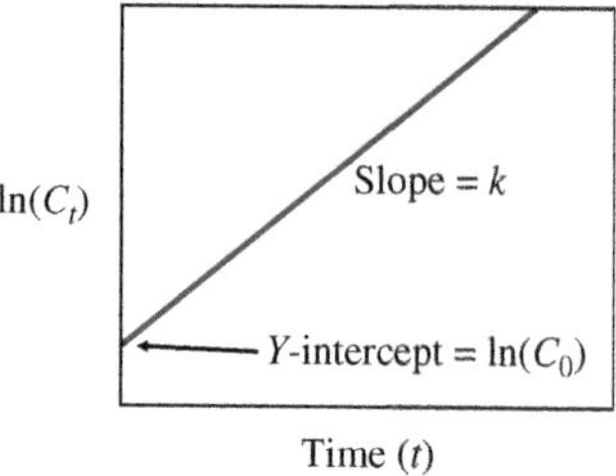

Figure 5.9 Semilog plot of first-order production reaction.

The **half-life**, $t_{1/2}$, is the time required for the quantity of a substance to decrease to 50% of its initial quantity. A generalized form of the half-life equation can be derived by substituting $t_{1/2}$ for t and $C_t = 0.5\ C_0$ into Equation (5.47):

$$0.5\ C_0 = C_0 e^{-kt_{1/2}} \tag{5.48}$$

Rearranging Equation (5.48) yield the half-life equation.

$$t_{1/2} = \frac{-\ln(0.5)}{k} = \frac{0.693}{k} \tag{5.49}$$

Thus, the half-life is easily calculated when k is known. The half-life for various radioactive isotopes is presented in Table 5.2.

Table 5.2 Half-lives of radioactive isotopes.

Radioisotope	Half-life	Radioisotope	Half-life
Radon-222	3.85 days	Americium-241	432 years
Iodine-131	8.04 days	Radium-226	1.6×10^3 years
Krypton-85	10.7 years	Carbon-14	5.73×10^3 years
Hydrogen-3	12.3 years	Plutonium-239	2.41×10^4 years
Strontium-90	28.8 years	Uranium-235	7×10^8 years
Cesium-137	30.2 years	Potassium-40	1.3×10^9 years

Source: Nazaroff and Alvarez-Cohen (2001) page 611. Reproduced by permission of John Wiley & Sons Ltd.

5.3.6 Second-order reactions

A second-order reaction proceeds at a rate that is proportional to the second power of a single reactant. An irreversible reaction involving the conversion of a single reactant to a single product is presented in Equation (5.50):

$$A \rightarrow P \tag{5.50}$$

The rate law equation can be written as Equation (5.51) for a second-order removal reaction:

$$r = -\frac{dC}{dt} = kC^2 \tag{5.51}$$

where C is the concentration of reactant A, mg/L or g/L.

The integral form of Equation (5.51), with the same integration limits for C, where C_0 is the concentration of A at time zero and C_t is the concentration of A at any time t, results in Equation (5.52).

$$\int_{C_0}^{C_t} \frac{-1}{C^2}\, dC = k \int_0^t dt \tag{5.52}$$

The integrated form of Equation (5.52) is presented below.

$$\frac{1}{C_t} = \frac{1}{C_0} + kt \tag{5.53}$$

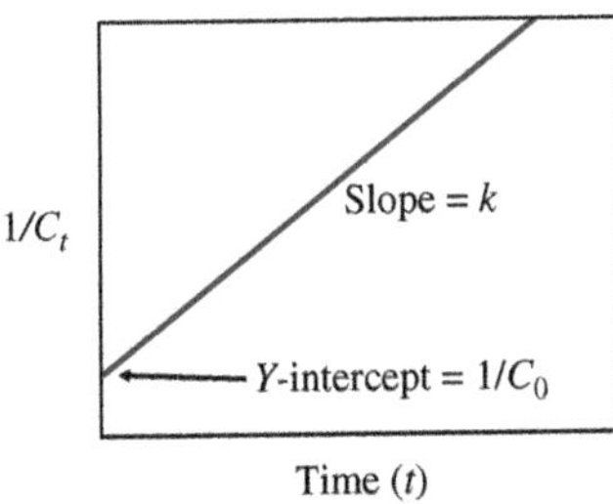

Figure 5.10 Arithmetic plot of first-order removal reaction.

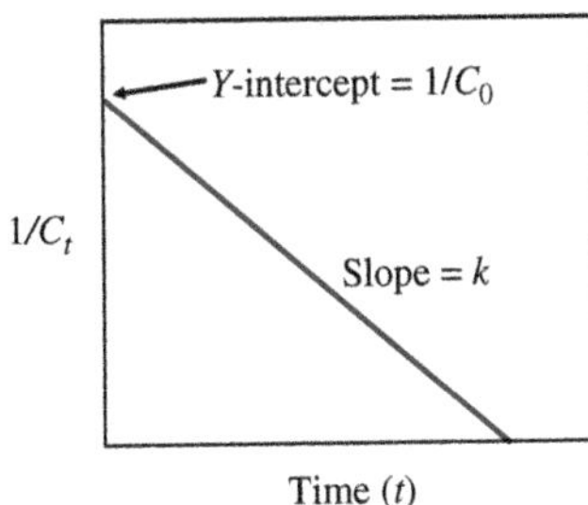

Figure 5.11 Plot of second-order production reaction.

A second-order **removal** reaction plots as a straight line with a positive slope of k and Y-intercept equal to $\frac{1}{C_0}$, as shown in Figure 5.10.

Figure 5.11 shows a second-order **production** reaction plot that produces a straight line with a negative slope. The absolute value of the slope is equal to k and Y-intercept equal to $\frac{1}{C_0}$.

Example 5.7 Reaction order and *k* determination

Bacto-peptone nutrient broth is added to a batch culture of microorganisms, and the concentration is measured in terms of chemical oxygen demand (COD) as a function of time. Evaluate the following data to determine whether the reaction is zero-, first-, or second-order removal. Also determine the magnitude of the rate constant k and list the appropriate units.

Time (minutes)	COD (mg/L)
0	300
5	234
10	182
15	142
20	110
30	67

Solution

First, make an arithmetic plot of the COD concentration (mg/L) plotted on the ordinate axis versus time (minutes) on the abscissa axis. If the plot produces a straight line, the reaction is zero-order and the slope of the line is equal to the value of the rate constant k.

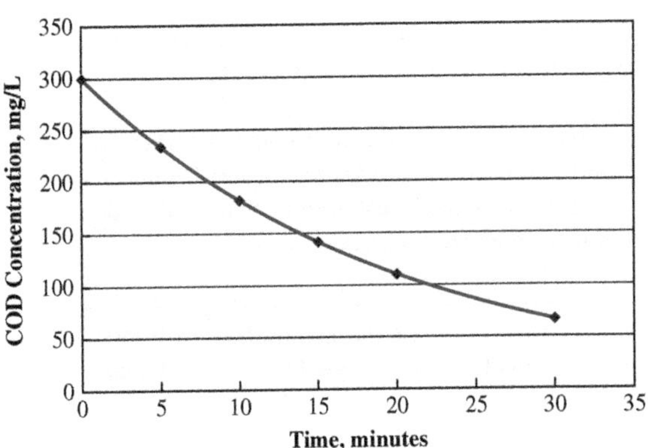

Figure E5.4

Since an arithmetic plot of the COD concentration versus time produced a curve, the reaction appears not to be of zero order. Therefore, make a plot of the natural logarithm (ln) of the COD concentration on the ordinate axis versus time (minutes) on the abscissa axis. If the plot produces a straight line, the reaction is first-order. The natural logarithm of the COD concentration must be determined before plotting as shown below. Microsoft *Excel* or similar spreadsheet software is recommended for performing these calculations and making the appropriate plots.

Time (minutes)	COD Concentration (mg/L)	ln [COD]
0	300	5.704
5	234	5.455
10	182	5.204
15	142	4.956
20	110	4.700
30	67	4.205

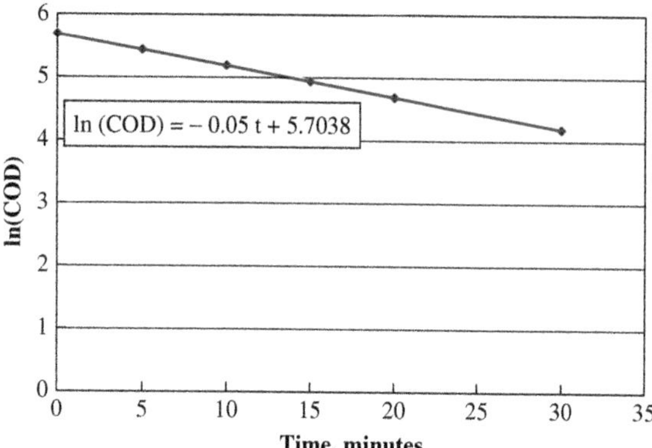

Figure E5.5

A semi-log plot of the COD concentration versus time produces a straight line, indicating the reaction is first-order. The value of k is equal to the absolute value of the slope of the line which, in this case, is $|-0.05| = 0.05\ \text{min}^{-1}$. The Y-intercept is equal to 5.7038, which is set equal to $\ln(\text{COD}_0)$. Therefore, the $\text{COD}_0 = \exp(5.7038) = 300$ mg/L.

Next, a plot of the reciprocal of the COD concentration (mg/L) on the ordinate axis versus time (minutes) on the abscissa axis is made, just to show that the data do not follow second-order kinetics.

Time (minutes)	COD concentration (mg/L)	1/[COD] (L/mg)
0	300	0.00333
5	234	0.00427
10	182	0.00549
15	142	0.00704
20	110	0.00909
30	67	0.0149

A plot of the reciprocal of COD versus time yields a curvilinear plot, so the reaction is not second order.

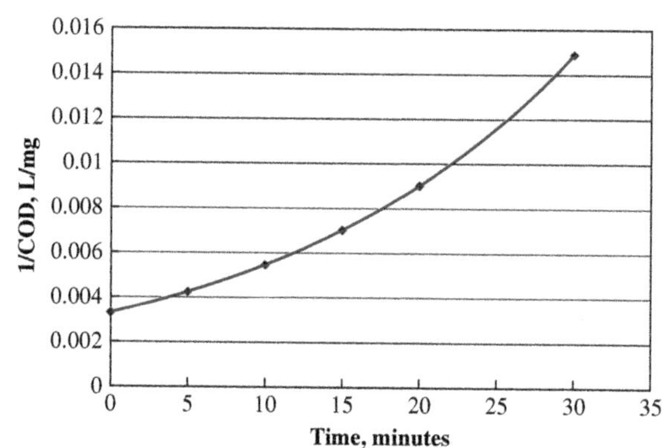

Figure E5.6

Should the above three types of plots fail to produce a straight line, the reaction may be fractional or variable order. Other texts, such as Levenspiel (1999) or Metcalf & Eddy (2003), should be consulted.

5.3.7 Temperature corrections

Biological and chemical reactions are affected by changes in temperature. The rule of thumb is that the reaction rate approximately doubles with a ten degree increase in temperature. The Arrhenius relationship (Equation (5.54)) shows that the reaction rate increases exponentially with an increase in temperature, as shown below:

$$k = Ae^{-(E_a/RT)} \tag{5.54}$$

where:

k = reaction rate constant, units dependent on order of the reaction
A = a constant that is independent of temperature for a specific reaction
E_a = activation energy, cal/mol
R = ideal gas constant, 1.98 cal/mol·K
T = reaction temperature, K.

If k is known at a given temperature, the Arrhenius relationship can be used to predict the rate constant at another temperature. The following is the derivation of the temperature correction coefficient (θ) used for applying temperature corrections to physical, chemical, and biological reactions.

Use the Arrhenius equation to development the temperature correction coefficient (θ) for two different temperatures identified as T_1 and T_2:

For temperature T_1:

$$k_1 = Ae^{-(E_a/RT_1)} \tag{5.55}$$

For temperature T_2:

$$k_2 = Ae^{-(E_a/RT_2)} \tag{5.56}$$

Dividing Equation (5.56) by Equation (5.55) yields Equation (5.57):

$$\frac{k_2}{k_1} = \frac{Ae^{-(E_a/RT_2)}}{Ae^{-(E_a/RT_1)}} \quad (5.57)$$

Taking the natural logarithm of both sides of Equation (5.57) and rearranging yields:

$$\ln\left(\frac{k_2}{k_1}\right) = \frac{-E_a}{RT_2} + \frac{E_a}{RT_1} \quad (5.58)$$

Simplifying the above equation yields Equation (5.59).

$$\ln\left(\frac{k_2}{k_1}\right) = \frac{E_a}{R}\frac{(T_2 - T_1)}{T_2T_1} \quad (5.59)$$

The term $\frac{E_a}{RT_2T_1}$ may be considered a constant (C) for most situations in environmental engineering and Equation (5.59) can be rearranged as follows:

$$\ln\left(\frac{k_2}{k_1}\right) = C(T_2 - T_1) \quad (5.60)$$

Taking the antilog of both sides of Equation (5.60) results in the following equation:

$$\frac{k_2}{k_1} = e^{C(T_2-T_1)} \quad (5.61)$$

Replacing e^C with the temperature correction coefficient (θ) results in the following equation that is commonly used for correcting physical, chemical, and biological reactions for temperature variations:

$$\frac{k_2}{k_1} = \theta^{(T_2-T_1)} \quad (5.62)$$

where:
k_2 = reaction rate constant at temperature T_2
k_1 = reaction rate constant at temperature T_1.

Absolute temperature values must be used in other forms of the Arrhenius equation, but the use of Celsius or Fahrenheit is acceptable in Equation (5.62), since only a difference in temperature is involved.

Example 5.8 Temperature effects on rate constants

A temperature correction coefficient (θ) of 1.047 is typically used for making temperature corrections when performing biochemical oxygen demand (BOD) analyses. If the BOD rate constant, k, at 20°C is 0.15 d^{-1}, determine the value of k for a temperature of 25°C.

Solution

Solve Equation (5.62) for k_2. Make appropriate substitutions and solve as follows.

$$k_2 = k_1\theta^{(T_2-T_1)}$$

$$k_{25°C} = 0.15\,d^{-1}(1.047)^{(25°C-20°C)}$$

$$k_{25°C} = 0.15\,d^{-1}(1.047)^{(5°C)} = 0.19\,d^{-1}$$

The value of the BOD rate constant k at a temperature of 20°C is equal to 0.15 d^{-1}. The rate of reaction for many systems is strongly dependent on temperature. In this example, increasing the temperature by 5°C increased the reaction rate by 27%.

5.4 Flow regimes and reactors

The detention time, tau (τ) and volumetric flow rate (Q) are key design parameters for sizing unit operations and processes in environmental engineering. A flow model or regime is used to evaluate the effects of detention time and flow rate on a given system.

Reactors are tanks or vessels in which physical, chemical, and biological reactions occur. The type of reactor used impacts the effectiveness of treatment or the degree of conversion of a given process.

Mathematically, the ideal or theoretical detention time (τ) is defined as follows:

$$\tau = \frac{\forall}{Q} \quad (5.63)$$

where:
τ = ideal hydraulic detention time, hr or d
$\forall$ = volume of the tank, lake, or reactor, ft^3, m^3, L, gal
Q = volumetric flow rate, $\frac{ft^3}{d}$, $\frac{m^3}{d}$, $\frac{L}{d}$, $\frac{gal}{d}$.

5.4.1 Flow regimes

There are three types of flow regimes related to reactor design and degree of mixing that are encountered in most environmental engineering applications.

In **continuous flow** systems, the flow regime will approach either ideal plug flow or ideal complete-mix flow.

For ideal **plug flow** to occur, all the elements of the fluid that enter the system (or control volume) at a given time pass through the system at the same velocity, remain in the system the same amount of time, and exit the system at the same time. The major aim of plug flow is to avoid mixing; the assumption is that no longitudinal mixing occurs between adjacent fluid elements. Plug flow occurs in long, narrow basins with length-to-width (L : W) ratios of 50 : 1 or greater. Water flowing through a pipe or hose is an example of plug flow.

Ideal **complete-mix flow** is approached when fluid elements entering the system are instantaneously and uniformly dispersed throughout the system. The concentration of the constituents within the reactor is identical to that discharged from the reactor. In actual reactor systems, the flow regime will be somewhere between these two extreme idealized cases and is often referred to as **dispersed-plug flow**. Levenspiel (1999), Reynolds and Richards (1996), and Metcalf & Eddy (2003) provide more details about these flow regimes.

5.4.2 Reactors

Reactors are the tanks or vessels in which physical, chemical, or biological reactions occur. Environmental engineers design reactor systems for treating water, wastewater, air, and solid and hazardous wastes. The principal types of reactors used in environmental engineering are batch, complete-mix, plug flow, packed-bed, and fluidized-bed. As implied by their name, a batch reactor processes materials on a "batch" basis, so there is no continuous flow of materials into and out of the reactor. Batch rectors are routinely used for mixing chemicals. There has been a resurgence in the use of batch rectors for accomplishing biological nutrient removal from wastewater by using sequencing batch reactors (SBR). Both complete-mix and plug flow reactors use continuous flow of materials into and out of the reactor.

Chemical and/or biological reactions typically occur within these systems. The oxidation and synthesis of organic carbon, as measured by five-day, biochemical oxygen demand (BOD_5), is an example of a biochemical reaction that takes place in a reactor. The addition of lime and soda ash to a rapid mixing basin is an example of a chemical reaction that takes place during hardness removal in the treatment of groundwater. From a kinetic and efficiency basis, batch and plug flow reactors will yield the same results.

Complete-mix reactors are less efficient than batch or plug flow reactors; however, using several complete-mix reactors in series simulates the plug flow regime. A packed-bed reactor contains some type of material, such as rock, slag, ceramic, or plastic. These systems may be operated either in the upflow or downflow mode. The trickling filter is an example of an attached growth, packed-bed reactor. Trickling filters will be discussed in the chapter on wastewater. Similar to the packed-bed reactor, a fluidized-bed reactor has its media fluidized by the incoming flow of air or water. The degree to which the bed is expanded is a function of the flow rate. An upflow anaerobic sludge blanket (UASB) reactor is an example of a fluidized-bed reactor. A detailed discussion of UASBs is found in Metcalf & Eddy (2003). It is difficult to achieve ideal complete-mix flow or ideal plug flow in full-scale reactors. Dispersed-plug flow is the type of flow regime that is somewhere between ideal plug flow and ideal complete-mix.

Presenting detailed procedures for the design of reactor systems is beyond the scope of this text. The reader should consult the following texts to learn more about reactor design: Levenspiel, 1999; Reynolds & Richards, 1996; and Metcalf & Eddy, 2003. A brief explanation of complete-mix batch, complete-mix flow, and plug flow reactors will be presented along with derivations of zero-, first-, and second-order reactions.

5.4.2.1 Ideal Complete-Mix Batch Reactor

Batch or fill-and-draw reactors are typically used in situations where the flow to the reactor is less than one million gallons per day (3,785 m^3/d) and operation involves a sequence of events. Since there is no continuous flow into and out of a batch reactor, the system operates under nonsteady-state conditions, i.e., the reaction occurring is unsteady. At any particular instant, the contents within the reactor are uniform throughout; however, the concentration of a particular species will decrease as reaction time increases.

First, the reactor is filled with the process stream containing the constituents to be processed. Next, the flow to the reactor is stopped and air or chemicals are added to the reactor, so that treatment may begin. The reactor is operated in this mode until the desired degree of treatment or conversion has been accomplished. Processing time may be several hours to several months, depending on the application. Once processing or treatment of the flow has been accomplished, the contents of the reactor are drained and a new batch of influent is added to the reactor for processing.

The advantage of using a batch reactor is that it allows flexibility of operation, since the reaction time can be varied. The main disadvantages are that multiple reactors may be required, and they are generally limited to small flows. A simple schematic diagram of a batch reactor is presented in Figure 5.12.

5.4.2.2 Complete-Mix Batch Reactor: Zero-Order Removal Reaction

This section shows the derivation of a zero-order removal reaction for a complete-mix, batch reactor (CMBR). There are no arrows to denote influent and effluent flows in Figure 5.12, since they are not continuous in a batch operation. Other important parameters used in the derivation are defined below.

Q = volumetric flow rate into and out of the batch reactor, $\frac{\text{volume}}{\text{time}}$

$\forall$ = liquid volume in the batch reactor, volume

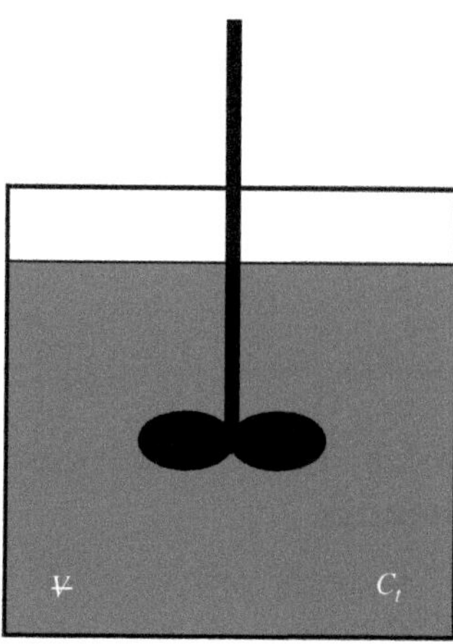

Figure 5.12 Schematic of a batch reactor.

C_0 = concentration of constituent in the influent, $\frac{\text{mass}}{\text{volume}}$
C_t = concentration of constituent at some specified time within the reactor, $\frac{\text{mass}}{\text{volume}}$.

Recall that a zero-order **removal** reaction takes the following form, Equation (5.38):

$$r = -\frac{dC}{dt} = kC^0 \text{ or } \frac{dC}{dt} = -k$$

To derive the **reaction time equation** for a complete-mixed, batch reactor, we begin with the materials balance equation, Equation (5.4).

$$[\text{accumulation}] = [\text{inputs}] - [\text{outputs}] + [\text{reaction}] \quad (5.4)$$

A materials balance will be performed on constituent "C", showing the mass of constituent "C" per unit of time as it enters, exits, accumulates, increases, or decreases within the reactor. Equation (5.4) is rewritten in differential equation form as Equation (5.64). The accumulation term $\left(\frac{dC}{dt}\right)_{accum}$ represents the concentration of "C" that accumulates within the reactor. It must be multiplied by the reactor volume (¥) to yield the proper units, mass/time:

$$\left(\frac{dC}{dt}\right)_{accum} ¥ = QC_0 - QC_t + \left(\frac{dC}{dt}\right) ¥ \quad (5.64)$$

Since batch reactors are operated intermittently, the input and output terms are zero. Since a zero-order **removal** reaction is occurring within the reactor, the reaction term is replaced with Equation (5.38), yielding the following:

$$\left(\frac{dC}{dt}\right)_{accum} ¥ = 0 - 0 + (-k)¥ \quad (5.65)$$

Equation (5.65) simplifies as follows:

$$\left(\frac{dC}{dt}\right)_{accum} = -k \quad (5.66)$$

The integral form of Equation (5.66) is presented below with integration limits for "t" varying from zero to some future time, t, and "C" having limits from C_0, the concentration of C at time zero, to C_t, the concentration of C at anytime, t.

$$\int_{C_0}^{C_t} dC = -k \int_0^t dt \quad (5.67)$$

The integrated form of Equation (5.67) is shown below.

$$C_t - C_0 = -kt \quad (5.68)$$

Rearranging Equation (5.68) and solving for "t" results in Equation (5.69), the reaction time equation for an ideal, complete-mix, batch reactor with a zero-order removal reaction taking place within.

Table 5.3 Reaction times for zero-, first-, and second-order removal and production reactions in a complete-mix batch reactor.

Order	Removal	Production
Zero	$t = \frac{C_0 - C_t}{k}$	$t = \frac{C_t - C_0}{k}$
First	$t = \frac{\ln(C_0) - \ln(C_t)}{k}$	$t = \frac{\ln(C_t) - \ln(C_0)}{k}$
Second	$t = \frac{(1/C_t - 1/C_0)}{k}$	$t = \frac{(1/C_0 - 1/C_t)}{k}$

$$t = \frac{C_0 - C_t}{k} \quad (5.69)$$

where:
t = reaction time, time
k = zero-order removal rate constant, $\frac{\text{mass}}{(\text{volume}\cdot\text{time})}$.

Knowing the reaction rate constant, along with the influent and effluent characteristics, the design engineer can determine the reaction time necessary for the desired degree of treatment. Table 5.3 lists the equations for reaction times for zero-, first-, and second-order reactions in a complete-mix batch reactor.

Example 5.9 Complete-mix batch reactor design

A CMBR is to be designed to pre-treat a food-processing wastewater that contains 500 mg/L of five-day biochemical oxygen demand (BOD_5). A treatability study performed on the food-processing wastewater determined the kinetics for BOD_5 removal to be zero-order with a rate constant of 40 mg/(L·h).

Determine:

a) Reaction or treatment time necessary to reduce the BOD_5 of the wastewater to 100 mg/L.
b) The volume of the reactor to treat 6,000 m^3 of wastewater daily.

Solution

Part A

Use Equation (5.69) to calculate the required reaction time since BOD_5 removal is assumed to follow zero-order kinetics. The required reaction time is 10.0 hours.

$$t = \frac{(500 - 100)\text{mg/L}}{40\ \text{mg/(L}\cdot\text{h)}} = \boxed{10.0\ \text{h}}$$

Part B

Assume that two batches of wastewater will be processed each day, with 3,000 m^3 of wastewater per batch. This will allow the reactor to be filled in one hour and drained in one hour, resulting in a total time of 12 hours for each batch of processed wastewater.

$$\boxed{V = 3000\ \mathrm{m}^3}$$

Use a reactor volume that is approximately 10% more than the required volume to allow additional depth for freeboard – in this case, 3,300 m^3 (1.10 × 3,000 m^3). **Freeboard** is extra height added to the depth of the reactor so that the reactor contents do not overflow.

5.4.2.3 *Ideal Complete-Mix Flow Reactor*

In an **ideal, complete-mix flow reactor** (CMFR), complete and instantaneous mixing of the fluid particles occurs on entering the reactor. The composition of the effluent is the same as the composition within the reactor. A CMFR is also known as a continuous flow stirred tank reactor (CFSTR). The influent concentration of a species (contaminant) is reduced to the final effluent concentration in a complete-mix reactor; this allows a CMFR to be resistant to upsets due to shock loadings of organics, toxic substances, or significant changes in influent pH. This is a tremendous advantage when using microorganisms to treat wastewater, since shock loadings lead to deterioration of effluent quality. Figure 5.13 shows a schematic of a complete-mix flow reactor.

Circular, square, and rectangular basins have been used successfully to achieve complete-mix flow. In practice, perfect mixing can be achieved by providing sufficient mixing and ensuring that the liquid is not too viscous (Reynolds & Richards, 1996, p 53).

CMFRs are not as efficient as plug flow reactors. To achieve the same degree of treatment or conversion in a CMFR requires a larger reactor volume compared to a plug flow or batch reactor.

5.4.2.4 *Complete-Mix Flow Reactor: First-Order Removal Reaction*

The derivation of the detention time equation for a first-order removal reaction in a CMFR is presented in this section.

Recall that a first-order **removal** reaction takes the following form, Equation (5.44):

$$r = -\frac{dC}{dt} = kC^1$$

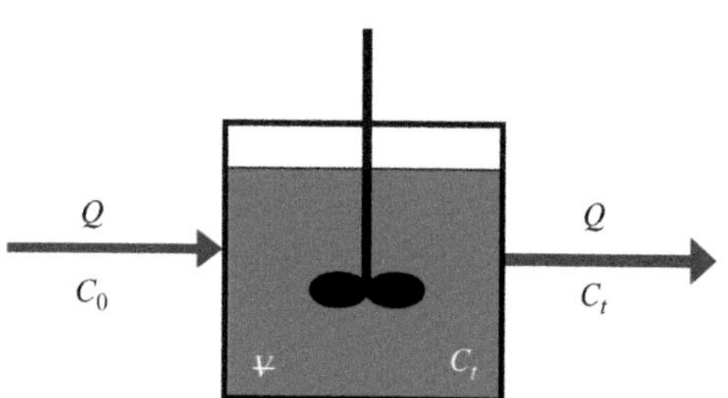

Figure 5.13 Schematic of a complete-mix flow reactor.

Table 5.4 Detention times for zero-, first-, and second-order removal and production reactions in a complete-mix flow reactor.

Order	Removal	Production
Zero	$\tau = \frac{C_0 - C_t}{k}$	$\tau = \frac{C_t - C_0}{k}$
First	$\tau = \frac{(C_0/C_t - 1)}{k}$	$\tau = \frac{(1 - C_0/C_t)}{k}$
Second	$\tau = \frac{1}{kC_t}\left(\frac{C_0}{C_t} - 1\right)$	$\tau = \frac{1}{kC_t}\left(1 - \frac{C_0}{C_t}\right)$

We start by performing a materials balance on constituent "C" equation using Equation (5.4).

$$[\text{accumulation}] = [\text{inputs}] - [\text{outputs}] + [\text{reaction}] \quad (5.4)$$

The material balance is written in differential form to show mass flow through the reactor.

$$\left(\frac{dC}{dt}\right)_{accum} V = QC_0 - QC_t + rV \quad (5.70)$$

The steady-state assumption is made, so the accumulation term goes to zero. Since we are assuming that a first-order removal reaction occurs in the reactor, the reaction term is replaced with Equation (5.44), $dC/dt = -k\,C_t$.

$$0 = QC_o - QC_t - kC_tV \quad (5.71)$$

Rearranging Equation (5.71) results in the following equation:

$$kC_tV = Q(C_0 - C_t) \quad (5.72)$$

The **detention time** (τ) is defined as the volume divided by the volumetric flow rate. When substituted into the equation above, we get:

$$\tau = \frac{V}{Q} = \frac{(C_0 - C_t)}{kC_t} = \frac{(C_0/C_t - 1)}{k} \quad (5.73)$$

Table 5.4 lists the detention time equations for continuous flow, complete-mix reactors for zero-, first-, and second-order removal and production reactions.

Example 5.10 Complete-mix flow reactor design

A complete-mix flow reactor is to be designed to treat an influent stream containing 200 mg/L of chemical oxygen demand (COD) at a flow rate of 200 gallons per minute (gpm). COD represents the total quantity of oxygen required to oxidize organic matter to carbon dioxide and water. COD removal follows a first-order removal reaction, with a rate constant k of $0.45\ h^{-1}$.

Determine:

a) The detention time in hours if the effluent is to contain 15 mg/L of COD.
b) The volume of the reactor in ft^3 if the effluent is to contain 15 mg/L of COD.

Solution

Part a

First, determine the required detention time by substituting the appropriate values into Equation (5.73). The detention time is equal to 27.4 hours.

$$\tau = \frac{(C_0 - C_t)}{kC_t} = \frac{(200\ \text{mg/L} - 15\ \text{mg/L})}{0.45\ \text{h}^{-1} \times 15\ \text{mg/L}} = \boxed{27.4\ \text{h}}$$

Part b

Next, calculate the volume of the reactor by rearranging the equation for detention time (τ). Recall that the definition of detention time is volume divided by volumetric flow rate.

$$\forall = \tau Q = 27.4\ \text{h} \times \left(\frac{200\ \text{gal}}{\text{min}}\right) \times \left(\frac{60\ \text{min}}{\text{h}}\right) \times \left(\frac{1\ \text{ft}^3}{7.48\ \text{gal}}\right)$$

$$= \boxed{43{,}957\ \text{ft}^3}$$

The reactor volume is approximately $44,000\ ft^3$.

Example 5.11 Complete-mix flow reactor design

A complete-mix flow reactor is to be designed to treat domestic wastewater with an influent five-day biochemical oxygen demand (BOD_5) concentration of 150 mg/L and a flow rate of $545\ m^3/d$. A 90% removal in BOD_5 is to be achieved. The concentration of microorganisms (X) in the reactor is maintained at 2,500 mg/L. Assuming that substrate utilization can be modeled by the following equation:

$$r = \frac{dS}{dt} = \frac{kXS}{K_s + S}$$

where:

k = maximum specific substrate utilization rate, $5\ d^{-1}$
K_s = half-velocity coefficient, 50 mg/L of BOD_5

Determine:

a) The equation for detention time by performing a mass balance on substrate.
b) The detention time in hours.

Solution

Start with Equation (5.4).

[accumulation] = [inputs] – [outputs] + [reaction]

The letter "S" is typically used to denote the substrate concentration in biological wastewater treatment systems. In this case, substrate is measured in terms of BOD_5. The substrate concentrations in the influent and effluent are denoted by S_0 and S_t, respectively.

$$\left(\frac{dS}{dt}\right)_{accum} \forall = QS_0 - QS_t - r\forall$$

Next, substituting $\frac{kXS}{K_s+S}$ for "r" yields the following:

$$\left(\frac{dS}{dt}\right)_{accum} \forall = Q(S_0 - S_t) - \left(\frac{kXS_t}{K_s + S_t}\right)\forall$$

At steady-state, the accumulation term is set equal to zero.

$$0 = Q(S_0 - S_t) - \left(\frac{kXS_t}{K_s + S_t}\right)\forall$$

$$\frac{Q(S_0 - S_t)}{\forall} = \left(\frac{kXS_t}{K_s + S_t}\right)$$

$$\frac{(S_0 - S_t)}{\tau} = \left(\frac{kXS_t}{K_s + S_t}\right)$$

The concentration of BOD_5 in the effluent is calculated as follows:

$$S_t = 150\ \frac{\text{mg}}{\text{L}}(1 - 0.9) = 15.0\ \frac{\text{mg}}{\text{L}}$$

$$\tau = \frac{(S_0 - S_t)(K_s + S_t)}{kXS_t} = \frac{(150 - 15\ \text{mg/L})(50 + 15\ \text{mg/L})}{5\ \text{d}^{-1} \times 2500\ \frac{\text{mg}}{\text{L}} \times \frac{15\ \text{mg}}{\text{L}}}$$

$$= 0.0468\ \text{d}$$

$$\tau = 0.0468\ \text{d} \times \left(\frac{24\ \text{h}}{\text{d}}\right) = \boxed{1.1\ \text{h}}$$

5.4.2.5 Complete-Mix Flow Reactors in Series

It is possible to improve the performance of a CMFR by replacing one large reactor with a series of smaller ones. The smaller reactors generally have the same volume and detention time, but it is possible to use reactors with different volumes. A series of ten CMFRs will approach the performance of a single plug flow reactor with a detention time equal to the total detention time of the 10 CMFRs (Reynolds & Richards, 1996, p 58). In this section, the derivation of "n" complete-mix reactors in series for a first-order removal reaction will be developed.

From Section 5.4.2.4, the detention time for CMFR with a first-order removal reaction yielded Equation (5.73). We will rearrange this equation and apply it to three CMFRs operating in series to derive an equation that can be used to calculate the effluent concentration from "n" complete-mixed reactors operating in series.

$$\tau = \frac{V}{Q} = \frac{(C_0 - C_t)}{kC_t} = \frac{(C_0/C_t - 1)}{k} \quad (5.74)$$

$$\tau k C_t = (C_0 - C_t) \quad (5.75)$$

$$\tau k C_t + C_t = C_0 \quad (5.76)$$

$$C_t(1 + k\tau) = C_0 \quad (5.77)$$

$$C_t = \frac{C_0}{(1 + k\tau)} \quad (5.78)$$

For three CMFRs operating in series as depicted in the figure below, Equation (5.78) can be applied to each separate reactor, resulting in Equations (5.79), (5.80), and (5.81). Combining all of these three equations results in Equation (5.82).

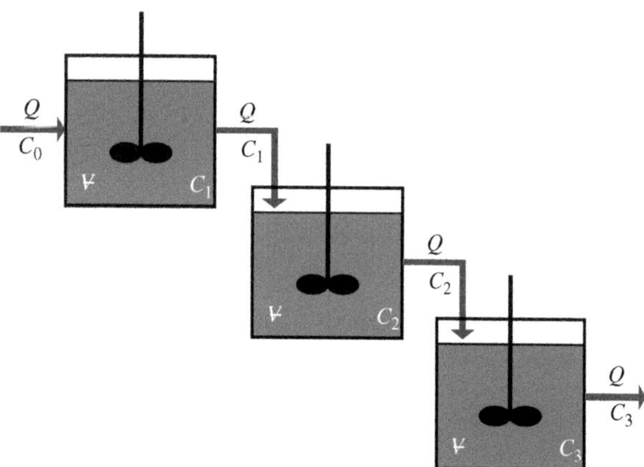

Figure E5.7

$$C_1 = \frac{C_0}{(1 + k\tau)} \quad (5.79)$$

$$C_2 = \frac{C_1}{(1 + k\tau)} \quad (5.80)$$

$$C_3 = \frac{C_2}{(1 + k\tau)} \quad (5.81)$$

$$C_3 = C_0\left(\frac{1}{1 + k\tau}\right)\left(\frac{1}{1 + k\tau}\right)\left(\frac{1}{1 + k\tau}\right) \quad (5.82)$$

A generalized equation for "n" number of rectors operating in series is presented as Equation (5.83):

$$C_n = C_0\left(\frac{1}{1 + k\tau}\right)^n \quad (5.83)$$

where:
C_0 = influent concentration to the first reactor, $\frac{\text{mass}}{\text{volume}}$
C_n = effluent concentration from the nth reactor, $\frac{\text{mass}}{\text{volume}}$
k = first-order reaction rate constant, time^{-1}
V = volume of each individual reactor in series, volume
Q = volumetric flow rate, $\frac{\text{volume}}{\text{time}}$
τ = detention time in each individual reactor, time, $\tau = \frac{V}{Q}$
n = number of individual reactors in series.

5.4.2.6 Ideal Plug Flow Reactor

In an **ideal plug flow reactor** (PFR), fluid particles enter at one end and exit at the other end in the same sequence as they entered. The composition varies along the length of the reactor. Plug flow is also called "pipe" or "tubular" flow. For a **removal** reaction, Figure 5.14 shows how the concentration of a constituent such as BOD_5 would decrease along the length of the reactor.

In plug flow, longitudinal or lateral mixing does not occur. Each element of fluid in a PFR is analogous to a complete-mix batch reactor that varies as a function of time along the length of the reactor. Ideal plug flow reactors are more efficient than completely mixed reactors from a kinetic analysis, and they yield the same results as batch reactors. This means that a smaller reactor volume is required, compared to a CMFR. A major drawback to PFRs when used in biological wastewater treatment is their susceptibility to shock loadings of organics, to toxic substances, or to dramatic pH shifts that can kill the microorganisms within the reactor.

5.4.2.7 Derivation of One-Dimensional Plug Flow

A derivation of the equation for one-dimensional, plug flow is presented in this section. The same nomenclature used in the previous derivations will be followed.

To derive the general form of the equation for one-dimensional plug flow, we must first use the materials balance equation, Equation (5.4).

$$[\text{accumulation}] = [\text{inputs}] - [\text{outputs}] + [\text{reaction}] \quad (5.4)$$

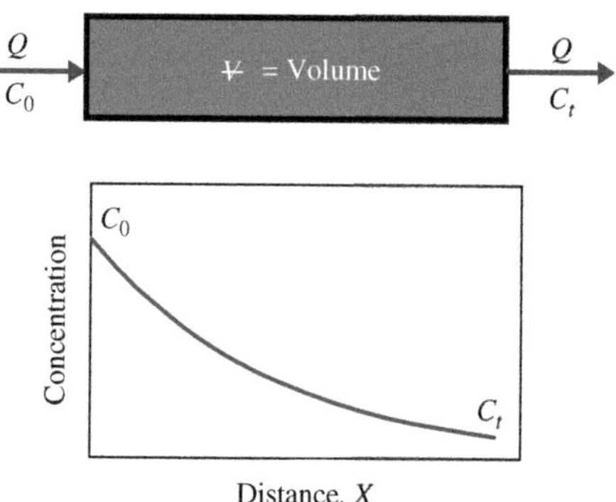

Figure 5.14 Schematic of plug flow reactor showing concentration decreasing as a function of length.

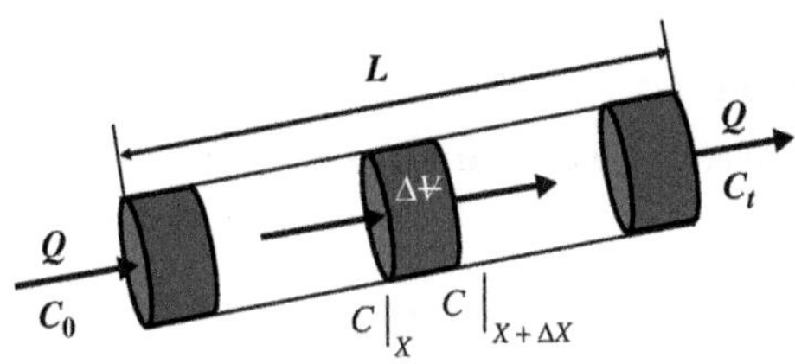

Figure 5.15 Schematic of a plug flow reactor.

Since the concentration varies both with time and distance, we must use a partial differential in the derivation. Figure 5.15 shows the slug or plug as it moves through the reactor.

$$\left(\frac{\partial C}{\partial t}\right)\Delta V = QC|_X - QC|_{X+\Delta X} + r\Delta V \tag{5.84}$$

where:
ΔV = elemental volume, L^3
A = cross-sectional area of the reactor, width times the depth, L^2
ΔX = width or thickness of the elemental area, L.

$$\Delta V = A\Delta X \tag{5.85}$$

$$\left(\frac{\partial C}{\partial t}\right)\Delta V = QC - Q\left(C + \frac{\Delta C}{\Delta X}\Delta X\right) + r\Delta V \tag{5.86}$$

$$\left(\frac{\partial C}{\partial t}\right)A\Delta X = QC - QC - \left(Q\frac{\Delta C}{\Delta X}\Delta X\right) + rA\Delta X \tag{5.87}$$

Dividing both sides of the above equation by $A\Delta X$ yields the following equation:

$$\left(\frac{\partial C}{\partial t}\right) = -\left(\frac{Q}{A}\frac{\Delta C}{\Delta X}\right) + r \tag{5.88}$$

Using the steady-state assumption, the time on the left side of the equation goes to zero, resulting in Equation (5.89) the general form of the one-dimensional plug flow equation. Remember that "r" is equal to the rate of reaction and may be a zero-, first-, or second-order reaction.

$$\left(\frac{Q}{A}\frac{\Delta C}{\Delta X}\right) = r \tag{5.89}$$

5.4.2.8 Plug Flow Reactor: Second-Order Removal Reaction

In this section, the detention time equation for a second-order **removal** reaction occurring within a continuous plug flow reactor is developed. The general form of the equation for one-dimensional plug flow is presented as Equation (5.89). Remember that "r" is equal to the rate of reaction. Either a production or removal reaction may take place. The sign will be positive if a substance is produced and negative if a reactant is removed. The reaction type may be zero-, first-, second-, or fractional order.

Recall that a second-order **removal** reaction takes the following form, Equation (5.51):

$$r = \frac{dC}{dt} = -kC^2$$

Table 5.5 Detention times for zero-, first-, and second-order removal and production reactions in a plug flow reactor.

Order	Removal	Production
Zero	$\tau = \frac{C_0 - C_t}{k}$	$\tau = \frac{C_t - C_0}{k}$
First	$\tau = \frac{\ln(C_0) - \ln(C_t)}{k}$	$\tau = \frac{\ln(C_t) - \ln(C_0)}{k}$
Second	$\tau = \frac{(1/C_t - 1/C_0)}{k}$	$\tau = \frac{(1/C_0 - 1/C_t)}{k}$

Substituting for "r" into Equation (5.89) and replacing $\Delta C/\Delta X$ in Equation (5.89) with dC/dX results in the following equation.

$$\left(\frac{Q}{A}\frac{dC}{dX}\right) = -kC^2 \tag{5.90}$$

Equation (5.91) is the integral form of Equation (5.90), with integration limits for C going from C_0, the initial concentration at the front of the reactor, where length is equal to zero, to C_t, the concentration at the end of the length of the reactor, L.

$$-\frac{A}{Q}\int_0^L dX = \frac{1}{k}\int_{C_0}^{C_t}\frac{1}{C^2}dC \tag{5.91}$$

$$-\frac{AL}{Q} = -\frac{V}{Q} = -\tau = \frac{1}{k}\left[-\left(\frac{1}{C_t}\right) - \left(-\frac{1}{C_0}\right)\right] \tag{5.92}$$

Solving Equation (5.92) for detention time (τ) yields:

$$\tau = \frac{1}{k}\left[\left(\frac{1}{C_t}\right) - \left(\frac{1}{C_0}\right)\right] \tag{5.93}$$

Table 5.5 lists the detention equations for continuous, plug flow reactors for zero-, first-, and second-order removal and production reactions. Note that the detention time equations shown in Table 5.5 for PFRs are identical to those shown in Table 5.3 for complete-mix batch reactors.

Example 5.12 Plug flow reactor design

A continuous PFR is to be designed to treat an influent stream containing 250 mg/L of acetic acid at a flow rate of 300 liters per minute (lpm). A second-order removal reaction is occurring, where the rate constant k is 0.0075 L/(mg·h).

Determine:

a) The detention time in hours to achieve 90% removal of acetic acid.
b) The volume of the reactor in m^3 to achieve 90% removal of acetic acid.

Solution

Part A

First, calculate the concentration of acetic acid in the effluent. This may be accomplished using the definition of percent removal (%), which is shown as Equation (5.94).

$$\text{percent removal (\%)} = \frac{(C_{in} - C_{out})\,100}{C_{in}} \qquad (5.94)$$

where:

C_{in} = Concentration of constituent in the influent, mass/volume

C_{out} = Concentration of constituent in the effluent, mass/volume.

$$90\%\ \text{removal} = \frac{(250\ \text{mg/L} - C_{out})\,100}{250\ \text{mg/L}}$$

$$C_{out} = 25\ \text{mg/L}$$

Next, the detention time necessary to achieve 90% removal of acetic acid for a second-order removal reaction can be determined using Equation (5.93).

$$\tau = \frac{1}{k}\left[\left(\frac{1}{C_t}\right) - \left(\frac{1}{C_0}\right)\right]$$

$$= \frac{1}{0.0075\ \text{L/mg}\cdot\text{h}}\left[\left(\frac{1}{25\ \text{mg/L}}\right) - \left(\frac{1}{250\ \text{mg/L}}\right)\right]$$

$$= \boxed{4.8\ \text{h}}$$

Part B

Finally, the volume of the plug flow reactor is determined by rearranging the equation for detention time (τ). Recall that detention time is equal to volume divided by the volumetric flow rate. Therefore, the volume of the reactor is equal to $\tau \times Q$.

$$\forall = \tau \times Q = 4.8\ \text{h}\left(\frac{300\ \text{L}}{\text{min}}\right)\left(\frac{60\ \text{min}}{\text{h}}\right)\left(\frac{1\ \text{m}^3}{1000\ \text{L}}\right)$$

$$= \boxed{86.4\ \text{m}^3}$$

Example 5.13 Derivation of τ equation for PFR with a first-order removal reaction

Derive an equation for calculating the detention time in a continuous flow, plug flow reactor which has a first-order **removal** reaction occurring within.

Solution

Recall that the differential equation for a first-order removal reaction has the following form, Equation (5.44).

$$r = -\frac{dC}{dt} = kC^1$$

Let us start with the one-dimensional equation previously derived for a continuous flow, plug flow reactor, Equation (5.89):

$$\left(\frac{Q}{A}\frac{\Delta C}{\Delta X}\right) = r$$

Since a first-order **removal** reaction is occurring, a negative sign precedes r.

$$\left(\frac{Q}{A}\frac{dC}{dX}\right) = -r = -kC$$

$$\frac{Q}{A}\frac{dC}{dX} = -kC$$

Develop the integral form of the equation above, with integration limits for C going from C_0, the initial concentration at the front of the reactor, where length is equal to zero, to C_t, the concentration at the end of the length of the reactor, L.

$$\int_{C_0}^{C_t} \frac{1}{C}\,dC = \frac{-kA}{Q}\int_0^L dX$$

$$\ln(C_t) - \ln(C_0) = \frac{-kAL}{Q}$$

$$\ln(C_t) - \ln(C_0) = \frac{-k\forall}{Q}$$

$$\ln\left({}^{C_t}/_{C_0}\right) = -k\tau$$

$$\boxed{\tau = \frac{\ln\left({}^{C_0}/_{C_t}\right)}{k}}$$

5.4.3 Tracer studies

In practice, the actual or measured hydraulic detention time of a process or system is seldom identical to the ideal detention time as calculated using Equation (5.63). For most cases, the measured detention time is much less than the theoretical value. Poor inlet and outlet construction to reactors, short circuiting, temperature gradients, dead zones, wind currents, and density currents may lead to decreased detention time within the process or system. On occasion, the actual detention time may exceed the theoretical one, but this is unusual.

A dye or tracer that is conservative (nonreactive) may be added to the influent of a reactor or process while the effluent concentration is monitored as a function of time. A plot of the dye effluent concentration versus time will indicate whether or

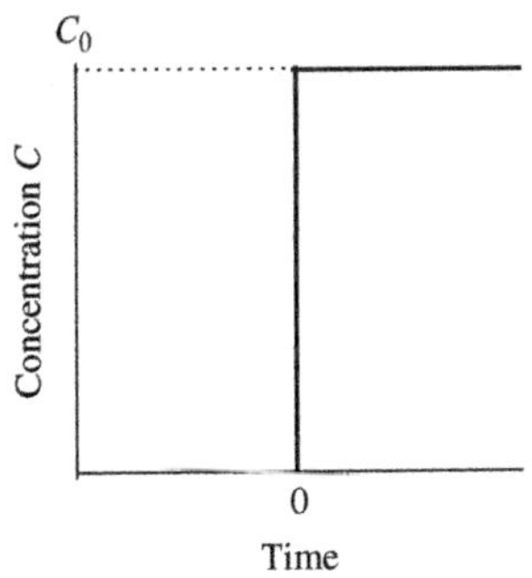

Figure 5.16 Batch reactor response to instantaneous or pulse addition of dye.

not the actual detention time (t) is the same as the ideal hydraulic detention time (τ). Rhodamine dye or sodium chloride is often used in tracer studies. It may be added instantaneously as a spike or slug, or on a continuous basis (step increase).

Figure 5.16 shows the concentration response in a complete-mix batch reactor after a spike or pulse addition of a dye is added. Notice that at time zero when the dye is added, the concentration of dye within the reactor reaches its maximum value. Recall that there are no flows entering or exiting a batch reactor, so the dye concentration remains constant, as it is non-reactive (i.e., it is conserved, so there is no increase or decrease in concentration).

A pulse addition of a dye or tracer to the influent of a plug flow reactor is shown in Figure 5.17. In this case, the residence time curve (RTC) is plotted versus the ratio of actual detention time to ideal detention time $\left(\frac{t}{\tau}\right)$. The residence time curve is the ratio of effluent concentration divided by the influent concentration $\left(\frac{C}{C_0}\right)$. If the actual and theoretical detention time are identical, the dye or tracer will reach the effluent when $\left(\frac{t}{\tau}=1\right)$.

Since complete-mix flow reactors are used extensively in several environmental applications, the development of the residence time curves for the addition of either a pulse or step increase in dose of tracer will be presented.

5.4.3.1 Pulse Addition of Dye to Complete-Mix Reactor

We will first perform a materials balance around a continuous flow, complete-mix reactor. Since the dye or tracer is added instantaneously, the rate at which it is introduced (input term) is set equal to zero. The reaction term is also set equal to zero, since the dye is conservative (i.e., nonreactive), so the concentration will not increase or decrease within the reactor.

$$[\text{accumulation}] = [\text{inputs}] - [\text{outputs}] + [\text{reaction}] \tag{5.4}$$

$$\left(\frac{dC}{dt}\right)\forall = QC_0 - QC + r\forall \tag{5.95}$$

$$\left(\frac{dC}{dt}\right)\forall = 0 - QC + 0 \tag{5.96}$$

$$\frac{dC}{dt} = \frac{-QC}{\forall} \tag{5.97}$$

$$\int_{C_0}^{C} \frac{1}{C}dC = -\frac{Q}{\forall}\int_0^t dt \tag{5.98}$$

$$\ln C - \ln C_0 = -\frac{t}{\tau} \tag{5.99}$$

$$C = C_0 e^{-(t/\tau)} \tag{5.100}$$

Figure 5.18 shows the effluent dye concentration (C) from the complete-mix reactor as a function of time. The figure is based on Equation (5.100) and assumes an influent dye concentration of 100 mg/L and an ideal detention time (τ) of 4 hours.

When the detention time within the reactor is four hours, approximately 36.8% of the dye is still in the reactor:

$$\frac{C}{C_0} = e^{-(t/\tau)} = e^{-(4.0/4.0)} = 0.368 \text{ or } 36.8\% \tag{5.101}$$

Alternatively, this means that 63.2% (100 – 36.8 = 63.2%) of the dye has exited the reactor in the effluent.

5.4.3.2 Step Addition of Dye to Complete-Mix Reactor

Start with a materials balance around a continuous flow, complete-mix reactor. The input term is not set equal to zero, as in the previous case, since the dye or tracer is continuously being added to the influent to the reactor. Since the dye is a conservative substance, the reaction term is set equal to zero.

$$[\text{accumulation}] = [\text{inputs}] - [\text{outputs}] + [\text{reaction}] \tag{5.4}$$

$$\left(\frac{dC}{dt}\right)\forall = QC_0 - QC + r\forall \tag{5.102}$$

$$\left(\frac{dC}{dt}\right)\forall = QC_0 - QC + 0 \tag{5.103}$$

$$\frac{dC}{dt} = \frac{Q(C_0 - C)}{\forall} \tag{5.104}$$

$$\int_0^C \frac{-1}{C_0 - C}dC = -\frac{Q}{\forall}\int_0^t dt \tag{5.105}$$

$$\ln(C_0 - C) - \ln(C_0 - 0) = -\frac{t}{\tau} \tag{5.106}$$

$$\ln C_0 - \ln C - \ln C_0 = -\frac{t}{\tau} \tag{5.107}$$

$$C = C_0(1 - e^{-(t/\tau)}) \tag{5.108}$$

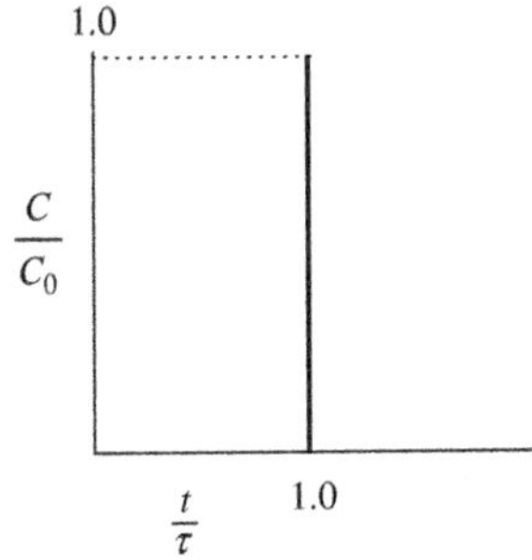

Figure 5.17 Residence time curve for pulse addition of dye to a plug flow reactor.

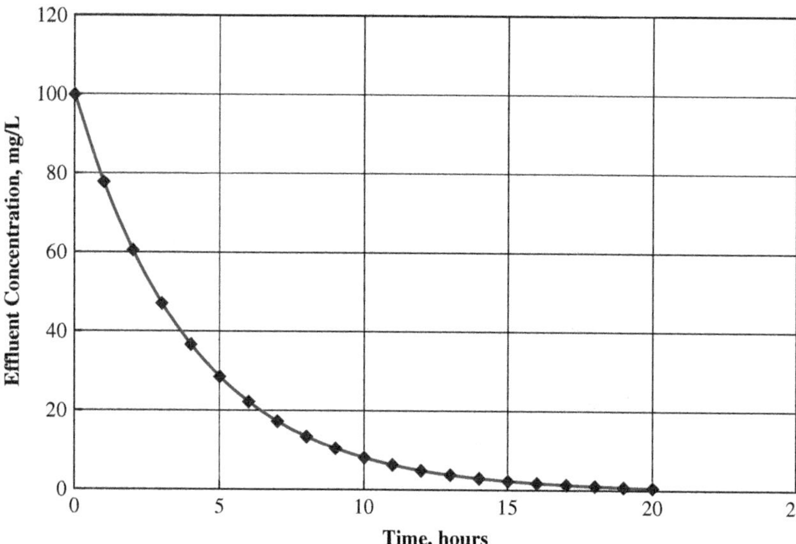

Figure 5.18 Complete-mix reactor response to instantaneous or pulse addition of dye.

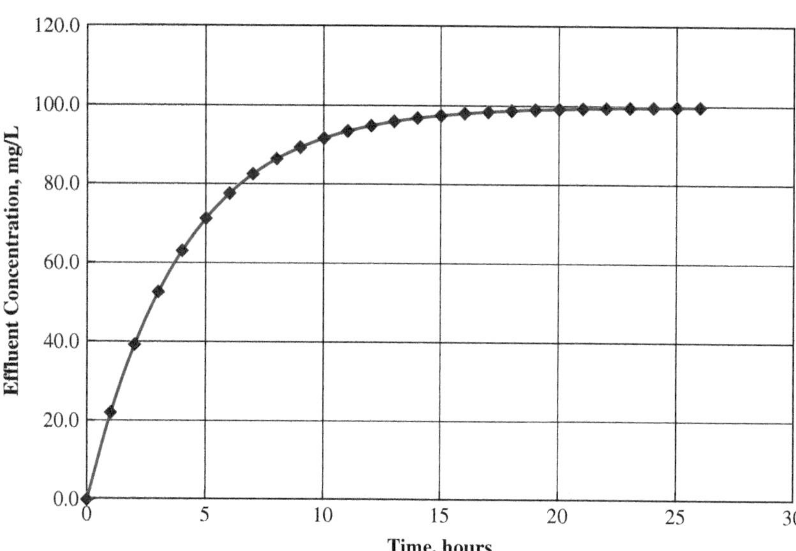

Figure 5.19 Complete-mix reactor response to step addition of dye.

Figure 5.19 shows the effluent dye concentration (C) from the complete-mix reactor as a function of time for a continuous addition of dye into the influent to the reactor. The figure is based on Equation (5.108) and assumes an influent dye concentration of 100 mg/L and an ideal detention time (τ) of four hours. When the detention time within the reactor is 4 hours, approximately 63.2% of the dye is still in the reactor:

$$\frac{C}{C_0} = (1 - e^{-(t/\tau)}) = (1 - e^{-(4.0/4.0)}) = 0.632 \text{ or } 63.2\% \quad (5.109)$$

The important point to note here is that, after three hydraulic detention times (actual detention time $= 3 \times \tau = 3 \times 4\,\text{h} = 12\,\text{h}$), the effluent concentration has reached approximately 95% of the influent concentration. In practical applications, environmental engineers approximate steady-state conditions when collecting data on treatment systems once three hydraulic detention times have been reached.

$$\frac{C}{C_0} = (1 - e^{-(t/\tau)}) = (1 - e^{-(3\times\tau/\tau)}) = 0.950 \text{ or } 95.0\% \quad (5.110)$$

5.5 Energy balances

It is important that an environmental engineer be able to perform energy balances, understand the various forms of energy, and convert from one form of energy to another. As the cost of energy continues to escalate and the world's supply of nonrenewable sources continues to dwindle, engineers are called upon to design efficient and sustainable processes and systems.

Fossil fuels and other nonrenewable sources of energy such as nuclear have served, and continue to serve, as the world's major sources of energy. According to the United States Department of Energy, approximately 84% of all US energy consumption comes from burning petroleum, natural gas, and coal, with 37% of that coming from petroleum (i.e., oil; EIA, 2008, p 37). The burning of coal, oil, peat, wood, natural gas, and gasoline not only contribute to the discharge of greenhouse gases but are in limited supply, so that efficiency of use also poses concerns.

In many cases, only 30% to 40% of the energy input to a power plant is converted to useful energy (Masters, 1998, Vesilind & Morgan, 2004). Vesilind & Morgan (2004) predict that the world's oil and gas supplies will run out in the next 50 years, based on modern reserve-to-production ratios. Without new discoveries of oil and gas deposits, or drastic changes in management of the world's energy resources, coal and nuclear power may likely be the major sources of energy by 2050 (Henry & Heinke, 1996). It is, therefore, imperative that alternative and sustainable sources of energy be identified, developed, and used throughout the world. Renewable sources, such as hydropower, biomass from plants, solar, geothermal, wind, and ocean energy from tides and waves will be used, and their individual contributions will have to increase substantially to off-set the use and/or lack of fossil fuels.

Although uranium is nonrenewable, the use of nuclear energy is also expected to increase, since it does not discharge carbon dioxide into the atmosphere. Several problems arise over the use of nuclear energy, however. According to Masters (1998), nuclear power plants have materials constraints, necessitating them to operate at lower temperatures than traditional coal-fired power plants, which results in efficiencies of approximately 33%. There is much concern over the disposal of hazardous wastes and the potential of terrorists stealing nuclear materials.

In the past, our natural resources were exploited by developing the most economical sources without concern for future generations or the impact on the environment. This same mentality has been applied to energy resources, too. As a consequence, future energy supplies will be more costly to develop, transport, and refine. Politics will continue to play a crucial role and, thus, governments must understand the need to manage the production and use of energy. Currently in the US for example, 28% of all energy consumption supports transportation alone and 95% of that energy for transportation comes from burning petroleum-based fuels (EIA, 2008, p 37). This dependency on a primary resource – petroleum – limits the ability to manage energy.

Utilizing secondary energy resources such as electricity and hydrogen (both clean energy), that can be reserved after being produced from a variety of primary resources, would permit fulfilling the energy needs for transportation from select resources such as coal or nuclear power or, more importantly, from renewable resources such as those mentioned above. Furthermore, the task of environmental engineers to design ways to reduce pollutants from the combustion of fossil fuels becomes more feasible considering a few large power-plant facilities, rather than millions of fossil-fuel burning vehicles.

Aside from managing the world's primary energy resources, it behooves us always to use our energy resources efficiently. Engineers must be able to analyze and improve the efficiency of power plants, as well as power-consuming devices. In this section, we introduce the two major laws of thermodynamics, explain the different forms of energy, and present several energy balances illustrating the flow of energy and materials through various processes.

5.5.1 Energy fundamentals

Thermodynamics is the study of energy and energy changes. Environmental engineers and scientists are primarily interested in chemical thermodynamics, which involves changes in energy associated with biological and chemical processes. Following the flow of energy through a coal-fired power plant or some environmental process is important, so that process performance can be increased and areas where there is a loss of energy can be identified. Energy losses are unavoidable and, in many cases, they occur as heat transfer is discharged into the environment. Recovery and reuse of this otherwise wasted energy to heat homes and businesses are ways to improve sustainability.

Animals, including humans, use the chemical energy content of food as their energy source. Some microorganisms use organic carbon and some inorganic forms for meeting their energy requirements. An understanding of the flow of energy through food chains and food webs is essential for maintaining vibrant ecosystems.

A brief review of some of the most important thermodynamic principles and terms are presented next. The reader should consult other texts on thermodynamics (e.g., Moran & Shapiro, 2004; Cengel & Boles, 2008) to learn more about these topics.

Energy is a concept of quantities representing a physical change or a potential for change. Physical change implies some form of motion, such as an object traveling at some speed, the internal energy of a substance associated with the vibration and rotation of molecules or atoms, or the motion of electrons, and even the motion of waves such as sound and light. These are generally classified as kinetic forms of energy. Potential forms of energy include gravitational, nuclear, or chemical; these provide a means for energy storage, in which energy remains available to create a later physical change.

Energy – both kinetic and potential forms – is observed to be a conserved quantity and, thus, is never created or destroyed, although it is connected to mass according to the **Theory of Relativity** which, basically, states that mass and energy are equivalent and transmutable (i.e., only the form is altered or changed). Mathematically, it is expressed as follows:

$$E = mc^2 \tag{5.111}$$

where:
E = energy equivalent of mass, J
m = mass of object at rest, kg
c = speed of light (3.0×10^8 m/s).

From a macroscopic perspective, three important forms of energy include internal, kinetic and potential energy.

Internal energy represents the sum of energy held within a substance, and accounts for all of the potential and kinetic forms of energy on the molecular level to include sensible, latent, chemical, and nuclear components. The sensible component is associated with the atoms and molecules of a substance. These energies include translational, rotational, and vibrational modes. The latent component refers to the energy associated with intermolecular forces related to the phase of substance. For example, steam contains more latent energy than liquid water at the same temperature and pressure. Energy stored in the chemical bonds between atoms that make up molecules is the chemical component of internal energy that is released during chemical reactions. Finally, a nuclear component accounts for the energy holding the nucleus together. Nuclear energy relates to kinetic energy and photons (particles of light or quantum of electromagnetic energy moving at the speed of light) that are released when radioactive elements disintegrate into more stable products or nuclei during the fission process. In nuclear reactions, the mass of the products formed is less than the original mass of the original reactions.

Kinetic energy (*KE*) is associated with motion and can produce energy by water flowing through turbines or windmills. The kinetic energy associated with a quantity of mass is calculated as follows:

$$KE = \frac{1}{2} m V^2 \quad (5.112)$$

where:
KE = kinetic energy associated with mass, BTU (kJ)
m = mass in system or control volume, lb_m (kg)
V = velocity of mass, fps (mps).

Potential energy (*PE*) is related to position, either in a gravitational or electric field. An object possesses potential energy only when the result of work done on the object results in a change in position. The water in an elevated storage tank possesses potential energy. The potential energy associated with a quantity of mass can be calculated as follows:

$$PE = mgz \quad (5.113)$$

where:
PE = potential of a quantity of mass, BTU (kJ)
g = acceleration of gravity, 32.2 ft/s^2 (9.81 m/s^2)
z = elevation difference, ft (m).

Internal energy (*U*) refers to the potential and kinetic energies associated with the atoms or molecules of a substance. All other properties being the same, a higher temperature implies more thermal energy. It represents the available or transferable energy in a substance that is not changing molecularly due to any chemical or nuclear reaction.

Equation (5.114) shows that the **total energy** (*E*) that a substance possesses is the sum of the internal, kinetic, and potential energies:

$$E = U + KE + PE = U + \frac{m V^2}{2} + mgz \quad (5.114)$$

The total energy of a system expressed on a unit mass basis is defined as *e*:

$$e = \frac{E}{m} \quad (5.115)$$

where:
E = total energy of a system, BTU (kJ)
m = total mass of a system, lb_m (kg)
e = total energy of a system per unit mass, $\frac{BTU}{lb_m}\left(\frac{kJ}{kg}\right)$.

The total internal energy of a system expressed on a unit mass basis is defined similarly as *u*:

$$u = \frac{U}{m} \quad (5.116)$$

where:
U = total internal energy of a system, BTU (kJ)
m = total mass of a system, lb_m (kg)
u = total internal energy of a system per unit mass, $\frac{BTU}{lb_m}\left(\frac{kJ}{kg}\right)$.

Equation (5.117) can be then be expressed on an energy per unit of mass basis as follows:

$$e = u + \frac{V^2}{2} + gz \quad (5.117)$$

The term **system** will refer to the object that is the subject of analysis. It may be a pipe, a reactor, a tank, or as complex as a coal-fired power plant. We will generally refer to the system as a control volume (C$\forall$). Everything external to the system or control volume is considered the system's surroundings, and the boundary separates the system from its surroundings. An open system allows mass and energy to be transferred into or out of it, whereas a closed system only allows the transfer of energy, not mass.

5.5.1.1 Heat and Work

Heat and work are not forms of energy but, rather, an energy interaction between a system and its surroundings, i.e., it involves physical actions that transfer (move) or change energy. Considering the total energy of a substance defined above, that sum of energy can only change if heat transfer and/or work occur. Thermal energy of a hot gas can be reduced, for example, if the gas performs work on the environment by expanding to lift a mass. Heat transfer from the hot gas to the cooler surroundings could also reduce the thermal energy.

Both heat transfer and work share common units with energy, although they are not properties of a substance, as is energy. Various units are used for measuring energy and, thus, also heat and work. These include British Thermal Unit (BTU), calorie (cal), and joule (J). The BTU is defined as the energy required to raise the temperature of one pound of water one degree Fahrenheit (°F). The basic unit of thermal energy, the calorie, is defined as the quantity of energy required to raise the temperature of one gram of water by one degree Celsius (°C). The joule is defined as the amount of work done by a force of one Newton to raise an object one meter.

Heat is the transfer of thermal energy from one location to another solely due to a temperature difference. Heat flows from a body of a higher temperature to another of a lower temperature. The symbol **Q** denotes the quantity of energy transferred across the boundary of a system. In this text, heat transfer into a system is considered to be positive, and heat transfer from a system is taken as negative.

Heat, like work, is not a property. Heat transfer depends on the details of the process, not just on the end states. The net rate of heat transfer is defined as $\dot{Q}$. The total quantity of heat transferred is expressed by Equation (5.118).

$$Q = \int_{t_1}^{t_2} \dot{Q} dt \qquad (5.118)$$

where:
Q = total quantity of heat transferred, BTU (kJ)
t_1 and t_2 = time at different periods, sec, min
dt = change in time, sec, min.

For some applications, the heat flux, $\dot{q}$, the heat transferred per unit area, is used. Mathematically, heat flux is defined as follows:

$$\dot{Q} = \int_A \dot{q} dA \qquad (5.119)$$

where:
A = total area across which heat is transferred, ft^2 (m^2)
dA = differential area element, ft^2 (m^2).

Work (W) is defined as transferring energy to an object by applying force and causing displacement. Energy is transferred and stored when work is done; however, work is not a property of the system or the surroundings. Work is often the desired end product and is the scalar product of force and distance.

$$W = \text{force} \times \text{distance} \qquad (5.120)$$

For a small displacement (ds), the change in work is the product of the force (F) and distance (ds).

$$W = \int_{s_1}^{s_2} F\,ds \qquad (5.121)$$

where:
W = work done on or by a system, BTU (kJ)
F = force applied to system, lb$_f$ (N)
F = $P \times A$, pressure times area
P = pressure, psi (kPa)
A = area, ft^2 (m^2)
s = distance, ft (m).

Work is either done on the system by the surroundings increasing the total energy in the system, or it is done by the system on the surroundings decreasing the total energy. We have adopted the sign convention that W is negative when work is done on the system by its surroundings, and positive when work is done by the system on its surroundings. The units of work are expressed in foot-pounds (ft·lb$_f$) or joules. One calorie of thermal energy is equivalent to 4.184 joules, and one BTU is equivalent to 778 ft·lb of work.

There are other types of work that involve less intuitive force and displacement scenarios such as the work to rotate shaft or create electrical current. A rotating paddle-wheel can do work on a gas or fluid. Power can be transmitted by a shaft according to the following equation:

$$\dot{W} = \Gamma \omega \qquad (5.122)$$

where:
$\dot{W}$ = electrical power, BTU/h (J/s or W)
Γ = torque, ft · lb$_f$ (N · m)
ω = angular velocity, radians per unit time.

Another important term is **Power**, which is defined as the rate of doing work. Power has units of energy per unit of time. Typical units for power include joules per second (J/s) or watts (W). One watt is equivalent to one J/s or 3.412 BTU/h.

The rate of energy transfer by work or power of an electrolytic cell is calculated using Equation (5.123).

$$\dot{W} = -\xi i \qquad (5.123)$$

where:
$\dot{W}$ = electrical power, BTU/h (J/s or W)
ξ = electrical potential difference, volts
i = electric current, amperes.

Similar to the connection between heat transfer and heat transfer rate, work is the accumulation of power over time, according to the equation:

$$W = \int_{t_1}^{t_2} \dot{W} dt \qquad (5.124)$$

where:
W = total work accumulated in the system, BTU (kJ)
$\dot{W}$ = rate of doing work, $\frac{\text{BTU}}{\text{h}} \left(\frac{\text{J}}{\text{s}}\right)$
t_1 and t_2 = time at different periods, sec, min
dt = change in time, sec, min.

Recall that energy is a conserved quantity and thus cannot be created nor destroyed, but merely moved and/or changed in form. Work and heat transfer both play a role in the dynamics of energy, which are always balanced. The First Law of Thermodynamics, discussed below, represents a mathematical expression for the conservation of energy. Furthermore, there are no processes that are 100% efficient at converting energy into useful work and, as a result, there will always be lost energy produced, usually in the form of waste heat. The Second Law of Thermodynamics identifies limits on the efficiency of a process and recognizes that not all energy is useful. For example, a power plant requires a source of heat to produce power and cannot extract any energy for the thermal energy from the surroundings to produce power. Thus, the thermal energy in the surrounding atmosphere of a power plant is effectively useless, although energy is present.

5.5.1.2 Enthalpy (H)

In many thermodynamic applications, the sum of the internal energy U and the product of pressure, P and volume, $\forall$, appears so frequently that a new property, called **enthalpy** (H), was developed. Enthalpy represents the total useful energy of a substance. It consists of the internal energy and the flow energy or work energy or $P \times \forall$ work. This thermodynamics property is a function of temperature, pressure, volume, and the composition of the material. The enthalpy of a substance is defined as follows:

$$H = U + P\forall \qquad (5.125)$$

where:
H = enthalpy of the substance, BTU (kJ)
U = the internal energy of the substance, BTU (kJ)
P = pressure of the system, lb_f/in^2 (kPa)
$\forall$ = volume of the system, ft^3 (m^3).

When working with customary US units, the value of $P\forall$ must be divided by the Joule's constant, 778.17 ft·lb_f/BTU, to obtain consistent units with Equation (5.125).

5.5.1.3 Specific Heat

The **specific heat** (c) of a substance is another important property involved in thermodynamics, and it is defined as the energy required to raise the temperature of a unit of mass of substance by one degree. In thermodynamics, the energy required in the process may occur under controlled conditions, such as constant volume or constant pressure.

When a process occurs without a change in volume, the change in internal energy can be calculated using Equation (5.126).

$$\Delta U = mc_v\Delta T \tag{5.126}$$

where:
ΔU = change in enthalpy, BTU (kJ)
m = mass of substance, lb_m (kg)
c_v = specific heat or heat capacity of the substance at constant volume, $\frac{BTU}{lb_m \cdot °F}\left(\frac{kJ}{kg \cdot K}\right)$
$\Delta T = T_2 - T_1$, temperature change, °F (K).

For a constant-volume process, the change in enthalpy (ΔH) is calculated as follows:

$$\Delta H = mc_p\Delta T \tag{5.127}$$

where:
ΔH = change in enthalpy, BTU (kJ)
m = mass of substance, lb_m (kg)
c_p = specific heat or heat capacity of the substance at constant pressure, $\frac{BTU}{lb_m \cdot °F}\left(\frac{kJ}{kg \cdot K}\right)$
$\Delta T = T_2 - T_1$, temperature change, °F (K).

Solids and liquids are considered "incompressible" under most normal environmental conditions. Therefore, the change in $P \times \forall$ is zero, resulting in $\Delta U = \Delta H$ and $c_v = c_p$ for solids and liquids. For an incompressible substance, it is unnecessary to distinguish between c_v and c_p, and both can be replaced by the symbol c. Equations (5.126) and (5.127) then become Equations (5.128) and (5.129), respectively.

$$\Delta U = mc\Delta T \tag{5.128}$$

$$\Delta H = mc\Delta T \tag{5.129}$$

where c = specific heat of the substance, $\frac{BTU}{lb_m \cdot °F}\left(\frac{kJ}{kg \cdot K}\right)$.

For most engineering applications, we are concerned with the rate of energy change across boundaries of systems.

Table 5.6 Specific heat, cp, of selected liquids and solids at 300 K.

Substance	Specific Heat, $c_p\left(\frac{BTU}{lb_m \cdot °F}\right)$	Specific Heat, $c_p\left(\frac{kJ}{kg \cdot K}\right)$
Aluminum	0.22	0.903
Ammonia	1.15	4.818
Concrete (stone mix)	0.21	0.880
Copper	0.092	0.385
Glass, plate	0.18	0.750
Iron	0.11	0.447
Lead	0.031	0.129
Mercury	0.033	0.139
Oil (unused engine)	0.46	1.909
Silver	0.056	0.235
Steel (AISI 302)	0.11	0.480
Water	1.0	4.179
Wood (fir, pine)	0.33	1.380

1 BTU/(lb_m · °F) = 4.1868 kJ/(kg · K)
Source: Moran and Shapiro (2004) p. 793. Reproduced by permission of John Wiley & Sons Ltd.

Therefore, Equation (5.129) is modified to account for the mass flow rate, $\dot{m}$, with units of mass per unit time:

$$[\text{rate of change in stored energy}] = \dot{m}\,c\Delta T \tag{5.130}$$

For thermal pollution problems in environmental engineering, the heat energy for a given water body is calculated by multiplying the mass of the water by the absolute temperature and by the specific heat of water. We assume that the specific heat of water does not vary significantly with temperature. At 15°C, the specific heat of water is 4.18 kJ/(kg·°C), 1.0 kcal/(kg·°C), or 1.0 BTU/(lb·°F). Table 5.6 lists the specific heat values of several substances.

5.5.1.4 Heat Transfer Modes

The transfer of heat energy into or out of a system may be accomplished by one of three modes: conduction, radiation, or convection. When two objects are at different temperatures, heat will be transferred from the object, with the higher temperature to the one at the lower temperature.

Conduction is the transfer of energy from more energetic particles of a substance to adjacent particles that are less energetic. The transfer of heat energy by direct physical contact between two bodies of different temperature is an example of conduction. The rate of heat transfer through a layer of thickness dx is quantified by Fourier's law, as expressed by Equation (5.131). Table 5.7 lists the thermal conductivity of several types of materials.

$$\dot{Q}_{cond} = -\kappa\frac{dT}{dx} \tag{5.131}$$

Table 5.7 Thermal Conductivities of Selected Substances at 300 K.

Substance	$k\left(\frac{BTU}{h \cdot ft \cdot °F}\right)$	$k\left(\frac{W}{m \cdot K}\right)$
Air	0.015	0.026
Aluminum	137	237
Blanket (glass fiber)	0.027	0.046
Concrete (stone mix)	0.81	1.4
Cork	0.023	0.039
Copper	232	401
Glass, plate	0.81	1.4
Lead	20	35.3
Steel (AISI 302)	8.7	15.1
Silver	248	429
Water	0.35	0.613

1 BTU/(h · ft · °F) = 1.731 W/(m · K)
Source: Moran and Shapiro (2004) pp. 793–794. Reproduced by permission of John Wiley & Sons Ltd.

where:
$\dot{Q}_{cond}$ = rate of heat conduction through a material of thickness (dx), $\frac{BTU}{h}\left(\frac{kJ}{h}\right)$
κ = thermal conductivity of material, $\frac{BTU}{(hr \cdot ft \cdot °F)}, \frac{W}{(m \cdot K)}$
$\frac{dT}{dx}$ = temperature change through distance dx, $\frac{°F}{ft}, \frac{K}{m}$.

Thermal radiation is emitted by matter as a result of changes in electronic configuration of the atoms or molecules within it. This energy is transported by electromagnetic waves or photons. Thermal radiation is electromagnetic radiation with a wavelength between 10^3 to 10^6 Å (1×10^{-7} to 1×10^{-4} m). The maximum rate of radiation that can be emitted from a surface at an absolute temperature, T_s, is determined by the Stefan-Boltzman law (Cengel & Boles, 2008) as expressed in Equation (5.132).

$$\dot{Q}_{emitt} = \sigma A T_s^4 \tag{5.132}$$

where:
$\dot{Q}_{emitt}$ = maximum rate of radiation emitted from the surface of a blackbody, $\frac{BTU}{h}$ (W)
σ = Stefan-Boltzman constant = 0.1713×10^{-8}, $\frac{BTU}{(hr \cdot ft^2 \cdot °R^4)}$
σ = Stefan-Boltzman constant = 5.67×10^{-8}, $\frac{W}{(m^2 \cdot K^4)}$
A = surface area of the body, ft^2 (m^2)
T_s = absolute temperature of at the surface of the body, °R (K).

The radiation emitted by surfaces other than an ideal blackbody must take into account the emissivity (ε) of the material. Equation (5.133) is a modification of Equation (5.132) to account for emissivity.

$$\dot{Q}_{emitt} = \varepsilon \sigma A T_s^4 \tag{5.133}$$

where ε = emissivity of the material, dimensionless.

All real bodies will absorb some portion of the radiation energy incident on their surface. For an ideal blackbody, the value of absorptivity (α) is set equal to a value of 1. The rate at which radiation is absorbed by the surface of a body is determined as follows:

$$\dot{Q}_{absorbs} = \alpha \dot{Q}_{incident} \tag{5.134}$$

where:
$\dot{Q}_{absorbs}$ = rate at which heat energy is absorbed by the surface of a body, $\frac{BTU}{h}\left(\frac{kJ}{h}\right)$
α = absorptivity of material at the surface of the body, dimensionless
$\dot{Q}_{incident}$ = rate at which radiation is incident on a body, $\frac{BTU}{h}\left(\frac{kJ}{h}\right)$.

The net rate of radiation heat transfer is the difference between the rate of radiation emitted by the surface and the radiation absorbed by the surface. Equation (5.135) is used for estimating the net rate of radiation heat transfer between the surface of a body and enclosed by a larger area that is separated by a gas such as air.

$$\dot{Q}_{radiation} = \varepsilon \sigma A (T_s^4 - T_{surr}^4) \tag{5.135}$$

where:
$\dot{Q}_{radiation}$ = net radiation heat transfer between two surfaces, $\frac{BTU}{h}$ (W)
T_s = absolute temperature of the surroundings, °R (K)
T_{surr} = absolute temperature of the surroundings, °R (K).

Convective heat transfer involves energy transfer between a solid surface at a temperature of T_s and an adjacent moving gas or liquid at another temperature, T_f. Mathematically, convective heat transfer is defined by Newton's Law of Cooling and is expressed as Equation (5.136).

$$\dot{Q}_{convective} = hA(T_s - T_f) \tag{5.136}$$

where:
$\dot{Q}_{convective}$ = heat transfer by convection, $\frac{BTU}{h}$ (W)
h = convection heat transfer coefficient, $\frac{BTU}{(h \cdot ft^2 \cdot °R)}\left(\frac{W}{(m^2 \cdot K)}\right)$
A = surface area through which heat transfer takes place, ft^2 (m^2)
T_s = temperature at surface of body, °R (K)
T_f = bulk temperature of gas or liquid flowing past the surface of the body, °R (K).

5.5.2 First law of thermodynamics

The conservation or balance of energy (also known as the **First Law of Thermodynamics**) is based purely on observation and is classified as a physical law. Energy has many forms and is inherently associated with mass and/or waves; it can be thought to have position and quantity. The first law states that energy can neither be created nor destroyed, and so can only relocate and/or change forms, so that the total energy in the "universe" is unchanging. In simple terms, during an interaction, the energy can change from one form to another, but the total amount of energy remains the same.

During an analysis of a system or process, all energy inputs and outputs, in addition to the accumulation of energy term, must balance. Just as in material balances, energy is conserved – only the form may change, through some sort of conversion process. Thus, engineered devices designed to produce a useful effect do not actually consume energy – for energy cannot be destroyed – but instead transform and relocate energy cleverly for a desired effect. Unlike mass, however, energy not only is transferred through the system by bulk flow, but also by heat transfer and work interactions with the surroundings.

Analogous to material or mass balances, energy balances can be performed on processes and systems. In this section, we are going to derive an equation for an overall energy balance around a control volume (C¥). As done earlier in the chapter, we start with a qualitative equation and transform it into a quantitative one.

$$[\text{accumulation}] = [\text{inputs}] - [\text{outputs}] \quad (5.1)$$

There is, however one main difference between an energy balance and a materials balance; there is no reaction term, since energy cannot be created or destroyed. Only its form is changed. The concept of an energy balance which applies at every instant in time can be stated as follows:

$$\begin{bmatrix}\text{net change of}\\ \text{energy in}\\ \text{the C¥}\\ \text{at time } t\end{bmatrix} = \begin{bmatrix}\text{net rate at which}\\ \text{energy is being}\\ \text{transferred in by}\\ \text{heat transfer}\\ \text{at time } t\end{bmatrix} - \begin{bmatrix}\text{net rate at which}\\ \text{energy is being}\\ \text{transferred out by}\\ \text{work}\\ \text{at time } t\end{bmatrix} + \begin{bmatrix}\text{total energy}\\ \text{entering C¥}\\ \text{at time } t\end{bmatrix} - \begin{bmatrix}\text{total energy}\\ \text{leaving C¥}\\ \text{at time } t\end{bmatrix} \quad (5.137)$$

Equation (5.137) can be simplified and re-stated as:

$$\begin{bmatrix}\text{the rate of change of}\\ \text{total energy}\\ \text{in C¥}\end{bmatrix} = \begin{bmatrix}\text{the rate at which}\\ \text{energy enters}\\ \text{C¥}\end{bmatrix} - \begin{bmatrix}\text{the rate at which}\\ \text{energy leaves}\\ \text{C¥}\end{bmatrix} \quad (5.138)$$

Expressing Equation (5.137) quantitatively yields:

$$\frac{dE_{C¥}}{dt} = \dot{Q} - \dot{W} + \dot{m}_i e_i - \dot{m}_e e_e \quad (5.139)$$

where:

$\frac{dE_{C¥}}{dt}$ = total energy accumulating in control volume, BTU (kJ)

$\dot{Q}$ = net rate of heat energy into or out off the control volume, BTU (kJ)

$\dot{W}$ = net rate of work done on or by the system, BTU (kJ)

$\dot{m}_i$, $\dot{m}_e$ = rate of mass transfer into and exiting the system, $\frac{\text{lb}_m}{\text{time}}\left(\frac{\text{kg}}{\text{time}}\right)$

e_i, e_e = total energy of the system per unit mass as defined by Equation (5.7), $\frac{\text{BTU}}{\text{lb}_m}\left(\frac{\text{kJ}}{\text{kg}}\right)$.

The net rate of work ($\dot{W}$) done on or by the system, with units of energy per unit of time, is made up of several components, as has been previously discussed. It consists of the flow energy or work energy ($\dot{W}_{flow}$) and all other transfers of energy, as work across the boundary of the control volume, such as electrical and shaft. Mathematically, the net rate of work is defined as:

$$\dot{W} = \dot{W}_{flow} + \dot{W}_{C¥} \quad (5.140)$$

Flow work is determined by multiplying force × distance as shown below:

$$\dot{W}_{flow} = (P_e A_e)V_e - (P_i A_i)V_i \quad (5.141)$$

where:

P_e, P_i = pressure at the exit and inlet of control volume, $\frac{\text{lb}_f}{\text{in}^2}$ (kPa)

A_e, A_i = cross section area of exit and inlet to control volume, ft^2 (m^2)

V_e, V_i = velocity at exit and inlet to control volume, $\frac{\text{ft}}{\text{s}}\left(\frac{\text{m}}{\text{s}}\right)$.

The mass rate of flow is calculated using onc of the following equations:

$$\dot{m} = \rho A V = \frac{1}{v} A V \quad (5.142)$$

where:

$\dot{m}$ = mass rate of flow, $\frac{\text{lb}_m}{\text{s}}\left(\frac{\text{kg}}{\text{s}}\right)$

ρ = density of mass, $\frac{\text{lb}_m}{\text{ft}^3}\left(\frac{\text{kg}}{\text{m}^3}\right)$

V = velocity of mass, $\frac{\text{ft}}{\text{s}}\left(\frac{\text{m}}{\text{s}}\right)$

v = specific volume of a substance, $\frac{\text{ft}^3}{\text{lb}_m}\left(\frac{\text{m}^3}{\text{kg}}\right)$.

Therefore, the mass flow rate times the specific volume is equal to the area times velocity as follows:

$$\dot{m}v = A V \quad (5.143)$$

Flow energy or flow work can now be expressed as follows by substituting Equation (5.143) into Equation (5.141).

$$\dot{W}_{flow} = \dot{m}_e v_e P_e - \dot{m}_i v_i P_i \quad (5.144)$$

Now, the net rate at which energy is transferred in or out of the control volume by work is determined from Equation (5.145) and we add the subscript C¥ to denote the heat transfer rate over the entire boundary of control volume.

$$\dot{W} = \dot{W}_{C¥} + \dot{m}_e v_e P_e - \dot{m}_i v_i P_i \quad (5.145)$$

Equation (5.145) is substituted into Equation (5.139) as follows:

$$\frac{dE_{C\forall}}{dt} = \dot{Q}_{C\forall} - \dot{W}_{C\forall} + \dot{m}_i e_i - \dot{m}_e e_e - \dot{m}_e v_e P_e + \dot{m}_i v_i P_i \quad (5.146)$$

Rearranging and collecting terms yields:

$$\frac{dE_{C\forall}}{dt} = \dot{Q}_{C\forall} - \dot{W}_{C\forall} + \dot{m}_i(e_i + v_i P_i) - \dot{m}_e(e_e + v_e P_e) \quad (5.147)$$

Replace the total energy per unit of mass (e) in Equation (5.147) with Equation (5.117) for the inlet and outlet results in the following equation:

$$\frac{dE_{C\forall}}{dt} = \dot{Q}_{C\forall} - \dot{W}_{C\forall} + \dot{m}_i\left(u_i + \frac{V_i^2}{2} + gz_i + v_i P_i\right) - \dot{m}_e\left(u_e + \frac{V_e^2}{2} + gz_e + v_e P_e\right) \quad (5.148)$$

From Equation (5.125), recall that the definition of enthalpy is $H = U + P\forall$. The specific enthalpy, h, can then be defined as follows:

$$h = \frac{H}{m} = u + Pv \quad (5.149)$$

The energy balance equation for an open system can then be shown as follows:

$$\frac{dE_{C\forall}}{dt} = \dot{Q}_{C\forall} - \dot{W}_{C\forall} + \dot{m}_i\left(h_i + \frac{V_i^2}{2} + gz_i\right) - \dot{m}_e\left(h_e + \frac{V_e^2}{2} + gz_e\right) \quad (5.150)$$

For several inlets and outlets to the control volume or system, the summation symbol is placed before each mass flow rate term. The general energy balance equation for an open system is then expressed as Equation (5.151).

$$\frac{dE_{C\forall}}{dt} = \dot{Q}_{C\forall} - \dot{W}_{C\forall} + \sum_i \dot{m}_i\left(h_i + \frac{V_i^2}{2} + gz_i\right) - \sum \dot{m}_e\left(h_e + \frac{V_e^2}{2} + gz_e\right) \quad (5.151)$$

Example 5.14 Thermal discharges

An industry in Macon, Georgia discharges approximately 19,000 liters per day (lpd) of industrial wastewater at a temperature of 35°C to the Rocky Creek WWTP. The Rocky Creek facility also receives 38,000 lpd of municipal wastewater at a temperature of 20°C. Determine the temperature of the industrial and municipal wastewater after mixing together at the headworks of the WWTP.

Solution

Assume that each wastewater has a density (1000 kg/m³) and specific heat $\left(\frac{4.18\ \text{kJ}}{\text{kg}\cdot{}^\circ\text{C}}\right)$.

A schematic of the three flow streams is presented below.

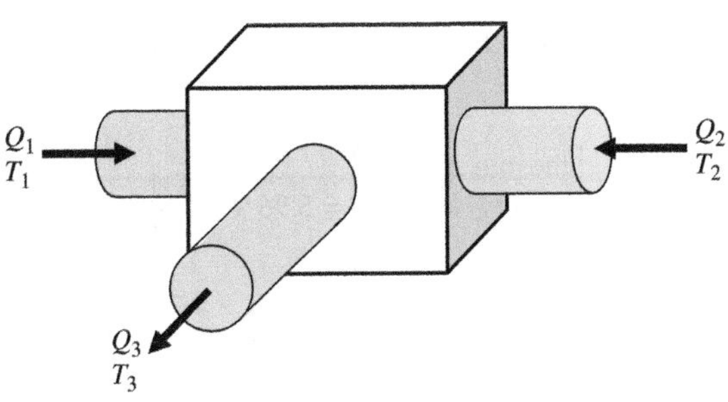

Figure E5.8

First, calculate the heat or thermal energy of the industrial and municipal wastewater streams using a modified form of Equation (5.130).

$$[\text{heat energy}] = \dot{m}cT$$

The heat energy in the industrial wastewater stream is calculated as:

$$[\text{heat energy}]_{\text{ind ww}} = 19{,}000\,\frac{\text{L}}{\text{d}}\left(4.18\,\frac{\text{kJ}}{\text{kg}\cdot{}^\circ\text{C}}\right)(35^\circ\text{C}) \times \left(\frac{1\ \text{m}^3}{1000\ \text{L}}\right)\left(\frac{1000\ \text{kg}}{\text{m}^3}\right)$$

$$[\text{heat energy}]_{\text{ind ww}} = 2.78 \times 10^6\,\frac{\text{kJ}}{\text{d}}$$

$$[\text{heat energy}]_{\text{mun ww}} = 38{,}000\,\frac{\text{L}}{\text{d}}\left(4.18\,\frac{\text{kJ}}{\text{kg}\cdot{}^\circ\text{C}}\right)(20^\circ\text{C}) \times \left(\frac{1\ \text{m}^3}{1000\ \text{L}}\right)\left(\frac{1000\ \text{kg}}{\text{m}^3}\right)$$

$$[\text{heat energy}]_{\text{mun ww}} = 3.18 \times 10^6\,\frac{\text{kJ}}{\text{d}}$$

We can write a simplified modification of Equation (5.138) to perform an energy balance at the head of the Rocky Creek WWTP as follows:

$$\begin{bmatrix}\text{the rate of change of}\\ \text{total energy}\\ \text{in } C\forall\end{bmatrix} = \begin{bmatrix}\text{the rate at which}\\ \text{energy enters}\\ C\forall\end{bmatrix} - \begin{bmatrix}\text{the rate at which}\\ \text{energy leaves}\\ C\forall\end{bmatrix}$$

Recall, at steady-state, the accumulation term is set equal to zero:

$$[0] = \left[2.78 \times 10^6 \frac{\text{kJ}}{\text{d}} + 3.18 \times 10^6 \frac{\text{kJ}}{\text{d}}\right] - \left[\begin{matrix}\text{the rate at which}\\ \text{energy leaves}\\ C\forall\end{matrix}\right]$$

The rate at which energy leaves $C\forall$ is equal to 5.96×10^6 kJ/d.

$$\left[\begin{matrix}\text{the rate at which}\\ \text{energy leaves}\\ C\forall\end{matrix}\right] = 5.96 \times 10^6 \frac{\text{kJ}}{\text{d}}$$

The temperature of the combined wastewater streams may then be calculated from the following equation:

$$[\text{rate of change in stored energy}] = \dot{m}cT$$

$$\left[5.96 \times 10^6 \frac{\text{kJ}}{\text{d}}\right] = \left(19{,}000 + 38{,}000 \frac{\text{L}}{\text{d}}\right)\left(4.18 \frac{\text{kJ}}{\text{kg} \cdot {}^\circ\text{C}}\right) \times \left(\frac{1\ \text{m}^3}{1000\ \text{L}}\right)\left(\frac{1000\ \text{kg}}{\text{m}^3}\right) T$$

$$\boxed{T = 25^\circ\text{C}}$$

Alternatively, a simpler way to estimate the temperature of two or more streams is by performing a materials balance on the streams using the following equation.

$$T_n = \frac{Q_1 T_1 + Q_2 T_2 + \ldots\ Q_n T_n}{Q_1 + Q_2 + \ldots\ Q_n} \qquad (5.152)$$

where:
Q_n = volumetric flow rate, $\frac{\text{gal}}{\text{d}}\left(\frac{\text{m}^3}{\text{d}}\right)$
n = number of streams that mix or combine together.

$$T = \frac{19{,}000 \frac{\text{L}}{\text{d}}(35^\circ\text{C}) + 38{,}000 \frac{\text{L}}{\text{d}}(20^\circ\text{C})}{19{,}000 \frac{\text{L}}{\text{d}} + 38{,}000 \frac{\text{L}}{\text{d}}} = \boxed{25^\circ\text{C}}$$

Example 5.15 Energy balance for nuclear power plant

Approximately two-thirds of the energy content of uranium fuel entering a 1,000 megawatt (MW) nuclear power plant is removed by condenser cooling water which is withdrawn from an adjacent stream. The flow upstream of the nuclear power plant is 100 m³/s and the temperature in the stream is 18°C. Perform an energy balance on the nuclear power plant to answer the following questions, and draw a simplified schematic of the process.

Determine:

a) Energy input to the nuclear power plant.
b) Necessary flow rate in the stream if the temperature in the cooling water is only allowed to rise 10°C in temperature.
c) Temperature in the stream after the heated cooling water is released back into the stream.

Solution

First, calculate the energy input to the power plant by dividing by the energy output by efficiency:

$$\text{energy input} = \frac{\text{energy output}}{\text{efficiency}} = \frac{1000\ \text{MW}_e}{1/3} = 3000\ \text{MW}_t$$

Total energy losses that must be dissipated by the cooling water are calculated as follows:

$$3000\ \text{MW} - 1000\ \text{MW} = 2000\ \text{MW}$$

Next, substitute into Equation (5.130) to determine the mass flow rate.

$$[\text{rate of change in stored energy}] = \dot{m}c\Delta T$$

$$\left[2000\ \text{MW}\left(\frac{10^6\ \text{W}}{1\ \text{MW}}\right)\left(\frac{1\ \text{J/s}}{1\ \text{W}}\right)\left(\frac{1\ \text{kJ}}{1000\ \text{J}}\right)\right] = \dot{m}\left(4.18 \frac{\text{kJ}}{\text{kg} \cdot {}^\circ\text{C}}\right)(10^\circ\text{C})$$

$$\dot{m} = 4.78 \times 10^4 \frac{\text{kg}}{\text{s}}$$

The stream flow rate is calculated as follows:

$$Q = \frac{\dot{m}}{\rho} = \frac{4.78 \times 10^4\ \text{kg/s}}{1000\ \text{kg/m}^3} = 47.8\ \text{m}^3/\text{s}$$

The temperature rise in the stream can be estimated as follows, assuming that the total flow in the stream remains 100 m³/s, i.e., 47.8 m³/s is withdrawn for cooling and 52.2 m³/s remains before the heated cooling water is discharged back into the river.

$$[\text{rate of change in stored energy}] = \dot{m}c\Delta T \qquad (5.130)$$

$$\Delta T = \frac{[\text{rate of change in stored energy}]}{\dot{m}c}$$

$$= \frac{\left[2000\ \text{MW}\left(\frac{10^6\ \text{W}}{1\ \text{MW}}\right)\left(\frac{1\ \text{J/s}}{1\ \text{W}}\right)\left(\frac{1\ \text{kJ}}{1000\ \text{J}}\right)\right]}{\left(100 \frac{\text{m}^3}{\text{s}}\right)\left(\frac{1000\ \text{kg}}{\text{m}^3}\right)\left(\frac{4.18\ \text{kJ}}{\text{kg} \cdot {}^\circ\text{C}}\right)}$$

$$\Delta T = 4.8^\circ\text{C}$$

The actual temperature in the stream will be:

$$4.8 + 18.0 = \boxed{22.8°C}$$

Alternatively, the temperature in the stream can be calculated using Equation (5.152).

$$T_n = \frac{Q_1T_1 + Q_2T_2 + \dots Q_nT_n}{Q_1 + Q_2 + \dots Q_n}$$

$$T = \frac{\left(100 - 47.8\frac{m^3}{s}\right)(18°C) + \left(47.8\frac{m^3}{s}\right)(18 + 10°C)}{100\ \frac{m^3}{s}}$$

$$= \boxed{22.8°C}$$

The schematic below shows the energy balance for the nuclear power plant.

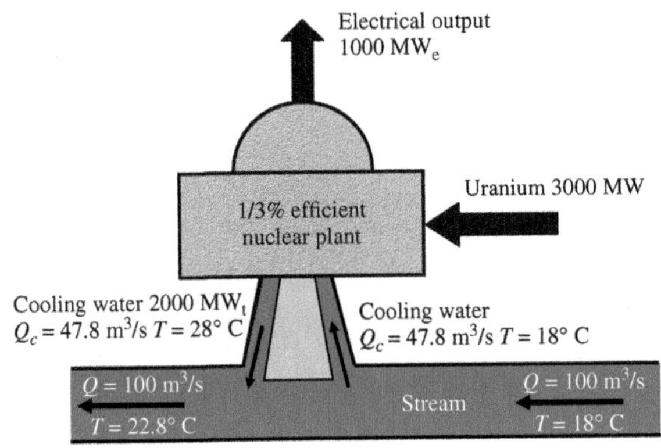

Figure E5.9

Example 5.16 Energy balance for coal-fired power plant

A 1000-MW$_e$ coal-fired power plant has an efficiency of 33%. 15% of the waste heat is released to the atmosphere through the stack and 85% is removed by the cooling water from an adjacent stream. The flow rate in the river upstream of the power plant is 3,500 cfs and has a temperature of 18.0°C. Perform an energy balance on the coal-fired power plant to answer the following questions and draw a simplified schematic of the process.

Determine:

a) quantity (mass) of coal required daily to generate 1000-MW of electricity if the energy content of the coal is 25 kJ/g.
b) required flow rate in the stream if the temperature in the cooling water is only allowed to rise 10°C in temperature.
c) temperature in the stream after the heated cooling water is released back into the stream.

Solution

First, calculate the energy input to the power plant by dividing by the energy output by efficiency.

$$\text{energy input} = \frac{\text{energy output}}{\text{efficiency}} = \frac{1000\ \text{MW}_e}{0.33} = 3030\ \text{MW}_t$$

$$\text{coal input} = 3030\ \text{MW}_t \left(\frac{10^6\ \text{W}}{1\ \text{MW}}\right)\left(\frac{1\ \text{J/s}}{1\ \text{W}}\right) \times \left(\frac{86{,}400\ \text{s}}{\text{d}}\right)\left(\frac{1\ \text{g coal}}{25\ \text{kJ}}\right)\left(\frac{1\ \text{kg}}{1000\ \text{g}}\right)$$

$$\text{coal input} = \boxed{1.05 \times 10^{10}\ \frac{\text{kg}}{\text{d}}}$$

Total energy losses that go to the stack and cooling water are calculated as follows:

$$3030\ \text{MW} - 1000\ \text{MW} = 2030\ \text{MW}$$

$$\text{stack losses} = 0.15 \times 2030\ \text{MW} = 305\ \text{MW}$$

$$\text{cooling water losses} = 0.85 \times 2030\ \text{MW} = 1725\ \text{MW}$$

Calculate the quantity of water ($\dot{m}$) required for cooling, starting with Equation (5.130):

$$[\text{rate of change in stored energy}] = \dot{m}c\Delta T$$

$$\left[1725\ \text{MW}\left(\frac{10^6\ \text{W}}{1\ \text{MW}}\right)\left(\frac{1\ \text{J/s}}{1\ \text{W}}\right)\left(\frac{1\ \text{kJ}}{1000\ \text{J}}\right)\right] = \dot{m}\left(4.18\frac{\text{kJ}}{\text{kg}\cdot°\text{C}}\right)(10°\text{C})$$

$$\dot{m} = 4.13 \times 10^4\ \frac{\text{kg}}{\text{s}}$$

The stream flow rate is calculated as follows:

$$Q = \frac{\dot{m}}{\rho} = \frac{4.13 \times 10^4\ \text{kg/s}}{1000\ \text{kg/m}^3} = 41.3\ \text{m}^3/\text{s}$$

$$Q = 41.3\ \frac{\text{m}^3}{\text{s}}\left(\frac{1000\ \text{L}}{1\ \text{m}^3}\right)\left(\frac{1\ \text{gal}}{3.785\ \text{L}}\right)\left(\frac{1\ \text{ft}^3}{7.48\ \text{gal}}\right)$$

$$= \boxed{1460\ \text{cfs}}$$

The temperature rise in the stream can be estimated as follows, assuming that the total flow in the stream remains 3500 cfs, i.e., 1460 cfs is withdrawn for cooling and 2040 cfs remains before the heated cooling water is discharged back into the river. The temperature change is calculated by rearranging Equation (5.130) as follows:

$$[\text{rate of change in stored energy}] = \dot{m}c\Delta T$$

$$\Delta T = \frac{[\text{rate of change in stored energy}]}{\dot{m}c}$$

$$\Delta T = \frac{\left[1725\ \text{MW}\left(\frac{10^6\ \text{W}}{1\ \text{MW}}\right)\left(\frac{1\ \text{J/s}}{1\ \text{W}}\right)\left(\frac{1\ \text{kJ}}{1000\ \text{J}}\right)\right]}{\left(3500\ \frac{\text{ft}^3}{\text{s}}\right)\left(\frac{7.48\ \text{gal}}{\text{ft}^3}\right)\left(\frac{3.785\ \text{L}}{\text{gal}}\right)\left(\frac{1\ \text{m}^3}{1000\ \text{L}}\right) \times \left(\frac{1000\ \text{kg}}{\text{m}^3}\right)\left(\frac{4.18\ \text{kJ}}{\text{kg}\cdot{}^\circ\text{C}}\right)}$$

$$\Delta T = 4.2^\circ\text{C}$$

The actual temperature in the stream will be:

$$4.2 + 18.0 = \boxed{22.2^\circ\text{C}}$$

Alternatively, the temperature in the stream can be calculated using Equation (5.152):

$$T_n = \frac{Q_1T_1 + Q_2T_2 + \dots Q_nT_n}{Q_1 + Q_2 + \dots Q_n}$$

$$T = \frac{(3500 - 1460\ \text{cfs})(18^\circ\text{C}) + (1460\ \text{cfs})(18 + 10^\circ\text{C})}{3500\ \text{cfs}}$$

$$= \boxed{22.2^\circ\text{C}}$$

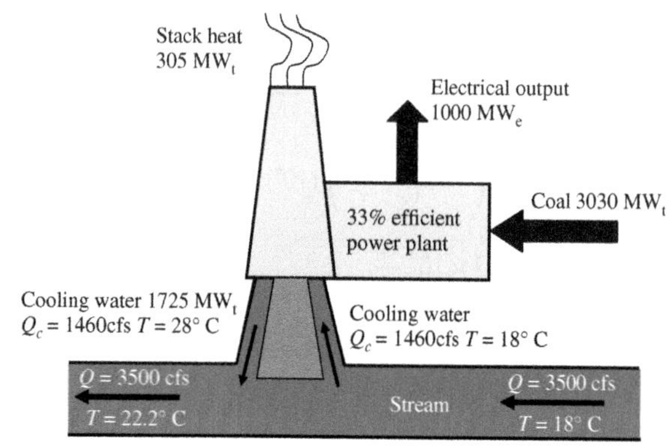

Figure E5.10

Example 5.17 Coal-fired power plant mass and energy balance

A 1,000 megawatt (MW) coal-fired power plant converts one-third of the coal's energy into electrical energy. In other words, for every three units of energy entering the power plant, approximately one unit is converted to electricity and two units are lost to the environment as waste heat. Assume that the coal has an energy content of 20 kJ/g and contains 55% carbon, 1.5% sulfur, and 8% ash. Assume that 65% of the ash is released as fly ash and 35% of the ash settles outside of the firing chamber and is collected as bottom ash. Approximately 15% of the waste heat is assumed to exit in the stack gases, and the cooling water is used to dissipate the remaining heat. Air emission standards restrict sulfur and particulate quantities to 260 g SO_2 per 10^6 kJ of heat input and 13 g particulates per 10^6 kJ of heat input into the coal-fired power plant. Perform a materials and energy balance around the coal-fired power plant to answer the following questions and draw a simplified schematic of the process.

Determine:

a) quantity of heat loss to the cooling water (MW).
b) quantity of cooling water (kg/s) and flow (m³/s) assuming a 10°C increase in the temperature of the cooling water.
c) efficiency of the sulfur dioxide removal system to meet air emission standards.
d) efficiency of the particulate removal system to meet air emission standards.

Solution

Part A

Perform the energy balance around the coal-fired power plant using Equation (5.138) as the basis.

$$\begin{bmatrix}\text{the rate of change of}\\ \text{total energy}\\ \text{in C}\forall\end{bmatrix} = \begin{bmatrix}\text{the rate at which}\\ \text{energy enters}\\ \text{C}\forall\end{bmatrix} - \begin{bmatrix}\text{the rate at which}\\ \text{energy leaves}\\ \text{C}\forall\end{bmatrix}$$

Recall that, at steady state, the energy accumulated is zero and the equation reduces to the following form:

$$\begin{bmatrix}\text{the rate at which}\\ \text{energy enters}\\ \text{C}\forall\end{bmatrix} = \begin{bmatrix}\text{the rate at which}\\ \text{energy leaves}\\ \text{C}\forall\end{bmatrix}$$

$$\begin{bmatrix}\text{energy in}\\ \text{coal}\end{bmatrix} = \begin{bmatrix}\text{energy out}\\ \text{in stack gases}\end{bmatrix} + \begin{bmatrix}\text{energy out}\\ \text{in cooling water}\end{bmatrix} + \begin{bmatrix}\text{energy out}\\ \text{useful electrial power}\end{bmatrix}$$

We estimate the energy in the coal by dividing the useful energy produced as electrical power by the efficiency of the coal-fired power plant, as follows:

$$\text{input power} = \frac{\text{output power}}{\text{efficiency}} = \frac{1000\ \text{MW}_e}{1/3}$$

$$= 3000\ \text{MW}_t$$

Determine the total energy losses in the system as follows:

$$\text{total losses} = \text{energy input} - \text{energy output}$$
$$= 3000 - 1000 = 2000 \text{ MW}_t$$

Estimate the stack losses assuming 15% of the total energy losses as follows:

$$\text{stack losses} = 0.15(2000 \text{ MW}_t) = 300 \text{ MW}_t$$

Calculate the energy loss in the cooling water.

$$\begin{bmatrix}\text{energy in}\\ \text{coal}\end{bmatrix} = \begin{bmatrix}\text{energy out}\\ \text{in stack gases}\end{bmatrix} + \begin{bmatrix}\text{energy out}\\ \text{in cooling water}\end{bmatrix} + \begin{bmatrix}\text{energy out}\\ \text{useful electrial power}\end{bmatrix}$$

$$[3000 \text{ MW}_t] = [300 \text{ MW}_t] + \begin{bmatrix}\text{energy out}\\ \text{in cooling water}\end{bmatrix} + [1000 \text{ MW}_e]$$

$$\begin{bmatrix}\text{energy out}\\ \text{in cooling water}\end{bmatrix} = \boxed{1700 \text{ MW}_t}$$

Part B

The mass flow rate ($\dot{m}$) of water required for cooling is calculated from Equation (5.130) and a specific heat (c) equal to 4.18 kJ/(kg·°C):

$$[\text{rate of change in stored energy}] = \dot{m}c\Delta T \quad (5.130)$$

$$1700 \text{ MW}_t = \dot{m}\left(4.18 \frac{\text{kJ}}{\text{kg}\cdot{}^\circ\text{C}}\right)(10^\circ\text{C}) \times \left(\frac{1 \text{ MW}}{10^6 \text{ J/s}}\right)\left(\frac{1000 \text{ J}}{\text{kJ}}\right)$$

$$\boxed{\dot{m} = 4.07 \times 10^4 \text{ kg/s}}$$

The volumetric flow rate of the cooling water is determined by dividing the mass flow rate by the density of the water (1,000 kg/m^3).

$$Q = \frac{\dot{m}}{\rho} = \frac{4.07 \times 10^4 \text{ kg/s}}{1000 \text{ kg/m}^3} = \boxed{40.7 \text{ m}^3/\text{s}}$$

Part C

Calculate the quantity of SO_2 and particulates that can be emitted per unit of heat input into the coal-fired power plant. First calculate the quantity of coal burned daily as follows:

$$3000 \text{ MW}\left(\frac{10^6 \text{ W}}{1 \text{ MW}}\right)\left(\frac{1 \text{ kW}}{1000 \text{ W}}\right)\left(\frac{24 \text{ h}}{\text{d}}\right)\left(\frac{1 \text{ kJ/s}}{\text{kW}}\right) \times \left(\frac{60 \text{ s}}{\text{min}}\right)\left(\frac{60 \text{ min}}{\text{h}}\right)\left(\frac{1 \text{ g coal}}{20 \text{ kJ}}\right)\left(\frac{1 \text{ kg}}{1000 \text{ g}}\right)$$

$$\text{quantity of coal burned daily} = 1.30 \times 10^7 \frac{\text{kg}}{\text{d}}$$

Next determine the quantity of SO_2 that is produced daily, knowing that the molecular weight of sulfur dioxide is 32 + 2(16) = 64.

$$SO_2 \text{ produced} = 1.30 \times 10^7 \frac{\text{kg coal}}{\text{d}}\left(0.015 \frac{\text{kg S}}{\text{kg coal}}\right)\left(\frac{64 \text{ kg } SO_2}{32 \text{ kg S}}\right)$$
$$= 3.89 \times 10^5 \frac{\text{kg}}{\text{d}}$$

The quantity of SO_2 that is permitted to be discharged daily to the atmosphere is calculated as:

$$SO_2 \text{ discharged} = \left(\frac{260 \text{ g } SO_2}{10^6 \text{ kJ}}\right)(3000 \text{ MW}_t)\left(\frac{10^6 \text{ W}}{1 \text{ MW}}\right) \times \left(\frac{1 \text{ J/s}}{1 \text{ W}}\right)\left(\frac{3600 \text{ s}}{\text{h}}\right)\left(\frac{24 \text{ h}}{\text{d}}\right)\left(\frac{1 \text{ kg}}{1000 \text{ g}}\right)\left(\frac{1 \text{ kJ}}{1000 \text{ J}}\right)$$

$$SO_2 \text{ discharged} = 6.74 \times 10^4 \frac{\text{kg } SO_2}{\text{d}}$$

The efficiency of air pollution control equipment can be calculated from the following equations:

$$\text{removal efficiency} = \frac{(C_{in} - C_{out})100}{C_{in}} = \frac{(M_{in} - M_{out})\,100}{M_{in}} \quad (5.94)$$

$$SO_2 \text{ removal efficiency} = \frac{(3.89 \times 10^5 - 6.74 \times 10^4 \text{ kg/d}) \times 100}{3.89 \times 10^5 \text{ kg/d}} = \boxed{82.7\%}$$

Part D

Determine the quantity of particulates or fly ash that enters the air pollution control equipment.

$$\text{fly ash produced} = 1.30 \times 10^7 \frac{\text{kg coal}}{\text{d}}\left(0.08 \frac{\text{kg ash}}{\text{kg coal}}\right)\left(0.65 \frac{\text{kg fly ash}}{\text{kg ash}}\right)$$
$$= 6.76 \times 10^5 \frac{\text{kg}}{\text{d}}$$

The quantity of fly ash or particulate matter that is permitted to be discharged daily is calculated as:

$$\left(\frac{13 \text{ g particulates}}{10^6 \text{ kJ}}\right)(3000 \text{ MW}_t)\left(\frac{10^6 \text{ W}}{1 \text{ MW}}\right)\left(\frac{1 \text{ J/s}}{1 \text{ W}}\right) \times \left(\frac{3600 \text{ s}}{\text{h}}\right)\left(\frac{24 \text{ h}}{\text{d}}\right)\left(\frac{1 \text{ kg}}{1000 \text{ g}}\right)\left(\frac{1 \text{ kJ}}{1000 \text{ J}}\right)$$
$$= 3.37 \times 10^3 \frac{\text{kg particulates}}{\text{d}}$$

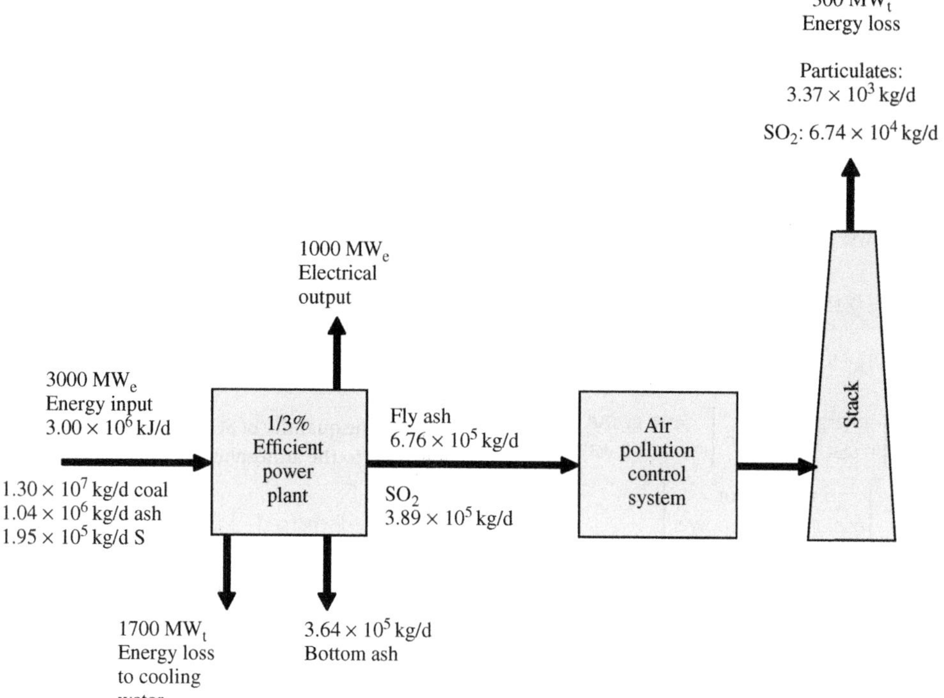

Figure E5.11

Estimate the particulate removal efficiency to meet air standards.

particulate removal efficiency

$$= \frac{(6.76 \times 10^5 - 3.37 \times 10^3 \text{kg/d})\ 100}{6.76 \times 10^5\ \text{kg/d}} = \boxed{99.5\%}$$

A simplified schematic of the materials and energy flow through the coal-fired power plant is presented above.

5.5.3 Second law of thermodynamics

A second observation of energy is that when its forms and spatial distribution are dynamic, it always moves and transforms towards spatial uniformity. As a result, all physical processes (changes) are irreversible, meaning that if one reverses the inputs to any real device, the device will not follow the same process backwards but must take a slightly different path.

This observation serves as the basis for another physical law available in many different mathematical expressions, all representing the **Second Law of Thermodynamics.**
The concept of entropy is based on the Second Law of Thermodynamics, which essentially states that all systems tend to move toward equilibrium or a steady-state condition. The significance of this law is that work can only be obtained from a system that is not at equilibrium.

Another way of stating this is that energy flows from a higher region of concentration to one of a lower concentration.

Physical chemists and thermo-dynamists state that systems go from one of order to one of disorder, i.e., the system becomes more random. **Entropy** (S) is defined by the following equation:

$$dS = \frac{dq_{rev}}{T} \tag{5.153}$$

where:
dS = change in entropy of the system, kJ
dq_{rev} = amount of heat delivered to or absorbed by the system in a reversible process, kJ
T = absolute temperature, K.

Engineers are interested in the change of entropy in a system, as evaluated below:

$$\Delta S = S_2 - S_1 = \int_1^2 \frac{dq_{rev}}{T} \tag{5.154}$$

where:
ΔS = change in entropy of the system, kJ
S_2 = entropy of system a state 2, kJ
S_1 = entropy of system a state 1, kJ
T = absolute temperature, K
ΔS = positive, meaning that a change will occur spontaneously
ΔS = negative, meaning that a change will occur in the opposite direction
ΔS = zero, the system is at equilibrium.

Thus, engineered devices designed to produce a useful effect do not actually consume energy, for energy cannot be destroyed, but instead transform and relocate energy, cleverly taking advantage of the First and Second Laws of Thermodynamics. Although not all devices are invented from this perspective, these two physical laws are powerful engineering analysis tools. Regardless of the process, energy conversion is never 100% efficient and a loss of useful energy occurs, normally through waste heat. The Second Law of Thermodynamics states that there will always be some waste heat released during energy conversions.

As an example, a gasoline engine converts the chemical energy stored in liquid fuel and converts it to kinetic energy in a rotating shaft, plus an unavoidable heat loss. Combustion of the fuel is an exothermic chemical reaction between the gasoline and oxygen, producing exhaust gases and heat. Much of the heat is absorbed by the gases trapped in a piston cylinder assembly. As the combustion proceeds, the temperature and pressure of the trapped gases increase, forcing the chamber to expand and thus moving the piston, which rotates a crank via the leverage of a crank arm. Since the engine cannot be insulated perfectly, much of the energy (typically 70–80%) leaves the engine through heat transfer (Q) rather than as work (W) of the moving piston. An engine is a clever means to convert chemical energy (or at least 20–30% of it) to the kinetic energy of a rotation shaft.

Sadi Carnot (1824), a French engineer, was one of the first to study the efficiency of a steam engine. The most efficient heat engine that can operate in a closed system is called a Carnot engine. It operates between two heat reservoirs, a hot reservoir at temperature T_h and a cold reservoir at temperature T_c. Figure 5.20 is a simplified schematic of the Carnot engine.

We are going to derive an equation to show the theoretical efficiency of a Carnot engine. First, we begin with Equation (5.139) and, realizing that there is no transfer of mass into or out of the system, this results in the following equation:

$$\frac{dE_{CV}}{dt} = \dot{Q} - \dot{W} \qquad (5.155)$$

At steady-state, the accumulation term, $\left(\frac{dE_{CV}}{dt}\right)$, is set equal to zero and Equation (5.155) can be expressed as:

$$0 = Q_h - Q_c - W \qquad (5.156)$$

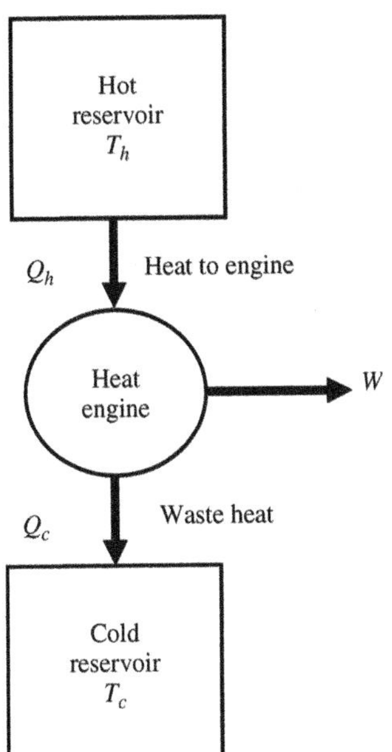

Figure 5.20 Simplified schematic of the Carnot engine.

where:
Q_h = heat energy into heat engine, BTU (kJ)
Q_c = heat energy out of heat engine, BTU (kJ)
W = work output from heat engine, BTU (kJ).

The thermal efficiency of a heat engine is the ratio of the work output (mechanical work) to the heat energy input to the heat engine. This is expressed mathematically as:

$$\eta = \frac{W}{Q_h} \qquad (5.157)$$

where η = thermal efficiency of heat engine, fraction.

Replacing W in Equation (5.157) with Equation (5.156) results in:

$$\eta = \frac{Q_h - Q_c}{Q_h} = 1 - \frac{Q_c}{Q_h} \qquad (5.158)$$

From Equation (5.153) the total amount of thermal energy transferred between the hot reservoir and the system is:

$$Q_h = T_h(S_c - S_h) \qquad (5.159)$$

where:
T_h = temperature of hot reservoir, °R (K)
S_c = entropy of the cold reservoir, $\frac{BTU}{lb_m \cdot °R}\left(\frac{kJ}{kg \cdot K}\right)$
S_h = entropy of the hot reservoir, $\frac{BTU}{lb_m \cdot °R}\left(\frac{kJ}{kg \cdot K}\right)$.

Similarly, the total amount of thermal energy transferred between the system and the cold reservoir is:

$$Q_c = T_c(S_c - S_h) \qquad (5.160)$$

where T_c temperature of cold reservoir, °R (K).

Then, according to Carnot, the most efficient heat engine can be determined as:

$$\eta = 1 - \frac{Q_c}{Q_h} = 1 - \frac{T_c(S_c - S_h)}{T_h(S_c - S_h)} = 1 - \frac{T_c}{T_h} \qquad (5.161)$$

Example 5.18 Ocean temperature energy conversion (OTEC) power plant

Ocean Temperature Energy Conversion (OTEC) Power Plants generate power by using the natural temperature gradient that occurs in oceans. In Florida, the ocean surface temperature is sometimes 80°F, while at a depth of 21,000 ft the temperature is 55°F. Determine the maximum thermal efficiency for any heat engine or power plant operating between these temperatures.

Solution

$$T_c = 55°\text{F} + 460 = 515°\text{R}$$

$$T_h = 80°\text{F} + 460 = 540°\text{R}$$

From Equation(5.161), calculate the maximum theoretical efficiency possible as:

$$\eta = 1 - \frac{T_c}{T_h} = 1 - \frac{515°\text{R}}{540°\text{R}} = 0.046$$

$$\boxed{\eta = 4.6\%}$$

The thermal efficiency of existing OTEC plants is only around 2.0%.

Summary

- Material or mass balances are used in modeling engineered and natural systems.
  - Mass flow rate $\dot{m}$ is determined by multiplying the volumetric flow rate Q of the stream by the density ρ of the flow stream.
  - Mass flow rate of a specific contaminant can be estimated by multiplying the volumetric flow rate Q of the stream by the concentration C of the contaminant and making appropriate conversions.
  - When systems operate at steady-state, wherein the influent and effluent concentrations of all parameters and flows remain constant, the accumulation term can be set equal to zero.
  - Dynamic systems are those that operate in a state of flux and where the mass balances yield differential equations that are more complex to solve.
- The rate of a reaction r is the rate of formation or disappearance of a chemical compound or species.
- A mathematical equation that expresses the relationship between the rate of reaction and the concentration of species involved in the reaction is called the rate law.
  - The order of the reaction is related to the exponent of the chemical species given in the rate law. The order of the reaction can be presented with respect to specific chemical species, or with respect to all of the chemical species that are involved in the reaction; either as reactants or products.
  - Most biological and chemical reactions involved in environmental engineering are either zero-, first-, or second-order reactions, with the majority of them being first-order or exponential.
- Complete-mix, plug, and dispersed-plug flow regimes are encountered when designing and modeling natural and engineered systems.
  - Complete-mix refers to systems in which their contents are uniform throughout.
  - Plug flow systems are those that are not well-mixed; the fluid particles pass through with little to no longitudinal mixing.
  - Dispersed-plug flow is the flow regime that lies between ideal complete-mix and ideal plug flow.
- Complete-mix batch reactors (CMBR) and complete-mix flow reactors (CMFR) are widely used in engineered systems for treating water and wastewater. They are less susceptible to changes in flow and influent characteristics compared to plug flow reactors (PFR).
  - A plug flow reactor is more efficient and requires a smaller volume than a complete-mix reactor.
  - Engineers often use a series of complete-mix reactors, since they offer the ability to resist shock loadings and approach the efficiency of a plug flow reactor. A minimum of three equally-sized complete-mix reactors should be used in series and preferably, ten, so that plug flow is approached.
- According to the Law of Conservation of Energy (First Law of Thermodynamics), energy can neither be created nor destroyed; only its form can be changed.
  - Heat and work involve energy interactions between the system and its surroundings.
  - The Second Law of Thermodynamics states that energy transformations are not 100% efficient. This inefficiency leads to the loss of useful energy in the form of work; primarily as lost thermal energy or heat.

Key Words

batch reactor
combustion
complete-mix
convection
control volume
detention time
dispersed-plug flow
emissivity
enthalpy
First Law of Thermodynamics
Fourier's law
half-life
heat
internal energy
kinetic energy
kinetics
mass balances
non-steady state
ocean thermal energy conversion (OTEC)
plug flow
potential energy
power
radiation
rate law
reaction order
reactor
Second Law of Thermodynamics
specific heat
steady-state
theory of relativity
thermodynamics
work

References

Cengel, Y.A., Boles, M.A. (2008). *Thermodynamics: An Engineering Approach*, p. 94. McGraw-Hill, New York, NY.

Eisenberg, D., Crothers, D. (1979). *Physical Chemistry with Applications to the Life Sciences*, pp. 212–261. The Benjamin/Cummings Publishing Company, Inc., Menlo Park, CA.

Energy Information Administration (2008). *Annual Energy Review 2008*, p 37. Report No. DOE/EIA, 0384, release date: June 26, 2009. Accessed on March 5, 2012 from: http://www.eia.gov/FTPROOT/multifuel/038408.pdf.

Henry, J.G., Heinke, G.W. (1996). *Environmental Science and Engineering*, pp. 60–64. Prentice Hall, Upper Saddle River, NJ.

Ladon, L. (2001). *Chemical Kinetics*. Accessed on March 5, 2012 from: http://pages.towson.edu/ladon/kinetics.html

Levenspiel, O. (1999). *Chemical Reaction Engineering*, 3rd edition. John Wiley and Sons, Inc., New York, NY.

Masters, G.M. (1998). *Introduction to Environmental Engineering and Science*, p. 25. Prentice Hall, Upper Saddle River, NJ.

Metcalf and Eddy (2003). *Wastewater Treatment and Reuse*, pp. 264–269, 279–282, 1005–1016. McGraw Hill, New York, NY.

Moran, M.J. and Shapiro, H.N. (2004). *Fundamentals of Engineering Thermodynamics*, p. 793. John Wiley & Sons, Inc., Hoboken, NJ.

Nazaroff, W.W., Alvarez-Cohen, L. (2001). *Environmental Engineering Science*, p. 611. John Wiley & Sons, Inc., New York, NY.

Reynolds, T.D., Richards, P.A. (1996). *Unit Operations and Processes in Environmental Engineering*, pp. 53, 58, 60–63, 427–435. PWS Publishing, Boston, MA.

Snoeyink, V., Jenkins, D. (1980). *Water Chemistry*, pp. 27–36, 59–60. John Wiley and Sons, Inc., New York, NY.

Vesilind, P.A., Morgan, S.M. (2004). *Introduction to Environmental Engineering*, 2nd Edition, pp 117–129. Thomson-Brooks/Cole, Belmont, CA.

Problems

1 The owner of a 1,000 ft^3 swimming pool desires to fill it up during a 24 hour period using one garden hose. Determine the following:

a. The necessary flow rate in gallons per minute (gpm) to fill the pool.

b. If two garden hoses flowing at 5 gpm are used, how long would it take to fill up the pool?

2 Four wastewater streams from a textile mill are blended in an equalization tank prior to further treatment. The flow rate and pH of each stream are as follows:

- grey water, 500 gallons at pH = 4.0;
- white water, 1000 gallons at pH = 7.3;
- dye waste, 1500 gallons at pH =11.0;
- kier waste, 500 gallons at pH = 11.8.

Perform a mass balance on the hydrogen ion concentration $[H^+]$ so that the pH of the equalized flow can be determined. Recall that pH = $-\log [H^+]$.

3 Three wastewater streams are combined at a food processing facility to equalize the pH prior to biological treatment. The flow rate and pH of each of the wastewater streams is presented in the following table. Perform a mass balance on flow and the hydrogen ion $[H^+]$ concentration so that the pH of the three combined streams may be determined. The pH of a solution is equal to the negative logarithm of the hydrogen ion concentration (pH = $-\log [H^+]$).

Wastewater stream	Flow (liters per minute)	pH
1	50	5.5
2	200	6.5
3	250	8.5

4 The recycle of various waste streams from unit operations and processes at a wastewater treatment plant (WWTP) can have a significant impact on the influent characteristics to the facility. Most recycle and waste streams from digesters and sludge handling operations are typically blended with the incoming raw wastewater at the headworks of the WWTP. Given the following schematic, determine the actual influent BOD_5 concentration (mg/L) that the facility must actually process.

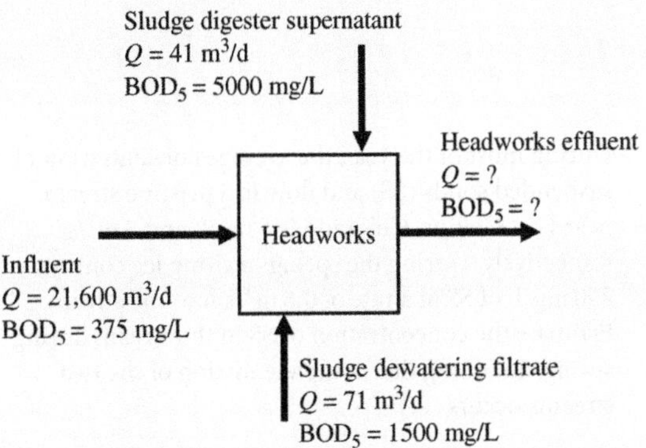

Figure E5.12

5 Raw primary sludge at a solids concentration of 5% is mixed with waste activated sludge (WAS) at a solids concentration of 1.0%. The primary and WAS flows are 29,000 gallons per day and 35,000 gallons per day, respectively. Assume that the specific gravity of both sludges is 1.0 so that a 1.0% solids concentration is equivalent to 10,000 mg/L. Determine the following:

a. The solids concentration (%) for the blended sludges.

b. The volume of thickened sludge and supernatant produced in gallons per day if the blended sludge in Part A is thickened to 8.0% solids. Assume that the thickener removes 100% of the solids that enter.

6 Reverse osmosis (RO) is being used more widely for treating groundwater for human consumption. RO treatment can produce water that essentially has no hardness. Hardness in water is attributed to divalent metallic cations primarily, calcium (Ca^{2+}) and magnesium (Mg^{2+}). A 3.785 mega-liter (ML) RO water treatment plant (WTP) treats groundwater containing 150 mg/L of hardness as $CaCO_3$. The product or

permeate from the RO WTP will be blended with untreated groundwater to achieve a final hardness of 50 mg/L as $CaCO_3$. How much of the untreated groundwater must be blended with permeate from the RO WTP? See the schematic diagram below to assist in the solution of the problem.

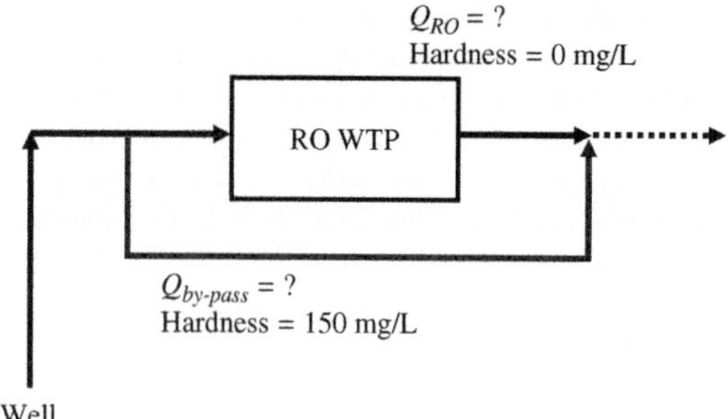

Figure E5.13

7 During most of the year, the average concentration of suspended solids (SS) and flow in a pristine stream near Fort Collins, Colorado is 5 mg/L and 3 m^3/s, respectively. During the spring, melting ice conveys 250 mg/L of SS at a rate of 0.6 m^3/s into the stream. Estimate the concentration of SS in the stream during spring, assuming that complete mixing of the two streams occurs.

8 Covington, Virginia, which is located in southwest Virginia, encounters inversions on a regular basis, as it is surrounded by mountains. Assume that the inversions typically limit a mixing depth of approximately 2,000 feet above the city. The city's land area is approximately 6.0 square miles and particulate emission rate is estimated at 10 lb of particulates per square foot per day. Determine:
   a. The concentration of particulates in the air above the city at the end of a 24 hour period when there is no wind. Assume that the length and width of the land area is approximately 2.45 miles = $\sqrt{6\ \text{mi}^2}$.
   b. The concentration of particulates in the air above the city at the end of a 24 hour period when the wind speed is 20 miles per hour. Assume there is no particulate matter in the air from outside of the city limits.

9 A company has been discharging a non-reactive pollutant at a concentration of 15 g/m^3 to a lagoon for several years. The wastewater flow rate is 0.1 mega-liters (ML) per day and the lagoon detention time is 10 days. The lagoon is assumed to be completely mixed. The overflow from the lagoon discharges into an adjacent river. Determine:
   a. The steady state concentration of the non-reactive pollutant in the effluent from the lagoon that overflows into the river.
   b. The concentration of the non-reactive pollutant in the effluent leaving the lagoon 10 days after the influent concentration is suddenly increased to 150 g/m^3.

10 A lagoon with a volume of 1500 m^3 receives an industrial wastewater flow of 100 m^3/day containing a conservative pollutant at a concentration of 20 mg/L.
   a. Assuming that complete mix conditions are achieved in the lagoon, determine the concentration of the conservative pollutant in the effluent from the lagoon at steady state.
   b. If the influent pollutant concentration is suddenly increased to 60 mg/L, what would the effluent concentration from the lagoon be 7 days later?

11 A lagoon with a volume of 1500 m^3 receives an industrial wastewater flow of 100 m^3/day containing a non-conservative or reactive pollutant at a concentration of 20 mg/L.
   a. Assuming that complete mix conditions are achieved in the lagoon and a reaction rate coefficient is $k = 0.20\ \text{d}^{-1}$, determine the concentration of the non-conservative pollutant in the effluent from the lagoon at steady state.
   b. If the influent pollutant concentration is suddenly increased to 120 mg/L, what would the effluent concentration from the lagoon be 7 days later?

12 Given the following reaction:

$$H_2 + D_2 \rightarrow 2\,HD$$

where D is deuterium or heavy hydrogen, and the rate law is given as follows:

$$r = k[H_2]^{0.38}[D_2]^{0.66}[Ar]$$

Determine:
   a. The reaction order with respect to each reactant.
   b. The total or overall reaction order.

13 Ammonia (NH_3) is often added to water treated with chlorine to form chloramines that provide a longer-lasting chlorine residual than hypochlorous acid (HOCl). Monochloramine (NH_2Cl) is produced according to the following reaction:

$$NH_3 + HOCl \rightarrow NH_2Cl + H_2O$$

The proposed rate law is:

$$r = -k[HOCl][NH_3]$$

Determine:

a. The reaction order with respect to each reactant.
b. The total or overall reaction order.

14 Consider the irreversible conversion of a single reactant (A) to a single product (P) for the following reaction: A → P. Evaluate the following data to determine whether the reaction is zero-, first-, or second-order. Also, determine the magnitude of the rate constant k and list the appropriate units.

Time (minutes)	Concentration of A (g/L)
0	1.00
11	0.50
20	0.25
48	0.10
105	0.05

15 During biological treatment of a synthetic wastewater, the concentration of microorganisms denoted by X was measured as a function of time. The observed microbe concentration at various time intervals is presented in the table below. Determine the reaction order and rate constant k and list the appropriate units. Are the microbes being produced or destroyed?

Time (hours)	Concentration of X (mg/L)
0	40
2.0	60
4.0	110
6.0	200
8.0	325

16 During biological treatment of a domestic wastewater, the concentration of chemical oxygen demand (COD) was measured as a function of time. The observed COD concentration at various time intervals is presented in the table below. Determine the reaction order and rate constant k and list the appropriate units. Is COD being produced or removed?

Time (hours)	COD Concentration (mg/L)
0	493
2.0	310
4.0	181
6.0	99
8.0	60
10.0	30

17 During a chemical reaction, the concentration of Species A was measured as a function of time. The observed concentration of Species A at various time intervals is presented in the table below. Determine the reaction order and rate constant k and list the appropriate units. Is Species A being removed or produced?

Time (min)	Concentration of A (mg/L)
0	0
10	18
20	41
30	59
40	82
50	105

18 During a chemical reaction, the concentration of Species A was measured as a function of time. The observed concentration of Species A at various time intervals is presented in the table below. Determine the reaction order and rate constant k and list the appropriate units. Is Species A being removed or produced?

Time (min)	Concentration of A (mg/L)
0	105
10	78
20	62
30	40
40	21
50	1

19 The concentration of Species D was measured as a function of time during a chemical reaction. The observed concentration of Species D at various time intervals is presented below. Determine the reaction order and rate constant, k. Is Species D being removed or produced?

Time (hr)	Concentration of D (mg/L)
0	200
1.0	160
3.0	105
5.0	65
7.0	45
9.0	30
11.0	20

20 During a chemical reaction, the concentration of Species B was measured as a function of time. The observed concentration of Species B at various time intervals is presented below. Determine the reaction order and rate constant, k. Is Species B being removed or produced?

Time (d)	Concentration of B (g/L)
0	50
2.0	60
4.0	75
6.0	90
8.0	110
10.0	140
12.0	170

21 During a chemical reaction, the concentration of Species C was measured as a function of time. The observed concentration of Species C at various time intervals is presented below. Determine the reaction order and rate constant, *k*. Is Species C being removed or produced?

Time (h)	Concentration of C (mg/L)
0	500
5.0	20
10.0	9.5
15.0	6.5
20.0	5.0
30.0	3.5
40.0	2.5
50.0	2.0
75.0	1.5
100	1.0

22 During a chemical reaction, the concentration of protein P was measured as a function of time. The observed concentration of P at various time intervals is presented below. Determine the reaction order and rate constant, *k*. Is P being removed or produced?

Time (h)	Concentration of P (mg/L)
0	0.50
10.0	0.55
15.0	0.58
20.0	0.62
30.0	0.70
50.0	0.95
100.0	10.00

23 Many biological reactions can be modeled as a mixed-order function commonly referred to as a Monod function, named after the French microbiologist. Monod found that microbial growth can be modeled with the follow equation:

$$\mu = \frac{\mu_{max} S}{K_s + S}$$

where:

μ = specific growth rate of microorganism, d^{-1}

μ_{max} = maximum specific growth rate of microorganism, d^{-1},

S = external substrate concentration, mg/L, and

K_s = half-velocity constant, mg/L.

During a biological reaction, the concentration of an amino acid denoted as *C* was measured as a function of time, along with the rate of production denoted as *r*. The observed concentration of *C* and rate of production of the amino acid at various time intervals is presented below. The biological reaction can be modeled by the following equation analogous to the equation proposed by Monod:

$$r = \frac{kC}{K_s + C}$$

where:

r = rate of amino acid production, d^{-1}

k = maximum rate of amino acid production, d^{-1}

C = amino acid concentration, mg/L

K_s = half-velocity constant of the amino acid, mg/L.

The coefficient k and K_s can be determined by performing a double reciprocal plot of $1/r$ versus $1/C$. Using the data in the table below, rearrange the equation presented above to determine the slope and y-intercept of the line of best fit through the data so that k and K_s

r (d^{-1})	C (mg/L)
0.85	10
1.50	20
2.50	50
3.00	75
3.75	150
4.00	200

24 If the half-life of a vinyl chloride in surface water is 5 days, estimate the concentration of vinyl chloride remaining in a water sample after setting in a laboratory for two weeks if the initial concentration was 20 mg/L.

25 Determine the half-life of a chemical compound if first-order removal kinetics were observed to be followed during a laboratory study. The initial and final concentration of the chemical at time 0 and 10 days later was 10 mg/L and 1.3 mg/L, respectively.

26 If the first-order removal rate constant k for a chemical compound is $0.5\ h^{-1}$, determine the half-life in days.

27 If the half-life of DDT in surface water is 56 days, determine the first-order removal rate constant k in d^{-1}.

28 Several reactor configurations are to be considered for reducing the influent substrate concentration from 200 mg/L to 20 mg/L at a design flow rate of

38,000 m^3/d. Assume that substrate removal follows first-order kinetics and the first-order rate constant k is 6.0 h^{-1}. Determine the reactor volume required for the following configurations operating at steady-state:

a. One ideal plug flow reactor.
b. One ideal complete-mix reactor.
c. Three ideal complete-mix reactors in series.
d. Ten ideal complete-mix reactors in series.

29 Several reactor configurations are to be considered for reducing the influent substrate concentration from 100 mg/L to 15 mg/L at a design flow rate of 5 million gallons per day (MGD). Assume that substrate removal follows first-order kinetics and the first-order rate constant k is 8.0 d^{-1}. Determine the reactor volume required for the following configurations operating at steady-state:

a. One ideal plug flow reactor.
b. One ideal complete-mix reactor.
c. Three ideal complete-mix reactors in series.
d. Ten ideal complete-mix reactors in series.

30 A complete-mix flow reactor is designed to treat an influent waste stream containing 130 mg/L of casein at a flow rate of 380 liters per minute (Lpm). Assume that casein removal follows first-order removal kinetics with a rate constant k of 0.5 h^{-1} and that the effluent should contain 13 mg/L of casein at steady state. Determine:

a. The detention time in hours.
b. The volume of the reactor in cubic meters.

31 A complete-mix flow reactor is designed to treat an influent stream containing 150 mg/L of total organic carbon (TOC) at a flow rate of 150 gallons per minute (gpm). Assume that TOC removal follows first-order removal kinetics with a rate constant k of 0.4 h^{-1}. The volume of the CMFR is 13,500 $ft^3$and steady-state conditions exist. Determine:

a. The detention time in hours.
b. The effluent TOC concentration.

32 A plug flow reactor (PFR) is designed to treat an influent stream containing 200 mg/L of acetic acid at a flow rate of 400 liters per minute (Lpm). The reactor has been in operation for several months and acetic acid removal is observed to follow second-order removal kinetics with a rate constant k of 0.0085 L/(mg·h). Ninety percent acetic acid removal is required. Determine:

a. The detention time in hours.
b. The volume of the plug flow reactor in m^3.

33 A plug flow reactor (PFR) is designed to treat 10 million gallons per day (MGD) of industrial wastewater containing contaminant A. Bench-scale studies indicate that contaminant A removal follows first-order removal kinetics with a reaction rate constant k of 9.0 d^{-1}. Steady-state conditions exist and 95% removal of contaminant A is required. Determine:

a. The detention time in hours.
b. The volume of the plug flow reactor in ft^3.

34 A 1,500 megawatt (MW) coal-fired power plant has a 33% efficiency of converting the energy of coal into electrical energy. The coal has an energy content of 24 kJ/ g and contains 55% carbon, 2.0% sulfur, and 7% ash. Also assume that 65% of the ash in the coal is released as fly ash with 35% of it settling outside of the firing chamber where it is collected as bottom ash. Approximately 15% of the waste heat is assumed to exit in the stack gases, and the cooling water dissipates the remaining heat. Assume that air emission standards restrict sulfur and particulate quantities to 260 g SO_2 per 10^6 kJ of heat input and 13 g particulates per 10^6 kJ of heat input into the coal-fired power plant. Perform a materials and energy balance around the coal-fired power plant to answer the following questions and draw a simplified schematic of the process. Determine:

a. The quantity of heat loss to the cooling water (MW).
b. The quantity of cooling water (kg/s) and flow (m^3/s), assuming a 10°C increase in the temperature of the cooling water.
c. The efficiency of the sulfur dioxide removal system to meet air emission standards.
d. The efficiency of the particulate removal system to meet air emission standards.

35 A solar collector panel with a surface area of 32 ft^2 receives energy from the sun at a rate of 160 BTU per hour per ft^2 of surface area. If 40% of the incoming energy is lost to the surroundings with the remainder going to heat water from 100 to 150°F, determine how many gallons of water at 150°F can be produced by eight solar collector panels in a 30 minute time period. There is a negligible pressure drop through the panel. Neglect potential and kinetic energy effects.

36 A bomb calorimeter is a device used to determine the heat energy value of materials when they are combusted. Typically, a sample is place into a stainless steel ball, to which oxygen under high pressure is added. The bomb is placed into an adiabatic water bath, so that heat cannot be transferred to the surroundings. Wires lead from the bomb to an electrical source so that a spark can be administered to combust the material. Determine the energy content of refuse derived fuel (RDF) if a 10 g sample is placed into a bomb calorimeter that holds 5 L of water. During combustion, the temperature in the water bath increases by 15°C. Ignore the mass of the bomb.

37 Three complete-mix reactors in series treat a municipal wastewater flow rate of 1.0 million gallons per day (MGD) containing 200 mg/L of BOD_5. If the volume of each reactor is 0.5 million gallons and the BOD removal rate coefficient, $k = 0.21\ h^{-1}$, calculate the effluent BOD_5 concentration from the third reactor.

38 A 5 ft × 10 ft solar collector is used to heat water flowing at 2.0 gallons per minute. Assume that 50% of

the sunlight is captured by the collector and the intensity of sunlight is 434 BTU/(ft^2·h). Determine the temperature of the water exiting the solar collector if the feed water temperature is 55°F.

39 Approximately two-thirds of the energy content of uranium fuel entering a 2,000 MW nuclear power plant is removed by condenser cooling water which is withdrawn from an adjacent stream. The flow upstream of the nuclear power plant is 200 m^3/s and the temperature in the stream is 19°C. Perform an energy balance on the nuclear power plant to answer the following questions and draw a simplified schematic of the process. Determine:

a. The energy input to the nuclear power plant.
b. The necessary flow rate in the stream if the temperature in the cooling water is only allowed to rise 10°C in temperature.
c. The temperature in the stream after the heated cooling water is released back into the stream.

Chapter 6

Design of water treatment systems

Richard O. Mines, Jr.

Learning Objectives

After reading this chapter, you should be able to:

- list the major objectives of water treatment;
- select appropriate unit operations and processes necessary to meet primary drinking water standards;
- design conventional unit operations and processes for public drinking water systems;
- design advanced unit operations and processes for public drinking water systems;
- design sludge thickening and dewatering systems for water treatment residuals;
- select the appropriate disinfectant for killing pathogens.

6.1 Drinking water standards

Standards for drinking water vary according to the type and size of water system. A community water system (CWS) is one that serves the same people year-round. Examples of CWS include those in cities and small towns that provide water to residences, homes, condominiums, and mobile home parks. A non-community water system (NCWS) serves the public but does not serve the same people on a year-round basis. A non-transient, non-community water system (NTNCWS) is defined as one which serves the same people for six months but not year round. A school that has its own water supply would be placed in this category. Rest areas and campgrounds are considered transient, non-community (TNCWS) because they serve the public but do not serve the same individuals for more than six months.

Drinking water standards for various contaminants have been established by the US Environmental Protection Agency (USEPA), as authorized by the Safe Drinking Water Act (SDWA), originally passed in 1974. The Act was amended in 1986 and again in 1996, and it does not regulate private wells that serve less than 25 people. The US EPA has the power to set the maximum contaminant level (MCL) for various contaminants and to enforce them if individual states give up their state primacy.

Each standard also has requirements for water systems to test for various contaminants in the water, and to make sure compliance is being achieved. Currently, community water systems and non-transient, non-community water systems monitor more than 83 contaminants. Maximum contaminant level goals (MCLGs) are also set by US EPA. MCLGs are established at a level of a contaminant below which there is not known or expected risk to health. In many cases, MCLGs have been set at zero. However, they are not-enforceable. As treatment technology evolves, resulting in higher levels of removal for various contaminants, MCLGs are often changed to MCLs.

For some contaminants, US EPA has established a treatment technique (TT) in lieu of a maximum contaminant level, because the best technology is not feasible, either economically or technically, and/or when the detection method for a specific contaminant is not reliable or economical. If research on a regulated contaminant indicates that it no longer poses a health hazard, it is removed from the list of primary drinking water standards. Drinking water standards are

Environmental Engineering: Principles and Practice, First Edition. Richard O. Mines, Jr.

continuously reviewed and updated by EPA. The most current standards may be found at the Office of Water at the US EPA website (www.epa.gov).

Drinking water standards are classified either as primary or secondary. Primary standards are established because they are directly related to health. States must adopt them or have regulations that are equal to or more stringent than the primary standard (this is known as State Primacy). Primary standards are enforceable. Secondary standards, however, are non-enforceable and are related to the aesthetics of the water, i.e., color, odor, and taste. Although, they are non-enforceable, most municipal water suppliers and purveyors meet secondary standards because they want to provide consumers with finished water that is aesthetically pleasing.

6.1.1 Primary drinking water standards

Primary standards are established to insure safe, finished water. Primary standards include microbiological organisms (M), disinfection by-products (DBP), organic chemicals (OC) – many of which are Priority Pollutants, inorganic chemicals (IOC), radionuclides (R), and turbidity. The contaminants listed in Tables 6.1A through 6.1F are currently regulated by the US EPA as Primary Drinking Water Standards. Table 6.2 is a list of treatment processes or recommendations made by the US EPA for removing various contaminants that are presented in Tables 6.1A–6.1F.

6.1.2 Secondary drinking water standards

Secondary Drinking Water Standards are not related to health and are, therefore, non-enforceable by the US EPA. Secondary contaminants cause a variety of problems that can be categorized as aesthetic, cosmetic, and technical effects. They relate to the aesthetics of the water, making it more desirable to the consumer. Table 6.3 is a listing of the secondary drinking water standards.

6.1.3 Other rules promulgated by EPA

Other rules promulgated by US EPA are described below, in order from the most recent. Regulations that are briefly discussed include: Ground Water Rule; Long Term 2 Enhanced Surface Water Treatment Rule; Stage 2 Disinfectants and Disinfection By-products Rule; Long Term 1 Enhanced Surface Water Treatment Rule; Filter Backwash Recycling Rule; Arsenic Rule; Stage 1 Disinfectants and Disinfection By-products Rule; Interim Enhanced Surface Water Treatment Rule; Lead and Copper Rule; Total Coliform Rule; and Surface Water Treatment Rule.

6.1.3.1 Ground Water Rule (GWR)

The Ground Water Rule was signed on October 11, 2006 and published in the Federal Register on November 8, 2006. Its

Table 6.1A Primary Drinking Water Standards.

Parameter	Category	MCL	MCLG	BAT
Acrylamide	OC	TT	Zero	LU
Alachlor	OC	0.002 mg/L	Zero	GAC
Alpha particles	R	15 pCi/L	Zero	NA
Antimony	IOC	0.006 mg/L	0.006 mg/L	C/F, RO
Arsenic	IOC	0.01 mg/L	Zero	AA, AX, MBIX, GF, O/C/F, LS, RO
Asbestos (fiber ≥ 10 1 10 μm)	IOC	7 million fibers/L (MFL)	7 MFL	C/F, CC, DF, DEF
Atrazine	OC	0.003 mg/L	0.003 mg/L	GAC
Barium	IOC	2 mg/L	2 mg/L	IX, RO, LS, ED
Benzene	OC	0.005 mg/L	Zero	GAC, PTA
Benzo(a)pyrene (PAHs)	OC	0.0002 mg/L	Zero	GAC
Beryllium	IOC	0.004 mg/L	0.004 mg/L	AA, C/F, IX, LS, RO
Beta particles and photon emitters	R	4 mRems/yr	Zero	C/F, RO, IX
Bromate	DBP	0.010 mg/L	Zero	MDP
Cadmium	IOC	0.005 mg/L	0.005 mg/L	C/F, IX, LS, RO
Carbofuran	OC	0.04 mg/L	0.04 mg/L	GAC
Carbon tetrachloride	OC	0.005 mg/L	Zero	GAC, PTA

Source: http://water.epa.gov/drink/contaminants/index.cfm#List. United States Environmental Protection Agency.

Table 6.1B Primary Drinking Water Standards.

Parameter	**Category**	**MCL**	**MCLG**	**BAT**
Chloramines (as Cl_2)	D	MRDL = 4.0 mg/L	MRDLG = 4.0 mg/L	RDD, CDTP
Chlordane	OC	0.002 mg/L	Zero	GAC
Chlorine (as Cl_2)	D	MRDL = 4.0 mg/L	MRDLG = 4.0 mg/L	RDD, CDTP
Chlorine dioxide (as ClO_2)	D	MRDL = 0.8 mg/L	MRDLG = 0.8 mg/L	RDD, CDTP
Chlorite	DBP	1.0 mg/L	0.8	MDP
Chlorobenzene	OC	0.1 mg/L	0.1 mg/L	GAC/ PTA
Chromium (total)	IOC	0.1 mg/L	0.1 mg/L	C/F, IX, RO, LS
Copper	IOC	TT, Action level=1.3 mg/L	1.3 mg/L	WQPM, PE, SOWT, CCT
Cryptosporidium	M	TT	Zero	NA
Cyanide (as free cyanide)	IOC	0.2 mg/L	0.2 mg/L	GAC, PTA
2,4-D	OC	0.07 mg/L	0.07 mg/L	GAC
Dalapon	OC	0.2 mg/L	0.2 mg/L	GAC
1,2-Dibromo-3-chloropropane (DBCP)	OC	0.0002 mg/L	Zero	GAC, PTA
o-Dichlorobenzene	OC	0.6 mg/L	0.6 mg/L	GAC & PTA
p-Dichlorobenzene	OC	0.075 mg/L	0.075 mg/L	GAC & PTA
1,2-Dichloroethane	OC	0.005 mg/L	Zero	GAC & PTA

Source: http://water.epa.gov/drink/contaminants/index.cfm#List. United States Environmental Protection Agency.

Table 6.1C Primary Drinking Water Standards.

Parameter	**Category**	**MCL**	**MCLG**	**BAT**
cis-1,2-Dichloroethylene	OC	0.07 mg/L	0.07 mg/L	GAC & PTA
trans-1,2-Dichloroethylene	OC	0.1 mg/L	0.1 mg/L	GAC & PTA
Dichloromethane	OC	0.005 mg/L	Zero	GAC & PTA
1,2-Dichloropropane	OC	0.005 mg/L	Zero	GAC & PTA
Di-(2-ethylhexyl) adipate	OC	0.4 mg/L	0.4 mg/L	GAC
Di-(2-ethylhexyl) phthalate	OC	0.006 mg/L	Zero	GAC
Dinoseb	OC	0.007 mg/L	0.007 mg/L	GAC
Dioxin (2,3,7,8-TCDD)	OC	3×10^{-8}	Zero	GAC
Diquat	OC	0.02 mg/L	0.02 mg/L	GAC
Endothall	OC	0.1 mg/L	0.1 mg/L	GAC
Endrin	OC	0.002 mg/L	0.002 mg/L	GAC
Epichlorohydrin	OC	TT	Zero	CUCA
Ethylbenzene	OC	0.7 mg/L	0.7 mg/L	GAC
Ethylene dibromide	OC	0.00005 mg/L	Zero	GAC
Fecal coliform and *E. coli*	M	MCL	Zero	NA
Fluoride	IOC	4.0 mg/L	4.0 mg/L	D, RO

Source: http://water.epa.gov/drink/contaminants/index.cfm#List. United States Environmental Protection Agency.

Table 6.1D Primary Drinking Water Standards.

Parameter	Category	MCL	MCLG	BAT
Giardia lamblia	M	TT	Zero	NA
Glyphosate	OC	0.7 mg/L	0.7 mg/L	GAC
Haloacetic acids	DBP	0.060 mg/L	N/A	OMR, MDP
Heptachlor	OC	0.0004 mg/L	Zero	GAC
Heptachlor epoxide	OC	0.0002 mg/L	Zero	GAC
Heterotrophic plate count (HPC)	M	TT	N/A	NA
Hexachlorobenzene	OC	0.001 mg/L	Zero	GAC
Hexachlorocyclopentadiene	OC	0.05 mg/L	0.05 mg/L	GAC & PTA
Lead	IOC	TT, Action level = 0.015 mg/L	Zero	WQPM, PE, SOWT, CCT, LSLM, LSLR
Legionella	M	TT	Zero	NA
Lindane	OC	0.0002 mg/L	0.0002 mg/L	GAC
Mercury (inorganic)	IOC	0.002 mg/L	0.002 mg/L	C/F, GAC, LS, RO
Methoxychlor	OC	0.04 mg/L	0.04 mg/L	GAC
Nitrate (as N)	IOC	10 mg/L	10 mg/L	IX, RO, ED
Nitrite (as N)	IOC	1 mg/L	1 mg/L	IX, RO
Oxamyl (Vydate)	OC	0.2 mg/L	0.2 mg/L	GAC

Source: http://water.epa.gov/drink/contaminants/index.cfm#List. United States Environmental Protection Agency.

Table 6.1E Primary Drinking Water Standards.

Parameter	Category	MCL	MCLG	BAT
Pentachlorophenol	OC	0.001 mg/L	Zero	GAC
Picloram	OC	0.5 mg/L	0.5 mg/L	GAC
Polychlorinated biphenyls (PCBs)	OC	0.0005 mg/L	Zero	GAC
Radium 226 and Radium 228 (combined)	R	5 pCi/L	Zero	MBIX, GF, LS, RO
Selenium	IOC	0.05 mg/L	0.05 mg/L	AA, C/F, LS, RO, ED
Simazine	OC	0.004 mg/L	0.004 mg/L	GAC
Styrene	OC	0.1 mg/L	0.1 mg/L	GAC & PTA
Tetrachloroethylene	OC	0.005 mg/L	Zero	GAC & PTA
Thallium	IOC	0.002 mg/L	0.0005 mg/L	AA, IX
Toluene	OC	1 mg/L	1 mg/L	GAC & PTA
Total coliform (including fecal coliform and *E. coli*)	M	5.0%	Zero	NA
Total trihalomethanes (TTHMs)	DBP	0.080 mg/L	N/A	EC/ES
Toxaphene	OC	0.003 mg/L	Zero	GAC
2,4,5-TP (Silvex)	OC	0.05 mg/L	0.05 mg/L	GAC
1,2,4-Trichlorobenzene	OC	0.07 mg/L	0.07 mg/L	GAC & PTA
1,1,1-Trichloroethane	OC	0.2 mg/L	0.2 mg/L	GAC & PTA

Source: http://water.epa.gov/drink/contaminants/index.cfm#List. United States Environmental Protection Agency.

Table 6.1F Primary Drinking Water Standards.

Parameter	Category	MCL	MCLG	BAT
1,1,2-Trichloroethane	OC	0.005 mg/L	0.003 mg/L	GAC & PTA
Trichloroethylene	OC	0.005 mg/L	Zero	GAC & PTA
Turbidity	M	TT	N/A	NA
Uranium	R	30 μg/L	Zero	AA, AX, MBIX, O/C/F, RO
Vinyl chloride	OC	0.002 mg/L	Zero	PTA
Viruses (enteric)	M	TT	Zero	NA
Xylenes (total)	OC	10 mg/L	10 mg/L	GAC & PTA

Best available technology = BAT
Control disinfectant treatment processes = CDTP
Controlled use of coagulant aids = CUCA
Disinfectants = D
Disinfection by products = DBP
Inorganic chemicals = IOC
Limit use = LU
Maximum contaminant level = MCL
Maximum contaminant level goal = MCLG
Maximum residual disinfection level goal = MRDLG
Microorganisms = M
Modification of Disinfection Practice = MDP
Not applicable = NA
Organic chemicals = OC
Organic matter removal = OMR
Radionuclides = R
Reduce disinfectant demand = RDD
Source: http://water.epa.gov/drink/contaminants/index.cfm#List. United States Environmental Protection Agency.

purpose is to reduce disease incidence associated with microbial contaminants in drinking water, and it applies to all systems that use ground water as a source of drinking water. The rule addresses risks through a risk-targeting approach that relies on four components:

- periodic survey of ground water systems that require evaluation of eight critical elements and identification of deficiencies;
- source water monitoring to test for the presence of *E. coli*, enterococci, or coliphage in the sample;
- corrective actions for systems with a significant deficiency or source water fecal contamination;
- compliance monitoring to ensure at least a 4-log (99.99%) inactivation or removal of viruses.

6.1.3.2 Long-Term 2 Enhanced Surface Water Treatment Rule (LT2)

The LT2 rule is to reduce illness linked with the contaminant *Cryptosporidium* and other disease-causing microorganisms. It was published in the Federal Register on January 5, 2006. Under this rule, systems will monitor their water sources to determine treatment requirements. An initial monitoring period of two years of monthly sampling for *Cryptosporidium* is required of large systems. Small filtered water systems can first monitor for *E. coli* and, should their results exceed specified levels, then *Cryptosporidium* monitoring is required. Systems must conduct a second round of monitoring six years after completing the initial round of monitoring. Filtered water systems will be classified in one of four treatment categories (bins) based on their monitoring results. Systems that store treated water in open reservoirs must either cover the reservoir or treat the reservoir discharge to accomplish inactivation of the following organisms: 4-log reduction of viruses, 3-log reduction of *Giardia lamblia*, and 2-log reduction of *Cryptosporidium*.

6.1.3.3 Stage 2 Disinfection and Disinfection By-products Rule (Stage 2 DBP)

The Stage 2 DBP rule focuses on public health protection by limiting exposure to disinfection by-products, specifically total trihalomethanes (TTHMs) and five haloacetic acids (HAA5). It was published in the Federal Register on January 4, 2006. This rule applies to all community water systems and non-transient, non-community water systems that add a primary or residual disinfectant other than ultraviolet (UV) light or deliver water that has been disinfected by a primary or residual disinfectant other than UV. Systems must conduct an evaluation of their distribution systems to identify locations with high disinfection by-product concentrations. Compliance with the MCLs for TTHM and HAA5 is determined for each monitoring location and is referred to as the locational running annual average (LRAA). The rule also requires each system to determine if they have exceeded an operational evaluation level and, if so, to submit a report identifying actions that may be taken to mitigate future high DBP levels.

Table 6.2 Best Available technologies and acronyms.

Treatment process or recommendation	Abbreviation
Activated alumina	AA
Anion exchange	AX
Corrosion control	CC
Corrosion control treatment	CCT
Coagulation filtration	C/F
Diatomaceous earth filtration	DEF
Direct filtration	DF
Distillation	D
Enhanced coagulation, enhanced softening	EC/ES
Electrodialysis	ED
Greensand filtration	GF
Ion exchange	IX
Lime softening	LS
Lead service line monitoring	LSLM
Lead service line replacement	LSLR
Mixed bed ion exchange	MBX
Oxidation, coagulation, filtration	O/C/F
Public education	PE
Packed tower aeration	PTA
Reverse osmosis	RO
Surface water treatment	SWT
Water quality parameter monitoring	WQPM

Source: United States Environmental Protection Agency.

6.1.3.4 Long Term 1 Enhanced Surface Water Treatment Rule (LT1)

The US EPA finalized the Long Term 1 Enhanced Surface Water Treatment Rule on January 14, 2002. The purposes of this rule are to improve the control of microbial pathogens in drinking water – specifically the protozoan *Cryptosporidium* – and address risk trade-offs with disinfection by-products. Systems that use surface water or ground water under the direct influence (GWUDI) of surface water, and serve fewer than 10,000 people, must comply with the rule.

There are four major provisions to the rule:

- Systems must achieve a 2-log reduction (99%) removal of *Cryptosporidium.*
- Filtered systems must comply with strengthened combined filter effluent turbidity performance requirements to assure a 2-log reduction in *Cryptosporidium*, and both conventional and direct filtration systems must continuously monitor the turbidity from each individual filter.
- Systems are required to develop a profile of microbial inactivation levels unless they perform monitoring which demonstrates their DBP levels are less than 80% of the MCLs established in the Stage 1 DBPR. They must determine their current lowest level of microbial inactivation and consult with the state before making significant changes to their disinfection practice.
- finished water reservoirs must be covered if construction begins 60 days after promulgation of the rule, and unfiltered water systems must comply with updated watershed control requirements that add *Cryptosporidium* as a pathogen of concern.

Table 6.3 Secondary Drinking Water Standards.

Parameter	Standard	Parameter	Standard
Aluminum	0.05–0.2 mg/L	Manganese	0.05 mg/L
Chloride	250 mg/L	Odor	3 TON
Color	15 c.u.	pH	6.5–8.5
Copper	1.0 mg/L	Silver	0.1 mg/L
Corrosivity	Noncorrosive	Sulfate	250 mg/L
Fluoride	2 mg/L	Total dissolved solids	500 mg/L
Foaming agents	0.5 mg/L	Zinc	5 mg/L
Iron	0.3 mg/L		

TON = threshold odor number
c.u. = color units based on platinum-cobalt
Source: http://water.epa.gov/drink/contaminants/index.cfm#List. United States Environmental Protection Agency.

6.1.3.5 Filter Backwashing Recycling Rule (FBRR)

In May 2001, US EPA released a rule governing the process of recycling wash water from the backwashing of drinking water filters. The Filter Backwash Recycling Rule (FBRR) is required by the Safe Drinking Water Act as a method of reducing potential risks to consumers from microbial contaminants. The rule requires that recycled filter backwash water, sludge thickener supernatant, and liquids from dewatering operations be returned to a location within the treatment train such that all processes including coagulation, flocculation, sedimentation, and filtration are used. If a system uses direct filtration, then sedimentation can be omitted.

6.1.3.6 Arsenic Rule

The US EPA adopted a new standard for arsenic on January 22, 2001, replacing the old standard of 50 parts per billion (ppb) with 10 ppb. The rule became effective on February 22, 2002. The rule requires monitoring for new systems and new drinking water sources, and also clarifies the procedures for compliance with MCLs established for inorganic contaminants (IOCs), synthetic organic chemicals (SOCs), and volatile organic chemicals (VOCs).

6.1.3.7 Stage I Disinfection and Disinfection By-products Rule (Stage I DBP)

On December 16, 1998, the US EPA published the final Stage 1 Disinfection and Disinfection By-products Rule (Stage 1 DBP). The maximum contaminant level for total trihalomethanes (TTHMs) was lowered from 0.10 to 0.080 mg/L. TTHMs are the sum of the concentration of chloroform, bromodicholoromethane, dibromochloromethane, and bromoform.

New standards were established for the following contaminants: 0.060 mg/L for five haloacetic acids (HAA5), 0.0101 mg/L for bromate, and 1.0 mg/L for chlorite. Haloacetic acids are the sum of monochloroacetic acid, dichloroacetic acid, trichloroacetic acid, monodibromoacetic acid, and dibromoacetic acid.

Maximum residual disinfection levels (MRDLs) were established for the following disinfectants: 4.0 mg/L for chlorine, 4.0 mg/L as total chlorine for chloramines, and 0.8 mg/L for chlorine dioxide.

Water systems that use surface water or ground water under the direct influence of surface water and use conventional filtration treatment are required to remove specific percentages of natural organic matter (NOM) as measured by total organic carbon (TOC). Removal is achieved by using enhanced coagulation or enhanced softening unless alternative criteria are met.

Enhanced coagulation is the term used to define the process of obtaining improved removal of DBP precursors by conventional treatment. Four processes are listed in the *Enhanced Coagulation and Enhanced Precipitative Softening Guidance Manual* (US EPA, 1999, p. A–24) as the most effective for NOM removal: coagulation/filtration, particularly at low pH; precipitative softening, particularly at high pH; granular activated carbon (GAC) adsorption; and membrane processes.

6.1.3.8 Interim Enhanced Surface Water Treatment Rule

The US EPA implemented the Interim Enhanced Surface Water Treatment Rule on December 16, 1998. This improves the control of microbial contaminants, particularly *Cryptosporidium*, in drinking water systems serving 10,000 persons or more and that use surface or ground water under the direct influence of surface water. A maximum contaminant goal (MCLG) of zero was established for *Cryptosporidium*, along with a 2-log reduction in *Cryptosporidium* for systems that use filtration.

6.1.3.9 Lead and Copper Rule

On June 7, 1992, US EPA published a regulation to control lead and copper in drinking water. The final revision to the Lead and Copper Rule was published on October 10, 2007. The rule requires all community water systems (CWS) and non-transient, non-community water systems (NTNCWS) to comply with the rule. An action level (AL) of 0.15 mg/L for lead and 1.3 mg/L for copper is based on 90th percentile level of tap water samples. An action level exceedance is not a violation, but it can trigger other requirements, such as corrosion control treatment (CCT), lead service line replacement (LSLR), public education, source water monitoring and treatment, and water quality parameter (WQP) monitoring.

6.1.3.10 Total Coliform Rule

The total coliform rule (TCR) was published in 1989 and became effective in 1990. The rule establishes both MCLGs and MCLs for the presence of total coliform in drinking water. All public water systems (PWS) must monitor for the presence of total coliforms in the distribution system at a frequency proportional to the number of people served. If any sample tests positive for total coliforms, the system must perform additional testing to include:

- determining the presence of either fecal coliforms or *Escherichia coli*;
- taking one set of 3–4 repeat samples at sites located within five or fewer sampling sites adjacent to the location of the routine positive sample within 24 hours;
- taking at least five routine samples during the next month of operation.

6.1.3.11 Surface Water Treatment Rule (SWTR)

The SWTR was promulgated on June 28, 1989 and seeks to prevent waterborne diseases caused by viruses *Legionella*, and *Giardia lamblia*. The rule requires public water systems that use surface water sources and groundwater sources, under the direct influence of surface water (GWUDI), to filter and disinfect the water. Under this rule, the US EPA established MCLGs of zero for *Giardia lamblia*, *Legionella*, and viruses.

6.2 Overview of typical processes used for contaminant removal

In addition to the best available technologies (BAT) listed with each Primary Drinking Water Standard in Table 6.1A–F, the following discussion presents some of the typical processes used for removing various contaminants from water and some of the parameters that may be adjusted by chemical addition.

6.2.1 Alkalinity

The alkalinity of the water may be insufficient to resist a change in pH when certain coagulants or coagulant aids are added to the water. Generally, lime (CaO), caustic (NaOH), or soda ash (Na_2CO_3) is added to the water to increase the alkalinity.

6.2.2 Arsenic

The US EPA website lists the following BATs for treating arsenic in water: lime softening (LS), anion exchange (AX), reverse osmosis (RO), granular activated carbon (GAC), granular ferric hydroxide (GFH), and activated alumina (AA). Arsenic dissolved in water is found as arsenite (As^{3+}) or arsenate (As^{5+}).

Coagulants such as alum, ferric chloride, ferric sulfate, and lime have been successfully used to remove arsenic from water

(MWH, 2005, p. 1561). However, arsenic must be in the form of As (V) for coagulation to be effective. If in the form of arsenite, As (III), first it must be oxidized to As (V) by adding chlorine or permanganate.

According to ASCE/AWWA (1990, p. 494), ferric sulfate coagulation at a pH from 6–8, alum coagulation at a pH from 6–7, and lime softening at a pH > 10.5, have been effective at removing As (V). The *Recommended Standards for Water Works* (GLUMRB, 2007, pp. 21–22) list several methods for arsenic removal:

- Adsorption onto filter media coated with iron, titanium, or aluminum.
- Chemical oxidation with chlorine, potassium permanganate, ozone, or manganese dioxide with filtration through greensand, anthracite, pyrolusite, or other proprietary filter media.
- Coagulation with ferric chloride, ferric sulfate, or alum followed by settling and filtration.
- anion exchange, electrodialysis, membrane filtration, and lime softening (LS).

6.2.3 Barium

Barium (Ba^{2+}) can be removed effectively by lime softening. Greater than 90% removals have been achieved in the pH range of 10–11 on groundwater containing 7–8.5 mg/L of barium (ASCE/AWWA, 1990, p. 496). Kawamura (1991, p. 566) indicates that barium can be removed either by lime softening at pH of 10–11 or by ion exchange with a sodium regenerate cycle.

6.2.4 Cadmium

Laboratory and pilot-scale studies have shown that greater than 98% removal of cadmium (Cd^{2+}) by lime softening can be achieved in the pH range of 8.5–11.3 for samples with an initial cadmium concentration of 0.3 mg/L. Ferric sulfate ($Fe_2(SO_4)_3$) and alum coagulation have also been used for cadmium removal, but their use is not as effective as lime softening (ASCE/AWWA, 1990, p. 496; Kawamura, 1991, p. 566).

6.2.5 Chromium

Chromium (Cr (III)) can be removed easily using alum or iron salts and by lime softening (ASCE/AWWA, 1990, p. 497). Aluminum and iron salts or lime softening are not effective at removing Cr (VI). Ferrous sulfate first must be used to reduce Cr (VI) to Cr (III) for these coagulants to be effective.

6.2.6 Color

Color removal traditionally has been accomplished by oxidation with chlorine, chlorine dioxide, oxygen, or potassium permanganate. Chlorine and chlorine dioxide are not now being used as often, due to the formation of trihalomethanes and other disinfection by-products such as chlorite (MWH, 2005, p. 513).

Granular activated carbon (GAC) and powdered activated carbon (PAC) can also be used for color removal. Coagulation of dissolved and colloidal species causing color may be accomplished with alum, copperas, or ferric salts.

6.2.7 Fluoride

Fluoride (F^-) can be removed during lime softening if the magnesium concentration is high, according to ASCE/AWWA (1990, p. 498). Alum coagulation at high dosages will remove fluoride, but is not considered economical. Ion exchange with activated alumina or lime coagulation at pH of 9.5–11.3 is recommended by Kawamura (1991, p. 566).

6.2.8 Gases

Dissolved gases that may be found in groundwater include hydrogen sulfide (H_2S), carbon dioxide (CO_2), methane (CH_4), and volatile organic chemicals (VOCs). Stripping of these gases is normally accomplished by tray aeration, packed tower aeration (PTA), cascade aeration, fountains, and spray nozzles. Diffused aeration is sometimes used instead of air stripping. Hydrogen sulfide can be oxidized with chlorine, ferrous sulfate ($FeSO_4 \cdot 7\,H_2O$), potassium permanganate ($KMnO_4$), or hydrogen peroxide (H_2O_2). Carbon dioxide is typically removed by air stripping when the concentration exceeds 10 mg/L; otherwise, lime may be added to neutralize the aqueous carbon dioxide which is in equilibrium with carbonic acid (H_2CO_3).

6.2.9 Iron

Dissolved iron (Fe (II)) must be oxidized to Fe (III) before it can be precipitated and removed by sedimentation and filtration. Oxygen (O_2), chlorine (Cl_2), ozone (O_3), chlorine dioxide (ClO_2), and potassium permanganate have all been used as the oxidizing agent.

6.2.10 Lead

The stable solid phase of lead depends on the pH and alkalinity of the water. In *Water Treatment: Principles and Design* (MWH, 2005, p. 1783), dissolved plumbic ion (Pb^{2+}) is the prevalent and stable lead form at low pH. Lead carbonate, $PbCO_3$, is favored at neutral pH, while lead hydroxycarbonate, $Pb_3(OH)_2(CO_3)_2$, or lead hydroxide, $Pb(OH)_2$, are formed at higher pH values. The solubility of lead increases dramatically as pH decreases below a pH of 8.

Lead can be removed by coagulation and lime softening (ASCE/AWWA, 1990, p. 499). Alum and ferric sulfate coagulation indicated >97% removal of 0.15 mg/L of lead within a pH range of 6–10. Lime softening experiments demonstrated >98% removal of lead at an initial concentration of 0.15 mg/L throughout a pH range of 8.5–11.3. Kawamura (1991, p. 566) also indicates that lead can be removed by ferrous sulfate addition at a pH of 7–9, alum coagulation at a pH of 6–9, and lime softening at a pH of 9.5 to 11.3.

6.2.11 Manganese

Manganese (Mn (II)) is the dissolved form; like iron, it must be oxidized (to Mn (IV)) in order to be precipitated and removed from water.

6.2.12 Mercury

Mercury is present in both inorganic and organic forms. Inorganic forms include mercuric chloride ($HgCl_2$), mercuric sulfide (HgS), and mercuric acetate ($Hg(C_2H_3O_s)_2$). 97% removal of inorganic mercury at a concentration of 0.05 mg/L was accomplished at a pH of 8, although only 66% removal of mercury occurred at pH 7 (ASCE/AWWA, 1990, p. 499). Kawamura (1991, p. 566) stated that inorganic mercury can be removed by coagulation with ferric sulfate at a pH of 7–8 and by lime softening at a pH > 11.

Organic mercury primarily exists as methyl mercury ion $(CH_3Hg)^+$ and ethyl mercury $(C_2H_5Hg)^+$. According to ASCE/AWWA (1990, p. 499), organic mercury in the form of methyl mercury ion is the more prevalent and toxic form, with <40% removal with alum and iron coagulation. Higher removals occur when higher levels of turbidity are present due to adsorption of the mercury onto solids. Granular and powdered activated carbons (PAC) are effective at removing organic mercury from water (ASCE/AWWA, 1990, p. 509).

6.2.13 Nitrate

Nitrate (NO_3^-) is relatively unaffected by conventional treatment consisting of coagulation/flocculation/sedimentation (ASCE/AWWA, 1990, p. 499). Ion exchange is the simplest and lowest-cost method of removing nitrate from contaminated groundwater (AWWA, 1999, p. 9.37). Nitrate may also be reduced by biological denitrification, a process commonly used in treating wastewater. Reverse osmosis will reduce nitrate to low levels, but this is generally used for high TDS and salt water (MWH, 2005, p. 254). Anion exchange, reverse osmosis, nanofiltration, and electrodialysis are treatment processes recommended for the removal of nitrate/nitrite by the Great Lakes-Upper Mississippi River Board of State and Provincial Public Health and Environmental Mangers (2007, p. 27).

6.2.14 pH

Often, the pH of the water must be adjusted by adding either an acid or base. Sulfuric acid (H_2SO_4) and phosphoric acid (H_3PO_4) can be added to lower the pH, whereas lime or caustic soda is added to raise the pH.

6.2.15 Radionuclides

The principal source of natural radionuclides is uranium ore (U_3O_8). Anthropogenic sources include nuclear power plants, research facilities, medical facilities that use nuclear medicine and X-rays services, TVs, smoke detectors, and nuclear weapons (MWH, 2005, p. 1627). In full-scale experiments, lime softening removed 70–95% of the influent radium concentration (ASCE/AWWA, 1990, p. 499).

The best available treatment for radon-222 is aeration, while combined radium-226 and radium-228 treatment consists of coagulation and flocculation. When radium-222 and radium-228 are removed separately, ion exchange, reverse osmosis (RO), and lime softening are the best technologies to use. Uranium is best removed by ion exchange.

6.2.16 Taste and odor

Taste and odor compounds may be related to organic compounds, secretions from algae, and/or inorganic compounds. Activated carbon is the most effective means of removing these compounds, whereas chlorine oxidation is the least expensive method. Chlorine, chlorine dioxide, PAC, GAC, copper sulfate, aeration, potassium permanganate, and ozone are treatment processes to be considered for the removal of taste and odor compounds in water, according to the Great Lakes-Upper Mississippi River Board of State and Provincial Public Health and Environmental Mangers (2007, pp 82–83).

6.2.17 Total organic carbon (TOC)

Investigators have found that TOC can be removed by lime softening and/or coagulation; however, coagulation is more effective than just using lime softening. Benefield & Morgan (1999, p. 10.51) make several generalizations regarding the removal of TOC, color, and disinfection by-product (DBP) precursors during water softening. They state that precipitation of calcium carbonate generally removes 10–30% of the TOC, DBP precursors, and color. Magnesium hydroxide precipitation removes approximately 30–60% of TOC and DBP precursors, and 50–80% of the color. Adding iron and aluminum coagulants during softening generally results in additional removals ranging from 5–15% for color, TOC, and DBP precursors for either calcium or magnesium precipitation.

6.3 Design flows and capacities

Engineers must establish various flow rates that will be used to design the unit operations and processes at the water treatment plant. It is essential to provide flexibility during design, such as including parallel treatment trains, redundant pumps and other equipment, and multiple injection points for chemicals and disinfectants. This is very important when the characteristics of

Table 6.4 Flow Ratios with respect to Average Daily Demand.

Flow ratio	Range	Average
Peak daily demand: average daily demand	1.5:1 to 3.5:1	2.0:1
Peak hourly demand: average daily demand	2:1 to 7.0:1	4.5:1
Minimum hourly demand: average daily demand*	0.25:1 to 0.5:1	N.A.

N.A. – Not available
Flow ratios from Shammas & Wang (2011) p. 133 and *McGhee (1991), p. 14.

the water source vary significantly, such as in a river or a lake. These sources may be seriously impacted by events such as the turn-over of a lake during the change of seasons, or significant rainfall events.

6.3.1 Water demand

According to the US Geological Survey, approximately 44,200 Mgal/d of water was withdrawn for public water supply serving an estimated 258 million people in the calendar year 2005 (Kenny *et al.*, 2009, p. 16). This converts to an average daily demand of 171 gallons per capita per day (647 Lpcd). Public water supply refers to water withdrawn by public and private water suppliers that provide water to at least 25 people or have a minimum of 15 connections. Public water supply water is delivered to users for domestic, commercial and industrial purposes, and is also used for public services and system losses.

Knowing the population served, one can estimate the average daily demand (ADD) by multiplying the population by the average daily demand of 171 gpcd. The peak daily demand (PDD), peak hourly demand (PHD), and minimum hourly demand (MHD) are important flow rates that must be considered during design. Once the average daily demand has been determined, the flow ratios in Table 6.4 can be used for estimating the other flow demands. The coincident demand is normally taken as the peak daily demand plus fire demand.

6.3.2 Fire demand

Preliminary estimates for fire flows can be estimated from the following equation proposed by the National Board of Fire Underwriters for communities with populations less than 200,000 people.

$$Q = C\sqrt{P}\left(1 - 0.01\sqrt{P}\right) \tag{6.1}$$

where:
Q = fire demand, gallons per minute when $C = 1{,}020$
Q = fire demand, m^3/min when $C = 3.86$
P = population in thousands.

Shammas & Wang (2011, pp. 135–137) describe three ways for calculating fire flow requirements. Depending on the population, the fire flow may have to be maintained up to ten hours.

Example 6.1 Calculating design flows

Estimate the water demand for a city of 200,000 people. Assume the annual average consumption rate is 171 gallons per capita per day (gpcd) and use the flow ratios given in Table 6.4 for estimating the following flow rates:

a) Average daily demand (MGD).
b) Peak daily demand (MGD).
c) Peak hourly demand (MGD).
d) Fire demand (MGD).
e) Coincident demand (MGD).

Solution part a

The average daily demand is estimated by multiplying the population of 200,000 people by the per capita water demand of 171 gallons per capita per day.

$$Q_{\text{Avg Day}} = 200{,}000 \text{ people} \times \frac{171 \text{ gal}}{\text{capita} \cdot \text{d}} \times \frac{\text{MG}}{10^6 \text{ gal}} = \boxed{34.2 \text{ MGD}}$$

Solution part b

From Table 6.4, assume a peak daily demand : average daily demand of 1.5:1. Calculate the maximum daily demand or flow by multiplying ADD by 1.5:1 as follows:

$$Q_{\text{PeakDay}} = Q_{\text{AvgDay}} \times 1.5 = 34.2 \text{ MGD} \times 1.5 = \boxed{51.3 \text{ MGD}}$$

Solution part c

From Table 6.4, assume a peak hourly demand : average hour demand of 2.5 : 1. Calculate the maximum daily demand or flow by multiplying ADD by 2.5 : 1 as follows:

$$Q_{\text{PeakHour}} = 2.5(Q_{\text{ADD}}) = 2.5(34.2 \text{ MGD}) = \boxed{85.5 \text{ MGD}}$$

Solution part d

Substituting into Equation (6.1):

$$Q_{\text{FireDemand}} = C\sqrt{P}\left(1 - 0.01\sqrt{P}\right) = 1020\sqrt{200}\left(1 - 0.01\sqrt{200}\right) = 12{,}385 \frac{\text{gal}}{\text{min}}$$

$$Q_{\text{FireDemand}} = 12{,}385 \frac{\text{gal}}{\text{min}}\left(\frac{60 \text{ min}}{\text{h}}\right)\left(\frac{24 \text{ h}}{\text{d}}\right)\left(\frac{\text{MG}}{10^6 \text{ gal}}\right) = \boxed{17.8 \text{ MGD}}$$

Solution part e

The coincidental flow is the fire demand added to the peak daily demand.

$$Q_{FireDemand} + Q_{MaxDay} = 17.8\,\text{MGD} + 51.3\,\text{MGD}$$
$$= \boxed{69.1\,\text{MGD}}$$

6.3.3 Design capacities

After establishing the various design flows, the capacity of the various unit operations and processes must be established – i.e., determine their dimensions, detention times, overflow rates, etc. Table 6.5 presents the capacity or flow rate which various conduits, pumps, and processes should be designed to handle.

Example 6.2 Calculating design capacities

Using the information in Example 6.1, determine the design capacities for the following:

a) Source (MGD).
b) Low-lift pumps (MGD).
c) Treatment plant (MGD).
d) High-service pumps (MGD).
e) Distribution system (MGD).

Solution part a

The source for the water supply must provide a peak daily demand of $\boxed{51.3\,\text{MGD.}}$

Solution part b

Low-lift pumps are sized to meet the peak daily demand plus 10–33% reserve. A minimum of two pumps should be used. Therefore, they should be designed with a capacity between 56.4 to 68.2 MGD.

$$51.3\,\text{MGD} \times (1.10) = \boxed{56.4\,\text{MGD}}$$

$$51.3\,\text{MGD} \times (1.33) = \boxed{68.2\,\text{MGD}}$$

Multiple pumps operating in parallel will be required to convey the PDD plus reserve flow. A redundant or stand-by pump should also be provided.

Solution part c

All unit processes at the water treatment plant should be designed to treat the peak daily demand of $\boxed{51.3\,\text{MGD.}}$

Solution part d

High-service pumps are sized to meet the peak hourly demand plus 10–33% reserve. Therefore, they should be designed with a capacity ranging from 94.1 to 114 MGD.

$$85.5\,\text{MGD} \times (1.10) = \boxed{94.1\,\text{MGD}}$$

$$85.5\,\text{MGD} \times (1.33) = \boxed{114\,\text{MGD}}$$

Several pumps operating in parallel will be required to provide this large flow to the distribution system. One or more redundant pumps should be included for emergency operation.

Solution part e

The water distribution system must be designed for a capacity to meet the peak hourly demand or the peak daily demand plus fire demand. Whichever flow is largest should be used for sizing the distribution systems; in this case, the peak hourly demand of 85.5 MGD controls the design of the water distribution system. See the calculations below.

$$\text{Peak hourly demand} = \boxed{85.5\,\text{MGD}}$$

$$\text{Peak daily demand} + \text{fire demand}$$
$$= 51.3\,\text{MGD} + 17.8\,\text{MGD} = \boxed{69.1\,\text{MGD}}$$

Table 6.5 Design capacities of various elements for treating water.

Element	Flow capacity
Water source	Peak daily demand
Conduit to WTP	Peak daily demand
Low-lift pumps	Peak daily demand plus reserve
High-lift pumps	Peak hourly demand plus reserve
Processes	Peak daily demand plus reserve
Distribution system	Peak hourly demand

Flow capacities from Shammas & Wang (2011) pp. 138–139, and Reynolds & Richards (1996, p. 78) recommend a reserve of 10–33%.

6.4 Preliminary treatment

Preliminary systems for treating drinking water are used to pre-treat the water prior to subsequent major unit operations and processes. Pre-treatment systems commonly used at drinking water plants include bar racks or coarse screens, fine screens, aeration, adsorption, and predisinfection. A brief discussion of each of these follows.

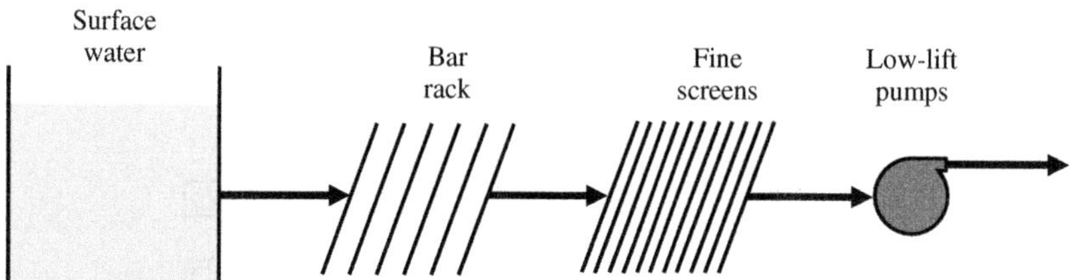

Figure 6.1 Schematic of pretreatment of surface water with bar racks and screens.

6.4.1 Screens

Screens are divided into two categories: coarse (racks) and fine screens. Screens are necessary for removing suspended materials such as leaves, logs, rags, and other floating debris when surface waters are used as the source. Screening also minimizes the impact of fish and other aquatic organisms from entering the treatment plant if properly designed. Figure 6.1 is a schematic showing the sequence of racks and screens used for treating surface water.

Coarse screens are called bar racks or trash racks with openings ranging from $\frac{3}{4}$ to 3 inches. Racks are generally made of $\frac{1}{2}$ - $\frac{3}{4}$ inch diameter (12.7–1.91 mm) metal bars. Typically, coarse screens are installed vertically or at an incline of approximately 30° from the vertical. Manual and automatic operation of the rack cleaning mechanism should be provided. The head loss through a bar rack is calculated using the following equation:

$$h_L = \frac{1}{C}\left(\frac{V_b^2 - V_a^2}{2g}\right) \tag{6.2}$$

where:
h_L = head loss, ft (m)
C = discharge coefficient, equal to 0.7 when clean and 0.6 when clogged
V_b = velocity of flow through bar rack, ft/s (m/s)
V_a = velocity of approach in channel upstream of bar rack, ft/s (m/s)
g = acceleration of gravity, 32.2 ft/s^2 (9.81 m/s^2).

Example 6.3 Bar screen design

A mechanical bar rack has bars that are $\frac{3}{8}$ inches in diameter with clear openings of 1 $\frac{1}{4}$ inches. If the velocity through the bars is 3 feet per second, determine the approach velocity and head loss through the rack assuming that the rack is clean.

Solution

Use the continuity equation to develop a relationship between the approach velocity and velocity through the bars.

$$Q = A_a V_a = A_b V_b$$

Select an arbitrary depth of D. The area of approach (A_a) is estimated as follows:

$$A_a = (1.25 \text{ in} + 3/8 \text{ in}) \times D = (1.625 \text{ in}) \times D$$

$$V_a = \frac{A_b V_b}{A_a} = \frac{(1.25 \text{ in} \times D)(3 \text{ fps})}{(1.625 \text{ in} \times D)} = \boxed{2.31 \text{ fps}}$$

The head loss through the rack is calculated using Equation (6.2).

$$h_L = \frac{1}{C}\left(\frac{V_b^2 - V_a^2}{2g}\right) = \frac{1}{0.7}\left(\frac{(3 \text{ fps})^2 - (2.31 \text{ fps})^2}{2 \times 32.2 \text{ ft/s}^2}\right)$$

$$= \boxed{0.08 \text{ ft} \cong 1 \text{ in}}$$

Traveling water-intake screens generally have openings of 0.24–0.35 inches (6–9 mm) and bucket screens have openings of $\frac{1}{8}$ - $\frac{3}{8}$ inches (3–10 mm) according to *Water Treatment Plant Design*, 4th Edition, 2005, pp. 4.31–4.32). The maximum velocity through the openings in the screen should be less than 2 ft/s (0.6 m/s). The head loss through fine screens is determined using the following equation:

$$h_L = \frac{1}{2g}\left(\frac{Q}{CA}\right) \tag{6.3}$$

where:
h_L = head loss, ft (m)
C = discharge coefficient, equal to 0.7 when clean and 0.6 when clogged
Q = discharge through screen, ft^3/s (m^3/s)
A = effective open area of submerged screen, ft^2 (m^2)
g = acceleration of gravity, 32.2 ft/s^2 (9.81 m/s^2).

6.4.2 Pre-sedimentation

Pre-sedimentation tanks are used for surface waters that may have high turbidities and coliform counts. Plain sedimentation involves settling of suspended solids without the aid of a coagulant. Two types of pre-sedimentation tanks or basins are used for river waters. The first type is for rivers that experience high turbidities and coliform counts on occasions of high rainfall events but normally have low to moderate turbidities and coliform counts. Reinforced concrete tanks with sludge removal mechanisms are used; these may be circular, rectangular, or square in plan-view, with detention times ranging from 0.5–1.0 hour. Overflow rates ranging from 1,000–3,000 gpd/ft^2 (40.7–122 m^3/d · m^2) are normally used. Surface waters with turbidities greater than 1,000 NTU are candidates for this type of pre-sedimentation (Reynolds & Richards, 1996, p. 128 and 255).

The second type of pre-sedimentation requires large tanks, primarily earthen basins with dikes being reinforced with riprap or concrete slabs. The detention times vary from 30–60 days. These long detention times are required for removing turbidities >10,000–40,000 NTU and coliform levels greater than 5,000/100 mL (Reynolds & Richards, 1996, p. 130).

6.4.3 Absorption and desorption

Aeration processes are primarily used to improve the quality of ground water which often contains dissolved gases such as NH_3, CH_4, H_2S, CO_2, volatile organic compounds (VOCs), and dissolved inorganic species such as Fe^{2+} and Mn^{2+}. Absorption is the process of adding a reactive gas such as oxygen (O_2), ozone (O_3), or chlorine (Cl_2) to water. Oxygen is added to oxidize and precipitate iron and manganese. Ozone is used for oxidation, color removal, taste and odor control, iron and manganese oxidation, algae removal, and disinfection (AWWA/ASCE, 2005, p. 10.55).

Air stripping or desorption involves the removal of nuisance gases such as carbon dioxide, hydrogen sulfide, and VOCs. Figure 6.2 shows the dispersion of a liquid such as water in a gas such as air. The concentration of the dissolved gas in the liquid at any time is denoted as C_t, and C_s represents the saturation concentration of the dissolved gas in the liquid.

Spray aerators or nozzles, multiple-tray aerators, cascade aeration, and packed column or packed tray aerators (PTA) are some of the usual methods used for dispersing water into air. Gas desorption and absorption are maximized by increasing the interfacial area per volume of water droplet. This is accomplished by minimizing the water droplet size. Using this approach works best for stripping gases from water. Figure 6.3 depicts the dispersion of a gas into a liquid. Diffused aeration or bubble systems are used for the absorption of gases such as O_2, O_3, and Cl_2. Mechanical surface and submerged turbine aerators have also been used in water treatment for adding oxygen to water (AWWA/ASCE, 2005, p. 5.7). Dispersing a gas into water works better for absorption than for desorption.

6.4.4 Air stripping or degasification

The major types of processes used for degasification are spray aerators or nozzles, multiple tray aerators, cascade aerators, and packet tower aerators. Detailed design information and theory for these systems is found in AWWA (1999, pp. 5.1–5.68), AWWA /ASCE (2005, pp. 5.1–5.25), and Metcalf & Eddy (2003, pp. 1162–1180). The US Environmental Protection Agency has identified packed tower aeration as a best available technology (BAT) for removing VOCs and radon from groundwater.

Packed tray aeration will be discussed briefly and an example is presented to introduce the equations used in design. PTA is used for removing dissolved gases and other chemical

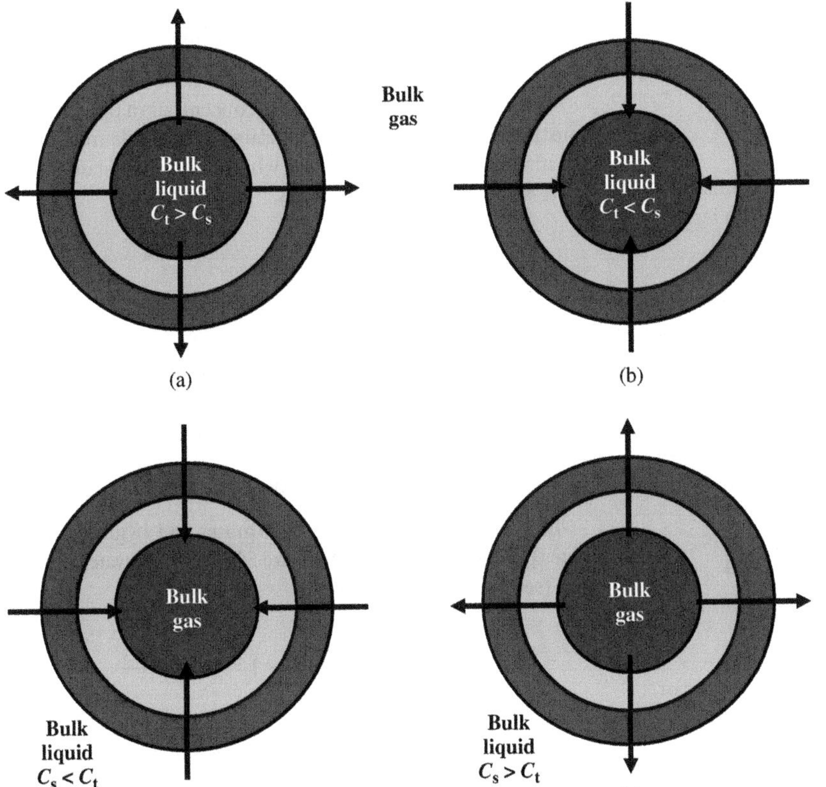

Figure 6.2 Dispersion of a liquid in gas: (a) Desorption and (b) Absorption.

Figure 6.3 Dispersion of a gas in liquid: (a) Desorption and (b) Absorption.

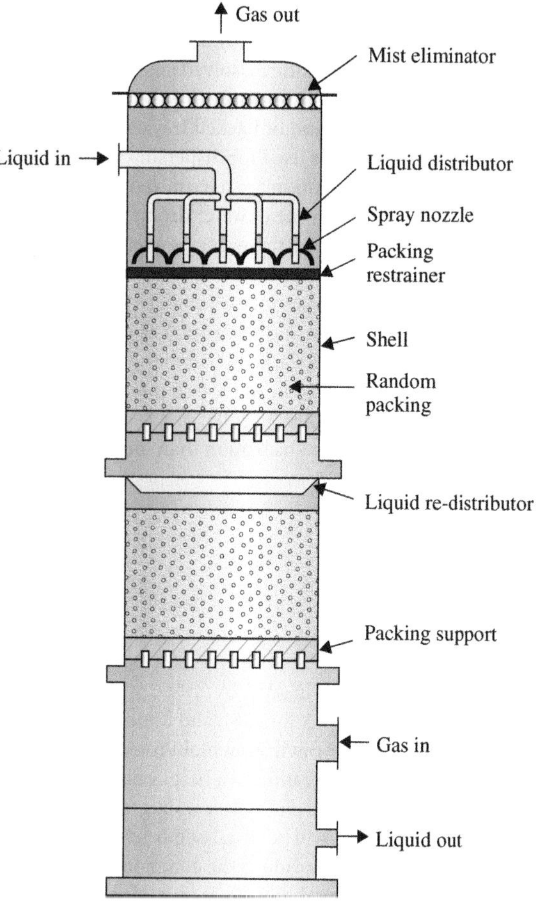

Figure 6.4 Packed tower for gas absorption or packed tray aeration. *Source:* U.S. EPA (2002), EPA Air Pollution Control Cost Manual, Chapter 1 Wet Scrubbers for Acid Gas, EPA/452/B-02-001, Research Triangle Park, NC 27709, p. 1–6. United States Environmental Protection Agency.

contaminants from water. A packed column typically consists of a cylindrical tower with packed media that provides a large surface area with force ventilation in counter-current flow. Figure 6.4 is a schematic of a PTA.

A chemical such as chlorine is often added to enhance oxidation, and an acid or base may also be added to achieve optimum efficiency at a specified pH. The quantity of air provided, (usually by a centrifugal blower) in relation to the water flow rate is called the air-to-liquid ratio ($G : L$), and this is an important design parameter. The exhaust gas from PTAs can create potential air quality problems, especially if it contains toxic VOCs. An effective method of treating the off-gas is by using a vapor-phase granular activated carbon system. The contaminated exhaust gas is first heated and then passed through a GAC contactor.

6.4.4.1 Packed Tray Aerator Design

To develop the equations used to design packed tray aerators or stripping towers, it is imperative to understand the following relationships: the equilibrium between a gas and liquid as represented by Henry's law, the material balance for a PTA, and the theory of gas mass transfer (discussed in Chapter 7, Section 6.2). Henry's law shows the relationship between the equilibrium aqueous concentration of a gas or constituent and the gas-phase concentration. There are several forms of Henry's law, depending on the units used for measuring Henry's constant. Henry's law is generally valid for low solute concentrations, <0.01 mole/L (MWH, 2005, p. 1168), but it has been reported to model solute concentrations as high as 0.1 mole/L. The various forms of Henry's law are discussed below. Equation (6.4) shows the relationship between the concentration of constituent A in the gas-phase measured in atmospheres and the concentration of constituent A in the liquid-phase measured in mole/L.

$$P_A = H_A^c C_A \tag{6.4}$$

where:
P_A = partial pressure of constituent A or concentration in gas-phase, atm
H_A^c = Henry's law constant based on liquid solute concentration, atm · L/mole
C_A = concentration of constituent A in liquid- or aqueous-phase, mole/L.

The concentration of constituent A in the gas-phase measured in atmospheres is determined using Dalton's law, Equation (6.5).

$$P_A = y_A P_T \tag{6.5}$$

where:
y_A = mole fraction of constituent A in gas-phase, dimensionless
y_A = mole of constituent A in gas-phase/(mole A + mole of air)
P_T = total pressure (typically 1 atm), atm.

Equation (6.6) indicates that the concentration of constituent A in the gas-phase is related to the mole fraction of constituent A in the liquid-phase when Henry's constant is expressed in units of atmospheres.

$$P_A = H_A^x x_A \tag{6.6}$$

where:
H_A^x = Henry's law constant based on liquid solute concentration expressed as mole fraction, atm
x_A = mole fraction of constituent A in liquid-phase, dimensionless
x_A = mole of A/(mole of A + mole of water).

In 1 liter of water, there are 1,000 grams and 18 grams of water per mole. Therefore, there are 55.6 moles of water per liter (see calculation below):

$$\left(\frac{1000\,g}{L}\right)\left(\frac{mole}{18\,gH_2O}\right) = 55.6\frac{moles}{L}$$

In most environmental applications, we are working with dilute solutions. Therefore, the moles of constituent A are much less than the moles of water present and can be neglected in calculating x_A. Equation (6.6) can be modified as follows for dilute solutions (solute concentration <0.01 moles/L):

$$P_A \cong H_A^x x_A = H_A^x \left(\frac{\text{moles A}}{\text{moles A} + \text{moles H}_2\text{O}}\right)$$

$$= H_A^x \left(\frac{\text{moles A/L}}{55.6\,\text{moles H}_2\text{O/L}}\right) \tag{6.7}$$

$$P_A = H_A^x \frac{C_A}{55.6} \tag{6.8}$$

The ideal gas law, as shown in Equation (6.9), is used for converting the Henry's constant to a dimensionless form (H):

$$P\forall = nRT \tag{6.9}$$

where:
P = absolute pressure, atm
$\forall$ = volume occupied by gas, L
n = moles of gas
R = universal gas law constant, 0.08206 atm · L/(mole · K)
T = absolute temperature, K (273.15 + °C).

At a pressure of 1 atmosphere and temperature of 0°C, the volume, $\forall$, occupied by an ideal gas or 1 mole of air is equal to 22.4 L. See the calculations below.

$$\forall = \frac{nRT}{P} = \frac{(1\text{ mole})\left(0.08206\,\frac{\text{atm}\cdot\text{L}}{\text{mole}\cdot\text{K}}\right)([273.15 + 0°\text{C}])}{(1\text{ atm})}$$

$$= 22.4\text{ L}$$

Equations (6.10) and (6.11) show how to convert to the dimensionless forms of Henry's constant (H) when Henry's constant is expressed as $H_A^c\left(\frac{\text{atm}\cdot\text{L}}{\text{mole}}\right)$ or H_A^x(atm), respectively.

$$H = \left(H_A^c \frac{\text{atm}\cdot\text{L}}{\text{mol}}\right)\left(\frac{1\text{ mole}\cdot\text{K}}{R\text{ atm}\cdot\text{L}}\right)\left(\frac{1}{T\text{ K}}\right) = \frac{H_A^c}{RT} \tag{6.10}$$

$$H = (H_A^x\text{ atm})\left(\frac{1\text{ mole}\cdot\text{K}}{R\text{ atm}\cdot\text{L}}\right)\left(\frac{1}{T\text{ K}}\right)\left(\frac{\text{L}}{55.6\text{ mole}}\right) = \frac{H_A^x}{RT(55.6)} \tag{6.11}$$

6.4.4.2 Mass Balance on Counter-Current Packed Tower

A material balance on solute in the gas and liquid streams entering the counter-current stripping tower must be performed to derive an equation for the required air-to-liquid ratio (Q_a/Q_l) for achieving effective removal of the solute.

The mass balance is first performed on the control volume of the lower portion of the stripping tower shown in Figure 6.5. At steady-state, no accumulation occurs, and the moles of solute entering the tower must equal the moles of solute leaving the tower, as presented in the following qualitative equation:

$$\begin{bmatrix}\text{moles of solute}\\\text{in influent}\\\text{liquid stream}\end{bmatrix} + \begin{bmatrix}\text{moles of solute}\\\text{in influent}\\\text{gas stream}\end{bmatrix}$$

$$= \begin{bmatrix}\text{moles of solute}\\\text{in effluent}\\\text{liquid stream}\end{bmatrix} + \begin{bmatrix}\text{moles of solute}\\\text{in efffluent}\\\text{gas stream}\end{bmatrix} \tag{6.12}$$

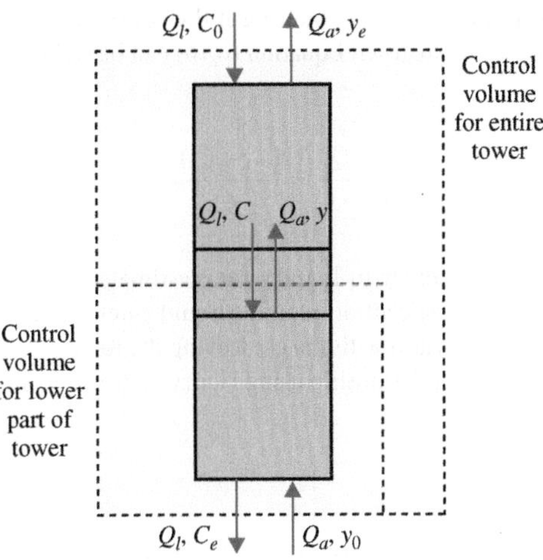

Figure 6.5 Schematic diagram for counter-current stripping tower used in mass balances.

Expressing Equation (0.1) mathematically results in:

$$Q_l C + Q_a y_0 = Q_l C_e + Q_a y \tag{6.13}$$

where:
Q_l = liquid (water) flow rate, m^3/s
C = solute concentration in liquid at a point within tower, mg/L
Q_a = gas (air) flow rate entering the bottom of tower, m^3/s
y_0 = solute concentration in gas entering the bottom of tower, mg/L
C_e = solute concentration in liquid leaving at the bottom of the tower, mg/L
y = solute concentration in gas at a point within the tower, mg/L.

Rearranging and solving (6.13) for the air-to-liquid ratio yields:

$$\frac{Q_a}{Q_l} = \frac{(C - C_e)}{(y - y_0)} \tag{6.14}$$

Performing a mass balance on the entire tower shown in Figure 6.5 results in the following equation:

$$Q_l C_0 + Q_a y_0 = Q_l C_e + Q_a y_e \tag{6.15}$$

where:
C_0 = solute concentration in liquid entering at the top of the tower, mg/L
y_e = solute concentration in gas leaving the top of the tower, mg/L.

Rearranging and combining terms in Equation (6.15) yields:

$$Q_a(y_0 - y_e) = Q_l(C_e - C_0) \tag{6.16}$$

Assuming that the air entering at the bottom of the tower contains no solute, then Equation (6.16) can be written as follows:

$$\frac{Q_a}{Q_l} = \frac{(C_0 - C_e)}{y_e} \tag{6.17}$$

The minimum air-to-liquid ratio is estimated by assuming that the solute concentration in the liquid entering the tower, C_0, is in equilibrium with the gas leaving the tower, y_e. Expressing this relationship using Henry's law results in:

$$y_e = HC_0 \tag{6.18}$$

where H = dimensionless form of Henry's constant.

The equation for calculating the minimum air-to-liquid ratio $\left(\frac{Q_a}{Q_l}\right)_{min}$ is derived by substituting Equation (6.18) into Equation (6.17) as follows:

$$\left(\frac{Q_a}{Q_l}\right)_{min} = \frac{(C_0 - C_e)}{y_e} = \frac{(C_0 - C_e)}{HC_0} \tag{6.19}$$

Alternatively, the minimum air-to-liquid ratio can be estimated from the following equation, where P_T is the total pressure in atmospheres:

$$\left(\frac{Q_a}{Q_l}\right)_{min} = \frac{P_T}{H_A^x}\frac{(C_0 - C_e)}{C_0} \tag{6.20}$$

When the effluent solute concentration is much less than the influent concentration ($C_e \ll C_0$), Equation (6.19) may be simplified as follows:

$$\left(\frac{Q_a}{Q_l}\right)_{min} = \frac{(C_0 - C_e)}{HC_0} = \frac{(C_0 - 0)}{HC_0} \cong \frac{1}{H} \tag{6.21}$$

In actual practice, the $\left(\frac{Q_a}{Q_l}\right)$ ratio will vary from 20 : 1 to 60 : 1 (Metcalf & Eddy, 2003, p. 1175).

Another important parameter is the stripping factor and is defined as follows:

$$S = \frac{(Q_a/Q_l)}{(Q_a/Q_l)_{min}} = \frac{(Q_a/Q_l)}{(1/H)} = H\left(\frac{Q_a}{Q_l}\right) \tag{6.22}$$

where:
S = stripping factor, dimensionless
Q_l = water flow rate, m^3/s
Q_a = air flow rate, m^3/s
H = Henry's constant, dimensionless.

When $S = 1$, the tower is operating at the minimum air-to-liquid ratio required for stripping. When $S<1$, the stripper cannot achieve the desire removal objective. When $S>1$, the stripper can remove the desired contaminant. The optimal air-to-liquid ratio is 3.5 times $\left(\frac{Q_a}{Q_l}\right)_{min}$ (MWH, 2005, p. 1194).

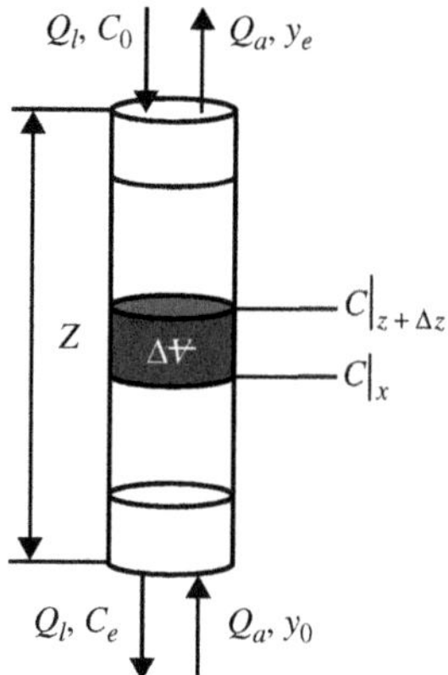

Figure 6.6 Diagram for mass transfer from the liquid phase to the gas phase.

6.4.4.3 Determination of Stripping Tower Height

To develop an equation for the height of the packing material in the stripping tower, a materials balance on the mass transfer of the solute into and out of a liquid differential volume within the tower must be performed (Figure 6.6).

A qualitative statement can be written as follows:

$$\begin{bmatrix}\text{accumulation}\\ \text{of solute}\\ \text{in tower}\end{bmatrix} = \begin{bmatrix}\text{inflow of}\\ \text{solute into}\\ \text{tower}\end{bmatrix} - \begin{bmatrix}\text{outflow of}\\ \text{solute from}\\ \text{tower}\end{bmatrix} - \begin{bmatrix}\text{volatilization}\\ \text{of solute}\end{bmatrix} + \begin{bmatrix}\text{generation}\\ \text{of solute}\\ \text{in tower}\end{bmatrix} \tag{6.23}$$

Equation (6.23) can be expressed mathematically as Equation (6.24) since the generation term is zero, i.e., there is no reaction occurring to increase the quantity of solute within the tower.

$$\frac{\partial C}{\partial t}\Delta V = Q_l C|_{(z+\Delta z)} - Q_l C|_{(z)} - r_V \Delta V + 0 \tag{6.24}$$

where:
$\frac{\partial C}{\partial t}$ = change in solute concentration with time, $g/(m^3{\cdot}s)$
ΔV = differential volume, m^3
$\Delta V = A\Delta z$
A = cross-sectional area, m^2
Δz = differential height, m
Q_l = liquid volumetric flow rate, m^3/s
C = solute concentration, g/m^3
r_v = rate of mass transfer of solute per unit volume per unit time, $g/(m^3{\cdot}s)$.

Replacing the differential volume with $A\Delta z$ and writing the differential form of $Q_l C|_{(z+\Delta z)}$ results in the following equation:

$$\frac{\partial C}{\partial t}A\Delta z = Q_l\left(C_z + \frac{\Delta C}{\Delta z}\Delta z\right) - Q_l C_z - r_V A\Delta z + 0 \tag{6.25}$$

Simplifying Equation (6.25) and taking the limit as Δz approaches zero yields:

$$\frac{\partial C}{\partial t}\frac{A\Delta z}{A\Delta z} = \frac{Q_l C_z}{A\Delta z} + \frac{\Delta C}{\Delta z}\frac{Q_l \Delta z}{A\Delta z} - \frac{Q_l C_z}{A\Delta z} - r_{\Psi}\frac{A\Delta z}{A\Delta z} \quad (6.26)$$

$$\frac{\partial C}{\partial t} = \frac{Q_l}{A}\frac{\partial C}{\partial z} - r_{\Psi} \quad (6.27)$$

The rate of mass transfer of the solute or gas due to volatilization as described in Chapter 7 is expressed by the following equation.

$$r_{\Psi} = K_L a(C_b - C_s) \quad (6.28)$$

where:

$K_L a$ = mass transfer coefficient of the specific solute or gas, s^{-1}

C_b = concentration of solute in the bulk liquid phase at time t, g/m^3

C_s = equilibrium concentration of solute in the liquid as determined by Henry's law, g/m^3.

At steady-state, the accumulation term $\frac{\partial C}{\partial t}$ is zero and substituting Equation (6.28) into Equation (6.27) results in:

$$\frac{\partial C}{\partial t} = 0 = \frac{Q_l}{A}\frac{dC_b}{dz} - K_L a(C_b - C_s) \quad (6.29)$$

Equation (6.29) can be rearranged and prepared for integration for the determination of the height of the tower:

$$\frac{dC_b}{dz} = \frac{AK_L a}{Q_l}(C_b - C_s) \quad (6.30)$$

$$\int_0^Z dz = \int_{C_e}^{C_0} \frac{Q_l}{AK_L a}\frac{dC_b}{(C_b - C_s)} \quad (6.31)$$

The equilibrium solute concentration is changing throughout the column as expressed by Henry's law, Equation (6.18). A relationship can be developed from the Henry's law equation and Equation (6.17), as shown by Equation (6.33).

$$y = HC_s \quad (6.18)$$

$$\frac{Q_a}{Q_l} = \frac{(C_0 - C_e)}{y_e} \quad (6.17)$$

$$\frac{Q_a}{Q_l} = \frac{(C_b - C_e)}{HC_s} \quad (6.32)$$

$$C_s = \frac{Q_l}{Q_a}\frac{(C_b - C_e)}{H} = \frac{Q_l}{Q_a}\frac{P_T}{H_A^{\times}}(C_b - C_e) \quad (6.33)$$

Hand *et al.* (1999, pp. 5.1–5.68) developed Equation (6.34) by substituting Equation (6.33) into Equation (6.31) and then integrating.

$$Z = \frac{Q_l}{AK_L a}\left[\frac{C_0 - C_e}{C_0 - C_e - C_0'}\right]\ln\left[\frac{C_0 - C_0'}{C_e}\right] \quad (6.34)$$

where:

$$C_0' = \frac{Q_l}{Q_a}\frac{(C_0 - C_e)}{H} = \frac{Q_l}{Q_a}\frac{P_T}{H_A^{\times}}(C_0 - C_e) \quad (6.35)$$

The pressure drop through a stripping tower can be estimated from Figure 6.7, developed by Eckert (1970, p. 40).

Equations (6.36) and (6.38) must be used for estimating the pressure drop in Newtons per m^2 per meter of packing height:

$$\text{value from x-axis} = \frac{L_m}{G_m}\left(\frac{\rho_g}{\rho_l - \rho_g}\right)^{0.5} \quad (6.36)$$

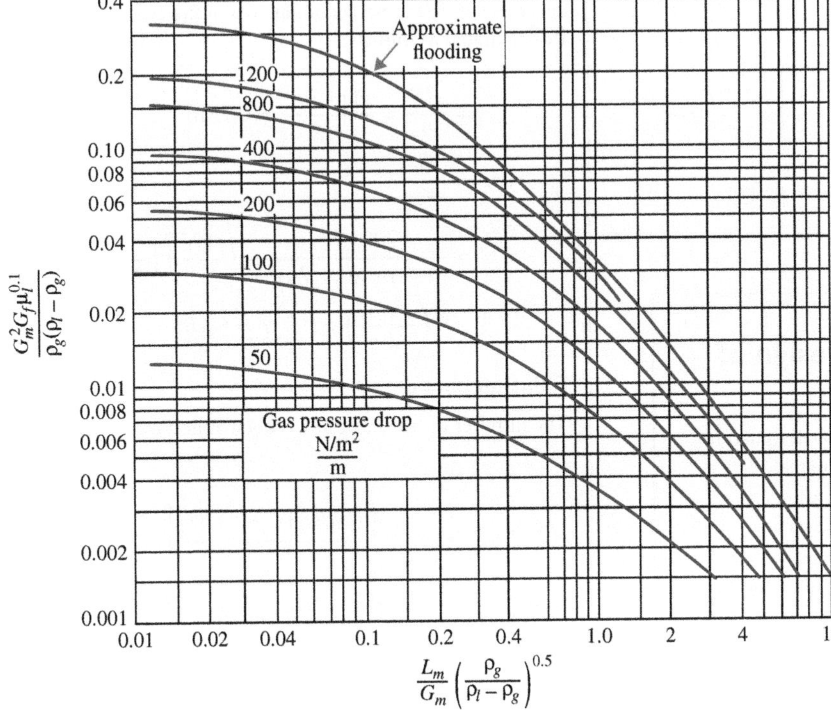

Figure 6.7 Generalized pressure drop for a random packed tower. *Source:* MWH (2005) p. 1205. Reproduced by permission of John Wiley & Sons Ltd.

$$\frac{G_m}{L_m} = \left(\frac{Q_a}{Q_l}\right)\left(\frac{\rho_g}{\rho_l}\right) \tag{6.37}$$

$$\text{value from } y\text{-axis} = \left[\frac{(G_m)^2(C_f)(\mu_l)^{0.1}}{\rho_g(\rho_l - \rho_g)}\right] \tag{6.38}$$

where:
L_m = liquid loading rate, kg/(m^2 · s)
G_m = air loading rate, kg/(m^2 · s)
C_f = packing factor, dimension, (See Table 6.7)
μ_l = absolute viscosity of water, kg/(m^2·s) or N·s/m^2
μ_l = 1.002 × 10^{-3} N·s/m^2 at 20°C
ρ_g = density of air or gas, kg/m^3
ρ_l = density of liquid or water, kg/m^3.

6.4.4.4 Design Equations for Stripping Towers

The equations used for designing stripping towers are presented in this section. The height of the packing material in the stripping tower Z is calculated by multiplying the height of a transfer unit (HTU) by the number of transfer units (NTU) as follows:

$$Z = \text{HTU} \times \text{NTU} \tag{6.39}$$

For practical designs, the value calculated using Equation (6.39) is increased by 20% and then rounded to the next whole number (LaGrega *et al.*, 1994, p. 457).

The height of a transfer unit is calculated using Equation (6.40):

$$\text{HTU} = \frac{Q_l}{AK_L a} \tag{6.40}$$

The number of transfer units is determined using the following equation:

$$\text{NTU} = \left[\frac{S}{S-1}\right] \ln\left[\frac{(C_0/C_e)(S-1)+1}{S}\right] \tag{6.41}$$

Table 6.6 provides typical parameters used to design stripping towers.

Example 6.4 Preliminary design of packed tray aerator

A packed tray aerator or air stripper is to be designed to lower the TCE concentration in a contaminated groundwater from 100 μg/L to 5 μg/L. The water volumetric flow rate is 2,000 m^3/d, the Henry's law coefficient for TCE is 307 atm, and the mass transfer coefficient ($K_L a$) for TCE is 0.0176 s^{-1}. Assume that the diameter of the stripping column is 1.0 m, a temperature of 20°C, and an air-to-liquid ratio of 20 : 1. Pall rings with a packing factor of 50 will be used in the design. The air density (ρ_{air}) at 20°C = 1.024 kg/m^3 and water density (ρ_{water}) at 20°C = 998.2 kg/ m^3.

Determine the following:

a) Dimensionless Henry's constant.
b) Stripping factor (S).
c) Height of transfer unit (HTU) in meters and feet.
d) Number of transfer units (NTU).
e) Height of packing in column (Z) in meters and feet.
f) Pressure drop through tower in Pascals.

Solution part a

Calculate the dimensionless Henry's constant using Equation (6.11):

$$H = (H_A^x\ \text{atm})\left(\frac{1\ \text{mole}\cdot\text{K}}{R\ \text{atm}\cdot\text{L}}\right)\left(\frac{1}{T\ \text{K}}\right)\left(\frac{\text{L}}{55.6\ \text{mole}}\right) = \frac{H_A^x}{RT(55.6)}$$

$$H = (307\ \text{atm})\left(\frac{1\ \text{mole}\cdot\text{K}}{0.08206\ \text{atm}\cdot\text{L}}\right)\left(\frac{1}{273.15+20\ \text{K}}\right) \times \left(\frac{\text{L}}{55.6\ \text{mole}}\right) = \boxed{0.23}$$

Solution part b

The stripping factor is determined from Equation (6.22):

$$S = \frac{(Q_a/Q_l)}{(Q_a/l)_{min}} = \frac{(Q_a/Q_l)}{(1/H)} = H\left(\frac{Q_a}{Q_l}\right)$$

$$S = 0.23\left(\frac{20}{1}\right) = \boxed{4.6}$$

Solution part c

The height of a transfer unit is determined from Equation (6.40). First, the cross-sectional area of the column must be calculated as follows.

$$A = \frac{\pi D^2}{4} = \frac{\pi(1.0\ \text{m})^2}{4} = 0.79\ \text{m}^2$$

$$\text{HTU} = \frac{Q_l}{AK_L a}$$

$$\text{HTU} = \frac{2000\ \text{m}^3/\text{d}}{(0.79\ \text{m}^2)(0.0176\ \text{s}^{-1})}\left(\frac{1\ \text{d}}{24\ \text{h}}\right)\left(\frac{1\ \text{h}}{60\ \text{min}}\right) \times \left(\frac{1\ \text{min}}{60\ \text{s}}\right) = \boxed{1.7\ \text{m} = 5.6\ \text{ft}}$$

Solution part d

The number of transfer units is determined from Equation (6.41):

$$\text{NTU} = \left[\frac{S}{S-1}\right] \ln\left[\frac{(C_0/C_e)(S-1)+1}{S}\right]$$

$$\text{NTU} = \left[\frac{4.6}{4.6-1}\right] \ln\left[\frac{(100\ \mu\text{g/L}/5\ \mu\text{g/L})(4.6-1)+1}{4.6}\right] = \boxed{3.5}$$

Table 6.6 Typical parameters used in the design of stripping towers for VOC removal.

Item	Nomenclature	Unit	Value
Gas loading rate	Q_a	lb/hr·ft^2 (kg/hr·m^2)	100 to 3,000 (489 to 14,700)
Liquid loading rate	Q_l	lb/hr·ft^2 (kg/hr·m^2)	400 to 15,000 (1955 to 73,300)
Allowable pressure drop	Δp	in of water/ft (N/m^2)/m	0.5 to 1.0 (409 to 818)
Column diameter	D	ft (m)	2 to 12 (0.6 to 3.6)
Packing depth	Z	ft (m)	4 to 12 (1.2 to 3.6)
Packing factor			
Berl saddles, ceramic 1 in (25 mm)	C_f	ft^{-1} (m^{-1})	110 (361)
Berl saddles, Ceramic 2 in (50 mm)	C_f	ft^{-1} (m^{-1})	45 (148)
Interlox saddles, ceramic 1 in (25 mm)	C_f	ft^{-1} (m^{-1})	98 (322)
Interlox saddles, ceramic 2 in (50 mm)	C_f	ft^{-1} (m^{-1})	40 (131)
Raschig rings, ceramic 1 in (25 mm)	C_f	ft^{-1} (m^{-1})	160 (525)
Raschig rings, ceramic 2 in (50 mm)	C_f	ft^{-1} (m^{-1})	65 (213)
Raschig rings, metal 1 in (25 mm)	C_f	ft^{-1} (m^{-1})	137 (449)
Raschig rings, metal 2 in (50 mm)	C_f	ft^{-1} (m^{-1})	57 (187)
Pall rings, plastic 1 in (25 mm)	C_f	ft^{-1} (m^{-1})	52 (171)
Pall rings, plastic 2 in (50 mm)	C_f	ft^{-1} (m^{-1})	25 (82)
Pall rings, metal 1 in (25 mm)	C_f	ft^{-1} (m^{-1})	48 (157)
Pall rings, metal 2 in (50 mm)	C_f	ft^{-1} (m^{-1})	20 (66)

Source: Values from Barbour *et al.* (1995) pp. 1-51 and 1-52.

Solution part e

The height of packing in column is calculated from Equation (6.39):

$$Z = \text{HTU} \times \text{NTU} = 1.7 \times 3.5 = \boxed{6.0\ \text{m}}$$

The actual height of the packing material will be increased by 20%.

$$Z = 6.0\ \text{m} \times 1.20 = 7.2 \cong \boxed{7.2\ \text{m} = 24\ \text{ft}}$$

Solution part f

The air density (ρ_{air}) at 20°C = 1.024 kg/m^3 and water density (ρ_{water}) at 20°C = 998.2 kg/ m^3 and μ_l = 1.002 × 10^{-3} N·s/m^2.

$$\frac{G_m}{L_m} = \left(\frac{Q_a}{Q_l}\right)\left(\frac{\rho_g}{\rho_l}\right) = \left(\frac{20}{1}\right)\left(\frac{1.024\ \text{kg/m}^3}{998.2\ \text{kg/m}^3}\right)$$

$$= 0.0205\ \frac{\text{kg air}}{\text{kg water}}$$

value for x-axis

$$= \frac{L_m}{G_m}\left(\frac{\rho_g}{\rho_l - \rho_g}\right)^{0.5}$$

$$= \frac{\text{kg water}}{0.0205\ \text{kg air}}\left(\frac{1.024\ \text{kg/m}^3}{998.2\ \text{kg/m}^3 - 1.024\ \text{kg/m}^3}\right)^{0.5}$$

value for x-axis = $\boxed{1.56}$

$$\frac{Q_a}{Q_l} = 20$$

$$Q_a = 20 \times Q_l = 20 \times 2000\ \frac{\text{m}^3}{\text{d}} = 40{,}000\frac{\text{m}^3}{\text{d}}$$

$$G_m = 40{,}000\frac{\text{m}^3}{\text{d}}\left(\frac{1.024\ \text{kg}}{\text{m}^3}\right)\left(\frac{1\ \text{d}}{24\ \text{h}}\right)\left(\frac{1\ \text{h}}{3600\ \text{s}}\right) \times \left(\frac{1}{0.79\ \text{m}^2}\right)$$

$$G_m = 0.60\ \frac{\text{kg}}{\text{m}^2 \cdot \text{s}}$$

value from y-axis

$$= \left[\frac{(G_m)^2(C_f)(\mu_l)^{0.1}}{\rho_g(\rho_l - \rho_g)}\right]$$

$$= \left[\frac{\left(0.60\ \frac{\text{kg}}{\text{m}^2 \cdot \text{s}}\right)^2(50)(1.002 \times 10^{-3}\text{N} \cdot \text{s/m}^2)^{0.1}}{1.024\ \text{kg/m}^3(998.2\ \text{kg/m}^3 - 1.024\ \text{kg/m}^3)}\right]$$

value from y-axis = $\boxed{0.0088}$

From Figure 6.7 using the x- and y- coordinates calculated above, the pressure drop in N/(m^2) per meter of packing is approximately 250.

$$\text{Pressure Drop} = 250\ \frac{\text{N}}{\text{m}^2 \cdot \text{m}}(7.2\ \text{m})$$

$$= \boxed{1800\ \frac{\text{N}}{\text{m}^2} = 1800\text{Pa}}$$

6.4.5 Adsorption

Adsorption is the process of collecting a substance at the surface of an adsorbent. It is a mass transfer operation, in which a constituent in the liquid phase (adsorbate) is removed and collected on the surface of the solid phase (adsorbent). Adsorption may be physical or chemical. Physical adsorption is primarily due to van der Waals forces and is reversible. In chemical adsorption, a chemical reaction occurs between the adsorbent and the adsorbate. Chemical adsorption is usually irreversible.

Activated carbon is the principal type of adsorbent used in water treatment. Reynolds & Richards (1996, p. 133) state that activated carbon is effective for removing organic compounds causing taste and odor or color problems, halogens, hydrogen sulfide, iron, manganese, and numerous other substances from water. Activated carbon is sold by a number of manufacturers either as powdered activated carbon (PAC) or granular activated carbon (GAC). In the past, PAC was more widely used by the water industry, but now GAC is being considered more seriously, as primary drinking water standards become more stringent (AWWA, 1999, p. 13.2). A brief discussion on the use of PAC and GAC is presented below.

6.4.5.1 Adsorption with PAC

Powdered activated carbon is added for the removal of taste and odor compounds and the removal of organics (AWWA/ASCE, 2005, p. 14.1). It may be added at the intake, prior to rapid mixing, at the rapid mixer, or filter influent. Better treatment occurs by providing longer contact times, so that equilibrium conditions can be reached. PAC is fed either as a powder or slurry using metering pumps.

Powdered activated carbon typically has a diameter less than 0.074 mm (200 sieve) and a total surface area of 800–1,800 m^2/g, and bulk density of 360–740 kg/m^3 (Metcalf & Eddy, 2003, p. 1139). PAC dosages range from 2–20 mg/L for nominal taste and odor compounds, but can exceed 100 mg/L for severe taste and odor problems or high concentrations of organic chemicals (AWWA/ASCE, 2005, p. 14.5).

6.4.5.2 Adsorption with GAC

Granular activated carbon is sometimes used as pre-filtration adsorption ahead of conventional filtration, but primarily as a post-filtration process, since it is more effective at this stage because much of the turbidity and organic constituents have already been removed. GAC has been used to retro-fit conventional sand filters. Existing filters can be converted to GAC filter adsorbers by replacing a portion or all of the existing media with GAC.

6.4.6 Pre-disinfection

It is not unusual to add some type of disinfectant to raw surface water if bacterial counts are high. Chlorine has been the disinfectant of choice in the past but, due to problems with the production of trihalomethanes (THMs) and other disinfection-by-products (DBPs), other oxidants/disinfectants such as ozone, potassium permanganate, and hydrogen peroxide should be considered. Chlorine is also added to ground water to suppress iron bacteria growth in pipes from the well field to the treatment plant and in the tray aerator (Viessman & Hammer, 2005, p. 344). If chlorine is used, the addition of ammonia to produce chloramines helps to prevent the formation of THMs and other DBPs.

6.5 Mixing, coagulation, and flocculation

6.5.1 Overview

Coagulation is a unit process consisting of the addition and mixing of a chemical reagent (coagulant or flocculant) for the destabilization of colloidal and fine solids that are suspended in

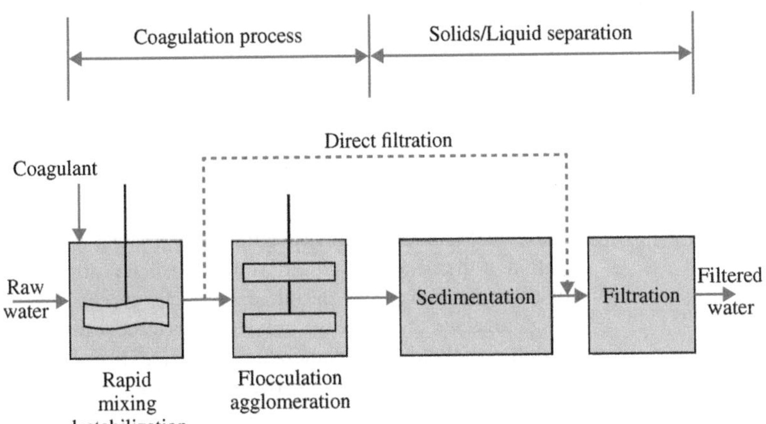

Figure 6.8 Schematic of coagulation and flocculation processes.

water. **Flocculation** is a unit process involving the slow stirring or gentile agitation to promote agglomeration of the destabilized particles from coagulation into heavy, rapidly settling flocs. The flocculated particles are subsequently removed by sedimentation and filtration. Figure 6.8 is a schematic showing the sequence of steps involved in coagulation and flocculation. Direct filtration of water occurs when sedimentation is omitted from the treatment sequence and filtration follows flocculation. Direct filtration is typically used if the raw water has low to moderate color ≤ 20 c.u., low turbidity ≤ 15 NTU and low TOC < 4 mg/L (MWH, 2005, p. 264).

6.5.2 Colloidal particles

Colloidal particles are small and have negligible mass and large surface area per unit volume. They are defined by their size, which ranges from 1 nanometer (10^{-9} m) to 1 μm (10^{-6} m). Due to their large surface areas, colloidal particles tend to acquire a negative surface charge. Colloidal constituents found in ground and surface water include microorganisms, trace organics, pathogens, algae, and inorganic particles such as clay and silt, as well as trace inorganics such as barium, strontium, and chromium. Turbidity in surface water is primarily caused by colloidal clay particles due to soil erosion. Color in water is usually attributed to colloidal forms of iron and manganese or from organic compounds (humic and fulvic acids).

Colloidal particles are classified as either hydrophilic or hydrophobic. Hydrophilic colloids have an affinity for water and contain water-soluble groups on the colloidal surface. These types of colloids promote hydration, causing a film of water to surround the colloid (water of hydration). Organic colloids, such as proteins and their degradation products, are hydrophilic. They may have ionization groups such as the amino group ($-NH_2$) and carboxyl ($-COOH$). The isoelectric point is the pH at which there is no surface charge on the colloid.

In most instances, colloidal particles can be removed more easily during coagulation and flocculation by maintaining the pH at the isoelectric point of the colloidal suspension. For the ionization groups listed above, the overall charge on a colloidal particle would be zero at the isoelectric point pH. At pH values below the isoelectric point, the amino group changes from $-NH_2$ to $-NH_3^+$ resulting in an overall positive charge on the colloid, whereas, at a pH value greater than that at the isoelectric point, $-COOH$ becomes $-COO^-$, resulting in a negatively charged colloidal particle.

Alternatively, hydrophobic colloids have little to no affinity for water. They do not have a water layer surrounding the colloid (no water of hydration). Inorganic colloids are typically clay particles.

Jar tests and pilot studies must be performed on the raw water to determine the optimum pH and specific type of coagulant that should be used for removing turbidity, color, and TOC.

6.5.2.1 *Jar Tests*

Jar tests are experiments conducted on the water to be treated to determine the proper dosage of coagulant and coagulant aid. These experiments are often performed using a six-gang stirring mechanism, as shown in Figure 6.9.

The coagulation/flocculation and sedimentation process sequence can be simulated using a jar test apparatus. Generally, 1- or 2-liter beakers or square containers are used with the jar test apparatus. The procedure for performing a jar test is presented below.

1 Fill six 1 L or 2 L beakers or square containers with a measured amount of the water to be evaluated.

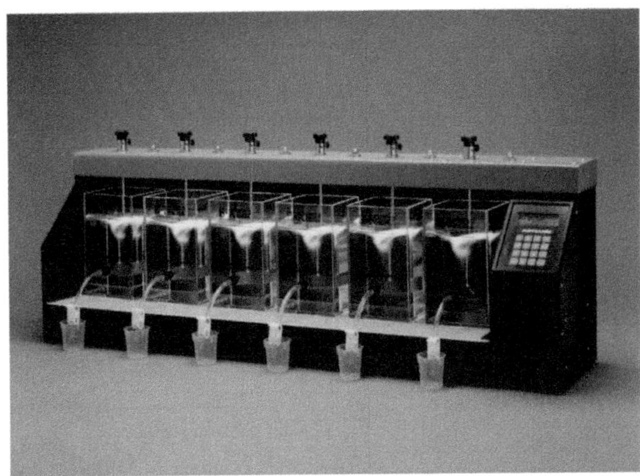

Figure 6.9 Photo of jar test apparatus. © 2012. Phipps and Bird, Richmond VA, with permission

2 Add a measured amount of coagulant and/or coagulant aid to all containers except for one of them, which serves as the control. For instance, add a measured amount of alum to result in dosages of 10, 50, 100, 150, and 200 mg/L, respectively in containers 1 through 5. Container 6 serves as the control, i.e., no coagulant or coagulant aid is added.

3 Simulate rapid mixing by operating the jar text apparatus at maximum speed (approximately 100 rpm) for 30 – 60 seconds.

4 Simulate flocculation by setting the operating speed of the apparatus to 20 – 70 rpm for 20 – 30 minutes to promote particle agglomeration, i.e., floc formation. Record the time at which floc appears in each container.

5 Raise the paddles out of the containers and allow the flocculated particles to settle for simulating the sedimentation process. Allow approximately 30 – 60 minutes for settling. Take samples of the supernatant and perform analytical tests such as turbidity, color, pH, TOC, etc.

Repeat the procedure using different coagulants, varying the pH, and/or dosages of coagulant added to determine best coagulant and optimum dose.

6.5.3 Electrical double layer theory

According to the electrical double layer (EDL) theory, a negative colloidal particle will have a layer of counter positive ions surrounding it known as the fixed or **Stern** layer (Figure 6.10).

Surrounding the fixed or Stern layer is another layer consisting primarily of counter-ions and co-ions called the **diffuse** layer. The electrical potential at the colloid surface is called the **Nernst potential** and at the surface of shear it is called the **zeta potential**. The greater the magnitude of the zeta potential, the greater the stability of the colloidal suspension. Zeta potential is measured using electrophoretic means. MWH (2005, p. 658), referencing Kruyt (1952), stated that rapid flocculation occurs when the absolute value of the zeta potential is reduced below approximately 20 mV.

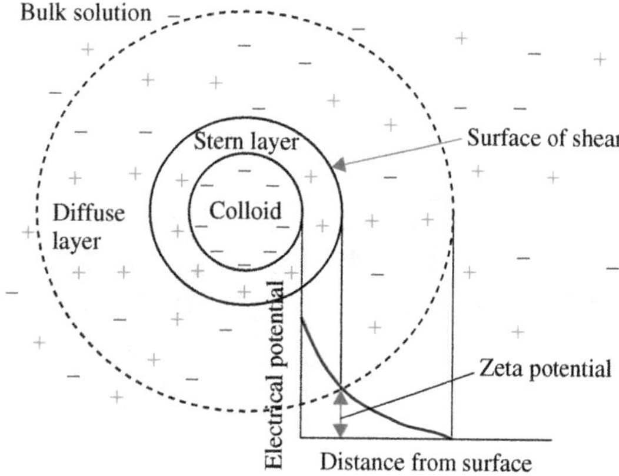

Figure 6.10 Diffuse double layer schematic of negatively charged colloid.

6.5.4 Colloidal destabilization

The stability of colloidal suspensions depends on both repulsive and attractive forces. **Repulsive** forces are associated with the electrostatic charge due to the double layer surrounding the colloid. **Attractive** forces primarily associated with van der Waals forces are essential for promoting particle agglomeration and eventual removal by gravity settling and granular media filtration. If the electrostatic charge on the particles can be reduced or neutralized, when colloidal particles get close enough together, the attractive van der Waals forces cause the destabilized colloids to agglomerate or coalesce. The random movement of colloidal particles (Brownian movement), due to bombardment by water molecules from the dispersion medium, also contributes to the destabilization of the sol. A representation of the interaction between the attractive and repulsive forces involved in colloidal destabilization is presented in Figure 6.11.

6.5.4.1 Destabilization Mechanisms

Coagulants and coagulant aids are added to water during rapid (flash) mixing to destabilize colloidal particles. In the literature, there are four mechanisms which have been hypothesized to describe how coagulants work:

1 Lowering the surface potential by compressing the electrical double layer.

2 Lowering the surface potential by adsorption and charge neutralization.

3 Enmeshment of the colloids in precipitate (sweep floc).

4 Attachment and interparticle bridging.

Several of the mechanisms may occur simultaneously to accomplish destabilization.

6.5.4.2 Compressing the Electrical Double Layer

By compressing the electrical double layer surrounding the colloid, the repulsive force can be reduced so that van der Waals forces of attraction can predominate to allow particles to agglomerate. This mechanism is only achieved by increasing the ionic strength or ionic concentration of the solution. Compression of the EDL is not a significant colloid destabilization mechanism at water treatment plants, since the coagulant dosages commonly used do not increase the ionic strength dramatically (Peavy *et al.*, 1985, p. 134).

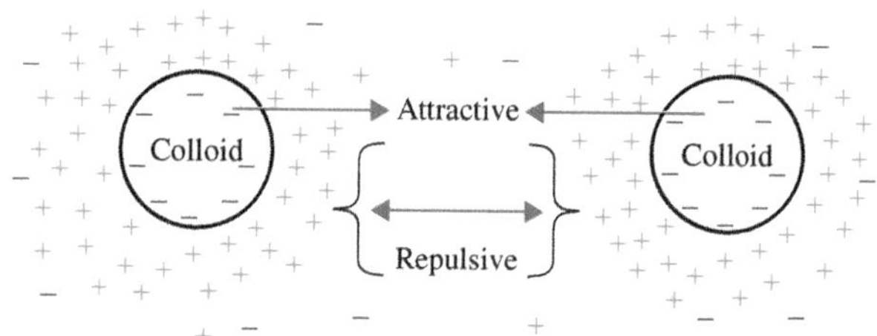

Figure 6.11 Attractive and repulsive force interaction on colloidal particles.

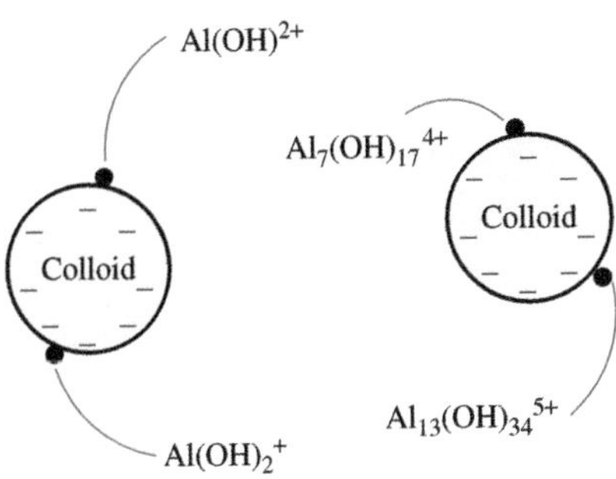

Figure 6.12 Schematic illustrating adsorption and charge neutralization of colloids.

6.5.4.3 Adsorption and Charge Neutralization

Colloidal particles can be destabilized by adsorption and charge neutralization of counterions that are produced when a coagulant is added to the water (Figure 6.12).

The selection of the type of coagulant is important, especially the charge of the counterions that are produced, in addition to how the ionic strength of the solution is impacted. Coagulants that produce multivalent versus monovalent counterions seem to be more effective. Aluminum and iron salts or cationic synthetic polymers have been used.

When aluminum sulfate (alum) is added to water, several hydroxometallic species are produced that have a positive charge, i.e., $Al(OH)^{2+}$, $Al(OH)_2^+$, $Al_7(OH)_{17}^{4+}$, and $Al_{13}(OH)_{34}^{5+}$ (Reynolds & Richards, 1996, p. 172; Peavy *et al.*, 1985, p. 134). These species are adsorbed to the negatively charged colloidal particle and reduce the zeta potential, allowing destabilization of the particles. Iron salts, such as ferric chloride and ferrous sulfate, act in a similar fashion.

6.5.4.4 Enmeshment of the Colloids in Precipitate

At coagulant dosages exceeding the solubility of the metal hydroxide, aluminum and iron salts form insoluble precipitates that enmesh and entrap suspended and colloidal particles in the precipitate (Figure 6.13). This mechanism is also referred to as *sweep floc* and predominates at a pH between 6 and 8.

6.5.4.5 Attachment and Interparticle Bridging

Another mechanism involved in the destabilization of colloidal particles is attachment and interparticle bridging (Figure 6.14). This destabilization mechanism often occurs when organic polymers are added to the water. Polymers are long-chain chemicals that have ionizable groups that adsorb or attach to the surface of the colloid. Because of this nature, several colloids may be bound to a single polymer molecule and a bridging structure formed. Polymer dosages typically range from 1 – 10 mg/L (MWH, 2005, p. 688).

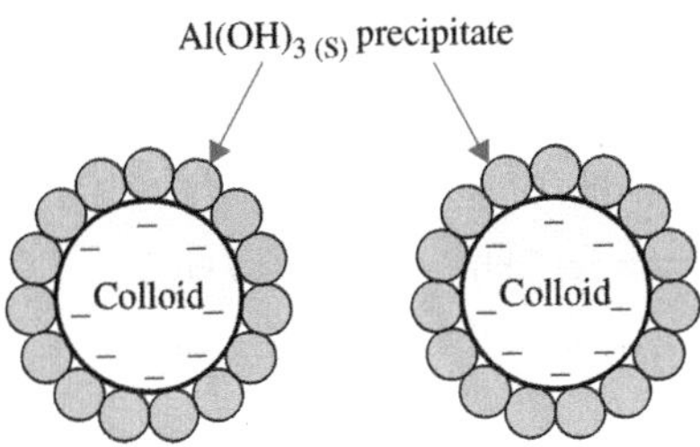

Figure 6.13 Schematic illustrating "sweep floc" removal of colloids.

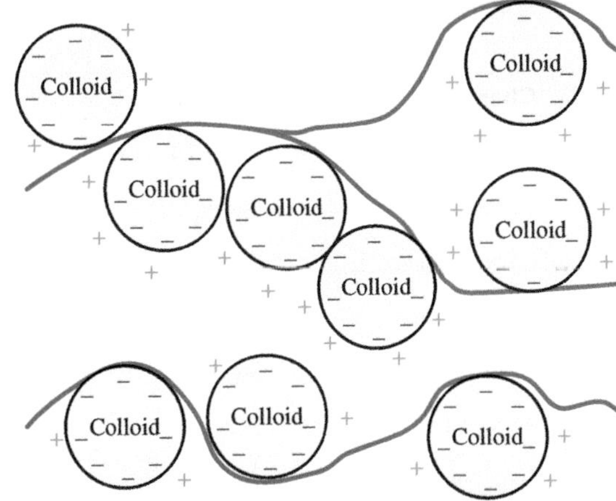

Figure 6.14 Schematic illustrating attachment and interparticle bridging removal of colloids.

6.5.5 Coagulants

Table 6.7 is a list of typical coagulants and chemicals that are used during water treatment. Coagulants are primarily added for the removal of turbidity, color, and TOC (DBP precursors). Other coagulants are added to aid the coagulation and flocculation processes. Some chemicals are added to provide alkalinity to the water or for pH adjustment.

Some of the most widely used coagulants and their stoichiometric reactions with alkalinity will be presented in this section. When coagulants are added to water, they react with numerous constituents in the water and the actual dosages will be greater than what is predicted from the reactions involving alkalinity. Temperature and pH affect the coagulant dosage. The only way to determine the actual dose of a chemical is to perform jar tests.

6.5.5.1 Aluminum Sulfate

Aluminum sulfate $[Al_2(SO_4)_3 \cdot 14H_2O]$, or alum, is the coagulant most widely used in the United States. It is an effective chemical used for coagulating turbidity, color, and TOC. It is more economical to use compared to iron salts; although the sludge or precipitate that is produced is amorphous and not as dense as that from iron coagulants. Typical alum dosages range from 10 – 150 mg/L (MWH, 2005, p. 683). Alum readily dissolves in water, with sulfate ions $SO_4{}^{2-}$ being dispersed throughout the liquid. The aluminum ions hydrolyze in water, with the simplest as $Al(H_2O)_6^{3+}$. Aluminum ions exist in water primarily as hydrated ions such as $Al(OH)^{2+}$, $Al_{17}(OH)_{17}^{4+}$, $Al_2(OH)_4^-$, and $Al_{13}(OH)_{34}^{5+}$, rather than as Al^{3+}. Equations (6.42) through (6.45) illustrate how some of these species are formed.

Note that adding alum does cause the pH of the water to decrease, due to the release of hydrogen ions. The optimum pH

Table 6.7 List of coagulants and chemicals used during water treatment.

Chemical	Formula	Molecular weight, g/mole	Application
Aluminum sulfate	$Al_2(SO_4)_3 \cdot 14H_2O$	594.4	Primary coagulant
Sodium aluminate	$Na_2Al_2O_4$	163.9	Primary coagulant
Ferric chloride	Fe_2Cl_3	162.2	Primary coagulant
Ferric sulfate	$Fe_2(SO_4)_3$	400	Primary coagulant
Cationic polymer	Specific to polymer	10^4-10^6	Primary coagulant
Activated silica	SiO_2	60.0	Coagulant aid
Bentonite (clay)	*See footnote	Varies	Coagulant aid
Anionic polymer	Specific to polymer	10^4-10^7	Coagulant aid
Nonionic polymer	Specific to polymer	10^5-10^7	Coagulant aid
Calcium oxide	CaO	56	Provide alkalinity, pH adjustment
Sodium hydroxide	NaOH	40	Provide alkalinity, pH adjustment
Soda ash	Na_2CO_3	106	Provide alkalinity, pH adjustment
Sulfuric acid	H_2SO_4	98.1	pH adjustment
Phosphoric acid	H_3PO_4	98	pH adjustment

*Bentonite is basically $Al_2O_3 \cdot 2SiO_2 \cdot H_2O$ that also contains Ca, Mg, Na, and/or K.

range for alum is from 4.5 to 8.0 (Reynolds & Richards, 1996, p. 174); however, the typical pH used in full-scale plants using it to remove turbidity, color, and TOC ranges from 6 to 8 (MWH, 2005, p. 679).

$$Al^{3+} + H_2O \rightleftharpoons Al(OH)^{2+} + H^+ \quad (6.42)$$

$$Al^{3+} + 2H_2O \rightleftharpoons Al(OH)_2^+ + 2H^+ \quad (6.43)$$

$$7Al^{3+} + 17H_2O \rightleftharpoons Al_7(OH)_{17}^{4+} + 17H^+ \quad (6.44)$$

$$Al^{3+} + 3H_2O \rightleftharpoons Al(OH)_3\downarrow + 3H^+ \quad (6.45)$$

Alum accomplishes coagulation in two primary ways: adsorption and charge neutralization, and sweep flocculation. Positively charged aluminum monomers and polymers are adsorbed onto negatively charged colloids, neutralizing the charge. The alum dose is less than the solubility limit of the metal hydroxide $Al(OH)_{3(s)}$. For the second case (sweep floc), a sufficient dose of alum is added such that the solubility limit of the metal hydroxide is exceeded, allowing the precipitation of $Al(OH)_{3(s)}$. This is a gelatinous, sticky precipitate that enmeshes and entraps colloids and other suspended solids as it settles.

Alum reacts with natural alkalinity in water according to the following equation.

$$Al_2(SO_4)_3 \cdot 14H_2O + 3Ca(HCO_3)_2 \rightleftharpoons 2Al(OH)_3\downarrow + 3CaSO_4 + 14H_2O + 6CO_2 \quad (6.46)$$

If the natural alkalinity of the water is insufficient to properly buffer the pH during coagulation, lime or caustic may be added to the water to increase alkalinity. Equations (6.47) and (6.48) show how alum reacts with each of these (MWH, 2005, p. 680).

$$Al_2(SO_4)_3 \cdot 14H_2O + 3Ca(OH)_2 \rightleftharpoons 2Al(OH)_3\downarrow + 3CaSO_4 + 14H_2O \quad (6.47)$$

$$Al_2(SO_4)_3 \cdot 14H_2O + 6NaOH \rightleftharpoons 2Al(OH)_3\downarrow + 3Na_2SO_4 + 14H_2O \quad (6.48)$$

6.5.5.2 Ferrous Sulfate

Ferrous sulfate or copperas ($FeSO_4 \cdot 7H_2O$) may be used as a primary coagulant. Oxygen is required for oxidizing the ferrous ion to ferric, since ferrous hydroxide is soluble. The pH must be raised to approximately 9.5, so that the ferric ions are precipitated as ferric hydroxide (Reynolds & Richards, 1996, p. 175). The simplified reactions are shown below:

$$FeSO_4 \cdot 7H_2O + Ca(OH)_2 \rightleftharpoons Fe(OH)_2 + CaSO_4 + 7H_2O \quad (6.49)$$

$$4Fe(OH)_2 + O_2 + 2H_2O \rightleftharpoons 4Fe(OH)_3\downarrow \quad (6.50)$$

A minimum dose of 50 mg/L of ferrous sulfate is recommended for estimating chemical requirements for removing suspended solids (O'Connell, 1982, p. 287).

6.5.5.3 Ferric Sulfate

Ferric sulfate ($Fe_2(SO_4)_3$) is another chemical used as a primary coagulant. It reacts with natural alkalinity in water as follows:

$$Fe_2(SO_4)_3 + 3\,Ca(HCO_3)_2 \rightleftharpoons 2\,Fe(OH)_3\downarrow + 3\,CaSO_4 + 6\,CO_2 \quad (6.51)$$

According to Reynolds & Richards (1996, p. 176), dense, rapid-settling flocs are produced, and the optimum pH range for ferric sulfate is from 4 to 12. For estimating chemical requirements when removing suspended solids, a minimum does of 25 mg/L should be considered (O'Connell, 1982, p. 288). Typical dosages of ferric sulfate range from 10 to 250 mg/L (MWH, 2005, p. 683).

6.5.5.4 Ferric Chloride

A dense, rapidly settling floc is produced when ferric chloride is used as the primary coagulant (Reynolds & Richards, 1996, p. 176). Ferric chloride ($FeCl_3$) reacts with natural alkalinity in water as follows:

$$2\,FeCl_3 + 3\,Ca(HCO_3)_2 \rightleftharpoons 2\,Fe(OH)_3\downarrow + 3\,CaCl_2 + 6\,CO_2 \quad (6.52)$$

If the natural alkalinity is insufficient for the coagulation process, lime or sodium hydroxide may be added to increase the alkalinity, as shown in the following equations.

$$2\,FeCl_3 + 3\,Ca(OH)_2 \rightleftharpoons 2\,Fe(OH)_3\downarrow + 3\,CaCl_2 \quad (6.53)$$

$$2\,FeCl_3 + 6\,NaOH \rightleftharpoons 2\,Fe(OH)_3\downarrow + 6\,NaCl \quad (6.54)$$

A typical dose of ferric chloride ranges from 5 to 150 mg/L (MWH, 2005, p. 683).

6.5.5.5 Polymers

Synthetic, high-molecular-weight organic polymers may be used as primary coagulants or coagulant aids, depending on the specific type of polymer used. Polymers are long-chain molecules with repeating chemical units, and they typically have an ionic group that imparts an overall electrical charge to the polymer chain. Cationic polymers possess a positive charge and have molecular weights ranging from 10^4 to 10^6 grams per mole. Their primary mechanism for removing colloids is by charge neutralization. They are strongly charged at low and neutral pH. Anionic polymers have a negative charge, and non-ionic polymers have a net charge of zero. Both types are only used as coagulant aids during coagulation. Interparticle bridging is the major mechanism involved when anionic polymers are used. Polymer dosages range from 1 – 10 mg/L for sedimentation (MWH, 2005, p. 688).

6.5.5.6 Coagulant Aids

Other chemicals or materials may be added to water to enhance coagulation. For some waters with low turbidity, bentonite or kaolinite clay has been added to increase interparticle collisions and optimize coagulation. Each coagulant has an optimum pH range and the addition of an acid or base may be required for adjusting the pH. Sulfuric and phosphoric acid are used to lower the pH, whereas lime, soda ash, and sodium hydroxide are used to raise the pH. Polymers often serve as coagulant aids with aluminum and iron salts to promote larger flocs. Viessman *et al.* (2009, p. 405) indicate that polymer dosages of 0.1 to 0.5 mg/L are used with alum.

Example 6.5 Coagulant dose

A surface water containing 50 mg/L of natural alkalinity as $CaCO_3$ is coagulated with an alum dose of 125 mg/L. Determine if there is sufficient alkalinity present in the source water to prevent inhibition of alum coagulation. If there is insufficient alkalinity present, calculate the amount of lime necessary to react with the alum.

Solution

MW of $CaCO_3$ is: $40 + 12 + 3(16) = 100$ g/mole

EW of $CaCO_3$ is:

$$\frac{100\ \text{g/mole}}{2\ \text{eq}} = 50\ \frac{\text{g}}{\text{eq}}.$$

MW of $Ca(HCO_3)_2$ is: $40 + (1+12+3\times16)\times2 =$ 162 g/mole

EW of $Ca(HCO_3)_2$ is:

$$\frac{162\ \text{g/mole}}{2\ \text{eq}} = 81\ \frac{\text{g}}{\text{eq}}.$$

MW of $Al_2(SO_4)_3{\cdot}14H_2O$ is: $2(27) + (32 + 4\times16)\times3 + 14(2\times1+16) = 594$ g/mole

From Equation (6.46) calculate the alkalinity required to react with the alum.

$$Al_2(SO_4)_3 \cdot 14H_2O + Ca(HCO_3)_2 \rightarrow 2\,Al(OH)_3\downarrow + 3\,CaSO_4 + 14H_2O + 6\,CO_2$$

Approximately 21 mg/L of alkalinity expressed as $Ca(HCO_3)_2$ is required to react with the alum.

$$125\ \frac{\text{mg Alum}}{\text{L}}\left(\frac{162\ \text{g}\ Ca(HCO_3)_2}{594\ \text{g Alum}}\right)\left(\frac{\text{eq}}{81\ \text{g}\ Ca(HCO_3)_2}\right)\times\left(\frac{50\ \text{g}\ CaCO_3}{\text{eq}}\right) = \boxed{21\ \frac{\text{mg}\ CaCO_3}{\text{L}}}$$

Since the actual alkalinity of the water exceeds that which is required to react with the alum, no additional adjustment to the alkalinity is necessary. In the event that there is insufficient alkalinity present in the source water, lime or sodium hydroxide could be added to increase alkalinity based on Equation (6.47) or (6.48).

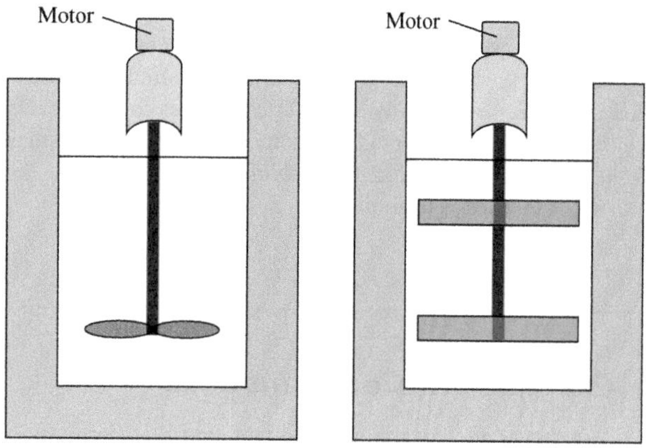

Figure 6.15 Rapid mixer: (a) Propeller and (b) Paddle.

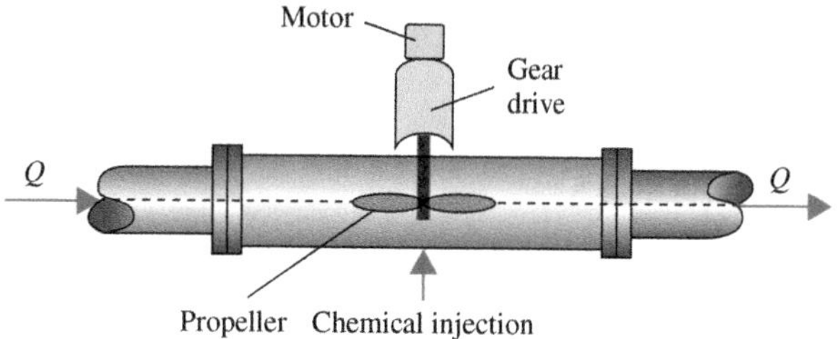

Figure 6.16 In-line mechanical mixer.

6.5.6 Mixing

Dispersing coagulants into the water is a critical step in the treatment process sequence. Mixing of chemicals may be accomplished by turbine and propeller mechanical mixers (Figure 6.15), over-and-under baffling, in-line mechanical mixers (Figure 6.16), in-line static mixers, and by injection into the suction intake of a pump.

Additional information on the design of mixing systems may be found in Metcalf & Eddy (2003, pp. 344–361) and Reynolds & Richards (1996, pp. 182–193). The velocity gradient (G) is one of the main parameters used in designing mixing systems and is determined using the following equation:

$$G = \left(\frac{W}{\mu}\right)^{0.5} = \left(\frac{P}{\mu \forall}\right)^{0.5} \tag{6.55}$$

where:

G = velocity gradient, fps/ft or s^{-1} (mps/m or s^{-1})

W = power imparted to the water per unit volume, $\frac{\text{ft·lb}_f}{\text{s·ft}^3}\left(\frac{\text{N·M}}{\text{s·m}^3}\right)$

P = power imparted to the water, $\frac{\text{ft·lb}_f}{\text{s}}\left(\frac{\text{N·M}}{\text{s·m}^3}\right)$

μ = absolute viscosity of the water, $\frac{\text{ft·lb}_f}{\text{s}}$

$\forall$ = volume of the basin, ft^3 (m^3).

In most instances, a minimum of two rapid or flash mixers are operated in parallel.

Rapid mix tanks are necessary for dispersing the coagulant and coagulant aid into the water. Either square or circular tanks may be used; however, stator blades and/or baffles along the sides of the walls should be used to prevent vortexes from forming.

Example 6.6 Rapid mix design

Determine the dimensions of a rapid-mixing basin to process 3.0 MGD of water. The basin will be square and the depth is assumed to be 1.5 times the width. Use a detention time of 30 s and velocity gradient of 750 fps/ft. The average minimum sustained temperature is 40°F and the absolute viscosity is 3.229×10^{-5} lb·s/ft². Two rapid-mixing basins will be constructed, each having the capacity to process the entire flow. Determine the following:

a) Dimensions of L, W, and D (ft).
b) Power input (hp).

Solution part a

Calculate the volume of the rapid-mixing basin by rearranging the detention time equation.

$$\forall = \tau \times Q = 30\text{s}\left(\frac{3.0 \times 10^6 \text{ gal}}{\text{d}}\right)\left(\frac{1 \text{ d}}{24 \text{ h}}\right)\left(\frac{1 \text{ h}}{3600 \text{ s}}\right) \times \left(\frac{1 \text{ ft}^3}{7.48 \text{ gal}}\right) = 139 \text{ ft}^3$$

$$\forall = L \times W \times D = 139 \text{ ft}^3 \quad L = W \text{ or } A = W^2$$

$$D = 1.5W$$

$$\forall = W \times W \times 1.5W = 139 \text{ ft}^3$$

$$W = \sqrt[3]{\frac{139 \text{ ft}^3}{1.5}} = \boxed{4.50 \text{ ft}} = L$$

$$D = 1.5 \times 4.50 \text{ ft} = \boxed{6.75 \text{ ft}}$$

Solution part b

The power input is calculated by rearranging Equation (6.55):

$$P = G^2 \times \mu \times \forall$$

$$P = (750 \text{ s}^{-1})^2\left(3.229 \times 10^{-5}\frac{\text{lb}\cdot\text{s}}{\text{ft}^2}\right)(4.5 \text{ ft} \times 4.5 \text{ ft} \times 6.75 \text{ ft})$$

$$P = 2.48 \times 10^3 \frac{\text{ft}\cdot\text{lb}}{\text{s}}\left(\frac{\text{hp}}{\frac{550\text{ft}\cdot\text{lb}}{\text{s}}}\right) = \boxed{4.5 \text{ hp}}$$

Therefore, construct two rapid-mixing basins with dimensions of 4.5 ft by 4.5 ft by 6.75 ft deep. Each basin would be provided with a nominal 4.5 hp mixer. The actual horsepower would be greater than this and would be estimated by dividing 4.5 hp by the combined efficiency of the mixer and motor.

6.5.7 Flocculation

Flocculation involves slow stirring/mixing of the destabilized colloids to promote particle aggregation and removal during subsequent settling operations. Mixing cannot be too intense, as the fragile flocs will shear into smaller fragments and not settle properly. Paddle-wheel flocculators and vertical turbine mixers are the primary types of devices used in water treatment. Figures 6.17 and 6.18 show the two types of flocculators.

Paddle-wheel flocculators rotate from 1 - 5 rpm, whereas vertical-shaft turbines rotate from 10 - 30 rpm (MWH, 2005, p. 743). Design criteria for both types of flocculators are shown in Table 6.8.

6.5.7.1 Paddle-Wheel Flocculators

Paddle-wheel flocculators may use horizontal or vertical shafts but, typically, horizontal shafts are preferred at conventional water treatment plants. In *Recommended Standards for Water Works* (2007, p. 61), engineers are encouraged to use tapered flocculation to produce large, dense floc. A detention time of at least 30 minutes is recommended, with a flow-through velocity no less than 0.5 ft/min (0.2 m/min) and no greater than 1.5 ft/min (0.46 m/min).

Tapered flocculation employing three separate compartments, each at a different velocity gradient, is normally practiced, so that a dense, good settling floc is produced. Typical G values used in the first, second, and third compartments respectively, are 50, 20, and 10 seconds^{-1}. The velocity gradient varies from 10 - 100 s^{-1} and the $G\tau$ parameter varies from 10^4 to 10^5. The power dissipated in the water by a paddle-wheel flocculator is calculated using the following equation:

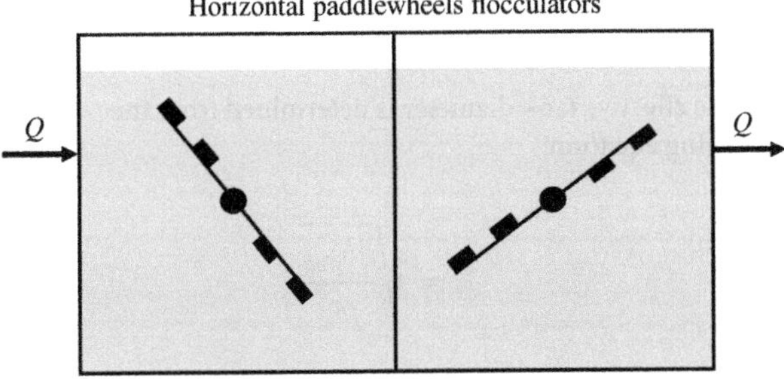

Figure 6.17 Horizontal paddle-wheel flocculators.

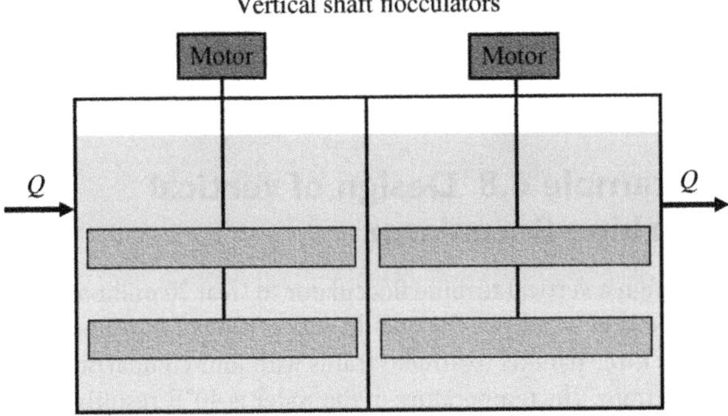

Figure 6.18 Vertical-shaft flocculators.

Table 6.8 Design criteria for horizontal-shaft paddle and vertical-shaft turbine flocculators.

Parameter	Units	Horizontal-shaft with paddles	Vertical-shaft turbine
Velocity gradient	s^{-1}	20–50	10–80
Maximum tip speed	m/s	1	2–3
Rotational speed	Rev/min (rpm)	1–5	10–30
Compartment dimensions (plan)			
Width	m	3–6	6–30
Length	m	3–6	3–5
Number of compartments	Number	2–6	4–6
Variable speed drives	–	Normally	Normally

Source: MWH (2012) p. 614. Reproduced by permission of John Wiley & Sons Ltd.

$$P = \frac{C_D A \rho v_r^3}{2} \quad (6.56)$$

where:

P = power dissipated in the water, ft·lb/sec (N·m/sec)
C_D = coefficient of drag, equal to 1.8 for flat blades, dimensionless
A = area of paddles, ft^2 (m^2)
ρ = density of water, lb·sec/ft^4 (kg/m^3)
v_r = velocity of the paddles relative to water, fps (mps).

Normally, v_r is 0.70 to 0.80 of the paddle velocity (v_p). The velocity of the paddle blade is calculated from the following equation:

$$v_p = 2\pi r N \quad (6.57)$$

where:

v_p = velocity of the paddle blade, fps (mps)
r = distance from the center of the shaft to the center of the blade, ft (m)
N = rotational speed, revolutions/second.

Example 6.7 Paddle-wheel flocculation basin design

A flocculator that is 30 m long, 15 m wide, and 5 m deep houses four, horizontal paddle-wheels. Each paddle-wheel has four symmetrical arms with fiber glass boards, 2.0 m from the shaft to the center of the board, rotating at 1.5 rpm. Paddles are 13 m long and 0.20 m wide. The water treatment plant processes 95,000 m^3/d of water. Assume that the mean water velocity is 25% of the paddle velocity and C_D equals 1.5, and μ and ρ are equal to 1.139×10^{-3} kg/(m·s) and 991.1 kg/m^3, respectively. Determine the following:

a) Velocity of paddle blades (m/s).
b) Total power input (W).
c) Velocity gradient (s^{-1}).
d) Detention time (min).
e) $G\tau$ (dimensionless).

Solution part a

$$v_p = 2\pi r N = 2 \times \pi \times (2.0\,\text{m})\left(\frac{1.5\,\text{rev}}{\text{min}}\right)\left(\frac{1\,\text{min}}{60\,\text{s}}\right)$$

$$= \boxed{0.31\,\frac{\text{m}}{\text{s}}}$$

Solution part b

$$v_r = (1 - 0.25)v_p = 0.75 \times 0.31\,\frac{\text{m}}{\text{s}} = 0.23\,\frac{\text{m}}{\text{s}}$$

$$A = (4\,\text{paddle-wheels})(4\,\text{arms})(13\,\text{m})(0.20\,\text{m}) = 41.6\,\text{m}^2$$

$$P = \frac{C_D A \rho v_r^3}{2} = \frac{(1.5)(41.6\,\text{m}^2)(999.1\,\text{kg/m}^3)(0.23\,\text{m/s})^3}{2}$$

$$= 379\,\frac{\text{kg}\cdot\text{m}^2}{\text{s}^3}$$

$$P = 379\,\frac{\text{kg}\cdot\text{m}^2}{\text{s}^3}\left(\frac{1\text{N}}{\frac{\text{kg}\cdot\text{m}}{\text{s}^2}}\right) = 379\,\frac{\text{N}\cdot\text{m}}{\text{s}}$$

$$P = 379\,\frac{\text{N}\cdot\text{m}}{\text{s}}\left(\frac{1\text{J}}{\text{N}\cdot\text{m}}\right)\left(\frac{1\text{W}}{\text{J/s}}\right) = \boxed{379\,\text{W}}$$

Solution part c

$$G = \left(\frac{W}{\mu}\right)^{0.5} = \left(\frac{P}{\mu V}\right)^{0.5}$$

$$= \left(\frac{379\,\frac{\text{N}\cdot\text{m}}{\text{s}}}{\left(1.139 \times 10^{-3}\,\frac{\text{kg}}{\text{m}\cdot\text{s}}\right)(30\,\text{m} \times 15\,\text{m} \times 5\,\text{m})}\right)^{0.5}$$

$$G = \boxed{12.2\,\text{s}^{-1}}$$

Solution part d

$$\tau = \frac{V}{Q} = \frac{(30\,\text{m} \times 15\,\text{m} \times 5\,\text{m})}{95{,}000\,\text{m}^3/\text{d}}\left(\frac{24\,\text{h}}{\text{d}}\right)\left(\frac{60\,\text{min}}{\text{h}}\right)$$

$$= \boxed{34\,\text{min}}$$

Solution part e

$$G\tau = (12.2\,\text{s}^{-1}) \times (34\,\text{min}) \times \left(\frac{60\,\text{s}}{\text{min}}\right) = \boxed{2.49 \times 10^4}$$

OK, since between 10^4 and 10^5

6.5.7.2 Vertical-Shaft Turbine Flocculators

In new water treatment plants and those that use direct filtration, vertical-shaft turbine flocculators are typically used. This type of flocculator has an impeller attached to a vertical shaft that is rotated by an electric motor. There are different types of impellers that produce either radial or axial flow mixing. Important design parameters for these types of flocculators include power number, effective tank diameter, and diameter to effective tank diameter ratio.

Details on the design of these systems can be found in *Water Treatment Principles and Design* (MWH, 2005, pp. 743–752). Typically, each compartment is designed as a cube.

The power number is calculated using the following equation:

$$N_p = \frac{P}{\rho N^3 D^5} \tag{6.58}$$

where:
N_p = power number, dimensionless
P = power input, hp (J/s or W)
ρ = fluid density, lb·s^2/ft^4 (kg/m^3)
N = rotational speed, rev/min (rpm)
D = impeller diameter, ft (m).

The effective tank diameter is determined from the following equation:

$$T_e = \sqrt{\frac{4A_{plan}}{\pi}} \tag{6.59}$$

where:
T_e = effective tank diameter, ft (m)
A_{plan} = plan area of the compartment, ft^2 (m^2).

The D/T_e ranges from 0.3–0.6, but a value of 0.4–0.5 is recommended (MWH, 2005, p. 749).

Example 6.8 Design of vertical turbine flocculator

Design a vertical turbine flocculator to treat 20 million gallons of water per day at a detention time of 20 minutes. Use three parallel treatment trains with four compartments per train. The temperature of the water is 40°F, resulting in values of 3.229×10^{-5} lb·s/ft^2 and 1.938 lb·s^2/ft^4 for μ and

ρ, respectively. The impeller diameter to effective tank diameter (T_e) ratio is 0.45. Assume a power number (N_p) of 0.25 for a three-pitch blade with camber, and a mean velocity gradient of 80 s^{-1}. Determine the following:

a) Dimensions of each compartment assuming they are cubes (ft).
b) Impellor diameter (ft).
c) Power input per compartment (hp).
d) Rotational speed of each turbine (rpm).

Solution part a

Determine the volume of each compartment by rearranging the detention time equation:

$$\forall = \tau \times Q = (20\,\text{min})\left(\frac{20 \times 10^6\,\text{gal}}{\text{d}}\right)\left(\frac{1\,\text{d}}{24\,\text{h}}\right) \times \left(\frac{1\,\text{h}}{60\,\text{min}}\right)\left(\frac{1\,\text{ft}^3}{7.48\,\text{gal}}\right) = 3.71 \times 10^4\,\text{ft}^3$$

$$\frac{\forall}{\text{compartment}} = \frac{3.71 \times 10^4\,\text{ft}^3}{(3\,\text{trains})(4\text{compartment/train})} = \frac{3090\,\text{ft}^3}{\text{compartment}}$$

$$L^3 = 3090\,\text{ft}^3 \qquad \boxed{L = 14.5\,\text{ft} = W = D}$$

Solution part b

Calculate the effective tank diameter using Equation (6.59):

$$T_e = \sqrt{\frac{4A_{plan}}{\pi}} = \sqrt{\frac{4 \times (14.5\,\text{ft} \times 14.5\,\text{ft})}{\pi}} = 16.4\,\text{ft}$$

$$\frac{D}{T_e} = 0.45 \qquad D = 0.45 \times T_e = (0.45)(16.4\,\text{ft}) = 7.38\,\text{ft}$$

Use:

$$\boxed{D = 7.5\,\text{ft}}$$

Solution part c

Determine the power input to each compartment by rearranging the mean velocity gradient Equation (6.55):

$$G = \sqrt{\frac{P}{\mu\forall}} \qquad P = G^2\mu\forall$$

$$P = (80\,\text{s}^{-1})^2\left(3.229 \times 10^{-5}\,\frac{\text{lb}\cdot\text{s}}{\text{ft}^2}\right)(14.5\,\text{ft})^3 = 630\,\frac{\text{ft}\cdot\text{lb}}{\text{s}}$$

$$P = \left(630\,\frac{\text{ft}\cdot\text{lb}}{\text{s}}\right)\left(\frac{\text{hp}}{550\,\frac{\text{ft}\cdot\text{lb}}{\text{s}}}\right) = \boxed{1.15\,\text{hp}}$$

Solution part d

Determine the rotational speed by rearranging the power number equation, as shown below:

$$N_p = \frac{P}{\rho N^3 D^5} \qquad N = \sqrt[3]{\frac{P}{\rho N_p D^5}}$$

$$N = \sqrt[3]{\frac{630\,\frac{\text{ft}\cdot\text{lb}}{\text{s}}}{\left(1.938\,\frac{\text{lb}\cdot\text{s}^2}{\text{ft}^4}\right)(0.25)(7.5\,\text{ft})^5}} = 0.38\,\frac{\text{rev}}{\text{s}}$$

$$N = 0.38\,\frac{\text{rev}}{\text{s}}\left(\frac{60\text{s}}{\text{min}}\right) = \boxed{22.8\,\text{rpm}}$$

6.6 Water softening

Water softening is the removal of polyvalent cations from water that cause hardness. Hardness is primarily attributed to the divalent metallic cations, calcium (Ca^{2+}) and magnesium (Mg^{2+}). Although ions of iron (Fe^{2+}), manganese (Mn^{2+}), strontium (Sr^{2+}), and aluminum (Al^{3+}) also produce hardness, they are generally present at insignificant concentrations.

The removal of ions that cause hardness can be accomplished by chemical precipitation, ion exchange, and membrane processes such as nanofiltration and reverse osmosis. Only chemical precipitation using stoichiometeric equations involving lime and soda ash will be discussed for determining chemical dosages for removing hardness. Equilibrium equations and Caldwell-Lawrence diagrams can also be used for calculating chemical dosages (Benefield *et al.*, 1982, pp. 269–303); Merrill, 1982, pp. 497–564). The reader should consult the following references to learn more about ion exchange and membrane processes used to soften water: *Water Quality and Treatment* (1999, pp. 9.1–9.87 and 11.1–11.67), *Water Treatment Plant Design* (2005, pp. 12.1–12.54 and 13.1–13.47), and *Water Treatment: Principles and Design* (2005, pp. 1429–1497).

Hard water does not lather well so, therefore:

- soap and detergent consumption increases;
- it produces a scum in the bathtub, shower stall, and on plumbing fixtures;
- it produces scale in water heaters and boilers.

Water hardness is primarily an aesthetics consideration. However, lime softening can effectively remove heavy metals (e.g. arsenic, chromium, lead, and mercury), iron, manganese, natural organic matter (NOM), total organic carbon, and turbidity (MWH, 2005, p. 1595).

Total hardness in water is defined as the sum of the calcium and magnesium ions, because these are the prevalent species present, especially in groundwater.

$$\text{TH} = Ca^{2+} + Mg^{2+} \tag{6.60}$$

where:
TH = total hardness, eq/l, meq/l, or mg/L as $CaCO_3$
Ca^{2+} = calcium ion concentration, eq/l, meq/l, or mg/L as $CaCO_3$
Mg^{2+} = magnesium ion concentration, eq/l, meq/l, or mg/L as $CaCO_3$.

Total hardness is subdivided into carbonate and non-carbonate hardness. Carbonate hardness refers to the portion of total hardness that is associated with the carbonate and bicarbonate anions (e.g., $CaCO_3$, $MgCO_3$, $Ca(HCO)_2$, and $Mg(HCO)_2$). Carbonate hardness is sometimes called temporary hardness because it can be removed by heating the water. Lime, either as CaO (quicklime) or $Ca(OH)_2$ (hydrated or slaked lime), is added to water for removing carbonate hardness and magnesium carbonate ($MgCO_3$).

Non-carbonate hardness refers to the hardness caused by ions that are associated with non-alkalinity anions such as chloride (Cl^-) and sulfate (SO_4^{2-}); typical species include magnesium sulfate ($MgSO_4$), magnesium chloride ($MgCl_2$), calcium chloride ($CaCl_2$), and calcium sulfate ($CaSO_4$).

Non-carbonate hardness cannot be removed by heating the water, so it is called permanent hardness. Soda ash (Na_2CO_3) is added for the removal of calcium non-carbonate hardness ($CaSO_4$ and $CaCl_2$). Sodium hydroxide (NaOH) can be used instead of lime and soda ash for water softening. The cost is much greater, but there is less sludge produced and caustic soda is easier to dose compared to slaking lime.

If the alkalinity in the water is less than the total hardness (TH), the carbonate hardness (CH) is equal to the alkalinity (ALK) and the water must be treated with both lime and soda ash. Mathematically, this is expressed as follows:

$$\text{If ALK} < \text{TH, then CH} = \text{ALK} \tag{6.61}$$

where:
ALK = total alkalinity of the water, meq/L or mg/L as $CaCO_3$
TH = total hardness of the water, meq/L or mg/L as $CaCO_3$
CH = carbonate hardness of the water, meq/L or mg/L as $CaCO_3$.

Non-carbonate hardness (NCH) is calculated below:

$$\text{NCH} = \text{TH} - \text{CH} \tag{6.62}$$

where NCH is the non-carbonate hardness expressed in meq/L or mg/L as $CaCO_3$.

When the alkalinity in the water is equal to or greater than the total hardness (TH), the carbonate hardness (CH) is equal to the total hardness (TH) and only lime addition is required for softening the water. Mathematically, this is expressed as follows:

$$\text{If ALK} \geq \text{TH, then CH} = \text{TH and NCH} = 0. \tag{6.63}$$

Table 6.9 shows one method of classifying water with regard to hardness. Most municipalities provide finished water with a hardness ranging from 50 – 150 mg/L as $CaCO_3$ (MWH, 2005, p. 1593).

Table 6.9 Typical classification of water hardness.

Category	Hardness, mg/L as $CaCO_3$
Soft	0 to <50
Moderately hard	50 to <100
Hard	100 to <150
Very hard	>150

Source: Hardness values from MWH (2005) p. 76. Reproduced by permission of John Wiley & Sons Ltd.

6.6.1 Chemical precipitation

Chemical precipitation of hardness-causing ions has traditionally been accomplished by adding lime and soda ash to water. Soda ash may be omitted when the magnesium ion concentration is less than 40 mg/L as $CaCO_3$ and when the non-carbonate hardness is insignificant. The precipitation of calcium (Ca^{2+}) and magnesium (Mg^{2+}) ions from solution is based on the solubility products of $CaCO_3$ and $Mg(OH)_2$, and LeChatelier's principle. The general form of a dissolution reaction is presented below:

$$A_xB_{y(s)} \rightleftharpoons xA^{y+} + yB^{x-} \tag{6.64}$$

The more familiar form of the solubility product expression follows.

$$K_{sp} = [A^{y+}]^x[B^{x-}]^y \tag{6.65}$$

where:
K_{sp} = solubility product constant
$[A^{y+}]$ = molar concentration of cation species, moles/L
$[B^{x-}]$ = molar concentration of anion species, moles/L
x = moles of cation species
y = moles of anion species.

Table 6.10 lists typical solubility product constants for several chemical species involved in environmental science and engineering applications.

The dissolution reactions and solubility expression for calcium carbonate and magnesium hydroxide at 25°C follow:

$$CaCO_{3(s)} \rightleftharpoons Ca^{2+} + CO_3^{2-} \tag{6.66}$$

$$K_{sp} = 4.6 \times 10^{-9} = [Ca^{2+}][CO_3^{2-}] \tag{6.67}$$

$$Mg(OH)_{2(s)} \rightleftharpoons Mg^{2+} + 2\,OH^- \tag{6.68}$$

$$K_{sp} = 2.0 \times 10^{-11} = [Mg^{2+}][OH^-]^2 \tag{6.69}$$

Precipitation of calcium and magnesium is a complex phenomenon related to other dissolution reactions involving Ca^{2+} and Mg^{2+}. The reader should consult Benefield & Morgan (1999, pp. 10.1 – 10.18) for a rigorous explanation of metal removal by chemical precipitation. The use of Equations (6.66) and (6.68) are for illustrative purposes only.

Table 6.10 Solubility Product Constants for Selected Compounds ($T = 25°C$).

Compound	pK_{sp}
Fe_2S_3	88
$Fe(OH)_3$ (amorphous)	38
$Fe_3(PO_4)_2$	33
$Al(OH)_3$ (amorphous)	33
PbS	27
$Ca_3(PO_4)_2$	26
ZnS	21.5
$AlPO_4$	21
$Cu(OH)_2$	19.3
$FePO_4$	17.9
FeS	17.3
$Zn(OH)_2$	17.2
$CaMg(CO_3)_2$ (dolomite)	16.7
$Fe(OH)_2$	14.5
$Pb(OH)_2$	14.3
$Mn(OH)_2$	12.8
$MgNH_4PO_4$	12.6
Ag_2CrO_4	11.6
$Mg(OH)_2$	10.7
CaF_2	10.3
$BaSO_4$	10
AgCl	10
$CaCO_3$ (calcite)	8.34
$CaCO_3$ (aragonite)	8.22
$PbSO_4$	7.8
$Ca(OH)_2$	5.3
$MgCO_3$	5.0
$PbCl_2$	4.8
Ag_2SO_4	4.8
$CaSO_4$	4.59
SiO_2 (amorphous)	2.8

Source: pK_{sp} values from Snoeyink and Jenkins (1980) p. 249. Reproduced by permission of John Wiley & Sons Ltd.

Based on Chatelier's principle, the addition of a common ion will cause a shift in equilibrium. If the concentration of carbonate in the water is increased by adding soda ash (Na_2CO_3), the concentration of carbonate in solution will be increased and the reaction will be shifted to the left, resulting in the precipitation of $CaCO_{3(s)}$. In other words, when the product of the Ca^{2+} ion concentration and the CO_3^{2-} ion concentration exceeds the solubility product of $CaCO_{3(s)}$, calcium carbonate will be precipitated from solution, resulting in a lower calcium ion concentration. The practical solubility limit of $CaCO_3$ is approximately 30 mg/L expressed as $CaCO_3$.

Example 6.9 Calculating the solubility of $CaCO_3$ at 25°C

Calculate the solubility of $CaCO_3$ in water at room temperature, given the pK_{sp} is 8.34.

Solution

$$\text{MW of } CaCO_3 = 40 + (12 + 3 \times 16) = 100\ \text{g/mole}$$

$$CaCO_{3(s)} \rightleftharpoons Ca^{2+} + CO_3^{2-}$$

$$K_{sp} = 10^{-8.34} = 4.6 \times 10^{-9}$$

$$K_{sp} = 4.6 \times 10^{-9} = [Ca^{2+}][CO_3^{2-}]$$

Assume that the concentration of $[Ca^{2+}] = [CO_3^{2-}] = X$.

$$(X)(X) = 4.6 \times 10^{-9}$$

$$X^2 = 4.6 \times 10^{-9} \qquad X = 6.8 \times 10^{-5}\,\frac{\text{moles } CaCO_3}{\text{L}}$$

The solubility of $CaCO_3$ expressed in mg/L is shown below:

$$X = 6.8 \times 10^{-5}\,\frac{\text{moles } CaCO_3}{\text{L}}\left(\frac{100\ \text{g}}{\text{mole}}\right)\left(\frac{1000\ \text{mg}}{\text{g}}\right)$$

$$= \boxed{6.8\,\frac{\text{mg } CaCO_3}{\text{L}}}$$

In practice, the solubility of $CaCO_3$ is approximately equal to 30 mg/L as $CaCO_3$, due to competing reactions, temperature, and process constraints.

Similarly, adding lime either as CaO or $Ca(OH)_2$ will increase the hydroxide ion (OH^-) and Equation (6.68) will be shifted to the left, resulting in the precipitation of $Mg(OH)_{2(s)}$. The removal mechanism involved in both lime and soda ash precipitation involves sweep coagulation (previously discussed in the coagulant section). The practical solubility limit of $Mg(OH)_2$ is 10 mg/L expressed as $CaCO_3$. Therefore, the lowest level of hardness that can be achieved by chemical precipitation with lime and soda ash is about 40 mg/L as $CaCO_3$.

Example 6.10 Calculating the solubility of $Mg(OH)_2$ at 25°C

Calculate the solubility of $Mg(OH)_2$ in water at room temperature, given the pK_{sp} is 10.7.

Solution

MW of $Mg(OH)_2 = 24.3 + (2 \times 17) = 58.3$ g/mole

$$Mg(OH)_{2(s)} \rightleftharpoons Mg^{2+} + 2\,OH^-$$

$$K_{sp} = 10^{-10.7} = 2.0 \times 10^{-11}$$

$$K_{sp} = 2.0 \times 10^{-11} = [Mg^{2+}][OH^-]^2$$

Assume that the concentration of $[Mg^{2+}] = X$ and $[OH] = 2X$.

$$(X)(2X)^2 = 2.0 \times 10^{-11}$$

$$4X^3 = 2.0 \times 10^{-11} \qquad X = 1.7 \times 10^{-4} \frac{\text{moles } Mg(OH)_2}{L}$$

The solubility of $Mg(OH)_2$ expressed in mg/L as $CaCO_3$ is shown below.

$$X = 1.7 \times 10^{-4} \frac{\text{moles } Mg(OH)_2}{L}\left(\frac{58.3\text{ g}}{\text{mole}}\right) \times \left(\frac{\text{eq}}{29.2\text{ g } Mg(OH)_2}\right)\left(\frac{50\text{ g } CaCO_3}{\text{eq}}\right)\left(\frac{1000\text{ mg}}{\text{g}}\right)$$

$$X = \boxed{17 \frac{\text{mg } CaCO_3}{L}}$$

In practice, the solubility of $Mg(OH)_2$ is approximately equal to 10 mg/L as $CaCO_3$ due to competing reactions, temperature, and process constraints.

When the carbonate hardness is adequate, precipitation of calcium carbonate and magnesium hydroxide can be accomplished by the addition of lime for pH adjustment. The pH of the water must be raised to about 9.5 to precipitate calcium carbonate, and to approximately 10.8 to precipitate magnesium hydroxide (Reynolds & Richards, 1996, p. 209). To raise the pH to 10.8 or higher typically requires an excess of between 30 and 70 mg/L of lime as $CaCO_3$ (MWH, 2005, p. 1597). Reynolds & Richards (1996, p. 209) and Viessman *et al.* (2009, p. 409) recommend a lime dosage of 1.25 meq/L in excess of the stoichiometric amount for raising the pH to 10.8 or higher.

Example 6.11 Excess lime impact on pH

Determine the pH of water after adding 1.25 meq/L of $Ca(OH)_2$ to water.

Solution

Calculate the molecular weight of calcium hydroxide.

$$\text{MW of } Ca(OH)_2 = 40 + (16 + 1) \times 2 = 74\text{ g/mole}$$

Calculate the equivalent weight of calcium hydroxide.

$$\text{EW of } Ca(OH)_2 = \frac{74\text{ g}}{2\text{ eq}} = \frac{37\text{ g}}{\text{eq}}$$

The moles of hydroxide ion are calculated as follows:

$$[OH^-] = 1.25\frac{\text{meq}}{L}\left(\frac{1\text{ eq}}{1000\text{ meq}}\right)\left(\frac{37\text{ g}}{\text{eq}}\right) \times \left(\frac{2 \times 17\text{ g}OH^-}{74\text{ g}Ca(OH)_2}\right)\left(\frac{1\text{ mole}}{17\text{ g}OH^-}\right) = 1.25 \times 10^{-3}\frac{\text{moles}}{L}$$

Calculate the pH after calculating the pOH of the solution. The pOH is determined as follows:

$$pOH = -\log[OH^-] = -\log[1.25 \times 10^{-3}] = 2.90$$

The pH is determined by rearranging the following equation:

$$14 = pH + pOH$$

$$pH = 14 - pOH = 14 - 2.90 = \boxed{11.1}$$

Ion exchange softening can remove all calcium and magnesium ions from the water (Viessman *et al.*, 2009, p. 478). In most applications, sodium ions from a strong cation exchange resin are exchanged for calcium and magnesium ions (AWWA/ASCE, 2005, pp. 12.4–12.5). Membrane softening can remove 80–95% of divalent metallic cations such as Ca^{2+} and Mg^{2+} (MWH, 2005, p. 1433). Nanofiltration membranes are used for softening and NOM removal (MWH, 2005, p. 958; Viessman *et al.*, 2009, p. 458).

6.6.2 Lime-soda ash softening stoichiometric equations

The stoichiometric equations involved in the lime-soda ash water softening process are presented below. Although the reactions are presented sequentially, in reality, they proceed simultaneously. In the following reactions, the arrow (↓) denotes the removal of a solid precipitate either as $CaCO_3$ or $Mg(OH)_2$. When quicklime (CaO) is added to water, it forms hydrated or slaked lime [$Ca(OH)_2$] according to Equation (6.70). This is an exothermic reaction.

$$CaO + H_2O \rightleftharpoons Ca(OH)_2 + \text{heat} \qquad (6.70)$$

Carbon dioxide may be present in the source water and is denoted as $CO_{2(aq)}$. This represents the quantity of gaseous carbon dioxide $CO_{2(g)}$ that is dissolved in the water $CO_{2(aq)}$ and is estimated using Henry's law.

$$CO_{2(g)} \rightleftharpoons CO_{2(aq)} \quad (6.71)$$

For ground waters, carbon dioxide is typically present, and at a much higher concentration than in most surface waters. It will be in equilibrium with carbonic acid (H_2CO_3), as shown in Equation (6.72):

$$CO_{2(aq)} + H_2O \rightleftharpoons H_2CO_3 \quad (6.72)$$

If carbon dioxide (carbonic acid) is present, it must be removed or neutralized before hardness-causing ions can be precipitated by adding chemicals. Equation (6.73) shows the precipitation of carbonic acid as $CaCO_{3(s)}$ by adding lime. There is no change in the hardness of the water. If the concentration of carbon dioxide is greater than 10 mg/L, air stripping is usually recommended as the most economical means for removal.

$$H_2CO_3 + Ca(OH)_2 \rightleftharpoons CaCO_3\downarrow + 2H_2O \quad (6.73)$$

Lime is added for the precipitation of calcium and magnesium carbonate hardness, and magnesium non-carbonate hardness. Equations (6.74) through (6.76) show reactions associated with the removal of carbonate hardness: $Ca(HCO_3)_2$, $Mg(HCO_3)_2$, and $MgCO_3$. Lime also reacts with magnesium non-carbonate hardness, as shown in Equations (6.77) and (6.78).

Equation (6.74) represents the removal of calcium carbonate hardness ($Ca(HCO_3)_2$) by converting the bicarbonate ions to carbonate ions and precipitating calcium as $CaCO_3$, by adding lime, $Ca(OH)_2$. This reaction shows that the addition of one equivalent of lime results in one equivalent of hardness being removed.

$$Ca\,(HCO_3)_2 + Ca\,(OH)_2 \rightleftharpoons 2\,CaCO_3\downarrow + 2H_2O \quad (6.74)$$

Equations (6.75) and (6.76) indicate that two equivalents of lime are required for the removal of each equivalent of magnesium bicarbonate hardness, $Mg(HCO_3)_2$. Adding lime converts magnesium bicarbonate to magnesium carbonate ($MgCO_3$) which is soluble, therefore, the hardness of the water does not change, Equation (6.75). The magnesium associated with $MgCO_3$ can then be precipitated as $Mg(OH)_2$ and the calcium associated with lime is precipitated as $CaCO_3$ according to Equation (6.76).

$$Mg\,(HCO_3)_2 + Ca\,(OH)_2 \rightleftharpoons CaCO_3\downarrow + MgCO_3 + 2H_2O \quad (6.75)$$

$$MgCO_3 + Ca\,(OH)_2 \rightleftharpoons CaCO_3\downarrow + Mg\,(OH)_2\downarrow \quad (6.76)$$

Equations (6.77) through (6.80) show how non-carbonate hardness is precipitated and, therefore, removed. Magnesium non-carbonate hardness ($MgSO_4$) can only be removed by adding both lime and soda ash. Equation (6.77) indicates that magnesium associated with $MgSO_4$ is converted to $Mg(OH)_2$, and that the calcium associated with lime is converted to $CaSO_4$, with the net result of no hardness being removed. Calcium sulfate can then be precipitated by adding soda ash, as represented in Equation (6.79). In summary, each equivalent of magnesium sulfate requires one equivalent of lime and one equivalent of soda ash, as shown in Equations (6.77) and (6.79), resulting in a net removal of hardness.

$$MgSO_4 + Ca\,(OH)_2 \rightleftharpoons CaSO_4 + Mg\,(OH)_2\downarrow \quad (6.77)$$

Magnesium non-carbonate hardness removal (in the form of $MgCl_2$) is accomplished by adding both lime and soda ash according to Equations (6.78) and (6.80). Equation (6.78) shows that the magnesium associated with $MgCl_2$ is converted to $Mg(OH)_2$ and that the calcium associated with lime remains in solution as $CaCl_2$. Therefore, the hardness of the water remains unchanged. Soda ash then must be added according to Equation (6.80), so that the calcium associated with $CaCl_2$ is precipitated as $CaCO_3$. In summary, one equivalent of lime and one equivalent of soda ash is required for each equivalent of magnesium chloride removed.

$$MgCl_2 + Ca\,(OH)_2 \rightleftharpoons Mg\,(OH)_2\downarrow + CaCl_2 \quad (6.78)$$

Calcium non-carbonate hardness in the form of $CaSO_4$ is only removed by adding soda ash (Na_2CO_3). For each equivalent of calcium sulfate, one equivalent of soda ash is required, resulting in a net removal of hardness as shown in Equation (6.79).

$$CaSO_4 + Na_2(CO)_3 \rightleftharpoons CaCO_3\downarrow + Na_2SO_4 \quad (6.79)$$

$$CaCl_2 + Na_2(CO)_3 \rightleftharpoons CaCO_3\downarrow + 2\,NaCl \quad (6.80)$$

Excess lime treatment is often practiced, wherein 1.25 milliequivalents per liter (meq/L) of lime is added above the amount predicted by the stoichiometric equations, to ensure that the pH of the water is raised above 10.8 to 11. Recarbonation is the process of adding carbon dioxide to lower the pH and stabilize the water. Equation (6.81) shows that excess lime, $Ca(OH)_2$, is converted to $CaCO_3$ by adding carbon dioxide. This reaction lowers the pH from around 11 to about 10.2 (Viessman *et al.*, 2009, p. 409).

$$Ca\,(OH)_2 + CO_2 \rightleftharpoons CaCO_3\downarrow + H_2O \quad (6.81)$$

Further recarbonation of the water, according to Equation (6.82), converts any remaining $CaCO_3$ that did not precipitate to $Ca(HCO_3)_2$. This reaction results in a final pH in the range of 9.5 to 8.5 (Viessman *et al.*, 2009, p. 409).

$$Ca\,CO_3 + CO_2 + H_2O \rightleftharpoons Ca\,(HCO_3)_2 \quad (6.82)$$

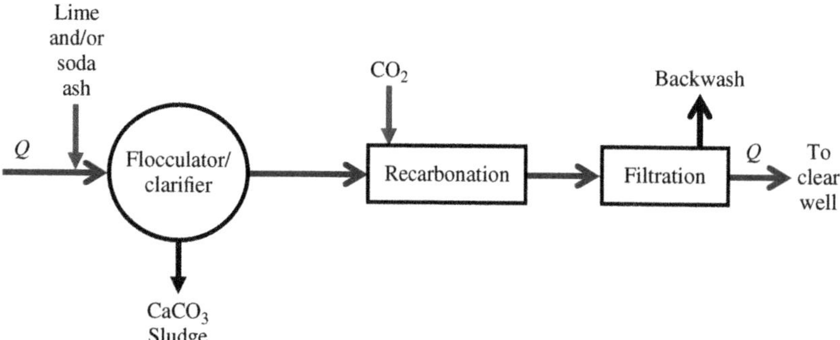

Figure 6.19 Single-stage softening schematic.

6.6.3 Process treatment trains for softening

There are three commonly used process treatment trains used to soften water. The actual process configuration selected will depend on the raw water characteristics and the desired finished water quality. The following three softening processes will be discussed and selected examples presented:

- single-stage, selective calcium removal treatment;
- two-stage excess lime-soda ash treatment; and
- split-flow treatment.

6.6.3.1 Single-Stage Softening

Conventional softening treatment has focused on the removal of calcium carbonate hardness from water. Single-stage softening, or selective calcium removal, can be used on waters with a magnesium concentration less than 40 mg/L as $CaCO_3$. The treatment sequence used consists of flash mixing with lime, flocculation, sedimentation, recarbonation, and filtration. A flocculator/clarifier or solids contactor is often used, rather than the conventional mixing/flocculation/sedimentation configuration. Lime is added for the removal of calcium carbonate hardness. If magnesium hardness removal is necessary, excess lime must be added. Soda ash must also be added if non-carbonate hardness is to be removed.

Figure 6.19 shows selective calcium removal with a flocculator/clarifier incorporated into the treatment sequence. To reduce chemical costs, sludge solids may be recirculated and mixed with the incoming raw water to provide nuclei for precipitation. Carbon dioxide (recarbonation) is added to stop $CaCO_3$ precipitation and to lower the pH. A pH of approximately 8.5 produces a stable water, i.e., one that does not produce scale in pipes downstream of conventional softening.

Example 6.12 Selective calcium removal

Calculate the lime, soda ash (if necessary), carbon dioxide dosages for recarbonation, and solids production for selective calcium removal given the following water analysis. Develop a bar graph showing the original theoretical species in the water.

Alkalinity (HCO_3^-) = 125 mg/L as $CaCO_3$
CO_2 = 13.2 mg/L Mg^{2+} = 12.2 mg/L Na^+ = 50.6 mg/L
Ca^{2+} = 94 mg/L SO_4^{2-} = 139.2 mg/L Cl^- = 88.75 mg/L

Solution

The following table presents concentrations, equivalent weights, and milliequivalents of all species analyzed.

Component	Concentration (mg/L)	Equivalent weight	Calculation	Concentration (meq/L)
CO_2/H_2CO_3	13.2	22	13.2/22 =	0.60
Ca^{2+}	94	20	94/20 =	4.70
Mg^{2+}	12.2	12.2	12.2/12.2 =	1.00
Na^+	50.6	23	50.6/23 =	2.20
Alkalinity (HCO_3^-)	125	50	125/50 =	2.50
SO_4^{2-}	139.2	48	139.2/48 =	2.90
Cl^-	88.75	35.5	88.75/35.5 =	2.50

The figure below is a bar graph showing the theoretical combination of chemical species in the raw water. Carbon dioxide is placed to the left of the bar graph if present. Cations are always placed at the top of the bar graph; calcium, magnesium, sodium, and potassium are shown in this order. Anions are shown at the bottom of the bar graph and should be sequenced as follows: hydroxide, carbonate, bicarbonate, sulfate, and chloride. The water should be electrically neutral so the concentration of cations must balance the anions. If not, the water analysis is suspect. Viessman *et al.* (2009, pp. 409–419) and McGhee (1991, pp. 246–253) provide details on bar graph development.

Since selective calcium hardness removal is being practiced, and the concentration of magnesium hardness is

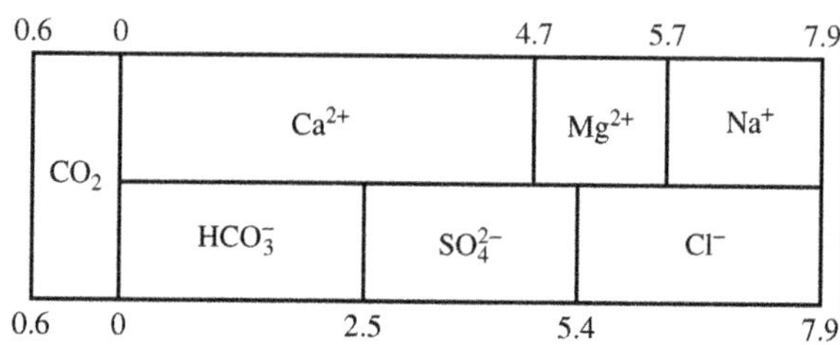

Figure E6.1 Bar graph of the raw water for Example 6.12.

relatively low, 50 mg/L as $CaCO_3$ (1.0 meq/L × 50 mg/meq), lime and soda ash will be required to lower the calcium hardness to approximately 30 mg/L as $CaCO_3$. Lime addition will be required for the neutralization of carbon dioxide according to Equation (6.73), and for the conversion of calcium bicarbonate to calcium carbonate according to Equation (6.74). Non-calcium carbonate hardness ($CaSO_4$) can only be removed by precipitating with sodium carbonate according to Equation (6.79).

Component (1)	Equation number (2)	Concentration, meq/L (3)	CaO, meq/L (4)	Na_2CO_3, meq/L (4)
CO_2/H_2CO_3	(6.73)	0.6	0.6	0
$Ca(HCO_3)_2$	(6.74)	2.5 – 0.6 = 1.9*	1.9	0
$CaSO_4$	(6.79)	2.2	0	2.2

Residual calcium in water is equal to 30 mg/L as $CaCO_3$ or 0.6 meq/L.

The final hardness of the water is calculated as follows and is related to the original magnesium concentration in the water plus the remaining calcium in the water, which is equal to the practical solubility of 30 mg/L as $CaCO_3$.

Lime Dosage

$$\left(0.6\,\frac{\text{meq}}{\text{L}} + 1.9\,\frac{\text{meq}}{\text{L}}\right) \times 28\,\frac{\text{mg CaO}}{\text{meq}} = \boxed{70\,\frac{\text{mg CaO}}{\text{L}}}$$

Soda Ash Dosage

$$\left(2.2\,\frac{\text{meq}}{\text{L}}\right) \times 53\,\frac{\text{mg Na}_2\text{CO}_3}{\text{meq}} = \boxed{117\,\frac{\text{mg Na}_2\text{CO}_3}{\text{L}}}$$

Carbon Dioxide Dosage

First-stage recarbonation will be required to stabilize the water. Assume a $CaCO_3$ concentration of 30 mg/L in the effluent from the settling basin, and that 50% of the carbonate is converted into bicarbonate, resulting in an equilibrium concentration of 15 mg/L of $CaCO_3$.

From Equation (6.81), the carbon dioxide required is:

$$\left(15\,\frac{\text{mg CaCO}_3}{\text{L}}\right)\left(\frac{\text{meq CaCO}_3}{50\text{ mg CaCO}_3}\right)\left(\frac{\text{meq CO}_2}{\text{meq CaCO}_3}\right) \times \left(\frac{22\text{ mg CO}_2}{\text{meq CO}_2}\right) = \boxed{6.6\,\frac{\text{mg CO}_2}{\text{L}}}$$

Solids production as $CaCO_3$

$$\left(0.6\,\frac{\text{meq}}{\text{L}} + 1.9\,\frac{\text{meq}}{\text{L}} + 2.2\,\frac{\text{meq}}{\text{L}} - 0.6\,\frac{\text{meq}}{\text{L}}\right) \times 50\,\frac{\text{mg CaCO}_3}{\text{meq}} = \boxed{205\,\frac{\text{mg CaCO}_3}{\text{L}}}$$

Bar graph of finished water is based on the following concentrations of ions:

$$Ca^{2+} = 0.6\,\frac{\text{meq}}{\text{L}} \qquad Mg^{2+} = 1.0\,\frac{\text{meq}}{\text{L}}$$

$$Na^{+} = 2.2\,\frac{\text{meq}}{\text{L}} + 2.2\,\frac{\text{meq}}{\text{L}} = 4.4\,\frac{\text{meq}}{\text{L}}$$

$$CO_3^{2-} = 0.5 \times \left(\frac{30\text{ mg CaCO}_3/\text{L}}{50\text{ mg CaCO}_3/\text{meq}}\right) = 0.3\,\frac{\text{meq}}{\text{L}}$$

$$HCO_3^{-} = 0.5 \times \left(\frac{30\text{ mg CaCO}_3/\text{L}}{50\text{ mg CaCO}_3/\text{meq}}\right) = 0.3\,\frac{\text{meq}}{\text{L}}$$

$$SO_4^{2-} = 2.9\,\frac{\text{meq}}{\text{L}} \qquad Cl^{-} = 2.5\,\frac{\text{meq}}{\text{L}}$$

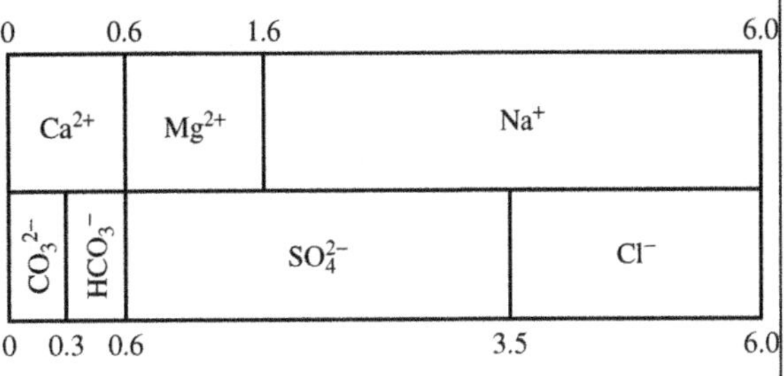

Figure E6.2 Finished Water Bar Graph

6.6.3.2 Two-stage Excess Lime-Soda Ash Treatment

When the magnesium concentration exceeds 40 mg/L as $CaCO_3$, then both calcium and magnesium must be removed from the water. Excess lime ranging from 30–70 mg/L as $CaCO_3$ is added to the water to raise the pH to above 10.5 for the precipitation of magnesium hydroxide (MWH, 2005, p. 1597). Figure 6.20 is a schematic of a two-stage, excess lime-soda ash softening process. This treatment alternative is used to remove both carbonate and non-carbonate hardness.

First-stage treatment consists of flash mixing of lime, followed by flocculation, sedimentation, recarbonation, and filtration. Flocculator/clarifiers or solids contact units are often used in place of conventional coagulation, flocculation, and sedimentation. Both types of these units provide all three functions in one compact unit. Lime is added in the first stage to meet the stoichiometric requirements for the neutralization of carbon dioxide (if present) and calcium carbonate hardness precipitation as $CaCO_{3(s)}$. An excess dosage of lime, typically 1.25 meq/L, is added above stoichiometric amounts to raise the pH above 11 for the precipitation of magnesium as magnesium hydroxide. first stage recarbonation then occurs to convert excess lime ($Ca(OH)_2$) to $CaCO_{3(s)}$ and to lower the pH to around 9.5–10. This also resolubilizes some of the suspended $Mg(OH)_{2(s)}$, resulting in a maximum magnesium concentration of 20 mg/L as $Mg(OH)_2$.

Second stage treatment involves flash mixing of soda ash, followed by flocculation, sedimentation, recarbonation, and filtration. Soda ash (Na_2CO_3) is added to the water for removing calcium non-carbonate hardness. Second stage recarbonation involves adding more carbon dioxide to convert

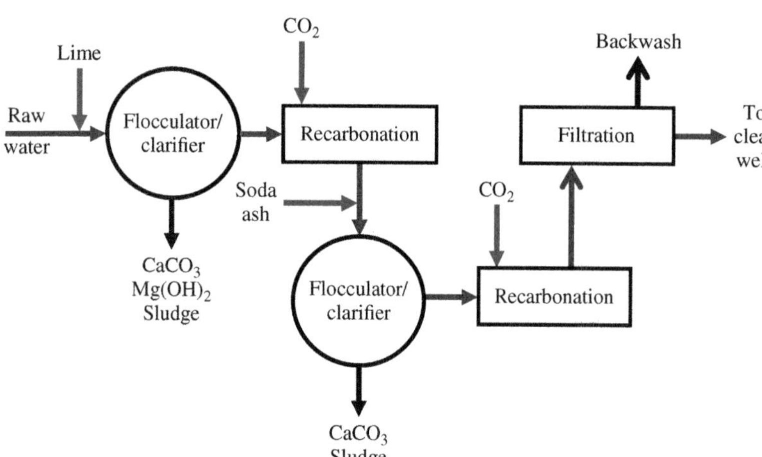

Figure 6.20 Two-stage lime-soda ash softening schematic.

some of the soluble $CaCO_3$ to $Ca(HCO_3)_2$, which lowers the pH to approximately 8.5. This stops the precipitation of $CaCO_{3(s)}$ and produces a stable water. The practical limits of hardness remaining after two-stage softening is approximately 10 mg/L of $Mg(OH)_2$ expressed as $CaCO_3$, and 30 mg/L of calcium expressed as $CaCO_3$, resulting in an overall hardness of 40 mg/L as $CaCO_3$ or 0.8 meq/L.

Example 6.13 Excess lime-soda ash softening problem

Calculate the lime and soda ash requirements to achieve the practical limits of hardness removal given the following water analysis. Assume that excess lime treatment in a two-stage system is being used. Develop a bar graph showing the original theoretical species in the water.

CO_2 = 10 mg/L HCO_3^- = 109.6 mg/L Mg^{2+} = 10 mg/L
K^+ = 7.0 mg/L Na^+ = 11.7 mg/L Ca^{2+} = 40 mg/L
SO_4^{2-} = 67.2 mg/L Cl^- = 11 mg/L

Solution

The following table presents concentrations, equivalent weights, and milliequivalents of all species analyzed.

Component	Concentration (mg/L)	Equivalent weight	Calculation	Concentration (meq/L)
CO_2/H_2CO_3	10	22	10/22 =	0.45
Ca^{2+}	40	20	40/20 =	2.00
Mg^{2+}	10	12.2	10/12.2 =	0.82
Na^+	11.7	23	11.7/23 =	0.51
K^+	7.0	39.1	7.0/39.1 =	0.18
HCO_3^-	109.6	61	109.6/61 =	1.80
SO_4^{2-}	67.2	48	67.2/48 =	1.40
Cl^-	11	35.5	11/35.5 =	0.31

A bar graph showing the theoretical combination of species in the raw water is presented below.

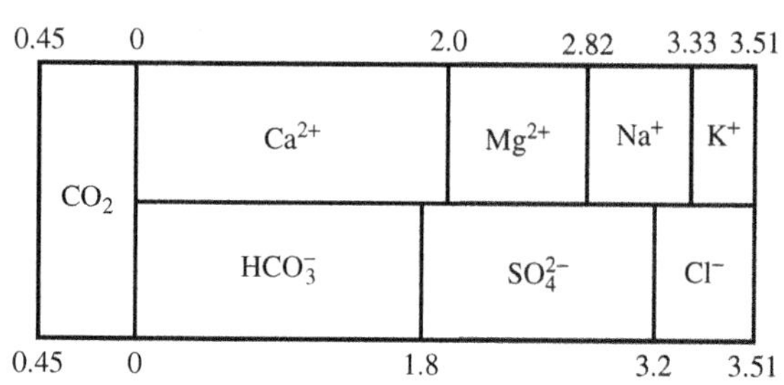

Figure E6.3 Bar graph of the raw water for Example 6.13.

The following table presents the theoretical species, along with the dosages of lime as calcium oxide (CaO) and soda ash as sodium carbonate (Na_2CO_3) that must be added to achieve a residual hardness of approximately 40 mg/L as $CaCO_3$. Excess lime treatment is assumed, which involves adding 1.25 meq/L of lime as CaO (equivalent to 35 mg/L as CaO) beyond the stoichiometric dosage. Both Equations (6.75) and (6.76) must be used to determine the quantity of lime required for precipitating magnesium bicarbonate. For each mole of $Mg(HCO_3)_2$, 2 moles of lime are required. Similarly, Equations (6.77) and (6.79) must be used to determine the quantity of lime and soda ash required for removing magnesium sulfate.

Component (1)	Equation (2)	Concen. (3) (meq/L)	CaO (4) (meq/L)	Na_2CO_3 (5) (meq/L)	$CaCO_3$ solids (6) (meq/L)	$Mg(OH)_2$ solids (7) (meq/L)
CO_2/H_2CO_3	(6.73)	0.45	0.45	0	0.45	0
$Ca(HCO_3)_2$	(6.74)	1.80	1.80	0	2 × 1.80 = 3.60	0
$CaSO_4$	(6.79)	0.20	0	0.20	0.20	
$MgSO_4$	(6.77) (6.79)	0.82	0.82	0.82	0.82	0.82
Excess lime CaO			1.25	0	0	0
Total			4.32	1.02	5.07	0.82

The total lime dosage as CaO assuming 90% purity is the sum of all species in Column 4:

$$4.32\,\frac{meq}{L} \times 28\,\frac{mg\ CaO}{meq} \times 8.34\,\frac{lb/MG}{mg/L} \times \frac{1}{0.90}$$

$$= \boxed{1120\,\frac{lb}{MG}}$$

where MG denotes millions of gallons.

The total soda ash dosage as Na_2CO_3, assuming 85% purity, is the sum of all species in Column 5:

$$1.02\,\frac{meq}{L} \times 53\,\frac{mg\ Na_2CO_3}{meq} \times 8.34\,\frac{lb/MG}{mg/L} \times \frac{1}{0.85}$$

$$= \boxed{530\,\frac{lb}{MG}}$$

First stage recarbonation

The carbon dioxide required in the first stage recarbonation process is determined by multiplying the sum of the excess lime added, plus the remaining magnesium hydroxide concentration (typically assumed to be 0.20 meq/L or 10 mg/L $Mg(OH)_2$ expressed as calcium carbonate), by the equivalent weight of carbon dioxide (22 mg/meq).

$$(1.25 + 0.2)\frac{meq}{L} \times 22\,\frac{mg\ CO_2}{meq} = \boxed{31.9\frac{mg\ CO_2}{L}}$$

Second stage recarbonation

The carbon dioxide required for second stage recarbonation is equal to the amount necessary to convert a portion of the remaining alkalinity associated with the residual hardness to calcium hardness. The residual hardness is normally 40 mg/L expressed as $CaCO_3$, with 30 mg/L of calcium and 10 mg/L as magnesium, both expressed in terms of calcium carbonate. For estimating CO_2 requirements in the second stage, approximately 50–75% of the remaining alkalinity is assumed to be converted to bicarbonate alkalinity. For this example, 50% conversion of the calcium carbonate is assumed to be converted to calcium bicarbonate.

$$(0.5) \times \left(0.8\,\frac{meq}{L}\right) \times 22\,\frac{mg\ CO_2}{meq} = \boxed{8.8\frac{mg\ CO_2}{L}}$$

Solids or sludge production

The solids or sludge produced during softening is estimated by summing up columns 6 and 7. Column 6 represents calcium carbonate solids, while Column 7 represents magnesium hydroxide solids. The quantity of calcium carbonate solids is reduced by 0.6 meq/L, since the practical solubility limit of calcium carbonate is assumed to be 30 mg/L as $CaCO_3$. The quantity of magnesium hydroxide solids is reduced by 0.2 meq/L, since the practical solubility limit of magnesium hydroxide is assumed to be 10 mg/L as $CaCO_3$. Since both are expressed in milli-equivalents per liter, the total sludge production will be expressed in terms of calcium carbonate solids, as shown below:

$$CaCO_3\ \text{solids} = (0.45 + 3.6 + 0.2 + 0.82 - 0.60)$$

$$\times 50\,\frac{mg\ CaCO_3}{meq} = \boxed{223.5\frac{mg\ CaCO_3}{L}}$$

$$CaCO_3\ \text{solids} = 223.5\frac{mg\ CaCO_3}{L} \times \frac{8.34\ lb}{MG/\frac{mg}{L}}$$

$$= \boxed{1864\frac{lb\ CaCO_3}{MG}}$$

$$Mg(OH)_2\ \text{solids} = (0.82 - 0.20) \times 50\,\frac{mg\ CaCO_3}{meq}$$

$$= \boxed{31\,\frac{mg\ CaCO_3}{L}}$$

$$Mg(OH)_2\ \text{solids} = 31\,\frac{mg\ CaCO_3}{L} \times \frac{8.34\ lb}{MG/\frac{mg}{L}}$$

$$= \boxed{259\frac{lb\ CaCO_3}{MG}}$$

$$\text{Total solids produce} = 1864\frac{lb\ CaCO_3}{MG} + 259\frac{lb\ CaCO_3}{MG}$$

$$= \boxed{2123\frac{lb\ CaCO_3}{MG}}$$

The bar graph after lime and soda ash addition is shown below. The excess lime (1.25 meq/L) is shown to the left of the graph. The practical solubility limits of calcium and magnesium are 30 mg/L as $CaCO_3$ (0.6 meq/L) and 10 mg/L $Mg(OH)_2$ as $CaCO_3$ (0.2 meq/L). The sodium concentration is equal to the original sodium concentration plus the increase due to soda ash addition (0.51 + 1.02 = 1.53 meq/L). The chloride, sulfate, and potassium concentrations remain unchanged. The total alkalinity remaining (0.8 meq/L) is assumed to be associated with the residual hardness of 0.8 meq/L, 0.2 meq/L being associated with hydroxide and the remaining 0.6 meq/L associated with carbonate.

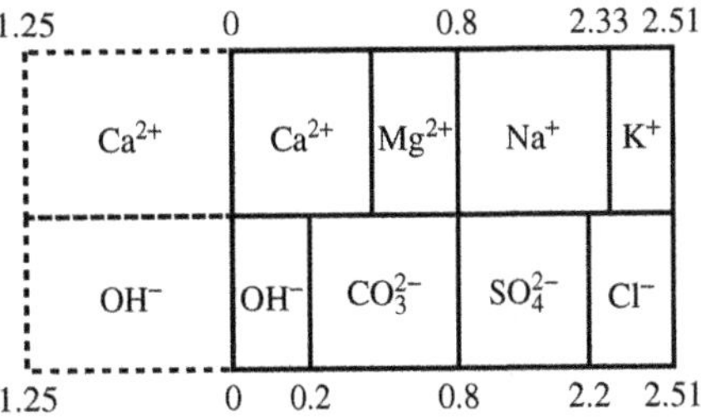

Figure E6.4 Bar graph after chemical addition for Example 6.13.

The bar graph of the finished water is shown below. The only difference in this and the bar graph after lime and soda ash addition is the conversion of 50% of the remaining alkalinity to carbonate and bicarbonate.

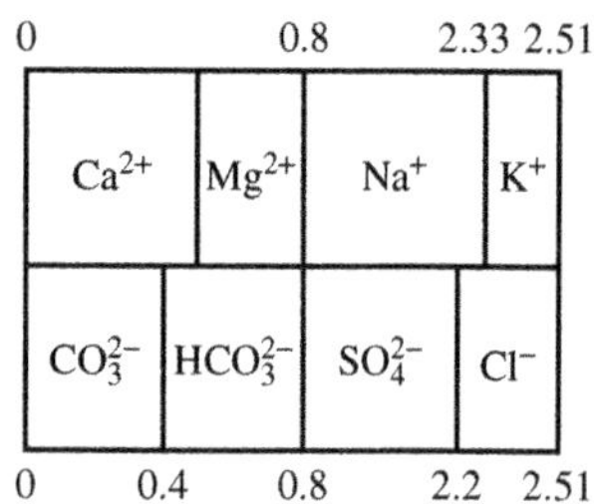

Figure E6.5 Bar graph of finished water for Example 6.13.

6.6.3.3 Split Treatment

Softening using the split treatment scheme involves treating a portion of the raw water by excess lime treatment, then blending this softened water with the remaining portion of raw water by-passed. Figure 6.21 is a schematic of the split treatment softening process.

Excess lime is added to the portion of raw water to be treated for neutralization of carbon dioxide, carbonate hardness removal, and some magnesium removal. Magnesium will not be removed to the lowest level. Using split treatment allows a municipality to achieve any desired water hardness above 40 mg/L as $CaCO_3$. Split treatment can lower overall treatment costs, due to smaller reactors and reduced chemical costs. The by-passed raw water contains alkalinity and carbon dioxide and, when blended with the chemically treated flow, this converts the alkalinity (in the by-passed flow) to carbonate. This is used for removing additional magnesium hardness, and the carbon dioxide (in the by-passed flow) is used for neutralizing the excess lime (in the treated flow). When the treatment objective is magnesium removal, a material balance on magnesium can be performed to determine the ratio of the raw water that must by-pass softening. The ratio of the by-pass flow to treated flow is determined using the following equation:

$$X = \frac{[Mg^{2+}]_{final} - [Mg^{2+}]_{soft}}{[Mg^{2+}]_{raw} - [Mg^{2+}]_{soft}} \qquad (6.83)$$

where:

X = ratio of by-passed flow, fraction

$[Mg^{2+}]_{raw}$ = magnesium concentration in raw water, mg/L as $CaCO_3$

$[Mg^{2+}]_{final}$ = magnesium concentration in final or finished water, mg/L as $CaCO_3$; typically, a concentration of 50 mg/L as $CaCO_3$ is used

$[Mg^{2+}]_{soft}$ = magnesium concentration in softened water, mg/L as $CaCO_3$; the practical solubility limit of magnesium, 10 mg/L as $CaCO_3$ is normally selected for this value.

6.6.4 Other softening methods

Ion exchange is another technique used for removing hardness from water, and it is extensively used in households. Calcium (Ca^{2+}) and magnesium (Mg^{2+}) ions are replaced by sodium ions (Na^{+}) by passing water through a kaolinite or montmorillonite bed, or a synthetic material such as a polymeric resin. Softening can also be accomplished when water is passed through a semi-permeable membrane such as those used in reverse osmosis, ultrafiltration, or other membrane processes. Information on these methods may be found in MWH (2005) and in Reynolds & Richards (1996).

6.7 Sedimentation

Surface water will contain turbidity in the form of colloidal and suspended solids that must be removed from the water. Gravity settling (sedimentation) is the major unit operation used for removing suspended solids. As previously discussed, coagulation is required to destabilize colloidal suspensions so that they can be removed by **sedimentation**. Sedimentation is considered a **unit operation** (physical treatment) involving solids-liquid separation using gravitational force. If the raw water contains turbidity greater than 1000 NTUs (Reynolds & Richards, 1996, p. 255), pre-treatment in the form of plain sedimentation is typically accomplished before proceeding to coagulation/flocculation processes.

During water treatment, there are two types of settling that are typically encountered: free and flocculent settling. Zone settling (discussed in Chapter 7) may occur during gravity

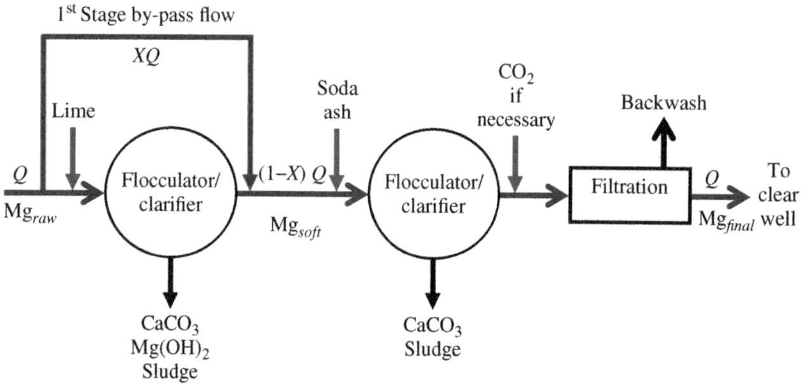

1st Stage by-pass flow
XQ
Lime
Soda ash
CO_2 if necessary
Backwash
Q
Flocculator/ clarifier
(1−X) Q
Flocculator/ clarifier
Filtration
Q
To clear well
Mg_{raw}
Mg_{soft}
Mg_{final}
$CaCO_3$ $Mg(OH)_2$ Sludge
$CaCO_3$ Sludge

Figure 6.21 Split treatment softening schematic.

thickening of residuals from water treatment. The theory behind free and flocculent settling will be discussed in the following sections.

6.7.1 Free settling

Free settling is observed when there are discrete, non-flocculent particles whose size, shape, and specific gravity do not change with time as they settle. Particles settle as individual entities, and there is no interaction between particles. Examples of free settling include grit and sand particles in grit removal systems and plain sedimentation of surface waters.

The terminal settling velocity of a particle (hereafter referred to as settling velocity) is derived in this section. There are three forces that act on a particle as it settles: gravitational force (F_g); buoyant force (F_b); and drag force (F_d). As a discrete particle settles, the particle will accelerate until the drag force equals the difference between the gravitational and buoyant forces, at which time a constant settling velocity (V_s) will be achieved.

$$F_g - F_b = F_d \tag{6.84}$$

$$F_g = \rho_p g Ɏ_p \tag{6.85}$$

$$F_b = \rho_w g Ɏ_p \tag{6.86}$$

$$F_d = \frac{C_d A_p \rho V_s^2}{2} \tag{6.87}$$

where:
F_g = gravitational force, (kg · m/s^2)
F_b = buoyant force, (kg · m/s^2)
F_d = force of drag, (kg · m/s^2)
A_p = cross-sectional area or projected area of particles in direction of flow, ft^2 (m^2)
V_s = settling velocity, fps (m/s)
g = acceleration of gravity, ft/s2 (m/s^2)
C_d = coefficient of drag (Dimensionless)
ρ_p = mass density of particle, lb·s2/ft^4 (kg/m^3)
ρ = mass density of liquid, lb·s2/ft^4 (kg/m^3).

For spherical particles, the volume and cross-sectional area are calculated using the following equations:

$$Ɏ_p = \frac{\pi d^3}{6} \tag{6.88}$$

$$A_p = \frac{\pi d^2}{4} \tag{6.89}$$

where:
d = diameter of particle, ft (m)
$Ɏ_p$ = volume of particle, ft^3 (m^3).

Substituting Equations (6.85) through (6.89) into Equation (6.84) and rearranging results in Newton's Law:

$$V_s = \left[\frac{4g}{3C_d}\left(\frac{\rho_p - \rho_w}{\rho_w}\right)d\right]^{0.5} \tag{6.90}$$

The coefficient of drag (C_d) is a function of the flow regime, which is estimated by calculating the Reynolds number (N_R):

$$N_R = \frac{\rho_w V_s d}{\mu} = \frac{V_s d}{\nu} \tag{6.91}$$

where:
μ = absolute or dynamic viscosity of water, lb·s/ft^2 (kg/(m·s))
ν = kinematic viscosity of water, ft^2/s (m^2/s).

When the Reynolds number is <1, laminar flow conditions exist and C_d is calculated using Equation (6.92):

$$C_d = \frac{24}{N_R} \tag{6.92}$$

During transitional flow between laminar and turbulent, N_R = 1 to 10^4 and C_d is determined from Equation (6.93):

$$C_d = \frac{24}{N_R} + \frac{3}{\sqrt{N_R}} + 0.34 \tag{6.93}$$

The coefficient of drag is assumed to be equal to 0.4 for turbulent flow when $N_R > 10^4$.

For laminar flow conditions, Newton's law simplifies to Stokes' law by substituting Equations (6.91) and (6.92) into Equation (6.90). This results in Equations (6.94) and (6.95):

$$V_s = \frac{g(\rho_p - \rho_w)d^2}{18\mu} \tag{6.94}$$

$$V_s = \frac{g(S_p - 1)d^2}{18\nu} \tag{6.95}$$

where S_p is the specific gravity of the particle (dimensionless).

6.7.2 Flocculent settling

Flocculent settling occurs in primary clarifiers and during settling of coagulated water and wastewater. As flocculent particles settle, their size, shape, density, and settling velocity will change with time. Newton's law or Stokes' law cannot be used for modeling flocculent settling. Settling column analyses must be performed with samples withdrawn at selected depths along the column. Percent removal of suspended solids (SS) is then plotted as a function of depth and time. Percent removal is calculated using Equation (6.96). This equation may be used to calculate the percent removal for any specific parameter, i.e., COD, BOD, TSS, etc.

$$\text{Percent removal}(\%) = \frac{(C_{initial} - C_{final}) \times (100)}{C_{initial}} \tag{6.96}$$

where:
$C_{initial}$ = initial concentration of species, mg/L
C_{final} = final concentration of species, mg/L.

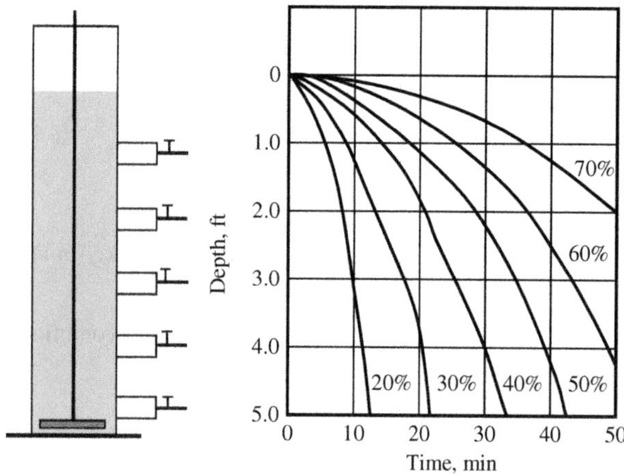

Figure 6.22 Settling column and settling trajectories for flocculent suspension.

Figure 6.22 shows a settling column with sampling ports. The curves in the plot represent the settling trajectories for a flocculent suspension and are referred to as isopercent removal lines.

Equation (6.97) is used for calculating the overall percent removal of solids that may be achieved at a given detention time:

$$R_T(\%) = R_0 + \frac{(\sum \Delta R \times Z_i)}{Z_0} \tag{6.97}$$

where:

R_T = total suspended solids removed at a given detention time, %
R_0 = isopercent removal line intersecting the time axis at specified settling column depth, %
ΔR = difference between adjacent isopercent removal lines, %
Z_i = distance from top of settling column or water level in column to the mid-point between two adjacent isopercent removal lines, ft (m)
Z_0 = total depth of settling column, ft (m).

6.7.2.1 Flocculent Settling Laboratory Procedure

Flocculent settling column analyses are typically performed in a clear, PVC column with sampling ports spaced approximately six inches (15 cm) apart. A column on the order of 5–6 feet (1.5 to 1.8 m) is used. The steps involved are listed below.

1. Fill settling column with coagulated water to be tested.
2. Take an initial sample for SS determination.
3. Take samples from each sampling port as a function of time.
4. Calculate the % SS removals and plot them as a function of time and depth.
5. Calculate overall % SS removals using Equation (6.97) for several detention times.
6. Plot overall % SS removal versus τ (detention time).

Example 6.14 illustrates how one uses settling column analysis data for designing a settling basin for coagulated surface water.

Example 6.14 Settling column analysis for flocculent settling

A batch settling column analysis is performed on a surface water sample containing an initial SS concentration of 597 mg/L. The settling column is 125 mm in diameter, 3.0 meters (11 ft) tall and has sampling ports located at 0.5, 1.0, 1.5, 2.0, and 2.5 meters from the water surface. SS concentrations at various sampling times are listed in the table below. Make a plot of overall % SS removal versus overflow rate.

SS concentration remaining versus depth and sampling time.

Depth		Suspended	Solids	Remaining	(mg/L)
(m)	10 min	20 min	30 min	40 min	50 min
0.5	394	298	227	179	119
1.0	465	358	299	239	179
1.5	490	400	340	281	221
2.0	507	418	358	316	263
2.5	513	436	376	328	287

The % SS removals as a function of depth and sampling time are presented in the following table. SS removal is calculated using Equation (6.96). The % SS removal at a depth of 0.5 meters and 10 minutes is calculated as follows:

$$\%\text{Removal} = \frac{(597\ \text{mg/L} - 394\ \text{mg/L}) \times (100)}{597\ \text{mg/L}}$$

$$= \boxed{34.0\ \%}$$

Percent SS removals versus depth and sampling time.

Depth		Suspended	Solids	Removals	(%)
(m)	10 min	20 min	30 min	40 min	50 min
0.5	34	50	62	70	80
1.0	22	40	50	60	70
1.5	18	33	43	53	63
2.0	15	30	40	47	56
2.5	14	27	37	45	52

Next, plot % SS removals as a function of depth and sampling time, and then trace isopercent removal lines as illustrated in Figure 6.22. Isopercent removals are lines of equal SS removal. Once the isopercent removals are drawn, Equation (6.97) can be applied at several detention times so that the overall SS removals can be determined as a function of overflow rate. At a detention time of 15 minutes, the overflow rate (V_0) is calculated as follows by rearranging the equation, Distance = Rate × time. The total depth of water in the column is 2.5 m.

$$V_0 = \frac{2.5\text{ meters}}{15\text{min}}\left(\frac{1000\text{ L}}{\text{m}^3}\right)\left(\frac{1\text{ m}^3}{1000\text{ L}}\right)\left(\frac{60\text{ min}}{\text{h}}\right)\left(\frac{24\text{ h}}{\text{d}}\right)$$

$$= \boxed{240\ \frac{\text{m}^3/\text{d}}{\text{m}^2}}$$

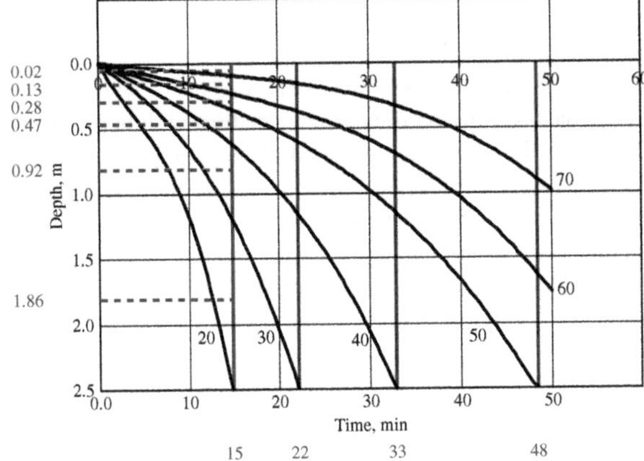

Figure E6.6 Settling column trajectories for flocculent suspension in Example 6.14.

Calculations for the determination of the overall % SS removal at a detention time of 15 minutes are presented below:

Determination of overall % removal (R_T) at a detention time of 15 minutes.

ΔR	Z_i	$\Delta R \times Z_i$
100 − 70 = 30	0.02	0.6
70 − 60 = 10	0.13	1.3
60 − 50 = 10	0.28	2.8
50 − 40 = 10	0.47	4.7
40 − 30 = 10	0.92	9.2
30 − 20 = 10	1.86	18.6
		$\sum(\Delta R \times Z_i) = 37.2$

The overall % SS removal at 15 minutes is estimated from Equation (6.97) as follows:

$$R_{T@15\text{min}} = R_0 + \frac{(\sum \Delta R \cdot Z_i)}{Z_0} = 20 + \frac{37.2}{2.5} = \boxed{34.9\ \%}$$

At a detention time of 22 minutes, the overflow rate (V_0) is calculated as follows:

$$V_0 = \frac{2.5\text{ meters}}{22\text{ min}}\left(\frac{1000\text{ L}}{\text{m}^3}\right)\left(\frac{1\text{ m}^3}{1000\text{ L}}\right)\left(\frac{60\text{ min}}{\text{h}}\right)\left(\frac{24\text{ h}}{\text{d}}\right)$$

$$= \boxed{164\ \frac{\text{m}^3/\text{d}}{\text{m}^2}}$$

Calculations for the determination of the overall % SS removal at a detention time of 22 minutes are presented below.

Determination of overall % removal (R_T) at a detention time of 22 minutes.

ΔR	Z_i	$\Delta R \times Z_i$
100 − 70 = 30	0.08	2.4
70 − 60 = 10	0.27	2.7
60 − 50 = 10	0.48	4.8
50 − 40 = 10	0.88	8.8
40 − 30 = 10	1.88	18.8
		$\sum(\Delta R \times Z_i) = 37.5$

The overall % SS removal at 22 minutes is estimated from Equation (6.97) as follows:

$$R_{T@22\text{min}} = R_0 + \frac{(\sum \Delta R \cdot Z_i)}{Z_0} = 30 + \frac{37.5}{2.5} = \boxed{45.0\ \%}$$

At a detention time of 33 minutes, the overflow rate (V_0) is calculated as follows:

$$V_0 = \frac{2.5\text{ meters}}{33\text{ min}}\left(\frac{1000\text{ L}}{\text{m}^3}\right)\left(\frac{1\text{ m}^3}{1000\text{ L}}\right)\left(\frac{60\text{ min}}{\text{h}}\right)\left(\frac{24\text{ h}}{\text{d}}\right)$$

$$= \boxed{109\ \frac{\text{m}^3/\text{d}}{\text{m}^2}}$$

Calculations for the determination of the overall % SS removal at a detention time of 33 minutes are presented below.

Determination of overall % removal (R_T) at a detention time of 33 minutes.

ΔR	Z_i	$\Delta R \times Z_i$
100 − 70 = 30	0.16	4.8
70 − 60 = 10	0.52	5.2
60 − 50 = 10	0.72	7.2
50 − 40 = 10	1.81	18.1
		$\sum(\Delta R \times Z_i) = 35.3$

The overall % SS removal at 33 minutes is estimated from Equation (6.97) as follows:

$$R_{T@33\text{min}} = R_0 + \frac{(\sum \Delta R \cdot Z_i)}{Z_0} = 40 + \frac{35.3}{2.5} = \boxed{54.1\ \%}$$

At a detention time of 48 minutes, the overflow rate (V_0) is calculated as follows:

$$V_0 = \frac{2.5\text{ meters}}{48\text{ min}}\left(\frac{1000\text{ L}}{\text{m}^3}\right)\left(\frac{1\text{ m}^3}{1000\text{ L}}\right)\left(\frac{60\text{ min}}{\text{h}}\right)\left(\frac{24\text{ h}}{\text{d}}\right)$$

$$= \boxed{75\ \frac{\text{m}^3/\text{d}}{\text{m}^2}}$$

Calculations for the determination of the overall % SS removal at a detention time of 48 minutes are presented below.

Determination of overall % removal (R_T) at a detention time of 48 minutes.

ΔR	Z_i	$\Delta R \times Z_i$
100 − 70 = 30	0.45	13.5
70 − 60 = 10	1.27	12.7
60 − 50 = 10	2.06	20.6
		$\Sigma(\Delta R \times Z_i) = 46.8$

The overall % SS removal at 48 minutes is estimated from Equation (6.97) as follows:

$$R_{T@48min} = R_0 + \frac{(\Sigma \Delta R \cdot Z_i)}{Z_0} = 50 + \frac{46.8}{2.5} = \boxed{68.7\ \%}$$

A plot of %SS removal versus overflow rate can then be made as shown below. Alternatively, %SS removal versus detention may also be plotted.

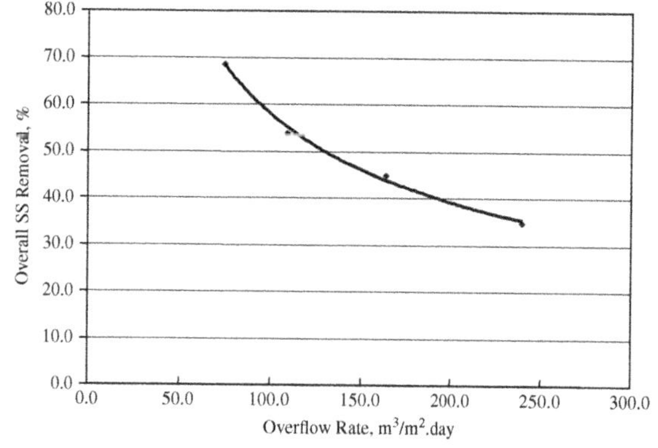

Figure E6.7 Plot of R_T versus overflow rate for Example 6.14.

6.7.3 Ideal settling basin

An ideal settling basin with an inlet, outlet, and sludge settling zone is presented in Figure 6.23.

Overflow rate serves as a design parameter used by engineers to determine the size of settling basins, whereas water treatment plant operators regulate the flow through the basin to maintain the proper detention time and overflow rate to achieve optimum solids removal. The overflow rate (V_0) is a different way of expressing the settling velocity (V_s) of a particle as calculated using Stokes' law.

Remember, settling velocity as shown in Equations (6.94) and (6.95) is a function of the diameter of the particle. Overflow rate is normally expressed in units of gpd/ft^2 or

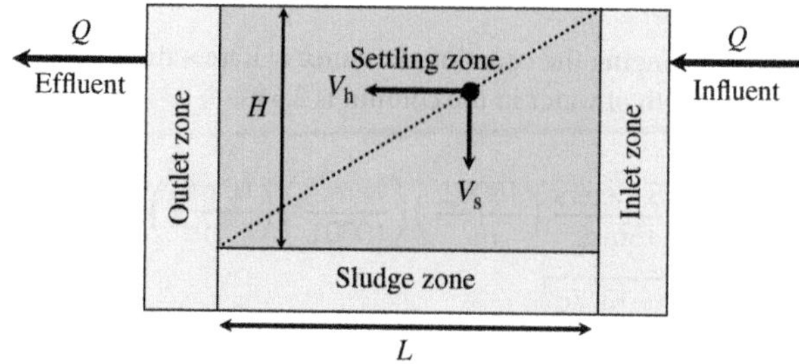

Figure 6.23 Schematic of ideal settling basin.

m^3/(d · m^2). When an engineer selects an overflow rate, they are actually specifying the particle size they wish to remove. Theoretically, all particles with a diameter equal to or greater than the particle size specified by overflow rate should be completely removed. Derivation of the overflow rate (V_0) equation is presented in this section.

The basis for design of a settling basin is to select a specific diameter particle that will settle by gravity during the time that it takes water to travel through the basin. Under ideal conditions, a particle will travel a vertical distance of H during time t (Equation (6.98)). Mathematically, the distance traveled is equal to the rate multiplied by time ($D = R \times t$).

$$t = \frac{H}{V_s} \tag{6.98}$$

where:
t = settling time, min, hours
H = depth of settling basin, ft (m)
V_s = settling velocity, fps (m/s).

This same particle will travel a distance of L in the horizontal direction during the same time (t), as shown in Equation (6.99).

$$t = \frac{L}{V_h} \tag{6.99}$$

where:
L = length of settling basin, ft (m)
V_h = horizontal flow-through velocity, fps (m/s)
W = width of settling basin, ft (m).

The horizontal flow-through basin velocity (V_h) (Equation (6.101)) is calculated by rearranging the continuity equation which is shown below:

$$Q = AV \tag{6.100}$$

where:
Q = flow rate through the basin, ft^3/s (m^3/s)
A = cross-sectional area = $W \times H$, ft^2 (m^2)
W = width of settling basin, ft (m)
V = average flow velocity, fps (m/s).

$$V_h = \frac{Q}{A} = \frac{Q}{W \times H} \tag{6.101}$$

The derivation of the equation for overflow rate (V_0) is presented as follows. We begin by replacing V_s with V_0 in Equation (6.98):

$$t = \frac{H}{V_s} = \frac{H}{V_0} \tag{6.102}$$

Then set Equation (6.102) equal to Equation (6.99) as follows:

$$t = \frac{H}{V_0} = \frac{L}{V_h}$$

Next, substitute Equation (6.101) into the equation above to yield Equation (6.103).

$$t = \frac{H}{V_0} = \frac{L}{Q/(W \times H)} \tag{6.103}$$

Rearranging:

$$V_0 = \frac{H \times Q}{L \times W \times H} = \frac{Q}{L \times W} = \frac{Q}{A_s} \tag{6.104}$$

The final form of the overflow rate equation is shown below.

$$V_0 = \frac{Q}{A_s} \tag{6.105}$$

where:
V_0 = overflow rate, gpd/ft^2 (m^3/(d · m^2))
A_s = surface area of settling basin = $L \times W$, ft^2 (m^2).

Another important design and operational parameter is detention time. Detention time, or retention time, is defined as the average time that the water remains in the settling basin. Typically, the Greek letter τ, rather than t, is used to denote detention time. The detention time equation is derived by substituting Equation (6.101) into Equation (6.99), resulting in the following:

$$\tau = t = \frac{L}{V_h} = \frac{L \times H \times W}{Q} = \frac{V\!\!\!/}{Q} \tag{6.106}$$

6.7.3.1 Weir Loading Rate

The **weir loading rate** (q) is defined as the volumetric flow rate (Q), divided by the length of the effluent weir. Regulatory agencies sometimes specify weir loading rates that must be maintained in settling basins. Mathematically, weir loading rate is given in Equation (6.107):

$$q = \frac{Q}{\text{weir length}} \tag{6.107}$$

where:
q = weir loading rate, gal/(d · m) (m^3/(d · m))
weir length = length of weir, ft (m).

6.7.3.2 Settling Basin Design Criteria

In water treatment, long rectangular or square settling basins are typically used, whereas circular basins or clarifiers are more widely used for wastewater applications. Solids contact basins, in which mixing, coagulation, flocculation, and settling occur all in the same unit, are now being used extensively in current water treatment practice. The actual design of a settling basin not only includes the ideal settling zone, but also:

- an inlet zone to distribute equally the incoming water and dissipate velocity currents;
- a sludge zone in which particles that settle out of solution accumulate and are subsequently removed for further processing; and
- an outlet zone in which baffles and weirs are used to ensure that no short circuiting occurs and quiescent conditions are maintained.

Overflow rate, detention time, and weir loading rate are the three main parameters used by environmental engineers in the design of settling basins. Table 6.11 provides typical design criteria for long rectangular, upflow radial, and reactor clarifier type settling basins.

Knowing the maximum daily flow to the water treatment plant and selecting an overflow rate, the surface area can be determined. Next, the detention time is selected and the depth is determined by dividing the volume by the surface area. The weir loading rate is calculated after laying out the weir configuration. Examples 6.14 and 6.15 illustrate the design procedures.

Table 6.11 Design criteria for settling basins.

Type of settling basin	Overflow rate	Depth	Detention time or settling time	Weir loading
Rectangular (horizontal flow)	0.34–1 gpm/ft^2 (0.83–2.5 m/h)	10–16 ft (3–5 m)	1.5–3.0 h	<15 gpm/ft (11 m/(min · h))
Circular or square radial upflow	0.5–0.75 gpm/ft^2 (1.3–1.9 m/h)	10–16 ft (3–5 m)	1.0–3.0 h	10 gpm/ft (7 m/(min · h))
Reactor clarifier	0.8–1.2 gpm/ft^2 (2.0–3.0 m/h)	–	1.0–2.0 h	10–20 gpm/ft (7.3–15 m/(min ·h))

Source: Adapted from Kawamura (1991) pp. 138–139. Reproduced by permission of John Wiley & Sons Ltd.

Figure 6.24 Photograph of long, rectangular settling basin.

Long, rectangular settling basins normally have minimum length-to-width ($L : W$) ratios ranging from 4 : 1 to 5 : 1, length-to-depth ($L : D$) ratios of 15 : 1, and bottom slope of 1 : 600 when mechanical sludge scraper mechanisms are used (MWH, 2005, p. 810). Where circular tanks are used, tank diameters of 125–150 feet (38–45 m) are recommended (Walker, 1982, p. 160). For rectangular basins with a length >100 ft (31 m) or circular basins with a 70 ft (21 m) diameter or greater, a side water depth (SWD) of 15–18 ft (4.5–5.5 m) is recommended (Walker, 1982, p. 157). Side water depth is the depth of water at the side of the tank from the top of the effluent weir to the bottom of the tank. An additional 1–2.5 feet (0.3–0.8 m) must be added to the SWD to act as a wind barrier and to account for flow variations. Figure 6.24 is a photo of a long rectangular settling basin.

Example 6.15 Circular settling basin design

Design two equally sized circular sedimentation basins for a total flow rate of 2.5 million gallons per day (MGD) using an overflow rate of 700 gpd/ft^2 and a detention time of four hours. Determine the diameter to the nearest five-foot interval.

What is the weir loading rate if the effluent flows radially outward to an effluent channel attached to the outside wall with a single peripheral weir (water flows into the channel from one side of the weir only)?

Solution

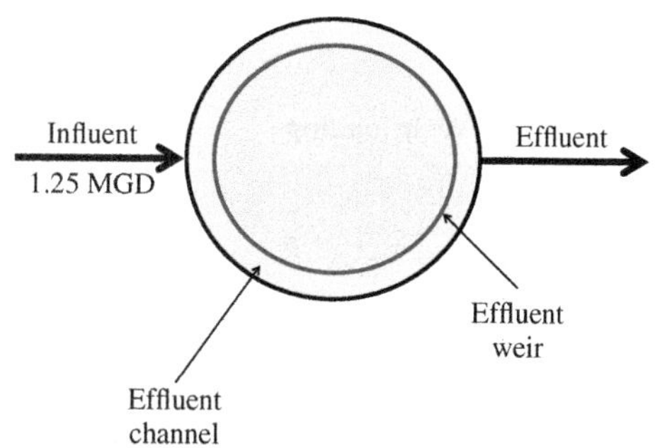

Figure E6.8

Calculate the surface area by rearranging Equation (6.105).

$$A_s = \frac{Q}{V_0} = \frac{1.25 \times 10^6\ \text{gpd}}{700\ \text{gpd/ft}^2} = 1786\ \text{ft}^2$$

Next, the diameter of the circular sedimentation basin is determined by using the equation for the area of a circle, $A = \pi D^2/4$. The surface area of each sedimentation basin is calculated below:

$$A_s = \frac{\pi D^2}{4} = 1786\ \text{ft}^2$$

As shown below, the final diameter of each sedimentation basin is 50 feet. Settling basins generally are constructed to the nearest five-foot interval.

$$D = \left(\frac{4 \times 1786\ \text{ft}^2}{\pi}\right)^{0.5} = 47.7\ \text{ft}\ \ \text{Use}\ \boxed{D = 50\ \text{ft}}$$

The weir loading rate is determined using Equation (6.107).

$$q = \frac{Q}{\text{weir length}} = \frac{1.25 \times 10^6\ \text{gpd}}{\pi(50\ \text{ft})} = \boxed{7960\ \frac{\text{gpd}}{\text{ft}}}$$

Example 6.16 Design of rectangular settling basin

Two rectangular settling basins are operated in parallel to settle 5,800 m^3/d of water. Each basin has the following dimensions: 28 m long, 7 m wide, and 13.5 m deep. The effluent weir length in each basin is equal to three times the tank width.

Determine the following:

a) The detention time (h).
b) Horizontal flow velocity (m/min).
c) Overflow rate (m/h).
d) Weir length (m).
e) Weir loading rate (m^3/(d · m)).

Solution

Determine the detention time by substituting into Equation (6.106).

$$\tau = \frac{\forall}{Q} = \frac{2 \times (28\ \text{m} \times 7\text{m} \times 3.5\ \text{m})}{5800\ \text{m}^3/\text{d}}\left(\frac{24\ \text{h}}{\text{d}}\right) = \boxed{5.68\ \text{h}}$$

Determine the horizontal flow-through velocity as follows:

$$V_h = \frac{Q}{A} = \frac{5800\ \text{m}^3/\text{d}}{2 \times (7\text{m} \times 3.5\ \text{m})}\left(\frac{1\ \text{d}}{24\ \text{h}}\right)\left(\frac{1\ \text{h}}{60\ \text{min}}\right)$$

$$= \boxed{0.082\ \text{m/min}}$$

The overflow rate (V_0) is determined from Equation (6.105).

$$V_0 = \frac{Q}{A_s} = \frac{5800\ \text{m}^3/\text{d}}{2 \times (28\ \text{m} \times 7\ \text{m})} = \boxed{14.8\ \text{m}^3/(\text{d} \cdot \text{m}^2)}$$

The weir length is calculated as follows:

$$\text{Weir length per basin} = 3 \times (7\ \text{m}) = \boxed{21\ \text{m}}$$

The weir loading rate is determined using Equation (6.107).

$$q = \frac{Q}{\text{weir length}} = \frac{5{,}800\ \text{m}^3/\text{d}}{2\ \text{basins} \times 21\ \text{m}} = \boxed{138\ \frac{\text{m}^3/\text{d}}{\text{m}}}$$

6.8 Filtration

Filtration of water is one of the most widely used unit operations at water treatment facilities. In the United States, the promulgation of the Surface Water Treatment Rule (SWTR) of May 1989 requires that all public water systems using surface water sources or ground water sources under the direct influence of surface water (GWUDI) be filtered and disinfected. The focus is placed on removing turbidity, *Giardia lamblia*, *Cryptosporidium parvum*, and viruses. Filtration is used to remove destabilized colloidal particles and other solids that remain after sedimentation. At most water treatment plants, granular media filters are used. However, membrane processes, including microfiltration (MF) and ultrafiltration (UF) are now being used.

Rapid filtration, which uses multiple layers of granular media such as anthracite, sand, granular activated carbon, ilmenite, and granite is typically used in the US. Slow sand filters are generally used at smaller treatment facilities; however, large metropolitan cities such as London and Amsterdam use this technology (Joslin, 1997. p. 295). Pre-coat pressurized filtration using diatomaceous earth (DE) is another method used to remove suspended solids from water. Details on the use and design of pre-coat filtration systems may be found in *Water Treatment: Principles and Design* (2005, pp. 878–880) and *Water Quality and Treatment* (1999, pp. 8.81–8.89).

In conventional water filtration, coagulation, flocculation, and sedimentation precede granular media filtration. Direct filtration of water is sometimes practiced, in which coagulated and flocculated water is applied directly to filters without having undergone sedimentation. In-line filtration is the process of applying coagulated water directly to the filter and omitting conventional flocculation and sedimentation. This is only used for raw water with relatively low turbidity.

6.8.1 Filter Classification

There are numerous ways of classifying filters. Filters that are operated using gravity flow are called gravity filters and operate in a downflow mode. Pressurized filtration is accomplished in pressurized containers. Another method of categorizing filters is according to the rate of filtration. The filtration rate is the flow rate applied to the filter divided by the surface area of the filter bed. Based on this system, filters are classified as slow sand filters, rapid filters, and high-rate filters.

The last classification method is based on the type of media used in the filter. Single-media or mono-media filters have only one type of media, which is typically sand or crushed anthracite coal. In most cases, mono-media filters have a uniform grain size and depths greater than those in multi-media applications, in order to achieve the same effluent quality. Dual-media filters normally consist of a layer of coarse anthracite coal above a layer of sand. Granular activated carbon is now being used in many applications, rather than anthracite, because it is effective in removing taste and odor compounds, color, and other organic compounds. Multi-media filters such as tri-media filters typically consist of a layer of coarse anthracite coal above a layer of sand and a layer of garnet or ilmenite below the layer of sand.

6.8.2 Suspended solids removal mechanisms

There are several mechanisms involved in removing suspended solids during granular media filtration. These include straining, sedimentation, flocculation, interception, biological growth, impaction, adhesion, and sorption (Metcalf & Eddy, 2003, p. 1048). The objective is to use the entire volume of media for removing particulate matter. This is known as *depth filtration*, and it is more effective than just the removal accomplished by surface straining.

Surface straining or mechanical screening of suspended solids occurs when the particles being removed are larger than the pore space through which the water flows. A cake of solids forms on the surface of the filter, increasing the overall efficiency of solids removal; however, the head loss also increases dramatically. Straining is the primary removal mechanism that occurs in slow sand filtration. The pore spaces or voids in the media act as miniature settling basins within the filter bed, allowing flocculation and sedimentation of solids to occur. Some particles are removed through interception as the flow streamlines bend around the media particles. In some cases, the particulate matter has sufficient mass that when the flow streamlines are altered, the trajectory of the particle cannot change and removal occurs because of inertial impaction.

Depending on the nature of the destabilized colloidal suspension, either physical or chemical adsorption to the media particles can also take place. Biological growth within and on the surface of the filter will reduce pore volume, resulting in enhanced removal of particles. The layer of cake solids and biological growth that occurs on the surface in filters is called

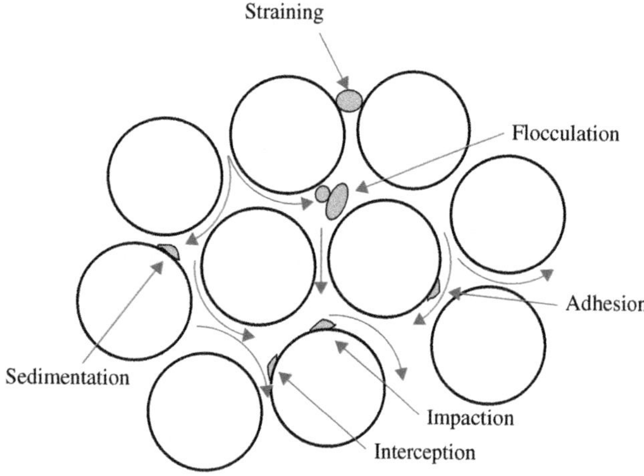

Figure 6.25 Illustration of six particle removal mechanisms during filtration.

"schmultzdecke." Figure 6.25 is an illustration of six of the particle removal mechanisms occurring in granular media filters. The reader is referred to the following sources to learn more about these mechanisms: Cleasby & Logsdon (1999, pp. 8.32–8.39), Viessman *et al.* (2009, pp. 343–345), and Metcalf & Eddy (2003, pp. 1047–1049).

6.8.3 Filter media

For drinking water systems in the US, the most commonly used media are sand, crushed anthracite coal, granular activated carbon (GAC), granite, and ilmenite. The **effective size (d_{10})** and **uniformity coefficient (UC)** are two key parameters in specifying filter media. The effective size is determined from a sieve analysis curve and represents the media size for which ten percent (d_{10}) by weight of the grains are smaller. The uniformity coefficient is a measure of the size range of the media and is defined as the ratio of d_{10}/d_{60} sizes that are determined from the sieve analysis curve. The media size for which 60 percent by weight of the grains are smaller is denoted as d_{60}.

Various properties of filter media are presented in Table 6.12. Silica sand and anthracite coal are the most widely used media types. The L/d_{10} ratio is used to estimate the depth of each layer of media once the effective size has been selected. Figure 6.26 is a schematic showing the media configurations of sand, dual-media, and mixed-media filters.

Gravity-type filters, consisting of dual- or multi-layers of coarse granular media, are the most widely used to treat water.

Table 6.12 Properties of filter media.

Property	Anthracite	Garnet	GAC	Ilmenite	Sand
Density (ρ_s), g/ml	1.4–1.8	3.6–4.2	1.3–1.7	4.5–5.0	2.65
Effective size (d_{10}), mm	0.8–2.0	0.2–0.4	0.8–2.0	0.2–0.4	0.4–0.8
Porosity (ε)	0.47–0.52	0.45–0.58	N.A.	N.A.	0.40–0.43
Specific gravity (S.G.)	1.4–1.8	3.6–4.2	1.3–1.7	4.5–5.0	2.65
[a]Shape factor (ϕ)	0.45–0.60	0.60	0.75	N.A.	0.7–0.8
Uniformity coefficient (UC)	1.3–1.7	1.3–1.7	1.3–2.4	4.5–5.0	1.3–1.7

N.A. = not available
[a]*Water Quality and Treatment* (1999), p. 8.10.
Source: Adapted from MWH (2005) p. 882. Reproduced by permission of John Wiley & Sons Ltd.

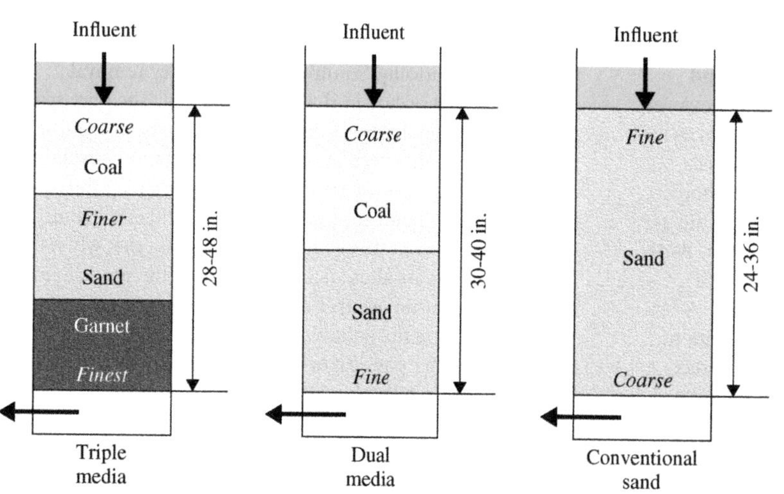

Figure 6.26 Schematic showing the media configurations of sand, dual-media, and mixed-media filters.

Table 6.13 Filter media specifications.

Filter type	Material	Effective size, d_{10} (mm)	Depth, L (in)	Uniformity coefficient, UC	L/d_{10}
Small dual-media	Anthracite and sand	1.0 0.5	20 10	1.45 1.30	1016
Intermediate dual-media	Anthracite and sand	1.48 0.5	30 15	1.50 1.20	1023
Large dual-media	Anthracite and sand	2.0 1.0	40 20	1.50 1.25	1016
Mixed-media	Anthracite, sand, and garnet	1.0 0.42 0.25	18 9 3	1.45 1.50 1.25	1306
Mono-media	Anthracite	1.0	40	1.4	1016

Source: Kawamura (1991) p. 200. Reproduced by permission of John Wiley & Sons Ltd.

The effective size, uniformity coefficient, depth of media (L), and depth of media to effective size (L/d_{10}) ratios for various types of filter media are presented in Table 6.13.

- Single-media filters typically consist of 24–36 inches (61–76 cm) of sand, supported by 15–24 inches (38–61 cm) of layered gravel and are commonly used in the United States.
- Dual-media filters consist of 18–30 inches (0.46–0.76 m) of anthracite (d_{10}: 0.8–1.22 mm) over 6–12 inches (0.1 5–0.30 m) of sand (d_{10}: 0.45–0.55 mm).
- Mixed-media filters consist of 18–24 inches (0. 5– 0.6 m) of anthracite (d_{10}: 0.8–1.2 mm), 6–9 inches (0.15–0.3 m) of silica sand (d_{10}: 0.35–0.5 mm), and 3–4 inches (5–10 cm) of garnet (d_{10}: 0.15–0.45 mm).
- Dual-media filters using GAC often use 12–48 inches (0.3–1.2 m) of GAC over 6–18 inches (0.14–0.5 m) of silica sand (d_{10}: 0.5–0.65 mm).
- Deep-bed, mono-media filters generally use silica sand or anthracite with an effective size of 0.5–6.0 mm, UC from 1.2–1.5, and a depth of 4–6 ft (1.2–1.8 m).
- Typical filtration rates for various types of granular media filters are presented in Table 6.14. According to Castro *et al.* (2005, p. 8.8), most regulatory agencies will not approve filtration rates exceeding 4–5 gpm/ft^2 (9.8–12.2 m/h).

Table 6.14 Typical filtration rates used in granular media filtration.

Filter type	Filtration rate
Slow sand filters	0.05–0.17 gpm/ft^2 (0.13–0.42 m/h)
Rapid sand filters (medium sand)	2.0–3.0 gpm/ft^2 (5.0–7.5 m/h)
High rate filters (coarse sand)	4.0–12.0 gpm/ft^2 (10.0–30.0 m/h)
Multimedia filters	4.0–10.0 gpm/ft^2 (10.0–25.0 m/h)
Granular activated carbon filters	3.0–6.0 gpm/ft^2 (7.5–15.0 m/h)

Source: Kawamura (1991), pp. 214–215. Reproduced by permission of John Wiley & Sons Ltd.

6.8.4 Granular media filtration

Figure 6.27 shows a cross-sectional view of a rapid sand filter during filtration. Coagulated water flows by gravity down through the filter (sand, gravel, and underdrains) and into the clear well.

The primary goal of granular media filtration is to use the entire filter bed for solids removal, rather than being strained at the surface. This is known as "depth filtration." Eventually, the head loss through the bed will build up to approximately 8–10 feet (2.4–3 m), at which time the filter must be backwashed to remove the particles and turbidity that has accumulated.

Backwashing involves reversing the flow of water back up through the bed, causing it to expand and thereby releasing the trapped particles. Auxiliary air and water jets may be used during backwashing to help scour the filter media releasing more particles. Disinfectant, typically chlorine, is added to the water as it enters the clear well, so that sufficient contact time is available for the disinfectant to inactivate any pathogens that may remain in the water. The flow control valve regulates the flow rate through the filter.

Typically, filters are operated to maintain a constant flow or production of water. This is accomplished by partially opening the flow control valve (FCV) and, as the filter run progresses, the FCV valve is opened wider to allow the same amount of flow through the filter as the head loss increased during a filter run. Although not shown, water from the clear well is pumped throughout the water distribution system using high-service pumps. Other valves shown in Figure 6.27 are either open or closed during the filtration and backwashing cycles. The underdrain system is approximately 1 foot (0.3 m) in depth. The static water head on the sand media generally varies from 3–4 feet (0.9–1.2 m).

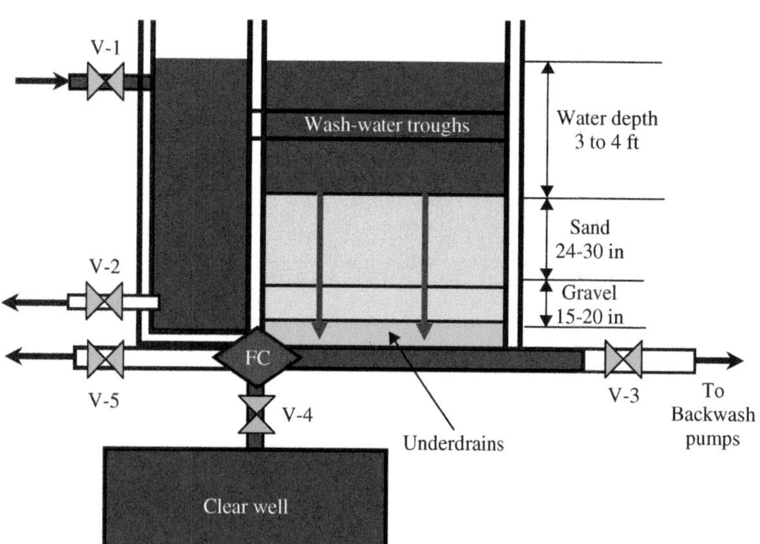

Figure 6.27 Cross-sectional view of a rapid sand filter during filtration.

6.8.5 Derivation of head loss through a clean filter

The derivation of the Carman-Kozeny equation for head loss across a clean filter bed is derived in this section.

We begin with the Darcy-Weisbach equation (Daugherty *et al.*, 1985, p. 208) as presented below:

$$h_L = f \frac{L}{D} \frac{V^2}{2g} \tag{6.108}$$

where:
h_L = head loss, ft (m)
f = dimensionless friction factor
L = length of pipe, ft (m)
D = pipe diameter, ft (m)
V = mean velocity of flow in pipe, fps, (mps)
g = acceleration of gravity, 32, ft/s^2 (9.81 m/s^2).

The hydraulic radius (R_H) is defined as the cross-sectional flow area ($A_{x\text{-section}}$) divided by the wetted perimeter (WP). As indicated in Equation (6.109) this is equivalent to the volume of voids ($\forall_v$) in the filter media, divided by the surface area of the media (A_s).

$$R_H = \frac{A_{x\text{-section}}}{WP} = \frac{\forall_v}{A_s} \tag{6.109}$$

For a pipe or conduit flowing full of fluid, the hydraulic radius is equal to $D/4$.

$$R_H = \frac{\pi D^2/4}{\pi D} = \frac{D}{4} \tag{6.110}$$

$$D = 4R_H \tag{6.111}$$

Substituting Equation (6.111) into Equation (6.108) for D yields:

$$h_L = f \frac{L}{4R_H} \frac{V^2}{2g} \tag{6.112}$$

The velocity of approach (V_A) is defined as the volumetric flow rate (Q) divided by the surface area of the filter (A_s), as follows:

$$V_A = \frac{Q}{A_s} \tag{6.113}$$

The velocity through the filter bed (V) is defined as the approach velocity divided by the fraction of the bed that is pores, also called bed porosity (ε).

$$V = \frac{V_A}{\varepsilon} \tag{6.114}$$

$$\varepsilon = \frac{\forall_v}{\forall_T} = \frac{\forall_v}{\forall_v + \forall_M} \tag{6.115}$$

where:
$\forall_v$ = volume of voids = total channel volume
$\forall_M$ = volume of media
$\forall_T$ = total volume (voids plus media).

$$\forall_v = \forall_T \times \varepsilon \tag{6.116}$$

$$\forall_T = \frac{\forall_v}{\varepsilon} \tag{6.117}$$

One minus the bed porosity is equal to the volume of media divided by the total volume as shown in Equation (6.118).

$$1 - \varepsilon = \frac{\forall_M}{\forall_T} \tag{6.118}$$

$$\forall_T = \frac{\forall_M}{(1-\varepsilon)} \tag{6.119}$$

Substituting Equation (6.117) into Equation (6.119) yields Equation (6.120).

$$\frac{V_V}{\varepsilon} = \frac{V_M}{(1-\varepsilon)} \quad (6.120)$$

The total volume of media is calculated as follows:

$$V_M = N V_p \quad (6.121)$$

where:
N = number of media particles
V_p = volume of an individual media particle.

Substituting Equation (6.121) into Equation (6.120) and rearranging yields:

$$V_V = \frac{\varepsilon N V_p}{(1-\varepsilon)} \quad (6.122)$$

The total surface area of the media is calculated using Equation (6.123).

$$A_s = N A_p \quad (6.123)$$

Where, A_p = surface area of each media particle.

Substituting Equations (6.122) and (6.123) into Equation (6.109) results in Equation (6.124) for the hydraulic radius:

$$R_H = \frac{V_V}{A_s} = \frac{\frac{\varepsilon N V_p}{(1-\varepsilon)}}{N A_p} = \frac{\varepsilon V_p}{(1-\varepsilon) A_p} \quad (6.124)$$

For spherical particles, the volume and surface area are given by Equations (6.125) and (6.126) respectively.

$$V_p = \frac{\pi d^3}{6} \quad (6.125)$$

$$A_p = \pi d^2 \quad (6.126)$$

where d = particle diameter, ft (m).

The ratio of V_p/A_p is equals to $d/6$ as follows:

$$\frac{V_p}{A_p} = \frac{\pi d^3/6}{\pi d} = \frac{d}{6} \quad (6.127)$$

For non-spherical particles, a dimensionless coefficient called shape factor (ϕ) must be incorporated into Equation (6.127):

$$\frac{V_p}{A_p} = \frac{\phi d}{6} \quad (6.128)$$

Substituting Equation (6.128) into Equation (6.124) yields:

$$R_H = \frac{V_V}{A_s} = \frac{\varepsilon \phi d}{(1-\varepsilon)6} \quad (6.129)$$

Substituting Equation (6.129) into Equation (6.112) results in Equation (6.130):

$$h_L = f\frac{L}{4R_H}\frac{V^2}{2g} = f\frac{L}{4\varepsilon}\frac{(1-\varepsilon)}{\phi d}\frac{6V^2}{2g} \quad (6.130)$$

Substituting V from Equation (6.114) into Equation (6.130) yields:

$$h_L = f\frac{L}{4\varepsilon}\frac{(1-\varepsilon)}{\phi d}\frac{6V^2}{2g} = f\frac{L}{4\varepsilon}\frac{(1-\varepsilon)}{\phi d}\frac{6V_A^2}{2\varepsilon^2 g}$$
$$= f\frac{L}{\phi d}\frac{(1-\varepsilon)}{\varepsilon^3}\frac{V_A^2}{g}\frac{6}{8} \quad (6.131)$$

Replacing $(6f)/8$ with a new friction factor, f', results in Equation (6.132):

$$h_L = f'\frac{L}{\phi d}\frac{(1-\varepsilon)}{\varepsilon^3}\frac{V_A^2}{g} \quad (6.132)$$

Equation (6.132) is the important Carman-Kozney Equation (Carman, 1937) for head loss for a bed of porous media of uniform particle size. Equation (6.132) may be modified to account for a mixture of different particle sizes according to Equation (6.133):

$$h_L = \frac{L}{\phi}\frac{(1-\varepsilon)}{\varepsilon^3}\frac{V_A^2}{g}\sum f'\frac{X_{ij}}{d_{ij}} \quad (6.133)$$

Equation (6.134) is used to calculate the friction factor, f':

$$f' = 150\left(\frac{1-\varepsilon}{N_R}\right) + 1.75 \quad (6.134)$$

where:
h_L = frictional head loss, ft (m)
L = depth of filter bed, ft (m)
ε = porosity of bed, dimensionless
ϕ = shape factor of media, dimensionless
V_A = filtration velocity or velocity of approach, total flow applied to the filter divided by the filter area, ft/s (m/s)
g = gravitational acceleration, ft/s^2 (m/s^2)
d_{ij} = particle size = geometric mean of adjacent sieve sizes, ft (m)
X_{ij} = weight fraction of media retained on adjacent sieve sizes
N_R = Reynolds number, dimensionless
f' = Carman-Kozeny friction factor, dimensionless.

6.8.5.1 Rose Equation

Rose (1945) developed an empirical equation for head loss through a clean filter bed that yields a similar result to the Carman-Kozeny equation. A derivation of the Rose equation for head loss across a clean granular media filter is found in Rich (1971, p. 143). Equations (6.135) and (6.136) are the head loss equations developed by Rose for uniform particle size and mixed particle size, respectively. Equation (6.137) is used to calculate the drag coefficient (C_D):

$$h_L = \frac{1.067}{\phi}\frac{C_D}{g}\frac{LV_A^2}{\varepsilon^4}\frac{1}{d} \quad (6.135)$$

$$h_L = \frac{1.067}{\phi}\frac{L}{g}\frac{V_A^2}{\varepsilon^4}\sum\frac{C_D X_{ij}}{d_{ij}} \quad (6.136)$$

$$C_D = \frac{24}{N_R} + \frac{3}{(N_R)^{0.5}} + 0.34 \quad (6.137)$$

Equation (6.138) is used to calculate the Reynolds number (N_R).

$$N_R = \frac{\phi \rho V d}{\mu} = \frac{\phi V d}{\nu} \quad (6.138)$$

where:

d = uniform particle or media diameter, ft (m)
C_D = coefficient of drag, dimensionless
ρ = density of water, lb·s²/ft⁴ (kg/m³)
d_{ij} = particle size = geometric mean of adjacent sieve sizes, ft (m)
μ = absolute or dynamic viscosity of water, lb·s/ft² (kg/m·s)
ν = kinematic viscosity of water, ft²/s (m²/s).

Example 6.17 Head loss across a granular media filter

Determine the head loss for a clean filter bed, using the Carman-Kozeny equation for a stratified bed with uniform porosity of 0.41. The rapid sand filter is made up of a 762 mm deep bed of sand, specific gravity of 2.65, shape factor (ϕ) = 0.80, temperature of 15°C, and filtration rate of 1.53 liters per second per square meter (Lpsm²). A sieve analysis is presented below.

Sieve (1)	% Sand retained (2) (X_{ij})	d_{ij} (3) ($m \times 10^{-4}$)	N_R (4)	f'_{ij} (5)	$f'_{ij}X_{ij}/d_{ij}$ (6)
14–20	0.87	10.006	1.08×10^{0}	8.41×10^{1}	731
20–28	8.63	7.111	7.64×10^{-1}	1.18×10^{2}	14,268
28–32	26.30	5.422	5.83×10^{-1}	1.54×10^{2}	74,524
32–35	30.10	4.572	4.91×10^{-1}	1.82×10^{2}	119,740
35–42	20.64	3.834	4.12×10^{-1}	2.17×10^{2}	116,578
42–48	7.09	3.225	3.47×10^{-1}	2.57×10^{2}	56,525
48–60	3.19	2.707	2.91×10^{-1}	3.06×10^{2}	36,057
60–65	2.16	2.274	2.44×10^{-1}	3.64×10^{2}	34,566
65–100	1.02	1.777	1.91×10^{-1}	4.65×10^{2}	26,702
	100				$\sum$ = 479,690

Solution

At a temperature of 15°C, $\nu = 1.139\times10^{-6}$ m²/s.

First calculate the approach velocity as follows:

$$V_A = 1.53\,\frac{\text{Lps}}{\text{m}^2}\left(\frac{1\text{ m}^3}{1000\text{ L}}\right) = 1.53\times10^{-3}\text{m/s}$$

For sieves 14–20, the Reynolds number is calculated as follows, using Equation (6.138):

$$N_R = \frac{\phi V d}{\nu} = \frac{0.80(1.53\times10^{-3}\text{m/s})(10.006\times10^{-4}\text{m})}{1.139\times10^{-6}\text{m}^2/\text{s}}$$
$$= 1.075$$

The new friction factor, f', is calculated using Equation (6.134) as follows:

$$f' = 150\left(\frac{1-\varepsilon}{N_R}\right) + 1.75 = 150\left(\frac{1-0.41}{1.075}\right) + 1.75$$
$$= 84.1$$

Calculate the following quantity:

$$\frac{f'_{ij}X_{ij}}{d_{ij}} = \frac{84.1(0.87/100)}{10.006\times10^{-4}\text{m}} = 731$$

The values in the table were calculated using a spreadsheet; hand calculations may differ slightly due to round-off error.

Next, calculate the head loss in the clean filter using Equation (6.133) as follows.

$$h_L = \frac{L(1-\varepsilon)V_A^2}{\phi\varepsilon^3 g}\sum\frac{f'_{ij}X_{ij}}{d_{ij}}$$

$$\sum\frac{f'_{ij}X_{ij}}{d_{ij}} = 479690$$

$$h_L = \frac{0.762\text{ m}(1-0.41)(1.53\times10^{-3}\text{m/s})^2}{(0.80)0.41^3(9.81\text{ m/s}^2)}(479690)$$
$$= \boxed{0.93\text{ m}}$$

The initial head loss of a clean filter at 15°C typically ranges from 0.8–2.5 feet (0.24–0.76 m) at filtration rates of 2 and 10 gpm/ft², respectively (Kawamura, 1991, p. 220). Viessman *et al.* (2009, p. 358) indicate that the head loss of a clean granular-media filter is typically less than 3 ft (0.9 m). Head loss is sensitive to both the bed porosity and temperature of the water (MWH, 2005, p. 892).

The head loss for a clean filter bed using the Rose equation is illustrated in the following example.

Example 6.18 Head loss across a granular media filter

Determine the head loss for a clean filter bed using the Rose equation for a stratified bed with uniform porosity of 0.41. The rapid sand filter is made up of a 762 mm deep bed of sand, specific gravity of 2.65, shape factor (ϕ) =

0.80, temperature of 15°C, and filtration rate of 1.53 Lpsm^2. A sieve analysis is presented below.

Sieve (1)	% Sand retained (2) (X_{ij})	d_{ij} (3) ($m \times 10^{-4}$)	N_R (4)	C_D (5)	$C_D X_{ij}/d_{ij}$ (6)
14–20	0.87	10.006	1.08×10^0	2.23×10^1	194
20–28	8.63	7.111	7.64×10^{-1}	3.14×10^1	3,812
28–32	26.30	5.422	5.83×10^{-1}	4.12×10^1	19,980
32–35	30.10	4.572	4.91×10^{-1}	4.88×10^1	32,159
35–42	20.64	3.834	4.12×10^{-1}	5.83×10^1	31,359
42–48	7.09	3.225	3.47×10^{-1}	6.93×10^1	15,224
48–60	3.19	2.707	2.91×10^{-1}	8.25×10^1	9,722
60–65	2.16	2.274	2.44×10^{-1}	9.82×10^1	9,329
65–100	1.02	1.777	1.91×10^{-1}	1.26×10^2	7,214
	100				$\sum = 128{,}993$

Solution

At a temperature of 15°C, $\nu = 1.139 \times 10^{-6}\ \mathrm{m^2/s}$.

First calculate the approach velocity as follows:

$$V_A = 1.53\,\frac{\mathrm{Lps}}{\mathrm{m^2}}\left(\frac{1\ \mathrm{m^3}}{1000\ \mathrm{L}}\right) = 1.53 \times 10^{-3}\mathrm{m/s}$$

For sieves 14–20, the Reynolds number is calculated as follows using Equation (6.138):

$$N_R = \frac{\phi V d}{\nu} = \frac{0.80(1.53 \times 10^{-3}\,\mathrm{m/s})(10.006 \times 10^{-4}\,\mathrm{m})}{1.139 \times 10^{-6}\,\mathrm{m^2/s}}$$

$$= 1.075$$

The coefficient of drag, C_D, is calculated by simplifying Equation (6.137) as follows for Reynolds numbers ≤ 1:

$$C_D = \frac{24}{N_R} = \frac{24}{1.075} = 22.3$$

Calculate the following quantity:

$$\frac{C_D X_{ij}}{d_{ij}} = \frac{22.3(0.87/100)}{10.006 \times 10^{-4}\,\mathrm{m}} = 194$$

Next, calculate the head loss in the clean filter using Equation (6.136) as follows:

$$h_L = \frac{1.067}{\phi}\frac{L}{g}\frac{V_A^2}{\varepsilon^4}\sum\frac{C_D X_{ij}}{d_{ij}}$$

$$h_L = \frac{1.067(0.762\,\mathrm{m})(1.53 \times 10^{-3}\mathrm{m/s})^2}{(0.80)(9.81\,\mathrm{m/s^2})(0.41^4)}(128993)$$

$$= \boxed{1.11\ \mathrm{m}}$$

In this example, the head loss estimated using the Rose equation is approximately 20% greater than that calculated using the Carman-Kozeny equation.

6.8.5.2 Ergun Equation

According to *Water Treatment: Principles and Design* (2005, p. 890), the most widely used empirical equation for estimating head loss in clean granular media filters is the one developed by Ergun (1952). Ergun's equation is based on an approach similar to that of Kozeny and is presented below.

$$h_L = \kappa_V\frac{(1-\varepsilon)}{\varepsilon^3}\frac{\mu L V}{\rho g d^2} + \kappa_I\frac{(1-\varepsilon)}{\varepsilon^3}\frac{L V^2}{g d} \qquad (6.139)$$

where:

κ_V = head loss coefficient due to viscous forces, unitless

κ_I = head loss coefficient due to inertial forces, unitless.

Head loss coefficients to use in Equation (6.139) are listed in Table 6.15.

6.8.6 Backwashing of filters

Periodically, granular media filters must be backwashed to remove particulates and turbidity that build up during the filter run. Typically, operators backwash a filter based on one of the following scenarios:

a) when the head loss through the filter reaches 8–10 feet (2.4–3 m);
b) when the turbidity in the effluent reaches a specified level such as 0.20 NTU;
c) when the filter has been in operation for a specified amount of time (24–96 hours, for example).

Filtered water used to backwash the filters is typically applied by gravity flow from an elevated storage tank or pumped from a ground-level or underground storage tank. Figure 6.28 is a cross-sectional view of a rapid sand filter during the backwashing process. Backwash water at a rate of around 15–20 gpm/ft^2 (611–815 Lpm/m^2) for 15–20 minutes is required to properly expand and clean a dirty filter. Upflow wash-water can be used with air scour or with surface wash; details on these methods are found in *Water Treatment; Principles and Design* (2005, pp. 937–940) and *Water Quality and Treatment* (1999, pp. 8.58–8.71).

Table 6.15 Head loss coefficients for use in Equation (6.139).

Medium	κ_V	κ_I	ε (%)
Sand	110–115	2.0–2.5	40–43
Anthracite	210–245	3.5–5.3	47–52

Source: MWH (2005) p. 891. Reproduced by permission of John Wiley & Sons Ltd.

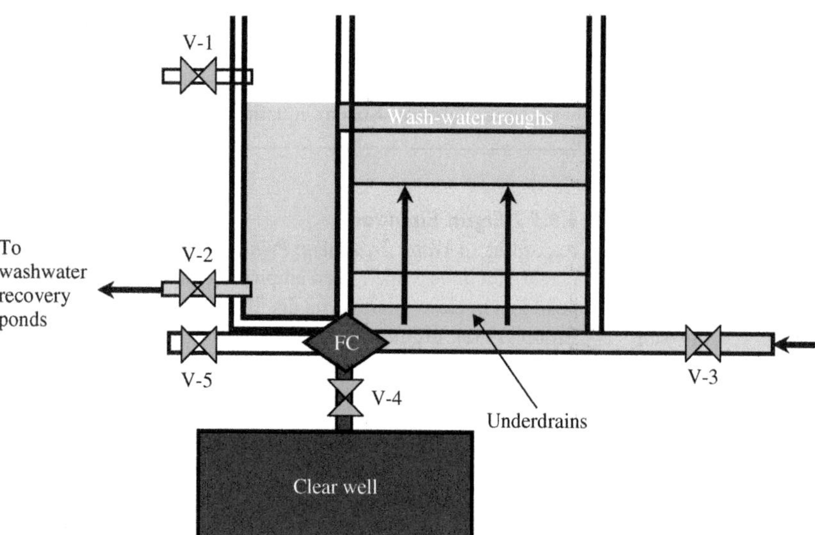

Figure 6.28 Cross-sectional view of a rapid sand filter during the backwashing process.

The entire backwashing operation may take up to 30 minutes before a filer is brought back on line. Approximately 1–5% of the filtered water is used during the backwashing process. The dirty wash-water is collected in the backwash water troughs and is transported to the wash-water recovery pond, where the solids are settled out of solution and the supernatant is returned to the head of the water treatment plant so that it can undergo coagulation/flocculation/sedimentation with the raw water entering the plant.

6.8.7 Backwash hydraulics

During **backwashing**, the **water flow through the filter bed is reversed**, thus expanding and fluidizing the bed. The frictional drag of the media particles suspended in the upward flowing water is equal to the pull of gravity or the buoyant weight of the media. This is expressed in the following equation:

$$\Delta p = \gamma h_L = \rho g h_L = (\rho_s - \rho)(1 - \varepsilon_e)(L_e)g \tag{6.140}$$

$$h_L = \left(\frac{\rho_s - \rho}{\rho}\right)(1 - \varepsilon_e)(L_e) \tag{6.141}$$

where:
Δp = pressure drop after fluidization, lb (N)
γ = specific weight of wash-water, lb/ft^3 (kN/m^3)
h_L = head loss as a water column height, ft (m)
ρ = density of wash-water, lb·s^2/ft^4 (kg/m^3)
ρ_s = density of media particles, lb·s^2/ft^4 (kg/m^3)
ε = porosity of unexpanded bed, dimensionless
ε_e = porosity of expanded bed, dimensionless
g = acceleration of gravity, ft/s^2 (m/s^2)
A_s = surface area of filter, ft^2 (m^2)
L = height of unexpanded bed, ft (m)
L_e = height of expanded bed, ft (m).

The volume of sand in an unexpanded bed is equal to the volume of sand in the expanded bed or performing a mass balance on the sand yields Equation (6.142):

$$(1 - \varepsilon)(A_s)(L) = (1 - \varepsilon_e)(A_s)(L_e) \tag{6.142}$$

$$L_e = \left(\frac{1 - \varepsilon}{1 - \varepsilon_e}\right)L \tag{6.143}$$

As referenced by Rich (1971, p. 149), Fair & Geyer (1954) found that the porosity of the expanded bed could be estimated using Equation (6.144):

$$\varepsilon_e = \left(\frac{V_b}{V_s}\right)^{0.22} \tag{6.144}$$

where:
V_s = settling velocity of a specified particle diameter or sieve size, ft/s (m/s)
V_b = upflow velocity of the backwash water, ft/s (m/s).

The settling velocity (V_s) for the largest diameter particle in the sieve analysis is used for calculating V_b by rearranging and substituting into Equation (6.144).

The depth of the expanded bed (L_e) can be calculated by substituting Equation (6.144) into Equation (6.143). This yields Equation (6.145), which only applies to bed expansions in which the bed consists of uniform media:

$$L_e = \left(\frac{1 - \varepsilon}{1 - (V_b/V_s)^{0.22}}\right)L \tag{6.145}$$

Equation (6.143) may be modified as Equation (6.146) for beds with non-uniform media, so that the expanded bed depth may be determined.

$$L_e = L(1 - \varepsilon)\sum \frac{X_{ij}}{(1 - \varepsilon_e)} \tag{6.146}$$

An equation must be developed for the settling velocity of a particle as a function of media diameter (d), viscosity (a function of temperature), and shape factor (ϕ) for the given media. Example 6.19 illustrates the procedure used for developing the appropriate relationship between settling velocity (V_s) and particle diameter. Newton's Law, as presented in Equation (6.90), is used in the derivation.

Example 6.19 Derivation of settling velocity as a function of particle diameter

Derive an equation relating the settling velocity of sand particles to particle diameter with the following assumptions: specific gravity (S.G.) of sand = 2.65, shape factor (ϕ) = 0.95, and temperature of 50°F (μ = 2.735×10^{-5} lb·s/ft² and ρ = 1.936 lb·s²/ft⁴). Assume that $C_D = 18.5/(N_R)^{0.6}$ for Reynolds number = 1.9 to 500 (Rich, 1971, p. 86).

Solution

$$C_D = \frac{18.5}{(N_R)^{0.6}}$$

Drag coefficient equation as proposed by Rich (1971).

Substituting Equation (6.138) into Rich's equation yields the following equation for C_D:

$$C_D = \frac{18.5}{(\phi\rho V_s d/\mu)^{0.6}}$$

Next, substitute the equation developed for C_D into Equation (6.90) to develop the relationship between settling velocity (V_s) and particle diameter.

$$V_s = \left[\frac{4}{3}\frac{g(\phi\rho V_s d)^{0.6}}{18.5\,\mu^{0.6}}(S.G._{sand} - 1)\,d\right]^{0.5}$$

$$V_s = \left[\frac{4}{3}\frac{32.2\text{ ft/s}^2(0.95 \times 1.936\,V_s d)^{0.6}}{18.5\,\mu^{0.6}}(2.65 - 1)\,d\right]^{0.5}$$

$$V_s = \left[\frac{5.52 V_s^{0.6}\, d^{1.6}}{\mu^{0.6}}\right]^{0.5}$$

$$V_s^2 = 5.52\, V_s^{0.6}\, d^{1.6}/\mu^{0.6}$$

$$V_s^{1.4} = \frac{5.52\, d^{1.6}}{\mu^{0.6}}$$

$$V_s = \left(5.52\, d^{1.6}/\mu^{0.6}\right)^{1/1.4} = \left(5.52\, d^{1.6}/\mu^{0.6}\right)^{0.714}$$

$$V_s = \left(5.52\, d^{1.6}/(2.735 \times 10^{-5})^{0.6}\right)^{0.714} = [3017.9\, d^{1.6}]^{0.714}$$

$$\approx 306\, d^{1.143}$$

$$\boxed{V_s = 306\, d^{1.143}}$$

This equation is easily used to calculate the particle settling velocity from its diameter.

Example 6.20 Calculating expanded bed depth in English units

Find the expanded bed depth in a rapid sand filter that is 24 inches deep and has a porosity of 0.40 for the sieve analysis given in the table below. Assume a water temperature of 55°F and a specific gravity (S.G.) for sand of 2.65. The shape factor (ϕ) is 0.95, μ = 2.55×10^{-5} lb·s/ft², and ρ = 1.94 lb·s²/ft⁴.

(1) Sieve #	(2) % Sand Retained (X_{ij})	(3) d_{ij} (ft × 10^{-3})	(4) V_S (fps)	(5) ε_e	(6) $\frac{X_{ij}}{(1-\varepsilon_e)}$
14–20	1.05	3.2883	0.457	0.40	0.0176
20–28	6.65	2.333	0.309	0.44	0.1188
28–32	15.7	1.779	0.227	0.47	0.2969
32–35	18.84	1.5	0.186	0.49	0.3707
35–42	18.98	1.258	0.152	0.51	0.3905
42–48	17.72	1.058	0.125	0.54	0.3826
48–60	14.25	0.888	0.102	0.56	0.3246
60–65	5.15	0.746	0.0839	0.59	0.1245
65–100	1.66	0.583	0.0633	0.62	0.0441
	100				$\sum$ **= 2.0703**

Solution

To calculate the settling velocities in Column 4 above, it is necessary to develop a settling velocity equation using Newton's Law, Equation (6.90), for a shape factor of 0.95, S.G. of 2.65, and viscosity at 55°F as shown in Example 6.19. For these parameters, the relationship between settling velocity and particle diameter is as follows:

$$V_s = 315\, d^{1.143}$$

The backwash velocity (V_b) is calculated by rearranging Equation (6.144) and substituting ε for ε_e as follows:

$$V_b = V_s \varepsilon^{4.5}$$

Next, substituting the settling velocity (V_s) for the largest diameter media particle that is retained between sieves 14–20, (V_s = 0.457 feet per second) yields a backwash velocity of 0.0074 fps:

$$V_b = V_s \varepsilon^{4.5} = 0.457\,(0.40)^{4.5} = 7.40 \times 10^{-3}\text{fps}$$

In Column 5, the expanded bed porosities (ε_e) are calculated using Equation (6.144):

$$\varepsilon_e = [V_b/V_s]^{0.22} = [7.40 \times 10^{-3}/4.57 \times 10^{-1}]^{0.22} = 0.404$$

For the largest sieve size (14–20), ε_e (0.404) should be approximately equal to ε, the unexpanded bed porosity, which is 0.40.

Column 6 is calculated using the following equation:

$$X_{ij}/(1-\varepsilon_e) = \frac{(1.05/100)}{(1-0.404)} = 0.0176$$

The expanded bed depth is calculated using Equation (6.146):

$$L_e = L(1-\varepsilon)\sum \frac{X_{ij}}{(1-\varepsilon_e)} = 2\ \text{ft}(1-0.40)(2.0703)$$

$$= \boxed{2.48\ \text{ft}}$$

Since the bed was two feet deep originally, note that the bed expanded 0.48 feet. Depending on the backwash rate utilized, it is not unusual for the bed to expand by 50% (Viessman & Hammer, 2005, p. 401).

At the beginning of the backwash process, the head loss is calculated using Equation (6.141).

$$h_L = [(S.G.)_P - 1](1-\varepsilon)L = [2.65-1](1-0.40)(2\ \text{ft})$$

$$= \boxed{1.98\ \text{ft}}$$

Example 6.21 Calculating expanded bed depth using SI units

Find the expanded bed depth in a rapid sand filter that is 0.610 m deep and has a porosity of 0.45 for the sieve analysis given below. Assume a water temperature of 10°C and a specific gravity (SG) for sand of 2.65. The shape factor (ϕ) is 0.82. $\nu = 1.3101 \times 10^{-6}\ \text{m}^2/\text{s}$.

(1) Sieve #	(2) % Sand Retained (X_{ij})	(3) d_{ij} (ft × 10^{-3})	(4) V_S (fps)	(5) ε_e	(6) $\frac{X_{ij}}{(1-\varepsilon_e)}$
14–20	0.87	10.006	0.127	0.454	0.0159
20–28	8.63	7.111	0.086	0.494	0.1707
28–32	26.3	5.422	0.0631	0.529	0.5586
32–35	30.1	4.572	0.0519	0.552	0.6724
35–42	20.64	3.834	0.0424	0.577	0.4883
42–48	7.09	3.225	0.0348	0.603	0.1786
48–60	3.19	2.707	0.0285	0.630	0.0863
60–65	2.16	2.274	0.0234	0.658	0.0632
65–100	1.02	1.777	0.0176	0.701	0.0341
	100				**$\sum = 2.2681$**

Solution

To calculate the settling velocities in Column 4 of the table, it is necessary to develop a settling velocity equation using Newton's Law, Equation (6.90) for a shape factor of 0.82, S.G. of 2.65, and viscosity at 10°C, which yields the following relationship. Example 6.19 illustrated the procedure that should be followed.

$$V_s = 340\, d^{1.143}$$

$$V_s = 340\,(10.006 \times 10^{-4}\ \text{m})^{1.143} = 0.127\ \text{m/s}$$

The backwash velocity (V_b) is calculated by rearranging Equation (6.144) as follows:

$$V_b = V_s \varepsilon^{4.5}$$

Next, substituting the settling velocity (V_s) of the largest diameter media particle that is retained between sieves 14–20 (V_s = 0.127 meters per second):

$$V_b = V_s \varepsilon^{4.5} = 0.127(0.45)^{4.5} = 3.49 \times 10^{-3}\ \text{m/s}$$

In Column 5, the expanded bed porosities (ε_e) are calculated using Equation (6.144):

$$\varepsilon_e = [V_b/V_s]^{0.22} = [3.49 \times 10^{-3}/0.127]^{0.22} = 0.454$$

For the largest sieve size (14–20), ε_e (0.454) should be approximately equal to ε (0.45), the unexpanded bed porosity.

Column 6 is calculated using the following equation:

$$\frac{X_{ij}}{(1-\varepsilon_e)} = \frac{(0.87/100)}{(1-0.454)} = 0.0159$$

The expanded bed depth is calculated using Equation (6.146):

$$L_e = L(1-e)\sum \frac{X_{ij}}{(1-\varepsilon_e)} = 0.61\ \text{m}(1-0.45)(2.2681)$$

$$= \boxed{0.76\ \text{m}}$$

Since the bed was 0.610 meter deep originally, note that the bed expanded 0.15 m.

At the beginning of the backwash process, the head loss is calculated using Equation (6.141):

$$h_L = [(S.G.)_p - 1](1-\varepsilon)L$$

$$= [2.65-1](1-0.45)(0.61\ \text{m}) = \boxed{0.55\ \text{m}}$$

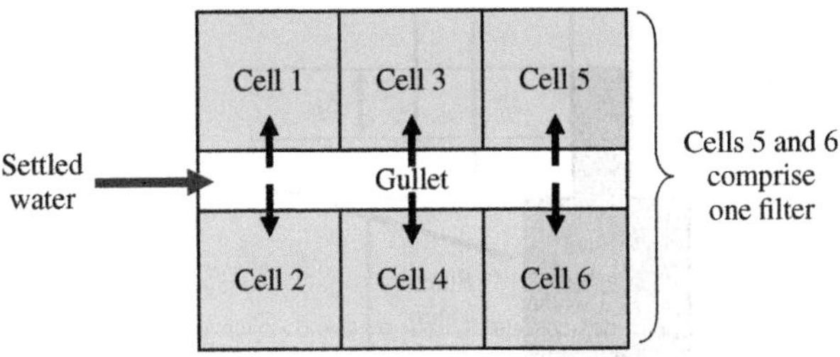

Figure 6.29 Filter layout along one or both sides of a pipe gallery or gullet.

6.8.8 Filter arrangement

Filters are typically placed adjacent to one another along one or both sides of a pipe gallery or gullet (See Figure 6.29). Clark *et al.* (1977, p. 402) state that rapid sand filters are usually rectangular in shape, with an average length : width ratio of about 1.25 and surface area ranging from 450–4,500 ft^2. According to Kawamura (1991, p. 210), conventional gravity filters use length to width (L:W) ratios of 2 : 1 to 4 : 1. Surface areas of gravity filters typically range from 250–1,000 ft^2 (25–100 m^2) and filter depths range from 12–20 feet (3.2–6 m).

For water treatment plants processing <2.0 MGD (88 L/s), a minimum of two filters should be used. If the capacity of the WTP exceeds 2.0 MGD, a minimum of four filters should be used. Equation (6.147) is used for estimating the number of filters (N) required as a function of flow.

$$N = 1.2\,Q^{0.5} \tag{6.147}$$

where:
N = total number of filters
Q = maximum daily flow rate, MGD.

For large water treatment plants processing over 20 MGD (75,700 m^3/d), filters are composed of two cells per filter, with a gullet running down the center.

Example 6.22 Calculating filter dimensions and backwash quantity

A water treatment plant has four rapid sand filters. Each filter is designed for a capacity of 3,800 m^3/d. Backwashing is accomplished for 15 minutes at a rate of 0.60 m/(min · m^2) once every 36 hours. The terminal head loss prior to backwashing averages 1.2–3 meters. Determine the following:

a) The filter dimensions if a filtration rate of 0.20 m/(min · m^2) is used.
b) The quantity of water required for backwashing and percentage of filtered water used in backwashing.

Solution part a

Determine the surface area (A_s) of the filters by dividing the design flow rate by the filtration rate (0.20 m/min · m^2) and making appropriate conversions.

$$A_s = \frac{3800\,\text{m}^3/\text{d}}{(0.20\,\text{m/min}\cdot\text{m}^2)(1440\,\text{min/d})} = 13.2\,\text{m}^2$$

For a square filter, $L = W$, therefore:

$$W^2 = 13.2\,\text{m}^2$$

$$W = \sqrt{13.2} = 3.6\,\text{m} \qquad \text{Use } W = L = \boxed{3.6\,\text{m}}$$

Solution part b

The backwash flow rate (Q_b) is estimated by multiplying the backwash velocity or rate by the filter surface area (A_s) times the backwash frequency (10 min/d).

$$Q_b = \left(\frac{0.60\,\text{m/min}}{\text{m}^2}\right)(3.6\,\text{m} \times 3.6\,\text{m})\left(\frac{10\,\text{min}}{\text{d}}\right) = \boxed{78\,\frac{\text{m}^3}{\text{d}}}$$

The percentage of water used in the backwashing process is estimated by dividing the backwash flow (Q_b) by the product flow of 3,800 m^3/d.

$$\text{Percentage} = \frac{78\,\text{m}^3/\text{d}}{3800\,\text{m}^3/\text{d}}(100\,\%) = \boxed{2.1\,\%}$$

The amount of water used for backwashing varies from 1% to 5% of the filtered water.

6.9 Membrane treatment

Filtration is the separation and removal of suspended solids (dispersed solids) from a liquid phase by passage through a porous medium. Membrane filtration extends the range of particle size to be removed down to 0.01 μm. The objective is the complete removal of particles such as sediment, algae, viruses, protozoa, bacteria, and colloids. Micro-filtration (MF) and ultra-filtration (UF) are examples.

Reverse osmosis (RO) is a membrane treatment process used to separate dissolved solutes and NOM from water. Both nanofiltration (NF) and RO membranes are used in RO applications. The objective of RO is to produce potable water from ocean or brackish waters, and to remove specific dissolved constituents. NF membranes are used for softening hard waters, to freshen brackish waters, and to reduce the concentration of DBP precursors. Both NF and RO membranes separate solutes by diffusion through a semi-permeable membrane as well as sieving action. Membranes are permeable to water but not to solutes or substances rejected and removed.

Table 6.16 Hierarchy of Pressure-Driven Membrane Processes.

Type of membrane	Types of materials rejected
Microfiltration, 0.1 μm pores	Particles, sediment, algae, protozoa, bacteria
Ultrafiltration, 0.01 μm pores	Small colloids, viruses
Nanofiltration, 0.001 μm pores	Dissolved organic matter, divalent ions
RO, nonporous	Monovalent ions plus all above

Source: Adapted from MWH (2005) p. 957. Reproduced by permission of John Wiley & Sons Ltd.

Typical constituents removed include salts, hardness, pathogens, color, turbidity, DBP precursors, synthetic organic compounds (SOCs), and pesticides. Dissolved gases such as H_2S and CO_2 and some pesticides will pass through RO membranes. The US EPA has designated RO as a best available treatment (BAT) for removing inorganic contaminants such as antimony, arsenic, barium fluoride, nitrate, nitrite, selenium, radionuclides (including radium-226), β emitters, photon emitters, and alpha emitters (MWH, 2005, p. 1433). Table 6.16 shows the hierarchy of pressure-driven membrane processes as delineated in *Water Treatment: Principles and Design* (2005, p. 957).

6.9.1 Diffusion

If a removable partition is gently removed from the tanks shown in Figure 6.30, which contain a concentrated salt solution and pure water, the tank contents would eventually reach equilibrium. The salt ions would diffuse throughout the entire tank until the concentration is the same throughout the entire volume. Conservation of mass requires a flux of water molecules from right to left. The salt ions would diffuse primarily from left to right.

6.9.2 Osmosis

If the removable partition in Figure 6.30 is replaced with a semi-permeable membrane and manometer tubes are installed in each tank, then osmotic flow occurs wherein water or solvent flows from right to left. The solvent flows from right to left, due to a drop in the solvent concentration (i.e., pure water to salty water). The semi-permeable membrane allows water or solvent to pass, but not the salt. Figure 6.31 illustrates this phenomenon.

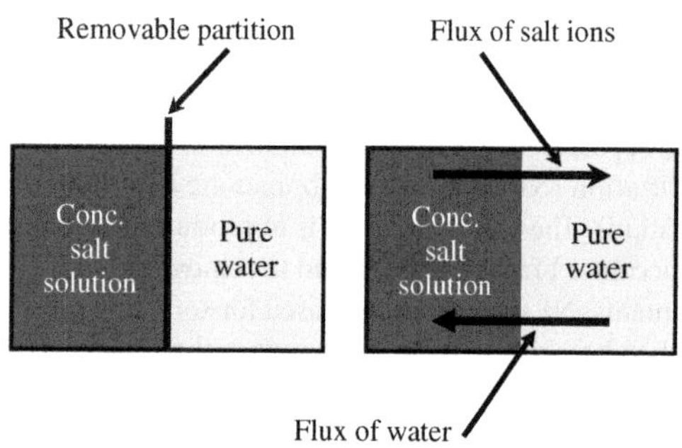

Figure 6.30 Diffusion of solute.

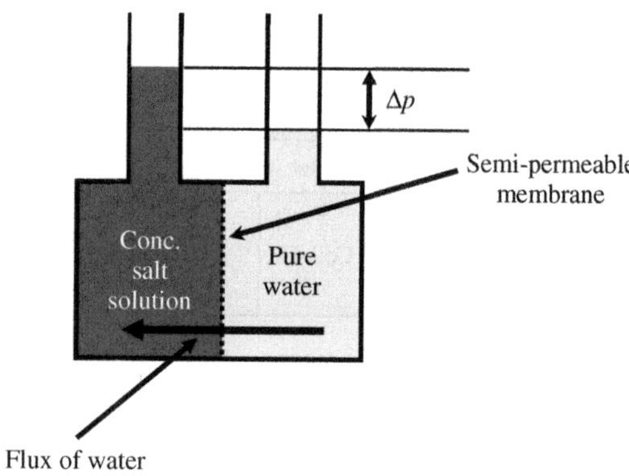

Figure 6.31 Osmotic flow.

6.9.3 Reverse osmosis

In reverse osmosis, an external force or hydrostatic pressure is applied which is greater than the osmotic pressure, causing the flux of water molecules going from left to right (i.e., producing pure water). See Figure 6.32 for illustration of reverse osmosis flow.

Reverse osmosis units may be run in three different modes and in either dead-end or cross-flow filtration. Figure 6.33 shows the three modes of operation: constant flux, constant pressure, and non-restricted flux and pressure. Transmembrane pressure is denoted by TMP. Traditionally, the constant flux mode of operation has been used. Newer studies indicate that allowing both the flux and TMP to vary is the best mode of operation.

Pressure-driven membrane processes are designed for dead-end (transverse) and cross-flow filtration operation modes, as illustrated in Figures 6.34 and 6.35. MF and UF systems that treat low-turbidity waters are designed to operate in the dead-end mode, where the waste retentate stream is

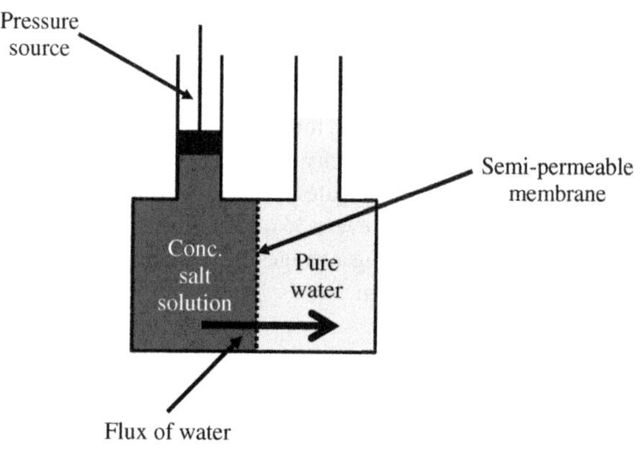

Figure 6.32 Reverse osmosis flow.

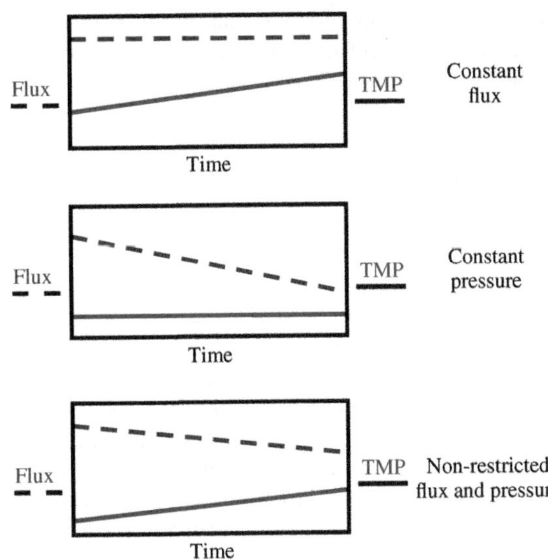

Figure 6.33 Reverse osmosis operational modes.

produced by an intermittent backwash. In the cross-flow mode of operation, the feed stream flows across the membrane, and the permeate or filtrate passes through the membrane tangential to the surface of the membrane. A continuous waste stream is generated during the cross-flow mode of operation.

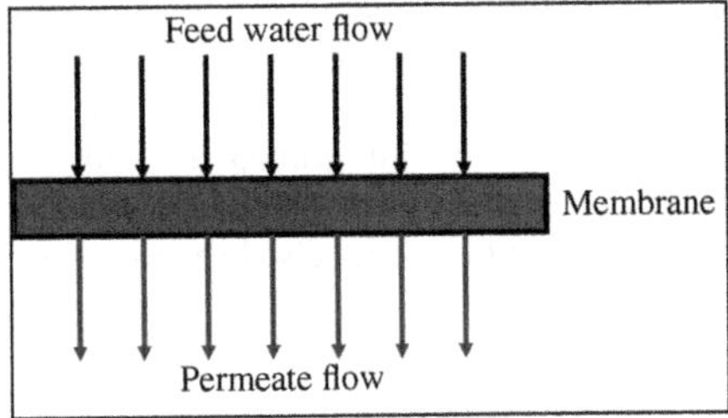

Figure 6.34 Dead-end flow mode of filtration.

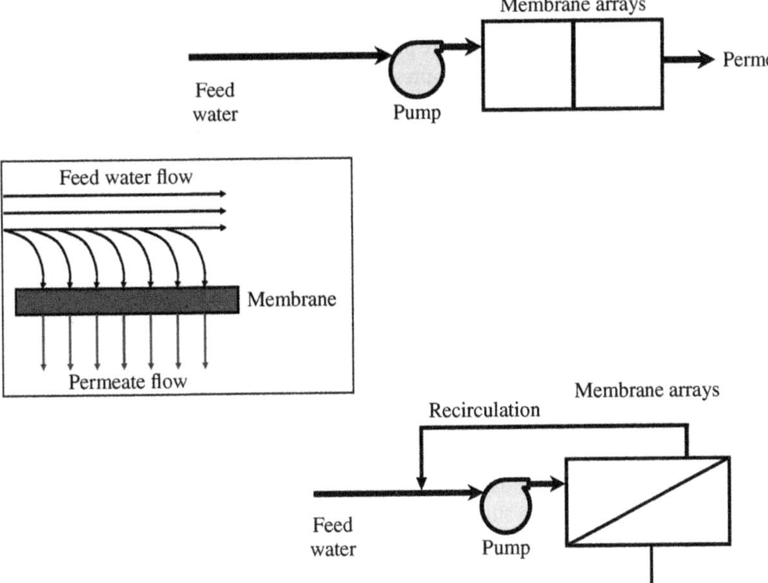

Figure 6.35 Cross-flow mode of filtration.

6.9.4 Membrane terminology

Figure 6.36 is a schematic showing the feed (*f*), permeate (*p*), and concentrate (*c*) streams for a membrane filtration system. The solute and mass transfer coefficients are represented by K_i and K_w, respectively. The rate at which permeate flows through a unit of membrane area is known as the water flux rate, F_W. The solute flux rate, F_i, is the rate at which a specific solute species passes through a unit of membrane area. The volumetric flow rate is represented by *Q*, the solute concentration by *C*, and the pressure by *P*.

The permeate stream exits at essentially atmospheric pressure and the concentrate stream remains close to the feed pressure. RO is a continuous separation process that does not

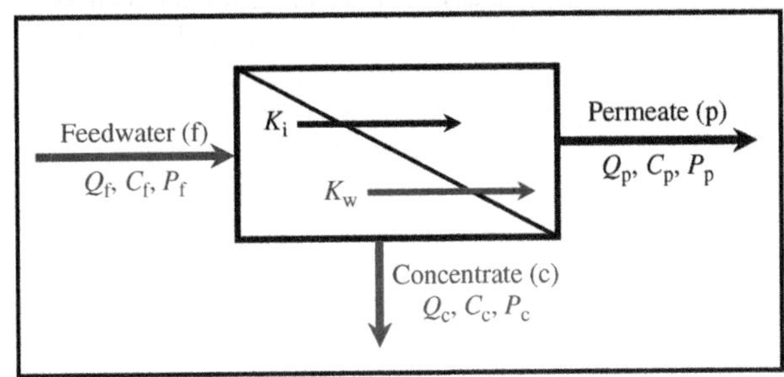

Figure 6.36 Schematic of membrane filtration system.

involve a backwash cycle. Periodically however, every three months to a year, the membranes must be chemically cleaned due to fouling of the membranes (Davis, 2011, p. 6–17). The equations used in the design of membrane systems will be presented in subsequent sections, along with typical units.

6.9.5 RO membranes

Membranes can be dense material without pores or voids as is the case for high-pressure RO membranes or porous materials used in NF membranes (MWH, 2005, p. 1435). RO membranes made from hydrophilic materials make them more vulnerable to physical, mechanical, and chemical degradation. The active membrane layer is normally between 0.1–2 μm in thickness and consists of several layers to provide structural support. They are operated at high pressure and must resist compaction. Typical operating pressures are between 700–1000 psi (47–67 bar). The osmotic pressure of seawater is approximately 360 psi (355 bar) or 10 psig/1000 mg/L of TDS.

RO membranes are typically made of cellulose acetate (CA) or polyamide (PA). Cellulose acetate membranes use asymmetric construction, i.e., they are formed from a single material that has active and support layers and they are usually hydrophilic. They maintain high flux rates and minimize fouling, are not tolerant of temperatures above 30°C, hydrolyze when pH is less than 3 or higher than 8, and are susceptible to biological degradation. These membranes compact over time due to high operating pressures, and they degrade in the presence of 1 mg/L chlorine concentration.

Polyamide membranes are chemically and physically more stable than CA membranes. They are generally thin-film composite membranes composed of two or more materials cast on top of each other. They are generally immune to bacterial degradation, are stable over a wide pH range (3–11), do not hydrolyze in water, produce higher water flux and higher salt rejection than CA membranes at similar operating pressures and temperatures, are more hydrophobic (water-hating) and susceptible to fouling, and are not tolerant of free chlorine at any concentration.

6.9.6 Membrane configurations

There are two primary membrane configurations: spiral wound and hollow fine fiber (HFF) modules (MWH, 2005, pp. 1438–1441). Spiral wound elements are 40–60 in (1–1.5 m) long, with diameters ranging from 8–12 in (0.2–0.3 m). Typically, 4–7 elements are arranged in series in a pressure vessel. The head loss across a membrane element is approximately 7 psi (0.5 bar) per element and the recovery per element ranges from 5–15%. The internal construction of a spiral-wound RO element is shown in Figure 6.37.

In the Hollow Fine Fiber (HFF) modules, feed water passes over the outside of the fiber and is collected in the inner annulus of the fiber. Figure 6.38 shows the internal construction of a HFF reverse osmosis membrane module. DuPont used to manufacturer HFF membranes with fibers having an inner diameter (ID) and outer diameter (OD) of 0.043 mm and 0.085 mm, respectively. Toyobo, a Japanese manufacture, is the only current manufacturer of HFF systems. They manufacture a HFF module which has a product recovery of about 30% per element (MWH, 2005, p. 1440).

6.9.7 Pre- and post-treatment of feed and finished water

Figure 6.39 shows a schematic of a typical Reverse Osmosis Water Treatment Plant. The feed or raw water to membrane treatment systems must undergo pre-treatment prior to membrane filtration. The major purpose of pre-treatment is to prevent scaling, i.e., the precipitation of various salts onto the membrane. Solute concentration increases in the feed stream as water passes through the semi-permeable membrane. Therefore, the solubility product of various salts such as calcium carbonate and calcium sulfate can be exceeded, causing them to precipitate.

Typically, pH adjustment with an acid (e.g., sulfuric or hydrochloric acid), and some type of antiscalant is also added to control scale formation. Conventional pre-treatment typically involves acid addition to lower pH and antiscalant addition to prevent precipitation of salts on the membrane. Antiscalants allow supersaturation of salts without precipitation occurring. Some antiscalants used include sodium hexametaphosphate (SHMP), polymeric compounds, and proprietary additives.

Feed water to membrane treatment systems must also be filtered. Pre-filtration is necessary for the removal of particulate matter that can accumulate on the membrane surface, since RO systems do not have a backwash cycle. Pre-filtration is usually accomplished using cartridge filters with 5 μm openings to remove particulates. Granular media filtration and membrane filtration using MF or UF membranes have also been used for pre-treating surface waters.

Pre-disinfection may also be included in pre-treatment, to help prevent biofouling of the membrane, which can be a significant problem when RO systems are taken off-line for servicing. Certain disinfectants, such as chlorine, may be incompatible with the type of membrane material being used.

High-pressure pumps are also required to pressurize the feed water. Low-pressure and brackish water RO systems typically require a pressure of 145–430 psi (10–30 bar). Seawater RO systems require a pressure of 800–1,200 psi (55–85 bar).

Post-treatment of the permeate stream involves the removal of dissolved gases (e.g., carbon dioxide and hydrogen sulfide) by aeration or gas stripping. The alkalinity and pH of the permeate require adjustment. Typically, a base such as sodium hydroxide or lime is added to the permeate to raise the pH, since the permeate is acidic from pre-treating the feed water. This increase in the alkalinity is necessary to help stabilize the finished water. Some type of disinfectant is added before the water is pumped into the distribution system. Corrosion inhibitors that are added to the degasified permeate include orthophosphates, polyphosphates, bimetallic phosphates, and sodium silicates.

The brine or concentrate must be properly disposed of. Disposal options include: discharge to brackish or surface waters, oceans, estuaries; discharge to municipal sewer; and deep well injection.

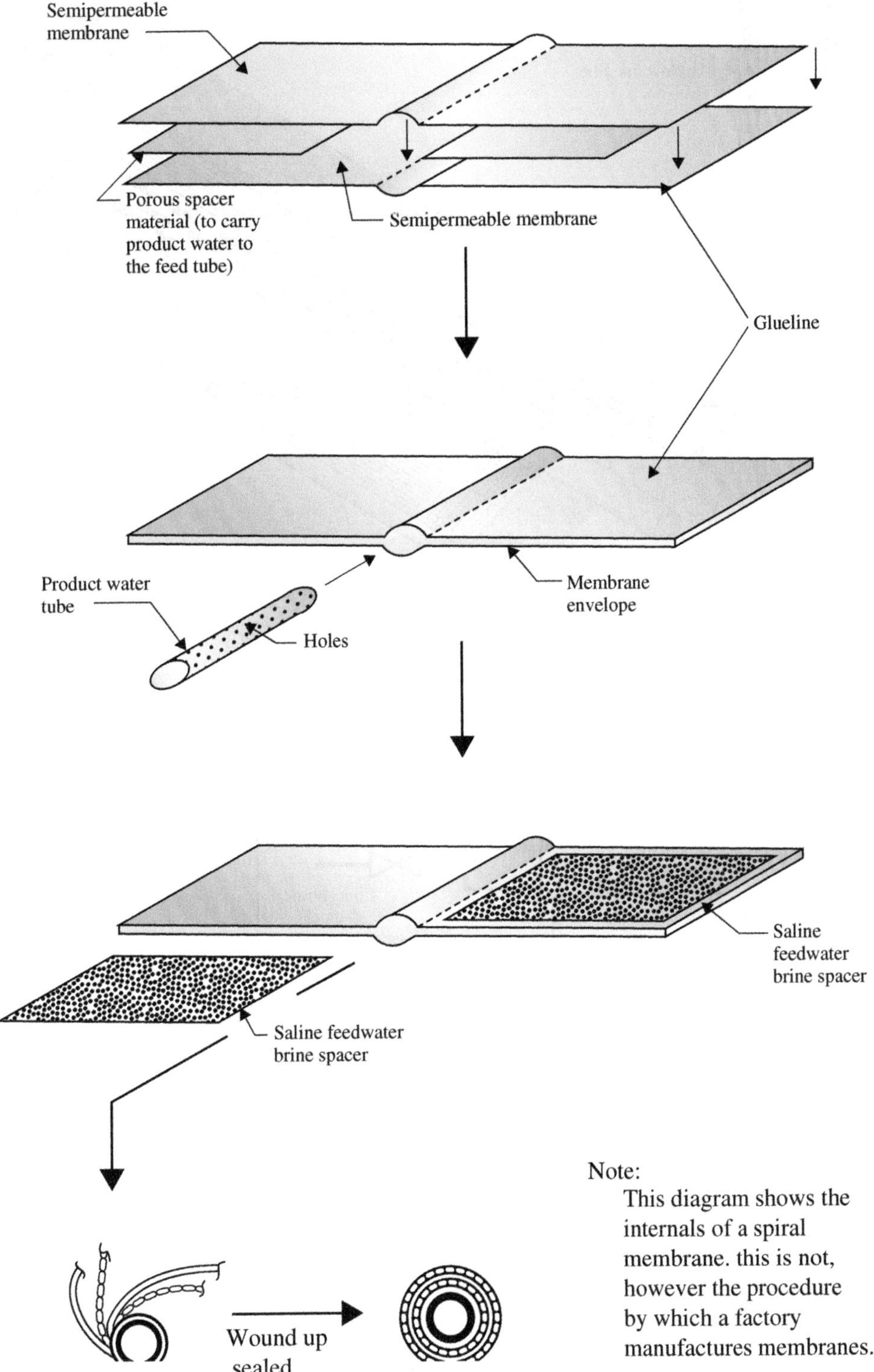

Figure 6.37 Internal construction of spiral-wound RO membrane. *Source*: Department of the Army, USA, (1986). *Technical Manual TM 5-813-8, Water Desalination*, p. 7–8.

A membrane element is the smallest physical production unit. Membrane elements are enclosed in pressure vessels mounted on skids which are connected by piping. A group of pressure vessels operating in parallel is called a stage. The concentrate from one stage can be fed to a second stage to increase water recovery. The number of pressure vessels decreases in each succeeding stage to maintain sufficient velocity in the feed channel as permeate is extracted from the feed water stream. This configuration is illustrated in Figure 6.40, and is called a multi-stage or brine-staged system.

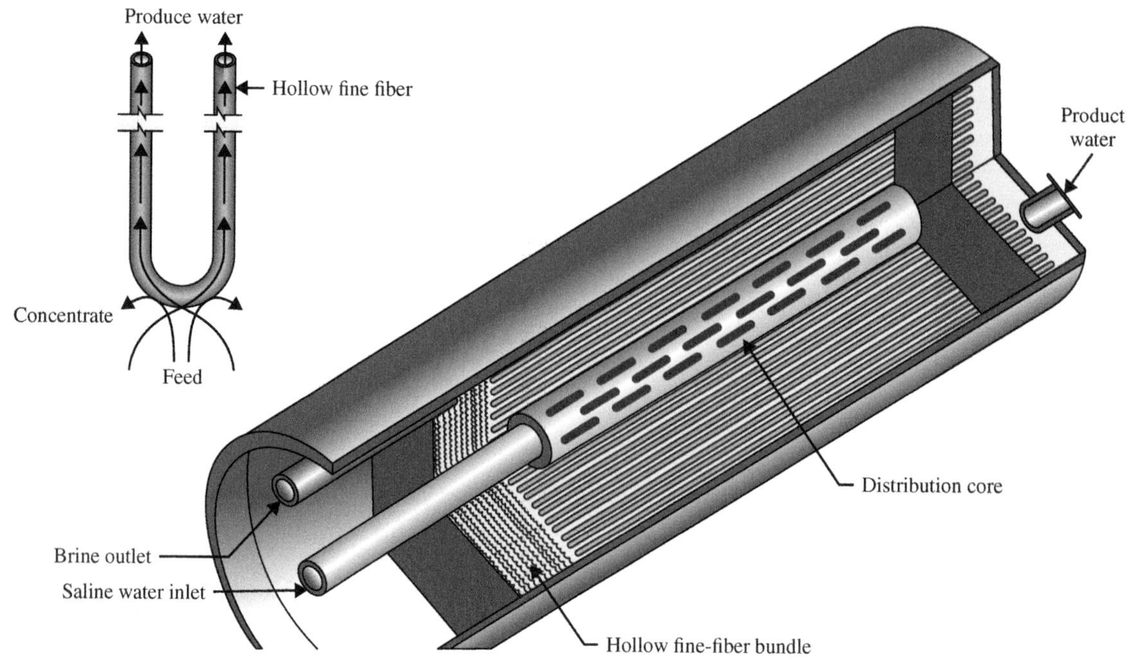

Figure 6.38 Internal construction of a hollow fine fiber reverse osmosis membrane module. *Source:* Department of the Army, USA, (1986). Technical Manual TM 5-813-8, Water Desalination, p. 7–9.

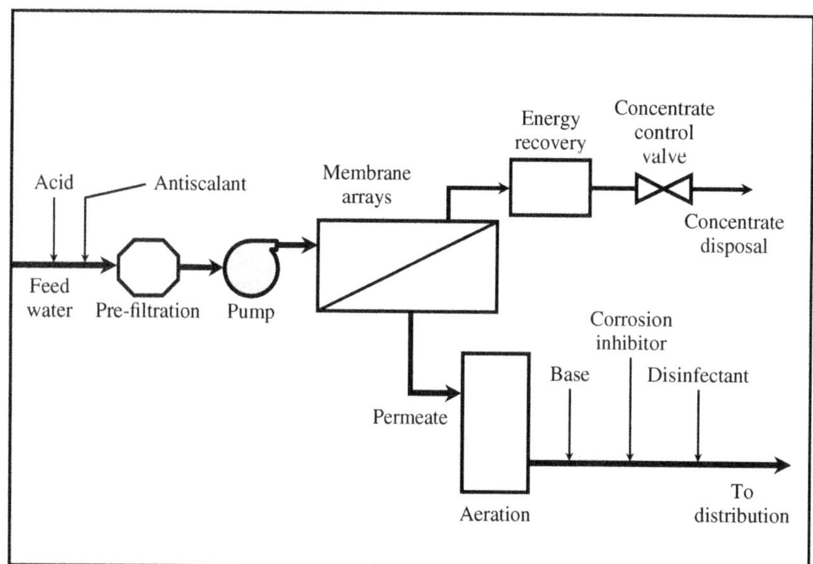

Figure 6.39 Schematic of typical reverse osmosis WTP. *Source:* MHW (2005) p. 1437. Reproduced by permission of John Wiley & Sons Ltd.

If the permeate from one stage is fed to a second stage to increase the rate of solute rejection, this type of operation is called a two-pass system or permeate-staged systems, and is illustrated in Figure 6.41. A unit of production capacity is known as an array, which may be made up of one or more stages.

6.9.8 Design equations for reverse osmosis

Several models have been proposed to describe the flux of water and solute through RO membranes. Brief descriptions of these can be found in the following references: *Water Treatment: Principles and Design* (2005, pp. 1450–1454) and Davis (2011, pp. 6-3 to 6-6). Design equations have been developed based on these models which relate mass flux of water or solute to a mass transfer coefficient and a driving force. These and similar equations may be found in the following references: MWH (2005) pp. 1454–1459; Davis (2011) pp. 6-3 to 6-6; Metcalf & Eddy (2003) pp. 1104–1116; Bergman (2005) pp. 13.1–13.49; and Taylor and Wiesner (1999) pp. 11.1–11.71. The water through a membrane can be calculated using the following equation:

$$F_W = K_W(\Delta P_a - \Delta\pi) = \frac{Q_p}{A} \tag{6.148}$$

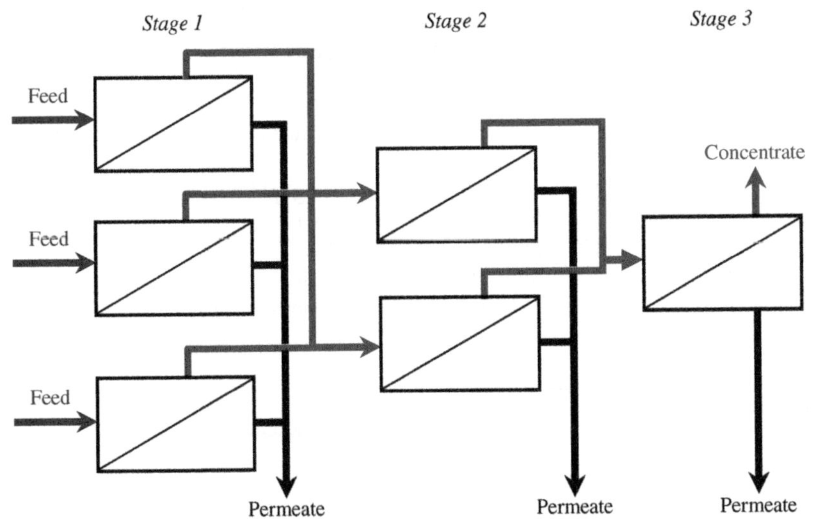

Figure 6.40 A 3×2×1 concentrate-staged array.

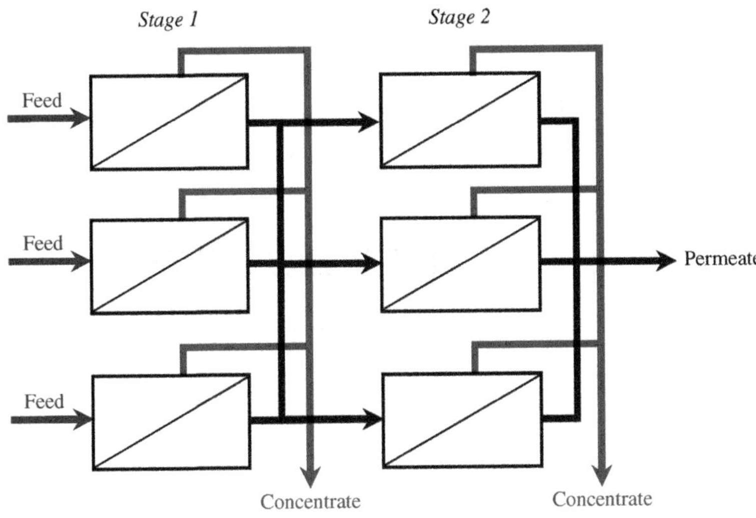

Figure 6.41 A two-pass, two-staged system.

where:
F_W = water flux rate, kg/(m^2 · s)
K_W = water mass transfer coefficient, s/m
ΔP_a =average imposed pressure gradient, kPa.

$$\Delta P_a = \left[\frac{P_f + P_c}{2}\right] - P_p \tag{6.149}$$

where:
P_f = inlet pressure of feed stream, kPa
P_c = pressure of concentrate steam, kPa
P_p = pressure of permeate steam, kPa
$\Delta\pi$ = osmotic pressure gradient, kPa.

$$\Delta\pi = \left[\frac{\pi_f + \pi_c}{2}\right] - \pi_p \tag{6.150}$$

where:
π_f, π_c, π_p = osmotic pressure in feed, concentrate, and permeate, kPa
Q_p = permeate stream flow, kg/s
A = membrane surface area, m^2.

The solute flux through the membrane can be calculated using the following equation:

$$F_i = K_i \Delta C_i = \frac{Q_p C_p}{A} \tag{6.151}$$

where:
F_i = flux of solute species i, kg/(m^2 · s)
K_i = solute mass transfer coefficient, m/s
ΔC_i = solute concentration gradient, kg/m^3.

$$\Delta C_i = \left[\frac{C_f + C_c}{2}\right] - C_p \tag{6.152}$$

where C_f, C_c, C_p = solute concentration in feed, concentrate, and permeate, kg/m^3.

The water recovery rate (r) is calculated as follows using Equation (6.153):

$$r(\%) = \frac{Q_p}{Q_f} \times 100 \tag{6.153}$$

Table 6.17 Summary of design and operational parameters of a reverse osmosis system.

Parameter	Value
Active membrane area/element[1]	300–378 ft^2/ft^3 (987–1241 m^2/m^3)
Influent pH[2]	4.5–5.5
Energy[2]	9–17 kWh/1,000 gal (2–5 kWh/m^3)
Pressure[3]	73–1,200 psig (5–85 bar)
Removal efficiency for specific impurities[3]	50–99%
Temperature[1]	35–40°C
Max temperature for CA membranes	35–40°C
Max temperature for PA membranes	
Water recovery[3]	
Water recovery (Seawater)	50%
Water recovery (Groundwater)	90%
Permeate flux[3]	0.6–30 gpd/ft^2 [1.0–50 L/($m^2 \cdot h$)]

CA = cellulose acetate; PA = polyamide.
Source: Values from [1]Bergman (2005), pp. 13.25 and 13.28; Qasim *et al.* (2000), p. 769, and MWH (2005), p. 959.

where:
Q_p = permeate flow, kg/s
Q_f = feed flow, kg/s.

The rate of solute rejection is calculated as follows using Equation (6.154).

$$R(\%) = \frac{C_f - C_p}{C_f} \times 100 \qquad (6.154)$$

Table 6.17 provides typical design and operational parameters for reverse osmosis systems. The use of the design equations and operating parameters will be illustrated in several example problems that follow.

Example 6.23 Calculating various RO design and operational parameters

Design an RO WTP to meet the following criteria. Influent-feed TDS = 1,000 mg/L. Q = 19,000 m^3/day is produced. TDS in treated water = 300 mg/L. Temperature of the feed = 27°C. Water recovery factor = 75% and the salt rejection = 95%. Design pressure = 4,140 kN/m^2 (600 psig) and flux rate = 0.82 $m^3/(m^2 \cdot d)$ (20 gpd/ft^2). Packing density = 820 m^2/m^3.

Calculate the following parameters:

a) TDS in the permeate (mg/L).
b) TDS in the concentrate stream (mg/L), assuming split treatment.
c) Membrane area (m^2) using split treatment.
d) Total module volume (m^3).
e) Total number of modules.
f) Number of pressure vessels.
g) Water power (W), brake power (W) and motor power (W).

Solution part a

Compute the TDS concentration in permeate:

$$R = \frac{95}{100} = \frac{C_f - C_p}{C_f}$$

$$C_p = C_f - \frac{95}{100}C_f = 1000\frac{mg}{L} - 0.95(1000) = \boxed{50\frac{mg}{L}}$$

Solution part b

Use split treatment to compute size.

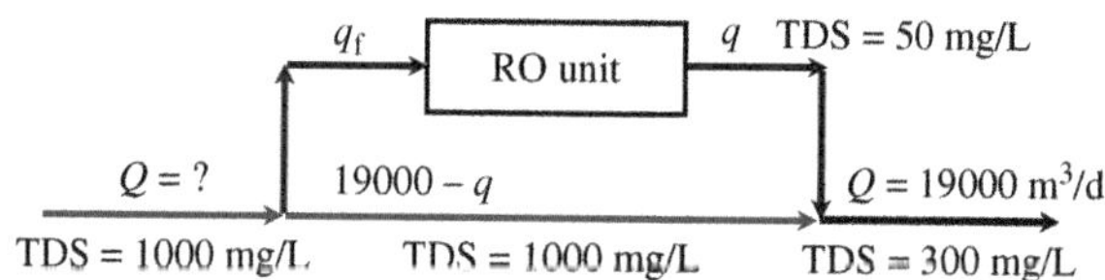

Figure E6.9

Perform mass balance on flows and TDS concentration around the system:

$$(19000 - q)\frac{m^3}{d}\left(1000\frac{mg}{L}\right) + (q)\frac{m^3}{d}\left(50\frac{mg}{L}\right) = \left(19000\frac{m^3}{d}\right)\left(300\frac{mg}{L}\right)$$

$$q = 14000 m^3/day$$

$$r = \frac{q}{q_f} = 0.75$$

$$q_f = \frac{q}{r} = \frac{14000\, m^3/d}{0.75} = 18{,}667\, m^3/d$$

Perform a flow balance:

$$Q = 18{,}667 + (19000 - 14000) = 23{,}667\, m^3/d$$

Brine or Concentrate Flow = 18667 – 14000 = 4667 m^3/d.

Calculate the TDS concentration in the brine or concentrate stream:

$$q_f\left(1000\frac{mg}{L}\right) = Q_c C_c + Q_p C_p$$

$$18667\,\frac{m^3}{d}\left(1000\,\frac{mg}{L}\right)$$

$$= 4667\,\frac{m^3}{d}(C_c) + 14000\,\frac{m^3}{d}\left(50\,\frac{mg}{L}\right)$$

$$\boxed{C_c = 3850\,\frac{mg}{L}}$$

Solution part c

Calculate the area of membranes.

$$A = \frac{Q_p}{F_w} = \frac{q}{F_w} = \frac{14000\,m^3/d}{0.82\,m^3/m^2\cdot d} = \boxed{17073\,m^2}$$

Solution part d

Calculate the total module volume.

$$\text{Total Module Volume} = \frac{A}{\text{Packing Density}}$$

Assume packing density = 820 m^2/m^3.

$$\text{Total Module Volume} = \frac{17073\,m^2}{820\,m^2/m^3} = \boxed{21\,m^3}$$

Solution part e

Calculate the total number of modules assuming 0.03 m^3/module. Bergman (2005, p. 13.28) states that typical elements are 8 inches in diameter and 40 in long, with an active surface area of 350 to 440 ft^2 (33–41 m^2). The volume per element or module then can be calculated as follows:

Volume of element or module

$$= \frac{\pi D^2}{4}\times L = \frac{\pi(8/12\,ft)^2}{4}\times 40\,in\left(\frac{ft}{12\,in}\right) = 1.16\,ft^3$$

Volume of element or module

$$= 1.16\,ft^3\times\frac{m^3}{(3.281\,ft)^3} = 3.3\times10^{-2}\,m^3 \cong \boxed{0.03\,m^3}$$

$$\text{Total \# of Modules} = \frac{21\,m^3}{0.03\,m^3/\text{module}} = \boxed{700}$$

Solution part f

Calculate the number of pressure vessels assuming 15 modules per pressure vessel.

$$\text{Total \# of Pressure Vessels} = \frac{700\text{ modules}}{15\text{modules/pressure vessel}}$$

$$= 46 \quad \boxed{\text{Go with 50}}$$

Solution part g

Compute Power Consumption:

$$\text{water Power} = \text{Design Pressure}\times\text{Flow Pressurized}$$

$$\text{water Power} = 4140\,\frac{kN}{m^2}\times 18667\,\frac{m^3}{d}\left(\frac{1\,d}{24\,h}\right)\left(\frac{1\,h}{60\,min}\right)$$

$$\times\left(\frac{1\,min}{60\,s}\right)\left(\frac{1000\,N}{1\,kN}\right)\left(\frac{1\,W}{1\,N\cdot m/s}\right)$$

$$\text{water Power} = \boxed{8.94\times10^5 W}$$

Assume a pump efficiency of 95%.

$$\text{Brake Power} = 8.94\times10^5/0.95 = \boxed{9.41\times10^5\,W}$$

Assume a motor efficiency of 88%.

$$\text{Motor Power} = 9.41\times10^5/0.88 = \boxed{1.07\times10^6\,W}$$

6.9.9 Water mass transfer coefficients units

The units for K_W may be expressed with units of s/m or m/(s · bar). We will examine these in the following examples.

Example 6.24 Calculating the value and units for K_W

Permeate flow = 3,520 m^3/day; membrane area=1,600 m^2; density of water = 1,000 kg/m^3; $(\Delta P_a - \Delta\pi)$ = 2,750 kPa. Express the units for K_W in s/m and m/(s · bar).

Solution

For K_W with units of s/m:

$$F_W = K_W(\Delta P_a - \Delta\pi) = \frac{Q_p}{A}$$

$$Q_p = 3520\,\frac{m^3}{d}\left(\frac{1000\,kg}{m^3}\right)\left(\frac{1\,d}{24\,h}\right)\left(\frac{1\,h}{60\,min}\right)\left(\frac{1\,min}{60\,s}\right)$$

$$= 40.74\,\frac{kg}{s}$$

$$(\Delta P_a - \Delta\pi) = 2750\,kPa\left(\frac{1000\,Pa}{kPa}\right)\left(\frac{1\,N/m^2}{1\,Pa}\right)$$

$$\times\left(\frac{1\,kg\cdot m/s^2}{1\,N}\right) = 2.75\times10^6\,\frac{kg}{m\cdot s^2}$$

$$K_W = \frac{Q_p}{(\Delta P_a - \Delta\pi)A} = \frac{40.74\,kg/s}{(2.75\times10^6\,kg/m\cdot s^2)(1600\,m^2)}$$

$$= \boxed{9.26\times10^{-9} s/m}$$

For K_W with units of m/(s · bar):

$$F_W = K_W(\Delta P_a - \Delta\pi) = \frac{Q_p}{A}$$

$$K_W = \frac{Q_p}{(\Delta P_a - \Delta\pi)A}$$

$$K_W = \frac{3520\,\text{m}^3/\text{day}}{(2750\,\text{kPa})(1600\,\text{m}^2)\left(0.01\,\frac{\text{bars}}{\text{kPa}}\right)}\left(\frac{1\,\text{d}}{24\,\text{h}}\right)\left(\frac{1\,\text{h}}{3600\,\text{s}}\right)$$

$$= \boxed{9.26 \times 10^{-7}\,\text{m/(s} \cdot \text{bar)}}$$

Example 6.25 Calculating various RO design and operational parameters

Determine the required membrane area, rejection rate, and concentration of TDS in the concentrate stream for the following conditions. If possible, reduce the membrane area by blending the feed water with the permeate stream.

Flow rate (feed) = 20,000 m^3/d
Influent TDS = 2,000 g/m^3 = 2.0 kg/m^3
Effluent TDS = 400 g/m^3 = 0.4 kg/m^3
Net Operating Pressure = 2,800 kPa
Recovery (%) = 89
$K_w = 1 \times 10^{-6}$ m/(s·bar)
$K_i = 6 \times 10^{-8}$ m/s

Solution part a

Calculate the water flux rate.

$$F_W = K_W(\Delta P_a - \Delta\pi) = \frac{Q_p}{A}$$

$$F_W = 1 \times 10^{-6}\,\frac{\text{m}}{\text{s} \cdot \text{bar}}(2800\,\text{kPa})\left(\frac{0.01\,\text{bars}}{\text{kPa}}\right)$$

$$= 2.8 \times 10^{-5}\,\text{m/s}$$

Calculate the membrane as follows:

$$A = \frac{Q_p}{F_W} = \frac{0.89(20000\,\text{m}^3/\text{d})}{2.8 \times 10^{-5}\,\text{m/s}}\left(\frac{1\,\text{d}}{24\,\text{h}}\right)\left(\frac{1\,\text{h}}{60\,\text{m}}\right)\left(\frac{1\,\text{m}}{60\,\text{s}}\right)$$

$$\boxed{A = 7358\,\text{m}^2}$$

Solution part b

Determine the TDS in the permeate assuming that the concentration of TDS in the concentration is approximately nine times that of the influent or $C_c = 9C_f$.

$$F_i = K_i \Delta C_i = \frac{Q_p C_p}{A} \quad (6.151)$$

$$\Delta C_i = \left[\frac{C_f + C_c}{2}\right] - C_p \quad (6.152)$$

Substitute Equation (6.152) into Equation (6.151) and solve for C_p:

$$C_p = \frac{K_i\left[\frac{C_f + C_c}{2}\right]A}{Q_p + K_i A} \qquad C_p = \frac{K_i\left[\frac{C_f + 9C_f}{2}\right]A}{Q_p + K_i A}$$

$$C_p = \frac{6 \times 10^{-8}\,\frac{\text{m}}{\text{s}}\left[\frac{2\,\text{kg/m}^3 + 9 \times 2\,\text{kg/m}^3}{2}\right]7358\,\text{m}^2}{\left(\begin{array}{c}0.89\left(20000\,\frac{\text{m}^3}{\text{d}}\right)\left(\frac{1\,\text{d}}{24\,\text{h}}\right)\left(\frac{1\,\text{h}}{3600\,\text{s}}\right)\\ +6 \times 10^{-8}\,\frac{\text{m}}{\text{s}}(7358\,\text{m}^2)\end{array}\right)}$$

$$\boxed{C_p = 0.0214\,\text{kg/m}^3} \le 0.4\,\text{kg/m}^3$$

Therefore, membrane area can be reduced by blending. Perform a mass balance on flows.

$$Q_f + Q_R = 20000\,\text{m}^3/\text{d} \qquad Q_f = 20000 - Q_R$$

$$r = \frac{Q_P}{(Q_f - Q_R)} \qquad Q_P = r(Q_f - Q_R)$$

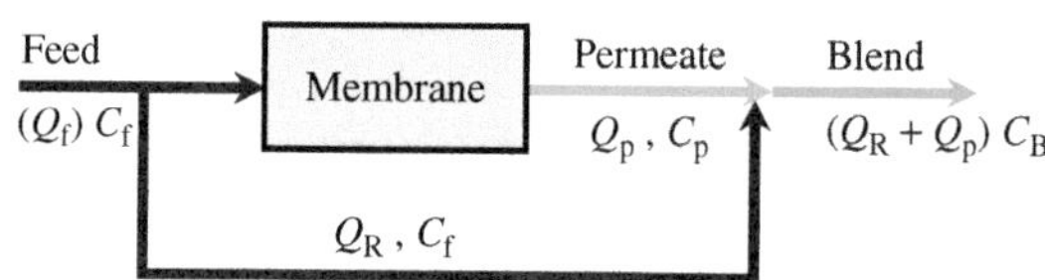

Figure E6.10

Develop an equation for calculating Q_R and substitute $r(Q_f - Q_R)$ for Q_p as follows:

$$Q_R C_f + Q_P C_P = (Q_R + Q_P)C_B$$

$$Q_R C_f + r(Q_f - Q_R)C_p = Q_R C_B + rQ_f C_B - rQ_R C_B$$

$$Q_R = \frac{(rQ_f C_p) - (rQ_f C_B)}{[C_B(1 - r) - C_f + rC_p]}$$

$$Q_R = \frac{(0.89 \times 20000 \times 0.0214) - (0.89 \times 20000 \times 0.4)}{[0.4(1 - 0.89) - 2 + 0.89 \times 0.0214]}$$

$$= 3479\,\frac{\text{m}^3}{\text{d}}$$

Calculate the new feed flow rate as follows:

$$20000\,\frac{\text{m}^3}{\text{d}} - 3479\,\frac{\text{m}^3}{\text{d}} = \boxed{16{,}521\,\frac{\text{m}^3}{\text{d}}}$$

Calculate the membrane area using the new feed flow rate:

$$A = \frac{Q_p}{F_W} = \frac{0.89(16521\,\text{m}^3/\text{d})}{2.8 \times 10^{-5}\,\text{m/s}}\left(\frac{1\,\text{d}}{24\,\text{h}}\right)\left(\frac{1\,\text{h}}{60\,\text{m}}\right)\left(\frac{1\,\text{m}}{60\,\text{s}}\right)$$

$$= \boxed{6078\,\text{m}^2}$$

Calculate the new permeate concentration using the new membrane area:

$$C_P = \frac{6\times10^{-8}\,\frac{\text{m}}{\text{s}}\left[\frac{2\,\text{kg/m}^3 + 9\times2\,\text{kg/m}^3}{2}\right]6078\,\text{m}^2}{\left(\begin{array}{c}0.89\left(16521\,\frac{\text{m}^3}{\text{d}}\right)\left(\frac{1\,\text{d}}{24\,\text{h}}\right)\left(\frac{1\,\text{h}}{3600\,\text{s}}\right)\\ +6\times10^{-8}\,\frac{\text{m}}{\text{s}}(6078\,\text{m}^2)\end{array}\right)}$$

$$\boxed{C_P = 0.0214\,\text{kg/m}^3} = 0.0214\,\text{kg/m}^3\text{OK}$$

Blending 3,479 m^3/d of feed with permeate will result in a blended product water with a TDS of approximately 400 g/m^3.

Estimate the rejection rate as follows:

$$R(\%) = \frac{C_f - C_p}{C_f}\times100$$

$$= \frac{(2.0\,\text{kg/m}^3 - 0.0214\,\text{kg/m}^3)}{2.0\,\text{kg/m}^3}\times100 = \boxed{98.9\,\%}$$

Solution part c

Next, estimate the concentrate stream TDS concentration.

$$Q_fC_f = Q_PC_P + Q_cC_c$$

$$C_c = \frac{Q_fC_f - Q_PC_P}{Q_c}$$

$$Q_c = (1-0.89)\,16{,}521\,\text{m}^3/\text{d} = 1{,}817\,\text{m}^3/\text{d}$$

$$C_c = \frac{\left(\begin{array}{c}16{,}521\,\text{m}^3/\text{d}\,(2\,\text{kg/m}^3)\\ -0.89(16{,}521\,\text{m}^3/\text{d})(0.0214\,\text{kg/m}^3)\end{array}\right)}{1817\,\text{m}^3/\text{d}}$$

$$= \boxed{18.0\,\text{kg/m}^3}$$

Check assumptions to ensure they are correct.

$$\frac{C_c}{C_f} = \frac{18.0\,\text{kg/m}^3}{2\,\text{kg/m}^3} = \boxed{9}$$

The TDS concentration in the concentrate is 9 times the value of the TDS concentration in the feed.

6.10 Fluoridation

Fluoride is added to drinking water for the prevention of dental cavities in children. Fluoride compounds are typically added following filtration. The most commonly added fluoride compounds are sodium fluoride (NaF), sodium fluorosilicate (Na_2SiF_6), and fluorosilic acid (H_2SiF_6). If added upstream of coagulation/flocculation and/or softening operations, significant quantities of fluoride will be removed.

Table 6.18 Recommended optimal fluoride levels for community public water supply systems.

Annual average of maximum daily air temperature (°F)	Annual average of maximum daily air temperature (°C)	Recommended fluoride concentration (mg/L)
50.0–53.7	10.0–12.0	1.2
53.8–58.3	12.1–14.6	1.1
58.4–63.8	14.7–17.7	1.0
63.9–70.6	17.8–21.4	0.9
70.7–79.2	21.5–26.2	0.8
79.2–90.5	26.3–32.5	0.7

Source: CDC (1995) p. 8. United States Department of Health and Human Services, Centers for Disease Control.

Ground water normally contains higher concentrations of dissolved fluoride ion from passing through geologic formations, whereas surface waters typically contain fluoride concentrations <0.3 mg/L (Viessman *et al.*, 2009, p. 273). When excess fluoride is present in the water supply, chemical precipitation with alum, lime softening, or ion exchange may be used to achieve safe levels (Kawamura, 1991, pp. 532–533). The major problem with ion exchange involves the permitting and disposal of the concentrated fluoride. Reeves (1999, pp. 15.1–15.19) gives an excellent overview of water fluoridation systems.

A fluoride concentration of approximately 1.0 mg/L is typically maintained at most drinking water facilities in the United States. The US EPA has established both primary and secondary drinking water standards for fluoride. The primary MCL and primary MCLG for fluoride is 4.0 mg/L to protect the public from bone disease (pain and tenderness of the bones) and mottling of teeth. The secondary standard is set at 2.0 mg/L to prevent "dental fluorosis" or mottling and discoloration of teeth. The recommended optimum fluoride concentration in drinking water is based on the annual average of the maximum daily air temperatures, since water consumption is affected by climate. Table 6.18 shows the recommended optimum fluoride concentration as a function of air temperature (CDC, 1995, p. 8).

6.11 Disinfection in water treatment

Disinfection practice at water treatment plants has traditionally been the use of chlorine gas as the primary means of killing pathogens in water. Chlorine is both an economical and effective disinfectant. However, there are some concerns with using chlorine, especially hazards associated with transporting a toxic gas to the treatment site and disinfection by-products produced such as trihalomethanes and haloacetic acids, among other chlorinated organics that may be potential carcinogens.

Some of the regulations promulgated concerning pathogens and disinfection by-products seem to be contradictory, and therefore a brief review of the most important ones will be presented. The mechanisms responsible for killing pathogens will be reviewed, and alternative disinfectants that are being used in the water utility industry will be presented. Since chlorine continues to be the primary disinfectant in use in the US, the chemistry of chlorination and use of various chlorine compounds will take precedence in the following paragraphs.

6.11.1 Goal of disinfection

The primary goal of disinfection is to inactivate pathogenic organisms such as bacteria, protozoa, and enteric viruses that may be present in the water. Microbial pathogens of greatest concern are protozoan cysts associated with *Giardia lamblia* and *Cryptosporidium parvum*, and various enteric viruses which may be resistant to specific types of disinfectants. Regardless of which disinfectant is selected, it is preferable to remove as much turbidity as possible before applying it. Microorganisms often attach to, or are enveloped in, suspended solids, which shield the organisms from the disinfectant, making it less effective. A lower disinfectant dose may be used when the turbidity has been removed.

A multiple barrier concept is used to achieve the necessary removals or inactivation of pathogens. Multiple barriers include proper watershed management, well-head protection programs, coagulation/flocculation/sedimentation, filtration, and disinfection. As will be discussed later, the US EPA gives credit for the removal of pathogens by using the multiple barrier concept in the treatment train and, therefore, the disinfection process need only achieve a 0.5 to 1.0 log reduction to meet regulations.

6.11.2 Disinfection Kinetics and C · t concept

The rate of kill or die-off of a microorganism is modeled as a first-order reaction known as Chick's law (1908). Under ideal conditions, it simulates how fast a microorganism will die when exposed to a specific type of disinfectant. The change in the number of organisms remaining in a unit of time is proportional to the number of organisms remaining, as shown in the following differential equation.

$$\frac{dN}{dt} = -kN \tag{6.155}$$

where:
$\frac{dN}{dt}$ = change in the number of microbes remaining with time, number/time
k = inactivation rate constant, time^{-1}
N = number of microorganisms.

Equation (6.155) may be rearranged and integrated from time is equal to zero to time "t" with limits of N varying from N_0 to N_t.

$$\int_{N_0}^{N_t} \frac{1}{N} dN = -k \int_0^t dt \tag{6.156}$$

$$\ln\left(\frac{N_t}{N_0}\right) = -kt \tag{6.157}$$

$$N_t = N_0 e^{-kt} \tag{6.158}$$

where:
N_0 = number of organisms at time zero
t = contact time, minutes
N_t = number of organisms at time t.

In separate disinfection studies, Herbert Watson (1908) found that the inactivation rate constant (k) was related to the disinfectant concentration (C) and die-off constant k' according to the following equation:

$$k = k' C^n \tag{6.159}$$

where:
k' = die-off constant
C = disinfectant concentration, mg/L
n = coefficient of dilution.

The rate of inactivation of a species of microorganisms is also a function of the disinfectant used, disinfectant concentration, pH, temperature, presence of suspended solids, turbidity, and contact time. A parameter in which the concentration of disinfectant (C) in mg/L multiplied by the contact time (t) in minutes or $C \cdot t$ is used in designing disinfection systems (Peavy *et al.*, 1985, p. 185, and Viessman & Hammer, 2005, p. 469). The following equation shows the relationship between these parameters:

$$C^n \cdot t = K \tag{6.160}$$

where:
C = disinfectant concentration, mg/L
t = contact time, min
n = experimental coefficient
K = a constant for a specific microorganism for a given disinfectant, pH, and temperature.

Several experiments at various disinfectant concentrations must be performed on a specific microorganism in order to determine the constants in Equation (6.160). Combining the Chick and Watson models allows one to develop a linear equation for estimating the constants (Haas & Kara, 1984).

Haas & Kara (1984) combined the proposed models of Chick and Watson as follows:

$$\frac{dN}{dt} = -k' C^n N \tag{6.161}$$

$$\int_{N_0}^{N_t} \frac{1}{N} dN = -k' C^n \int_0^t dt \tag{6.162}$$

$$\ln\left(\frac{N_t}{N_0}\right) = -k' C^n t \tag{6.163}$$

$$N_t = N_0 e^{-k' C^n t} \tag{6.164}$$

The linear form of Equation (6.161) is shown below:

$$\ln(C) = -\frac{1}{n}\ln(t) + \frac{1}{n}\ln\left[\frac{1}{k'}\left(-\ln\left(\frac{N_t}{N_0}\right)\right)\right] \quad (6.165)$$

A plot of ln(*C*) versus ln(*t*) will produce a straight line with a slope of *n*.

When $n = 1$, the $C \cdot t$ value remains constant. Viessman & Hammer (2005, p. 469), referencing Haas & Kara (1984), indicate that *n* varies from 0.5–2.0, with *n* averaging close to 1.0. In most engineering applications, $C \cdot t$ values are based on assuming an *n* equal to 1.0. The value of *n* is dependent on the type of disinfectant selected. When *n* is greater than 1, disinfection action depends on the disinfectant concentration. Conversely, when *n* is less than 1, the disinfection action is dependent on the contact time. Reynolds & Richards (1996, p. 742) state that the microorganism concentrations encountered in water treatment applications has very little effect on Equation (6.160).

6.11.2.1 C · t Values Used in Surface Water Treatment

The US EPA has published $C \cdot t$ data for various types of microorganisms and disinfectant chemicals. In most cases, the results are presented on a log reduction basis. For example, a 2-log, 3-log, and 4-log reduction equates to a 99%, 99.9%, and 99.99% inactivation or kill, respectively. $C \cdot t$ values for a 2-log reduction in the concentration of various types of microorganisms for specific disinfectants at 5°C are presented in Table 6.19. Table 6.20 presents the $C \cdot t$ values for achieving a 1-log reduction or 90% inactivation of *Giardia lamblia* cysts for four different disinfectants. $C \cdot t$ values required to achieve various log-removals/inactivation of viruses using four different disinfectants are shown in Table 6.21.

Engineers must provide flexible designs for disinfection systems because a reactor's size is fixed upon construction; the only variable that can change is the disinfectant dosage. Therefore, disinfectant feed systems must be capable of achieving a wide range of dosages over a range of anticipated flows.

Table 6.20 $C \cdot t$ values in (mg/L) · min for 90% Inactivation of *Giardia lamblia* cysts by disinfectants at various temperatures.

		Water temperature			
Disinfectant	**pH**	**≤ 1.0°C**	**5°C**	**10°C**	**15°C**
Free chlorine[1]	6	*49	35	26	18
Free chlorine[1]	7	*70	50	37	25
Free chlorine[1]	8	*101	72	54	36
Free chlorine[1]	9	*146	104	78	52
Preformed chloramines	6–9	1,270	735	615	500
Chlorine dioxide	7	21	8.7	7.7	6.3
Ozone	7	*0.97	0.63	0.48	0.32

Free chlorine[1] values are based on a residual of 1.0 mg/L.
*At $T \leq 0.5$°C
Source: Adapted from Guidance Manual for Compliance with the Filtration and Disinfection Requirements for Public Water Systems using Surface Water Sources (Environmental Protection Agency, 1991) Appendix E and F. United States Environmental Protection Agency.

6.11.2.2 Ground Water Disinfection

Ground water not under the direct influence of surface water is covered under the Ground Water Rule promulgated in 2006 (See Section 6.1.3). This rule targets ground water systems that are susceptible to fecal contamination. Systems that identify a positive sample during monitoring for the Total Coliform Rule are required to monitor for the presence of *E. coli*, enterococci, or coliphage and implement corrective actions for systems with significant deficiencies. Corrective actions may include providing an alternate water source, eliminating the source of the contamination, and providing treatment to achieve a 4-log reduction/inactivation in viruses. Chlorination of ground water supplies typically involves adding chlorine to provide a "protective" residual in the range of 0.2–0.6 mg/L of free chlorine (Viessman & Hammer, 2005, p. 477).

Table 6.19 $C \cdot t$ values [(mg/L) · min] for 99% inactivation of various microorganism by disinfectants at 5°C.

	Disinfectant			
Microorganism	**Free chlorine pH 6–7**	**Preformed chloramines pH 8–9**	**Chlorine dioxide pH 6–7**	**Ozone pH 6–7**
E. coli	0.034–0.05	95–180	0.4–0.75	0.02
Poli I	1.1–2.5	768–3740	0.2–6.7	0.1–0.2
Rotavirus	0.01–0.05	3806–6476	0.2–2.1	0.006–0.06
Bacteriophage t_2	0.08–0.18	–	–	–
G. lamblia cysts	47 ≥ 150	–	–	0.5–0.6
G. muris cysts	30–630	–	7.2–18.5	1.8–2.0

Source: EPA/600/S2-86/067 (September 1986), p. 5. United States Environmental Protection Agency.

Table 6.21 $C \cdot t$ values[1] in (mg/L) · min for inactivation of viruses by disinfectants at various temperatures and pH 6–9.

Disinfectant	Log inactivation	Water temperature ≤1.0°C	5°C	10°C	15°C	20°C
Free chlorine	2	*6	4	3	2	1
Free chlorine	3	*9	6	4	3	2
Free chlorine	4	*12	8	6	4	3
Preformed chloramines	2	1,243	857	643	428	321
Preformed chloramines	3	2,063	1,423	1,067	712	534
Chlorine dioxide	2	8.4	5.6	4.2	2.8	2.1
Chlorine dioxide	3	25.6	17.1	12.8	8.6	6.4
Ozone	2	*0.9	0.6	0.5	0.3	0.25
Ozone	3	*1.4	0.9	0.8	0.5	0.4

$C \cdot t$ values[1] for free chlorine, ozone, and chlorine dioxide include safety (uncertainty) factors. No safety factor was applied to the chloramines $C \cdot t$ values since chloramination conducted in the field is more effective than using preformed chloramines in the laboratory.
*$T = 0.5$°C
Source: Adapted from Guidance Manual for Compliance with the Filtration and Disinfection Requirements for Public Water Systems using Surface Water Sources 1991, Appendix E and F. United States Environmental Protection Agency.

6.11.3 Major disinfectants

The major disinfectants used in the water industry include: chlorine gas (Cl_2); combined chlorine compounds such as sodium hypochlorite (NaOCl), calcium hypochlorite [$Ca(OCl)_2$], and chloramines (NH_2Cl-monochloramine, $NHCl_2$-dichloramine, and NCl_3-trichloramine); chlorine dioxide (ClO_2); ozone (O_3); and ultraviolet radiation (UV).

Haas (2002, p. 189) reported that in 1992, over 90% of water supplies in the US used chlorine or chlorine compounds for disinfection, and approximately 20% used chloramines. The advantages and disadvantages of selected disinfectants are discussed below.

6.11.3.1 Free Chlorine and Combined Chlorine Compounds

Chlorine is delivered to water treatment plants in pressurized containers ranging in size from 100 lb (45 kg) to one ton (908 kg) containers (Metcalf & Eddy, 2003, p. 1231). At large installations, railroad tank cars with capacities up to 55 tons (49.9 Mg) have been used. The chlorine is liquefied under high pressure and occupies approximately 85% of the container volume, with the remaining volume occupied by chlorine gas. It may be withdrawn either as a gas or liquid, depending on how the chlorination system is designed. Figure 6.42 is a schematic of a one-ton cylinder showing that the chlorine will be liquid when withdrawn from the bottom tap and gaseous when withdrawn from the top tap.

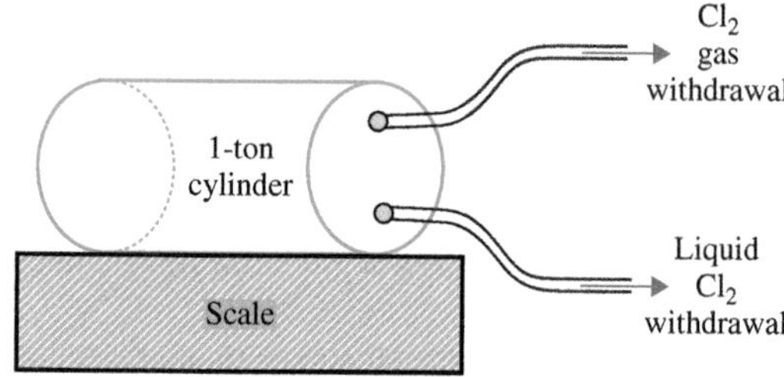

Figure 6.42 One-ton chlorine cylinder showing gaseous and liquid chlorine withdrawal.

Use of liquid chlorine requires conversion to gas using an evaporator. Evaporators typically are used when chlorine dosages approach 1,500 lb/d (680 kg/d) (Metcalf & Eddy, 2003, p. 1266). Chlorine gas is seldom injected directly into a water pipe or process, due to potential gas leaks in the chlorine piping system.

Chlorine gas is a greenish-yellow poisonous gas and is approximately 2.5 times heavier than air. It is very corrosive, and it is normally dissolved into a side stream of treated water using a vacuum pressure system and regulator. The concentrated chlorine solution is then dispersed through a diffuser at the desired point of treatment.

Multiple chlorine injection points should be provided along the treatment train, such as before and after filtration, and directly into the clear well. Pre-disinfection and oxidation with chlorine at the front end of the treatment train is not as widely practiced, due to concerns of forming disinfection by-products that may be potentially carcinogenic and mutagenic.

Chemistry of Chlorination Chlorine gas is soluble in water and hydrolyzes into hypochlorous acid (HOCl) as follows:

$$Cl_2 + H_2O \rightleftharpoons HOCl + H^+ + Cl^- \qquad (6.166)$$

Hypochlorous acid dissociates or ionizes into a hydrogen ion (H^+) and a hypochlorite ion (OCl^-) according to the following reaction:

$$HOCl \rightleftharpoons H^+ + OCl^- \qquad (6.167)$$

The ionization constant for hypochlorous acid at 0°C and 25°C is 1.5×10^{-8} and 2.9×10^{-8} moles/L, respectively (Metcalf & Eddy, 2003, p. 1235).

Figure 6.43 shows the distribution of hypochlorous acid as a function of pH and temperature. At pH values less than 6, the predominate species is hypochlorous acid. Hypochlorous acid is approximately 40–80 times more efficient in killing pathogens than the hypochlorite ion. The sum of the concentration of hypochlorous acid and hypochlorite ion represents the free chlorine residual. Free chlorine residual has a higher oxidizing capacity than combined chlorine residual.

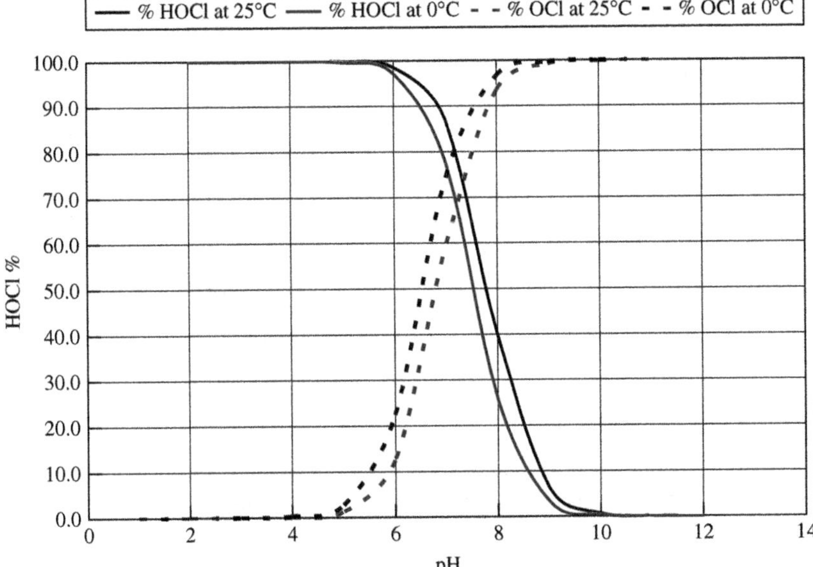

Figure 6.43 Distribution of HOCl as a function of pH.

As shown in Equations (6.166) and (6.167), adding chlorine gas to water results in a decrease in pH due to the release of hydrogen ions.

The effectiveness of compounds containing chlorine can be evaluated by comparing the percent actual chlorine and percent available chlorine. The percent actual chlorine is calculated as follows:

$$\text{Actual chlorine (\%)} = \frac{\text{(Weight of chlorine in compound)}}{\text{(Molecular weight of compound)}} \times 100 \quad (6.168)$$

The percent available chlorine is calculated by multiplying the actual chlorine percentage by the chlorine equivalent. The oxidizing power of chlorine is based on the value of the valence of the chloride in the compound that is reduced to a valence of −1. Percent available chlorine is determined as:

$$\text{Available chlorine (\%)} = [\text{Actual chlorine (\%)}] \times (\text{Chlorine equivalent}) \quad (6.169)$$

Example 6.26 Calculating the percent actual and percent available chlorine

Determine the percent actual and percent available chlorine for chlorine dioxide. The half-reactions for chlorine dioxide (ClO_2) and chlorite ion (ClO_2^-) are given below.

$$ClO_2 + e^- \rightleftharpoons ClO_2^-$$

$$ClO_2^- + 2H_2O + 4\,e^{-1} \rightleftharpoons Cl^- + 4\,OH^-$$

Solution

Calculate the percent actual chlorine as follows:

$$\text{Actual chlorine (\%)} = \frac{\text{(Weight of chlorine in compound)}}{\text{(Molecular weight of compound)}} \times 100 \quad (6.168)$$

$$\text{Actual chlorine (\%)} = \frac{(35.5)}{(35.5 + 2 \times 16)} \times 100 = \boxed{52.6\%} \quad (6.169)$$

Calculate the percent available chlorine as follows:

$$\text{Available chlorine (\%)} = [\text{Actual chlorine (\%)}] \times (\text{Chlorine equivalent})$$

The valence of the chlorine species in chlorine dioxide changes from a +4 to a −1, i.e., a chlorine equivalent of 5.

$$\text{Available chlorine (\%)} = [52.6\ (\%)] \times (5) = \boxed{263\%}$$

Percent actual and percent available chlorine for chlorine and other compounds containing chlorine that are used for disinfection are provided in Table 6.22.

Calcium hypochlorite ($Ca(OCl)_2$) or HTH and sodium hypochlorite (NaOCl), also known as "bleach", are two other combined chlorine forms that may be added to water, producing similar results to that of chlorine gas. These compounds are typically used at small installations and for disinfecting swimming pools. Both compounds are more expensive than chlorine gas, but do not cause the pH to

Table 6.22 Percent actual and percent available chlorine for selected chlorine compounds.

Compound	Molecular weight	Chlorine equiva-lent	Actual chlorine (%)	Available chlorine (%)
ClO_2	67.5	5	52.6	263
$NHCl_2$	86	2	82.6	165
NH_2Cl	51.5	2	68.9	138
HOCl	52.5	2	67.6	135
Cl_2	71	1	100	100
$Ca(OCl)_2$	143	2	49.7	99.4
NaOCl	74.5	2	47.7	95.4

drop. The reactions showing the dissociation of calcium and sodium hypochlorite are presented below. The hypochlorite ion is in equilibrium with the hydrogen ion to form hypochlorous acid as shown in Equation (6.167).

$$Ca(OCl)_2 \rightleftharpoons Ca^{+2} + 2\,OCl^- \tag{6.170}$$

$$NaOCl \rightleftharpoons Na^+ + OCl^- \tag{6.171}$$

6.11.3.2 Chloramination

Some municipalities practice chloramination of their drinking water to provide a combined chlorine residual throughout the water distribution system. Combined chlorine residual is more stable and longer-lasting than is free chlorine residual, but it is not as powerful as an oxidant or disinfectant. Using chloramines reduces the rate of disinfection by-product formation (Viessman & Hammer, 2005, p. 467). Chlorine residual is necessary in the distribution system to ensure that microbial regrowth does not occur.

Chloramination involves the addition of ammonia to chlorinated water to produced chloramines. The type of chloramine produced depends on the temperature, pH, and the chlorine-to-nitrogen ratio, i.e., the quantity of chlorine added. The reactions that occur when ammonia is added to chlorinated water are shown below:

$$HOCl + NH_3 \rightleftharpoons H_2O + NH_2Cl \text{ (monochloramine)} \tag{6.172}$$

$$HOCl + NH_2Cl \rightleftharpoons H_2O + NHCl_2 \text{ (dichloramine)} \tag{6.173}$$

$$HOCl + NHCl_2 \rightleftharpoons H_2O + NCl_3 \text{ (trichloramine)} \tag{6.174}$$

Mono- and di-chloramine are the major chloramine species that are produced between a pH of 4.5 to 8.5. Trichloramine or nitrogen trichloride (NCl_3) occurs at pH values less than 4.4. Disadvantages of using chloramines include: less powerful disinfectants compared to free chlorine residuals; nitrification problems can occur in the distribution system; and the formation of N-nitrosodimethyl-amine (NDMA), a suspected carcinogen (MWH, 2005, p. 1542).

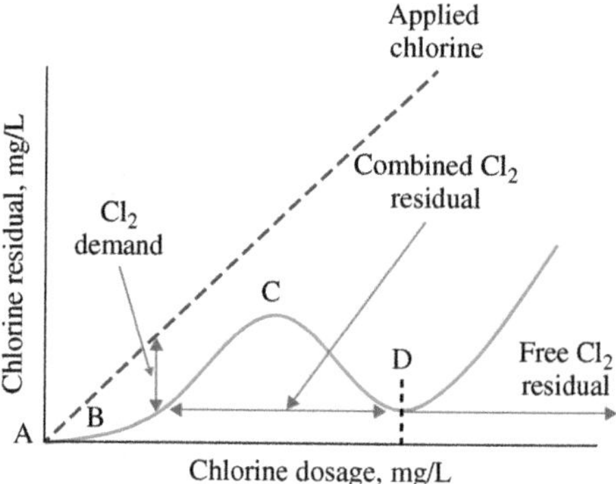

Figure 6.44 Breakpoint chlorination curve.

6.11.3.3 Breakpoint Chlorination

Some municipalities prefer to provide free chlorine residual in the distribution system. Free chlorine residual is comprised of hypochlorous acid and the hypochlorite ion. These species are more powerful disinfectants, although they are less stable and do not provide a long-lasting residual. To achieve breakpoint chlorination requires high dosages of chlorine to react with reduced species such as nitrites, ferrous ions, manganous ions, and sulfides as well as ammonia and organic nitrogen.

Figure 6.44 shows the chlorine residual curve for breakpoint chlorination. The applied chlorine dosage is shown as a straight line at 45°. The difference between the applied dose line and chlorine residual curve represents the chlorine demand. When chlorine is added to water, it first reacts with reducing agents, and minimal chlorine residual is developed, as shown from Point A to B. The addition of more chlorine past Point B results in the production of chloramines, primarily mono- and di-chloramine, which results in an increase in the combined chlorine residual as shown on the curve going from Point B to Point C. In this case, the chlorine is reacting with ammonia to form chloramines (chloramination). Adding more chlorine beyond Point C results in the formation of free chlorine residual, which oxidizes the chloramines, resulting in a decrease in the chlorine residual as exhibited from Point C to Point D. Oxidized nitrogen species such as nitrous oxide (N_2O), nitrogen gas (N_2), and nitrogen trichloride (NCl_3) are formed (Metcalf & Eddy, 2003, p. 1238). Point D on the curve is called the breakpoint; adding chlorine beyond this point creates free chlorine residual.

6.11.3.4 Chlorine Dioxide

Chlorine dioxide (ClO_2) is a powerful oxidizing agent that has been used primarily to remedy taste and odor problems in drinking water. It is also being used as a primary disinfectant, but not to the extent of chlorine gas, because it is more costly to produce. Chlorine dioxide must be generated on-site by mixing solutions of aqueous chlorine with sodium chlorite, as shown below.

$$2\,NaClO_2 + Cl_2 \rightleftharpoons 2\,ClO_2 + 2\,NaCl \tag{6.175}$$

Chlorine dioxide is a stronger disinfectant than hypochlorous acid and produces a residual. It does not form trihalomethanes or chloramines when added to water. The major disadvantages include the production of both chlorite and chlorate, which are regulated at MCLs of 1.0 mg/L and 0.8 mg/L, respectively. Advantages and disadvantages of using chlorine dioxide are presented below (Chlorine Chemistry Council, 2003, p. 23):

Advantages

1. Effective disinfectant against *Cryptosporidium*.
2. Up to 5 times faster than chlorine at inactivating *Giardia*.
3. ClO_2 disinfection is only moderately affected by pH.
4. Does not form chlorinated by-products such as THMs or HAA5s.
5. Does not oxidize bromide to bromine.
6. More effective than chlorine in treating some taste and odor problems.
7. A selective oxidant used for manganese oxidation and targeting some chlorine resistant organics.

Disadvantages

1. Chlorite and chlorate by-products are formed.
2. Highly volatile residuals that can be easily stripped form solution.
3. Requires on-site generation equipment and handling of chlorine and sodium chlorite.
4. A high level of technical competence is required to operate and monitor product and residuals.
5. Poses unique odor and taste problems occasionally.
6. High operating cost since sodium chlorite cost is high.

6.11.3.5 Ozone

Ozone is one of the strongest oxidants and disinfectants available. It is an allotropic form of oxygen that must be generated on-site, typically by passing clean dry air or oxygen through a high-strength electrical field. It is highly reactive and does not maintain a residual, decomposing into dissolved oxygen and the hydroxyl radical (OH·). There are two modes of reaction: direct oxidation, which is rather slow and extremely selective, and auto-decomposition into the hydroxyl radical (Hesby, 2005, p. 10.40). Advantages and disadvantages of using ozone are listed below (Chlorine Chemistry Council, 2003, p. 23).

Advantages

1. Strongest oxidant/disinfectant available.
2. Does not produce chlorinated by-products such as THMs or HAA5s.
3. An effective disinfectant against *Cryptosporidium* at high concentrations.
4. May be used with advanced oxidation processes (AOPs) to oxidize refractory organics.

Disadvantages

1. A high-level or technical competence required for process operation and maintenance.
2. Does not provide a residual for disinfection.
3. Forms brominated by-products such as bromate and brominated species.
4. Forms non-halogenated by-products such as ketenes, organic acids, and aldehydes.
5. Breaks down complex organic matter into smaller compounds that may result in microbial re-growth in the distribution systems in addition to increased disinfection by-product formation during secondary disinfection.
6. Operating and capital costs are higher than those for chlorination.
7. Difficult to control and monitor under variable loading conditions.

6.11.3.6 Ultraviolet Radiation

Ultraviolet radiation (UV) generated by mercury-vapor arc lamps is a non-chemical means of disinfection. UV radiation in the range of 245–285 nm is effective against bacteria and viruses. UV lamps are placed inside of quartz sleeves, which are then placed into the water. Modules of lamps are placed either vertically or horizontally to the flow. The number of UV lamps per module is 2, 4, 8, 12 or 16 (Metcalf & Eddy, 2003, p. 1302). To be effective, suspended solids, turbidity, and scale-forming chemicals must be removed prior to UV exposure. UV radiation damages the genetic material of organisms, preventing them from reproducing. Advantages and disadvantages of using UV are presented below (Chlorine Chemistry Council, 2003, p. 23).

Advantages

1. Effective at inactivating most viruses, spores, and cysts
2. Does not require chemical generation, storage, or handling
3. Effective disinfectant against *Cryptosporidium*
4. No known by-products generated at levels of concern

Disadvantages

1. Does not provide disinfection residual.
2. Low inactivation of some viruses such as retroviruses and rotaviruses.
3. Difficult to monitor the efficacy of the process.
4. Some irradiated organisms can repair and reverse the destructive effects of UV through photo-reactivation.
5. May require additional treatment steps to maintain high-clarity water.
6. High cost of providing backup/emergency capacity.
7. Mercury lamps may pose a potable water and environmental toxicity risk.

The UV dose is calculated using Equation (6.176). It is analogous to the $C \cdot t$ concept previously discussed. A typical dose applied to drinking water is 24,000 mW · s/cm².

$$D = I \times t \qquad (6.176)$$

where:
D = UV dose, mJ/cm² (1 mJ/cm² = mW · s/cm²)
I = UV intensity, mW · s/cm²
t = exposure time, s.

6.12 Residuals, solids, and quantities of sludge

The residuals, solids, or sludges that are generated at water treatment plants have two primary sources:

1. solids that are removed from settling basins which consist of chemical precipitates, inorganic and organic solids, and microorganisms, and
2. solids that have been removed from filters that end up in the spent backwash.

Other sources of residuals include debris from screens and concentrated streams from membrane processes that must be properly disposed. In general, residuals from WTPs are less offensive than those from wastewater treatment plants. There is concern, however, over potential pathogens, including the cysts from *Giardia lamblia* and oocysts from *Cryptosporidium*, in addition to toxic compounds such as arsenic that may accumulate in the sludge.

Water treatment plant sludges are tested using the toxicity characteristic leaching procedure (TCLP) or waste extraction test (WET) to determine their suitability for land application and/or disposal in a sanitary landfill. If the leachate from these types of tests exceeds the concentration of regulated compounds, found in the List of Inorganic Persistent and Bioaccumulative Toxic Substances and Their Soluble Threshold Limit Concentration (US EPA, 1992, p. 7–5), then the sludge is considered a hazardous waste and land application cannot be used as a disposal method. Hazardous waste or sludge can only be placed in a hazardous waste landfill.

Sludge quantities can be estimated from stoichiometric equations or from empirical equations developed by researchers and from plant operations. Lime-soda ash softening sludge quantities can be estimated from the softening equations presented in Section 6.62. Example 6.13 illustrated how to estimate the quantity of calcium carbonate and magnesium hydroxide sludge produced per million gallons of flow each day. Other equations have been used for estimating sludge quantities from water treatment plants.

Viessman *et al.* (2009, p. 630) recommend the following equation for estimating the dry solids production from alum coagulation, surface water treatment plants:

$$M_s = 8.34 \times Q(0.44 \times \text{alum dose} + 0.74 \times \text{turbidity}) \qquad (6.177)$$

where:
M_s = mass of dry solids produced, lb/d
Q = volumetric flow rate, million gallons per day
alum dose = aluminum sulfate dosage expressed in, mg/L
turbidity = turbidity expressed in nephelometric turbidity units (NTU).

An alternative equation presented by Davis (2011, p. 11–7) when using alum as the primary coagulant is shown below:

$$M_s = 86.4\, Q(0.44\, A + SS + M) \qquad (6.178)$$

where:
M_s = mass of dry solids produced, kg/d
A = aluminum sulfate dose, mg/L
Q = volumetric flow rate, m³/s
SS = suspended solids concentration in raw water, mg/L
M = other chemical additions including clay, polymer, and carbon, mg/L.

When an iron coagulant such as ferric chloride or ferrous sulfate is used as the coagulant, Davis (2011, p. 11–7) recommends the following equation for calculating the dry mass of sludge solids produced from adding iron.

$$M_s = 86.4\, Q(2.9\, \text{Fe} + SS + M) \qquad (6.179)$$

where Fe is the mg/L of iron coagulant added to the water.

Example 6.27 Sludge produced from alum coagulation

Estimate the quantity of sludge produced from adding 50 mg/L of alum if the raw water turbidity is 15 NTU and contains 25 mg/L of SS. No other chemicals are added to the water. The volumetric flow rate is 5.0 MGD.

Determine:

a) The quantity of dry solids produced using the equation recommended by Viessman *et al.* in both pounds per day and kilograms per day.
b) The quantity of dry solids produced using the equation recommended by Davis in both pounds per day and kilograms per day.

Solution

Solving Equation (6.177) recommended by Viessman *et al.* (2009, p. 630):

$$M_s = 8.34 \times Q(0.44 \times \text{alum dose} + 0.74 \times \textit{turbidity})$$

$$M_s = 8.34 \times 5.0 \frac{\text{MG}}{\text{d}} \left(0.44 \times 50 \frac{\text{mg}}{\text{L}} + 0.74 \times 15\ \text{NTU}\right)$$

$$= \boxed{1380 \frac{\text{lb}}{\text{d}}}$$

$$M_s = 1380 \frac{\text{lb}}{\text{d}} \times 454 \frac{\text{g}}{\text{lb}} \times \frac{\text{kg}}{1000\ \text{g}} = \boxed{627 \frac{\text{kg}}{\text{d}}}$$

Solving Equation (6.178) recommended by Davis (2011).

$$M_s = 86.4\,Q(0.44A + SS + M)$$

$$Q = 5.0 \times 10^6 \frac{\text{gal}}{\text{d}} \times 3.785 \frac{\text{L}}{\text{gal}} \times \frac{1\ \text{m}^3}{1000\ \text{L}} \times \frac{1\ \text{d}}{24\ \text{h}} \times \frac{1\ \text{h}}{60\ \text{min}} \times \frac{1\ \text{min}}{60\ \text{s}}$$

$$Q = 0.219 \frac{\text{m}^3}{\text{s}}$$

$$M_s = 86.4 \times 0.219 \frac{\text{m}^3}{\text{s}} \left(0.44 \times 50 \frac{\text{mg}}{\text{L}} + 25 \frac{\text{mg}}{\text{L}} + 0\right)$$

$$M_s = \boxed{889 \frac{\text{kg}}{\text{d}}}$$

$$M_s = 889 \frac{\text{kg}}{\text{d}} \times \frac{1000\ \text{g}}{1\ \text{kg}} \times \frac{1\ \text{lb}}{454\ \text{g}} = \boxed{1{,}958 \frac{\text{lb}}{\text{d}}}$$

There is a significant difference in the quantity of sludge produced using each equation. Best design practice would be to use the largest sludge quantity.

6.12.1 Sludge mass-volume relationships

Sludge mass-volume relationships are important when designing solids handling and treatment systems. The specific gravity of volatile (organic) solids and fixed (mineral) solids is typically assumed to be 1.0 and 2.5, respectively. Using these values and knowing the solids content (P_s) or moisture content (P_w) of the sludge and the specific gravity of the dry solids (S_s), the specific gravity of the sludge (S_{sl}) can then be determined. The water or moisture content of sludge is determined from the following equation:

$$P_w = \left(\frac{M_w}{M_w + M_s}\right) \tag{6.180}$$

where:
M_s = mass of dry solids, lb_m (kg)
M_w = mass of water, lb_m (kg)
P_w = water content or moisture content of sludge expressed as a fraction.

The solids concentration of a sludge expressed as a fraction is calculated from Equation (6.181).

$$P_s = (1.0 - P_w) = \left(\frac{M_s}{M_s + M_w}\right) \tag{6.181}$$

where P_s = solids content of sludge expressed as a fraction.

Using the definition of specific gravity (S), the volume occupied by the total mass of solids V_s can be determined from Equation (6.184):

$$S_s = \frac{\rho_s}{\rho} = \frac{\gamma_s}{\gamma} \tag{6.182}$$

$$S_s = \frac{\rho_s}{\rho} = \frac{M_s/V_s}{\rho} \tag{6.183}$$

$$V_s = \frac{M_s}{S_s \rho} \tag{6.184}$$

where:
Ss = specific gravity of dry solids, dimensionless
ρ_s = density of dry solids, lb_m/ft^3 (kg/m^3)
ρ = density of water, 62.4 lb_m/ft^3 (1,000 kg/m^3)
γ_s = specific weight of solids, lb_f/ft^3 (N/m^3)
γ = specific weight of water, 62.4 lb_f/ft^3 (9,810 N/m^3).

Since the total mass of solids is composed of a fixed and organic fraction, Equation (6.184) can be expressed as follows:

$$\frac{M_s}{S_s \rho} = \frac{M_v}{S_v \rho} + \frac{M_f}{S_f \rho} \tag{6.185}$$

where:
M_v = mass of volatile solids, lb_m (kg)
M_f = mass of fixed solids, lb_m (kg).

The specific gravity of sludge (S_{sl}) is estimated using the following equation which is based on Equation (6.184):

$$\frac{M_{sl}}{S_{sl} \rho} = \frac{M_w}{S_w \rho} + \frac{M_s}{S_s \rho} \tag{6.186}$$

where:
S_{sl} = specific gravity of sludge, dimensionless
S_{sl} = 1.005 to 1.05 (Metcalf & Eddy, 2003, p. 1456)
M_{sl} = mass of sludge, lb_m (kg)
S_w = specific gravity of water, which is equal to 1.

The specific gravity of sludge can also be defined as follows using Equation (6.183) as the basis:

$$S_{sl} = \frac{\rho_{sl}}{\rho} = \frac{M_{sl}/V_{sl}}{\rho} \tag{6.187}$$

Rearranging Equation (6.187) yields:

$$V_{sl} = \frac{M_{sl}}{S_{sl} \rho} \tag{6.188}$$

The mass of dry solids is determined by dividing the mass of sludge by the solids concentration as follows:

$$M_s = \frac{M_{sl}}{P_s} \tag{6.189}$$

Substituting Equation (6.189) into Equation (6.188) provides the equation for calculating the volume occupied by the sludge:

$$V_{sl} = \frac{M_s}{S_{sl} \rho P_s} \tag{6.190}$$

where: V_s = volume of sludge with mass M_{sl}, ft^3 (m^3).

Example 6.28 illustrates how to calculate the specific gravity of dry solids, the specific gravity of sludge, and the volume occupied by the sludge.

Example 6.28 Sludge mass-volume relationship

Alum sludge is thickened to a solids density of 2.0% in a gravity thickener. 20% of the solids are volatile. Determine:

a) The specific gravity (S_s) of the dry solids.
b) The specific gravity (S_{s1}) of the sludge.
c) The volume of the thickened sludge, if the mass of alum produced daily is 1,000 lb of dry solids.

Solution part a

Estimate the specific gravity of the dry solids knowing that 20% are volatile and 80% are fixed using Equation (6.185). Assume that the specific gravity of the volatile (organic) and fixed (mineral) solids are 1.0 and 2.5, respectively, and the total mass of the dry solids is 100 lb. Therefore, 0.20 × 100 = 20 lb of volatile solids and 80 lb of fixed solids. Substitute the appropriate values into Equation (6.185):

$$\frac{M_s}{S_s\rho} = \frac{M_v}{S_v\rho} + \frac{M_f}{S_f\rho} \tag{6.185}$$

$$\frac{100\text{ lb}}{S_s\rho} = \frac{80\text{ lb}}{2.5\,\rho} + \frac{20\text{ lb}}{1.0\,\rho} \qquad \boxed{S_s = 1.92}$$

Solution part b

Calculate the specific gravity of the sludge using Equation (6.186) and a value of 1.92 for the specific gravity of the dry solids. Assume that the total mass of the sludge is 100 lb; therefore, the mass of dry solids is 100 lb × 0.02 = 2 lb and the mass of water is 100 lb – 2 lb = 98 lb.

$$\frac{M_{s1}}{S_{s1}\rho} = \frac{M_w}{S_w\rho} + \frac{M_s}{S_s\rho} \tag{6.186}$$

$$\frac{100\text{ lb}}{S_{s1}\rho} = \frac{98\text{ lb}}{1\,\rho} + \frac{2\text{ lb}}{1.92\,\rho} \qquad \boxed{S_{s1} = 1.01}$$

Solution part c

Estimate the volume of the sludge after thickening to 2% solids (dry weight basis) by gravity thickening. Use Equation (6.190). The density of water is equal to 62.4 lb/ft^3 (1.94 slugs/ft^3) at 40°F and is the value generally used in sludge calculations.

$$\forall_{s1} = \frac{M_s}{S_{s1}\rho P_s} \tag{6.190}$$

$$\forall_{sl} = \frac{1000\text{ lb}}{(1.01)\left(\frac{62.4\text{ lb}}{\text{ft}^3}\right)(0.02)} = \boxed{793\text{ ft}^3}$$

6.12.2 Residuals processing options

Not only is the reduction in volume and mass of water treatment plant residuals of major concern in designing residuals management systems but, more importantly, the recycling and beneficial use of these residuals are the major goals that engineers now focus on. Figure 6.45 shows five generic steps that may be involved in handling, processing, and disposing of water treatment residuals. There are a number of unit operations and processes that may be used in each step.

The following factors must be considered when selecting the proper residuals processing scheme:

1 Characteristics of the sludge or residuals.
2 Quantity of solids generated.
3 The size or capacity of the water treatment facility.
4 Desired end-product and/or ultimate disposal of the residuals.

At most water treatment plants, the residuals are thickened, conditioned, dewatered, and then disposed of. A recovery step may be included at larger facilities to recalcinate lime or process the sludge to recover alum or a specific chemical. The reuse and recovery of wash-water is almost universally expected at water treatment facilities in operation today. Table 6.23 lists some of the options that are available for each of the steps shown in Figure 6.45.

Figure 6.46 is a schematic of the residual treatment scheme for the Amerson Water Treatment Plant, operated by the Macon Water Authority, which uses a surface water supply. The following references should be consulted for details on specific residuals handling and processing options: MWH, 2005; Davis, 2011; Peck & Russell, 2005; and Cornwell, 1999.

6.12.3 Thickening

Thickening of water treatment plant residuals is primarily accomplished by mechanical gravity thickening in concrete basins. The objective of thickening is to increase the solids concentration of the residuals before furthering processing or disposal. Thickening systems generally produce a final solids concentration of <15%. A polymer or coagulant may be added to the sludge to enhance thickening.

Lined basins constructed from earth (e.g., sludge lagoons) have also been used for increasing the solids content of residuals. Dissolved air flotation, in which air under high pressure (60–80 psi) is injected into the sludge feed to the unit, results in the release of small bubbles, which attach themselves to suspended solids, causing them to rise to the surface. A scraper mechanism is then used to skim the floating solids from the surface. Heavier solids settle to the bottom and must be scraped to one end of the unit before being pumped out for additional treatment.

6.12.3.1 Gravity Sludge Thickeners

Gravity sludge thickeners are similar to secondary clarifiers used for solids-liquid separation at wastewater treatment plants. They consist of a concrete basin with a sludge scraper

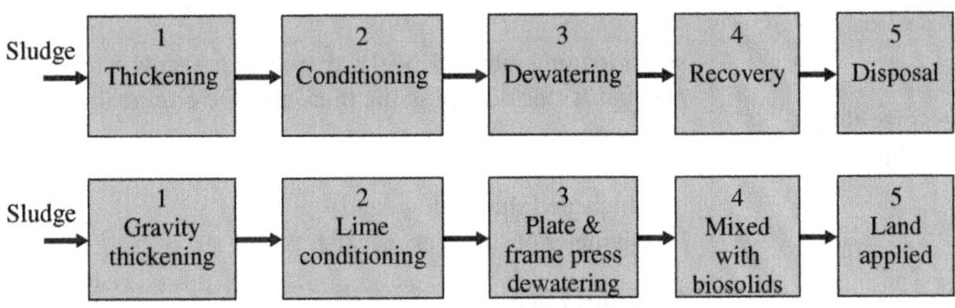

Figure 6.45 Five generic steps involved in residuals handling and processing.

Figure 6.46 Schematic of Amerson WTP sludge handling operations, Macon, Georgia.

Table 6.23 Residuals processing options.

Step #	Option 1	Option 2	Option 3	Option 4	Option 5	Option 6
Thickening	Gravity thickening	Lagoon thickening	Dissolved air flotation			
Conditioning	Chemical addition	Freeze-thawing	Heat	Inerts addition		
Dewatering	Belt filter press	Vacuum filtration	Plate & frame press	Drying beds	Sludge lagoons	Heat drying
Recovery	Alum recovery	Lime calcining	Acid treatment	Washwater recovery		
Disposal/reuse	Landfill: sanitary, monofill, or hazardous waste	Land application to crop land	Wastewater collection system	Land reclamation	Composting with biosolids	

Source: Adapted from MWH (2012) p. 1671. Reproduced by permission of John Wiley & Sons Ltd.

mechanism and steep bottom slopes. The objective is to accomplish consolidation of the water treatment residuals, i.e., increase the solids concentration of the residuals while producing a supernatant with relatively low solids concentration. Figure 6.47 shows a cross-sectional view of a gravity thickener. Sludge may be fed to the unit either continuously or on a batch basis.

Gravity thickeners are used for concentrating the solids from sedimentation basins and the spent wash water from backwashing filters. The design of a gravity thickener considers both clarification and thickening operations. In most instances, thickening operations will control the design of the system. Settling column analyses may be performed on the specific type of sludge to determine the limiting solids flux that is used to design the thickener. When settling column analyses cannot be performed, appropriate solids loading rates and overflow rates can be selected from the literature

When applying the overflow rate during design, the general practice is to use the supernatant flow coming out of the thickener instead of the sludge feed rate. Hindered settling and

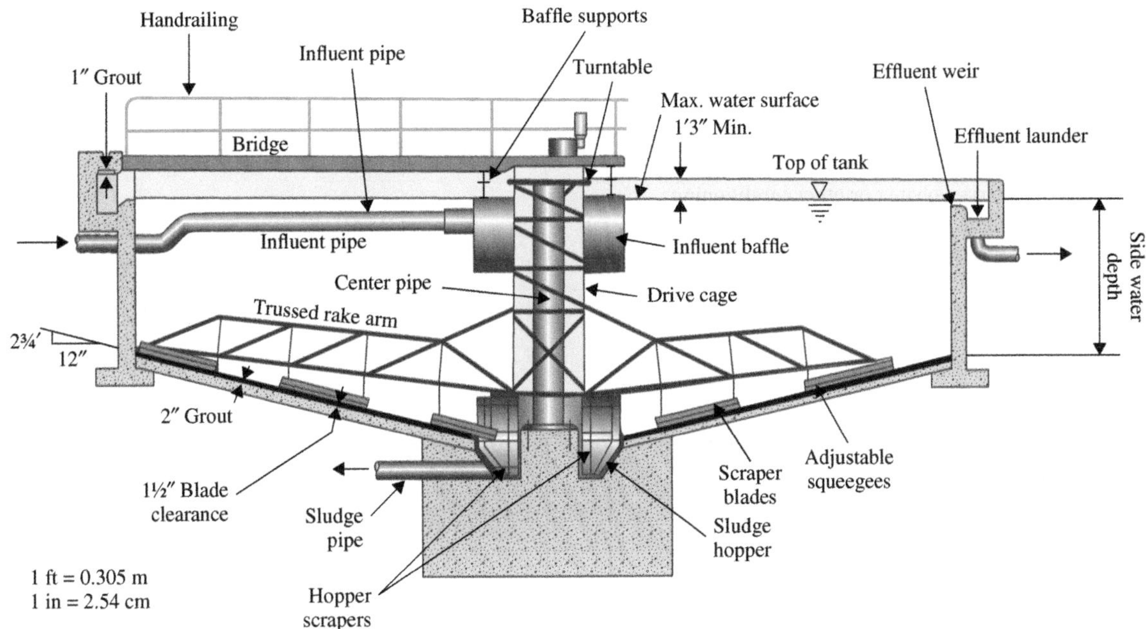

Figure 6.47 Schematic of gravity thickener. *Source*: EPA (1979) Sludge Treatment and Disposal Manual, p. 5–6. United States Environmental Protection Agency.

Table 6.24 Performance and design criteria for mechanical gravity thickeners.

Parameter	Units	Coagulant sludge	Lime softening sludge
Feed solids	%	0.2–1	1–4
Thickened solids	%	2–3	>5
Solids recovery	%	80–90	80–90
Solids loading	$lb/(ft^2 \cdot d)$	4–16	20–40
Solids loading	$kg/(m^2 \cdot d)$	20–80	100–200

Source: MWH (2012) p. 1674 (2012) p. 1671. Reproduced by permission of John Wiley & Sons Ltd.

the use of solids flux analysis for designing secondary clarifiers is presented in Chapter 7. The equations used for determining the thickener area are presented below.

$$A_c = \frac{Q_{supernatant}}{V_0} \tag{6.191}$$

where:

A_c = area required for clarification, ft^2 (m^2)
$Q_{supernatant}$ = supernatant flow from the gravity thickener, gpd (m^3/d)
V_0 = overflow rate, $gpd/ft^2[m^3/(d \cdot m^2)]$.

$$A_t = \frac{Q_{sludge}}{SLR} = \frac{Q_{sludge}C_{sludge}}{G_L} \tag{6.192}$$

where:

A_t = area required for thickening, ft^2 (m^2)
Q_{sludge} = sludge flow into the gravity thickener, gpd (m^3/d)
SLR = solids loading rate, $lb/(d \cdot ft^2)$ $[kg/(d \cdot m^2)]$
G_L = limiting solids flux from settling column analysis, $lb/(d \cdot ft^2)$ $[kg/d \cdot m^2]$.

Table 6.24 shows typical solids concentrations that can be achieved by mechanical gravity thickeners along with design and performance criteria. Polymer or other conditioning chemical addition will increase solids settling rates. Qasim *et al.*, (2000, p. 643) indicate that gravity thickeners typically have a depth of 15–21 ft (4.5–6.5 m) and detention times ranging from 8 to 24 hours.

Example 6.29 Design of a gravity thickener

A mechanical gravity thickener is used to thicken alum sludge from a surface water treatment facility. The sludge flow is 300,000 gallons per day at a solids concentration of 1% and specific gravity of 1.002. The thickened solids concentration is 2.5%, with a specific gravity of 1.005, and the supernatant contains 1,000 mg/L of solids. Assume a solids loading rate of $5\ lb/(ft^2 \cdot d)$ and an overflow rate of $600\ gpd/ft^2$ for designing the thickener. Determine the following:

a) Thickened sludge flow (gpd)
b) Diameter of the thickener (ft).

Solution part a

All sludge processing schemes require material balances be performed on the solids entering and exiting the thickener. Calculate the quantity of sludge that enters the gravity thickener.

$$Q_{sludge} \times \text{Concentration}\left(\frac{mg}{L}\right) \times 8.34 \frac{lb}{MG \times \frac{mg}{L}}$$

$$300{,}000\frac{gal}{d}\left(\frac{1\ MG}{10^6\ gal}\right) \times 1\%\ \text{solids}\left(\frac{10{,}000\ mg/L}{1\%\ \text{solids}}\right)$$

$$\times 8.34 \frac{lb}{MG \times \frac{mg}{L}} \times 1.002 = 25{,}070\frac{lb}{d}$$

A flow balance can be written as follows:

$$Q_{feed} = Q_{supernatant} + Q_{underflow}$$

$$300{,}000\frac{gal}{d}\left(\frac{1\ MG}{10^6\ gal}\right)$$

$$= 1.00\,Q_{supernatant} + 1.00\,Q_{underflow}$$

$$0.300\frac{MG}{d} = 1.000\,Q_{supernatant} + 1.000\,Q_{underflow}$$

A solids balance is developed below; recall that mass is equal to $Q \times C$.

$$m_{feed} = m_{supernatant} + m_{underflow}$$

$$m_{supernatant} = Q_{supernatant} \times \left(1000\frac{mg}{L}\right) \times 8.34 \frac{lb}{MG \times \frac{mg}{L}}$$

$$= 8340\,Q_{supernatant}$$

$$m_{underflow} = Q_{underflow} \times 2.5\%\text{solids}\left(\frac{10{,}000mg/L}{1.0\%\text{solids}}\right)$$

$$\times 8.34 \frac{lb}{MG \times \frac{mg}{L}} \times 1.005 = 209{,}543\,Q_{underflow}$$

$$25{,}070\frac{lb}{d} = 8340\,Q_{supernatant} + 209{,}543\,Q_{underflow}$$

$$25{,}070\frac{lb}{d} = 8340 \times (0.300 - 1.000\,Q_{underflow})$$

$$+ 209{,}543Q_{underflow}$$

$$25{,}070\frac{lb}{d} = 2502 - 8340\,Q_{underflow} + 209{,}543\,Q_{underflow}$$

$$201203\,Q_{underflow} = 22{,}568$$

$$Q_{underflow} = \boxed{0.112\,MGD} = 112{,}000\,\frac{gal}{d}$$

$$0.3000\,\frac{MG}{d} = 1.000\,Q_{supernatant} + 1.000\,Q_{underflow}$$

$$0.3006\,\frac{MG}{d} = 1.005 \times 0.112\frac{MG}{d} + 1.000\,Q_{supernatant}$$

$$Q_{supernatant} = \boxed{0.188\,MGD} = \boxed{118{,}000\,\frac{gal}{d}}$$

Solution part b

$$A_t = \frac{Q_{sludge}}{SLR} = \frac{Q_{sludge}C_{sludge}}{G_L} = \frac{25{,}070\,lb/d}{5\,ppd/ft^2} = \boxed{5014\,ft^2}$$

$$A_c = \frac{Q_{supernatant}}{V_0} = \frac{0.188 \times 10^6\,gpd}{600\,\frac{gpd}{ft^2}} = 313\,ft^2$$

Thickening considerations govern the design of the thickener, so use an area of 5,014 ft^2.

Determine the diameter of the circular gravity thickener as follows:

$$A = 5014\,ft^2 = \frac{\pi D^2}{4} \qquad D = \sqrt{\frac{5014\,ft^2 \times 4}{\pi}} = \boxed{79.9\,ft}$$

Use two 80 ft diameter gravity thickeners, with one of them on standby.

6.12.3.2 Dissolved Air Flotation Thickener

Thickening using dissolved air flotation has been used in both the water and wastewater industry for separating fats, oils, grease, and suspended solids from a liquid phase. Flotation is advantageous over sedimentation, because it more completely removes small particles and those that settle slowly, and accomplishes this in a shorter time.

Figure 6.48 shows a schematic of a typical dissolved air flotation system. Generally, air is injected into the effluent recycle stream at a pressure between 40 to 70 psi (280 to 489 kPa) (Reynolds & Richards, 1996, p. 633). This stream combines with the sludge feed before entering the flotation tank. The recycle stream ranges from 15–150% of the sludge feed (Metcalf & Eddy, 2003 p. 421 and Reynolds & Richards, 1996, p. 633).

Upon entering the flotation tank which is open to the atmosphere, small bubbles on the order of 50–100 μm are formed, which attach or become adsorbed to the sludge solids, causing them to rise to the surface. A layer of solids known as "float" is formed at the surface, and this is removed by a skimmer mechanism. In some instances, the entire sludge feed stream is pressurized, rather than just the recycle stream. Table 6.25 presents performance and design criteria for dissolved air flotation thickening.

Arnold *et al.* (1995, p. 329) report that dissolved air flotation technology may be used to treat most surface water with turbidities up to 100 NTU. They suggest that DAF systems offer

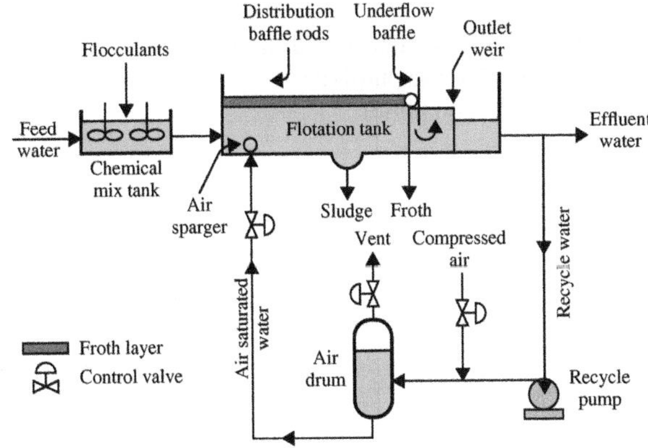

Figure 6.48 Schematic of dissolved air flotation unit. *Source:* http://upload.wikimedia.org/wikipedia/commons/7/75/DAF_Unit.png Accessed June 21, 2012. Courtesy Milton Beychok.

Table 6.25 Performance and Design Criteria for Dissolved Air Flotation Thickening.

Parameter	Units	Coagulant sludge	Lime softening sludge
Feed solids	%	0.5–1	0.5–1
Thickened solids	%	3–5	3–5
Solids recovery	%	80–90	80–90
Solids loading	lb/(ft^2 · d)	10–24	10–24
Solids loading	kg/(m^2 · d)	48–120	48–120
Volumetric loading	gpd/ft^2	2,800–3,600	2,800–3,600
Volumetric loading	m^3/(m^2 · d)	110–150	110–150

Source: MWH (2012) p. 1675 (2012) p. 1671. Reproduced by permission of John Wiley & Sons Ltd.

significant advantages over other technologies when the water quality includes the following: algae, high color, low turbidity-low alkalinity, water supersaturated with air, and cold waters.

DAF technology used by the town of New Castle, New York (Arnold *et al.*, 1995, p. 334) reduced influent turbidity from 2 to <0.5 NTU after adding 4 mg/L of polyaluminum hydroxychlorosulfate at a flotation surface loading rate up to 4 gpm/ft^2. Filtered effluent turbidity values were <0.1 NTU.

Example 6.30 Design of dissolved air flotation unit

A 30 ft diameter dissolved air flotation thickener treats 100 gpm of backwash water containing 1,000 mg/L of TSS. Air is added at 5.0 ft^3/min and assume that the density of air is 0.075 lb/ft^3. Determine the following:

a) Hydraulic loading rate (gal/(d · ft^2))
b) Solids loading rate (lb/(d · ft^2))
c) Air : solids ratio (lb/lb)

Solution part a

$$A = \frac{\pi D^2}{4} = \frac{\pi(30\text{ ft})^2}{4} = 707\text{ ft}^2$$

$$\text{Hydraulic loading rate} = \frac{100\text{ gpm}}{707\text{ ft}^2} = \boxed{0.14\frac{\text{gpm}}{\text{ft}^2}} = \boxed{202\frac{\text{gpd}}{\text{ft}^2}}$$

Solution part b

$$\text{Solids loading} = 100\frac{\text{gal}}{\text{min}}\left(\frac{8.34\text{ lb}}{\text{MG}\cdot\text{mg/L}}\right)\left(\frac{1\text{ MG}}{10^6\text{ gal}}\right) \times \left(1000\frac{\text{mg}}{\text{L}}\right) = 0.834\frac{\text{lb}}{\text{min}}$$

$$\text{Solids loading rate} = \frac{0.834\text{ lb/min}}{707\text{ ft}^2}\left(\frac{60\text{ min}}{\text{h}}\right)\left(\frac{24\text{ h}}{\text{d}}\right) = \boxed{1.7\frac{\text{lb}}{\text{d}\cdot\text{ft}^2}}$$

Solution part c

$$\text{Air mass} = 5.0\frac{\text{ft}^3}{\text{min}}\left(0.075\frac{\text{lb}}{\text{ft}^3}\right) = \boxed{0.375\frac{\text{lb}}{\text{min}}}$$

$$A:S = \frac{0.375\text{ lb air/min}}{0.834\text{ lb solids/min}} = \boxed{0.45\frac{\text{lb}}{\text{lb}}}$$

6.12.3.3 Lagoons

Lagoons may be used for thickening, dewatering, and drying water treatment plant residuals where land is readily available on-site. Most lagoons are lined earthen basins, but some are more complex, involving concrete basins with overflow structures for supernatant removal. According to Qasim *et al.* (2000, p. 642), alum and iron coagulated sludges thicken to approximately 4–6% solids in a month with continuous decanting. For design purposes, assume that a 5–10% solids concentration can be achieved in two or three months of thickening. Lagoons used for dewatering residuals will be discussed in more detail in Section 6.12.5.

6.12.4 Conditioning

Chemical conditioning of water treatment residuals is most commonly accomplished by adding polymers to the sludge prior to mechanical thickening and dewatering operations. As stated in *Water Treatment Principles and Design* (2005, p. 1689), the major objectives of conditioning the sludge are to improve the physical and structural characteristics so that water will be easily released and allow free drainage of the released water.

Water existing in residuals consists of free water and bound water. Normura *et al.*, (2007, p. 34) state that free water is easily removed from the residuals by mechanical means, because it is not associated with or influenced by suspended solids. Veslind (1994) divides bound water into three subcategories: interstitial water; vicinal water on the surface of solids; and water of hydration. Bench-scale and pilot-scale tests are typically conducted to select the right polymer or conditioning chemical for a specific application. If possible, full-scale tests should be performed. Polymer dosages generally range from 1–10 mg per gram of sludge solids (Qasim *et al.*, 2000, p. 644).

Flotation thickening requires adding polymer or other conditioning agent at a dosage of approximately 4.4 lb/dry ton (2.2g/kg), (Peck & Russell, 2005, p. 17.30). Polymer is required as a conditioning agent for dewatering sludge using bowl centrifuges. Dewatering alum sludge requires 1–2 lb polymer per dry ton (0.5–1.0 g/kg), with a 98% solids capture at 30% cake solids (Peck & Russell, 2005, p. 17.47). For some low turbid waters, up to 4 lb polymer per dry ton of solids (2g/kg) is required for dewatering alum sludge producing a cake containing 15% solids (Peck & Russell, 2005, p. 17.47).

In colder climates, residuals are allowed to freeze in lagoons or storage beds during the winter; during the summer, the residuals thaw, producing a dewatered sludge. Natural freezing of alum sludge at 2% solids can produce a dewatered sludge ranging from 20–30% solids (MWH, 2005, p. 1691; Qasim *et al.*, 2000, p. 644). Davis (2011, p. 11.35) states that the freeze-thaw process reduces the sludge volume by freezing the water molecules and is followed by dehydration of the solids. Upon thawing, the solid mass forms granular particles that easily dewater. Freeze-thaw treatment is only economical where natural freezing occurs.

Inert material, such as diatomaceous earth or fly ash, is often added to residuals that are applied to pressure filters such as plate and frame presses or diaphragm presses. Lime may also be added to residuals to enhance dewatering in pressure filters. Perlite, cellulose, or some combination thereof must be applied to rotary drum vacuum filters when direct filtration cannot be practiced and when precoat filtration is required.

Heat treatment has been used to condition and stabilize wastewater sludge (Metcalf & Eddy, 2003, p. 1557). Heating the residuals under pressure for short periods of time coagulates the solids, destroys the gelatinous structure, and reduces the water affinity of the solids. This conditioning method is not economically attractive, because of high capital and energy costs (Metcalf & Eddy, 2003, p. 1557; MWH, 2005, p. 1691).

6.12.5 Dewatering

Dewatering of water treatment plant residuals may be accomplished by mechanical or non-mechanical means. Dewatering processes are used to remove free water from sludge beyond what can be accomplished by thickening systems. Dewatering systems typically produce a final solids concentration above 15%. The major mechanical dewatering processes used include vacuum filtration, belt filter presses (BFP), dewatering centrifuges, plate and frame presses, and diaphragm presses. Non-mechanical dewatering processes include sludge lagoons and dewatering sand beds.

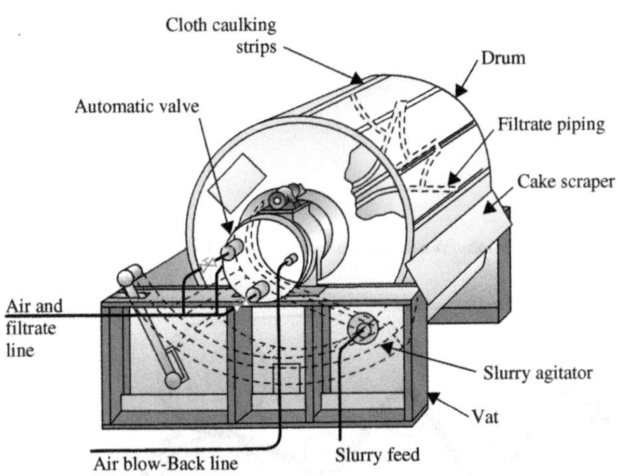

Figure 6.49 Schematic of rotary drum vacuum filter. *Source*: EPA (1979) Sludge Treatment and Disposal Manual, p. 9–29. United States Environmental Protection Agency.

6.12.5.1 Vacuum Filtration

Rotary drum vacuum filters (RDVFs) are used for dewatering coagulant and softening sludges. The system consists of a cylindrical drum that is partially submerged in the sludge or slurry. RDVFs operate continuously, and a typical revolution of the drum involves cake formation, cake washing if necessary, drying, and cake discharge.

As the drum rotates, a vacuum draws the liquid through the filtering material (cloth or stainless steel coils) and the solids are retained on the surface of the drum. Vacuum filters operate continuously and the cake solids are typically removed from the filter medium by a rubber or plastic scrapper blade. For some types of residuals, an inert material such as diatomaceous earth, perlite, cellulose, or a combination thereof, must be applied to the drum surface before applying slurry or residuals. The precoat solids thickness may vary between 4 to 6 inches (10–15 cm) (Komline-Sanderson, 1996).

Figure 6.49 is a schematic of a rotary drum vacuum filter. Alum and ferric hydroxide sludges require chemical conditioning with lime and/or polymer prior to vacuum filtration. Filtering areas of rotary drum filters vary from 9.4 to 1,526 ft^2 (0.87 to 142 m^2), and drum diameters vary from 3–13.5 ft (0.9–4.1 m) (Komline-Sanderson, 1996). Table 6.26 provides performance and design data for precoat RDVFs.

Example 6.31 Design of a vacuum filter

30,000 liters per day of lime-softening sludge at 5% solids is dewatered on a rotary drum vacuum filter. If the RDVF has a dry solids yield of 1.25 kg/(m^2·h), determine the required filter drum area (m^2).

Solution

Calculate the mass of solids that are applied to the rotary drum vacuum filter per hour, assuming that the specific gravity of the sludge is approximately 1.0.

$$30{,}000\,\frac{\text{L}}{\text{d}}(5\,\%)\left(\frac{10{,}000\text{ mg/L}}{1\%\text{ solids}}\right)\left(\frac{1\text{ g}}{1000\text{ mg}}\right)\times\left(\frac{1\text{ kg}}{1000\text{ g}}\right)\left(\frac{1\text{ d}}{24\text{ h}}\right)=62.5\,\frac{\text{kg}}{\text{h}}$$

The required drum area is calculated below:

$$A=\frac{62.5\text{ kg/h}}{\dfrac{1.25\text{ kg}}{\text{m}^2\cdot\text{h}}}=\boxed{50\text{ m}^2}$$

6.12.5.2 Belt Filter Presses

Belt filter presses use both gravity drainage along with shear and mechanical dewatering of conditioned residuals. Figure 6.50 shows the three basic stages of a belt filter press.

Belt filter presses require 2–5 lb polymer/ton (1–2.5 g/kg) for conditioning residuals. Polymer between 0.25–0.50% by weight is applied to the feed sludge (Cornwell, 1999, p. 16.34). Performance and design data for belt filter presses dewatering water treatment plant residuals are reported in Table 6.27.

Table 6.26 Performance and design data for precoat rotary drum vacuum filters for dewatering water treatment plant residuals.

Parameter	Units	Range of values
Feed solids	%	2–6
Feed rate	gal/(ft^2 · h)	2–6
Feed rate	L/(m^2 · h)	0.7–2.1
Solids recovery	%	96–99+
Dry solids yield	lb/(ft^2 · h)	1.0 - 1.5
Dry solids yield	kg/(m^2 · h)	0.2–0.3
Thickened alum sludge	%	15–25
Thickened lime sludge	%	20–40
Filtrate suspended solids	mg/L	10–20
Precoat recovery	%	30–35
Precoat rate	lb/(ft^2 · h)	0.1–0.2
Precoat rate	kg/(m^2 · h)	0.02–0.04
Precoat thickness	in	1.5–2.5
Precoat thickness	mm	38.1–63.5
Drum speed	rev/min	0.2– 0.3
Operating vacuum	in of Hg	5–20
Operating vacuum	mm of Hg	127–508

Source: MWH (2012) p. 1682 (2012) p. 1671. Reproduced by permission of John Wiley & Sons Ltd.

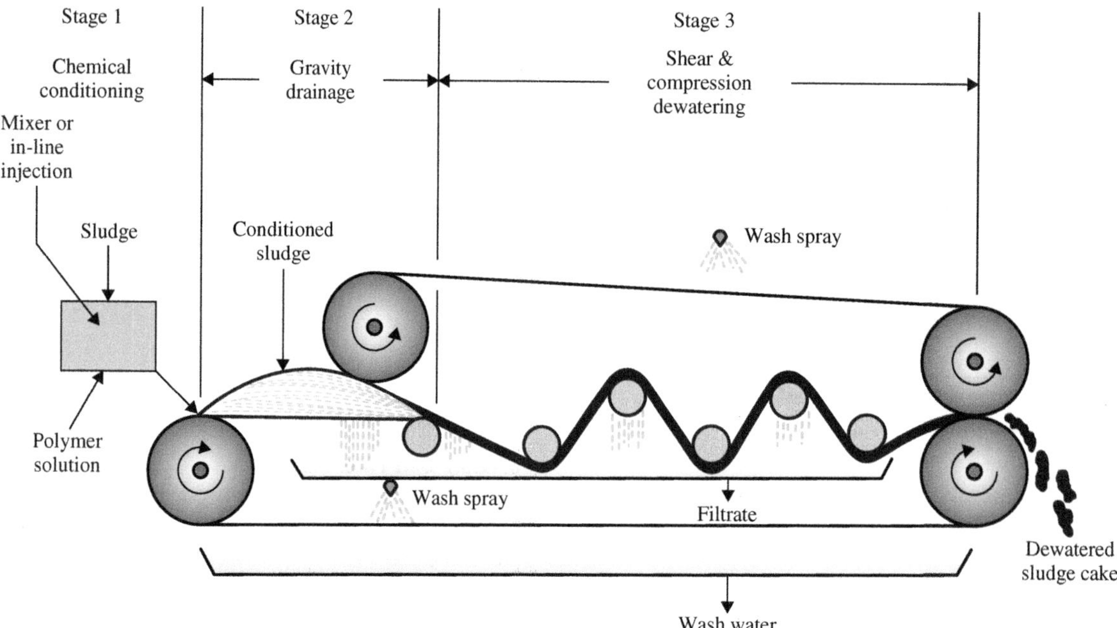

Figure 6.50 Schematic of belt filter press. *Source*: EPA (1979) Sludge Treatment and Disposal Manual, p. 9–46. United States Environmental Protection Agency.

Table 6.27 Performance and design data for belt filter presses dewatering water treatment plant residuals.

Parameter	Unit	Range of values
Feed solids	%	4–30
Thickened alum sludge	%	15–30
Thickened lime sludge	%	25–60
Solids recovery	%	95–99+
Cake yield	$lb/(ft^2 \cdot h)$	4–20
Cake yield	$kg/(m^2 \cdot h)$	0.8–4.0
Filtrate solids	mg/L	950–1,500
Operating pressure	lb/in^2	80–120
Operating pressure	kPa	550–830

Source: MWH (2012) p. 1685 (2012) p. 1671. Reproduced by permission of John Wiley & Sons Ltd.

Belt filter presses are manufactured in belt widths of 1.0, 1.5, 2.0, and 3.0 meters. Design of belt filter presses is based on solids loading rate and hydraulic loading rate.

6.12.5.3 Dewatering Centrifuges

Centrifuges are used for thickening and dewatering water treatment plant residuals. There are two basic types: solid bowl and basket centrifuges. Solid bowl centrifuges, also known as scroll or decanter centrifuges, are the most widely used.

A centrifuge uses centrifugal force to separate solids from the liquid (centrate). The residuals must be conditioned with polymer to accomplish effective dewatering. Polymer dosages typically range from 2–4 lb/ton (1–2 g/kg) of dry solids (MWH, 2005, p. 1697).

Table 6.28 Performance and design data for centrifuges thickening water treatment plant residuals.

Parameter	Unit	Coagulant sludge	Lime softening sludge
Feed solids	%	1–6	10–25
Thickened solids	%	12–15*	35–50
Solids recovery	%	90–96	90–97
Polymer dosage	lb/ton	2–4	N.A.
Polymer dosage	g/kg	1–2	N.A.

N.A. = not applicable.
*Up to 25% has been achieved with conditioning chemicals.
Source: MHW (2012) p. 1687 (2012) p. 1671 Reproduced by permission of John Wiley & Sons Ltd.

Table 6.28 provides performance and design data for centrifuge thickening of water treatment plant residuals.

A continuous countercurrent solid bowl centrifuge is shown in Figure 6.51 and a continuous concurrent solid bowl centrifuge is shown in Figure 6.52.

6.12.5.4 Plate and Frame Filter Presses

There are two categories of filter presses: plate and frame presses and diaphragm presses. Plate and frame presses consist of a series of circular or rectangular frames. A filter cloth covers the edges of the frame. During dewatering, the frames are pressed together, either electromechanically or hydraulically, between a fixed end and moving end. The residuals are pumped into the cavities or recessed areas. Next, a pressure between 100–225 psi (690–1,550 kPa) is applied within each chamber forcing the liquid through the filter cloth and plate outlet (Peck & Russell, 2005, p. 17.39). At the end of the cycle, the plates are separated and the sludge is removed.

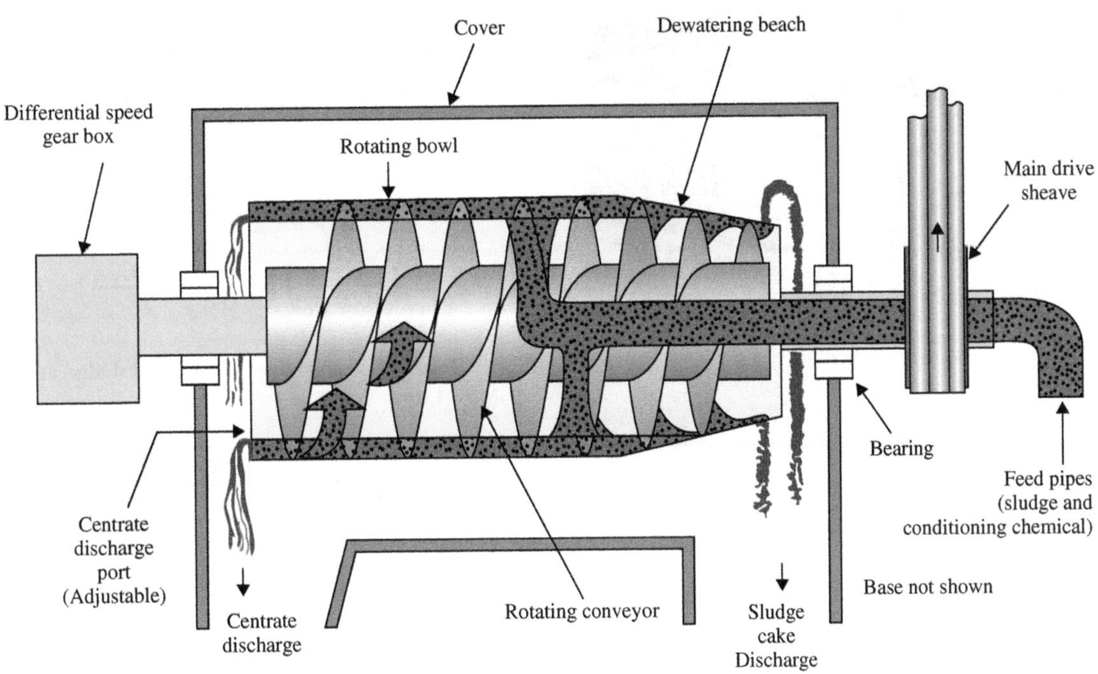

Figure 6.51 Schematic of continuous countercurrent solid bowl centrifuge. *Source*: EPA (1982) Dewatering Municipal Wastewater Sludges, p. 14. United States Environmental Protection Agency.

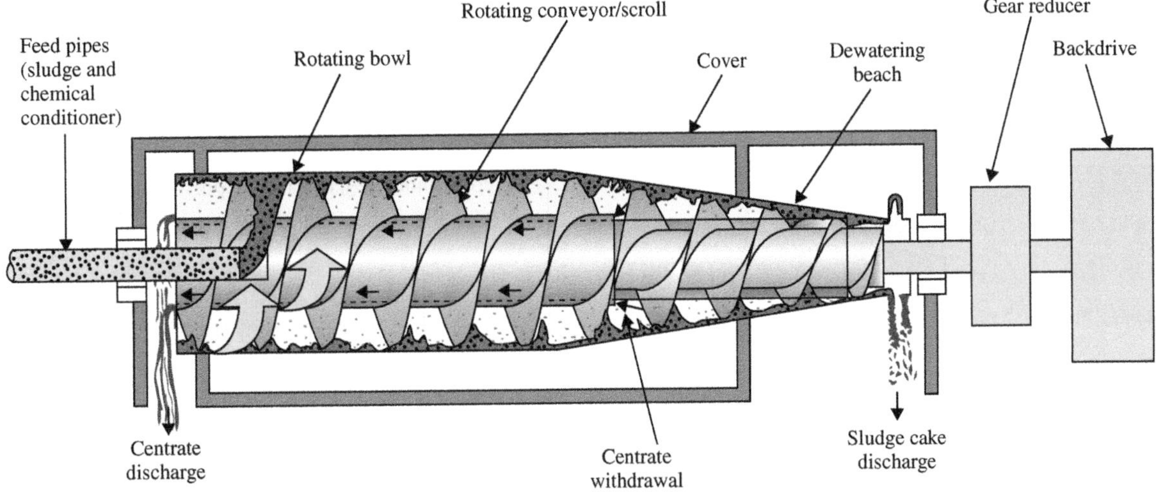

Figure 6.52 Schematic of continuous concurrent solid bowl centrifuge. *Source*: EPA (1982) Dewatering Municipal Wastewater Sludges, p. 15. United States Environmental Protection Agency.

Diaphragm presses are similar to plate and frame presses, but they offer the advantage of having a variable diaphragm within the plates that is expanded by air or water to achieve further dewatering of the residuals. Chemicals such as lime, ferric compounds, and polymers are used to condition the residuals. Diatomaceous earth and fly ash have also been added as conditioning agents. According to Peck & Russell (2005, p. 17.43), filter presses can achieve a cake solids concentration of 20–50% with a pressing cycle time of 2.5–22 hours with an eight-hour average. Figure 6.53 shows a filter press. Table 6.29 provides filter press manufacturer's specifications.

Example 6.32 Design of a filter press

A filter press (plate & frame press) is to be selected for dewatering a coagulation sludge that has been thickened to 2.8% solids. The design sludge flow to the filter press is 280 m^3/d. The press will be operated on a cycle time of three hours and the dewatered solids concentration is 28%. Determine the required filter press size based on manufacturer's data presented in Table 6.29.

Figure 6.53 Filter press (Courtesy of Macon Water Authority).

Table 6.29 Filter press manufacturer specifications.

Press	Units	Minimum	Maximum
1200 mm	Volume (ft^3)	50	125
Height: 116 in	Volume (L)	1,416	3,540
Height: 2,946 mm	Chambers	39	98
Width: 97 in	Length (in)	223	368
Width: 2,464 mm	Length (mm)	5,664	9,347
1500 mm	Volume (ft^3)	125	275
Height: 137 in	Volume (L)	3,540	7,788
Height: 3,840 mm	Chambers	64	141
Width: 107 in	Length (in)	310	514
Width: 2,718 mm	Length (mm)	7,874	13,056
1.5 × 2.0 m	Volume (ft^3)	200	350
Height: 166 in	Volume (L)	5,664	9912
Height: 4,214 mm	Chambers	74	129
Width: 107 in	Length (in)	348	494
Width: 2,718 mm	Length (mm)	8,839	12,548
2.0 × 2.0 m	Volume (ft^3)	300	500
Height: 153 in	Volume (L)	8496	14,160
Height: 3,886 mm	Chambers	89	148
Width: 156 in	Length (in)	482	647
Width: 3,962 mm	Length (mm)	12,243	16,434

Source: http://www.water.siemens.com/en/products/sludge_biosolids_processing/filter_press/Pages/dewatering_systems_generic_overhead_press_specs.aspx. [accessed June 20, 2012]. Dimensions and data are for general reference only. Reproduced with permission of Siemens Water Technologies © 2013.

Solution

Calculate the mass of solids that are applied to the filter press, recalling that $M_s = Q \times C$.

$$M_s = \left(280\ \frac{\text{m}^3}{\text{d}}\right)(2.8\ \%)\left(\frac{10{,}000\ \text{mg/L}}{1\%\text{solids}}\right) \times \left(\frac{1\ \text{g}}{1000\ \text{mg}}\right)\left(\frac{1\ \text{kg}}{1000\ \text{g}}\right)\left(\frac{1000\ \text{L}}{\text{m}^3}\right) = 7840\ \frac{\text{kg}}{\text{d}}$$

Determine the volume of dewatered sludge, assuming $S_{s1} = 1.0$.

$$V_{s1} = \frac{M_s}{S_{s1}\rho P_s} = \frac{7480\ \text{kg/d}}{(1.0)\left(\frac{1000\ \text{kg}}{\text{m}^3}\right)(0.28)} = 28.0\ \frac{\text{m}^3}{\text{d}}$$

$$V_{s1} = 28.0\ \frac{\text{m}^3}{\text{d}}\left(\frac{1000\ \text{L}}{\text{m}^3}\right) = 2.80 \times 10^4\ \frac{\text{L}}{\text{d}}$$

Calculate the volume of the press based on an operating cycle of three hours for filling, pressing, and discharging dewatered solids:

$$\frac{\text{volume}}{\text{cycle}} = \left(2.80 \times 10^4\ \frac{\text{L}}{\text{d}}\right)\left(\frac{1\ \text{d}}{24\ \text{h}}\right)\left(\frac{3\ \text{h}}{\text{cycle}}\right) = \boxed{3500\ \frac{\text{L}}{\text{cycle}}}$$

From Table 6.29, at least two possibilities exist for meeting the design requirements:

Press: 2946 mm × 2464 mm with 98 chambers and volume = 3540 L

Press: 3480 mm × 2718 mm with 64 chambers and volume = 3540 L

Best design practice would be to go with the second press. Although it is longer, it has fewer chambers and plates that are less likely to cause operational and maintenance problems. This press can also be expanded by adding plates.

6.12.5.5 Dewatering Lagoons

Lagoons are one of the oldest techniques used for handling water treatment residuals. They can be used for storage, dewatering, drying, and final disposal of residuals. Lagoon depths vary from 4–20 ft (1.2–6.1m) and have surface areas from 0.5–15 acres (0.2–6.1 ha) (Peck & Russell, 2005, p. 17.35).

Lagoon design is based either on the solids loading rate or maximum residual design flow rate. According to *Water Treatment: Principles and Design* (2005, p. 1686), lagoons are typically operated on a three-month filling cycle and a three-month drying cycle. A solids loading rate of 8.2 and 16.4 lb/ft^2 (40 and 80 kg/m^2) of effective lagoon area is used in designing lagoons for wet and dry regions, respectively. The actual lagoon area required is estimated by multiplying the effective lagoon area by a factor of 1.5 to account for the area required for berms and access roads. The depth at which sludge is applied ranges from 8–30 in (20–75 cm) for coagulant sludge, and 12–48 in (30–122 cm) for lime sludge (Peck & Russell, 2005, p. 17.37).

In the 2007th *Edition of Recommended Standards for Water Works*, GLUMRB (2007, pp. 134–136), several suggestions are made when designing lagoons for handling softening and alum residuals. Temporary storage lagoons for softening residuals should be designed on the basis of 0.7 ac/(MGD · 100 mg/L of hardness removed) (0.75 m^2/(m^3/d· 100 mg/L of hardness removed)). For alum sludge, a minimum of two lagoons should be used, with a minimum usable depth of 5 ft (1.5 m) and adequate freeboard of at least 2 ft (0.61 m). The effective surface area of a lagoon based on solids loading rate can be determined as follows:

$$A_s = \frac{(M_s)(t)}{SLR} \tag{6.193}$$

where:

A_s = effective surface area of lagoon, ac (m^2)
M_s = mass of solids or residuals to be dewatered, lb/d (kg/d)
t = filling cycle duration, d
SLR = solids loading rate, lb/(d·ac) [kg/(d·m^2)].

Example 6.33 Design of a dewatering lagoon

Design a sludge dewatering lagoon system for dewatering lime softening residuals. The lime sludge flow rate is 100 m^3/d and the solids concentration is 10%. Assume that the climate is dry and that four lagoons will be operated on a six-month cycle (three months for filling and three months for drying). Calculate the effective lagoon surface area, based on the lime sludge flow rate and the solids loading rate. Assume sludge is applied at a depth of 50 cm.

Solution

The lagoon surface area based on solids loading rate is calculated in the following steps.

Recall that the filling cycle duration is three months and a solids loading rate is assumed to be 80 kg/m^2 for the dry climate. Assume that sludge can be applied to each lagoon for 100 days during the fill cycle.

$$M_s = 100\,\frac{\text{m}^3}{\text{d}}\left(1000\frac{\text{L}}{\text{m}^3}\right)(10\%\text{ solids}) \times \left(\frac{10{,}000\text{ mg/L}}{1\,\%\text{ solids}}\right)\left(\frac{1\text{ kg}}{10^6\text{ g}}\right) = 10{,}000\,\frac{\text{kg}}{\text{d}}$$

$$A_s = \frac{(M_s)(t)}{SLR} = \frac{(10{,}000\text{ kg/d})(100\text{ d})}{\left(\dfrac{80\text{ kg}}{\text{m}^2}\right)} = 12{,}500\text{ m}^2$$

The actual lagoon area

$$= 12{,}500\text{ m}^2 \times 1.5 = 1.88 \times 10^4\text{ m}^2\left(\frac{1\text{ ha}}{10{,}000\text{ m}^2}\right)$$

$$= \boxed{1.88\text{ ha}}$$

or

$$\frac{1.88\text{ ha}}{4\text{ lagoons}} = \boxed{0.47\,\frac{\text{ha}}{\text{lagoon}}}$$

6.12.5.6 Drying Beds

Drying beds are another widely used method of dewatering water treatment residuals. They are classified as a natural or non-mechanical dewatering process and there are several variations: conventional sand drying beds, solar drying beds, wedge wire beds, and vacuum assisted drying beds (Peck & Russell, 2005, p. 17.32). Residuals are typically applied at a much lower depth of loading: 1–3 ft (0.3–0.9 m) for drying beds, compared to 5–20 ft (1.5–6.1 m) for lagoons. Details on the design of these systems are provided in Davis (2011, p. 11–28) and Peck & Russell (2005, p. 17.32–17.39).

A simplified example is presented below to illustrate one type of design methodology.

Example 6.34 Calculation of conventional sand drying bed area

Two thousand pounds of alum residuals are produced per million gallons of water treated. The surface water treatment plant has a design capacity of 3.0 MGD. The alum sludge has a solids concentration of 2.5%, and 20 sand drying beds will be used for dewatering the alum residuals. Determine the surface area required for each drying bed, assuming that the alum sludge will be applied at a depth of 16 inches.

Solution

First, calculate the mass of solids to be dewatered each year.

$$M_s = (3.0\text{ MGD})\left(2000\,\frac{\text{lb}}{\text{MG}}\right) = 6000\,\frac{\text{lb}}{\text{d}}$$

The volume of thickened sludge to be dewatered each year is calculated below using Equation (6.190):

$$\forall_{sl} = \frac{M_s}{S_{sl}\,\rho P_s} = \frac{6000\text{ lb/d}}{(1.0)\left(62.4\,\dfrac{\text{lb}}{\text{ft}^3}\right)(0.025)} = 3850\,\frac{\text{ft}^3}{\text{d}}$$

$$\text{Total volume} = \left(3850\,\frac{\text{ft}^3}{\text{d}}\right)\left(365\,\frac{\text{d}}{\text{y}}\right) = 1.41 \times 10^6\text{ ft}^3$$

The area of each sludge drying bed is calculated as follows:

$$A = \frac{V}{\text{Depth} \times \text{Number of sand beds}}$$

$$= \frac{1.41 \times 10^6\ \text{ft}^3}{(16\ \text{in})\left(\frac{1\ \text{ft}}{12\ \text{in}}\right)(20)} = \boxed{53{,}000\ \text{ft}^2}$$

6.12.6 Ultimate disposal

There are several options available for ultimate disposal of water treatment plant residuals in addition to the beneficial reuse of these residuals. Table 6.30 lists some of the options, along with the solids concentration required for the specific disposal method. Other residuals, such as reject or concentrated brine from membrane processes, may be injected into deep wells. In some cases, regulatory agencies will allow these types of wastes to be discharged into the ocean. A brief description of the major methods of disposing or using water treatment plant residuals is presented below.

6.12.6.1 Beneficial Reuse

It is most prudent to recycle and recover as much of the water in the residuals as possible and then select an ultimate disposal method if the residuals are not going to be reused. Land application of the residuals is a beneficial reuse method. Davis (2011, p. 11–49), referencing Novak (1993), stated that coagulant sludges, lime softening sludges, slow sand filtration washings, and nanofiltration concentrate have all been land applied. The addition of a lime softening sludge is advantageous, because it raises the pH of an acid soil. Other beneficial uses include: composting with biosolids and/or municipal solids waste; using the residuals in brick and cement manufacturing; and using the residuals as landfill cover material or as a subgrade for road construction (Peck & Russell (2005), p. 17.11–17.12; Davis (2011), p. 11–51).

Table 6.30 Disposal Options Available Based on Solids Content of Residuals.

Disposal option	Required solids concentration (%)
Land application	<1 to 15
Landfilling (codisposal)	<8 to 25
Landfilling (monofill)	>25
Discharge to sewer	<1 to 8
Direct stream discharge	<1 to 8
Residuals reuse	<1 to >25

Solids concentration values from EPA (1996), pp. 57–58.

6.12.6.2 Landfilling

Water treatment residuals may be disposed of in a sludge only landfill (monofill) or co-disposed with municipal solid waste in a sanitary landfill. Should the residuals not meet the TCLP or WET tests, they would have to be buried in a hazardous waste landfill under the Resource Conservation and Recovery Act (RCRA).

6.12.6.3 Direct Stream Discharge

Disposal of water treatment residuals by direct surface water discharge is not a viable option in today's environmentally conscious society. All discharges to waterways require a National Pollution Discharge Elimination System (NPDES) permit. Some states have banned the discharge of residuals to streams, while other states have established limits based on pH and solids concentration (Peck & Russell, 2005, p. 17.13).

6.12.6.4 Discharge to Sanitary Sewer

Peck & Russell (2005, p. 17.10) state that the practice of discharging water treatment plant residuals to sanitary sewers has become increasingly more common, since it allows a municipality to combine water and wastewater residuals handling at one site. This also reduces the inorganic content of the resulting sludge, making it more amenable to land application.

Summary

- Drinking water standards have been established by the US EPA, as authorized by the Safe Drinking Water Act. Primary standards (based on health concerns) and secondary standards (based on esthetics) have been developed.
  - Maximum contaminant levels (MCLs) are concentrations or levels for specific contaminants that cannot be exceeded.
  - Maximum contaminant level goals (MCLGs) are concentrations for specific contaminants that are established below which there is no known or expected health risk.
  - For some contaminants, a treatment technique (TT) rather than an MCL is established, because the best technology is not feasible, either economically or technically, or the detection method is not reliable or economical.
- US EPA has developed a list of best available technologies (BATs) for each primary drinking water standard.
- The average daily demand for water in the United States in 2005 was 171 gallons per capita per day (647 liters per capita per day).
- Design capacity of various elements in the water treatment train is based on peak flow rates. Peak daily demand (PDD), peak hourly demand (PHD), and peak daily demand plus fire flow demand are the flow rates of most concern to design engineers. The capacity of a water treatment plant (WTP) is based on the peak daily demand flow rate.

- Preliminary systems for treating drinking water include bar racks, fine screens, aeration, adsorption, and pre-disinfection. With the exception of pre-disinfection, these are all unit operations which physically remove specific contaminants.
- Coagulation is a unit process involving the addition and mixing of a chemical reagent for the destabilization of colloidal particles and fine solids.
- Flocculation is a unit process consisting of slow stirring to promote the agglomeration of the destabilized particles following the coagulation process.
- Colloidal destabilization is accomplished by reducing repulsive forces between colloidal particles, so that attractive forces – primarily van der Waals forces – can predominate and cause particle agglomeration.
- There are four destabilization mechanisms presented in the literature to describe how coagulants work. They include: lowering the surface potential by compressing the electrical double layer; lowering the surface potential by adsorption and charge neutralization; enmeshment of the colloids in precipitate; and attachment and interparticle bridging.
- Coagulants are added to water for removing turbidity, color, and TOC. Major coagulants used during water treatment are aluminum sulfate, ferrous sulfate, ferric sulfate, ferric chloride, and polymers.
- Chemicals or materials added to water to assist in coagulation are called coagulant aids. Major coagulant aids include: bentonite and kaolinite clay; acids such as sulfuric and phosphoric; lime and caustic for pH adjustment and alkalinity addition; and polymers.
- Mechanical mixers installed in rapid mixing basins are typically used for dispersing chemicals during water treatment. In-line mechanical and in-line static mixers are also used, in addition to over-and-under baffling, and by adding the coagulant to the suction intake of a pump.
  - The velocity gradient (G) is the main design parameter used in designing mixing systems.
- Flocculation basin design traditionally has used paddle-wheel flocculators. Present design makes use of vertical-shaft turbine flocculators.
- Water softening is the removal of polyvalent cations (consisting primarily of calcium and magnesium ions) from water.
  - Water softening is typically accomplished by adding lime and soda ash for precipitating calcium and magnesium as calcium carbonate and magnesium hydroxide.
  - Ion exchange and membrane processes are two other means of removing hardness from water.
- Sedimentation is a unit operation involving solids-liquid separation using gravitational force. Settling basin design is based on free settling if the particles are discrete, or flocculant settling for waters that have been coagulated.
  - Settling basin design is primarily based on the overflow rate (V_0). Both detention time (τ) and weir loading rate (q) are also important design parameters.
- Filtration of water is required for all public water supplies using surface water sources or groundwater sources under the direct influence of surface water (GWUDI).
  - Filtration is primarily accomplished using multi-media granular filters.
  - Microfiltration and ultrafiltration, using semi-permeable membranes, are also used.
  - The head loss through a clean filter can be estimated using the Rose, Carman-Kozeny, or Ergun equations.
- Granular media filters are generally backwashed when finished water turbidity levels reach some specified level, or when the head loss through the filter reaches 8 – 10 ft, or when the filter has been in operation for 24 – 48 hours.
- Reverse osmosis (RO) is a membrane treatment process used to separate dissolved solutes and NOM from water. Both nanofiltration (NF) and RO membranes are used in RO applications.
  - The objective of RO is to produce potable water from ocean or brackish waters and to remove specific dissolved constituents.
  - NF membranes are used for softening hard waters, freshen brackish waters, and reduce the concentration of DBP precursors. Both NF and RO membranes separate solutes by diffusion through a semi-permeable membrane, as well as sieving action.
  - Membranes are permeable to water, but not to solutes or substances rejected and removed.
- Fluoride is added to drinking water at a dose of approximately 1.0 mg/L to prevent dental cavities in children.
- The goal of disinfection of water is to inactivate pathogenic organisms such as bacteria, protozoa, and enteric viruses.
  - Principal disinfectants used in the water industry include: chlorine gas; sodium hypochlorite, calcium hypochlorite, chloramines, chlorine dioxide, ozone, and ultraviolet radiation.
- Water treatment plant residuals are produced from settling basins and from solids removed during backwashing.
- Thickening of water treatment plant residuals may be accomplished by gravity thickening, dissolved air flotation thickening, and lagoons.
- Conditioning of residuals is traditionally accomplished by adding chemicals, primarily polymers, before thickening and dewatering operations. In some cases, lime and inert materials such as fly ash and/or diatomaceous earth may be added to condition the sludge.
- Dewatering water treatment plant residuals increases the solids concentration beyond that which can be accomplished by thickening. Primary dewatering processes include: vacuum filtration, belt filter presses, dewatering centrifuges, plate and frame presses, dewatering lagoons, and drying beds.
- Ultimate disposal of water treatment plant residuals is accomplished through beneficial reuse, landfilling, direct stream discharge, or discharge to sanitary sewer.

Key Words

alkalinity
arsenic
backwashing
barium
BAT
belt filter press
beneficial reuse
breakpoint chlorination
cadmium
Carman-Kozeny equation
centrifuge
chloramines
chlorine dioxide
chloramination
chromium
coagulant
color
conditioning
Cryptosporidium
CWS
DBP
dewatering
dissolved air flotation
drying beds
electrical double layer
Ergun equation
flocculation
fluoridation
gas
Giardia
GWR
HAA5
hollow fiber membrane
IOC
iron
Lead and Copper Rule
Legionella
LRAA
LT1
LT2
manganese
MCL
MCLG
mercury
MRDL
NCWS
nitrate
NTNCWS
ozone
packed tray aeration
pH
plate and frame presses
primary standards
PWS
Radionuclides
rapid mix
reverse osmosis
Rose equation
secondary standards
SOC
spiral wound membrane
Stage 1 DBP
Stage 2 DBP
state primacy
sweep floc
SWTR
taste and odor
thickening
TLCP
TNCWS
TOC
Total Coliform Rule
TT
TTHMs
vacuum filtration
VOC
WET
spiral wound membrane

References

Arnold, S.R., Grubb, T.P., Harvey, P.J. (1995). Recent Applications of Dissolved Air Flotation Pilot Studies and Full Scale Design. *Wat Sci Tech* **31**, 3-4, 327-340.

AWWA (1999). *Water Quality and Treatment: A Handbook of Community Water Supplies*, 5th Edition, pp. 5.1-5.68, 8.10, 8.58-8.71, 8.81-8.89, 9.1-9.87, 9.37, 11.1-11.67, 13.2. McGraw-Hill, New York, NY.

AWWA/ASCE (2005). *Water Treatment Plant Design*, 4th Edition, pp. 4.31-4.32, 5.1-5.25, 5.7, 10.55, 12.1-12.54, 12.4-12.5, 13.1-13.47, 14.1, 14.5. McGraw-Hill, New York, NY.

ASCE/AWWA (1990). *Water Treatment Plant Design*, 2nd Edition, pp. 494, 496, 497, 498, 499, 509. McGraw-Hill, New York, NY.

Barbour, Wiley, Oommen, Roy, Shareer, Gunseli S., Vataguk, W.M. (1995). *Wet Scrubbers for Acid Gas*, EPA/452/B-02-001, pp. 1-51 and 1-52. US EPA, Research Triangle Park, NC.

Benefield, L.D., Morgan, J.M. (1999). Chemical Precipitation in *Water Quality and Treatment: A Handbook of Community Water Supplies*, pp. 10.1-10.18, 10.51. McGraw-Hill, New York, NY.

Benefield, L.D., Judkins, J.F., Weand, B.L. (1982). *Process Chemistry for Water and Wastewater Treatment*, pp. 269-303. Prentice-Hall, Englewood Cliffs, New Jersey.

Bergman, R.A. (2005). Membrane Processes in *Water Treatment Plant Design*, 4th Edition, AWWA/ASCE, pp. 13.1-13.49 and 13.25-13.28. McGraw-Hill, New York, NY.

Carman, P.C. (1937). Fluid Flow Through Granular Beds. *Transactions of Institute of Chemical Engineers (London)* **15**, 150.

Castro, K., Logsdon, G., Martin, S.R. (2005). High-Rate Granular Media Filtration in *Water Treatment Plant Design*, 4th Edition, p. 8.8. AWWA/ASCE, McGraw-Hill, New York, NY.

CDC (1995). *Engineering and Administrative Recommendations for Water Fluoridation, 1995*, p. 8. US Department of Health and Human Services, Centers for Disease Control, Atlanta, GA.

Chick, H. (1908). Investigation of the Law of Disinfection, *Journal of Hygiene* **8**, 92.

Chlorine Chemistry Council (2003). *Drinking Water Chlorination: A Review of Disinfection Practices and Issues*, p. 23. Arlington, VA.

Clark, J.W., Viessman, Jr., W., Hammer, M.J. (1977). *Water Supply and Pollution Control*, 3rd Edition, p. 402. IEP-Dun-Donnelley, New York, NY.

Cleasby, J.L., Logsdon, G.S. (1999). Granular Bed and Precoat Filtration in *Water Quality and Treatment: A Handbook of Community Water Supplies*, 5th Edition, pp. 8.32-8.39. McGraw-Hill, New York, NY.

Cornwell, D.A. (1999). Water Treatment Plant Residuals Management in *Water Quality and Treatment: A Handbook of Community Water Supplies*, 5th Edition, pp. 16.1-16.51, 16.34. McGraw- Hill, New York, NY.

Davis, M.L. (2011). *Water and Wastewater Engineering: Design, Principles, and Practice*, pp. 6-3 to 6-6, 6-17, 11-7, 11-28, 11-35, 11-49, and 11-51. McGraw-Hill, New York, NY.

Daugherty, R.L., Franzini, J.B., Finnemore, E.J. (1985). *Fluid Mechanics with Engineering Applications*, 8th Edition, p. 208. McGraw-Hill, New York, NY.

Department of the Army, USA, (1986). *Technical Manual TM 5-813-8, Water Desalination*, pp. 7-8 and 7-9.

Eckert, J.S. (1970). Selecting the Proper Distillation Column Packing. *Chemical Engineering Progress* **66**(3), 39-44.

Ergun, S. (1952). Fluid Flow through Packed Columns. *Chemical Engineering Progress* **48**(2), 89-94.

Fair, G.M. and Geyer, J.C. (1954). *Water Supply and Waste-Water Disposal*, John Wiley and Sons, New York.

GLUMRB (2007). *Recommended Standards for Water Works*, Great Lakes-Upper Mississippi Board of State and Provincial Public Health and Environmental Managers, Health Education services, pp. 21-22, 27, 54, 61, 82-83, and 134-136. Albany, New York, NY.

Haas, C.N., Kara, S. (1984). Kinetics of *Microbial Inactivation by Chlorine-I Review of Results in Demand-Free Systems Water Research* **18**(11), 1443-1449.

Haas, C.N. (2002). Chlorine and Chloramines, in *Control of Microorganisms in Drinking Water*, p. 189. ASCE, Reston, VA.

Hand, D.W., Hokanson, D.R., Crittenden, J.C. (1999). Air Stripping and Aeration in *Water Quality and Treatment: A Handbook of Community Water Supplies*, 5th Edition, pp. 5.1-5.68. McGraw-Hill, New York, NY.

Hesby, J.C. (2005). Oxidation and Disinfection in *Water Treatment Plant Design*, 4th Edition, p. 10.40. AWWA/ASCE, McGraw-Hill, New York, NY.

Hoff, J.C. (1986). Project Summary, Inactivation of Microbial Agents by Chemical Disinfectants, EPA/600/S2-86/067 (September 1986), p. 5. Cincinnati, OH.

Joslin, W.R. (1997). Slow Sand Filtration: A Case Study in the Adoption and Diffusion of a New Technology. *Journal New England Water Works Association* **111**(3), 294-303.

Kawamura, S. (1991). *Integrated Design of Water Treatment Facilities*, pp. 138-139, 200, 210, 214-215, 220, 532-533, 548, and 566. John Wiley & Sons, Inc., New York, NY.

Kenny, J.F., Barber, N.L., Hutson, S.S., Linsey, K.S., Lovelace, J.K., Maupin, M.A. (2009). Estimated Use of Water in the United States in 2005. US Geological Survey Circular **1344**, p. 16.

Komline-Sanderson (1996). The Komline-Sanderson Rotary Drum Vacuum Filter for process filtration, wastewater clarification and sludge dewatering, pp. 1–8. Peapack, NewYork, NY.

Kruyt, H.R. (1952). *Colloid Science*. Elsevier, New York, NY.

LaGrega, M.D., Buckingham, P.L, Evans, J.C. (1994). *Hazardous Waste Management*, pp. 447–554, p. 457. McGraw-Hill, New York, NY.

McGhee, T.J. (1991). *Water Supply and Sewerage*, p. 14, 16, and pp. 246–253. McGraw-Hill, New York, NY.

Merill, D.T. (1982). Chemical Conditioning for Water Softening and Corrosion Control in *Water Treatment Plant Design For the Practicing Engineer*, pp. 497–564. Ann Arbor Science, Ann Arbor, Michigan.

Metcalf & Eddy (2003). *Wastewater Engineering: Treatment and Reuse*, pp. 344–361, 421, 1048, 1047–1049, 1104–1116, 1139, 1162–1180, 1175, 1231, 1235, 1237, 1238, 1266, 1302, 1456, and 1557. McGraw-Hill, New York, NY.

MWH (2005). *Water Treatment Principles and Design*, 2nd Edition, pp. 76, 254, 264, 513, 658, 679, 680, 683, 688, 743–752, 749, 810, 878–880, 882, 890, 891, 892, 937–940, 957, 958, 959, 1168, 1194, 1205, 1429–1497, 1433, 1435, 1437, 1438–1441, 1450–1454, 1454–1459, 1542, 1561, 1593, 1595, 1597, 1627, 1686, 1689, 1691, 1697, and 1783. John Wiley and Sons, Inc., Hoboken, New Jersey.

MWH (2012). *Water Treatment Principles and Design*, 3rd Edition, pp. 614, 1075, 1345, 1671, 1674, 1675, 1682, 1685, and 1687. John Wiley and Sons, Inc., Hoboken, New Jersey.

Normura, Toshiyuki, Araki, Shunsuke, Magao, Takanori, Konishi, Yasuhiro (2007). Resource Recovery Treatment of Waste Sludge Using a Solubilizing Reagent. *J. Mater Cycles Waste Manag* **9**, 34–39.

Novak, J.T. (1993). *Demonstration of Cropland Application of Alum Sludges*. American Water Works Association Research Foundation, Denver, Colorado.

O'Connell, R.T. (1982). Suspended Solids Removal in *Water Treatment Plant Design For the Practicing Engineer*, pp. 283-298, 287, and 288. Ann Arbor Science, Ann Arbor, Michigan.

Peavy, H.S., Rowe, D.R., Tchobanoglous, G. (1985). *Environmental Engineering*, pp. 134 and 185. McGraw-Hill, New York, NY.

Peck, B.E., Russell, J.S. (2005). Process Residuals in *Water Treatment Plant Design*, 4th Edition, AWWA/ASCE, pp. 17.1–17.70, 7.10, 7.11–17.12, 17.13,17.30, 17.32, 17.32–17.39, 17.35, 17.37, 17.39, 17.43, 17.47, and 17.59. McGraw-Hill, New York, NY.

Qasim, S.R., Motley, E.M., Zhu, G. (2000). *Water Works Engineering: Planning, Design, and Operation*, pp. 634-683, 642, 643, 644 and 769. Prentice Hall PTR, Upper Saddle River, New Jersey.

Reeves, T. G. (1999). Water Fluoridation in *Water Quality and Treatment: A Handbook of Community Water Supplies*, 5th Edition, pp. 15.1–15.19. McGraw-Hill, New York, NY.

Reynolds, T.D., Richards, P.A. (1996). *Unit Operations and Processes in Environmental Engineering*, 2nd Edition, pp. 77, 78, 128, 130, 133, 172, 174, 175, 176, 182–193, 209, 255, 633, and 742. PWS Publishing Company, Boston, Massachusetts.

Rich, L. (1971). *Unit Operations of Sanitary Engineering*, pp. 86, 143, and 149. John Wiley & Sons, Inc., New York, NY.

Rose, H.E. (1945). An Investigation of the Laws of Flow of Fluids through Beds of Granular Materials. *Proceedings Institute of Mechanical Engineers* **153**, 141.

Shammas, N.K., Wang, L.K. (2011). *Water Supply and Wastewater Removal*, pp. 133, 135–139. John Wiley and Sons, Inc., Hoboken, NJ.

Snoeyink, V.L., Jenkins, D. (1980). *Water Chemistry*, NY, p. 249. John Wiley & Sons, Inc., New York, NY.

Taylor, J.S., Wiesner, M. (1999). Membranes in *Water Quality and Treatment: A Handbook of Community Water Supplies*, 5th Edition, pp. 11.1–11.71. AWWA, McGraw-Hill, New York, NY.

US EPA (1979). *Process Design Manual for Sludge Treatment and Disposal*, EPA 625/1-79-011, pp. 5-6, 9-29, and 9-46. Municipal Environmental Research Laboratory, Cincinnati, Ohio.

US EPA (1982). *Design Manual for Dewatering Municipal Wastewater Sludge*, pp. 14–15. Center for Environmental Research Information, Cincinnati, Ohio.

US EPA (1991). *Guidance Manual for Compliance with the Filtration and Disinfection Requirements for Public Water Systems Using surface Water Sources*, Office of Water, EPA/CN-68-01-6989. March 1991, Washington, D.C., Appendix E and F.

US EPA (1992). *Test Methods for Evaluating Solid Waste, Physical/Chemical Methods*, pp. 7–5. EPA Publication SW-846, 3rd Edition, September 1986, amended by Update I (July 1992), US Environmental Protection Agency, Washington, D.C.

US EPA (1996). *Management of Water Treatment Plant Residuals Enhanced*, pp. 57–58. Office of Research and Development, EPA/625/R-95/008.

US EPA (1999). *Enhanced Coagulation and Enhanced Precipitative Softening Guidance Manual*, pp. A-24. Office of Water, EPA 815-R-99-012.

US EPA (2002). Air Pollution Control Cost Manual, Barbour, Wiley, Oommen, R., Shareef, G., Vatavuk, W., Chapter 1 *Wet Scrubbers for Acid Gas*, pp. 1–6. EPA/452/B-02-001, Research Triangle Park, NC 27709.

Vesilind, P.A. (1994). The Role of Water in Sludge Dewatering. *Water Environment Research* **66**(1), 4–11.

Viessman, Jr., W., Hammer, M.J., Perez, E.M., Chadik, P.A. (2009). *Water Supply and Pollution Control*, 8th Edition, pp. 273, 343–345, 344, 358, 405, 409–419, 458, 467, 477, 478, 630, and 644. Pearson Prentice Hall, Upper Saddle River, New Jersey.

Viessman, Jr., W., Hammer, M.J. (2005). *Water Supply and Pollution Control*, 7th Edition, pp. 344, 401, 467, 469, and 477. Pearson-Prentice Hall, Upper Saddle River, New Jersey.

Walker, J.D. (1982). Sedimentation in *Water Treatment Plant Design For the Practicing Engineer*, pp. 149–182, pp. 157 and 160. Ann Arbor Science, Ann Arbor, Michigan.

Watson, H.E. (1908). A Note on the Variation of the Rate of Disinfection with Change in the Concentration of the Disinfectant. *Journal of Hygiene* **8**.

Problems

1 Estimate the water demand for a city of 100,000 people. Assume the annual average consumption rate is 647 liters per capita per day (Lpcd) and use the flow ratios given in Table 6.4 for estimating the following flow rates:

a. Average daily demand (m^3/d).
b. Peak daily demand (m^3/d).
c. Peak hourly demand (m^3/d).
d. Fire demand (m^3/d).
e. Coincident demand (m^3/d).

2 Using the information in Problem 6.1, determine the design capacities for the following:

a. Source (m^3/d).
b. Low-lift pumps (m^3/d).
c. Treatment plant (m^3/d).
d. High-service pumps (m^3/d).
e. Distribution system (m^3/d).

3 A mechanical bar rack has bars that are 20 mm in diameter, with clear openings of 25 mm. If the velocity of approach is 0.5 m/s, determine the head loss through the rack, assuming that the rack is clean.

4 A stripping tower will be designed to reduce the toluene (C_7H_8) concentration from a contaminated well from 100 μg/L to 5 μg/L. The water flow rate is 3,000 m^3/d, the mass transfer coefficient (K_La) for toluene is 0.0206 s^{-1}, and the dimensionless Henry's constant for toluene is 0.268. Assume a stripping factor of 2 and packing factor of 50, and a pressure drop of 200 N/(m^2·m). The air and water temperature are 20°C, resulting in $\rho_l = 998.2$ kg/m^3, $\rho_g =$ 1.204 kg/m^3, and $\mu_l = 1.002 \times 10^{-3}$ N·s/m^2. Determine the following:
a. Theoretical air flow rate (m^3/min).
b. Diameter of the stripping tower (m).
c. Height of the stripping tower (m).

5 An air stripping tower is to be designed to lower the ethylbenzene (C_8H_{10}) concentration in a contaminated groundwater from 5 mg/L to 30 μg/L. The water flow rate is 700 m^3/d, the mass transfer coefficient (K_La) for ethylbenzene is 0.015 s^{-1}, and the dimensionless Henry's constant for toluene is 0.27. Assume a stripping factor of 3 and packing factor of 40 and a pressure drop of 100 N/(m^2·m). The air and water temperature are 20°C, resulting in $\rho_l = 998.2$ kg/m^3, $\rho_g = 1.204$ kg/m^3, and $\mu_l = 1.002 \times 10^{-3}$ N·s/m^2. Determine the following:
a. Theoretical air flow rate (m^3/min).
b. Diameter of the stripping tower (m).
c. Height of the stripping tower (m).

6 A surface water containing 100 mg/L of natural alkalinity as $CaCO_3$ is coagulated with a ferric chloride dose of 150 mg/L. Determine if there is sufficient alkalinity present in the source water to prevent inhibition of ferric chloride coagulation. If there is insufficient alkalinity present, calculate the amount of sodium hydroxide necessary to react with the ferric chloride.

7 Determine the dimensions of a rapid-mixing basin to process 12,000 m^3/d of water. The basin will be square and the depth is assumed to be 1.5 times the width. Use a detention time of 30 s and velocity gradient of 700 s^{-1}. The average minimum sustained temperature is 15°C and the absolute viscosity is 1.139×10^{-3} kg/(m·s). Two rapid-mixing basins will be constructed, with each having the capacity to process the entire flow. Determine the following:
a. Dimensions of L, W, and D (m).
b. Power input (W).

8 A flocculator that is 100 ft long, 50 ft wide, and 20 ft deep houses four, horizontal paddle-wheels. Each paddle-wheel has four symmetrical arms with fiber glass boards, 6.5 ft from the shaft to the center of the board rotating at 1.5 rpm. Paddles are 42 ft long and 8 inches wide. The water treatment plant processes 25 MGD of water. Assume that the mean water velocity is 25% of the paddle velocity and C_D equals 1.8, and μ and ρ are equal to 2.735×10^{-5} lb·s/ft^4 and 1.936 lb·s/ft^2, respectively. Determine the following:
a. Velocity of paddle blades (ft/s).
b. Total power input (hp).
c. Velocity gradient (s^{-1}).
d. Detention time (min).
e. $G\tau$ (dimensionless).

9 Design a vertical turbine flocculator to treat 75,700 m^3/d of water per day at a detention time of 30 minutes. Use three parallel treatment trains with four compartments per train. The temperature of the water is 20°C, resulting in values of 1.002×10^{-3} kg/(m·s) and 998.2 kg/m^3 for μ and ρ, respectively. The impeller diameter (D) to effective tank diameter (T_e) ratio is 0.4. Assume a power number (N_p) of 0.25 for a three pitch blade with camber, and a mean velocity gradient of 70s^{-1}. Determine the following:
a. Dimensions of each compartment assuming they are cubes (m).
b. Impeller diameter (m).
c. Power input per compartment (W).
d. Rotational speed of each turbine (rpm).

10 Calculate the solubility of $Al(OH)_3$ in water at 25°C, given the K_{sp} is 1.0×10^{-31}.

11 Calculate the solubility of $MgCO_3$ in water at 25°C, given the K_{sp} is 4.0×10^{-5}.

12 Calculate the lime, soda ash (if necessary), and carbon dioxide dosages for recarbonation, and solids production for selective calcium removal, given the following water analysis. Develop a bar graph showing the original theoretical species in the water.

Alkalinity (HCO_3^-) = 115 mg/L as $CaCO_3$
CO_2 = 10 mg/L Mg^{2+} = 9.7 mg/L Na^+ = 6.9 mg/L
Ca^{2+} = 70 mg/L SO_4^{2-} = 96 mg/L Cl^- = 10.6 mg/L

13 Calculate the lime and soda ash requirements to achieve the practical limits of hardness removal, given

the following water analysis. Assume excess lime treatment in a two-stage system is being used. The purity of the lime and soda ash are 85% and 90%, respectively. Develop a bar graph showing the original theoretical species in the water.

CO_2 = 22 mg/L HCO_3^- = 185 mg/L as $CaCO_3$
Mg^{2+} = 15.8 Na^+ = 8 mg/L Ca^{2+} = 60 mg/L
SO_4^{2-} = 28.6 mg/L Cl^- = 12.4 mg/L

14 The results of a batch settling column analysis are plotted below. Determine the overall removal efficiency that can be achieved at a detention time of 26 minutes and a depth of 8.0 ft.

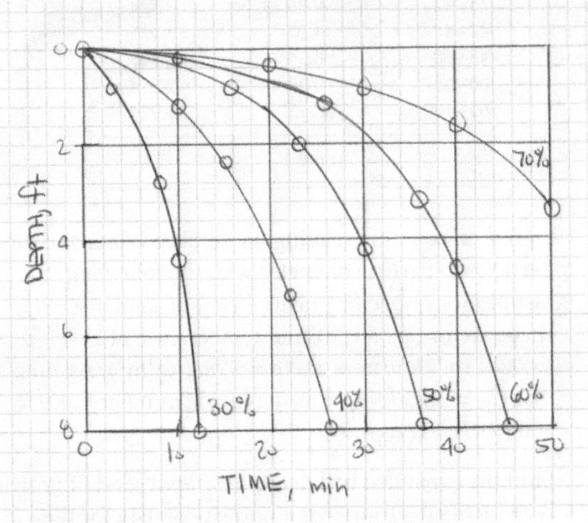

Figure E6.11

15 Design two equally sized rectangular settling basins for treating a total flow rate of 2.5 MGD using an overflow rate of 500 gpd/ft^2 and a detention time of four hours. Use a length : width ratio of 4 : 1 and assume that the basins are operating in parallel and that the effluent weir length in each basin is three times the basin width. Determine:

a. The dimensions of each basin (ft).
b. The weir loading rate (gpd/ft).

16 Four rectangular settling basins operating in parallel are to be sized for treating 37,850 m^3/d of water. Use a length : width ratio of 3 : 1 and assume a detention time of 3.0 hours. The effluent weir length in each basin is equal to three times the tank width. Determine the following:

a. The dimensions of each tank (m).
b. The weir loading rate [m^3/(m·d)].

17 Determine the head loss for a clean filter bed using the Carman-Kozeny equation for a stratified bed with uniform porosity of 0.45. The rapid sand filter is made up of a 610 mm deep bed of sand, specific gravity of 2.65, shape factor (ϕ) = 0.82, temperature of 10°C, and filtration rate of 1.70 liters per second per square meter (Lpsm^2). A sieve analysis is presented below.

Sieve # (1)	% Sand retained (X_{ij}) (2)	d_{ij} (m × 10^{-4}) (3)
14–20	0.87	10.006
20–28	8.63	7.111
28–32	26.30	5.422
32–35	30.10	4.572
35–42	20.64	3.834
42–48	7.09	3.225
48–60	3.19	2.707
60–65	2.16	2.274
65–100	1.02	1.777
	100	

18 Determine the head loss for a clean filter bed using the Rose equation for a stratified bed with uniform porosity of 0.45. The rapid sand filter is made up of a 610 mm deep bed of sand, specific gravity of 2.65, shape factor (ϕ) = 0.82, temperature of 10°C, and filtration rate of 1.70 Lpsm^2. A sieve analysis is presented below.

Sieve # (1)	% Sand retained (X_{ij}) (2)	d_{ij} (m × 10^{-4}) (3)
14–20	0.87	10.006
20–28	8.63	7.111
28–32	26.30	5.422
32–35	30.10	4.572
35–42	20.64	3.834
42–48	7.09	3.225
48–60	3.19	2.707
60–65	2.16	2.274
65–100	1.02	1.777
	100	

19 Determine the head loss for a clean filter bed using the Carman-Kozeny equation for a stratified bed with uniform porosity of 0.45. The rapid sand filter is made up of a 24-in deep bed of sand, specific gravity of 2.65, shape factor (ϕ) = 0.82, temperature of 50°F, and filtration rate of 2.5 gpm/ft^2. A sieve analysis is presented below.

Sieve # (1)	% Sand retained (X_{ij}) (2)	d_{ij} (ft × 10^{-4}) (3)
14–20	0.87	32.8300
20–28	8.63	23.33
28–32	26.30	17.79
32–35	30.10	15.000
35–42	20.64	12.58
42–48	7.09	10.58
48–60	3.19	8.88
60–65	2.16	7.46
65–100	1.02	5.83
	100	

20 Determine the head loss for a clean filter bed using the Rose equation for a stratified bed with uniform porosity of 0.45. The rapid sand filter is made up of a 24 in deep bed of sand, specific gravity of 2.65, shape factor (ϕ) = 0.82, temperature of 50°F, and filtration rate of 2.5 gpm/ft^2. A sieve analysis is presented below.

Sieve # (1)	% Sand retained (X_{ij}) (2)	d_{ij} (ft × 10^{-4}) (3)
14–20	0.87	32.8300
20–28	8.63	23.33
28–32	26.30	17.79
32–35	30.10	15.000
35–42	20.64	12.58
42–48	7.09	10.58
48–60	3.19	8.88
60–65	2.16	7.46
65–100	1.02	5.83
	100	

21 Determine the head loss for a clean filter bed using the Rose equation for a stratified bed with uniform porosity of 0.40. The rapid sand filter is made up of a 760 mm deep bed of sand, specific gravity of 2.65, shape factor (ϕ) = 0.80, temperature of 10°C, and filtration rate of 1.70 Lpsm^2. A sieve analysis is presented below.

Sieve # (1)	% Sand retained (X_{ij}) (2)	d_{ij} (m × 10^{-4}) (3)
14–20	0.44	10.006
20–28	14.33	7.111
28–32	43.22	5.422
32–35	27.07	4.572
35–42	9.76	3.834
42–48	4.22	3.225
48–60	0.54	2.707
60–65	0.29	2.274
65–100	0.13	1.777
	100	

22 Determine the head loss for a clean filter bed using the Carman-Kozeny equation for a stratified bed with uniform porosity of 0.40. The rapid sand filter is made up of a 760 mm deep bed of sand, specific gravity of 2.65, shape factor (ϕ) = 0.80, temperature of 10°C, and filtration rate of 1.70 liters per second per square meter (Lpsm^2). A sieve analysis is presented below.

Sieve # (1)	% Sand retained (X_{ij}) (2)	d_{ij} (m × 10^{-4}) (3)
14–20	0.44	10.0060
20–28	14.33	7.111
28–32	43.22	5.422
32–35	27.07	4.572
35–42	9.76	3.834
42–48	4.22	3.225
48–60	0.54	2.707
60–65	0.29	2.274
65–100	0.13	1.777
	100	

23 Derive an equation relating the settling velocity of sand particles to particle diameter with the following assumptions: specific gravity (S.G.) of sand = 2.65, shape factor (ϕ) = 0.90, and temperature of 20°C (μ = 1.002×10^{-3} kg/(m·s) and ρ = 998.2 kg/m^3). Assume that $C_D = 18.5/(N_R)^{0.6}$ for Reynolds number = 1.9 to 500 (Rich, 1971, p. 86).

24 Find the expanded bed depth in a rapid sand filter that is 28 inches deep and has a porosity of 0.45 for the sieve analysis given in the table below. Assume a water temperature of 68°F and a specific gravity (S.G.) for sand of 2.65. The shape factor (ϕ) is 0.90, μ = 2.11×10^{-5} lb·s/ft^2, and ρ = 1.932 lb·s^2/ft^4.

Sieve # (1)	% Sand retained (X_{ij}) (2)	d_{ij} (ft × 10^{-4}) (3)
14–20	1.05	3.2883
20–28	6.65	2.333
28–32	15.7	1.779
32–35	18.84	1.5
35–42	18.98	1.258
42–48	17.72	1.058
48–60	14.25	0.888
60–65	5.15	0.746
65–100	1.66	0.583
	100	

25 Find the expanded bed depth in a rapid sand filter that is 0.710 m deep and has a porosity of 0.45 for the sieve analysis given below. Assume a water temperature of 20°C and a specific gravity (SG) for sand of 2.65. The shape factor (ϕ) is 0.90, $\nu = 1.003 \times 10^{-6}$ m^2/s, ρ = 998.2 kg/m^3, and $\mu = 1.002 \times 10^{-3}$ kg/(m·s).

Sieve # (1)	% Sand retained (X_{ij}) (2)	d_{ij} ($m \times 10^{-4}$) (3)
14–20	1.05	10.006
20–28	6.65	7.111
28–32	15.7	5.422
32–35	18.84	4.572
35–42	18.98	3.834
42–48	17.72	3.225
48–60	14.25	2.707
60–65	5.15	2.274
65–100	1.66	1.777
	100	

26 A water treatment plant is designed for a maximum daily flow rate of 11.5 MGD at a filtration rate of 5 gpm/ft^2. The filters are backwashed once every 48 hours at a backwash rate of 20 gpm/ft^2 for approximately for 15 minutes. Determine the following:

a. The number of rapid sand filters required.
b. The dimensions if square filters are used.
c. The quantity of water required for backwashing.

27 Design an RO WTP to meet the following criteria. Influent feed TDS = 3000 mg/L. Finished water produced flow rate = 1.5 MGD. TDS in treated water = 300 mg/L. Temperature of the feed = 20°C. Water recovery factor (r) = 90% and the salt rejection = 95%. Design pressure = 600 psi and water flux rate = 15 gpd/ft^2. Packing density = 250 ft^2/ft^3. Calculate the following parameters:

a. TDS in the permeate (mg/L).
b. TDS in the concentrate stream (mg/L) assuming split treatment.
c. Membrane area (ft^2) using split treatment.
d. Total module volume (ft^3).
e. Total number of modules.
f. Number of pressure vessels.
g. Water power (hp), brake power (hp) and motor power (hp).

28 Given the following data, determine the water flux rate coefficient (K_W) in units of s/ft and the solute mass transfer coefficient (K_i) in units of ft/s.

Conversion factor: 1 psi = 6.8948 kPa
Feed flow rate = 1.0 MGD
Feed TDS concentration = 2,500 mg/L
Permeate TDS concentration = 20 mg/L
Net operating pressure = 406 psi
Membrane area =18,300 ft^2
Recovery (r) = 90%

29 Given the following data, determine the water flux rate coefficient (K_W) in units of s/m and the solute mass transfer coefficient (K_i) in units of m/s.

Feed flow rate = 15,000 m^3/d
Feed TDS concentration = 4,500 g/m^3
Permeate TDS concentration = 50 g/m^3
Net operating pressure = 2,900 kPa
Membrane area =8,500 m^2
Recovery (r) = 87%

30 Determine the required membrane area, rejection rate, and concentration of TDS in the concentrate stream for the following conditions. If possible, reduce the membrane area by blending the feed water with the permeate stream.

Flow rate = 25,000 m^3/d
Influent TDS = 2,200 g/m^3 = 2.2 kg/m^3
Effluent TDS = 400 g/m^3 = 0.4 kg/m^3
Net operating pressure = 2,500 kPa
Recovery (%) = 85
$K_w = 1 \times 10^{-6}$ m/(s·bar)
$K_i = 6 \times 10^{-8}$ m/s

31 Determine the percent actual and percent available chlorine for calcium hyplochlorite.The half reactions for calcium hypochlorite ($Ca(OCl)_2$) and hypochlorite ion (OCl^-) are given below.

$$Ca(OCl)_2 + 2\,H^+ + 4\,e^- \leftrightarrow 2\,Cl^- + CaO + H_2O$$

$$OCl^- + H_2O + 2\,e^{-1} \leftrightarrow Cl^- + 2\,OH^-$$

32 Determine the percent actual and percent available chlorine for sodum chlorite.The half reactions for sodium chlorite (NaOCl) and chlorite ion (ClO_2^-) are given below.

$$NaOCl + H^+ + 2e^- \leftrightarrow Cl^- + NaOH$$

$$ClO_2^- + 2H_2O + 4\,e^{-1} \leftrightarrow Cl^- + 4\,OH^-$$

33 A surface water treatment plant processes 37,850 m^3 of water daily. Estimate the quantity of sludge produced from adding 125 mg/L of alum if the raw water turbidity is 20 NTU and contains 20 mg/L of SS. No other chemicals are added to the water. Determine:

a. The quantity of dry solids produced using the equation recommended by Viessman *et al.* (2009) in both kilograms per day and pounds per day.
b. The quantity of dry solids produced using the equation recommended by Davis (2011) in both kilograms per day and pounds per day.

34 A surface water treatment plant processes 37,850 m^3 of water daily. The raw water contains 25 mg/L of SS and the turbidity is 20 NTU. Calculate the ferric chloride dosage using the equation recommended by Davis (2011) if 12,000 kg/d of iron sludge solids are produced. No other chemicals are added during coagulation.

35 Lime sludge thickens to 5.0% solids in a gravity thickener. Assume that 25% of these solids are volatile and that the specific gravity of the fixed and volatile fractions in the solids is 2.5 and 1.0, respectively. Determine:

a. The specific gravity (S_s) of the dry solids.

b. The specific gravity (S_{s1}) of the sludge.
c. The volume of the thickened sludge, if the mass of alum produced daily is 2000 kg of dry solids.

36 The water content of a solids slurry is reduced from 99% to 96% by gravity thickening. Assume that the dry solids contain 20% organic matter with a specific gravity of 1.0 and 80% inorganic or mineral matter with a specific gravity of 2.0. Lime sludge thickens to 5.0% solids in a gravity thickener. Assume that 25% of these solids are volatile and that the specific gravity of the fixed and volatile fractions in the solids is 2.5 and 1.0, respectively. Determine:
a. The specific gravity (S_s) of the dry solids in the slurry.
b. The specific gravity (S_{s1}) of the 99% solids slurry.
c. The specific gravity (S_{s1}) of the 96% solids slurry.
d. The percent reduction in volume when the water content is reduced from 99% to 96%.

37 A lime softening sludge flow of 100 m^3/d is gravity thickened in two circular tanks with a diameter of 10 m. The solids concentration for the un-thickened and thickened lime sludge is 5 % and 8%, respectively. The suspended solids capture rate is 85%. Determine the following:
a. Solids loading rate ($kg/(m^2{\cdot}d)$).
b. The quantity of thickened lime sludge (m^3/d).

38 A lime softening sludge flow of 30,000 gpd is gravity thickened from 2.0% to 5.0% in a mechanical gravity thickener. The gravity thickener is to be designed using a solids loading rate of 10 $lb/(ft^2{\cdot}d)$ and a solids capture of 85%. Assume that the thickener has a side water depth (SWD) of 10 ft, an overflow rate of 500 gpd/ft^2 will be used in design, and that the specific gravity of all sludge streams is approximately 1. Determine the following:
a. The thickened sludge flow rate (gpd) if 800 mg/L of SS are contained in the supernatant.
b. The diameter (ft) of the gravity thickener considering both the solids loading rate and overflow rate.

39 A DAF processes 250 m^3 of coagulated softening sludge from 1% solids to 4% in a 14-hour period. The solids recovery is 88% and solids loading rate is 50 $kg/(m^2{\cdot}d)$. Determine the following:
a. Surface area (m^2) of the DAF.
b. The quantity of float (m^3) produced in 14 hours.
c. The hydraulic or volumetric loading rate [$m^3/(d \cdot m^2)$]

40 Lime softening sludge is thickened from 0.5% to 4% solids with a solids recovery of 95% using a dissolved air flotation thickener. A polymer dose of 35 mg/L is used to enhance thickening in the 10 ft diameter thickener. The lime sludge feed rate is 40,000 gallons per day. Determine the following:
a. Solids loading rate [$lb/(ft^2{\cdot}d)$].
b. Volume of float solids (ft^3/d).
c. Hydraulic or volumetric loading rate [$ft^3/(ft^2 \cdot d)$].
d. Polymer addition rate (lb polymer/dry ton of solids removed).

41 30,000 liters per day of lime-softening sludge at 5% solids is dewatered on a rotary drum vacuum filter. If the RDVF has a dry solids yield of 1.25 $kg/(m^2{\cdot}h)$, determine the required filter drum area (m^2).

42 A filter press (plate & frame type) is to be selected for dewatering a lime softening sludge that has been thickened to 5.0% solids. The design sludge flow to the filter press is 300 m^3/d. The press will be operated on a cycle time of three hours and the dewatered solids concentration is 25%. Determine the appropriate size filter press required, based on manufacturer's data presented in Table 6.29.

43 Design a sludge dewatering lagoon system for dewatering alum residuals. The alum sludge flow rate is 26,500 gallons per day and the solids concentration is 5%. Assume that the climate is wet and that four lagoons will be operated on a 6 month cycle (three months for filling and three months for drying). Sludge will be applied for 100 days during the fill cycle. Calculate the effective lagoon surface area, based on the alum sludge flow rate and the solids loading rate. Assume that sludge is applied at a depth of 24 in.

44 A softening water treatment plant treats 40,000 m^3/d of groundwater. 1,000 kilograms of lime sludge are produced per 1000 m^3 of water treated. If the lime sludge has a solids concentration of 5% and there are ten sand drying beds, each with a surface area of 10,000 m^2, determine the depth (cm) of lime sludge applied to each bed on an annual basis.

Chapter 7

Design of wastewater treatment systems

Richard O. Mines, Jr.

Learning Objectives

After reading this chapter, you should be able to:

- list the major objectives of domestic wastewater treatment;
- select appropriate unit operations and processes necessary to meet secondary wastewater treatment standards;
- design conventional unit operations and processes for treating domestic wastewater;
- design biological systems using microorganisms to treat wastewater;
- understand the relationship between sludge volume and mass;
- design sludge thickening and dewatering systems;
- design sludge stabilization processes;
- understand the basic biological nutrient removal (BNR) processes used to remove nitrogen and phosphorus from wastewater;
- defend the use of indicator organisms as surrogate parameters for pathogens;
- select appropriate chemicals for disinfecting wastewater effluent.

7.1 Wastewater standards

Designing domestic wastewater treatment plants (WWTPs) requires knowledge of the influent wastewater characteristics and flows in addition to the effluent requirements for proper selection of the treatment system. The treatment scheme that is selected will consist of a series of unit operations and processes. A unit operation is defined as a physical treatment such as screening, grit removal, sedimentation, and sludge dewatering. Any treatment that involves a biological or chemical reaction is called a unit process. Typical unit processes include activated sludge, trickling filters, rotating biological contactors (RBCs), membrane bioreactors (MBRs), chlorination, and phosphorus removal by chemical addition.

Each WWTP has two treatment trains: one for treating the liquid portion of the wastewater; and the other for the residuals or sludge that is generated during treatment. Sludge or residuals that have been properly stabilized at domestic WWTPs is called "biosolids." This chapter presents the design of conventional unit operations and processes for treating domestic wastewater. A brief introduction to biological nutrient removal (BNR) systems will also be discussed in a later section.

7.1.1 Secondary treatment

Figure 7.1 is a schematic of the liquid treatment train for a typical secondary wastewater treatment facility. Secondary wastewater treatment implies that a biological process is used to treat the wastewater.

The primary objectives of secondary treatment are to remove organic matter (as measured by five-day biochemical oxygen demand, BOD_5) and total suspended solids (TSS), and

Environmental Engineering: Principles and Practice, First Edition. Richard O. Mines, Jr.

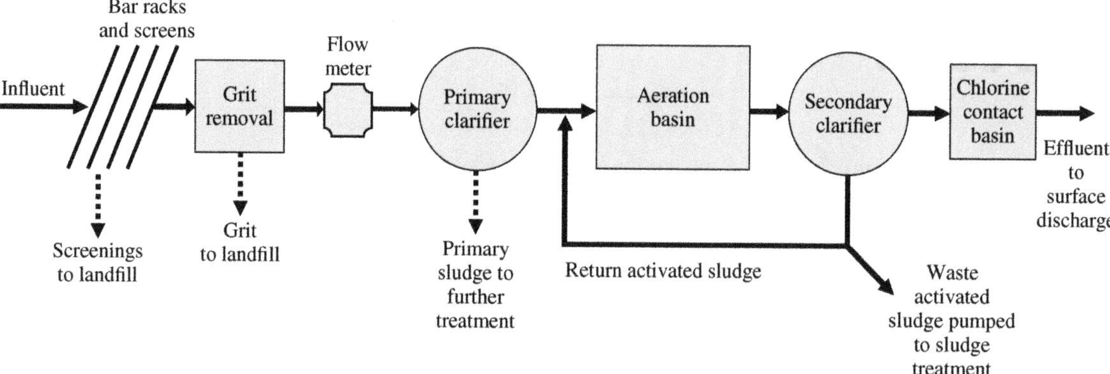

Figure 7.1 Liquid treatment train of a secondary WWTP.

kill pathogens. Most wastewater treatment facilities have parallel treatment trains, so that unit operations and processes may be taken off-line for maintenance. Effluent from the process depicted in Figure 7.1 should be capable of meeting effluent limits of 30 mg/L for both BOD_5 and TSS on an average monthly basis, or 85% removal for each parameter.

As shown in Figure 7.1, the influent or untreated wastewater first passes through bar racks for the removal of large objects and other debris, followed by screens with smaller openings for removing paper, plastics, hair, and rags. The material removed from bar racks and screens is called "screenings." These residuals are usually disposed of in a sanitary landfill. Screening is followed by grit removal, so that inorganic particles such as sand, silt, grit, coffee grinds, and eggshells are removed. Grit is transported to a landfill for disposal.

Flow measurement by a Parshall flume or some other type of flow meter (ultrasonic or magnetic) typically occurs during preliminary treatment. Flow measurement is necessary to meet regulatory permit requirements and for pacing chemical feed systems. If necessary, primary clarification follows preliminary treatment. At average flow, primary clarifiers typically remove 35% of the incoming BOD_5 and 50% of the TSS.

The aeration basin and secondary clarifier comprise the activated sludge process. Here, a heterogeneous culture of microorganisms (primarily bacteria) consumes soluble, colloidal, and particlulate organic matter. Most of these substances are oxidized to carbon dioxide and water to produce energy for the microbes, and a portion of the organics is used in the synthesis of biomass (growth of new microorganisms).

The secondary clarifier separates the biological and inert solids from the wastewater to produce a clarified effluent and a thickened sludge stream (return activated sludge) that is recycled back to the front of the aeration basin. Secondary effluent then enters a chlorine contact basin (CCB), where chlorine is added to kill any remaining pathogens in the wastewater. The chlorine must be removed by some type of dechlorination process before the effluent is discharged to a water body. Sulfur dioxide is often added for dechlorinating the wastewater.

Waste-activated sludge (WAS) produced during biological treatment (activated sludge process) is processed in a sludge treatment train, as depicted in Figure 7.2. A typical sludge processing scheme for WAS would consist of an aerobic digester, followed by thickening with a gravity belt thickener (GBT) prior to land application of the sludge. Aerobic digestion is a stabilization process that reduces the quantity of sludge that must be disposed of, and also kills pathogens.

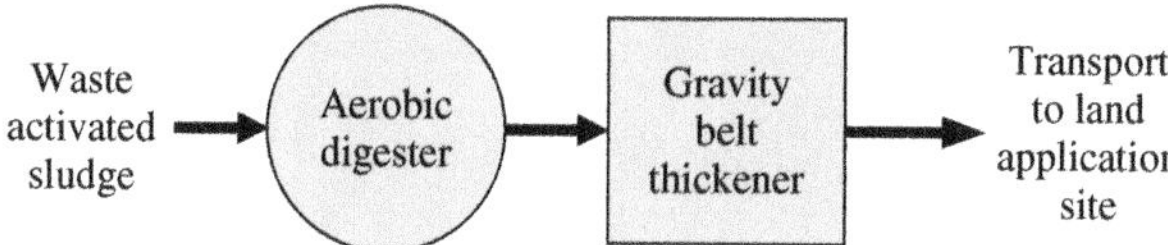

Figure 7.2 Sludge treatment train for secondary WWTP.

Thickening with a GBT reduces the volume of sludge that eventually may be applied to an agricultural site for beneficial reuse of the biosolids.

7.1.2 Advanced wastewater treatment (AWT)

Advanced wastewater treatment (AWT) systems are those used for producing a higher quality effluent than can be accomplished with secondary treatment. AWT facilities are designed to remove nitrogen and phosphorus through biological and/or chemical means. Additional BOD_5 and TSS removals beyond those achieved during secondary treatment are possible. Most AWT facilities include some type of effluent filtration prior to disinfection, which results in lower effluent BOD_5 and TSS concentrations. Effluent from AWT facilities can yield the following concentrations, depending on the specific unit operations and processes used: $BOD_5 \leq 5$ mg/L, TSS ≤ 5 mg/L, total nitrogen (TN) ≤ 3 mg/L, and total phosphorus ≤ 1 mg/L.

7.1.3 Water quality standards

All surface waters in the United States are classified by a system based on water quality standards often called "stream standards." Each state develops its own classification system to specify appropriate water uses to be achieved and protected. Water quality standards serve as the foundation of the water quality-based pollution control program mandated by the

Clean Water Act. A water quality standard consists of four basic elements:

1 Designated use (e.g., recreation, water supply, aquatic life, agriculture).
2 Water quality criteria (numeric pollutant concentrations and narrative requirements).
3 An antidegradation policy (to maintain and protect existing uses and high quality waters).
4 General policies addressing implementation issues (e.g., low flows, variances, mixing zones).

The State of Georgia classifies its water bodies on the following scale:

a) Drinking water supplies.
b) Recreation.
c) Fishing, propagation of fish, shellfish, game and other aquatic life.
d) Wild river.
e) Scenic river.
f) Coastal fishing.

Georgia's water quality standards (*Rules and Regulations for Water Quality Control*, Chapter 391-3-6, 2005) were established on the following basis: "to provide enhancement of water quality and prevention of pollution; to protect the public health or welfare in accordance with the public interest for drinking water supplies, conservation of fish, wildlife and other beneficial aquatic life, and agricultural, industrial, recreational, and other reasonable and necessary uses and to maintain and improve the biological integrity of the waters of the State."

7.1.4 NPDES permits and effluent standards

To ensure that water quality standards are achieved, the US Environmental Protection Agency (EPA) oversees the National Pollution Discharge Elimination System (NPDES). Each wastewater treatment plant that discharges to surface waters must have an NPDES permit which quantifies the level of treatment that must be achieved to meet effluent requirements.

Table 7.1 lists typical effluent parameters and requirements that must be achieved for compliance with NPDES permits issued to secondary WWTPs. Whichever condition is most stringent must be met (i.e., the effluent concentration or percent removal). Percent removal requirements may be waived on a case-by-case basis. The average 30-day percent removal for both five-day biochemical oxygen demand (BOD_5) and total suspended solids (TSS) must be less than 85%. At the option of the NPDES permitting authority, the five-day carbonaceous biochemical oxygen demand ($CBOD_5$) may be substituted in lieu of BOD_5.

NPDES permit requirements for AWT facilities are more stringent, with typical effluent concentrations set at 5 mg/L for BOD_5, 5 mg/L for TSS, 3 mg/L for total nitrogen (TN), and 1 mg/L for total phosphorous (TP). Many of the advanced wastewater treatment facilities operating in Florida must achieve effluent limits of 5, 5, 3, 1 mg/L or less for BOD_5, TSS, TN, and TP, respectively (Thabaraj, 1993).

Table 7.1 Typical secondary effluent NPDES permit requirements.

Parameter	30-day average (% removal)	30-day average (mg/L)	7-day average (mg/L)
BOD_5	85	30	45
$CBOD_5$	85	25	40
TSS	85	30	45
pH	–	6.5–8.5	6.5–8.5

Source: Federal Register (1988) and Federal Register (1989).

7.1.5 Total maximum daily loads (TMDLs)

In some cases, even when the effluent standards from a wastewater treatment facility are met, water quality or stream standards are still not achieved. Under section 303(d) of the Clean Water Act, states, territories, and authorized Indian tribes are required to develop lists of impaired waters. The law requires that these jurisdictions establish priority rankings for these impaired waters and to develop total maximum daily loads (TMDLs). The TMDL rule was promulgated in 2000 (US EPA, 2000). A TMDL is the maximum amount of a pollutant that a water body can receive and still safely meet water quality standards. It represents the sum of the waste-load allocation from individual wastewater treatment plants (point sources of pollution), waste-load allocation from non-point sources (agricultural and stormwater runoff), natural background levels in the water body, and a safety margin (US EPA, 2000).

7.2 Design flows and loadings

The design of a WWTP is based upon volumetric flow rate (Q) and mass loading rate ($\dot{m}$) at the end of the design period. Where existing systems are in place, influent flow rates and wastewater characteristics can be monitored so that probability plots and/or regression analysis can be performed to estimate design flows and loadings. For new treatment facilities, population data and projections must be made. Wastewater characteristics and wastewater generation rates are then assumed for estimating design flows and loadings. Population data may be obtained from the US Census Bureau (www.census.gov). Other sources to consider include local census bureaus, utility companies, chamber of commerce, and planning commissions. Several states have population data and population projects, i.e., State of Georgia (http://www.opb.state.ga.us/media/3016/georgia_population_projections_reduced_web_5_25_05.pdf).

The annual average daily flow (ADF), as discussed below, may be estimated by multiplying 120 gallons of wastewater generated per person per day (460 L/capita · d) by the estimated design population at the end of the design period (Viessman & Hammer, 2005, page 542). This value includes nominal infiltration into the sewerage system. Infiltration is primarily groundwater that enters the wastewater collection system due to defective pipes and joints. Metcalf & Eddy (2003, page 165) state that infiltration may range from 100 to 10,000 gallons per day per in · mi $\left(0.01 \text{ to } 1.0 \frac{\text{m}^3}{\text{d·mm·km}}\right)$.

Maximum month, peak hour, and minimum design flows can then be determined by assuming appropriate peaking factors. The following sources provide detailed information regarding design flows: Reynolds & Richards (1996), Viessman & Hammer (2005), and Metcalf & Eddy (2003).

7.2.1 Design flows and capacities

A WWTP must be designed to accommodate a range of flows to ensure that proper treatment is achieved at all times. The most critical design flows to be considered are peak hour flow (PHF) and minimum hour flow (MHF). The design capacity of a wastewater treatment plant is based either on the annual average daily flow (ADF) or maximum month flow (MMF). Definitions of various design flows and their importance and impact on design are presented below.

7.2.1.1 Average Daily Flow (ADF)

The annual average daily flow denoted as (ADF) is computed as the total flow during a year divided by 365. Essentially, it is the average flow rate occurring in a 24-hour period based on annual flow data. If the ADF is calculated for some period other than a calendar year, the period that was used should be specified. ADF should be considered when designing unit processes. However, maximum month flow (MMF) is usually what governs their design.

The ADF is used for estimating chemical quantities. Sludge production at ADF is often used for determining the ultimate disposal requirements for sludge. Pumping stations in the collection system and at the WWTP should be designed such that an effective and efficient combination of pumps can handle this flow.

The average dry weather flow (ADWF) is similar to ADF; however, it is based on sustained periods when low rainfall and limited infiltration occurs. Similarly, the average wet weather flow (AWWF) is based on sustained periods when high rainfall and infiltration occur.

7.2.1.2 Maximum Month Flow (MMF)

The maximum month flow (MMF) represents the highest monthly average flow for a given reporting period (it may also be calculated as a 30-day moving average rather than on a calendar month basis). For instance, if flow records are for a calendar year, the month with the highest average daily flow would be designated as the maximum month flow. NPDES permit requirements usually stipulate that wastewater treatment facilities report the MMF on a calendar month basis and, typically, this represents the design plant capacity in the permit.

Unit processes are typically designed to treat maximum month flows. Maximum month wastewater flows should be considered in sizing most liquid treatment unit processes, including primary clarifiers, biological processes (activated sludge, trickling filters, rotating biological contactors, biological nutrient removal processes), and secondary clarifiers. It is recommended that the design of sludge thickening, stabilization, and dewatering processes be based on the maximum month flow of sludge that is generated.

7.2.1.3 Peak Hour Flow (PHF)

The highest short-term flow that is recorded or expected in a one-hour period is called the peak hour flow (PHF). From a design viewpoint, the WWTP must be able to handle this flow without having to bypass unit operations or processes. This also means that the wastewater should not overflow any of the basins or processes. Wastewater pumping stations on-site and within the collection system must be designed to handle the PHF with the largest pump out of service. The peak hour flow is used for sizing pumps, pipes, channels, and conduits. In addition to these, all unit operations (physical treatment units), such as screens, grit removal systems, sedimentation tanks, effluent launders, filters, and chlorine contact basins, are designed to treat the PHF.

7.2.1.4 Other Peak Flows

Two other important peak flows to consider are peak week flow (PWF) and peak day flow (PDF). Peak week flow is the highest seven-day average flow for a specified reporting period. It is generally calculated as a seven-day moving average. The highest daily flow in a 24-hour period is called the peak day flow. Sludge pumping, thickening, and stabilization facilities are typically based on sludge production at the PWF. Sometimes, the PDF is used for sizing the hydraulic capacity of structures. However, in most instances, the peak hour flow is used.

7.2.1.5 Minimum Daily Flow (MDF)

Minimum flows must be considered when designing WWTPs, because they can have a significant impact on plant operations during the initial start-up of a facility or when minimum flow occurs. Solids deposition in channels and pipes may occur at low flows, when velocities are less than 2 feet per second (Viessman *et al.* 2009, page 193). Pumping stations, aeration systems, and flow control devices are particularly vulnerable at low flows, which can lead to an inability to operate properly and result in reduced removal efficiencies. The minimum daily flow (MDF) is the average lowest flow in a 24-hour period. Often, the minimum hour flow (MHF) is used, since it is used to size the turndown capacity of pumps and flow meters. It represents the average lowest flow in a one-hour period.

7.2.2 Peaking factors

Peaking factors (PF) are used for comparing various flows, concentrations, and mass loadings to those at average conditions. A peaking factor is the ratio of the peak or minimum condition to the average condition. Commonly used peaking factors for flow include maximum month, peak hour, peak day, minimum day, and minimum hour. Peaking factors

Table 7.2 Flow ratios for residential wastewater flows.

Description of flow	Ratio to average
Peak day	2.0 : 1
Peak hour	3 : 1
Minimum day	0.67 : 1
Minimum hour	0.33 : 1

Ratio values from Fair, Geyer, and Okun (1968a), Vol. 1, pp. 5–24.

for mass loadings are often used too. These include maximum month, peak week, peak day, and minimum month.

Peaking factors for both flows and mass loadings should be developed from historical flow and loading data when possible. When this is not possible, PF values from the literature must be used. Peaking factors for flows and loads should not be combined. One should assume that peak concentrations will not occur when a peak flow condition exists. Typically, peak flows result from infiltration and inflow into the sewerage system resulting in a low concentration condition.

Table 7.2 lists typical peaking factors for residential wastewater flows. The WEF Manual of Practice #8 (1998a) states that the ratio of the maximum month flow to the average daily flow may range from 0.9 to 1.2, whereas the minimum month flow to average daily flow may range from 0.9 to 1.1. Peaking factors for maximum month concentration to average concentration and minimum month concentration to average concentration for both BOD and TSS are provided in Table 7.3.

7.2.3 Design parameters and loadings

It is imperative to know the influent wastewater characteristics, so that correct unit operations and processes can be selected and designed to meet effluent requirements.

When historical data are available, average and maximum concentrations for BOD_5, TSS, total Kjeldahl nitrogen (TKN), ammonia nitrogen (NH_3-N)) and total phosphorus (TP) should be estimated, in addition to the average, minimum, and maximum influent pH values. Average, minimum, and maximum ambient air temperatures should be estimated from weather data for the proposed WWTP site. Climatic data are available from the National Oceanic and Atmospheric Administration (www.noaa.gov). When historical wastewater influent data are not available, wastewater characteristics from the literature must be used. Table 7.4 presents typical influent characteristics for untreated domestic wastewater.

Once influent characteristics and flows have been established, design loadings for major influent parameters are developed. Wastewater loadings ($\dot{m}$) refer to the mass concentration of a specific constituent multiplied by the

Table 7.3 Peaking factors for maximum and minimum month BOD and TSS concentrations.

Parameter	$\frac{\text{maximum month concentration}}{\text{average concentration}}$	$\frac{\text{minimum month concentration}}{\text{average concentration}}$
BOD	1.1 : 1	0.9 : 1
TSS	1.1 : 1 to 1.3 : 1	0.9 : 1

Values from WEF *Manual of Practice* (1998a), Vol. 1, pp. 3–19.

Table 7.4 Typical influent domestic wastewater characteristics and concentrations.

Parameter	Units	US[a]	US[b]	Manchester, UK[c]	Nairobi, Kenya[c]
Biochemical oxygen demand (BOD)	mg/L	200	190	240	520
Chemical oxygen demand (COD)	mg/L	–	430	520	1120
Total solids (TS)	mg/L	800	720	–	–
Suspended solids (TSS)	mg/L	240	210	210	520
Total nitrogen (as N)	mg/L	35	40	–	–
Free ammonia (as N)	mg/L	–	25	22	33
Total phosphorus (as P)	mg/L	10	7	–	–
Soluble phosphorus (as P)	mg/L	7	–	–	–
pH	standard	–	–	7.4	7.0

[a] Values from Hammer (1986), p. 324.
[b] Values from Metcalf and Eddy (2003), p. 186 medium strength.
[c] Values from Horan (1990) p. 27.

Table 7.5 Peaking factors for wastewater loading rates.

Parameter	1-d Sustained Peak / Average	1-d Sustained Minimum / Average	7-d Sustained Peak / Average	7-d Sustained Minimum / Average	30-d Sustained Peak / Average	30-d Sustained Minimum / Average
BOD	2.6	0.12	1.6	0.59	1.3	0.82
TSS	2.8	0.14	1.6	0.71	1.3	0.71
TKN	2.3	0.18	1.2	0.88	1.3	0.94
NH_3	1.7	0.18	1.2	0.82	1.3	0.88
TP	1.9	0.18	1.2	0.82	1.3	0.88

Adapted from Metcalf & Eddy (2003), p. 195.

volumetric flow rate. Equations (7.1) and (7.2) are used to calculate mass loading rates in English and S.I. units, respectively.

$$\dot{m}\left(\frac{\text{lb}}{\text{d}}\right) = Q(\text{MGD}) \times C\left(\frac{\text{mg}}{\text{L}}\right) \times 8.34 \frac{\text{lb}}{\text{MG} \times \text{mg/L}} \quad (7.1)$$

$$\dot{m}\left(\frac{\text{kg}}{\text{d}}\right) = Q\left(\frac{\text{m}^3}{\text{d}}\right) \times C\left(\frac{\text{mg}}{\text{L}}\right) \times \left(\frac{1\ \text{kg}}{10^6\ \text{mg}}\right) \times \left(\frac{1000\ \text{L}}{\text{m}^3}\right) \quad (7.2)$$

where:

Q = volumetric flow rate, MGD $\left(\frac{\text{m}^3}{\text{d}}\right)$

C = concentration of constituent (BOD, TSS, TKN), mg/L

$\dot{m}$ = mass loading rate of constituent, $\frac{\text{lb}}{\text{d}}\left(\frac{\text{kg}}{\text{d}}\right)$.

When mass loading peaking factors cannot be determined from historical records, peaking factors must be taken from the literature. Table 7.5 is a list of mass loading peaking factors that can be used for estimating peak and minimum mass loading rates for BOD, TSS, total Kjeldahl nitrogen (TKN), ammonia (NH_3), and TP. The data in Table 7.5 have been adapted from data presented in Metcalf & Eddy (2003, page 195).

Example 7.1 illustrates how to calculate design flows and mass loading rates using values from the literature.

Example 7.1 Calculating design flows and mass loadings using literature values

A new residential development in Blacksburg, Virginia is expected to have a build-out population of 5000 people by the year 2020. Assume a domestic wastewater with characteristics from column 1 in Table 7.4 for the United States. Using the values in Table 7.2, estimate the following for the year 2020:

a) Average daily flow (m^3/d).
b) Peak hour flow (m^3/d).
c) Minimum hour flow (m^3/d).
d) Average daily BOD_5 mass loading (kg/d).
e) Average daily TSS mass loading (kg/d).
f) Average daily TKN mass loading (kg/d).

Solution Part A

The average daily flow in 2020 is estimated by multiplying the build-out population of 5000 people by the per capita wastewater generation rate of 460 liters per capita per day.

$$\text{ADF} = 5000\ \text{people} \times \frac{460\ \text{L}}{\text{capita} \times \text{d}} \times \frac{1\ \text{m}^3}{10^3\ \text{L}}$$

$$= \boxed{2.3 \times 10^3 \frac{\text{m}^3}{\text{d}}}$$

Solution Part B

From Table 7.2, assume a PHF : ADF peaking factor of 3 : 1. Calculate the peak hour flow for 2020 by multiplying ADF by 3 : 1 as follows:

$$\text{PHF} = \text{ADF} \times \text{PF}$$

$$\text{PHF} = \text{ADF} \times \frac{\text{PHF}}{\text{ADF}} = 2.3 \times 10^3 \frac{\text{m}^3}{\text{d}} \times \frac{3}{1}$$

$$= \boxed{6.9 \times 10^3 \frac{\text{m}^3}{\text{d}}}$$

Solution Part C

The minimum hour flow (MHF) is calculated similarly. From Table 7.2, assume a MHF : ADF peaking factor of 0.33 : 1. Calculate the minimum hour flow for 2020 by multiplying ADF by 0.33 : 1 as follows:

$$\text{MHF} = \text{ADF} \times \text{PF}$$

$$\text{MHF} = \text{ADF} \times \frac{\text{MHF}}{\text{ADF}} = 2.3 \times 10^3 \frac{\text{m}^3}{\text{d}} \times \frac{0.33}{1}$$

$$= \boxed{7.59 \times 10^2 \frac{\text{m}^3}{\text{d}}}$$

Solution Part D

Average daily mass loading rates are calculated using Equation (7.2). The average daily BOD_5 mass loading rate is calculated by multiplying the ADF by the assumed BOD_5 concentration in Table 7.4:

$$\dot{m}\left(\frac{kg}{d}\right) = Q\left(\frac{m^3}{d}\right) \times C\left(\frac{mg}{L}\right) \times \left(\frac{1\ kg}{10^6\ mg}\right) \times \left(\frac{1000\ L}{m^3}\right)$$

$$\dot{m}\left(\frac{kg}{d}\right) = 2.3 \times 10^3 \frac{m^3}{d} \times 200 \frac{mg}{L} \times \left(\frac{1\ kg}{10^6\ mg}\right) \times \left(\frac{1000\ L}{m^3}\right)$$

$$= \boxed{4.60 \times 10^2\ \frac{kg\ BOD_5}{d}}$$

Solution Part E

Using Equation (7.2) and the assumed TSS concentration in Table 7.4, the average daily TSS mass loading is calculated as follows:

$$\dot{m}\left(\frac{kg}{d}\right) = Q\left(\frac{m^3}{d}\right) \times C\left(\frac{mg}{L}\right) \times \left(\frac{1\ kg}{10^6\ mg}\right) \times \left(\frac{1000\ L}{m^3}\right)$$

$$\dot{m}\left(\frac{kg}{d}\right) = 2.3 \times 10^3 \frac{m^3}{d} \times 240 \frac{mg}{L} \times \left(\frac{1\ kg}{10^6\ mg}\right) \times \left(\frac{1000\ L}{m^3}\right)$$

$$= \boxed{5.52 \times 10^2\ \frac{kg\ TSS}{d}}$$

Solution Part F

Using Equation (7.2) and the assumed TN concentration in Table 7.4 of 35 mg/L, the average daily TN mass loading is calculated as follows:

$$\dot{m}\left(\frac{kg}{d}\right) = Q\left(\frac{m^3}{d}\right) \times C\left(\frac{mg}{L}\right) \times \left(\frac{1\ kg}{10^6\ mg}\right) \times \left(\frac{1000\ L}{m^3}\right)$$

$$\dot{m}\left(\frac{kg}{d}\right) = 2.3 \times 10^3 \frac{m^3}{d} \times 35 \frac{mg}{L} \times \left(\frac{1\ kg}{10^6\ mg}\right) \times \left(\frac{1000\ L}{m^3}\right)$$

$$= \boxed{8.05 \times 10^1\ \frac{kg\ TKN}{d}}$$

Example 7.2 demonstrates the procedure used for estimating design flows and mass loadings when historical data are available.

Example 7.2 Calculating design flows and mass loadings from historical data

Tabulated below are the average monthly influent values from an actual full-scale operating WWTP. Estimate the following:

a) Annual average daily, maximum month daily, and minimum month daily flows in MGD.
b) Peaking factors for maximum month and minimum month daily flows.
c) Average influent BOD, maximum month BOD, and minimum month BOD concentrations in mg/L.
d) Peaking factors for maximum month and minimum month BOD concentrations.
e) Average influent TSS, maximum month TSS, and minimum month TSS concentrations in mg/L.
f) Peaking factors for maximum month and minimum month TSS concentrations.
g) Average influent BOD, maximum month BOD, and minimum month BOD mass loadings in lb/d.
h) Peaking factors for maximum month and minimum month BOD mass loadings.
i) Average influent TSS, maximum month TSS, and minimum month TSS mass loadings in lb/d.
j) Peaking factors for maximum month and minimum month TSS mass loadings.

Date	Influent flow, MGD (1)	Influent BOD concentration, mg/L (2)	Influent TSS concentration, mg/L (3)
Aug 08	3.83	182	244
Sep 08	4.06	158	166
Oct 08	4.50	161	173
Nov 08	4.03	172	193
Dec 08	4.03	187	212
Jan 09	4.62	191	223
Feb 09	4.71	152	172
Mar 09	5.03	166	203
Apr 00	4.78	136	150
May 09	3.66	160	205
Jun 09	3.95	130	197
Jul 09	3.97	144	183

Solution

For this type of problem it is best to use some type of spreadsheet to perform the calculations. The following table shows the results of the spreadsheet analysis. At the bottom of the table the peaking factors for flows, concentrations, and mass loadings are listed. Sample calculations are provided beneath the table. Although the complete data set was not provided in the table above, the bottom row is provided to show the yearly averages (based on 365 individual values for each of the five categories presented).

Date	Influent flow, MGD (1)	Influent BOD concentration, mg/L (2)	Influent TSS concentration, mg/L (3)	Influent BOD mass loading, lb/d (4)	Influent TSS mass loading, lb/d (5)
Aug 08	3.83	182	**244**	5803	7751
Sep 08	4.06	158	166	5341	**5612**
Oct 08	4.50	161	173	6007	6472
Nov 08	4.03	172	193	5781	6482
Dec 08	4.03	187	212	6292	7154
Jan 09	4.62	**191**	223	**7350**	**8604**
Feb 09	4.71	152	172	5970	6774
Mar 09	**5.03**	166	203	6971	8482
Apr 00	4.78	136	**150**	5397	5953

Date	Influent flow, MGD (1)	Influent BOD concentration, mg/L (2)	Influent TSS concentration, mg/L (3)	Influent BOD mass loading, lb/d (4)	Influent TSS mass loading, lb/d (5)
May 09	**3.66**	160	205	4887	6241
Jun 09	3.95	**130**	197	**4265**	6435
Jul 09	3.97	144	183	4749	6066
Average (A)	**4.26**	**162**	**193**	**5734**	**6835**
Max. month (B)	5.03	191	244	7350	8604
Min. month (C)	3.66	130	150	4265	5612
Max. month PF (D)	1.18	1.18	1.26	1.28	1.26
Min. month PF (E)	0.86	0.80	0.78	0.74	0.82
Yearly average (all data) (F)	4.30	162	194	5734	6844

Solution Part A

The solution to Part A is determined as follows. The annual average daily flow is the arithmetic average or mean of the 12 months of data shown in Column 1 (Column 1, Row A lists the answer). For this data set, the annual average daily flow rate is 4.26 MGD.

$$\text{ADF} = \frac{\left(\begin{array}{c}3.83 + 4.06 + 4.50 + 4.03 + 4.03 + 4.62 \\ + 4.71 + 5.03 + 4.78 + 3.66 + 3.95 + 3.97\end{array}\right)}{12}$$

$$\text{ADF} = \boxed{4.26\ \text{MGD}}$$

The maximum month flow for this data set is the maximum average month flow value, which is 5.03 MGD for March 2009, whereas the minimum month flow is the minimum average month flow, which corresponds to 3.66 MGD for May 2009.

Solution Part B

The peaking factors for flow are calculated as follows. The maximum month flow to average daily flow is determined by dividing the maximum month flow by the average daily flow, as shown below (See Column 1, Row D).

$$PF_{\text{Max month flow}} = \frac{\text{maximum month flow}}{\text{average daily flow}}$$

$$= \frac{5.03\ \text{MGD}}{4.26\ \text{MGD}} = \boxed{1.18}$$

The WEF MOP #8 (1998a) indicates this PF should range from 0.9 to 1.2.

The minimum month flow to average daily flow is determined by dividing the minimum month flow by the average daily flow as shown below (See Column 1, Row E).

$$PF_{\text{Min month flow}} = \frac{\text{minimum month flow}}{\text{average daily flow}}$$

$$= \frac{3.66\ \text{MGD}}{4.26\ \text{MGD}} = \boxed{0.86}$$

The WEF MOP #8 (1998a) indicates this PF should range from 0.9 to 1.1.

Solution Part C

The annual average BOD concentration is the arithmetic average of the 12 months of data shown in Column 2 (Column 2, Row A gives the answer). For this data set, the annual average BOD concentration is 162 mg/L.

$$\text{Average BOD} = \frac{\left(\begin{array}{c}182 + 158 + 161 + 172 \\ +187 + 191 + 152 + 166 \\ +136 + 160 + 130 + 144\end{array}\right)}{12}$$

$$\text{Average BOD} = \boxed{162\ \text{mg/L}}$$

The maximum month BOD concentration for this data set is the maximum average month BOD value, which is 191 mg/L for January 2009, whereas the minimum average month BOD concentration is the minimum average month value, which corresponds to 130 mg/L for June 2009.

Solution Part D

The peaking factors for BOD concentrations are calculated as follows. The maximum month BOD concentration to average BOD concentration is determined by dividing the maximum month BOD concentration by the average BOD concentration, as shown below (See Column 2, Row D).

$$PF_{\text{Max month conc}} = \frac{\text{maximum month BOD concentration}}{\text{average BOD concentration}}$$

$$= \frac{191\ \text{mg/L}}{162\ \text{mg/L}} = \boxed{1.18}$$

Table 7.3 lists a PF of 1.1 for typical domestic WWTPs.

The minimum month BOD concentration to average BOD concentration is determined by dividing the minimum month BOD value by the average BOD value, as shown below (See Column 2, Row E).

$$PF_{\text{Minimum month}} = \frac{\text{minimum month BOD concentration}}{\text{average daily BOD concentration}}$$

$$= \frac{130\ \text{mg/L}}{162\ \text{mg/L}} = \boxed{0.80}$$

This value is slightly less than the range shown in Table 7.3, which is 0.9 to 1.0 for a typical domestic WWTP.

Solution Part E

The annual average TSS concentration is the arithmetic average or mean of the 12 months of data shown in Column 3 (Column 3, Row A gives the answer). For this data set, the annual average TSS concentration is 193 mg/L.

$$\text{Average TSS} = \frac{\begin{pmatrix} 244+166+173+193 \\ +212+223+172+203 \\ +150+205+197+183 \end{pmatrix}}{12}$$

$$\text{Average TSS} = \boxed{193\ \text{mg/L}}$$

The maximum month TSS concentration for this data set is the maximum average month TSS value which is 244 mg/L for August 2008, whereas the minimum average month TSS concentration is the minimum average month value, which corresponds to 150 mg/L for April 2009.

Solution Part F

The peaking factors for TSS concentrations are calculated as follows. The ratio of the maximum month TSS concentration to average TSS concentration is as shown below (See Column 3, Row D).

$$PF_{\text{Max month conc}} = \frac{\text{maximum month TSS concentration}}{\text{average TSS concentration}} = \frac{244\ \text{mg/L}}{193\ \text{mg/L}} = \boxed{1.26}$$

This values lies between 1.0 and 1.3, as listed in Table 7.3 for a typical domestic WWTP.

The ratio of the minimum month TSS concentration to average TSS concentration is determined as follows (See Column 3, Row E):

$$PF_{\text{Minimum month}} = \frac{\text{minimum month TSS concentration}}{\text{average daily TSS concentration}} = \frac{150\ \text{mg/L}}{193\ \text{mg/L}} = \boxed{0.78}$$

This value is slightly less than the range shown in Table 7.3, which is 0.9 to 1.0.

Solution Part G

The annual average BOD mass loading rate is the arithmetic average of the 12 months of data shown in Column 4 (Column 4, Row A gives the answer). For this data set, the annual average BOD mass loading rate is 5,734 lb/d.

$$\text{Average BOD mass loading} = \frac{\begin{pmatrix} 5803+5341+6007+5781 \\ +6292+7350+5970+6971 \\ +5397+4887+4265+4749 \end{pmatrix}}{12}$$

$$\text{Average BOD mass loading} = \boxed{5734\ \text{lb/d}}$$

The maximum month BOD concentration for this data set is the maximum average month BOD value, which is 7,350 lb/d for January 2009, whereas the minimum average month BOD concentration is the minimum average month value, which corresponds to 4,265 lb/d for June 2009.

Solution Part H

The peaking factors for BOD mass loadings are calculated as follows. The ratio of the maximum month BOD mass loading rate to average BOD mass loading rate is determined as (See Column 4, Row D):

$$PF_{\text{Max month load}} = \frac{\text{maximum month BOD loading}}{\text{average BOD loading}} = \frac{7350\ \text{lb/d}}{5734\ \text{lb/d}} = \boxed{1.28}$$

The ratio of the maximum month BOD mass loading rate to average BOD mass loading rate in Table 7.5 is 1.3.

The ratio of the minimum month BOD mass loading rate to average BOD mass loading rate is shown below (See Column 4, Row E):

$$PF_{\text{Minimum month load}} = \frac{\text{minimum month BOD loading}}{\text{average BOD loading}} = \frac{4265\ \text{lb/d}}{5734\ \text{lb/d}} = \boxed{0.74}$$

The ratio of the minimum month BOD mass loading rate to average BOD mass loading rate in Table 7.5 is 0.82, which is slightly greater than our observed ratio.

Solution Part I

The annual average TSS mass loading rate is the arithmetic average of the 12 months of data shown in Column 5 (Column 5, Row A gives the answer). For this data set, the annual average TSS mass loading rate is 6,835 lb/d.

$$\text{Average TSS mass loading} = \frac{\begin{pmatrix} 7751+5612+6472+6482 \\ +7154+8604+6774+8482 \\ +5953+6241+6435+6066 \end{pmatrix}}{12}$$

$$\text{Average TSS mass loading} = \boxed{6835\ \text{lb/d}}$$

The maximum month TSS mass loading rate for this data set is the maximum average month TSS mass loading value, which is 8,604 lb/d for January 2009, whereas the minimum average month TSS mass loading rate is the minimum average month TSS mass loading value, which corresponds to 5,612 lb/d for September 2008.

Solution Part J

The peaking factors for TSS mass loadings are calculated as follows. The ratio of the maximum month TSS mass loading rate to average TSS mass loading rate is shown below (See Column 5, Row D):

$$PF_{\text{Max month load}} = \frac{\text{maximum month TSS loading}}{\text{average TSS loading}}$$

$$= \frac{8604\ \text{lb/d}}{6835\ \text{lb/d}} = \boxed{1.26}$$

The ratio of the maximum month TSS mass loading rate to average TSS mass loading rate in Table 7.5 is 1.3.

The ratio of the minimum month TSS mass loading rate to average TSS mass loading rate is calculated as follows (See Column 5, Row E):

$$PF_{\text{Minimum month load}} = \frac{\text{minimum month TSS loading}}{\text{average TSS loading}}$$

$$= \frac{5612\ \text{lb/d}}{6835\ \text{lb/d}} = \boxed{0.82}$$

The ratio of the minimum month TSS mass loading rate to average TSS mass loading rate in Table 7.5 is 0.71, which is slightly less than our calculated value of 0.82.

7.3 Preliminary treatment

Untreated wastewater or sewage that enters a WWTP is usually treated in a number of preliminary steps before subsequent treatment in major unit operations and processes. Preliminary treatment of wastewater focuses primarily on physical treatment schemes involving screening, grit removal, flow measurement, pumping, flow equalization, and pre-aeration. The purpose of preliminary treatment is to remove large debris, paper, plastics, hair, grit, eggshells, coffee grinds, and sand. This is necessary to prevent accumulation of these materials in aeration basins, settling tanks, and digesters. Grit is abrasive to pump impellers, pipes, and process equipment. Rags and hair can be of particular concern, since they can clog pumps, flow meters, and valves.

In many instances, chemicals such as chlorine or ozone are added to pre-disinfect and oxidize inorganic compounds such as hydrogen sulfide. Other chemical species, used to minimize odor and corrosion problems at WWTPs, include chlorine, ferrous sulfate, oxygen, hydrogen peroxide, and potassium permanganate (US EPA, 1985a, pp. 35–68). These compounds may be added at the headworks of WWTPs, but they are more effective if added at various points throughout the wastewater collection system. Where hydrogen sulfide is a concern, all equipment and components should be made of stainless steel or other material resistant to corrosion.

Detailed design criteria for preliminary treatment systems can be found in Metcalf & Eddy (2003) and WEF Manual of Practice #8 (1998b). Only screening and grit removal design will be addressed in the following sections.

7.3.1 Screens

Screening of wastewater is normally the first unit operation encountered in wastewater treatment. Often, mechanically or manually cleaned bar racks or trash racks are used first, followed by bar screens with smaller openings. Bar racks have clear openings ranging from 1.5–6 inches (38–150 mm). Bar racks are used for removing large objects such as logs, tires, or other debris from entering the WWTP.

Mechanically cleaned coarse screens with openings from 1–2 inches (25–50 mm) are used for removing rags, paper, and other debris. Bar racks and bar screens are made up of parallel bars or rods while fine screens consisting of wires, grating, wire mesh, or perforated plates. Figure 7.3 is a photo of a coarse bar screen.

Table 7.6 presents typical design criteria for coarse bar screens.

Figure 7.3 Photograph of a bar screen.

Table 7.6 Design criteria for bar screens.

Parameter	Manually cleaned	Mechanically cleaned
Acceptable head loss	6 in (150 mm)[a]	6–24 in (150–600 mm)[b]
[b]Bar size depth	1.0–1.5 in (25–38 mm)	1.0–1.5 in (25–38 mm)
[b]Bar size width	0.2–0.6 in (5–15 mm)	0.2–0.6 in (5–15 mm)
[a]Screen openings	0.25–1.0 in (6–30 mm)	0.25–0.5 in (6–13 mm)
[b]Slope from vertical	30–45	0–30
[a]Velocity of approach, V_A	1.3–2.5 ft/s (0.4–0.8 m/s)	3.0 ft/s (0.9 m/s)
[a]Velocity thru openings, V_B	1.2–2.5 ft/s (0.3–0.6 m/s)	2.0–4.0 ft/s (0.6–1.2 m/s)

[a] Values from Veslind (2003), p. 4–12.
[b] Values from Metcalf & Eddy (2003), p. 316.

The head loss through a bar screen can be estimated from the following equation:

$$h_L = \frac{1}{C}\left(\frac{V_{BS}^2 - V_A^2}{2g}\right) \quad (7.3)$$

where:
h_L = head loss through bar screen, ft (m)
C = discharge coefficient, 0.6 for clogged screen and 0.7 for a clean screen
V_{BS} = velocity of flow through bar screen, fps (mps)
V_A = velocity of approach in upstream channel, fps (mps)
g = acceleration due to gravity, 32.2 ft/s^2 (9.81 m/s^2).

Example 7.3 Bar screen design

A mechanical bar screen with 1 inch openings and $\frac{5}{8}$ inch bars is installed in a rectangular channel where the approach velocity should not exceed 2.0 ft per second. Estimate:

a) The velocity between the bars.
b) The head loss through the screen, assuming it is clean.

Solution Part A

Assume that the width and depth of flow in the rectangular channel are W and D, respectively. Estimate the net area of the openings in the bar screen by multiplying the cross-sectional area of the rectangular channel by the ratio of the width of the bar screen openings to the width of opening plus width of the bar as:

$$\text{net area of openings} = WD\left[\frac{1.0}{1.0 + 5/8}\right] = 0.615\,WD$$

From the continuity equation:

$$Q = V_A A_A = V_{BS} A_{BS}$$

where:
Q = volumetric flow rate, L^3/time
A_A = cross-sectional area of approach channel, WD
A_{BS} = net cross-sectional area of openings in bar screen, $0.615WD$.

Calculate the velocity of flow through the bar screen, V_{BS}, as:

$$V_{BS} = \frac{V_A A_A}{A_{BS}} = \frac{2.0\text{ fps}(WD)}{(0.615\,WD)} = \boxed{3.25\text{ fps}}$$

Note that the velocity through the bar screen (3.25 fps) is significantly larger than the velocity of approach (2.0 fps).

Solution Part B

Estimate the head loss through the bar screen using Equation (7.3).

$$h_L = \frac{1}{C}\left(\frac{V_{BS}^2 - V_A^2}{2g}\right) = \frac{1}{0.7}\left(\frac{(3.25\text{ fps})^2 - (2.0\text{ fps})^2}{2 \times 32.2\text{ ft/s}^2}\right)$$
$$= \boxed{0.15\text{ ft}}$$

This is not an appreciable loss of energy or pressure drop. The head loss through mechanically cleaned coarse screens is typically 6 inches (150 mm) of water.

7.3.2 Grit removal

Grit removal is another unit operation that is normally performed during preliminary treatment and follows screening. Grit consists of sand, silt, small gravel, cinders, coffee grounds, egg shells, and other inert materials that typically have a specific gravity around 2.65. These materials are abrasive, cause pump impellers to wear excessively, and accumulate in tanks, digesters, and pipes. The three major types of grit removal systems in use today are aerated chambers, horizontal flow through basins, and vortex removal systems (Metcalf & Eddy, 2003). Horizontal velocity of flow through aerated grit chambers is selected so that all particles retained on a 65 mesh screen (> 0.21 mm diameter particle) will be removed by gravity, whereas vortex-type systems use centrifugal and gravitational forces to separate the grit from the wastewater using proprietary equipment.

Vortex grit removal systems are frequently installed at newly constructed WWTPs. There are two primary manufacturers of such systems. Smith & Loveless makes the PISTA® system and Eutek® makes the Teacup. Figures 7.4 and 7.5 show typical

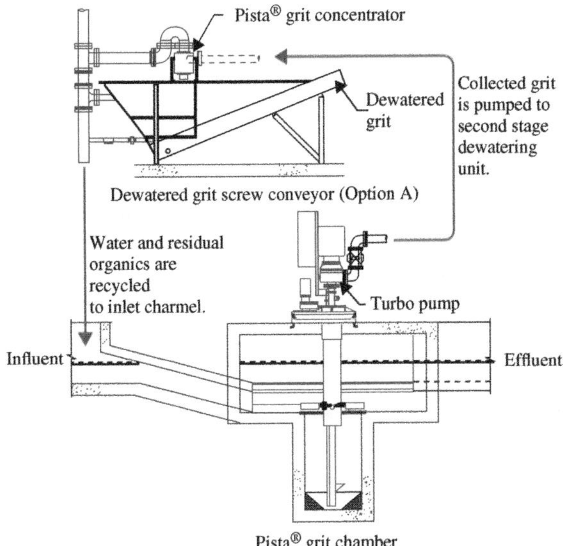

Figure 7.4 Pista Grit removal system. *Source:* Courtesy of Smith & Loveless, Inc.

Figure 7.5 High performance Eutek SlurryCup™ grit washing/sludge degritting unit mounted on a Eutek Grit Snail quiescent dewatering Escalator. *Source*: Courtesy of Hydro International.

Table 7.7 Design criteria for vortex type grit removal chambers.

Parameter	Value
[a]Detention time at peak flow	20–30 s
[a]Influent channel velocity	2–3 ft/s (0.6–0.9 m/s)
[b]Diameter	
Upper chamber	4.0–24.0 ft (1.2–7.2 m)
Lower chamber	3.0–6.0 ft (0.9–1.8 m)
[b]Height	9.0–16.0 ft (2.7–4.8 m)
[a]Head loss	0.25 in (6 mm)
[a]Removal efficiency	Up to 73% of 140 mesh (0.11 mm diameter)

[a] WEF (1998b) Vol 2, pp. 9-29, 9-31.
[b] Metcalf & Eddy (2003), p. 394

vortex removal systems. Table 7.7 lists design criteria for vortex grit removal systems.

Once grit has been separated from the wastewater, it must be washed to remove organic matter that may be attached to the grit particles. Both inclined rake and inclined screw conveyor washer systems are used. In some cases, hydrocyclones are used prior to the grit washer, to enhance separation of the grit and organics. Detailed design information for these systems can be found in Metcalf & Eddy (2003), WEF MOP #8 (1998b), and Reynolds & Richards (1996).

7.3.2.1 Grit Basin Design: Type I or Discrete Particle Settling

Grit basin design is based on Type I settling, i.e., discrete settling. Type I settling involves discrete, non-flocculent particles whose size, shape, and specific gravity do not change with time as they settle. Particles settle as individual entities, and there is no interaction between particles. Examples of discrete settling include grit and sand particles in grit removal systems.

The settling velocity of a particle is derived by equating the gravitational force minus the buoyant force on the particle to the frictional resistance or drag on the particle. There are three forces acting on the particle as it settles: gravitational force (F_g), buoyant force (F_b), and drag force (F_d). The units for force are either pound of force (lb_f) or Newtons (N). As a discrete particle settles, the particle will accelerate until the drag force equals the gravitational force minus the buoyant force at which time a constant settling velocity (V_s) will be achieved.

$$F_g - F_b = F_d \tag{7.4}$$

$$F_g = \rho_p g \forall_p \tag{7.5}$$

$$F_b = \rho g \forall_p \tag{7.6}$$

$$F_d = \frac{C_D A \rho V_s^2}{2} \tag{7.7}$$

For spherical particles, the volume and cross-sectional area are calculated using the following equations:

$$\forall_p = \frac{\pi d^3}{6} \tag{7.8}$$

$$A = \frac{\pi d^2}{4} \tag{7.9}$$

Substituting Equations (7.5) through (7.9) into Equation (7.4) and rearranging, results in Newton's Law:

$$V_s = \left[\frac{4g}{3C_D}\left(\frac{\rho_p - \rho}{\rho}\right)d\right]^{0.5} \tag{7.10}$$

where:
V_S = settling velocity, fps (m/s)
V_p = volume of particle, $ft^3 (m^3)$
g = acceleration of gravity, ft/s^2 (m/s^2)
C_D = coefficient of drag (Dimensionless)
ρ_p = mass density of particle, $lb_m \cdot s^2/ft^4$ (kg/m^3)
ρ = mass density of liquid, $lb_m \cdot s^2/ft^4$ (kg/m^3)
d = diameter of particle, ft (m)
μ = absolute or dynamic viscosity, $lb \cdot s/ft^2$ ($kg/m \cdot s$)

The coefficient of drag (C_D) is a function of the flow regime, which is estimated by calculating the Reynolds number (N_R).

$$N_R = \frac{\rho V_S d}{\mu} = \frac{V_s d}{\nu} \tag{7.11}$$

When the Reynolds number is < 1, laminar flow conditions exist and C_D is calculated using Equation (7.12):

$$C_D = \frac{24}{N_R} \tag{7.12}$$

During transitional flow between laminar and turbulent, $N_R = 1$ to 10^4 and C_D is determined from Equation (7.13):

$$C_D = \frac{24}{N_R} + \frac{3}{(N_R)^{0.5}} + 0.34 \quad (7.13)$$

The coefficient of drag is assumed to be equal to 0.4 for turbulent flow when $N_R > 10^4$.

Stokes' law, which is used in several water, wastewater, and air pollution applications, is derived below. For laminar flow conditions, the coefficient of drag is determined by substituting Equation (7.11) into Equation (7.12), which yields the following:

$$C_D = \frac{24}{N_R} = \frac{24\mu}{\rho V_s d} \quad (7.14)$$

Substituting Equation (7.14) into Equation (7.10) and simplifying results in Equation (7.15).

$$V_s = \left[\frac{4g\rho V_s d}{3 \times 24\mu}\left(\frac{\rho_p - \rho}{\rho}\right) d\right]^{0.5} = \left[\frac{gV_s}{18\mu}\left(\rho_p - \rho\right) d^2\right]^{0.5} \quad (7.15)$$

Squaring both sides of Equation (7.15) results in the following equation:

$$V_s^2 = \left[\frac{gV_s}{18\mu}\left(\rho_p - \rho\right) d^2\right] \quad (7.16)$$

Next, divide both sides of Equation (7.16) by V_s and simplify to yield Equation (7.17), which is one form of Stokes' law:

$$\frac{V_s^2}{V_s} = \left[\frac{gV_s}{18\mu V_s}\left(\rho_p - \rho\right) d^2\right]$$

$$V_s = \frac{g(\rho_p - \rho)d^2}{18\mu} \quad (7.17)$$

From fluids (Houghtalen *et al.*, 2010, pg 6), the kinematic viscosity (ν) is defined as:

$$\nu = \frac{\mu}{\rho} \quad (7.18)$$

The specific gravity of a substance is defined as:

$$S_{sub} = \frac{\rho_{sub}}{\rho} \quad (7.19)$$

where:

S_{sub} = specific gravity of substance (dimensionless)

ρ_{sub} = density of substance, $\frac{lb_m \cdot s^2}{ft^4}\left(\frac{kg}{m^3}\right)$.

Therefore, the specific gravity of a particle (S_p) is calculated as follows:

$$S_p = \frac{\rho_p}{\rho} \quad (7.20)$$

Substituting Equations (7.18) and (7.20) into Equation (7.17) yields another form of Stokes' law.

$$V_s = \frac{g(\rho_p - \rho)\, d^2}{18\mu} \quad (7.17)$$

Equation (7.18) is rearranged to solve for absolute viscosity, μ.

$$\mu = \rho\nu \quad (7.21)$$

Substituting Equation (7.21) into Equation (7.17) yields:

$$V_s = \frac{g(\rho_p - \rho)d^2}{18\rho\nu} \quad (7.22)$$

Separating variables and dividing by density results in Equation (7.23):

$$V_s = \frac{g}{18\nu}\left(\frac{\rho_p}{\rho} - \frac{\rho}{\rho}\right) d^2 \quad (7.23)$$

An alternative form of Stokes' law is presented as Equation (7.24) after substituting Equation (7.20) for $\frac{\rho_p}{\rho}$:

$$V_s = \frac{g}{18\nu}(S_p - 1)d^2 \quad (7.24)$$

In grit chambers, the flow regime is turbulent so the value of C_D is 0.4 and Equation (7.10) reduces to Equation (7.25):

$$V_s = \left[\frac{4g}{3 \times 0.4}\left(S_p - 1\right) d\right]^{0.5}$$

$$V_s = [3.3g(S_p - 1)d]^{0.5} \quad (7.25)$$

7.3.2.2 Horizontal-flow Grit Chamber

Horizontal-flow type grit chambers are designed to remove discrete particles with diameters of 0.008 in (0.2 mm) and specific gravity of 2.65. A horizontal-flow through velocity of 1.0 fps (0.3 m/s) is typically used in design. Detailed design procedures for sizing horizontal-flow through grit chambers are presented by Reynolds & Richards (1996, pp. 137–156). Metcalf & Eddy (1991, page 458) recommend a settling velocity of 3.8 fpm (1.15 m/min) for 0.21 mm diameter grit particles with a specific gravity of 2.65.

Example 7.4 Design of a horizontal-type grit chamber

A horizontal-flow type of grit chamber is designed to remove grit particles with a diameter of 0.2 mm and specific gravity of 2.65. A flow-through velocity of 0.3 m/s will be maintained by a proportioning weir. The average daily wastewater flow is 5,000 m^3/d. The PHF : ADF ratio is 2.0 : 1.0. Determine the channel dimensions for the PHF.

Solution

First, calculate the PHF as follows:

$$\text{PHF} = \text{ADF} \times 2.0 = 5000\,\frac{\text{m}^3}{\text{d}} \times 2.0 = 10{,}000\,\frac{\text{m}^3}{\text{d}}$$

Assume that a rectangular cross-section for the grit chamber will be used and that the depth of the chamber is 1.5 × width at maximum flow. The cross-sectional area is determined using the continuity equation, Equation (7.26):

$$Q = AV \qquad (7.26)$$

where:
Q = volumetric flow rate, $\frac{\text{ft}^3}{\text{s}}\left(\frac{\text{m}^3}{\text{s}}\right)$
A = cross-sectional area, ft² (m²)
V = velocity of flow, fps (m/s).

$$A = \frac{Q}{V} = \frac{10{,}000\,\text{m}^3/\text{d}}{0.3\,\text{m/s}}\left(\frac{1\,\text{d}}{24\,\text{h}}\right)\left(\frac{1\,\text{h}}{3600\,\text{s}}\right) = 0.39\,\text{m}^2$$

Next, determine the width (W) and depth (D) of the channel as follows:

$$A = W \times D = 0.39\,\text{m}^2$$

Recall that $D = 1.5W$

$$A = W \times (1.5W) = 0.39\,\text{m}^3$$

$$1.5W^2 = 0.39\,\text{m}^2$$

$$W = \sqrt{\frac{0.39\,\text{m}^2}{1.5}} = \boxed{0.51\,\text{m}}$$

$$D = 1.5\,W = 1.5 \times 0.51\,\text{m} = \boxed{0.77\,\text{m}}$$

Estimate the settling velocity of a 0.2 mm diameter particle with a specific gravity of 2.65 using Equation (7.25):

$$V_s = [3.3g(S_p - 1)d]^{0.5}$$

$$= \left[3.3 \times 9.81\,\text{m/s}^2\,(2.65 - 1)\,0.2\,\text{mm}\left(\frac{1\,\text{m}}{1000\,\text{mm}}\right)\right]^{0.5}$$

$$= 0.103\,\frac{\text{m}}{\text{s}}$$

The detention time, τ, is calculated by dividing the depth by the particle settling velocity as follows:

$$\tau = \frac{D}{V_s} = \frac{0.77\,\text{m}}{0.103\,\text{m/s}} = 7.5\,\text{s}$$

Length of the grit chamber is equal to the detention time multiplied by the horizontal flow-through velocity as follows:

$$L = \tau \times V_h = 7.5\,\text{s} \times 0.3\,\frac{\text{m}}{\text{s}} = \boxed{2.25\,\text{m}}$$

Metcalf & Eddy (2003) recommend that the theoretical length be increased by 50% to account for influent and effluent turbulence. Therefore, the overall length should be equal to:

$$1.5 \times 2.25\,\text{m} = \boxed{3.38\,\text{m}}.$$

7.3.2.3 Aerated Grit Chambers

Aerated grit removal chambers offer the advantages of increasing the dissolved oxygen concentration of the incoming wastewater, oxidizing hydrogen sulfide, stripping volatile organic compounds (VOCs), and producing a well-washed grit which contains little organic matter. Rectangular tanks with air introduced along the side of the bottom produces a spiral flow pattern. Design criteria for aerated grit chambers are presented in Table 7.8.

Example 7.5 Design of aerated grit removal chamber

Design an aerated grit chamber for an average daily wastewater flow rate of 30,000 m³/d. Assume two grit chambers are operating in parallel. The peak hour flow rate

Table 7.8 Aerated grit chamber design criteria.

Parameter	Range
[a]Air supply per unit length	3.2–7.8 ft[c]/(min × ft) (0.3–0.72 m[c]/(min · m))
[a] Depth	7–16 ft (2–5 m)
[b]Length	25–65 ft (7.5–20 m)
[b]Width	8–23 ft (2.5–7 m)
[c]Minimum detention time at peak flow	2–5 min
[b]Grit quantity	0.5–27 ft³/Mgal (0.004–0.20 m³/(10³ m³))
[c]Type of diffuser	Medium to coarse bubble
[d] Width : length ratio	1:3 to 1:5

[a] WEF (1998b) Vol 2, pp. 9–28.
[b] Metcalf and Eddy (2003) p. 389.
[c] Vesilind (2003) p. 4–16.
[d] Reynolds and Richards (1996) p. 154.

is three times the ADF. Use the design criteria in Table 7.8 to determine the following:

a) The dimensions of each grit chamber.
b) The total air required (m^3/d).

Solution Part A

The peak flow passing through the grit chamber must be calculated first.

$$\text{PHF} = 3 \times \text{ADF} = 3 \times 30{,}000\ \frac{m^3}{d} = 90{,}000\ \frac{m^3}{d}$$

The volume of the grit chambers is calculated using the detention time equation:

$$\tau = \frac{V}{Q}$$

where:
τ = detention time, min
V = volume of the grit chambers, m^3
Q = volumetric flow rate, m^3/d.

Assuming a detention time of 3.0 min at PHF, the volume of the grit chambers is 187.5 m^3.

$$V = \tau \times Q$$
$$= 3.0\,\text{min}\left(\frac{1\,\text{h}}{60\,\text{min}}\right)\left(\frac{1\,\text{d}}{24\,\text{h}}\right) \times 90{,}000\ \frac{m^3}{d}$$
$$= 187.5\ m^3$$
$$\frac{V}{\text{grit chamber}} = \frac{187.5\ m^3}{2} = 94\ m^3$$

Assuming a length : width ratio of 4 : 1 and a width : depth ratio of 1.5 : 1 from Table 7.8, calculate the width as follows:

$$\frac{L}{W} = \frac{4}{1} \qquad L = \frac{4W}{1} = 4W$$
$$\frac{W}{D} = \frac{1.5}{1} \qquad D = \frac{1W}{1.5}$$

Recall the definition of volume:

$$V = L \times W \times D$$
$$94\ m^3 = 4W \times W \times \frac{W}{1.5}$$
$$W^3 = \frac{94\ m^3 \times 1.5}{4}$$
$$W = \sqrt[3]{\frac{94\ m^3 \times 1.5}{4}} = \boxed{3.3\ m}$$
$$L = 4W = 4 \times 3.3\ m = \boxed{13.2\ m}$$
$$D = \frac{1W}{1.5} = \frac{3.3\ m}{1.5} = \boxed{2.2\ m}$$

Solution Part B

From Table 7.8, the quantity of air required per unit of length is assumed to be 0.5 $m^3/(min \cdot m)$. The total quantity of air required is determined as follows:

$$\text{air required} = 0.5\ \frac{m^3}{\text{min}\cdot m}\left(\frac{60\ \text{min}}{h}\right)\left(\frac{24\ h}{d}\right) \times 2\ \text{chambers} \times 13.2\ m$$

$$\text{air required} = \boxed{19{,}000\ \frac{m^3}{d}}$$

7.4 Primary sedimentation

Primary sedimentation or clarification follows preliminary treatment and is a unit operation that involves the separation of settleable solids and the removal of oil, grease, and scum that float to the surface of the wastewater. Settleable solids are the portion of suspended solids that will settle out of solution under the force of gravity, since their specific gravity is greater than 1.0. Oil, grease, scum, and other floating materials rise to the surface since their specific gravities are less than 1.0. A skimmer mechanism that rotates at the same speed as the sludge removal mechanism forces these materials to a scum box attached the wall of the clarifier. A diaphragm pump or progressing cavity pump is typically used to convey scum and other materials to digestion for treatment.

Primary sedimentation does not remove soluble or colloidal organic materials, so the only organic removal that is accomplished is due to settling of particulate organic matter to the bottom of the clarifier and the lightweight organics that float to the surface and are skimmed off. The sludge that accumulates at the bottom of a primary clarifier is normally stabilized by anaerobic digestion before disposal. Primary sludge is called "raw" sludge and it is objectionable, since it contains pathogens, organics, and produces odors, especially if it becomes anaerobic. Typical removal for BOD and TSS range from 25–40% and 50–70%, respectively (Metcalf & Eddy, 2003, page 396). Figure 7.6 gives the percent removal for BOD and TSS as a function of overflow rate for primary clarifiers treating domestic wastewater.

Primary clarifiers or primary settling basins are normally circular by design. However, long, rectangular basins are often used in situations where land space is limited. Figures 7.7 and 7.8 show photographs of rectangular and circular primary clarifiers, respectively. Table 7.9 presents typical dimensions of primary clarifiers and Table 7.10 gives typical design criteria.

The design of primary clarifiers is based on the overflow rate, detention time, and weir loading rate. The overflow rate (V_o) or surface loading rate is defined as the flow rate divided by the surface area of the clarifier, as expressed in Equation (7.27).

$$V_o = \frac{Q}{A_s} \tag{7.27}$$

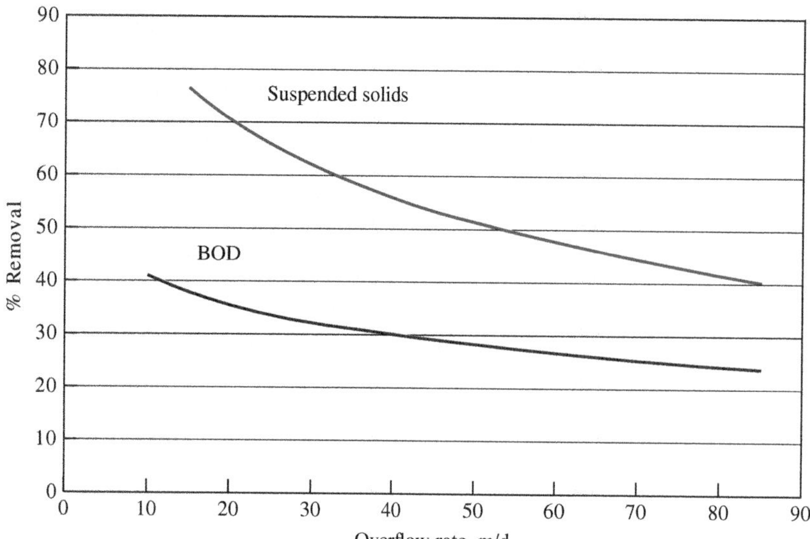

Figure 7.6 BOD and suspended solids removal versus overflow rate. Adapted from McGhee (1991), pp. 421–422.

Figure 7.7 Rectangular primary clarifiers.

Figure 7.8 A circular primary clarifier.

where:
V_o = overflow rate gpd/ft^2 (m^3/d · m^2)
Q = design flow rate, MGD (m^3/d)
A_S = surface area of the clarifier, ft^2 (m^2).

The surface area of the clarifier or settling basin is determined by dividing the design flow rate by the overflow rate. The surface area that is calculated is translated into either a circular or rectangular area. The depth of the clarifier may be determined once the detention time is selected and the total volume of the clarifier is determined. Detention time (τ) is the average unit of time that the wastewater remains in the clarifier. It is determined using Equation (7.28):

$$\tau = \frac{\forall}{Q} \tag{7.28}$$

where:
$\forall$ = volume of the primary clarifier in ft^3 (m^3)
Q = design flow rate, MGD (m^3/h or m^3/d).

The clarifier depth is estimated by dividing the volume of the clarifier by the surface area of the clarifier.

Weir loading rate (q) is the third parameter that must be satisfied when designing primary clarifiers. Mathematically, q is defined as the design flow rate divided by the length of the weir, as shown in Equation (7.29):

$$q = \frac{Q}{\text{weir length}} \tag{7.29}$$

where:
q = weir loading rate, gpd/ft (m^3/d · m)
Q = design flow rate, gpd (m^3/d)
weir length = length of primary clarifier effluent weir, ft (m).

The weir loading rate is the last parameter to be checked. For circular clarifiers, a peripheral weir that extends around the entire circumference of the clarifier is used. For rectangular clarifiers, the design engineer uses inboard box weirs to provide

Table 7.9 Typical dimensions for rectangular and circular primary sedimentation tanks.

Circular:	
Depth	8–13 ft (2.4–4.0 m)
Diameter	10–300 ft (3–90 m)
[a]Slope of bottom	$\frac{3}{4}$–2 in/ft (1/16–1/6 mm/mm)
Rectangular:	**Value**
Depth	10–16 ft (3.0–4.9 m)
Length	50–300 ft (15–90 m)
Width	10–80 ft (3.0–24 m)
Diameter	10–200
Bottom slope	0.75–2 inches/ft

WEF (1998b) Vol 2, pp. 10-4 to 10-7.
[a] Metcalf & Eddy (2003), p. 398

Table 7.10 Design criteria for primary sedimentation tanks.

Parameter	**Value**
[a]Detention time	1.5–2.0 h at average flow
[b]Average overflow rate	600–800 gpd/ft^2 [25–33 m^3/(d·m^2)]
[b]Peak overflow rate	1,200–1,500 gpd/ft^2 [49–61 m^3/(d·m^2)]
[b]Depth	12–15 ft (3.7–4.6 m)
[c]Average overflow rate	800–1,200 gpd/ft^2 [33–49 m^3/(d·m^2)]
[c]Peak overflow rate	2,000–3,000 gpd/ft^2 [82–122 m^3/(d·m^2)]
[c]Depth	10–12 ft (3.0–3.7 m)

[a] Metcalf & Eddy (2003), p. 407
[b] Primary settling with waste-activated sludge, EPA (1975b) pp. 7–14.
[c] Primary settling followed by secondary treatment, EPA (1975b) pp. 7–14.
Source: United States Environmental Protection Agency.

the necessary weir length to meet the weir loading rate. Example 7.6 illustrates the procedure for designing primary clarifiers.

Example 7.6 Primary clarifier design

A municipal WWTP receives an average daily wastewater flow of 5.0 MGD. Two, rectangular, primary clarifiers operating in parallel will treat the flow. The peak hour flow anticipated is 2.5 times the average daily flow. Use the design criteria in Tables 7.9 and 7.10, and assume that the effluent weir in each clarifier is eight times the clarifier width. Determine:

a) The dimension of each primary clarifier.
b) The detention time in each clarifier.
c) The weir loading rate (gpd/ft) for each clarifier at PHF.
d) The BOD and suspended solids removal efficiencies at ADF.

Solution Part A

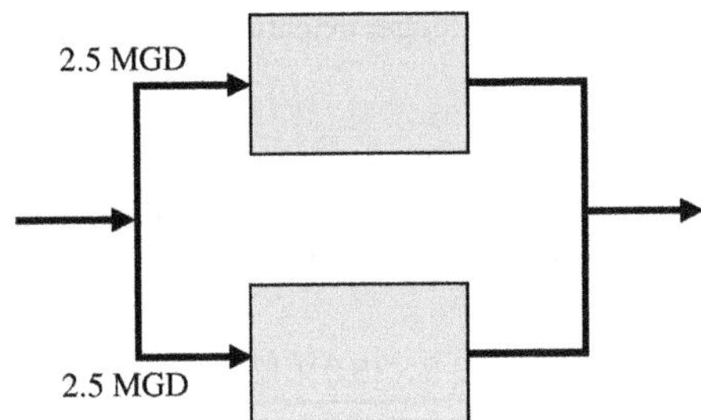

Figure E7.1

From Table 7.10, assume overflow rates of 800 and 2,500 gpd/ft^2 at average and peak flows and then calculate the surface areas of the clarifiers. At ADF, the surface area is:

$$A_S = \frac{Q}{V_o} = \frac{5.0 \times 10^6 \text{ gpd}}{800 \text{ gpd/ft}^2} = 6250 \text{ ft}^2$$

At PHF, the surface area is:

$$A_S = \frac{Q}{V_o} = \frac{2.5 \times (5.0 \times 10^6) \text{gpd}}{2500 \text{ gpd/ft}^2} = 5000 \text{ ft}^2$$

Use the larger of the two areas which is 6250 ft^2 or 3125 ft^2 per clarifier. Using a length : width ratio, calculate the length and width of each clarifier.

$$\frac{L}{W} = \frac{4}{1} \quad \text{or} \quad L = 4 \times W$$

$$A = L \times W = 4W^2 = 3125 \text{ ft}^2$$

$$W = 27.9 \cong \boxed{28 \text{ ft}}$$

$$L = 4W = 4 \times 28 \text{ ft} = \boxed{112 \text{ ft}}$$

From Table 7.9, select a side water depth (SWD) of 10 ft. Add two feet of freeboard to the SWD. **Freeboard** is the distance from the water surface to the top of the wall of the clarifier.

$$SWD = \boxed{10 \text{ ft}} \quad \text{Overall depth} = 10 \text{ ft} + 2 \text{ ft} = \boxed{12 \text{ ft}}$$

Solution Part B

Calculate the detention time by dividing the clarifier volume by the flow rate. The detention time at average flow in a primary clarifier at average flow should range from 45 minutes to 2.0 hours (Reynolds & Richards (1996), p. 257).

$$\tau = \frac{V}{Q} = \frac{112\text{ ft}\times 28\text{ ft}\times 10\text{ ft}}{2.5\times 10^6\text{gpd}}\left(\frac{7.48\text{ gal}}{1\text{ ft}^3}\right)\left(\frac{24\text{ h}}{1\text{ d}}\right)$$
$$= \boxed{2.25\text{ h}}$$

The detention time is greater than what is required; however, this will provide a longer detention time at peak flow too.

Solution Part C

The weir loading rate is the volumetric flow rate divided by the weir length; it is determined as follows:

$$q = \frac{Q}{\text{weir length}} = \frac{2.5\times(5.0\times 10^6\text{ gal/d})}{8\times 28\text{ ft}\times 2\text{ basins}}$$
$$= \boxed{27{,}900\ \frac{\text{gpd}}{\text{ft}}} < 30{,}000\ \frac{\text{gpd}}{\text{ft}}\quad \text{OK}$$

Peak weir loading rates generally should not exceed $30{,}000\ \frac{\text{gpd}}{\text{ft}}\left(248\ \frac{\text{m}^3}{\text{d}\cdot\text{m}}\right)$ (Reynolds & Richards (1996), p. 258).

Solution Part D

The efficiency of removal for BOD and suspended solids in a primary clarifier is estimated from Figure 7.6. To use the diagram, the overflow rate at ADF must be calculated and converted to m/d. Surface overflow rate at ADF is calculated as:

$$V_o = \frac{Q}{A_S} = \frac{(5.0\times 10^6)\text{gpd}}{2\times(112\text{ ft}\times 28\text{ ft})} = 797\ \frac{\text{gpd}}{\text{ft}^2}$$
$$V_o = 797\ \frac{\text{gpd}}{\text{ft}^2}\left(\frac{3.785\text{ L}}{\text{gal}}\right)\left(\frac{3.281^2\text{ ft}^2}{\text{m}^2}\right)\left(\frac{1\text{ m}^3}{1000\text{ L}}\right)$$
$$= 32.5\ \frac{\text{m}}{\text{d}}$$

From Figure 7.6, the % removal for BOD and suspended solids is 32% and 61%, respectively.

7.5 Secondary wastewater treatment

Secondary wastewater treatment implies that the major mechanism used for treating the wastewater is a biological process. A biological process uses a heterogeneous culture of microorganisms to treat the wastewater. The biological process selected may contain microbes that are suspended in the wastewater or attached to some type of media. The former type of process is called "suspended growth", while the later type is called "attached growth." Most suspended growth processes use the activated sludge process or some modification of it. Attached growth processes, in which the media remains stationary, include trickling filters and some types of packed beds. Rotating biological contactors (RBCs) use attached microbial growth on media that is rotated through the wastewater.

Both attached and suspended growth systems essentially involve growing microorganisms indigenous to the process. The microbes in the wastewater use organic carbon, along with the nutrients nitrogen and phosphorus, and other trace elements, to grow more microorganisms, primarily bacteria. Heterotrophic bacteria use the organic matter as measured by biochemical oxygen demand (BOD) or chemical oxygen demand (COD) for their energy source and carbon source. Oxidation of the organic matter produces energy that is captured in the microbe's biochemical pathways, while a portion of the organic matter is used to synthesize new biomass. Most bacteria reproduce by binary fission, which means the cell divides into two new cells. Equation (7.30) shows how organic matter is oxidized for energy, whereas Equation (7.31) shows that some of the organic matter is synthesized into new microbial cells, which is represented by the following formula ($C_{60}H_{87}O_{23}N_{12}P$).

$$\text{Organics} + O_2 \rightarrow CO_2 + H_2O + \text{Energy} \tag{7.30}$$

$$\text{Organics} + O_2 + N + P \xrightarrow{\text{Microorganisms}} C_{60}H_{87}O_{23}N_{12}P \tag{7.31}$$

The time required for each division is called the generation or doubling time, which may vary from days to less than 20 minutes (Metcalf & Eddy, 2003). For example, if a bacterium has a generation time of 20 minutes, this one bacterium would yield 4.72×10^{21} (2^{72}) bacteria in a 24-hour period. Similarly, one bacterium with a generation time of 1 hour would yield 1.68×10^7 (2^{24}) bacteria in one day.

In simplistic terms, domestic wastewater is treated by having a heterogeneous culture of microorganisms convert organic carbon, nitrogen, phosphorus, and trace elements into biological solids or biomass. These biological solids then must be removed from the wastewater to produce a high-quality effluent. Removal of the biological solids is usually accomplished in a secondary or final clarifier. Membrane filters are now being used at many WWTPs to eliminate the need for clarifiers and improve the quality of the effluent. Secondary clarification is discussed in Section 7.9.

7.5.1 Microbial growth

Bacteria are ubiquitous in nature and are found in water, air, and soil. Hoover & Porges (1952) proposed the following formula as the composition of a bacterium, $C_5H_7O_2N$. Another well known formula includes phosphorus, $C_{60}H_{87}O_{23}N_{12}P$ (McCarty, 1970). Based on the composition that includes phosphorus, a bacterial cell has a total formula weight of 1,374 and consists of approximately 52.4% carbon, 12.2% nitrogen, and 2.3% phosphorus by dry weight. It is easy to recognize that bacteria and other microbes require carbon, nitrogen, and phosphorus so that their growth will not be limited. Other trace nutrients, such as sodium, iron, and potassium, are also required.

To flourish, all microorganisms require acceptable environmental conditions, including proper moisture, a pH in the range of 6–8.5, and a temperature ranging from 15–30°C. Some species of microorganisms can withstand extreme pH and temperature conditions, and even the lack of moisture. However, this is not routinely the case.

When a pure culture of bacteria are grown in the laboratory in a batch reactor under proper environmental conditions, a growth curve as depicted in Figure 7.9 will result. This figure shows the logarithm of the mass of microbes as a function of time.

Normally, environmental engineers and scientists quantify the mass of organisms grown as a concentration in milligrams of dry mass per liter of solution, while microbiologists measure the number of organisms grown. Suspended solids and volatile solids analyses are used in most environmental engineering applications for measuring the concentration of microorganisms (biomass).

In Figure 7.9, Phase 1 of the growth curve is called the "lag phase" that results as the bacteria become acclimated to their new environment. The lag phase occurs because it takes time for organisms to develop the proper enzymes necessary to synthesize compounds from the media.

Phase 2 is called the "exponential growth phase"; the bacteria are growing at their maximum growth rate. Excess substrate (food) and nutrients exist, so there is nothing to limit growth.

The "declining growth phase" occurs in Phase 3; the growth rate starts to slow down and the bacterial death rate increases. As substrate becomes limiting, the growth rate declines and metabolic waste products accumulate which also inhibit growth.

Phase 4 is called the "stationary phase", where growth rate equals death rate. The mass or concentration of bacteria begins to decline at the end of this phase.

Finally, Phase 5, the "endogenous phase" occurs, in which the death rate exceeds the growth rate. Bacteria that are still alive oxidize their own cellular components and feed on the remains of the dead bacteria. The concentration of biomass decreases at an exponential rate during this phase. Exogenous or external substrate has been exhausted by this time.

7.5.2 Microbial growth in batch reactor

Microbial growth in a batch reactor, as depicted in Figure 7.9 can be expressed by the following equation:

$$\left(\frac{dX}{dt}\right)_G = \mu X \tag{7.32}$$

where:

$\left(\frac{dX}{dt}\right)_G$ = microorganism growth rate, mass/(volume · time)

μ = specific growth rate of microorganism, time^{-1}

X = microorganism concentration, mass/volume.

The specific growth rate of a microorganism (μ) is associated with a particular species of microorganism. In most biological wastewater systems, heterogeneous cultures of microbes are used and, therefore, biokinetic coefficients for the overall heterogeneous culture are used in design.

The French microbiologist Monod (1949) found that the specific growth rate of a microorganism is dependent upon some growth-limiting substrate or nutrient. He developed Equation (7.33), which indicates that the microorganism specific growth rate is a function of both the maximum specific growth rate and the concentration of the limiting substrate:

$$\mu = \frac{\mu_{max} S}{K_S + S} \tag{7.33}$$

where:

S = growth limiting substrate or nutrient concentration, mass/volume

μ_{max} = maximum specific growth rate, time^{-1}

K_s = half-saturation constant, concentration of limiting substrate or nutrient at which half the maximum specific growth rate occurs, mass/volume.

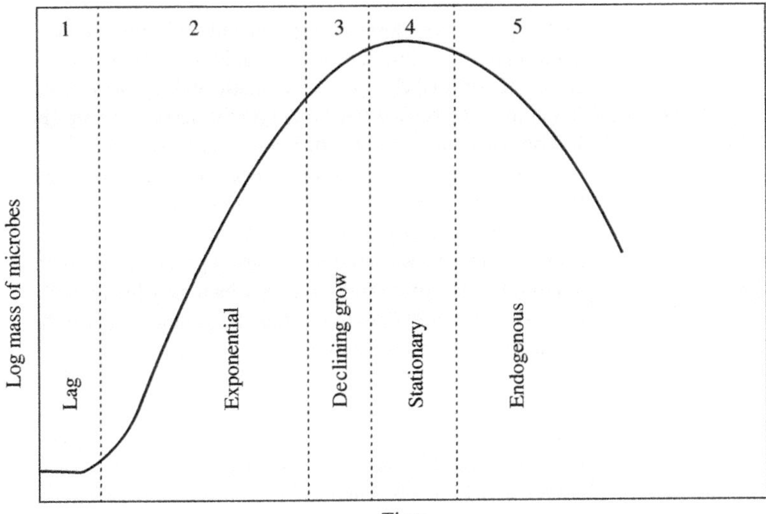

Figure 7.9 Bacterial growth curve in a batch reactor.

Combining Equations (7.32) and (7.33) yields Equation (7.34), which represents microbial growth rate under batch operating conditions:

$$\left(\frac{dX}{dt}\right)_G = \frac{\mu_{max} XS}{K_S + S} \tag{7.34}$$

7.5.3 Microbial growth in continuous flow reactor

Most full-scale engineered biological treatment systems use continuous flow reactors rather than a batch reactor. Equation (7.32) must be modified to account for the portion of biomass or microorganisms that are lost through death and decay (endogenous decay). Endogenous decay is represented by Equation (7.35).

$$\left(\frac{dX}{dt}\right)_{ED} = -k_d X \tag{7.35}$$

where:

$\left(\frac{dX}{dt}\right)_{ED}$ = endogenous decay rate, mass/(volume · time)

k_d = endogenous decay rate constant, time^{-1}.

The net growth rate of microorganisms in a continuous flow biological reactor can then be expressed as Equation (7.36):

$$\left(\frac{dX}{dt}\right)_{NG} = \left(\frac{dX}{dt}\right)_G + \left(\frac{dX}{dt}\right)_{ED} \tag{7.36}$$

Substituting Equations (7.34) and (7.35) into the above expression yields:

$$\left(\frac{dX}{dt}\right)_{NG} = \frac{\mu_{max} XS}{K_s + S} - k_d X \tag{7.37}$$

The cell yield or yield coefficient of a microbe is another useful biological term. Qualitatively, it is defined as the quantity of biomass produced per unit of substrate oxidized. Mathematically, the yield coefficient is expressed as follows:

$$Y = \frac{(dX/dt)_G}{(dS/dt)_U} \tag{7.38}$$

where:

Y = microbial yield coefficient, mass of biomass produced/mass of substrate utilized

$\left(\frac{dX}{dt}\right)_G$ = microbial growth rate, mass/(volume · time)

$\left(\frac{dS}{dt}\right)_U$ = substrate utilization rate, mass/(volume · time).

The specific substrate utilization rate (U) with units of inverse time (time^{-1}) is defined by the following equation:

$$\left(\frac{dS}{dt}\right)_U = UX \tag{7.39}$$

where

$\left(\frac{dS}{dt}\right)_U$ = substrate utilization rate, mass/(volume · time).

Equation (7.40) may be developed by substituting Equations (7.32) and (7.39) into Equation (7.38):

$$Y = \frac{(dX/dt)_G}{(dS/dt)_U} = \frac{\mu}{U} \tag{7.40}$$

Equation (7.41) is another way of expressing net microbial growth rate in a continuous flow reactor. It is obtained by substituting Equations (7.35) and (7.38) into Equation (7.36).

$$\left(\frac{dX}{dt}\right)_{NG} = Y\left(\frac{dS}{dt}\right)_U - k_d X \tag{7.41}$$

The net growth rate of a heterogeneous culture is normally expressed as Equation (7.41). This derived equation has been successfully demonstrated in actual studies performed by Heukelikian *et al.*, (1951) and is used in many environmental texts: Peavy *et al.*, 1985, p. 233; Mihelcic, 1999, p. 235; Reynolds & Richards, 1996, p. 475; Metcalf & Eddy, 2003, p.584; and Viessman *et al.*, 2009, p. 508. Equation (7.41) is used in the design and operation of activated sludge treatment processes.

7.5.4 Activated sludge

7.5.4.1 Overview

The activated sludge process is a biological process that has a long and successful history in treating domestic and industrial wastewaters. Biological treatment processes occur in nature, i.e., organic matter undergoes biological oxidation by microorganisms in water bodies. The activated sludge process is an engineered process wherein oxygen is added to a reactor to speed up the process that naturally occurs in rivers. Aeration is the major energy-consuming operation during the activated sludge process. Ardern & Lockett (1914) are credited with developing the process in England. The process is an aerobic, suspended growth, biological process used primarily to remove dissolved and colloidal organic matter from wastewater.

There are two major steps that characterize the process. The first step involves substrate sorption and utilization in the aeration basin, followed by solids/liquid separation in a secondary or final clarifier. A schematic of the process is shown in Figure 7.10. Wastewater flows into the aeration basin where it is brought into contact with a heterogeneous culture of microbes, primarily heterotrophic bacteria. The liquid inside the aeration basin is called "mixed liquor" or activated sludge.

Microorganisms growing in the aeration basin are indigenous to the wastewater and, given proper environmental conditions, they grow and proliferate by binary fission with generation times of days to less than 20 minutes (Metcalf & Eddy, 2003, p. 566). Microorganisms use the organic matter (substrate) for synthesis and energy. Synthesis involves the production of carbohydrates, lipids, proteins, etc. for cell maintenance and for reproduction by binary fission. Substrate is also oxidized through respiration for the production of energy to drive the biochemical reactions required for synthesizing biomass and for motility.

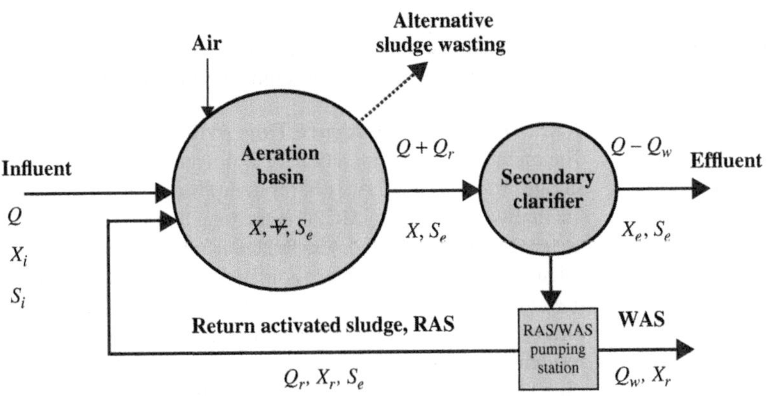

Figure 7.10 Activated sludge process schematic with complete-mix reactor.

Given the generic formula of a microorganism as $C_{60}H_{87}O_{23}N_{12}P$, it is easy to see that the major constituents required for growth are carbon (52%), nitrogen (12%), and phosphorus (2.3%). A qualitative equation representing the growth and utilization of organics and nutrients from the wastewater is presented as Equation (7.42). Simply stated, organics and nutrients in the wastewater are removed by the indigenous microorganisms and are transformed into new microorganisms, carbon dioxide, water, and energy.

$$\begin{aligned}&\text{organics} + \text{N} + \text{P} + O_2 + \text{microbes}\\&\quad\rightarrow \text{new microbes} + CO_2 + H_2O + \text{energy}\end{aligned} \tag{7.42}$$

The active heterogeneous culture of microbes living in the activated sludge process is also known as biomass. Bacteria are the predominant microbial species present, with most of them being aerobes. Some, however, are facultative, meaning they can survive in an anaerobic environment. Protozoans also make up a small portion of the biomass. Some feed on organics, while others are predators of bacteria. Generally, their presence indicates that good treatment is being accomplished and that high levels of DO and low levels of organic matter exist. Fungi may be present within the biomass, but are undesirable, since they produce poor settling sludge, and poor sludge settleability may lead to NPDES permit violations. The next highest stage of life form beyond protozoans is the rotifers. These have minimal impact on treatment, but indicate a healthy biological process.

Biomass that enters the clarifier is separated from the liquid portion of the wastewater by gravity. Supernatant or clarified wastewater that overflows the effluent weir is called "secondary effluent." Biomass that settles to the bottom forms a layer of sludge called the "sludge blanket." This thickened sludge, also known as underflow, normally flows by gravity from the bottom of the clarifier to a pumping station that contains a wet well and pumps. Thickened biomass pumped back to the head of the aeration basin is called the recycle flow (Q_r) or return activated sludge (RAS) flow.

The quantity of biomass recycled to the aeration basin, along with that wasted from the system, determines the concentration of active biomass that can be maintained in the aeration basin. The active biomass or microorganism concentration in the aeration basin (denoted by X) is typically measured as mixed liquor suspended solids (MLSS) or mixed liquor volatile suspended solids (MLVSS). Using either of these parameters has one major deficiency, in that active microbial cells cannot be differentiated from non-proliferating cells and inert suspended solids. MLVSS is actually preferable, since it relates to the organic fraction of the suspended solids concentration that is more likely to be active cells.

Droste (1997) reported that several researchers indicate that adenosine triphosphate (ATP) and dehyrogenase activity are better indicators of active biomass than MLSS or MLVSS. Benefield & Randall (1980, p. 190), referencing (Weddle & Jenkins, 1971), state that MLVSS is just as accurate a measurement as cellular deoxyribonucleic acid (DNA), ATP, dehyrogenase activity, or organic nitrogen. Mean cell residence time (MCRT) is the average unit of time that the biomass remains in the system; it serves both as a design and operational parameter.

To maintain steady-state operating conditions, the quantity of biomass that is grown within the system is equal to the quantity of biomass that must be wasted from the system. The sludge wasted (Q_w) from the system is called "waste-activated sludge" or WAS. Sludge wasting from the process is normally accomplished by wasting biomass from the sludge return line or from the RAS/WAS pumping station. The TSS concentration of WAS ranges from 5,000 – 10,000 mg/L. Alternatively, sludge may be wasted directly from the aeration basin (TSS concentration of 1,500 – 3,000 mg/L), providing a better means for controlling the process. However, it has a high water content (lower solids concentration than thickened sludge), resulting in higher costs associated with thickening, stabilizing, and disposing of the waste sludge, compared to wasting thickened sludge from the secondary clarifier.

The rate at which sludge settles in the secondary clarifier is dependent on the MLSS concentration entering the clarifier, characteristics of the wastewater, MCRT, and the sludge return rate. Sludge volume index (SVI) is often used by WWTP operators to determine how well sludge settles and for estimating the RAS flow rate. SVI is determined by filling a 1 L graduated cylinder with a sample of mixed liquor and allowing the solids to settle in a 30-minute period. The volume (in mL) occupied by the settled sludge is used for calculating the SVI. SVI values ranging from 50 – 150 mL/g indicate a good settling sludge, whereas values greater than 150 indicate poor settling sludge. The sludge volume index is determined according to Equation (7.43):

$$\text{SVI} = \frac{\text{sludge volume after settling (mL/L)} \times 1000}{\text{MLSS (mg/L)}} \tag{7.43}$$

where
SVI = sludge volume index after settling 30 minutes in a 1 L graduated cylinder, mL/g.

An estimate of Q_r can be determined from Equation (7.44) and knowing the sludge volume after settling as determined from the SVI analysis:

$$\frac{Q_r}{(Q+Q_r)} = \frac{\text{sludge volume after settling(mL/L)}}{1000\,\text{mL}} \tag{7.44}$$

where:
Q = influent wastewater flow rate, MGD (m^3/d)
Q_r = return activated sludge (RAS) flow rate, MGD, (m^3/d).

An alternative means of estimating the RAS flow is by performing a materials balance on solids around the secondary clarifier and assuming that the solids concentration in the secondary effluent is negligible. Equations (7.45) and (7.46) are developed from Figure 7.11.

$$(Q+Q_r)\text{MLSS} = (Q_r)X_r + (Q)0 \tag{7.45}$$

$$\frac{Q_r}{Q} = \frac{\text{MLSS}}{X_r - \text{MLSS}} \tag{7.46}$$

Knowing the SVI of the sludge, it is possible to estimate the maximum suspended solids concentration $(X_r)_{max}$ in the return sludge flow using Equation (7.47):

$$(X_r)_{max} = \frac{10^6}{\text{SVI}} \tag{7.47}$$

where
$(X_r)_{max}$ = maximum TSS concentration in RAS flow or underflow, mg/L.

Often, solids concentrations of sludge are expressed as % solids, where, 1% solids=10,000 mg/L assuming that the specific gravity of the wet sludge is the same as water, equal to 1.

7.5.5 Design and operational parameters

Five design and operational parameters are presented in this section. These parameters are used by process engineers for designing activated sludge processes. Wastewater treatment plant operators use these same parameters for operating and controlling the process to meet effluent requirements.

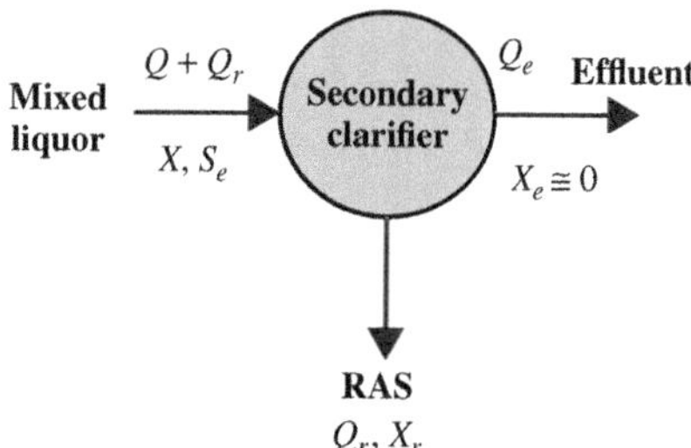

Figure 7.11 Materials balance on solids around secondary clarifier.

7.5.5.1 Mean Cell Residence Time (MCRT)

The main design and operational parameter for the activated sludge process is the mean cell residence time (MCRT). A qualitative definition of MCRT is the total biomass in the system divided by the biomass wasted from the system. When calculating the MCRT value, a consistent set of unit must be used for X, i.e., either TSS or VSS. MCRT represents the average unit of time that the biomass remains in the system. Mean cell residence time is also called sludge age or solids retention time (SRT) and is denoted as θ_c. Mathematically, MCRT is expressed as Equation (7.48), using the nomenclature presented in Figure 7.10.

$$\theta_c = \frac{X\forall}{(Q-Q_w)X_e + Q_w X_r} \tag{7.48}$$

where:
θ_c = mean cell residence time, days
X = active biomass concentration in aeration basin measured as MLSS orMLVSS concentration, mg/L
$\forall$ = volume of the aeration basin, ft^3 (m^3)
X_e = secondary effluent TSS or VSS concentration, mg/L
X_r = TSS or VSS concentration in return activated sludge, mg/L
Q = influent wastewater flow rate, MGD (m^3/d)
Q_w = sludge wastage flow rate, MGD (m^3/d).

MCRT generally varies from 5–30 days and determines the overall substrate removal efficiency of the process. The longer the MCRT, the lower the effluent soluble substrate concentration (S_e) as measured by biochemical oxygen demand (BOD) or chemical oxygen demand (COD) – i.e., the organic matter becomes oxidized. The remaining organic matter is less amenable to biological degradation, and an accumulation of microbial end-products and the release of secondary substrate products occurs at long MCRTs (Droste, 1997). MCRT is much longer than the hydraulic detention time (τ). Typically, the magnitude of θ_c is 10–40 times the value of τ.

7.5.5.2 Hydraulic Detention Time

The time that the wastewater resides in a particular unit operation or process is called the hydraulic detention time (τ). Detention time is mathematically defined as volume divided by flow rate, as follows:

$$\tau = \frac{\forall}{Q} \tag{7.49}$$

where:
τ = hydraulic detention time, d or h
$\forall$ = volume of unit operation or process, ft^3 (m^3)
Q = influent flow to unit operation or process excluding recycles, gal/d (m^3/d).

Design engineers use τ to determine the volume of aeration basins, secondary clarifiers, etc. Wastewater treatment plant operators monitor the detention time of various processes to ensure that proper treatment is obtained.

7.5.5.3 Food-to-Microorganism Ratio (F/M)

Another design and operational parameter is the food-to-microorganism ratio (F/M) which is defined as follows:

$$F/M = \frac{QS_i}{X\forall} \quad (7.50)$$

where:
F/M = food-to-microorganism ratio, d^{-1}
S_i = influent substrate concentration prior to mixing with sludge recycle expressed as BOD or COD, mg/L
$\forall$ = volume of aeration basin, ft^3 (m^3)
Q = wastewater flow rate prior to mixing with the RAS flow, gal/d (m^3/d).

S_i is normally expressed in terms of total BOD (TBOD) or total COD (TCOD), i.e., the sum of the soluble and particulate organic fractions. The volume of the aeration basin can be determined by assuming values for the F/M ratio, X, Q, and S_i. Operationally, there is an inverse relationship between the F/M ratio and θ_c. A large value for the F/M ratio results in a small value for θ_c, whereas a small F/M ratio results in a large value for θ_c.

7.5.5.4 Specific Substrate Utilization Rate (U)

Specific substrate utilization rate (U) is used similar to the F/M ratio. The only difference is that the effluent soluble substrate concentration is taken into account. Mathematically, U is defined as follows:

$$U = \frac{Q(S_i - S_e)}{X\forall} \quad (7.51)$$

where:
U = specific substrate utilization rate, d^{-1}
S_e = effluent soluble substrate concentration expressed as soluble BOD (SBOD) or soluble COD (SCOD), mg/L.

7.5.5.5 Efficiency of Treatment (E)

The efficiency of treatment is generally based on substrate removal. However, the efficiency of treatment, often called the removal efficiency, can be calculated for any analytical parameter using the following equation:

$$E = \frac{(C_i - C_e)}{C_i} \times 100 \quad (7.52)$$

where:
E = efficiency of treatment expressed as a percentage
C_i = concentration of parameter in the influent to process, mg/L
C_e = concentration of parameter in the effluent from process, mg/L.

Example 7.7 Calculating activated sludge operating parameters

A complete-mix activated sludge process treats 5.0 MGD of wastewater containing 225 mg/L BOD_5 to 20 mg/L BOD_5. The volume of the aeration basin is 111,400 ft^3 and the MLSS concentration is 2500 mg/L. Determine the following:

a) Detention time in the aeration basin.
b) F/M ratio.
c) Specific substrate utilization rate.
d) Substrate removal efficiency.

Solution Part A

Calculate the detention time using Equation (7.49):

$$\tau = \frac{\forall}{Q} = \frac{111{,}400\ ft^3}{5 \times 10^6\ gal/d}\left(\frac{7.48\ gal}{ft^3}\right)\left(\frac{24\ h}{d}\right) = \boxed{4.0\ h}$$

Solution Part B

F/M ratio is calculated using Equation (7.50). First, convert the volume of the aeration basin to millions of gallons (MG):

$$\forall = 111{,}400\ ft^3\left(\frac{7.48\ gal}{ft^3}\right)\left(\frac{MG}{10^6\ gal}\right) = 0.83\ MG$$

$$F/M = \frac{QS_i}{X\forall} = \frac{5.0\ MGD \times 225\ mg/L}{2500\ mg/L \times 0.83\ MG}$$

$$= \boxed{0.54\ \frac{mg\ BOD_5}{mg\ TSS \cdot d}}$$

Solution Part C

U is calculated using Equation (7.51), as follows:

$$U = \frac{Q(S_i - S_e)}{X\forall} = \frac{5.0\ MGD(225\ mg/L - 20 mg/L)}{2500 mg/L \times 0.83\ MG}$$

$$= \boxed{0.50\ \frac{mg\ BOD_5}{mg\ TSS \cdot d}}$$

Solution Part D

Treatment efficiency is calculated using Equation (7.52):

$$E = \frac{(C_i - C_e)}{C_i} \times 100$$

$$= \frac{(225\ mg/L - 20\ mg/L)}{225\ mg/L} \times 100 = \boxed{91\%}$$

7.5.6 Biochemical kinetics of complete-mix activated sludge (CMAS) systems

Kinetic equations based on the growth rate of microbes in pure cultures have been used to model heterogeneous microbial cultures such as activated sludge systems. This section presents the development of kinetic equations used for designing complete-mix activated sludge (CMAS) systems. Material balances and the assumptions made in developing the kinetic equations are presented. The nomenclature used is shown in Figure 7.10.

The following assumptions are made in developing the CMAS kinetic equations:

1 Steady-state conditions prevail throughout the system; flows, biomass concentrations, and substrate concentrations do not vary with time.
2 Complete mixing is achieved in the aeration basin; therefore, the concentration of all species within the aeration basin is the same as the concentration of all species in the effluent from the aeration basin.
3 Biological activity only occurs in the aeration basin, not in the secondary clarifier.
4 The concentration of microorganisms in the influent wastewater is negligible.
5 The mean cell residence time is based only on the biomass in the aeration basin and does not include biomass in the secondary clarifier.
6 No sludge accumulation occurs in the secondary clarifier.
7 The substrate concentration is measured as SBOD or SCOD.
8 Sludge is purposely wasted from the sludge return line.

7.5.6.1 Net Growth Rate Equation

A relationship between net growth rate and MCRT is derived in the section below.

The net growth rate of the biomass is defined by the following equation:

$$\left(\frac{dX}{dt}\right)_{NG} = Y\left(\frac{dS}{dt}\right)_U - k_d X \quad (7.53)$$

where:

$\left(\frac{dX}{dt}\right)_{NG}$ = net microbial growth rate, mass/(volume · time)

$\left(\frac{dS}{dt}\right)_U$ = substrate utilization rate, mass/(volume · time)

Y = biomass yield coefficient, biomass produced per unit of substrate utilized, mass/mass

k_d = endogenous decay coefficient, time^{-1}

X = active biomass concentration expressed as MLSS or MLVSS, mass/volume.

Perform a materials balance on biomass (X) around the entire system (see Figure 7.12).

A qualitative expression for a materials balance is presented as Equation (7.54):

$$[\text{accumulation}] = [\text{inputs}] - [\text{outputs}] + [\text{reaction}] \quad (7.54)$$

$$\left(\frac{dX}{dt}\right)_{accumulation} \forall = QX_i - (Q - Q_w)X_e - Q_w X_r + \left(\frac{dX}{dt}\right)_{NG} \forall \quad (7.55)$$

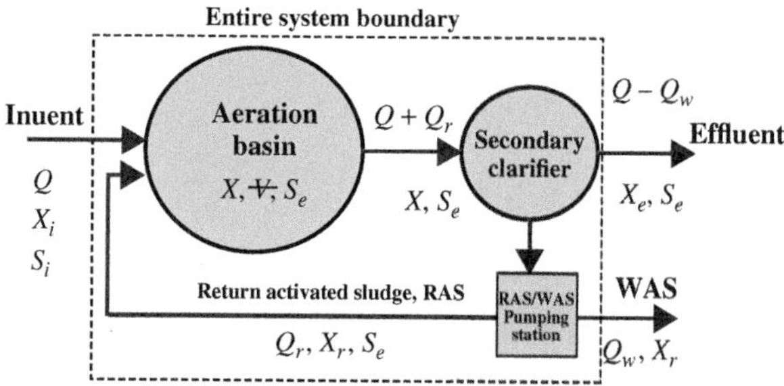

Figure 7.12 System boundary for materials balance on biomass for complete-mix reactor.

The accumulation term goes to zero during steady-state conditions, and the influent biomass concentration, X_i, is assumed to be negligible. Therefore, Equation (7.55) can be written as:

$$\left(\frac{dX}{dt}\right)_{NG} = \frac{(Q - Q_w)X_e + Q_w X_r}{\forall} \quad (7.56)$$

Substituting Equation (7.53) into Equation (7.56) results in Equation (7.57):

$$\left(\frac{dX}{dt}\right)_{NG} = \frac{(Q - Q_w)X_e + Q_w X_r}{\forall} = Y\left(\frac{dS}{dt}\right)_U - k_d X \quad (7.57)$$

Dividing Equation (7.57) by X yields Equation (7.58):

$$\frac{(dX/dt)_{NG}}{X} = \frac{(Q - Q_w)X_e + Q_w X_r}{X\forall} = \boxed{\frac{1}{\theta_c} = Y\frac{(dS/dt)_U}{X} - k_d} \quad (7.58)$$

7.5.6.2 Effluent Soluble Substrate Equations

Substrate removal in activated sludge systems has been modeled using various expressions. Substrate utilization using a Michaelis-Menten type equation was proposed by Lawrence & McCarty (1970) and is one of the most widely accepted equations.

$$\left(\frac{dS}{dt}\right)_U = \frac{kXS_e}{K_S + S_e} \quad (7.59)$$

where:

k = maximum specific substrate utilization rate, d^{-1}

K_s = half-velocity constant, substrate concentration at one-half the maximum specific substrate utilization rate, mg/L.

The expressions in Equations (7.60) and (7.61) have also been used to model substrate utilization. Equation (7.60) follows first-order removal kinetics with respect to S_e, while Equation (7.61), proposed by Grau *et al.*, (1975), is recommended where significant variations in influent substrate concentration occur. Each of these equations may be used to derive a specific equation for calculating S_e for a complete-mix activated sludge process.

$$\left(\frac{dS}{dt}\right)_U = KXS_e \tag{7.60}$$

$$\left(\frac{dS}{dt}\right)_U = K_1 X \frac{S_e}{S_i} \tag{7.61}$$

where:
K = pseudo-first-order reaction rate coefficient, time^{-1}
K_1 = specific substrate utilization rate coefficient, time^{-1}.

We will now derive the widely used S_e equation developed by Lawrence & McCarty (1970). Begin by substituting Equation (7.59) into Equation (7.58):

$$\frac{1}{\theta_c} = Y\frac{(dS/dt)_U}{X} - k_d = Y\frac{\left(\frac{kXS_e}{K_S + S_e}\right)}{X} - k_d \tag{7.62}$$

$$\frac{1}{\theta_c} = Y\frac{kS_e}{K_S + S_e} - k_d \tag{7.63}$$

$$\frac{1}{\theta_c} + k_d = Y\frac{kS_e}{K_S + S_e} \tag{7.64}$$

$$\left(\frac{1}{\theta_c} + k_d\right)(K_S + S_e) = YkS_e \tag{7.65}$$

$$\frac{K_S}{\theta_c} + K_S k_d + \frac{S_e}{\theta_c} + S_e k_d = YkS_e \tag{7.66}$$

$$\frac{K_S}{\theta_c} + K_S k_d = YkS_e - \frac{S_e}{\theta_c} - S_e k_d \tag{7.67}$$

$$K_S\left(\frac{1}{\theta_c} + k_d\right) = S_e\left(Yk - \frac{1}{\theta_c} - k_d\right) \tag{7.68}$$

$$K_S \times \theta_c\left(\frac{1}{\theta_c} + k_d\right) = S_e\left(Yk - \frac{1}{\theta_c} - k_d\right) \times \theta_c \tag{7.69}$$

$$K_S(1 + k_d\theta_c) = S_e[Yk\theta_c - (1 + k_d\theta_c)] \tag{7.70}$$

$$S_e = \frac{K_S(1 + k_d\theta_c)}{[Yk\theta_c - (1 + k_d\theta_c)]} \tag{7.71}$$

The final form of the effluent soluble substrate concentration from a CMAS process as developed by Lawrence & McCarty is presented as Equation (7.72):

$$\boxed{S_e = \frac{K_S(1 + k_d\theta_c)}{\theta_c(Yk - k_d) - 1}} \tag{7.72}$$

7.5.6.3 Minimum Mean Cell Residence Time

Biological wastewater treatment systems should not be operated at the minimum mean cell residence time, $(\theta_c)_{min}$. This is known as the washout point, because the biomass cannot reproduce fast enough, and therefore the effluent substrate concentration (S_e) will be equal to the influent substrate concentration (S_i). Engineers select a design mean cell residence time $(\theta_c)_{design}$ by multiplying $(\theta_c)_{min}$ by some safety factor, typically ranging from 2 to 10.

Lawrence & McCarty (1970) indicate that the safety factor in several types of activated sludge processes ranges from 4 to more than 70. They also state that high BOD removal can be accomplished at θ_c values less than one day; however, they recommend that MCRTs of three days or greater be used, so that bioflocculation of the sludge will occur, which is necessary to produce a clarified effluent.

The minimum MCRT is calculated by replacing S_e with S_i in Equation (7.63) as follows:

$$\frac{1}{(\theta_c)_{min}} = Y\frac{kS_i}{K_S + S_i} - k_d \tag{7.73}$$

where
$(\theta_c)_{min}$ = minimum mean cell residence time or washout point, days.

Equation (7.74) is used to calculate the design MCRT:

$$(\theta_c)_{design} = (\theta_c)_{min} \times SF \tag{7.74}$$

where:
$(\theta_c)_{design}$ = design mean cell residence time, days
SF = a safety factor typically ranging from 2 to 10.

Figure 7.13 is a plot of effluent soluble substrate concentration calculated using Equation (7.72) and efficiency of treatment (substrate removal) using Equation (7.52) as a function of mean cell residence time. The minimum mean cell residence time as shown in the figure is approximately 0.21 days, which can be calculated from Equation (7.73).

7.5.6.4 Total Effluent Substrate Concentration

Regulatory agencies and NPDES permits generally list effluent BOD_5 requirements in terms of the total BOD_5 ($TBOD_5$). Total BOD_5 consists of particulate BOD_5 ($PBOD_5$) and soluble BOD_5 ($SBOD_5$), as expressed in Equation (7.75):

$$TBOD_5 = PBOD_5 + SBOD_5 \tag{7.75}$$

where:
$TBOD_5$ = total five-day biochemical oxygen demand, mg/L
$PBOD_5$ = particulate five -day biochemical oxygen demand, mg/L
$SBOD_5$ = soluble five -day biochemical oxygen demand, mg/L.

The soluble or filtered BOD_5 can be measured in the laboratory or may be calculated from one of the kinetic equations for S_e. Particulate BOD_5 can be approximated using Equation (7.76):

$$PBOD_5 = (1.42VSS_e)\left(\frac{2}{3}\right) \tag{7.76}$$

where:
VSS_e = VSS concentration in the effluent from the secondary or final clarifier following aeration, mg/L
1.42 = theorectical COD of biomass, lb O_2/lb biomass (kg O_2/kg biomass)
$\frac{2}{3}$ = factor for approximating the conversion from COD to BOD_5.

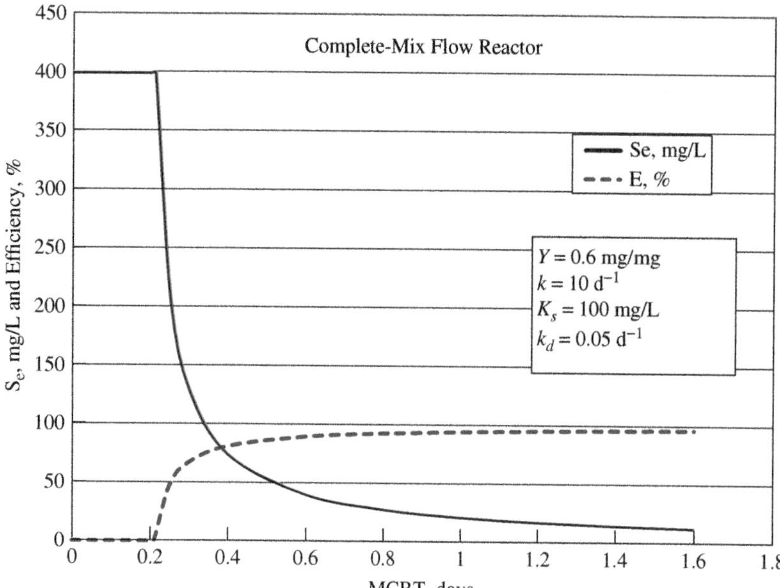

Figure 7.13 Effluent substrate concentration and treatment efficiency versus MCRT for a complete-mix reactor.

Example 7.8 Estimating particulate BOD_5 concentration

NPDES permits typically require that activated sludge WWTPs meet an annual average effluent BOD_5 concentration of 20 mg/L. If a plug flow reactor activated sludge plant produces an effluent containing 5 mg/L of soluble BOD_5 and 15 mg/L of TSS, will the effluent BOD_5 standard be met? Assume that the volatile fraction of the effluent suspended solids is 70%.

Solution

Estimate the particulate BOD_5 in the effluent using Equation (7.76):

$$VSS_e = (0.70)\left(\frac{15\ \text{mg}\ TSS_e}{L}\right) = \frac{10.5\ \text{mg}}{L}$$

$$PBOD_5 = (1.42VSS_e)\left(\frac{2}{3}\right)$$

$$= (1.42)\left(\frac{10.5\ \text{mg VSS}}{L}\right)\left(\frac{2}{3}\right) = \frac{9.94\ \text{mg}}{L}$$

The total BOD_5 in the effluent can be calculated from Equation (7.75):

$$TBOD_5 = PBOD_5 + SBOD_5 = \frac{9.94\ \text{mg}}{L} + \frac{5\ \text{mg}}{L}$$

$$= \boxed{\frac{14.9\ \text{mg}}{L}}$$

Since the calculated $TBOD_5$ is less than 20 mg/L, the WWTP will meet the effluent standard for BOD_5. It is important to recognize the impact that biological solids have on meeting effluent requirements. At many WWTPs, effluent filters consisting of coarse media are added onto liquid treatment trains following secondary clarification, in order to achieve higher removals of BOD_5 and TSS so that strict effluent limits can be met.

7.5.6.5 Aeration Basin Biomass Concentration Equation

An equation for calculating the active biomass concentration (X) will now be derived.

A materials balance on substrate entering and leaving the aeration basin in Figure 7.12 can be written to develop an equation for calculating the active microorganism concentration inside the aeration basin. We begin with our qualitative equation for a materials balance, Equation (7.54):

$$[\text{accumulation}] = [\text{inputs}] - [\text{outputs}] + [\text{reaction}] \quad (7.54)$$

$$\left(\frac{dS}{dt}\right)_{accumulation} \forall = QS_i + Q_rS_e - (Q + Q_r)S_e - \left(\frac{dS}{dt}\right)_U \forall \quad (7.77)$$

During steady-state conditions, the substrate accumulation term becomes zero and Equation (7.77) becomes:

$$0 = QS_i + Q_rS_e - QS_e - Q_rS_e - \left(\frac{dS}{dt}\right)_U \forall \quad (7.78)$$

Dividing both sides of the above equation by V results in:

$$\left(\frac{dS}{dt}\right)_U = \frac{Q(S_i - S_e)}{\forall} \quad (7.79)$$

Substituting Equation (7.79) into Equation (7.58) and rearranging gives the active biomass or microorganism concentration (X) inside the aeration basin of a complete-mix reactor:

$$\left(\frac{dX}{dt}\right)_{NG} = \frac{(Q - Q_w)X_e + Q_w X_r}{X\forall}$$

$$= \boxed{\frac{1}{\theta_c} = Y\frac{(dS/dt)_U}{X} - k_d} \qquad (7.58)$$

$$\frac{1}{\theta_c} + k_d = Y\left(\frac{Q(S_i - S_e)}{\forall X}\right) \qquad (7.80)$$

$$X = \frac{YQ(S_i - S_e)}{\forall\left(\frac{1}{\theta_c} + k_d\right)} \qquad (7.81)$$

$$X = \frac{Y(S_i - S_e)}{\left(\frac{1}{\theta_c} + k_d\right)}\frac{Q}{\forall}\frac{\theta_c}{\theta_c} \qquad (7.82)$$

Recall the definition of detention time from Equation (7.49).

$$\tau = \frac{\forall}{Q} \qquad (7.49)$$

$$\boxed{X = \frac{Y(S_i - S_e)}{(1 + k_d\theta_c)}\frac{\theta_c}{\tau}} \qquad (7.83)$$

The volume of a complete-mix aeration basin can be determined by rearranging Equation (7.83) and substituting Equation (7.49) for τ:

$$\boxed{\forall = \frac{YQ(S_i - S_e)}{(1 + k_d\theta_c)}\frac{\theta_c}{X}} \qquad (7.84)$$

7.5.6.6 Sludge Production Equation

Equations for calculating excess biomass production (P_x) and the total quantity of sludge (SP) that must be wasted from the activated sludge process are developed in this section.

The observed growth yield (Y_{obs}) is defined as the net microbial growth rate divided by the substrate utilization rate as follows:

$$Y_{obs} = \frac{(dX/dt)_{NG}}{(dS/dt)_U} \qquad (7.85)$$

Equation (7.89) is derived in the following steps as it is a more useful definition of the observed yield coefficient. Begin by substituting Equation (7.57) into Equation (7.85):

$$Y_{obs} = \frac{Y\left(\frac{dS}{dt}\right)_U - k_d X}{(dS/dt)_U} \qquad (7.86)$$

Next, rearrange Equation (7.58) to solve for $\left(\frac{dS}{dt}\right)_U$ as shown in Equation (7.87):

$$\frac{1}{\theta_c} = Y\frac{(dS/dt)_U}{X} - k_d \qquad (7.58)$$

$$\left(\frac{dS}{dt}\right)_U = \left(\frac{1}{\theta_c} + k_d\right)\frac{X}{Y} \qquad (7.87)$$

Replace $\left(\frac{dS}{dt}\right)_U$ in the denominator of Equation (7.86) with Equation (7.87)and multiply the right side of the equation by $\frac{\theta_c}{\theta_c}$, as shown below in Equation (7.88):

$$Y_{obs} = \frac{Y\left(\frac{dS}{dt}\right)_U - k_d X}{\left(\frac{1}{\theta_c} + k_d\right)\frac{X}{Y}} = \frac{Y\frac{(dS/dt)_U}{X} - \frac{k_d X}{X}}{\left(\frac{1}{\theta_c} + k_d\right)\frac{1}{Y}}$$

$$= \frac{Y\left(\frac{1}{\theta_c}\right)}{\left(\frac{1}{\theta_c} + k_d\right)}\left(\frac{\theta_c}{\theta_c}\right) \qquad (7.88)$$

$$Y_{obs} = \frac{Y}{1 + k_d\theta_c} \qquad (7.89)$$

Sherrard & Schroeder (1972) proposed that the net microbial growth rate could be modeled by the following equation:

$$\left(\frac{dX}{dt}\right)_{NG} = Y_{obs}\left(\frac{dS}{dt}\right)_U \qquad (7.90)$$

On a finite time basis, generally one day, Equation (7.90) can be expressed as:

$$P_x = Y_{obs}Q(S_i - S_e) \qquad (7.91)$$

where
P_x = quantity of excess biomass produced on a dry weight basis, $\frac{\text{lb TSS}}{\text{d}}\left(\frac{\text{kg TSS}}{\text{d}}\right)$.

Substituting Equation (7.89) into Equation (7.91) yields the final equation for calculating the quantity of excess biomass produced:

$$\boxed{P_x = \frac{YQ(S_i - S_e)}{1 + k_d\theta_c}} \qquad (7.92)$$

The total quantity of sludge that is produced at a WWTP not only includes the excess biomass produced as a result of substrate removal, but also from a buildup of non-degradable volatile suspended solids (NDVSS) and fixed suspended solids (FSS) that are present in the influent to the aeration basin. Equation (7.92) must be modified to account for these additional suspended solids as follows:

$$SP = P_x + Q(\text{NDVSS} + \text{FSS}) \qquad (7.93)$$

where:
SP = total quantity of sludge produced on a dry weight basis, $\frac{\text{lb TSS}}{\text{d}}\left(\frac{\text{kg TSS}}{\text{d}}\right)$
Q = flow rate to the aeration basin excluding RAS flow, MGD (m^3/d)

NDVSS = non-degradable VSS in aeration basin (excluding RAS flow), mg/L

FSS = fixed suspended solids concentration in aeration basin influent, excluding RAS flow, mg/L.

The total quantity of sludge that is produced daily is equal to the quantity of solids that are accidentally lost in the effluent (Q_eX_e), plus the quantity of solids that are purposely wasted from the system, as presented in Equation (7.94):

$$SP = Q_eX_e + Q_wX_r \tag{7.94}$$

where:

Q_eX_e = quantity of sludge solids accidentally lost in the effluent, $\frac{lb}{d}\left(\frac{kg}{d}\right)$.

Q_wX_r = quantity of sludge solids purposely wasted from the system, $\frac{lb}{d}\left(\frac{kg}{d}\right)$.

The actual quantity of sludge solids that must be purposely wasted from the activated sludge process can be calculated by rearranging Equation (7.94) and solving for Q_wX_r as follows:

$$Q_wX_r = SP - Q_eX_e \tag{7.95}$$

Either knowing or assuming the TSS concentration X_r in the RAS flow or underflow, the actual WAS flow rate (Q_w) can be estimated as follows:

$$Q_w = \frac{Q_wX_r}{X_r} \tag{7.96}$$

7.5.6.7 Nitrifying Activated Sludge Systems

Nitrification is an aerobic biological process mediated by autotrophic bacteria collectively called the "nitrifiers." These microorganisms use inorganic carbon (CO_2 and HCO_3^-) for synthesizing cellular components and derive energy from the oxidation of ammonia to nitrite and, ultimately, to nitrate. Nitrification is traditionally modeled as a two-step sequential reaction, in which the genus *Nitrosomonas* oxidizes ammonium to nitrite followed by oxidation of nitrite to nitrate by the genus *Nitrobacter*. Equations (7.97) through (7.100) show the energy equations involved with nitrification:

$$2NH_4^+ + 3O_2 \xrightarrow{Nitrosomonas} 2NO_2^- + 4H^+ + 2H_2O \tag{7.97}$$

$$2NO_2^- + O_2 \xrightarrow{Nitrobacter} 2NO_3^- \tag{7.98}$$

$$2NH_4^+ + 4O_2 \xrightarrow{Nitrifiers} 2NO_3^- + 4H^+ + 2H_2O \tag{7.99}$$

$$NH_4^+ + 2O_2 \xrightarrow{Nitrifiers} NO_3^- + 2H^+ + H_2O \tag{7.100}$$

Based on the overall nitrification reaction shown as Equation (7.100), the following conclusions can be ascertained:

1

$$\left(\frac{2\times 32\,gO_2}{14\,gNH_4^+-N}\right) = \boxed{\frac{4.57\,gO_2\text{ are required}}{gNH_4^+-N\text{ oxidized}}}$$

2

$$\left(\frac{2\times 1\,gH^+}{14\,gNH_4^+-N} \times \frac{1\,eqH^+}{1\,gH^+} \times \frac{50\,gCaCO_3}{eq\,CaCO_3}\right) = \boxed{\frac{7.14\text{ g alkalinity as }CaCO_3\text{ are required}}{gNH_4^+-N\text{ oxidized}}}$$

In addition to requiring a minimum dissolved oxygen (DO) concentration of 2.0 mg/L to ensure complete nitrification, a pH in the range of 7.5–8.0 is recommended. According to Metcalf & Eddy (2003, p. 615), MCRT ranges from 10–20 days at 10°C and from 4–7 days at 20°C to achieve nitrification. At a DO level of < 0.5 mg/L, nitrification is inhibited and nitrite accumulation or build-up will occur since *Nitrobacter* are inhibited (Mines & Sherrard, 1983, p. 843; Metcalf & Eddy, 2003, p. 615).

Painter (1970) indicated that other autotrophic bacteria genera, *Nitrosococcus*, *Nitrosospira*, *Nitrosolobus*, and *Nitrosorobrio* are also capable of oxidizing ammonia to nitrite. *Nitrococcus*, *Nitrospira*, *Nitrospina*, and *Nitroeystis* have also been found to oxidize nitrite to nitrate (Metcalf & Eddy, 2003, p. 612). The oxidation of ammonia to nitrite can be considered to be the rate-limiting step in the overall nitrification process (Metcalf & Eddy, 2003, p. 614) at temperatures below 28°C. Therefore, the biokinetic coefficients for *Nitrosomonas* may be used to design nitrification systems. Typical biokinetic coefficients for *Nitrosomonas* treating domestic wastewater are presented in Table 7.11.

Table 7.11 Typical biokinetic coefficients at 20°c for suspended growth nitrification processes.

Biokinetic coefficient	Basis	Range	Typical
^aY_n	$g VSS/gNH_4^+-N$	0.10–0.15	0.12
^bY_n	$g VSS/gNH_4^+-N$	0.04–0.29	0.15
^cY_n	$g COD/gNH_4^+-N$	–	0.24
^dY_n	$g VSS/gNH_4^+-N$	–	0.10
$^a\mu_{mn}$	$g VSS/g VSS\cdot d$	0.20–0.90	0.75
$^c\mu_{mn}$	h^{-1}	–	0.032
^aK_n	gNH_4^+-N/m^3	0.5–1.0	0.74
^bK_n	gNH_4^+-N/m^3	0.2–5.0	1.4
^cK_n	gNH_4^+-N/m^3	–	1.0
^dK_n	gNH_4^+-N/m^3	–	1.0
$^ak_{dn}$	$g VSS/g VSS\cdot d$	0.05–0.15	0.08
$^bk_{dn}$	d^{-1}	0.03–0.06	0.05
$^ck_{dn}$	h^{-1}	–	0.004
$^dk_{dn}$	d^{-1}	–	0.04

[a] Metcalf & Eddy (2003), p. 705.
[b] WEF (1998b) Vol 2, pp. 11–19.
[c] Grady *et al.* (2011), p. 202.
[d] Henze *et al.* (2008), p. 90.

7.5.6.8 Biokinetic Coefficients for Activated Sludge

Equations (7.72), (7.83), and (7.92) were developed for calculating the effluent soluble substrate concentration, heterotrophic biomass concentration, and quantity of excess heterotrophic biomass produced during biological treatment. To use these equations, biokinetic coefficients for heterotrophic microorganisms must be assumed or determined through wastewater treatability studies. These same equations are applicable for other types of microorganisms, such as autotrophic nitrifying bacteria and heterotrophic denitrifying bacteria. Typical biokinetic coefficients obtained from laboratory and field studies are shown in Table 7.12 for activated sludge systems. Activated sludge biokinetic coefficients for removal of carbonaceous material (based on biodegradable COD, bCOD) by heterotrophic bacteria are presented in Table 7.13.

7.5.6.9 Temperature Corrections

Variations in temperature can have a significant impact on the design and operation of biological treatment systems. In general, Benefield *et al.* (1975) found that temperature variations have a minimal effect on the yield coefficient (Y) in the temperature range between 15–25°C. However, the endogenous decay coefficient (k_d) varied within this temperature range. The maximum specific substrate utilization rate (k), maximum specific growth rate (μ_{max}), and half-velocity coefficient (K_s) should be corrected for temperature (Metcalf & Eddy, 2003, pp. 704–705).

Temperature corrections are made using the modified Arrhenius relationship, as shown below:

$$K_2 = K_1(\theta)^{(T_2-T_1)} \quad (7.101)$$

where:

K_2 = reaction rate coefficient at temperature T_2
K_1 = reaction rate coefficient at temperature T_1
Θ = temperature correction coefficient, dimensionless
Θ = ranges from 1.02 to 1.09, depending upon the application
T_1 and T_2 = temperature at two different conditions, °C, (°F).

Values for all of the following temperature correction coefficients are found in Metcalf & Eddy (2003, pp. 704–705). The temperature correction coefficient (θ) for k_d varies from 1.03 to 1.08, with 1.04 being typically used for both heterotrophic and nitrification reactions. A theta value of 1.0 is used to make temperature corrections for K_s values involving heterotrophic reactions. Theta values ranging from 1.03–1.123 are used for making temperature corrections for K_s values involving nitrification reactions, with a typical value of 1.053. Temperature correction coefficients for the heterotrophic μ_{max} coefficient range from 1.03–1.08, with a typical value of 1.07, while for nitrification reactions, θ varies from 1.06–1.123, with a typical value of 1.07. The temperature correction coefficients applied to μ_{max} should also be applicable to the maximum specific substrate utilization rate k since temperature has a minimal effect on the yield coefficient and $\mu_{max} = Yk$.

7.5.6.10 Nutrient Requirements

For biological wastewater treatment systems to operate properly, sufficient concentrations of nitrogen and phosphorus must be available for microbial synthesis. Domestic wastewater generally has excess nitrogen (N) and phosphorus (P). However, a deficiency may exist for various types of industrial wastewaters.

A $BOD_5 : N : P$ ratio of 100 : 5 : 1 is commonly considered to be acceptable for accomplishing biological treatment.

Table 7.12 Typical biokinetic coefficients at 20°C for activated sludge treating domestic wastewater at 20°C.

Biokinetic coefficient	Basis	Range	Typical value
[a]Y	g VSS/g BOD	0.4–0.8	0.6
[a]Y	g VSS/g bsCOD	0.3–0.6	0.4
[b]Y	g VSS/g COD	0.25–0.4	0.4
[a]k	g bsCOD/g VSS · d	2–10	5
[a]K_s	g BOD/m^3	25–100	60
[a]K_s	g bsCOD/m^3	10–60	40
[a]k_d	g VSS/g VSS · d	0.06–0.15	0.10
[b]k_d	d^{-1}	0.004–0.075	0.06
[c]μ_{max}	g VSS/g VSS · d	0.6–6	2

[a] Metcalf & Eddy (2003) p. 585; bsCOD = biodegradable chemical oxygen demand.
[b] WEF(1998b) Vol 2, pp. 11–18.
[c] Calculated using $\mu_{max} = Yk$.

Table 7.13 Typical activated sludge biokinetic coefficients for heterotrophic bacteria at 20°C.

Biokinetic coefficient	Basis	Range	Typical value
[a]Y	g VSS/g bCOD	0.3–0.5	0.40
[b]Y	g TSS/g COD	–	0.50
[c]Y	g COD/g COD	–	0.67
[c]Y	g VSS/g COD	–	0.45
[d]k	g VSS/g VSS · d	10–26.4	15
[a]K_s	g bCOD/m^3	5–40	20
[b]K_s	g COD/m^3	–	20
[a]k_d	g VSS/g VSS · d	0.06–0.20	0.12
[b]k_d	d^{-1}	–	0.18
[a]μ_{max}	g VSS/g VSS · d	3.0–13.2	6.0
[b]μ_{max}	d^{-1}	–	6.0

[a] Metcalf & Eddy (2003), p. 704; bCOD = biodegradable chemical oxygen demand.
[b] Grady *et al.* (2011), p. 410.
[c] Henze *et al.* (2008), p. 58.
[d] Calculated using: $k = \frac{\mu_{max}}{Y}$.

As MCRT increases, less biomass is wasted from activated sludge systems, so the quantity of N and P required by the microorganisms decreases, too. Sherrard & Schroeder (1976) found that the BOD_5/N/P ratio varies from approximately 100 : 5.4 : 1 at $\theta_c = 3$ days to 200 : 5.4 : 1 at $\theta_c = 20$ days.

Using the molecular formula of $C_{60}H_{87}O_{23}N_{12}P$ to represent the composition of biomass (microorganisms), the following equations may be used to estimate the quantity of nitrogen and phosphorus required as a function of MCRT. The molecular weight is 1374. Nitrogen represents 12.2% $\left(\frac{12\times14}{1374}\times100\%\right)$ and phosphorus represents 2.3% $\left(\frac{1\times31}{1374}\times100\%\right)$ of the dry weight fraction of biomass.

$$\text{N required} = 0.122P_x \quad (7.102)$$

where:

N required = quantity of nitrogen required for synthesis, $\frac{\text{lb}}{\text{d}}\left(\frac{\text{kg}}{\text{d}}\right)$

0.122 = decimal fraction of nitrogen content of biomass.

Recall that P_x is quantity of excess biomass produced on a dry weight basis:

$$\text{P required} = 0.023P_x \quad (7.103)$$

where:

P required = quantity of phosphorus required for synthesis, $\frac{\text{lb}}{\text{d}}\left(\frac{\text{kg}}{\text{d}}\right)$

0.023 = decimal fraction of phosphorus content of biomass.

7.5.6.11 Oxygen Requirements

The heterogeneous culture of microorganisms in the aeration basin primarily consists of aerobes; therefore, they require oxygen to meet metabolic requirements. Oxygen serves as the electron acceptor in aerobic processes, with the ultimate end product being water. In most activated sludge systems, air (which contains 21% oxygen by volume or 23% oxygen by weight) is supplied either by diffused aeration or mechanical aerators. Diffused aeration systems are more efficient on a mass transfer basis than mechanical aerators. Diffusers should be used in colder climates, where freezing conditions can prevail for extended time periods. Heat is added to the air as the blowers compress the air for transport to and through the diffuser system. This helps to maintain a higher temperature in the mixed liquor than can be achieved with mechanical surface aerators.

Typically, a minimum of 2.0 mg/L of dissolved oxygen (DO) is maintained in the aeration basin to meet process oxygen requirements. In high purity oxygen activated sludge processes, the DO concentration may range from 4–10 mg/L (Reynolds & Richards, 1996, p. 450). Sufficient oxygen must be transferred to the mixed liquor to meet both the carbonaceous and nitrogenous oxygen demands. The carbonaceous oxygen demand relates to the oxygen required by heterotrophic organisms for stabilizing the organic matter of the wastewater as measured by BOD or COD. The nitrogenous oxygen demand (NOD) refers to the quantity of oxygen required during the nitrification process.

The total quantity of oxygen required to meet process requirements (metabolic requirements) is calculated using Equation (7.104). If nitrification is not desired, a MCRT less than five days is typically used and the NOD term in the equation is set equal to zero.

$$O_2 = Q(S_i - S_e)(1 - 1.42Y) + 1.42k_d X\forall + \text{NOD} \quad (7.104)$$

$$\text{NOD} = Q(\text{TKN}_o)(4.57) \quad (7.105)$$

where:

O_2 = total quantity of oxygen required to meet process requirements, $\frac{\text{lb}}{\text{d}}\left(\frac{\text{kg}}{\text{d}}\right)$

NOD = nitrogenous oxygen demand that occurs during nitrification, $\frac{\text{lb}}{\text{d}}\left(\frac{\text{kg}}{\text{d}}\right)$

TKN_o = total Kjeldahl nitrogen concentration entering the aeration basin excluding RAS flow, mg/L

4.57 = quantity of oxygen required to oxidize ammonia to nitrate, $\frac{\text{lb }O_2}{\text{lb }NH_3\text{–N}}\left(\frac{\text{kg }O_2}{\text{kg}NH_3\text{–N}}\right)$.

Example 7.9 Activated sludge design without nitrification

A complete-mix activated sludge process (CMAS) is to be designed to treat 6.0 million gallons per day (MGD) of domestic wastewater having a BOD_5 of 250 mg/L. The NPDES permit requires that the effluent BOD_5 and TSS concentrations should be 20 mg/L or less on an annual average basis. The following biokinetic coefficients obtained at 20°C will be used in designing the process: $Y =$ 0.6 mg VSS/mg BOD_5, $k = 5\ d^{-1}$, $K_s = 60$ mg/L BOD_5, and $k_d = 0.06\ d^{-1}$. Assume that the MLVSS concentration in the aeration basin is maintained at 3,000 mg/L and the VSS : TSS ratio is 0.80. The temperature of the wastewater during the winter months is expected to remain at 18°C for extended periods. During the summer, the wastewater temperature may reach 25°C for several weeks. Determine the following:

a) Effluent soluble BOD_5 ($SBOD_5$) concentration necessary to meet the effluent total BOD_5 ($TBOD_5$) requirement of 20 mg/L.
b) Mean cell residence time necessary to meet the NPDES permit during the winter months.
c) Volume of the aeration basin in cubic feet.
d) Mean cell residence time necessary to meet the NPDES permit during the summer months.
e) Oxygen requirements assuming no nitrification during the summer months.
f) The quantity of excess biomass produced in terms of TSS at the shortest MCRT that the facility will be operated.

Solution Part A

First, the effluent particulate BOD_5 ($PBOD_5$) concentration is calculated from Equation (7.76):

$$PBOD_5 = (1.42VSS_e)\left(\frac{2}{3}\right) \quad (7.76)$$

$$VSS_e = (0.80)\left(\frac{20\,mg}{L}\right) = \frac{16\,mg}{L}$$

$$PBOD_5 = 1.42\left(\frac{16\,mg}{L}\right)\left(\frac{2}{3}\right) = \frac{15.1\,mg}{L}$$

The effluent soluble BOD_5 ($SBOD_5$) concentration is estimated from Equation (7.75):

$$TBOD_5 = PBOD_5 + SBOD_5 \quad (7.75)$$

$$SBOD_5 = TBOD_5 - PBOD_5 = 20 - 15.1 = \boxed{4.9\,\frac{mg}{L}}$$

Solution Part B

Before the MCRT can be determined, the biokinetic coefficients k, K_s, and k_d must be corrected from a temperature of 20°C to 18°C. It is unnecessary to correct the yield coefficient for temperature variations. Temperature correction coefficients (θ) of 1.07, 1.00, and 1.04 will be used to correct k, K_s, and k_d, respectively. Equation (7.101) is used to correct for temperature variations:

$$K_2 = K_1(\theta)^{(T_2 - T_1)} \quad (7.102)$$

$$(k)_{18°C} = 5\,d^{-1}(1.07)^{(18-20°C)} = 4.4\,d^{-1}$$

$$(K_s)_{18°C} = 60\,\frac{mg}{L}(1.00)^{(18-20°C)} = 60\,\frac{mg}{L}$$

$$(k_d)_{18°C} = 0.06\,d^{-1}(1.04)^{(18-20°C)} = 0.06\,d^{-1}$$

Calculate the MCRT using Equations (7.58) and (7.59):

$$\left(\frac{dX}{dt}\right)_{NG} = \frac{(Q - Q_w)X_e + Q_w X_r}{X\forall} = \frac{1}{\theta_c} = Y\frac{(dS/dt)_U}{X} - k_d \quad (7.58)$$

$$\left(\frac{dS}{dt}\right)_U = \frac{kXS_e}{K_S + S_e} \quad (7.59)$$

$$\frac{1}{\theta_c} = Y\frac{(dS/dt)_U}{X} - k_d = Y\left(\frac{k\,S_e}{K_S + S_e}\right) - k_d \quad (7.62)$$

$$\frac{1}{\theta_c} = \frac{0.6\,mg\,VSS}{mg\,BOD_5}\left(\frac{4.4\,d^{-1} \times 4.9\,mg/L}{60\,mg/L + 4.9\,mg/L}\right) - 0.06\,d^{-1}$$

$$= 0.14\,d^{-1}$$

$$\theta_c = \frac{1}{0.14\,d^{-1}} = \boxed{7.1\,d}$$

Solution Part C

The volume of the aeration basin is calculated from Equation (7.84):

$$\forall = \frac{YQ(S_i - S_e)\,\theta_c}{(1 + k_d\theta_c)\,X} \quad (7.84)$$

$$\forall = \frac{\frac{0.60\,mg\,VSS}{mg\,BOD_5}\left(6.0 \times 10^6\,\frac{gal}{d}\right)\left(\frac{250\,mg}{L} - \frac{4.9\,mg}{L}\right)}{(1 + 0.06\,d^{-1} \times 7.1\,d)} \times \frac{7.1\,d}{3000mg/L}$$

$$= 1.46 \times 10^6\,gal$$

$$\forall = 1.46 \times 10^6\,gal\left(\frac{1\,ft^3}{7.48\,gal}\right) = \boxed{1.95 \times 10^5\,ft^3}$$

Solution Part D

Calculate the MCRT necessary to meet effluent requirements during the summer months. First, the biokinetic coefficients must be corrected from 20°C to 25°C. Use the same temperature correction coefficients that were used in Part B:

$$K_2 = K_1(\theta)^{(T_2 - T_1)} \quad (7.101)$$

$$(k)_{18°C} = 5\,d^{-1}(1.07)^{(25-20°C)} = 7.0\,d^{-1}$$

$$(K_s)_{18°C} = 60\,\frac{mg}{L}(1.00)^{(25-20°C)} = 60\,\frac{mg}{L}$$

$$(k_d)_{18°C} = 0.06\,d^{-1}(1.04)^{(25-20°C)} = 0.07\,d^{-1}$$

Use Equations (7.63) to determine MCRT:

$$\frac{1}{\theta_c} = Y\frac{(dS/dt)_U}{X} - k_d = Y\left(\frac{k\,S_e}{K_S + S_e}\right) - k_d \quad (7.63)$$

$$\frac{1}{\theta_c} = \frac{0.6\,mg\,VSS}{mg\,BOD_5}\left(\frac{7.0\,d^{-1} \times 4.9\,mg/L}{60\,mg/L + 4.9\,mg/L}\right) - 0.07d^{-1}$$

$$= 0.25\,d^{-1}$$

$$\theta_c = \frac{1}{0.25\,d^{-1}} = \boxed{4.0\,d}$$

Solution Part E

The actual biomass concentration X at a MCRT of 4.0 days is determined from Equation (7.83). Recall that the volume of the aeration basin is 1.46×10^6 gal = 1.46 MG. The detention time τ is determined below:

$$\tau = \frac{\forall}{Q} = \frac{1.46\,MG}{6.0\,MGD} = 0.24\,d \quad (7.49)$$

$$X = \frac{Y(S_i - S_e)\,\theta_c}{(1 + k_d\,\theta_c)\,\tau} \quad (7.83)$$

$$X = \frac{\frac{0.60\,mg\,VSS}{mg\,BOD_5}\left(\frac{250\,mg}{L} - \frac{4.9\,mg}{L}\right)}{(1 + 0.07\,d^{-1} \times 4.0\,d)}\,\frac{4.0\,d}{0.24\,d}$$

$$= \boxed{\frac{1915\,mg\,VSS}{L}}$$

Finally, the oxygen requirements are determined from Equation (7.104):

$$O_2 = Q(S_i - S_e)(1 - 1.42Y) + 1.42k_d X\mathcal{V} + \text{NOD} \quad (7.104)$$

The nitrogenous oxygen demand (NOD) will be assumed to be zero since nitrification was assumed not to occur at a MCRT of 4.0 days.

$$O_2 = \left[\frac{6.0\,\text{MG}}{\text{d}}(250 - 4.9)\frac{\text{mg}}{\text{L}}\left(1 - 1.42 \times \frac{0.60\,\text{mg VSS}}{\text{mg BOD}_5}\right) + 1.42 \times 0.07\,\text{d}^{-1} \times \frac{1915\,\text{mg}}{\text{L}} \times 1.46\,\text{MG} + 0\right] \times \frac{8.34\,\text{lb}}{\text{MG} \times \frac{\text{mg}}{\text{L}}}$$

$$O_2 = \boxed{4130\,\frac{\text{lbO}_2}{\text{d}}}$$

Solution Part F

The largest quantity of excess biomass produced occurs at the shortest MCRT, which in this case is 4.0 days. Excess biomass is determined from Equation (7.92):

$$P_x = \frac{YQ(S_i - S_e)}{1 + k_d\theta_c} \quad (7.92)$$

$$P_x = \frac{YQ(S_i - S_e)}{1 + k_d\theta_c} = \frac{\frac{0.60\,\text{mg VSS}}{\text{mg BOD}_5}(6.0\,\text{MGD})\left(\frac{250\,\text{mg}}{\text{L}} - \frac{4.9\,\text{mg}}{\text{L}}\right)}{1 + 0.07\,\text{d}^{-1} \times 4.0\,\text{d}} \times \left(\frac{8.34\,\text{lb}}{\text{MG} \cdot \frac{\text{mg}}{\text{L}}}\right)$$

$$P_x = 5750\,\frac{\text{lb VSS}}{\text{d}}$$

$$P_x = 5750\,\frac{\text{lb VSS}}{\text{d}} \times \left(\frac{1\,\text{lb TSS}}{0.80\,\text{lb VSS}}\right) = \boxed{7190\,\frac{\text{lb TSS}}{\text{d}}}$$

7.5.7 Biokinetic coefficient determination

To use the kinetic equations developed in previous sections of this chapter, one must conduct wastewater treatability studies to determine the various biokinetic coefficients, or assume typical values from the literature. When possible, it is prudent to perform bench-scale and/or pilot-scale studies for 6–12 months or more on the particular wastewater to be evaluated. Normally, several CMAS reactors are operated in parallel at different MCRTs. Various analyses are performed on the influent, effluent, and mixed liquor within the reactor. A schematic of a complete-mix, bench-scale reactor used for such studies is shown in Figure 7.14.

In addition to measuring the influent, effluent, and sludge wastage rates, the parameters listed in Table 7.14 are typically monitored once steady-state conditions have been reached. Data are normally collected daily for a minimum period of 1–2 weeks once reaching steady-state. Recall from Chapter 6 that the reactors must be operated a minimum of three MCRTs, e.g., 3 × MCRT value, before the system will reach steady-state conditions. For instance, if the CMAS bench-scale reactor is to be operated at an MCRT of ten days, the system should be operated for a minimum of 30 days prior to collecting data.

7.5.7.1 Yield and Endogenous Decay Coefficients

The yield and endogenous decay coefficients are determined by plotting the reciprocal of MCRT on the y-axis versus specific substrate utilization rate (U) on the x-axis according to

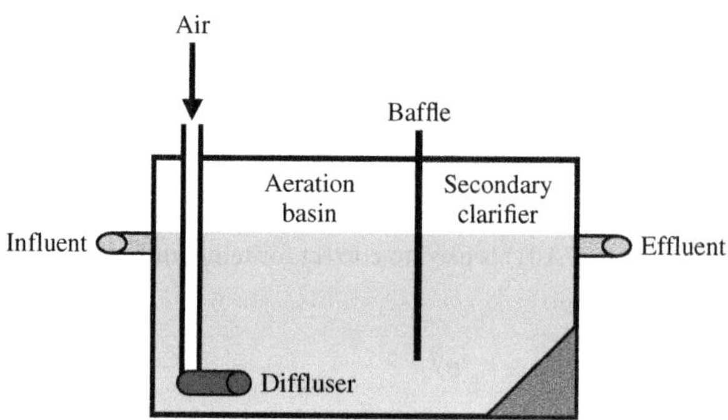

Figure 7.14 Bench-scale CMAS reactor.

Table 7.14 Parameters to monitor for evaluating biokinetic coefficients.

Parameter	Locations
Flow rate	Influent, effluent, wasted sludge
BOD_5 or COD (total and soluble)	Influent, effluent, wasted sludge
TKN	Influent, effluent, wasted sludge
$NH_3 - N$	Influent, effluent, wasted sludge
$NO_3^- - N$	Influent, effluent
$NO_2^- - N$	Effluent
Dissolved oxygen (DO)	Influent, effluent, mixed liquor
Oxygen uptake rate (OUR)	Mixed liquor
Total phosphorus (TP)	Influent, effluent, wasted sludge
Orthophosphorus	Influent, effluent, wasted sludge
TSS or VSS	Influent, effluent, wasted sludge

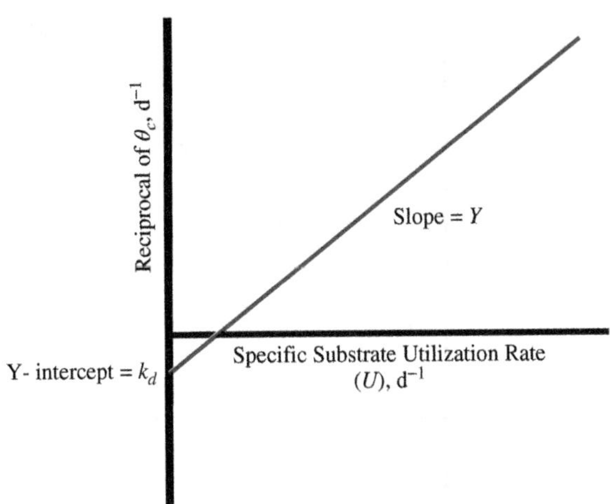

Figure 7.15 Determination of Y and k_d biokinetic coefficients.

Equation (7.106):

$$\frac{1}{\theta_c} = YU - k_d \quad (7.106)$$

As shown in Figure 7.15, the slope of the line of best fit is equal to the yield coefficient (Y) and the y-intercept is equal to k_d. Equation (7.48) is used to calculate the MCRT or θ_c values.

Specific substrate utilization rate may also be defined as follows:

$$\left(\frac{dS}{dt}\right)_U = UX \quad (7.107)$$

For complete-mix activated sludge processes, substrate utilization was defined by Equation (7.79):

$$\left(\frac{dS}{dt}\right)_U = \frac{Q(S_i - S_e)}{\forall} \quad (7.108)$$

Equation (7.109) is used to calculate the value of U:

$$U = \frac{(dS/dt)_U}{X} = \frac{Q(S_i - S_e)}{X\forall} \quad (7.109)$$

The specific substrate utilization rate is required in several of the plots for estimating biokinetic coefficients.

7.5.7.2 Substrate Removal Biokinetic Coefficients

Substrate removal kinetics may be zero-, first-, pseudo-first, hyperbolic, or fractional order. Equations (7.59) through (7.61) give the most common kinetic reactions encountered in biological treatment systems. Typically, either a pseudo-first-order or Michaelis-Menten type equation is used to model substrate removal kinetics.

Assuming that substrate removal follows pseudo-first order removal kinetics according to Equation (7.60), Equation (7.109) can be derived to calculate the value of K:

$$\left(\frac{dS}{dt}\right)_U = KXS_e \quad (7.60)$$

$$\frac{(dS/dt)_U}{X} = U = KS_e \quad (7.109)$$

Plotting U on the y-axis versus the effluent soluble substrate concentration S_e on the x-axis yields a line with the slope equal to the valuc of K. If a pseudo-first-order substrate removal reaction occurs, Equation (7.60) will plot as Figure 7.16 when BOD_5 is used for measuring the substrate concentration, whereas Figure 7.17 will result when substrate concentration is measured as COD. The x-intercept represents the non-degradable COD.

If Michaelis-Menten type kinetics is assumed to exist, as presented in Equation (7.59), then Equation (7.112) is plotted to determine the value of k and K_s:

$$\left(\frac{dS}{dt}\right)_U = \frac{kXS_e}{K_S + S_e} \quad (7.59)$$

Dividing both sides of Equation (7.59) by X yields:

$$\frac{(dS/dt)_U}{X} = U = \frac{kS_e}{K_s + S_e} \quad (7.110)$$

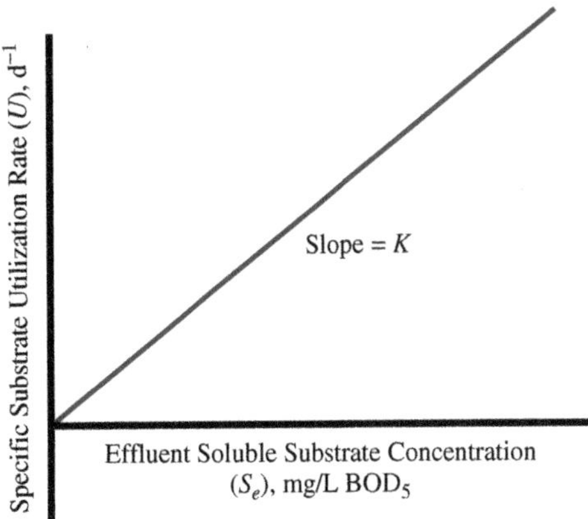

Figure 7.16 Determination of K biokinetic coefficient using BOD_5.

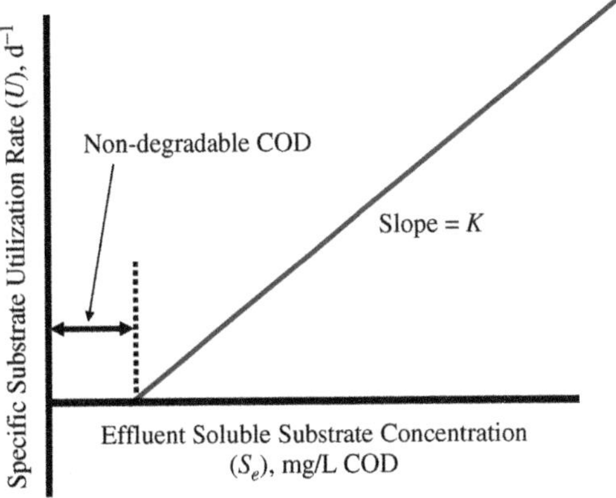

Figure 7.17 Determination of K biokinetic coefficient using COD.

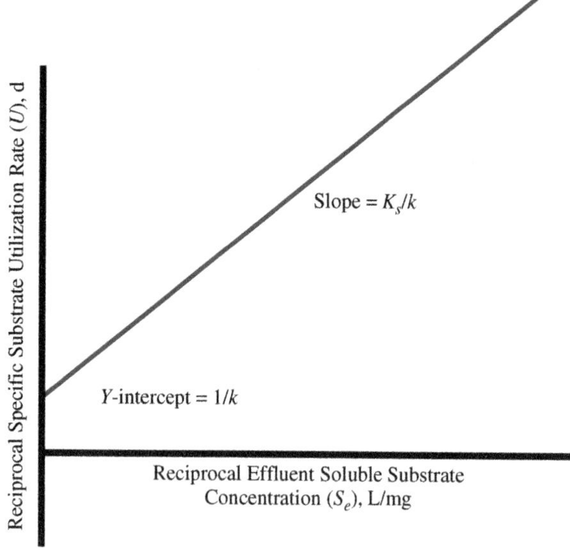

Figure 7.18 Determination of biokinetic coefficients k and Ks.

Taking the reciprocal of both sides of Equation (7.110) results in Equation (7.111):

$$\frac{1}{U} = \frac{K_S + S_e}{kS_e} \tag{7.111}$$

Separating the variables on the right side of Equation (7.111) results in Equation (7.112):

$$\frac{1}{U} = \frac{K_S}{k}\frac{1}{S_e} + \frac{1}{k} \tag{7.112}$$

A plot of the reciprocal of U on the y-axis versus the reciprocal of the effluent soluble substrate concentration S_e on the x-axis will produce a straight line, as shown in Figure 7.18.

The slope of the line will be equal to $\frac{K_s}{k}$ and the y-intercept will be equal to $\frac{1}{k}$.

7.5.7.3 Oxygen Use Coefficients

In lieu of using Equation (7.104) for estimating oxygen requirements for activated sludge systems, Equation (7.113) may be used when treatability studies are conducted on the specific wastewater to be treated in the full-scale system:

$$\frac{dO_2}{dt} = a\left(\frac{dS}{dt}\right)_U + bX \tag{7.113}$$

where:

$\frac{dO_2}{dt}$ = oxygen utilization rate, $\frac{\text{mass}}{(\text{volume}\cdot\text{time})}$

a = oxygen use coefficient for synthesis, $\frac{\text{mgO}_2}{(\text{mg substrate used})}$

b = oxygen use coefficient for energy of maintenance, $\frac{\text{mgO}_2}{(\text{mg biomass}\cdot\text{d})}$.

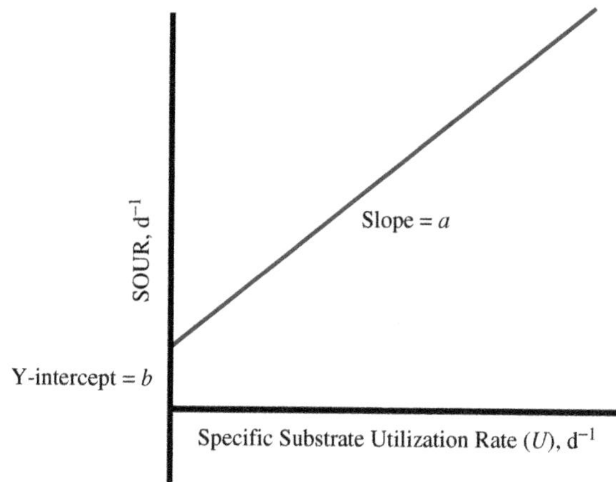

Figure 7.19 Determination of oxygen use coefficients a and b.

The oxygen utilization rate (OUR), also known as oxygen uptake rate, is determined by placing a sample of mixed liquor in a BOD bottle, saturating with oxygen, and then measuring the DO concentration versus time (*Standard Methods* (1998), p. 2-79). The slope of the linear portion of the graph is equal to OUR.

Dividing the OUR by the active biomass concentration X, yields the specific oxygen utilization rate (SOUR) as shown in Equation (7.114):

$$\text{SOUR} = \frac{\text{OUR}}{X} \tag{7.114}$$

where:

OUR = oxygen utilization rate, $\frac{\text{mass}}{(\text{volume}\cdot\text{time})}$

SOUR = specific oxygen utilization rate, time^{-1}.

Equation (7.113) is divided by the active biomass concentration X so that the oxygen use coefficients a and b may be determined. A plot of SOUR on the y-axis versus U on the x-axis should produce a straight line. The slope of the line of best fit yields the value of a and the y-intercept is equal to b (Figure 7.19).

$$\frac{dO_2/dt}{X} = a\left(\frac{(ds/dt)_U}{X}\right) + \frac{bX}{X}$$

$$\text{SOUR} = a\,(U) + b \tag{7.115}$$

Example 7.10 illustrates how to estimate the biokinetic coefficients Y, k, K_s, and k_d.

Example 7.10 Determination of biokinetic coefficients from treatability studies

A wastewater treatability study was performed on an industrial wastewater and the data in the table below were collected on four, separate bench-scale CMAS reactors.

Reactor #	MLVSS (mg/L)	τ (d)	S_i (mg/L)	S_e (mg/L)	OUR (mg/L · h)	$(dX/dt)_{NG}$ (mg MLVSS/L · d)
1	920	1.0	885	60	30	310
2	1350	1.0	885	45	35	340
3	2070	1.0	885	30	40	315
4	3880	1.0	885	15	60	140

Solution

A spreadsheet is an excellent resource for solving these types of problems and for graphing. The values in the table below are used to make the necessary plots to determine the biokinetic coefficients.

Reactor #	θ_c (d)	$1/\theta_c$ (d^{-1})	U (d^{-1})	1/U (d)	SOUR (d^{-1})
1	2.97	0.34	0.90	1.11	0.78
2	3.97	0.25	0.62	1.61	0.62
3	6.57	0.15	0.41	2.44	0.46
4	27.7	0.04	0.22	4.55	0.37

Sample Calculations

Calculations showing how each value in the above table for reactor #1 will be presented.

Mean cell residence time or θ_c is calculated using Equation (7.58):

$$\frac{(dX/dt)_{NG}}{X} = \frac{(Q - Q_w)X_e + Q_w X_r}{X\forall} = \frac{1}{\theta_c}$$

$$= Y\frac{(dS/dt)_U}{X} - k_d \qquad (7.58)$$

$$\frac{(dX/dt)_{NG}}{X} = \frac{1}{\theta_c}$$

$$\theta_c = \frac{920 \text{ mg MLVSS/L}}{(310 \text{ mg MLVSS/L} \cdot \text{d})} = \boxed{2.97 \text{ d}}$$

$$\frac{1}{\theta_c} = \frac{1}{2.97 \text{ d}} = \boxed{0.34 \text{ d}^{-1}}$$

The specific substrate utilization rate U is determined using Equation (7.51):

$$U = \frac{Q(S_i - S_e)}{X\forall} \qquad (7.51)$$

Recall the definition of detention time from Equation (7.49):

$$\tau = \frac{\forall}{Q} \qquad (7.49)$$

Substituting Equation (7.49) into Equation (7.51) results in the following equation for U:

$$U = \frac{(S_i - S_e)}{X\tau} = \frac{(885 \text{ mg/L} - 60 \text{ mg/L})}{920 \text{ mg/L} \times 1.0 \text{ d}} = \boxed{0.90 \text{ d}^{-1}}$$

$$\frac{1}{U} = \frac{1}{0.90 \text{ d}^{-1}} = \boxed{1.11 \text{ d}}$$

The specific oxygen utilization rate is determined using Equation (7.116):

$$\text{SOUR} = \frac{\text{OUR}}{X} \qquad (7.116)$$

$$\text{SOUR} = \frac{30 \text{ mg } O_2/(\text{L} \cdot \text{h})}{920 \text{mg MLVSS/L}}\left(\frac{24 \text{ h}}{\text{d}}\right) = \boxed{0.78 \text{ d}^{-1}}$$

Plot Equation (7.106) to determine Y and k_d:

$$\frac{1}{\theta_c} = YU - k_d \qquad (7.106)$$

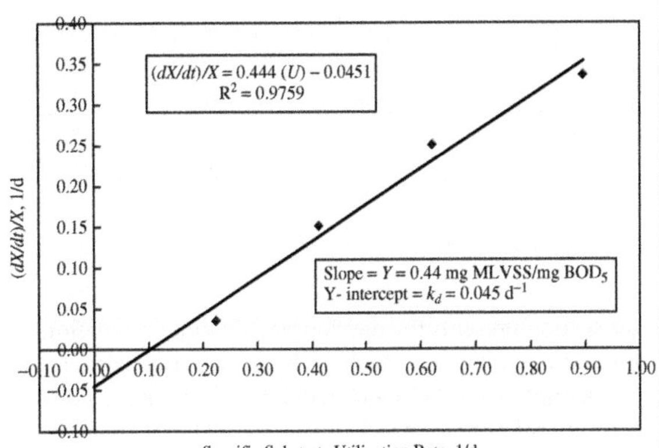

Figure E7.2

Plot Equation (7.112) to determine k and K_s.

$$\frac{1}{U} = \frac{K_s}{k}\frac{1}{S_e} + \frac{1}{k} \qquad (7.112)$$

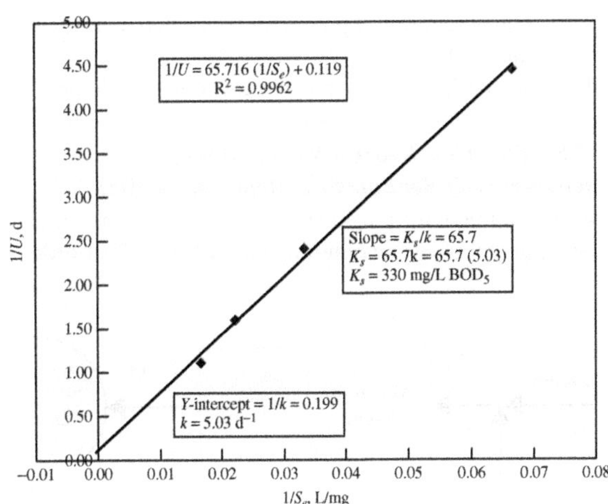

Figure E7.3

Plot Equation (7.115) to determine a and b.

$$\text{SOUR} = a(U) + b \tag{7.115}$$

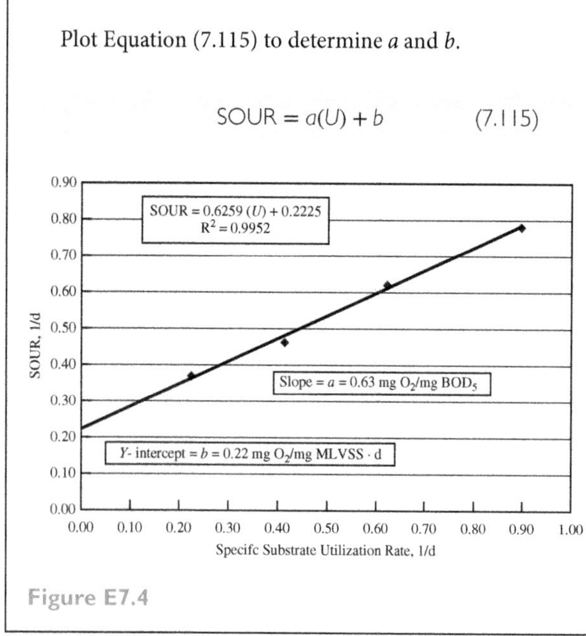

Figure E7.4

7.5.8 Biochemical kinetics of activated sludge in plug flow systems

This section presents the development of kinetic equations used for designing plug flow reactors (PFR). Material balances and the assumptions used in the model development are presented. The nomenclature used in Figure 7.20 will be used.

As previously developed in Chapter 5, the general steady-state equation for a plug flow reactor for component A in the wastewater is presented below:

$$Q dC_A = r_A d\forall \tag{7.117}$$

where:

dC_A = change in the concentration of component A in the wastewater, mass/volume

r_A = rate of removal or production of component A within the reactor

$d\forall_r$ = change in the elemental reactor volume, volume.

7.5.8.1 Plug Flow Reactor Volume Based on Pseudo-First-Order Substrate Removal Kinetics

The derivation of the plug flow reactor volume is presented, assuming that substrate removal follows pseudo-first-order removal kinetics. It is imperative to keep in mind that the equations developed in this section are based on the total flow entering the reactor $(Q + Q_r)$ and the substrate concentration S_o after the influent and recycle streams combine at the front of the aeration basin.

Assuming that the rate of substrate removal within a plug flow reactor is a pseudo-first-order removal reaction and that component A is the substrate concentration measured as BOD or COD, the following equation represents substrate removal. The minus sign denotes that substrate is being removed during the process.

$$r_s = \frac{dS}{dt} = -K\overline{X}S \tag{7.118}$$

where:

r_s = rate of substrate removal in plug flow reactor, mass/(volume · time)

K = rate constant for substrate removal, volume/(mass · time)

$\overline{X}$ = average microorganism concentration in a plug flow reactor, mass/volume

S = soluble substrate concentration surrounding the biomass, mass/volume.

The total flow rate into a plug flow reactor with recycle is actually $(Q + Q_r)$. This sum must be substituted into Equation (7.117) and the substrate concentration at the beginning of the aeration basin is S_o:

$$(Q + Q_r)\, dS = r_s\, d\forall_r = -K\overline{X}S\, d\forall_r \tag{7.119}$$

$$\int_{S_o}^{S_e} \frac{1}{S} dS = \frac{-K\overline{X}}{(Q + Q_r)} \int_0^{\forall_r} d\forall_r \tag{7.120}$$

$$\ln(S_e) - \ln(S_o) = -\frac{K\overline{X}\forall_r}{(Q + Q_r)} = -K\overline{X}\tau' \tag{7.121}$$

$$\tau' = \frac{\ln(S_e) - \ln(S_o)}{-K\overline{X}} = \frac{\ln(S_o) - \ln(S_e)}{K\overline{X}} \tag{7.122}$$

The effluent soluble substrate concentration from a plug flow reactor may be determined by taking the antilogarithm of both sides of Equation (7.121), resulting in Equation (7.123).

$$\frac{S_e}{S_o} = e^{-K\overline{X}\tau'} \tag{7.123}$$

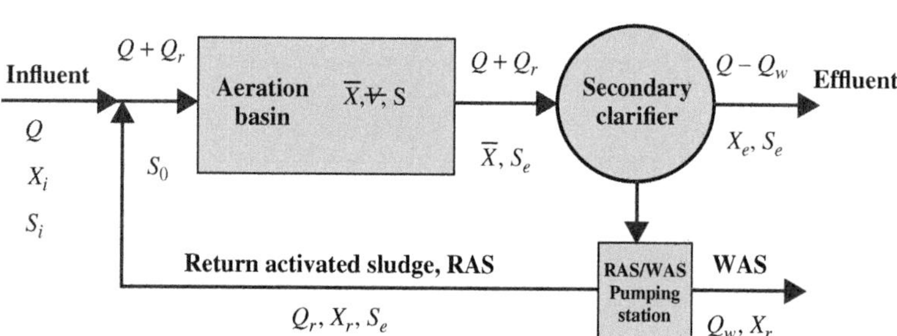

Figure 7.20 Schematic of plug flow activated sludge process.

where
τ' = actual liquid residence or treatment time in reactor, d or h, and is not equal to the nominal hydraulic detention time, τ, which is based only on the influent flow rate, Q.

The development of these kinetic equations is based on the total flow $(Q + Q_r)$ entering the aeration basin and the substrate concentration S_o, which is the value after the raw influent (S_i) and recycle streams meet. Therefore, it is imperative to remember that the actual detention time in plug flow and dispersed-flow reactors is calculated as follows:

$$\tau' = \frac{\forall}{(Q + Q_r)} \quad (7.124)$$

The volume of a plug flow reactor assuming pseudo, first-order substrate removal is calculated as follows:

$$\forall = \frac{(Q + Q_r)[\ln(S_o) - \ln(S_e)]}{K\overline{X}} \quad (7.125)$$

From Figure 7.20, a materials balance on substrate must be performed to calculate the substrate concentration in the combined influent to the head of the plug flow reactor as follows:

$$QS_i + Q_rS_e = (Q + Q_r)S_o \quad (7.126)$$

$$S_o = \frac{QS_i + Q_rS_e}{Q + Q_r} = \frac{QS_i + rQ}{Q + rQ} \quad (7.127)$$

where:
S_i = influent substrate concentration prior to mixing with the return activated sludge, mass/volume
S_o = Substrate concentration of combined influent and return activated sludge flows that is applied to the plug flow reactor, mass/volume
r = Recycle ratio or Q_r/Q, dimensionless.

7.5.8.2 Plug Flow Reactor Volume Based on Michaelis-Menten Substrate Removal Kinetics

The model assumes that the biomass concentration in the effluent from the aeration basin is not significantly different than the influent concentration. Therefore, an average concentration of microorganisms or biomass $(\overline{X})$ is used in the derivation. This is only applicable if $\frac{MCRT}{\tau}$ is greater than 5. The derivation of the plug flow reactor volume is presented on the assumption that substrate removal follows Michaelis-Menten kinetics. It is imperative to keep in mind that the equations developed in this section are based on the total flow entering the reactor $(Q + Q_r)$ and the substrate concentration S_o after the influent and recycle streams combine at the front of the aeration basin.

We begin with Equation (7.117), substituting S for C_A:

$$Q\,dS = r_s\,d\forall \quad (7.117)$$

The derivation of the plug flow reactor volume is presented assuming that substrate removal follows pseudo-first-order removal kinetics.

$$\frac{dS}{dt} = -\frac{k\overline{X}S}{K_s + S} \quad (7.128)$$

$$(Q + Q_r)\,dS = r_s\,d\forall = -\frac{k\overline{X}S}{K_s + S} \quad (7.129)$$

$$\int_{S_0}^{S_e} \frac{(K_s + S)}{S}\,dS = \frac{-k\overline{X}}{(Q + Q_r)}\int_0^{\forall} d\forall \quad (7.130)$$

$$K_s\int_{S_0}^{S_e} \frac{1}{S}\,dS + \int_{S_0}^{S_e} dS = \frac{-k\overline{X}\forall}{(Q + Q_r)} \quad (7.131)$$

$$K_s \ln\frac{S_e}{S_0} + (S_e - S_0) = \frac{-k\overline{X}\forall}{Q(1 + r)} \quad (7.132)$$

Perform a materials balance on biomass around the entire system.

$$[\text{accumulation}] = [\text{inputs}] - [\text{outputs}] + [\text{reaction}] \quad (7.133)$$

$$\left(\frac{dX}{dt}\right)_{accumulaton} \forall = QX_i - (Q - Q_w) - Q_wX_r + \left(\frac{dX}{dt}\right)_{NG} \forall \quad (7.134)$$

At steady-state, the accumulation term is set equal to zero and the influent biomass concentration (X_i) is assumed to be negligible and is also set equal to zero.

$$0 = 0 - (Q - Q_w) - Q_wX_r + \left(\frac{dX}{dt}\right)_{NG} \forall \quad (7.135)$$

$$\left(\frac{dX}{dt}\right)_{NG} \forall = (Q - Q_w) + Q_wX_r \quad (7.136)$$

Divide both sides of Equation (7.136) by V as follows:

$$\left(\frac{dX}{dt}\right)_{NG} \frac{\forall}{\forall} = \frac{(Q - Q_w) + Q_wX_r}{\forall} \quad (7.137)$$

Recall that the net growth rate of a microorganism is expressed as Equation (7.41):

$$\left(\frac{dX}{dt}\right)_{NG} = Y\left(\frac{dS}{dt}\right)_U - k_d\overline{X}$$

Equating Equations (7.137) and (7.41) and dividing both sides by $\overline{X}$ yields the following equation.

$$\frac{(dX/dt)_{NG}}{\overline{X}} = \frac{(Q - Q_w) + Q_wX_r}{\overline{X}\forall} = \frac{Y\left(\frac{dS}{dt}\right)_U - k_d\overline{X}}{\overline{X}} \quad (7.138)$$

The reciprocal of MCRT or θ_c is substituted into Equation (7.138) to produce Equation (7.139):

$$\frac{1}{\theta_c} = Y\left(\frac{(dS/dt)_U}{\overline{X}}\right) - k_d \quad (7.139)$$

The specific substrate utilization rate for a plug flow reactor is calculated by replacing X in Equation (7.51) with $\overline{X}$:

$$\frac{(dS/dt)_U}{\overline{X}} = \frac{Q(S_i - S_e)}{\forall \overline{X}} \tag{7.140}$$

Substituting Equation (7.140) into Equation (7.139) yields Equation (7.141):

$$\frac{1}{\theta_c} = Y\left(\frac{Q\,(S_i - S_e)}{\overline{X}\forall}\right) - k_d \tag{7.141}$$

Rearranging Equation (7.141) and solving for $\overline{X}\forall$ yields:

$$\overline{X}\forall = \frac{YQ\,(S_i - S_e)}{\dfrac{1}{\theta_c} + k_d} \tag{7.142}$$

Equation (7.142) is then inserted into Equation (7.132) and rearranged in the following steps to produce an equation for MCRT:

$$K_s \ln\frac{S_e}{S_0} + (S_e - S_0) = \frac{-k\overline{X}\forall}{Q(1+r)} = \frac{-k}{Q(1+r)}\left(\frac{YQ\,(S_i - S_e)}{\dfrac{1}{\theta_c} + k_d}\right) \tag{7.143}$$

$$K_s \ln\frac{S_e}{S_0} + (S_e - S_0) = \frac{-k}{(1+r)}\left(\frac{Y\,(S_i - S_e)}{\dfrac{1}{\theta_c} + k_d}\right) \tag{7.144}$$

Replace S_0 in Equation (7.144) with Equation (7.127):

$$K_s \ln\frac{S_e}{S_0} + \left(S_e - \frac{S_i + rS_e}{1+r}\right) = \frac{-k}{(1+r)}\left(\frac{Y\,(S_i - S_e)}{\dfrac{1}{\theta_c} + k_d}\right) \tag{7.145}$$

$$K_s \ln\frac{S_e}{S_0} + \left(S_e - \frac{S_i + rS_e}{1+r}\right) = \frac{-k}{(1+r)}\left(\frac{Y\,(S_i - S_e)}{\dfrac{1}{\theta_c} + k_d}\right) \tag{7.146}$$

Multiply both sides of Equation (7.146) by $(1 + r)$:

$$\left[K_s \ln\frac{S_0}{S_e} + \left(\frac{S_i + rS_e}{1+r} - \frac{(1+r)}{(1+r)}S_e\right) = \frac{k}{(1+r)}\left(\frac{Y\,(S_i - S_e)}{\dfrac{1}{\theta_c} + k_d}\right)\right] \times (1+r) \tag{7.147}$$

$$(1+r)K_s \ln\frac{S_0}{S_e} + S_i + rS_e - S_e - rS_e = k\left(\frac{Y\,(S_i - S_e)}{\dfrac{1}{\theta_c} + k_d}\right) \tag{7.148}$$

Multiply both sides of Equation (7.148) by $\left(\frac{1}{\theta_c} + k_d\right)$:

$$\left[(1+r)K_s \ln\frac{S_0}{S_e} + (S_i - S_e) = k\left(\frac{Y\,(S_i - S_e)}{\dfrac{1}{\theta_c} + k_d}\right)\right] \times \left(\frac{1}{\theta_c} + k_d\right) \tag{7.149}$$

$$\left(\frac{1}{\theta_c} + k_d\right) \times \left[(1+r)K_s \ln\frac{S_0}{S_e} + (S_i - S_e)\right] = Yk\,(S_i - S_e) \tag{7.150}$$

Rearranging Equation (7.150) and solving for reciprocal of θ_c as follows:

$$\boxed{\frac{1}{\theta_c} = \frac{Yk(S_i - S_e)}{\left[(S_i - S_e) + (1+r)K_s \ln\dfrac{S_0}{S_e}\right]} - k_d} \tag{7.151}$$

Equation (7.151) is similar to Equation (7.63) for complete-mix activated sludge systems; however, the main difference is that the MCRT in a plug flow reactor is dependent on the influent wastewater concentration S_i. Plug flow reactor activated sludge systems are theoretically more efficient in oxidizing soluble organic wastes than CMAS systems but, in practice, it is extremely difficult to truly achieve plug flow in a reactor. As discussed in Chapter 5, a series of complete-mix reactors can be used to approximate plug flow. Design engineers often use this approach to increase process efficiency and benefit from the dilution effect that reduces harmful effects associated with shock loadings of organics and toxics that otherwise disrupt plug flow reactors.

Figure 7.21 is a plot of effluent soluble substrate concentration calculated using Equation (7.151) and efficiency of treatment (substrate removal) using Equation (7.52) as a function of mean cell residence time. The minimum mean cell residence time as shown in the figure is approximately 0.21 days, which can be calculated from Equation (7.73).

7.5.9 Activated sludge modifications

A brief description of the various types of activated sludge processes and their modifications is presented in this section. Typical design parameters for various types of activated sludge processes are listed in Table 7.15.

7.5.9.1 Conventional Activated Sludge

The conventional activated sludge process consists of primary clarifiers followed by biological treatment in long, rectangular basins with air diffusers along one side to approach plug flow. Recall that plug flow or pipe flow results when fluid particles exit the reactor in the same sequence in which they enter; there is no longitudinal mixing. As wastewater flows through the reactor, the organic carbon concentration varies both as a function of time and length along the basin.

Figure 7.22 illustrates the conventional activated sludge process using plug flow. Thickened sludge, known as return

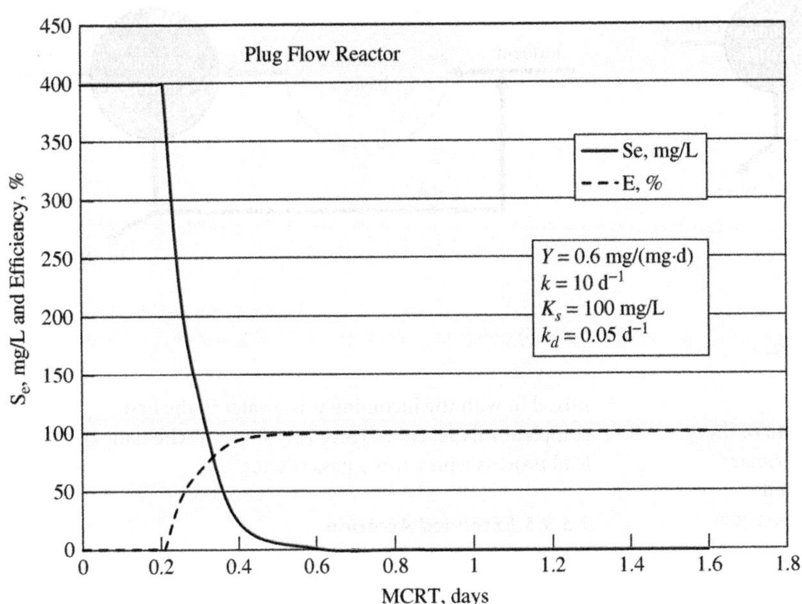

Figure 7.21 Effluent substrate and treatment efficiency versus MCRT for a plug flow reactor.

Table 7.15 Typical design parameters for various activated sludge processes.

Process	**Reactor type**	**MCRT (d)**	**F/M** $\left(\frac{\text{kg BOD}}{\text{kg MLVSS} \cdot \text{d}}\right)$	**Volumetric loading** $\left(\frac{\text{lb BOD}}{1000\ \text{ft}^3 \cdot \text{d}}\right)$	**Volumetric loading** $\left(\frac{\text{kg BOD}}{\text{m}^3 \cdot \text{d}}\right)$	**MLSS (mg/L)**	**Nominal detention time, τ (d)**	**RAS (Q_r/Q)**
[a]Conventional	PFR	3–8	0.2–0.4	20–40	0.3–0.6	1200–3000	4–8	0.25–0.75
[d]Tapered aeration	PFR	5–15	0.2–0.4	20–40	0.3–0.6	1500–3000	4–8	0.25–1.00
[b]Complete-mix	CMFR	3–15	0.2–0.6	20–100	0.3–1.6	1500–4000	3–5	0.25–1.00
Step feed	PFR	5–15	0.2–0.4	40–60	0.7–1.0	1500–3500	3–5	0.25–0.75
[a]Extended aeration	PFR	20–30	0.05–0.15	10–25	0.2–0.4	1500–5000	18–36	0.50–1.50
[a]Oxidation ditch	PFR	10–30	0.05–0.30	5–30	0.1–0.5	1500–5000	8–36	0.75–1.50
[b]High-purity oxygen	PFR	1.3–4	0.5–1.0	80–200	1.3–3.2	2000–5000	1–3	0.25–0.50
[b]Contact stabilization	PFR	5–10	0.2–0.6	60–75	1.0–1.3			0.50–1.50
Contact basin						1000–3000	0.5–1.0	
Stabilization basin						6000–10000	2–4	
[b]Sequencing batch reactor	CMBR	10–30	0.04–0.10	5–15	0.1–0.3	2000–5000	15–40	NA

PFR = plug flow reactor, CMFR = complete-mix flow reactor, CMBR = complete-mix batch reactor.
[a] Metcalf & Eddy (1991) p. 550.
[b] Metcalf & Eddy (2003) p. 747.
[c] Grady *et al.* (2011) p. 383.
[d] Reynolds and Richards (1996) p. 429.

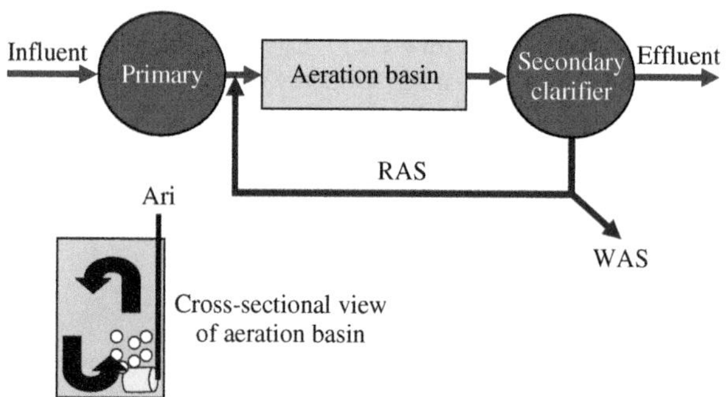

Figure 7.22 Conventional activated sludge process.

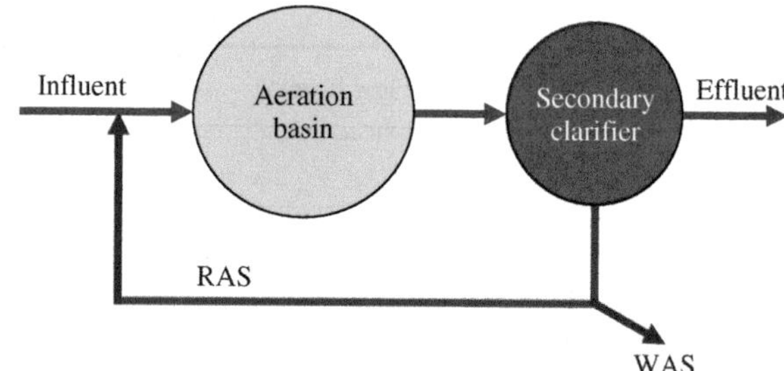

Figure 7.23 Schematic of complete-mix activated sludge (CMAS) process.

activated sludge (RAS), is recycled back to the front of the aeration basin, where it is mixed with incoming primary effluent. Two major deficiencies encountered with the conventional activated sludge process are that the oxygen demand of the wastewater at the front of the aeration tank exceeds the capacity of the aeration system, and that shock loadings due to toxic and/or high-strength waste entering the aeration tank often result in process failure.

7.5.9.2 Tapered Aeration

To overcome deficiencies with conventional activated sludge processes, engineers place more diffusers at the front of the aeration basin and increase diffuser spacing near the end of the basin. Using this design allows more oxygen to be provided at the front of the tank to meet the wastewater loading, and some longitudinal mixing, which helps to reduce the detrimental effects associated with shock loadings. The flow schematic shown in Figure 7.22 is applicable to tapered aeration activated sludge processes, with the exception of the design of the aeration system as noted above.

7.5.9.3 Complete-Mix

Complete-mix or completely-mixed activated sludge processes are the most widely used, since the concentration of constituents within the reactor is the same as that in the effluent. In complete mix reactors, the influent wastewater concentration is reduced instantaneously to the effluent concentration upon entering the reactor. This precludes any problems with shock loadings due to toxic and high-strength wastes entering the reactor.

Diffused aeration or mechanical aerators may be used to achieve the complete-mix flow regime. The contents of the reactor are uniform throughout creating a uniform oxygen demand and uniform biological solids concentration. Primary clarifiers are typically omitted from the process treatment train when complete mix reactors are used. A typical flow schematic for the complete-mix activated sludge process is shown in Figure 7.23. Return activated sludge (RAS) is either recycled back to the influent pipeline to blend in with the incoming wastewater, or is returned directly to the aeration basin.

7.5.9.4 Step Feed

The step feed activated sludge process uses a series of plug flow reactors, with the incoming wastewater feed at the front of each aeration basin to help equalize the oxygen demand incurred from the wastewater. Return activated sludge is recycled and mixed in with the incoming wastewater in the first complete-mix reactor. Figure 7.24 shows a schematic of the step feed process with a three-pass reactor.

7.5.9.5 Extended Aeration

Extended aeration activated sludge plants use plug flow reactors with long hydraulic detention time, ranging from 18–36 hours, but typically 24 hours. Primary clarifiers are omitted and the objective is to minimize the amount of sludge or biosolids produced. The process is operated at long mean cell residence times (MCRTs), so that most of the substrate (organic carbon as measured by BOD or COD) is oxidized for energy and maintenance of cell functions, rather than the synthesis of new microbes. Theoretically, the process was developed so that excess sludge production would not occur. However, non-biodegradable solids and some excess biomass must be wasted from extended aeration facilities. The flow schematic presented for the oxidation ditch (Figure 7.25) is applicable for extended aeration.

7.5.9.6 Oxidation Ditch

The oxidation ditch is a version of the extended aeration process, but an oval or racetrack configuration type of reactor is used. Oxygen is supplied to the process by either mechanical aerators or brush rotors. The flow regime approaches plug flow. Figure 7.25 shows a schematic of an oxidation ditch with mechanical aerators at each end of the ditch.

7.5.9.7 High-Purity Oxygen Activated Sludge

Enclosed activated sludge systems with a high-purity oxygen environment have been used at many installations. Typically,

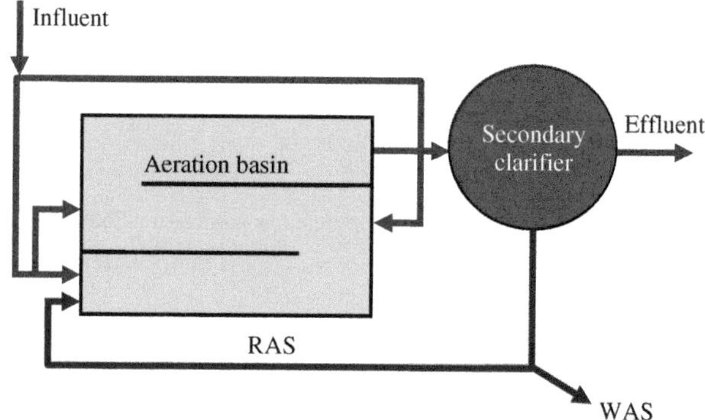

Figure 7.24 Schematic of step feed activated sludge process.

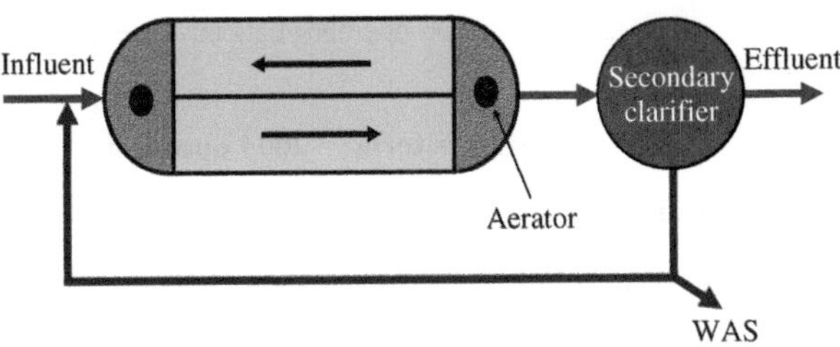

Figure 7.25 Schematic of oxidation ditch.

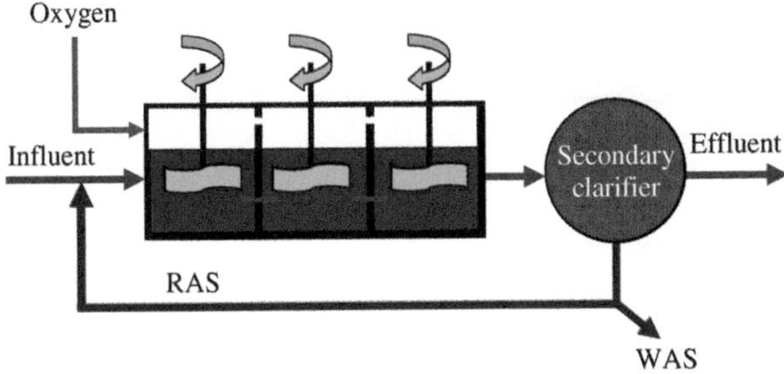

Figure 7.26 Schematic of high-purity oxygen activated sludge process.

pure oxygen systems can treat higher strength wastewaters and require smaller reactor volumes than traditional air-fed systems. The complete-mix flow regime is used, and mechanical mixers entrain the oxygen from the high-purity oxygen atmosphere into the wastewater. Figure 7.26 shows a schematic of a high-purity oxygen process.

7.5.9.8 Contact Stabilization

The contact stabilization modification of the activated sludge process was developed primarily to treat wastewaters that contain complex colloidal/soluble substrates. Two aeration basins are used, along with a secondary clarifier (Figure 7.27). The first aeration basin is called the contact basin, since it has a short detention time of 0.5–1.0 hour.

Soluble substrate is used by the microbes and the complex colloidal matter is adsorbed by the microorganism. From the contact basin, the activated sludge is separated from the wastewater in a secondary clarifier. The thickened sludge is recycled to a stabilization basin, which has a detention time of 3–6 hours. Here, the microbes have sufficient time to stabilize or oxidize the organic matter that was adsorbed in the contact basin. The contents from the stabilization basin then pass back into the contact basin to react with incoming wastewater. Plug flow reactors are used for both the contact and the stabilization basins.

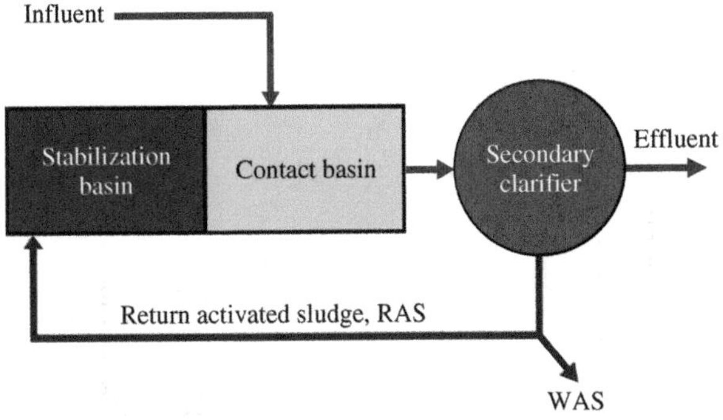

Figure 7.27 Schematic of contact-stabilization activated sludge process.

7.5.9.9 Single-Sludge Systems

Modification of the traditional aerobic activated sludge process has lead to the development of biological nutrient removal (BNR) processes in a single-sludge process. Single-sludge activated sludge processes use different environments to grow various types of bacteria to accomplish various treatment objectives. Figure 7.28 depicts a single-sludge process with an anaerobic, anoxic, and oxic (aerobic) zone for removing nitrogen and phosphorus from wastewater biologically. These systems will be explained in more detail later in the chapter.

The anaerobic zone is required for fermentation of the incoming wastewater, so that volatile fatty acids can be produced for enhancing the growth of phosphorus-accumulating organisms (PAOs) such as *Acinetobacter*. An anoxic environment is required for growing heterotrophic, denitrifying bacteria that reduce nitrates to nitrogen gas, i.e., accomplishing nitrogen removal biologically. Finally, the aerobic or oxic zone is necessary for removing most of the organic matter, as measured by BOD or COD, and accomplishing biological nitrification, i.e., conversion of ammonia to nitrates. Excess phosphorus removal uptake by the PAOs also occurs in the oxic zone.

7.5.10 Sequencing batch reactor (SBR)

Sequencing batch reactors (SBRs) are a departure from the continuous flow treatment processes discussed above. In addition to being a batch process versus a continuous flow process, SBRs do not require a secondary clarifier. A single tank or reactor is used to accomplish biological treatment of the wastewater and solids/liquid separation. SBRs are suspended growth biological processes similar to the activated sludge process but, instead of being a continuous flow process, they are operated on a batch basis.

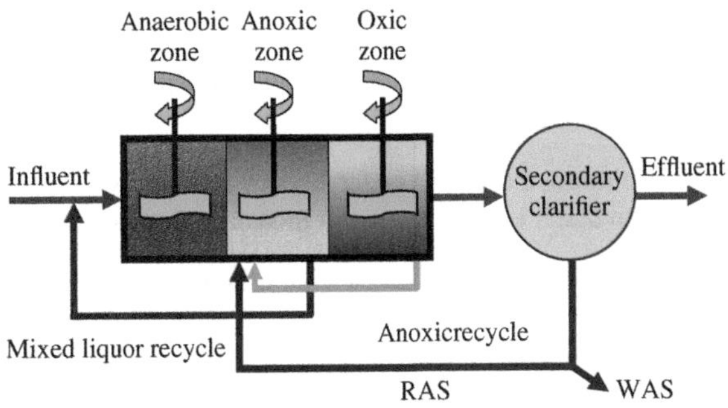

Figure 7.28 Schematic of single-sludge process.

Figure 7.29 Camp Branch WWTP in Calera, Alabama. *Source:* Courtesy of Aqua-Aerobic Systems, Inc.

The same treatment mechanisms used in the activated sludge process are applicable to SBRs. Air is provided either by a diffused aeration system or mechanical aerator. Retrievable coarse- or fine-bubble diffuser systems are typically used. A mechanical mixer may be used during various phases of operation to keep the biomass in suspension when aeration is not provided. Since SBRs are fill-and-draw reactors, they are designed and operated on the number of batches that can be processed daily.

Typically, an SBR is designed to complete 4–6 cycles per day (CPD). Fewer cycles are used for treating wastewaters that require a higher level of treatment to meet effluent requirements. Alum or other chemicals may be added during the react phase to accomplish phosphorus removal. Figure 7.29 is a photograph of the Camp Branch WWTP in Calera, Alabama that uses SBRs.

Major advantages of the process are:

1 A single basin is used to grow the microorganisms and separate the biomass from the wastewater without the need of a secondary clarifier.

2 The time needed to achieve a specified level of treatment can be varied to accommodate changes in influent wastewater characteristics.

3 It is possible to produce anaerobic, anoxic, and aerobic phases of operation with an SBR, so that biological nitrogen and phosphorus removal can be accomplished.

Major disadvantages of the process are:

1 SBRs are usually limited to wastewater flows that are 5.0 MGD (18,900 m^3/d) or less.

2 A minimum of two, and preferably three, SBRs should be used so that it is possible to treat incoming wastewater flows.

3 During the decanting phase, a large volume of wastewater must be removed from the SBR within 45–60 minutes. Therefore, post-equalization (holding tank) is required so that hydraulic loading rates on downstream unit operations, and processes such as effluent filters and chlorine contact basins, are not exceeded.

Table 7.16 Design criteria and operating data for the Peurifoy WWTP.

	Design criteria		2004 operating data	
Parameter	**Influent**	**Effluent**	**Influent**	**Effluent**
ADF (MGD)	1.5	–	0.639	0.577
MDF (MGD)	1.9	–	–	–
BOD_5 (mg/L)	250	20	255	2.9
TSS (mg/L)	250		222	1.6
TKN (mg/L)	40	–	–	–
NH_3–N (mg/L)	–	1.3	14.1	0.38
TP (mg/L)	5.0	0.3	7.4	0.15

Source: Humphries (2006).

Effluent quality from sequencing batch reactors is similar to that from activated sludge processes or even better, when operated to accomplish biological nutrient removal. Table 7.16 shows the effluent quality data for the Peurifoy WWTP in Stockbridge, Georgia, which accomplishes biological nutrient removal. Design criteria and operating data for 2004 are presented.

7.5.10.1 SBR Phased Operation for Conventional Biological Treatment

SBRs are typically operated on 4–6 cycles per day, meaning that each cycle may vary from six to four hours. For conventional biological treatment, 4–5 sequenced phases are used per cycle and include: fill, react, settle, decant/sludge waste, and idle. Sludge wasting may also be accomplished during the react phase to ensure that a uniform discharge of solids occurs.

Typically, sludge is wasted from the SBR for 3–4 minutes during each cycle. More often than not, the idle phase is omitted. Figure 7.30 shows the main phases used to treat domestic wastewater in an SBR. A brief description of each phase is presented below, along with some of the options available. Wastewater that enters the SBR has undergone preliminary treatment, consisting of screening and grit removal and, in some cases, primary clarification.

Fill Phase During the fill phase, wastewater that has undergone pretreatment enters the reactor. The reactor is filled

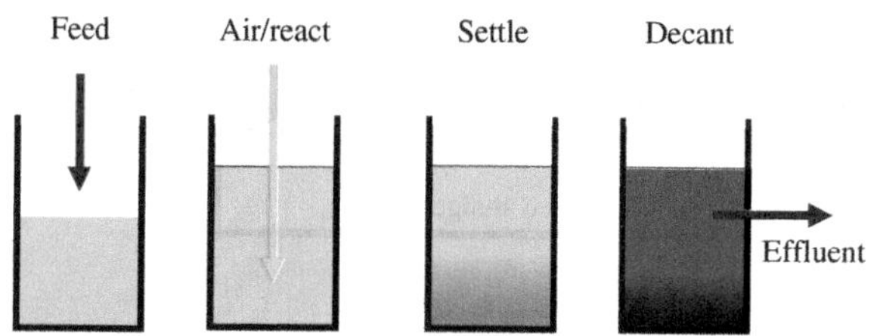

Figure 7.30 Operating phases for conventional biological treatment in an SBR.

to its maximum operating volume. Mixing of the wastewater during the fill phase is normally accomplished by a mechanical mixer. Air may or may not be added during this phase depending on the treatment objectives. Approximately 25% of the total cycle time is consumed during the fill phase.

React Phase In this phase, air is added and, normally, the mechanical mixer is turned off. The objective here is to allow the heterogeneous culture of microorganisms within the reactor sufficient time to adsorb and utilize the substrate (as measured by BOD or COD). It is also possible to accomplish nitrification (oxidation of ammonia to nitrate) during this phase. The dissolved oxygen concentration is generally maintained at 2.0 mg/L. Sludge wasting sometimes occurs during the react period, so that a uniform concentration of solids can be wasted from the rector. The react phase usually represents 35% of the total cycle time.

Settle Phase The reactor serves as a secondary clarifier during the settle phase. Aeration and mixing are terminated during this phase to allow the biomass to settle by the force of gravity. The solids settle to the bottom, producing a clarified supernatant that is discharged as effluent during the decant phase. Approximately 20% of the total cycle time is devoted to the settling phase.

Decant Phase Once the settling phase has been completed, clarified supernatant is removed from the reactor by some type of decanting mechanism. Quiescent conditions must continue to be maintained during this phase, so aeration and mixing are turned off. There are various types of decanting mechanisms used; typically, a floating decanting mechanism that withdraws supernatant below the surface to avoid scum and other floating debris that may result in a permit violation. Figure 7.31 is a photograph of a floating decanting mechanism.

Some facilities use adjustable weirs during the decant phase. The clarified supernatant must be withdrawn in a short time period, usually in 45–60 minutes. At many SBR plants, a holding tank follows the reactor to equalize the flow, thereby eliminating high hydraulic loading rates on unit operations and processes downstream of the SBR. Normally, near the end of the decant phase, thickened sludge is pumped from the bottom of the reactor to waste excess biomass from the system. This excess sludge is typically stabilized in an aerobic digester. The decant phase represents approximately 15% of the total cycle time.

Idle Phase The idle phase may be implemented where multiple SBRs are used to treat a continuous flow. When one reactor is being filled, another one may be in react mode, while a third reactor may be in idle. Air may be turned off and on periodically during the idle phase to keep the biomass aerobic. Five percent of the total cycle time is typically devoted to the idle phase.

SBR Phased Operation for Biological Nutrient Removal To accomplish biological nitrogen and phosphorus removal in a sequencing batch reactor, five or six sequenced phases are used, consisting of: mixed-fill, react-fill, react, settle, decant/sludge waste, and idle. Figure 7.32 shows five operating phases for an AquaSBR®.

Figure 7.31 Aqua-Aerobics Decanter (hose design). *Source:* Courtesy of Aqua-Aerobic Systems, Inc.

Mixed-Fill Phase Pretreated wastewater enters the reactor during the mixed-fill phase. As the reactor is filled, a mechanical mixer is used to stir the contents to accomplish complete mixing. An anaerobic environment may be created to ferment organic matter in the incoming wastewater, to produce volatile fatty acids which are required for the growth and proliferation of phosphorus-accumulating organisms (PAOs) such as *Acinetobacter*. The reactor is filled to its maximum operating volume. Mixing of the wastewater during the fill phase is normally accomplished by a mechanical mixer. Approximately 16–28% of the total cycle time is allotted to the mixed-fill phase.

React-Fill Phase As wastewater continues to fill the reactor, aeration is initiated to allow BOD and nitrification to occur. Timers are used to allow the aeration system to be turned on and off so that both aerobic and anoxic conditions can be maintained at various times during this phase. An anoxic environment is described as one in which a low DO concentration exists (< 0.5 mg/L), however, bound oxygen in the form of nitrites (NO_2^-) and nitrates (NO_3^-) are acceptable. When anoxic conditions exist, heterotrophic denitrifying microorganisms oxidize organic matter (as measured by BOD or COD) and use nitrates (that were generated during nitrification) as electron acceptors, reducing them to nitrogen gas (N_2). The mixer is normally turned on when aeration is

Phase 1 Mixed fill

Phase 2 React fill

Phase 3 React

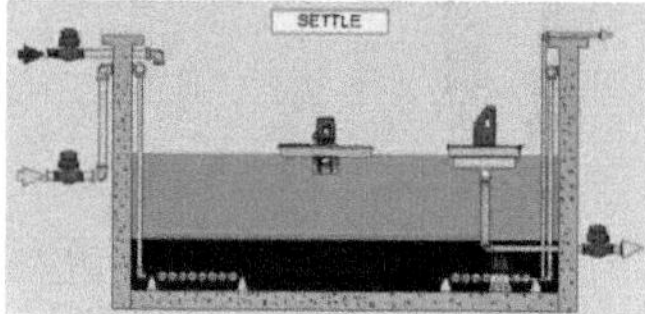
Phase 4 Settle

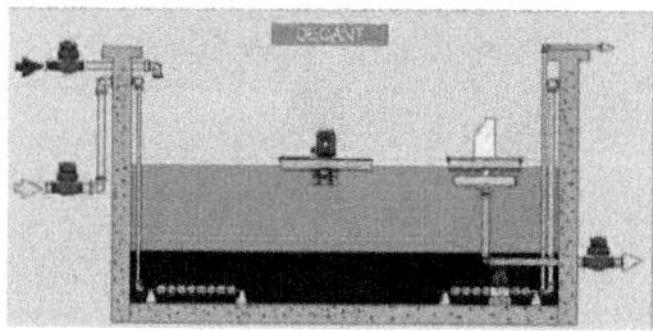
Phase 5 Decant/sludge waste

Figure 7.32 Phases for an AquaSBR® Sequencing Batch Reactor System. *Source*: Courtesy of Aqua-Aerobic Systems, Inc.

turned off, to keep the contents of the reactor well-mixed. The react-fill phase accounts for 22–33% of the total cycle time.

React Phase The react phase serves more as a polishing step when biological nutrient removal (BNR) is being achieved in an SBR. At the beginning of this phase, an anoxic environment exists to complete denitrification that was initiated during the react-fill phase. Aeration is then turned back on to keep the biomass aerobic at the start of the decant phase. Typically, 16–21% of the total cycle time is expended in the react phase.

Settle Phase Separation of the biomass from the wastewater to produce a clarified supernatant occurs during the settle phase. Aeration and mixing are terminated to allow the biomass to settle under quiescent conditions. Approximately 16–17% of the total cycle time is devoted to this phase.

Decant Phase/ Sludge Waste Clarified supernatant is decanted or withdrawn from the reactor by some type of decanting mechanism. Sludge is normally pumped off the bottom of the SBR for 3–4 minutes during each cycle. The large volume of wastewater that is decanted must be equalized in a holding tank or a post-equalization basin, as discussed previously. The decant phase represents approximately 12–16% of the total cycle time.

Idle Phase An idle phase may be implemented when multiple SBRs are used to treat a continuous flow. One reactor is being filled, while a second reactor will be in the idle mode. Air may be cycled on and off to keep the biomass aerobic. Typically, the idle phase is omitted when SBRs are operated to accomplish BNR.

Example 7.11 Design of a sequencing batch rector

Design a sequencing batch reactor WWTP to treat an average daily flow of 1 MGD and a peak flow of 2.4 MGD. Use two, square SBRs, a four cycle per day operating scheme, and a low-water-level (LWL) of ten feet. Determine:

a) The volume of each SBR.
b) Dimensions of each SBR.

Solution Part A

SBRs must be sized to treat the peak wastewater flow rate. The design of the aeration system should be based on the average daily flow rate. A simplified procedure will be used in this example. Metcalf & Eddy (2003) provide a detailed kinetic approach for designing SBRs.

First, calculate the volume necessary to treat the average daily flow using the following equation:

$$\forall = \frac{Q}{\text{CPD} \times \#\text{ of SBRs}} \tag{7.152}$$

$$\forall = \frac{1 \times 10^6\ \text{gal/d}}{\dfrac{4\ \text{cycles}}{\text{d}} \times 2} = 125{,}000\ \text{gal}$$

$$\forall = 125{,}000\text{gal}\left(\frac{\text{ft}^3}{7.48\ \text{gal}}\right) = 16{,}700\ \text{ft}^3$$

where:

$\forall$ = volume of SBR required to treat a specified flow, ft^3 (m^3)
CPD = cycles per day (4 to 6)
Q = volumetric flow rate, gpd (Lpd)
\# of SBRs = number of sequencing batch reactors in operation.

The volume of 16,700 ft^3 must be allocated between the LWL and the average water level (AWL), as shown in the figure provided in the Part B solution.

Next, the volume necessary to accommodate the peak wastewater flow is calculated using Equation (7.152):

$$\forall = \frac{Q}{\text{CPD} \times \#\text{ of SBRs}} = \frac{2.4 \times 10^6\ \text{gal/d}}{\dfrac{4\ \text{cycles}}{\text{d}} \times 2}\left(\frac{1\ \text{ft}^3}{7.48\ \text{gal}}\right)$$

$$= 40{,}100\ \text{ft}^3$$

The volume of 40,100 ft^3 must be allocated between the AWL and the high-water-level (HWL).

Solution Part B

Determine the dimensions of the square reactors. Assume that the low water level is 10 feet deep and each reactor is 100 ft by 100 ft. Typically, the LWL ranges from 9–13 feet. The surface area (A_s) of each reactor is:

$$A_s = L \times W = 100\,\text{ft} \times 100\,\text{ft} = 10{,}000\,\text{ft}^2$$

The depth required to handle the average daily flow is calculated as follows:

$$D = \frac{\forall}{A_s}$$

where:
D = depth of wastewater, ft (m)
A_s = surface area of SBRs, ft^2 (m^2).

$$D = \frac{\forall}{A_s} = \frac{16{,}700\,\text{ft}^3}{10{,}000\,\text{ft}^2} = 1.67\,\text{ft}$$

Therefore, the AWL is 10 ft + 1.67 ft = 11.67 ft.

The depth required to handle the peak flow is calculated as follows:

$$D = \frac{\forall}{A_s} = \frac{40{,}100\,\text{ft}^3}{10{,}000\,\text{ft}^2} = 4.01\,\text{ft}$$

Therefore, the HWL is 11.67 ft + 4.01 ft = 15.68 ft. Make the total depth of each SBR 18 feet; this allows 2.32 feet of freeboard. Freeboard is additional depth to allow for unexpected flows. The figure below shows the cross-sectional area of one of the SBRs.

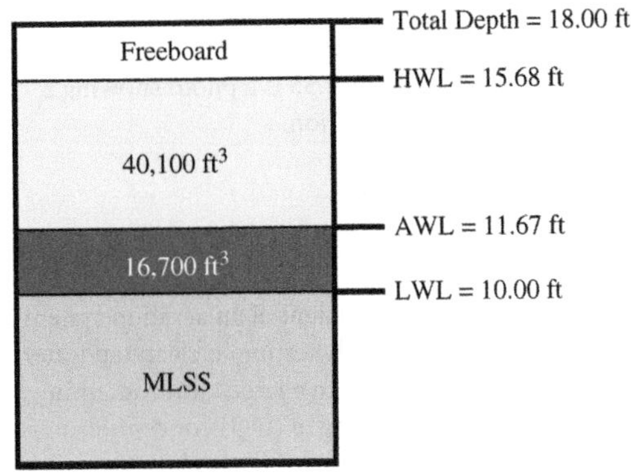

Figure E7.5

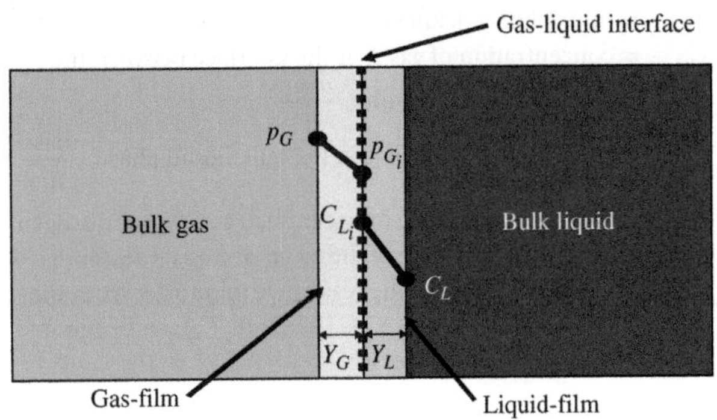

Figure 7.33 Schematic of two-film theory model.

7.6 Oxygen transfer and mixing

7.6.1 Theory

The design of aeration systems for the activated sludge process is based on the two-film theory by Lewis & Whitman (1924) and Fick's Law of Diffusion. Gas molecules are transported to the outer face of the gas-film by mixing and diffusion. Then they diffuse across the gas-film to the gas-liquid interface, where the gas molecules dissolve in the liquid-film. Next, the dissolved gas diffuses through the liquid-film to the boundary between the film and bulk liquid phase. Finally, the dissolved gas molecules are transported throughout the bulk liquid phase by mixing. Figure 7.33 illustrates the two-film theory model.

7.6.2 Derivation of mass transfer equation

The rate of transfer of gas A diffusing through a stagnant gas-film and stagnant liquid-film at steady-state conditions can be expressed by the following equation:

$$\frac{dm_A}{dt} = K_G A(p_G - p_{G_i}) = K_L A(C_{L_i} - C_L) \tag{7.153}$$

where:
$\frac{dm_A}{dt}$ = rate of mass transfer of gas A as it diffuses through gas-film and liquid-film, mass/time
D_G = diffusion coefficient through the gas-film
D_L = diffusion coefficient through the liquid-film
K_G = mass transfer coefficient of the gas in the gas phase, $\frac{\text{mass}}{\text{h·ft}^2\text{·atm}}$
A = area across which diffusion takes place, ft^2
K_L = mass transfer coefficient of gas in the liquid phase, $\frac{\text{ft}}{\text{h}}$
p_G = partial pressure of gas A in the bulk gas phase, atm
p_{G_i} = partial pressure of gas A at the bulk gas, gas-film interface, atm
Y_G = gas-film thickness

Y_L = liquid-film thickness
C_{L_i} = concentration of gas A at the interface between the gas-film and liquid-film, $\frac{\text{mass}}{\text{ft}^3}$
C_L = concentration of gas A in the bulk liquid phase, $\frac{\text{mass}}{\text{ft}^3}$.

For gases slightly soluble in water, such as oxygen, nitrogen, and carbon dioxide, the rate of diffusion or mass transfer of these gases is controlled by the resistance to transfer from the liquid-film, meaning ($K_G \gg K_L$). Therefore, K_G can be ignored and Equation (7.154) reduces to:

$$\frac{dm_A}{dt} = K_L A(C_{L_i} - C_L) \quad (7.154)$$

The concentration of the dissolved gas at the liquid-gas interface (C_{L_i}) is equal to the saturation concentration C_s, as determined from Henry's law ($C_s = Hp_G$). Replacing C_{L_i} with C_s in Equation (7.154) yields the following:

$$\frac{dm_A}{dt} = K_L A(C_s - C_L) \quad (7.155)$$

Dividing both sides of Equation (7.155) by the volume of the reactors results in the following:

$$\frac{d(m_A/V)}{dt} = K_L \frac{A}{V}(C_s - C_L) \quad (7.156)$$

Rearranging Equation (7.156) yields the overall mass transfer equation for oxygen transfer:

$$\frac{dC}{dt} = K_L a(C_s - C_L) \quad (7.157)$$

where:
$\frac{dC}{dt}$ = overall oxygen transfer rate, $\frac{\text{mass}}{\text{volume} \cdot \text{time}}$
$K_L a$ = overall oxygen transfer coefficient, time^{-1} $K_L a = K_L \frac{A}{V}$
$\frac{A}{V}$ = a interfacial surface area, L^2
C_s = dissolved oxygen saturation concentration for a given temperature and pressure, mass/volume
C_L = actual dissolved oxygen concentration in the aeration basin or tank, mass/volume.

7.6.3 Types of aeration systems

There are two major types of aeration systems: mechanical and diffused. Mechanical aerators or surface aeration systems consist of mixers or brush rotors that transfer oxygen into the wastewater by spraying the wastewater into the atmosphere. Figure 7.34 is a photo of a mechanical aerator.

Diffused aeration systems are similar to fish aquarium tanks, in which oxygen diffuses through a diffuser stone or membrane, thereby transferring oxygen into the wastewater. Diffused aeration systems are categorized by the size of bubbles they produce, ranging from coarse-bubble (large), through medium- to fine-bubble diffusers. Coarse-bubble diffusers are usually made of stainless steel pipes with orifices, while fine-bubble diffusers may consist of ceramic material with small pore sizes. In general, diffused aeration systems are more efficient in transferring oxygen into the wastewater. However, they are more costly from a capital and operational and maintenance perspective. Figure 7.35 is a photo showing a diffused aeration system in operation.

Figure 7.34 Mechanical aerator in operation.

Figure 7.35 Diffused aeration system in operation.

7.6.4 Aerator testing

The overall oxygen transfer coefficient of an aeration system is typically determined by nonsteady testing in clean tap water. For instance, the aerator is tested in a large tank containing clean tap water. The dissolved oxygen (DO) concentration is measured at several points throughout the tank, and an average DO is calculated and used to determine the quantity of sodium sulfite that is added to the water. Sodium sulfite reacts with oxygen according to the following stoichiometric reaction. It is used to deoxygenate the water prior to performing the nonsteady-state oxygen transfer test.

$$Na_2SO_3 + 0.5\,O_2 \rightarrow Na_sSO_4 \qquad (7.158)$$

Theoretically, approximately 7.9 mg/L of sodium sulfite are required for each mg/L of DO present. However, during actual aerator testing, it is common practice to add between 1.5–2.0 times the theoretical quantity of sodium sulfite to achieve complete deoxygenation of the water. Cobalt chloride (CoCl) is added at a dose of 0.05 mg/L to serve as a catalyst.

Rearranging Equation (7.157) for integration from the initial DO concentration (C_0) at time zero to the DO concentration at time t yields:

$$\int_{C_0}^{C_t} \frac{-1}{(C_s - C_L)}\,dC = -K_L a \int_0^t dt \qquad (7.159)$$

Integrating results in the following:

$$\ln(C_s - C_t) - \ln(C_s - C_0) = -K_L a(t) \qquad (7.160)$$

A plot of the $\ln(C_s - C_t)$ versus time t should yield a straight line with a negative slope equal to the value of $K_L a$ and the y-intercept equal to $\ln(C_s - C_0)$. Figure 7.36 is an example of such a plot.

Aeration manufacturers evaluate aeration equipment at standard conditions which include using clean tap water at 20°C containing zero mg/L of dissolved oxygen (initially) and at one atmosphere of pressure. Aeration equipment must transfer the desired quantity of oxygen to the microorganisms in the wastewater. To translate the oxygen transfer rate under standard conditions to those under process conditions (i.e., in the wastewater), three coefficients must be applied: α, β, and θ. Each coefficient is defined by the following equations:

$$\alpha = \frac{(K_L a)_{ww}}{(K_L a)_{tap\ water}} \qquad (7.161)$$

$$\beta = \frac{(C_s)_{ww}}{(C_s)_{tap\ water}} \qquad (7.162)$$

$$(K_L a)_{T°C} = (K_L a)_{20°C}(\theta)^{(T°C-20°C)} \qquad (7.163)$$

where:

$(K_L a)_{ww}$ = overall oxygen transfer coefficient at process conditions in wastewater, time^{-1}

$(K_L a)_{tap\ water}$ = overall oxygen transfer coefficient at standard conditions in water, time^{-1}

$(C_s)_{ww}$ = dissolved oxygen saturation concentration at specified temperature and pressure in wastewater, mass/volume

$(C_s)_{tap\ water}$ = dissolved oxygen saturation concentration at standard conditions in water, mass/volume

T = actual temperature of wastewater under process conditions, °C

θ = temperature correction factor, normally 1.02 to 1.024

$(K_L a)_{T°C}$ = overall oxygen transfer coefficient at temperature T, time^{-1}

$(K_L a)_{20°C}$ = overall oxygen transfer coefficient at temperature of 20°C, time^{-1}.

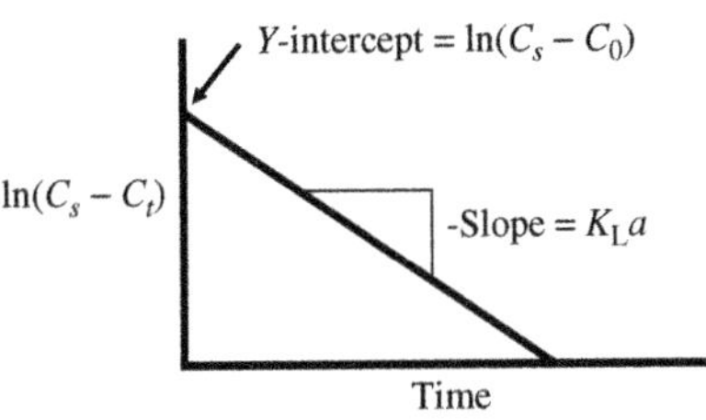

Figure 7.36 Plot used for determining the value of $K_L a$.

7.6.5 Derivation of oxygen transfer equation

The actual oxygen transfer rate represents the quantity of oxygen that the aerator or aeration system can provide at process conditions. Therefore, the overall oxygen transfer coefficient for the aeration device must be corrected to account for the constituents in the wastewater that affect the transfer rate (α) and dissolved oxygen saturation concentration (β). Temperature also affects the $K_L a$ value and the C_s concentration. The theta value (θ) is used for correcting $K_L a$ for temperature variations. The actual oxygen transfer rate expressed in equation form is presented below:

$$\left(\frac{dC}{dt}\right)_{Actual} = \alpha(K_L a)_{20°C}(\beta C_s - C_t)(\theta)^{(T-20°C)} \qquad (7.164)$$

The standard oxygen transfer rate is the rate at which the aeration device transfers oxygen in clean tap water at standard conditions of 20°C and 1 atmosphere of pressure, and is shown as Equation (7.165):

$$\left(\frac{dC}{dt}\right)_{Standard} = (K_L a)_{20°C}(C_s)_{20°C} \qquad (7.165)$$

$$\frac{(dC/dt)_{Actual}}{(dC/dt)_{Standard}} = \frac{\alpha(K_L a)_{20°C}(\beta C_s - C_t)(\theta)^{(T-20°C)}}{(K_L a)_{20°C}(C_s)_{20°C}} \qquad (7.166)$$

$$\frac{(dC/dt)_{Actual}}{(dC/dt)_{Standard}} = \frac{\alpha(\beta C_s - C_t)(\theta)^{(T-20°C)}}{(C_s)_{20°C}} \qquad (7.167)$$

DO saturation concentration at 20°C and 1 atmosphere of pressure is approximately 9.17 mg/L.

$$\left(\frac{dC}{dt}\right)_{Actual} = \left(\frac{dC}{dt}\right)_{Standard} \frac{\alpha(\beta C_s - C_t)(\theta)^{(T-20°C)}}{9.17} \qquad (7.168)$$

The oxygen transfer rate can be expressed in mass of oxygen per unit of time (mass/time) by multiplying the oxygen transfer coefficient by the volume ($\forall$) of the basin or tank.

$$\text{Oxygen transfer rate} = OTR = \left(\frac{dC}{dt}\right)\forall \qquad (7.169)$$

$$SOTR = \left(\frac{dC}{dt}\right)_{Standard} ¥ = (K_L a)_{20°C}(C_s)_{20°C}¥ \quad (7.170)$$

$$AOTR = \left(\frac{dC}{dt}\right)_{Actual} ¥ \quad (7.171)$$

The actual oxygen transfer rate equation under process conditions (transferring oxygen into wastewater) is developed by multiplying both sides of Equation (7.168) by volume and substituting Equations (7.170) and (7.171).

$$AOTR = (SOTR)\frac{\alpha\left(\beta C_s - C_t\right)(\theta)^{(T°C-20°C)}}{9.17} \quad (7.172)$$

where:

$AOTR$ = actual oxygen transfer rate at process conditions, lb/d (kg/d)

$SOTR$ = standard oxygen transfer rate at standard conditions, lb/d (kg/d).

7.6.6 Mixing requirements

Sufficient oxygen must be supplied to the microorganisms so that biological processes are not limited. Also, mixing requirements must be maintained so that the microorganisms remain suspended in the aeration basin. If diffused aeration systems are used, then a specified amount of air is required for mixing. The mixing requirement for a mechanical aerator is based on the amount of energy supplied to the water, in terms of power per unit volume. Aeration systems must be designed to meet the following two conditions:

1 Process oxygen requirements (oxygen required to keep microorganisms alive):

$$O_2 = Q(S_i - S_e)(1 - 1.42Y) + 1.42k_d X¥ + NOD \quad (7.173)$$

$$NOD = Q(TKN_0)(4.57) \quad (7.174)$$

where

NOD = nitrogenous oxygen demand due to nitrification, $\frac{lb}{d}\left(\frac{kg}{d}\right)$.

2 Mixing requirements: diffused aeration: 20 to 30 scfm/1,000 ft^3 $\left(20 \text{ to } 30 \frac{m^3}{min \cdot 1000\ m^3}\right)$; mechanical aerators: 0.75–1.5 hp/1,000 ft^3 (20–40 kW/1,000 m^3)

The design of a mechanical aerator system for a completely-mixed activated sludge process is presented in Example 7.12.

Example 7.12 Design of mechanical aerator system

Given the following information for a complete-mix activated sludge process: $Q = 5.0$ MGD, influent soluble COD = 280 mg/L, $Y = 0.5$ g VSS/ g COD, $k_d = 0.1\ d^{-1}$, $K_S = 60$ mg/L COD, $k = 5\ d^{-1}$, MLVSS = 0.8 MLSS, TKN_0 = 40 mg/L, MLSS = 3,000 mg/L, and DO = 2 mg/L. Determine:

a) Volume of the aeration basin (MG) at an SRT = 10 days.
b) Oxygen required (ppd), SRT = 10 days.
c) Number and hp of mechanical aerators assuming SOTR/aerator = 3.2 lb O_2/hp · h.

Solution Part A

First, find the effluent soluble COD concentration at a 10-day SRT using Equation (7.72).

$$S_e = \frac{K_s(1 + k_d\theta_c)}{\theta_c(Yk - k_d) - 1}$$

$$= \frac{60\ mg/L(1 + 0.1\ d^{-1} \times 10\ d)}{10\ d(0.5\ g/g \times 5\ d^{-1} - 0.1\ d^{-1}) - 1}$$

$$= 5.2\ \frac{mg\ COD}{L}$$

Determine the volume of the aeration basin (MG) using Equation (7.84).

$$¥ = \frac{YQ(S_0 - S_e)\,\theta_c}{1 + k_d\theta_c}\,\frac{}{X}$$

$$= \frac{(0.5\,g\ VSS/g\ COD)(5\,MGD)(280 - 5.2\,mg/L)(10\,d)}{(1 + 0.1\ d^{-1} \times 10\ d)(3000\ mg\ SS/L \times 0.8\ VSS/TSS)}$$

$$¥ = \boxed{1.43\ MG}$$

Solution Part B

Determine the oxygen process requirements using Equations (7.104) and (7.105):

$$NOD = Q(TKN_0)4.57$$

$$= (5\ MGD)\left(40\ \frac{mg}{L}\right)(4.57)\left(8.34\ \frac{lb/MG}{mg/L}\right)$$

$$= 7620\ \frac{lb\ O_2}{d}$$

$$O_2 = Q(S_0 - S_e)(1 - 1.42Y) + 1.42k_d X¥ + NOD$$

$$O_2 = (5\ MGD)\left(280 - 5.2\ \frac{mg}{L}\right)(1 - 1.42 \times 0.5) \times \left(\frac{8.34\ lb/MG}{mg/L}\right) + 1.42(0.1\ d^{-1}) \times \left(3000\ \frac{mg\ TSS}{L} \times 0.8\ \frac{g\ VSS}{g\ TSS}\right)(1.43\ MG) \times \left(\frac{8.34\ lb/MG}{mg/L}\right) + 7620$$

$$O_2 = \left(3320\ \frac{lb\ O_2}{d}\right) + \left(4060\ \frac{lb\ O_2}{d}\right) + 7620\ \frac{lb\ O_2}{d}$$

$$= \boxed{15{,}000\ \frac{lb\ O_2}{d}}$$

Solution Part C

Design a mechanical aerator system for summer conditions, the worst case scenario assuming that the temperature of the wastewater rises to 30°C. Assume the alpha and beta coefficients for the mechanical aerators, as supplied by the aeration manufacturer, are 0.85 and 0.95, respectively, and use a temperature correction factor (θ) of 1.024. The dissolved oxygen (DO) saturation concentration of water is 7.54 mg/L at sea level and 30°C. Typically, a DO concentration (C_t) of 2.0 mg/L is maintained in the aeration basin, so that oxygen does not limit the biological process.

Calculate the actual oxygen transfer rate (*AOTR*) of the mechanical aerators using Equation (7.172).

$$AOTR = (SOTR)\frac{\alpha(\beta C_s - C_t)(\theta)^{(T°C-20°C)}}{9.17}$$

$$AOTR = \left(3.2\,\frac{\text{lb O}_2}{\text{hp}\cdot\text{h}}\right) \times \frac{0.85(0.95 \times 7.54 - 2.0\text{ mg/L})(1.024)^{(30°C-20°C)}}{9.17}$$

$$= 1.9\,\frac{\text{lb O}_2}{\text{hp}\cdot\text{h}}$$

The total horsepower required for the mechanical aerators is estimated by dividing the total oxygen requirements determined in Part B by the *AOTR* determined above.

$$\text{total HP} = \left(\frac{15{,}000\text{ ppd O}_2}{1.9\text{ pphO}_2/\text{hp}}\right)\left(\frac{1\text{ d}}{24\text{ h}}\right) = 330$$

Assume there are four aerators in the aeration basin. Therefore, the hp per aerator is determined by dividing the total hp by four:

$$\frac{\text{hp}}{\text{aerator}} = \frac{330\text{ hp}}{4}$$

$$= 82.5\text{ hp Therefore, size up to } \boxed{85\text{ hp}} \text{ per aerator.}$$

Next, it is necessary to check the mixing requirements to ensure that the microorganisms remain in suspension. Recall that mixing requirements dictate that the hp per 1,000 ft^3 of basin should range from 0.75 to 1.5.

$$\frac{\text{Actual hp}}{1000\text{ ft}^3} = \frac{4 \times 85\text{ hp}}{1.43\text{ MG}}\left(\frac{1\text{ MG}}{10^6\text{ gal}}\right) \times \left(\frac{7.48\text{ gal}}{\text{ft}^3}\right)\left(\frac{1000\text{ ft}^3}{1000\text{ ft}^3}\right)$$

$$= 1.8$$

There is sufficient power to meet the mixing requirements. Mechanical aerators are not as efficient as diffused aeration systems and should not be used where icing might be a problem. Diffused aeration systems are more efficient but more costly to build and operate.

7.6.7 Design of diffused aeration systems

Diffused aeration systems use either coarse-bubble or fine-bubble diffusers. The oxygen transfer rate is higher for fine-bubble diffusers, but they are more likely to foul or clog. Centrifugal or positive displacement blowers are used to compress ambient air to provide the required air flow to diffused aeration systems. The oxygen transfer equation must be modified slightly to account for the change in the solubility of oxygen as a function of depth, i.e., as depth increases, the pressure and saturation of oxygen increase.

$$AOTR = (SOTR)\frac{\alpha(C_M - C_t)(\theta)^{(T°C-20°C)}}{9.17} \qquad (7.175)$$

where

C_M = dissolved oxygen saturation concentration at mid-depth, mg/L.

The DO saturation at mid-depth is calculated from the following equation:

$$C_M = \beta C_s\left(\frac{Pr}{C} + \frac{O_e}{42}\right) \qquad (7.176)$$

where:

C_s = dissolved oxygen saturation concentration at the surface of the basin at a specified temperature and elevation or pressure, mg/L

Pr = absolute pressure at the depth of bubble release, psia (kPa)

O_e = percent oxygen content in the exit air flow, %

C = 29.4 for US customary units and 203 for SI units.

The power required for adiabatic compression of a gas is given in Equation (7.177):

$$P_w = \frac{wRT_1}{C_1 ne}\left[\left(\frac{p_2}{p_1}\right)^{0.283} - 1\right] \qquad (7.177)$$

where:

P_w = brake or shaft power required of each blower, hp (kw)

w = mass air flow rate, lb/s (kg/s)

w = $G_{air} \times \rho_{air}$

G_{air} = air flow rate, $\frac{\text{ft}^3}{\text{s}}\left(\frac{\text{m}^3}{\text{s}}\right)$

ρ_{air} = density of air, $\frac{\text{lb}}{\text{ft}^3}\left(\frac{\text{kg}}{\text{m}^3}\right)$

R = engineering gas constant for air, $\frac{53.3\text{ ft}\cdot\text{lb}}{(\text{lb air})}\left(\frac{8.314\text{ kJ}}{k\cdot\text{mole}\cdot\text{K}}\right)$

T_1 = absolute inlet air temperature, °R (K)
C_1 = $\frac{550\text{ ft·lb}}{\text{s·hp}}\left(\frac{29.7\text{ m·N}}{\text{s·kw}}\right)$
p_1 = absolute inlet pressure, psia (atm)
p_2 = absolute outlet pressure, psia (atm)
e = efficiency of the compressor expressed as a fraction (usually 0.7 to 0.9)
n = $(k - 1)/k = 0.283$ for air
k = 1.395 for air.

To determine the actual power required by the motor, the power calculated by Equation (7.177) must be divided by the efficiency of the motor. The motor efficiency typically varies from 95 to 98% (Reynolds & Richards, 1996, p. 511).

Example 7.13 Design of diffused aeration system

Use the data and answers from Example 7.12 to design a coarse-bubble diffused aeration system. Aeration basin volume = 1.43 MG; assume the alpha and beta coefficients for the diffusers are 0.40 and 0.95, respectively and use a temperature correction factor (θ) of 1.024. The dissolved oxygen (DO) saturation concentration of water is 7.54 mg/L at sea level and 30°C. A DO concentration of 2.0 mg/L is maintained in the aeration basin. The oxygen transfer efficiency is assumed to be 10% since coarse-bubble diffusers are being used. The side water depth of the aeration basin is 15 ft with diffusers placed 1 foot from the bottom of the tank. The barometric pressure is 760 mm Hg and the specific weight of water is 62.4 lb/ft³. Assume the density of air at seal level at 100°F is $\rho_{air} = 0.070\text{ lb/ft}^3$. Determine:

a) Volume of the air (scfm) to meet oxygen requirement of 15,000 lb per day.
b) Horsepower required for centrifugal blowers.

Solution Part A

The dissolved oxygen saturation concentration at mid-depth must be calculated using Equation (7.176). We first must estimate the oxygen concentration (%) in the off-gas. The concentration of oxygen in the atmosphere is approximately 21% by volume and 23% by weight. The concentration of nitrogen in the atmosphere is approximately 79% by volume. Assuming 100 moles of air to be compressed, there would be 21 moles of oxygen (n_{O_2}) and 79 moles of nitrogen (n_{N_2}). The moles of oxygen in the off-gas are estimated as follows:

$$n_{O_2} = (21\text{ moles }O_2)(1 - \text{OTE})$$

$$n_{O_2} = (21\text{ moles }O_2)\left(1 - \frac{10\%}{100\%}\right) = 19\text{ moles}$$

The percentage of oxygen in the off-gas may now be calculated.

$$O_e = \frac{n_{O_2}}{n_{O_2} + n_{N_2}}(100) = \frac{19}{19 + 79}(100) = 19\%$$

The hydrostatic pressure due to water is calculated using Equation (7.178)

$$p = \gamma h \qquad (7.178)$$

where:
p = hydrostatic pressure, $\frac{\text{lb}}{\text{ft}^2}\left(\frac{\text{N}}{\text{m}^2}\right)$
γ = specific weight of water, 62.4 $\frac{\text{lb}}{\text{ft}^3}\left(9800\frac{\text{N}}{\text{m}^3}\right)$, Houghtalen *et al.* (2010)
h = depth of water, ft (m).

The pressure at the point of air release (Pr) at the bottom of the aeration basin is calculated below and must be expressed in terms of absolute pressure. Absolute pressure (p_{abs}) is gauge pressure (p_{gauge}) plus barometric pressure (p_{bar}):

$$\text{Pr} = (15\text{ ft} - 1\text{ ft})\left(\frac{62.4\text{ lb}}{\text{ft}^3}\right)\left(\frac{1\text{ ft}^2}{144\text{ in}^2}\right) + 760\text{ mm Hg}\left(\frac{14.7\text{ psi}}{760\text{ mm Hg}}\right) \qquad (7.179)$$

Now, calculate the dissolved oxygen saturation concentration at mid-depth.

$$C_M = \beta C_s\left(\frac{\text{Pr}}{C} + \frac{O_e}{42}\right) \qquad (7.176)$$

$$C_M = (0.95)\left(\frac{7.54\text{ mg}}{\text{L}}\right)\left(\frac{21\text{ psia}}{29.4} + \frac{19\%}{42}\right) = \frac{8.4\text{ mg}}{\text{L}}$$

Use Equation (7.175) to determine the quantity of oxygen required at standard conditions.

$$AOTR = (SOTR)\frac{\alpha(C_M - C_t)(\theta)^{(T°C-20°C)}}{9.17} \qquad (7.175)$$

$$15{,}000\,\frac{\text{lb}}{\text{d}} = (SOTR)\frac{0.40(8.4\text{ mg/L} - 2.0\text{ mg/L}) \times (1.024)^{(30°C-20°C)}}{9.17}$$

$$SOTR = 4.2 \times 10^4\,\frac{\text{lb}}{\text{d}}$$

The standard cubic feet of air required per minute is calculated by dividing the *SOTR* by the specific weight of air $\left(\frac{0.0175\text{ lb air}}{\text{ft}^3\text{ air}}\right)$ at standard conditions (20°C and 1 atm of pressure) and by the concentration of oxygen in the atmosphere by weight $\left(\frac{0.23\text{ lb }O_2}{\text{lb air}}\right)$.

$$\text{scfm} = 4.2 \times 10^4 \frac{\text{lb } O_2}{\text{d}}\left(\frac{1 \text{ ft}^3 \text{ air}}{0.0175 \text{ lb air}}\right)\left(\frac{1 \text{ lb air}}{0.23 \text{ lb } O_2}\right)\left(\frac{1 \text{ d}}{24 \text{ h}}\right)\left(\frac{1 \text{ h}}{60 \text{ min}}\right) = \boxed{7300}$$

Check to see if enough air has been provided to ensure adequate mixing of the wastewater in the aeration basin, i.e., divide the standard cubic feet per minute of air required by the volume of the aeration basin.

$$\text{Mixing air} = \frac{7300 \text{ scfm}}{1.43 \times 10^6 \text{ gal}}\left(\frac{7.48 \text{ gal}}{\text{ft}^3}\right)\left(\frac{1000 \text{ ft}^3}{1000 \text{ ft}^3}\right) = \boxed{\frac{38 \text{ scfm}}{1000 \text{ ft}^3}}$$

The amount of air required for mixing typically ranges from 20 – 30 scfm/1,000 ft^3. We have more than enough to meet mixing requirements, but cannot reduce this amount, since 7,300 scfm are required to meet process air requirements. As a design engineer, other types of diffuser systems could be evaluated that have a higher transfer efficiency, which would reduce the amount of air required.

Solution Part B

Next, calculate the power required by the centrifugal blower for compressing the air using Equation (7.177). We will assume that the air temperature may get up to 100°C during the summer months.

$$\text{Air temp.} = 100°\text{F} + 460 = 560°\text{R}$$

$$\text{Std. temp.} = 68°\text{F} + 460 = 528°\text{R}$$

The combined gas law must be used to compute the air required at a different temperature and pressure:

$$\frac{P_1V_1}{T_1} = \frac{P_2V_2}{T_2} \tag{7.180}$$

$$V_2 = \frac{P_1V_1T_2}{P_2T_1}$$

Assume that the barometric pressure remains 1 atmosphere at both temperatures, which is equivalent to 760 mm Hg.

$$7300 \text{ scfm} \times \left(\frac{760 \text{ mm}}{760 \text{ mm}}\right)\left(\frac{560°\text{R}}{528°\text{R}}\right) = 7700 \text{ cfm}$$

$\rho_{air} = 0.070 \text{ lb/ft}^3$ at sea level and 100°F (Reynolds & Richards, 1996)

$$w = (7700 \text{ cfm})\left(0.070\frac{\text{lb}}{\text{ft}^3}\right)\left(\frac{1 \text{ min}}{60 \text{ s}}\right) = 9.0\frac{\text{lb}}{\text{s}}$$

Assume that the inlet pressure (p_1) = 14.7 psia (760 mm Hg) at sea level and that the outlet pressure (p_2) = 21 psia = Pr, T_1 = 100°F + 460 = 560°R and T_2 = 68°F + 460 = 528°R. Also assume the efficiency of the blower, e = 75%, efficiency of the electrical motor, e_m = 95%, R = 53.5 and n = 0.283 for air:

$$P_w = \frac{wRT_1}{C_1 ne}\left[\left(\frac{p_2}{p_1}\right)^{0.283} - 1\right] \tag{7.177}$$

$$P_w = \frac{9.0 \text{ lb/s}\left(53.3\frac{\text{ft}\cdot\text{lb}}{\text{lb air}}\right)(528°\text{R})}{\left(550\frac{\text{ft}\cdot\text{lb}}{\text{s}\cdot\text{hp}}\right)(0.283)(0.75)}\left[\left(\frac{21 \text{ psia}}{14.7 \text{ psia}}\right)^{0.283} - 1\right] = \boxed{230 \text{ hp}}$$

$$\text{bhp} = 230$$

$$\text{mhp} = \text{motor horsepower}$$

$$\text{mhp} = \frac{\text{bhp}}{e_m}$$

$$\text{mhp} = \frac{230 \text{ hp}}{0.95} = 242 \cong \boxed{245 \text{ hp}}$$

7.7 Attached-growth biological systems

In attached-growth processes, some type of media or inert packing material is used, on which the microorganisms grow. Media that have been used include rock, gravel, slag, plastics, nylon, redwood, and other synthetics. For most attached growth processes, the media remains stationary and the wastewater is applied to the media or flows through a packed bed of inert material. The alternative is to have the media rotate through the wastewater. In both types of systems, the microbial growth that forms on the media is known as a biofilm.

The heterogeneous culture of organisms in the biofilm convert organic carbon and nutrients into new biomass, with a portion oxidized to meet energy needs, similar to the activated sludge process. Attached growth systems may be operated as aerobic, anoxic, or anaerobic processes. The media or packing material may be completely submerged in liquid, partially submerged, or not submerged.

This section will present the design of two types of attached growth processes: trickling filters and rotating biological contactors (RBCs).

7.7.1 Trickling filters

An alternative biological process that was often used before activated sludge systems had gained wide acceptance is the aerobic, attached-growth process called the trickling filter. With trickling filters, wastewater is applied to some type of filter media, on which a heterogeneous culture of microorganisms is growing. In the past, rock or slag was the media of choice, but these have been replaced with plastic or synthetic inert media.

Trickling filters do not filter the wastewater like a traditional sand filter; instead, the biomass growing on the media uses the organic matter, along with a portion of the nitrogen and phosphorus, to grow new microorganisms. Primary clarifiers must precede trickling filters to remove suspended solids that can clog the openings (orifices) on the distribution arms of trickling filters. The pressure of the wastewater discharging from the distributor arms (nozzles and/or orifices) provides the driving force causing the arms to rotate. Some small trickling filters use electric motors to drive the distribution arms.

Figure 7.37 is a schematic diagram of a conventional trickling filter system. Figure 7.38 shows a trickling filter with rock media that contains no biological growth on it since it has been taken out of service. Figure 7.39 shows a trickling filter consisting of plastic media with biological growth.

Trickling filters that use plastic media are called "biotowers", since their depths (20–30 ft) are much greater than traditional rock media filters (6–10 ft). The design of trickling filter processes has primarily been based on organic and hydraulic loading rates. Several kinetic equations have been proposed for designing trickling filters or biotowers that use plastic media.

7.7.1.1 Design of Trickling Filters with Rock Media

Traditional designs of trickling filters using rock media have been based on the National Research Council equations (NRC, 1946). Equations for single- and two-stage trickling filters were developed and are presented below. Similar to the activated sludge process, most trickling filters employ the recycle of effluent back through the units to keep them moist at all times and to increase removal of organics from the wastewater. The recycle is different from return activated sludge, since liquid wastewater, rather than thickened sludge solids, is recycled back through the filters.

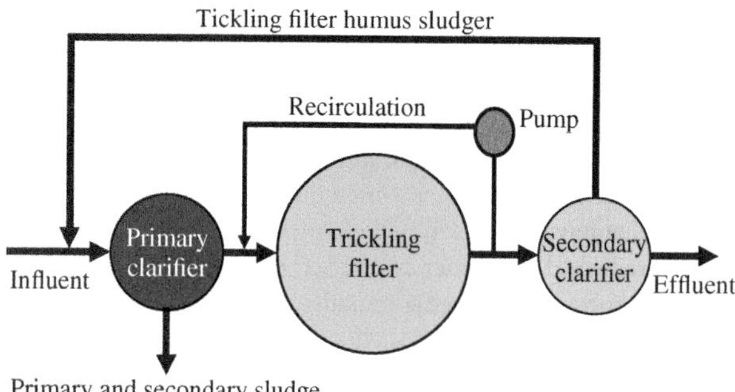

Figure 7.37 Schematic of a trickling filter wastewater treatment plant.

Figure 7.38 Photograph of conventional trickling filter with rock media.

Figure 7.39 Photograph of trickling filter with plastic media.

The recirculation factor (F) accounts for the recycled flow and reflects the effective number of times that the organic matter passes through the filter. The efficiency that is obtained using the NRC equations reflects the BOD removal through the trickling filter and secondary clarifier. The equations were developed for a temperature of 20°C, and a temperature correction coefficient (θ) of 1.035 may be used to correct efficiency to other temperatures. The efficiency of a first-stage trickling filter and secondary clarifier can be determined from Equation (7.181).

$$E_1 = \frac{100}{1 + C\left(\frac{W_1}{\forall F}\right)^{0.5}} \tag{7.181}$$

where:

E_1 = BOD removal efficiency for first-stage trickling filter and clarifier at 20°C, %

C = 0.0561 for USCS units and 0.443 for SI units

W_1 = BOD loading applied to the first-stage trickling filter, lb/d (kg/d)

$\frac{W_1}{\forall}$ = BOD loading to first-stage, $\frac{\text{lb}}{(\text{d}\cdot 1000\ \text{ft}^3)}\left(\frac{\text{kg}}{\text{m}^3}\right)$

F = recirculation factor.

$$F = \frac{1 + R}{(1 + 0.1R)^2} \tag{7.182}$$

where:

R = recirculation ratio, Q_r/Q

Q = influent wastewater flow rate excluding recycle flow, MGD (m^3/d)

Q_r = recycle flow rate, MGD(m^3/d).

The removal efficiency of the second-stage is calculated as follows:

$$E_2 = \frac{100}{1 + \left[\frac{C}{(1 - E_1)}\right]\left(\frac{W_2}{\text{Ұ}F}\right)^{0.5}} \tag{7.183}$$

where:
W_2 = BOD loading applied to the second-stage trickling filter, lb/d (kg/d)
$\frac{W_2}{\text{Ұ}}$ = BOD loading to second-stage, $\frac{\text{lb}}{(\text{d} \cdot 1000\ \text{ft}^3)}$ (kg/m^3)
E_2 = BOD removal efficiency for second-stage trickling filter at 20°C, %.

Equations (7.181) and (7.183) yield the BOD removal efficiencies for first- and second-stage trickling filters at 20°C. Equation (7.184) is used to correct removal efficiency for temperature variations.

$$E_{T^\circ C} = E_{20^\circ C}(1.035)^{(T^\circ C - 20^\circ C)} \tag{7.184}$$

where:
$E_{T^\circ C}$ = BOD removal efficiency of trickling filter at temperature T°C, %
$E_{20^\circ C}$ = BOD removal efficiency for second-stage trickling filter at 20°C, %.

Example 7.14 Design of trickling filters with rock media

A two-stage high-rate trickling filter with 100% recycle will be used to treat 1.2 million gallons per day (MGD). The influent BOD to the primary clarifier is 250 mg/L and the primary removes 35% of the incoming BOD. The effluent BOD should be 20 mg/L or less. Assume that the design BOD loading to the trickling filter system is 45 lb BOD/(d·1000 ft^3) and a depth of six feet will be used, along with a recycle ratio of 1.0. Use the NRC Equations for designing the trickling filter system. Determine:

a) The diameter of the high-rate trickling filter system.
b) The effluent BOD concentration.

Solution Part A

First, calculate the actual BOD loading to the trickling filter system.

$$W_1 = (1.2\ \text{MGD})(1 - 0.35)(250\ \text{mg BOD/L})\left(\frac{8.34\ \text{lb/MG}}{\text{mg/L}}\right) = 1630\ \text{ppd}$$

The factor (1 – 0.35) accounts for the BOD that is removed in the primary clarifier.

Next, calculate the volume of the trickling filter media by dividing the actual BOD loading by the design BOD loading:

$$\text{Ұ} = \frac{1630\ \text{ppd}}{45\ \text{lb}/(\text{d} \cdot 1000\ \text{ft}^3)} = 36{,}200\ \text{ft}^3$$

Volume per trickling filter is calculated by dividing the total volume by 2, since this is a two-stage trickling filter.

$$\frac{\text{Ұ}}{\text{trickling filter}} = \frac{36{,}200\ \text{ft}^3}{2} = 18{,}100\ \text{ft}^3$$

The area of each trickling filter is determined by dividing the volume by the depth of 6 ft.

$$A = \frac{18{,}100\ \text{ft}^3}{6\ \text{ft}} = 3017 \cong 3020\ \text{ft}^2$$

$$A = 3020\ \text{ft}^2 = \pi r^2 \qquad r = \sqrt{3020 ft^2/\pi} = 31\ \text{ft}$$

Diameter = 2r = 2(31 ft) = 62 ft, therefore use a diameter of 65 ft. Wastewater equipment normally is manufactured in increments of 5 ft.

Next, calculate the new area based on a diameter of 65 ft or radius of 32.5 ft:

$$A = \pi r^2 = \pi(32.5\ \text{ft})^2 = 3318\ \text{ft}^2$$

Actual volume of each trickling filter is calculated by multiplying the area by the 6 ft depth.

$$\text{Ұ} = A \times Depth = 3318\ \text{ft}^2 \times 6\ \text{ft} = 19{,}908 \cong 19{,}910\ \text{ft}^3$$

Determine the recirculation factor (F) using Equation (7.182).

$$F = \frac{1 + R}{(1 + 0.1R)^2} = \frac{1 + 1}{(1 + 0.1 \times 1)^2} = 1.65$$

Solution Part B

Estimate the efficiency of the first-stage trickling filter using Equation (7.181):

$$E_1 = \frac{100}{1 + 0.0561\left(\frac{W_1}{\text{Ұ}F}\right)^{0.5}} = \frac{100}{1 + 0.0561\left(\frac{1630\ \text{ppd}}{(19{,}910/1000)\ 1.65}\right)^{0.5}} = 71.7\%$$

Next, calculate the BOD loading to the second-stage trickling filter (W_2).

$$W_2 = (1 - 0.717)(1630\ \text{ppd}) = 461\ \text{ppd}$$

Now, it is possible to calculate the efficiency of the second-stage using Equation (7.183).

$$E_2 = \frac{100}{1 + \left[\frac{0.0561}{(1 - E_1)}\right]\left(\frac{W_2}{VF}\right)^{0.5}}$$

$$E_2 = \frac{100}{1 + \left[\frac{0.0561}{(1 - 0.717)}\right]\left(\frac{461 \text{ ppd}}{(19{,}910/1000)\, 1.65}\right)^{0.5}}$$

$$= 57.4\%$$

Estimate the final BOD in the effluent to see if it meets the standard.

$$\text{Effluent BOD}_e = \left(250 \frac{mg}{L}\right)(1 - 0.35)(1 - 0.717)(1 - 0.574)$$

$$= \boxed{19.6 \frac{mg}{L}} \leq 20 \frac{mg}{L}$$

A trickling filter cannot achieve the same effluent limits as that of an activated sludge system. However, trickling filters are lower in cost and require less operator attention.

7.7.1.2 Design of Trickling Filters with Plastic Media

Various kinetic equations have been proposed for designing trickling filters and biotowers containing plastic media (Schultze, 1960; Eckenfelder, 1960; Germain, 1966; Balarishnan *et al.*, 1969.) BOD removal through a trickling filter or biotower containing plastic media can be modeled by a first-order equation, as shown below (Germain, 1966):

$$\frac{L_e}{L_o} = e^{-kD/q^n} \quad (7.185)$$

where:

L_o = BOD_5 of primary effluent fed to the filter, excluding recirculation flow, mg/L
L_e = BOD_5 of settled effluent, mg/L
k = wastewater treatability and media coefficient, $(gpm)^{0.5}/ft^2$ $((L/s)^{0.5}/m^2)$ based on $n = 0.5$
q = hydraulic loading rate of primary effluent, excluding recirculation flow, gpm/ft^2 (Lps/m^2)
q = Q/A
Q = primary effluent flow rate, gpm (Lps)
A = filter cross section area, ft^2 (m^2)
n = constant for type of media, typically 0.5 for plastic and rock media.

The value of k can be corrected for temperature variations using the following equation (Metcalf & Eddy, 2003, p. 913):

$$k_{T°C} = k_{20°C}(1.035)^{(T°C-20°C)} \quad (7.186)$$

where:

$k_{T°C}$ = wastewater treatability and media coefficient at temperature, T, $(gpm)^{0.5}/ft^2$ $[(L/s)^{0.5}/m^2]$ based on $n = 0.5$
$k_{20°C}$ = wastewater treatability and media coefficient at 20°C, $(gpm)^{0.5}/ft^2$ $[(L/s)^{0.5}/m^2]$ based on $n = 0.5$
T = wastewater temperature, °C.

Equation (7.185) can be rewritten by substituting the soluble BOD_5 concentration (S_e/S_o) for (L_e/L_o) as follows (WEF, 1998b):

$$\frac{S_e}{S_o} = e^{-kD/q^n} \quad (7.187)$$

where:

S_o = soluble BOD_5 of primary effluent fed to the filter, excluding recirculation flow, mg/L
S_e = soluble BOD_5 of settled effluent, mg/L.

The value of k must also be normalized to a specified depth and effluent concentration, using the following equation (Metcalf & Eddy, 2003, p. 918; WEF, 1998b):

$$k_2 = k_1\left(\frac{D_1}{D_2}\right)^{0.5}\left(\frac{S_1}{S_2}\right)^{0.5} \quad (7.188)$$

where:

S_1 = 150 mg BOD_5/L
S_2 = site specific influent BOD_5 concentration, mg/L
D_1 = media depth, 20 ft (6.1 m)
D_2 = site specific media depth, ft (m)
k_2 = normalized value of k for site specific media depth and influent BOD_5 concentration
k_1 = k value at depth of 20 ft (6.1 m) and influent BOD_5 concentration of 150 mg/L.

The units for k are $(gpm)^{0.5}/ft^2$ or $(L/s)^{0.5}/m^2$. The value of k_1 at 20°C for domestic wastewater, as reported in Metcalf & Eddy (2003, p. 918), is 0.078 $(gpm)^{0.5}/ft^2$ or 0.210 $(L/s)^{0.5}/m^2$. Values for k_1 for other types of wastewater are also presented in Metcalf & Eddy (2003, p. 918). When possible, it is preferable to conduct pilot-scales studies on the specific wastewater to be treated in order to determine the wastewater treatability and media coefficient, k. A minimum wetting rate of 0.75 gpm/ft^2 (0.5 Lps/m^2) is recommended to keep the biomass moist and, therefore, trickling filters and biotowers are designed to have wastewater recycle systems (Metcalf & Eddy, 2003, p. 918).

Example 7.15 Design of trickling filters with plastic media

Two biotowers operating in parallel are to be designed to treat a flow of 4.0 million gallons per day (MGD). The BOD_5 in the effluent from the primary clarifier is 135 mg/L and the effluent BOD_5 from the secondary clarifier is 20 mg/L. The minimum wastewater temperature is expected to be 16°C. The value of k is assumed to be 0.078 $(gpm)^{0.5}/ft^2$ at 20°C. Biotower depth = 20 ft. Determine:

a) The radius of each biotower, ft.
b) The volume of media, ft^3.

c) The recycle flow around the biotowers if the minimum wetting rate is 0.75 gpm/ft² and the recycle ratio, *R*.

Solution

Normalize *k* for site specific depth and influent BOD_5 concentration using Equation (7.188):

$$k_2 = k_1\left(\frac{D_1}{D_2}\right)^{0.5}\left(\frac{S_1}{S_2}\right)^{0.5}$$

$$= 0.078\,(\text{gpm})^{0.5}/\text{ft}^2\left(\frac{20\text{ ft}}{20\text{ ft}}\right)^{0.5}\left(\frac{150\text{ mg/L}}{135\text{ mg/L}}\right)^{0.5}$$

$$k_2 = 0.082\,(\text{gpm})^{0.5}/\text{ft}^2$$

Correct k_2 for a temperature of 16°C using Equation (7.186):

$$k_{T°C} = k_{20°C}(1.035)^{(T°C-20°C)}$$

$$= 0.082(\text{gpm})^{0.5}/\text{ft}^2\,(1.035)^{(16°C-20°C)}$$

$$k_{20°C} = 0.071\,(\text{gpm})^{0.5}/\text{ft}^2$$

The hydraulic loading rate, *q*, is determined by substituting the appropriate values into Equation (7.187):

$$\frac{S_e}{S_o} = e^{-kD/q^n}$$

$$\frac{20\text{ mg/L}}{135\text{ mg/L}} = e^{-0.071\,(\text{gpm})^{0.5}/\text{ft}^2(20\text{ ft})/q^{0.5}}$$

$$q^{0.5} = \frac{-0.071(\text{gpm})^{0.5}/\text{ft}^2(20\text{ ft})}{\ln\left(\frac{20\text{ mg/L}}{135\text{ mg/L}}\right)}$$

$$q = \left(\frac{-0.071(\text{gpm})^{0.5}/\text{ft}^2(20\text{ ft})}{\ln\left(\frac{20\text{ mg/L}}{135\text{ mg/L}}\right)}\right)^{(1/0.5)} = 0.55\,\frac{\text{gpm}}{\text{ft}^2}$$

The surface area of the biotowers can now be determined by dividing the volumetric flow rate by the hydraulic loading rate as follows:

$$A_{\text{cross section}} = \frac{Q}{q} = \frac{4.0\times10^6\text{ gal/d}}{0.55\text{ gpm/ft}^2}\left(\frac{1\text{ d}}{24\text{ h}}\right)\left(\frac{1\text{ h}}{60\text{ min}}\right)$$

$$= 5.05\times10^3\text{ft}^2$$

$$\frac{A_{\text{cross section}}}{\text{biotower}} = \frac{5.05\times10^3\text{ft}^2}{2} = 2.53\times10^3\text{ft}^2$$

$$A_{\text{cross section}} = 2.53\times10^3\text{ft}^2 = \pi r^2$$

$$r = \sqrt{\frac{2.53\times10^3\text{ft}^2}{\pi}} = 28\text{ ft} \cong \boxed{30\text{ ft}}$$

The volume of the biotower media is determined by multiplying the cross sectional area by the depth or height:

$$V = \pi r^2 \times D = \pi(30\text{ ft})^2 \times 20\text{ ft} = 5.65\times10^4\text{ ft}^3$$

$$V_{\text{Total}} = (5.65\times10^4\text{ ft}^3)\times 2\text{ biotowers} = \boxed{1.13\times10^5\text{ ft}^3}$$

The minimum wetting rate has been specified as 0.75 gpm/ft², so the sum of the hydraulic loading rate (q) and recycle loading rate (q_r) must be at least equal to or greater than the minimum wetting rate:

$$q + q_r = 0.75\text{ gpm/ft}^2$$

$$q_r = 0.75\text{ gpm/ft}^2 - 0.55\text{ gpm/ft}^2 = \boxed{0.20\text{ gpm/ft}^2}$$

The recycle ratio is calculated as follows:

$$R = \frac{q_r}{q} = \frac{0.20\text{ gpm/ft}^2}{0.55\text{ gpm/ft}^2} = \boxed{0.36}$$

7.7.2 Rotating biological contactors

Another type of attached-growth process that produces effluent equivalent to secondary treatment uses rotating biological contactors (RBCs). The process was first used in Europe, and came to the United States in the early 1970s. BOD removal and nitrification can be accomplished in a staged RBC system. High-density corrugated polystyrene or polyvinyl chloride circular disks approximately 12 ft (3.7 m) in diameter are placed on horizontal shafts. The typical length for a shaft is 27 ft (8.23 m). The surface area of the disks with these standard dimensions is approximately 100,000 ft² (9,300 m²) for low-density media and 150,000 ft² (13,900 m²) for high-density media (Metcalf & Eddy, 2003, page 931; WEF, 1998b).

The disks are typically submerged 40% into the wastewater and rotate from 1 – 2 revolutions per minute (RPM). As the disks rotate through the wastewater, a biological slime or biofilm grows on the media. The biofilm adsorbs organics and nutrients as it passes through the wastewater and the microorganisms utilize these for synthesizing new biomass and oxidizing some of the organics for energy. Oxygen is transferred to the biofilm as it rotates out of the wastewater and is exposed to air. An electric motor, typically 5 or 7.5 hp (3.7 or 5.6 kW) per shaft provides the power necessary to rotate the disks (WEF, 1998b).

Typically, fiberglass covers are provided to protect the RBCs from UV light, inclement weather, and exposure to sunlight which causes problems with algal growth. In some cases, the RBCs are enclosed in a building, which must be well ventilated, and the off-gas is collected and treated for hydrogen sulfide and other potential volatile organic carbons that may be in the wastewater. Figure 7.40 shows a schematic of an RBC disk partially immersed in wastewater.

Table 7.17 lists the hydraulic loadings, soluble and total organic loadings, and anticipated effluent BOD and NH_3 concentrations from rotating biological wastewater treatment facilities.

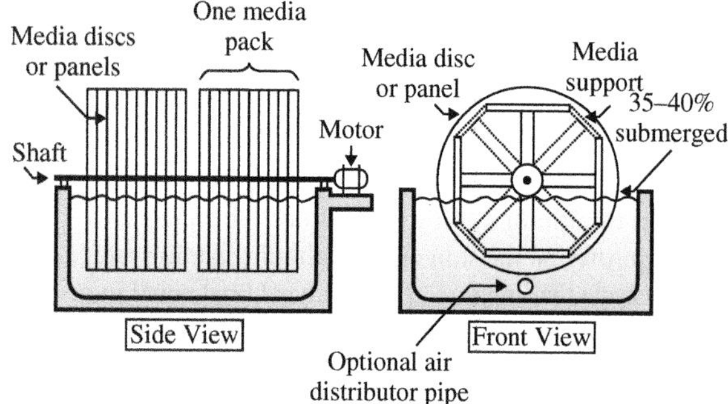

Figure 7.40 Schematic of a rotating biological contactor. *Source:* http://en.wikipedia.org/wiki/File:Rotating_Biological_Contactor.png. Courtesy Milton Beychok.

Table 7.17 Design criteria for rotating biological contactors.

Parameter	BOD removal and/or nitrification
Maximum total organic loading on first stage, lb BOD_5/1000 $ft^2 \cdot d$ (g BOD_5/ $m^2 \cdot d$)	6.4 (31.2)
Maximum soluble organic loading on first stage, lb $sBOD_5$/1000 $ft^2 \cdot d$ (g $sBOD_5$/ $m^2 \cdot d$)	2.5–4.0 (12.2–19.5)
Maximum inter-stage total organic loading, lb BOD_5/1000 $ft^2 \cdot d$ (g BOD_5/ $m^2 \cdot d$)	6 (29)
Maximum inter-stage soluble organic loading, lb $sBOD_5$/1000 $ft^2 \cdot d$ (g $sBOD_5$/ $m^2 \cdot d$)	2.5 (12.2)

Source: EPA (1993) *Manual Nitrogen Control*, pp. 187–188. United States Environmental Protection Agency.

According to the US EPA design report (1984a, p. 22), RBC treatment facilities should be staged. Table 7.18 presents the recommended number of stages recommended by RBC manufacturers.

Table 7.18 Number of RBC Stages Recommended by Manufacturers.

Target soluble effluent BOD_5 concentration (mg/L)	Recommended minimum number of stages
>25	1
15–25	1 or 2
10–15	2 or 3
< 10	3 or 4

Source: US EPA (1984a), p. 22. United States Environmental Protection Agency.

The design of RBC treatment systems has primarily relied on manufacturer's operating literature and has been based on BOD loading rate. A temperature correction factor is applied to increase the amount of disk surface area required when the temperature drops below 55°C. Figure 7.41 shows the temperature correction factor as a function of temperature as advocated by the Walker Process (2009).

The media surface area is adjusted by dividing the media surface area by the temperature correction factor, which increases the total surface area. Kornegay (1975) and Eckenfelder (1989, pp. 239–245) have developed kinetic models for designing RBCs. A second-order removal model developed by US EPA (1984b), and modified by Grady *et al.* (1999, p. 927), has been successfully used to design RBCs. The model is used to calculate the soluble BOD_5 from each stage for a given disk surface area and flow rate. Equation (7.189) presents the form of the model promoted by Grady *et al.* (1999, p. 927):

$$S_n = \frac{-1 + \sqrt{1 + (4)(0.00974)(A_s/Q)S_{n-1}}}{(2)(0.00974)(A_s/Q)} \tag{7.189}$$

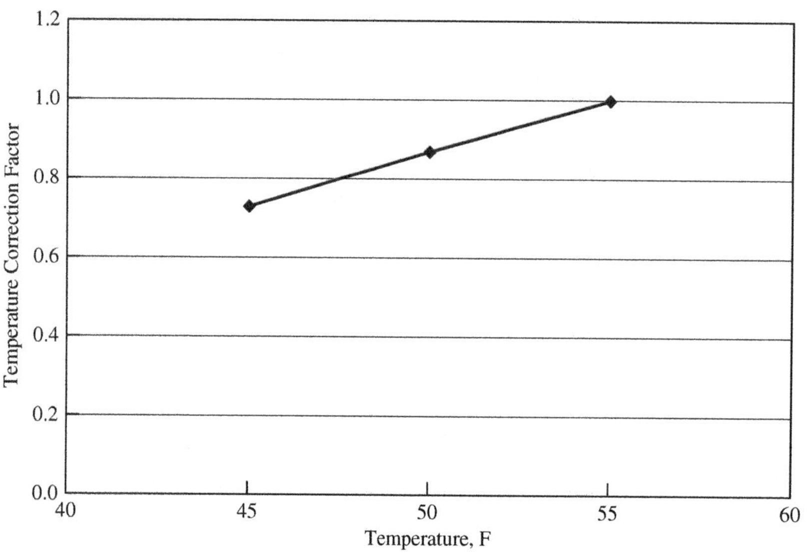

Figure 7.41 Temperature correction factor for RBC surface area.

where:
S_n = effluent soluble BOD_5 from stage n, g/m^3
A_s = disk surface area for stage n, m^2
Q = flow rate, m^3/d.

A design example using the EPA method is now presented.

Example 7.16 Design of rotating biological contactors for BOD removal

Design a multi-stage RBC WWTP for the following wastewater for BOD removal. Influent flow = 6,340 m^3/d, influent BOD_5 = 250 g/m^3, primary clarifier removes 34% of BOD_5, total effluent BOD_5 and effluent TSS = 20 g/m^3, effluent soluble BOD_5 ($sBOD_5$) = 10 g/m^3, and T = 20°C. Determine:

a) The number of stages and number of trains.
b) The soluble BOD_5 in the effluent from each stage.
c) The organic loading on the first stage.
d) The overall organic loading to the RBC WWTP.
e) The overall hydraulic loading rate to the RBC WWTP.

Solution Part A

Determine the number of RBCs shafts for the first stage, assuming that the primary effluent soluble BOD_5 is 50% of the primary effluent total BOD_5, and that the total soluble BOD_5 loading rate is 15 g $sBOD_5$/(m^2 · d).

First, calculate the total BOD_5 in the primary effluent:

$$\text{Total BOD}_5 \text{ in primary effluent} = (1 - 0.34) \times 250 \frac{g}{m^3}$$
$$= 165 \frac{g}{m^3}$$

Calculate the soluble BOD_5 in the primary effluent:

$$\text{Soluble BOD}_5 \text{ in primary effluent} = 165 \frac{g}{m^3} \times 0.5$$
$$= 82.5 \frac{g}{m^3}$$

Calculate the soluble BOD_5 loading to the first stage:

$$\text{Soluble BOD}_5 \text{ Loading} = \left(6.34 \times 10^3 \frac{m^3}{d}\right) \times \left(82.5 \frac{g}{m^3}\right) = 5.23 \times 10^5 \frac{g}{d}$$

Determine the number of shafts in the first stage:

$$\text{Media Surface Area} = \frac{5.23 \times 10^5 \text{ gsBOD}_5/d}{\frac{15.0 \text{ gsBOD}_5}{(m^2 \cdot d)}}$$
$$= 3.49 \times 10^4 m^2$$

$$\text{Number of Shafts} = \frac{3.49 \times 10^4 m^2}{9{,}300 \text{ m}^2/\text{shaft}}$$
$$= 3.8 \cong 4.0 \text{ shafts per stage}$$

Assume there are three treatment trains and four stages with 4.0 shafts per stage.

Solution Part B

Next, calculate the flow per train:

$$\text{Flow per train} = \frac{6.34 \times 10^3 \text{ m}^3/d}{3 \text{ trains}} = \frac{2.11 \times 10^3 m^3/d}{\text{train}}$$

Determine the ratio of surface area to the flow rate as follows:

$$\frac{A_s}{Q} = \frac{9300 \text{ m}^2}{2.11 \times 10^3 \text{ m}^3/d} = \frac{4.4 \text{ d}}{\text{m}}$$

Calculate the soluble BOD_5 from each stage using Equation (7.189):

$$S_1 = \frac{-1 + \sqrt{1 + (4)(0.00974)(A_s/Q)S_0}}{(2)(0.00974)(A_s/Q)}$$

$$S_1 = \frac{-1 + \sqrt{1 + (4)(0.00974)(4.4 \text{ d/m})(82.5 \text{ g/m}^3)}}{(2)(0.00974)(4.4 \text{ d/m})}$$

$$= \boxed{33.7 \frac{g}{m^3}}$$

$$S_2 = \frac{-1 + \sqrt{1 + (4)(0.00974)(A_s/Q)S_1}}{(2)(0.00974)(A_s/Q)}$$

$$S_2 = \frac{-1 + \sqrt{1 + (4)(0.00974)(4.4 \text{ d/m})(33.7 \text{ g/m}^3)}}{(2)(0.00974)(4.4 \text{ d/m})}$$

$$= \boxed{18.7 \frac{g}{m^3}}$$

$$S_3 = \frac{-1 + \sqrt{1 + (4)(0.00974)(A_s/Q)S_2}}{(2)(0.00974)(A_s/Q)}$$

$$S_3 = \frac{-1 + \sqrt{1 + (4)(0.00974)(4.4 \text{ d/m})(18.7 \text{ g/m}^3)}}{(2)(0.00974)(4.4 \text{ d/m})}$$

$$= \boxed{12.3 \frac{g}{m^3}}$$

$$S_4 = \frac{-1 + \sqrt{1 + (4)(0.00974)(A_s/Q)S_3}}{(2)(0.00974)(A_s/Q)}$$

$$S_4 = \frac{-1 + \sqrt{1 + (4)(0.00974)(4.4 \text{ d/m})(12.3 \text{ g/m}^3)}}{(2)(0.00974)(4.4 \text{ d/m})}$$

$$= \boxed{8.9 \frac{g}{m^3}} < 10.0 \frac{g}{m^3} \text{ Good!}$$

Solution Part C

The soluble organic loading on the first-stage is calculated below:

$$\text{First-stage soluble BOD}_5 \text{ loading} = \frac{(6.34 \times 10^3\,\text{m}^3/\text{d})(82.5\,\text{g/m}^3)}{(4\text{shafts})(9300\,\text{m}^2/\text{shaft})} = \boxed{14.0\,\frac{\text{g sBOD}_5}{(\text{m}^2\cdot\text{d})}}$$

$$\text{Within range}\left(12 - 15\,\frac{\text{g sBOD}_5}{(\text{m}^2\cdot\text{d})}\right)$$

Solution Part D

The total organic loading on the RBC WWTP is calculated as follows.

$$\text{Total BOD}_5 \text{ loading} = \frac{(6.34 \times 10^3\,\text{m}^3/\text{d})(165\,\text{g/m}^3)}{(3\,\text{trains})(4\,\text{shafts/stage})(9300\,\text{m}^2/\text{shaft})} = \boxed{9.4\,\frac{\text{g BOD}_5}{(\text{m}^2\cdot\text{d})}}$$

$$\text{Within range}\left(8 - 20\,\frac{\text{g BOD}_5}{(\text{m}^2\cdot\text{d})}\right)$$

Solution Part E

The hydraulic loading on the RBC WWTP is calculated as follows.

$$\text{Hydraulic loading} = \frac{(6.34 \times 10^3\,\text{m}^3/\text{d})}{(3\,\text{trains})(4\,\text{shafts/stage})(9300\,\text{m}^2/\text{shaft})} = \boxed{0.057\,\frac{\text{m}^3}{(\text{m}^2\cdot\text{d})}}$$

$$\text{Slightly below range}\left(0.08 - 0.16\,\frac{\text{m}^3}{(\text{m}^2\cdot\text{d})}\right)$$

but should work.

7.8 Advanced biological treatment systems

Advanced wastewater treatment (AWT) systems are designed to remove nitrogen and phosphorus, along with additional BOD and TSS, from the wastewater. AWT systems use both chemical and biological unit processes, along with physical unit operations such as dual-media filters. Biological nutrient removal (BNR) processes use various types of microorganisms to remove nitrogen and phosphorus beyond that which can be accomplished in conventional activated sludge systems. Both attached-growth and suspended-growth biological processes are used, which may be operated in anaerobic, anoxic, or oxic (aerobic) environments. By manipulating a microorganism's environment, engineered systems can be designed to remove only nitrogen or only phosphorus, or to accomplish both nitrogen and phosphorus removal.

All BNR processes should be equipped with chemical feed systems in order to have the ability to add coagulants such as alum or lime to precipitate phosphorus when an upset occurs to the biological process. Higher quality effluent can also be achieved with the addition of filters following BNR treatment and clarification.

AWT systems may also use chemical systems for removing nitrogen and phosphorus. Ammonium nitrogen can be converted to ammonia gas (NH_3) by adding lime or sodium hydroxide to raise the pH above 11 and then passing the wastewater through a stripper to remove the ammonia. Phosphorus, in the form of orthophosphate (PO_4^{3-}), is easily removed from wastewater by adding alum or lime for precipitation as aluminum phosphate or calcium phosphate. Advanced biological and chemical processes for removing nutrients from wastewater can be found in the following references: WEF (1998b, 1998d); Metcalf & Eddy (2003); and US EPA (1993). The mechanisms involved in biological nitrogen and phosphorus removal systems will be presented, along with some of the most widely used processes.

7.8.1 Biological nitrogen removal

Removing nitrogen from wastewater biologically is accomplished by nitrification, an aerobic process by which nitrifying autotrophic bacteria oxidize ammonia to nitrite, and ultimately to nitrate, is followed by denitrification, an anoxic process wherein heterotrophic bacteria reduce nitrates produced during nitrification to the end product, nitrogen gas.

In the past, separate, two-staged reactor systems were often used to accomplish nitrogen removal. Successful nitrogen removal was achieved using a nitrifying activated sludge process, with secondary clarification in the first stage followed by dentrification and final clarification in the second stage.

Figure 7.42 is a schematic of a two-stage, nitrification-denitrification process. Using two separate stages offers the advantages of optimization of each stage and better control of the process. The major disadvantages include large reactor volumes, which increase capital costs, and also methanol or other type of carbon source must be added to the denitrification process, leading to increased operating costs. Most suspended-growth BNR systems in use today are called single-sludge systems, since they use one reactor with multiple staged zones. Each zone is designed to create the proper environment to grow specific types of microorganisms.

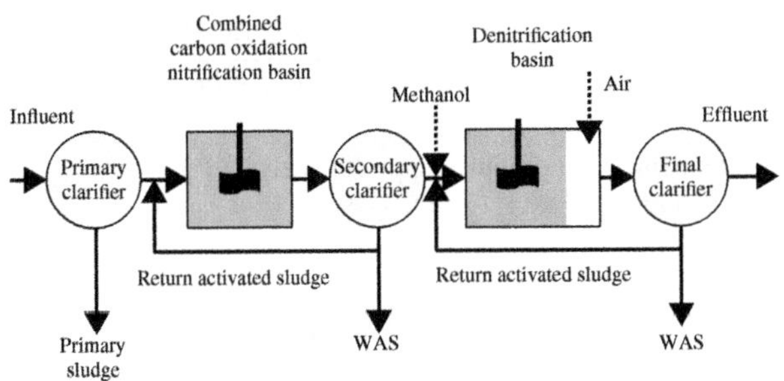

Figure 7.42 Two-stage nitrification – denitrification process.

Table 7.19 lists design criteria for some suspended-growth, nitrification-denitrification systems.

7.8.1.1 Nitrification

As stated previously, nitrification and denitrification must be accomplished in order to remove nitrogen from wastewater. Nitrification was discussed in detail in Section 7.5.6.7 of this chapter. During nitrification, organic nitrogen is deaminated and transformed into ammonia (NH_3) and ammonium (NH_4^+). The form of the nitrogen is pH dependent but, when the pH is in the range of 7–8.5, it is primarily in the ammonium ion form. At pH values above 9.3, ammonia is the predominant species. In the past, a commonly used process for removing nitrogen from wastewater was to add lime (CaO) or caustic soda (NaOH) to raise the pH above 10.8, and then pass through a stripper to release ammonia gas to the atmosphere.

Nitrification is principally accomplished by two autotrophic bacterial genera, *Nitrosomonas* and *Nitrobacter*. Autotrophs expend energy to fix and reduce inorganic carbon, which leads to lower yields. The nitrifiers also obtain less energy from their electron donors (NH_4^+ and NO_2^-). This makes them slow growers in comparison to heterotrophic microbes that compete with them in activated sludge systems. Equations (7.97) through (7.100) show the energy equations used to model nitrification. Rittman and McCarty (2001, p. 473) give the following equation, which represents the overall nitrification reaction in activated sludge operating at a MCRT of 15 days:

$$NH_4^+ + 1.815\,O_2 + 0.1304\,CO_2 \rightarrow 0.0261\,C_5H_7O_2N + 0.973\,NO_3^- + 0.091\,H_2O + 1.973\,H^+ \quad (7.190)$$

Table 7.19 Design criteria for nitrification-denitrification systems.

Parameter	Pre-denitrification	Post-dentrification	Pre- and Post-Denitrification
System MCRT, d	6–30	6–33	6–40
HRT, h			
First anoxic zone	2–5	–	2–5
First oxic zone	4–12	4 –12	4–12
Second anoxic zone	–	2–5	2–5
Reaeration zone	–	–	0.5–1.0
MLSS, mg/L	1,500–4,000	1,500–4,000	2,000–5,000
Mixed liquor recycle	(2–4) Q_{design}	–	(4–6) Q_{design}
RAS flow	(0.5–1.0) Q_{design}	(0.5–1.0) Q_{design}	(0.5–1.0) Q_{design}
DO, mg/L			
Anoxic zones	0	0	0
Oxic zones	1–2	1–2	1–2
Mixing requirements			
Anoxic zones, kw/10^3 m^3	4–10	4–10	4–10
Oxic zones, kw/10^3 m^3	20–40	20–40	20–40
Oxic zones, m^3 of air/(min ≅ 10^3 m^3)	10–30	10–30	10–40

Source: WEF Biological and Chemical Systems for Nutrient Removal (1998d).

Based on the above equation the following conclusions can be reached:

1 Nitrification requires approximately 4.15 $gO_2/g\,NH_4^+ - N$ when synthesis is incorporated in the overall reaction.

$$\left(\frac{1.815 \times 32\,g/\text{mole}}{14\,gNH_4^+ - N/\text{mole}}\right) = 4.15\,\frac{gO_2}{gNH_4^+ - N}$$

2 Nitrification consumes approximately 7.05 g alkalinity as $CaCO_3/g\,NH_4^+ - N$ when synthesis is incorporated into the overall reaction.

$$\left(\frac{1.973 \times 1\,gH^+/\text{mole}}{14\,gNH_4^+ - N/\text{mole}}\right)\left(\frac{\text{eq}}{1\,gH^+}\right)\left(\frac{50\,gCaCO_3}{\text{eq}}\right) = 7.05\,\frac{\text{g of akalinity as } CaCO_3}{g\,NH_4^+ - N}$$

When designing and operating nitrifying activated sludge processes, one typically uses 4.57 $gO_2/g\,NH_4^+ - N$ and 7.14 g alkalinity as $CaCO_3/g\,NH_4^+ - N$ instead of the values calculated above. By doing this, more conservatism is built into the engineering design.

7.8.1.2 Denitrification

To remove the nitrate that is formed during nitrification, a second biological process known as denitrification, must be used. An anoxic environment is required (i.e., one void of dissolved oxygen), but bound oxygen in the form of nitrites and nitrates is necessary. Heterotrophic, denitrifying microorganisms reduce nitrate to nitrogen gas that is released into the atmosphere. Since denitrifiers are heterotrophs, they require organic matter to serve as the electron donor, and nitrites and nitrates serve as the electron acceptor rather than oxygen.

The first denitrification systems (post-denitrification) followed nitrifying activated sludge systems, so there was little soluble organic matter remaining to drive the denitrification reaction. Therefore, methanol was typically used in these systems. The following stoichiometric equation shows what happens during denitrification when methanol (CH_3OH) serves as the carbon source (electron donor).

$$6NO_3^- + 5CH_3OH \xrightarrow{\text{Denitrifiers}} 3N_2 + 6OH^- + 5CO_2 + 7H_2O \quad (7.191)$$

Using a denitrification process decreases the overall quantity of oxygen required to stabilize or oxidize organic matter, since it occurs in an anoxic environment. Denitrification also generates alkalinity, which helps to replenish some of the alkalinity that is consumed during the nitrification process. According to Equation (7.191), the following conclusions can be made.

1 The methanol to nitrate feed ratio is 1.9 g $CH_3OH/gNO_3^- - N$.

$$\frac{5(32\,gCH_3OH)}{6(14\,gNO_3^- - N)} = \frac{1.9\,g\,CH_3OH}{g\,NO_3^- - N}$$

2 Denitrification produces approximately 3.57 g alkalinity as $CaCO_3/g\,NO_3^- - N$.

$$\frac{6(17\,gOH^-)}{6(14\,gNO_3^- - N)}\left(\frac{\text{eq}}{17\,gOH^-}\right)\left(\frac{50\,gCaCO_3}{\text{eq}}\right) = 3.57\,\frac{\text{g of akalinity as } CaCO_3}{g\,NO_3^- - N}$$

Most BNR systems in operation today have denitrification processes that precede the combined nitrification and organic removal step in an oxic zone; these are called pre-denitrification systems, and they use the soluble organics in the incoming wastewater as the electron donor. Equation (7.192) represents the denitrification reaction when wastewater is used:

$$10NO_3^- + C_{10}H_{19}O_3N \xrightarrow{\text{Denitrifiers}} 5N_2 + 10OH^- + 10CO_2 + 3H_2O + NH_3 \quad (7.192)$$

Based on Equation (7.192), 3.57 g alkalinity as $CaCO_3/gNO_3^- - N$ is produced during pre-denitrification using the biodegradable organics in the wastewater as electron donors.

$$\frac{10(17\,gOH^-)}{10(14\,gNO_3^- - N)}\left(\frac{\text{eq}}{17\,gOH^-}\right)\left(\frac{50\,gCaCO_3}{\text{eq}}\right) = 3.57\,\frac{\text{g of akalinity as } CaCO_3}{g\,NO_3^- - N}$$

7.8.1.3 Post-Denitrification

Wuhrmann was one of the earliest to investigate post-dentrification suspended growth systems (US EPA, 1993, p. 251). Figure 7.43 is a schematic of the Wuhrmann process.

Combined carbon oxidation and nitrification in an aerobic zone precedes the denitrification process that occurs in an anoxic reactor. Since the exogenous carbon source has been exhausted after aerobic treatment, the denitrification reaction is driven by endogenous decay, i.e., the oxidation of intracellular components. According to the *Nitrogen Control Manual* (US EPA, 1993, p. 251), the Wuhrmann process, when tested at pilot-scale, was able to achieve an 88% reduction in total nitrogen.

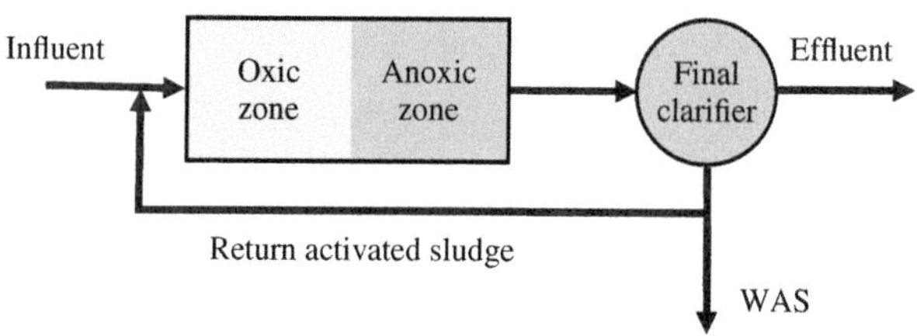

Figure 7.43 Wuhrmann denitrification process.

The Wuhrmann process, followed by post-aeration, is used in several BNR configurations that will be discussed later. An external organic substrate, such as methanol, is often added to post-denitrification systems to provide electron donors for the reaction. Typically, the methanol to nitrate dose ratio is 3 : 1.

7.8.1.4 Pre-Denitrification

The Ludzack-Ettinger (LE) and Modified Ludzack-Ettinger (MLE) processes have anoxic zones preceding the combined organic removal/nitrification step. When the denitrification step is placed up-stream of the aerobic zone, organic matter in the incoming wastewater serves as the carbon source, i.e., electron donor for the reaction. Figures 7.44 and 7.45 are schematics of each of these processes. The MLE process works better than the LE process, since internal mixed liquor recycle, ranging from 100–400% of the design flow, is recycled from the aerobic zone to the anoxic zone. This recycles nitrates produced in the aerobic zone to the anoxic zone, allowing them to be reduced by the denitrifiers.

7.8.1.5 Four-Stage Bardenpho Process

A schematic of the Four-stage Bardenpho process is presented in Figure 7.46. This is a proprietary process developed by Barnard and marketed in the United States by EIMCO (US EPA, 1993, p. 259). The process uses a pre- and post-anoxic zone and two aerobic zones. The Four-stage Bardenpho process has achieved an effluent total nitrogen (TN) concentration of 3.0 mg/L and 90% TN removal. An approach that may be used to design a Four -stage Bardenpho process is presented below.

This section presents design criteria, design equations, a design example, and a case history for a combined pre- and post-denitrification system. Figure 7.46 is a schematic of a typical combined pre-and post-denitrification system. This layout is typical of a modified Bardenpho system.

The first-stage anoxic zone precedes the first-stage aerobic or oxic zone. The first-stage anoxic zone generally removes approximately 60–75% of the nitrate nitrogen. The

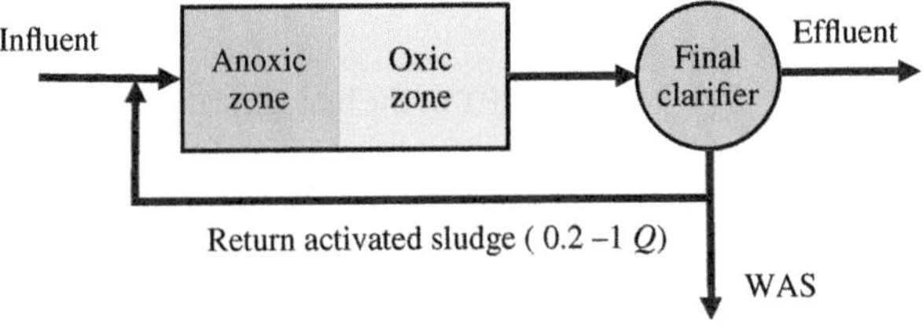

Figure 7.44 Ludzack-Ettinger denitrification process.

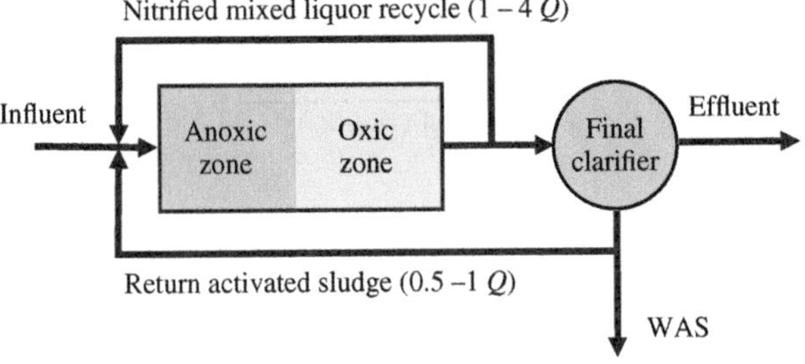

Figure 7.45 Modified Ludzack-Ettinger dentrification process.

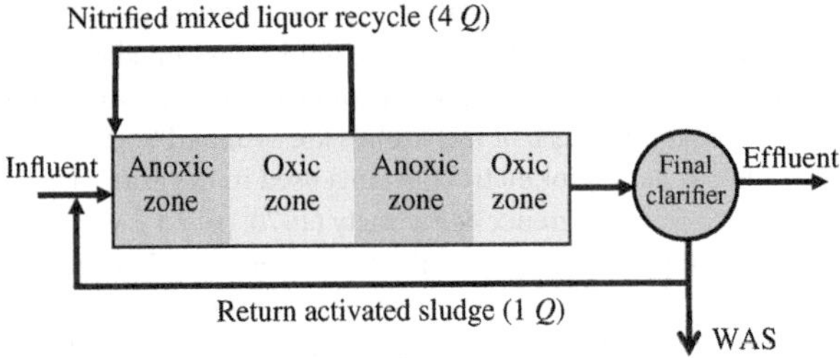

Figure 7.46 Four-stage Bardenpho denitrification process.

carbonaceous and nitrogenous oxygen demand of the wastewater is primarily met in the aerobic zone. Mixed liquor recycle (MLR) is pumped from the end of the aerobic zone back to the head of the first-stage anoxic zone to allow biological denitrification to occur.

The second-stage anoxic zone follows the first-stage aerobic zone. The second-stage anoxic zone is required to consistently meet effluent requirements of less than 3 mg/L of total nitrogen. This zone is sized to remove the remaining nitrates, resulting in an 80–90% removal of nitrates.

A second-stage oxic or aerobic zone generally follows the second-stage anoxic zone, with a detention time of about 30–60 minutes. The aeration is necessary to prevent denitrification in the secondary clarifiers and to convert any residual ammonia to nitrate. Methanol, acetate, raw wastewater, or primary effluent can be added to the second-stage anoxic zone to enhance biological denitrification, although it is seldom practiced.

Design Criteria Design criteria used in sizing combined pre- and post- denitrification system are presented in Table 7.19. For combined systems, the total anoxic volume usually represents 30–50% of the total volume of the biological process. Mixed liquor recycle is normally returned at a rate of about 4 : 1 up to 6 : 1 of the incoming wastewater flow rate to the plant. Submerged vertical turbine mixers on platforms or submerged propeller mixers are normally used to keep the biomass suspended in the anoxic zone. Either diffused aeration or surface aerators are utilized in the first-stage aerobic zone. Generally, fine-bubble diffusers are used in the second-stage aerobic zone for re-aerating the mixed liquor prior to clarification. As mentioned previously, the MLR pump intakes should be located such that minimal DO is recycled back to the first stage anoxic zone.

Design Equations There are two procedures that can be used for designing pre-nitrification-denitrification systems. Each method uses a similar approach for sizing the aerobic treatment zone, whereas the sizing of the anoxic zone uses different approaches. The first method is that presented in the US EPA *Phosphorus Removal Manual* (1987a, pp. 42–43), and is similar to the procedure presented in the WEF *Manual of Practice* No. 8 (1992), whereas the second approach bases the sizing of the anoxic zone on matching the respiration requirements (oxygen equivalents) of the denitrifiers with the oxygen equivalents supplied by the nitrates produced during nitrification which are recycled back to the anoxic zone. The details of the second approach are presented elsewhere (Mines *et al.*, 1992).

The procedure used in sizing the aerobic or oxic portion of the biological process is based on the model developed by Lawrence & McCarty (1970). In the design examples presented in this chapter, 60 mg/L of inert solids are assumed to enter the biological system. Biokinetic constants used in the examples were taken from Lawrence & McCarty (1970, p. 763), Metcalf & Eddy (1991, pp. 394, 701), and the WEF *Manual of Practice* No. 8 (1998b). Temperature correction coefficients (θ) were taken from WEF MOP 8 (1998b). The procedure for the conventional approach is presented below. A spreadsheet should be used to perform the calculations so that multiple scenarios can be evaluated at different flows and loadings.

First Oxic Zone Calculations

1 Select minimum weekly temperature, minimum pH, and minimum DO level to be maintained.

2 Calculate the maximum growth rate of *Nitrosomonas* using the following equation by multiplying its yield coefficient (Y_{NS}) by its maximum specific substrate utilization rate (k_{NS}) corrected for pH and DO as presented in the EPA Nitrogen Control Manual (US EPA, 1975a):

$$(\mu_{MAX})_{NS} = Y_{NS}k_{NS}\left[\frac{DO}{K_{DO}+DO}\right][1-0.833\,(7.2-pH)] \quad (7.193)$$

where:
T = mixed liquor temperature, °C
$(\mu_{max})_{NS}$ = maximum growth rate of *Nitrosomonas*, d^{-1}
Y_{NS} = *Nitrosomonas* yield coefficient, g VSS/g NH_4^+-N, mg/L
k_{NS} = maximum specific substrate utilization rate for *Nitrosomonas*, d^{-1}
pH = pH of mixed liquor, standard pH units
DO = DO level in aeration basin or oxic zone, mg/L
K_{DO} = half-saturation coefficient for DO, mg/L
K_{DO} = 0.5 mg/L (Metcalf & Eddy, 2003, page 705).

3 Calculate the minimum mean cell residence time (MCRT) required for a given temperature, DO, pH, and influent ammonia concentration, using the following equation:

$$\frac{1}{(\theta_c)_{min}} = \frac{(\mu_{max})_{NS}(NH_4^+-N)_0}{K_{NS}+(NH_4^+-N)_0} - (k_d)_{NS} \quad (7.194)$$

where:
$(\theta_c)_{min}$ = minimum MCRT required for nitrification based on the *Nitrosomonas* growth rate, d^{-1}
$(NH_4^+\text{-}N)_0$ = influent ammonium nitrogen concentration, mg/L
$(k_d)_{NS}$ = endogenous decay coefficient for *Nitrosomonas*, d^{-1}
K_{NS} = half-saturation coefficient for ammonium nitrogen, mg/L.

4 Calculate the design MCRT based on a safety factor and peaking factor as follows:

$$(\theta_c)_{design} = (\theta_c)_{min}(SF)(PF) \quad (7.195)$$

where:
$(\theta_c)_{design}$ = design MCRT for first oxic reactor, days
SF = safety factor based on uncertainty of performance, usually 1.2 to 2.0
PF = peaking factor based on the peak nitrogen loading to average loading for the treatment unit, usually 1.1 to 1.2.

5 Estimate the multiplication factor (MF) to account for the anoxic and re-aeration volumes which will increase the MCRT of the system. MF is estimated using Equation (7.196):

$$MF = \frac{1}{[(1-Z_1)(1-Z_2)(1-Z_3)]} \quad (7.196)$$

where:
Z_1 = pre-anoxic zone fraction of total reactor volume,
Z_2 = post-anoxic zone fraction of total reactor volume,
Z_3 = re-aeration zone fraction of total reactor volume,
MF = multiplication factor to account for anoxic and re-aeration zones.

For most BNR systems, the multiplication factor will vary from approximately 1.4–2.0. Alternatively, the MF value in Equation (7.197) can be replaced with a value from 1.4–2, rather than using Equation (7.196).

6 Estimate the overall MCRT of the biological system using Equation (7.197):

$$(\theta_c)_{overall} = (\theta_c)_{design}(MF) \quad (7.197)$$

where
$(\theta_c)_{overall}$ = overall MCRT of the biological system, days.

7 For a given set of heterotrophic biokinetic coefficients, calculate the effluent soluble substrate concentration (S_e). The substrate concentration may be in terms of BOD_5, COD, or TOC. Therefore, make sure that the biokinetic coefficients selected are appropriate for the biokinetic equation used. S_e is typically measured as five-day carbonaceous biochemical oxygen demand ($CBOD_5$).

$$S_e = \frac{K_s\,[1+k_d\,(\theta_c)_{overall}]}{(\theta_c)_{overall}[(Yk-k_d)]-1} \quad (7.198)$$

where:
S_e = effluent soluble substrate concentration, mg/L (g/m^3)
k_d = endogenous decay coefficient for heterotrophs, d^{-1}
K_s = half-saturation coefficient for organic substrate, mg/L

Y = biomass yield coefficient for heterotrophic bacteria, $\frac{\text{mg VSS}}{\text{mg substrate}}$
k = maximum substrate utilization rate, d^{-1}.

8 For a given set of autotrophic biokinetic coefficients, calculate the effluent ammonium nitrogen concentration using the following equation:

$$(NH_4^+ - N)_e = \frac{K_{NS}\left[1 + (k_d)_{NS}(\theta_c)_{design}\right]}{(\theta_c)_{design}\left[Y_{NS}\,k_{NS} - (k_d)_{NS}\right] - 1} \quad (7.199)$$

where:
$(NH_4^+ - N)_e$ = effluent ammonium nitrogen concentration, mg/L
k_{NS} = maximum specific ammonium utilization rate, d^{-1}.

9 The nitrogen utilized in synthesis neglecting the small amount of nitrogen synthesized by the nitrifiers is calculated using Equation (7.200):

$$N_{syn} = \frac{Y(S_i - S_e)F_N}{\left[1 + k_d(\theta_c)_{overall}\right]} + (X_e)F_N \quad (7.200)$$

where:
N_{syn} = nitrogen used in synthesizing biomass, mg/L
S_i = influent substrate concentration excluding internal recycle flows (usually total BOD or total COD is used), mg/L
X_e = effluent VSS concentration, mg/L
F_N = fraction of nitrogen in volatile suspended solids, fraction
F_N = 5 to 12%.

10 Determine the amount of nitrogen to be oxidized using Equation (7.201):

$$NO = TKN_0 - (NH_4^+ - N)_e - N_{syn} \quad (7.201)$$

where:
NO = influent nitrogen converted to oxidized nitrogen, mg/L
TKN_0 = influent total Kjeldhal nitrogen concentration, mg/L.

11 Calculate the volume of the first oxic zone necessary to achieve nitrification for a given temperature, pH, and DO as follows:

$$\forall_{oxic_1} = \frac{Q(\theta_c)_{design}}{X}\left[\frac{Y(S_i - S_e)}{1 + k_d(\theta_c)_{design}} + X_L\right] \quad (7.202)$$

where:
Q = influent flow rate excluding internal recycle flows, m^3/d
$\forall_{oxic_1}$ = volume of the first oxic zone, m^3,
X_L = inert solids in influent (FSS plus non-degradable VSS), mg/L
X = active biomass concentration measured in VSS, mg/L.

First-Stage Anoxic Zone Calculations

1 Estimate the nitrate nitrogen concentration in the mixed liquor recycle returned to the first-stage anoxic zone using the following equation:

$$N = \frac{NO(Q)}{(Q_{mlr} + Q_r + Q)} \quad (7.203)$$

where:
N = nitrate nitrogen concentration returned to anoxic zone, mg/L
Q_{mlr} = nitrified mixed liquor recycle flow, m^3/d
Q_r = return activated sludge flow, m^3/d

2 Estimate the nitrate equivalence of dissolved oxygen in the mixed liquor recycle as follows:

$$(NO_3^- - N)_{eq_1} = (DO)_{mlr}\left(0.35\frac{g\,NO_3^- - N}{g\,O_2}\right)(Q_{mlr}) \times \left(\frac{1\,kg}{1000\,g}\right) \quad (7.204)$$

where:
$(NO_3^- - N)_{eq_1}$ = nitrogen equivalence of DO in the mixed liquor recycle, kg/day
$(DO)_{mlr}$ = DO concentration in nitrified mixed liquor recycle, mg/L
0.35 = $g\,NO_3^- - N/g$ of DO.

3 Calculate the mass of nitrates to be removed in the first stage anoxic zone using the following equation:

$$NOR_1 = [Q_r(NO_3^-)_e + Q_{mlr}(N)]\left(\frac{1kg}{1000g}\right) \quad (7.205)$$

where
NOR_1 = mass of nitrates to be removed in the anoxic zone, kg/day.

4 The total mass of nitrates to be removed in the first stage anoxic zone is calculated as follows:

$$TNOR_1 = NOR_1 + (NO_3^- - N)_{eq_1} \quad (7.206)$$

where
$TNOR_1$ = total mass of nitrates to be removed in the first-stage anoxic zone, kg/day.

5 Estimate the specific denitrification rate in the first-stage anoxic zone, corrected for ambient temperature, using the following equation (Burdick *et al.*, 1982):

$$SDNR_1 = [0.03\,(F/M) + 0.029](1.06)^{(T-20°C)} \quad (7.207)$$

where
$SDNR_1$ = specific denitrification rate in the first-stage anoxic zone, d^{-1}.

6 The food-to-microorganism ratio is defined as:

$$F/M = \frac{Q(S_i)}{X(\text{V}_{anoxic_1})} \quad (7.208)$$

where:
F/M = food-to-microorganism ratio, d^{-1}
V_{anoxic_1} = volume of first-stage anoxic zone, m^3.

7 Calculate the specific denitrification rate in the first-stage anoxic zone as a function of the volume as given in the following equation:

$$SDNR_1 = \frac{TNOR_1(1000\ g/kg)}{\text{V}_{anoxic_1}(X)} \quad (7.209)$$

8 The volume of the first-stage anoxic zone can be determined by equating Equations (7.207) and (7.209) and replacing the food-to-microorganism ratio by Equation (7.208) as follows:

$$\text{V}_{anoxic_1} = \frac{TNOR_1(1000\,g/kg) - 0.03(Q)(S_i)(1.06)^{(T-20°C)}}{0.029(X)(1.06)^{(T-20°C)}} \quad (7.210)$$

Second-Stage Anoxic Zone Calculations

1 Calculate the quantity of nitrates to be removed in the second-stage anoxic zone using Equation (7.211).

$$NOR_2 = \frac{Q\left[NO - \left(NO_3^- - N\right)_e\right]}{1000\ g/kg} - NOR_1 \quad (7.211)$$

where
NOR_2 = mass of nitrates to be removed in the second-stage anoxic zone, kg/day.

2 The nitrate equivalence of dissolved oxygen in the mixed liquor from the first-stage oxic zone is calculated as:

$$(NO_3^- - N)_{eq_2} = (Q + Q_r)(DO_{ml})\left(0.35\frac{g\,NO_3^- - N}{g\,O_2}\right) \times \left(\frac{1\,kg}{1000\,g}\right) \quad (7.212)$$

where:
$(NO_3^- - N)_{eq_2}$ = nitrate equivalence of DO in the mixed liquor from first-stage oxic zone, kg/day
$(DO)_{ml}$ = DO concentration in the mixed liquor from first-stage oxic zone, mg/L.

3 The total mass of nitrates to be removed in the second-stage anoxic zone is calculated using Equation (7.213):

$$TNOR_2 = NOR_2 + (NO_3^- - N)_{eq_2} \quad (7.213)$$

where
$TNOR_2$ = total mass of nitrates to be removed in the second-stage anoxic zone, kg/day.

4 The specific denitrification rate in the second-stage anoxic zone is calculated using Equation (7.214) corrected for ambient temperature (Burdick *et al.*, 1982):

$$SDNR_2 = 0.12(\theta_c)_{overall}^{-0.706}[1.02]^{(T-20°C)} \quad (7.214)$$

where:
$SDNR_2$ = specific denitrification rate in the second-stage anoxic zone, d^{-1}
T = wastewater temperature, °C.

5 The volume of the second-stage anoxic zone is calculated as:

$$\text{V}_{anoxic_2} = \frac{TNOR_2(1000\ g/kg)}{(X)(SDNR_2)} \quad (7.215)$$

where
V_{anoxic_2} = volume of the second-stage anoxic zone, m^3.

Second Oxic Zone or Re-aeration Zone Calculations
The second-stage oxic zone is sized according to detention time. Generally, a detention time of 30–60 minutes is maintained at the design influent flow excluding internal recycle flows. The second-stage oxic zone volume is calculated as:

$$\text{V}_{oxic_2} = (\tau)\left(\frac{1\ hr}{60\ min}\right)\left(\frac{1\ d}{24\ hr}\right)(Q) \quad (7.216)$$

where:
V_{OXIC_2} = volume of the second-stage-stage oxic zone, m^3
τ = detention time in second-stage-stage oxic zone, min.

Check the Overall Design The overall volume of the biological system must be calculated and the overall MCRT compared with the initial assumption.

1 First, calculate the total volume of the biological system using the following equation:

$$\text{V}_{total} = \text{V}_{anoxic_1} + \text{V}_{oxic_1} + \text{V}_{anoxic_2} + \text{V}_{oxic_2} \quad (7.217)$$

Where
V_{total} = total volume of the biological system, m^3.

2 Determine the total quantity of sludge produced as follows:

$$P_x = \left[\frac{Y\,(S_i - S_e)}{1 + k_d\,(\theta_c)_{overall}} + X_L\right](Q)\left(\frac{1\ kg}{1000\ g}\right) \quad (7.218)$$

where:
P_x = waste sludge production, kg/day
X_L = inert solids entering biological system, mg/L (g/m^3).

3 Compute the new overall MCRT using Equation (7.219):

$$(\theta_c)_{overall} = \frac{(V_{total})(X)}{P_x(1000\ g/kg)} \quad (7.219)$$

Where
$(\theta_c)_{overall}$ = overall MCRT of the biological system, days.

If the calculated $(\theta_c)_{overall}$ is not within 5% of the assumed $(\theta_c)_{overall}$ in Step 5 of the first oxic zone calculations, the entire design procedure must be repeated from that point forward.

Oxygen Requirements Oxygen is required to meet both the carbonaceous and nitrogenous demand. The total kilograms of oxygen required can be estimated using the following equations:

$$O_2 = CBOD + NOD - DOC \quad (7.220)$$

$$CBOD = \left\{Q\left[(1 - 1.42\,Y)\,(S_i - S_e)\right] + 1.42\,(k_d)(X)\,V_{oxic_1}\right\} \times \left(\frac{1\ kg}{1000\ g}\right) \quad (7.221)$$

$$NOD = Q\,(4.57)\,(NO)\left(\frac{1\ kg}{1000\ g}\right) \quad (7.222)$$

$$DOC = Q\left(2.86\,\frac{gO_2}{g\,NO_3^- - N}\right)[NO - (NO_3^-)_e]\left(\frac{1\ kg}{1000\ g}\right) \quad (7.223)$$

where:
O_2 = total quantity of oxygen that must be supplied to the first oxic zone, kg/day
CBOD = carbonaceous oxygen demand, kg/day
NOD = nitrogenous oxygen demand, kg/day
DOC = denitrification oxygen credit, kg/day.

Best design practice is to ignore the denitrification oxygen credit.

Alkalinity Requirements Sufficient alkalinity must be maintained so that the pH does not drop during nitrification, thereby inhibiting the process. Normally, 50–100 mg/L of alkalinity as $CaCO_3$ are maintained in the effluent from BNR systems. The effluent alkalinity can be calculated as follows for the four-stage anoxic/oxic process:

$$ALK_e = ALK_o - 7.14\,(NO) + 3.57\,[NO - (NO_3^- - N)_e] \quad (7.224)$$

where:
ALK_e = effluent alkalinity, mg/L as $CaCO_3$
ALK_o = influent alkalinity, mg/L as $CaCO_3$.

Example 7.17 Preliminary design of a 4-Stage Bardenpho process

A four-stage Bardenpho process is to be designed to meet a 5/5/3/ mg/L effluent standard for BOD_5, TSS, and TN, respectively. The influent wastewater characteristics, mixed liquor parameters, and biokinetic coefficients used in the example are presented in the tables below. Assume that the influent ammonium concentration is equal to the TKN concentration, RAS flow = 1Q, MLR flow = 3Q, DO in the RAS and MLR recycle is 0.5 mg/L and 1.0 mg/L entering the second anoxic zone.

Wastewater characterization for design example

Influent		Mixed liquor	
Q	38,000 m^3/d	DO	2.0 mg/L
pH	7.1	MLSS	3,500 mg/L
Minimum temperature	15°C	MLVSS	0.75 MLSS
TKN	30 mg/L	K_{DO}	0.5 mg/L
Alkalinity	200 mg/L as $CaCO_3$	MLVSS	2,625 mg/L
BOD_5	200 mg/L		
Inerts (X_L)	60 mg/L		

Heterotrophic biokinetic coefficients at 20°C

Coefficient		Coefficient	
Y	0.6 g VSS/g BOD_5	k_d	0.06 d^{-1}
k	5 d^{-1}	K_s	60 mg/L BOD_5

***Nitrosomonas* biokinetic coefficients at 20°C**

Coefficient		Coefficient	
Y_{NS}	0.15 g VSS/g $NH_4^+ - N$	$(k_d)_{NS}$	0.05 d^{-1}
k_{NS}	3 d^{-1}	$(K_s)_{NS}$	0.74 mg/L $NH_4^+ - N$

Heterotrophic biokinetic coefficients at 15°C

Coefficient		Coefficient	
Y	0.6 g VSS/g BOD_5	k_d	0.049 d^{-1}
k	3.6 d^{-1}	K_s	60 mg/L BOD_5

***Nitrosomonas* biokinetic coefficients at 15°C**

Coefficient		Coefficient	
Y_{NS}	0.15 g VSS/g $NH_4^+ - N$	$(k_d)_{NS}$	0.041 d^{-1}
k_{NS}	2.14 d^{-1}	$(K_s)_{NS}$	0.57 mg/L $NH_4^+ - N$

A temperature correction coefficient (θ) of 1.04 was used for correcting both the heterotrophic and *Nitrosomonas* k_d values, while a θ of 1.07 was used for correcting both the heterotrophic and *Nitrosomonas* k

values. The heterotrophic K_s value was corrected with a θ of 1.00 and the *Nitrosomonas* K_s was corrected with a θ value of 1.053. All temperature correction coefficients were obtained from Metcalf & Eddy (2003, pp. 704–705).

Solution

First oxic zone calculations

1 Calculate the maximum growth rate of *Nitrosomonas* for the ambient temperature of 15°C.

$$(\mu_{MAX})_{NS} = Y_{NS}k_{NS}\left[\frac{DO}{K_{DO}+DO}\right] \times [1-0.833(7.2-pH)] \quad (7.193)$$

$$(\mu_{MAX})_{NS} = \left(\frac{0.15\text{g VSS}}{\text{g NH}_4^+-\text{N}}\right) \times 2.14\text{d}^{-1}\left[\frac{2\text{ mg/L}}{0.5\text{ mg/L}+2\text{ mg/L}}\right] \times [1-0.833(7.2-7.1)] = 0.23\text{ d}^{-1}$$

2 Correct the *Nitrosomonas* decay coefficient for temperature.

$$(k_d)_{NS(T°C)} = (k_d)_{NS(20°C)}(1.04)^{(T-20°C)} \quad (7.101)$$

$$(k_d)_{NS(15°C)} = (0.05\text{ d}^{-1})_{NS(20°C)}(1.04)^{(15°-20°C)} = 0.041\text{ d}^{-1}$$

3 Calculate the minimum MCRT required for nitrification.

$$\frac{1}{(\theta_c)_{min}} = \frac{(\mu_{max})_{NS}(NH_4^+-N)_0}{K_{NS}+(NH_4^+-N)_0} - (k_d)_{NS} \quad (7.194)$$

$$\frac{1}{(\theta_c)_{min}} = \frac{(0.23\text{ d}^{-1})(30\text{ mg/L})}{(0.57\text{mg/L})+(30\text{ mg/L})} - 0.041\text{ d}^{-1}$$

$$(\theta_c)_{min} = \frac{1}{0.185\text{ d}^{-1}} = 5.41\text{ d}$$

4 Calculate the design MCRT based on a safety factor of 1.25 and peaking factor of 1.2 as follows:

$$(\theta_c)_{design} = (\theta_c)_{min}(SF)(PF) \quad (7.195)$$

$$(\theta_c)_{design} = 5.41\text{ d}(1.25)(1.2) = 8.12\text{ d}$$

5 Estimate the multiplication factor (MF) to account for the anoxic and re-aeration volumes that will increase the MCRT of the system. MF is estimated using Equation (7.196). Assume that the pre-anoxic, post-anoxic, and re-aeration volumes occupy 27%, 30%, and 6% of the total reactor volume, respectively.

$$MF = \frac{1}{[(1-Z_1)(1-Z_2)(1-Z_3)]} \quad (7.196)$$

$$MF = \frac{1}{[(1-0.27)(1-0.30)(1-0.06)]} = 2.08$$

6 Estimate the overall MCRT of the biological system using Equation (7.197):

$$(\theta_c)_{overall} = (\theta_c)_{design}(MF) \quad (7.197)$$

$$(\theta_c)_{overall} = (8.12\text{ d})(2.08) = 16.9\text{ d}$$

7 For a given set of heterotrophic biokinetic coefficients, calculate the effluent soluble substrate concentration (S_e):

$$S_e = \frac{K_s[1+k_d(\theta_c)_{overall}]}{(\theta_c)_{overall}[(Yk-k_d)]-1} \quad (7.198)$$

$$S_e = \frac{60\frac{\text{mg}}{\text{L}}[1+0.049\text{ d}^{-1}(16.9\text{ d})]}{(16.9\text{ d})\left[\left(0.6\frac{\text{g VSS}}{\text{g BOD}_5}\times 3.6\text{ d}^{-1} - 0.049\text{ d}^{-1}\right)\right]-1} = 3.20\frac{\text{mg}}{\text{L}}$$

8 For a given set of autotrophic biokinetic coefficients, calculate the effluent ammonium nitrogen concentration using the following equation:

$$(NH_4^+-N)_e = \frac{K_{NS}\left[1+(k_d)_{NS}(\theta_c)_{design}\right]}{(\theta_c)_{design}\left[Y_{NS}k_{NS}-(k_d)_{NS}\right]-1} \quad (7.199)$$

$$(NH_4^+-N)_e = \frac{0.57\frac{\text{mg}}{\text{L}}[1+0.041\text{ d}^{-1}(8.12\text{ d})]}{(8.12\text{ d})\left[\left(0.15\frac{\text{g VSS}}{\text{gNH}_4^+-\text{N}}\right)\times 2.14\text{d}^{-1}-0.041\text{ d}^{-1}\right]-1}$$

$$(NH_4^+-N)_e = 0.60\frac{\text{mg}}{\text{L}}$$

9 The nitrogen utilized in synthesis, neglecting the small amount of nitrogen synthesized by the nitrifiers, is calculated using Equation (7.200) and assuming $F_N = 0.12$:

$$N_{syn} = \frac{Y(S_i-S_e)F_N}{[1+k_d(\theta_c)_{overall}]} + (X_e)F_N \quad (7.200)$$

$$N_{syn} = \frac{0.60\frac{\text{g VSS}}{\text{g BOD}_5}\left(200\frac{\text{mg}}{\text{L}}-3.20\frac{\text{mg}}{\text{L}}\right)0.12}{[1+0.049\text{ d}^{-1}(16.9\text{ d})]} + \left(5\frac{\text{mg TSS}}{\text{L}}\right)\left(0.75\frac{\text{mg VSS}}{\text{mg TSS}}\right)0.12$$

$$N_{syn} = 8.2\,\frac{\text{mg N}}{\text{L}}$$

10 Determine the amount of nitrogen to be oxidized using Equation (7.201).

$$NO = TKN_0 - (NH_4^+ - N)_e - N_{syn} \qquad (7.201)$$

$$NO = 30\,\frac{\text{mg}}{\text{L}} - 0.60\,\frac{\text{mg}}{\text{L}} - 8.2\,\frac{\text{mg}}{\text{L}} = 21.2\,\frac{\text{mg}}{\text{L}}$$

11 Calculate the volume of the first oxic zone necessary to achieve nitrification for a given temperature, pH, and DO as follows:

$$\text{V}_{oxic_1} = \frac{Q(\theta_c)_{design}}{X}\left[\frac{Y(S_i - S_e)}{1 + k_d(\theta_c)_{design}} + X_L\right] \qquad (7.202)$$

$$\text{V}_{oxic_1} = \frac{38{,}000\,\frac{\text{m}^3}{\text{d}}(8.12\text{d})}{3500\,\frac{\text{mg}}{\text{L}}} \times \left[\frac{\left(0.6\,\frac{\text{mg VSS}}{\text{mg BOD}_5}\left(\frac{\text{g TSS}}{0.75\text{ g VSS}}\right) \times\left(200\frac{\text{mg}}{\text{L}} - 3.20\frac{\text{mg}}{\text{L}}\right)\right)}{1 + 0.049\text{d}^{-1}(8.12\text{ d})} + 60\frac{\text{mg}}{\text{L}}\right]$$

$$\text{V}_{oxic_1} = 15{,}200\text{ m}^3$$

First-stage anoxic zone calculations

1 Estimate the nitrate nitrogen concentration in the mixed liquor recycle returned to the first-stage anoxic zone using the following equation:

$$N = \frac{NO(Q)}{(Q_{mlr} + Q_r + Q)} \qquad (7.203)$$

$$N = \frac{\left(21.2\,\frac{\text{mg}}{\text{L}}\right)\left(38{,}000\,\frac{\text{m}^3}{\text{d}}\right)}{\left(38{,}000\,\frac{\text{m}^3}{\text{d}} \times 3 + 38{,}000\,\frac{\text{m}^3}{\text{d}} + 38{,}000\,\frac{\text{m}^3}{\text{d}}\right)} = 4.24\frac{\text{mg}}{\text{L}}$$

2 Estimate the nitrate equivalence of dissolved oxygen in the mixed liquor recycle as follows:

$$(NO_3^- - N)_{eq_1} = (DO)_{mlr}\left(0.35\,\frac{\text{g NO}_3^- - \text{N}}{\text{g O}_2}\right) \times (Q_{mlr})\left(\frac{1\text{ kg}}{1000\text{ g}}\right) \qquad (7.204)$$

$$(NO_3^- - N)_{eq_1} = \left(0.5\,\frac{\text{mg}}{\text{L}}\right)\left(0.35\,\frac{\text{g NO}_3^- - \text{N}}{\text{g O}_2}\right) \times\left(38{,}000\,\frac{\text{m}^3}{\text{d}} \times 3\right)\left(\frac{1\text{ kg}}{1000\text{ g}}\right)$$

$$(NO_3^- - N)_{eq_1} = 20.0\,\frac{\text{kg}}{\text{d}}$$

3 Calculate the mass of nitrates to be removed in the first-stage anoxic zone using the following equation and assuming that the nitrate nitrogen concentration in the effluent is 1.0 mg/L as N:

$$NOR_1 = [Q_r\,(NO_3^-)_e + Q_{mlr}(N)]\left(\frac{1\text{ kg}}{1000\text{ g}}\right) \qquad (7.205)$$

$$NOR_1 = \left[38{,}000\,\frac{\text{m}^3}{\text{d}}\left(1\,\frac{\text{mg}}{\text{L}}\right) + 38{,}000\,\frac{\text{m}^3}{\text{d}} \times 3\left(4.24\,\frac{\text{mg}}{\text{L}}\right)\right] \times\left(\frac{1\text{ kg}}{1000\text{ g}}\right)$$

$$NOR_1 = 521\,\frac{\text{kg}}{\text{d}}$$

4 The total mass of nitrates to be removed in the first-stage anoxic zone is calculated as follows:

$$TNOR_1 = NOR_1 + (NO_3^- - N)_{eq_1} \qquad (7.206)$$

$$TNOR_1 = 521\,\frac{\text{kg}}{\text{d}} + 20.0\frac{\text{kg}}{\text{d}} = 541\,\frac{\text{kg}}{\text{d}}$$

5 The volume of the first-stage anoxic zone can be determined using Equation (7.210):

$$\text{V}_{anoxic_1} = \frac{TNOR_1(1000\text{ g/kg}) - 0.03(Q)(S_i)(1.06)^{(T-20^\circ\text{C})}}{0.029(X)(1.06)^{(T-20^\circ\text{C})}} \qquad (7.210)$$

$$\text{V}_{anoxic_1} = \frac{\left((541\text{ kg/d})(1000\text{ g/kg}) - 0.03(38{,}000\text{ m}^3/\text{d}) \times\left(200\,\frac{\text{mg}}{\text{L}}\right)(1.06)^{(15-20^\circ\text{C})}\right)}{0.029\left(3500\,\frac{\text{mg}}{\text{L}}\right)(1.06)^{(15-20^\circ\text{C})}}$$

$$\text{V}_{anoxic_1} = 4890\text{ m}^3$$

Second-stage anoxic zone calculations

1 Calculate the quantity of nitrates to be removed in the second-stage anoxic zone using Equation (7.211):

$$NOR_2 = \frac{Q[NO - (NO_3^- - N)_e]}{1000\text{ g/kg}} - NOR_1 \qquad (7.211)$$

$$NOR_2 = \frac{38{,}000\frac{m^3}{d}\left[21.2\frac{mg}{L} - 1.0\frac{mg}{L}\right]}{1000\,g/kg} - 521\frac{kg}{d}$$

$$= 247\frac{kg}{d}$$

2 The nitrate equivalence of dissolved oxygen in the mixed liquor from the first-stage oxic zone (1.0 mg/L) is calculated as:

$$(NO_3^- - N)_{eq_2} = (Q + Q_r)(DO_{ml})\left(0.35\frac{g\,NO_3^- - N}{g\,O_2}\right) \times \left(\frac{1\,kg}{1000\,g}\right) \quad (7.212)$$

$$(NO_3^- - N)_{eq_2} = \left(38{,}000\frac{m^3}{d} + 38{,}000\frac{m^3}{d}\right)\left(1.0\frac{mg}{L}\right) \times \left(0.35\frac{g\,NO_3^- - N}{g\,O_2}\right)\left(\frac{1\,kg}{1000\,g}\right)$$

$$(NO_3^- - N)_{eq_2} = 26.6\frac{kg}{d}$$

3 The total mass of nitrates to be removed in the second-stage anoxic zone is calculated using Equation (7.213):

$$TNOR_2 = NOR_2 + (NO_3^- - N)_{eq_2} \quad (7.213)$$

$$TNOR_2 = 247\frac{kg}{d} + 26.6\frac{kg}{d} = 274\frac{kg}{d}$$

4 The specific denitrification rate in the second-stage anoxic zone is calculated using Equation (7.214) corrected for ambient temperature (Burdick *et al.*, 1982):

$$SDNR_2 = 0.12(\theta_c)_{overall}^{-0.706}[1.02]^{(T-20°C)} \quad (7.214)$$

$$SDNR_2 = 0.12(16.9\,d)^{-0.706}[1.02]^{(15-20°C)}$$

$$= 0.0148\,d^{-1}$$

5 The volume of the second-stage anoxic zone is calculated as:

$$\forall_{anoxic_2} = \frac{TNOR_2(1000\,g/kg)}{(X)(SDNR_2)} \quad (7.215)$$

$$\forall_{anoxic_2} = \frac{(274\,kg/d)(1000\,g/kg)}{\left(3500\frac{mg}{L}\right)(0.0148\,d^{-1})} = 5290 m^3$$

Second oxic zone or re-aeration zone calculations

The second-stage oxic zone volume is calculated as follows assuming a detention time of 45 minutes:

$$\forall_{oxic_2} = (\tau)\left(\frac{1\,hr}{60\,min}\right)\left(\frac{1\,d}{24\,hr}\right)(Q) \quad (7.216)$$

$$\forall_{oxic_2} = (45\,min)\left(\frac{1\,hr}{60\,min}\right)\left(\frac{1\,d}{24\,hr}\right)\left(38{,}000\frac{m^3}{d}\right)$$

$$= 1190\,m^3$$

Check the overall design

1 First, calculate the total volume of the biological system using the following equation:

$$\forall_{total} = \forall_{anoxic_1} + \forall_{oxic_1} + \forall_{anoxic_2} + \forall_{oxic_2} \quad (7.217)$$

$$\forall_{total} = 4890\,m^3 + 15{,}200\,m^3 + 5290\,m^3 + 1190\,m^3$$

$$= 26{,}570\,m^3$$

2 Determine the total quantity of sludge produced as follows:

$$P_x = \left[\frac{Y(S_i - S_e)}{1 + k_d(\theta_c)_{overall}} + X_L\right](Q)\left(\frac{1\,kg}{1000\,g}\right) \quad (7.218)$$

$$P_x = \left[\frac{\left(0.60\frac{g\,VSS}{g\,BOD_5} \times \frac{g\,TSS}{0.75\,g\,VSS} \times \left(200\frac{mg}{L} - 3.20\frac{mg}{L}\right)\right)}{1 + 0.049\,d^{-1}(16.9\,d)} + 60\frac{mg}{L}\right] \times \left(38{,}000\frac{m^3}{d}\right)\left(\frac{1\,kg}{1000\,g}\right)$$

$$P_x = 5550\frac{kg}{d}$$

3 Compute the new overall MCRT using Equation (7.219):

$$(\theta_c)_{overall} = \frac{(\forall_{total})(X)}{P_x(1000\,g/kg)} \quad (7.219)$$

$$(\theta_c)_{overall} = \frac{(26{,}570\,m^3)(3500\,g/m^3)}{\left(5550\frac{kg}{d}\right)\left(1000\frac{g}{kg}\right)} = 16.8\,d$$

$$\%Difference = \frac{(16.9\,d - 16.8\,d) \times 100\%}{16.9\,d}$$

$$= 0.6\% < 5.0\%\ \text{OK}$$

Since the difference between the original and the calculated values for the overall MCRT is within 5%, it is not necessary to repeat the procedure.

Oxygen requirements

Oxygen is required to meet both the carbonaceous and nitrogenous demand. The total kilograms of oxygen required can be estimated using the following equations:

$$O_2 = CBOD + NOD - DOC \quad (7.220)$$

$$CBOD = \{Q[(1 - 1.42\,Y)(S_i - S_e)] + 1.42(k_d)(X)\forall_{oxic_i}\}$$

$$\times\left(\frac{1\ \text{kg}}{1000\ \text{g}}\right) \quad (7.221)$$

$$CBOD = \left\{38{,}000\frac{\text{m}^3}{\text{d}}\left[\left(1 - 1.42 \times 0.6\frac{\text{g VSS}}{\text{g BOD}_5}\right)\right.\right.$$

$$\left.\times\left(200\frac{\text{mg}}{\text{L}} - 3.20\frac{\text{mg}}{\text{L}}\right)\right] + 1.42(0.049\ \text{d}^{-1})$$

$$\left.\left(2625\frac{\text{mg}}{\text{L}}\right)15{,}200\ \text{m}^3\right\}\left(\frac{1\ \text{kg}}{1000\ \text{g}}\right)$$

$$CBOD = 3880\frac{\text{kg}}{\text{gd}}$$

$$NOD = Q(4.57)(NO)\left(\frac{1\ \text{kg}}{1000\ \text{g}}\right) \quad (7.222)$$

$$NOD = 38{,}000\frac{\text{m}^3}{\text{d}}(4.57)\left(21.2\frac{\text{mg}}{\text{L}}\right)\left(\frac{1\ \text{kg}}{1000\ \text{g}}\right)$$

$$= 3680\frac{\text{kg}}{\text{gd}}$$

$$DOC = Q\left(2.86\frac{\text{g}O_2}{\text{g}\,NO_3^- - N}\right)[NO - (NO_3^-)_e]\left(\frac{1\ \text{kg}}{1000\ \text{g}}\right) \quad (7.223)$$

$$DOC = 38{,}000\frac{\text{m}^3}{\text{d}}\left(2.86\frac{\text{g}O_2}{\text{g}\,NO_3^- - N}\right)$$

$$\times\left[21.2\frac{\text{mg}}{\text{L}} - 1.0\frac{\text{mg}}{\text{L}}\right]\left(\frac{1\ \text{kg}}{1000\ \text{g}}\right) = 2200\frac{\text{kg}}{\text{d}}$$

$$O_2 = CBOD + NOD - DOC$$

$$= 3880\frac{\text{kg}}{\text{d}} + 3680\frac{\text{kg}}{\text{d}} - 2200\frac{\text{kg}}{\text{d}} = 5360\frac{\text{kg}}{\text{d}}$$

Alkalinity Requirements

The effluent alkalinity from the 4-stage anoxic/oxic process can be calculated as follows:

$$ALK_e = ALK_o - 7.14(NO) + 3.57[NO - (NO_3^- - N)_e] \quad (7.224)$$

$$ALK_e = 200\frac{\text{mg}}{\text{L}} - 7.14\left(21.2\frac{\text{mg}}{\text{L}}\right)$$

$$+ 3.57\left[21.2\frac{\text{mg}}{\text{L}} - 1.0\frac{\text{mg}}{\text{L}}\right] = 121\frac{\text{mg}}{\text{L}}$$

7.8.2 Phosphorus removal systems

The following is an over-simplification of enhanced biological phosphorus removal (EBPR) systems. To accomplish EBPR, an anaerobic zone is necessary for the fermentation of incoming organic matter into volatile fatty acids (VFAs), which are required for the growth and proliferation of phosphorus-accumulating organisms (PAOs) such as *Acinetobacter*. PAOs are capable of removing excess quantities of phosphorous (e.g., 7 to 14% by weight; Randall *et al.* (1992), p.103; WEF (1998d), p. 210).

Recall that the composition of most microbes can be represented by $C_{60}H_{87}O_{23}N_{12}P$, which indicates that a microbe usually contains 2.3% phosphorous by weight. The anaerobic zone must be followed by an oxic zone, in which the PAOs actually take up soluble phosphorus in excess of their normal biochemical requirements. Bernard (1975) used the term *Phoredox* to describe processes that use an alternating anaerobic/oxic environment for accomplishing EBPR. Ultimately, phosphorus is removed by wasting excess biomass from the system.

EBPR systems typically require an influent BOD_5/TP ratio of greater than 15–20 (Metcalf & Eddy, 2003, p. 802). To achieve proper fermentation of the incoming wastewater, an anaerobic zone with a detention time of 0.5–2.0 hours is required (Metcalf & Eddy, 2003, p. 814). Typical design criteria for three major types of biological removal processes are presented in Table 7.20.

7.8.2.1 A/O

The anaerobic/oxic (A/O) process is a proprietary process that was originally patented by Air Products and is currently

Table 7.20 Design criteria for mainstream biological phosphorus removal systems.

Parameter	A/O	Sequencing batch reactor
System MCRT, d	2–6	–
HRT, h		
Anaerobic zone	0.5–1.5	1.8–3.0
Aerobic zone	1–3	0.4–1.0
F : M ratio, lb BOD/(lb MLVSS · d)	0.2–0.7	0.16–0.42
MLSS, mg/L	2,000–4,000	–
RAS flow	(0.25–0.40) Q	–
Parameter	**PhoStrip**	
Feed, % of influent flow	20–30	
Solids detention time, h	5–20	
Elutriation flow, % of stripper feed	50–100	
Underflow, % of influent flow	10–20	
Reactor clarifier		
Overflow rate, $m^3/(m \cdot d)$	48	
pH	9–9.5	
Lime dosage, mg/L	100–300	

MCRT = mean cell residence time.
HRT = hydraulic retention time.
Source: *Design Manual Phosphorus Removal* (1987a) EPA/625/1-87/001, pp. 21–23. United States Environmental Protection Agency.

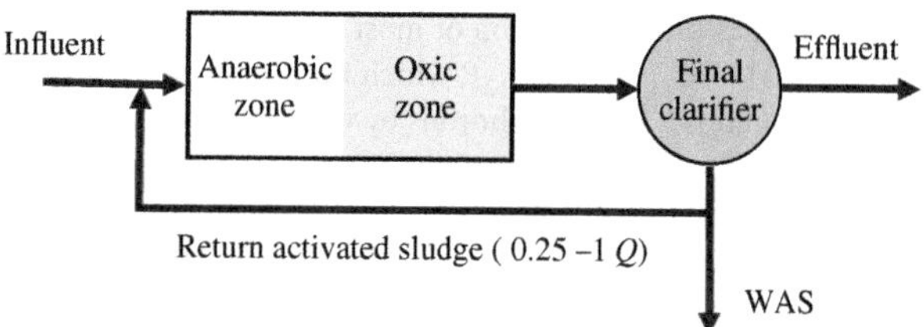

Figure 7.47 Schematic of the A/O biological phosphorus removal process.

licensed to and marketed by I. Krueger of Cary, North Carolina. The process is capable of achieving an effluent total phosphorus concentration of less than 1.0 mg/L (US EPA, 1987a, p. 29). Figure 7.47 is a schematic of the A/O process.

7.8.2.2 PhoStrip

The PhoStrip process uses anaerobic/oxic treatment of thickened waste-activated sludge. The WAS is held in a gravity thickener to obtain anaerobic conditions, which releases phosphorus stored in the biomass. The US EPA Design Manual for Phosphorus Removal (1987a, p. 19) states that a detention time of approximately 8 – 12 hours is necessary to achieve anaerobic conditions to produce a phosphorus rich supernatant. A chemical – typically, lime – is then added to the supernatant for precipitation of the phosphorus in a reactor/clarifier as calcium hydroxyapatite ($Ca_5(PO_4)_3OH$) and calcium carbonate ($CaCO_3$). A final effluent TP concentration of < 1.0 mg/L is possible (US EPA, 1987a, pp. 25 – 26). Alternatively, aluminum and ferric salts may be used rather than lime for precipitating orthophosphate. A schematic of the PhoStrip process is shown in Figure 7.48.

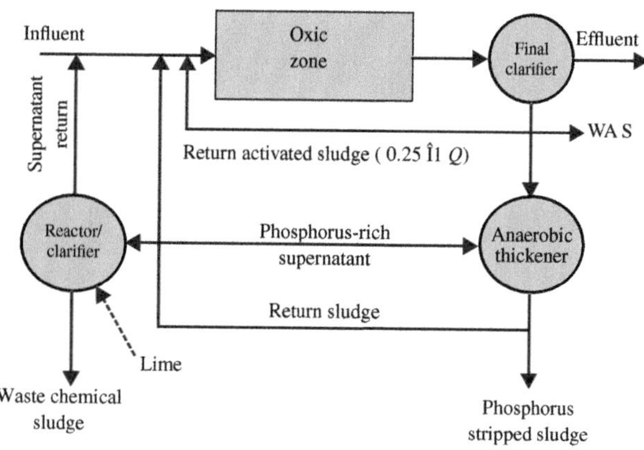

Figure 7.48 Schematic of the PhoStrip biological phosphorus removal process.

7.8.3 Combined biological nitrogen and phosphorus removal processes

Several biological processes (most of which are proprietary) that remove both nitrogen and phosphorus include: A^2/O, University of Cape Town (UCT), Virginia Initiative Plant (VIP) process, 5-Stage Bardenpho, and sequencing batch reactors. SBRs were discussed in Section 7.5.10. Table 7.21 provides design criteria for each of these processes. Additional information about them can be found in the following references: Metcalf & Eddy (2003), US EPA (1993), WEF (1998b, 1998d), Soap and Detergents Manufacturer's Association (1989), and Randall *et al.* (1992).

Table 7.21 Design criteria for combined biological nitrogen and phosphorus removal systems.

Parameter	A^2/O**	UCT**	VIP**	5-Stage Bardenpho*	SBR*
System MCRT, d	5–10	5–10	5–10	10–40	20–40
HRT, h					
Anaerobic zone	0.5–1.0	1–2	1–2	1.0–2.0	1.0–3.0
First anoxic zone	0.5–1.0	1–2	1 – 2	2–4	0–1.6
First aerobic zone	3.5–6.0	2.5–4	2.5–4	4–12	0.5–1.0
Second anoxic zone	–	–	–	2–4	0–0.3
Second aerobic zone	–	–	–	0.5–1.0	0–0.3
MLSS, mg/L	3,000–5,000	1,500–3,000	1,500–3,000	2,000–5,000	600–5,000
Nitrified recycle	(1–2) Q	(2–4) Q	(2–4) Q	(4–6) Q	–
Oxic recycle	NA	(0.5–2) Q	(0.5–2) Q	–	–
RAS flow	(0.25–0.5) Q	(0.5–1.0) Q	(0.5–1.0) Q	(0.8–1.0) Q	–

*WEF (1998b) Vol 2, pp. 15–60.
**EPA *Nitrogen Control Manual* (1993), p. 254.
MCRT = mean cell residence time.
HRT = hydraulic retention time.

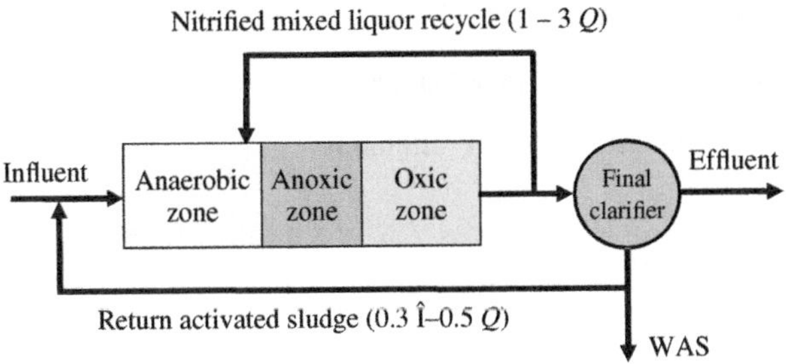

Figure 7.49 Schematic of the A^2/O process.

7.8.3.1 A^2/O

The anaerobic/anoxic/oxic (A^2/O) process (Figure 7.49) is a combined biological nitrogen and phosphorus removal process licensed and marketed by I. Krueger. This process can achieve effluent TN concentrations ranging from 6–12 mg/L and average monthly effluent TP concentrations from 0.5–4.6 mg/L. (Soap and Detergent Association, 1989, pp. 153, 167).

7.8.3.2 University of Cape Town (UCT) Process

Effluent quality from the UCT biological nutrient removal process should be similar to that of the A^2/O and VIP process or even better. The UCT process has not been used in the United States so there are no data available for assessment purposes. A schematic of the UCT process is shown in Figure 7.50.

7.8.3.3 Virginia Initiative Plant (VIP) Process

The VIP Process (Figure 7.51) was developed and patented by the Hampton Roads Sanitation District and CH_2M Hill. According to the Soap and Detergent Association (1989, p. 171) the original design goals were to achieve 67% phosphorus removal on a year-round basis and 70% nitrogen removal when the temperature of the wastewater is above 20°C.

Actual operating data from July 2003 at an average monthly flow rate of 30 MGD are presented in Table 7.22.

Based on the above data, the facility achieved the following removals: > 98% BOD_5, 98% TSS, 95% TKN, 69% TN, and 95% TP.

7.8.3.4 Five-Stage Bardenpho Process

EIMCO Process Equipment of Salt Lake City markets the 4-stage and 5-stage Bardenpho Processes. The 5-stage Bardenpho process configuration is presented in Figure 7.52.

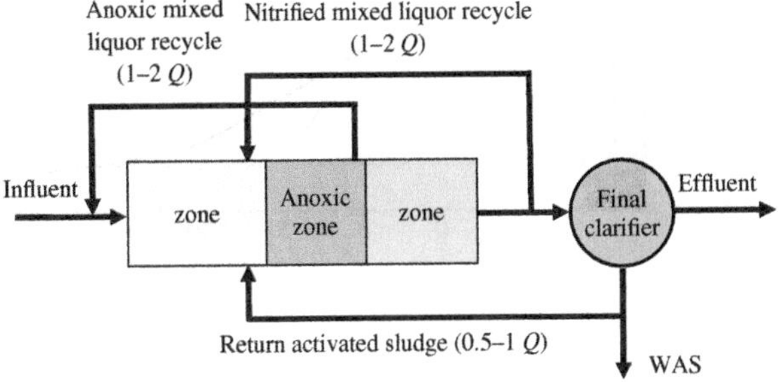

Figure 7.50 Schematic of the University of Cape Town (UCT) process.

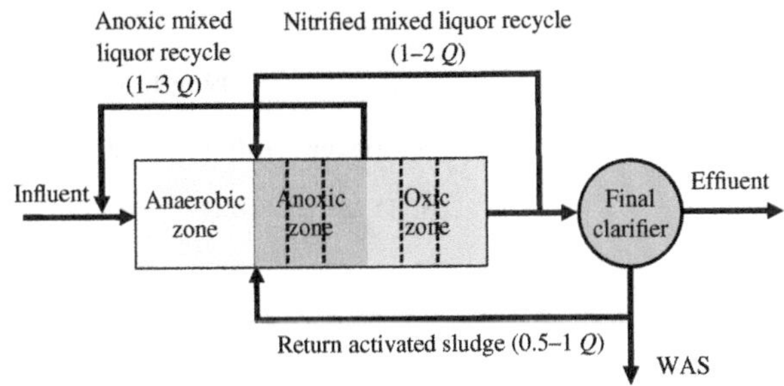

Figure 7.51 Schematic of the Virginia Initiative Plant (VIP) process.

Table 7.22 VIP process operating data for July 2003.

Parameter	Influent	Effluent
BOD_5	174	< 2.0
TSS	122	2.0
TKN	25.5	1.1
TN	–	7.8
TP	4.2	0.21
pH	6.8	7.1

Source: Monthly Operating Report VIP July 2003.

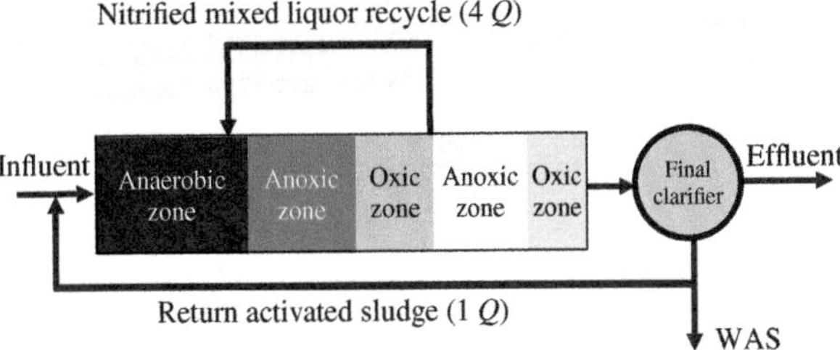

Figure 7.52 Schematic of the 5-stage Bardenpho process.

This BNR configuration is capable of meeting effluent TN concentrations < 3.0 mg/L and TP concentrations < 1.0 mg/L when effluent filtration is included in the process train (US EPA, 1987a, p. 26). Operating data from three 5-stage Bardenpho plants in Florida are presented in Table 7.23 (Mines, 1996, p. 607). High quality effluent is achieved with the process. Each of the facilities presented in Table 7.23 have effluent sand filters to enhance effluent quality.

7.9 Secondary clarification

If biological processes are to operate properly, it is essential that secondary clarifiers be designed correctly. For suspended-growth systems, such as activated sludge, secondary clarifiers must be designed to accomplish both clarification and thickening of solids.

Table 7.23 Annual operating data from three 5-Stage Bardenpho plants in Florida.

	Northeast		East		Marshall St	
Parameter	**Influent**	**Effluent**	**Influent**	**Effluent**	**Influent**	**Effluent**
$CBOD_5$	244	2.9	186	2.8	229	1.1
TSS	190	1.3	131	1.3	181	1.2
TKN	30	–	26	–	29	–
TN	–	1.9	–	1.3	–	2.1
TP	6.1	2.1	4.8	1.9	5.1	0.25

Source: Mines (1996), p. 607.

Figure 7.53 Photograph of secondary clarifier.

When mixed liquor from the activated sludge process enters the secondary clarifier, two process streams are produced. One is the clarified secondary effluent, and the other a more concentrated stream called the underflow. Clarification relates to the removal of suspended solids that enter the clarifier so that effluent standards can be met. Thickening of the biomass at the bottom of the clarifier to produce an underflow, or thickened sludge, at solids concentrations ranging from 5,000 to 15,000 mg/L, is also desired. Increasing the solids concentration of the biomass reduces the volume of biomass that must be handled during sludge treatment. Figure 7.53 is a photograph of a secondary clarifier.

7.9.1 Type II and Type III settling

The biological solids generated during secondary treatment from processes such as activated sludge are flocculent in nature. Depending on the concentration, they will exhibit either Type II or Type III settling characteristics. Type II settling involves flocculent particles at dilute solids concentrations $< 1{,}000$ mg/L (Peavy *et al.*, 1985, p. 268). As these particles settle, the particles flocculate and they thus increase in size, causing them to settle at a faster velocity. Type III setting is called hindered or zone settling; it involves flocculent particles at solids concentrations ranging from 1,000–4,000 mg/L. These flocculent particles are so close together that interparticle forces cause them to remain in a fixed position relative to each other, and the mass of particles settle as a zone. The settling rate of these particles is called the zone settling velocity (*ZSV*).

7.9.2 Solid flux theory

The design of secondary clarifiers for activated sludge processes and gravity thickeners for processing sludge is based on solids flux theory, as proposed by Dick (1970). Batch settling column analyses must be conducted to obtain the limiting solids flux, G_L, that is used in sizing the clarifier. Solids flux (G) is defined as the rate at which solids thickening occurs per unit of area in plan-view. The units for G are mass of solids per unit of time per unit of area.

Reynolds & Richards (1996, pp. 239–247) present a graphical procedure for estimating the limiting solids flux which is used to size secondary clarifiers. Batch- or pilot-scale settling column analyses are performed to obtain *ZSV* at various solids concentrations. Figure 7.54 shows a settling

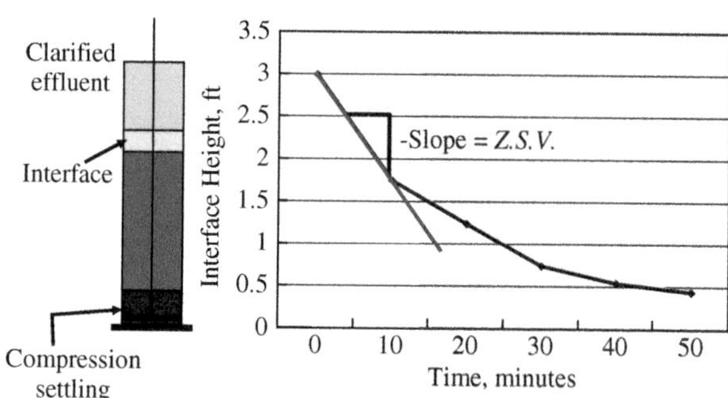

Figure 7.54 Diagram of settling column apparatus and interface height plot.

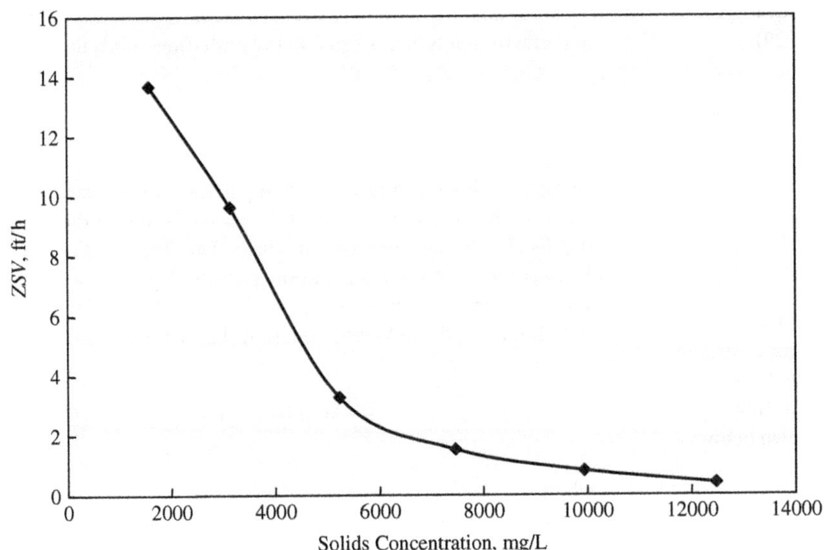

Figure 7.55 Plot of zone settling velocity versus solids concentration.

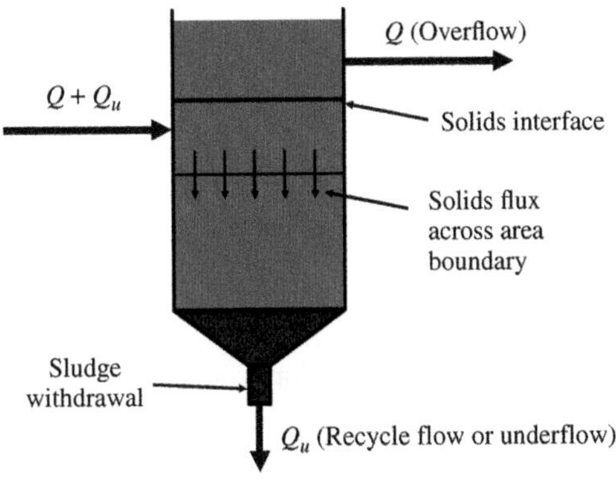

Figure 7.56 Schematic of a secondary clarifier operating at steady-state.

column and a plot of interface height versus time. The slope of the line of best fit through the linear portion of the curve is equal to the zone settling velocity.

Figure 7.55 shows what the relationship between *ZSV* and solids concentration should look like.

The equations used to describe solid flux theory are developed below. Figure 7.56 shows a schematic of an ideal continuous flow, secondary clarifier or settling tank.

At any level in the settling tank, the solids flux due to gravity settling (G_s) is calculated as follows:

$$G_s = C_i \times ZSV_i \tag{7.225}$$

where:

G_s = solids flux due to gravity, $\frac{\text{lb}}{\text{ft}^2\cdot\text{h}}\left(\frac{\text{kg}}{\text{m}^2\cdot\text{h}}\right)$

C_i = solids concentration, mg/L

ZSV_i = zone setting velocity at solids concentration C_i, $\frac{\text{ft}}{\text{s}}, \left(\frac{\text{m}}{\text{s}}\right)$.

There is also solids flux due to the withdrawal of solids from the bottom of the settling basin in the underflow, G_u.

$$G_u = C_i \times V_u \tag{7.226}$$

where:

G_u = solids flux due to underflow withdrawal, $\frac{\text{lb}}{\text{ft}^2\cdot\text{h}}\left(\frac{\text{kg}}{\text{m}^2\cdot\text{h}}\right)$

V_u = underflow withdrawal velocity, $\frac{\text{ft}}{\text{s}}, \left(\frac{\text{m}}{\text{s}}\right)$.

The total solids flux (G_t) is the sum of the gravity flux and underflow withdrawal:

$$G_t = G_s + G_u = C_i \times ZSV_i + C_i \times V_u \tag{7.227}$$

At steady-state, the mass rate of solids that enter the settling tank must be equal to the mass rate of solids that are settling.

$$G_t = Q_0 C_0 = Q_u C_u \tag{7.228}$$

where:

Q_0 = influent flow rate into settling tank or clarifier, $\frac{\text{gal}}{\text{d}}, \left(\frac{\text{m}^3}{\text{d}}\right)$

Q_u = underflow withdrawal rate from settling tank or clarifier, $\frac{\text{gal}}{\text{d}}, \left(\frac{\text{m}^3}{\text{d}}\right)$

C_0 = influent solids concentration, mg/L

C_u = solids concentration in the underflow, mg/L.

There is a limiting solid flux rate (G_L), which can be handled by a settling basin with a plan or cross-sectional area (A). At solid flux rates above G_L, the sludge blanket rises and solids escape in the effluent from the clarifier. Mathematically, the limiting solid flux rate is defined by Equation (7.229):

$$G_L = \frac{G_t}{A} = \frac{Q_0 C_0}{A} = \frac{Q_u C_u}{A} \quad (7.229)$$

where:
G_L = limiting solids flux, $\frac{\text{lb}}{\text{ft}^2\cdot\text{h}}\left(\frac{\text{kg}}{\text{m}^2\cdot\text{h}}\right)$
A = settling tank or clarifier plan area, ft^2 (m^2).

The plan area of the settling tank or clarifier can be calculated knowing the influent flow rate, solids concentration, and the value for G_L. The bulk downward velocity due to underflow withdrawal can be calculated by substituting $Q_u = AV_u$ into Equation (7.229) and rearranging as shown below:

$$G_L = \frac{Q_u C_u}{A} = \frac{A V_u C_u}{A} \quad (7.230)$$

$$V_u = \frac{G_L}{C_u} \quad (7.231)$$

Figure 7.57 illustrates the relationships among the various parameters that were derived above.

The limiting solids flux (G_L) is estimated by selecting an underflow solids concentration (C_u) and drawing a line tangent to the solids flux curve. The y-intercept value for this tangent line is G_L. The slope of the tangent line is equal to the underflow withdrawal velocity (V_u). At the point of tangency, the gravity solids flux value (G_s) is equal to the y-coordinate and the limiting solids concentration (C_L) is equal to the x-coordinate. The difference between the limiting solids flux and gravity solids flux is equal to the underflow solids flux.

7.9.3 Type IV settling

At high solids concentrations, when particles contact each other, settling can only happen if the mass of solids in the lower depths of a clarifier compress or consolidate. This is called Type IV or compression settling, and the particles have only a slight velocity. Compression settling is modeled as a first-order decay function using the following equation (Lin, 2007):

$$\frac{dH}{dt} = i(H - H_\infty) \quad (7.232)$$

where:
H = height of sludge at time t, ft (m)
H_∞ = height of sludge after long settling period (24 hours), ft (m)
i = constant for a given sludge suspension, time^{-1}.

Integrating Equation between the limits of H_2 at t_2 and H_t at time t results in Equation (7.234):

$$\int_{H_2}^{H_t} \frac{1}{(H - H_\infty)} dH = i \int_0^t dt \quad (7.233)$$

$$H_t - H_\infty = (H_2 - H_\infty)e^{-i(t-t_2)} \quad (7.234)$$

A plot of $\ln(H_t - H_\infty) - \ln(H_1 - H_\infty)$ versus $(t - t_2)$ yields a straight line, having a negative slope equal to the value of i.

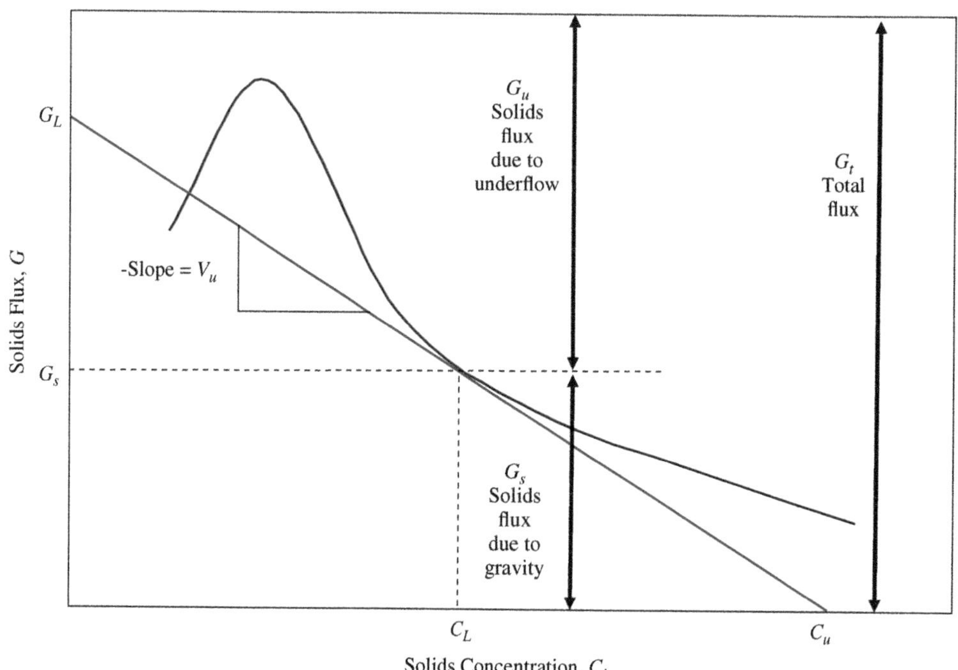

Figure 7.57 Solids flux versus solids concentration.

7.9.4 Secondary clarifier design for suspended growth systems

As stated at the beginning of this section, proper separation of biological solids from the wastewater must be accomplished during secondary clarification to ensure successful operation of activated sludge or any biological process. For suspended biological growth processes, both clarification and thickening of the biological solids is desired.

During the design of most municipal WWTPs, batch settling column studies are typically not conducted and, therefore, alternative procedures must be used to design clarifiers. In some instances, studies will be conducted, especially on industrial wastewaters. When settling column data are not available, solids loading rates (*SLR*) must be used for sizing secondary clarifiers in lieu of having a limiting solids flux value.

Secondary clarifier design protocol requires calculation of the surface area or plan area of the clarifier based on clarification and also thickening considerations. Once the areas are calculated, the larger of the two areas controls the final design. The surface area of the secondary clarifier, based on clarification, is determined by dividing the design flow rate by the overflow rate according to the following equation:

$$A_C = \frac{Q}{V_o} \tag{7.235}$$

where:
A_C = area of secondary clarifier based on clarification, ft^2 (m^2)
Q = wastewater design flow rate applied to the secondary clarifier, excluding the return activated sludge flow rate, MGD (m^3/d)
V_o = overflow rate or surface loading rate, gpd/ft^2 ($m^3/d \cdot m^2$).

The surface area, based on solids thickening considerations, is calculated using either Equation (7.236) or (7.237). Equation (7.236) is used when settling column data are available for ascertaining the limiting solids flux value. Otherwise, a solids loading rate is selected from the literature, and Equation (7.237) is used for calculating the secondary clarifier area:

$$A_T = \frac{(Q + Q_r)SS}{G_L} \tag{7.236}$$

where:
A_T = secondary clarifier surface area based on thickening considerations, ft^2 (m^2)
Q = wastewater design flow rate applied to the secondary clarifier excluding the return activated sludge flow rate, MGD (m^3/d)
Q_r = return activated sludge flow rate, MGD (m^3/d)
SS = suspended solids concentration in the aeration basin
G_L = limiting solids flux value determined from settling column analysis, $\frac{lb}{ft^2 \cdot d}\left(\frac{kg}{m^2 \cdot d}\right)$.

$$A_T = \frac{(Q + Q_r)SS}{SLR} \tag{7.237}$$

where SLR = solids loading rate design criteria, typically 25 ppd/ft^2 (122 kg/d $\cdot$ m^2) at average daily flow and 50 ppd/ft^2 (244 kg/d $\cdot$ m^2) at peak flow.

Table 7.24 presents design criteria for secondary clarifiers. Secondary settling basins are normally rectangular or circular in plan-view. Standard sizes for circular clarifiers come in 5 ft (1.6 m) intervals. The design of a secondary clarifier for a suspended-growth process is presented in Example 7.18.

Example 7.18 Secondary clarifier design for suspended growth

Design a secondary clarifier system for an activated sludge WWTP that treats 38,000 m^3/d of domestic wastewater at average daily flow and 76,000 m^3/d at peak flow. At ADF, the MLSS concentration is 3,000 mg/L and return activated sludge (RAS) flow = 19,000 m^3/d. It is anticipated that, at peak flow, the MLSS concentration will drop to 2,000 mg/L and the RAS flow will be 38,000 m^3/d. State regulations require overflow rates of 20 and 50 $m^3/(d \cdot m^2)$ at average

Table 7.24 Design criteria for secondary clarifiers.

	V_o gpd/ft^2 ($m^3/d \cdot m^2$)		SLR ppd/ft^2 (kg/d $\cdot$ m^2)		
Parameter	**Average**	**Peak**	**Average**	**Peak**	**Depth ft (m)**
Air activated sludge (except extended aeration)	400–800 (16.3–32.6)	1,000–2,000 (40.8–81.6)	20–30 (98–244)	50 (244)	12–15 (3.7–4.6)
Extended aeration	200–400 (8.15–16.3)	800 (32.6)	20–30 (98–147)	50 (244)	12–15 (3.7–4.6)
Oxygen activated sludge (with primary settling)	400–800 (16.3–32.6)	1,000–2,000 (40.8–81.6)	25–35 (122–177)	50 (244)	12–15 (3.7–4.6)
Trickling filters	400–600 (16.3–24.5)	1,000–2,000 (40.8–81.6)	–	–	10–12 (3.0–3.7)

Source: U.S. EPA (1975b) *Suspended Solids Removal Manual*, pg. 7–16. United States Environmental Protection Agency.

and peak flow rates and solids loading rates of 5 and 8 kg/(h · m^2) at average and peak loadings, respectively. Assume a minimum of two circular clarifiers will be used. Determine the diameter of the clarifiers.

Solution

First, determine the surface area of the secondary clarifiers based on clarification considerations using Equation (7.235) for both the average and peak flow conditions. At ADF, the total clarifier area required is 1,900 m^2.

$$A_C = \frac{Q}{V_o} = \frac{3.8 \times 10^4\,\text{m}^3/\text{d}}{20\,\text{m}^3/(\text{d} \cdot \text{m}^2)} = 1.90 \times 10^3\,\text{m}^2 \quad (7.235)$$

At peak flow, the total clarifier area required is 1,520 m^2.

$$A_C = \frac{Q}{V_o} = \frac{7.6 \times 10^4\,\text{m}^3/\text{d}}{50\,\text{m}^3/(\text{d} \cdot \text{m}^2)} = 1.52 \times 10^3\,\text{m}^2$$

Next, determine the surface area of the secondary clarifiers based on thickening considerations using Equation (7.237) at average and peak loading conditions. Under average loading conditions, the total clarifier area required is 1,430 m^2.

$$A_T = \frac{(Q + Q_r)SS}{SLR}$$

$$= \frac{\left(\begin{array}{c} \left(3.8 \times 10^4 + 1.9 \times 10^4\right) \dfrac{\text{m}^3}{\text{d}} \\ \times \left(3000\,\dfrac{\text{mg}}{\text{L}}\right)\left(\dfrac{\text{kg}}{10^6\,\text{mg}}\right)\left(1000\,\dfrac{\text{L}}{\text{m}^3}\right) \end{array} \right)}{5\,\dfrac{\text{kg}}{\text{h} \cdot \text{m}^2}\left(\dfrac{24\,\text{h}}{\text{d}}\right)}$$

$$A_T = 1.43 \times 10^3\,\text{m}^2 \quad (7.237)$$

Total clarifier surface area required for peak loading conditions is calculated below:

$$A_T = \frac{(Q + Q_r)SS}{SLR}$$

$$= \frac{\left(\begin{array}{c} \left(7.6 \times 10^4 + 3.8 \times 10^4\right) \dfrac{\text{m}^3}{\text{d}} \\ \times \left(2000\,\dfrac{\text{mg}}{\text{L}}\right)\left(\dfrac{\text{kg}}{10^6\,\text{mg}}\right)\left(1000\,\dfrac{\text{L}}{\text{m}^3}\right) \end{array} \right)}{8\,\dfrac{\text{kg}}{\text{h} \cdot \text{m}^2}\left(\dfrac{24\,\text{h}}{\text{d}}\right)}$$

$$A_T = 1.19 \times 10^3\,\text{m}^2$$

In this example, clarification rather than thickening controls the design of the secondary clarifiers. Therefore, the total surface area that is necessary is 1,900 m^2.

Finally, the diameter of each secondary clarifier may be determined by equating the total surface area required to the area of a circle. Since there are two circular clarifiers, the surface area must be divided by 2. The radius and diameter of the clarifiers can then be determined.

$$\frac{A}{\text{clarifier}} = \frac{1900\,\text{m}^2}{2} = \pi r^2$$

$$r = \sqrt{950\,\text{m}^2/\pi} = \boxed{17.4\,\text{m}} \quad \text{Diameter} = \boxed{34.8\,\text{m}}$$

Example 7.19 Design of secondary clarifier based on settling column analyses

Batch settling column analyses have been performed using activated sludge with the results provided in the following table:

Run number	Solids concentration (mg/L)	Zone settling velocity (ft/h)	Solids flux $\left(\frac{\text{lb}}{\text{d} \cdot \text{ft}^2}\right)$
1	12, 500	0.410	7.67
2	9, 950	0.818	12.2
3	7, 450	1.55	17.3
4	5, 250	3.30	25.9
5	3, 140	9.70	45.6
6	1, 590	13.5	32.1

The design flow to the secondary clarifier is 3.65 MGD ($Q + Q_r$) and the suspended solids concentration (mixed liquor suspended solids) is 2,500 mg/L. The underflow solids concentration is 12,000 mg/L (1.2% solids as TSS). Determine the diameter of the secondary clarifier if the manufacturer sells clarifier equipment in 5 ft increments of tank diameter.

Solution

A plot of solids flux G versus suspended solids concentration is made. Then a line is drawn tangent to the solids flux curve starting at the underflow solids concentration of 12,000 mg/L as shown in the figure below. The y-intercept of the tangent line provides the value of the limiting solids flux $G_L = 43\,\text{lb}(\text{d} \cdot \text{ft}^2)$ which is used to determine the clarifier diameter.

Calculate the areas of the secondary clarifier, based on thickening considerations, using the limiting solids flux value and Equation (7.236):

$$A_T = \frac{(Q + Q_r)SS}{G_L}$$

$$= \frac{(3.65\,\text{MGD})(2500\,\text{mg/L})\left(\dfrac{8.34\,\text{lb}}{\text{MG/mg/L}}\right)}{43\,\dfrac{\text{lb}}{\text{d} \cdot \text{ft}^2}}$$

$$A_T = 1770\,\text{ft}^2$$

The diameter of the clarifier is determined as follows:

$$A_T = 1770\,\text{ft}^2 = \frac{\pi D^2}{4}$$

$$D = \sqrt{\frac{4 \times 1770\,\text{ft}^2}{\pi}} = 47.5\,\text{ft} \cong \boxed{50\,\text{ft}}$$

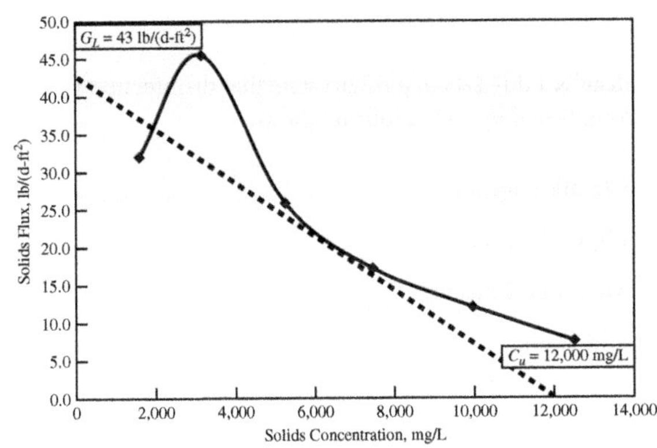

Figure E7.6

7.9.5 Secondary clarifier design for attached-growth systems

The design of secondary or final clarifiers for attached-biological growth systems is similar to that for primary clarification. For this situation, clarification of the wastewater is the main objective, whereas thickening is not. Biomass produced from trickling filters or bio-towers is called humus, and it has settling characteristics similar to those of discrete particles. The size of the secondary clarifier must be increased if recirculation flows, typically 1Q to 4Q pass, through the clarifier. If direct recirculation around the trickling filter is used, clarifier size is reduced. Sludge that is collected in secondary clarifiers from attached-growth systems is often pumped back to the primary clarifiers for concentration with primary sludge; otherwise, it is removed from the secondary clarifier and pumped directly to sludge processing facilities. Example 7.20 illustrates how the size of a secondary clarifier is impacted depending on how recirculation flows are handled around a trickling filter.

Example 7.20 Design of secondary clarifier for a trickling filter

A secondary clarifier system is to be designed to clarify wastewater from a high-rate trickling filter process.

The trickling filters are designed to treat 2.5 million gallons of wastewater per day (MGD) at average daily flow (ADF) and 6.25 MGD at peak flow. Two clarifiers operating in parallel will be used. Recirculation pumps will be sized to provide a recirculation flow of 10 MGD (4Q at ADF). The state regulatory agency requires overflow rates of 500 and 1,100 gpd/ft^2 at ADF and peak flow, respectively. Determine the following:

a) The size of the clarifiers when the recirculation flow enters the secondary clarifier.
b) The size of the clarifiers when direct recirculation around the trickling filters is used.

Solution Part A

Determine the area of the secondary clarifiers at ADF. Recall that the surface area is calculated by dividing the flow rate by the overflow rate. In this case, the total flow to the clarifiers is 12.5 MGD at average flow conditions.

$$A_s = \frac{Q}{V_o} = \frac{(2.5 + 10.0)\,\text{MG/d}}{500\,\text{gpd/ft}^2}\left(\frac{10^6\,\text{gal}}{1\,\text{MG}}\right)$$

$$= 25{,}000\,\text{ft}^2$$

$$\frac{A_s}{\text{clarifier}} = \frac{25{,}000\,\text{ft}^2}{2} = 12{,}500\,\text{ft}^2$$

$$A_s = \frac{\pi D^2}{4} = 12{,}500\,\text{ft}^2$$

$$D = \sqrt{\frac{4 \times 12{,}500\,\text{ft}^2}{\pi}} = \boxed{126\,\text{ft}}$$

Determine the area of the secondary clarifiers at peak flow:

$$A_s = \frac{Q}{V_o} = \frac{(6.25 + 10.0)\,\text{MG/d}}{1100\,\text{gpd/ft}^2}\left(\frac{10^6\,\text{gal}}{1\,\text{MG}}\right)$$

$$= 14{,}800\,\text{ft}^2$$

$$\frac{A_s}{\text{clarifier}} = \frac{14{,}800\,\text{ft}^2}{2} = 7{,}400\,\text{ft}^2$$

$$A_s = \frac{\pi D^2}{4} = 7{,}400\,\text{ft}^2$$

$$D = \sqrt{\frac{4 \times 7{,}400\,\text{ft}^2}{\pi}} = 97\,\text{ft}$$

Two 126 ft diameter secondary clarifiers would be necessary to meet the design criteria for Part A.

Solution Part B

The schematic of the process for direct recirculation of flow around the trickling filters was presented in Figure 7.37.

Determine the area of the secondary clarifiers at ADF. The average flow to the clarifiers is 2.5 MGD, since direct recirculation around the trickling filters is being used.

$$A_s = \frac{Q}{V_o} = \frac{2.5\ \text{MG/d}}{500\ \text{gpd/ft}^2}\left(\frac{10^6\ \text{gal}}{1\ \text{MG}}\right) = 5{,}000\ \text{ft}^2$$

$$\frac{A_s}{\text{clarifier}} = \frac{5{,}000\ \text{ft}^2}{2} = 2{,}500\ \text{ft}^2$$

$$A_s = \frac{\pi D^2}{4} = 7{,}400\ \text{ft}^2$$

$$D = \sqrt{\frac{4 \times 2{,}500\ \text{ft}^2}{\pi}} = 56\ \text{ft}$$

Determine the area of the secondary clarifiers at peak flow. The peak flow to the clarifiers is 6.25 MGD, since direct recirculation around the trickling filters is being used.

$$A_s = \frac{Q}{V_o} = \frac{6.25\ \text{MG/d}}{1100\ \text{gpd/ft}^2}\left(\frac{10^6\ \text{gal}}{1\ \text{MG}}\right) = 5{,}680\ \text{ft}^2$$

$$\frac{A_s}{\text{clarifier}} = \frac{5{,}680\ \text{ft}^2}{2} = 2{,}840\ \text{ft}^2$$

$$A_s = \frac{\pi D^2}{4} = 2{,}840\ \text{ft}^2$$

$$D = \sqrt{\frac{4 \times 2{,}840\ \text{ft}^2}{\pi}} = \boxed{60\ \text{ft}}$$

Two 60 ft diameter secondary clarifiers will meet design criteria if direct recirculation around the trickling filers is practiced.

The size of the secondary clarifiers is dramatically impacted when recirculation flows are passed through them. In this example, the diameter of the clarifiers is more than doubled, i.e., $\frac{126\ \text{ft}}{60\ \text{ft}} = 2.1$. Ultimately, the capital cost for each flow process will determine which one is selected. For both scenarios, pumps will be required to circulate 10.0 MGD of flow around the trickling filters. It may be more cost effective to increase the size of the clarifiers and place recirculation pumps on a concrete pad rather than building a separate pumping station that contains a large wet well and pumps.

7.10 Disinfection

The purpose of disinfecting wastewater is to kill and inactivate disease-causing organisms (pathogens) such as bacteria, protozoa, helminths, and viruses. Wastewater must be disinfected in situations where humans may come in contact with it. Disinfection is not the same as sterilization, which is the destruction of all organisms.

Effluent from WWTPs is disinfected before it is released back into the environment, either discharged to a receiving water body or used for irrigation of crops, golf courses, or parks. Historically, chlorine has been the disinfectant of choice, since it is effective at killing pathogens and is economical compared to other disinfection methods. However, chlorine must be removed from the final effluent prior to discharging to a water body, because it is toxic to aquatic organisms.

In some cities, municipalities have stopped using chlorine, due to safety concerns during transport to the site. In some states, disinfection must be preceded by filtration for the removal of suspended solids, which can inhibit the effectiveness of the disinfectant. This section provides a brief overview of methods of disinfection and major disinfectants, mechanisms of inactivation, chlorination equipment, and design of chlorine contact basins.

7.10.1 Methods of disinfection

Metcalf & Eddy (2003, p. 1220) state that disinfection is accomplished by one of four methods:

1 Chemical agents.
2 Physical agents.
3 Mechanical means.
4 Radiation.

The main chemical agents used include chlorine and its various forms, such as chlorine dioxide (ClO_2), chloramines (NCl_3, $NHCl_2$, and NH_2Cl), sodium hypochlorite (NaOCl), and calcium hypochlorite (Ca_2OCl), along with ozone (O_3). Ozone is a powerful oxidizing agent and effectively kills bacteria, viruses, and cysts. The problem with ozone, however, is that no residual is formed, and chlorine or other disinfectant must be added following ozonation to provide a disinfectant residual if the treated wastewater is to be distributed in a reuse distribution system. Reynolds & Richards (1996, p. 749), referencing Rice *et al.* (1979), state that ozone has a half-life of 20–30 minutes in distilled water at 20°C. In the presence of oxidant-demanding materials in solution, this half-life will be shortened.

Physical agents that are used include heat, light, and sound waves. Heating wastewater to boiling point is, of course, not practical. However, the ultraviolet (UV) radiation spectrum of sunlight is effective at inactivating pathogens. Low-pressure and medium-pressure UV lamps are now being used to disinfect wastewater. These lamps typically generate monochromatic radiation at a wavelength of 254 nm, which is effective for microbial inactivation (Metcalf & Eddy, 2003, p. 1298). Using UV radiation as the major disinfectant requires a low turbidity or suspended solids concentration in the wastewater, since these species absorb the UV light and shield microorganisms from exposure. UV radiation systems are expensive, and the light bulbs are housed in quartz glass sleeves that must be periodically cleaned to prevent the build-up of scale and biological growth that may occur. As with ozonation, an auxiliary disinfectant must be added if the wastewater is to be reused.

Mechanical means are those methods that remove potential pathogens from the wastewater as it passes through screens, grit chambers, or settling basins. Gamma radiation and electron beam technology have also been used to disinfect wastewater and sludge.

Chemical addition with chlorine or ozone and UV treatment are the primary means of disinfecting wastewater today. Both ozonation and UV radiation offer the advantage of not having to remove chlorine if the effluent is discharged to surface waters. Ozone and UV treatment also eliminate concerns over the production of triohalomethanes and haloacetic acids, which are chlorinated organic species, and suspected carcinogens that result when chlorine is added to water containing organic compounds.

7.10.2 Proposed disinfection mechanism

Asano *et al.* (2007, p. 604) list five mechanisms that describe how disinfectants inactivate and kill pathogens. The proposed mechanisms include:

1 Damage to the cell wall.
2 Alteration of cell permeability.
3 Alteration of the colloidal nature of the cell's protoplasm.
4 Alteration of the cell's nucleic acid (DNA or RNA).
5 Inhibition of enzyme activity.

The effectiveness of a specific disinfectant is primarily related to the targeted organism, i.e., bacteria, virus, and protozoan, dosage of disinfectant, and contact time.

7.10.2.1 Contact Time

An effective disinfectant should act quickly (fast rate of kill) and provide a residual. The rate of kill is modeled as a first-order removal equation known as Chick's Law, Equation (7.240):

$$\frac{dN}{dt} = -kN \tag{7.238}$$

$$\int_0^t \frac{1}{N} dN = -k \int_0^t dt \tag{7.239}$$

$$N_t = N_0 e^{-kt} \tag{7.240}$$

where:
N_0 = number of microorganisms initially present
N_t = number of microorganisms remaining at time t
k = rate constant, time^{-1}
t = contact time, time.

Figure 7.58 shows a plot of chlorine concentration versus contact time for three forms of chlorine to achieve a 99% kill of *Escherichia coli.*

7.10.2.2 $C_R t$ Concept

The rate constant and contact time necessary to achieve a specified kill is a function of the type of pathogen and the type of disinfectant used. The $C_R t$ parameter is often used for selecting the proper disinfectant and used in the design of tanks and reactors to accomplish a specified level of kill.

$$C_R t = C_R \times t \tag{7.241}$$

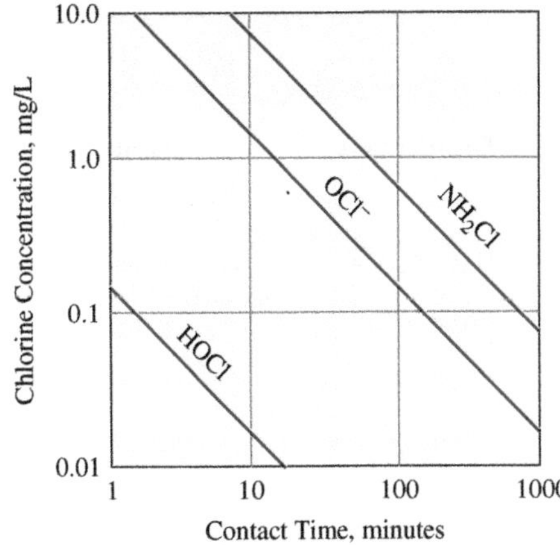

Figure 7.58 Chlorine concentration versus contact time for 99% kill of E. coli by various forms of chlorine. *Source*: Adapted from Reynolds and Richards, 1996, pg. 742. Reproduced with permission of PWS Publishing Company, Boston, MA.

where:
$C_R t$ = $C_R t$ concept or parameter, mg · min/L
C_R = residual disinfection concentration, mg/L
t = contact time, minutes.

Table 7.25 lists the $C_R t$ value required for various types of disinfectants to achieve a specific level of inactivation (1-log to 4-log reduction), i.e., 90% to 99.99%.

7.10.3 Chlorination equipment

Chlorine is delivered to wastewater treatment facilities as a liquefied gas at high pressure in containers ranging in size from 150 lb (68 kg) cylinders, through one ton (908 kg) cylinders, all the way up to railroad tank cars, which may contain up to 55 tons (49.9 Mg) of chlorine (Metcalf & Eddy, 1991, p. 495). Chlorine may be applied either as a gas or in aqueous solution.

At most WWTPs, chlorine gas is dissolved in non-potable effluent, which is then pumped to the chlorine contact chamber. Since chlorine is liquefied at high pressure, it may be withdrawn from the container either as a gas or a liquid, depending on the withdrawal rate. At large WWTPs, chlorine is usually withdrawn as a liquid and passed through an evaporator to be converted to a gas. Evaporators are recommended where the quantity of chlorine supplied approaches 2,000 lb/d (908 kg/d). The chlorine gas is then dissolved into a side stream of water, using a vacuum pressure system and regulator. Due to safety concerns with potential gas leaks (chlorine is toxic and corrosive), dissolving chlorine gas directly into the wastewater is seldom practiced. Chlorine dosage rates are dependent upon the type of treatment process used. Typical chlorine dosages are listed in Table 7.26 for various applications.

7.10.4 Chlorine contact basin design

Chlorine contact basins (CCB) generally consist of long, rectangular serpentine channels to allow chlorine sufficient time to react with pathogens in the wastewater. Chlorine is added at the front of the CCB, and the detention time of the

Table 7.25 $C_R t$ values for various disinfectants to achieve specific levels of inactivation for *Giardia* cysts and viruses.

Disinfectant	Units	$C_R t$ value for 1-log inactivation	$C_R t$ value for 2-log inactivation	$C_R t$ value for 3-log inactivation	$C_R t$ value for 4-log inactivation
		***Giardia* cysts**			
Free chlorine	mg · min/L	35	69	104	
Chloramine	mg · min/L	615	1,230	1,850	
Chlorine dioxide	mg · min/L	7.7	15	23	
Ozone	mg · min/L	0.48	0.95	1.43	
		Viruses			
Free chlorine	mg · min/L		3	4	6
Chloramine	mg · min/L		643	1,067	1,491
Chlorine dioxide	mg · min/L		4.2	12.8	25.1
Ozone	mg · min/L		0.5	0.8	1.0
UV	mJ/cm^2		21	36	N.A.

Source: EPA (1999) *Guidance Manual Alternative Disinfectants and Oxidants*, pg. 2–26. United States Environmental Protection Agency

Table 7.26 Probable chlorine dosages for various types of wastewater to produce a Cl_2 residual of 0.5 mg/L after 15 minutes of contact.

Application	Chlorine dosage range, mg/L
Raw wastewater	6–24
Primary effluent	3–18
Chemical precipitation effluent	3–12
Trickling filter effluent	3–9
Activated sludge effluent	3–9
Filtered activated sludge effluent	1–6

Source: Fair, Geyer, and Okun (1968b) pg. 31–26. Reproduced by permission of John Wiley & Sons Ltd.

basins is set at 30 minutes at the average daily flow rate or 15 minutes at the peak hourly flow. The flow condition resulting in the largest volume is selected for the final design. A three-pass CCB with a length-to-width (L : W) ratio of 10 : 1 to 40 : 1 is used in design to prevent short circuiting (Metcalf & Eddy, 1991, p. 502) and to approach plug flow. The design of a chlorine contact basin is illustrated in the following example.

Example 7.21 Chlorine contact basin design

The design of a chlorine contact basin system will be based on maintaining a detention time of 30 minutes at the average daily flow (ADF) or 15 minutes at peak hourly flow (PHF). The CCB will be designed to treat effluent from an extended aeration activated sludge wastewater treatment plant that treats an average daily flow of 38,000 cubic meters per day. The peak hour flow to average daily flow ratio is 3 : 1. Design a three-pass, serpentine chlorine contact basin assuming a length-to-width ratio of 10 : 1. Assume there will be two CCB operating in parallel, with each basin capable of treating the design flows with one unit out of service.

Solution

The volume of the CCB is determined for each flow condition. Volume is calculated by rearranging the detention time equation.

Volume of CCB at ADF:

$$\forall = \tau Q = 30\,\text{min}\left(\frac{1\,\text{h}}{60\,\text{min}}\right)\left(\frac{1\,\text{d}}{24\,\text{h}}\right)\left(\frac{38{,}000\,\text{m}^3}{\text{d}}\right)$$

$$= 792\,\text{m}^3$$

Volume of CCB at PHF:

$$\text{PHF} = 3\left(\frac{38{,}000\,\text{m}^3}{\text{d}}\right) = \frac{1.14\times10^5\,\text{m}^3}{\text{d}}$$

$$\forall = \tau Q = 15\,\text{min}\left(\frac{1\,\text{h}}{60\,\text{min}}\right)\left(\frac{1\,\text{d}}{24\,\text{h}}\right)\left(\frac{1.14\times10^5\,\text{m}^3}{\text{d}}\right)$$

$$= 1.19\times10^3\,\text{m}^3$$

The volume of the CCB is controlled by the PHF and should be 1.19×10^3 m^3.

Assume the depth of the equally sized chlorine contact basins is 3.0 meters. The volume per basin and the surface area are calculated below:

$$\frac{\text{volume}}{\text{basin}} = 1190\,\text{m}^3$$

$$\frac{\text{area}}{\text{basin}} = \frac{1190\,\text{m}^3}{3.0\,\text{m}} = 397\,\text{m}^2 \cong 400\,\text{m}^2$$

Assuming three passes per basin and an $L : W$ ratio of $10 : 1$ or $L = 10W$, the area of the basin per pass is calculated as:

$$A = L \times W = 10\,W^2 = 400\,\text{m}^2$$

$$W = \sqrt{400\,\text{m}^2/10} = 6.3\,\text{m}$$

Use a width of 6.3 m and length of 63 m. Each CCB basin will have a plan view as seen below.

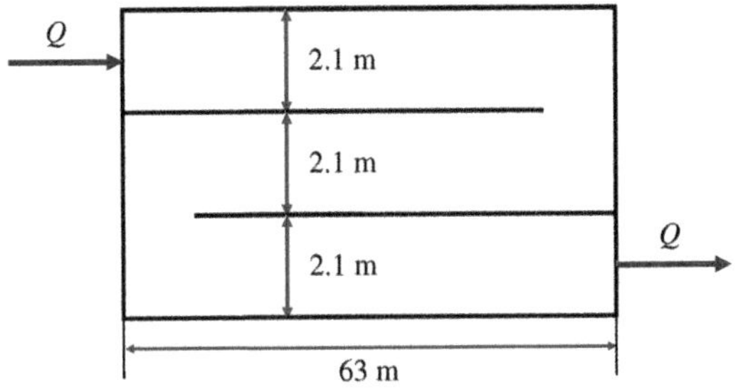

Figure E7.7

The actual dimensions of each chlorine contact basin is 7.3 m by 63.5 m, if the wall thickness of the basins and baffles is assumed to be 0.25 m thick.

7.10.5 Dechlorination

Sulfur dioxide (SO_2), sodium sulfite (Na_2SO_3), and sodium bisulfate ($NaHSO_3$) are some of the chemicals used for removing chlorine from wastewater (Reynolds & Richards, 1996, p. 748). Municipalities typically use SO_2 for dechlorinating effluents. Sulfur dioxide reacts with chlorine compounds according to the following reactions (Metcalf & Eddy, 1991, p. 344):

$$SO_2 + H_2O \rightarrow HSO_3^- + H^+ \quad (7.242)$$

$$HOCl + HSO_3^- \rightarrow Cl^- + SO_4^{2-} + 2H^+ \quad (7.243)$$

$$SO_2 + HOCl + H_2O \rightarrow Cl^- + SO_4^{2-} + 3H^+ \quad (7.244)$$

Sulfur dioxide reacts with monochloramine according to the following reaction (Metcalf & Eddy, 2003, p. 1262):

$$SO_2 + NH_2Cl + 2H_2O \rightarrow Cl^- + SO_4^{2-} + NH_4^+ + 2H^+ \quad (7.245)$$

The above reactions are nearly instantaneous and do not require a separate contact chamber. In most situations, the sulfur dioxide is added at the end of the chlorine contact basin.

Sodium sulfite reacts with chlorine as follows (Metcalf & Eddy, 2003, p. 1262):

$$Na_2SO_3 + Cl_2 + 2H_2O \rightarrow Na_2SO_4 + 2HCl \quad (7.246)$$

Sodium bisulfate reacts with chlorine as shown below (Metcalf & Eddy, 2003, p. 1263):

$$NaHSO_3 + Cl_2 + H_2O \rightarrow NaHSO_4 + 2HCl \quad (7.247)$$

7.11 Solids handling and treatment systems

7.11.1 Solids and sludge sources and quantities of sludge

The primary sources of solids or residuals at wastewater treatment plants include screens, grit chambers, primary clarifiers, secondary clarifiers, tertiary clarifiers, and digesters. The solids produced from screening the wastewater consists of paper, plastics, hair, rocks, rags, leaves, tree limbs, and other debris that must be disposed of in a sanitary landfill. The volume of screenings varies from 0.5–10 ft^3/million gallons or 4–74 L/1000 m^3 of wastewater flow (Metcalf & Eddy, 2003, p. 329).

Grit chambers remove sand, gravel, egg shells, coffee grounds, glass, and bone chips. According to Metcalf & Eddy (2003, p. 395) approximately 0.53–5.0 ft^3 of grit/million gallons (0.004–0.037 m^3 grit/1,000 m^3) of wastewater is generated in separate wastewater collection systems. Typically, grit separators and grit washers are used in conjunction with grit removal chambers. Solids from grit chambers are transported to a landfill for disposal.

Primary clarifiers generate primary sludge that contains organic matter and pathogens. Primary sludge generally has a solids content ranging from 5–9%. Due to the high organic content, anaerobic digestion is a prime candidate for stabilization of primary sludge. The concentration of anaerobically digested sludge ranges from 1.5–5.0% solids.

Secondary clarifiers generate secondary sludge. If attached biological systems are used, the sludge is called "humus" with solids concentrations ranging from 1–3%, while suspended growth processes generate waste-activated sludge at solids concentrations from 0.5–1.5%. Secondary sludge consists primarily of microorganisms or biomass, and aerobic digestion is often used for stabilizing waste-activated sludge.

Aerobic digestion generally produces a sludge at a solids concentration ranging from 0.8–7.0%, depending on the type of sludge fed to the digester. Chemical sludge or precipitate is produced when chemicals such as aluminum sulfate ($Al_2(SO_4)_3 \cdot 18\,H_2O$) or lime ($CaO$) is added to precipitate

Table 7.27 Quantities and characteristics of sludge from various unit operations and processes.

Operation or process	Specific gravity of solids	Specific gravity of sludge	lb of dry solids per 10^3 gal Range	kg of dry solids per 10^3 m^3 Range
[a]Primary sedimentation	1.4	1.02	0.8–2.5	96–300
[a]Activated sludge	1.08	1.0002	0.5–0.8	60–96
[b]Trickling filter	1.45	1.025	0.5–0.8	60–100
[b]Filtration	1.20	1.005	0.1–0.2	12–24

[a]EPA (1979) pp. 4-1, 4-8, 4-21 to 4-28.
[b]Metcalf & Eddy (2003) p. 1456.

phosphorus from wastewater. Sometimes this is accomplished during primary sedimentation or secondary clarification.

Chemical addition may also take place after secondary treatment, in a tertiary clarifier. Chemical sludge is often handled separately, since it normally does not contain a lot of organic matter. At large WWTPs, it may prove economical to reclaim and reuse chemicals from a chemical sludge. Sludge from municipal wastewater treatment facilities that has been properly stabilized is called "biosolids." Biosolids that do not contain high concentrations of metals or pathogens can be safely applied to the land, which promotes the beneficial reuse of sludge solids. Table 7.27 provides the anticipated quantities of solids generated from various unit operations and processes.

7.11.2 Sludge mass-volume relationships

It is necessary to be familiar with sludge mass-volume relationships when designing solids handling and treatment systems. The specific gravity of the dry solids and sludge are important values to know. Typically, the specific gravity of volatile (organic) solids and fixed (mineral) solids is assumed to be 1.0 and 2.5, respectively. Using these values and knowing the solids content (P_s) or moisture content (P_w) of the sludge, and the specific gravity of the dry solids (S_s), the specific gravity of the sludge (S_{sl}) can then be determined. The water or moisture content of sludge is determined from the following equation:

$$P_w = \left(\frac{M_w}{M_w + M_s}\right) \tag{7.248}$$

where:
Ms = mass of dry solids, lb_m (kg)
M_w = mass of water, lb_m (kg)
P_w = water content or moisture content of sludge expressed as a fraction.

The solids concentration of a sludge expressed as a fraction is calculated from Equation (7.249):

$$P_s = (1.0 - P_w) = \left(\frac{M_s}{M_s + M_w}\right) \tag{7.249}$$

where P_s = solids content of sludge expressed as a fraction.

Using the definition of specific gravity (S), the volume occupied by the total mass of solids ($\forall_s$) can be determined from Equation (7.252):

$$S_s = \frac{\rho_s}{\rho} = \frac{\gamma_s}{\gamma} \tag{7.250}$$

$$S_s = \frac{\rho_s}{\rho} = \frac{M_s/\forall_s}{\rho} \tag{7.251}$$

$$\forall_s = \frac{M_s}{S_s\rho} \tag{7.252}$$

where:
S_s = specific gravity of dry solids, dimensionless,
ρ_s = density of dry solids, $lb_m/ft^3 (kg/m^3)$,
ρ = density of water, 62.4 lb_m/ft^3 (1,000 kg/m^3) at 4°C,
γ_s = specific weight of solids, lb_f/ft^3 (N/m^3),
γ = specific weight of water, 62.4 lb_f/ft^3 (9,810 N/m^3) at 4°C.

Since the total mass of solids is composed of a fixed and organic fraction, Equation (7.252) can be expressed as follows:

$$\frac{M_s}{S_s\rho} = \frac{M_v}{S_v\rho} + \frac{M_f}{S_f\rho} \tag{7.253}$$

where:
M_v = mass of volatile solids, lb_m (kg)
M_f = mass of fixed solids, lb_m (kg).

The specific gravity of sludge (S_{sl}) is calculated using the following equation, which is based on Equation (7.251):

$$\frac{M_{sl}}{S_{sl}\rho} = \frac{M_w}{S_w\rho} + \frac{M_s}{S_s\rho} \tag{7.254}$$

where:
S_{sl} = specific gravity of sludge, dimensionless
S_{sl} = 1.005 to 1.05 (Metcalf & Eddy, 2003, p. 1456)
M_{sl} = mass of sludge, lb_m (kg)
S_w = specific gravity of water, which is equal to 1.

The specific gravity of sludge can also be defined as follows, using Equation (7.250) as the basis:

$$S_{sl} = \frac{\rho_{sl}}{\rho} = \frac{M_{sl}/\forall_{sl}}{\rho} \tag{7.255}$$

Rearranging Equation (7.255) yields:

$$\forall_{sl} = \frac{M_{sl}}{S_{sl}\rho} \quad (7.256)$$

The mass of dry solids is determined by multiplying the mass of sludge by the solids concentration, as follows:

$$M_s = M_{sl} \times P_s \quad (7.257)$$

Substituting Equation (7.257) into Equation (7.256) provides the equation for calculating the volume occupied by the sludge:

$$\forall_{sl} = \frac{M_s}{S_{sl}\rho P_s} \quad (7.258)$$

where V_{sl} = volume of sludge with mass M_{sl}, ft^3 (m^3).

Example 7.22 illustrates how to calculate the specific gravity of dry solids, the specific gravity of sludge, and the volume occupied by the sludge.

Example 7.22 Sludge mass-volume relationship

Primary sludge contains 5% dry solids that are 70% volatile. Determine:

a) The specific gravity (S_s) of the dry solids.
b) The specific gravity (S_{sl}) of the sludge.
c) The volume of the thickened sludge, if 1,000 kg of dry solids are thickened to 6% in a gravity belt thickener.

Solution Part A

Estimate the specific gravity of the dry solids, knowing that 70% are volatile and 30% are fixed using Equation (7.253). Assume that the specific gravities of the volatile (organic) and fixed (mineral) solids are 1.0 and 2.5, respectively and the total mass of the dry solids is 100 kg. Therefore, 0.70 × 100 = 70 kg of volatile solids and 30 kg of fixed solids. Substitute the appropriate values into Equation (7.253):

$$\frac{M_s}{S_s\rho} = \frac{M_v}{S_v\rho} + \frac{M_f}{S_f\rho} \quad (7.253)$$

$$\frac{100\text{ kg}}{S_s\rho} = \frac{70\text{ kg}}{1.0\rho} + \frac{30\text{ kg}}{2.5\rho} \qquad \boxed{S_s = 1.22}$$

Solution Part B

Calculate the specific gravity of the sludge using Equation (7.254) and the specific gravity of the dry solids of 1.22. Assume that the total mass of the sludge is 100 kg; therefore, the mass of dry solids is 100 kg × 0.05 = 5 kg and the mass of water is 100 kg − 5 kg = 95 kg.

$$\frac{M_{sl}}{S_{sl}\rho} = \frac{M_w}{S_w\rho} + \frac{M_s}{S_s\rho} \quad (7.254)$$

$$\frac{100\text{ kg}}{S_{sl}\rho} = \frac{95\text{ kg}}{1\rho} + \frac{5\text{ kg}}{1.22\rho} \qquad \boxed{S_{sl} = 1.01}$$

Solution Part C

Estimate the volume of the sludge after thickening to 6% solids (dry weight basis) by gravity belt thickening. Use Equation (7.258). The density of water is equal to 1000 kg/m^3 at 4°C and is the value generally used in sludge calculations.

$$\forall_{sl} = \frac{M_s}{S_{sl}\rho P_s} \quad (7.258)$$

$$\forall_{sl} = \frac{1000\text{ kg}}{(1.01)\left(\frac{1000\text{ kg}}{m^3}\right)(0.06)} = \boxed{16.5\text{ m}^3}$$

Thickening sludge has a significant impact on costs at WWTPs. Increasing the sludge solids concentration from 1% to 2% results in a 50% decrease in sludge volume! This translates into tremendous cost savings, both capital and operational, since the size of sludge treatment and handling systems can be reduced.

7.11.3 Sludge unit operations and processes

The major goals of processing sludge are:

1. Reduce the volume of the sludge.
2. Kill pathogenic organisms in the sludge.
3. Stabilize the sludge by furthering reducing the organic content.
4. Produce a product that can be used beneficially or safely disposed.

There are numerous options available for sludge processing, and the treatment scheme used depends on the type of sludge processed and the ultimate disposal method selected. Figure 7.59 illustrates a simplified schematic of sludge treatment alternatives.

Table 7.28 provides a list of alternatives for each category. Additional details for designing sludge handling and treatment systems are available elsewhere (US EPA, 1979; Metcalf & Eddy, 2003; WEF, 1998c; Droste, 1997; and Reynolds & Richards, 1996).

7.11.4 Sludge thickening systems

Thickening of sludge is very important, since it reduces the volume of the sludge to be handled, thereby reducing the size of

Table 7.28 Sludge processing alternatives.

Thickening	Stabilization	Conditioning	Dewatering	Heat drying and other	Thermal reduction	Reuse and disposal
Centrifuge	Aerobic digestion	Chemical	Belt filter press	Alkaline stabilization pasteurization	Co-incineration	Land application
Dissolved air flotation	Alkaline stabilization	Freeze-thaw	Centrifuge	Composting	Incineration	Surface disposal site
Gravity	Anaerobic digestion	Heat	Drying beds	Direct drying		Landfill
Gravity belt			Lagoons	Indirect drying		Monofill
Rotary drum			Plate and frame press			

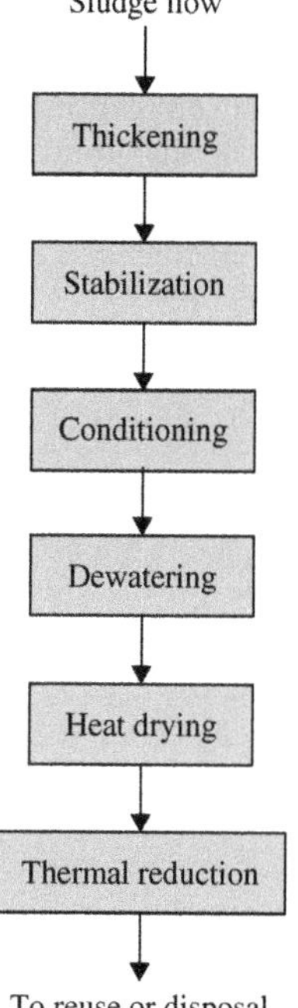

Figure 7.59 A generalized sludge-processing flow schematic.

subsequent solid handling processes, such as sludge stabilization. Increasing the solids content of sludge from 1% to 3% results in a 67% reduction in the sludge volume. Typical sludge thickening operations, with approximate solids concentration associated with each, include gravity thickening (2% to 10% solids), gravity belt thickeners (3% to 6% solids), dissolved air flotation (3% to 6% solids), and thickening centrifuges (4% to 8% solids).

Recall that a 1% solids concentration is approximately equal to 10,000 mg/L, assuming that the specific gravity of the sludge is 1.0. Designing sludge thickening unit operations involves performing mass balances on the sludge flows and sludge (solids) quantities into and out of the unit operation. Example 7.23 illustrates the design of a gravity belt thickener.

7.11.4.1 Gravity Thickening

Gravity thickening of sludge is similar to the clarification and thickening of mixed liquor that occurs during secondary clarification, and it is typically accomplished in circular basins. Sludge is fed to a center-feed well and allowed to thicken and compact. Design of gravity thickeners is based on solids loading rate (*SLR*) and overflow rate (V_0). In most cases, the solids loading rate (thickening), rather than overflow rate (clarification), will control the design. Recommended solids loading rates and overflow rates for designing gravity thickeners are listed in Table 7.29. Figure 7.60 is a schematic of a gravity thickener.

Table 7.29 Design criteria for gravity thickeners.

Type of sludge	Solids loading rate lb/(ft^2 · d)	Solids loading rate kg/(m^2 · d)	Overflow rate gal/(ft^2 · d)	Overflow rate m^3/(m^2 · d)
Primary	20–30	100–150	380–760	15.5–31
Waste activated	4–8	20–40	100–200	4–8
Combined primary and WAS	5–14	25–70	150–300	6–12

Adapted from EPA (1979), pp. 5–7 and 5–8; Metcalf & Eddy (2003), pp. 1491–1492.

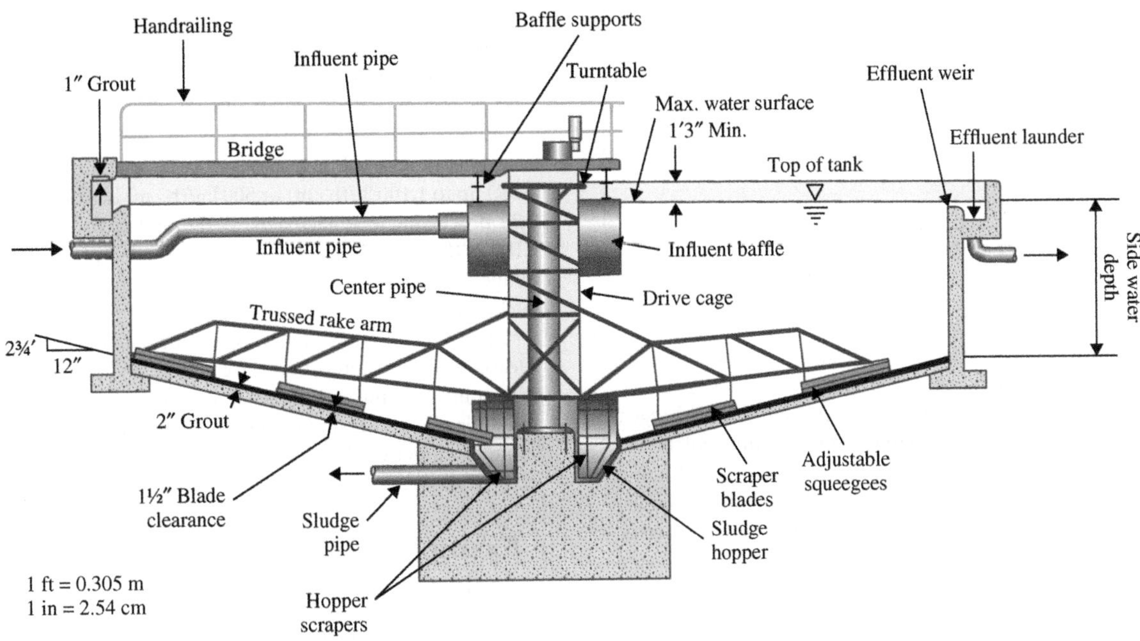

Figure 7.60 Schematic of a gravity thickener. *Source:* U.S. EPA (1979) pg. 5–6. United States Environmental Protection Agency.

Example 7.23 Design of a gravity thickener

Gravity thickening is used to thicken primary sludge at a municipal WWTP. The sludge flow is 950 m^3/d at a solids concentration of 20,000 mg/L and specific gravity of 1.02. The solids concentration in the supernatant is 800 mg/L and a 5.0% solids concentration is anticipated in the thickened sludge at a specific gravity of 1.05. Determine the following:

a) Diameter of the thickener, m.
b) Thickened sludge flow, m^3/d.

Solution Part A

Calculate the actual solids loading to the thickener. Recall that 1 mg/L = 1 g/m^3.

$$950\,\frac{m^3}{d}\times 20{,}000\,\frac{g}{m^3}\times\frac{1\,kg}{1000\,g}\times 1.02 = 1.94\times 10^4\,\frac{kg}{d}$$

From Table 7.29, use a solids loading rate of 125 $kg/(m^2 \cdot d)$ to determine the area of the thickener:

$$A_s = \frac{Q\times C}{SLR} = \frac{1.94\times 10^4\ kg/d}{125\ kg/(m^2\cdot d)} = 155\,m^2$$

The diameter of the thickener is calculated as follows:

$$A_s = \frac{\pi D^2}{4} = 155\,m^2$$

$$D = \boxed{14\,m}$$

Solution Part B

Perform materials balance on flows as shown below:

$$Q_{in} = Q_{underflow} + Q_{supernatant} = 950\,\frac{m^3}{d}$$

$$Q_{supernatant} = 950\,\frac{m^3}{d} - Q_{underflow}$$

Perform materials balance on solids as follows:

$$Q_{in}\times Conc\times S = Q_{underflow}\times Conc\times S + Q_{supernatant}\times Conc\times S$$

S = specific gravity of each solids stream. Recall that 1% solids = 10,000 mg/L.

$$\frac{1\,kg}{1000\,g}\left[950\,\frac{m^3}{d}\times 20{,}000\,\frac{g}{m^3}\times 1.02 = Q_{underflow}\times\left(5\times 10{,}000\,\frac{g}{m^3}\right)\times 1.05 + Q_{supernatant}\times 800\,\frac{g}{m^3}\times 1.00\right]$$

$$1.94\times 10^4\,\frac{kg}{d} = 52.5\,Q_{underflow} + 0.8\,Q_{supernatant}$$

$$1.94\times 10^4\,\frac{kg}{d} = 52.5\,Q_{underflow} + 0.8\left(950\,\frac{m^3}{d} - Q_{underflow}\right)$$

$$\boxed{Q_{underflow} = 360\,\frac{m^3}{d}}$$

$$Q_{supernatant} = 950\ \frac{m^3}{d} - Q_{underflow} = 950\ \frac{m^3}{d} - 360\ \frac{m^3}{d}$$

$$= \boxed{590\ \frac{m^3}{d}}$$

7.11.4.2 Gravity Belt Thickening

Gravity belt thickeners (GBTs) are mechanical devices that consist of a fabric-mesh belt that is moved over a series of rollers by a variable speed drive. Polymer-conditioned sludge is applied to a distribution box at one end of the unit, which evenly distributes the liquid sludge across the entire width of the belt. As the belt moves, water drains through the porous belt, increasing the solids concentration of the sludge. A series of plow blades along the longitudinal axis of the unit produces ridges and furrows, enhancing the draining of water from the sludge. An adjustable pitch ramp is located at the discharge end of the GBT. Thickened sludge travels over the ramp and drops into a hopper, to be transported for further processing. The belt is then rinsed with high-pressure water spray to remove solids and polymer from the pores of the porous fabric.

Figure 7.61 Photo of a gravity belt thickener.

The effective width of the belts used typically range from 1.0–3.0 meters in increments of 0.5 m. Design and hydraulic loading rates for GBTs are presented in Table 7.30.

A hydraulic loading of 200 gallons per minute/m (800 Liters per minute/m) is recommended when no pilot-plant data are available (Metcalf & Eddy, 2003, p. 1497). A solids loading rate (SLR) up to 1,100 lb/(h · m) or 500 kg/(h · m) has been used on GBT processing waste-activated sludge and up to 1,100 lb/(h · m) or 770 kg/(h · m) when processing digested sludge (WEF, 1998c, pp. 20–82). Solids capture for gravity belt thickeners range from 90 to 98%.

Operating a GBT is a messy operation, and off-gases released from the sludge must be collected and processed for odor control. At the end of each operating day, the GBT unit and floor around the unit is rinsed with water. Figure 7.61 is a photo of a gravity belt thickener.

Example 7.24 Gravity belt thickener design

Design a gravity belt thickener (GBT) to thicken waste-activated sludge (WAS) from 1% solids to 5% solids. 10,000 kilograms of WAS are produced daily. The GBT will operate five days per week, for eight hours each day. The SS concentration in the filtrate is 1,000 mg/L. Use the following design criteria to design the GBT: hydraulic loading rate = 140–450 gpm/m (6.7 to 47 L/s · m) (Metcalf & Eddy, 2003, p. 1497, recommend a hydraulic loading rate of 200 gpm/m or 800 L/(m · min); solids loading rate = 370–1,200 lb/(m · h) or 200–600 kg/(m · h); and washwater rate = 25 gpm/meter of belt (95 Lpm/m). Although not shown on the schematic below, GBT operation requires polymer addition to the sludge feed. Typical polymer dosages range from 6–14 lb dry polymer per ton of dry solids (3–7 kg of dry polymer per Mg of dry solids).

Solution

Calculate the daily solids loading rate to the GBT:

$$10{,}000\ \frac{kg}{d}\left(\frac{7\ d}{wk}\right)\left(\frac{1\ wk}{5\ d}\right)\left(\frac{1\ d}{8\ h}\right)$$

$$= 1750\ \frac{kg}{h} \quad \text{or} \quad 14{,}000\ \frac{kg}{d}$$

Table 7.30 Typical operational data for gravity belt thickeners.

Type of sludge	Feed solids (%)	Solids loading rate (kg/m · h)	Thickened sludge solids (%)	Polymer dosage (g/kg)
[a]Aerobically digested WAS	1–2.5	500–700	5–6	3–5
[a]Anaerobically digested WAS	1.5–3.5	500–700	5–7	4–6
[b]Waste activated	0.5–2.0	270–910	5–8	2–4
[b]Anaerobically digested primary and WAS	2.5–5	910–1360	6–10	2.5–5

Vesilind (2003), pp. 14–15.
Viessman et al. (2009), p. 650.

Determine the sludge volume entering the GBT using Equation (7.258) and assuming the specific gravity of the wet sludge (S_{sl}) is equal to 1.0:

$$\forall_{sl} = \frac{M_S}{S_{sl}\,\rho\,P_s} = \frac{14{,}000\ \text{kg/d}}{(1.00)(1000\ \text{kg/m}^3)(0.01)}$$
$$= 1400\ \text{m}^3/\text{d or } 2916\ \text{Lpm}$$

Draw a schematic diagram so that a flow and materials balance can be performed:

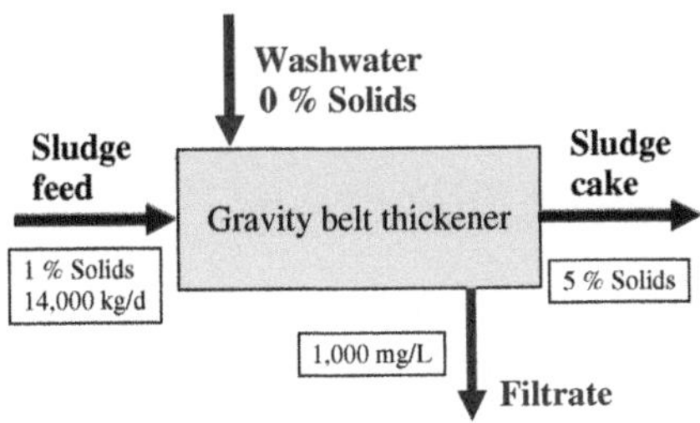

Figure E7.8

Determine the belt width of the GBT. Belt widths come in increments of 0.5 meter, ranging from 0.5–3 meters. Typical GBTs in service use 2.0 m belts. Pick a hydraulic loading rate of 800 Lpm/m from the design criteria given.

$$\text{belt width} = \frac{2916\ \text{Lpm}}{800\ \text{Lpm/m}} = 3.6\ \text{m}$$

Use two 2.0 m belts.

Check the actual solids loading rate (SLR) to see if it meets the design criteria.

$$\text{solids loading rate} = \frac{1750\ \text{kg/h of solids}}{2 \times 2\ \text{m}} = 438\ \frac{\text{kg}}{\text{m}\cdot\text{h}}$$

This is acceptable since the SLR design criteria specify 200–600 kg/(m · h).

Determine the sludge cake and filtrate flows by performing a flow and materials balance around the GBT. Cake and filtrate flows are denoted as Q_C and Q_F, respectively.

$$Q_{\text{sludge}} + Q_{\text{washwater}} = Q_F + Q_C$$
$$2916\ \text{Lpm} + 2 \times 2\ \text{m} \times 95\ \text{Lpm/m} = Q_F + Q_C$$
$$3296\ \text{Lpm} = Q_F + Q_C$$
$$3296\ \frac{\text{L}}{\text{min}}\left(\frac{60\ \text{min}}{\text{h}}\right)\left(\frac{8\ \text{h}}{\text{d}}\right)\left(\frac{\text{m}^3}{1000\ \text{L}}\right) = 1582\ \frac{\text{m}^3}{\text{d}}$$
$$1582\ \frac{\text{m}^3}{\text{d}} = Q_F + Q_C$$
$$Q_C = 1582\ \frac{\text{m}^3}{\text{d}} - Q_F$$

Perform a materials balance on solids. Recall that a 1% solids concentration is equal to 10,000 mg/L, assuming the specific gravity is equal to 1.0.

$$M_{\text{sludge}} = M_C + M_F$$
$$14{,}000\ \text{kg/d} = Q_C\left(5 \times 10{,}000\ \frac{\text{g}}{\text{m}^3}\right)\left(\frac{1\ \text{kg}}{1000\ \text{g}}\right)$$
$$+\, Q_F\left(1000\ \frac{\text{g}}{\text{m}^3}\right)\left(\frac{1\ \text{kg}}{1000\ \text{g}}\right)$$
$$14{,}000\ \text{kg/d} = 50Q_C + 1.0Q_F$$
$$14{,}000\ \text{kg/d} = 50\left(1582\ \frac{\text{m}^3}{\text{d}} - Q_F\right) + 1.0Q_F$$
$$49Q_F = 65{,}100 \qquad Q_F = 1329\ \frac{\text{m}^3}{\text{d}} \text{ or } 2769\ \text{Lpm}$$
$$Q_C = 1582\ \frac{\text{m}^3}{\text{d}} - Q_F = 1582\ \frac{\text{m}^3}{\text{d}} - 1329\ \frac{\text{m}^3}{\text{d}}$$
$$= 253\ \frac{\text{m}^3}{\text{d}} \quad \text{or} \quad 527\ \text{Lpm}$$

Now calculate the solids in the filtrate.

$$1329\ \frac{\text{m}^3}{\text{d}}\left(1000\ \frac{\text{g}}{\text{m}^3}\right)\left(\frac{1\ \text{kg}}{1000\ \text{g}}\right) = 1329\ \frac{\text{kg}}{\text{d}}$$

The percent capture through the GBT can be calculated using the following equation:

$$\%\ \text{Capture} = \frac{[\text{solids in feed} - \text{solids in filtrate}] \times 100}{\text{solids in feed}} \tag{7.259}$$

$$\%\ \text{Capture} = \frac{[14{,}000\ \text{kg/d} - 1329]\text{kg/d} \times 100}{14{,}000\ \text{kg/d}}$$
$$= 90.5\ \%$$

A schematic diagram of the complete materials balance is shown below:

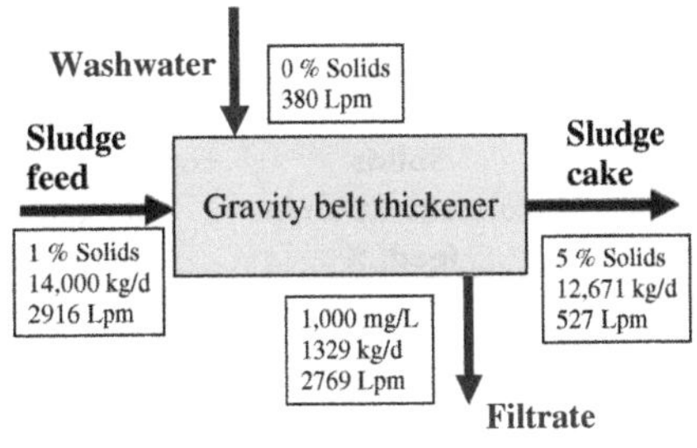

Figure E7.9

7.11.4.3 Centrifuges

Centrifuges are used for thickening and dewatering sludge. Waste-activated sludge may be thickened using a solid-bowl centrifuge without having to condition the sludge with polymer. To improve performance, however, it is recommended that standby polymer addition systems be provided. According to Metcalf & Eddy (2003, p. 1496), operational and maintenance costs can be substantial for centrifugal thickening, and it is only recommended for facilities larger than 5 MGD (0.2 m^3/s).

A centrifuge uses centrifugal force to speed up the separation of solids from the liquid wastewater. The bowl speed for horizontal solid-bowl centrifuges ranges from 1,000 – 2,600 rpm, generating forces 1,000 – 4,000 times the force of gravity (WEF, 1998c, pp. 21 – 6 and Viessman *et al.*, 2009, p. 683). Polymer dosages for thickening WAS range from 0 – 8 lb dry polymer per ton of dry solids (0 – 4 kg dry polymer per Mg of dry solids (Metcalf & Eddy, 2003, p. 1496). Figure 7.62 is a schematic of a countercurrent, solid-bowl centrifuge. Operating results from solid-bowl centrifuges are reported in Table 7.31.

The solids capture or recovery may also be determined by Equations (7.260) or (7.261):

$$\text{solids capture} = \frac{\text{mass of dry solids in cake}}{\text{mass of dry solids in feed}} \times 100 \quad (7.260)$$

$$\text{solids capture} = \frac{C_{cake}(C_{feed} - C_{filtrate})}{C_{feed}(C_{cake} - C_{filtrate})} \times 100 \quad (7.261)$$

where:

C_{cake} = solids concentration in cake (thickened sludge), mg/L or % solids

C_{feed} = solids concentration in feed (unthickened sludge), mg/L or % solids

$C_{filtrate}$ = solids concentration in filtrate, mg/L or % solids.

An example on using a centrifuge for dewatering sludge is presented in Section 7.11.7.1 of this chapter.

7.11.5 Sludge stabilization

The objectives of sludge stabilization are to reduce the liquid volume and quantity of sludge solids, prevent nuisance odors, and reduce the pathogen content. Two primary means of stabilizing wastewater sludge include aerobic and anaerobic digestion. A brief description of aerobic digestion will be presented, followed by a discussion of anaerobic digestion.

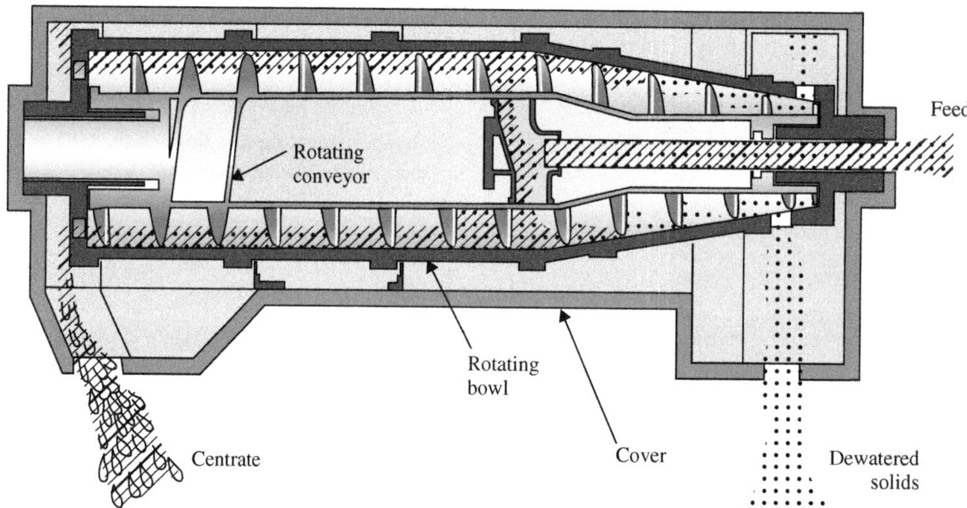

Figure 7.62 Schematic of solid-bowl centrifuge. *Source*: U. S. EPA (1979) pg. 5–50. United States Environmental Protection Agency.

Table 7.31 Reported operating results for horizontal solid-bowl centrifuges.

Type of activated sludge	Solids concentration in feed, %	Solids concentration in thickened sludge, %	Solids capture, %	Polymer use, g of active polymer/kg of dry solids	Type of centrifuge
Air	0.48–0.60	3.6–6.0	77–96	0.2–2.2	Cocurrent
Air	0.48–0.60	1.7–8.2	57–97	0.4–1.4	Countercurrent
High-purity oxygen	0.50	7	66	6	Countercurrent
Air	0.60–0.80	3–5.5	92–93	–	Cocurrent

Source: WEF (1998c) pg. 20–62. McGraw-Hill.

7.11.5.1 Aerobic Digestion

Aerobic digestion is actually a continuation of the activated sludge process operating in the endogenous phase. Typically, waste-activated sludge is aerated for extended periods, so that microorganisms oxidize their own protoplasm since exogenous (external) carbon has been depleted. Equation (7.262) is a simplified qualitative reaction showing what happens during aerobic digestion:

$$\text{organic matter(biomass)} + O_2 \xrightarrow{\text{aerobic microbes}} \text{new cells} + \text{energy} + CO_2 + H_2O + \text{endproducts} \quad (7.262)$$

If the microbial solids (biomass) are represented by the formula $C_5H_7O_2N$, a quantitative reaction for aerobic digestion can be developed as shown in Equation (7.263):

$$C_5H_7O_2N + 7O_2 \rightarrow 5CO_2 + 3H_2O + H^+ + NO_3^- \quad (7.263)$$

Equation (7.263) indicates that 1.98 lb of oxygen per lb of cells (7 × 32/113), or 1.98 kg of oxygen per kg of cells, are required during digestion. For design purposes, engineers typically assume 2.0 kg of oxygen per kg of volatile suspended solids. It is also important to note that hydrogen ions are produced during aerobic digestion, resulting in 0.44 kg of alkalinity as $CaCO_3$ being consumed per kg of cells oxidized.

Aerobic digesters may be operated either on a continuous flow basis, or as a batch operation. Aerobic digestion works best on waste-activated sludge, since the exogenous carbon has been used up and, therefore, synthesis of new biomass does not occur. Blended primary sludge and secondary sludge may be treated by aerobic digestion. However, the digestion time must be increased, since primary sludge contains organic matter that must be oxidized. Advantages and disadvantages of aerobic digestion are listed below.

Advantages:

1. Volatile solids reduction approximately equal to that obtained anaerobically.
2. Lower BOD concentrations in the supernatant.
3. Production of humus-like, biologically stable end-product.
4. Recovery of the basic fertilizer value.
5. Ease of operation.
6. Lower capital cost as compared to anaerobic digestion.

Disadvantages:

1. High power cost associated with aeration.
2. Digested sludge has poor mechanical dewatering characteristics.
3. Process is significantly affected by temperature.
4. Cannot recover methane.
5. Generally, a lower solids content in digested sludge.
6. May have trouble meeting the vector attraction reduction rule (twelve options available. Option 1, 38% reduction in volatile solids, is the most widely accepted option).

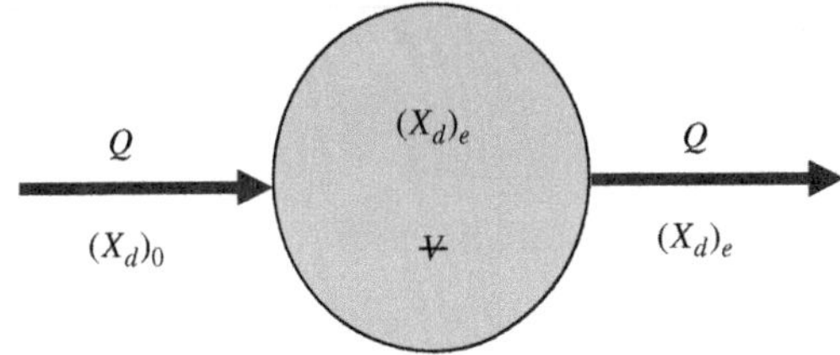

Figure 7.63 Continuous flow, complete-mix aerobic digester schematic.

Aerobic Digestion Detention Time The design of aerobic digesters has typically been based on a first-order removal reaction, as shown below.

$$\frac{dX_d}{dt} = -K_d X_d \quad (7.264)$$

where:
X_d = concentration of degradable VSS, mg/L
K_d = reaction rate or degradation constant, time^{-1}.

Performing a materials balance on degradable VSS around a complete-mix, aerobic digester as shown in Figure 7.63, will yield Equation (7.265), which is used to calculate the required digestion time:

$$\left(\frac{dX_d}{dt}\right)_{\text{accumulation}} \forall = Q(X_d)_0 - Q(X_d)_e + \left(\frac{dX_d}{dt}\right) \forall \quad (7.265)$$

At steady-state, the accumulation term is set equal to zero, and substituting Equation (7.264) into Equation (7.265) results in the following equation:

$$0 = Q(X_d)_0 - Q(X_d)_e - K_d(X_d)_e \forall \quad (7.266)$$

Rearranging Equation (7.266) and solving for digestion time yields Equation (7.269):

$$Q(X_d)_0 - Q(X_d)_e = K_d(X_d)_e \forall \quad (7.267)$$

$$\frac{\forall}{Q} = t_d = \frac{(X_d)_0 - Q(X_d)_e}{K_d(X_d)_e} \quad (7.268)$$

$$t_d = \frac{[(X_0 - X_n) - (X_e - X_n)]}{K_d[(X_e - X_n)]} \quad (7.269)$$

where:
t_d = detention time or digestion time, d
$(X_d)_0$ = degradable VSS in sludge fed to digester, mg/L
$(X_d)_0 = X_0 - X_n$
X_0 = VSS concentration in sludge fed to digester, mg/L

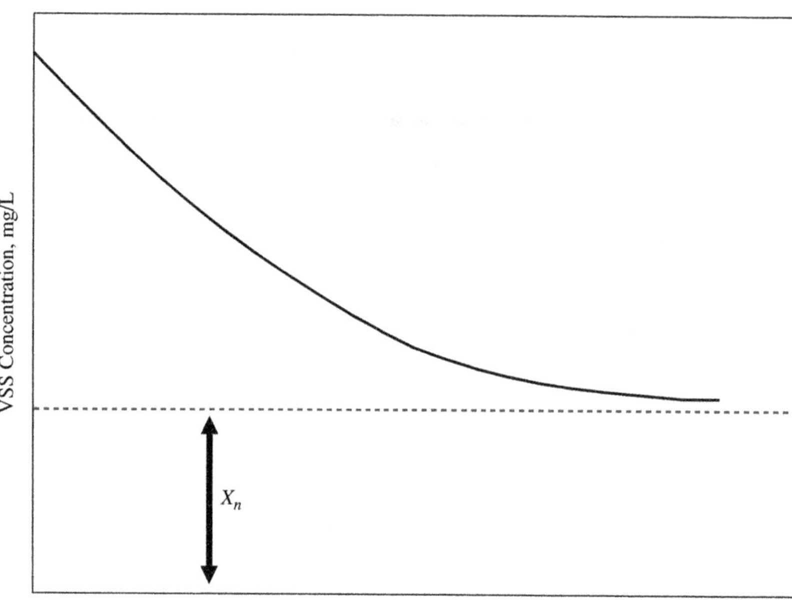

Figure 7.64 Plot of VSS concentration remaining versus digestion time.

X_n = nondegradable VSS concentration in sludge that is assumed to remain constant through the digestion process, mg/L
$(X_d)_e$ = degradable VSS concentration in digested sludge, mg/L
$(X_d)_e = X_e - X_n$
X_e = VSS concentration in digested sludge, mg/L
Q = volumetric flow rate of sludge, volume/time
$\forall$ = aerobic digester volume, L.

Batch studies for a particular sludge must be performed to determine K_d and X_n. A plot of the VSS concentration versus digestion time, as shown in Figure 7.64, is used for estimating the value of X_n.

Once X_n has been determined, a plot of the natural log of degradable VSS concentration versus time should yield a straight line with a negative slope (Figure 7.65).

The magnitude of the slope of the line of best fit is equal to the degradation constant, K_d. The degradable VSS concentration must be determined by subtracting X_n from each VSS concentration prior to making the semi-log plot. Benefield & Randall (1980, p. 490) indicate K_d varies from 0.10–0.12 d^{-1} for waste-activated sludge at 20°C.

When bench-scale studies cannot be performed, aerobic digester design may be based on detention time values taken from the literature. Reynolds & Richards (1996, p. 615) show that hydraulic detention times at 20°C range from 12–16 days for the digestion of waste-activated sludge, and 18–22 days for combined primary and waste-activated sludge or trickling filter humus. They also recommend that the minimum design mean cell residence time should range from 10–15 days for WAS and 15–20 days for combined primary and waste-activated sludge. The detention time may be corrected for temperatures other than 20°C by using the following equation:

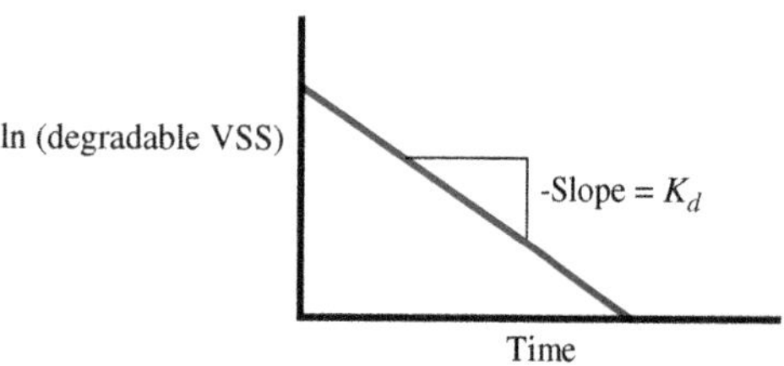

Figure 7.65 Plot of ln(Degradable VSS) versus digestion time.

$$t_{T^\circ C} = t_{20^\circ C}(\theta)^{(20^\circ - T^\circ C)} \tag{7.270}$$

where:

$t_{T^\circ C}$ = hydraulic detention time in aerobic digester at temperature T, d
$t_{20^\circ C}$ = hydraulic detention time in aerobic digester at 20°C, d
T = actual digester operating temperature, °C
Θ = temperature correction factor varies from 1.02 to 1.11, typically 1.065 is used.

Air and Mixing Requirements for Aerobic Digestion

Similar to the activated sludge process, sufficient oxygen must be supplied to the aerobic digester so that oxygen does not limit the biological reactions that are occurring. For design purposes, 2.0 lb of oxygen per lb of solids destroyed (2.0 kg/kg) must be supplied to meet process oxygen requirements. This value is used for both waste-activated sludge and trickling filter humus. Approximately 1.9 lb of oxygen per lb of BOD (1.9 kg/kg) are used for computing the oxygen required for stabilizing primary sludge (Reynolds & Richards, 1996, p. 615). Also, air or mechanical energy must be input to the digester to keep the microorganisms in suspension. Mechanical mixers need 0.75–1.25 hp/1,000 ft^3 (20 to 40 kW/1,000 m^3) to meet the mixing requirement. Diffused aeration systems must supply 20–40 cfm/1,000 ft^3 (25 to 40 m^3/min·1,000 m^3) for digesters treating WAS and 60 cfm/1,000 ft^3 (60 m^3/min·1,000 m^3) for aerobic digesters treating a combination of primary sludge and WAS (Reynolds & Richards, 1996, p. 615). Table 7.32 presents other design criteria for aerobic digesters.

Table 7.32 Design criteria for aerobic digesters.

Parameter	Value
[a]MCRT for PSRP	40 d at 20°C 60 d at 15°C
[b]MCRT at 20°C	
Primary and WAS or Trickling Filter Humus	18–22 d
Waste Activated Sludge without Primaries	16–18 d
Waste Activated Sludge	12–16 d
[c]Dissolved oxygen concentration, mg/L	1–2
[c]Energy for mixing	
Diffused aeration	20–40 $ft^3/(10^3\ ft^3 \cdot min)$ [0.02–0.04 $m^3/(m^3 \cdot min)$]
Mechanical aerators	0.75–1.5 $hp/(10^3\ ft^3)$ [20–40 $kw/(10^3\ m^3)$]
[c]Oxygen requirements	1.6–2.3 lb O_2/lb VS [1.6–2.3 kg O_2/kg VS]
[c]Volatile Solids Destruction	38–50%
[b]Volatile Solids Loading Rate	0.04–0.20 lb $VS/(ft^3 \cdot d)$ [0.64– 3.20 kg $VS/(m^3 \cdot d)$]

PSRP = Process for the significant reduction of pathogens
[a]US EPA (2003a), p. 45; to meet pathogen reduction requirements for PSRP.
[b]Reynolds & Richards (1996), p. 615.
[c]Metcalf & Eddy (2003), p. 1536.

The design of an aerobic digestion system is presented in Example 7.25.

Example 7.25 Aerobic digester design

An aerobic digester is to be designed to stabilize 3,300 kg/d of combined primary and waste-activated sludge. The volume of the combined sludge that will be fed to the digester is 104 m^3/d. The volatile solids concentration of the combined sludge is 70% and the minimum design operating temperature is 16°C. A temperature correction coefficient (θ) of 1.06 should be used to correct the detention or digestion time for temperature. The primary sludge contains 940 kg of BOD, with an oxygen demand of 1.9 kg of oxygen per kg of BOD, while the WAS consists of 865 kg of VSS, with an oxygen demand of 2.0 kg of oxygen per lb of VSS destroyed.

Determine the following:

a) Design hydraulic detention time or digestion time in days.
b) Aerobic digester volume in m^3.
c) Volatile solids loading rate to the aerobic digester in kg/d.
d) Quantity of oxygen required to stabilize the primary sludge in $kg/(m^3 \cdot d)$.
e) Quantity of oxygen required to stabilize the secondary sludge (WAS) in kg/d.
f) Total quantity of oxygen required to stabilize the combined sludge in kg/d.
g) Total air required in m^3/min if air contains 0.28 kg of oxygen per cubic meter (0.075 lb O_2/ft^3) and the diffusers have a transfer efficiency of 5.0%.

Solution Part A

Select a hydraulic detention time of 20 days from Table 7.32 and correct it for the minimum design operating temperature of 16°C using Equation (7.270):

$$t_{T°C} = t_{20°C}(1.06)^{(20°-T°C)} = 20\,d(1.06)^{(20°-16°)}$$
$$= 25.2\,d$$

Solution Part B

The aerobic digester volume is calculated by multiplying the hydraulic detention time by the volume of sludge fed to the digester as:

$$V = tQ_{sludge} = 25.2\,d\left(104\,\frac{m^3}{d}\right) = 2620\,m^3$$

Solution Part C

Calculate the volatile solids loading to the digester:

$$\text{VS loading} = 3300\,kg/d(0.70) = 2310\,kg/d$$

The volatile solids loading rate to the digester is determined as:

VS loading to digester $= \frac{2310\ kg/d\ VS}{2620\ m^3} = 0.88\ \frac{kg\ VS}{m^3 \cdot d}$.
Meets criteria in Table 7.32.

Solution Part D

Oxygen required to stabilize the BOD in the primary sludge is estimated as:

$$O_2 \text{ to stabilize primary BOD}$$
$$= \left(940\,\frac{kg\ BOD}{d}\right)\left(1.9\,\frac{kg\ O_2}{kg\ BOD}\right) = 1790\,\frac{kg\ O_2}{d}$$

Solution Part E

Oxygen required to stabilize the VS in the secondary sludge (WAS) is estimated as:

$$O_2 \text{ to stabilize VS in WAS}$$
$$= \left(865\,\frac{kg\ VS}{d}\right)\left(2.0\,\frac{kg\ O_2}{kg\ VS}\right) = 1730\,\frac{kg\ O_2}{d}$$

Solution Part F

The total oxygen required to stabilize the combined primary and secondary sludge is determined below:

$$\text{total } O_2 \text{ required} = 1790 + 1730 = 3520\ \frac{\text{kg } O_2}{\text{d}}$$

Solution Part G

Finally, the total air required ($m^3/(\text{min}\cdot 1{,}000\ m^3)$) to stabilize the combined sludge is estimated as follows, assuming that air contains 0.28 kg O_2/m^3 and the diffuser transfer efficiency is 5%:

$$\text{air required} = \frac{3520\ \text{kg } O_2/\text{d}}{[(0.28\ \text{kg } O_2/m^3)(0.05)]}\left(\frac{1\ \text{d}}{24\ \text{h}}\right)\left(\frac{1\ \text{h}}{60\ \text{min}}\right)$$

$$= 175\ \frac{m^3}{\text{min}}$$

Check to see if the mixing requirements have been met.

$$\frac{\text{air required}}{1000\ m^3} = \frac{175\ m^3/\text{min}}{2620\ m^3}\left(\frac{1000\ m^3}{1000\ m^3}\right)$$

$$= \frac{66.7\ m^3/\text{min}}{1000\ m^3} \geq \frac{60\ m^3/\text{min}}{1000\ m^3}$$

This does meet the mixing requirements. Therefore, no additional air must be supplied beyond process requirements.

7.11.5.2 Anaerobic Digestion

Anaerobic digestion is the decomposition of organic matter and inorganic matter such as sulfate in the absence of oxygen. It is modeled as a three-step process: hydrolysis, acidogenesis, and methanogenesis. During **hydrolysis**, complex organic solids are hydrolyzed by bacteria through secretion of extra-cellular enzymes. Carbohydrates, proteins, and fats are converted to simple carbohydrates, amino acids, and fatty acids.

Acidogenesis involves the conversion of soluble carbon from hydrolysis into organic acids and H_2 by facultative and anaerobic bacteria. The pH will drop, and other fermenting bacteria partially oxidize organic acids to acetic acid and hydrogen gas.

Methanogenesis is the bacterial conversion of fatty acids and hydrogen into CH_4 and CO_2 by strict anaerobes. This gasification of organics is accomplished by acid-splitting, methane-forming bacteria known as methanogens.

Anaerobic digestion produces a stable sludge and a valuable by-product, methane gas, which can be combusted to provide heat for the digesters or used for generating electricity. Only small quantities of organic matter and cellular protoplasm remain in the digested sludge. Anaerobic digesters are used extensively at WWTPs that have primary clarifiers. Most of the energy needs for plant operation can be provided with sufficient methane production. A disadvantage of using anaerobic digestion is that large reactors are required to provide the long detention times necessary to accomplish biological degradation under anaerobic conditions. Anaerobic digesters are sensitive to operate and are prone to biological upsets. Anaerobically digested sludge is usually difficult to dewater by mechanical means.

Table 7.33 Design criteria for conventional and high-rate anaerobic digesters.

Parameter	Low-rate	High-rate
MCRT, d	30–60	10–20
Mixing	None	Yes
Volatile solids loading rate, lb VS/($ft^3\cdot$d)	0.04–0.10	0.15–0.40
[a]Feed concentration of primary and secondary sludge, %	2–4	4–6
Volume criteria, ft^3/capita		
Primary sludge	2–3	1.3
Primary sludge and trickling filter humus	4–5	2.7–3.3
Primary sludge and WAS	4–6	2.7–4.0
[a]Digester underflow concentration, %	4–6	4–6

EPA (1979), pp. 6–19.
[a]Vesilind (2003), p. 15–12.

Two-stage anaerobic digestion is normally practiced where undigested sludge is fed to a mixed, heated, anaerobic digester. The next step involves a second digester that is unmixed and unheated. Figure 7.66 shows a schematic of a two-stage anaerobic digester.

Design criteria for conventional and high-rate anaerobic digesters are presented in Table 7.33.

The volume of a low- or standard-rate anaerobic digester can be determined using Equation (7.271):

$$\text{Ꝟ} = \frac{\text{Ꝟ}_1 + \text{Ꝟ}_2}{2} \times T_1 + \text{Ꝟ}_2 T_2 \qquad (7.271)$$

where:

Ꝟ = total digester capacity, ft^3 (m^3)
Ꝟ_1 = volume of raw sludge fed daily, cfd (m^3/d)
T_1 = time required for digestion, days (approximately 25 days)
Ꝟ_2 = volume of daily digested sludge accumulation, cfd (m^3/d)
T_2 = digestion storage period, 20–120 days.

Equation (7.272) is used for calculating the volume of a high-rate anaerobic digester:

$$\text{Ꝟ}_I = \text{Ꝟ}_1 \times T \qquad (7.272)$$

The volume of a second-stage, low-rate anaerobic digester is determined as follows:

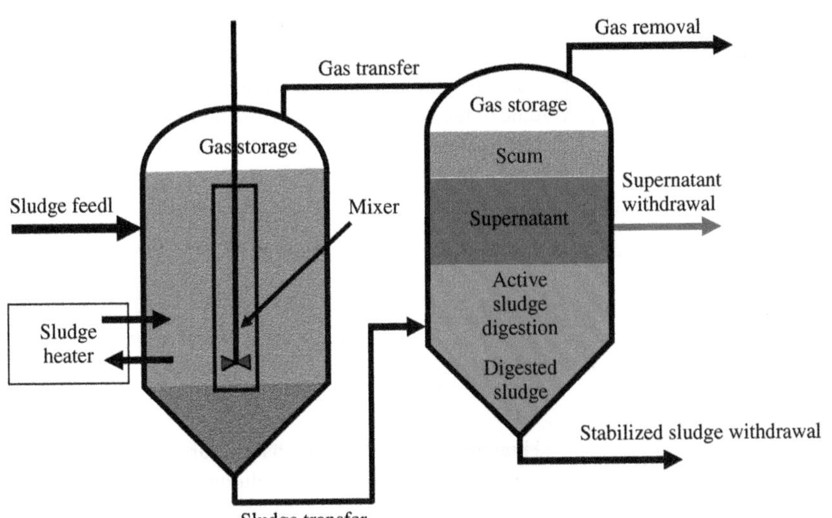

Figure 7.66 Two-stage anaerobic digestion.

$$\forall_{II} = \frac{\forall_I + \forall_2}{2} \times T_I + V_2 T_2 \quad (7.273)$$

where:

$\forall_{II}$ = digester capacity required for second stage, ft^3 (m^3)
$\forall_I$ = volume of average daily raw sludge feed = volume of digested sludge feed, cfd (m^3/d)
$\forall$ = volume of daily digested sludge accumulation in tank, cfd (m^3/d)
T_I = period required for thickening, days
T_2 = period required for digested sludge storage, days.

Example 7.26 illustrates the design of a single-stage conventional anaerobic digester.

Example 7.26 Conventional anaerobic digester design

Calculate the anaerobic digester capacity (ft^3) required for conventional single-stage anaerobic digestion, given the following data: raw sludge production = 630 lb/d; VS in raw sludge = 70%; moisture content of raw sludge = 95%; solids content of raw sludge = 5% solids; digestion period = 30 days; VS Reduction = 50%; moisture content of digested sludge = 93%; storage time required = 90 days; percent solids in thickened digested sludge = 7%, and specific gravity of digested sludge = 1.03.

Solution

Calculate the volume of raw sludge fed to the digester using Equation (7.258), assuming S_{sl} is 1.0. The density of water ρ is 1000 kg/m^3 = 62.4 lb$_m$/ft^3 = 8.34 lb$_m$/gal:

$$\forall_{sl} = \frac{M_S}{S_{sl}\rho P_s} = \frac{630\ \text{lb/d}}{(1.0)(62.4\ \text{lb}_m/\text{ft}^3)(0.05)} = 202\ \frac{\text{ft}^3}{\text{d}}$$

Perform a materials balance on the mass of solids entering and exiting the digester. First, determine the volatile and fixed solids in the raw sludge fed to the digester. Recall that **total solids = volatile solids + fixed solids.**

Mass of volatile solids in raw sludge

$$= 0.70(630\ \text{lb/d}) = 441\ \text{lb/d}$$

Mass of fixed solids in raw sludge

$$= \text{Total solids} - \text{Volatile solids}$$

$$= 630 - 441 = 189\ \text{lb/d}$$

Next, calculate the volatile and total solids quantities remaining after digestion. Digestion of the solids causes a 50% reduction in the volatile solids and the fixed solids are assumed to remain unchanged.

Mass of volatile solids remaining after digestion

$$= 0.50 \times 441\ \text{lb/d} = 220.5\ \text{lb/d}$$

Mass of total solids remaining after digestion

$$= \text{VS} + \text{FS} = 220.5 + 189 = 409.5\ \text{lb/d}$$

Next, determine the volume of digested sludge that accumulates daily in the tank, assuming that the sludge thickens to 7% solids using Equation (7.258) and that the specific gravity of the digested sludge is 1.03:

$$\forall_{sl} = \frac{M_S}{S_{sl}\rho P_s} = \frac{409.5\ \text{lb/d}}{(1.03)(62.4\ \text{lb}_m/\text{ft}^3)(0.07)} = 91\ \frac{\text{ft}^3}{\text{d}} = \forall_2$$

Finally, the volume of the conventional, single-stage anaerobic digester is determined from Equation (7.271).

$$\forall = \frac{\forall_1 + \forall_2}{2} \times T_1 + \forall_2 T_2$$

$$= \frac{(202\ \text{ft}^3/\text{d} + 91\ \text{ft}^3/\text{d})}{2} \times 30\ \text{d} + \frac{91\ \text{ft}^3}{\text{d}} \times (90\text{d})$$

$$= 1.26 \times 10^4\ \text{ft}^3$$

Example 7.27 First-stage high-rate anaerobic digester design

The average daily quantity of thickened raw waste sludge produced at a municipal WWTP is 15,000 gallons containing 10,000 lb of solids as TS. The solids are 70% volatile. Determine:

a) % moisture in the thickened sludge.
b) Volume required for a first-stage, high-rate digester, based on a maximum loading of 100 lb VS/1000 cfd and a detention time of 15 days.

Solution Part A

First calculate the % solids in the thickened raw sludge using Equation (7.258), assuming that the specific gravity of the thickened sludge is 1.0:

$$\forall_{sl} = \frac{M_s}{S_{sl}\rho P_s}$$

$$P_s = \frac{M_s}{(\forall_{sl})\rho S_{sl}} = \frac{10{,}000\ \text{lb/d}}{\left(\frac{15{,}000\ \text{gal}}{\text{d}}\right)\left(\frac{8.34\ \text{lb}_m}{\text{gal}}\right)(1.0)} = 0.08$$

The moisture content of the sludge is determined by rearranging Equation (7.249):

$$P_s = (1.0 - P_w) = \left(\frac{M_s}{M_s + M_w}\right)$$

$$P_w = (1.0 - P_s) = (1.0 - 0.08) = 0.92 \text{ or } 0.92 \times 100\% = \boxed{92\%} \qquad (7.249)$$

Solution Part B

Next calculate the volatile solids (VS) loading rate to the digester as follows:

$$\text{VS loading} = 10{,}000\ \text{ppd TS} \times (0.70) = 7000\ \text{ppd}$$

The volume of the first stage high-rate digester is calculated by dividing the actual VS loading rate by the VS loading rate design criteria of 100 lb VS/1,000 ft^3 · d.

$$\forall = \frac{7000\ \text{ppd}}{100\ \text{lb VS}/1000\ \text{ft}^3 \cdot \text{d}} = 7.00 \times 10^4\ \text{ft}^3$$

Check the volume of the first-stage, high-rate digester based on detention time. Detention time is defined as the volume of the reactor divided by the flow rate to the reactor. Rearrange the detention time equation to solve for the reactor volume:

$$\tau = \frac{\forall}{Q}$$

$$V = \tau Q = 15\ \text{d}\left(\frac{15{,}000\ \text{gal}}{\text{d}}\right)\left(\frac{\text{ft}^3}{7.48\ \text{gal}}\right)$$

$$= 3.01 \times 10^4\ \text{ft}^3$$

Final design would dictate that the larger of the two volumes be used; therefore, the volume of the digester should be 70,000 ft^3.

7.11.5.3 Lime Stabilization

Lime stabilization is a process in which lime or some other type of alkaline material, such as cement kiln dust or lime kiln dust, is added to raise the pH of the untreated sludge above 12. At high pH, bacteria, viruses, and other microorganisms are inactivated. The principal advantages include low cost, simple process, and easy to implement. Major disadvantages include higher quantity of solids to dispose, no direct reduction in organic matter occurs, and when pH drops below 11, odors and biological degradation of the sludge will resume.

Quicklime (CaO) or hydrated lime ($Ca(OH)_2$) are the most common forms of lime added. When quicklime is added to water or reacts with water in sludge, the following reaction occurs:

$$CaO + H_2O \rightarrow Ca(OH)_2 + 64\ \frac{\text{kJ}}{\text{mole}} (\text{Heat}) \qquad (7.274)$$

When lime is added to sludge prior to dewatering operations, it is called "lime pretreatment." After dewatering, it is called "lime post-treatment."

Lime stabilization is an approved Process to Significantly Reduce Pathogens (PSRP), 40CFR Part 503, and can be used to produce Class B biosolids that can be applied to agricultural land (US EPA, 2003a, p. 48). To meet pathogen requirements, sufficient lime must be added to the sewage sludge to raise the pH to 12 for two or more hours of contact. Vector attraction reduction requirements stipulate that sufficient alkali be added to raise the pH of the sludge to at least 12 at 25°C (77°F) and maintain a pH ≥ 12 for two hours and a pH ≥ 11.5 for 22 additional hours.

7.11.6 Conditioning

Conditioning of sludge or biosolids following stabilization is required for most dewatering operations. Chemical conditioning of the sludge results in coagulation and release of absorbed water. Lime, alum, ferric chloride, and polymers are typically added in liquid form. Heat treatment and freeze-thaw methods have also been used for conditioning sludge. Other sources should be consulted for details about sludge conditioning (Metcalf & Eddy (2003), pp. 1554–1558; WEF (1998c), pp. 19-1 to 19-61; and US EPA (1979), pp. 8-1 to 8-33).

7.11.7 Sludge dewatering systems

Dewatering is a unit operation primarily accompanied by mechanical means. The objective of dewatering is to increase the solids content of the sludge, thereby reducing the volume of sludge that must be handled. The major types of dewatering processes commonly used include centrifuges, belt filter presses (BFP), plate and frame presses, drying beds, and lagoons. Some typical dewatering operations and their anticipated solids concentration in the dewatered sludge are centrifuges (5–35% solids), belt filter presses (12–32% solids), plate and frame presses (30–52% solids) or sand drying beds (≅ 40% to > 50% solids) (Metcalf & Eddy (2003), pp. 1562, 1566, 1568, 1572). In most instances, the sludge must be conditioned by adding a polymer or coagulant aid such as lime or ferric chloride to increase the effectiveness of the dewatering operations. Material balances on flow and solids must be performed when designing any type of solids handling equipment and process.

7.11.7.1 Dewatering Centrifuges

As stated previously, centrifuges are used for thickening and dewatering sludge. Centrifuges are primarily used for dewatering stabilized sludge or biosolids. Solid-bowl centrifuges are typically used with anticipated solids concentrations ranging from 5–35% in the dewatered sludge (Metcalf & Eddy, 2003, p. 1562). Table 7.34 provides performance data for solid-bowl centrifuges for various types of sludge.

Metcalf & Eddy (2003, p. 1561) indicate that polymer dosages range from 2–15 lb of dry polymer per ton of dry solids (1.0–7.5 kg of dry polymer per 1,000 kg of dry solids).

Example 7.28 Dewatering centrifugation design

Lime is added to secondary wastewater effluent for the removal of phosphorus. A centrifuge is used for dewatering this lime sludge. The plant capacity is 31 million gallons per day (MGD); 800 lb of dry solids are produced per MG of flow. The solids content of the sludge is 10% and the cake solids from the centrifuge is 50%. The solids capture rate is 90% and the specific gravity of the aqueous slurry is 1.06.

Determine the following:

a) The gallons of wet sludge produced daily.
b) The pounds of dry cake produced daily.
c) The pounds of moist cake produced daily.
d) The gallons of centrate produced daily.
e) The solids content of the centrate, mg/L.

Solution Part A

First, draw a schematic of the process:

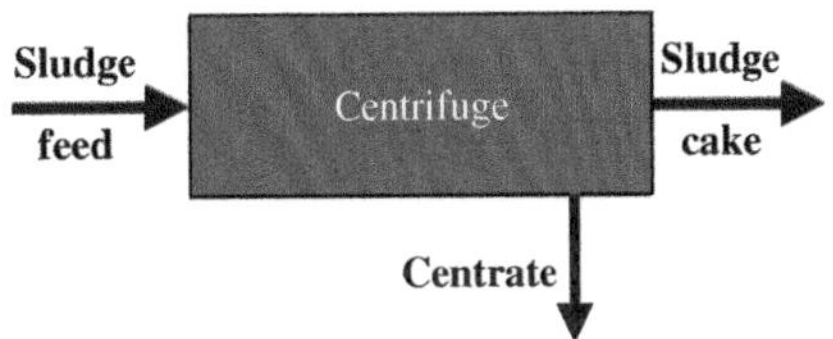

Figure E7.10

Next, determine the quantity of dry sludge that enters the centrifuge:

$$31\ \text{MGD}\left(800\ \frac{\text{lb}}{\text{MG}}\right) = 24{,}800\ \text{ppd of dry solids}$$

Table 7.34 Dewatering performance data for solid-bowl centrifuges for various types of sludge.

Type of sludge	Feed solids %	Average cake solids, %	[a]Polymer required lb/ton	[b]Recovery, %
Untreated primary	5–8	25–36	1–5	90–95
Untreated primary	5–8	28–36	0	70–90
Untreated primary and WAS	4–5	18–25	3–7	90–95
WAS	0.5–3	8–12	10–15	85–90
Anaerobically digested primary sludge	2–5	28–35	6–10	98+
Anaerobically digested primary sludge	9–12	30–35	0	65–80
Aerobically digested WAS	1–3	8–10	3–6	90–95

Source: U.S. EPA (1979) pg. 9– 24. United States Environmental Protection Agency. US EPA (1979) pp. 9–24.
Pound of dry polymer per ton of dry feed solids.
Recovery based on centrate.

The volume of the wet sludge entering the centrifuge is determined as follows, using Equation (7.258). Recall that the density of water $\rho = 1{,}000\,\text{kg/m}^3 = 62.4\,\text{lb}_m/\text{ft}^3 = 8.34\,\text{lb}_m/\text{gal}$.

$$\text{¥}_{sl} = \frac{M_S}{S_{sl}\rho P_s} = \frac{24{,}800\ \text{ppd}}{(1.06)(8.34\ \text{lb}_m/\text{gal})(0.10)} = 28{,}053\ \text{gpd}$$

Solution Part B

The quantity of dry solids in the cake is estimated as:

$$24{,}800\ \text{ppd}(0.90) = 22{,}320\ \text{ppd}$$

Solution Part C

The quantity of moist solids in the cake is estimated by dividing the dry cake mass by the fraction of solids in the cake.

$$\frac{22{,}320\ \text{ppd}}{0.50} = 44{,}640\ \text{ppd mass of moist cake}$$

Solution Part D

The volume occupied by the dewatered cake is estimated using Equation (7.258) and assuming $S_{sl} = 1.0$:

$$\text{¥}_{sl} = \frac{M_S}{S\rho P_s} = \frac{22{,}320\ \text{ppd}}{(1.0)(8.34\ \text{lb}_m/\text{gal})(0.50)} = 5353\ \text{gpd}$$

Solution Part E

Perform a mass balance on flows to determine the centrate flow.

$$Q_{\text{centrate}} = Q_{\text{sludge}} - Q_{\text{cake}} = 28{,}053 - 5353 = 22{,}700\ \text{gpd}$$

Calculate the mass of solids in the centrate (M_{centrate}) by performing a mass balance on solids around the centrifuge.

$$M_{\text{sludge}} = M_{\text{centrate}} + M_{\text{cake}}$$

$$M_{\text{centrate}} = M_{\text{sludge}} - M_{\text{cake}} = 24{,}800 - 22{,}320 = 2480\ \text{ppd}$$

The concentration of solids in the centrate is determined by dividing the mass of solids in the centrate by the centrate flow rate. This is accomplished by rearranging Equation (7.252) and assuming $S_s = 1.0$:

$$\text{¥}_s = \frac{M_S}{S_s\rho P_s}$$

$$P_s = \frac{M_S}{(\text{¥}_s)\rho S_s} = \frac{2480\ \text{ppd}}{(22{,}700\ \text{gpd})\left(\frac{8.34\ \text{lb}_m}{\text{gal}}\right)(1.0)}$$

$$= 0.013\ \text{solids or}\ \boxed{1.3\%\ \text{solids}}$$

Waste streams such as filtrate or centrate from sludge processing are normally recycled back to the headworks of the WWTP. These recycled flow streams may have a significant impact on process performance, especially if effluent limits are stringent for BOD, TSS, TN, and TP. Material balances should be performed around the entire treatment plant to determine the effect on performance. Although the TSS concentration is high in this case, the low volume of wastewater (22,700 gpd), compared to the influent flow (31 MGD), produces a negligible effect.

7.11.7.2 Belt-Filter Presses

Belt-filter presses (BFPs) are similar to gravity-belt thickeners. In addition to having a gravity drainage section for thickening the sludge, a low-pressure and a high-pressure section follows where the sludge passes between two opposing belts. This pressure squeezes additional water from the sludge achieving a high concentration of solids, on the order of 12–32%. The sludge must be conditioned with polymer before being applied to the belt press. Figure 7.67 is a drawing of a belt-filter press. Operational data showing typical hydraulic and solids loading rates is presented in Table 7.35.

7.11.7.3 Filter Presses

Filter presses accomplish dewatering by forcing water from the sludge under high-pressure. There are two primary types of filter presses in use for dewatering sludge: fixed-volume recessed chamber, and variable-volume recessed chamber or diaphragm filter. The advantages and disadvantages of using filter presses are listed below (WEF, 1998c, p. 21–39).

Advantages:

1 Generally produces a cake that is drier than other dewatering alternatives.
2 Can consistently achieve a cake solids concentration of 35%.
3 Can adapt to a wide range of feed solids concentrations.
4 Have acceptable mechanical reliability.
5 Comparable energy requirements to vacuum-filter dewatering systems.
6 Produce high filtrate quality that lowers recycle stream treatment requirements.

Disadvantages:

1 High capital cost.
2 Substantial quantities of conditioning chemicals or precoat materials are required.
3 Periodic adherence of cake to the filter medium that requires manual removal.
4 Relatively high operation and maintenance costs.

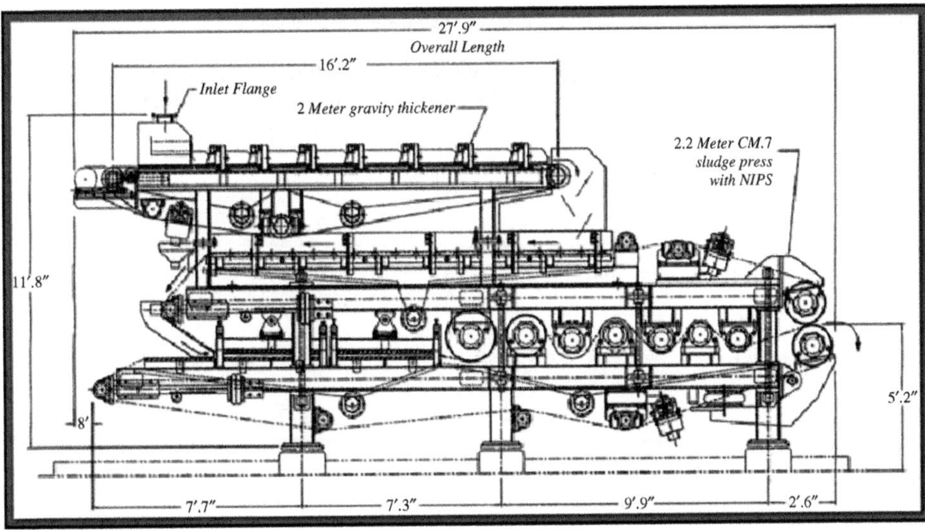

Figure 7.67 Drawing of a belt-filter press. *Source*: www.presstechnologies.com/gravity/

Table 7.35 Operational data for belt-filter presses.

Type of sludge	Feed solids (%)	Hydraulic loading ($m^3/m \cdot h$)	Solids loading rate ($kg/m \cdot h$)	Cake solids (%)	Polymer dosage (kg/tonne)
Anaerobically digested primary	4–6	9–11	450–750	25–35	2–3
Anaerobically digested primary and WAS	2–5	9–11	250–450	15–26	3–6
Aerobically digested WAS	1–3	7–10	100–250	11–22	4–7
Raw primary and waste activated	3–6	9–11	350–550	16–25	2–5
Thickened WAS	3–5	9–11	350–450	14–20	3–4
Extended aeration WAS	1–3	7–10	100–250	11–22	4–7

Source: Hammer (1986) pg. 454. Reproduced by permission of John Wiley & Sons Ltd.

Sludge must be conditioned prior to being pumped to a filter press. In the USA, ferric chloride ($FeCl_3$) and lime (CaO) are the two most widely used chemicals added to enhance dewatering. Ash and various types of polymer have been used, as well (WEF, 1998c, p. 21–51). Tables 7.36 and 7.37 provide anticipated performance for filter presses along with chemical dosages typically used.

7.11.7.4 Fixed-Volume Recessed Chamber

The fixed-volume recessed chamber filter press consists of series of recessed rectangular plates that are positioned face-to-face in a vertical position on a frame. Figure 7.68 shows a fixed-volume recessed chamber filter press.

At one end of the frame there is a fixed head, while at the other there is a movable head. The plates are held together by a hydraulic ram or a powered screw during the filtration phase. Recessed chamber filter presses are operated on a batch basis with chemically conditioned sludge pumped to the units and a pressure ranging from 100–225 lb_f/in^2gauge (690–1550 kPa) is applied from 1–3 hours, which forces liquid through the filter medium (typically cloth), leaving a concentrated solids cake remaining between the filter media and recessed plate (WEF, 1998c, p. 21–44; Metcalf & Eddy (2003), p. 1568).

The filtrate drains to internal conduits and collects at the end of the press for discharge. The plates are then separated and the dewatered cake is removed. The filtrate is returned to the influent of the WWTP. The filtration cycle time varies from 2–5 hours and includes the following steps (Metcalf & Eddy (2003), p. 1568):

1. fill the press;
2. maintain the press under pressure;
3. open the press;
4. wash and discharge the cake; and
5. close the press.

The cross-sectional view of a fixed-volume recessed plate filter assembly is presented in Figure 7.69. Table 7.36 lists the expected performance for fixed-volume, recessed plate pressure filters.

7.11.7.5 Variable-Volume Recessed Chamber

The variable-volume recessed chamber filter press is similar to the fixed-volume recessed plate filter, except that a flexible,

Table 7.36 Expected dewatering performance for fixed-volume recessed plate pressure filters.

Type of sludge	Feed solids (%)	$FeCl_3$ dose (lb/ton dry solids)	CaO dose (lb/ton dry solids)	Ash dose (lb/ton dry solids)	Cake solids with chemicals (%)	Cake solids without chemicals (%)
Raw primary (P)	5–10	100	200		45	39
Raw P	5–10			2,000	50	25
Raw P with < 50% WAS	3–6	100	200		45	39
Raw P with < 50% WAS	3–6			3,000	50	20
Raw P with > 50% WAS	1–4	120	240		45	38
Raw P with > 50% WAS	1–4			4,000	50	17
Anaerobically digested mixture of P and WAS < 50% WAS	6–10	100	200		45	39
Anaerobically digested mixture of P and WAS < 50% WAS	6–10			2,000	50	25
Anaerobically digested mixture of P and WAS > 50% WAS	2–6	150	300		45	37
Anaerobically digested mixture of P and WAS > 50% WAS	2–6			4,000	50	17
WAS	1–5	150	300		45	37
WAS	1–5			5,000	50	14

P = primary sludge; WAS = waste activated sludge
Source: U.S. EPA (1979) pg. 9–56. United States Environmental Protection Agency.

Table 7.37 Typical dewatering performance for variable-volume recessed plate pressure filters.

Type of sludge	Feed solids (%)	$FeCl_3$ dose (lb/ton dry solids)	CaO dose (lb/ton dry solids)	Cake solids with chemicals (%)	Cake solids without chemicals (%)
Anaerobically digested (AND): 60% P and 40% WAS	3.8	120	320	37	30
AND: 60% P and 40% WAS	3.2	180	580	36	25
AND: 40% P and 60% WAS	3.8	120	340	40	32
AND: 40% P and 60% WAS	2.5	180	500	42	30
AND: 50% P and 50% WAS	6.4	80	220	45	39
AND: 60% P and 40% WAS	3.6	160	320	50	40
Raw WAS	4.3	180	460	34	25
Raw 60% P and 40% WAS	4.0	100	300	40	33
Thermal conditioned: 50% P and 50% WAS	14.0	0	0	60	60

P = primary sludge; WAS = waste activated sludge.
Source: U.S. EPA (1979) pg. 9–57. United States Environmental Protection Agency.

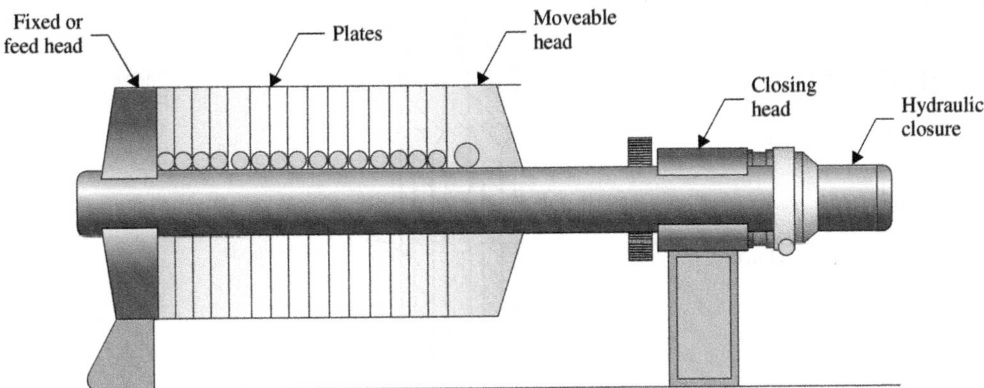

Figure 7.68 Schematic of a recessed plate pressure filter. *Source:* U.S. EPA (1979) pg. 9–53. United States Environmental Protection Agency.

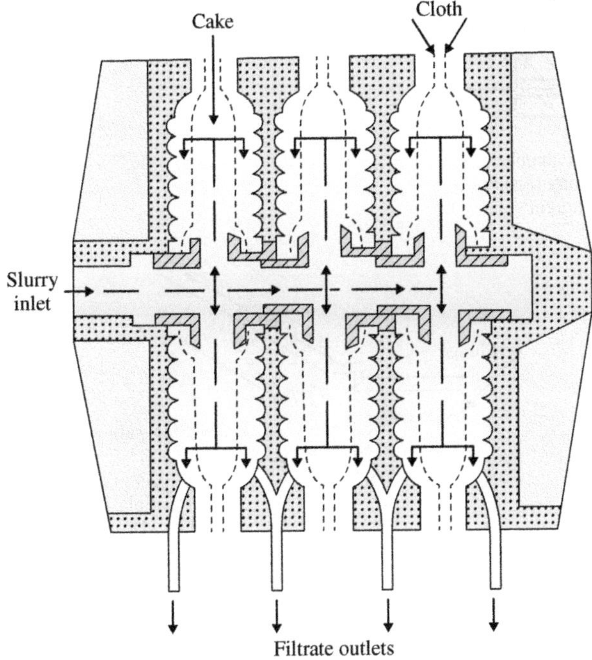

Figure 7.69 Cross-sectional view of fixed-volume recessed plate filter assembly. *Source:* U. S. EPA (1979) pg. 9–53. United States Environmental Protection Agency.

rubber diaphragm is placed between the cloth filter media and recessed plate. By expanding the diaphragm, pressure is created further reducing the water content of the sludge. According to Metcalf & Eddy (2003, p. 1569), approximately 10–20 minutes are required to fill the press, and an additional 20–30 minutes of constant pressure are required to achieve the desired solids concentration in the dewatered sludge.

During the initial dewatering stage, the pressure applied to the filter chamber ranges from 100–125 lb_f/in^2gauge (690–860 kPa). The pressure necessary to achieve final compression of the cake varies from 200–300 lb_f/in^2gauge (1380–2070 kPa). This type of press requires a considerable amount of maintenance (Metcalf & Eddy, 2003, p. 1569). Cross-sectional views of a variable-volume recessed plate filter are provided in Figure 7.70. Typical dewatering performance for variable-volume recessed plate pressure filters is presented in Table 7.37.

7.11.8 Sludge drying beds

Sludge drying beds are the most widely used methods of dewatering sludge in the USA. (Metcalf & Eddy (2003), p. 1570; US EPA (1987b), pp. 135–142). Dewatering of sludge in drying beds is accomplished by two mechanisms. The main mechanism for removing water is through drainage and the secondary mechanism is through evaporation.

The major advantages of using drying beds include:

1 low cost;
2 less operator attention;
3 low energy consumption; and
4 high cake solids concentration.

Major disadvantages to sludge drying beds include:

1 large land requirements;
2 impact of climate on performance;
3 labor-intensive sludge removal; and
4 real and perceived odor and visual nuisances.

Five types of drying beds that have been used for dewatering sludge are:

1 Conventional sand beds.
2 Vacuum-assisted drying beds (VADB).
3 Paved beds.
4 Wedgewire beds.
5 Solar beds.

Detailed information on these systems may be found in the following references: Metcalf & Eddy (2003, pp. 1570–1579); WEF (1998c, pp. 21–89 to 21–114); US EPA (1987b, pp. 135 to 142); US EPA (1979, pp. 9-1 to 9-14).

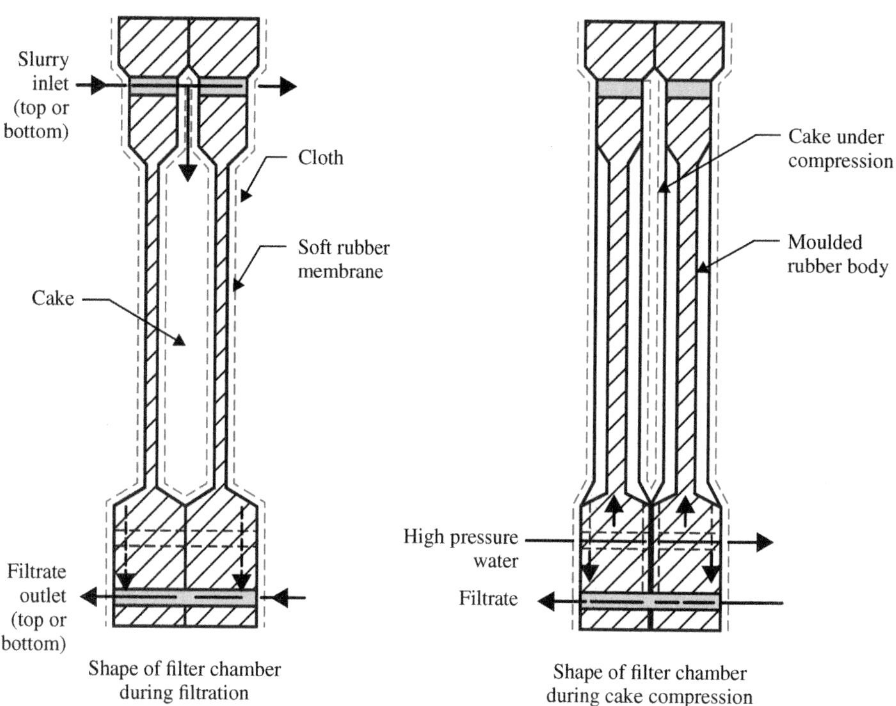

Figure 7.70 Cross-sectional views of a variable volume recessed plate filter. *Source:* U. S. EPA (1979) pg. 9–55. United States Environmental Protection Agency.

7.11.8.1 Sand Drying Beds

Conventional sand beds have been traditionally used for small- and medium-sized communities. At small facilities, it is not unusual to enclose the sand beds within a greenhouse to enhance dewatering during inclement weather and provide year-round operation. Typical sand beds used for dewatering sludge are rectangular, with dimensions of 20 ft wide by 20 - 100 ft long (6 m wide by 6 - 30 m long). Drying beds are normally sized to accommodate the quantity of sludge that is removed from digesters in one drawing.

The filter bed material consists of a layer of sand 9 - 12 inches (230 - 300 mm) deep, overlying an 8 - 18 inches (20 - 46 cm) layer of graded gravel (Metcalf & Eddy(2003), pp. 1570 - 1573; US EPA (1979), p. 9-5). An under-drain system beneath the gravel collects the filtrate. A cross-sectional view of a typical sand bed is depicted in Figure 7.71.

Stabilized sludge is normally applied so that an 8 - 12 in (200 - 300 mm) layer forms on the sand. the sludge is then allowed to dry, to form a dewatered cake which has a black to dark-brown color. The final solids concentration of the dewatered sludge is typically between 30 - 40% solids (Peavy *et al.*, 1985, p. 293; Reynolds & Richards, 1996, p. 653) in two weeks of drying under favorable conditions.

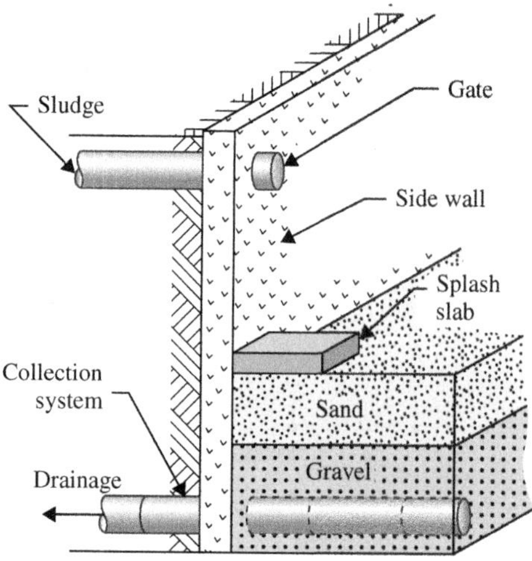

Figure 7.71 Cross-sectional view of typical sand drying bed. *Source:* U.S. EPA (1979) pg. 9–6. United States Environmental Protection Agency.

7.11.9 Sludge reuse and disposal

The sludge or residuals that are generated during wastewater treatment must be properly disposed of. Recall that sewage sludge that has been stabilized and meets federal and state standards is referred to as "biosolids." As required by the Clean Water Act Amendments of 1987, the US Environmental Protection Agency (EPA) developed a new sludge regulation to protect the public health and the environment from any potential adverse effects of certain pollutants that might be present.

This regulation, entitled "The Standards for Use or Disposal of Sewage Sludge" (Title 40 of the Code of Federal Regulations (CFR), Part 503), was published in the Federal Register (58 FR248 to 9404) in February 19, 1993 (US EPA, 1994, p. 1) and became effective on March 22, 1993. The Part 503 rule established requirements for the final use or disposal of biosolids that are:

1 applied to land to condition the soil or fertilize crops or other vegetation;
2 placed on a surface site for final disposal; or
3 fired in a biosolids incinerator.

Ultimate disposal of biosolids and residuals such as screenings and grit is typically handled by burial in a landfill or incinerated if land application is not selected or applicable. Other disposal methods used include dedicated land disposal and ocean dumping.

7.11.9.1 Land Application

Land application of biosolids is considered a beneficial use by the US EPA and is encouraged because biosolids contain beneficial plant nutrients (e.g., nitrogen, phosphorus, and potassium) and soil conditioning properties. Bastain (1997) reported that 54% of the sewage sludge generated in the US was applied to land in 1995. Biosolids that are applied to land must meet specific concentrations for ten metals, pathogen densities, and requirements for vector reduction. If these standards are not met, then land application of the biosolids is not allowed. In such cases, they are typically co-disposed with municipal solid waste in a sanitary landfill. Subpart B of the Part 503 regulation covers the requirements for land application of biosolids. The requirements for placing biosolids on a surface for final disposal are presented in Subpart C. Requirements for pathogen reduction and vector attraction reduction (VAR) are covered in Subpart D. Specific requirements for each of these are presented below.

Table 7.38 Alternatives for meeting Class A pathogen requirements.

Name of Alternative	Description
Alternative 1: Thermally treated sewage sludge for time and temperature.	There are four time-temperature regimes. See Table 7.39.
Alternative 2: Sludge treated in a high pH-high temperature process.	Elevate and maintain at pH > 12 for 72 hr and temperature above 52°C for at least 12 hr, air dry to 50% after the 72 hr period at elevated pH.
Alternative 3: Sewage sludge treated in other processes.	Density of fecal coliform < 1,000 MPN per g of TS or *Salmonella* sp. < 3 MPN per g TS at the time of disposal. Density of enteric viruses after pathogen treatment < 1 PFU per 4 grams of TS. Density of viable helminth ova in the sewage sludge after pathogen treatment < 1 per 4 grams of TS.
Alternative 4: Sewage sludge treated in unknown processes.	Density of fecal coliform < 1,000 MPN per g of TS or *Salmonella* sp. < 3 MPN per g TS at the time of disposal. Density of enteric viruses after pathogen treatment < 1 PFU per 4 grams of TS. Density of viable helminth ova in the sewage sludge after pathogen treatment < 1 per 4 grams of TS.
Alternative 5: Use of process to further reduce pathogens (PFRP). See Table 7.40.	Density of fecal coliform < 1,000 MPN per g of TS or *Salmonella* sp. < 3 MPN per g TS at the time of disposal. PFRP processes: composting, heat drying, heat treatment, thermophilic aerobic digestion, beta ray irradiation, gamma ray irradiation, and pasteurization.
Alternative 6: Use of process equivalent to PFRP.	Density of fecal coliform < 1,000 MPN per g of TS or *Salmonella* sp. < 3 MPN per g TS at the time of disposal. Use of a process that is equivalent to PFRP.

Source: U.S. EPA (2003a) pp. 28 to 33. United States Environmental Protection Agency.

Pathogen Reduction Requirements Pathogen reduction requirements were implemented to reduce pathogen levels in sewage sludge, including pathogenic bacteria, enteric viruses, protozoa, viable helminth ova, and other disease causing organisms. The requirements are divided into two categories: Class A and Class B.

Class A Pathogen Requirements Class A requirements must be met if:

1 biosolids are sold or given away in a bag or other container for application to land;

2 bulk biosolids are applied to a lawn or home garden;

3 bulk biosolids are applied to other types of land and if site restrictions are not met.

Since the goal of Class A pathogen reduction requirements is to reduce the levels of pathogens to below detectable levels, six alternative processing methods are available. The six alternative sludge processing methods are presented in Table 7.38.

Table 7.39 Four time-temperature regimes for alternative 1 to meet Class A pathogen requirements.

Regime	Sludge type	Required time-temperature relationship
A	Sewage sludge with at least 7% solids (except those covered by Regime B)	$D = 131{,}700{,}000/10^{0.1400 \cdot t}$ $t \geq 50°C$ $D \geq 0.0139$ d
B	Sewage sludge with at least 7% solids that are small particles heated by contact with either warmed gases or an immiscible liquid	$D = 131{,}700{,}000/10^{0.1400 \cdot t}$ $t \geq 50°C$ $D \geq 1.74 \times 10^4$ d
C	Sewage sludge with less than 7% solids treated in processes with less than 30 minutes contact time	$D = 131{,}700{,}000/10^{0.1400 \cdot t}$ 1.74×10^4 d $\leq D \leq 0.021$ d
D	Sewage sludge with less than 7% solids treated in processes with at least 30 minutes contact time	$D = 50{,}070{,}000/10^{0.1400 \cdot t}$ $t \geq 50°C$ $D \geq 0.021$ d

D = time in days; t = temperature in °C

Source: U.S. EPA (2003a) pg. 29. United States Environmental Protection Agency.

Table 7.40 Processes to further reduce pathogens (PFRP).

Name of Process	Description
1 Composting	Using either the within-vessel or static aerated pile composting method, the temperature of the sewage sludge is maintained at 55°C or higher for 3 days. Using the windrow composting method, the temperature of the sewage sludge is maintained at 55°C or higher for 15 days or longer. During the period when the compost is maintained at 55°C or higher, there shall be a minimum of five turnings of the windrow.
2 Heat drying	Sewage sludge is dried by direct or indirect contact with hot gases to reduce the moisture content to 10% or lower. Either the temperature of the sewage sludge particles exceeds 80°C or the wet bulb temperature of the gas in contact with the sludge as it leaves the dryer exceeds 80°C. Flash dryers, spray dryers, rotary dryers, and the Carver-Greenfield Process are commonly used for heat drying sewage sludge.
3 Heat treatment	Liquid sewage sludge is heated to a temperature of 180°C or higher for 30 minutes. The Porteous and Zimpro processes have been used for heat treatment of sewage sludge.
4 Thermophilic aerobic digestion	Liquid sewage sludge is agitated with air or oxygen to maintain aerobic conditions and the mean cell residence time of the sewage sludge is 10 days at 55°C to 60°C.
5 Beta ray radiation	Sewage sludge is irradiated with beta rays from an accelerator at dosages of at least 1.0 megarad at room temperature (ca. 20°C).
6 Gamma ray radiation	Sewage sludge is irradiated with gamma rays from certain isotopes, such as Cobalt 60 and Cesium 137 (at dosages of at least 1.0 megarad) at room temperature (ca. 20°C).
7 Pasteurization	The temperature of the sewage sludge is maintained at 70°C or higher for 30 minutes or longer.

Source: U.S. EPA (1994, pg. 116), U.S. EPA (2003a) pp. 51 to 57. United States Environmental Protection Agency.

The four time-temperature regimes that may be used to meet Alternative 1 for meeting Class A Pathogen Requirements are presented in Table 7.39. Table 7.40 lists and describes processes to further reduce pathogens (PFRP).

In addition to using one of the alternative processes, Class A biosolids must meet one of the following requirements at the time when the biosolids are ready to be used or disposed:

1. the density of fecal coliforms in the sewage sludge must be less than 100 most probable number (MPN) per gram of total solids (TS) on a dry weight basis; or
2. the density of *Salmonella* species bacteria in the sewage sludge must be less than 3 MPN per 4 grams of TS on a dry weight basis.

Class B Pathogen Requirements Biosolids that are applied to agricultural land, a forest, a public contact site, or a reclamation site must meet Class B requirements. Class B biosolids cannot be applied to a lawn or home garden. Application of Class B biosolids must be conducted in compliance with site restrictions. Class B pathogen requirements can be met in three different ways:

- Alternative 1: Seven samples of biosolids are collected, and the geometric mean of these samples must be less than 2×10^6 colony forming units (CFUs) or MPN per gram of TS on a dry weight basis.
- Alternative 2: Biosolids that have been treated in one of the "Processes to Significantly Reduce Pathogens" (PSRP) which are listed in Table 7.41.

Table 7.41 Processes to significantly reduce pathogens (PSRP).

Name of Process	Description
Aerobic digestion	Aerobically digest for 40 days at 20°C or 60 days at 15°C.
Air drying	Sewage sludge is dried on sand beds or paved or unpaved basins for a minimum of three months. Ambient temperature must be above 0°C for two of the months.
Anaerobic digestion	Anaerobically digest between 15 days at 35–55°C or 60 days at 20°C.
Composting	Using either within-vessel, static aerated pile, or windrow composting methods, the temperature of the sewage sludge is raised to 40°C or higher and remains at 40°C or higher for five days. For four hours during the five-day period, the temperature in the compost pile exceeds 55°C.
Lime stabilization	Sufficient lime is added to raise the pH of the sewage sludge to 12 for ≥ 2 hours of contact.
Use of processes equivalent to PSRP	Processes must be approved by EPA's Pathogen Equivalency Committee.

Source: U.S. EPA (2003a) pp. 43 to 50. United States Environmental Protection Agency.

Table 7.42 Options for meeting vector attraction reduction criteria.

Name of option
Option 1: 38% reduction in volatile solids (VS) during sewage sludge treatment.
Option 2: Demonstrate < 17% additional VS reduction in bench-scale anaerobic digestion of sludge for 40 additional days at 30–37°C.
Option 3: Demonstrate < 15% additional VS reduction in bench-scale aerobic digestion of sludge for 30 additional days at 20°C.
Option 4: Meet a specific oxygen uptake (SOUR) rate ≤ 1.5 mg oxygen per hr per gram of TS.
Option 5: Aerobic treatment of sewage sludge for at least 14 days at over 40°C with an average temperature of 45°C.
Option 6: Addition of alkali sufficient to raise pH of sludge to at least 12 at 25°C and maintain at pH ≥ 12 for two hours and a pH of ≥ 11.5 for 22 more hours.
Option 7: Percent of dry solids ≥ 75% prior to mixing with other materials. Includes sludges that contain no unstabilized solids.
Option 8: Percent of dry solids ≥ 90% prior to mixing with other materials. Includes sludges with unstabilized solids.
Option 9: Sewage sludge that is injected into the soil so that no significant sludge is present after one hour of injection. Class A sludge must be injected within 8 hr after the pathogen reduction process.
Option 10: Sewage sludge is incorporated into the soil within 6 hr after application to land or placement on a surface disposal site. Class A must be applied or placed on the land surface within eight hours after the pathogen reduction process. Also, domestic septage applied to agricultural land, forest, reclamation site, or surface disposal site is included.
Option 11: Sewage sludge placed on a surface disposal site must be covered with soil or other material at the end of each operating day (includes septage).
Option 12: The pH of domestic septage must be raised to ≥ 12 at 25°C by alkali addition and maintained ≥ 12 for 30 minutes without adding more alkali.

Source: U.S. EPA (1994, pg. 121), U.S. EPA (2003a) pp. 58 to 64. United States Environmental Protection Agency.

Table 7.43 Management practices for surface disposal sites.

Type of management practice
Biosolids placed on a disposal unit must not harm or threaten endangered species.
The active biosolids unit must not restrict base flood flow.
The active biosolids unit must be located in a geologically stable area.
The active biosolids unit cannot be located in wetlands.
Runoff must be collected from the surface disposal site with a system capable of handling a 25-year, 24-hour storm event.
Only where there is a liner must leachate be collected and must the owner/operator maintain and operate a leachate collection system.
Only where there is a cover, must there be limits on concentrations of methane gas in air in any structure on the site and in air at the property line of the surface disposal site.
The owner/operator cannot grow crops on site (unless allowed by the permitting authority).
The owner/operator cannot graze animals on site (unless allowed by the permitting authority).
The owner/operator must restrict public access.
The biosolids placed in the active biosolids unit must not contaminate an aquifer.

Source: U.S. EPA (1994, pg. 63). United States Environmental Protection Agency.

- Alternative 3: Biosolids that have been treated by a process determined to be equivalent to PSRP. The regulatory authority has responsibility for determining process equivalency.

Vector Attraction Reduction (VAR) Requirements
Vector reduction requirements were developed to reduce the potential of the biosolids to attract vectors such as birds, rodents, insects, and other organisms that can transport pathogens. The Subpart B requirements provide twelve options to demonstrate vector attraction reduction. Table 7.42 summarizes the twelve options that are available.

7.11.9.2 Incineration

Requirements for incineration of biosolids only are delineated in Subpart E of the 503 regulation. Incineration of biosolids is actually a volume reduction method, since the remaining ash must be disposed of. Incinerators are increasingly becoming more difficult to permit by regulatory agencies, because of stringent air pollution requirements. Rotary multiple hearth incinerators and fluidized bed furnaces are used for incinerating biosolids. A dewatered sludge with a solids content of approximately 30–40% is necessary to maintain self-sustaining combustions (e.g., no auxiliary fuel necessary).

Pollutant limits for seven metals – arsenic, beryllium, cadmium, chromium, lead, mercury, and nickel – were established for sewage sludge fired in a biosolids incinerator. The monthly average concentration for total hydrocarbon (THC) emissions as propane (corrected for 0% moisture and 7% oxygen) may not exceed 100 ppm on a volumetric basis (US EPA, 2003b).

7.11.9.3 Sanitary Landfill

Screenings, grit, and ash are typically co-disposed with municipal solids waste in a sanitary landfill. Most municipalities require that biosolids must have a solids concentration of approximately 20–25% to be accepted for disposal in a landfill.

Monofills, which are sludge-only landfills, are regulated under Subpart C of the rule. Management practices for surface disposal of biosolids which include monofills are outlined in Table 7.43.

7.11.9.4 Ocean Dumping

Pumping biosolids through pipelines for disposal in the ocean or barging them out to sea for dumping is no longer practiced in the United States. In 1988, Congress passed the Federal Ocean Dumping Ban Act, which was implemented December 31, 1991 (Liptak (1991), pp. 60–67). Prior to that time, cities such as New York and Boston disposed their biosolids in the ocean. Boston stopped disposing biosolids in 1991 (DeCocq *et al.* (1998), p. 2) and New York in 1992 (US EPA (1992), p. 1).

Summary

- Wastewater treatment plants (WWTPs) consist of a series of unit operations and unit processes for treating the liquid wastewater and the residuals or sludge that is generated during treatment.

- The design engineer must know the influent wastewater characteristics and effluent requirements in order to select and design the proper unit operations and processes.

- Unit operations involve some type of physical treatment.
  - Major unit operations discussed in this chapter in sequential order include: screens; grit removal; primary sedimentation; aeration; secondary clarification; sludge thickening; and sludge dewatering.

- Unit processes involve biological or chemical reactions.
  - Major unit processes discussed in this chapter include the activated sludge process and associated modifications, sequencing batch reactors, and trickling filters.

- The capacity of a WWTP is based on both a hydraulic flow and process loading.
  - The design of unit operations such as screening, grit removal, sedimentation, and disinfection is typically based on treating the peak hourly flow (PHF).
  - Unit processes such as activated sludge, BNR processes, RBCs, and trickling filters are typically designed to treat the maximum month organic loading.

- Preliminary treatment of wastewater consists of bar racks, screens, and grit removal. These unit operations remove debris, trash, plastics, tree limbs, gravel, sand and other coarse materials that may accumulate in settling basins and digesters, or that can damage equipment such as pumps.

- The design of a primary clarifier is based on overflow rate and detention time.

- Biological processes, either suspended-growth or attached-growth systems, use a heterogeneous culture of microorganism to treat wastewater.
  - Biological wastewater treatment is the removal of organics and nutrients from wastewater by microorganisms, which transform them into new microorganisms, carbon dioxide, water, and other end products.
  - Organic matter measured as BOD, COD, or TOC is primarily used by heterotrophic microbes as a carbon and energy source.
  - Nutrients such as nitrogen and phosphorus are used in synthesizing new biomass.
  - Autotrophic organisms such as the genera *Nitrosomonas* and *Nitrobacter* oxidize ammonium to nitrite and, ultimately, to nitrate.

- Suspended-growth biological processes include the following activated sludge processes: conventional or plug flow; complete-mix; extended aeration; oxidation ditch; tapered aeration; high-purity oxygen activated sludge; and contact stabilization. Sequencing batch reactors (SBRS), which are fill-and-draw reactors, were discussed.

- The design of mechanical and diffused aeration systems was presented.
  - Biological processes must have sufficient oxygen present to keep the biomass alive, and enough energy imparted to the wastewater so that the microorganisms remain in suspension.

- Attached-growth biological processes presented include trickling filters and rotating biological contactors.

- Biological processes are only as good as the design of the secondary clarification system. Biomass produced from biological processes must undergo clarification and thickening in a secondary clarifier.

- The design of a secondary clarifier is based on the overflow rate and solids loading rate. The area of the clarifier is calculated using each of the above design parameters, with the one yielding the largest area being used for sizing the clarifier.

- Biological nutrient removal (BNR) systems typically use single-sludge processes or staged-reactors with engineered environments, such that specific types of microorganisms are grown to accomplish either nitrogen removal or phosphorus removal or both.
  - Biological nitrogen removal systems use the nitrification process and the denitrification process.
  - Nitrification is an aerobic process carried out by autotrophic organisms, collectively known as the "nitrifiers."
  - Denitrification is an anaerobic process in which, heterotrophic organisms oxidize organic compounds to carbon dioxide and water, with nitrate serving as the electron acceptor rather than oxygen.
  - BNR systems for nitrogen removal include: Wuhrmann process, Ludzack-Ettinger process, modified Ludzack-Ettinger (MLE) process, and 4-stage Bardenpho process.
  - Enhanced biological phosphorus removal (EBPR) systems are those that are configured to have alternating anaerobic-aerobic treatment. The anaerobic zone is required for the fermentation of the incoming wastewater for the proliferation of phosphate accumulating organisms (PAOs) such as the genus, *Acinetobacter*. PAOs are capable of taking up excess quantities of phosphorus from the wastewater, resulting in a phosphorus content of 7–14% on a dry weight basis.
  - Mainstream biological phosphorus removal systems discussed were anaerobic/oxic (A/O) and the PhoStrip process.
  - Combined BNR processes for nitrogen and phosphorus removal presented include: anaerobic/anoxic/oxic (A^2/O), Virginian Initiative Plant (VIP), University of Cape Town (UCT), and 5-stage Bardenpho process.

- Disinfection of the treated effluent is required to kill any remaining pathogens before the effluent is discharged or reused.
  - Primary disinfectants include chlorine, chlorine dioxide, chloramines, and ultraviolet radiation.

- Sludge or residuals produced during wastewater treatment must be handled, treated, and ultimately disposed of.

- A sludge treatment train typically consists of thickening, stabilization, dewatering, and disposal.

- Sludge thickening operations and processes include: gravity thickening, gravity belt thickeners (GBTs), and thickening centrifuges.

- Sludge stabilization systems are necessary for reducing the quantity of sludge, killing pathogens, and oxidizing the remaining organics.
  - Three types of sludge stabilization systems include aerobic digestion, anaerobic digestion, and lime stabilization.

- Sludge conditioning involves adding chemicals such as lime, ferric chloride, or polymers to enhance the removal of water during the dewatering process.

- Dewatering operations are required to increase the solids content beyond that accomplished during thickening.
  - Sludge dewatering processes discussed were belt filter presses (BFP), dewatering centrifuges, and sand drying beds.

- Ultimate disposal of the residuals or biosolids is accomplished primarily by application to agricultural land or burial in a sanitary landfill.

Key Words

α
activated sludge
ADF
aerobic digestion
aerated grit chamber
aeration basin
anaerobic digestion
anaerobic zone
anoxic zone
AOTR
AWT
β
bar rack
bar screens
Bardenpho process
batch reactor
belt-filter press
biokinetic coefficient determination
biokinetic equations
biological nutrient removal
biomass
biosolids
biotower
centrifuge
Chick's law
chlorine contact basin
CMAS
complete-mix
compression settling
contact stabilization
contact time
conventional activated sludge
$C_r \cdot t$
decant phase
dechlorination
denitrification
design flows
design loadings
detention time
dewatering
diffused aeration
discrete settling
disinfection
drying beds
endogenous decay
extended aeration
F/M
fill phase
flocculent settling
decant phase
gravity belt thickener
half-velocity coefficients
high-purity oxygen activated sludge
horizontal-flow grit chamber
humus
idle phase
$K_L a$
land application of sludge
lime stabilization
Ludzack-Ettinger process
gravity belt thickener
mean cell residence time
mechanical aerators
Michaelis-Menten kinetics
microorganism concentration
mixing requirements
moisture content
Monod kinetics
nitrification
nitrogenous oxygen demand
NPDES permits
ocean dumping
OUR
overflow rate
oxic zone
oxidation ditch
oxygen requirements
oxygen transfer
oxygen use coefficients
PAOs

pathogen reduction criteria
pathogens
peaking factors
PFRP
PHF
Pista grit
plate and frame press
plug flow
post-denitrification
power requirements
pre-denitrification
preliminary treatment
primary clarifier
primary sedimentation
PSRP
θ
RAS
react phase
rotating biological contactors
screening
secondary clarifier
sequencing batch reactors
settle phase
single-sludge
sludge production
solid flux theory
solids content
solids flux
solids loading rate
SOTR
SOUR
specific substrate utilization rate
stabilization
step feed activated sludge
sterilization
Stokes' law
SVI
$TBOD_5$
Teacup
temperature correction
thickening
TMDLs
trickling filters
two-film theory
unit operation
unit process
vector attraction reduction criteria
VFA
VIP process
vortex grit removal systems
WAS
wastewater characteristics
weir loading rate
yield coefficient
zone settling

References

Ardern, E., Lockett. W.T. (1914). Experiments on the Oxidation of Sewage without the Aid of Filters. *Journal Society Chemical Industry* **22**, 523–539, 1122–1124.

Asano, T., Burton, F.L., Leverenze, H.L., Tsuchihashi, R., Tchobanoglous, G. (2007). *Water Reuse: Issues, Technologies, and Applications*, pg. 604. McGraw Hill, New York, NY.

Balarishnan, S., Eckenfelder, Jr., W.W., Brown, C. (1969). Organics Removal by a Selected Trickling Filter Media. *Water and Wastes Engineering* **6**(1), A22–A25.

Bastain, R.K. (1997). The Biosolids (Sludge) Treatment, Beneficial Use, and Disposal Situation in the USA. *European Water Pollution Control Journal* **7**(2), 62–79.

Benefield, L.D., Randall, C.W., King, P.H. (1975). *Temperature Considerations in the Design and Control of Completely-Mixed Activated Sludge Plants.* Paper presented at 2nd Annual National Conference on Environmental Engineering Research, Development and Design, ASCE, University of Florida, Gainesville, FL.

Benefield, L.D., Randall, C.W. (1980). *Biological Process Design for Wastewater Treatment*, pp. 190, 490. Prentice Hall, Inc., Englewood Cliffs, NJ.

Bermard, J.L. (1975). Biological Nutrient Removal without the Addition of Chemicals, *Water Research* **9**, 485–490.

Burdick, C. R., Refling, D.R. and Stensel, H.D., and H. D. Stensel (1982). Advanced Biological Treatment to Achieve Nutrient Removal. *J. Water Pollution Control Federation*, **54**(7): 1078–1086.

DeCocq, J., Gray, K., Churchill, R. (1988). Sewage Sludge Pelletization in Boston: Moving Up the Pollution Prevention Hierarchy. *National Pollution Prevention Center for Higher Education.* October 1988, p. 2.

Dick, R.I. (1970). Role of Activated Sludge Final Settling Tanks, *Journal Sanitary Engineering Division.* **96**, SAE2, 423.

Droste, R.L. (1997). *Theory and Practice of Water and Wastewater Treatment*, pp. 550–551. John Wiley and Sons, Inc., New York, NY.

Eckenfelder, Jr., W.W.. (1960). Trickling Filter Design and Performance, *Journal Sanitary Engineering Division* **87**(SA6), 87.

Eckenfelder, Jr., W.W. (1989). *Industrial Water Pollution Control* pp. 239–245. McGraw-Hill, New York, NY.

Eckenfelder, Jr., W.W. Ford, D.L., Englande, Jr., A. (2009). *Industrial Water Quality.* McGraw Hill, New York, NY.

Fair, G. M., Geyer, J. C., and Okun, D. A. (1968a). *Water and Wastewater Engineering, Volume 1: Water Supply and Wastewater Removal*, pg. 5–24. John Wiley & Sons, Inc., New York, NY.

Fair, G. M., Geyer, J. C., and Okun, D. A. (1968b). *Water and Wastewater Engineering, Volume 2: Water Purification and Wastewater Treatment and Disposal*, John Wiley & Sons, Inc., New York, pg. 31–26.

Federal Register (1988). 40 CFR Part 133, Secondary Treatment Regulations.

Federal Register (1989). 40 CFR Part 133, Amendments to the Secondary Treatment Regulations: Percent Removal Requirements During Dry Weather Periods for Treatment Works Served by Combined Sewers.

Germain, J.E. (1966) Economical Treatment of Domestic Waste by Plastic-Medium Trickling Filters. *J Water Pollution Control Federation* **38**(2), 192–203.

Grady, C.P.L., Daigger, G.T., Lim, H.D. (1999). *Biological Wastewater Treatment*, 2nd Edition, pg. 927. Marcel Deker, New York, NY.

Grady, C.P. Lesle, D., Glen, T., Love, N.G., Felipe, C.D.M. (2011). *Biological Wastewater Treatment*, pg. 202, 383, 410, 3rd Edition. IWA Publishing, London, UK.

Grau, P. Dohanyos, M., Chudoba, J. (1975). Kinetics of Multicomponent Substrate Removal by Activated Sludge. *Water Research* **9**, 637–642.

Hammer, Mark J. (1986). *Water and Wastewater Technology, SI Version*, pp. 324, 454. John Wiley & Sons, New York, NY.

Henze, M., Van Loosdrecht, M. C. M., Ekama, G. A. and Brdjanovic, D. (2008). *Biological Wastewater Treatment: Principles, Modeling and Design*, IWA Publishing, London, UK, pp. 58, 90.

Heukelekian, H., Oxford, H.E., Manganelli, R. (1951). Factors Affecting the Quantity of Sludge Production in the Activated Sludge Process. *Sewage and Industrial Wastes* **23**, 945.

Hoover, S.R., Porges, N. (1952). Assimilation of Dairy Wastes by Activated Sludge, *Sewage and Industrial Wastes* **24**, 306–312.

Horan, N. J. (1990a). *Biological Wastewater Treatment Systems: Theory and Operation*, p. 27. John Wiley & Sons, Ltd, West Sussex, England.

Houghtalen, R.J., Akan, A.O., Hwang, N.H.C. (2010). *Fundamentals of Hydraulic Engineering Systems*, 4th Edition, pg. 6. Prentice Hall, Upper Saddle River, NJ.

Humphries, S. (2006). *Presentation on Sequencing Batch Reactor (SBR) Technology given to the EVE 405 Design and Analysis of Wastewater Systems class*, Mercer University, April 6, 2006.

Kornegay, B.H. (1975). Modeling and Simulation of Fixed Film Biological Reactors. In Keinath, T.M., Wanielista, M. (eds.) *Mathematical Modeling of Water Pollution Control Processes.* Ann Arbor Science Publications Inc., Ann Arbor, Michigan.

Lawrence, A.W., McCarty, P.L. (1970). Unified Basis for Biological Treatment and Design and Operation. *Journal of Sanitary Engineering Division, ASCE* **96**, 757–778.

Lewis, W.K., Whitman, W.C. (1924). Principles of Gas Adsorption. *Journal Industrial and Engineering Chemistry* **16**, 1215–1220.

Lin, S.D. (2007). *Water and Wastewater Calculations.* McGraw-Hill, New York, NY.

Liptak, B.G. (1991). *Municipal Waste Disposal in the 1990s*, pp. 60–67. Chilton Book Company, Radnor, PA.

McCarty, P.L. (1970). Phosphorus and Nitrogen Removal by Biological Systems, *Proceedings, Wastewater Reclamation and Reuse Workshop*, Lake Tahoe, CA, June 25–27, 1970.

McGhee, T.J. (1991). *Water Supply and Sewerage*, pp. 421–422. McGraw-Hill, Inc., New York, NY.

Metcalf & Eddy (1991). *Wastewater Engineering: Treatment, Disposal, and Reuse*, pp. 344, 394, 458, 495, 502, 550, 701. McGraw Hill, New York, NY.

Metcalf & Eddy (2003). *Wastewater Engineering: Treatment and Reuse*, pp. 165, 186, 195, 315–396, 389, 393, 398, 407, 566, 584, 585, 612, 614, 615, 704, 705, 720–723, 747, 802, 814, 913, 918, 931, 1220, 1262–1263, 1298, 1456, 1491–1492, 1496, 1497, 1536, 1554–1558, 1561, 1562, 1566, 1568, 1569, 1570-1579, McGraw Hill, New York, NY.

Mihelcic, J.R. (1999). *Fundamentals of Environmental Engineering*, pg. 235. John Wiley and Sons, Inc., Hoboken, NJ.

Mines, R.O. and Sherrard, J.H. (1983). Activated Sludge Treatment of A High Strength Nitrogenous Waste. *Proceedings, 40th Industrial Waste Conference*, Purdue University, 837–846.

Mines, R.O., Smith, D G., Dahl, B.W. Holcomb, S.P. (1992). Bionutrient Removal Using a Modified Pure Oxygen Process. *Paper presented at 65th Annual Conference Water Pollution Control Federation*, New Orleans, LA.

Mines, Jr., R.O. (1996). Assessment of AWT Systems in Tampa Bay Area. *Journal of Environmental Engineering* **122**(7), 605–611.

Monod, J. (1949). The Growth of Bacterial Cultures. *Annual Review of Microbiology* **3**, 371–394.

NRC (1946). National Research Council 'Trickling Filters in Sewage Treatment at Military Installations'. *Sewage Works Journal* **18**(5), 787–982.

Painter, H.A. (1970). A Review of Literature of Inorganic Nitrogen Metabolism in Microorganisms. *Water Research*, 4, 393-350.

Peavy, H.S., Rowe, D.R., Tchobanoglous, G. (1985). *Environmental Engineering*, pp. 233, 268, 293. McGraw-Hill, New York, NY.

Randall, C.W., Barnard, J.L, Stensel, H. D. (1992). *Design and Retrofit of Wastewater Treatment Plants for Biological Nutrient Removal*, p. 103. Technomic Publishing Company, Lancaster, PA.

Rice, R.G., Miller, G.W., Robson, C.M, Hill, A.G. (1979). Ozone Utilization in Europe, *AIChE, 8th Annual Meeting*, Houston, TX.

Reynolds, T.D., Richards, P.A. (1996). *Unit Operations and Processes in Environmental Engineering*, pp. 137–156, 239–247, 257, 258, 429, 450, 475, 511, 615, 653, 742, 748, 749. PWS Publishing Company, Boston, MA.

Rittman, B.E., McCarty, P.L. (2001). *Environmental Biotechnology: Principles and Application*, p. 473. McGraw Hill, New York, NY.

Schultze, K.L. (1960). Load and Efficiency of Trickling Filters. *Journal Water Pollution Control Federation* **32**(3), 245–261.

Sherrard, J.H., Schroeder, E.W. (1972). Relationship between the Observed Yield Cell Coefficient and Mean Cell Residence Time in the Completely Mixed Activated Sludge Process. *Water Research* **6**, 1039–1049.

Sherrard, J.H., Schroeder, E.W. (1976). Stoichiometry of Industrial Biological Wastewater Treatment. *Journal Water Pollution Control Federation* **48**(4), 742–747.

Soap and Detergent Association (1989). *Principles and Practices of Nutrient Removal for Municipal Wastewater*, pp. 153, 167, 171. New York, NY.

Standard Methods (1998). *Standard Methods for the Examination of Water and Wastewater*, p. 2–79. American Public Health Association, 1015 Fifteenth Street, NW, Washington, D.C.

Thabaraj, G.J. (1993). *AWT Processes in Florida*, 1–19. Florida Department of Environmental Regulation, Tampa, FL.

US EPA (1975a). *Process Design Manual for Nitrogen Control*, EPA/625/1-75-007, U.S. EPA, Washington, D.C.

US EPA (1975b). *Suspended Solids Removal, Process Design Manual*, pp. 7–14, 7–16. Office of Research and Development, Office of Water, Washington, D.C.

US EPA (1979). *Process Design Manual: Sludge Treatment and Disposal*, EPA/625/1–87/014, pp. 4–1, 4–8. 4–21 to 4–28, 5–6 to 5–9, 5–50, 6–19, 9–1 to 9–14, 9–24, 9–50, 8–1 to 8–33, 9–53, 9–55, 9–56, 9–57, 19–1 to 19–61. Center for Environmental Research Information, Cincinnati, OH.

US EPA (1984a). *Summary of Design Information on Rotating Biological Contactors*, p. 22. EPA/430/9-84-008/, Cincinnati, OH.

US EPA (1984b). *Design Information on Rotating Biological Contactors*, pp. 5–12. EPA-600/2-84-106, Cincinnati, OH.

US EPA (1985a). *Design Manual: Odor and Corrosion Control in Sanitary Sewerage Systems and Treatment Plants*, pp. 35–68. US Environmental Protection Agency, EPA 625/1-85/018, Cincinnati, OH.

US EPA (1987a). *Design Manual: Phosphorus Removal*, EPA/625/1-87/001, pp. 19, 21–23, 25–26, 29, 42–43. Water Research Laboratory, Cincinnati, OH.

US EPA (1987b). *Design Manual: Dewatering Municipal Wastewater Sludges*, EPA/625/1-79/011, pp. 135–142. Center for Environmental Research Information, Cincinnati, OH.

US EPA (1992). *Reilly in New York to Mark End of Sewage Sludge Dumping*. accessed from EPA website on July 7, 2012 at: http://www.epa.gov/aboutepa/history/topics/mprsa/03.html, p. 1.

US EPA (1993). *Manual Nitrogen Control*, EPA/625/R-93/010, pp. 187–188, 251, 254, 259, 261. Office of Research and Development, Office of Water, Washington, DC.

US EPA (1994). *A Plain English Guide to the EPA Part 503 Biosolids Rule*, pp. 1, 63, 116, 121. EPA/832/R-93/003, Office of Wastewater Management, Washington, DC.

US EPA (1999). *Guidance Manual Alternative Disinfectants and Oxidants*, pp. 2–26. EPA/815/R-99/014, Office of Water, Washington, D.C.

US EPA (2000). *Total Maximum Daily Load (TMDL)*. US Environmental Protection Agency, EPA 8410-F-00-009, Washington, DC.

US EPA (2003a). *Environmental Regulations and Technology: Control of Pathogens and Vector Attraction in Sewage Sludge*, pp. 28–33, 51–57, 43–50, 58–64. EPA/625/R-92-013, Center for Environmental Research Information, Cincinnati, OH.

US EPA (2003b). *Biosolids Technology Fact Sheet: Use of Incineration for Biosolids Management*, Office of Water EPA 832-F-03-013, June 2003, pg. 8. Municipal Technology Branch, Washington, DC.

Veslind, P.A. (2003). *Wastewater Treatment Plant Design*, pp. 4–12, 4–15, 4–16, 6–13, 15–12. IWA Publishing, London, UK and WEF, Alexandria, VA.

Viessman, Jr., W., Hammer, M.J. (2005). *Water Supply and Pollution Control*, p. 542. Pearson/Prentice Hall, Upper Saddle River, NJ.

Viessman, Jr., W., Hammer, M.J., Perez, E.M., Chadik, P.A. (2009). *Water Supply and Pollution Control*, 8th Edition, pp. 193, 508, 650, 683. Pearson/Prentice Hall, Upper Saddle River, NJ.

Walker Process Equipment Brochure (2009). *EnviroDisc™ Rotating Biological Contactor*. Aurora, IL.

Water Environment Federation (1992). *Design of Municipal Wastewater Treatment Plants, Manual of Practice No 8*. Alexandria, VA.

Weddle, C.L., Jenkins, D.L. (1971).The Viability and Activity of Activated Sludge. *Water Research* **5**, 621.

WEF (1998a). *Design of Municipal Wastewater Treatment Plants, Volume I: Planning and Configuration of Wastewater Treatment Plants*, 4th Edition, pg. 3–19. Water Environment Federation, Reston, VA.

WEF (1998b). *Design of Municipal Wastewater Treatment Plants, Volume II: Liquid Treatment Processes*, 4th Edition, pp. 9-1 to 9-58,

10-4 to 10-7, 11-18, 11-19, 11-27, 12-43, 12-56, 12-121, 12-122, 15-60. Water Environment Federation, Reston, VA.

WEF (1998c). *Design of Municipal Wastewater Treatment Plants, Volume III: Solids Processing and Disposal*, 4th Edition, pp. 19-1 to 19-61, 20-62, 20-82, 21-6, 21-39, 21-44, 21-51, 21-89 to 21-114. Water Environment Federation, Reston, VA.

WEF (1998d). *Biological and Chemical Systems for Nutrient Removal*, pp. 158, 210. Water Environment Federation, Alexandria, VA.

Bibliography

Cooper, C.D., Dietz, J.D., Reinhart, D.R. (1990). *Foundations of Environmental Engineering*. Waveland Press, Inc., Prospect Heights, IL.

Crites, R., Tchobanoglous, G. (1998). *Small and Decentralized Wastewater Management Systems*. McGraw-Hill, New York, NY.

Davis, M.L., Cornwell, D.A. (2008). *Introduction to Environmental Engineering*. McGraw-Hill, New York, NY.

Drury, D.D., Carmona, J., Delgadillo, A. (1986). Evaluation of High Density Cross Flow Media for Rehabilitating an Existing Trickling Filter. *Journal Water Pollution Control Federation* **58**(5), 364–367.

Eckenfelder, Jr., W.W. (1963). Trickling Filter Design and Performance. *Transactions American Society of Civil Engineers* **128** Part 3, 371.

Eckenfelder, Jr., W.W. (1980). *Principles of Water Quality Management*, pg. 275. CBI Publishing Company, Inc., Boston, MA.

Gaudy, Jr., A.F. and Gaudy, E.T. (1988). *Elements of Bioenvironmental Engineering*. Engineering Press, Inc., San Jose, CA.

Horan, N.J. (1990b). *Biological Wastewater Treatment Systems: Theory and Operation*. John Wiley and Sons, Inc., West Sussex, England.

Lue-Hing, C., Zenz, D.R., Kuchenrither, R. (1992). *Municipal Sewage Sludge Management: Processing, Utilization, and Disposal*. Technomic Publishing Co, Inc., Lancaster, PA.

Masters, G.M. (1998). *Introduction to Environmental Engineering and Science*. Prentice Hall, Upper Saddle River, NJ.

Mines, Jr., R.O., Thomas, IV, W.C. (1996). Biological Nutrient Removal Using the VIP Process. *Journal Environmental Science and Health* **A31**(10), 2557–2575.

Mines, Jr., R.O. (1997a). Design and Modeling of Post-Denitrification Single-Sludge Activated Sludge Processes. *Journal Water Air Soil Pollution* **100**(1–2), 79–88.

Mines, R.O., Mohamed, S.R. (1997). Oxygen Uptake Rate at the Valrico Advanced Wastewater Treatment Plant. *Advances in Environmental Research* **1**(2), 166–177.

Mines, Jr., R.O. (1997b). Nutrient Removal Using a Modified Pure Oxygen Activated Sludge Process. *Advances in Environmental Research* **1**(1), 15–26.

Mines, Jr., R.O., Sherrard, J.H. (1997). Biological Treatment of A High Strength Nitrogenous Wastewater. *Journal Environmental Science and Health* **A32**(5), 1353–1375.

Mines, Jr., R.O. (1997). Design and Modeling of 4-Stage Single-Sludge Systems. *Advances in Environmental Research* **1**(3), 323–332.

Mines, Jr., R.O. (1998). Design and Modeling of Pre-Denitrification Single-Sludge Activated Sludge Processes. *Journal Environmental Science and Health* **A33**(1), 111–128.

Mines, Jr., R.O., Milton, G.D. (1998). Bionutrient Removal with A Sequencing Batch Reactor. *Journal Water Air Soil Pollution* **197**, 81–89.

Mines, R.O., Sherrard, J.H. (1999). Temperature Interactions in the Activated Sludge Process. *Journal Environmental Science and Health* **A34**(2), 329–340.

Mines, Jr., R.O., Vilagos, J.L., Echelberger, Jr., W.F., Murphy, R.J. (2001). Conventional and AWT Mixed-Liquor Settling Characteristics. *Journal Environmental Engineering Division, ASCE* **127**, 249–258.

Mines, Jr., R.O., Robertson, R.R. (2003). Treatability Study of a Seafood Processing Waster. *Journal Environmental Science and Health*, **A38**(10), 1927–1937.

Mines, Jr., R.O., Behrend, G., Bell, H.G. (2004). Assessment of AWT Systems in the Metro Atlanta Area. *Journal of Environmental Management* **70**, 309–314.

Mines, Jr., R., Lackey, L., Behrend, G. (2006). The Impact of Flow and Loading on Wastewater Treatment Plant Performance. *Journal Water, Air, and Soil Pollution*, published on-line, 8 August 2006, DOI: 10.1007/s11270-006-9220-0.

Mines, Jr., R.O., Lackey, L.W., Behrend, G.R. (2006b). Performance Assessment of Major Wastewater Treatment Plants (WWTPs) in the State of Georgia. *Journal Environmental Science and Health*, Part A **41**(10) 2175–2198.

Mines, Jr., R.O., Northenor, C. B., Murchison, M. (2008). Oxidation and Ozonation of Waste Activated Sludge. *Journal Environmental Science and Health*, Part A **43**(6) 610–618.

Mines, Jr., R.O., Lackey, L.W. (2009). Bench-Scale Ozonation Study of Waste Activated Sludge. *Journal Environmental Science and Health*, Part A **44**, 38–47.

Mines, R.O., Lackey, L.W. (2009). *Introduction to Environmental Engineering*. Prentice Hall, Upper Saddle River, NJ.

Qasim, S.R. (1994). *Wastewater Treatment Plants: Planning, Design, and Operation*. Technomic Publishing Company, Inc., Lancaster, PA.

Painter, H.A. (1970). A Review of Literature of Inorganic Nitrogen Metabolism in Microorganisms. *Water Research* **4**, 393–450.

Sherrard, J.H., Schroeder, E.W. (1973). Cell Yield and Growth Rate in Activated Sludge. *Journal Water Pollution Control Federation* **5**, 1189.

Tchobanoglous, G., Schroeder, E.D. (1985). *Water Quality Characteristics, Modeling, and Modification*. Addison-Wesley Publishing Company, Reading, MA.

US EPA (1985). *Review of Current RBC Performance and Design Procedures*. US Environmental Protection Agency, EPA 600/2-85/033, Cincinnati, OH.

Vesilind, P.A., Peirce, J.J., Weiner, R.F. (1994). *Environmental Engineering*. Butterworth-Heinemann, Newton, MA.

Vesilind, P.A. (1997). *Introduction to Environmental Engineering*. PWS Publishing Company, Boston, MA.

Vesilind, P.A. (2003). *Wastewater Treatment Plant Design*. Water Environment Federation, Alexandria, VA.

Yoshioka, N. *et al.* (1957). *Continuous Thickening of Homogenous Flocculated Slurries*. Chem Eng 21, Tokyo.

Problems

1 A developer in Lexington, Virginia is proposing a new retirement community called "Keydet Acres", which is expected to have a build-out population of 3,000 people by the year 2025. Assume a domestic wastewater with weak characteristics and use the values in Tables 7.2 and 7.4 for estimating the following for 2025:

a. Average daily flow (MGD).
b. Peak hour flow (MGD).
c. Minimum hour flow (MGD).

d. Average daily COD mass loading (lb/d).
e. Average daily TN mass loading (lb/d).
f. Average daily TP mass loading (lb/d).

2 Tabulated below are the average monthly influent flow and wastewater characteristics data for a wastewater treatment plant near Orlando, Florida. Estimate the following:
a. Annual average daily, maximum month daily, and minimum month daily flows in MGD.
b. Peaking factors for maximum month and minimum month daily flows.
c. Average influent COD, maximum month COD, and minimum month COD concentrations in mg/L.
d. Peaking factors for maximum month and minimum month COD concentrations.
e. Average influent TP, maximum month TP, and minimum month TP concentrations in mg/L.
f. Peaking factors for maximum month and minimum month TP concentrations.
g. Average influent COD, maximum month COD, and minimum month COD mass loadings in lb/d.
h. Peaking factors for maximum month and minimum month COD mass loadings.
i. Average influent TP, maximum month TP, and minimum month TP mass loadings in lb/d.
j. Peaking factors for maximum month and minimum month TP mass loadings.

Date	Influent flow MGD (1)	Influent COD concentration, mg/L (2)	Influent TP concentration, mg/L (3)
Jan 09	3.72	362	7.00
Feb 09	4.48	416	6.35
Mar 09	4.73	386	7.29
Apr 09	4.65	419	7.09
May 09	4.63	427	6.70
Jun 09	5.41	394	7.15
Jul 09	5.69	365	7.15
Aug 09	5.59	353	7.16
Sep 09	4.87	292	7.00
Oct 09	5.70	325	6.52
Nov 09	5.95	333	8.11
Dec 09	5.28	283	9.05

3 A mechanical bar screen with $\frac{1}{2}$ inch openings and $\frac{3}{8}$ inch bars is installed in a rectangular channel where the approach velocity should not exceed 2.0 ft per second. Estimate:
a. The velocity (fps) between the bars.
b. The head loss (ft) through the screen assuming it is clean.

4 Estimate the head loss through a coarse screen before and after the accumulation of solids occurs. Assume the following conditions for solving the problem. The approach velocity and velocity through the screen are 0.5 meters per second and 0.9 meters per second, respectively. The open area through the clean bar screen is 0.20 m^2.
a. Estimate the head loss (m) through the clean bar screen assuming a discharge coefficient of 0.7 for a clean screen.
b. Estimate the head loss (m) through the clogged bar screen assuming a discharge coefficient of 0.6 for a clogged screen and that 50% of the flow area has been blocked by debris.

5 Two horizontal-flow type grit chambers are designed to remove grit particles with a diameter of 0.15 mm (100 mesh) and specific gravity of 2.65. A flow-through velocity of 1.0 ft/s will be maintained by a proportioning weir. The average daily wastewater flow is 2.5 MGD. The PHF : ADF ratio is 2.5 : 1.0. Determine the channel dimensions (ft) for the PHF.

6 Design an aerated grit chamber for an average daily wastewater flow rate of 10 MGD. Assume two grit chambers are operating in parallel. The peak hour flow rate is three times the ADF. Use the design criteria in Table 7.8 to determine the following:
a. The dimensions (ft) of each grit chamber.
b. The total air required (ft^3/d).

7 A municipal WWTP receives an average daily wastewater flow of 10.0 MGD. Two rectangular primary clarifiers, operating in parallel, will treat the flow. The peak hour flow anticipated is 2.75 times the average daily flow. Use the design criteria in Tables 7.9 and 7.10 and assume that the effluent weir length in each clarifier is 12 times the clarifier width. Determine:
a. The dimension (ft) of each primary clarifier.
b. The detention time (h) in each clarifier.
c. The weir loading rate (gpd/ft) for each clarifier at PHF.
d. The BOD and suspended solids removal efficiencies (%) at ADF.

8 An industrial WWTP receives an average daily wastewater flow of 38,000 m^3/d. Two circular primary clarifiers, operating in parallel, will treat the flow. The peak hour flow anticipated is 1.5 times the average daily flow. Use the design criteria in Tables 7.9 and 7.10, and assume that a peripheral effluent weir is used for each clarifier. Determine:
a. The diameter (m) of each primary clarifier.
b. The detention time (h) in each clarifier.
c. The weir loading rate ($m^3/d \cdot m$) for each clarifier at PHF.
d. The BOD and suspended solids removal efficiencies (%) at ADF.

9 A conventional activated sludge process treats 3,785 m^3/d of wastewater, containing 250 g/m^3 BOD_5, and produces an effluent containing 20 g/m^3 BOD_5. The nominal detention time in the aeration basin excluding the return activated sludge flow is six hours and the MLSS concentration is 3,000 g/m^3. Determine the following:

a. The aeration basin volume (m^3).
b. F/M ratio (d^{-1}).
c. Specific substrate utilization rate (d^{-1}).
d. Substrate removal efficiency (%).

10 A step-aeration activated sludge process treats 38,000 m^3/d of wastewater containing 220 g/m^3 BOD_5. The F/M ratio based on VSS in the aeration basin is 0.30 d^{-1} and the MLVSS concentration is 3,000 g/m^3. Determine the following:
a. The aeration basin volume (m^3).
b. The detention time (h).
c. The effluent substrate concentration if the specific substrate utilization rate is 0.28 d^{-1}.
d. Substrate removal efficiency (%).

11 NPDES permits typically require that activated sludge WWTPs meet an annual average effluent BOD_5 concentration of 20 mg/L. If an oxidation ditch type of activated sludge process produces an effluent containing 20 mg/L of TSS, estimate the soluble BOD_5 (mg/L) necessary to meet the effluent BOD_5 standard. Assume that the volatile fraction of the effluent suspended solids is 65%.

12 A complete-mix activated sludge process (CMAS) is to be designed to treat 5.0 million gallons per day (MGD) of primary effluent having a BOD_5 of 180 mg/L. The NPDES permit requires that the effluent BOD_5 and TSS concentrations should be 20 mg/L or less on an annual average basis. The following biokinetic coefficients obtained at 20°C will be used in designing the process: $Y = 0.6$ mg VSS/mg BOD_5, $k = 4\ d^{-1}$, $K_s =$ 70 mg/L BOD_5, and $k_d = 0.05\ d^{-1}$. Assume that the MLVSS concentration in the aeration basin is maintained at 2,500 mg/L and the VSS : TSS ratio is 0.75. The temperature of the wastewater during the winter months is expected to remain at 15°C for extended periods. During the summer, the wastewater temperature may reach 30°C for several weeks. Determine the following:
a. Effluent soluble BOD_5 ($SBOD_5$) concentration in mg/L necessary to meet the effluent total BOD_5 ($TBOD_5$) requirement of 20 mg/L.
b. Mean cell residence time (d) necessary to meet the NPDES permit during the winter months.
c. Volume of the aeration basin in cubic feet.
d. Mean cell residence time (d) necessary to meet the NPDES permit during the summer months.
e. Oxygen requirements (lb/d) assuming complete nitrification during the summer months and a $TKN_0 = 35$ mg/L.
f. The quantity of excess biomass produced (lb/d), in terms of TSS, at the shortest MCRT that the facility will be operated.

13 A complete-mix activated sludge process (CMAS) is to be designed to treat 19,000 m^3/d of raw wastewater having a BOD_5 of 200 mg/L. The NPDES permit requires that the effluent BOD_5 and TSS concentrations should be 20 mg/L or less on an annual average basis. The following biokinetic coefficients obtained at 20°C will be used in designing the process: $Y = 0.6$ mg VSS/mg BOD_5, $k = 4\ d^{-1}$, $K_s = 50$ mg/L BOD_5, and $k_d = 0.05\ d^{-1}$. Assume that the MLVSS concentration in the aeration basin is maintained at 2800 mg/L and the VSS : TSS ratio is 0.70. The temperature of the wastewater during the winter months is expected to remain at 17°C for extended periods. During the summer, the wastewater temperature may reach 28°C for several weeks. Determine the following:
a. Effluent soluble BOD_5 ($SBOD_5$) concentration in mg/L necessary to meet the effluent total BOD_5 ($TBOD_5$) requirement of 20 mg/L.
b. Mean cell residence time (d) necessary to meet the NPDES permit during the winter months.
c. Volume of the aeration basin in cubic meters.
d. Mean cell residence time (d) necessary to meet the NPDES permit during the summer months.
e. Oxygen requirements (kg/d), assuming complete nitrification during the summer months and a $TKN_0 = 30$ mg/L.
f. The quantity of excess biomass produced (kg/d) in terms of TSS at the shortest MCRT that the facility will be operated.

14 A wastewater treatability study was performed on a municipal wastewater and the data in the table below were collected in a bench-scale CMAS reactor. Determine the following biokinetic coefficients: Y, k, k_d, K_s, a, and b.

MCRT (days)	MLVSS (mg/L)	τ (d)	S_i (mg/L)	S_e (mg/L)	OUR (mg/L · h)
5	2,280	0.167	200	7.00	48
10	3,590	0.167	200	4.30	59
15	4,420	0.167	200	3.50	66
20	5,000	0.167	200	3.10	71
25	5,420	0.167	200	2.80	75

15 A wastewater treatability study was performed on a nitrogenous wastewater with an influent ammonia concentration of 300 mg/L as N and the data in the table below were collected in a bench-scale CMAS reactor. Determine the following biokinetic coefficients for nitrification: Y, k, k_d, and K_s.

MCRT (days)	MLVSS (mg/L)	τ (d)	S_i (mg/L)	S_e (mg/L)
6	875	0.167	300	0.36
12	1,322	0.167	300	0.21
16	1,515	0.167	300	0.17
20	1,661	0.167	300	0.16
24	1,774	0.167	300	0.14

16 A wastewater treatability study was performed on a soluble synthetic wastewater and the data in the table

below were collected in a bench-scale CMAS reactor. The substrate concentration was measured as COD. The reactor volume was 5 liters and the volumetric flow rate averaged 20 liters per day. Determine the following biokinetic coefficients: Y, k, k_d, K_s, a, and b.

MCRT (days)	MLVSS (mg/L)	S_i (mg/L)	S_e (mg/L)	OUR (mg/L · h)
3.9	1,030	300	20.00	50
4.2	1,100	300	18.00	60
8.5	1,940	300	10.00	65
13.8	2,650	300	7.00	70
21.0	3,320	300	6.00	75

17 Design a sequencing batch reactor WWTP to treat an average daily flow of 11,400 m^3/d and a peak flow of three times the ADF. Use three circular SBRs, and six cycles per day operating scheme, and a low-water-level (LWL) of 3 meters. Determine:
a. The volume (m^3) of each SBR.
b. Dimensions (m) of each SBR.

18 Given the following information for an extended aeration activated sludge process: $Q = 3.0$ MGD, influent soluble $BOD_5 = 220$ mg/L, $Y = 0.7$ g VSS/ g BOD_5, $k_d = 0.05d^{-1}$, $K_S = 60$ mg/L $BOD_5 k = 5\ d^{-1}$, MLVSS = 0.75 MLSS, $TKN_0 = 35$ mg/L, MLSS = 3,000 mg/L, DO = 2 mg/L, $\alpha = 0.80$, and $\beta = 0.98$. Assume the biokinetic coefficients have been corrected for a wastewater temperature of 30°C, which is expected during the summer months. Determine:
a. Volume of the aeration basin (MG) at an SRT = 25 days.
b. Oxygen required (lb/d), SRT = 25 days.
c. Number and hp of mechanical aerators assuming SOTR/aerator = 3.0 lb O_2/hp · h.

19 Given the following information for an oxidation ditch activated sludge process: $Q = 11{,}400\ m^3/d$, influent soluble $BOD_5 = 250$ mg/L, $Y = 0.6$ g VSS/g BOD_5, $k_d = 0.06d^{-1}$, $K_S = 70$ mg/L BOD_5, $k = 6\ d^{-1}$, MLVSS = 0.70 MLSS, $TKN_0 = 40$ mg/L, MLSS = 3,500 mg/L, DO = 2 mg/L. $\alpha = 0.88$, and $\beta = 0.95$. Assume the biokinetic coefficients have been corrected for a wastewater temperature of 28°C, which is expected during the summer months, and there are three aeration basins with two aerators per basin. Determine:
a. Volume of the aeration basin (m^3) at an SRT = 30 days.
b. Oxygen required (kg/d), SRT = 30 days.
c. Number and kw of mechanical aerators assuming SOTR/aerator = 3.0 kg O_2/kw · h.

20 Design a coarse-bubble diffused aeration system to meet an oxygen demand of 10,000 pounds per day. The plant is located at an elevation of 1,000 feet. There are two aeration basins with a total volume equal to 1.43 MG. Assume the alpha and beta coefficients for the diffusers are 0.45 and 0.98, respectively and use a temperature correction factor (θ) of 1.024. The dissolved oxygen (DO) saturation concentration of water is 7.54 mg/L at sea level and 30°C. A DO concentration of 2.0 mg/L is maintained in the aeration basins. The oxygen transfer efficiency is assumed to be 10%, since coarse-bubble diffusers are being used. The side water depth of the aeration basin is 15 ft with diffusers placed one foot from the bottom of the tank. The atmospheric pressure at an elevation of 1,000 feet is 733 mm Hg. Assume that the specific weight of water is 62.4 lb/ft^3. Determine:
a. Volumetric flow rate of the air (scfm) to meet oxygen requirement of 10,000 lb per day.
b. Horsepower required for centrifugal blowers.

21 Design a fine-bubble diffused aeration system to meet an oxygen demand of 5,000 kg/d. The plant is located at an elevation of 610 meters (2,000 ft). There are two aeration basins with a total volume equal to 5,400 m^3. Assume the alpha and beta coefficients for the diffusers are 0.35 and 0.95, respectively and use a temperature correction factor (θ) of 1.024. The dissolved oxygen (DO) saturation concentration of water is 7.81 mg/L at sea level and 28°C. A DO concentration of 2.0 mg/L is maintained in the aeration basins. The oxygen transfer efficiency is assumed to be 30%, since fine-bubble diffusers are being used. The side water depth of the aeration basin is 4.9 m with diffusers placed 0.5 m from the bottom of the tank. The atmospheric pressure at an elevation of 610 meters is 706 mm Hg. Assume that the specific weight of water is 9,810 N/m^3. Determine:
a. Volumetric flow rate of the air (m^3/min) to meet oxygen requirement of 5,000 kg per day.
b. Power required (kw) for centrifugal blowers.

22 A single-stage trickling filter is to be designed to treat primary effluent containing a BOD_5 concentration of 120 g/m^3. The minimum wastewater temperature anticipated is 15°C. Using a recycle ratio of 1.0, determine the minimum allowable BOD_5 loading rate (kg/m^3 · d) for a stone-media filter, based on the NRC equation, if an average effluent BOD_5 concentration of 30 g/m^3 is to be achieved.

23 A single-stage trickling filter is to be designed to treat primary effluent containing a BOD_5 concentration of 170 mg/L. The minimum wastewater temperature anticipated is 16°C. Using a recycle ratio of 0.5, determine the minimum allowable BOD_5 loading rate (lb/1,000 ft^3 · d) for a stone-media filter, based on the NRC equation, if an average effluent BOD_5 concentration of 30 mg/L is to be achieved.

24 A 3.0 MGD trickling filter (TF) plant consists of a primary clarifier, followed by a single-stage trickling filter containing stone media, which is followed by a secondary clarifier. The influent to the WWTP contains 150 mg/L of BOD_5 and the primary clarifier removes 35% of the BOD_5 and the temperature of the wastewater during the winter averages 14°C.

Determine the effluent BOD_5 concentration (mg/L) if the diameter of the TF is 80 ft and the depth is 8 ft. Use a recycle ratio of 1.0 and assume there are two parallel treatment trains.

25 A two-stage high-rate trickling filter with 200% recycle will be used to treat 5.0 million gallons per day (MGD). The influent BOD to the primary clarifier is 225 mg/L and the primary removes 30% of the incoming BOD. The effluent BOD should be 20 mg/L or less. Assume that the design BOD loading to the trickling filter system is 40 lb BOD/(d·1,000 ft^3) and a depth of 6 ft will be used. Use the NRC equations for designing the trickling filter system, and assume that there are two treatment trains operating in parallel. Determine:
a. The diameter (ft) of the high-rate trickling filter system.
b. The effluent BOD concentration (mg/L).

26 A two-stage high-rate trickling filter with 100% recycle will be used to treat 5,000 m^3/d. The influent BOD to the primary clarifier is 220 mg/L and the primary removes 33% of the incoming BOD. The effluent BOD should be 20 mg/L or less. Assume that the design BOD loading to the trickling filter system is 1.0 kg BOD/(d · m^3), and a depth of 1.83 m will be used. Use the NRC equations for designing the trickling filter system and assume that the operating temperature is 18°C. The coefficient 0.0561 in Equation (7.181) can be replaced with 0.443, and the organic loadings must be expressed in kg/d and the volume of media in cubic meters.
Determine:
a. The diameter (m) of the high-rate trickling filter system.
b. The effluent BOD concentration (mg/L).

27 Two biotowers operating in parallel are to be designed to treat a flow of 6.0 million gallons per day (MGD). The BOD_5 in the influent to the primary clarifier is 200 mg/L and the clarifier removes 35% of the BOD_5. The minimum wastewater temperature is expected to be 15°C. The value of k is assumed to be 0.078 $(gpm)^{0.5}/ft^2$ at 20°C and an effluent BOD_5 concentration of 20 mg/L is to be achieved. The biotower depth = 20 ft. Determine:
a. The radius (ft) of each biotower.
b. The volume (ft^3) of biotowers.
c. The recycle flow (gpm/ft^2) around the biotowers if the minimum wetting rate is 0.75 gpm/ft^2.
d. The recycle ratio (R).

28 Two biotowers operating in parallel are to be designed to treat a flow of 16,000 m^3/d. The BOD_5 in the influent to the primary clarifier is 200 mg/L and the clarifier removes 35% of the BOD_5. The minimum wastewater temperature is expected to be 14°C. The value of k is assumed to be 0.210 $(Lps)^{0.5}/m^2$ at 20°C and an effluent BOD_5 concentration of 20 mg/L is to be achieved. The biotower depth = 6.1 m. Determine:
a. The radius (m) of each biotower.
b. The volume (m^3) of biotowers.
c. The recycle flow (Lps/ft^2) around the biotowers if the minimum wetting rate is 0.5 Lps/ft^2.
d. The recycle ratio (R).

29 Design a multi-stage RBC WWTP to treat the following wastewater for BOD removal. Influent flow = 6,000 m^3/d, influent BOD_5 = 260 g/m^3, primary clarifier removes 35% of BOD_5, total effluent BOD_5 and effluent TSS = 20 g/m^3, effluent soluble BOD_5 ($sBOD_5$) = 10 g/m^3, and T = 10°C. Determine:
a. The number of stages and number of trains.
b. The soluble BOD_5 in the effluent from each stage (g/m^3).
c. The organic loading on the first stage (g/m^2 · d).
d. The overall organic loading to the RBC WWTP (g/m^2 · d).
e. The overall hydraulic loading rate to the RBC WWTP (m^3/m^2 · d).

30 A modified Ludzack-Ettinger (MLE) biological nutrient removal process is to be designed to meet a 20/20/10 mg/L effluent standard for BOD_5, TSS, and TN, respectively. The influent wastewater characteristics, mixed liquor parameters, and biokinetic coefficients to be used in the design are presented in the tables below. Assume that the influent ammonium concentration is equal to the TKN concentration, RAS flow = 1Q and MLR flow = 3Q, DO in the RAS and MLR recycle is 1.0 g/m^3.

Wastewater characterization

Influent		Mixed liquor	
Q	38,000 m^3/d	DO	2.0 mg/L
pH	7.0	MLSS	3500 mg/L
Minimum temperature	13°C	MLVSS	0.75 MLSS
TKN	30 mg/L	K_{DO}	0.5 mg/L
Alkalinity	200 mg/L as $CaCO_3$	MLVSS	2,625 mg/L
BOD_5	200 mg/L		
TSS	167 mg/L		
Inerts (X_L)	50 mg/L		

Heterotrophic biokinetic coefficients at 20°C

Coefficient		Coefficient	
Y	0.6 g VSS/g BOD_5	k_d	0.06 d^{-1}
k	5 d^{-1}	K_s	60 mg/L BOD_5

***Nitrosomonas* biokinetic coefficients at 20°C**

Coefficient		Coefficient	
Y_{NS}	0.15 g VSS/g $NH_4^+ - N$	$(k_d)_{NS}$	0.05 d^{-1}
k_{NS}	3 d^{-1}	$(K_s)_{NS}$	0.74 mg/L $NH_4^+ - N$

Heterotrophic biokinetic coefficients at 13°C

Coefficient		Coefficient	
Y	0.6 g VSS/g BOD_5	k_d	0.046 d^{-1}
k	3.1 d^{-1}	K_s	60 mg/L BOD_5

***Nitrosomonas* biokinetic coefficients at 13°C**

Coefficient		Coefficient	
Y_{NS}	0.15 g VSS/g $NH_4^+ - N$	$(k_d)_{NS}$	0.038 d^{-1}
k_{NS}	1.9 d^{-1}	$(K_s)_{NS}$	0.52 mg/L $NH_4^+ - N$

A temperature correction coefficient (θ) of 1.04 should be used for correcting both the heterotrophic and *Nitrosomonas* k_d values, whereas a θ of 1.07 should be used for correcting both the heterotrophic and *Nitrosomonas* k values. The heterotrophic K_s value is to be corrected with a θ of 1.00, and the *Nitrosomonas* K_s is corrected with a θ value of 1.053. All temperature correction coefficients were obtained from Metcalf & Eddy (2003). Use a safety factor of 1.25 and peaking factor of 1.2 and assume that $F_n = 0.10$.

31 A modified Ludzack-Ettinger (MLE) biological nutrient removal process is to be designed to meet a 20/20/10 mg/L effluent standard for BOD_5, TSS, and TN, respectively. The influent wastewater characteristics, mixed liquor parameters, and biokinetic coefficients to be used in the design are presented in the tables below. Assume that the influent ammonium concentration is equal to the TKN concentration, RAS flow = 1Q and MLR flow = 3Q, DO in the RAS and MLR recycle is 1.0 g/m^3.

Wastewater characterization

Influent		Mixed liquor	
Q	38,000 m^3/d	DO	2.0 mg/L
pH	7.1	MLSS	3500 mg/L
Minimum temperature	10°C	MLVSS	0.75 MLSS
TKN	30 mg/L	K_{DO}	0.5 mg/L
Alkalinity	225 mg/L as $CaCO_3$	MLVSS	2625 mg/L
BOD_5	200 mg/L		
TSS	167 mg/L		
Inerts (X_L) =	60 mg/L		

Heterotrophic biokinetic coefficients at 20°C

Coefficient		Coefficient	
Y	0.6 g VSS/g BOD_5	k_d	0.06 d^{-1}
k	5 d^{-1}	K_s	60 mg/L BOD_5

***Nitrosomonas* biokinetic coefficients at 20°C**

Coefficient		Coefficient	
Y_{NS}	0.15 g VSS/g $NH_4^+ - N$	$(k_d)_{NS}$	0.05 d^{-1}
k_{NS}	3 d^{-1}	$(K_s)_{NS}$	0.74 mg/L $NH_4^+ - N$

Heterotrophic biokinetic coefficients at 10°C

Coefficient		Coefficient	
Y	0.6 g VSS/g BOD_5	k_d	0.041 d^{-1}
k	2.5 d^{-1}	K_s	60 mg/L BOD_5

***Nitrosomonas* biokinetic coefficients at 10°C**

Coefficient		Coefficient	
Y_{NS}	0.15 g VSS/g $NH_4^+ - N$	$(k_d)_{NS}$	0.034 d^{-1}
k_{NS}	1.5 d^{-1}	$(K_s)_{NS}$	0.44 mg/L $NH_4^+ - N$

A temperature correction coefficient (θ) of 1.04 should be used for correcting both the heterotrophic and *Nitrosomonas* k_d values, whereas a θ of 1.07 should be used for correcting both the heterotrophic and *Nitrosomonas* k values. The heterotrophic K_s value is to be corrected with a θ of 1.00, and the *Nitrosomonas* K_s is corrected with a θ value of 1.053. All temperature correction coefficients were obtained from Metcalf & Eddy (2003). Use a safety factor of 1.25 and peaking factor of 1.2 and assume that $F_n = 0.10$.

32 A 4-stage Bardenpho process is to be designed to meet a 5/5/3/ mg/L effluent standard for BOD_5, TSS, and TN, respectively. The influent wastewater characteristics, mixed liquor parameters, and biokinetic coefficients to be used in the design are presented in the tables below. Assume that the influent ammonium concentration is equal to the TKN concentration, RAS flow = 1Q and MLR flow = 3Q, DO in the RAS and MLR recycle is 0.5 mg/L and 1.0 mg/L entering the second anoxic zone.

Wastewater characterization

Influent		Mixed liquor	
Q	38,000 m^3/d	DO	2.0 mg/L
pH	7.1	MLSS	3,500 mg/L
Minimum temperature	15°C	MLVSS	0.75 MLSS
TKN	30 mg/L	K_{DO}	0.5 mg/L
Alkalinity	200 mg/L as $CaCO_3$	MLVSS	2,625 mg/L
BOD_5	200 mg/L		
TSS	167 mg/L		
Inerts (X_L)	50 mg/L		

Heterotrophic biokinetic coefficients at 20°C

Coefficient		Coefficient	
Y	0.6 g VSS/g BOD_5	k_d	0.06 d^{-1}
k	5 d^{-1}	K_s	60 mg/L BOD_5

***Nitrosomonas* biokinetic coefficients at 20°C**

Coefficient		Coefficient	
Y_{NS}	0.15 g VSS/g $NH_4^+ - N$	$(k_d)_{NS}$	0.05 d^{-1}
k_{NS}	3 d^{-1}	$(K_s)_{NS}$	0.74 mg/L $NH_4^+ - N$

Heterotrophic biokinetic coefficients at 15°C

Coefficient		Coefficient	
Y	0.6 g VSS/g BOD_5	k_d	0.05 d^{-1}
k	3.6 d^{-1}	K_s	60 mg/L BOD_5

***Nitrosomonas* biokinetic coefficients at 15°C**

Coefficient		Coefficient	
Y_{NS}	0.15 g VSS/g $NH_4^+ - N$	$(k_d)_{NS}$	0.04 d^{-1}
k_{NS}	2.1 d^{-1}	$(K_s)_{NS}$	0.57 mg/L $NH_4^+ - N$

A temperature correction coefficient (θ) of 1.04 should be used for correcting both the heterotrophic and *Nitrosomonas* k_d values, whereas a θ of 1.07 should be used for correcting both the heterotrophic and *Nitrosomonas* k values. The heterotrophic K_s value is to be corrected with a θ of 1.00, and the *Nitrosomonas* K_s is corrected with a θ value of 1.053. All temperature correction coefficients were obtained from Metcalf & Eddy (2003). Use a safety factor of 1.25 and peaking factor of 1.2 and assume that F_n =0.12.

33 A 4-stage Bardenpho process is to be designed to meet a 5/5/3/ mg/L effluent standard for BOD_5, TSS, and TN, respectively. The influent wastewater characteristics, mixed liquor parameters, and biokinetic coefficients to be used in the design are presented in the tables below. Assume that the influent ammonium concentration is equal to the TKN concentration, RAS flow = 1Q and MLR flow = 4Q, DO in the RAS and MLR recycle is 1.0 mg/L and 1.0 mg/L entering the second anoxic zone.

Wastewater characterization

Influent		Mixed liquor	
Q	56,775 m^3/d	DO	2.0 mg/L
pH	7.2	MLSS =	3,500 mg/L
Minimum temperature	18°C	MLVSS	0.70 MLSS
TKN	30 mg/L	K_{DO}	0.5 mg/L
Alkalinity	250 mg/L as $CaCO_3$	MLVSS	2,625 mg/L
BOD_5	200 mg/L		
TSS	167 mg/L		
Inerts (X_L)	50 mg/L		

Heterotrophic biokinetic coefficients at 20°C

Coefficient		Coefficient	
Y	0.6 g VSS/g BOD_5	k_d	0.055 d^{-1}
k	4.4 d^{-1}	K_s	60 mg/L BOD_5

***Nitrosomonas* biokinetic coefficients at 20°C**

Coefficient		Coefficient	
Y_{NS}	0.15 g VSS/g $NH_4^+ - N$	$(k_d)_{NS}$	0.046 d^{-1}
k_{NS}	2.6 d^{-1}	$(K_s)_{NS}$	0.67 mg/L $NH_4^+ - N$

Heterotrophic biokinetic coefficients at 18°C

Coefficient		Coefficient	
Y	0.6 g VSS/g BOD_5	k_d	0.05 d^{-1}
k	3.6 d^{-1}	K_s	60 mg/L BOD_5

***Nitrosomonas* biokinetic coefficients at 18°C**

Coefficient		Coefficient	
Y_{NS}	0.15 g VSS/g $NH_4^+ - N$	$(k_d)_{NS}$	0.04 d^{-1}
k_{NS}	2.1 d^{-1}	$(K_s)_{NS}$	0.57 mg/L $NH_4^+ - N$

A temperature correction coefficient (θ) of 1.04 should be used for correcting both the heterotrophic and *Nitrosomonas* k_d values, whereas a θ of 1.07 should be used for correcting both the heterotrophic and *Nitrosomonas* k values. The heterotrophic K_s value is corrected with a θ of 1.00, and the *Nitrosomonas* K_s is corrected with a θ value of 1.053. All temperature correction coefficients were obtained from Metcalf & Eddy (2003). Use a safety factor of 2.5 and peaking factor of 1.2 and assume that F_n = 0.12.

34 A batch settling column analysis on sludge from an extended aeration activated sludge process is presented below. Determine the diameter (ft) of two secondary clarifiers operating in parallel if the design flow (Q) is 10 MGD, return activated sludge (RAS) flow (Q_r) is 10 MGD, the MLSS concentration is 3,000 mg/L, and the underflow solids concentration is 12%.

SS(mg/L)	V(fph)
1,000	15.00
3,000	10.50
4,500	7.00
5,500	5.00
7,600	1.60
10,000	0.50
12,500	0.25
15,000	0.13

35 A batch settling column analysis on sludge from an oxidation ditch is presented below. Determine the diameter (m) of two secondary clarifiers operating in parallel if the total flow ($Q + Q_r$) to the clarifiers is 56,775 m^3/d at a MLSS concentration of 3,500 g/m^3, and the underflow solids concentration is 10,000 g/m^3.

SS(mg/L)	V(fph)
1,000	2.5
3,000	1.5
4,500	0.6
5,500	0.25
7,500	0.11
10,000	0.05
12,500	0.03

36 A circular secondary clarifier processes a total wastewater flow ($Q + Q_r$) of 7,570 m^3/d at a MLSS concentration of 2,800 g/m^3. Given the following batch settling column analysis and a limiting solids flux of 7.0 kg/(m$^2 \cdot$ h), determine the following:

a. Solids underflow concentration (g/m^3)
b. Diameter of the secondary clarifiers (m)
c. Clarifier overflow rate (m/h)

SS(mg/L)	V(fph)
1,500	5.00
2,600	3.25
3,950	2.00
5,420	1.00
6,950	0.55
9,000	0.25
12,000	0.13

37 A batch settling column analysis on sludge from a complete-mix activated sludge process is presented below. A circular secondary clarifier processes a total flow of 2.0 MGD ($Q + Q_r$) at a MLSS concentration of 3,200 mg/L. Determine the following, assuming a limiting solids flux value of 1.2 lb/(ft$^2 \cdot$ h):

a. Solids concentration in the underflow (mg/L)
b. Underflow rate (gpd)
c. Diameter of the secondary clarifiers (ft)
d. Overflow rate (gpd/ft^2)

SS(mg/L)	V(fph)
1,000	19.5
1,200	19
2,000	12
3,800	3.5
6,100	1.1
8,200	0.5
10,000	0.3
11,000	0.2

38 Two circular secondary clarifiers are to be designed to treat an average daily flow (ADF) of 10.0 MGD from a CMAS process. The peak : average flow ratio is 2.5. The MLSS at ADF is 3,500 mg/L and the return activated sludge (RAS) flow is 6.0 MGD. During the peak flow condition, the MLSS concentration will be lowered to 2,500 mg/L and the RAS flow increased to 10.2 MGD. Use overflow rates of 600 and 1,500 gpd/ft^2 at average and peak flows. Solids loading rates of 25 and 50 ppd/ft^2 are to be used at average and peak loading conditions, respectively. Determine the diameter of the clarifiers (ft).

39 Two circular secondary clarifiers will be used to treat an average daily flow (ADF) of 5,000 m^3/d from a high-purity oxygen activated sludge process with a mixed liquor suspended solids concentration of 6,000 mg/L. At peak flow (10,000 m^3/d), the MLSS concentration is reduced to 4,500 g/m^3. The RAS flow at average and peak flows are 3,000 and 6,000 m^3/d, respectively. Additional design criteria include overflow rates of 30 and 70 m^3/(d $\cdot$ m^2) and solids loading rates of 100 and 240 kg/(d $\cdot$ m^2), respectively at average and peak loading conditions. Determine the diameter (m) of the clarifiers.

40 Two circular secondary clarifiers are designed to treat wastewater from a high-rate trickling filter system. The average and peak flow rates are 10 and 25 MGD, respectively. A recirculation flow of 2Q, based on average daily flow, will be used at all times. The overflow rates at average and peak flow will be 400 and 1,000 gpd/ft^2, respectively. Determine:

a. The diameter (ft) of the clarifiers if the total flow is applied to the secondary clarifiers.
b. The diameter (ft) of the clarifiers when direct recirculation around the clarifiers is used.

41 Two existing chlorine contact basins (CCBs) at a secondary WWTP have the following dimensions: 7 ft × 70 ft × 15 ft. The average daily flow (ADF) and peak hour flow (PHF) to the basins are 5 MGD and 10 MGD, respectively. Will the CCBs meet a detention time of 30 minutes at ADF or 15 minutes at PHF with one unit out of service? If not, what could be done?

42 Primary sludge contains 4% dry solids that are 75% volatile. Determine:

a. The specific gravity (S_s) of the dry solids.
b. The specific gravity (S_{sl}) of the sludge.
c. The volume (ft^3) of the thickened sludge if 1,000 lb of dry solids are thickened to 7% in a gravity belt thickener.

43 Waste activated sludge contains 1% dry solids that are 80% volatile. Determine:

a. The specific gravity (S_s) of the dry solids.
b. The specific gravity (S_{sl}) of the sludge.
c. The volume (m^3) of the thickened sludge if 60,000 kg of dry solids is thickened to 6% in a gravity belt thickener.

44 Ten thousand gallons of primary sludge at 5% solids is blended with 5,500 gallons of WAS at 1.2% solids. Determine:

a. The solids concentration (%) of the blended sludge.
b. The final volume (ft^3) of the blended sludge if it is thickened to 8% using a thickening centrifuge. Assume the specific gravity of the blended sludge is 1.0

45 Five thousand cubic meters of primary sludge at 6% solids is blended with 3,500 m^3 of WAS at 0.8% solids. Determine:

a. The solids concentration (%) of the blended sludge.
b. The final volume (m^3) of the blended sludge if it is thickened to 7% using a GBT. Assume the specific gravity of the blended sludge is 1.03.

46 One thousand cubic meters of primary sludge at 3% solids concentration and specific gravity of 1.03 is to be

thickened to 5% solids in a gravity thickener. The solids concentration in the supernatant is estimated to be 700 g/m^3. Determine:

a. The diameter of the thickener (m).
b. Thickened sludge flow (m^3/d).

47 Fifty thousand gallons of primary sludge at 3.5% solids concentration and specific gravity of 1.05 is to be thickened to 5.5% solids in a gravity thickener. The solids concentration in the supernatant is estimated to be 600 mg/L. Determine:

a. The diameter of the thickener (ft).
b. Thickened sludge flow (gpd).

48 A gravity belt thickener (GBT) is designed to thicken 10,000 pounds per day of anaerobically digested primary and WAS at a 3.0% solids concentration to 6% solids. The GBT will operate 5 days per week for 8 hours each day. The SS concentration in the filtrate is 900 mg/L. Use the following design criteria to design the GBT: hydraulic loading rate = 100 gpm/m; solids loading rate = 1,100 lb/(m · h); and washwater rate = 25 gpm/meter of belt.

49 A gravity belt thickener (GBT) is designed to thicken 600 m^3/d of WAS at a 0.8% solids concentration to 5% solids. The GBT will operate 6 days per week for 7 hours each day. The SS concentration in the filtrate is 1,200 g/m^3. Use the following design criteria to design the GBT: hydraulic loading rate = 800 Lpm/m; solids loading rate = 200 – 600 kg/(m · h) ; and washwater rate = 95 Lpm/meter of belt.

50 57,000 gallons of waste activated sludge at a 1.2 % solids concentration is to be processed daily by an aerobic digester. The volatile solids content is 75% and the minimum design operating temperature is 12°C. A temperature correction coefficient (θ) of 1.06 will be used for adjusting the digestion time for temperature variations. Assume an oxygen demand of 2.0 pounds of oxygen per pound of volatile solids destroyed and a 50% reduction in volatile solids. Determine the following:

a. Design hydraulic detention time (τ) or digestion time (days).
b. Aerobic digester volume (ft^3).
c. Volatile solids loading rate ($lb/d \cdot ft^3$) to the aerobic digester.
d. Quantity of oxygen required (lb/d) to stabilize the waste activated sludge.
e. Total air required (ft^3/min), assuming 0.075 lb air/ft^3, 0.23 lb O_2/lb air and the diffusers have a transfer efficiency of 6.0%.

51 An aerobic digester is to be designed to stabilize 600 kg/d of primary sludge at 5% solids and 75% volatiles, along with 400 kg/d of waste-activated sludge at 1% solids and 75% volatiles. The volume of the combined sludge that will be fed to the digester is 104 m^3/d. The minimum design operating temperature is 18°C. A temperature correction coefficient (θ) of 1.06 should be used to correct the detention or digestion time for temperature. The primary sludge contains 210 kg of BOD, with an oxygen demand of 1.9 kg of oxygen per kg of BOD, whereas use an oxygen demand of 2.0 kg of oxygen per lb of VS destroyed for the WAS. Assume that 90% of the BOD contained in the primary sludge is destroyed and that 50% of the volatiles solids in the WAS is destroyed. Determine the following:

a. Design hydraulic detention time (τ) or digestion time (days).
b. Aerobic digester volume (m^3).
c. Volatile solids loading rate ($kg/d \cdot m^3$) to the aerobic digester.
d. Quantity of oxygen required (kg/d) to stabilize the primary sludge.
e. Quantity of oxygen required (kg/d) to stabilize the secondary sludge (WAS).
f. Total quantity of oxygen required (kg/d) to stabilize the combined sludge.
g. Total air required (m^3/min), if there are 1.2 kg of air per cubic meter of air and the atmosphere consists of 23% oxygen by weight (0.23 kg O_2/kg air) and the diffusers have a transfer efficiency of 4.0%.

52 Calculate the anaerobic digester capacity (m^3) required for conventional single-stage anaerobic digestion given the following data: raw sludge production = 1,000 kg/d; VS in raw sludge = 72%; moisture content of raw sludge = 94%; solids content of raw sludge = 5% solids; digestion period = 30 days; VS Reduction = 52%; storage time required = 90 days; percent solids in thickened digested sludge = 5.5%; and specific gravity of digested sludge = 1.02.

53 Calculate the anaerobic digester capacity (ft^3) required for conventional single-stage anaerobic digestion given the following data: 40,000 gallons per day of combined primary sludge and WAS at 3.0% solids; volatile solids content = 75%; digestion period = 30 days; VS Reduction = 55%; storage time required = 100 days; percent solids in thickened digested sludge = 6.5%; and specific gravity of digested sludge = 1.04.

54 Determine the volume (m^3) of a first-stage, high-rate anaerobic digester based on a maximum volatile solids loading rate of 2.5 kg VS/($m^3 \cdot d$) and a detention time of 18 days. 700 cubic meters per day of combined primary and waste activated sludge, at 3.6% solids and a volatile content of 71%, are fed to the digester.

55 Determine the volume (ft^3) of a first-stage, high-rate anaerobic digester based on a maximum volatile solids loading rate of 0.15 lb VS/($ft^3 \cdot d$) and a detention time of 16 days. 20,000 gallons of raw primary sludge at 5% solids and a volatile content of 73% are fed to the digester.

56 An activated sludge WWTP produces an average of 100,000 gallons per day of primary sludge at 5% solids

and 250,000 gallons per day of WAS at 1.1% solids. If the WAS is thickened to 6.0% solids by dissolved air flotation and then combined with the primary sludge prior to dewatering, what is the quantity (lb/d) and solids concentration of the blended sludge (% solids)? Determine the belt width (m) of a belt filter press (BFP) if the BFP produces a sludge cake of 22% solids. Assume that the BFP operates 5 days a week for 8 hours each day. The solids in the filtrate is assumed to be 1,800 mg/L and the addition of polymer for conditioning contributes negligible weight. Assume a washwater rate of 25 gpm/meter of belt and hydraulic loading rate = 45 gpm/meter of belt.

57 A 2.0 m belt filter press dewaters 100 gpm of anaerobically digested sludge at a solids content of 6.5%. Four pounds of polymer are added per ton of sludge, and the liquid polymer flow rate is 6.4 gallons per minute. The washwater flow rate is 30 gpm per meter of belt. The solids concentrations in the filtrate and cake are 1,800 mg/L and 30%, respectively. Perform a materials balance around the BFP and calculate the following:

a. Hydraulic loading rate, gpm/m.
b. Solids loading rate, lb/(m · h).
c. Solids capture (%).

Chapter 8

Municipal solid waste management

Philip T. McCreanor

Learning Objectives

After reading this chapter, you should be able to:

- identify and describe the primary federal regulation addressing solid waste disposal in the United States;
- explain how a waste is identified as "hazardous";
- discuss solid waste generation and disposal trends in the United States;
- describe the composition of generated, disposed, and recycled solid waste streams in the United States;
- perform mass balance calculations on the composition of generated, disposed, and recycled solid waste streams;
- calculate "bulk" material properties of a solid waste sample;
- determine the approximate size of a landfill required to meet a community's solid waste disposal needs;
- design the basic configuration of a landfill's leachate collection system;
- calculate potential leakage rates from surface impoundments and liner systems;
- predict the volume and rate of landfill gas production.

8.1 Introduction

Solid waste management became an issue for humankind when we transitioned from a nomadic, mobile culture to a society that revolved around a single geographic position. In the nomadic lifestyle, the group moved away from solid waste materials as they were generated. Public health and security issues made it necessary for stationary cultures to develop plans for the management of solid waste materials. Allowing solid waste to accumulate on the streets of a municipality attracts rodents and other vermin, and the diseases associated with them. On the other hand, dumping of waste materials directly over and/or outside of a city's defensive walls will eventually create a slope that could be used by enemies to enter the city. Municipal authorities were therefore motivated to develop regulations and systems for the management of solid waste.

A key component of the municipal response would be a central location for the ultimate placement of the generated solid waste materials. Archeologists often refer to the locations used by ancient civilizations as 'mounds'. These mounds provide a wealth of information on how a community lived and operated. Modern communities refer to these locations as landfills. The municipality would typically identify a nearby depression to 'fill in' with their solid waste materials. This method of disposal was acceptable as long as the materials disposed consisted of degradable organic matter and non-degradable inerts.

With the industrial revolution, the materials placed in the municipal landfill began to include potentially hazardous materials generated both by industry and residences. This resulted in significant contamination of groundwater resources, as indicated by the presence of 250 municipal solid waste

Environmental Engineering: Principles and Practice, First Edition. Richard O. Mines, Jr.

landfills on the US EPA's Superfund National Priority List (Denison & Ruston, 1997). According to the US EPA, there were approximately 1,900 active landfills in the USA in 2009 (US EPA, 2010), and the industry generated in excess of 56 billion dollars in yearly revenues in 2009 (*Waste Business Journal*, 2009).

8.2 Regulations

The primary federal regulation in the United States addressing solid waste management is the Resource Conservation and Recovery Act (RCRA). This regulation has several subtitles which address a variety of topics, including underground storage tanks and the management of hazardous and non-hazardous solid wastes. Subtitles C and D of RCRA regulate solid, hazardous wastes and solid, non-hazardous wastes, respectively. The flow chart presented in Figure 8.1 can be useful in determining whether or not a waste meets the regulatory definition of a solid waste and is therefore regulated under RCRA.

A RCRA hazardous waste meets at least one of the following criteria:

- The material is specifically listed by the US EPA.
- The material meets hazardous waste physical property characteristics (ignitability, corrosivity, reactivity, and toxicity).
- The material is declared hazardous by the generator.

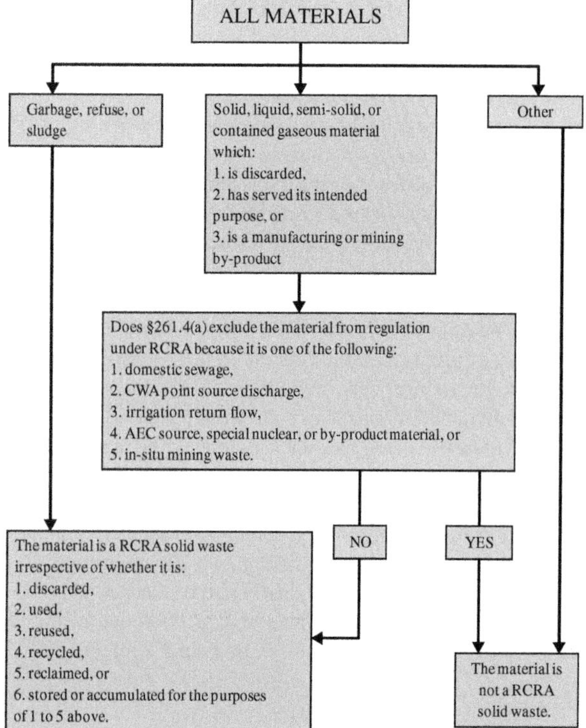

Figure 8.1 Identification of a RCRA solid waste. *Source*: United States Environmental Protection Agency.

This chapter will focus on RCRA subtitle-D wastes and municipal solid waste (MSW) in particular. Examples of common RCRA subtitle-D and municipal solid wastes are presented in Tables 8.1 and 8.2, respectively.

Sources for RCRA subtitle-D wastes include: residential, commercial, institutional, industrial, agricultural, treatment plants, and open areas such as streets and parks. These wastes may be characterized as infectious, and can include medical waste material from disposable equipment, instruments, utensils, or fomites from the rooms of patients who have been diagnosed or are suspected of having a communicable disease.

Table 8.1 Common RCRA subtitle-D waste materials.

Residential wastes
Commercial (business) wastes
Household hazardous wastes
Municipal wastewater treatment sludge
Non-hazardous industrial wastes
Municipal combustion ash
Small quantity generator's hazardous waste
Construction and demolition debris materials
Agricultural wastes
Oil and gas wastes
Mining wastes

Table 8.2 Common municipal solid waste materials.

Durable goods having a lifespan of greater than three years, including: • appliances, also known as white goods, consisting of inoperative and discarded refrigerators, ranges, water heaters, freezers, and other similar large appliances; • electronics, sometimes referred to as beige goods, including computers, televisions, audio equipment, etc.; and • furniture, tires, and oversize, bulky items.
Non-durable goods: newspaper, clothing, paper towels, cups
Containers and packaging
Food wastes
Yard wastes
Miscellaneous inorganics including stones, concrete, soil, ashes, residues.
Clothing
Household hazardous wastes

Laboratory wastes such as tissues, blood specimens, excreta, and secretions from patients or lab animals are also regulated by RCRA subtitle-D, as are surgical room pathological specimens and other materials from outpatient areas and emergency rooms.

8.3 Waste generation – international perspectives

One of the most important parameters of solid waste planning and decision making is the rate at which waste is generated. Waste generation rates (millions of tons and lb/capita/day) and gross national products (total and per capita) for a variety of countries are presented in Figure 8.2 and Table 8.3, respectively. Various resources suggest that there is a strong correlation between per capita Gross National Product (GNP) and the *per capita* waste generation of a country. Figure 8.2 and Table 8.3 also demonstrate that the waste materials generated by households are only a portion of the waste generated by a municipality.

8.4 Waste generation in the United States

Total municipal waste generation and disposal quantities in the United States from 1960 to 2009 are shown in Figure 8.3.

Refuse generation has grown steadily since the 1960s, but has shown a decline since 2007, which may be associated with the slowdown in the US economy. Material recovery for recycling and composting were relatively insignificant throughout the 1960s and 1970s. The difference between generation and disposal reflects recovery and recycling activities. The rate of recycling in 1960 was only 6.4%, with minimal change in this observed rate through 1980, when the recycle rate was 9.6%. In the late 1980s, a renewed interest in recycling pushed the recovery rate up to 16.2% by 1990, and it was measured at 33.7% in 2009. Figure 8.4 presents total and per capita waste generation rates in the US.

8.5 Waste composition

Another important waste parameter is the composition of the generated waste. Composition studies require the manual sorting of waste components into predefined categories. Knowledge of individual components is important for calculating waste physical properties, projecting the potential impact of recycling, performing landfill calculations, and designing waste incineration facilities. Composition studies should be performed seasonally to define equipment needs, management programs, and trends for future planning. Seasonal variations in composition include yard waste increase in the summer and gift wrapping/packaging increase during Christmas and other holidays. Well defined composition trends include an increase in food percentage with a decrease in

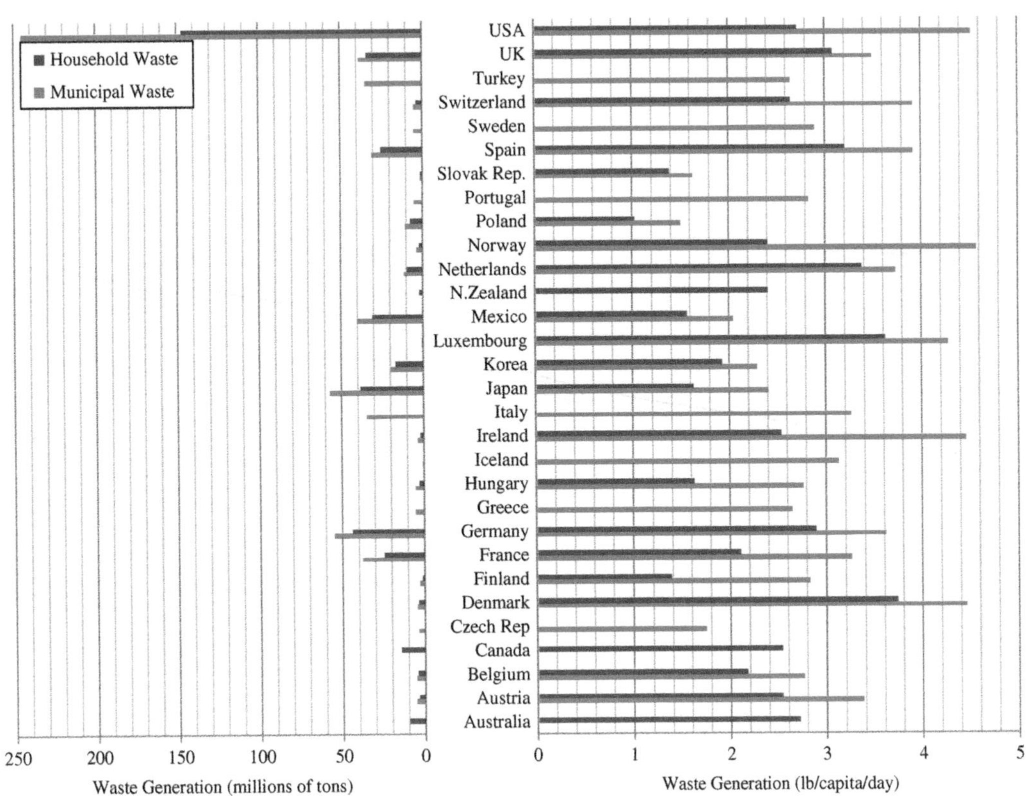

Figure 8.2 International municipal waste generation rates. Source data: Organization for Economic Co-operation and Development, 2007.

Table 8.3 International Gross National Products and waste generation rates.

	Gross National Product		Total waste generation millions of tons		per capita waste generation lb/capita/day	
Country	**$/10⁹**	**$/capita**	**MSW**	**Household**	**MSW**	**Household**
Australia	444	24	–	9.79	–	2.71
Austria	226	27	5.05	3.76	3.38	2.53
Belgium	264	26	5.33	4.10	2.77	2.17
Canada	760	24	–	14.71	–	2.53
France	1,543	26	37.36	24.20	3.25	2.11
Germany	2,242	27	54.52	43.87	3.62	2.89
Italy	1,260	22	34.84	–	3.25	–
Japan	4,852	38	56.77	38.28	2.41	1.63
Mexico	578	6	39.70	30.56	2.05	1.57
Netherlands	429	27	11.20	10.01	3.74	3.38
Poland	188	$5	10.29	7.15	1.51	1.02
Spain	651	16	30.35	25.01	3.92	3.19
Sweden	275	31	4.78	–	2.89	–
Switzerland	286	39	5.34	3.56	3.92	2.65
Turkey	212	3	34.49	–	2.65	–
USA	10,533	38	245.15	147.09	4.52	2.71

" – "= data not available
Source data: Organization for Economic Co-operation and Development, 2007.

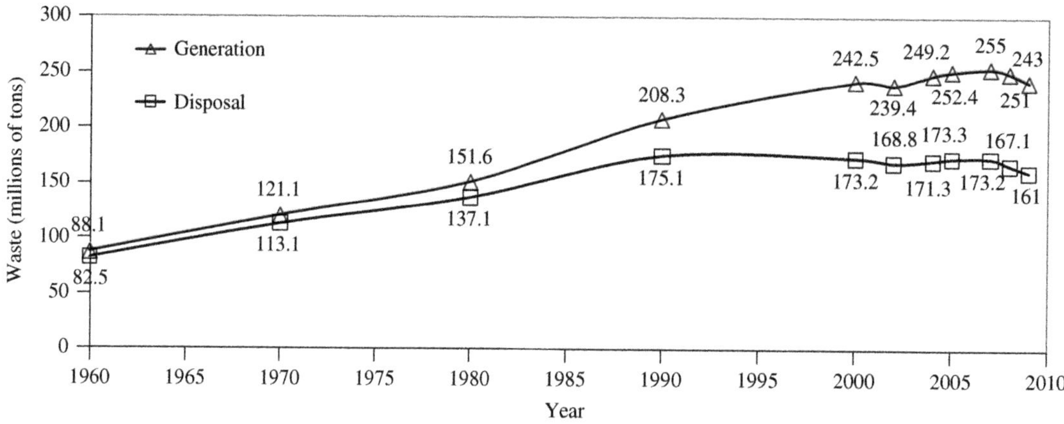

Figure 8.3 MSW generation and disposal in the USA. Source Data: US EPA, 2010.

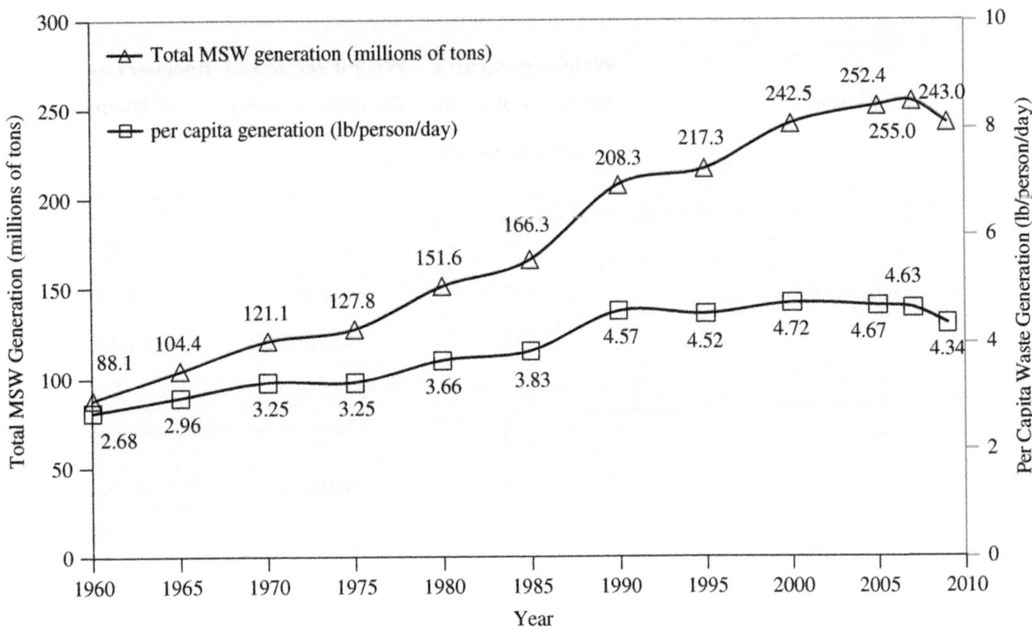

Figure 8.4 Total and per capita MSW generation in the USA. Source Data: US EPA, 2010.

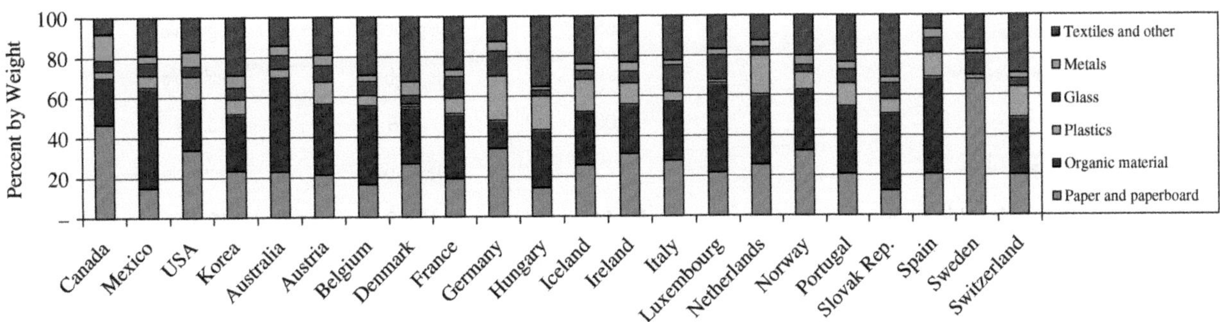

Figure 8.5 The composition of municipal waste in major countries. Source data: Organization for Economic Co-operation and Development, 2007.

income, paper waste increase with increasing income, and paper waste decrease with an increase in recycling.

Figure 8.5 provides a snapshot of the composition of MSW of selected major countries. The waste organic content is the primary difference between MSW generated in industrialized and developing nations. Industrialized countries tend to produce more organics than developing countries. Countries located in humid, tropical and semitropical areas typically generate waste characterized by a high concentration of plant debris. Areas that undergo seasonal changes in temperature often experience an increase in ash production during the winter months if wood or coal is primarily used for cooking and heating.

Figure 8.6 and Table 8.4 describe the waste composition in the United States on a materials basis and provides insight on the impact of recycling on waste composition. Historically, paper and paperboard dominated the US waste generation materials category and accounted for 36.4% and 28% of the materials generated in 1980 and 2009, respectively (US EPA, 2010). The second largest materials component of US MSW is yard trimmings, which accounted for 22.7% and 14% of generation in 1980 and 2009, respectively. The notable decrease in yard trimmings that occurred during this 45-year period has been attributed to local legislation prohibiting disposal of this type of refuse by landfilling and an increased emphasis on residential composting.

Table 8.4 Generation and recovery of MSW materials in the United States in 2009.

Material	Weight generated (millions of tons)	Weight recovered (millions of tons)	Recovery as percent of generation
	Durable goods		
Steel	13.34	3.72	27.9%
Aluminum	1.35	Negligible	Negligible
Other non-ferrous metals*	1.89	1.30	68.8%
Glass	2.12	Negligible	Negligible
Plastics	10.65	0.40	3.8%
Rubber and leather	6.43	1.07	16.6%
Wood	5.76	Negligible	Negligible
Textiles	3.49	0.44	12.6%
Other materials	1.61	1.23	76.4%
Total durable goods	**46.64**	**8.16**	**17.5%**
	Non-durable goods		
Paper and posterboard	33.48	17.43	52.1%
Plastics	6.65	Negligible	Negligible
Rubber and leather	1.06	Negligible	Negligible
Textiles	9.00	1.46	16.2%
Other materials	3.25	Negligible	Negligible
Total non-durable goods	**53.44**	**18.89**	**35.3%**
	Containers and packaging		
Steel	2.28	1.51	66.2%
Aluminum	1.84	0.69	37.5%
Glass	9.66	3.00	31.1%
Paper and paperboard	34.94	25.07	71.8%
Plastics	12.53	1.72	13.7%
Wood	10.08	2.23	22.1%
Other materials	0.24	Negligible	Negligible
Total containers and packaging	**71.57**	**34.22**	**47.8%**
	Other wastes		
Food, other**	34.29	0.85	2.5%
Yard trimmings	33.2	19.9	59.9%
Miscellaneous inorganic wastes	3.82	Negligible	Negligible
Total other wastes	**71.31**	**20.75**	**29.1%**
Total municipal solid waste	**242.96**	**82.02**	**33.8%**

Source Data: US EPA, 2010.

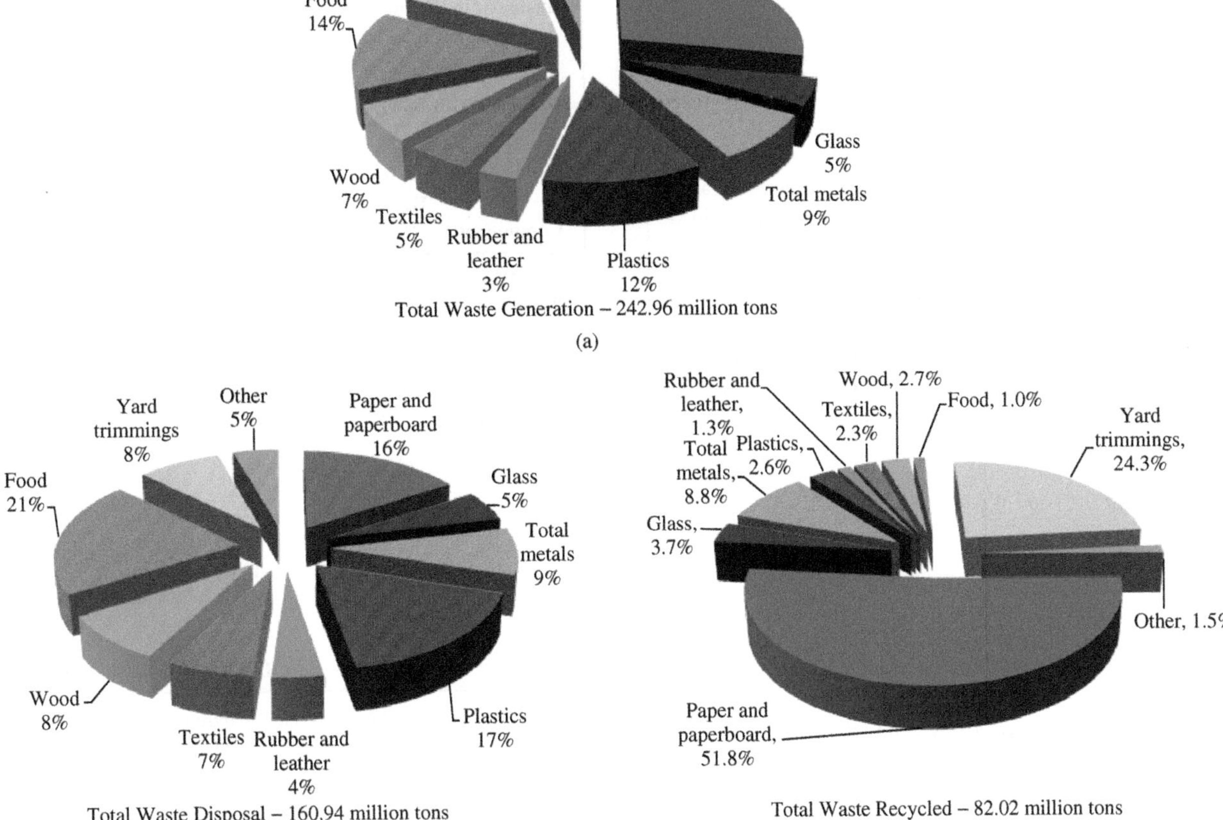

Figure 8.6 Composition of waste generated (a) disposed (b) and recycled (c) in the United States in 2009. *Source:* United States Environmental Protection Agency.

Example 8.1 Calculating the composition of generated waste

Use the following data to calculate the composition of the generated waste for the outlined conditions.

Use of food waste grinders:

- 20% of the homes have them
- 25% of the food waste generated in homes with food grinders is ground and discharged to the sewer system

Waste recycling:

- 11% of the generated MSW is recycled
- The recycled MSW has the following weight-based composition
  - Paper: 50%
  - Cardboard: 10%
  - Plastic: 6%
  - Yard wastes: 8%
  - Tin cans: 4%
  - Glass: 18%
  - Aluminum: 1%
  - Nonferrous metal: 3%

Disposed waste compositions for use in Example Problem 8.1:

Component	Disposed waste composition (% by weight)
Food waste	9
Paper	34
Cardboard	6
Plastic	7
Textile	2
Rubber	0.5
Leather	0.5
Yard waste	18.5
Wood	2
Glass	8
Tin	6
Aluminum	0.5
Other metals	3
Dirt, ash, etc.	3
Total	**100**

Solution

The basic solid waste mass balance is:

$$\text{Generated Waste} = \text{Disposed Waste} + \text{Recycled Materials} + \text{Diverted Materials} \quad (8.1)$$

In this problem, the diverted material is the food waste processed via the grinder. The problem statement provided data detailing food waste grinder use, disposed waste composition, and composition of recycling.

Step 1

To simplify the problem solution, assume a 100 lb sample of disposed waste. This enables the individual component data to be treated as weights (lb) rather than percentage (%).

Step 2

Account for the Food Waste (FW) ground up:

$$FW_{generated} = \frac{FW_{disposed}}{(\text{Fraction of generated waste which is disposed})}$$

From Table, $FW_{disposed} = (0.09) \times (100\,\text{lb}) = 9.0\,\text{lb}$

$$\text{Fraction of } FW_{ground} = \left(\frac{20}{100}\text{homes have grinders}\right) \times \left(\frac{25\,\text{lb}}{100\,\text{lb}}\text{ of FW is disposed by grinding}\right) + \left(\frac{80}{100}\text{ homes without grinder}\right) \times \left(\frac{0\,\text{lb}}{100\,\text{lb}}\text{ of FW is ground}\right)$$

$$\text{Fraction of } FW_{ground} = 0.05$$

$$FW_{generated} = \frac{9\,\text{lb}}{(1 - 0.05)} = 9.5\,\text{lb}$$

Therefore,

$$\text{Ground FW} = (9.5 - 9.0)\text{lb} = 0.5\,\text{lb (Diverted)}$$

Step 3

Now, recalling the solid waste mass balance (Equation 8.1), if 11% of the generated waste is recycled then, 89% of the generated waste must be diverted and disposed.

$$\text{Disposed} + \text{Diverted} = (100 + 0.5)\,\text{lb} = 100.5\,\text{lb}$$

$$\text{Generated Waste} = (100.5\,\text{lb}/0.89) = 113\,\text{lb}$$

$$\text{Recycled Material} = (113 - 100.5)\,\text{lb} = 12.5\,\text{lb}$$

Step 4

Construct a computation table to calculate the generated waste composition

Example Problem 8.1: Computations for generated waste composition.

Component	Disposed Waste (lb)	Ground Food Waste (lb)	Recycled Material (lb)[1]	Generated Waste (lb)	Generated Waste (% by wt.)
Food Waste	9	0.5		9.5	8.4
Paper	34		6.25	40.25	35.6
Cardboard	6		1.25	7.25	6.4
Plastic	7		0.75	7.75	6.9
Textile	2			2	1.8
Rubber	0.5			0.5	0.4
Leather	0.5			0.5	0.4
Yard Waste	18.5		1.0	19.5	17.3
Wood	2			2	1.8
Glass	8		2.25	10.25	9.1
Tin	6		0.5	6.5	5.8
Aluminum	0.5		0.13	0.63	0.6
Other Metals	3		0.38	3.38	3.0
Dirt, Ash, etc.	3			3	2.7
Totals	**100**		**12.5**	**113.01**	**100.0**

[1]Individual Recycled Materials calculated from total weight recycled (12.5 lb) multiplied by the percentage of the individual material (example: Paper = 50 × 12.5/100)

$$\text{Generated paper waste} = 34\,\text{lb} + 6.25\,\text{lb} = 40.25\,\text{lb}$$

8.6 Properties of municipal solid waste

Physical properties important to the characterization of MSW include specific weight, particle size and distribution, moisture content, field capacity and hydraulic conductivity.

8.6.1 Specific weight (γ)

The specific weight is often used in waste volume calculations; it is defined as the weight of material per unit volume and is typically given in units of lb/ft^3 or lb/yd^3. Literature values for specific weight possess little uniformity and are frequently reported as loose, found in containers, uncompacted, compacted, etc, so care should be taken when using and reporting these values. Variables such as geographic location, season, storage time, and processing equipment used (compaction, shredding, etc.) all affect the specific weight. MSW transferred in compaction vehicles are characterized by specific weights ranging from 300 to 700 lb/yd^3, as delivered. A value of 500 lb/yd^3 is typically assumed for waste collection

Table 8.5 Specific weight estimates and moisture content values for components of MSW.

	Specific weight (lb/ft^3)		MC_{wet} (%)	
Waste component	**Range**	**Typical**	**Range**	**Typical**
Food	220–810	490	50–80	70
Paper	70–220	150	4–10	6
Cardboard	70–135	85	4–8	5
Plastics	70–220	110	1–4	2
Textiles	70–170	110	6–15	10
Rubber	170–340	220	1–4	2
Leather	170–440	270	8–12	10
Yard Waste	220–540	170	30–80	60
Wood	270–810	400	15–40	20
Glass	85–270	330	1–4	2
Aluminum	110–405	270	2–4	3
Dirt, ash, etc.	540–1685	810	6–12	8

Source data: Tchobanoglous, et *al.* (1993) and Vesilind et *al.* (2002).

calculations while the specific weight of the landfill waste mass can range from 1,000 to 2,000 lb/yd^3. Table 8.5 lists specific weight values for several common residential wastes.

8.6.2 Moisture content

Properly assessing moisture content is crucial when processing refuse into fuel, directly combusting, or degrading a waste mass. When MC is defined on a weight basis, it may be presented as either wet or dry such that:

$$MC_{wet} = \frac{\text{weight of water}}{\text{initial wet weight of sample}} = \frac{w-d}{w} \tag{8.2}$$

$$MC_{dry} = \frac{\text{weight of water}}{\text{dry weight of solids}} = \frac{w-d}{d} \tag{8.3}$$

where:

MC_{wet} = moisture content defined on wet basis, fraction or percentage
MC_{dry} = moisture content defined on dry basis, fraction or percentage
w = the initial wet weight of the sample, lb or kg
d = the weight of the sample after drying at 105°C, lb or kg.

Note that the terms "wet" and "dry" do not describe the condition of the sample, but rather identify the denominator against which the water content is compared.

Table 8.5 provides the MC_{wet} on an uncompacted, wet weight basis for several common household items. Typically, the MC_{wet} for MSW in the United States will range between 15–40%, depending on the waste composition, season, humidity and precipitation. Data from the table highlight that the MC_{wet} of material can vary widely. Example 8.2 shows how literature values can be used to estimate the weight based MC of a specified residential waste mixture.

Moisture content can also be expressed on a volumetric basis, such that:

$$MC_{vol} = \frac{\text{Volume of water}}{\text{Volume of sample}} \tag{8.4}$$

where:

MC_{vol} = moisture content defined on a volume basis, fraction or percentage

Volume of water and volume of sample are typically expressed in units of ft^3 or liters.

Finally, moisture content can be expressed as the portion (percent or fractional percentage) of the pore space filled with water. This expression is referred to as saturation (S_w) (Equation 8.5).

$$S_w = \frac{\text{Volume of water}}{\text{Volume of voids in sample}} \tag{8.5}$$

where:

S_w = moisture content expressed as saturation, fraction or percentage

Volume of water and volume of voids in a sample are typically expressed in units of ft^3 or liters.

Example 8.2 Estimating moisture content of a sample of MSW

A residential waste can be characterized as shown in the table below. Use typical values for MC_{wet} found in Table 8.5 to estimate the moisture content of the sample on both a wet and dry weight basis.

Disposed waste compositions for use in Example Problem 8.2:

Component	Disposed waste composition (% by weight)
Food waste	10
Paper	35
Cardboard	6
Plastic	8
Textile	3
Rubber	1
Leather	1
Yard waste	20.5
Wood	3
Glass	8
Aluminum	0.5
Dirt, ash, etc.	4
Totals	**100**

Solution:

Step 1

Assume an initial wet disposed waste weight of 100 lb and create a calculation table.

Example Problem 8.2: Data and computations.

Component	Disposed waste (lb)	Typical moisture content (wet basis)	Dry weight[a] (lb)
Food Waste	10	70	3
Paper	35	6	32.9
Cardboard	6	5	5.7
Plastic	8	2	7.84
Textile	3	10	2.7
Rubber	1	2	0.98
Leather	1	10	0.9
Yard Waste	20.5	60	8.2
Wood	3	20	2.4
Glass	8	2	7.84
Aluminum	0.5	3	0.485
Dirt, Ash, etc.	4	8	3.68
Totals	**100 lb wet**		**76.6 lb dry**

[a]Calculation based on 100 lb. For example, the dry weight for food waste is determined as 10 lb(1 − 0.70) = 3 lb.

Step 2

Use Equation (8.2) to determine the moisture content on a wet weight basis:

$$MC_{wet} = \frac{w-d}{w}(100) = \frac{100-76.6}{100}(100) = \boxed{23.4\%}$$

For MC_{dry}, use Equation (8.3).

$$MC_{dry} = \frac{w-d}{d}(100) = \frac{100-76.6}{76.6}(100) = \boxed{30.5\%}$$

8.6.3 Field capacity

The field capacity (FC) is the total amount of water a sample can hold under free (unrestricted) drainage conditions and can be expressed as a MC_{dry}, MC_{wet}, MC_{vol}, or percentage saturation. This concept is important in solid waste management as, theoretically, no leachate (water that has contacted waste) will be produced by a landfill until the FC is exceeded. FC can be used to predict leachate production and the potential leachate storage available when designing leachate recirculation systems.

Typical values for FC for uncompacted, commingled residential and commercial waste range from 50–60% by volume (Tchobanoglous *et al.*, 1993). For compacted wastes, FC can be expected to range between 30–55% (Qian *et al.*, 2002) by volume or approximately 42.5% which corresponds to 425 ft^3 liquid/1,000 ft^3 of MSW. When adjusted for moisture inherently found in the waste (~27.5% by volume), the leachate storage capacity is approximately 15% on a volume basis or 150 ft^3 liquid storage available/1,000 ft^3 refuse (Vesilind *et al.*, 2002).

8.6.4 Hydraulic conductivity (K)

The hydraulic conductivity (K) is an important parameter, because it describes how both liquids and gases move in porous media. Hydraulic conductivity can be used to predict leachate movement and production, to evaluate landfill liner system designs, and to estimate the leakage rate from landfill liners. Hydraulic conductivity is the coefficient relating the hydraulic gradient, $\frac{dh}{dL}$, to the discharge velocity, v. This relationship (Equation 8.6) was first postulated by Henri Darcy in 1856 and is commonly referred to as Darcy's Law.

$$v = K\frac{dh}{dL} \tag{8.6}$$

where:

v = discharge velocity, in./h, cm/s, m/d
K = hydraulic conductivity, in./h, cm/s, m/d
$\frac{dh}{dL}$ = hydraulic gradient, in./in., cm/cm, m/m.

Hydraulic conductivity values for MSW range from 10^{-3} to 10^{-6} cm/s. For comparison, the hydraulic conductivities of gravel, sand, and clay are approximately 10^{-1}, 10^{-2}, and 10^{-6} cm/s; respectively.

8.6.5 Chemical composition

Knowledge of the chemical composition of MSW is primarily used for combustion and waste to energy (WTE) calculations but can also be used to estimate biological and chemical behaviors. Generally, waste is considered to consist of combustible (e.g. paper) and non-combustible materials (e.g. glass). If it is to be used as fuel, it is important to chemically characterize the waste using four common properties including:

1 proximate analysis;
2 fusing point of ash;
3 ultimate analysis (major elements); and
4 energy content.

8.6.5.1 Proximate Analysis

The organic component of MSW is subject to proximate analysis that is used to estimate the fraction of volatile organics and fixed carbon in the fuel. Proximate analysis includes the following tests:

1 Loss of moisture when the temperature is increased to and sustained at 105°C for 1 h.
2 Volatile Combustible Matter (VCM) – measurement of additional weight loss caused by ignition at 950°C in a closed crucible.
3 Fixed Carbon – the measured weight of the residue from VCM analyses.
4 Ash – the measured weight of residue remaining after ignition at 950°C in an open crucible.

Table 8.6 provides proximate analysis and energy data for components found in municipal and commercial refuse. Caution should be taken when using these tabulated values because of the heterogeneity associated with MSW.

8.6.5.2 Fusing Point of Ash

The fusing point of ash is defined as the temperature where the ash, or incombustible residue from burning waste will form a solid (clinker) of carbon and metal by agglomeration and fusion. Typical fusing temperatures for MSW range from 2,000 to 2,200°F.

8.6.5.3 Ultimate Analysis (major elements)

The ultimate analysis of a waste involves determining its elemental components on a percent basis. Elements commonly reported include C, H, N, O, S, and P. If chlorinated compounds are suspected in the waste mixture, halogenated compounds may also be reported in the ultimate analysis. Table 8.7 provides ultimate analysis data for individual combustible materials commonly associated with MSW.

8.6.6 Energy content

The energy content of MSW can be determined:

1 through laboratory experiments using calorimeters;
2 by using a full-scale boiler as a calorimeter (this method is often inconvenient); and
3 by calculations based on the waste elemental composition.

Energy content values are often characterized as being on an as-collected basis, moisture-free, or moisture- and ash-free. Converting between reported values is easily accomplished using Equations (8.7) and (8.8). Table 8.6 provides energy

Table 8.6 Proximate analysis and heating values for household and commercially generated waste components.

	Proximate analysis (% by weight)				Energy content (BTU/lb)		
Waste component	Moisture	Volatile matter	Fixed carbon	Ash	As collected	Moisture-free	Moisture- and ash-free
Food (mixed)	70.0	21.4	3.6	5.0	1,797	5,983	7,180
Paper (mixed)	10.2	75.9	8.4	5.4	6,799	7,571	8,056
Cardboard	5.2	77.5	12.3	5.0	7,042	7,428	7,842
Plastics (mixed)	0.2	95.8	2.0	2.0	14,101	14,390	16,024
Textiles	10.0	66.0	17.5	6.5	7,960	8,844	9,827
Rubber	1.2	83.9	4.9	9.9	10,890	11,022	12,250
Leather	10.0	68.5	12.5	9.0	7,500	8,040	8,982
Yard waste	50.0	30.0	9.5	0.5	2,601	6,503	6,585
Wood (mixed)	20.0	68.1	11.3	0.6	6,640	8,316	8,383
Dirt, ash, etc.	8	–	–	70	3,000	–	–

Source data: Tchobanoglous *et al.* (1993) and US DoD (2004).

Table 8.7 Ultimate analysis of components commonly associated with residential and commercial MSW.

Waste component	C	H	O	N	S	Ash
	% by weight, dry basis					
Food (mixed)	48.0	6.4	37.6	2.6	0.4	5.0
Paper (mixed)	43.4	5.8	44.3	0.3	0.2	6.0
Cardboard	43.0	5.9	44.8	0.3	0.2	5.0
Plastics (mixed)	60.0	7.2	22.8	–	–	10.0
Textiles	48.0	6.4	40.0	2.2	0.2	3.2
Rubber	69.7	8.7	–	–	1.6	20.0
Leather	60.0	8.0	11.6	10.0	0.4	10.0
Yard waste	46.0	6.0	38.0	3.4	0.3	6.3
Wood (mixed)	49.5	6.0	42.7	0.2	< 0.1	1.5

Source data: Tchobanoglous *et al.*, 1993.

content data for individual waste components commonly found in MSW.

$$\text{Energy Content (moisture free)}\frac{\text{BTU}}{\text{lb}} = \frac{\text{BTU}}{\text{lb}}(\text{as collected})\left(\frac{100}{100 - \%\ \text{moisture}}\right) \quad (8.7)$$

$$\text{Energy Content (ash and moisture free}\frac{\text{BTU}}{\text{lb}} = \frac{\text{BTU}}{\text{lb}}(\text{as collected})\left(\frac{100}{100 - \%\ \text{moisture} - \%\ \text{ash}}\right) \quad (8.8)$$

When BTU values are not available for a given material, approximate values may be obtained by calculating the bulk chemical compositions and then applying Equation (8.9), known as the Dulong Formula (Tchobanoglous *et al.*, 1993).

$$\text{Energy Content}\frac{\text{BTU}}{\text{lb}} = 145C + 610\left(H - \frac{1}{8}O\right) + 40S + 10N \quad (8.9)$$

where C, H, O, S, and N represent the percent by weight of each element or compound in the individual waste material.

Example 8.3 Estimating the energy content of a waste

Consider a residential waste composition characterized as shown in the first two columns of the table below. Estimate the as-collected energy content of the waste.

Example Problem 8.3 data and calculation of total energy

Given data		Collected values	Calculated values
Component	Disposed waste (% by weight or lb/100 lb)	Energy content (BTU/lb)[a]	Total energy (BTU)
Food waste	10	1,797	17,970
Paper	35	6,799	237,965
Cardboard	6	7,042	42,252
Plastic	8	14,101	112,808
Textile	3	7,960	23,880
Rubber	1	10,890	10,890
Leather	1	7,500	7,500
Yard waste	20.5	2,601	53,320.5
Wood	3	6,640	19,920
Glass	8	60	480
Aluminum	0.5	–	0
Dirt, ash, etc.	4	3,000	12,000
Totals	**100**	**n/a**	**538985.5**

Solution:

Step 1

Using Table 8.6, list the energy value for each compound (shown in the third column in the table above).

Step 2

Assume a 100 lb sample of disposed waste (Column 2) and calculate the total energy available from each compound (shown in last column in the table above).

Step 3

Calculate the bulk energy content of the disposed waste.

$$\text{Disposed waste bulk energy content} = \frac{538{,}985.5\ \text{BTU}}{100\ \text{lb}} = \boxed{5390\frac{\text{BTU}}{\text{lb}}}$$

Example 8.4 Estimating the energy content of MSW using the Dulong Formula

Assume the chemical composition of a MSW sample can be represented as $C_{650}H_{1700}O_{757}N_{11}S$. Estimate the energy content based using the Dulong formula.

Solution:

Step 1

Compose a calculations table to determine the percent weight contribution made by each element in the representative compound.

Calculation table for Example Problem 8.4

Component	Number of atoms per mole	Atomic weight	Elemental weight contribution per mole	% Weight Contribution
C	650	12	7800	35.8
H	1700	1	1700	7.8
O	757	16	12112	55.6
N	11	14	154	0.7
S	1	32	32	0.1
Totals			**21798**	**100.0**

Step 2

Apply the Dulong Formula (Equation 8.9).

$$\text{Energy Content} = [145(35.8) + 610\left(7.8 - \frac{1}{8}(55.6)\right) + 40(0.1) + 10(0.7)]\ \text{BTU/lb}$$

$$\text{Energy Content} = \boxed{5721\ \text{BTU/lb}}$$

8.6.7 Biological properties

The organic fraction of refuse is often equated with both the volatile solids content, as determined from ignition at 550°C, and the biodegradability of the MSW. However, the use of volatile solids to estimate the waste biodegradability is often misleading, as not all organic materials (e.g., newsprint containing high lignin content) are easily biodegradable. Approximately 53% of MSW is amenable to biodegradation (Verma, 2002). Components degraded biologically are converted into gases, organic detritus, and inorganics. The primary gases produced through the decomposition of MSW are methane (CH_4) and carbon dioxide (CO_2). Methane and carbon dioxide are produced in approximately equal quantities and account for almost 100% of the volume of landfill gas produced. Other components of landfill gas include odorous compounds such as hydrogen sulfide (characterized by the smell of rotten eggs), methyl mercaptans, and aminobutyric acid.

8.7 Municipal landfill design

Modern landfills are significantly different than the local 'dump' many people envision. RCRA subtitle-D facilities are engineering structures designed and operated to keep waste, as well as the liquid emissions associated with its storage and decomposition, isolated from the environment. Systems are also designed to control the release of gaseous emissions associated with the waste mass.

8.7.1 Sizing a landfill

The airspace a landfill occupies is approximately the shape of a truncated pyramid (Figure 8.7). This airspace is occupied by the waste disposed as well as daily and intermediate cover materials, gas collection wells, and leachate recirculation devices. It is therefore necessary first to estimate the amount (volume) of waste that will be landfilled over the design life of the facility. Then, an estimate should be made of the percentage (usually 10–20%) of the airspace that will be occupied by materials other than waste. The landfill volume is then increased accordingly. When estimating the volume of landfilled waste, it is important to account for density changes due to compaction occurring during waste placement.

Once the total volume of the landfill has been estimated, the areal dimensions and final height of the landfill may be estimated. The final configuration of the landfill will depend on site geography and environmental conditions. A simplified approach to sizing a landfill is to assume a truncated pyramid design as shown in Figure 8.7. When the pyramid has a square base, an equation for the landfill volume may be developed as follows:

$$V_{LF} = \left(\frac{A_{base} + A_{top}}{2}\right) H \tag{8.10}$$

where:

V_{LF} = total landfill volume, [L^3]
A_{base} = area of the landfill base, [L^2]
L = width and length of the landfill base, [L]
A_{top} = area of the landfill top, [L^2]
L' = width and length of the landfill top, [L]
H = final height of landfill, [L].

Using the definition of a slope $= \frac{rise}{run} = \frac{1}{N}$ combined with simple geometry, we know that:

$$L' = L - 2HN \tag{8.11}$$

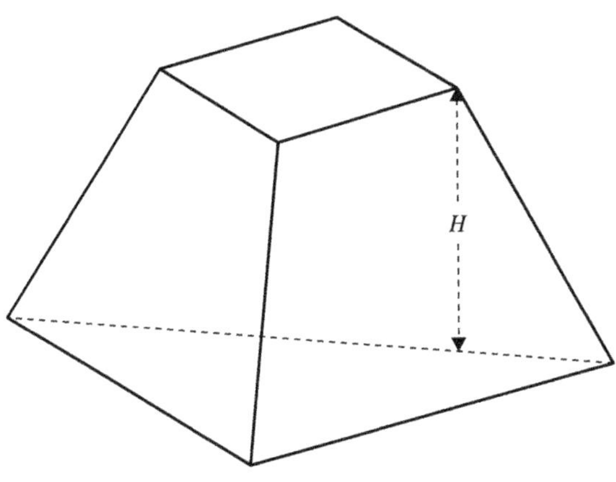

Figure 8.7 The typical landfill shape, a truncated pyramid.

Making these substitutions into the landfill volume equations provides:

$$V_{LF} = (L^2 - 2HNL + 2H^2N^2)H \tag{8.12}$$

Example 8.5 Sizing a landfill

Determine the size of the landfill required to serve a community with a population of 100,000 for ten years. The community has a waste disposal rate of 4.67 lb/capita/day and a by weight recycling rate of 33.8%. Assume that the population, disposal rate, and recycling rate will remain constant and waste density of 1,500 lb/yd^3.

Solution:

Waste Generation Data

Assume generation rate ~ constant @ 4.67 lb/cap/day

Waste Disposal Data

Recycling rate 33.8 %

Disposal rate 66.2 %

Waste Volume Calculation Table

Year	Population	Waste Generation Rate (lb/person/day)	Waste Disposal Rate (lb/person/day)	Yearly Waste Disposal (lb)
1	100,000	4.67	3.09	1.13E+08
2	100,000	4.67	3.09	1.13E+08
3	100,000	4.67	3.09	1.13E+08
4	100,000	4.67	3.09	1.13E+08
5	100,000	4.67	3.09	1.13E+08
6	100,000	4.67	3.09	1.13E+08
7	100,000	4.67	3.09	1.13E+08
8	100,000	4.67	3.09	1.13E+08
9	100,000	4.67	3.09	1.13E+08
10	100,000	4.67	3.09	1.13E+08
		Total waste disposed =		1.13E+09

$$\text{Yearly waste disposal rate} = \left(4.67\frac{\text{lb generated}}{\text{person}\cdot\text{day}}\right)$$

$$\times\left(\frac{66.2\text{ lb disposed}}{100\text{ lb generated}}\right) = 3.09\frac{\text{lb generated}}{\text{person}\cdot\text{day}}$$

Yearly waste disposal

$$= \left(3.09\frac{\text{lb disposed}}{\text{person}\cdot\text{day}}\right)(100{,}000\text{ persons})(365\text{ days})$$

$$= 1.13\times10^8\text{ lb}$$

$$\text{Total volume of waste disposed} = 1.13\times10^8\text{ lb}$$

$$\times\left(\frac{\text{yd}^3}{1500\text{ lb}}\right)\left(\frac{27\text{ ft}^3}{\text{yd}^3}\right) = 2.03\times10^7\text{ ft}^3$$

Assume that the waste occupies 90% of the landfill volume.

$$\text{Landfill volume} = 2.03\times10^7\text{ ft}^3\left(\frac{100\text{ ft}^3\text{ landfill}}{90\text{ ft}^3\text{ waste}}\right)$$

$$= 2.26\times10^7\text{ ft}^3$$

Assume a landfill height of 100 ft and side slopes with N=3 (3 run : 1 rise), then solve Equation (8.12) for L, length and width of base, using the Quadratic Equation or a root solver program. Solution using Microsoft *Excel's* Solver tool yields L = 669 ft.

Final Landfill Design:

- Side slopes (run:rise) = 3 : 1
- Landfill height (H) = 100 ft
- Length and width of landfill base (L) = 700 ft
- Area required for base = 490,000 ft^2 = 11.25 acres
- Landfill volume (airspace) = 2.5 × 10^7 ft^3

8.7.2 Leachate Production and Removal

Leachate is defined as liquid that has contacted waste or leachate. Leachate may contain dissolved or suspended materials leached from the waste mass. Most landfill leachate is associated with water entering the landfill from infiltrating rainfall. Leachate is also produced as the waste material is biologically degraded. Leachate production quantity can be estimated empirically, or by using a water mass balance technique that considers precipitation, evapotranspiration, surface runoff and moisture storage. A landfill water balance can be written as:

$$P = F + R + ET + \Delta S + \Delta M \tag{8.13}$$

where:

P = water from rainfall or snowmelt, L

F = water from rainfall or snowmelt that infiltrates the landfill, L

R = runoff water from rainfall or snowmelt that does not enter the landfill, L

ET = water removed from the landfill through the evapotranspiration process, L

ΔS = change in water storage within the landfill, L

ΔM = water consumed or produced in the landfill by microbial reactions, L.

Climate conditions have been shown to significantly affect leachate generation rates. Leachate generation rates in the Northeastern United States may be as high as 1,500 gallons per acre per day (gpad) while, in more arid regions, only 1–7 gpad may be generated (Veslind *et al.*, 2002 and Qian *et al.*, 2002).

The chemical composition of leachate changes as the landfill matures. In general, for a new landfill that is less than two years old, the pH value will be relatively low (pH ~6) (Tchobanoglous *et al.*, 1993), causing the heavy metals concentration to be relatively high. The five-day biochemical oxygen demand (BOD_5), total organic carbon (TOC), chemical oxygen demand (COD) and nutrients are also relatively high in new landfills. For example, the typical BOD_5 and COD concentrations in new landfills are 20,000 and 18,000 mg/L (McBean *et al.*, 1995 and Tchobanoglous *et al.*, 1993), respectively.

As the landfill matures, the average leachate pH tends to increase to approximately pH = 7, causing the apparent metals concentration to decrease (metals tend to be less soluble at neutral pH values). The BOD_5, COD and TOC also show a decrease in mature landfills (greater than ten years). Typical BOD_5 and COD values for mature landfill leachate are 50 and 300 mg/L, respectively (McBean *et al.*, 1995 and Tchobanoglous *et al.*, 1993).

The biodegradability of the leachate measured by the BOD_5/COD ratio also changes as the landfill matures. Initial BOD_5/COD ratios may be greater than 0.5, indicating that the organic matter in the leachate is readily degradable. In a mature landfill, this ratio drops to approximately 0.1, indicating that the remaining organics (typically humic and fulvic acids) in the leachate are not readily degradable.

8.7.3 Leachate collection and recovery system

All subtitle-D landfills are designed with leachate collection and recovery systems (LCRS). Design considerations for the system include a barrier layer, rigid liner, and collection system. A composite liner system, a geomembrane above a clay barrier layer (Figure 8.8), is more effective at limiting leachate migration into the subsoil than either a clay layer or a geomembrane liner alone, and is required by RCRA subtitle-D for MSW landfills.

Geomembranes, also called flexible membrane liners (FML), are often constructed from high-density polyethylene (HDPE) and are considered to be not only strong but impermeable to water. The use of this material helps to minimize the transfer of leachate from the landfill to the environment. Federal law requires that the thickness of geomembranes used in landfill liner construction have a minimum thickness of 60 mm. RCRA subtitle-D minimum design standard for the compacted clay layer is a 60 cm thickness, with a hydraulic conductivity of no more than 10^{-7} cm/s.

A leachate collection and recovery system, consisting of a lateral drainage layer (sand, gravel, and/or a geonet) and a series of perforated leachate collection pipes (Figure 8.8), is used to convey leachate off of the geomembrane and out of the landfill. Once out of the LCRS, leachate is routed via gravity flow or pumps, through solid piping, to holding tanks or ponds for storage prior to treatment.

The potential for LCRS damage associated with waste placement is reduced by placing a protective layer above the lateral drainage layer (Figure 8.8.). This protective layer has traditionally been composed of soil, sand, and gravel, but many landfills now use a layer of soft refuse, such as paper, organic refuse, shredded tires, and rubber. The design of a leachate collection system consists of: estimating the leachate loading rate; specifying the lateral pipe spacing, drainage media characteristics, and slope of liner system; and predicting the resulting depth of leachate on the liner system. RCRA subtitle-D specifies the maximum allowable depth of leachate on the liner system to be 30.5 cm.

Various researchers have developed equations to predict the head on the liner as a function of these design parameters and leachate loading rate. The majority of these equations were developed based on a hydraulic flow scheme similar to the one shown in Figure 8.9.

McEnroe (1993) developed a set of Equations (8.14a, b, and c) for use in the prediction of the maximum depth of leachate head above the impervious sloped bed. The equations were developed using Darcy's Law and the Dupuit-Theim Assumption. The upstream boundary condition was assumed to be either a drainage divide or an impervious side wall which would result in no flow laterally across it. A free drainage condition, with the water level in the drain trench beneath the top of the barrier layer and a hydraulic gradient of − 1, was assumed at the downstream boundary condition. It is important to recognize that, if the drain is not working properly, the free drainage condition may not be met and the actual leachate depth would exceed the predicted level for the specified design conditions.

for $R < \frac{1}{4}$:

$$Y_{max} = \sqrt{R - RS + R^2S^2}\left[\frac{(1 - A - 2R)(1 + A - 2R)}{(1 + A - 2R)(1 - A - 2R)}\right]^{1/2A} \tag{8.14a}$$

for $R = \frac{1}{4}$:

$$Y_{max} = \frac{R(1 - 2RS)}{(1 - 2R)} = \exp\left(\frac{2R(S - 1)}{(1 - 2RS)(1 - 2R)}\right) \tag{8.14b}$$

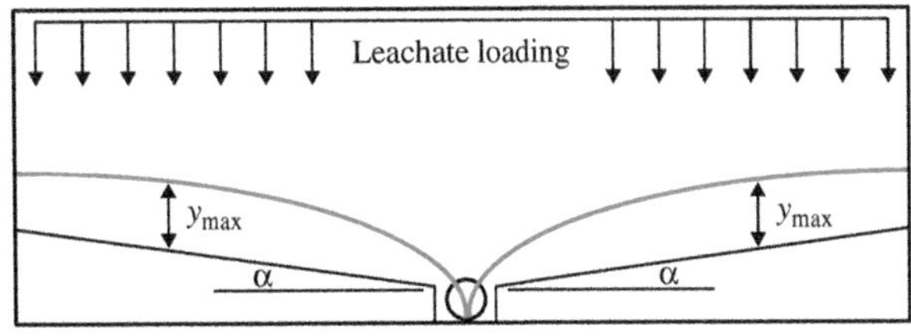

Figure 8.9 Definition sketch for landfill drainage systems.

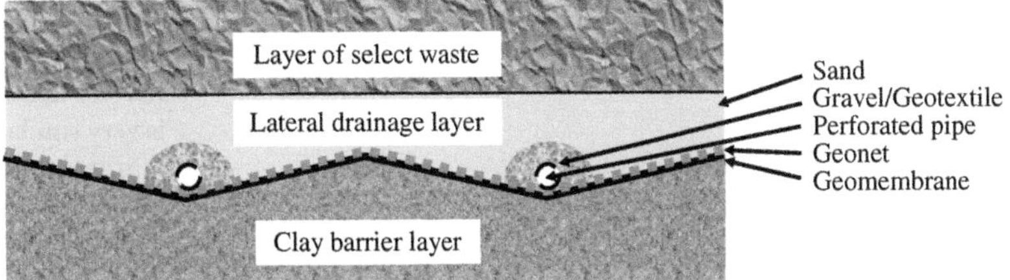

Figure 8.8 Example of a composite liner system.

for $R > \frac{1}{4}$:

$$Y_{max} = \sqrt{R - RS + R^2S^2} \times \exp\left(\frac{1}{B}\tan^{-1}\left(\frac{2RS-1}{B}\right) - \frac{1}{B}\tan^{-1}\left(\frac{2R-1}{B}\right)\right) \quad (8.14c)$$

where:

R = $\frac{r}{K\sin^2\alpha}$, unitless
A = $\sqrt{1-4R}$, unitless
B = $\sqrt{4R-1}$, unitless
S = $\tan(\alpha)$, slope of liner, unitless
Y_{max} = $\frac{y_{max}}{L}$, dimensionless maximum head on liner
y_{max} = maximum head on line, [L]
L = horizontal drainage distance, [L]
$2L$ = lateral pipe spacing, [L]
α = inclination of liner from horizontal, degrees
K = hydraulic conductivity of the drainage layer, $[LT^{-1}]$
r = leachate loading rate, $[LT^{-1}]$.

The use of a geonet above the geomembrane liner will significantly increase the allowable pipe spacing. The lateral distance between collection pipes in a LCRS using a geonet between the liner and drainage media may be calculated using Equation (8.15) (US EPA, 1989).

$$\theta_{reqd} = \frac{qL^2}{4h_{max} + 2L\sin\alpha} \quad (8.15)$$

where:

θ_{reqd} = transmissivity of geonet, $[L^2T^{-1}]$
L = distance between collection pipes, [L]
h_{max} = maximum head on liner, [L]
q = leachate loading rate, $[LT^{-1}]$
α = slope of drainage system, degrees.

8.7.4 Leakage through a clay barrier layer

Darcy's Law (Equation 8.6) can be used to develop an equation for the leakage rate through a soil liner without a geomembrane liner (Equation 8.16). The development of Equation (8.16) includes the assumptions that the hydraulic conductivity of the material above the soil barrier is much greater than the hydraulic conductivity of the soil barrier, and that the zone beneath the soil barrier is unsaturated and has a hydraulic conductivity much greater than the soil barrier's.

$$Q = kiA = k\left(1 + \frac{h}{D}\right)A \quad (8.16)$$

where:

Q = soil barrier leakage rate, $[L^3T^{-1}]$
k = hydraulic conductivity of the liner, $[LT^{-1}]$
A = surface area of the liner, $[L^2]$
i = hydraulic gradient, [L/L]
h = liquid depth above the soil barrier, [L]
D = thickness of the soil barrier, [L].

8.7.5 Leakage through perforations in a geomembrane liner

Giroud *et al.* (1994) presented empirical equations for the calculation of leakage rates through a variety of configurations of natural and geosynthetic layers. The leakage rate through perforations in the geomembrane component of a traditional composite liner (Figure 8.8) was found to be a function of the perforation's size, the hydraulic conductivity of the barrier layer, the liquid depth above the liner, and contact between the geomembrane and the barrier layer. Good contact conditions correspond to a geomembrane installed with few wrinkles placed on top of a well-compacted, low-permeability layer with a smooth surface. Poor contact conditions are characterized by a geomembrane installed with a number of wrinkles or placement on a poorly-compacted, low-permeability layer with an uneven surface.

Equations (8.17a, b, and c) can be used to calculate leakage rates for good, average, and poor contact conditions, respectively, when the hydraulic conductivity of the barrier layer is less than 10^{-6} m/s and the thickness of the barrier layer is greater than the hydraulic head. The average contact equation was formulated by averaging the coefficients from the poor and good contact equations.

$$Q = 1.15a^{0.1}h^{0.9}k^{0.74} \quad \text{(poor contact)} \quad (8.17a)$$

$$Q = 0.6a^{0.1}h^{0.9}k^{0.74} \quad \text{(average contact)} \quad (8.17b)$$

$$Q = 0.21a^{0.1}h^{0.9}k^{0.74} \quad \text{(good contact)} \quad (8.17c)$$

where:

Q = leakage rate through a liner perforation, $[m^3/s]$
a = defect size, $[m^2]$
h = depth of liquid above the defect, [m]
k = hydraulic conductivity of the barrier layer, [m/s].

8.7.6 Vapor diffusion across a geomembrane liner

In addition to pressure driven flow through pinholes and defects, leachate can move across an intact geomembrane liner by vapor diffusion. The USEPA HELP Model's documentation (Schroeder *et al.*, 1994) describes a calculation process (Equations 8.18a, b) for vapor diffusion across an intact geomembrane, based on work by Giroud & Bonaparte (1989). The equations were developed by combining concepts from Fick's Law for concentration gradient driven flow, and Darcy's Law for hydraulic gradient driven flow. The resulting equation is very similar to Equation 8.16 (clay barrier leakage) and utilizes a term called the equivalent hydraulic conductivity (Table 8.8) for diffusion calculations which combines the effect of the diffusivity and thickness of the geomembrane and the density of the liquid layer.

Table 8.8 Geomembrane characteristics for vapor diffusion calculations.

Material	Equivalent hydraulic conductivity for vapor diffusion calculations, cm/sec
Neoprene	3×10^{-12}
Nitrile Rubber	3×10^{-11}
Low-Density Polyethylene (LDPE)	4×10^{-13}
High-Density Polyethylene(HDPE)	2×10^{-13}
Polyvinyl Chloride (PVC)	4×10^{-11}
Saran Film	6×10^{-14}

Source: Schroeder *et al.*, 1994.

$$Q_g = 0 \quad \text{for } h = 0 \qquad (8.18a)$$

$$Q_g = K_g\left(1 + \frac{h}{T}\right)A \quad \text{for } h > 0 \qquad (8.18b)$$

where:

Q_g = diffusion drive leakage rate, [L^3/T]
K_g = equivalent hydraulic conductivity for diffusion calculation, [L/T]
h = average depth of liquid above geomembrane liner, [L]
T = geomembrane thickness, [L]
A = surface area of the liner, [L^2].

8.7.7 Stormwater control

Stormwater control strategies regulated through Subtitle D of RCRA are used to reduce leachate production. During the active phase of landfilling, intermediate cover layers consisting of compacted soil are used to limit infiltration, and the associated production of leachate, by increasing surface runoff. Intermediate cover is thicker and more permanent than daily cover, in which typically a six-inch thick soil layer is used to limit vectors, odor, and litter. Intermediate cover is typically no less than 12 inches thick and consists of soil that is readily available on-site or can be obtained from nearby borrow pits.

The final cap or final cover layer on a landfill is primarily used to limit infiltration (leachate production) and to control gas emissions. The typical configuration of a final cap can be seen in Figure 8.10.

Cap design is based on:

1 hydrologic principles that maximize runoff to limit infiltration;
2 geotechnical principles to prevent slope failures (if the drainage layer is not designed properly, pore pressure in the soil at the geomembrane surface will increase and the soil above the liner will slide down); and
3 gas collection system requirements.

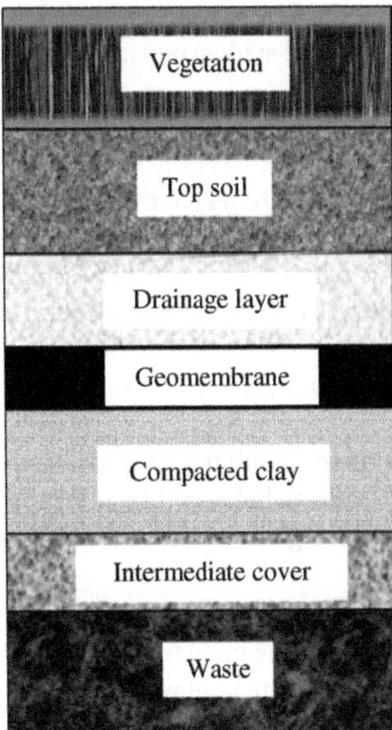

Figure 8.10 Schematic of a typical RCRA subtitle-D cap used for closing landfills.

8.7.8 Gas generation and control

Microbially mediated oxidation-reduction reactions that occur as organic material is decomposed are principally responsible for gas generation in landfills. Knowledge of the gas production volume over the lifetime of the landfill, and the chemical composition, are important to owners, operators, and the public, as the methane production is a potential energy source, a potent greenhouse gas, and an explosive danger. Thus, there are safety and public health concerns about the potential migration of landfill gases away from the landfill to nearby. Gases produced may also cause an odor nuisance.

The stoichiometric equation used to describe the breakdown of organic waste (Barlaz & Ham, 1993) can be approximated as:

$$C_aH_bO_cN_d + \left(\frac{4a - b - 2c + 3d}{8}\right)H_2O \rightarrow \left(\frac{4a + b - 2c - 3d}{8}\right)CH_4 + \left(\frac{4a - b + 2c + 3d}{8}\right)CO_2 + dNH_3 \qquad (8.19)$$

where a, b, c, and d are the number of atoms of carbon, hydrogen, oxygen, and nitrogen, respectively.

The composition of the gas produced is typically 40–60% methane, with the balance primarily being carbon dioxide. Higher percentages of methane can be realized as CO_2 is absorbed by the leachate.

Landfill stabilization occurs as a result of microbial metabolism or the anaerobic fermentation process. Stabilization can be considered a three-step process that begins with enzyme-mediated hydrolysis reactions that convert high molecular weight compounds such as lipids and proteins into fatty acids, amino acids and monosaccharides.

Products from the first step are converted through a microbial process called acidogenesis, which involves acidogens or acid formers.

Products from the second step include lower molecular weight compounds such as acetic acid that are converted by methanogens (strict anaerobes) in the final step into methane and carbon dioxide.

A synotrophic relationship exists between the methanogens and acidogens. While the acidogens produce products that the methanogens use as reactants, the methanogenic bacteria remove compounds from the system that inhibit the growth of acidogens.

Gas production in a landfill is often characterized by identifying five sequential phases that occur throughout the life of a landfill (Tchobanoglous *et al.*, 1993 and Vesilind *et al.*, 2002).

- *Initial Adjustment* – the microbial decomposition that occurs just after placement of waste into the landfill. This phase is characterized by the aerobic decomposition of waste that continues until the available oxygen is utilized and a transition is made to anaerobic conditions. Moisture may accumulate and some lag time may be observed until favorable conditions for stabilization are developed.
- *Transition Phase* – the transition that occurs as oxygen is depleted and anaerobic conditions develop. The oxidation/reduction potential (ORP) drops to values ranging between −150 to −300 mV. A drop in pH is also observed due to the formation of organic acids.
- *Acid Phase* – acidogens become active and the formation of organic acids are realized. Heavy metals are mobilized due to the decrease in pH. If leachate is not recycled, essential nutrients may be lost in the leachate stream.
- *Methane Fermentation Phase* – In this stage, methanogenic organisms become more predominant and methane and acid production proceed simultaneously. A reduced production rate of acid may be noticed. As the organic acids present are converted into methane and carbon dioxide, the pH begins to rise to a more neutral value. This is followed by metal complexation and the production of landfill gas peaks. The BOD, COD and conductivity of the leachate will decrease.
- *Maturation Phase* – As the landfill matures, the lack of available nutrients begin to limit microbial growth and conditions shift from active degradation to relative dormancy. Gas production drops and aerobic conditions may reappear depending on the capping technique used. With continued degradation, a humic-like substance is produced.

Example 8.6 Calculating Landfill Gas Generation

A landfill is expected to receive waste with the composition described in Example 8.3 at a rate of 100 tons per day, 5 days per week. The landfill has a design life of 5 years. Determine the volume and rate of landfill gas production.

Solution:

Microsoft *Excel* was used to perform the calculations associated with this solution.

Basic data for Example Problem 8.6

Component	Disposed waste composition (% by wt.)	Decay rate	Wet moisture Content (%)	Chemical composition of individual waste components (% by weight on dry basis)					
				Carbon	Hydrogen	Oxygen	Nitrogen	Sulfur	Ash
Food waste	10.0	Rapid	70	48	6.4	37.6	2.6	0.4	5
Paper	35.0	Rapid	6	43.5	6	44	0.3	0.2	6
Cardboard	6.0	Rapid	5	44	5.9	44.6	0.3	0.2	5
Plastics	8.0	n/a	2	60	7.2	22.8	0	0	10
Textiles	3.0	Slow	10	55	6.6	31.2	4.6	0.15	2.5
Rubber	1.0	Slow	2	78	10	0	2	0	10
Leather	1.0	Slow	10	60	8	11.6	10	0.4	10
Yard wastes	20.5	50/50	60	47.8	6	38	3.4	0.3	4.5
Wood	3.0	Slow	20	49.5	6	42.7	0.2	0.1	1.5
Glass	8.0	n/a	2	0.5	0.1	0.4	0.1	0	98.9
Aluminum	0.5	n/a	2	4.5	0.6	4.3	0.1	0	90.5
Dirt, ash, etc.	4.0	n/a	8	26.3	3	2	0.5	0.2	68
sum:	100.0								

Steps 1 and 2 from Table 8.9:

Table 8.9 Process for calculating the volume and rate of landfill gas generation.

1 Separate waste materials (by weight) into RAPIDLY-, SLOWLY-, and NON-degradable components.

2 Calculate the chemical composition (in lb·mole) without sulfur of the RAPIDLY and SLOWLY degradable materials. Normalize the chemical formula.

3 Use a stoichiometric equation (Equation 8.19) to calculate the CH_4 and CO_2 produced by the decomposition of the RAPIDLY and SLOWLY degraded material in moles.

4 Calculate ft^3 of methane and carbon dioxide produced by SLOW and RAPID decomposition. Review the examples below paying particular attention to the units. Be sure you understand the different components of the equations:

- CH_4 (ft^3) = (moles of methane, RAPID or SLOW, from STEP 3)(16 lb/lb·mole)(Dry weight of RAPID or SLOW material)/(MW of substrate in stoichiometric equation)/(0.0448 lb/ft^3 CH_4)
- CO_2 (ft^3) = (moles of carbon dioxide, RAPID or SLOW, from STEP 3)(44 lb/lb·mole)(Dry weight of RAPID or SLOW material)/ (MW of substrate in stoichiometric equation)/(0.1235 lb/ft^3 CO_2).

5 Determine the total theoretical gas production from RAPID and SLOW decomposition per lb of total sample.

- Example: RAPID = (CH_4 + CO_2)/total sample weight *(generally 100 lb)*.

6 Calculate the total RAPID and SLOW gas produced by the first year of waste disposal using the data calculated in Step 5) and the total amount of waste disposed.

7 Determine the gas production profile:

- Step A. Make assumptions regarding the RAPID and SLOW gas production profiles.
  - Example assumptions: Rapid production occurs over five years, with the peak occurring at year 1. Slow production occurs over 15 years, with the peak occurring at year 5.
- Step B. Calculate the peak RAPID and SLOW gas generation rates.
  - Note: The area under the gas production vs. time curve is the total volume of gas produced.

8 Calculate the yearly gas production rates and volumes from the first year of waste disposal.

- The results from a single year would plot roughly as shown below:

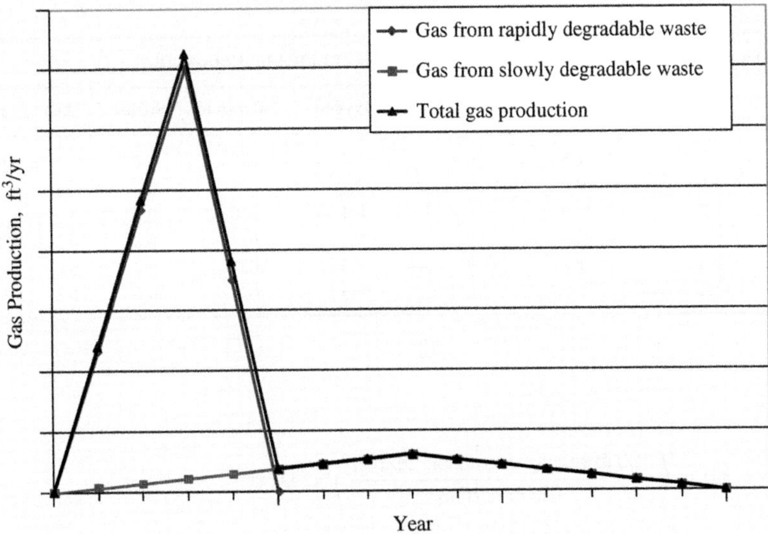

9 Repeat Steps 6) through 9) for every year waste is disposed.

10 Use superposition to combine yearly production rates.

11 Plot this data to create the gas production characteristics for the entire landfill life.

Rapidly degradable materials

Component	Wet weight (lb)	Dry weight (lb)	Weights of individual atoms (lb)						
			Carbon	Hydrogen	Oxygen	Nitrogen	Sulfur	Ash	*sum:*
Food waste	10.0	3.0	1.4	0.2	1.1	0.1	0.0	0.2	3.0
Paper	35.0	32.9	14.3	2.0	14.5	0.1	0.1	2.0	32.9
Cardboard	6.0	5.7	2.5	0.3	2.5	0.0	0.0	0.3	5.7
Yard wastes	10.25	4.1	2.0	0.2	1.6	0.1	0.0	0.2	4.1
sum:	61.3	45.7	20.2	2.7	19.7	0.3	0.1	2.6	

Example calculations:

$$\text{Wet weight of food waste} = 10.00\ \text{lb}\left(\frac{100\ \text{lb rapidly degradable food waste}}{100\ \text{lb food waste}}\right) = 10.0\ \text{lb}$$

$$\text{Dry weight of food waste} = 10.0\ \text{lb}\left(1 - \frac{70\ \text{lb water}}{100\ \text{lb food waste}}\right) = 3.0\ \text{lb}\left(\frac{48\ \text{lb carbon}}{100\ \text{lb food waste}}\right) = 1.4\ \text{lb}$$

$$\text{Carbon content of food waste} = 3.0\ \text{lb}$$

$$\text{Wet weight of food waste} = 20.5\ \text{lb}\left(\frac{50\ \text{lb rapidly degradable food waste}}{100\ \text{lb food waste}}\right) = 10.25\ \text{lb}$$

$$\text{Dry weight of food waste} = 10.25\ \text{lb}\left(1 - \frac{60\ \text{lb water}}{100\ \text{lb food waste}}\right) = 4.1\ \text{lb}\left(\frac{6\ \text{lb carbon}}{100\ \text{lb food waste}}\right) = 0.2\ \text{lb}$$

$$\text{Hydrogen content of yard waste} = 4.1\ \text{lb}$$

Slowly Degradable Materials

Component	Wet Weight (lb)	Dry Weight (lb)	Weights of Individual Atoms (lb)						
			Carbon	Hydrogen	Oxygen	Nitrogen	Sulfur	Ash	*sum:*
Textiles	3.0	2.7	1.5	0.2	0.8	0.1	0.0	0.1	2.7
Rubber	1.0	1.0	0.8	0.1	0.0	0.0	0.0	0.1	1.0
Leather	1.0	0.9	0.5	0.1	0.1	0.1	0.0	0.1	0.9
Yard wastes	10.25	4.1	2.0	0.2	1.6	0.1	0.0	0.2	4.1
Wood	3.0	2.4	1.2	0.1	1.0	0.0	0.0	0.0	2.4
sum:	18.3	11.1	5.9	0.7	3.5	0.4	0.0	0.5	

Example Calculations:

$$\text{Wet weight of textiles} = 3.0\ \text{lb}\left(\frac{100\ \text{lb rapidly degradable textiles}}{100\ \text{lb textiles}}\right) = 3.0\ \text{lb}$$

$$\text{Dry weight of textiles} = 3.0\ \text{lb}\left(1 - \frac{10\ \text{lb water}}{100\ \text{lb food waste}}\right) = 2.7\ \text{lb}$$

$$\text{Carbon content of textiles} = 2.7\ \text{lb}\left(\frac{55\ \text{lb carbon}}{100\ \text{lb food waste}}\right) = 1.5\ \text{lb}$$

$$\text{Wet weight of yard waste} = 20.5\ \text{lb}\left(\frac{100\ \text{lb rapidly degradable yard waste}}{100\ \text{lb food waste}}\right) = 10.25\ \text{lb}$$

$$\text{Dry weight of yard waste} = 10.0\ \text{lb}\left(1 - \frac{60\ \text{lb water}}{100\ \text{lb yard waste}}\right) = 4.1\ \text{lb}$$

$$\text{Hydrogen content of yard waste} = 4.1\ \text{lb}\left(\frac{6\ \text{lb carbon}}{100\ \text{lb food waste}}\right) = 0.2\ \text{lb}$$

	Carbon	Hydrogen	Oxygen	Nitrogen
Molecular weight (lb/mole)	12.0	1	16	14
Molar composition of rapidly degradable waste	1.68	2.75	1.23	0.02
Molar composition of slowly degradable waste	0.49	0.74	0.22	0.03

Normalized Molar Compositions

	Carbon 'a'	Hydrogen 'b'	Oxygen 'c'	Nitrogen 'd'	Molecular Weight (lb/lb·mole)
Rapidly degradable waste	70.8	115.5	51.7	1.0	1807
Slowly degradable waste	18.3	27.3	8.2	1.0	392

Example Calculations:

Normalization is the process of adjusting a chemical composition based on 1 mole of the least prevalent atom. In the case above, the least prevalent atom is nitrogen. The normalized composition of the rapidly degradable waste is

$$C_{\frac{1.68}{0.02}}H_{\frac{2.75}{0.02}}O_{\frac{1.23}{0.02}}N = C_{70.8}H_{115.5}O_{51.7}N$$

The molecular weight of the rapidly degradable waste was calculated as follows:

$$(70.8 \text{ mole C})\left(\frac{12 \text{ lb}}{\text{lb mole}}\right) + (115.5 \text{ mole H})\left(\frac{1 \text{ lb}}{\text{lb mole}}\right) + (51.7 \text{ mole O})\left(\frac{16 \text{ lb}}{\text{lb mole}}\right) + (1 \text{ mole N})\left(\frac{14 \text{ lb}}{\text{lb mole}}\right)$$
$$= 1807 \frac{\text{lb}}{\text{lb mole}}$$

Step 3 from Table 8.9:

	Coefficient for each molecule			
	H_2O	CH_4	CO_2	NH_3
Rapidly degradable waste	16.8	36.5	34.3	1.0
Slowly degradable waste	8.2	10.2	8.2	1.0

Step 4 and 5 from Table 8.9:

Gas production from rapidly degradable waste

CH_4	329.7	ft^3/100 lb disposed waste
CO_2	345.2	ft^3/100 lb disposed waste
sum:	675.1	ft^3/100 lb disposed waste

Gas production from slowly degradable waste

CH_4	91.8	ft^3/100 lb disposed waste
CO_2	82.2	ft^3/100 lb disposed waste
sum:	174.0	ft^3/100 lb disposed waste

Example calculation:

$$CH_4 \text{ produced from rapid waste decomposition} = \left(\frac{45.7 \text{ lb rapidly degradable waste}}{100 \text{ lb waste disposed}}\right)$$
$$\times \left(\frac{\text{mole rapidly degradable waste}}{1807 \text{ lb rapidly degradable waste}}\right)\left(\frac{36.5 \text{ mole } CH_4 \text{ produced}}{\text{mole rapidly degradable waste}}\right)\left(\frac{16 \text{ lb } CH_4}{\text{mole } CH_4}\right)$$
$$\times \left(\frac{ft^3\ CH_4}{0.0448 \text{ lb } CH_4}\right) = 329.7 \frac{ft^3\ CH_4}{100 \text{ lb waste disposed}}$$

Steps 6, 7, and 8 from Table 8.9:

Now, 100 tons are disposed per day, 5 days per week for a total yearly waste disposal rate of 52×10^6 lb. The analysis will be conducted for 5 years of waste disposal and the analysis will use a time step of 1 year.

Rapid waste degradation profile: 5 year degradation time with a peak at year 3

Slow waste degradation profile: 15 year degradation time with a peak at year 8

Total gas production per year		
Rapidly degradable waste	3.51E+08	ft^3
Slowly degradable waste	9.05E+07	ft^3
Peak production rates		
Rapidly degradable waste	1.40E+08	ft^3/yr
Slowly degradable waste	1.21E+07	ft^3/yr

Example calculations:

$$\text{Yearly gas production from rapidly degradable wastes} = (52 \times 10^6 \text{ lb of yearly waste disposal}) \times \left(\frac{675.1 \text{ ft}^3 \text{ of gas production}}{100 \text{ lb waste disposed}}\right) = 3.51 \times 10^8 \text{ ft}^3$$

The gas production rate profile is assumed to have a triangular shape with the height of the triangle equal to the peak gas production rate and the base of the triangle equal to the total time over which gas is produced. The area under the gas production rate curve is equal to the volume of gas produced. The area of the triangular gas production profile and the volume of gas produced are equal and an equation for the peak gas production rate can be written as follows:

$$\text{Peak gas production rate} = 2\frac{\text{Volume of gas produced}}{\text{Duration of gas production}}$$

In this problem, the rapidly and slowly degradable materials were assumed to decompose completely over 5 and 15 years, respectively.

$$\text{Peak production rate of gas produced from rapidly degradabe wastes} = \frac{2(3.51 \times 10^8 \text{ ft}^3}{5 \text{ years}} = 1.40 \times 10^8 \frac{\text{ft}^3}{\text{yr}}$$

$$\text{Peak production rate of gas produced from slowly degradabe wastes} = \frac{2(9.05 \times 10^7 \text{ ft}^3}{15 \text{ years}} = 1.21 \times 10^7 \frac{\text{ft}^3}{\text{yr}}$$

	Gas production rate (ft^3/yr)		
Year	**Rapidly degradable waste**	**Slowly degradable waste**	**Total**
0	0	0	0
1	4.7E+07	1.5E+06	4.8E+07
2	9.4E+07	3.0E+06	9.7E+07
3	1.4E+08	4.5E+06	1.4E+08
4	7.0E+07	6.0E+06	7.6E+07
5	0	7.5E+06	7.5E+06
6		9.0E+06	9.0E+06
7		1.1E+07	1.1E+07
8		1.2E+07	1.2E+07
9		1.0E+07	1.0E+07
10		8.6E+06	8.6E+06
11		6.9E+06	6.9E+06
12		5.2E+06	5.2E+06
13		3.4E+06	3.4E+06
14		1.7E+06	1.7E+06
15		0	0

Example calculations:

For the rapidly degradable waste, the peak gas production rate, 1.4x10^8 ft^3/yr, was assumed to occur in year 3 of the five-year decomposition period. The degradation rates at years 1, 2, and 4 can be calculated based on the three-year rise to the peak gas production rate and the two-year fall from the peak production rate to the end of gas production from rapidly degradable waste.

Year 1 gas production rate from rapidly degradable waste

$$= \left(\frac{1}{3}\right)\left(1.4 \times 10^8 \frac{ft^3}{yr}\right) = 4.7 \times 10^7 \frac{ft^3}{yr}$$

Year 2 gas production rate from rapidly degradable waste

$$= \left(\frac{2}{3}\right)\left(1.4 \times 10^8 \frac{ft^3}{yr}\right) = 9.4 \times 10^7 \frac{ft^3}{yr}$$

Year 4 gas production rate from rapidly degradabe waste

$$= \left(\frac{1}{2}\right)\left(1.4 \times 10^8 \frac{ft^3}{yr}\right) = 7.0 \times 10^7 \frac{ft^3}{yr}$$

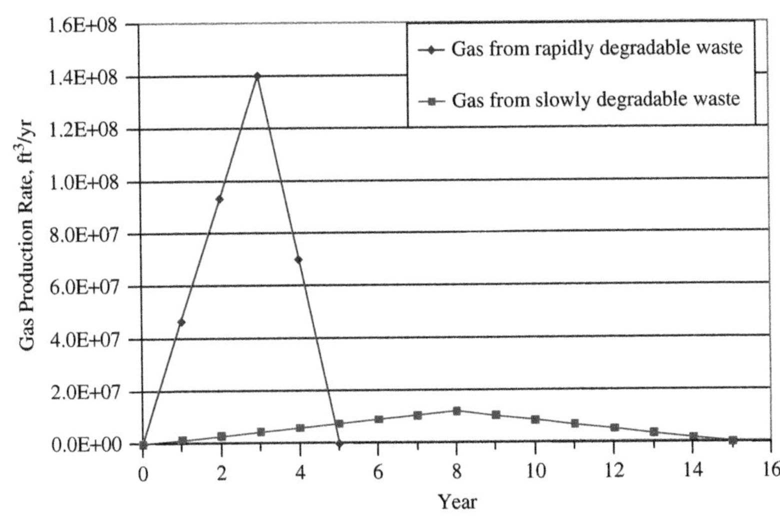

Steps 9, 10, and 11 from Table 8.9:

Rapid Gas Production Rates, ft^3/yr

Year	1	2	3	4	5	sum:
0	0					0.0E+00
1	4.7E+07	0				3.9E+07
2	9.4E+07	4.7E+07	0			1.2E+08
3	1.4E+08	9.4E+07	4.7E+07	0		2.3E+08
4	7.0E+07	1.4E+08	9.4E+07	4.7E+07	0	2.9E+08
5	0	7.0E+07	1.4E+08	9.4E+07	4.7E+07	2.9E+08
6		0	7.0E+07	1.4E+08	9.4E+07	2.5E+08
7			0	7.0E+07	1.4E+08	1.7E+08
8				0	7.0E+07	5.8E+07
9					0	0.0E+00

Slow Gas Production Rates, ft^3/yr

Year	1	2	3	4	5	sum:
0	0					0.0E+00
1	1.5E+06	0				1.2E+06
2	3.0E+06	1.5E+06	0			3.7E+06
3	4.5E+06	3.0E+06	1.5E+06	0		7.4E+06
4	6.0E+06	4.5E+06	3.0E+06	1.5E+06	0	1.2E+07
5	7.5E+06	6.0E+06	4.5E+06	3.0E+06	1.5E+06	1.9E+07
6	9.0E+06	7.5E+06	6.0E+06	4.5E+06	3.0E+06	2.5E+07
7	1.1E+07	9.0E+06	7.5E+06	6.0E+06	4.5E+06	3.1E+07
8	1.2E+07	1.1E+07	9.0E+06	7.5E+06	6.0E+06	3.7E+07
9	1.0E+07	1.2E+07	1.1E+07	9.0E+06	7.5E+06	4.1E+07
10	8.6E+06	1.0E+07	1.2E+07	1.1E+07	9.0E+06	4.1E+07
11	6.9E+06	8.6E+06	1.0E+07	1.2E+07	1.1E+07	4.0E+07
12	5.2E+06	6.9E+06	8.6E+06	1.0E+07	1.2E+07	3.5E+07
13	3.4E+06	5.2E+06	6.9E+06	8.6E+06	1.0E+07	2.8E+07
14	1.7E+06	3.4E+06	5.2E+06	6.9E+06	8.6E+06	2.1E+07
15	0	1.7E+06	3.4E+06	5.2E+06	6.9E+06	1.4E+07
16		0	1.7E+06	3.4E+06	5.2E+06	8.5E+06
17			0	1.7E+06	3.4E+06	4.2E+06
18				0	1.7E+06	1.4E+06
19					0	0.0E+00

Summary Production Rates, ft^3/yr

Year	Rapid	Slow	Total
0	0.0E+00	0.0E+00	0.0E+00
1	3.9E+07	1.2E+06	4.0E+07
2	1.2E+08	3.7E+06	1.2E+08
3	2.3E+08	7.4E+06	2.4E+08
4	2.9E+08	1.2E+07	3.0E+08
5	2.9E+08	1.9E+07	3.1E+08
6	2.5E+08	2.5E+07	2.8E+08
7	1.7E+08	3.1E+07	2.1E+08
8	5.8E+07	3.7E+07	9.5E+07
9	0.0E+00	4.1E+07	4.1E+07
10		4.1E+07	4.1E+07
11		4.0E+07	4.0E+07
12		3.5E+07	3.5E+07
13		2.8E+07	2.8E+07
14		2.1E+07	2.1E+07
15		1.4E+07	1.4E+07
16		8.5E+06	8.5E+06
17		4.2E+06	4.2E+06
18		1.4E+06	1.4E+06
19		0.0E+00	0.0E+00

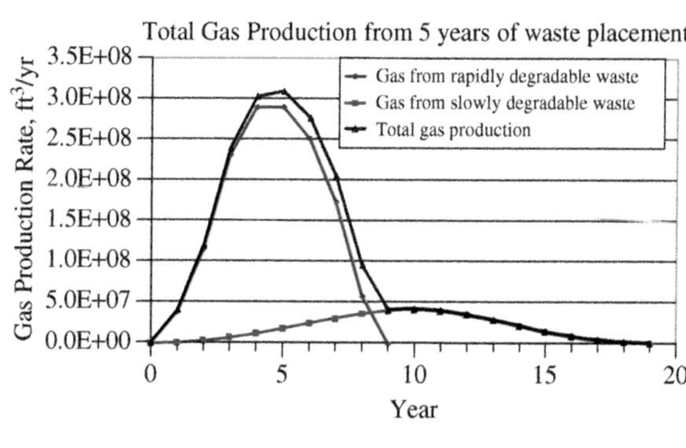

8.7.9 Monitoring and post-closure care

Post-closure care of a landfill requires both groundwater monitoring and methane monitoring (40CFR258). After closure, groundwater and methane monitoring must continue until the owner or operator of the municipal solid waste landfill (MSWLF) can demonstrate that there is no potential for the migration of hazardous constituents derived by the landfill. This period of monitoring is typically required to last for 30 years after the landfill has been closed. This demonstration must be made through use of site-specific field collected measurements, sampling, and analysis of physical, chemical and biological processes affecting contaminant fate and transport. Contaminant fate and transport predictions that maximize contaminant migration and consider impacts on human health and the environment are also to evaluate the potential impact of closed landfills.

Owners or operators of all MSWLF units must implement a routine methane monitoring program to ensure that the concentration of methane gas generated by the facility does not exceed 25% of the lower explosive limit (LEL) for methane in facility structures (excluding gas control or recovery system components), and that the concentration of methane gas does not exceed the LEL for methane at the facility property boundary. The sampling routine must be designed on the basis of factors such as soil, hydrogeologic, and hydraulic conditions surrounding the facility with the minimum frequency of monitoring being quarterly. Methane is explosive when present in a range of 5 (lower explosive limit, LEL) to 15 (upper explosive limit (UEL)) percent by volume in air. Methane is not explosive when present in concentrations greater than 15%, but fire and asphyxiation are still a threat at these levels. Also, any dilution due to mixing with ambient air could bring the mixture back into the explosive range.

Air criteria requirements are becoming a bigger issue for landfills as the control of greenhouse gases (GHGs) becomes more of a priority. Both CH_4 and CO_2 are greenhouse gases, and CH_4 is ten times more potent than CO_2. A potential benefit of aerobic landfilling is a reduction in the CH_4 emissions; however, aerobic processes produce NO_X, another green-house gas, which is more difficult to treat than methane. Landfill gas flares convert CH_4 to CO_2 and H_2O, thus reducing the severity of GHGs emitted from the LF.

8.8 Case study – Mecklenburg County, North Carolina (NC), USA

Prepared by Gregory D. Boardman, Professor, Department of Civil and Environmental Engineering, Virginia Tech, Blacksburg, VA 24061

Certainly, a critical dimension of sustainability relates to the use and reuse of materials. For years, the United States "enjoyed" being a throwaway type of society; raw materials were readily available and affordable. Given the way we think now in the 21st Century, it might shock you to learn that the first curbside recycling programs in the US were begun in the 1980s (surely, we should have started before then?). We are still a bit "throw-away" minded, but the mindset is changing, and the importance and effectiveness of recycle programs are being realized.

There are many exemplary solid waste management programs in the USA, but the activities and progress in Mecklenburg County, NC, serve as a nice model. The solid waste management plan for Mecklenburg County encompasses one city (Charlotte) and six towns (Cornelius, Davidson, Huntersville, Matthews, Mint Hill, and Pineville).

Naturally, there were a host of logistical and political issues associated with developing a plan for several communities that extend over a large area and produce a lot of solid waste. A ten-year plan (hereafter referred to as the Plan) for the county, released in 2009, was developed by a Steering Committee, consisting of one person from Charlotte and from each of the six towns, and an Advisory Board of 12 citizen members.

In FY07/08, the population of the county was about 860,000, and the solid waste produced was in the area of 1.44 million tons, which yielded a per capita rate of 1.67 tons/person/year (tpy). According to the goals of the Plan, the per capita rate will decrease to 1.27 tpy for a population of 1.14 million people in FY18/19. Using FY98/99 as a baseline, the production rate in FY07/08 was 15% lower; the goal is to reduce the production rate by 35% with respect to FY07/08 in FY18/19.

Table 8.10 was constructed from data given in the Plan to show how the county's overall goals will be achieved in the three major components of their solid waste stream: residential, commercial, and construction and demolition (C&D). In the actual Plan, the goals for each year are described.

Note from Table 8.10 that the greatest production of solid waste is in the commercial sector and that the greatest reduction rate in FY18/19 is expected for the C&D sector. Although there was a bit of a set-back in reduction rate for residential in FY07/08 (−1%), it is known that current recycling efforts, begun in 2010, are making a significant difference. In FY11/12, it is expected that the rate reduction over FY98/99 will be 19%.

With respect to reduction and reuse for residential waste, the county now has a number of successful programs:

1. Don't Dispose – Donate it: links are provided to non-profit charities, Freecycle (citizen website for giving and accepting items) and NC Waste Trader (reuse of materials by businesses and industry).
2. Enviroshopping: citizens can look online for information to help make more environmentally friendly choices in what they buy (e.g., less packaging, non-toxic products, etc.).
3. Junk mail reduction: references to websites where citizens can opt out of mailings.
4. Household hazardous waste reduction: online information (www.wipeoutwaste.com) about household hazardous waste, its disposal, and friendly alternatives.
5. Holiday waste reduction: educational program (public announcements, cinema ads, presentations, etc.) dedicated to reducing waste over Thanksgiving and Christmas holiday season (mid-November to New Year's day).
6. America Recycles Day: national event on November 15; encourage people to recycle and purchase recycled products.

7 PLANT (Piedmont Landscaping and Naturescaping Training): workshops on soil testing, composting, erosion control, landscaping with native plants, grasscycling, beneficial insects, organic and habitat gardening, and vermicomposting; over 100 participants each year and many more on waiting lists.

8 Composting classes and instruction for schools.

Among the materials recycled in most of the county's towns are PET/HDPE plastics, glass, spiral cans, steel cans, aluminum cans, mixed paper, newsprint, old corrugated cardboard (OCC), magazines and phone books. Three towns are not now recycling spiral cans, and one is not collecting OCC. Once per week, materials are collected from 16–22 gallon containers (depends on the community; 18-gallon is the most common size) using a dual stream method, i.e., paper is separated from the other materials.

The County operates 14 recycling centers, five full-service centers and nine self-service centers. Household hazardous waste (HHW) is collected full time at each of the full-service centers, but HHW is managed by a private contractor. The HHW waste is categorized as paint, flammables, cleaners, batteries, and other. The county recycled and/or disposed of 392 tons of HHW at a cost of $0.58 per pound in FY08.

Table 8.10 Waste production goals for each of the three major solid waste components.

	Baseline FY98/99	FY07/08	FY18/19
Population	618,853	863,147	1,144,371
Residential			
– Waste disposed with proposed programs [a], tons	N/A [b]	364,458	349,033
– Rate, tons/person/year	0.42	0.42	0.305
– Reduction rate, %	N/A	–1	27
Commercial			
– Waste disposed with proposed programs, tons	N/A	752,550	785,429
– Rate, tons/person/year	1.04	0.87	0.686
– Reduction rate, %	N/A	16	34
C&D			
– Waste disposed with proposed programs, tons	N/A	325,979	317,912
– Rate, tons/person/year	0.51	0.38	0.278
– Reduction rate, %	N/A	26	45

[a] Source reduction and reuse
[b] Not applicable; baseline data

Table 8.11 Solid waste management rate structure for Charlotte and selected communities.

Entity	County tax/yr/ household	Single household fees/yr	Multi-family structure fees/yr
Charlotte	$15	$45	$27
Cornelius	$15	Solid waste: $91.80 Recycling: $41.16 Yard waste: $51.48	–
Davidson	$15	$200	–
Huntersville	$15	Solid waste: $18 Recycling: $18 Yard waste: $18	–
Pineville	$15	Solid waste: $167.16 Recycling: $36.96	–

Electronic scrap can also be taken to the full-service centers free of charge at any time. In FY08, the cost of recycling 211 tons of electronics was $19,344. The county also provides for the disposal of scrap tires and white goods (primarily appliances).

In FY08, the county disposed of 18,134 tons of tires at a cost of $1,325,150 ($1,075,371 was reimbursed by the NC Scrap Tire Disposal Fund) and 1,243 tons of white goods. Funding for managing the white goods in FY08 was obtained from the sale of scrap metal to a contractor (1820 tons) and from the NC White Goods Disposal Fund (in the amount of $155,012).

Due to the county's efforts to recover monies through recycling, the cost to residents for solid waste management services has been reduced to low levels. Charlotte and each of the participating towns are responsible for setting their own rates, but the fee charged by the county to each household is $15 per year. This fee is assessed as a county tax. Table 8.11 provides the additional fees which are charged by Charlotte and a few towns in the county (FY2010-11). Thus, for example, a Cornelius resident in a single family home would pay a total of $199.44/yr ($15 + $91.80 + $41.16 + $51.48) for solid waste management (curb service, recycling, yard waste, disposal).

As mentioned, the county has outsourced some of the services at their staffed recycling centers, which has increased efficiency and effectiveness. The names of companies that are working with the county to manage various materials are provided in Table 8.12.

In their efforts to increase recycling rates, the county conducted a survey and learned that the primary reasons for not recycling are:

1 Don't have a bin: 18%
2 Inconvenient to clean and separate: 16%
3 Don't know: 16%
4 Forgot, lazy, no time: 12%
5 Accidently put in trash: 6%
6 Other: 28%

Table 8.12 Contractors involved in managing various materials for the county.

Material	Contractor
Garbage	Republic Waste
Tires	US Tire
HHW	Ecoflo
Electronics	Computel
Cooking oil/grease	Valley Protein
Oil filters	Clean Green
Lead-acid batteries	Interstate Battery
Propane cylinders	Heritage Recycling
Donated items	Goodwill Industry
Motor oil/antifreeze	Safety Kleen

For commercial reduction and reuse, the county feels that local governments have little influence on the behavior of businesses, given the global economy. However, the county determined in 2005 that too much recoverable material was being sent to landfills from the commercial sector (in units of tons):

- OCC: 54,450
- mixed recyclable paper: 29,960
- high grade office paper: 21,520
- newsprint: 15,762
- food waste: 63,000
- untreated wood: 3,853
- wood pallets: 11,128
- other ferrous metals: 30,900

Thus, the county will do what it can to reduce disposal of these materials in landfills and continues to work on educational programs for businesses and supporting legislation that will reduce packaging materials.

There are many dimensions and issues associated with solid waste management and recycling. An overview of the situation and ten-year plan for Mecklenburg County was provided to illustrate the breadth and complexity of recycling programs. Mecklenburg's systems are working and moving in the right direction. Among the lessons learned (and surely more will come) at Mecklenburg are:

- communicate with the public;
- offer training programs, make services convenient;
- keep good records;
- develop a business plan;
- develop partnerships with the private sector;
- take advantage of state programs;
- and look to the future.

Of critical importance is the role that we all play in making recycling programs successful. We cannot allow the excuses listed above to decide the fate of recycling programs.

Summary

The primary federal regulation addressing municipal solid waste management in the United States is RCRA subtitle-D. This regulation addresses the transport, storage, and ultimate disposal of non-hazardous solid wastes. In 2006, international waste generation rates ranged from 1.51 to 4.52 lb/capita/day, where the higher generation rates were associated with higher GNPs. For the 1990 to 2009 time period, the average per capita generation rate in the United States was approximately 4.6 lb/day. In 2009, a total of 243 million tons of waste were generated and 161 million tons were disposed in either landfills or waste incineration operations. The primary components of waste generated in the United States in 2009 – 68% of the total sample composition – were paper and paperboard, plastics, food, and yard trimmings. Some of the most important physical parameters associated with solid waste calculations are specific weight, moisture content, field capacity, hydraulic conductivity, chemical composition, and energy content.

The primary mode of ultimate disposal for solid wastes is landfilling. The physical design of a landfill includes calculating the total volume of waste to be placed in the facility over its design life and then estimating the total volume of landfill space required to accommodate the waste and cover materials. The footprint required for the landfill is then calculated, based on the expected geometry of the final landfill. The environment is protected from leachate releases through two methods:

1. controlling the formation of leachate by increasing stormwater runoff and thereby limiting infiltration; and
2. a leachate collection and removal system at the bottom of the landfill, which rapidly removes leachate from the landfill via a series of perforated pipes, a plastic membrane liner, and a low permeability soil barrier.

The process of anaerobic waste degradation will produce CO_2 and CH_4 in approximately equal quantities. Methane is an extremely potent GHG, and its impact can be reduced by flaring the evolving gas stream or using it as an energy source in a combustion process.

Additional topics that fall under the study of municipal solid waste management include:

- Waste collection and transport operations,
- Composting facilities for the management of solid waste and/or yard debris,
- Advanced landfill design and operation paradigms including leachate recirculation and bioreactor landfills,

- Landfill gas to energy projects, and
- Incineration (waste to energy) facilities.

It is the author's sincere hope that the reader will find solid waste management as interesting as he does and will thus pursue a more in-depth study of this complex topic.

Key Words

airspace
closure
composition
daily cover
Darcy's Law
dry moisture content
Dulong Formula
energy content
field capacity
final cap
flexible membrane liner
gas migration
gas production
generation
geomembrane
geonet
hazardous waste
hydraulic conductivity
intermediate cover
landfill
leachate
leachate collection and removal system
leakage
methane
moisture content
municipal solid waste
proximate analysis
RCRA subtitle-C
RCRA subtitle-D
recycling
saturation
specific weight
ultimate analysis
vapor diffusion
volumetric moisture content
waste disposal
wet moisture content

References

Barlaz, M.A., Ham, R.K. (1993).Leachate and Gas Generation. in Daniel, D. (Ed.) *Geotechnical Practice for Waste Disposal*, Chapter 6, pp. 113–136. Chapman and Hall, New York, NY.

Denison, R.A., Ruston, J.F. (1997). Recycling is Not Garbage. *Technology Review* **100**(7), 55–60.

Giroud, J.P., Bonaparte, R. (1989). Leakage through liners constructed with geomembrane liners – parts I and II and technical note, *Geotextiles and Geomembranes* **8**(1), 27–67, **8**(2), 71–111, **8**(4), 337–340.

Giroud, J.P., Gadu-Tweneboah, K., Soderman, K.L. (1994). Evaluation of Landfill Liners, *Proceedings of the Fifth International Conference on Geotextiles, Geomembranes, and Related Products*, pp. 981–986. Singapore.

McBean, E.A., Pohland, R., Rovers, F.A., Crutcher, A.J. (1982). Leachate Collection Design for Containment Landfills. *Journal of Environmental Engineering Division, Proceedings of the American Society of Civil Engineers* **108**(EE1), 204–209.

McEnroe, B.M. (1993). Maximum Saturated Depth Over Landfill Liner. *J Env Eng* **119**(2), 262–270.

Organization for Economic Co-operation and Development (OECD) (2007). *Environmental Data Compendium Report 2006/2007.*

Qian, X., Koerner, R.M., Gray, D.H. (2002). *Geotechnical aspects of landfill design and construction.* Prentice-Hall, Upper Saddle River, NJ.

Schroeder, P.R., Dozier, T.S., Zappi, P.A., McEnroe, B.M., Sjostrom, J.W., Peyton, R.L. (1994). *The Hydrologic Evaluation of Landfill Performance (HELP) Model: Engineering Documentation for Version 3*, EPA/600/R-94/168b, September 1994, US Environmental Protection Agency Office of Research and Development, Washington, DC.

Tchobanoglous, G., Theisen, H., Vigil, S. (1993). *Integrated Solid Waste Management Engineering Principles and Management Issues.* McGraw-Hill, Inc.

US DoD (2004). *Unified Facilities Criteria (UFC) – Solid Waste Incineration.* UFC 3-240-05A.

US EPA (1989). *Requirements for Hazardous Waste Landfill Design, Construction, and Closure.* EPA/625/4-89-022.

US EPA (2010). *Municipal Solid Waste in the United States – 2009 Facts and Figures.* EPA530-R-10-012.

Vesilind, A.P., Worrell, W.A., Reinhart, D.R. (2002). *Solid Waste Engineering.* Brooks/Cole, California.

Verma, S. (2002). *Anaerobic Digestion of Biodegradable Organics in Municipal Solid Wastes.* Department of Earth & Environmental Engineering, Columbia University.

Waste Business Journal (2009). *Waste Market Overview and Outlook 2009.*

Problems

1 Review Figure 8.4: why does total waste generation continue to increase when per capita generation has become almost constant?

2 Determine the yearly waste generation, disposal, and recycling for your city, county, or state in tons.

3 Determine the yearly waste generation, disposal, and recycling for your city, county, or state in cubic yards.

4 Use the information provided in the table below to calculate the volume reduction and weight reduction in percent due to recycling.

Component	Disposed waste composition (% by wt.)	Recycling efficiency (%)	Density (lb/ft^3)
Mixed paper	37	10	5.6
Metals	3	80	20.0
Plastics	43	30	4.1
Organics	17	15	18.1
sum:	100	N/A	N/A

5 The moisture content (dry basis) of a sample is 0.45 g/g. Calculate the moisture content (wet basis) of this sample.

6 Calculate the chemical composition of the waste disposed in your city, county, or state.

7 A community disposes of its waste stream via a Waste to Energy (WTE) Facility. The implementation of a recycling program will decrease waste disposal by 20% (weight basis) and change the energy content of the disposed waste from 5,000 Btu/lb to 4,500 Btu/lb.

Determine the percentage reduction in the WTE Facility's energy production potential.

8 Develop an equation for the volume of a truncated pyramid with a rectangular base.

9 Rework Example Problem 8.5, assuming that the population grows at a rate of 5% per year.

10 Determine the footprint (acres) of a landfill that can service your city, county, or state for a ten-year period. Use a height of 150 feet and side slopes of 3.5:1 (run:rise).

11 A recycling program will reduce the volume of waste disposal by 15%. How much will this extend the design life of the landfill? Express your answer in %.

12 Size a leachate collection and recovery system (drainage slope, and pipe spacing) for a landfill in your state. The drainage layer will be sand, with a hydraulic conductivity of 10^{-2} cm/s. Assume open cell conditions and use the average yearly rainfall for your state as the leachate loading rate.

13 Size a leachate collection and recovery system (drainage slope, and pipe spacing) for a landfill in your state. The drainage layer will be sand, with a hydraulic conductivity of 10^{-3} cm/s. Assume open cell conditions and use the average yearly rainfall for your state as the leachate loading rate.

14 A one-acre surface impoundment holds an average liquid depth of 30 cm. The barrier layer consists of 30 cm of a 10^{-6} cm/s clay. Calculate the leakage rate in cm/yr and m^3/yr.

15 Calculate the leakage rate from a 1 cm^2 perforation in the liner component of traditional composite liner system. The soil barrier is a 5×10^{-8} cm/s clay with a thickness of 45 cm. The average leachate head above the perforation is 20 cm.

16 Calculate the vapor diffusion (cm/yr and m^3/yr) across a one-acre, 60-mil HDPE geomembrane liner. The average leachate head on the liner is 30 cm.

17 Rework Example Problem 8.6, assuming that the landfill has a ten-year design life.

18 Rework Example Problem 8.6, assuming that the rapidly degrading waste degrades over six years, with a peak at year 3, and the slowly degrading waste degrades over 12 years, with a peak at year 6.

Chapter 9

Air pollution

Arthur B. Nunn, III

Learning Objectives

After reading this chapter, you should be able to:

- understand the different types, sources, and effects of air pollutants, including local and global impacts;
- understand the fundamentals of how meteorology impacts the evaluation of air pollutant emissions, and the basis of atmospheric dispersion modeling;
- understand the basic design and function of different types of air pollution control technologies for particulate and gaseous air pollutants.

9.1 Principal air pollutants

Air pollutants are generally categorized as either Criteria Pollutants or Hazardous Air Pollutants (HAPs). Criteria pollutants are regulated under Title 40 of the Code of Federal Regulations, Part 50. This regulatory framework presents maximum allowable ambient air concentrations for six compounds, for different averaging periods ranging from one-hour average to annual average. The six pollutants identified as Criteria Pollutants are:

- sulfur dioxide (SO_2), PM_{10} (particulate matter greater than 2.5 micrometers (μm) and less than or equal to 10 μm in size);
- $PM_{2.5}$ (particulate matter less than or equal to 2.5 μm);
- carbon monoxide (CO);
- ozone (O_3)
- nitrogen dioxide (NO_2);
- lead (Pb).

In order to limit exposure to these six compounds, EPA has developed National Ambient Air Quality Standards (NAAQS), which consist of primary and secondary standards. Primary standards provide public health protection, including protecting the health of "sensitive" populations such as asthmatics, children, and the elderly. Secondary standards provide public welfare protection, including protection against decreased visibility and damage to animals, crops, vegetation, and buildings. In many cases, the primary and secondary standards are the same.

A full listing of criteria air pollutants and their respective NAAQS are presented in Table 9.1. A summary of the health and environmental impacts of these pollutants is presented in Table 9.2.

In addition to the six criteria pollutants, hydrocarbons or volatile organic compounds (VOCs) are regulated as source level emissions. There are no specific ambient air quality standards for VOCs, but their emissions are regulated because, when present in the ambient air, they react to form the criteria pollutant ozone.

To achieve attainment of the NAAQS, emissions of criteria pollutants are regulated on the federal level by New Source Performance Standards (NSPS), which are contained in Part 60 of Title 40 of the Code of Federal Regulations. These regulations limit emissions from a broad spectrum of new, modified, or reconstructed industrial facilities, by requiring them to install Best Available Control Technology (BACT). BACT consists of the technology that will reduce emissions to the lowest possible level, taking into consideration the capital and operating costs of the available options for emissions reduction.

Emissions from existing facilities are regulated by individual state regulatory agencies, whose programs are contained in State Implementation Plans (SIPs), which must be approved by EPA. In addition, permit requirements for new major facilities in areas that have attained compliance with NAAQS are required to obtain Prevention of Significant Deterioration

Environmental Engineering: Principles and Practice, First Edition. Richard O. Mines, Jr.

Table 9.1 National Ambient Air Quality Standards for criteria pollutants.

Pollutant	Standard type	Maximum allowable concentration	Averaging period	Regulatory citation
Sulfur dioxide (SO_2)	Primary	0.14 ppm (365 μg/m^3)	24-hour	40 CFR 50.4(b)
	Primary	0.03 ppm (80 μg/m^3)	Annual	40 CFR 50.4(a)
	Secondary	0.50 ppm (1,300 μg/m^3)	3-hour	40 CFR 50.5(a)
PM_{10}	Primary and secondary	150 μg/m^3	24-hour	40 CFR 50.6(a)
$PM_{2.5}$	Primary and secondary	35 μg/m^3	24-hour	40 CFR 50.7(a)
	Primary and secondary	15 μg/m^3	Annual	40 CFR 50.7(a)
Carbon monoxide (CO)	Primary	35 ppm (40 mg/m^3)	1-hour	40 CFR 50.8(a)(2)
	Primary	9 ppm (10 mg/m^3)	8-hour	40 CFR 50.8(a)(1)
Ozone (O_3)	Primary and secondary	0.12 ppm (235 μg/m^3)	1-hour	40 CFR 50.9(a)
	Primary and secondary	0.075 ppm (150 μg/m^3)	8-hour	40 CFR 50.10(a)
Nitrogen dioxide (NO_2)	Primary and secondary	0.075 ppm (150 μg/m^3)	Annual	40 CFR 50.10(a) and (b)
Lead (Pb)	Primary and secondary	0.15 μg/m^3	Rolling 4 months	40 CFR 50.12

Based on: http://www.epa.gov/air/criteria.html

(PSD) permits, which insure that such facilities do not deteriorate air quality beyond a level that will exceed the maximum allowable impact levels contained in the PSD regulations, or result in attainment areas becoming non-attainment. These PSD regulations, in some instances, will result in proposed new, modified, or reconstructed facilities having to meet lower emission limits than those contained in the NSPS.

Similarly, new facilities wishing to locate in an area that is not in attainment with NAAQS may be allowed to do so under available permitting regulations. This consists of a two-phased approval process. First, the facility will be required to install and operate the level of emissions control equipment known as Lowest Achievable Emission Rate (LAER). The difference between this and BACT is that it must achieve the lowest possible level of emissions, regardless of capital and operating cost. The second criterion is that the proposed new facility must also facilitate emissions reductions in the same air quality basin by an amount greater than the quantity of emissions to be generated by the facility. Such emission reductions can occur from other facilities in the air quality basin, as well as at the owner's facility, or from other emission sources in the air quality basin. The purpose of these non-attainment permitting regulations is to allow economic expansion in non-attainment areas, while providing that such expansion results in a net reduction of overall emissions in the area, thereby enhancing progress toward achievement of attainment of the NAAQS.

Hazardous Air Pollutants (HAPS) comprise a list of 187 compounds, or classes of compounds, contained in Title I, Section 112 of the Clean Air Act. Provisions requiring control of the emissions of these compounds were added to the Clean Air Act as of the amendments to the act enacted in 1990. There are no federal ambient air quality standards for these compounds, though some state agencies have regulations in place which establish such limits, with such standards typically based upon some fraction of OSHA worker exposure standards for the same compounds. In addition, there is very little monitoring of ambient air quality concentrations of HAP pollutants. This is, in part, because the ambient impact of HAPS emissions are closely located to the sources from which they are released, rather than being generally regional, as is the case with criteria pollutants.

On the federal level, the Clean Air Act Amendments of 1990 required EPA to develop emissions standards for these compounds, for major emission sources. For the purpose of HAPS regulations, a major source is defined as any facility which emits ten tons per year or more of any single HAP, or 25 tons per year or more of any combination of HAPS.

Federal regulations limiting emissions of HAPS are contained in Part 63 of Title 40 of the Code of Federal Regulations. Such regulations require implementation of Maximum Achievable Control Technology (MACT). MACT is defined as a level of control that must not be less stringent than the average emission level achieved by controls on the best-performing 12% of existing sources of similar industrial and utility sources. MACT regulations apply to both new and existing facilities, but economic and technical considerations may result in lesser levels of stringency for existing facilities.

Table 9.2 Health and environmental impacts of criteria pollutants.

Pollutant	Type of impact	Pollutant effects
Sulfur dioxide (SO_2)	Primary	Increased asthma symptoms Bronchoconstriction Worsening of emphysema Worsening of bronchitis
	Secondary	Visibility reduction Acid rain
PM_{10} and $PM_{2.5}$	Primary	Increased respiratory symptoms, such as irritation of the airways, coughing, or difficulty breathing, e.g., decreased lung function Aggravated asthma Development of chronic bronchitis Irregular heartbeat Nonfatal heart attacks Premature death in people with heart or lung disease
	Secondary	Visibility reduction Acid rain Aesthetic damage to stone and other building materials
Carbon monoxide (CO)	Primary	Reduced oxygen delivery to the body's organs
Ozone (O_3)	Primary	Airway irritation, coughing, and pain when taking a deep breath Wheezing and breathing difficulties during exercise or outdoor activities Inflammation, which is much like a sunburn on the skin Aggravation of asthma and increased susceptibility to respiratory Illnesses like pneumonia and bronchitis Permanent lung damage with repeated exposures
	Secondary	Interfering with the ability of sensitive plants to produce and store food Damaging the leaves of trees and other plants Reducing forest growth and crop yields
Nitrogen dioxide (NO_2)	Primary	Increased asthma symptoms Airway inflammation Worsening of emphysema Worsening of bronchitis
	Secondary	Visibility reduction Acid rain
Lead (Pb)	Primary	Adversely affects the nervous system, kidney function, immune system, reproductive and developmental systems and the cardiovascular system. Restricts blood's ability to carry oxygen. Neurological effects in children and cardiovascular effects (e.g., high blood pressure and heart disease) in adults
	Secondary	Contamination of food supply Health impacts on animals

Based on: http://www.epa.gov/air/criteria.html

9.2 Air pollution sources

Air pollutants are emitted into the atmosphere from a variety of industrial, commercial, natural, and transportation related sources. Some emission sources are referred to as point sources, because they are released from fixed stacks or vents, whereas fugitive emissions are generated from widespread areas, such as fields, roads, construction sites, or quarries, or from specific source operations from which emissions are released over the area of operations, rather than through a stack or vent. Transportation sources include on-road and off-road vehicles, as well as other means of transportation, including trains and aircraft.

Criteria pollutants tend to be generated from a variety of point sources and area sources across the spectrum of

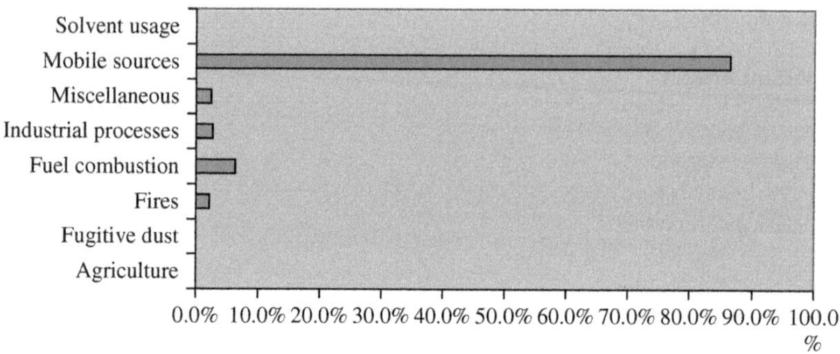

Figure 9.1 Sources of CO emissions. Based on EPA National Emissions Inventory – http://www.epa.gov/air/emissions/

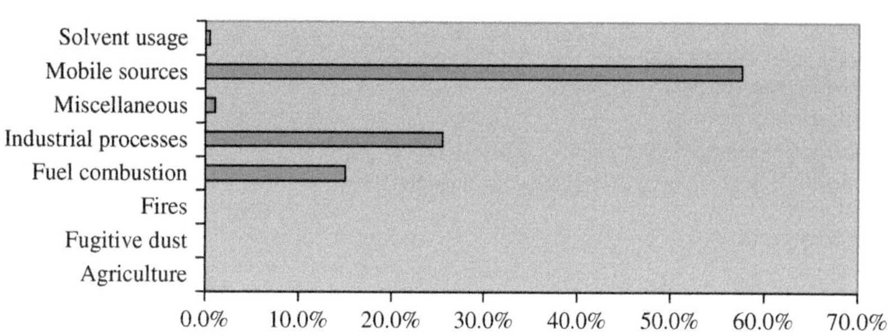

Figure 9.2 Sources of lead emissions. Based on EPA National Emissions Inventory – http://www.epa.gov/air/emissions/

industrial, commercial, natural, and transportation related sources. Carbon monoxide is generated as a product of incomplete combustion, and is released therefore, exclusively from combustion processes. Presented in Figure 9.1 is a graphical depiction of the primary sources of nationwide CO emissions. As presented therein, the vast majority of carbon monoxide emissions are generated by mobile sources, primarily on-road vehicles.

Lead emissions, as shown in Figure 9.2, are generated from a mix of transportation sources, industrial sources, and fuel combustion sources.

On-road transportation sources were, at one time, a major source of lead emissions, but the advent of unleaded gasoline eliminated those emissions. The single largest source of lead emissions is still from transportation sources, but it is now primarily from aircraft engines. Industrial lead emissions are primarily generated by ferrous and non-ferrous metals production, and fuel combustion lead emissions are primarily from utility power plants.

As presented in Figure 9.3, the majority of nitrogen oxides emissions are generated by transportation sources, largely by on-road vehicles, and to a lesser degree from off-road vehicles, locomotives, and commercial marine operations.

Secondary to transportation sources is utility fuel combustion. Industrial sources constitute a minority of overall NO_x emissions and are released from a variety of different sources, including oil production, cement plants, chemical production, and pulp and paper production.

Volatile organic compounds are emitted from a variety of different sources, as shown in Figure 9.4.

The largest overall source is transportation related, from on-road and off-road vehicles. Emissions from these sources are primarily related to the evaporation of fuel, as well as from hydrocarbons contained in the vehicle engine exhaust. Solvent usage sources include a broad spectrum of consumer and commercial product usage, industrial surface coating, graphic arts, degreasing, and dry cleaning operations. Industrial process VOC emissions are generated primarily from oil and gas production operations. There are also a number of miscellaneous sources of VOCs, including gas stations and bulk gasoline terminal operations.

Emissions of particulate matter (PM) are generated, as shown in Figure 9.5, from a wide and diverse number of sources. The largest source of emissions overall is fugitive dust from unpaved roads, construction operations, and paved roadways. Second to that is emissions from fuel combustion, with these emissions split almost evenly between residential heating and utility power plants. Other sources of PM emissions include on-road and off-road vehicle engines, waste disposal, cooking operations, and numerous industrial operations.

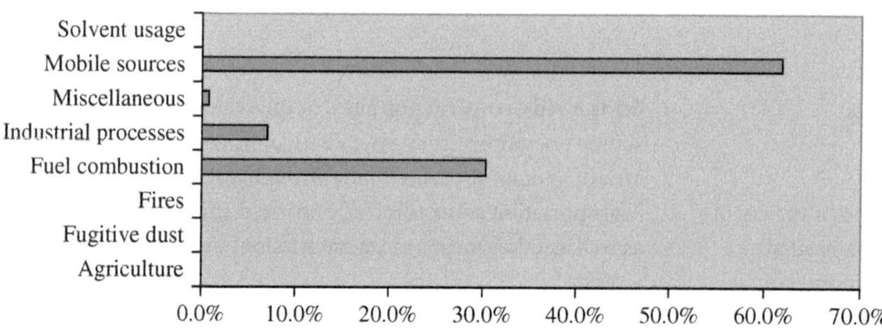

Figure 9.3 Sources of NO_x emissions. Based on EPA National Emissions Inventory – http://www.epa.gov/air/emissions/

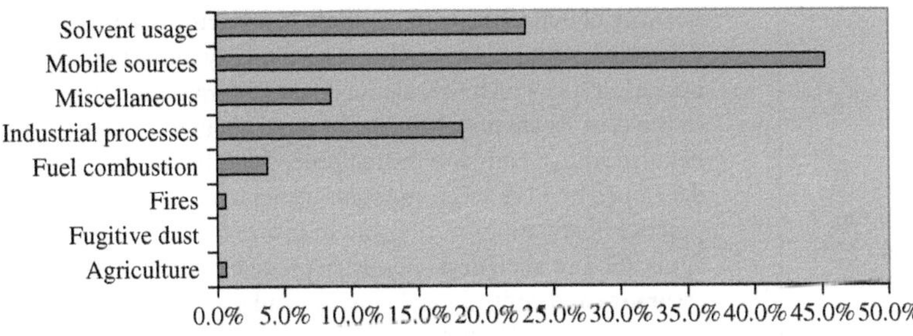

Figure 9.4 Sources of VOC emissions. Based on EPA National Emissions Inventory – http://www.epa.gov/air/emissions/

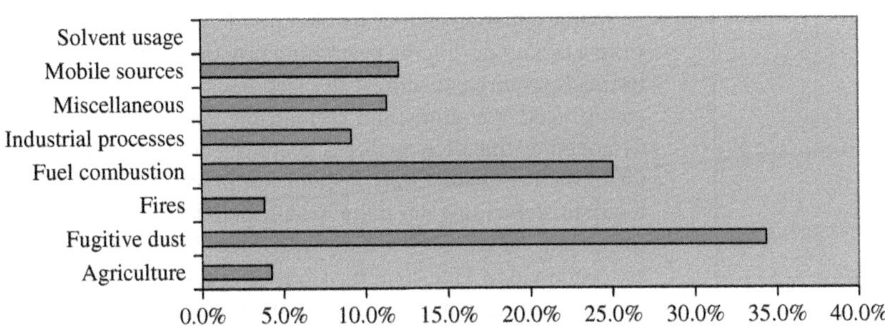

Figure 9.5 Sources of PM emissions. Based on EPA National Emissions Inventory – http://www.epa.gov/air/emissions/

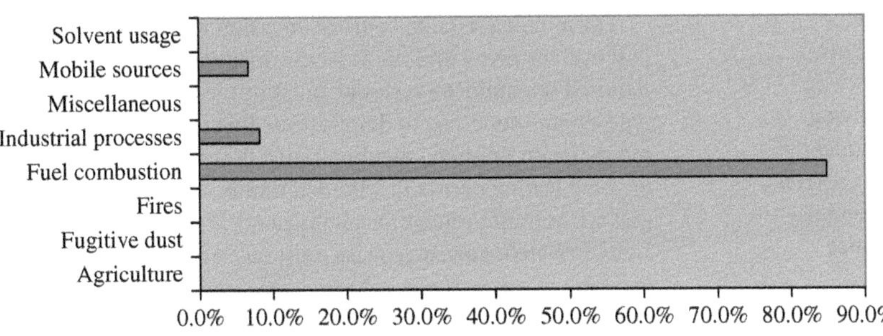

Figure 9.6 Sources of SO_2 emissions. Based on EPA National Emissions Inventory – http://www.epa.gov/air/emissions/

Sulfur dioxide emissions are generated by the oxidation of sulfur in fuels in different combustion processes. As shown in Figure 9.6, the vast majority of SO_2 emissions are produced from utility power plants, with lesser amounts generated by industrial and commercial boilers, and combustion of sulfur-containing diesel fuels.

Hazardous air pollutants are generated primarily from industrial and transportation related sources. HAPS are, in general, process-specific, with some HAP compounds generated by a limited number of different industrial processes. EPA has issued rules covering over 96 categories of major industrial sources, such as chemical plants, oil refineries, aerospace manufacturers, cement plants, waste incineration facilities, utility power plants, and steel mills, as well as categories of smaller sources, such as dry cleaners, commercial sterilizers, secondary lead smelters, industrial boilers, and chromium electroplating facilities. These, and other sources of HAP emissions, vary significantly in size and location, and are inclusive of continuous and batch operations, as well as including transportation related emissions. Industrial HAP emissions are primarily point source in nature, though there are fugitive HAP emissions associated with some types of processes.

Air pollutant emissions from a large number of sources can be estimated utilizing air pollutant emission factors developed by EPA and various industrial organizations. Emission factors provide a mechanism for calculating emissions on the basis of different production information that may be available regarding a facility. In their simplest form, air pollutant emission factors follow the form as shown in Equation (9.1).

$$E = P \times F \times (1 - ER/100) \tag{9.1}$$

where:
E = emissions, lb/h
P = product production rate, tons/h
F = emission factor, lb/ton
ER = overall emission reduction efficiency of air pollution control equipment.

In other cases, the factor is based upon the content of a given substance in the raw material or fuel used in an industrial process.

An example of this is the calculation of emission of sulfur oxides (SO_x) from a coal fired boiler. This is calculated as shown below.

$$E = P \times 38(S) \tag{9.2}$$

where:
E = emissions, lb/h
P = coal firing rate, tons/h
S = coal sulfur content, %
38 = EPA emission factor for SO_x emissions.

Example 9.1 Calculation of air pollutant emissions

For example, a boiler burning 50 tons per hour of coal containing 1.5% sulfur, the total SO_x emissions from the boiler would be calculated as follows using Equation (9.2):

Solution

$$E = P \times 38(S) \tag{9.2}$$

$$E = 50\,\frac{\text{ton}}{\text{h}} \times 38(1.5\%)$$

$$E = \boxed{2{,}850\,\frac{\text{lb}}{\text{h}}}$$

Many emission factors are contained in the EPA Compilation of Emission Factors, referred to as AP-42, which is contained on the EPA website at http://www.epa.gov/ttn/chief/ap42/index.html. This consists of a living compilation of emission factors that is updated frequently, and it contains hundreds of emission factors in 15 different broad source categories, ranging from fossil fuel combustion, through chemical plant operations, to ordinance detonation.

AP-42 emission factors are assigned an emission factor rating ranging from A to E. Emission factors assigned a rating of A are considered to be highly accurate and reliable, as they are based upon a large database of measured emission rates from a large spectrum of emission sources throughout the United States. Emission factors with a rating of E, on the other hand, are considered rough estimates only, and are subject to significant variability from source to source. Based upon the emission factor rating, environmental regulatory agencies may or may not use the published emission factors in developing air pollutant emission limits for industrial or commercial facilities for which air permits are granted.

In the absence of published emission factors, and in the case of many HAPs, emission rates are estimated using mass balances coupled with engineering knowledge about the process in question. For example, a mercury balance across a chemical process can be employed to estimate the quantity of mercury emission from a facility if enough information is available about the mercury content of the raw materials, the quantity of raw materials utilized, the quantity of product produced, the mercury content of the final product, and the mercury content and quantity of any solid waste and/or byproduct materials generated in the production process. By knowing the total quantity of mercury entering a process and subtracting from that the total quantity leaving a process in the form of the final product, and all wastes and byproducts, the remainder is assumed to have been released to the atmosphere.

Emission factors and mass balances, no matter how well defined, are still just estimates of the quantities of emissions from varying emission sources. The only way to know the quantity of emissions with certainty is to perform actual emissions measurements. This is particularly important for operations for which well-defined factors (emission factor rating A or B) are not available. In order to fill this need, detailed source emission testing procedures have been developed by EPA and some state agencies, to standardize the practice of emissions testing and to insure the proper application of analytical chemistry procedures and quality assurance and control mechanisms to all source test data generated in the United States.

While emission factors are often used to estimate emissions from a facility during the permitting process, source emissions testing is required, in almost all cases, once a facility has commenced operations, and periodically thereafter, in order to demonstrate that each facility is operating in compliance with established emission limits. Specific test procedures are required, depending upon the specific pollutants to be quantified. All emission testing procedures developed by EPA are contained in Title 40 of the Code of Federal Regulations, Part 60, Appendix A. All of these test procedures are also accessible at http://www.epa.gov/ttn/emc.

For mobile emission sources, emission testing is not practical for every unit. In that case, manufacturers are required to conduct a series of emissions tests on every vehicle type, engine style, etc. to demonstrate that each such vehicle meets the established emission limits. Beyond that, the only ongoing testing occurs in a limited number of geographic areas that are non-attainment for photochemical oxidants. In those areas, privately owned vehicles must take and pass an annual emission test as part of a vehicle inspection and maintenance program, designed to reduce overall vehicle emissions in the area.

The general category of emission sources that remains essentially unregulated or tested is that of household emissions, including fireplaces, and home heating systems. The same is true for many office and educational buildings. Some individual burners must meet certain design criteria but, once sold and installed, they are not controlled, quantified, or subjected to regulated preventive maintenance activities.

9.3 Pollutant affects on humans and environment

The impact of air pollutants on humans varies by pollutant and location. Impacts on humans have been discussed in Section 9.1, and they are typically closely located to the emission source. Other nearby impacts are as discussed relative to secondary impacts, and as referenced by the NAAQS secondary standards. These include such items as etching of masonry structures and dust accumulation.

Large area impacts are generally more related to the cumulative effects of a broad spectrum of emission sources located throughout a region, or even throughout the world. Some of these impacts include global climate change, acid rain, tropospheric ozone, and stratospheric ozone. Each of these environmental impacts is discussed in the following sections.

9.3.1 Global climate change

The term "climate change" is often used interchangeably with the term "global warming". According to the National Academy of Sciences, however, "the phrase 'climate change' is growing in preferred use to 'global warming', because it helps convey that there are other changes in addition to rising temperatures." Climate change refers to any significant change in measures of climate (such as temperature, precipitation, or wind) lasting for an extended period (decades or longer).

Climate change may result from:

- natural factors, such as changes in the sun's intensity, slow changes in the Earth's orbit around the sun, or changes in activity levels of the sun itself;
- natural processes within the climate system (e.g., volcanic eruptions or changes in ocean circulation);
- human activities that change the atmosphere's composition (e.g., increases in concentrations of greenhouse gases resulting from burning fossil fuels) and the land surface (e.g., deforestation, reforestation, urbanization, desertification, etc.).

Climate change is not a recent phenomenon, given that the Earth's climate has changed constantly for millions of years. As presented in Figure 9.7, there have been significant changes in global temperatures over the last 1,000 years, including such peaks as the high temperatures of the medieval period, and the "little ice age" of the 1600s.

There are numerous natural phenomena that can be related to global climate change, including slight changes in the Earth's axis, the presence or absence of volcanic eruptions, and sunspots. Presented in Figure 9.8, for example, is a graph depicting the rate of occurrence of sunspots versus global temperatures.

Although the Earth's climate has changed many times throughout its history, recent more-rapid warming cannot be explained only by natural processes. Human activities are increasing the amount of greenhouse gases in the atmosphere. Some amount of greenhouse gases is necessary for life to exist

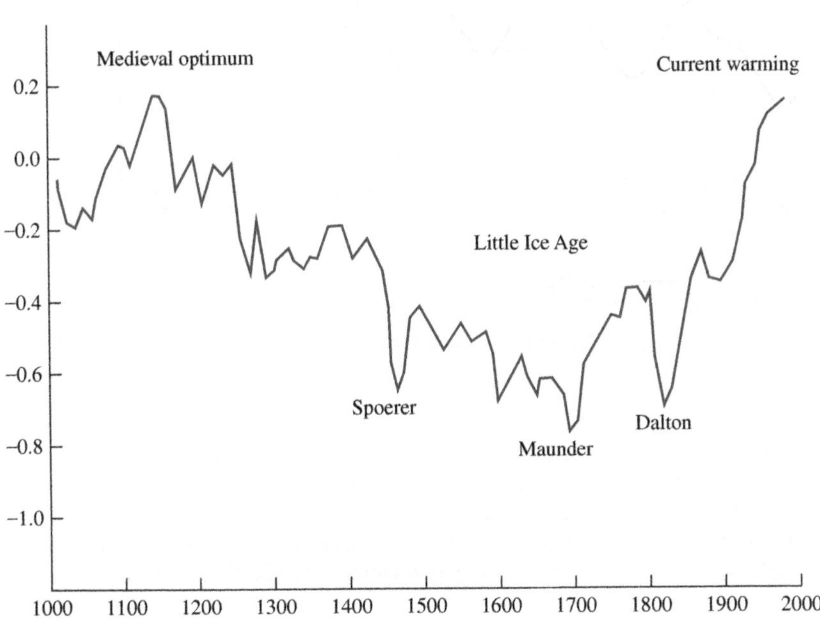

Figure 9.7 Climate history. Based on González-Rouco F. *et al.* (2003). Deep Soil Temperature As Proxy For Surface Air-Temperature In A Coupled Model Simulation Of The Last Thousand Years. *Geophysical Research Letters* **30**, 2116.

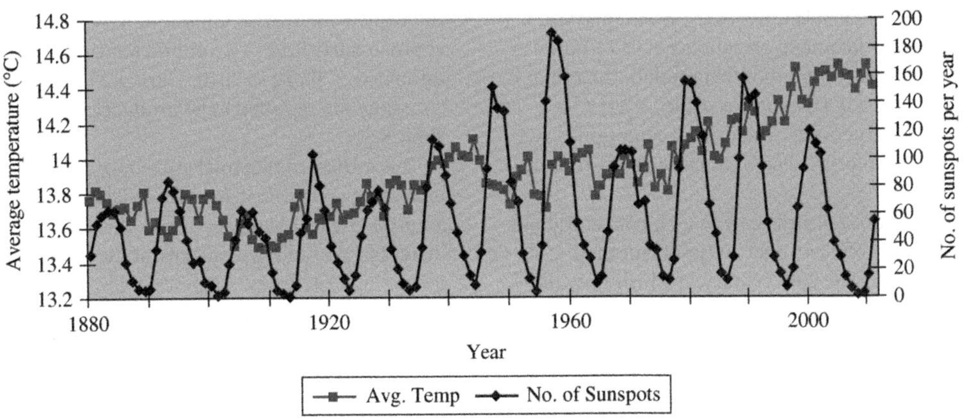

Figure 9.8 Global temperature versus number of sunspots. Based on National Oceanic and Atmospheric Administration (2011) http://www.noaa.gov; and Solar Influences Data Analysis Center (2011) http://sidc.oma.be/sunspot-data.

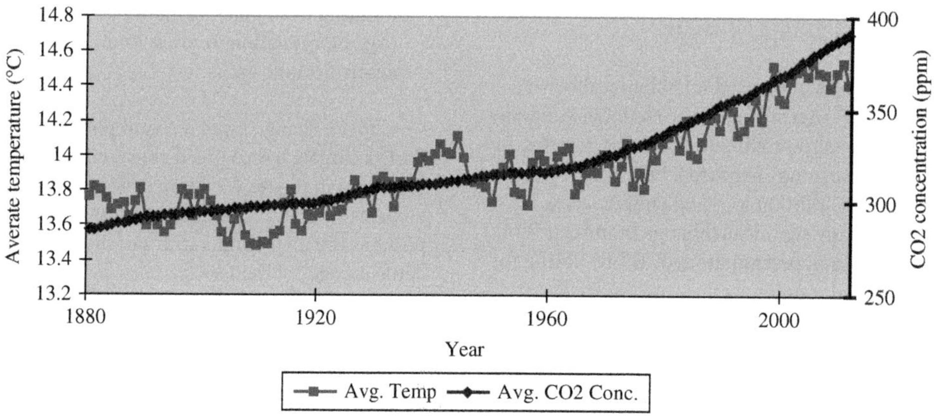

Figure 9.9 Global temperature versus carbon dioxide concentration. Based on National Oceanic and Atmospheric Administration (2011) http://www.noaa.gov; and Carbon Dioxide Information Analysis Center (2011) http://www.cdiac.gov.

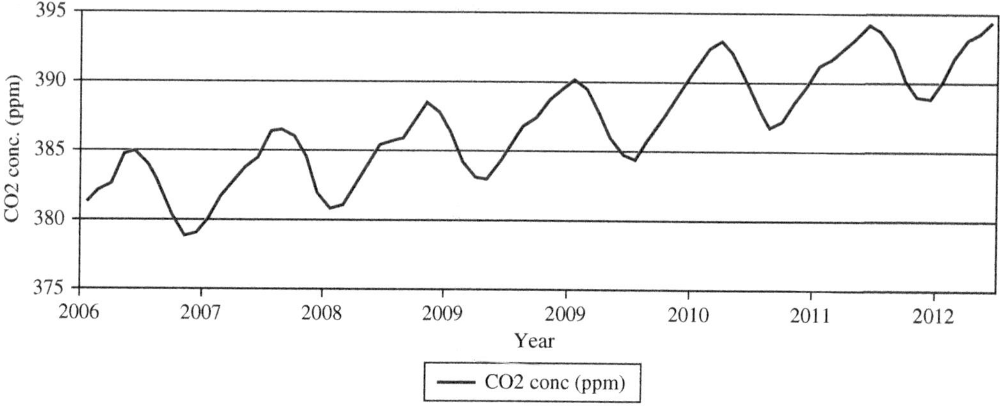

Figure 9.10 Carbon dioxide concentration variability. *Source*: Based on National Oceanic and Atmospheric Administration. 2012, http://www.esrl.noaa.gov.

on Earth, as they provide for photosynthesis of plant life and they trap heat in the atmosphere, keeping the planet warm and in a state of equilibrium. This natural greenhouse effect is being strengthened by human activities (combustion of fossil fuels), which are adding more of these gases to the atmosphere, resulting in a shift in the Earth's equilibrium. Presented in Figure 9.9 is a depiction of the comparative increases in global temperature and carbon dioxide levels over the past 200 years.

The most commonly identified greenhouse is CO_2, however there are numerous other compounds that produce the same effect, including methane (CH_4), nitrous oxide (N_2O), halocarbons, ozone, and aerosols. These other compounds are, however, minor contributors to global climate change, relative to carbon dioxide.

As presented in Figure 9.9, atmospheric CO_2 concentrations in the late 19th century were less than 300 ppm, but they increased gently until the late 20th century, at which time the concentration started to increase at a more rapid rate. Though most of this increase is believed to be attributable to the burning of fossil fuels, it has been estimated that approximately 35% of the increase over the last 300 years is due to changes in agricultural land use (Foley, 2005).

Measurements of atmospheric CO_2 concentrations at the Mauna Lua monitoring station reveal a saw tooth pattern each year, with peak concentrations in the winter, and lower concentrations in the summer (Figure 9.10). This tends to link measured concentrations with increased heating activities for homes, businesses, etc. during the winter, as well as increased vegetative activity resulting in CO_2 collection during summer months.

The Environmental Protection Agency reports that, within the United States, CO_2 emissions have increased from approximately 5,100 Tg (million metric tons) in 1990 to an estimated 5,706 Tg in 2010. During 2010, the majority of these emissions are reported to be produced by the sources shown in Table 9.3.

It is estimated that total anthropogenic CO_2 emissions are increasing approximately twice as fast as is the average global atmospheric CO_2 concentration. This tends to indicate an increasing assimilative capacity of the Earth to absorb increasing CO_2 levels. Analyses of phytoplankton in the North Sea and the Northeast Atlantic indicate growth from 1948 to 2003, with levels continuing to increase (Raitsos, 2005).

Clearly, the impact of greenhouse gas emissions on the Earth's atmosphere is a subject of significant study and debate. Greenhouse gas concentrations continue to increase in the Earth's atmosphere. However, the real impact of this on the overall climate of the Earth is not fully understood, and it

Table 9.3 CO_2 emissions in the United States.

Emission Source	Tg of CO_2 emissions
Electricity generation	2,228.4
Transportation	1,745.5
Industrial fuel combustion	777.8
Residential fuel combustion	340.2
Commercial fuel combustion	224.2

Based on EPA National Emissions Inventory Worldwide; the Carbon Dioxide Information Analysis Center (http://www.cdiac.gov) estimates total 2010 CO_2 emissions to be approximately 30,000 Tg.

influences numerous areas ranging from average global temperature to ocean vegetation.

9.3.2 Tropospheric ozone

Tropospheric, or ground level ozone, is a concern because of the direct human health effects discussed previously, as well as the fact that it is a greenhouse gas and a contributor to visibility impairment. It is also a strong oxidizing compound, which reacts with other pollutants in the atmosphere to form fully or partially oxidized chemical products, some of which may be toxic air contaminants.

Ozone is not emitted directly into the atmosphere, but is created by chemical reactions between oxides of nitrogen (NO_x) and volatile organic compounds (VOCs) emitted by a wide variety of fuel combustion and associated industrial processes. Ozone most often reaches unhealthy levels on hot sunny days in urban environments. It can also, however, be transported long distances by wind. For this reason, even rural areas can experience high ozone levels. Ozone is also the primary constituent of smog, occurring predominantly in urban areas.

Ozone can affect human health, even at relatively low levels. People with lung disease, children, older adults, and people who are active outdoors, may be particularly sensitive to ozone. Children are at greatest risk, because their lungs are still developing. Children are also more likely than adults to have asthma. Ozone also impacts sensitive vegetation, including trees and plants during the growing season.

9.3.3 Stratospheric ozone

The region of the Earth's atmosphere from 6–30 miles (10–50 kilometers) above the Earth's surface is known as the stratosphere. In this region, ozone plays a vital role in absorbing harmful ultraviolet radiation from the sun. During the past 20 years, concentrations of ozone have been threatened by human-made gases released into the atmosphere, including those known as chlorofluorocarbons (CFCs). These chemical compounds, as well as meteorological conditions in the stratosphere, affect the concentration of stratospheric ozone.

The ozone layer absorbs a portion of the radiation from the sun, preventing it from reaching the Earth's surface. The most important aspect of this is that it absorbs the portion of ultraviolet light called UVB, which has been linked to skin cancer and cataracts. UVB is a band of ultraviolet radiation produced by the sun, with wavelengths of 280–320 nanometers.

Ozone molecules are constantly formed and destroyed in the stratosphere. The total amount, however, remains relatively stable. While ozone concentrations vary naturally with sunspots, the seasons, and latitude, these processes are well understood and predictable. Scientists have established records spanning several decades that detail normal ozone levels during these natural cycles. Each natural reduction in ozone levels has been followed by a recovery. Recently, however, scientific evidence has shown that the ozone shield is being depleted well beyond changes due to natural processes.

Chlorofluorocarbons (CFCs) have been used for decades for various purposes because they are very stable, non-toxic, and relatively inexpensive to produce. Evaluation of the depletion of stratospheric ozone, however, has lead to the conclusion that it is these CFCs that are depleting the ozone layer. As stated, CFCs are very stable compounds, which allow them to exist for long periods of time in the atmosphere, and they are eventually carried up into the stratosphere. There, these CFCs can be broken down by very strong UV radiation. Upon breakdown, chlorine atoms are released, and it is these chlorine atoms that react with ozone molecules, thereby depleting the ozone concentration. One chlorine atom can destroy over 100,000 ozone molecules. The net effect is to destroy ozone faster than it can be naturally created (http://www.epa.gov/ozone/science/sc_fact.html).

There are natural sources of chlorine that can be transported to the troposphere. These include large fires, some forms of marine life, and volcanic eruptions. It is estimated, however, that these natural sources comprise only approximately 15% of the chlorine contribution to stratospheric ozone depletion.

In addition to chlorine, from CFCs or other sources, other compounds have been linked to stratospheric ozone depletion. These include nitric oxide (NO), nitrous oxide (N_2O), and compounds with hydroxyl (OH^-) radicals. While these compounds will react with, and thereby deplete, ozone, neither their potency nor their prevalence in the troposphere is believed to approach that of CFCs.

Reductions in ozone levels will increase the amount of UVB reaching the Earth's surface. The sun's output of UVB does not change, but less ozone means less protection, and hence more UVB reaching the Earth. Laboratory and epidemiological studies have demonstrated that UVB causes nonmelanoma skin cancers and plays a major role in malignant melanoma development.

9.3.4 Acid rain

Acid rain is a term referring to the formation and ultimate deposition of sulfuric and nitric acids onto the Earth's surface and waterways. The pollutants that are involved in the formation of acid rain are sulfur dioxide (SO_2) and nitrogen oxides (NO_x). These compounds are generated by natural sources such as volcanic eruptions and decaying vegetation, and from anthropogenic sources such as combustion of fossil fuels. Acid rain occurs when these gases react in the atmosphere with water, oxygen, and other chemicals to form acidic compounds.

This may occur hundreds of miles from the source of emissions of the precursor compounds, because these compounds can be carried long distances by the prevailing winds in the area. Deposition upon the surface of the Earth can be either liquid form, from reaction with rain droplets, or dry deposition, via adsorption of acidic materials onto dust particles.

Acid rain can cause acidification of lakes and streams and can contribute to damage to trees at high elevations (e.g., red spruce trees above 2,000 feet). Damage to some sensitive forest soils can also occur. In addition, acid rain accelerates the decay of building materials and paints. This produces an economic impact to building owners and structural infrastructure. There is also a potential for cultural impact by degradation of statues and other works of art.

9.3.5 Visibility

One of the most readily identifiable forms of air pollution is haze, which degrades visibility in many cities and scenic areas, such as national parks. Haze is caused when sunlight encounters tiny pollution particles in the air, which reduces the clarity and color of areas and objects viewed through the haze. Some light is absorbed by particles, while other light is scattered by particles. There is a direct relationship between the number of particles in the air and the degree of obstruction of visibility. Some types of particles, such as sulfates, scatter more light, particularly during humid conditions.

Air pollutants which obstruct visibility come from a variety of natural and anthropogenic sources. Natural sources can include windblown dust and soot from wildfires. Anthropogenic sources can include motor vehicles, electric utility and industrial fuel burning, and manufacturing operations. Some haze-causing particles are directly emitted to the air. Others are formed when gases, which are emitted to the air, form particles as they are carried many miles from the source of the pollutants. Examples of such pollutants are sulfur dioxide, which may form sulfate particles, and nitrogen oxides, which may form nitrates. This is the same mechanism that can result in the formation and deposition of acid rain.

9.4 Air pollution meteorology

The impact of air pollutant emissions on nearby and distant surrounding communities is as much a function of local and regional meteorology as it is a source of emissions. Meteorology dictates the pattern and degree of dispersal of pollutants after they are released from the process from which they are generated.

The atmosphere is divided into four separate layers: the troposphere, the stratosphere, the mesosphere, and the thermosphere. The lowest layer, called the troposphere, accounts for roughly three-quarters of the mass of the atmosphere and contains nearly all of the water associated with the atmosphere (vapor, clouds and precipitation); it is the portion of the atmosphere in which all air masses, fronts, and storms are located. The depth of the troposphere varies with latitude and season. The top of the troposphere is typically about 54,000 ft (16,500 m) over the equator and about 28,000 ft (8,500 m) over the poles. Seasonal changes affect the thickness of the troposphere, causing it to be thicker in summer than in winter.

Air pollutants are emitted into the troposphere. Air pollution transport is controlled by the speed and direction of the winds. The rate of pollutant dispersion is influenced by the thermal characteristics of the atmosphere, as well as by the mechanical agitation of the air mass as it moves over the different surface features of the Earth.

The thermal characteristics of the atmosphere are governed by a number of different factors, including the intensity of solar radiation, the amount of cloud cover, and a surface feature known as the albedo. Solar radiation varies with the seasons and the area of the globe, and cloud cover varies with normal changes in day-to-day surface weather conditions. These changes can all be measured and recorded at meteorological monitoring stations around the world. Albedo, on the other hand, is a fixed factor determined by the features of the Earth's surface at one location or another. Albedo is defined as the fraction (or percentage) of incoming solar energy that is reflected back to space. Different surfaces, such as soil, water, or snow, have different reflective properties; hence, they have different albedo values. Presented in Table 9.4 is a listing of different albedo factors for different surface conditions.

Wind is the most fundamental element in the general circulation of the atmosphere. Winds are always named by the direction from which they blow. For example, a "north wind" is a wind blowing from the north toward the south, and a wind that blows from west to east is a "westerly wind". When wind blows primarily from one direction, that direction is known as the prevailing wind.

Wind speed increases rapidly with height above the ground level, because frictional drag decreases. Wind is not a steady current, but is comprised of a succession of gusts. Close to the Earth, this is caused by irregularities of the surface, which create eddies. Eddies are variations from the main current of wind flow. Larger irregularities are caused by vertical movement of heat known as convection. These, and other forms of turbulence, contribute to the movement of heat, moisture, and pollutants into the upper atmosphere.

The physical characteristics of the Earth's surface are referred to as terrain features or topography. Topographical features influence the way the Earth and its surrounding air heat up, and they also affect the way air flows. This effect primarily impacts air flow relatively close to the Earth's surface. These features can be grouped into four categories: flat, mountain/valley, land/water, and urban. Topography plays a significant impact upon pollutant dispersion.

Vertical motion is important in air pollution meteorology, because vertical motion helps to determine how much air is available for pollutant dispersal. Atmospheric temperature and pressure influence the buoyancy of air. A parcel of air that becomes warmer than the surrounding air rises, or is buoyant. As the parcel rises, it expands, thereby decreasing its pressure and, therefore, it cools. The initial cooling of an air parcel has the opposite effect. Warm air rises and cools, whereas cool air descends and warms. The extent to which an air parcel rises or falls depends on the relationship of its temperature to that of the surrounding air. The point at which the air parcel reaches its maximum rise is known as the mixing height. The air below the mixing height is the mixing layer. The higher the mixing height, the deeper the mixing layer, and thus the greater the volume of air into which pollutants can be dispersed.

Table 9.4 Albedo values of various surfaces.

Surface	Albedo (as percentage of incoming solar radiation)
Black soil – dry	14
Black soil moist	8
Plowed soil – moist	14
Sand, bright, fine	37
Dense, dry and clean soil	86–95
Sea ice, slightly porous, milky bluish	36
Ice sheet, covered by a water layer of 15–20 cm	26
Woody farm covered with snow	33–40
Deciduous forest	17
Tops of oak	18
Pine forest	40
Desert shrub land	20–29
Swamp	10–14
Prairie	12–13
Winter wheat	16–23
Heather	10
Yuma, Arizona	20
Washington, DC (September)	12–13
Winnipeg, Manitoba (July)	13–16
Great Salt Lake, Utah	3

Based on: US EPA (2005). *Basic Air Pollution Meteorology*, APTI Self Instructional Course 409, Air Pollution Training Institute, Environmental Research Center, MD, Table 2-2.

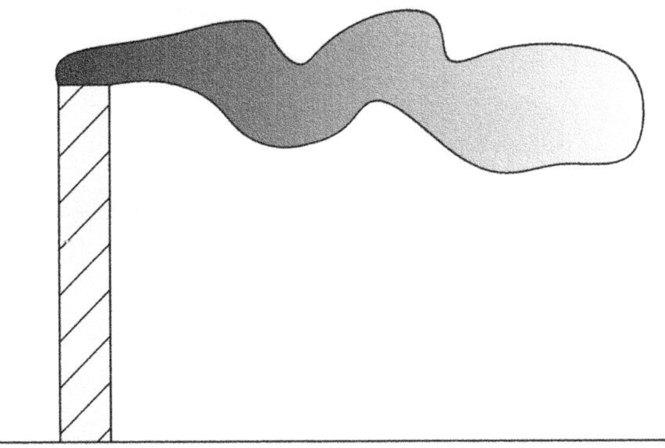

Figure 9.11 Looping plume. *Source*: Reproduced by permission of The Air Compliance Group, LLC.

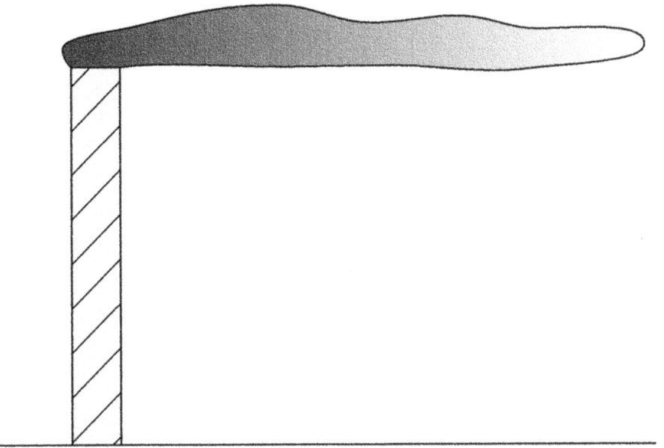

Figure 9.12 Fanning plume. *Source*: Reproduced by permission of The Air Compliance Group, LLC.

The degree of stability of the atmosphere is determined by the difference in the temperature of the air parcel, and that of the air surrounding it. This difference can cause the parcel to move vertically. In stable conditions, this vertical movement is discouraged, whereas in unstable conditions, there is significant vertical movement. The air parcel tends to move upward or downward and to continue that movement. When conditions present no vertical movement, conditions are considered neutral. When conditions are extremely stable, cooler air near the surface can be trapped by a layer of warmer air above it. This condition, called an inversion, allows virtually no vertical air motion. Different stability conditions are directly related to pollutant dispersion and resultant concentrations in the ambient air.

The combination of horizontal and vertical air mass movement dictates the form and extent of pollutant dispersion. On a typical day, with unstable conditions, a looping plume will result from stack emissions. A diagram of a normal looping plume is presented in Figure 9.11.

During a period of stable atmospheric conditions, an emissions plume will transport pollutants downwind in a steadier manner. This is referred to as a fanning plume, as depicted in Figure 9.12.

A coning plume develops during neutral periods when there is little in the way of air movement to direct the exhaust from a stack. A diagram of this is presented in Figure 9.13.

A major influencing factor for pollutant dispersion is an inversion layer, because it acts as a barrier to vertical mixing. The height of a stack in relation to the height of the inversion layer will influence ground-level pollutant concentrations during an inversion. When the stack top protrudes above the inversion layer, emissions are trapped above the inversion and may not make it down to the ground surface. This is known as lofting, and is shown in Figure 9.14.

If the top of the stack is under an inversion layer, downwind pollutant concentrations can increase rapidly. As the ground warms in the morning, air below an inversion layer becomes unstable. When this occurs, the pollutants can be rapidly transported down toward the ground. This is known as fumigation, which is shown in Figure 9.15. This is a potentially dangerous condition in some instances, depending on the severity and duration of the conditions and the nature of the pollutants that are emitted.

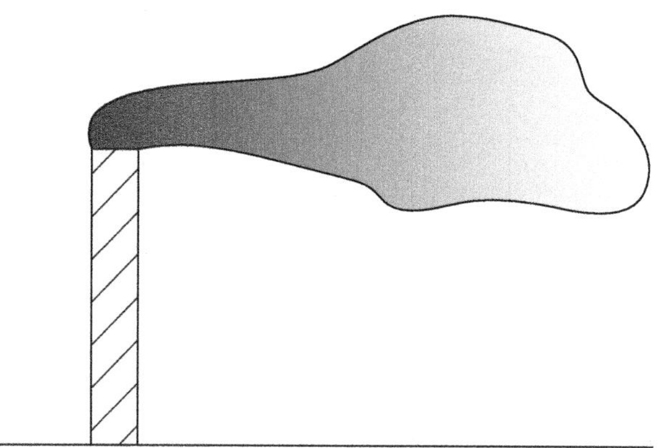

Figure 9.13 Coning plume. *Source:* Reproduced by permission of The Air Compliance Group, LLC.

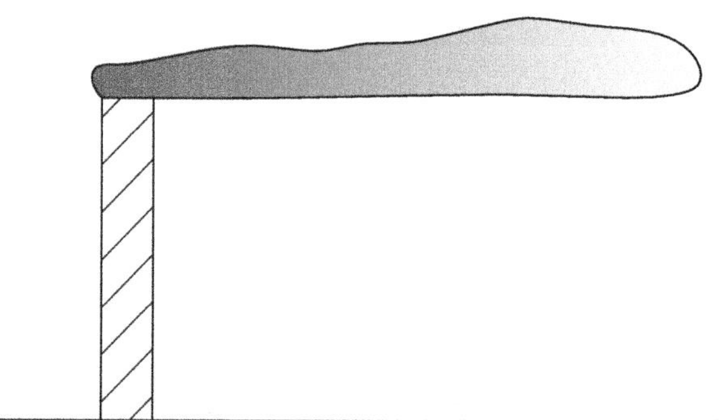

Figure 9.14 Lofting plume. *Source:* Reproduced by permission of The Air Compliance Group, LLC.

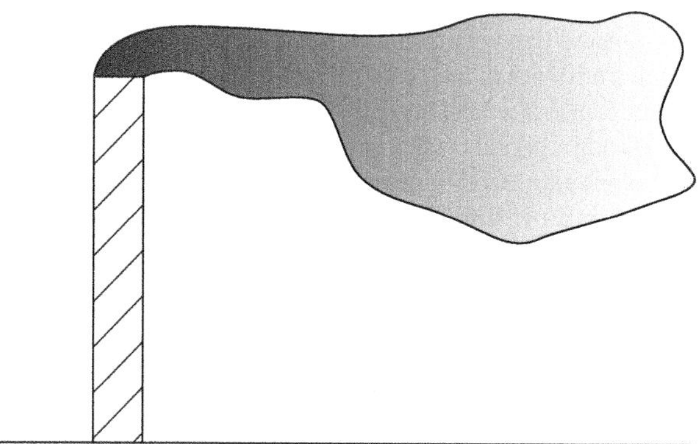

Figure 9.15 Fumigation. *Source:* Reproduced by permission of The Air Compliance Group, LLC.

9.5 Dispersion modeling

Estimating and evaluating the downwind impacts of air pollutant emissions is performed by using dispersion modeling. There are numerous computer based models available for different types of applications and analyses. However, they all share a fundamental basis in terms of plume rise and basic dispersion algorithms.

The first part of any modeling analysis is to calculate the height to which the exhaust plume rises after release from the stack, and from which downwind transport begins. This is a critical step, because the concentration of pollutants, as they impact the ground, is directly related to the height of the plume. The higher the height of the plume, the longer it will have to travel before it ultimately reaches ground level.

Plume height is the sum of the physical height of the stack, the momentum plume rise, and the buoyancy plume rise. Momentum plume rise is that which is created by the velocity of the exhaust gases as they exit the top of the stack. Buoyancy plume rise (F), on the other hand, is also known as thermal plume rise, and is the result of hot air rising, due to the temperature of the exhaust gas being higher than the temperature of the air mass into which it is released.

For dispersion modeling purposes, plume rise is calculated by the Briggs Plume Rise Equation, as presented in Equations (9.3) and (9.4). (USEPA APTI Course 409) The final plume height (H) is the sum of the physical stack height (H_s) plus the plume rise (ΔH).

$$\Delta H = \frac{1.6\,F^{1/3}\,X^{2/3}}{\bar{u}} \tag{9.3}$$

where:
ΔH = plume rise above the stack, ft (m)
F = buoyancy flux, ft^3/s^3 (m^3/s^3)
x = downwind distance from the stack, ft (m)
$\bar{u}$ = wind speed ft/s (m/s)
g = acceleration due to gravity, 32.2 ft/s^2 (9.81m/s^2)
V = volumetric flow rate to stack gas, ft^3/s (m^3/s)
T_s = temperature of stack gas, °R (K)
T_a = temperature of ambient air, °R (K).

$$F = \frac{g}{\pi} V \left[\frac{T_s - T_a}{T_a}\right] \tag{9.4}$$

After emissions are released from a stack, and rise to their maximum plume height, they move downwind and spread out in the horizontal (y-axis) and vertical (z-axis) directions. The degree to which dispersion progresses is strongly dependent upon the level of atmospheric stability, as discussed in the previous section. A summary of the different atmospheric stability classes used in dispersion modeling, labeled A through F, is presented in Table 9.5.

With the knowledge of the plume height, the wind speed, and the atmospheric stability class, downwind ambient pollutant concentrations can be calculated using the Gaussian dispersion equation presented in Equation (9.5). (USEPA APTI Course 409)

$$\chi_{(x,y,z)} = \frac{Q}{2\pi\,\sigma_y \sigma_z\,u}\, e^{-\frac{1}{2}\left(\frac{y}{\sigma_y}\right)^2} \left\{ e^{-\frac{1}{2}\left(\frac{z-H}{\sigma_z}\right)^2} + e^{-\frac{1}{2}\left(\frac{z+H}{\sigma_z}\right)^2} \right\} \tag{9.5}$$

where:
$\chi_{(x,y,z)}$ = downwind pollutant concentration, g/m^3,
Q = pollutant mass emission rate, g/s,

Table 9.5 Atmospheric stability classes.

	Daytime incoming solar radiation			Night-time cloud cover	
Surface wind speed (m/s)	**Strong**	**Moderate**	**Slight**	**$\geq\frac{1}{2}$ Cloud cover**	**$\leq\frac{3}{8}$ Cloud cover**
<2	A	A–B	B	–	–
2–3	A–B	B	C	E	F
3–5	B	B–C	C	D	E
5–6	C	C–D	D	D	D
>6	C	D	D	D	D

Based on: US EPA (2005). *Basic Air Pollution Meteorology*, APTI Self Instructional Course 409, Air Pollution Training Institute, Environmental Research Center, MD, Table 6-1.

H = effective plume height ($H_s + \Delta H$), m,
u = wind speed, m/s,
x = downwind distance, km,
y = horizontal distance from plume centerline, m,
σ_y = standard deviation of pollutant concentration in the y (horizontal) direction,
z = vertical distance from plume centerline, m,
σ_z = standard deviation of pollutant concentration in the z (vertical) direction.

In order to utilize this equation, it is necessary to calculate the σ_y and σ_z values. These are calculated as presented in Equations (9.6) through (9.8). (USEPA APTI Course 409)

$$\sigma_y = 465.11628(x)(\tan\theta) \quad (9.6)$$

$$\theta = 0.017453293(c - d\ln(x)) \quad (9.7)$$

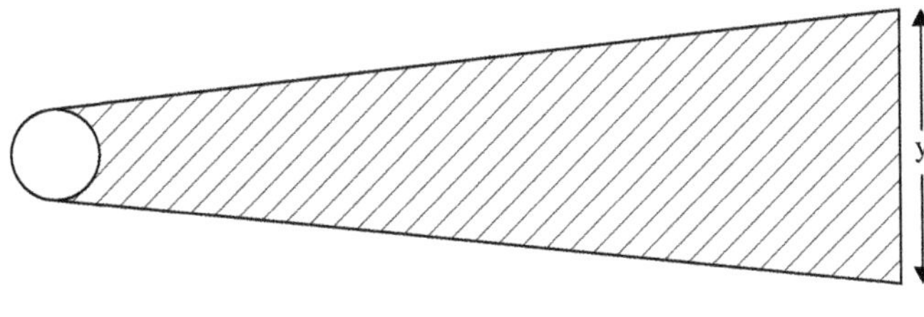

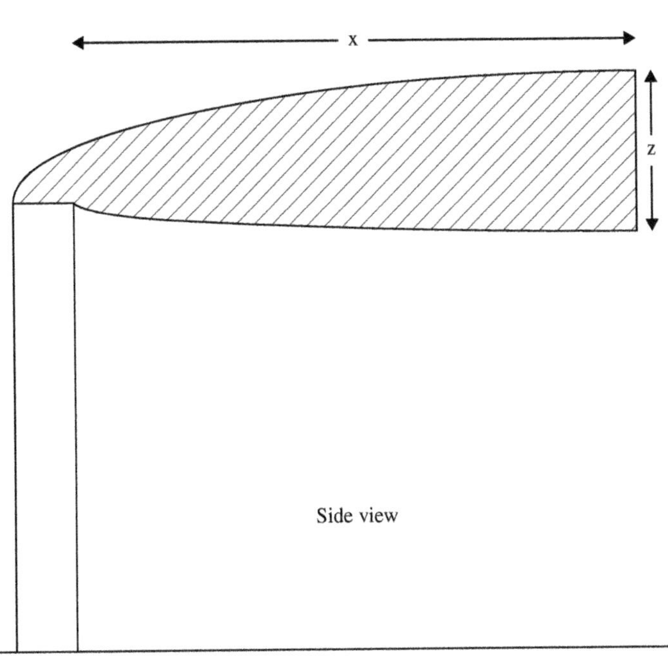

Figure 9.16 Plume dispersion pattern. *Source*: Reproduced by permission of The Air Compliance Group, LLC.

where:
θ = angle, radians
x = downwind distance, km
c = see Table 9.7
d = see Table 9.7.

$$\sigma_z = a\,x^b \tag{9.8}$$

where:
x = downwind distance, km,
a = see Table 9.6
b = see Table 9.6.

Figure 9.16 is a plan and profile view of a plume dispersion pattern.

Example 9.2 Calculation of downwind pollutant concentrations

Calculate the ambient pollutant concentration, at a point 0.5 km downwind, at the plume centerline ($z=0$ and $y=0$) of a source with a plume height (H) of 50 meters and a pollutant emission rate of 10 grams per second. Wind speed is 6.0 m/s with an atmospheric stability class D.

Solution

From Table 9.6 at $x=0.5$ km and for class D stability, the values of a and b are 32.093 and 0.81066, respectively.

Table 9.6 Calculation values for σ_z.

Stability Category	x (km)	a	b
A	<0.10	122.800	0.94470
	0.10–0.15	158.080	1.05420
	0.16–0.20	170.220	1.09320
	0.21–0.25	179.520	1.12620
	0.26–0.30	217.410	1.26440
	0.31–0.40	258.890	1.40940
	0.41–0.50	346.750	1.72830
	0.51–3.11	453.850	2.11660
	>3.11	*	*
B	<0.20	90.673	0.93198
	0.21–0.40	98.483	0.98332
	>0.40	109.300	1.09710
C	All	61.141	0.91465
D	<0.30	34.459	0.86974
	0.31–1.00	32.093	0.81066
	1.01–3.00	32.093	0.64403
	3.01–10.00	33.504	0.60486
	10.01–30.00	36.650	0.56589
	>30.00	44.053	0.51179
E	<0.10	24.260	0.83660
	0.10–0.30	23.331	0.81956
	0.31–1.00	21.628	0.75660
	1.01–2.00	21.628	0.63077
	2.01–4.00	22.534	0.57154
	4.01–10.00	24.703	0.50527
	10.01–20.00	26.970	0.46713
	20.01–40.00	35.420	0.37615
	>40.00	47.618	0.29592
F	<0.20	15.209	0.81558
	0.21–0.70	14.457	0.78407
	0.71–1.00	13.953	0.68465
	1.01–2.00	13.953	0.63227
	2.01–3.00	14.823	0.54503
	3.01–7.00	16.187	0.46490
	7.01–15.00	17.836	0.41507
	15.01–30.00	22.651	0.32681
	30.01–60.00	22.074	0.27436
	>60.00	32.219	0.21716

*σ_z is set to 5000 m whenever the calculated value equals of exceeds 5,000 m
**σ_z is equal to 5,000 m
Source: US EPA (1995) *User's Guide for the Industrial Source Complex (ISC3) Dispersion Models – Volume II – Description of Model Algorithms*, US EPA, Research Triangle Park, NC, EPA-454/B-95-003b, pp 117–118.

From Table 9.7 for class D stability, the values of c and d are 8.3330 and 0.72382, respectively.

Determine the value of θ from Equation (9.7):

$$\theta = 0.017453293(c - d\ln(x))$$

$$\theta = 0.017453293(8.3330 - 0.72382\ln(0.5))$$

$$\theta = 0.15420 \text{ radians} = 8.835°$$

Table 9.7 Theta Calculation Values for σ_y.

Stability Category	c	d
A	24.1670	2.5334
B	18.3330	1.8096
C	12.5000	1.0857
D	8.3330	0.72382
E	6.2500	0.54287
F	4.1667	0.36191

Source: U.S. EPA (1995) User's Guide for the Industrial Source Complex (ISC3) Dispersion Models – Volume II – Description of Model Algorithms, U.S. EPA, Research Triangle Park, NC, EPA-454/B-95-003b, p 116. United States Environmental Protection Agency.

Using the value of θ, calculate σ_y using Equation (9.6):

$$\sigma_y = 465.11628(x)(\tan\theta)$$

$$\sigma_y = 465.11628(0.5)(\tan(8.835))$$

$$\sigma_y = 36.15$$

Next, determine σ_z from Equation (9.8):

$$\sigma_z = a\,x^b$$

$$\sigma_z = 32.093(0.5)^{0.81066} = 18.3$$

Finally, substitute the appropriate coefficients into Equation (9.5) as follows:

$$\chi_{(x,y,z)} = \frac{Q}{2\pi\sigma_y\sigma_z u} e^{-\frac{1}{2}\left(\frac{y}{\sigma_y}\right)^2} \times \left\{ e^{-\frac{1}{2}\left(\frac{z-H}{\sigma_z}\right)^2} + e^{-\frac{1}{2}\left(\frac{z+H}{\sigma_z}\right)^2} \right\}$$

$$\chi_{(0.5,0,0)} = \frac{10}{2\pi(36.15)(18.3)(6)} e^{-\frac{1}{2}\left(\frac{0}{36.15}\right)^2} \times \left\{ e^{-\frac{1}{2}\left(\frac{0-50}{18.3}\right)^2} + e^{-\frac{1}{2}\left(\frac{0+50}{18.3}\right)^2} \right\}$$

$$\boxed{\chi_{(0.5,0,0)} = 1.92\times10^{-5}\ \text{g/m}^3}$$

Example 9.2A Calculation of downwind pollutant concentrations

Using the same emission source and meteorology for the above example, calculate the ambient pollutant concentration, at a point 1.0 km downwind, at a point 100 m (0.1 km) off the plume centerline in the horizontal direction ($z = 0$ and $y = 0.1$)

Solution

The calculation of θ and σ_z are not repeated here, because they are the same as above.

Using the value of θ (from above), calculate σ_y using Equation (9.6):

$$\sigma_y = 465.11628(x)(\tan\theta)$$

$$\sigma_y = 465.11628(1.0)(\tan(8.835))$$

$$\sigma_y = 72.30$$

Finally, substitute the appropriate coefficients into Equation (9.5) as follows:

$$\chi_{(x,y,z)} = \frac{Q}{2\pi\sigma_y\sigma_z u} e^{-\frac{1}{2}\left(\frac{y}{\sigma_y}\right)^2} \times \left\{ e^{-\frac{1}{2}\left(\frac{z-H}{\sigma_z}\right)^2} + e^{-\frac{1}{2}\left(\frac{z+H}{\sigma_z}\right)^2} \right\}$$

$$\chi_{(1.0,0.5,0)} = \frac{10}{2\pi(72.30)(18.3)(6)} e^{-\frac{1}{2}\left(\frac{0.1}{72.30}\right)^2} \times \left\{ e^{-\frac{1}{2}\left(\frac{0-50}{18.3}\right)^2} + e^{-\frac{1}{2}\left(\frac{0+50}{18.3}\right)^2} \right\}$$

$$\boxed{\chi_{(1.0,0.1,0)} = 9.60\times10^{-6}\ \text{g/m}^3}$$

Computer-based dispersion modeling programs are available for a wide variety of applications. The EPA SCREEN model is a simple, automated worst-case modeling analysis that is quick and easy to perform. For more complex regulatory determinations, the EPA AERMOD model is a complex application of the equations presented herein, modified for a variety of factors, including surface roughness, albedo, stack and building downwash, and other factors impacting air currents and resulting pollutant dispersion characteristics in differing topographical and meteorological conditions.

These models are typically run with five years of recorded meteorological data from the National Weather Service location closest to the emission source, in order to be able to evaluate the full range of weather conditions found in the area of the source. EPA Models can be found, in their basic form, at http://www.epa.gov/ttn/scram/. More user-friendly versions of these models are available for several different commercial suppliers.

9.6 Air pollution control technologies

Air pollution control technologies are generally divided into those that are used to control particulate emissions, and those that are used to control gaseous emissions. Within each category of control equipment, control of toxic air pollutants can also be obtained.

9.6.1 Particulate emissions control

9.6.1.1 Particulate Emissions

Particulate emissions, including toxic metals and/or other pollutants contained on or within particles, are controlled by physical mechanisms including gravitational settling, cyclonic separation, electrostatic collection, and filtration. Gravitational settling chambers are the most rudimentary and least expensive type of particulate control systems. Particulate laden gas streams flow through ductwork designed to keep gas stream velocities high enough to keep particles entrained, in order to keep particles from settling out in the ducts.

Settling chambers are designed for the opposite goal. Settling chambers are, in essence, simply open chambers (boxes) in which the gas flow rate slows quickly, thereby allowing particles to settle out of the gas stream and to be collected into hoppers below. An example of a simple settling chamber is presented in Figure 9.17. Settling chambers can be modified by the inclusion of baffle plates, which are designed to improve particulate removal performance by causing some particles to be removed by impaction upon the baffles. Also, by forcing the gas stream to follow other than a straight path from entry to exit, baffles can increase the time spent in the chamber.

Settling chambers are limited in their particulate removal efficiency to particles greater than 40–60 μm in diameter; hence, they are of little value in meeting many current emissions limitations. Sometimes, however, settling chambers are used as an initial scalping device to remove large particles from a gas stream inexpensively, in order to lessen the burden on more effective control devices downstream.

Settling chambers are designed largely on the basis of Stokes' law, as presented in Equation (9.9). (USEPA APTI Course 413)

$$v_t = \frac{g(\rho_p - \rho_g)d^2}{18\mu_g} \qquad (9.9)$$

where:
v_t = settling velocity, m/s
ρ_p = particle density, kg/m^3
ρ_g = gas stream (air) density, kg/m^3
g = gravitational acceleration, m/s^2
d = particle diameter, m
μ_g = gas stream (air) viscosity, kg/(m·s).

As shown, this equation, and hence the settling velocity, is heavily dependent upon the diameter of the particles being separated from the gas stream by the settling chamber.

In order to calculate the efficiency of a settling chamber for a given particle size, it is necessary to know the velocity of the gas stream as it moves through the chamber. This is calculated using Equation (9.10) (USEPA APTI Course 413).

$$v = \frac{Q}{HW} \qquad (9.10)$$

where:
v = linear velocity, m/s
Q = gas stream volumetric flow rate, m^3/s
H = settling chamber height, m
W = settling chamber width, m.

The efficiency (E) of the settling chamber is determined by modifying Equation (9.9) to incorporate the gas stream velocity, as presented in Equation (9.11). (USEPA APTI Course 413)

$$E = \frac{g(\rho_p - \rho_g)d^2}{18\mu_g v}(100) \qquad (9.11)$$

Example 9.3 Calculation of settling chamber efficiency

For a settling chamber measuring 5 meters in length, 3 meters in width, and 3 meters in height, treating a gas volume of 15 m^3/s, calculate the chamber efficiency for 40 μm and 80 μm diameter particles. Assume a gas density

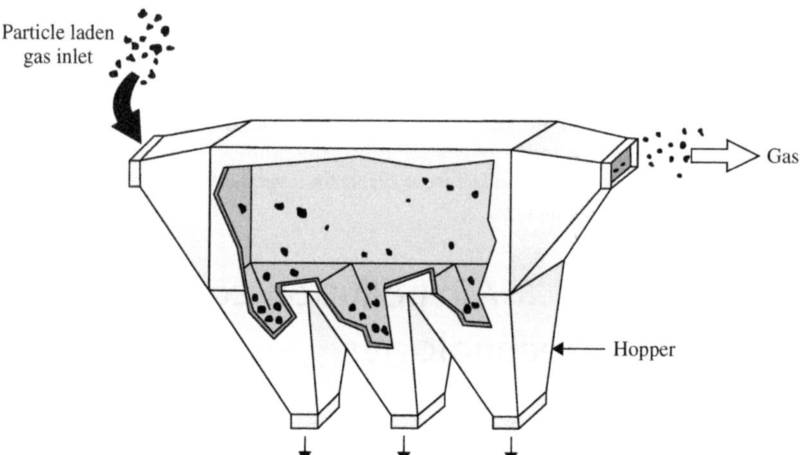

Figure 9.17 Settling chamber. *Source:* U.S. EPA (1999b) Control of Particulate Matter Emissions – Student Manual, APTI Course 415, Air Pollution Training Institute, Research Triangle Park, NC, Chapter 5 – page 2. United States Environmental Protection Agency.

of 1.2kg/m^3, a particle density of 1,500 kg/m^3, and a gas viscosity of 1.85×10^{-5} kg/(m·s).

Solution

First use Equation (9.10) to calculate the gas stream velocity:

$$v = \frac{Q}{LW} = \frac{15\,m^3/s}{(3\,m)(3\,m)} = 1.67\,m/s$$

Next, use Equation (9.11), with a particle diameter of 40 μm (4.0×10^{-5} m), to calculate the settling chamber removal efficiency.

$$E = \frac{\left(9.81\frac{m}{s^2}\right)\left(1500\frac{kg}{m^3} - 1.2\frac{kg}{m^3}\right)(4.0 \times 10^{-5}\,m)^2}{18\left(1.85 \times 10^{-5}\frac{kg}{mg \cdot s}\right)\left(1.67\frac{m}{s}\right)}(100)$$

$\boxed{E = 4.2\%}$

Finally, repeat the same equation substituting the particle size with the smaller 80 μm (8.0×10^{-5}m) to calculate the settling chamber removal efficiency.

$$E = \frac{\left(9.81\frac{m}{s^2}\right)\left(1500\frac{kg}{m^3} - 1.2\frac{kg}{m^3}\right)(8.0 \times 10^{-6}m)^2}{18\left(1.85 \times 10^{-5}\frac{kg}{mg \cdot s}\right)\left(1.67\frac{m}{s}\right)}(100)$$

$\boxed{E = 16.9\%}$

The above example shows how the efficiency of a settling chamber is strongly influenced by particle size, and how such chambers provide minimal overall particulate removal efficiencies.

9.6.1.2 Cyclones

Another form of mechanical collector that is more efficient than a simple settling chamber is a cyclone, a general diagram of which is presented in Figure 9.18. Cyclones use centrifugal force to separate particles from a gas stream. Basically, the gas stream enters the cone-shaped device tangentially, and begins to spin inside the cyclone body. As the gas stream spins, particles are thrown to the outer wall of the cyclone, where they collect and fall to the collection hopper below.

There are two main types of cyclones:

1 large diameter cyclones; and

2 small diameter multi-cyclones.

Large diameter cyclones range in size from < 0.3 m to > 4 m in diameter and are used for collection of large diameter particulate matter. Multi-cyclone collectors are groups of small diameter cyclones, typically 0.15 to 0.3 m in diameter. Such

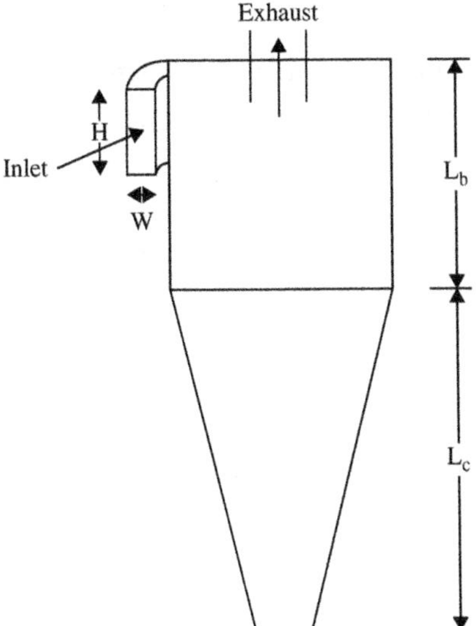

Figure 9.18 Air pollution control cyclone. *Source*: The Air Compliance Group, LLC, Courtesy of The Air Compliance Group, LLC, Roanoke, VA, USA.

collections of small cyclones have better particulate removal capability than large diameter cyclones.

Several factors impact upon the performance of a cyclone collector, the most important of which include the size and density of the particles, the gas velocity through the unit, and the cyclone diameter. Since inertial forces are used to separate the particles from the gas stream, collection efficiency increases as the size and density of the particle increases and as the gas velocity through the unit increases. The centrifugal force increases as the radius of the turns within the cyclone decrease. Because of this, smaller diameter cyclones are more efficient than larger diameter cyclones.

The particulate removal efficiency of a cyclone can be estimated, for a given particle size, using Equation (9.12) (USEPA APTI Course 413):

$$E = \frac{\pi N \rho_p d_p^2 V_g}{9\,\mu W} \tag{9.12}$$

where:
E = Efficiency, %
N = number of turns within the cyclone
ρ_p = particle density, kg/m^3
d_p = particle diameter, m
V_g = gas velocity, m/s
μ = gas stream (air) viscosity, kg/(m·s)
W = width of the cyclone inlet, m.

The number of turns (N) is calculated using Equation (9.13) (USEPA APTI Course 413):

$$N = \frac{1}{H}\left(L_b + \frac{L_c}{2}\right) \tag{9.13}$$

where:
H = height of cyclone inlet, m (See Figure 9.18)
L_b = length of cyclone upper chamber, m (See Figure 9.18)
L_c = length of cyclone lower chamber, m. (See Figure 9.18).

Example 9.4 Calculate the particulate removal efficiency of a cyclone

A cyclone is used to remove dust from a grain processing facility. The height of the cyclone inlet (H) is 0.5 m, the width of the cyclone inlet (W) is 0.3 m, the length of the upper chamber (L_b) is 1.5 m, and the length of the lower chamber (L_c) is 3.0 m. The inlet gas velocity is 15.0 m/s, and the particle density is 2000 kg/m^3. Assume gas viscosity of 1.85×10^{-5} kg/(m·s). Calculate the cyclone removal efficiency over a range of particle sizes from 40 μm to 80 μm.

Solution

First, calculate the number of turns within the cyclone, using Equation (9.13):

$$N = \frac{1}{0.5\,\text{m}}\left(1.5\,\text{m} + \frac{3.0\,\text{m}}{2}\right)$$

$$\boxed{N = 6}$$

Next, use Equation (9.12) to calculate the cyclone efficiency for the smallest particle size of 40 microns:

$$E = \frac{\pi(6.0)(2000\,\text{kg/m}^3)(4.0\times10^{-5}\text{m})^2(15.0\,\text{m/s})}{9\left(1.85\times10^{-5}\dfrac{\text{kg}}{\text{m}\cdot\text{s}}\right)(0.3\,\text{m})}$$

$$\boxed{E = 18.1\%}$$

Similarly, use Equation (9.12) to calculate the cyclone efficiency for the largest particle size of 80 microns:

$$E = \frac{\pi(6.0)(2000\,\text{kg/m}^3)(8.0\times10^{-5}\text{m})^2(15.0\,\text{m/s})}{9\left(1.85\times10^{-5}\dfrac{\text{kg}}{\text{m}\cdot\text{s}}\right)(0.3\,\text{m})}$$

$$\boxed{E = 72.5\%}$$

This example shows how the efficiency of a cyclone is also strongly influenced by particle size, as well as by the number of turns within the device. It also demonstrates that cyclones are capable of much higher particulate removal efficiencies than simple settling chambers. However, they are still fairly low-efficiency devices that rarely meet stringent environmental standards alone.

9.6.1.3 Venturi Scrubbers

Wet scrubbers are rarely used for particulate control, because standard scrubbers offer little efficiency. One type of scrubber, referred to as a venturi scrubber, is often used in small industrial applications because it is more efficient than settling chambers or cyclones. Venturi scrubbers can be used for wet gas streams that would present plugging problems for mechanical collectors, and because they offer a level of absorptive capacity for removal of certain gaseous air pollutants from the same gas stream (see section 9.6.2.1).

Venturi scrubbers are examples of high-energy wet scrubbers, although they can also be operated as a medium-energy scrubber. The fixed throat venture scrubber, shown in Figure 9.19, is the most common design.

The gas stream entering the converging section of the venturi is accelerated to a velocity between 60–200 meters per second at the throat inlet. Liquid (water) is injected into the throat and atomized into droplets with a typical size of 50–75 μm by the rapid flow rate of the gas stream into which it has been introduced. These droplets are initially moving relatively slowly compared to the gas stream. Impaction occurs

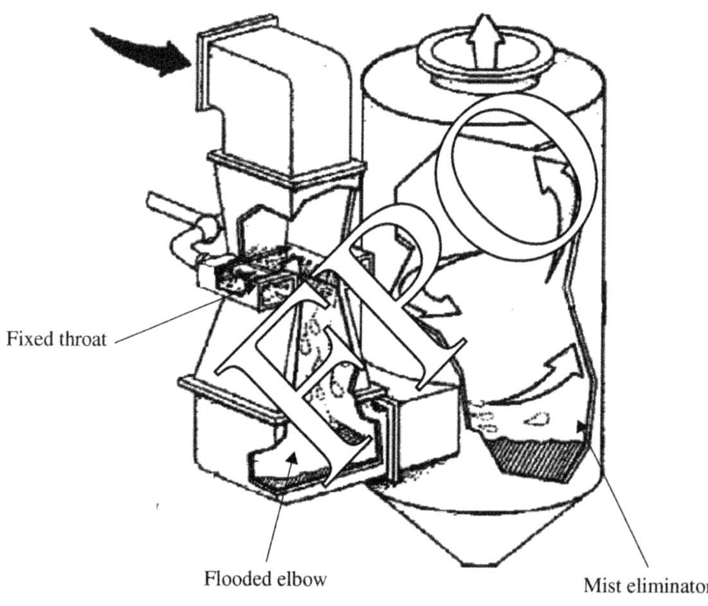

Figure 9.19 Venturi scrubber. *Source*: EPA (1999b) Control of Particulate Matter Emissions – Student Manual, APTI Course 415, Air Pollution Training Institute, Research Triangle Park, NC, Chapter 8 – page 32. United States Environmental Protection Agency.

on the droplets due to the large difference in the gas stream velocity and the velocity of the water droplets.

The effectiveness of a venturi scrubber is primarily related to the maximum difference in the droplet and gas stream velocities. Because the fixed throat has a constant open area, the actual gas velocity in the throat section is dependent upon the gas flow rate. Particle collection efficiency is, therefore, gas flow rate dependent. Fixed throat venturi scrubbers are used on sources where the gas flow rate is relatively constant, or where the particle size distribution is sufficiently large that some variation in gas velocity is tolerable. When this is not the case, variable throat venturi scrubbers can be used. As the name implies, the geometry of the converging section of the scrubber can be varied mechanically to provide more or less restriction and, hence, higher or lower gas velocity through the venture throat.

Proper liquid distribution is essential in obtaining optimum performance in a venturi scrubber. Because of the high gas velocities, the residence time of the gas stream in the venturi throat, where most collection occurs, is only 0.001 – 0.005 s. Most of the particles that make it through the venturi throat will not be collected in the latter stages of the scrubber. Obviously, if liquid is not distributed completely across the area of the throat, those particles passing through a section in which water has not been injected will pass through uncollected.

The venturi scrubber system shown in Figure 9.19 includes a flooded section in the elbow directly under the venture throat. This elbow leads from the diverging section to the mist eliminator. This section is termed a "flooded elbow", and it provides abrasion protection. Droplets that have accelerated to a high velocity in the venturi will erode the bottom of this duct if it is not protected. From this point, the gas stream passes through a mist eliminator to trap the water droplets and the particles contained therein, and carry them away for disposal.

Calculation of the design and operating efficiency for venturi scrubbers is largely empirical, though there are several procedures available. The process must start with the calculation of the liquid droplet size created when liquid is injected into the high gas velocity throat of the scrubber, using Equation (9.14). This equation is generally accepted in spite of the mixed units involved. Following this, two intermediate parameters (the Cunningham Correction Factor and the Inertial Impaction Factor) are calculated using Equations (9.15) and (9.16). Finally, the estimated particulate collection efficiency can be calculated using Equation (9.17) (USEPA APTI Course 413):

$$d_d = \frac{16,400}{v_g} + 1.45\left(\frac{Q_l}{Q_g}\right)^{1.5} \tag{9.14}$$

where:
d_d = liquid droplet diameter, μm
v_g = gas velocity, ft/s
Q_l = liquid flow rate, gal/min
Q_g = gas flow rate (1,000 cubic feet per minute).

$$C_c = 1 + \frac{6.21 \times 10^{-4} T}{d_p} \tag{9.15}$$

where:
C_c = Cunningham correction factor, (dimensionless)
T = gas temperature (K)
d_p = particle diameter, μm.

$$\psi_I = \frac{C_c d_p^2 \rho_p V_r}{18 \mu_g d_d} \tag{9.16}$$

where:
ψ_I = inertial impaction factor, dimensionless
d_p = particle diameter, cm
ρ_p = particle density, g/cm^3
V_r = relative velocity between particle and droplet, cm/s
μ_g = gas stream (air) viscosity, g/(cm·s)
d_d = droplet diameter, cm

$$E = 1 - e^{-k\sqrt{\psi_I}\frac{Q_l}{Q_g}} \tag{9.17}$$

where:
E = particulate removal efficiency, fractional
K = constant, 1,000 ft^3/gal.

The constant in equation (9.17) (k) has been developed empirically, and is typically 0.1 – 0.2 (1,000 ft^3/gal) (USEPA APTI Course 413).

In addition to the particulate removal performance of a venturi scrubber, it is necessary to evaluate the pressure drop associated with the scrubber operation. High efficiency venturi scrubbers often operate at high pressure drops, thereby necessitating a powerful blower, with the associated operating costs. The anticipated pressure drop of a venturi scrubber is largely dependent upon the velocity of the gas stream through the scrubber throat, which is dictated largely by the degree of constriction provided by this fixed or variable throat. Pressure drop can be estimated using Equation (9.18) (Air Pollution Engineering Manual):

$$\Delta P = \frac{v^2 \rho_g A^{0.133} (L/G)^{0.78}}{3870} \tag{9.18}$$

where:
ΔP = scrubber pressure drop, cm H_2O
V = gas velocity in the venturi throat, cm/s
ρ_g = gas density, g/cm^3
A = the throat cross-sectional area, cm^2
L/G = the liquid/gas flow rate ratio, L/m^3.

Example 9.5 Calculation of venturi scrubber efficiency and pressure drop

A venture scrubber is used to control particulate emissions from a medical waste incinerator exhaust stack. The incinerator exhaust is cooled to 350°F (≅ 450 K), and the scrubber throat measures 30 cm × 45 cm. The gas velocity through the throat is 300 ft/s (9,144 cm/s), and the liquid to gas ratio is 8 gal/1,000 ft^3 (1.07 L/m^3). Assume a "k" constant of 0.15 1,000ft^3/gal. Assume also, a particle density of 1.5 g/cm^3, a gas density of 0.0015 g/cm^3, and gas viscosity of 1.8×10^{-4}g/(cm·s).

Solution

First calculate the liquid droplet size upon injection into the venture throat, using Equation (9.14):

$$d_d = \frac{16,400}{300\ \text{ft/s}} + 1.45(8)^{1.5}$$

$$\boxed{d_d = 87.5\ \mu\text{m}}$$

With the droplet size determined, it is possible to work through Equations (9.15), (9.16), and (9.17) to calculate the venture scrubber efficiency at removing a 1.0 μm size particle.

$$C_c = 1 + \frac{(6.21 \times 10^{-4})(450\ \text{K})}{1\ \mu\text{m}} = 1.28$$

$$\psi_I = \frac{(1.28)(1 \times 10^{-4}\ \text{cm})^2 \left(1.5\ \frac{\text{g}}{\text{cm}^3}\right) \times \left(300\ \frac{\text{ft}}{\text{s}}\right) 2.54 \frac{\text{cm}}{\text{in}} \times 12 \frac{\text{in}}{\text{ft}}}{18\left(1.8 \times 10^{-4} \frac{\text{g}}{\text{cm} \cdot \text{s}}\right)(8.75 \times 10^{-3}\text{cm})} = 6.193$$

$$E = 1 - e^{-0.15\sqrt{6.193}(8.0)}$$

$$\boxed{E = 0.95 \times 100 = 95\%}$$

In order to calculate the scrubber pressure drop associate with this scrubber, which will achieve a 95% removal of 1.0 micron particles, apply Equation (9.18):

$$\Delta P = \frac{(9144\ \text{cm/s})^2(0.0015\ \text{g/cm}^3) \times (1350\ \text{cm}^2)^{0.133}(1.07\ \text{L/m}^3)^{0.78}}{3870}$$

$$\boxed{\Delta P = 89.1\ \text{cm} - H_2O}$$

9.6.1.4 Electrostatic Precipitator

Electrostatic precipitators (ESPs) are used in many industries for the high-efficiency collection of particulate matter. During the 1940s, precipitators began to be used for particulate matter control at coal-fired boilers, cement kilns, and Kraft recovery boilers. The applications of precipitators have steadily increased since the 1940s, due to their ability to efficiently collect particles from a variety of sources without generating restrictions to gas flow resulting in high pressure drops. Also, ESP units are adaptable to very large gas flow rates, and gas streams of varying temperatures.

Electrostatic precipitators contain discharge and collection electrodes aligned in rows within the internal chamber of the unit. Discharge electrodes are connected to power sources and are used to create a field, through which particles pass and, in doing so, become negatively charged. Collection electrodes consist of grounded plates, toward which the charged particles migrate and collect. Physical impact devices, known as rappers, periodically pound on these collection plates to create vibration, which dislodges the collected particles, allowing them to fall into hoppers below the ESP chambers. An example of an electrostatic precipitator is presented in Figure 9.20.

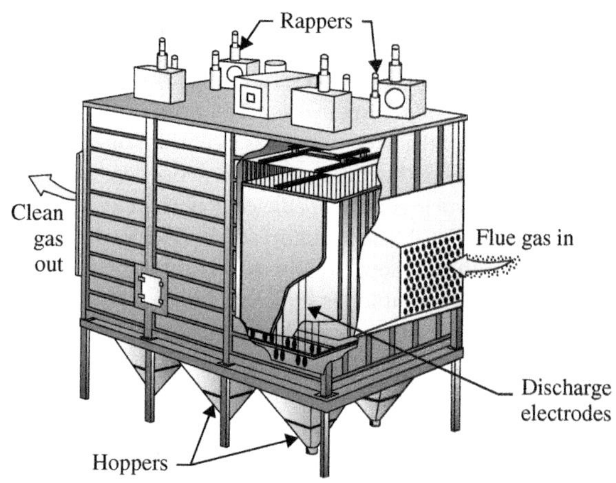

Figure 9.20 Typical electrostatic precipitator. *Source:* U.S. EPA (1999b) Control of Particulate Matter Emissions – Student Manual, APTI Course 415, Air Pollution Training Institute, Research Triangle Park, NC, Chapter 9 – page 15. United States Environmental Protection Agency.

The electrical discharges from the precipitator discharge electrodes are termed corona discharges and are needed to charge the particles electrostatically. Within the negative corona discharge, electrons are accelerated by the very strong electrical field, and strike and ionize gas molecules. The corona discharges are often described as an electron avalanche, since large numbers of electrons are generated during multiple electron-gas molecule collisions. The ionized gas molecules, in turn, transfer the negative charge to solid particles in the gas stream.

The magnitude of the charge picked up by each particle is dependent upon the particle size. Small particles have a low charge, because the gas ions have only a small surface on which to deposit. The charge increases with surface area or with the square of the particle diameter. Large particles accumulate higher electrical charges on their surface and are, therefore, more strongly affected by the applied electrical field generated between the discharge electrode and the collection electrode.

Once the particles have attached negative ions, they are influenced by the electrical field between the discharge electrode and the grounded collection plate. As a result, the charged particles begin to migrate toward the grounded plates. At the same time, opposing ballistic forces, which depend on the particle mass or diameter and the gas velocity, attempt to keep the particles moving straight through the precipitator. As a result, the smaller μm-sized particles are deposited near the inlet and progressively larger particles are deposited farther into the precipitator. Because of this, ESP designers will consider the particle size distribution of the subject gas stream when designing the length of the ESP chambers.

The ability of collected dust to adhere to, and to be released from, the collection electrodes (plates) is influenced by the resistivity of the collected dust, which is a measure of the resistance of the dust to ion flow through the dust to the electrode. If the resistivity is too low, particles can be re-dispersed into the gas stream during rapping, causing a short-term spike in emissions. As the resistivity increases into the moderate range, the dust cake is dislodged from the collection plate as cohesive sheets or clumps that are large enough to fall rapidly into the hopper below. If the voltage drop across the dust layers becomes too high (high resistivity), the dust that is collected on the plate can produce an insulating effect, thereby reducing the ability of the precipitator to collect particles.

Electrostatic precipitators work best when the dust resistivity is in the moderate range. It should provide some resistance to current flow, but not too much. The measure of dust resistivity is referred to in units of ohm-centimeters, which is simply the ohms of resistance created by each centimeter of dust in the dust layer. High resistivity is generally considered to be equal to or above 5×10^{10} ohm-cm. Low resistivity is generally considered to be equal to or below 5×10^{8} ohm-cm. The region between 5×10^{8} and 5×10^{10} ohm-cm is, therefore, the preferred range.

Electrostatic precipitators consist of a large number of parallel gas passages, with discharge electrodes mounted in the center and grounded collection plates on either side. The discharge electrodes are spaced 6–15 cm away from each of the collection plates. The gas stream flows laterally between these plates. The gas stream enters the precipitator in a transition area that slows and distributes the gas flow between the plates. At this point, the gas flow rate is reduced to approximately 1.0–2.0 m/s.

Precipitators are divided into fields. These are portions of the precipitator controlled by separate power supplies. Fields are arranged in series, so that the gas stream flows through each of the fields sequentially. Typically, electrostatic precipitators have from three to five fields.

The particulate removal efficiency of each field is dependent, in large part, upon the migration velocity of the particles within the area between the discharge electrodes and collection plates. Migration velocities have been developed empirically for a variety of different types of particles, over many years of study. Presented in Table 9.8 is a summary of migration velocities that have been established for use in estimating ESP efficiencies. This is performed using Equation (9.18) (USEPA APTI Course 413).

$$E = 1 - e^{-\omega \frac{A}{Q}} \qquad (9.18)$$

where:
ω = migration velocity, m/s
A = total collection plate area, m^3
Q = total gas flow rate, m^3/s.

The total collection plate area (A) is calculated using Equation (9.19) (USEPA APTI Course 413). This is presented graphically in Figure 9.21.

$$A = 2(n - 1)(H)(L) \qquad (9.19)$$

where:
A = Total Surface Area of Collection Plates (m^2)
n = number of collection plates
H = height of collection plate, m
L = length of collection plate, m.

Table 9.8 Effective migration velocities for various industries.

	Effective Migration Velocity	
Application	**ft/s**	**cm/s**
Utility coal-fired boiler	0.13–0.67	4.0–20.4
Pulp and paper mill	0.21–0.31	6.4–9.5
Sulfuric acid mist	0.19–0.25	5.8–7.6
Cement (wet process)	0.33–0.37	10.7–11.3
Cement (dry process)	0.19–0.23	5.8–7.0
Gypsum	0.52–0.64	15.8–19.5
Open-hearth furnace	0.16–0.19	4.9–5.8
Blast furnace	0.20–0.46	6.1–14.0

Source: U.S. EPA (1999a) Control of Gaseous Emissions – Student Manual, APTI Course 413, Air Pollution Training Institute, Research Triangle Park, NC, Table 9-1. United States Environmental Protection Agency.

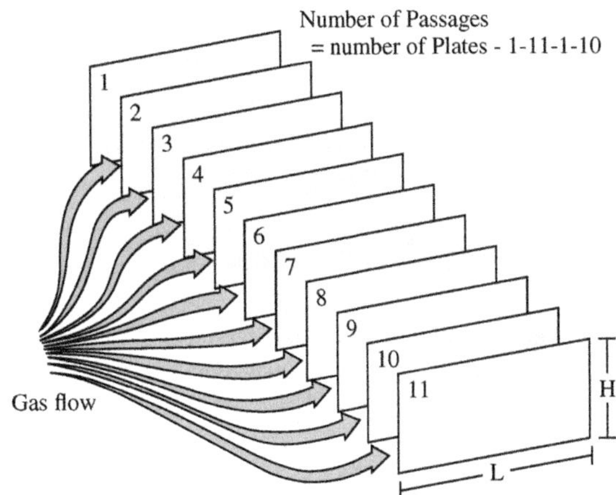

Figure 9.21 ESP collector plate arrangement. *Source:* U.S. EPA (1999b) Control of Particulate Matter Emissions – Student Manual, APTI Course 415, Air Pollution Training Institute, Research Triangle Park, NC, Chapter 9 – page 33. United States Environmental Protection Agency.

As presented in Equations (9.18) and (9.19), the efficiency of any field in an ESP is the same for all fields with the same physical design, treating the same gas stream. In some designs, however, the first field will be designed larger than the subsequent fields, in order to handle the higher particulate loading.

The calculation of the estimate particulate concentration at the exhaust of an ESP is performed using Equation (9.20). (USEPA APTI Course 413)

$$\text{Emissions} = \text{conc}\left(1-\frac{\text{eff}_1}{100}\right)\left(1-\frac{\text{eff}_2}{100}\right) \times\left(1-\frac{\text{eff}_3}{100}\right)\left(1-\frac{\text{eff}_4}{100}\right) \quad (9.20)$$

where:
conc = inlet particulate concentration, mg/m^3
eff_1 = efficiency of first field, %
eff_2 = efficiency of second field, %
eff_3 = efficiency of third field, %
eff_4 = efficiency of fourth field, %.

Example 9.6 Calculation of ESP efficiency and emissions

Calculate the particulate removal efficiency of each field of a four-field precipitator servicing a coal fired boiler. All four fields have the same design. The gas flow rate is $120\ m^3/s$. Each field contains 11 collection plates measuring 12 m high by 18 m in length. The inlet particulate concentration is $4{,}000\ mg/m^3$. Assume a migration velocity of 0.05 m/s.

Solution

First calculate the plate collection area (A) for each field of the precipitator using Equation (9.19).

$$A = 2(11-1)(12\ m)(18\ m) = 4320\ m^2$$

Using this value, and the data presented for this precipitator, calculate the particulate removal efficiency for each identically designed field:

$$E = 1 - e^{-(0.05\ m/s)\frac{4320\ m^2}{120\ m/s}}$$

$$E = 0.835 \qquad \boxed{E = 0.835 \times 100 = 83.5\%}$$

Using the inlet particulate concentration of $4{,}000\ mg/m^3$, the above calculated efficiency, and Equation (9.20), calculate the particulate concentration of the controlled gas stream.

$$\text{Emissions} = \text{conc}\left(1-\frac{\text{eff}_1}{100}\right)\left(1-\frac{\text{eff}_2}{100}\right) \times\left(1-\frac{\text{eff}_3}{100}\right)\left(1-\frac{\text{eff}_4}{100}\right)$$

$$\text{Emissions} = 4000\ mg/m^3\left(1-\frac{83.5}{100}\right)\left(1-\frac{83.5}{100}\right) \times\left(1-\frac{83.5}{100}\right)\left(1-\frac{83.5}{100}\right)$$

$$\boxed{\text{Emissions} = 2.96\ mg/m^3}$$

Electrostatic precipitators are very popular control devices because they have a long history of solid performance, they can be designed to be highly efficient, and they are readily adaptable to large gas flow rates with minimum pressure drop. Their primary limitations are that they are large in size, and they are not very adaptable to small emission sources.

9.6.1.5 Fabric Filters

Fabric filters, or baghouses, are extremely popular particulate emission control devices because they are moderately priced, they are adaptable to a wide range of gas flow rates, and they are highly efficient.

As the name implies, fabric filters utilize fabrics, in the form of long cylindrical bags, to trap particles as the gas stream passes through the bags. In the simplest form of comparison, baghouses operate similarly to vacuum cleaners.

Multiple mechanisms are responsible for particle capture within dust layers and fabrics in a baghouse. Impaction is an inertial mechanism that is most effective on particles larger than about 1 μm. It is effective in fabric filters because there are many sharp changes in flow direction as the gas stream moves around the various particles and bag fibers.

Impaction is a predominant collection mechanism in baghouses, because there are multiple opportunities for particle impaction due to the numbers of individual dust cake particles and fabric fibers in the gas stream path.

Brownian diffusion is moderately effective for collecting sub-micrometer particles, because of the close contact between the gas stream and the dust cake. The particle does not have to be displaced a long distance in order to come into contact with a dust cake particle or fiber. Furthermore, the displacement of sub-μm particles can occur over a relatively long time as the gas stream moves through the dust cake and fabric.

Electrostatic attraction is another particle collection mechanism. Particles can be attracted to the dust layer and fabric due to the moderate electrical charges that accumulate on the fabrics, the dust layers, and the particles. Both positive and negative charges can be generated, depending on the chemical make-up of the materials. Particles are attracted to the dust layer particles or fabric fibers when there is a difference in charge polarity or when the particle has no electrical charge.

One important factor to understand about baghouses is that the fabric itself is not highly effective in particulate removal. It is the dust cake built up on the bags that provides the highest degree of filtration. When bags are new, particles will permeate the fabric material but, as it does so, many particles become strongly lodged within the fibers. As this continues, a permanent "cake" is built on the bags, significantly increasing particulate removal efficiency. Ultimately, a fully caked bag becomes a very high efficiency collection device. This cake is not removed during the bag cleaning operations that baghouses undergo as part of their operational design.

One of the most important criteria in the design of a baghouse is known as the gas-to-cloth ratio, which is simply a measure of the amount of gas flow to fabric area in the baghouse. The *G/C* ratio is calculated as presented in Equation (9.21) (USEPA APTI Course 413):

$$G/C = \frac{Q}{A} \quad (9.21)$$

where:
G/C = gas-to-cloth ratio, $m^3/(s \cdot m^2)$
Q = gas flow rate, m^3/s
A = total fabric area, m^2.

The proper gas-to-cloth ratio varies with different types of fabrics and different baghouse designs. For any application however, if the ratio is too high, small particles may be pushed through the fabric, when they would otherwise be captured. Also, high gas flow rates, above the desired G/C ratio, can open pores in the bags as fibers are stretched, decreasing the life of the bags and decreasing particulate removal efficiency.

There are three primary types of baghouses. The oldest form is know as a shaker baghouse, as depicted in Figure 9.22.

In this type of baghouse, bags are connected to a cell plate (also known as a tube sheet) at the bottom of the structure, immediately above the dust collection hopper. The tops of the bags are secured at the top of the structure, to connections that provide a shaking motion when the bags are cleaned. The dirty gas enters the bags at the cell plate and moves up through the bags; the filtered air then exits near the top of the structure. When the pressure drop across the baghouse increases to a point where flow is restricted, the shaker mechanism activates, one row of bags at a time, to shake the collected dust off of the bags (the cake remains). The collected dust falls to the hopper below. When the shaker stops, particle collection resumes.

Shaker baghouses are cleaned one row at a time, so that the filtration process is never interrupted. In larger, multi-compartment baghouses, dampers may close to isolate the entire compartment. The shakers are then activated to clean the

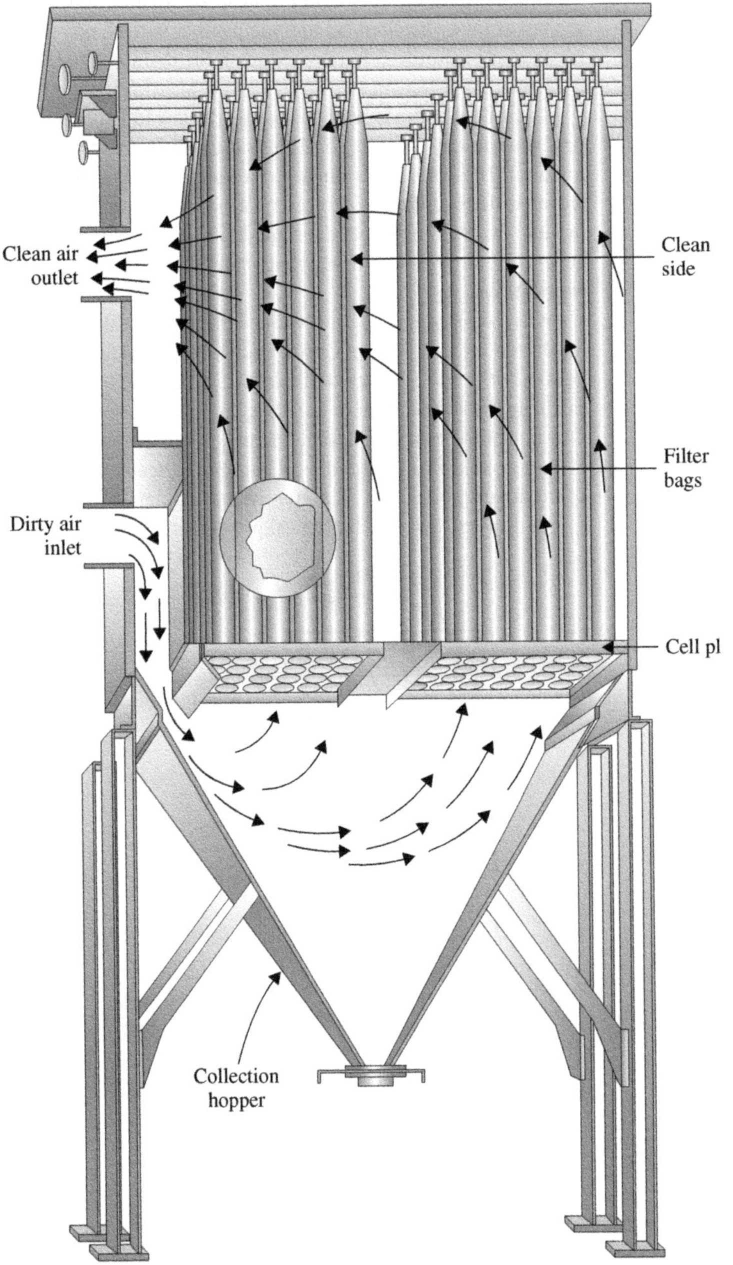

Figure 9.22 Shaker baghouse. *Source*: U.S. EPA (1999b) Control of Particulate Matter Emissions – Student Manual, APTI Course 415, Air Pollution Training Institute, Research Triangle Park, NC, Chapter 7 – page 10. United States Environmental Protection Agency.

bags, then the compartment dampers are re-opened and the compartment is placed back on line.

Shaker baghouses have been in existence for decades, in large and small applications. They typically operate with gas-to-cloth ratios of 0.1 – 0.3 $m^3/(sec \cdot m^2)$. Because of the violent nature of the shaker cleaning process, bag life is less than that of other types of baghouses.

The next generation of baghouses developed is the reverse air type of unit (Figure 9.23).

The design is similar to that of a shaker, in that the bags are connected to a cell plate at the bottom and the dirty gas stream flows up into and through the bags. The difference is in the cleaning process. Reverse air baghouses are typically large, multi-compartment units. In these systems, cleaning occurs when a compartment is taken off-line by closure of dampers. When this is done, a fan is activated which blows air through the baghouse in the reverse direction, thereby dislodging the collected dust from the inside of the bags and sending it down to the hopper below. The bags are constructed with anti-collapse rings to keep the bags from crimping together, thereby preventing the collected dust from falling down to the hopper. Because cleaning occurs off-line, reverse air baghouses must have enough compartments to be able to handle the entire process gas flow adequately with one unit off-line.

Reverse air cleaning is a very gentle method of cleaning, which extends the lifetime of the bags. These units typically operate with a gas-to-cloth ratio of 0.05 – 0.2 $m^3/(sec \cdot m^2)$, so they are somewhat larger than shaker baghouses. Reverse air baghouses are very popular with coal-fired utility power plants, cement plants, and other large industrial operations.

Pulse jet baghouses are the most recent development in baghouse designs. These units are significantly different in that the cell plate is located at the top of the structure (Figure 9.24) and the bags extend down toward the bottom.

The contaminated air flow still enters at the bottom but, in this design, filtration occurs on the outside of the bags, because the gas moves from the outside to the inside of the bags, then exits through a cell plate (tube sheet) at the top. The bags are supported by tubular wire structures, known as cages, that are inserted into the bags to insure that they keep their shape during the filtration process. Across the top of each row of bags is a pulse pipe. This pipe directs a short, high-intensity blast of compressed air down the center of the bags to dislodge collected particles from the bag exterior surface. One row of bags is pulsed at a time, which allows on-line cleaning.

Because of the strong nature of the pulse cleaning, felted bag materials can be used, rather than simple woven materials. These provide a larger basis upon which cake is developed, and allow more gas flow through the bags without some of the bleed through fabric weave distortion possible with normal woven bags. As a result, much higher gas-to-cloth ratios can be obtained with pulse jet baghouses, (up to 15 : 1), thereby making them much smaller in physical size. In addition, there are not as many mechanical moving parts as with shaker or reverse air baghouses, thereby reducing the maintenance level required.

Regardless of the type of baghouse, the key to making the unit work in any given application is the selection of the bag material. These materials range from simple woven cotton, to woven or felted fiberglass, to chemical-resistant synthetic fibers or Teflon coated materials. Presented in Tables 9.9 and 9.10 are general summaries of different bag materials and their applications.

When designing a baghouse, once the appropriate material is selected for a given application and the appropriate gas-to-cloth ratio is selected (Table 9.11), then one has only to calculate the surface area of each bag, and use the volumetric flow rate of the gas stream to be treated, to calculate the number of bags needed in a unit. Pulse jet bags are typically approximately 0.15 m in diameter by approximately 3 meters in length. Reverse air bags are typically approximately 0.3 m in

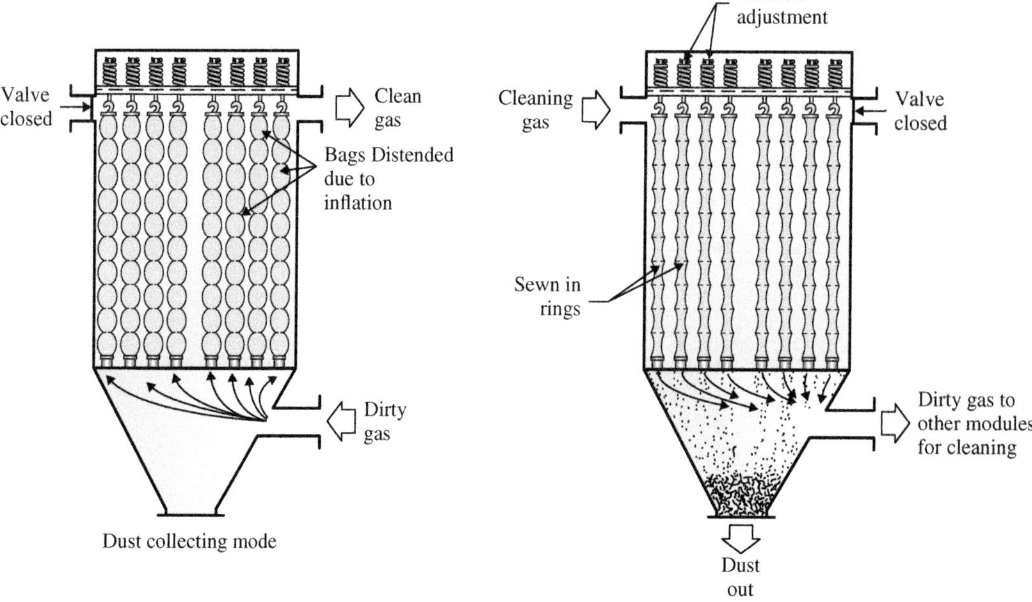

Figure 9.23 Reverse air baghouse. *Source*: The Air Compliance Group, LLC, In-House Drawing Courtesy of The Air Compliance Group, LLC, Roanoke, VA, USA.

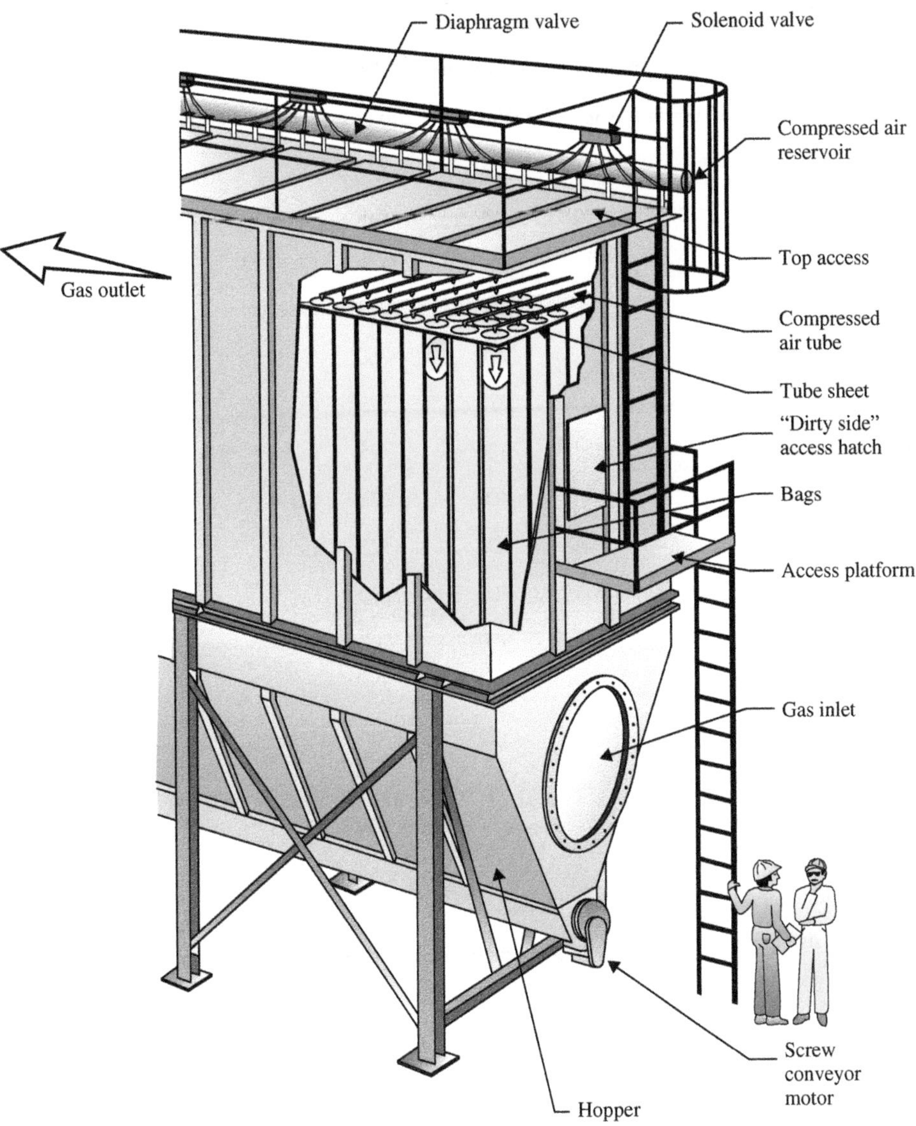

Figure 9.24 Pulse jet baghouse. *Source*: U.S. EPA (1999b) Control of Particulate Matter Emissions – Student Manual, APTI Course 415, Air Pollution Training Institute, Research Triangle Park, NC, Chapter 7 – page 15. United States Environmental Protection Agency.

diameter, with lengths up to 8 m. Shaker bags are more like reverse air bags, but usually shorter.

To calculate the number of bags needed in a baghouse, individual bag surface area is calculated using Equation (9.23). Using the gas-to-cloth ratio and the exhaust gas flow rate, the minimum required filtration filter surface area is calculated using Equation (9.24). The number of bags required is then simply calculated using Equation (9.25) (USEPA APTI Course 413).

$$A = \pi DL \tag{9.23}$$

where:
A = bag surface area, m^2
D = bag diameter, m
L = bag length, m.

$$F = \frac{Q}{G/C} \tag{9.24}$$

where:
F = total fabric area, m^2
Q = gas flow rate, m^3/min
G/C = gas-to-cloth ratio, $(m^3/s)/m^2$.

$$Bags = \frac{F}{A} \tag{9.25}$$

where $Bags$ = number of bags required.

Example 9.7 Calculation of the number of bags needed in a pulse jet baghouse

Calculate the number of bags required for a pulse jet baghouse servicing a foundry gas flow of 50 m^3/s. Assume a gas-to-cloth ratio of $0.5(m^3/s)/m^2$, and bag dimensions of 0.3 m diameter by 2.5 m in length.

Table 9.9 Temperature and acid resistance characteristics of different fabrics.

Generic name	Common or trade name	Maximum temperature (°F) Continuous	Surges	Acid resistance
Natural fiber, Cellulose	Cotton	180	225	Poor
Polyolefin	Polyolefin	190	200	Good to excellent
Polypropylene	Polypropylene	200	225	Excellent
Polyamide	Nylon®	200	225	Excellent
Acrylic	Orlon®	240	260	Good
Polyester	Dacron®	275	325	Good
Aromatic polyamide	Nomex®	400	425	Fair
Polyphenylene sulfide	Ryton®	400	425	Good
Polyimide	P-84®	400	425	Good
Fiberglass	Fiberglass	500	550	Fair
Fluorocarbon	Teflon®	400	500	Excellent
Stainless steel	Stainless steel	750	900	Good
Ceramic	Nextel®	1300	1400	Good

Source: U.S. EPA (1999a) Control of Gaseous Emissions – Student Manual, APTI Course 413, Air Pollution Training Institute, Research Triangle Park, NC, Table 7-1. United States Environmental Protection Agency.

Solution

First, calculate the surface area of each bag using Equation (9.23):

$$A = (\pi)(0.3\text{ m})(2.5\text{ m}) = 2.36\text{ m}^2$$

Next, use Equation (9.24) to calculate the total required surface area of filtration area required.

$$F = \frac{50\text{ m}^3/\text{s}}{0.5(\text{m}^3/\text{s})/\text{m}^2} = 100\text{ m}^2$$

Lastly, calculate the number of bags needed for this application using Equation (9.25).

$$Bags = \frac{100\text{ m}^2}{2.36\text{ m}^2}$$

$$\boxed{Bags = 42.4 = 42\text{ bags}}$$

Table 9.10 Fabric resistance to abrasion and flex.

Generic name	Common or trade name	Resistance to abrasion and flex
Natural fiber, cellulose	Cotton	Good
Polyolefin	Polyolefin	Excellent
Polypropylene	Polypropylene	Excellent
Polyamide	Nylon®	Excellent
Acrylic	Orlon®	Good
Polyester	Dacron®	Excellent
Aromatic polyamide	Nomex®	Excellent
Polyphenylene sulfide	Ryton®	Excellent
Polyimide	P-84®	Excellent
Fiberglass	Fiberglass	Fair
Fluorocarbon	Teflon®	Fair
Stainless steel	Stainless steel	Excellent
Ceramic	Nextel®	Fair

Source: U.S. EPA (1999a) Control of Gaseous Emissions – Student Manual, APTI Course 413, Air Pollution Training Institute, Research Triangle Park, NC, Table 7-2. United States Environmental Protection Agency.

Bags are typically arranged in rows of equal numbers of bags. In the above example, the small baghouse would typically be constructed with seven rows of six bags each.

Something that cannot be calculated with respect to a baghouse is the particulate removal efficiency, which is a

Table 9.11 Typical gas-to-cloth ratios for selected industries.

Industry	Gas-to-cloth ratio (ft^3/min)/ft^2 Shaker	Reverse air	Pulse jet
Basic oxygen furnaces	2.5–3.0	1.5–2.0	6.0–8.0
Brick manufacturers	2.5–3.2	1.5–2.0	9.0–10.0
Coal-fired boilers	1.5–2.5	.0–2.0	3.0–5.0
Electric arc furnaces	2.5–3.0	1.5–2.0	6.0–8.0
Ferroalloy plants	2.0	2.0	9.0
Grey iron foundries	2.5–3.0	1.5–2.0	7.0–8.0
Lime kilns	2.5–3.0	1.5–2.0	8.0–9.0
Municipal waste incinerators	1.5–2.5	1.0–2.0	2.5–4.0
Phosphate fertilizer producers	3.0–3.5	1.8–2.0	8.0–9.0
Portland cement kilns	2.0–3.0	1.2–1.5	7.0–10.0

Source: U.S. EPA (1999a) Control of Gaseous Emissions – Student Manual, APTI Course 413, Air Pollution Training Institute, Research Triangle Park, NC, Table 7-3. United States Environmental Protection Agency.

function of the individual gas stream characteristics and the resulting bag materials employed. In general, efficiencies well in excess of 99.5% are readily obtained with any baghouse with an acceptable gas-to-cloth ratio. If the gas-to-cloth ratio becomes too large, efficiencies may be reduced.

In terms of operation, baghouses are very forgiving units; however, their efficiency and cost of operation can be closely tied to proper cleaning. Baghouse cleaning is typically tied to measured pressure drop across a given baghouse compartment. When the pressure drop reaches a pre-set level, a cleaning sequence is started. When on-line, row-by-row cleaning is utilized, the cleaning sequence continues, one row at a time, until the measured pressure drop is reduced to a preset level. In the case of off-line cleaning, the compartment is isolated by dampers and the entire compartment is cleaned, then placed back on line.

If pressure drops are allowed to increase too high before cleaning, the cost of operation can increase, as more electrical power will be required to push the gas stream through the baghouse. On the other hand, if bags are excessively cleaned to a point where the pressure drop is reduced to a level below the manufacturer's recommendation, the cake on the bags has been reduced, resulting in an increase in particulate emissions until such time as the cake is re-established. Cleaning also places physical stress on the bags themselves, so it is desirable to clean only as often as necessary. Cleaning too often may not only increase emissions, but will also reduce the lifespan of the bags.

Overall, baghouses have become the most popular form of particulate control for many applications, and they are also replacing electrostatic precipitators for many applications. This is not only because of their smaller size, especially for pulse-jet units, but also because of the higher efficiency that can be obtained, especially for sub-μm particles. They do have limitations, however. For one thing, baghouses cannot process a wet gas stream, because the moisture would wet the bags, creating a mud cake which would be impermeable to gas flow. They also have limited applicability to applications with oily, sticky particles, though some pre-treatments are available to help them work in such environments. Lastly, bag fabrics have a limited temperature range in which they can operate. Other technologies, including electrostatic precipitators, can be designed to operate at considerably higher gas stream temperatures, and would therefore be the technology of preference for such applications.

9.6.2 Gaseous emissions control

Gaseous emissions include acid gases such as SO_2 and HCl, organic compounds, and products of combustion such as CO and NO_x. Control technologies include such techniques as absorption in wet scrubbers, activated carbon adsorption, oxidation, and catalytic reduction. These techniques are discussed in the following sections.

9.6.2.1 Absorption

Absorbers, or scrubbers, use aqueous scrubbing liquids to remove gaseous pollutants from exhaust gas streams. Absorption refers to the transfer of a gaseous component from the gas phase to a liquid phase. Absorption can occur into liquid droplets dispersed in the gas stream, sheets of liquid covering packing material, or jets of liquid sprayed into the gas stream. The liquid surface area available for mass transfer and the time available for diffusion of the gaseous molecules into the liquid are important factors affecting scrubber performance.

Absorption can be divided into two broad classifications: straight dissolution of absorbate (the pollutant) into absorbent (scrubber liquid); and dissolution, accompanied by irreversible chemical reaction. Most scrubbers are built and operated on the basis of straight dissolution. More advanced systems used for flue gas desulfurization incorporate chemical reactions into the process in order to enhance the efficiency of pollutant removal and, in some cases, to create a marketable by-product.

In order to work properly, the gaseous contaminant being absorbed (absorbate) must be at least slightly soluble in the scrubbing liquid (absorbent). Mass transfer of the gas into the liquid will continue until the liquid becomes saturated with the pollutant. At saturation, equilibrium is established between the gas and liquid phases. As a result, the solubility of the pollutant in the liquid limits the amount of pollutant removal that can occur with a given quantity of liquid. This solubility limit can be overcome by providing reactants in the liquid phase that react with the dissolved gas contaminant, forming a dissolved compound that will not be released from the liquid. This is the case in flue gas desulfurization, where a compound such as CaO is added to the scrubbing liquid to react irreversibly with SO_2. Similarly, when scrubbing HCl from a gas stream, the addition of NaOH to the scrubbing liquid will force a reaction between the HCl and the NaOH to form NaCl (salt).

Some of the most popular types of scrubber systems are spray tower absorbers, tray tower absorbers, and packed bed absorbers. Venturi scrubbers, as discussed previously for particulate control, are rarely used for the purpose of gas absorption, since their strength of design is much more directed toward particulate removal. All types of scrubbers are followed by mist-eliminators.

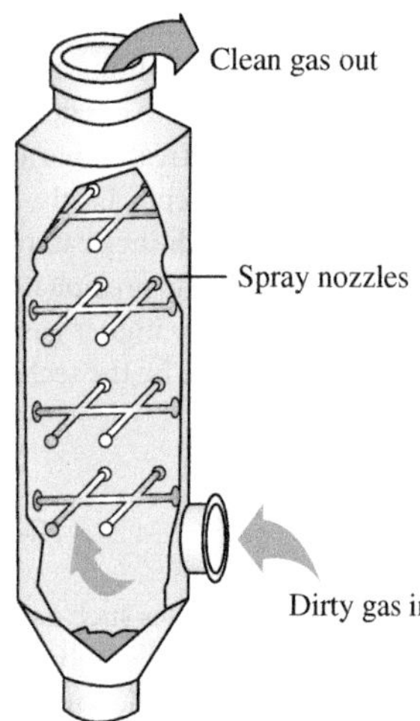

Figure 9.25 Spray tower absorber. *Source:* U.S. EPA (1999a) Control of Gaseous Emissions – Student Manual, APTI Course 413, Air Pollution Training Institute, Research Triangle Park, NC, page 5-2. United States Environmental Protection Agency.

Spray towers are the simplest devices used for gas absorption. They consist of an open vessel and one or more sets of liquid spray nozzles that distribute the scrubbing liquid into the gas stream. Typically, the flow is countercurrent, with the gas stream entering near the bottom of the tower and flowing upward, while the liquid enters near the top and flows downward. This maximizes the contact between the liquid droplets and the pollutant to be removed. An example of a typical spray tower scrubber is presented in Figure 9.25. These types of scrubbers range in size from 5 to 100,000 actual cubic feet per minute (ACFM) (0.14 to 2,800 m^3/min).

A key design criteria for any type of absorber, including spray towers, is the quantity of liquid compared to the quantity of gas flow through the scrubber. This is expressed as the liquid-to-gas ratio (L/G). The (L/G) is expressed in units of gallons per minute of liquid divided by the gas flow rate in units of 1,000 ACFM. Typical (L/G) ratios for spray-tower absorbers vary from 5 to more than 50 gallons per 1,000 ACF, depending upon the solubility of the pollutant(s) in the liquid and by the mass transfer characteristics in the spray tower. Pollutant scrubbing efficiency increases with increasing (L/G).

Because of limited contact between the liquid droplets and the gas stream, spray-tower absorbers are most often used in applications where the gases are extremely soluble in the scrubbing liquid, where high pollutant removal efficiency is not required, or where the chemical reactions in the absorbing liquid could result in salts which could cause plugging in other types of absorbers. Spray towers are also used in a number of flue gas desulfurization systems. The main advantage of spray-tower absorbers is that they are completely open internally. They have no internal components, except for the spray nozzles and connecting piping. As a result, they have a very low gas-stream pressure drop.

The next level of complexity in scrubbers is the tray-tower absorber, which is a vertical column with one or more trays mounted horizontally inside to promote gas-liquid contact. The gas stream enters at the bottom and flows upward, passing through openings in the trays. Liquid enters at the top of the

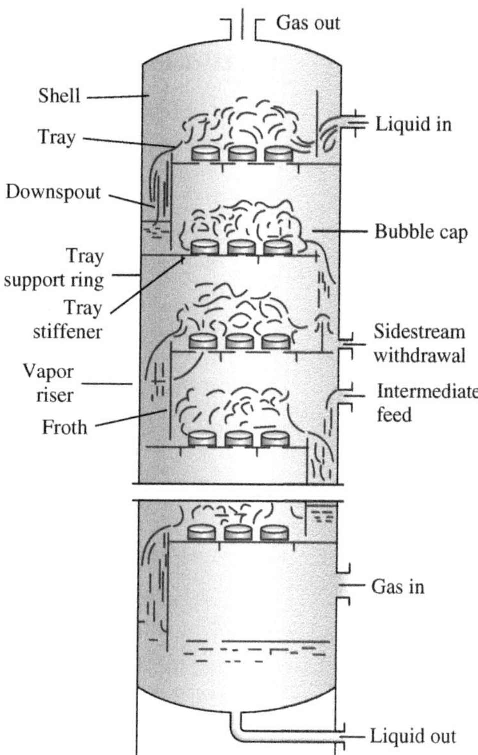

Figure 9.26 Tray tower absorber. *Source:* U.S. EPA (1999a) Control of Gaseous Emissions – Student Manual, APTI Course 413, Air Pollution Training Institute, Research Triangle Park, NC, page 5-9. United States Environmental Protection Agency.

tower and travels across each tray, then through a downcomer to the tray below, until it reaches the bottom of the tower. Mass transfer occurs in the liquid spray created by the gas velocity through the openings in the tray, and the gas stream forces itself essentially upstream, against the flow of the downward pouring liquid. Presented in Figure 9.26 is an illustration of a typical bubble cap-tray tower unit.

There are numerous different designs for the trays of a tray-tower scrubber. Figure 9.26 shows a bubble cap tray, but other tray designs include sieve trays, float valve™ trays, and impingement trays. Different scrubber manufacturers have different proprietary designs of these and other tray designs. High pollutant removal efficiencies are possible in all properly designed tray towers, because they offer a high degree of gas-liquid contact that can be achieved on a tray. The use of several trays in series also ensures that gas-liquid distribution is maximized more than would be the case on a single tray.

Perhaps the most popular form of scrubber in use today for air pollution control purposes is the packed tower scrubber. In this design, the absorbing liquid is dispersed over a bed of packing material, which provides a large surface area for gas-liquid contact. The most common packed-bed absorber is the countercurrent-flow tower shown in Figure 9.27.

The gas stream enters the bottom of the tower, flows upward through the packing material and exits from the top. Liquid is introduced at the top of the packed bed by sprays or weirs, and flows downward over the packing. With this design, the most dilute gas contacts the least saturated absorbing liquid, near the top of the bed, and the concentration difference

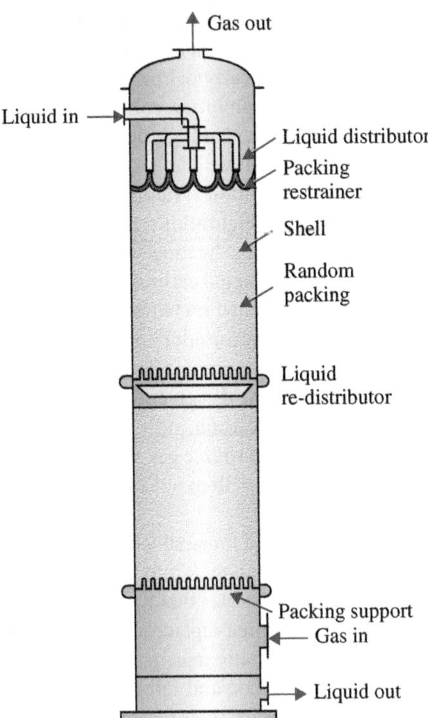

Figure 9.27 Packed tower absorber. *Source:* U.S. EPA (1999a) Control of Gaseous Emissions – Student Manual, APTI Course 413, Air Pollution Training Institute, Research Triangle Park, NC, page 5-5. United States Environmental Protection Agency.

between the liquid and gas phases (which is necessary for mass transfer) is reasonably constant throughout the column length. The maximum (*L/G*) is limited by a condition known as flooding, which occurs when the upward force exerted by the gas is sufficient to prevent the liquid from flowing downward.

Scrubber manufactures design the *L/G* to insure that there is a thin film of liquid over the surfaces of all packing materials, without reaching the state of flooding. The primary purpose of the packing material is to provide a large surface area for mass transfer. Figure 9.28 illustrates some of the most commonly used packing materials. These materials are usually made of plastic (polyethylene, polypropylene, or polyvinylchloride), but can also be ceramic or metal.

Packed-bed absorbers are most suited to applications where high gas removal efficiency is required, and the feed gas stream is relatively free from particulate matter. They are also used to control odors in rendering plants, petroleum refineries, and wastewater treatment plants. For odor control applications, the packed bed scrubbing liquor usually contains an oxidizing reagent such as sodium hypochlorite. The gas flow rate through packed towers can vary from 5 to 30,000 ACFM (0.14 to 850 m^3/min).

Absorption systems generate liquid droplets that tend to be entrained in the gas stream leaving the scrubber. A mist eliminator is used to remove these entrained droplets prior to discharge from the stack. The droplet sizes generated in absorbers range from approximately 200–1,000 μm. Though the larger droplets tend to settle out of the gas stream quickly, the small droplets are easily entrained and must be removed by means of a mist eliminator.

Common types of mist eliminators used on absorbers include chevrons and mesh pads. Chevrons are simply zig-zag baffles that force the gas to make several sharp turns while passing through the mist eliminator. Water droplets collect on the chevron blades and coalesce into larger droplets that fall downward into the absorber. Chevron mist eliminators are generally limited to gas velocities of less than approximately 20 ft/s (6 m/s). Mesh pads are formed from woven or randomly interlaced metal or plastic fibers that serve as impaction targets, and these pads can be up to six inches thick. This maximum gas velocity is usually in the range of 12 ft/s (3.7 m/s).

Sizing of wet scrubbers of all types has evolved into a series of empirical databases held by scrubber manufacturers, for their specific detailed designs. In many cases, manufacturers maintain pilot scale units of their scrubbers, in which they will test their designs for a given application in order to confirm the size needed for that application.

Wet scrubbers, or absorption systems, are widely and effectively used for air pollution control activities. They are effective for a wide variety of water-soluble compounds, and have undergone proprietary modifications over the years to create highly efficient flue gas desulfurization systems.

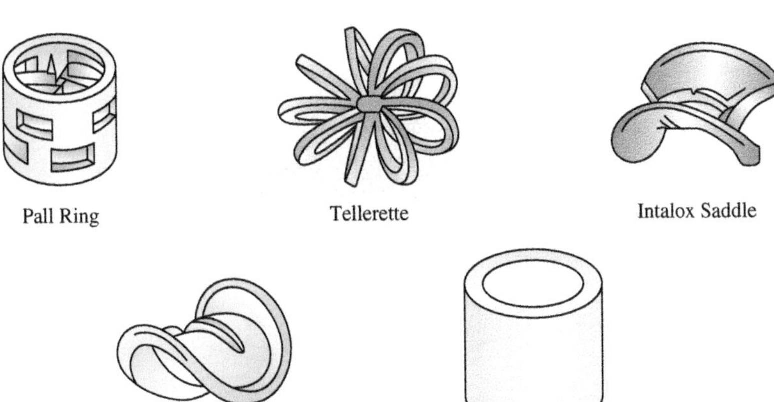

Figure 9.28 Tower absorber packing materials. *Source:* U.S. EPA (1999a) Control of Gaseous Emissions – Student Manual, APTI Course 413, Air Pollution Training Institute, Research Triangle Park, NC, page 5-7. United States Environmental Protection Agency.

9.6.2.2 Adsorption

Adsorption is a process in which a gas stream passes through a bed of solid sorbent material. Pollutants contained in the gas stream are attracted to, and adhere to, the solid surface of the sorbent material, thereby being removed from the gas stream. These systems are used primarily for solvents comprised of volatile organic compounds (VOCs). Adsorption systems that are designed for odor control and low pollutant concentration applications (< 10 ppm) are relatively simple. In these systems, the adsorbent bed is designed to be easily discarded and replaced when it approaches saturation with the pollutant being controlled. These systems are termed non-regenerative, because the absorbent material is not reused.

Adsorption processes are also used on large-scale applications having solvent vapor concentrations in the range of 10 to 10,000 ppm. Because of the large quantities of adsorbent needed, it is not practical to discard the adsorbent material. Instead, systems are included that recover the adsorbed materials and regenerate the adsorbent material. This type of system not only removes air pollutants from a gas stream, but it also provides an economic benefit of recovering solvent material for re-use.

Adsorber systems that operate continuously must have either multiple fixed beds of adsorbent, fluidized bed contactors with separate adsorption and desorption vessels, or rotary bed adsorbents that cycle continuously between adsorption and desorption operations. Because the adsorbent material is desorbed and placed back into service, these adsorption processes are termed regenerative adsorbers.

During adsorption, the gas stream passes through a bed or layer of highly porous material called the adsorbent. The compound or compounds to be removed, termed the adsorbate(s), diffuse to the surface of the adsorbent and are retained in the pores of the adsorbent, while the carrier gas passes through the bed without being adsorbed. Adsorption occurs on the internal surfaces of the materials, as shown in Figure 9.29.

There are numerous adsorbent materials, including numerous synthetic materials that are used in the chemical process industry. Activated carbon, however, is by far the most widely used sorbent for air pollution control purposes. Activated carbon operates with a high degree of efficiency and can be regenerated many times.

Non-regenerative adsorption systems are manufactured in a wide variety of physical configurations, including thin flat or pleated filters, similar to home heating system filters. These adsorbent beds contain activated carbon, sandwiched between sheets of cellulose or perforated metal. These types of small adsorbers are used primarily for odor control in very small applications. Larger, more common non-regenerative adsorption systems are in the design and size of a 55-gallon drum containing activated carbon, sitting on a perforated plate, through which the emission source gas stream passes upward, through the activated carbon, then out to the atmosphere (See Figure 9.30).

These are frequently used for small emission sources, where it is convenient to sit a carbon tank into place, utilize it until a periodic inspection demonstrates that hydrocarbon vapors are starting to break through, then replace it with a new carbon drum. The old drum is typically returned to the manufacturer, who will regenerate the carbon and return it to their working inventory of carbon drums for their customers. Many of these applications are on industrial or commercial facilities whose emissions are too small to be regulated by permits, but from which odorous emissions, if not controlled, could result in an undesirable odor being released from the facility into the surrounding community. Small carbon drums work very well in preventing these odors.

For larger VOC control systems, large fixed tanks of carbon are used, and these are equipped with on-site regeneration and

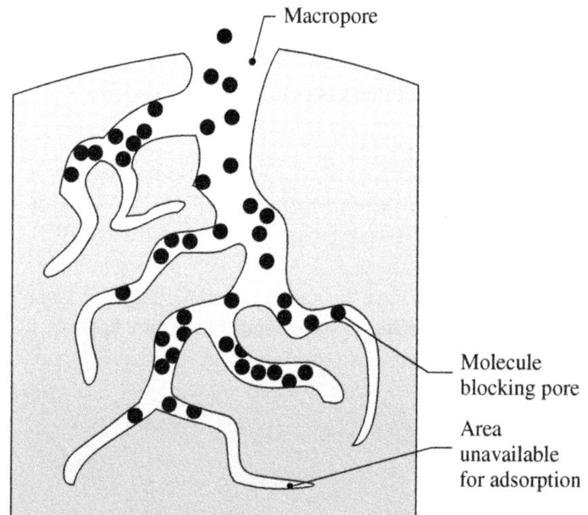

Figure 9.29 Vapor adsorption into carbon pores. *Source:* U.S. EPA (1999a) Control of Gaseous Emissions – Student Manual, APTI Course 413, Air Pollution Training Institute, Research Triangle Park, NC, page 4-2. United States Environmental Protection Agency.

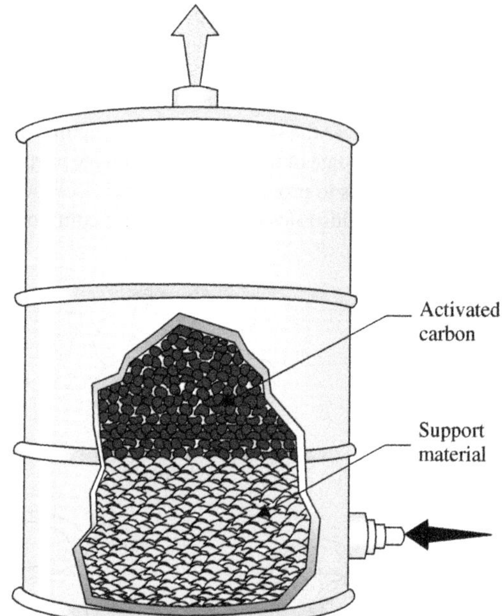

Figure 9.30 Non-regenerative carbon adsorption tank. *Source:* U.S. EPA (1999a) Control of Gaseous Emissions – Student Manual, APTI Course 413, Air Pollution Training Institute, Research Triangle Park, NC, page 4-9. United States Environmental Protection Agency.

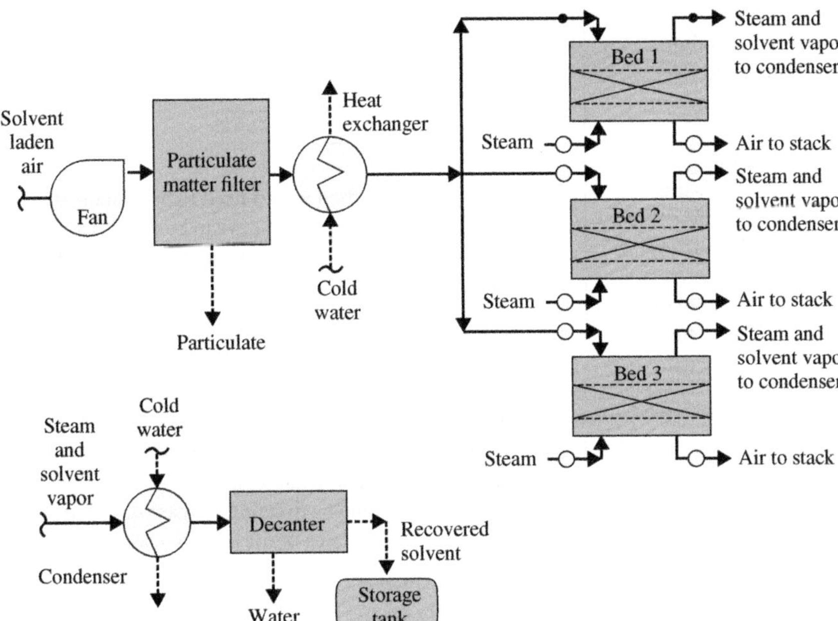

Figure 9.31 Regenerative multi-tank carbon adsorption system. *Source*: U.S. EPA (1999a) Control of Gaseous Emissions – Student Manual, APTI Course 413, Air Pollution Training Institute, Research Triangle Park, NC, page 4-12. United States Environmental Protection Agency.

recovery of the adsorbed solvents. A typical example of such a system is presented in Figure 9.31.

These systems consist of at least two, and typically three or more, tanks containing large quantities of activated carbon sitting upon perforated plates, through which the untreated gas stream enters. In these systems, the contaminated gas stream typically enters through the top and works its way down through the bed, to the point where it enters a duct at the bottom of the bed prior to release to the atmosphere. One carbon bed (tank) is typically being regenerated, while the others are in service treating emissions. Regeneration is accomplished by passing steam through the beds. The steam volatilizes the trapped hydrocarbon compounds, stripping them off of the carbon and carrying them away from the carbon bed. The stream of steam carrying the stripped hydrocarbons is cooled in a condenser, then separated in a decanter.

The decanted water is either disposed of or returned to the boiler. The decanted solvent is either sold or re-used in the original operation. Re-use is, by far, the most common practice, because it saves a great deal of money for the manufacturer. In many cases, the hydrocarbon stream consists of either a single compound, or a simple mix of compounds, such as is the case with petroleum hydrocarbon solvent industrial dry cleaners or degreasing operations. In some cases where the solvent mix is of critical proportions, the recovered solvent is sent back to the solvent producer for re-refining and re-use.

When hydrocarbon vapors are adsorbed, in any type of a fixed carbon bed, the adsorption process occurs in an area of the bed known as the mass transfer zone (see Figure 9.32).

The depth of the mass transfer zone is dependent upon a number of factors, including the compound(s) being adsorbed, the temperature of the gas stream, and the gas flow rate through the bed. Hydrocarbon vapors begin to be adsorbed at the top of the mass transfer zone, and are fully adsorbed at the bottom of the zone. As adsorption continues over time, the mass transfer zone moves downward, leaving fully saturated carbon above it.

When the mass transfer zone reaches the bottom of the bed, breakthrough begins to occur. The concentration of the

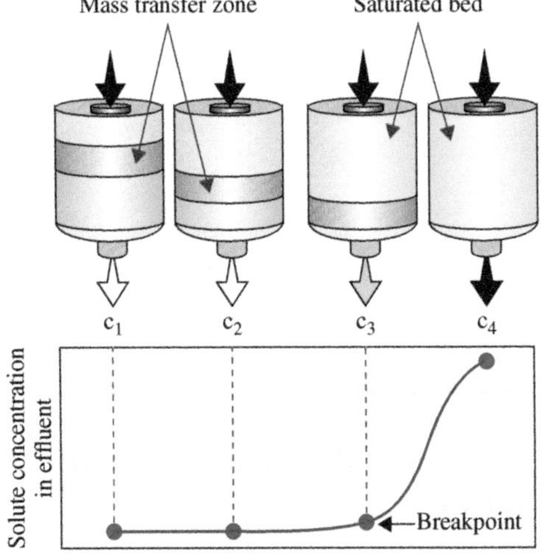

Figure 9.32 Carbon adsorption mass transfer zone. *Source*: U.S. EPA (1999a) Control of Gaseous Emissions – Student Manual, APTI Course 413, Air Pollution Training Institute, Research Triangle Park, NC, page 4-31. United States Environmental Protection Agency.

hydrocarbon vapors in the exhaust stream will increase as adsorption continues, and the mass transfer zone continues to move downward until the entire bed is saturated, at which time the hydrocarbon concentration in the exhaust gas equals the concentration in the inlet gas.

For carbon adsorption units with on-site regeneration, the regeneration process will be automatically started, in a given bed, when a hydrocarbon vapor sensor begins to detect that breakthrough has begun to occur. At that point, that bed will be taken off-line for regeneration. In a system containing three or more tanks (beds), one tank would have already been regenerated and would have been sitting idle. This cleaned tank will be placed back into service at the same time that the other tank is taken off-line for regeneration.

The most important factor in the design of a carbon adsorption unit is that of the adsorptive capacity of the carbon. This is a factor which varies significantly, depending upon the compound(s) to be adsorbed, the overall porosity of the carbon particles, and the pore structures within the particles. This cannot be effectively estimated by calculation, because no two manufacturing processes create the same carbon with the same characteristics. Because of this, manufacturers maintain their own empirical databases for their carbon, used in a wide variety of applications.

Utilizing manufacturer's data, it is still necessary to calculate the amount of carbon needed for any application. This can be performed using saturation capacity and working capacity data from the manufacturer, and Equations (9.26) and (9.27) (USEPA APTI Course 415).

$$M_{CS} = \frac{((M_C)(T))}{W_S} \tag{9.26}$$

where:
M_{cs} = mass of carbon at the saturation capacity, kg
M_c = mass rate of pollutant, kg/h
T = time (duty) cycle, h
W_s = saturation capacity of carbon (fractional).

$$M_{CW} = \frac{M_{CS}}{W_C} \tag{9.27}$$

where:
M_{cw} = mass of carbon at working capacity, kg
W_c = working capacity of carbon (fractional)

Example 9.8 Calculation of carbon quantity

Calculate the amount of carbon needed to operate a carbon adsorption unit treating an emission rate of 100kg/h of carbon tetrachloride (CCl_4). The adsorber is to be designed for a four-hour duty cycle, and the carbon provider reports a saturation adsorptive capacity for CCl_4 of 45% (0.45). Assume a working capacity of 50% of the saturation capacity.

Solution

First, calculate the capacity of carbon needed to achieve saturation in 4 hours, using Equation (9.26):

$$M_{CS} = \frac{((100\ \text{kg/h})(4\ \text{h}))}{0.45}$$

$$\boxed{M_{CS} = 889\ \text{kg}}$$

Once the saturation capacity is know, use Equation (9.27) to calculate the mass of carbon needed for a design of 50% working capacity:

$$M_{CW} = \frac{889\ \text{kg}}{0.5}$$

$$\boxed{M_{CW} = 1778\ \text{kg}}$$

This quantity of carbon would need to be contained in a single tank system, or it would have to be divided among the number of tanks on-line together at any one time, for a multiple tank system.

In order to achieve 90% or more capture efficiency, most carbon adsorption systems are designed for a maximum gas velocity of 100 ft/min (30 m/min) through the adsorber. A lower limit of at least 20 ft/min (6 m/min) is maintained to avoid flow problems such as channeling. Gas velocity through the adsorber is determined by dividing the gas volumetric flow rate by the cross-sectional area of the adsorber, as presented in Equation (9.28):

$$V = \frac{Q}{A} \tag{9.28}$$

Properly designed, carbon adsorption units are very popular air pollution control units which can provide in excess of 90% removal efficiency, and they can also provide a potentially significant cost savings to the operating entity, in terms of solvent recovery and re-use.

9.6.2.3 Oxidation

Oxidation systems are used to destroy organic compounds classified as volatile organic compounds (VOCs) and/or air toxic compounds. At sufficiently high temperatures and adequate residence times, essentially all organic compounds can be oxidized to form carbon dioxide and water vapor. High-temperature thermal oxidation uses temperatures in the range of 1,000–2,000°F (540–1,100°C). Catalytic oxidation processes, on the other hand, operate at a lower temperature range of 400–1,000°F (200–540°C).

In many cases, thermal oxidizers are equipped with either recuperative or regenerative heat exchangers to capture heat and reduce operating costs. The terms "recuperative" and "regenerative" refer to the type of heat exchanger used to increase system efficiency. A recuperative heat exchanger is a tubular or plate heat exchanger, where heat is transferred through the metal surface. Captured heat is used to pre-heat the incoming gas stream, thereby reducing the amount of energy that needs to be added to bring the gas temperature up to the desired temperature range to complete the oxidation process. A regenerator uses a set of refractory packed beds that store heat. This is not an add-on piece of equipment, but is part of the integral design of the oxidizer. Both types of heat recovery reduce the amount of supplement fuel needed to oxidize the contaminants in the combustion chamber. With sufficiently high organic concentration, the energy released during oxidation may be sufficient to maintain the necessary temperature without the addition of supplemental fuel.

The key to effective oxidizer design is to maximize the three "Ts" of combustion-time, temperature, and turbulence. Large residence times, high temperatures, and highly turbulent flow

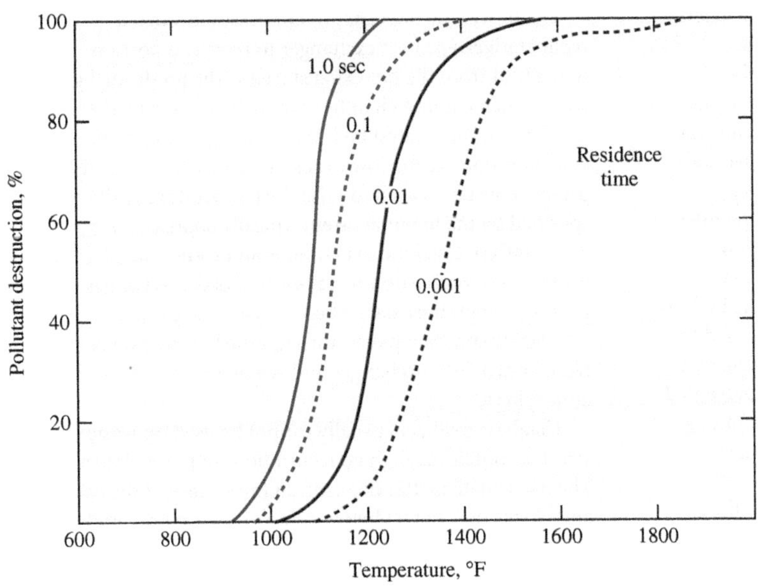

Figure 9.33 Relationship of temperature vs. residence time in a thermal oxidizer. *Source:* U.S. EPA (1999a) Control of Gaseous Emissions – Student Manual, APTI Course 413, Air Pollution Training Institute, Research Triangle Park, NC, page 6-3. United States Environmental Protection Agency.

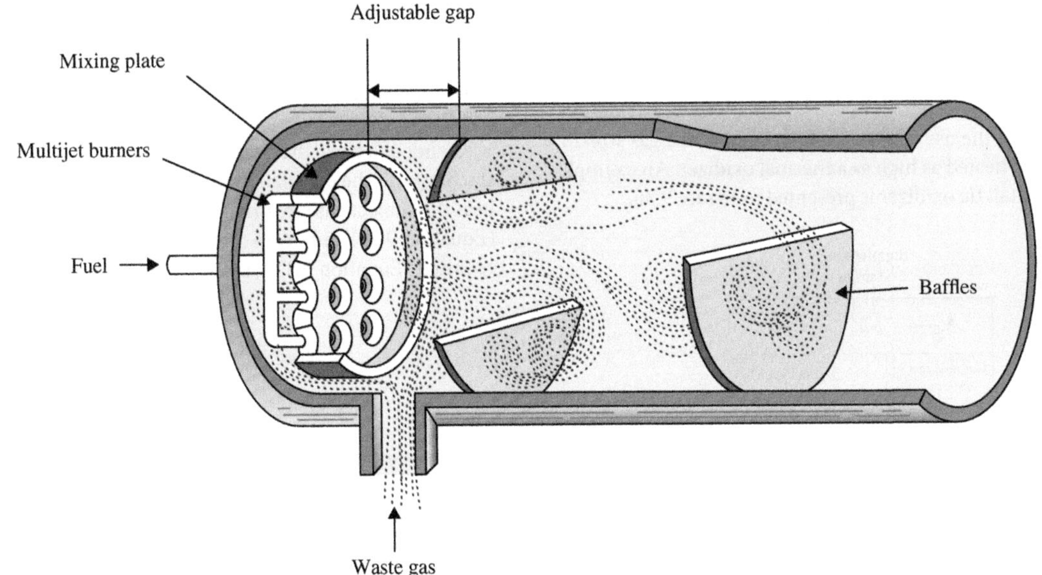

Figure 9.34 Basic thermal oxidizer with mixing baffles. *Source:* U.S. EPA (1999a) Control of Gaseous Emissions – Student Manual, APTI Course 413, Air Pollution Training Institute, Research Triangle Park, NC, page 6-3. United States Environmental Protection Agency.

all contribute to the complete destruction of the organic pollutant. Figure 9.33 shows the interplay between time and temperature for a hypothetical compound. With a residence time of one second, 100% destruction can be achieved at a temperature of about 1,200°F. However, with a residence time of 0.01 s, a temperature of about 1,500°C is required. Some state environmental agencies require a minimum temperature of 1,500°F with a residence time of 0.5 s.

Turbulent flow is necessary to ensure that the oxidizer feed gas stream is well-mixed with the products from the supplemental fuel burners, and that none of the pollutants are allowed to bypass the zone of maximum temperature. As shown in Figure 9.34, mixing can be enhanced by the proper use of baffles within the oxidization chamber.

The combustion chamber of a thermal oxidizer is sized to provide sufficient residence time to complete the oxidation reactions, typically from approximately 0.3 to more than 2 seconds. The time needed for high-efficiency destruction is strongly dependent on the operating temperature and the mixing within the chamber. Current design is typically based on a residence time of 0.5 s. Residence time is calculated using Equation (9.29) (USEPA APTI Course 415):

$$t = \frac{V}{Q} \tag{9.29}$$

where:

t = residence time, s
V = oxidizer volume, m^3
Q = gas flow rate, m^3/s.

Regenerative thermal oxidizers have heat recovery efficiencies as high as 95%. Because of the high pre-heated inlet gas temperatures created by the heat recovery, burner fuel is required only if the organic vapor concentrations in the gas stream are very low. At moderate-to-high concentrations, the heating value of the organic pollutant is sufficient to maintain the necessary temperatures in the combustion chamber. High-efficiency heat recovery is achieved by passing the inlet gas stream through a large packed bed containing ceramic packing that has been previously pre-heated by passing the outlet gases from the combustion chamber through the bed. At least two beds are required, and gas flow dampers are used to switch the inlet and outlet gas streams to the appropriate beds.

An example of a two-bed regenerative oxidizer is presented in Figure 9.35. One of the beds is used to pre-heat the inlet gas stream, while the second is heated with the post combustion exhaust stream. When the second bed comes to a pre-set temperature, the valves are switched to reverse the gas flow pattern. Bed #2 then becomes the pre-heat bed, while the exhaust gas stream heats bed #1. This pattern of gas stream reversal continues as part of the operation of the system.

Catalytic oxidizers are designed much like thermal oxidizers, except that they contain a catalyst bed through which the polluted gas stream passes after being heated to a specified level by a gas-fired burner. The catalyst completes the oxidation process without the use of as much fuel, because the gas stream does not to be heated as high as a thermal oxidizer. An example of a typical catalytic oxidizer is presented in Figure 9.36.

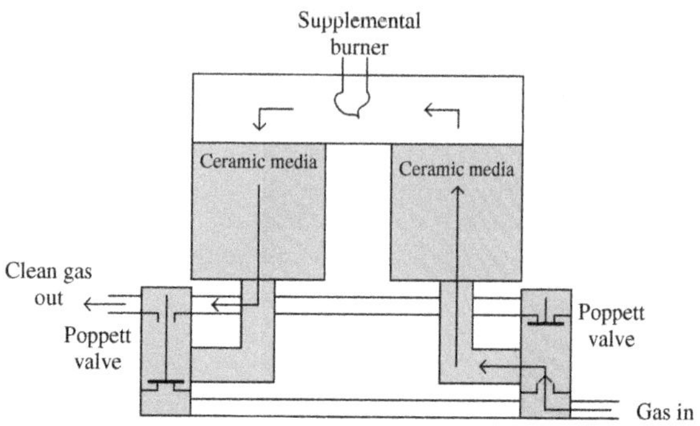

Figure 9.35 Regenerative thermal oxidizer with two cells. *Source:* The Air Compliance Group, LLC, in-house drawing.

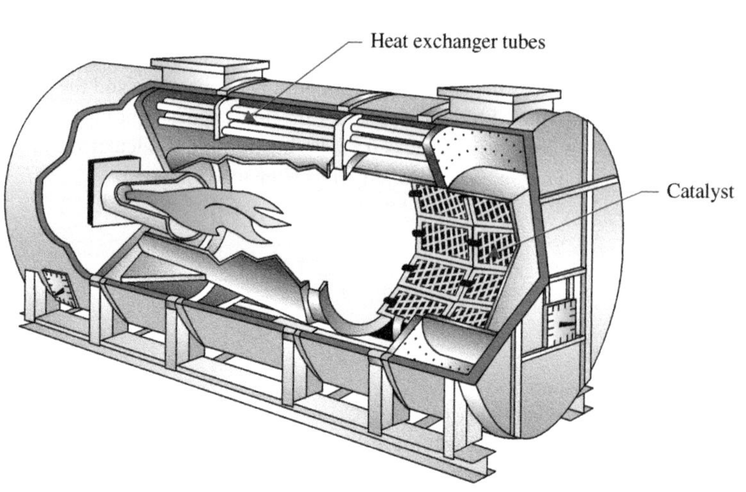

Figure 9.36 Catalytic oxidizer. *Source:* US EPA (1999a) Control of Gaseous Emissions – Student Manual, APTI Course 413, Air Pollution Training Institute, Research Triangle Park, NC, p. 6-15. Source Environmental Protection Agency.

As shown, the inlet gas stream passes through a recuperative-type heat exchanger to recover a portion of the heat from the hot exhaust gases. The pre-heated gas stream then enters a chamber, where it is distributed across the inlet face of the catalyst bed. If the inlet gas temperature and concentration are too low to sustain the catalytic reactions, a pre-heat burner is used to raise the temperature to the range specified by the manufacturer (typically approximately 250 – 500°C). Catalytic oxidation is an exothermic reaction. The exothermic combustion reactions that occur as the gas stream passes through the catalyst bed increase the gas temperature. The gas stream then passes through the hot gas side of the recuperative heat exchanger and is exhausted to the atmosphere.

Catalytic oxidizers usually cannot be used on waste gas streams containing high concentrations of particulate matter. The particulate matter deposits on the surface of the catalyst and blocks the contact between the organic compounds and the catalyst surface. Oil droplets can also block access to the catalyst bed, unless they are vaporized in the pre-heat section. Periodically cleaning and washing the catalyst allows the catalyst activity to be restored. The catalyst can also be poisoned by exposure to many heavy metals in the gas stream. A poisoned catalyst cannot be cleaned back into working condition.

Sizing of a basic thermal oxidizer is a four-step process, starting with the calculation of the amount of energy needed to heat the subject gas stream to the desired temperature (Equation 9.30) and, from there, the quantity of natural gas required (Equation 9.31) (USEPA APTI Course 415):

$$Q_{net} = (V_{in})(60)(h_i - h_c) \tag{9.30}$$

where:

Q_{net} = quantity of energy required, BTU/h
V_{in} = oxidizer inlet gas volume, scfm
h_i = enthalpy of gas stream at inlet temp, BTU/scf
h_c = enthalpy of gas stream at oxidizer temp, BTU/scf.

$$F = \frac{Q_{net}}{h_f} \tag{9.31}$$

where:
F = quantity of natural gas required, scfm
h_f = enthalpy of natural gas, BTU/scf.

With that information, the volumetric flow rate of the processed gas stream through the thermal oxidizer is calculated (Equation 9.32). The oxidizer chamber volume is then calculated for the targeted residence time, using Equation (9.33) (USEPA APTI Course 415).

$$V_{total} = \left(V_{in} + \frac{F}{60}\right)\left(\frac{T_c + 460}{T_s + 460}\right) \quad (9.32)$$

where:
V_{total} = total gas flow rate at oxidizer temperature, ft^3
T_c = oxidizer temperature ,°F
T_s = standard temperature ,°F.

$$V_c = \left(\frac{V_{total}}{60}\right)(RT) \quad (9.33)$$

where:
V_c = oxidizer internal volume, ft^3
RT = oxidizer residence time, s.

Example 9.9 Calculation of thermal oxidizer size

Calculate the size of a thermal oxidizer that will be used to control VOC emissions from a gas 4,000 scfm gas stream. The temperature of the gas stream entering the oxidizer is 300°F and the oxidizer design operating temperature is 1,500°F. For these two temperatures, use an enthalpy of 4.24 BTU/ft^3 at 300°F and 28.24 BTU/ft^3 at 1,500°F. Use a standard temperature of 68°F and a heating value of natural gas of 1,030 BTU/ft^3. Also use a target oxidizer residence time of 0.5 second.

Solution

First, calculate the energy required to heat the incoming gas stream from 300°F to 1,500°F using Equation (9.30).

$$Q_{net} = (4000\ \text{scfm})(60\ \text{min/h}) \times (28.24\ \text{BTU/scf} - 4.42\ \text{BTU/scf})$$

$$Q_{net} = 5.72 \times 10^6\ \text{BTU/h}$$

Then use that value to calculate the natural gas usage necessary to accomplish the desired heat rise, using Equation (9.31):

$$F = \frac{5.72 \times 10^6\ \text{BTU/h}}{1030\ \text{BTU/ft}^3}$$

$$F = 5553\ \text{ft}^3/\text{h}$$

After the quantity of natural gas is calculated, use Equation (9.32) to calculate the total volumetric flow rate of the gas stream moving through the oxidizer chamber at the desired temperature of 1,500°F:

$$V_{total} = \left((4000\ \text{ft}^3/\text{min}) + \left(\frac{5553\ \text{ft}^3/\text{h}}{60\ \text{min/h}}\right)\right) \times \left(\frac{1500 + 460}{68 + 460}\right)$$

$$V_{total} = 15192\ \text{ft}^3/\text{min}$$

Use this value and the specified residence time to calculate the oxidizer volume necessary to treat the gas stream (Equation 9.33):

$$V_c = \left(\frac{15192\ \text{ft}^3/\text{min}}{60\ \text{s/min}}\right)(0.5\text{s})$$

$$\boxed{V_c = 126.6\ \text{ft}^3}$$

9.6.2.4 NO_x Control

Nitrogen oxides emissions are produced during the combustion process of any fuel. Very little of the NO_x that is generated is derived from any nitrogen in the fuel. Most of it is what is known as thermal NO_x, meaning that it is formed from the nitrogen in ambient air. Nitrogen oxides are not highly soluble in most scrubber liquids, so therefore liquid scrubbing is of little value in terms of emissions control. As a result, NO_x is controlled either by combustion modifications, or by reaction with ammonia to form elemental nitrogen.

Combustion modifications typically consist of the use of low-NO_x burners in boilers or incinerators. Thermal NO_x is formed most rapidly at temperatures in excess of 1,800°F. Below that, NO_x formation drops off significantly, although the peak flame temperature of any burner, regardless of fuel, is typically much higher than that – hence, the thermal NO_x formation. Low-NO_x burners are designed to extend the flame pattern. With the extended pattern, the same amount of heat is released into the boiler or incinerator overall although, with the dispersed flame, the peak temperature is lowered. This will not eliminate NO_x formation, but it can reduce it by as much as 40% in some cases. Actual efficiency is a case-by-case situation that varies with every different application because of differing boiler geometries, fuel flow rates, and other considerations.

Reduction of NO_x to N_2 and H_2O can be accomplished either with or without a catalyst bed, using the following reaction (EPA Control Cost Manual):

$$2NO + 2NH_3 + \frac{1}{2}O_2 \rightarrow 2N_2 + 3H_2O \quad (9.34)$$

Reduction can be accomplished in some boilers using a technique known as Selective Non-Catalytic Reduction (SNCR). The SNCR process occurs within the combustion unit, which acts as the reaction chamber. Ammonia is injected into the flue gas through nozzles mounted on the wall of the combustion unit. The injection nozzles are generally located in the post-combustion area, where heat is recovered. The injection causes mixing of the reagent and flue gas. The heat of the boiler provides the energy for the reduction reaction. The NO_x molecules are reduced, as shown in Equation (9.34), and the reacted flue gas then carried out of the boiler.

The NO_x reduction reaction occurs within a specific temperature range, in the SNCR process, where adequate heat is available to drive the reaction. At lower temperatures, the reaction kinetics are slow, and ammonia passes through the boiler (ammonia slip). At higher temperatures, the ammonia oxidizes and additional NO_x is generated. The optimum operating range for peak SNCR effectiveness is between 1,700°F and 1,900°F. Effectiveness varies widely, depending upon the specifics of each boiler application; however, NO_x removal efficiencies of up to 68% can be realized.

When SNCR cannot be implemented, due to the unavailability of a location of ammonia to be injected in the proper temperature range, and with adequate mixing, NO_x reduction can be accomplished using Selective Catalytic Reduction (SCR). As the name implies, this version of the process involves passing the gas stream through a catalyst bed, which fosters the reduction reaction. These types of systems operate at much lower temperatures, making them applicable to many different types of installations. The optimum temperature range for effective SCR operations varies by manufacturer and the precise nature of their catalyst materials, but typically ranges from 480–800°F. Maintenance of the temperature within the specified range, for the specific catalyst, is of significant importance in maximizing SCR NO_x removal efficiency. A well-designed and well-operated SCR system should, however, be capable of achieving NO_x reduction of greater than 90%.

9.7 Indoor Air Pollution

Indoor pollution sources that release gases or particles into the air are the primary cause of indoor air quality problems in homes, offices, and other buildings. Inadequate ventilation can increase indoor pollutant levels by not bringing in enough outdoor air to dilute emissions from indoor sources and by not carrying indoor air pollutants out of the building. High temperature and humidity levels can also increase concentrations of some pollutants.

There are many sources of indoor air pollution. These include: combustion sources such as fossil fuels or wood used for heating purposes; tobacco products; asbestos-containing insulation; wet or damp carpet; and cabinetry or furniture made of certain pressed wood products. In addition, products for cleaning and maintenance, personal care, or hobbies, as well as central heating and cooling systems and humidification devices; can impact indoor air quality. Outdoor air pollutants can also enter and be trapped inside a building.

The relative importance of any single source depends on how much of a given pollutant it emits and how hazardous those emissions are. In some cases, factors such as how old the source is, and whether it is properly maintained, are significant. For example, an improperly adjusted gas stove can emit significantly more carbon monoxide than one that is properly adjusted.

Some sources, such as building materials, furnishings, and household products like air fresheners, release pollutants more or less continuously. Other sources, related to activities carried out in the home, release pollutants intermittently. These include: smoking; the use of unvented or malfunctioning stoves, furnaces, or space heaters; the use of solvents in cleaning and hobby activities; the use of paint strippers in redecorating activities; and the use of cleaning products and pesticides in house-keeping. High pollutant concentrations can remain in the air for long periods after some of these activities.

If too little outdoor air enters a home, pollutants can accumulate to levels that can pose health and comfort problems. Unless they are built with special mechanical means of ventilation, buildings that are designed and constructed to minimize the amount of outdoor air that can "leak" into and out of the building may have higher pollutant levels than other buildings. However, because some weather conditions can drastically reduce the amount of outdoor air that enters a home, pollutants can build up even in buildings that are normally considered not to be overly tight.

Summary

- The study of air pollutants, their impacts, and their control is a complex mixture of scientific and engineering principles.
- Criteria pollutants and hazardous air pollutants impact differently upon the surrounding and global communities and are regulated differently because of those different impacts.
- Maximum ambient air quality levels of criteria pollutants have been established by EPA in the form of National Ambient Air Quality Standards (NAAQS).
- Evaluation of the impact of air pollutants released from a given source or group of sources is performed by dispersion modeling techniques which incorporate micro-scale and macro-scale meteorological conditions.
- Dispersion modeling is based upon a Gaussian dispersion algorithm, and EPA has developed computer models that can be used to perform these functions.
- Control of criteria and hazardous air pollutants is based upon their physical and chemical characteristics: solid vs. gaseous; acidic vs. organic, etc.
- Particulate matter can be controlled by gravitation settling chambers, cyclones, venturi scrubbers, electrostatic precipitators, and baghouses.
- Gravitational settling chambers have very low efficiency, particularly for smaller particles, but they are still used as scalping techniques, to remove large particles prior to gas entry into a more efficient control device.
- Cyclones are more efficient than settling chambers, and are still used as stand-alone control devices for some sources,

but they are also used as scalping devices for more efficient systems.

- Venturi scrubbers can remove particles at high efficiencies, but are limited to small facilities. Their efficiency increases with increasing pressure drop.
  - Venturi scrubbers are good selections for control of very wet gas streams.
- Electrostatic precipitators (ESPs) are very popular with large installations such as coal-fired power plants. They are very efficient and can be designed to handle very large gas flow rates at varying temperatures.
  - ESPs are limited in some cases because they are physically very large.
- Baghouses are extremely efficient particulate control devices that are adaptable to a wide variety of emission sources and sizes.
  - Baghouses are designed as either shaker, reverse-air, or pulse-jet designs, with pulse-jet becoming the most popular in recent years.
  - Pulse-jet baghouses can be designed with much higher gas-to-cloth ratios, meaning that they can be considerably smaller than the other two designs.
  - Baghouse application can be limited by gas stream moisture content and temperature.
- Gaseous emissions can be controlled, depending upon their physical and chemical makeup, by absorbers, adsorbers, and oxidizers.
- Absorbers (scrubbers) are primarily used for inorganic and acid gases.
  - Absorbers can be designed as spray towers, plate towers, or packed towers.
  - Packed tower scrubbers are the most efficient, but they are not always needed if the pollutant being absorbed is of sufficient solubility for spray tower or plate tower designs.
  - Spray tower absorbers can be modified to serve as flue gas desulfurization applications.
- Adsorbers, primarily activated carbon adsorbers, are used for control of organic compounds.
  - Carbon adsorbers provide a high level of efficiency, and they recover the collected organic compounds for re-use, thereby potentially saving significant money for some industrial installations.
- Oxidizers can destroy organic compounds at a very high level of efficiency.
  - Oxidizers can be classified as either thermal oxidizers, catalytic oxidizers, or regenerative thermal oxidizers. The primary difference is the amount of energy (natural gas) required to operate the units.
- Nitrogen oxides are formed in the combustion process; they can be controlled by low-NO_x burners, selective catalytic reduction (SCR), or selective non-catalytic reduction (SNCR).
  - Low-NO_x burners disperse the flame pattern, thereby reducing the peak flame temperature, which reduces the formation of nitrogen oxides.
  - SCR and SNCR involve reaction, under specified conditions, of the NO_x in the gas stream with ammonia, to form N_2 and H_2O.
- Indoor air pollutants can be generated by sources such as building materials, cleaning supplies, smoking, etc., inside the building. They are trapped inside if the building does not offer adequate ventilation.

Key Words

absorption
acid rain
activated carbon
adsorption
albedo
Baghouse
Best Available Control; Technology (BACT)
Briggs plume rise
carbon dioxide (CO_2)
carbon monoxide
catalyst bed
catalytic oxidizer
chlorofluorocarbons (CFCs)
criteria pollutants
Cyclones
dispersion modeling
electrostatic precipitator (ESP)
emission factor
fabric filter
gaussian dispersion
global climate change
gravitational settling chambers
hazardous air pollutants (HAPS)
indoor air pollution
low-NO_x burner
Maximum Achievable Control Technology (MACT)
meteorology
methane (CH_4)
mixing height
National Ambient Air Quality Standards (NAAQS)
New Source Performance Standards (NSPS)
nitrogen oxides
nitrous oxide (N_2O)
oxidation
ozone
packed tower
PM_{10}
$PM_{2.5}$
Prevention of Significant Deterioration (PSD)
primary standards
pulse jet
regenerative thermal oxidizer (RTO)
reverse air
scrubber
secondary standards
Selective Catalytic Reduction (SCR)
Selective Non-Catalytic Reduction (SNCR)
shaker
spray tower
State Implementation Plan (SIP)
stratospheric ozone
sulfur dioxide
thermal oxidizer
tray tower
tropospheric ozone
venture scrubbers
visibility
volatile organic compounds (VOCs)
wind direction
wind speed

References

Code of Federal Regulations, Title 40, Part 60, Appendix A. http://www.epa.gov/ozone/science/sc_fact.html

Foley, J. *et al.* (2005). Global Consequences of Land Use. *Science* **309**, 570–573

González-Rouco F. *et al.* (2003). Deep Soil Temperature As Proxy For Surface Air-Temperature In A Coupled Model Simulation Of The Last Thousand Years. *Geophysical Research Letters* **30**, 2116.

Raitsos D. *et al.* (2005). Extending the SeaWifs chlorophyll data set back 50 years in the northeast Atlantic. *Geophysical Research Letters*, L06603.

Uriarte, Anton (2011). *Earth's Climate History*. http://web.me.com/uriarte/Earths_Climate/Earths_Climate_History.html

US EPA (1995). *User's Guide for the Industrial Source Complex (ISC3) Dispersion Models – Volume II – Description of Model Algorithms*. US EPA, Research Triangle Park, NC, EPA-454/B-95-003b.

US EPA (1999a). *Control of Gaseous Emissions – Student Manual*. APTI Course 413, Air Pollution Training Institute, Research Triangle Park, NC.

US EPA (1999b). *Control of Particulate Matter Emissions – Student Manual*. APTI Course 415, Air Pollution Training Institute, Research Triangle Park, NC.

US EPA (2002). *Air Pollution Control Cost Manual – Office of Air Quality Planning and Standards*. US EPA, Research Triangle Park, NC, EPA-452/B-02-001

US EPA (2005). *Basic Air Pollution Meteorology*. APTI Self Instructional Course 409, Air Pollution Training Institute, Environmental Research Center, MD.

US EPA (Updated 2011). *Compilation of Air Pollutant Emission Factors – Fifth Edition*. US EPA, Research Triangle Park, NC, AP-42.

Useful Websites

- http://www.cdiac.gov
- http://epa.gov/ozone/science/sc_fact.html
- http://www.epa.gov/iaq/
- http://www.epa.gov/air/criteria.html

Problems

1 Utilizing on-line resources from EPA, determine the NAAQS compliance status for all criteria pollutants in your home town.

2 Utilizing the EPA on-line emission inventory, determine the total annual emissions of sulfur dioxide over the time frame 1995–2000.

3 Go on-line to the EPA AP-42 emission factor website (http://www.epa.gov/ttn/chief/efpac/index.html), find the emission factors for stationary gas turbines, and calculate the uncontrolled emission rates for NO_x and CO from a 250 MMBtu/hr, gas fired turbine.

4 Describe the difference between tropospheric and stratospheric ozone, and their role in the global environment.

5 Describe the role of chlorofluorocarbons in the depletion of the ozone layer.

6 Calculate the plume height of a boiler exhaust plume, at a distance of 1,000 meters downwind. Assume a physical stack height of 40 meters, a wind speed of 6.0 m/s, an exhaust temperature of 160°C, and an ambient temperature of 18.0°C.

7 Calculate the downwind ambient concentration of SO_2 emissions of a boiler emission source. Use a downwind distance of 1,000 meters and the wind speed and plume height from Problem 6. Use Stability Class C, and an emission rate of 5.5 grams per second. Calculate for the plume centerline.

8 Calculate the particulate matter removal efficiency of a settling chamber measuring 5 m in length, 4 m in width, and 3 m in height. The gas stream volumetric flow rate entering the chamber is 20 m^3/s, and the average particle size is 80 μm. Assume a particle density of 2,000 kg/m^3 and a gas stream viscosity of 1.85×10^{-5} kg/(m·s).

9 A cyclone is used as a scalping unit on a coal fired boiler. The height of the cyclone inlet (H) is 0.75 m, the width of the cyclone inlet (W) is 0.6 m, the length of the upper chamber (L_b) is 2.0 m, and the length of the lower chamber (L_c) is 5.0 m. The inlet gas velocity is 20.0 m/s and the particle density is 2,000 kg/m^3. Assume gas viscosity of 1.85×10^{-5} kg/(m·s). Calculate the cyclone removal efficiency over a range of particle sizes equal to or greater than to 80 μm.

10 Calculate the particulate removal efficiency of a three-field precipitator servicing a Portland cement kiln. All three fields have the same design. The gas flow rate is 100 m^3/s. Each field contains 11 collection plates measuring 12 m high by 12 m in length. The inlet particulate concentration is 4,500 mg/m^3. Assume a migration velocity of 0.05 m/s. Then calculate the increase in overall particulate removal efficiency that can be achieved by adding a fourth identical field.

11 Compare the number of bags required for a pulse jet baghouse versus a reverse air baghouse servicing a foundry gas flow of 75 m^3/ s. Assume a gas-to-cloth ratio of 0.5 $(m^3/s)/m^2$, and bag dimensions of 0.3 m diameter by 2.5 m in length for the pulse jet unit and a gas-to-cloth ratio of 0.1 $(m^3/s)/m^2$, and bag dimensions of 0.6 m diameter by 7.0 m in length for the reverse air baghouse. With this information, calculate the appropriate number of rows and columns of bags for each baghouse type.

12 For a gas stream containing a high quantity of solids, would a spray tower or a packed tower be the technology of preference for an acid gas removal scrubber? Explain your rationale.

13 An activated carbon adsorber contains 2,000 kg of carbon. It is treating a process exhaust with an emission rate of 115 kg/h of toluene. The adsorber is to be designed for a four-hour duty cycle, and the carbon provider reports a saturation adsorptive capacity for toluene of 60%. At what working capacity will the adsorber be operating?

14 A state agency is requiring a source to install a thermal oxidizer to control VOC emissions from their facility. The oxidizer is required to operate at a temperature of 1,500°F to insure a high efficiency of destruction. The inlet gas stream temperature is 125°F, and the flow rate is 15,000 ft^3/min. Calculate the amount of natural gas required to meet the oxidizer temperature requirement. Calculate also the volume of the oxidizer chamber needed to meet the additional requirement of a residence time of 0.5 s.

15 What is the most important building design criteria in order to avoid excessive buildup of indoor air contaminants?

Chapter 10

Environmental sustainability

John C. Little and Zhe Liu

Learning Objectives

After reading this chapter, you should be able to:

- explain the causes and consequences of global climate change;
- describe the environmental effects of rapid human development;
- estimate ecological footprints;
- evaluate the potential for use of different forms of renewable energy;
- understand the use of indicators of human development that include economic output, sustainability and well-being.

10.1 Overview

Concern for the environment, and a realization that humanity needs to develop in a sustainable way, is not new. In 1919, Svante Arrhenius, the Director of the Nobel Institute, wrote (Arrhenius, 1926):

> *"Engineers must design more efficient internal combustion engines capable of running on alternative fuels such as alcohol, and new research into battery power should be undertaken ... Wind motors and solar engines hold great promise and would reduce the level of CO_2 emissions. Forests must be planted ... To conserve coal, half a tonne of which is burned in transporting the other half tonne to market ... The building of power plants should be in close proximity to the mines ... All lighting with petroleum products should be replaced with more efficient electric lamps."*

He also called for a reduction in waste generated by industry, so that future generations would also be able to meet their needs. This call is echoed in the Brundtland Report (Brundtland, 1987), whose frequently quoted definition of sustainable development is: "Development that meets the needs of the present without compromising the ability of future generations to meet their own needs."

The Brundtland Report, also known as "Our Common Future", was published in 1987 by the World Commission on Environment and Development. The report highlighted three fundamental components of sustainable development: environmental protection, economic growth and social equity.

The former Chief Economist for the World Bank, Herman Daly, outlined three simple operational rules to help define environmental sustainability:

1. Renewable resources such as fish, soil, and groundwater must be used no faster than the rate at which they regenerate (for example, fish are harvested unsustainably if their rate of capture is greater than the rate of growth of the remaining population).
2. Nonrenewable resources such as minerals and fossil fuels must be used no faster than renewable substitutes for them can be put into place.
3. Pollution and wastes must be emitted no faster than natural systems can absorb them, recycle them, or render them harmless.

These rules can be considered as providing a system for making operational the Brundtland definition of sustainability. The Brundtland definition provides the ethical goal of non-depletion of natural capital, whereas the Daly rules define how this goal can be achieved.

Environmental Engineering: Principles and Practice, First Edition. Richard O. Mines, Jr.

There has been a growing public awareness that humans have the capacity to have a severe impact on the balance of nature; that natural resources are not infinite; that humanity, in order to survive, needs to ensure that resources are used sustainably; and that waste products and pollution do not irreversibly change the environment. In her book *Silent Spring*, published in 1962, Rachel Carson brought home the consequences of pollution and pesticide use to the general public. The most important legacy of *Silent Spring* was this new public awareness that nature was vulnerable to human intervention, and that, to prevent catastrophic loss of diversity and life, attention has to be paid to the consequences of technological advances.

In a seminal work published in 1972 titled *The Limits to Growth* (Meadows, Meadows, Randers & Behrens III, 1972), the authors modeled how rapid population growth, together with unchecked consumption, would lead to a point where humanity was using more resources than were being replaced, given that the Earth has only finite resources. In this work, the authors created a computer model called *World3* that was used to simulate a number of scenarios varying population growth, available resources, agricultural productivity and environmental protection. From these simulations, the authors predicted that a crisis point would be reached around 2030, assuming exponential population growth. In the updated edition published in 2004, *Limits to Growth. The 30-Year Update* (Meadows, Randers, & Meadows, 2004), the authors show that the original models were surprisingly accurate.

Possibly the most pressing environmental issue is that of climate change. People have long suspected that human activity could change the local climate. The ancient Greeks wondered what effect cutting down forests would have on the rainfall in a region. In 1896, Arrhenius speculated that changes in the levels of carbon dioxide in the atmosphere could alter the surface temperature of the Earth. This was called the "greenhouse effect", and the law that he formulated is:

> *"If the quantity of carbonic acid increases in geometric progression, the augmentation of the temperature will increase nearly in arithmetic progression."*

This "greenhouse effect" was only one of many speculations about climate change. With more data available, and better climate models evolving, the Intergovernmental Panel on Climate Change (IPCC) in 2007 reported that scientists were more confident than ever that humans were changing the climate (IPCC, 2007). The primary cause for this change is the amount of carbon dioxide being emitted into the atmosphere. The link between carbon dioxide and temperature is demonstrated using data from the Vostok ice cores, as shown in Figure 10.1. This figure shows historical CO_2 (right axis) and reconstructed temperature (as a difference from the mean temperature for the last 100 years) records, based on Antarctic ice cores, providing data for the last 800,000 years. Temperatures can be reconstructed by using proxies such as tree rings, pollen counts, and the isotopic composition of snow and coral.

As will be shown in the following section, temperatures are indeed rising, and are expected to rise further this century. The effect of this rise on the climate is also discussed.

With the knowledge that human demand on nature is not sustainable, metrics to quantify the demand are needed to reveal which policy interventions are most urgently required. The Ecological Footprint (EF), first proposed by Wackernagel and Rees at the University of British Columbia (Rees, 1992), was developed to provide this knowledge, based on a global perspective. The Ecological Footprint measures the amount of land and water required to supply humanity with the resources

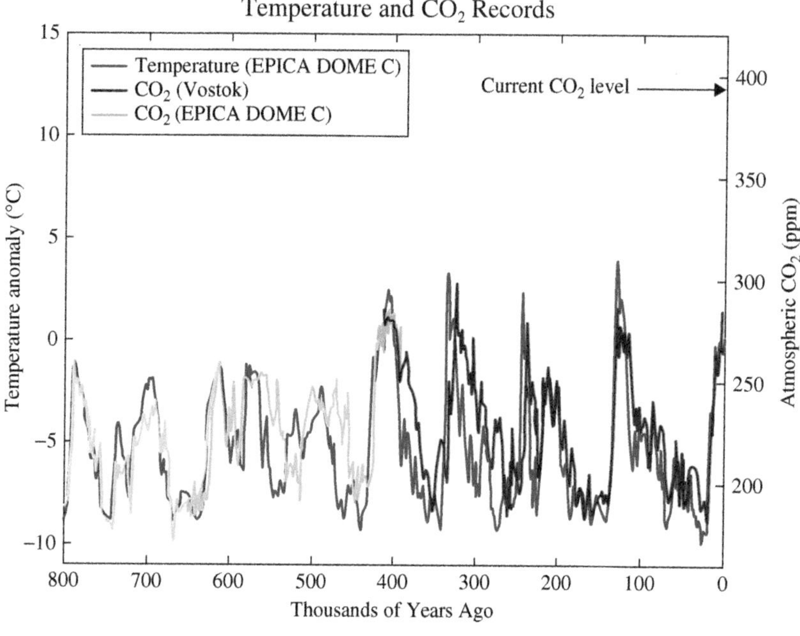

Figure 10.1 Historical CO_2 and temperature for the last 800,000 years. *Source*: http://en.wikipedia.org/wiki/File:Co2-temperature-plot.svg by Leland McInnes.

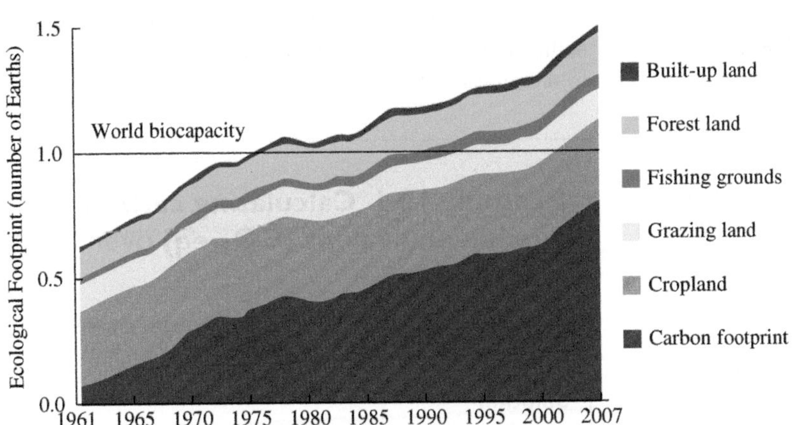

Figure 10.2 Humanity's ecological footprint, 1961–2007. *Source:* Global Footprint Network (2010). Courtesy of The Global Footprint Network.

it needs and to absorb its wastes. The land is measured in Global Hectares, defined as hectares with world-average productivity. Figure 10.2 shows the Ecological Footprint from 1961 to 2007, broken down by different demands.

Figure 10.2 shows that the Ecological Footprint has roughly doubled between 1961 and 2007, with the biggest growth being due to the increase in carbon emissions. The Ecological Footprint shows that, sometime in the early 1970s, humanity went into "overshoot", meaning that we are using more resources than can sustainably be replaced. By 2007, using the EF as a measure, humanity was using the equivalent of 1.5 Earths. This is obviously not sustainable.

Since levels of carbon dioxide in the atmosphere are regarded as being primarily responsible for increasing temperatures, an obvious response would be for humanity to reduce emissions of this gas. Burning of fossil fuels is the major source of carbon dioxide emissions, so finding alternatives for these fuels is imperative. Many alternatives are already available, but they are not cost effective in the prevailing economic system, which does not incorporate the costs of pollution and environmental degradation in the pricing of goods and services. These alternative energy technologies are presented in Section 10.3.

Sustainable development can only be achieved if society, economics and the environment are all simultaneously taken into consideration. In Section 10.4, we therefore look at various measures of sustainability.

10.2 Unsustainable earth

10.2.1 Climate change

Climate can be defined simply as "average weather." The World Meteorological Organization (WMO) defines the averaging period to be 30 years. A more rigorous statistical definition of climate is the mean and variability of appropriate factors over a period of time, which could range from months to millions of years. The factors used in describing climate include temperature, precipitation and wind. There is an even broader definition of climate in which climate is the state and statistical description of the climate *system*. This system is powered by solar radiation, with changes in the incoming or reflected radiation causing a corresponding change in climate.

The climate *does* change in response to changes in factors that affect climate. The almost constant variables affecting climate include latitude, altitude, proximity to oceans and mountains. These variables can change, but do so only over periods of millions of years, such as during shifts in the Earth's continental plates. In addition, slight changes in Earth's position and orientation relative to the Sun bring the Earth closer to the Sun and take it further away from the Sun in predictable cycles (called Milankovitch cycles). Variations in these cycles are believed to be the cause of Earth's ice ages (Weart, 2011).

Other factors affecting the climate are more dynamic. The thermohaline circulation of the ocean results in a 5°C warming of the northern Atlantic Ocean. Extent of ice and snow cover determines how much of the Sun's energy gets reflected back into the atmosphere, and the type and density of vegetation affects solar heat absorption, water retention, and rainfall. Alterations to the quantity of greenhouse gases emitted result in changes to the greenhouse effect.

10.2.1.1 Greenhouse Effect

The greenhouse effect is a natural phenomenon that insulates the Earth from the cold of space. As incoming solar radiation is absorbed and re-emitted back from the Earth's surface in the form of infrared energy, greenhouse gases (GHGs) in the atmosphere prevent a portion of this heat from escaping into space, instead reflecting the energy back to warm the surface further. Without this greenhouse effect, the average temperature of the Earth would be approximately 30°C cooler.

External causes of change in the climate are known as drivers or forcings. These factors can be natural phenomena, such as volcanic eruptions, or human-induced change to the atmosphere, such as the release of greenhouse gases. Detecting climate change is the process of showing that the climate has changed in some demonstrable, statistical sense. Attribution is the process of finding the most likely causes of the detected climate change with some level of confidence. The factors or variables determining climate are numerous, and their interactions are complex. Since it is not possible to run planet-sized experiments, detection and attribution rely on observed data and climate models.

Climate scientists try to understand how climate changes occur. For example, they consider whether the balance of solar radiation is affected by a natural phenomenon such as changes in the Earth's orbit or the Sun itself, or by changes in the amount of radiation reflected, or by changes in the atmosphere causing radiation to be reflected back to Earth. Feedback phenomena are also observed in climate studies. For example, if the global temperature increases, ice will melt, leaving less ice to reflect radiation. Therefore, more warming occurs, perhaps causing permafrost melt, thus releasing methane, which is a strong greenhouse gas. Detecting, understanding, and quantifying these feedbacks are the focus of a great deal of climate science research.

10.2.1.2 Radiative Forcing

Radiative forcing (RF) is a measure of the influence that some climatic factor (e.g., ice albedo or tropospheric aerosol) has in disturbing the balance between incoming solar radiation and outgoing infrared radiation of the Earth's atmosphere. Radiative forcing is expressed in W/m^2. A positive forcing tends to warm the surface of the Earth, while a negative forcing tends to cool it. The concept of RF is useful, because a linear relationship has been found between the global mean equilibrium surface temperature and the amount of RF.

The radiative forcing of different greenhouse gases is not the same, so a common metric, called carbon dioxide-equivalent (CO_2-eq) emissions and concentrations, is used. This is based on the radiative forcing of CO_2. A CO_2-eq for the emission of a long-lived greenhouse gas or mixture of gases is obtained by multiplying the emission of a greenhouse gas by its Global Warming Potential (GWP) (Table 10.1). The CO_2-eq for a mix of gases is obtained by summing the CO_2-eq of each gas. The GWP is a measure of the relative effectiveness of a greenhouse gas to trap the Earth's heat. The GWP of a gas is the warming caused over a 100-year period by the emission of a certain mass of the gas, relative to the warming caused over the same period by the emission of the same mass of CO_2.

Example 10.1 Calculating carbon dioxide equivalent (CO_2-eq) over 100 years

Calculate the CO_2-eq of 12 tonnes of methane emitted from a landfill for a 100-year period.

Solution

$$CO_2 - eq_{methane} = \text{Mass of methane} \times \text{GWP of methane}$$

$$CO_2 - eq_{methane} = 12\ \text{tonnes} \times 25$$

$$CO_2 - eq_{methane} = \boxed{300\ \text{tonnes}}$$

If the landfill is next to a cement factory that emits 12 tonnes of CO_2 and 5 tonnes of N_2O, calculate the total CO_2-eq of these three gases for a 100-year period.

$$CO_2 - eq_{carbon\ dioxide} = 12\ \text{tonnes}$$

$$CO_2 - eq_{nitrous\ oxide} = 5\ \text{tonnes} \times 298$$

$$CO_2 - eq_{nitrous\ oxide} = 1490\ \text{tonnes}$$

The total CO_2-eq of these three greenhouse gases (GHGs) is:

$$CO_2 - eq_{methane} + CO_2 - eq_{carbon\ dioxide} + CO_2 - eq_{nitrous\ oxide} = 300 + 12 + 1490 = \boxed{1802\ \text{tonnes}}$$

Table 10.1 The Global Warming Potential and atmospheric lifetimes of various greenhouse gases relative to CO_2.

Gas	Chemical Formula	Lifetime (years)	Global Warming Potential (GWP) 20-year period	100-year	500-year
Carbon dioxide	CO_2		1	1	1
Methane	CH_4	12	72	25	7.6
Nitrous oxide	N_2O	114	289	298	153
CFC-12	CCl_2F_2	100	11,000	10,900	5,200
HCFC-22	$CHClF_2$	12	5,160	1,810	549
Tetrafluoromethane	CF_4	50,000	5,210	7,390	11,200
Hexafluoroethane	C_2F_6	10,000	8,630	12,200	18,200

The table shows that the GWP of methane is 25 times higher than carbon dioxide over a one hundred year period.

It is possible to use radiative forcing to determine whether global warming is a result of human activities or of natural influences such as volcanic activity and solar irradiance. This is explored further in the next section.

Example 10.2 Calculating radiative forcing and temperature rise

Calculate the radiative forcing and temperature rise for a doubling of CO_2.

Solution

Theoretically, the radiative forcing for a given greenhouse gas, such as CO_2, should be determined by examining each spectral line. For simplification, empirical equations have been developed for CO_2, CH_4, N_2O, CFC-11, and CFC-12 by IPCC. For example, radiative forcing for CO_2, ΔF_{CO_2}, can be calculated as follows:

$$\Delta F_{CO_2} = 5.35 \times \ln(C/C_0)$$

where C is the CO_2 concentration in ppm and C_0 is the reference concentration, 278 ppm.

Therefore, radiative forcing caused by doubling the CO_2 concentration is:

$$\Delta F_{CO_2} = 5.35 \times \ln(C/C_0)$$
$$= 5.35 \times \ln(2 \times 278\ \text{ppm}/278\ \text{ppm})$$
$$\Delta F_{CO_2} = 5.35 \times \ln(2) = \boxed{3.7\ \frac{\text{W}}{\text{m}^2}}$$

The resulting change in surface temperature, ΔT, caused by the radiative forcing can be calculated by:

$$\Delta T_s = \lambda \times \Delta F$$

where λ is the climate sensitivity, which is about 0.8 $K/(W/m^2)$.

So the temperature change due to doubling the CO_2concentration is:

$$\Delta T_s = \lambda \times \Delta F = 0.8\ \frac{\text{K}}{\text{W/m}^2} \times 3.7\ \frac{\text{W}}{\text{m}^2} = \boxed{3.0\ \text{K}}$$

10.2.1.3 Intergovernmental Panel on Climate Change (IPCC)

The United Nations Environment Program (UNEP) and the World Meteorological Organization (WMO) established the IPCC in 1988. The original mandate was:

1. Identification of uncertainties and gaps in our present knowledge with regard to climate change and its potential impacts, and preparation of a plan of action over the short-term in filling these gaps.
2. Identification of information needed to evaluate policy implications of climate change and response strategies.
3. Review of current and planned national/international policies related to the greenhouse gas issue.
4. Scientific and environmental assessments of all aspects of the greenhouse gas issue and the transfer of these assessments and other relevant information to governments and intergovernmental organizations, to be taken into account in their policies on social and economic development and environmental programs.

Thousands of scientists from around the world voluntarily contribute to the work of the IPCC in reviewing and assessing the latest scientific, technical and socio-economic information relevant to the understanding of climate change. The First Assessment Report was published by the IPCC in 1990. This led to the creation of the United Nations Framework Convention on Climate Change (UNFCCC), which is the key international treaty to reduce global warming and cope with the consequences of climate change.

The Second Assessment Report (1995) was used in preparing the Kyoto Protocol, which set binding targets for reducing greenhouse gas emissions. The Third Assessment Report was published in 2001, and the Fourth in 2007. The Fifth is due in 2014. The IPCC was awarded the Nobel Peace Prize in 2007, in recognition of their contribution to the world in the form of a thorough, dispassionate and comprehensive study of the issues associated with climate and climate change over the previous two decades. In addition to the Assessment Reports, the IPCC has also delivered several special reports on topics such as "Carbon Dioxide Capture and Storage" and "Methodological and Technological Issues in Technology Transfer".

The definition of climate change from the Fourth Assessment Report (IPCC, 2007) is stated as follows:

> *"Climate change in IPCC usage refers to a change in the state of the climate that can be identified (e.g. using statistical tests) by changes in the mean and/or the variability of its properties, and that persists for an extended period, typically decades or longer. It refers to any change in climate over time, whether due to natural variability or as a result of human activity. This usage differs from that in the United Nations Framework Convention on Climate Change (UNFCCC), where climate change refers to a change of climate that is attributed directly or indirectly to human activity that alters the composition of the global atmosphere and that is in addition to natural climate variability observed over comparable time periods."*

10.2.1.4 Observed Changes in Climate

The 2007 IPCC report documented the following climate trends. Warming of the climate system is evident, as is widespread melting of snow and ice (particularly in the northern hemisphere), and an increase in the global sea level.

Temperature Instrumental records of global surface temperatures exist for many regions of the globe for the period since 1850. The Third Assessment Report showed a linear heating trend of 0.6°C for the period 1901 to 2000. The Fourth Assessment Report showed a linear trend of 0.74°C for the period 1906 to 2005.

Sea Level The global average sea level rose at an average rate of 1.8 mm/year from 1961 to 2003, and by 3.1 mm/year from 1993 to 2003. This trend is consistent with warming. Since 1993, the contributions to sea level rise have been attributed as follows:

- 57% due to the thermal expansion of the oceans.
- 28% due to the decrease in glaciers and ice caps.
- 15% due to melting of the polar ice sheets.

This rise is significantly larger than the rate averaged over the last several thousand years.

Snow and Ice Extent Snow cover in the northern hemisphere has been consistently below average since 1987. Since 1966, there has been a 10% decrease in snow cover extent, mostly due to a decrease in spring and summer snow in the northern hemisphere. Measurement of sea ice extent has only been possible since readings could be taken by satellite. Using these data, it has been determined that September Arctic sea ice has decreased between 1973 and 2007 at a rate of about 10% ± 0.3% per decade. In September 2007, the sea ice extent was 23% below the previous record low. Sea ice extent in the Antarctic has shown very little trend either way for the same period.

Other Observations Trends have been observed in the amount of precipitation, as well as an increase in intensity of tropical cyclone activity. Many natural systems are being affected by regional climate change, in particular by the increased temperature. The next Assessment Report from the IPCC, due out in 2014, will focus more closely on these regional impacts.

10.2.1.5 Attribution of These Observed Climate Changes

The rise in sea level, loss of snow and ice extent, and the rise in ocean temperature, can all be attributed to the observed rise in global temperatures. To explain these changes, we need to look for some factor, or factors, that hold up to rigorous testing. In addition, we need to eliminate alternative explanations for these observed changes by subjecting them to the same rigorous examination.

There are a number of factors that might cause a global rise in temperature: an increase in solar radiation; a change in the Earth's position relative to the Sun; changes in land cover; or a change in the composition of the atmosphere. We consider each of these possibilities briefly.

Solar Radiation Natural attributions to warming have been observed. Satellite observations since the late 1970s has shown that solar energy has been variable. Together with paleoclimatic reconstructions of solar radiation, a trend of about +0.12 W/m^2 since 1750 is suggested. Although there is a great deal of uncertainty in these estimates, the contribution of direct solar irradiance forcing is small compared to the forcing attributed to greenhouse gases.

Earth's Position Relative to the Sun The Earth's orbit does vary slightly, bringing us closer to and further away from the Sun in predictable cycles. Although these Milankovitch cycles have great value for explaining ice ages and long-term changes in the climate, they are unlikely to have an impact on predictions of climate change in the immediate future, since orbital changes occur over thousands of years.[1]

Change in Land Cover There are two primary functions of land cover that directly affect global temperatures. Snow and ice tend to reflect incoming radiation, and vegetation (particularly forests) removes carbon dioxide from the atmosphere, thus reducing greenhouse gases. The reduced snow and ice extent as a result of warming results in less reflection of radiation as the darker surfaces are exposed, which feeds back into further warming. As the Earth's population increases, and deforestation occurs, less CO_2 is removed from the atmosphere, causing an increase in global CO_2. As the Earth warms, permafrost starts to melt, releasing methane into the atmosphere, which, as we saw above, has a GWP 25 times larger than CO_2.

Change in the Composition of the Atmosphere There are ten primary greenhouse gases, four of which are naturally occurring, with the remaining six being due to industrial emissions. The four naturally occurring are water vapour (H_2O), carbon dioxide (CO_2), methane (CH_4) and nitrous oxide (N_2O). The other six are perfluorocarbons (CH_4, C_2F_6), hydrofluorocarbons (CHF_3, CF_3CH_2F, CH_3CHF_2), and sulfur hexafluoride (SF_6).

Water vapour is the most abundant greenhouse gas. Its concentration depends on temperature and other meteorological conditions, and is not directly related to human activity. Using paleoclimatic records and more recent direct measurements, the atmospheric concentrations of CO_2, CH_4, and N_2O were shown to be relatively stable from 10,000 years ago until the mid-nineteenth century. As shown in Figure 10.3, all three of these gases have shown a marked increase, with CH_4 increasing 150%, and CO_2 by even more. CO_2 is, by volume, the primary anthropogenic greenhouse gas. Between 1970 and 2004, its annual emissions have grown by 80%. In 2004, CO_2 comprised 77% of all the greenhouse gases. The rate at which CO_2-eq greenhouse gases increased in the decade preceding 2005 was almost twice that for the period 1970 to 1994.

As mentioned earlier, radiative forcing can be used to determine the effects that greenhouse gases, aerosols, and clouds have on climate change.

Figure 10.4 shows the global average radiative forcing estimates and ranges in 2005 for anthropogenic CO_2, CH_4,

[1] For an excellent history of scientific enquiry into the causes of the ice ages and an explanation of the Milankovic cycles, see "Past Climate Cycles: Ice Age Speculations" at http://www.aip.org/history/climate/cycles.htm

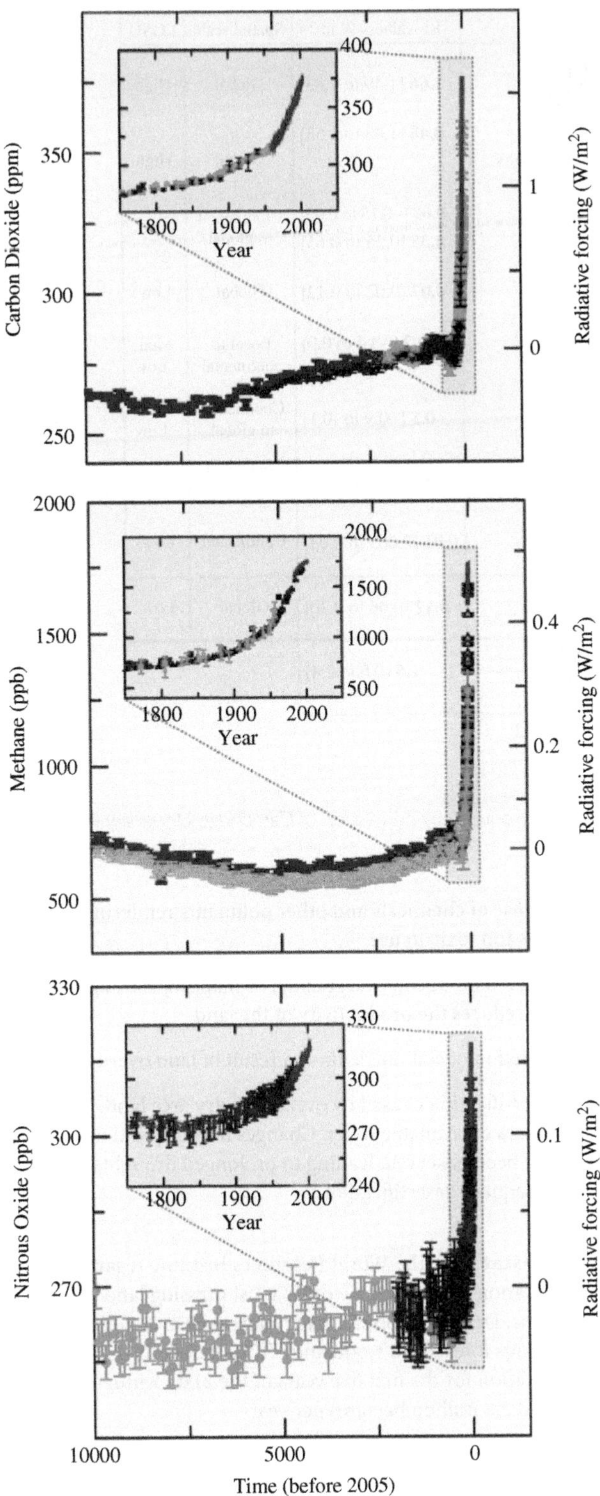

Figure 10.3 Atmospheric concentrations of CO_2, CH_4 and N_2O over the last 10,000 years, and since 1750 (inset panels). Measurements are shown from ice cores (symbols with different colors for different studies) and atmospheric samples (red lines). The corresponding radiative forcings relative to 1750 are shown on the right hand axes of the large panels. *Source*: IPCC WGI Figure SPM.1 (Figure 2.3 pg 38 IPCC Synthesis report). USGS.

N_2O and other important agents and mechanisms, together with the typical geographical extent (spatial scale) of the forcing and the assessed level of scientific understanding (LOSU). The net anthropogenic radiative forcing and its range are also shown. These require summing asymmetric uncertainty estimates from the component terms, and cannot be obtained by simple addition. Additional forcing factors not included here are considered to have a very low LOSU. Volcanic aerosols contribute an additional natural forcing, but are not included due to their episodic nature. The range for linear contrails does not include other possible effects of aviation on cloudiness.

The net effect of human activities since 1750 has been a warming trend (to which the IPCC attributes very high confidence), with an average radiative forcing of $+1.6\ W/m^2$. In comparison, the radiative forcing due to natural solar irradiance for the same period is $+0.12\ W/m^2$.

10.2.2 Environmental depletion and degradation

Environmental degradation is the deterioration of the environment through depletion of resources such as air, water and soil, the destruction of ecosystems and the extinction of wildlife. It is defined as any change or disturbance to the environment perceived to be deleterious or undesirable. Some of the more important areas of resource degradation are briefly discussed here.

Water

Water is a renewable resource because, unlike fossil fuels, it has a sustainable yield. It is also sometimes referred to as a "flow" resource, meaning that while regional precipitation patterns may change, the overall quantity does not change and it is essentially renewable on a global scale. Figure 10.5 shows the global distribution of the Earth's water.

Fresh Water Resources Fresh water resources currently face the dual problems of quality and quantity. These problems will become more acute over time, with population growth and climate-driven changes in the water cycle. In many parts of the world, there is already a stress on fresh water due to population growth, economic development, and changes in precipitation patterns causing desertification in some areas and flooding in others.

Air Pollution It is estimated that more than two million people die prematurely each year due to both indoor and outdoor air pollution. Indoor air pollution has many sources, though the most severe levels occur where open fires are used indoors for heating and cooking without adequate ventilation. Cooking with oils at high temperature is another source of potentially dangerous pollutants. In addition, many building materials and furnishings contain harmful chemicals that are released into the indoor environment. Outdoor air pollution arises from many sources, including industrial processes, energy generation, vehicle emissions, wildfires and volcanoes. Outdoor air pollution has successfully been reduced in many countries through laws regulating emissions.

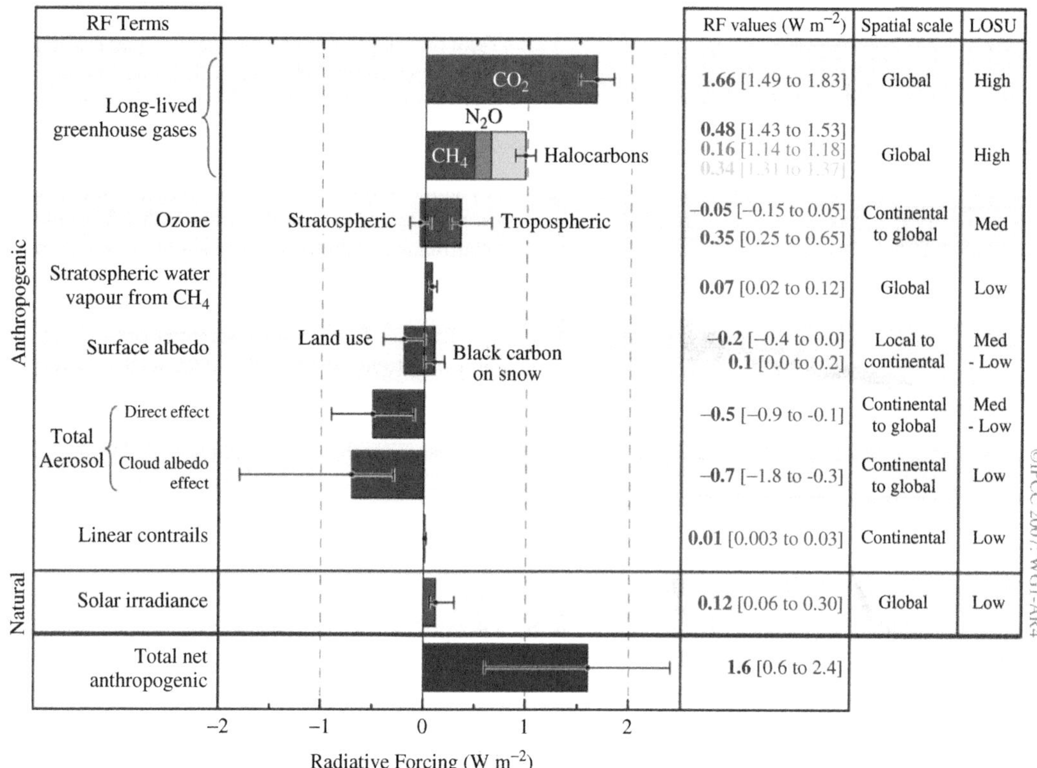

Figure 10.4 Radiative forcing components. *Source*: (IPCC 2007 WGI AR4). Reproduced by permission of Cambridge University Press.

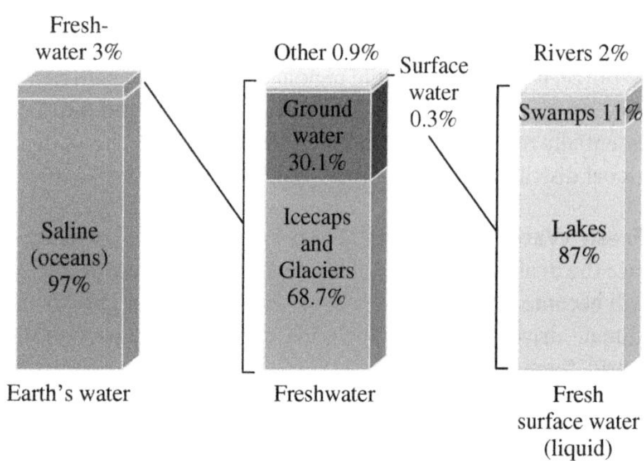

Figure 10.5 Global distribution of Earth's water. *Source*: USGS, http://ga.water.usgs.gov/edu/waterdistribution.htm.

Ozone Depletion The ozone layer in the stratosphere protects life on Earth from harmful ultraviolet radiation. In the 1970s, the first papers linking CFC emissions with ozone depletion were published. After 1985, when a study showed that the ozone hole was much bigger than expected, the Montreal Protocol on Substances that Deplete the Ozone Layer was signed. Since then, emissions of ozone depleting compounds have decreased but, due to their long lifetimes, it is estimated that the damage will continue until 2050.

Land Use Unsustainable land use has led to land degradation. This is caused by:

- The use of chemicals and other pollutants rendering some areas too toxic to use.
- Soil erosion due to overgrazing, or inappropriate agriculture that reduces the productivity of the land.
- Depletion of soil nutrients as a result of land overuse.
- Desertification caused by overuse of dry area land, and stresses on available water. Changes in precipitation patterns have been observed, leading to prolonged droughts and subsequent desertification.

Deforestation The World Resources Institute regards deforestation as one of the world's most pressing land-use problems. Deforestation is the permanent destruction of indigenous forests and woodlands. The rate of global deforestation for the first five years of the 21st Century averaged 7.3 million hectares per year.

Biodiversity The United Nations declared 2010 the "Year of Biodiversity" to try to increase awareness of the importance of biodiversity for a sustainable future. Biodiversity, or biological diversity, is defined as "the variability among living organisms from all sources, including, terrestrial, marine and other aquatic ecosystems, and the ecological complexes of which they are part: this includes diversity within species, between species and of ecosystems."

Biodiversity decline is more rapid now than at any time in human history, and the associated loss of ecosystem services is a constant threat to future development. Ecosystems such as forests and wetlands are being transformed and, in some cases, irrevocably lost, with concurrent loss of species.

Waste Disposal In 2006, US residents, businesses, and institutions produced more than 220 million tonnes of municipal solid waste, which is approximately 2.1 kg of waste per person per day (EPA, 2011). In addition, American industrial facilities generate and dispose of approximately 6.9 billion tons of industrial solid waste each year. In the industrialized countries, it is becoming more and more costly to dispose of this waste, and many countries are simply dumping their waste on the less developed parts of the world. The problem is particularly acute for hazardous wastes.

10.2.3 Ecological footprint (EF)

A core aspect of sustainability is the protection of natural capital, in particular the renewable resources constituting the ecosystem. Humans can live without non-renewable resources such as metals and fossil fuels, but cannot live without the services that the ecosystem provides. Some method is, therefore, necessary to measure the supply of this natural capital, taking into account its ability to regenerate itself. The Ecological Footprint was developed for this purpose, as well as to provide the ability to track progress and set targets for sustainability. The Ecological Footprint measures how much land and water area a human population requires to produce the resources it consumes and to absorb its wastes, using prevailing technology, and compares this to the "carrying capacity" of the Earth. The term, "Ecological Footprint" was first used in an article by William Rees in 1992 (Rees, 1992).

Humanity has, until recently, been able to live within the capacity of the Earth to regenerate the resources used and to reabsorb the wastes generated. However, in the late 1970s, human demand on ecological systems began to exceed the Earth's ability to regenerate. It is now estimated that humanity is using the equivalent of one and a half of Earth's renewable resources per year. The difference between demand and supply is known as ecological overshoot.

Since crossing that threshold three decades ago, the ecological overshoot has grown steadily each year, as shown in Figure 10.2. The implications of this are obvious, and the results are already being seen in climate change, shrinking forests, biodiversity loss, stress on water supplies from overuse, and change in rainfall patterns, among other things.

The advantages of using the Ecological Footprint as a measure of sustainability are that it is easily understandable; it provides a clear reference system that is updated annually; it provides a way to assign responsibilities; and, importantly, it is peer-reviewed. The Ecological Footprint does not tell us anything about, for example, human health, standards of living, and literacy rates, but these shortfalls can be addressed by supplementing the Ecological Footprint with an indicator of human wellbeing, such as the Human Development Index (HDI) from the UNDP.

Countries meet their demands for the resources they need by using their own biocapacity and the biocapacity of other countries. With population growth, and increasing consumption in many countries, competition for resources is increasing rapidly. As shortages develop and prices rise, economies will suffer, with wealthier countries being able to obtain resources at the expense of the poorer. In 2007, the Food and Agriculture Organization of the United Nations (UN FAO) began warning about absolute food shortages, with resulting "biocapacity grabs", where one country buys cropland biocapacity in another. For example, South Korea has leased land in Tanzania to grow food, and other countries have similar arrangements with African countries (Reuters, 2010).

Countries that are large emitters of greenhouse gases also make demands on the biocapacity of other countries, because the emissions disperse through the atmosphere and biocapacity somewhere else on Earth is required to sequester them.

In 1961, most countries could support themselves, but by 2006 the situation had changed dramatically, with less than 20% of the world's population living in countries that can keep up with their own demands. The Ecological Footprint accounting method shows that the world is currently living unsustainably. In Section 10.4, we will consider how the Footprint is determined, and what it can tell us about moving toward a more sustainable future.

10.3 Addressing climate change

Recognizing that the climate is changing as a result of an anthropogenic driver, we now consider various ways to reduce our footprint.

10.3.1 Greenhouse gases

The previous section described the greenhouse effect, and how greenhouse gases perform the function of keeping the planet warm. Data collected from direct measurements, ice cores, and other paleoclimatic sources show that concentrations of the major greenhouse gases have increased significantly since the mid-18th century, with the growth being exponential over the past half century. Despite international agreements, such as the Kyoto Protocol, to reduce the rate of greenhouse gas emissions, there has been no significant drop in the amount of these gases entering the atmosphere.

A gas's contribution to the greenhouse effect is based on its molecular structure as well as its abundance. For example, methane has a GWP 25 times greater than carbon dioxide, but is present in lower concentrations, so its effect is smaller. CO_2 and other poly-atomic gas molecules absorb infrared radiation, and it is this property of the various greenhouse gases that causes the Earth's temperature to rise. CO_2 is the primary anthropogenic greenhouse gas, accounting for approximately 77% of the human contribution to the greenhouse effect. As discussed in the previous section, scientists agree that if emissions of CO_2 continue to increase, then global temperatures will also continue to increase, leading to potentially catastrophic climate change. To stabilize atmospheric CO_2, we need to reduce emissions.

10.3.2 Reducing CO_2

To reduce the amount of CO_2 being discharged into the atmosphere, the obvious route is to reduce the amount of energy we use, and to replace the burning of fossil fuels with other forms of energy. We consider some of the solar alternatives (solar heating, photovoltaic cells, wind, biofuels) as well as geothermal energy. We also need to consider energy efficiency to reduce CO_2 emissions, and evaluate the much-discussed topic of carbon sequestration as a viable alternative. The two primary sources of CO_2 emissions in the United States are power generation and transport. Renewable energy sources need to be found to replace the old ways of doing these things. Power, in the form of electricity, can be generated to replace fossil fuel generators, and alternatives to petroleum for powering vehicles are being developed.

Another reason for finding alternatives to fossil fuels, particularly oil, is energy security. Since fossil fuels are a non-renewable resource, and the demand for energy is only going to increase, it also makes sense immediately to try to replace their use through efficiency measures, and with carbon-free renewable alternatives.

10.3.2.1 Efficiencies

Efficiency technologies, from better insulation to more efficient engines, are improving quickly. Using a compact fluorescent light bulb in place of an incandescent bulb reduces energy usage by a factor of four for a given light output (Energy Star, 2011). Many automobile companies have developed, or have prototypes of, cars with fuel economy of 65 miles per US gallon (3.6 L/100 km) or more. A conservative estimate shows that the USA could halve its energy usage simply by using currently available efficiency technologies.

10.3.3 Solar power

Solar power is the term generally used to describe technologies that use solar energy directly to provide heat, light, hot water, electricity, and even cooling, for homes, businesses, and industry. Although wind power, wave power and the power generated from bio-fuels are also technically forms of solar power, they are usually discussed separately, because the principles used to generate power are different, as are their applications. Solar water heating, passive solar design for space heating and cooling, and solar photovoltaics for electricity are the most commonly used solar technologies for homes and businesses. Utilities and independent developers are also using solar photovoltaics and concentrated solar power to generate electricity on a larger scale.

10.3.3.1 Principles

The Sun is the star at the center of the Solar System, about 1.5×10^{11} m away from the Earth. It is a near-perfect sphere with a diameter of about 1.39×10^9 m, or 109 times that of the Earth, and a mass of about 2×10^{30} kg, or 330,000 times that of the Earth. About three-quarters of the Sun's mass is hydrogen, with the remaining one-quarter primarily helium. From the center outward, the Sun can be divided into a core, a radiative zone, a convective zone, a photosphere (the visible surface), and the solar atmosphere.

The core of the Sun extends from the center to about 0.2 to 0.25 of the solar radius. It has a density of up to 1.5×10^5 kg/m^3 and a temperature of about 1.5×10^7 K. For comparison, the Sun's surface temperature is about 5,800 K, which is often considered to be the effective blackbody temperature of the Sun. The core is the only region of the Sun where thermal energy is generated through nuclear fusion reactions, which convert hydrogen into helium. About 3.846×10^{26} W of thermal energy are generated within the core with heat transferred outward from the core through the other layers to the solar photosphere, and then into space in the form of sunlight (solar electromagnetic radiation) or kinetic energy of particles (NASA, 2007). Because the effective blackbody temperature of the Sun is 5,800 K, the total energy radiated per unit time from the Sun, P_{sun}, can be estimated by the Stefan-Boltzmann law, or:

$$P_{sun} = A_{sun}\sigma T_{sun}^4 \quad (10.1)$$

where:

A_{sun} = the surface area of the Sun, which can be calculated from its diameter

σ = the Stefan-Boltzmann constant, whose value is 5.67×10^{-8} Js^{-1}m^{-2}K^{-4}

T_{sun} = the effective blackbody temperature of the Sun.

Therefore, P_{sun} can be calculated to be 3.89×10^{26} W. Considering an imaginary sphere with the Sun at the center and the average Sun-Earth distance as the radius, the solar electromagnetic radiation cast on the inner surface of the imaginary sphere per unit area can be calculated by dividing P_{sun} by the surface area, or:

$$\frac{P_{sun}}{4\pi D^2_{sun\text{-}earth}} = \frac{3.89 \times 10^{26}\text{W}}{4\pi(1.5 \times 10^{11}\text{m})^2} = 1377\text{W/m}^2 \quad (10.2)$$

The result shows that, at the average Sun-Earth distance from the Sun, the solar electromagnetic radiation received by a plane perpendicular to the rays is about 1,377 W/m^2, which is called the solar constant. While this calculation is an estimate, the solar constant measured on the outer surface of the Earth's atmosphere equals 1,367 W/m^2 (Fröhlich & Brusa, 1981). The actual direct solar electromagnetic radiation received on the outer surface of Earth's atmosphere fluctuates somewhat, due to the Earth's varying distance from the Sun, as well as the variance of the solar electromagnetic radiation itself. Employing the solar constant, the total radiation energy received by the Earth at the top of the atmosphere can be estimated as the product of the solar constant and the cross sectional area of the Earth (the projected area of the Earth on the inside of the imaginary sphere) or about 1.74×10^{17} W.

The solar electromagnetic radiation covers a large spectrum of wavelengths, as shown in Figure 10.6 (which represents sunlight at the top of the atmosphere). About 48% of the total radiation energy is in the form of visible light, with wavelength

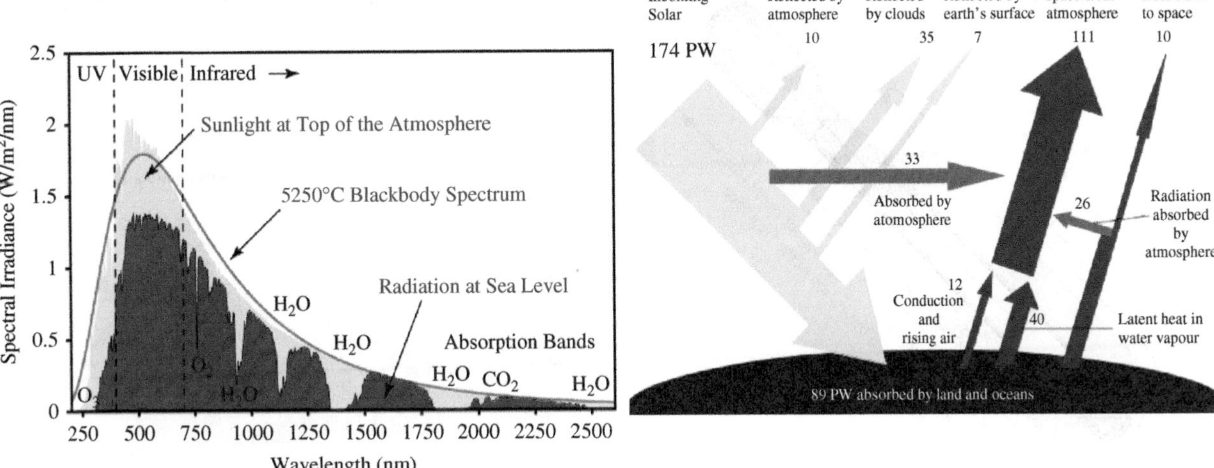

Figure 10.6 Solar radiation spectrum and the breakdown of the incoming solar energy.
Image 1: *Source*: USGS, http://ga.water.usgs.gov/edu/waterdistribution.htm. Robert A. Rohde as part of the Global Warming Art project;
Image 2: *Source*: http://en.wikipedia.org/wiki/File:Breakdown_of_the_incoming_solar_energy.svg. Frank van Mierlo.

from 380–780 nm, and about 45% falls in the infrared range with wavelength larger than 780 nm. The remaining radiation energy is in the ultraviolet (UV) range with wavelength smaller than 380 nm (Pelegrini *et al.*, 2007). However, when the sunlight passes through the Earth's atmosphere, a significant fraction is reflected (about 30%), scattered or absorbed by the atmosphere, and only part of the sunlight reaches the surface of the Earth. Because the atmosphere has a different absorption capacity for specific ranges of wavelengths, the total energy is reduced when reaching the Earth's surface and the spectrum is also changed. In Figure 10.6, the radiation spectrum at sea level and absorption bands of water, oxygen and ozone molecules are shown.

It is estimated that the total solar energy reaching the Earth's surface is 8.9×10^{16} W – about half of the total radiation energy received at the top of the atmosphere. The energy keeps the Earth's surface at an average temperature of 14°C, drives atmospheric circulation and the hydrologic cycle, and powers the growth of photosynthetic plants. Solar energy is therefore the ultimate energy source for several forms of energy used by humans, including fossil fuel, biofuel, biomass, hydroelectricity and wind power.

10.3.3.2 Applications

Humans use sunlight in a wide variety of ways, from passive use for heating and growing crops to modern solar photovoltaic electricity generation. Broadly speaking, the various applications can be classified into thermal use and electricity generation, with representative techniques or applications introduced below.

Solar Thermal **Water heating**: sunlight can be used to heat water for domestic use with solar water heaters. The most common types of solar water heaters include evacuated tube collectors, glazed flat plate collectors and unglazed flat plate collectors.

As shown in Figure 10.7, evacuated tube collectors have a series of transparent glass tubes, each of which has a copper heat pipe (called the absorber tube) and a vacuum zone between the outer and inner tubes. The inner absorber tube contains a working fluid (usually methanol) and is covered with a heat-absorbent material to collect the heat, while the vacuum prevents absorbed heat from escaping. When the working fluid in the absorber tube is heated and vapourized, it rises to the top of the absorber tube, where cooler water flowing through a surrounding manifold is heated. The cold vapour liquefies and returns to the bottom of the absorber tube to repeat the cycle (to allow the fluid to return to the bottom, the tubes must be mounted with a minimum tilt angle of around 25°).

An important feature of this technology is the vacuum, which helps to retain absorbed heat within the absorber tube and therefore achieves higher efficiency compared to traditional collectors. This enables the production of water with higher temperatures, in the range of 70–170°C. Although more efficient and versatile in an unfavorable climate, such as cold and cloudy weather, evacuated tube collectors are generally more expensive than other types of collectors.

Common glazed flat plate collectors (Figure 10.8) comprise: a network of flow tubes, where water or another heating fluid flows; a dark color absorber plate; a glazed glass cover which protects the absorber plate and prevents loss of heat; and insulation. The heat is absorbed by the absorber plate and is then transferred to the fluid circulating through the collectors in the flow tubes. If a heating fluid other than water is used (e.g., glycol prevents freezing of water in the flow tubes during cold seasons), a heat exchanger is required to transfer heat from the heating fluid to the water. Glazed flat plate collectors can produce water with temperatures of up to 70°C. Unglazed collectors, which do not have a glazed cover, are a simpler and cheaper version of the glazed flat plate collectors. These generate water at lower temperature, and are mainly used for heating swimming pools and homes.

Solar cooling: active solar cooling systems use solar thermal collectors to provide thermal energy to drive absorption chillers. In contrast to a regular refrigerator, which requires a compressor to condense the refrigerant, absorption chillers use heat to drive a circulation system. Currently, most absorption

Figure 10.7 Evacuated tube collector. *Source*: http://www.sunmaxxsolar.com/how-evacuated-tube-solar-collectors-work.php. © Hickory Ridge Solar.

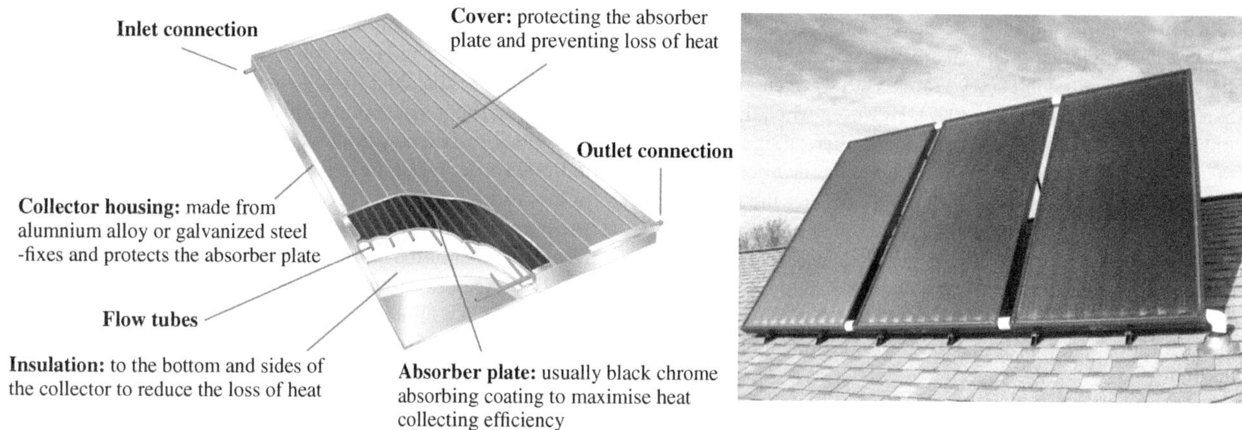

Figure 10.8 Glazed flat plate collectors. *Source*: http://www.greenspec.co.uk/solar-collectors.php and http://www.htproducts.com/graphics.html

chilling systems use a lithium-bromide or ammonia solution as the refrigerant. Neither of these would deplete ozone and are, therefore, preferable to compressor refrigerators that typically use ozone-depleting hydrochlorofluorocarbons (HCFCs). Hot water (or some other working fluid) can thus be produced from solar thermal collectors and used as a heat source to power the absorption systems. Efficient absorption chillers require a heat source of high temperature, so glazed flat plate collectors or some other techniques are needed.

Solar cooker: Concentrated solar heat can be used for cooking. Although solar cookers vary widely, they all have similar basic components: reflective devices used to concentrate sunlight into a small cooking area; solar absorbing devices such as dark surfaces used to convert sunlight into heat; and heat trapping devices such as a glass cover. Figure 10.9 shows a simple but effective solar cooker design. The foldable polished metal surface reflects and concentrates sunlight, while the black steel converts sunlight into heat, which is trapped in the air-tight glass pot so that the pot reaches a temperature high enough for cooking.

Electricity Generation Using solar energy to generate electricity has received increasing attention in recent years, and several promising technologies have been developed. Solar energy can be converted to electricity directly based on the photovoltaic effect, or indirectly using concentrated solar power (CSP), which is also a solar thermal application.

Solar photovoltaics: the photovoltaic effect allows direct conversion of solar energy into electricity using a photovoltaic or solar cell (IEA, 2010).

Figure 10.10 briefly shows how a solar cell works. When sunlight is cast onto the solar cell, which is essentially a large series of semiconductor P-N junctions, photons with sufficient energy can knock electrons loose from their atoms. Due to the special composition of solar cells, these electrons can only move in one direction. When an external load is connected to the terminals, free electrons flow through the external circuit, generating an electric current. Solar cells are connected together to form photovoltaic modules or solar panels and further to form solar panel arrays, which are able to provide a usable amount of direct current (DC) electricity. Various

Figure 10.9 Solar cooker. *Source*: http://atlascuisinesolaire.free.fr

semiconductor materials can be used for solar cells, including monocrystalline silicon, polycrystalline silicon, amorphous silicon, cadmium telluride, and copper indium selenide/sulfide. The different materials have different efficiencies and costs. Solar panels can be used in a wide variety of ways to provide electricity.

Example 10.3 Calculating electricity production from a solar cell

At the 37th IEEE Photovoltaic Specialist Conference in Seattle, researchers presented a thin-film gallium-arsenide solar cell product which could convert 27.6% of the solar energy reaching the cell surface into electricity. A household in California is going to use this product on the roof of their house, about 20 m^2. Determine the total solar energy received by the solar cell and the total electricity produced.

Solution

Assuming average solar energy received at the roof is 200 W/m^2 during day time (10 hours), the total solar energy received by the solar cell during a day is calculated as follows:

$$200\ \text{W/m}^2 \times 20\ \text{m}^2 \times 10\ \text{h} = 40\ \text{kilowatt} - \text{h(kWh)}$$

$$40\ \text{kWh} \times \left(3.6 \times 10^6 \frac{\text{J}}{\text{kWh}}\right) = \boxed{1.44 \times 10^8\ \text{J}}$$

Total electricity the household could get from the solar cell is:

$$\frac{27.6\%}{100\%} \times 40\ \text{kWh} = \boxed{11\ \text{kWh}}$$

$$11\ \text{kWh} \times \left(3.6 \times 10^6 \frac{\text{J}}{\text{kWh}}\right) = \boxed{3.96 \times 10^7\ \text{J}}$$

Concentrated solar power (CSP) systems use lenses or mirrors to concentrate a large area of sunlight into a small area (Aringhoff *et al.*, 2005). The concentrated light is converted to a heat source with a very high temperature, which drives a heat engine to generate electricity. In contrast to solar photovoltaics, which directly convert solar energy into electricity, CSP systems generate heat sources that are then used to generate electricity with a conventional power plant. There are several concentrating technologies available to achieve the required high temperatures. Three techniques that are commonly used are the solar trough, parabolic dish, and solar power tower. The techniques are discussed below and Figure 10.11 shows all three.

A *solar trough* has a linear parabolic reflector and a receiver positioned along the reflector's focal line. The reflector concentrates sunlight onto the receiver, which is a tube, and the working fluid flowing through the receiver is heated. The parabolic trough is made to follow the Sun during the day by tracking along a single axis, and this increases the efficiency. The working fluid can be heated to between 150–350°C for subsequent use as a heat source in a power plant.

A *stand-alone parabolic dish* has a parabolic reflector that concentrates sunlight onto a receiver positioned at the

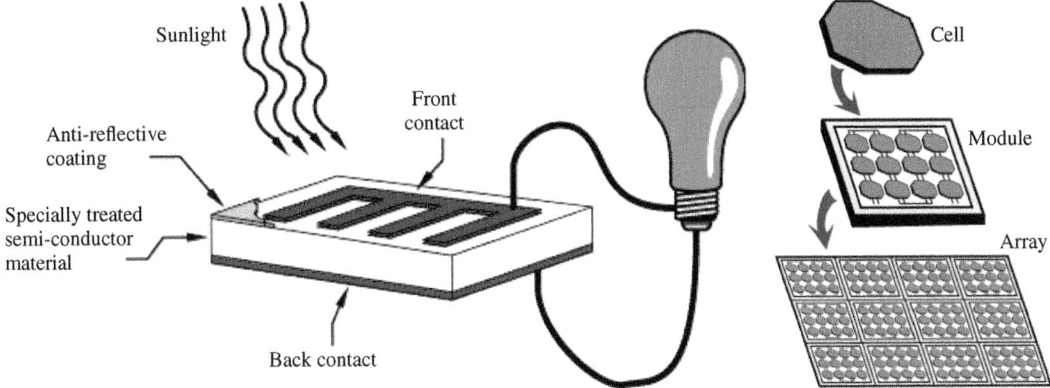

Figure 10.10 Schematic of a solar cell and the composition of a solar array. *Source*: http://www.dynaspede.net/PowerSystems_StandAloneSolarHomeLightingLEDLamps.htm and http://www.zeh.ca/SolarPanels/tabid/57/Default.aspx

Figure 10.11 Solar trough, parabolic dish, and solar tower. *Source:* http://commons.wikimedia.org/wiki/File:Solar_troughs_in_the_Negev_desert_of_Israel.jpg; http://www.psa.es/webesp/instalaciones/discos.php; http://en.wikipedia.org/wiki/File:Th%C3%A9mis.jpg. (a) David Shankbone (b) Plataforma Solar de Almería (c) David66

parabolic reflector's focal point. Better than a linear solar trough, a parabolic dish can track the Sun along two axes to receive more solar energy. The working fluid in the receiver is heated to between 250–700°C, which can be directly used as a heat source by a Stirling engine placed together with the receiver. By receiving more solar energy and eliminating the energy loss during the flow of working fluid as in the solar trough, a parabolic dish can achieve higher efficiency.

A *solar power tower* consists of an array of dual-axis tracking reflectors that concentrate sunlight onto a central receiver placed on a tower. The working fluid in the receiver (molten nitrate salt) can be heated up to 1,000°C and the system has higher energy storage capacity.

10.3.3.3 Pros and Cons

As one of the main sources of renewable energy, the advantage of solar energy is obvious: no matter whether using solar energy for heating, cooling or generating electricity, there is no emission of greenhouse gases and no fossil fuel consumption; it is renewable and there is a practically unlimited supply available. There are a great many technologies that use solar energy in various environments, and many of them are well developed and convenient to deploy.

However, the Sun only shines during the day and does not shine consistently, depending on the weather. Energy storage and other complementary systems are usually required to provide a consistent power supply. Because solar energy is a diffuse source, the energy conversion efficiency is low and large collection areas are needed. For solar photovoltaic and CSP systems, the initial investment is high, so that the average cost for solar electricity is sometimes not competitive, compared to conventional thermal power system and other forms of renewable energy such as wind power.

10.3.4 Wind energy

10.3.4.1 Principles

Whenever a pressure gradient exists, air flows spontaneously from the high-pressure region to the low-pressure region, and the resulting large-scale horizontal flow of air is wind. The basic cause of all wind, whether global-scale atmospheric circulation or a local-scale sea breeze, can be traced to temperature differences, which occur because of differential heating and cooling of the Earth and the atmosphere. Solar radiation is, therefore, the ultimate power source driving the wind. It is estimated by meteorologists that about 1% of the total solar radiation energy received by the Earth is converted into wind (Musgrove, 1987).

Wind power is the conversion of the kinetic energy of wind, into a useful form of energy, such as mechanical or electric power. To estimate the energy in wind, we can imagine holding up a hoop with an area A, facing the wind whose velocity is v. The mass of air (m) passing through the hoop in time interval t is the product of the density of air ρ and the volume of the parcel of air, which is the product of v, t and A.

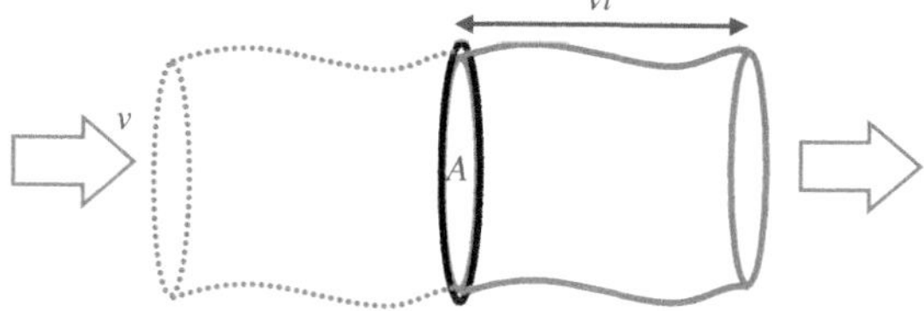

The kinetic energy (KE) of the parcel of air in motion is determined as:

$$KE = \frac{1}{2}mv^2 = \frac{1}{2}(\rho\, vt\, A)v^2 = \frac{1}{2}\rho\, Atv^3 \qquad (10.3)$$

Thus, the power available in the wind, or the kinetic energy passing through the area A per unit time, is given by

$$\text{Power} = \frac{\frac{1}{2}mv^2}{t} = \frac{1}{2}\rho\, Av^3 \qquad (10.4)$$

However, not all of the kinetic energy can be extracted and converted into useful forms of energy. When passing through a windmill, kinetic energy is lost and the wind velocity (and hence kinetic energy) decreases. If all the kinetic energy was extracted from the wind, the wind would have to come to a complete stop.

The German physicist Albert Betz calculated the maximum fraction of the incoming energy that can be extracted by a windmill or wind turbine. According to Betz, the ideal or maximum fraction of total wind energy that can be captured by a windmill or a turbine is 59.3%. This is referred to as Betz's limit or Betz's law (Betz, 1966).

According to the wind power equation, the larger the surface area of a wind power device facing the wind, the greater the power. Eliminating the impact of the size of the device by dividing power by A, the expression of $1/2\rho v^3$ is defined as Wind Power Density (WPD), with units of W/m^2. WPD depends solely on wind speed and the air density, which decreases with increasing altitude and varies with temperature and humidity. Therefore, WPD provides an estimate of the effective power of the wind at a particular location. Because wind speed and air density vary substantially across the Earth, the resulting WPD is an important factor when determining whether a location is suitable for wind power utilization. The National Renewable Energy Laboratory (NREL) of the US Department of Energy investigates and provides data on wind resources for the US, based on mean WPD, as found in the Wind Energy Resource Atlas of the United States (NREL, 1986).

Generally, wind speed is reduced near ground level (because of friction at the Earth's surface) and increases with altitude, although the exact functional relationship is complicated and depends on several environmental conditions, such as the roughness of the surrounding terrain. As a simple rule of thumb, doubling the altitude typically increases wind speed by 10%. There are also several standard formulas that can be used to estimate wind speed as a function of height. For example, according to NREL, wind speed at a height of z (m) can be calculated from wind speed at a height of 10 m using a power law, or:

$$v(z) = v_{10}\left(\frac{z}{10}\right)^{\alpha} \quad (10.5)$$

where:
v_{10} = the wind speed at 10 m
α = 1/7 (NREL, 1986).

For comparison, the Danish Wind Industry Association uses a logarithmic law, or:

$$v(z) = v_{ref}\frac{\log(z/z_0)}{\log(z_{ref}/z_0)} \quad (10.6)$$

where:
z_0 = roughness length
v_{ref} = wind speed at a reference height z_{ref}.

The roughness length for agricultural land with some houses and sheltering hedgerows is z_0=0.1 m (Danish Wind Industry Association, 2003).

10.3.4.2 Applications

Historical Use Wind power has a very long history, with wind being used to propel sailboats for thousands of years. Taking advantage of wind power via windmills dates back to the 7th century AD, when people in Afghanistan, Iran and Pakistan used vertical-axis windmills for pumping irrigation water and milling grain (al-Hassan & Hill, 1986). Windmills then spread to China and India through the Middle East and Central Asia (Hill, 1991). The use of windmills in Europe began around 1200 AD, and they were extensively used for milling, with some horizontal-axis windmills still in existence. The Industrial Revolution brought steam engines, which provided a more stable source of power for machines, leading to a gradual decline in the use of windmills. However, small windmills for pumping water have remained in use in many parts of the world (Dodge, 2006).

Using wind power to generate electricity has a much shorter history. The first windmill for electricity production was built in Scotland in 1887 and, in the 1890s, wind turbines were built by the Danish scientist Poul la Cour (Price, 2005). In the first part of the 20th century, small wind turbines for lighting or isolated rural buildings were common. For example, by the 1930s, windmills were widely used to generate electricity on farms where power distribution systems had not yet been installed in the United States. The modern wind power industry began in 1979 with the serial production of wind turbines by several Danish manufacturers. Since then, both wind power technology and commercialization have developed quickly, due to the 1970s oil crisis and, later, the rising concerns over global warming, energy security and fossil fuel depletion. Government policies have accelerated the development in many countries.

Modern Technologies Modern wind turbines can be categorized into two basic types according to whether the rotation is about a horizontal or a vertical axis (EWEA, 2004).

Horizontal Axis Wind Turbines (HAWTs) are the older and more common type. A horizontal axis wind turbine, as shown in Figure 10.12, has blades that look like a propeller that spin on the horizontal axis.

Figure 10.12 Horizontal axis wind turbines. *Source:* http://en.wikipedia.org/wiki/Wind_turbine#Horizontal_axis, http://holland.portfoliocms.com/Brix?pageID=111A) Hans Hillewaert.

Horizontal axis wind turbines have the main rotor shaft, gearbox and electrical generator in a nacelle mounted at the top of a tower. The airfoils, or blades, intercept the airflow that drives torque that turns the shaft. In fact, the wind exerts two forces on the airfoils: aerodynamic lift and drag. The lift force is perpendicular to the direction of airflow, and this causes rotation about the hub, while the drag force is parallel to the direction of airflow and this impedes rotation. To be efficient, blades should therefore have a relatively high lift-to-drag ratio.

The gearbox connected to the shaft adjusts the rotational speed to drive an electrical generator. To enable the airfoils to intercept the airflow most effectively, the nacelle should be pointed into the wind. If the wind speed exceeds the designed range, the nacelle must be yawed out of the wind to prevent damage to the generator. Small turbines are adjusted into or out of the wind by a simple wind vane placed square with the rotor (blades), while large turbines generally use a wind sensor coupled with a servo-motor. Because a tower produces turbulence in the wind's wake, which may lead to fatigue failures, the turbine is usually positioned upwind of its supporting tower.

Since HAWTs are installed on a tall tower, they can access stronger winds and, therefore, achieve a higher power output. Because the blades of HAWTs always move perpendicularly to the wind, receiving power through the whole rotation, the efficiency is high compared to vertical axis wind turbines (VAWTs). However, to construct the supporting tower and install and maintain the nacelle on the top of the tower costs more. Furthermore, HAWTs require a yaw control mechanism to turn the blades into and out of the wind.

Example 10.4 Calculating the electricity output of a HAWT

Calculate the electricity output of a HAWT whose nacelle is 25 m high and airfoils are 12 m long. The HAWT is installed at a flat rural area where average wind speed is 6 m/s at 10 m height.

Solution

The wind velocity at the height of the HAWT nacelle is determined from Equation (10.5):

$$V_{25m} = V_{10m}\left(\frac{25}{10}\right)^{1/7} = 6\,\text{m/s} \times \left(\frac{25}{10}\right)^{1/7} = 6.84\,\text{m/s}$$

The power available in the wind is determined from Equation (10.4).

$$\text{Power} = \frac{1}{2}\rho AV^3 = \frac{1}{2} \times (1.237\,\text{kg/m}^3) \times (\pi \times 12^2\text{m}^2) \times (6.84\text{m/s})^3 = \boxed{89540.68\,\text{W}}$$

Considering Betz's limit, the power that can be extracted by the HAWT is:

$$59.3\% \times 89540.68\,\text{W} = 53097.62\,\text{W}$$

Assuming the electricity generator in the HAWT has an overall efficiency of 35% and ignoring other energy losses, the electricity generated by the HAWT is:

$$\frac{35\%}{100\%} \times 53097.62\,\text{W} = 18584.17\text{W}$$

$$18584.17\text{W} \times \left(\frac{\text{kW}}{1000\,\text{W}}\right) \approx \boxed{18.6\,\text{kW}}$$

Vertical Axis Wind Turbines (VAWTs), as shown in Figure 10.13, have the main rotor shaft arranged vertically. In contrast to HAWTs, which require yaw control mechanisms to turn the nacelle into the wind, VAWTs are always aligned with the wind. With a vertical axis, the generator and gearbox can be placed at ground level making installation and service easier.

Figure 10.13 Vertical axis wind turbines. *Source:* http://science.howstuffworks.com/environmental/green-science/wind-power2.htm, http://news.cnet.com/8301-11128_3-9956965-54.html, http://windturbinezone.com/wind-turbine/vertical-wind-turbine. All reproduced with permission.

However, without a tall supporting tower, the elevation of VAWTs is lower, where wind speed is low, and turbulence may be higher. Finally, a VAWT cannot start to move spontaneously, because the starting torque is very low, and they require a boost from an electrical system to get started. For these reasons, VAWTs are generally less efficient than HAWTs.

Wind farms: to capture wind power and generate electricity on a large scale, a group of wind turbines (from a few dozen to several hundred) can be installed in the same location to form a wind farm. As of November 2010, the world's largest wind farm is the Roscoe Wind Farm (780 MW) in Texas (O'Grady, 2009). Because VAWTs require a larger footprint, HAWTs are generally used in wind farms. Individual turbines are connected to a medium voltage power collection system and communications network. The medium-voltage electrical current is increased in voltage with a transformer for connection to a high voltage transmission system. A wind farm may cover an area of hundreds of square kilometers, because individual turbines cannot be placed too close together or the upwind ones will cast wind-shadows on the downwind ones (a wake effect). To minimize loss of efficiency, turbines should be placed at least three to five diameters apart perpendicular to the prevailing wind, and five to nine diameters apart in the direction of the prevailing wind (Wagner & Mathur, 2009). The land between the turbines may be used for agriculture or other purposes.

Depending on their location, wind farms can be classified as being onshore, nearshore or offshore. Onshore wind farms are those in hilly or mountainous regions, generally three kilometers or more inland from the nearest shoreline. Nearshore wind farms are on land within three kilometers of a shoreline, or on water within ten kilometers of land. Offshore wind farms are generally ten kilometers or more from land. Currently, most wind farms around the world are either onshore or nearshore, because the construction and maintenance costs are lower and they are easier to connect to power grids. However, onshore wind farms require large areas of land. Although they are compatible with some land uses, such as agriculture, they may create adverse aesthetic problems, including noise and visual impacts, and they may disrupt wildlife and birds. In contrast, offshore wind farms can take advantage of steadier and stronger ocean winds and have lower noise and visual impacts. However, they cost more for construction and maintenance.

10.3.4.3 Pros and Cons

Except for the energy consumption and environmental impact caused by the manufacture of wind turbines, operation does not require additional energy or cause additional emissions. Wind power is renewable and abundant. For example, the global onshore and nearshore wind power potential is about 72 TW, or over five times the world's current energy use in all forms (Archer & Jacobson, 2005). In addition to being clean, renewable and abundant, wind power is more cost effective for electricity generation than other forms of renewable energy. Even though offshore wind power is much more expensive than onshore higher speed wind power, it remains cheaper than solar photovoltaic (Solar PV) and concentrated solar power (Solar CSP) at present.

Unfortunately, wind resources are quite area specific, and the best sites are often located in sparsely populated areas far away from the population centers they need to serve. Extra transmission lines are often required, so that costs are increased, while efficiencies are reduced from transmission losses. The wind is intermittent and unpredictable, so that the efficiencies of wind turbines are reduced. Furthermore, wind turbines may cause noise and visual problems and impact adversely on birds and wildlife.

10.3.5 Geothermal energy

10.3.5.1 Principles

The literal meaning ("geo" means Earth and "thermal" means heat) for geothermal energy is the thermal energy stored and generated in the Earth. The heat in the Earth primarily has two sources: the residual heat left over from the Earth's formation (about 20%), and the decay of radioactive isotopes embedded in the Earth (80%), mainly comprising potassium (^{40}K), thorium (^{232}Th), and uranium (^{235}U and ^{238}U) (Turcotte & Schubert, 2002).

In addition to the internal heat deep in the Earth, the top 10 m of the ground accumulates solar energy during summer and releases the heat during winter, which also contributes to geothermal energy. Naturally, the heat moves from the Earth's interior towards the surface via conduction, with an average gradient of temperature through the crust of about 25–30°C/km of depth (Fridleifsson *et al.*, 2008). The average heat flow through the continental and oceanic crusts is about 65 mW/m^2 and 101 mW/m^2, respectively, with a total global heat flux of over 4.4×10^{13} W (Pollack *et al.*, 1993).

Although the total output is larger than the current energy consumption of all primary sources, not all of the heat can be reasonably extracted for use at costs competitive with other forms of energy, simply because geothermal energy is too diffuse and the average heat flux is too small. Nevertheless, the natural heat flux at some locations of the Earth is much higher than the average value, such as in regions near tectonic plate boundaries.

The geological structure at these boundaries is not stable, and geological phenomena, including earthquakes and volcanic activity, often occur – the so-called "Pacific ring of fire" is a good example. In these locations, fluids such as magma, and water penetrating through the crust, can enhance the heat flow substantially by convection, which brings heat to the surface directly in the form of hot springs, fumaroles, steam vents and geysers. Due to the higher heat flux and a heat source that is closer to the surface, it is possible to take economic advantage of the geothermal energy in these areas.

To demonstrate how geothermal energy can be extracted, an ideal geothermal field with hydrothermal system is described here. Such a system generally consists of a heat source, a reservoir covered by impermeable rocks, a recharge area and connecting paths through which cool superficial water can penetrate into the reservoir and escape back to the surface. The heat source can be either a very high temperature magmatic intrusion (600–900°C, typically at a depth of 7–15 km) or the Earth's normal temperature gradient, which increases with depth. The reservoir gains heat from the heat source via conduction within an aquifer, where flowing water is heated via convection. After heating, water rises back to the surface as a liquid (hot springs) or vapour (steam vents). Cool water (such as rainwater) collects in the recharge area and penetrates into the reservoir to replace the escaping heated water. As a hot spring or geyser, a hydrothermal system can be a natural formation, through which geothermal energy (in the form of hot water or vapour) can be extracted directly at rather low cost for utility use, such as electricity generation and space heating.

Aside from the heat source, other components of a hydrothermal system can be engineered. For example, cool water can be pumped via a geothermal well into the deep Earth to reach hot rocks, which serve as both the heat source and the reservoir. The heated water is then pumped back to the surface through another well. Natural hydrothermal systems can be roughly divided into three types: hot water, wet steam, and dry steam.

About 30% of commercially exploitable hydrothermal systems are hot water fields, which generally have rather small heat sources and produce water with temperature less than 100°C.

Wet steam fields contain pressurized water at temperatures exceeding 100°C. These are the most common types of hydrothermal systems. Approximately 60% of commercially exploitable hydrothermal systems and more than 90% of currently exploited hydrothermal reservoirs on an industrial scale are wet steam fields.

Dry steam fields contain dry saturated or slightly superheated steam at pressures above atmospheric pressure. They are quite rare, comprising only 10% of commercially exploitable hydrothermal systems, although about 50% of the geothermal electricity in the world comes from six large dry steam fields.

10.3.5.2 Applications

The geothermal energy in hot springs has been used for bathing, space heating and even cooking for much of history. For modern applications, the Lindal diagram, initially proposed by Icelandic engineer Baldur Lindal, is a commonly accepted reference illustrating the potential use of geothermal fluids at different temperatures. Although geothermal energy can be used for various purposes, as shown in the modified Lindal diagram (Figure 10.14), we will consider only two applications: electricity generation and direct heat.

Electricity Generation There are four main types of geothermal power plants: dry steam, flash steam, binary cycle, and dry hot rock power plants (EERE, 2011).

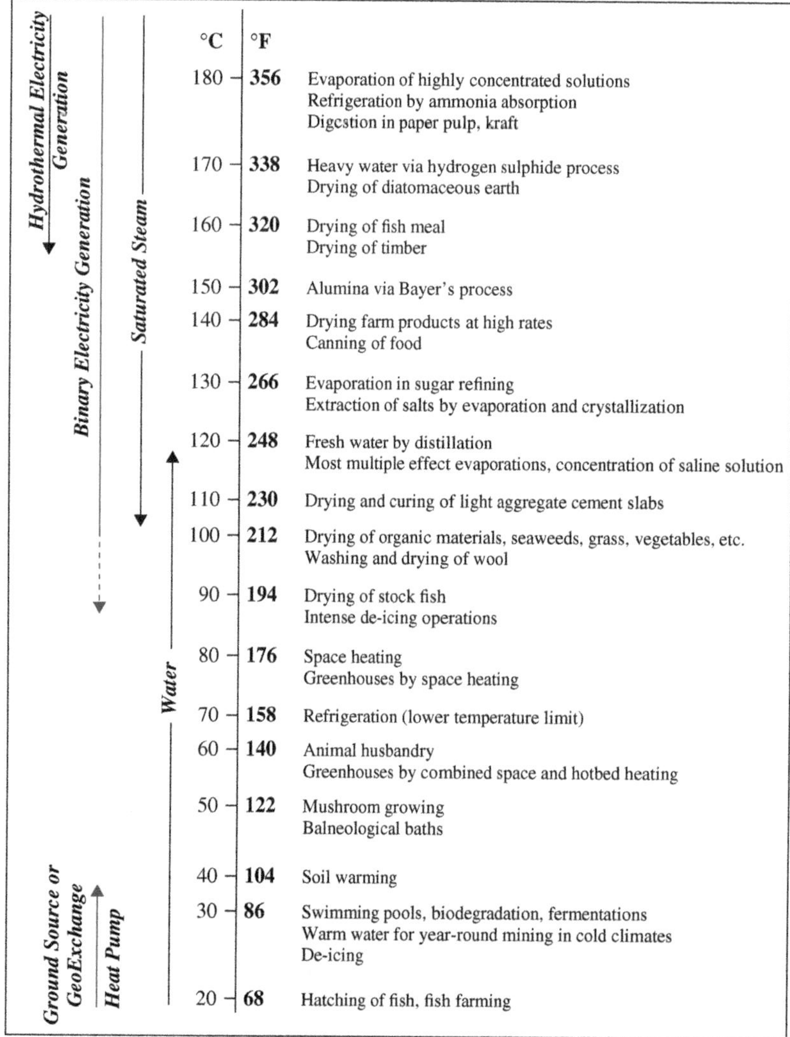

Figure 10.14 The Lindal diagram. *Source:* http://geosurvey.state.co.us/energy/renewables/Geothermal/Pages/GeothermalDirect%20Use.aspx. Colorado Department of Natural Resources.

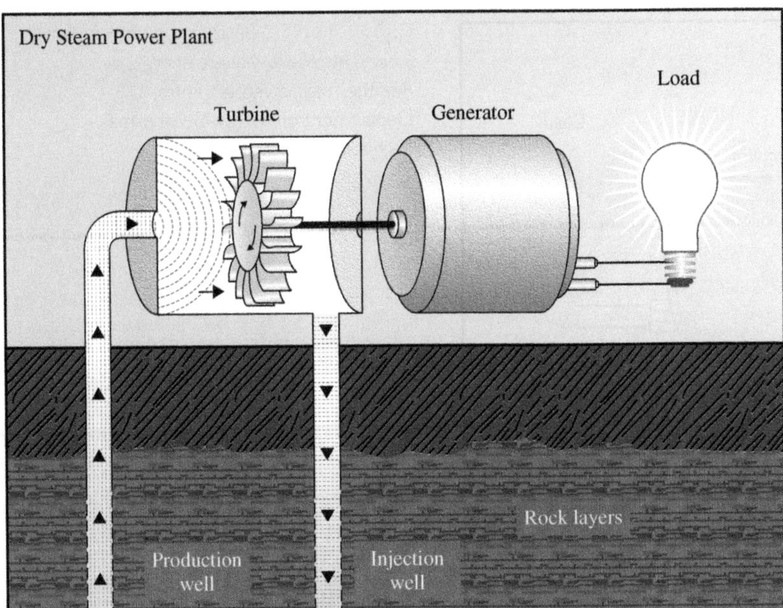

Figure 10.15 Dry steam power plant. *Source*: http://www1.eere.energy.gov/geothermal/powerplants.html.

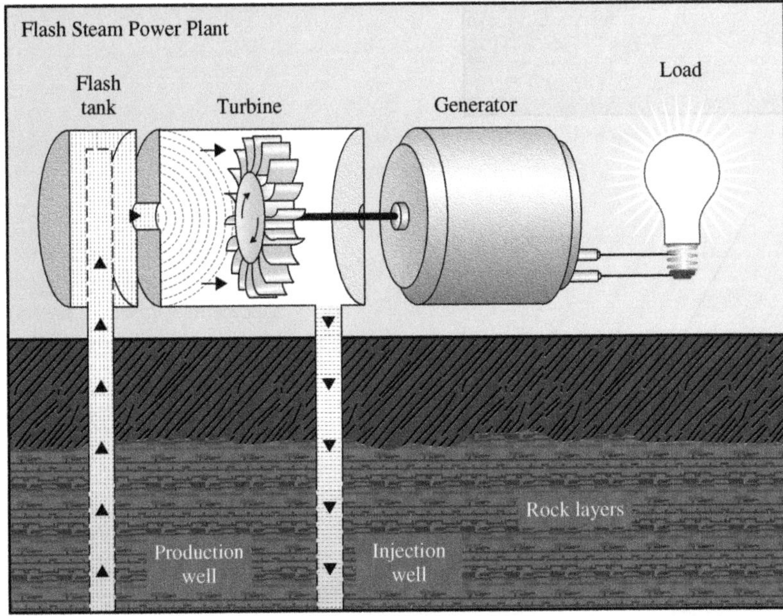

Figure 10.16 Flash steam power plant. *Source*: http://www1.eere.energy.gov/geothermal/powerplants.html. U.S. Department of Energy, Geothermal Technologies Office.

Dry steam power plants (Figure 10.15): can be used in dry steam fields and some wet steam fields. The steam from the geothermal well is passed directly through a turbine that drives a generator that produces electricity. The steam exiting the turbine can be discharged or condensed with cold air or cold water and reintroduced into the geothermal reservoir through an injection well. The dry steam power plant is the simplest and oldest type of geothermal power plant, but it is very efficient. It requires dry steam of 150°C or more and is only feasible at a limited number of hydrothermal system sites.

Flash steam power plants (Figure 10.16): when the temperature is not high enough to directly vapourize a large amount of water from the geothermal well into steam, hot water is first sprayed into lower-pressure flash tanks to produce steam and the resulting steam is used to drive turbines. Flash steam power plants are the most common type of geothermal power plants in operation today.

Binary cycle power plants (Figure 10.17): if the water temperature in the geothermal well is not high enough for a flash steam power plant, binary cycle power plants can extract heat from moderate temperature water to generate electricity. Hot geothermal water is passed through a heat exchanger, together with a secondary, low boiling-point working fluid (typically butane or pentane), so that heat from the hot water causes the secondary working fluid to vapourize. The vapour of the secondary working fluid is then used to drive the turbine to generate electricity. The vapour leaving the turbine is condensed and cycled back through the heat exchanger for

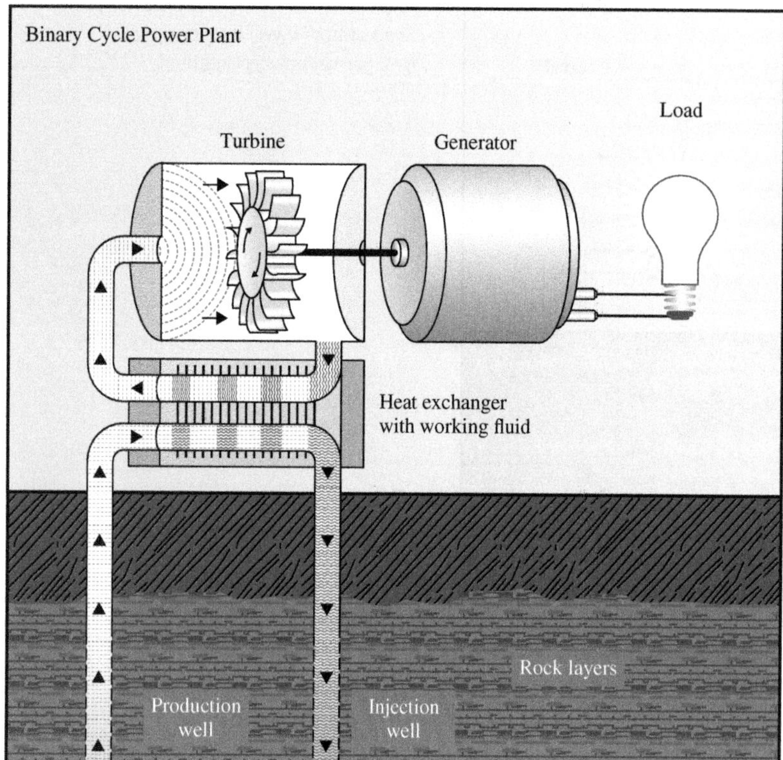

Figure 10.17 Binary cycle power plant. *Source*: http://www1.eere.energy.gov/geothermal/powerplants.html. U.S. Department of Energy, Geothermal Technologies Office.

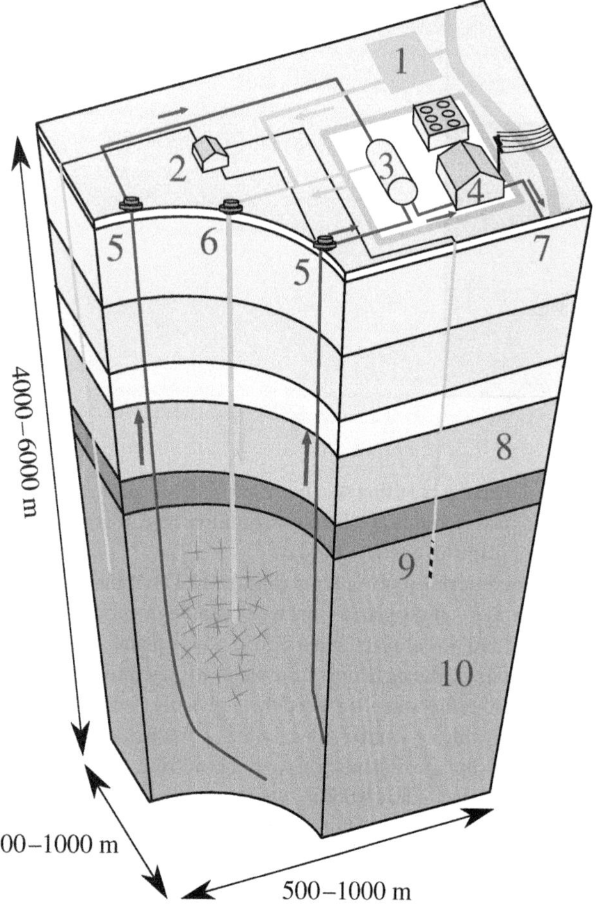

Figure 10.18 Dry hot rock power plant. *Source*: http://en.wikipedia.org/wiki/File:EGS_.Geothermie_Prinzip01.jpg. Modifications made by Ytrottier.

a new cycle. Since moderate temperature water is the most common geothermal resource, most geothermal power plants in the future will be the binary cycle type.

Dry hot rock power plants (Figure 10.18): although the three types of power plant described above can operate at different geothermal water temperatures, they all rely on natural hydrothermal systems. A dry hot rock power plant, also called an enhanced geothermal system (EGS), is developed to produce geothermal electricity from previously unusable sites, where the naturally occurring water and rock porosity is insufficient to carry heat to the surface. In dry hot rock power plants, wells are drilled a few kilometers deep into the earth, reaching hot dry rocks. Cold water is then pumped into the earth through the wells and the pressure creates fissures in the deep rocks so that the water flows into the fissures, creating an artificial geothermal reservoir where the water is heated. The heated water is then pumped back up to the surface and used to generate electricity using one of the three previously described procedures. The exhaust water is pumped back into the deep earth for a new cycle. Although more expensive to construct than the three previous types of plant, dry hot rock power plants provide a promising approach for geothermal electricity in areas without natural hydrothermal systems.

Direct Applications **Direct heating**: low-temperature geothermal resources not viable for electricity are typically used for direct heating. Water or steam from natural hot springs or artificial geothermal wells can be captured and piped directly into radiators, or passed through heat exchangers to heat a separate working fluid. Waste heat, such as hot water and steam from the turbine of a geothermal power plant, can also be used for heating. In addition, due to the greater thermal inertia of the shallow ground, the seasonal variation in ground temperature is smaller than in air, and it disappears completely below a depth of 10 m. Therefore, heat in the warmer Earth can sometimes be extracted by circulating water through tubes that are used for direct heating.

Geothermal heat pump: a heat pump is a device that pumps heat from a low-temperature region to a high-temperature region, normally powered by electricity. Based on the same principles as an ordinary air conditioner with a compressor (which is a common type of heat pump), a heat pump can be used to move heat from the indoor environment to outdoors in summer for cooling, and from outdoors to indoors in winter for heating. When used for heating in winter, heat pumps can have a much higher efficiency than an electric heater because, not only is heat moved from the heat source into the region which needs to be heated, but the electricity used to power the heat pump is also converted into heat.

A geothermal heat pump uses the Earth as a heat source in the winter or a heat sink in the summer. In comparison, an ordinary heat pump uses the outdoor air as the heat source in the winter or heat sink in the summer. Therefore, geothermal heat pumps are often called ground source heat pumps. As mentioned above, the Earth is warmer than the surrounding air in winter and cooler than the air in summer. Depending on the operating mechanism of a heat pump, the efficiency is higher for either a higher-temperature heat source or a lower-temperature heat sink. Geothermal heat pumps are thus more efficient than ordinary air conditioners, or air source heat pumps.

Example 10.5 Calculating space heating costs

Compare the annual cost for space heating using baseboard electric heater, geothermal heat pump, and furnace burning gas in a household. The annual energy consumption for space heating by the household is 20,000 kilowatt-hours.

Solution

The efficiency of the baseboard electric heater is 100% (all the electricity is converted to heat) and electricity cost is \$0.1/kilowatt-hours. So the annual cost would be:

$$20,000 \times \frac{100\%}{100\%} \times \$0.1/\text{kWh} = \boxed{\$2000}$$

The coefficient of performance (COP, the ratio of heating output to the supplied electricity) of a geothermal heat pump is generally 2.5 to 4 and here is assumed to be 3. So the annual cost would be:

$$20,000\ \text{kWh} \times \frac{1}{3} \times \$0.1/\text{kWh} = \boxed{\$667}$$

The efficiency of a forced air furnace burning gas is 78% and the gas cost is \$0.035/ kilowatt-hours (calculated from its heating value and cost). Thus the annual cost would be:

$$20,000\ \text{kWh} \times \left(\frac{1}{78\%/100\%}\right) \times \$0.035/\text{kWh} = \boxed{\$897}$$

In summary, the operation cost using geothermal heat pump saves about \$1,330 per year compared to the electric heater and about \$230 compared to the furnace. But the initial investment to install the system and the operating life should be taken into account when evaluating the overall economics.

10.3.5.3 Pros and Cons

Geothermal energy is generally renewable and causes little impact on the environment. It is reliable and consistent compared to wind and solar energy. Geothermal energy is also cost-effective and does not require much land. However, large-scale use of geothermal energy, especially for electricity generation, has been limited to areas with known geothermal resources. New technologies, such as enhanced geothermal systems and geothermal heat pumps, have dramatically expanded the range and size of viable resources. However, there are still some problems that need to be considered, such as local depletion, which occurs when the extraction rate exceeds the regeneration rate in the heat reservoir. Furthermore, enhanced geothermal systems, and some traditional geothermal power plants, may affect land stability, due to underground work such as drilling wells, pumping underground water to the surface and injecting water into the deep ground, so that subsidence and even earthquakes may be triggered (Lund, 2007).

10.3.6 Carbon sequestration

"The term carbon sequestration is used to describe both natural and deliberate processes by which CO_2 is either removed from the atmosphere or diverted from emission sources and stored in the ocean, terrestrial environments (vegetation, soils, and sediments), and geologic formations" (USGS, 2008).

The carbon cycle describes natural sequestration of CO_2, whereby the oceans and plants act as sinks for CO_2. Until the advent of the Industrial Revolution, CO_2 sources generally produced less CO_2 than the sinks were capable of handling, but after the invention of the steam engine, and the development of coal fired power plants, this was no longer possible. The rate of deforestation over the past century has only exacerbated this imbalance. The oceans have historically been the primary sink for atmospheric CO_2, absorbing around a third of anthropogenic CO_2. However recent studies are showing a decreasing trend in the oceans uptake ability (McKinley *et al.*, 2011).

In this section, we look only at deliberate efforts to capture and sequester carbon, which comprises four steps, although capture and compression usually go together:

1. CO_2 capture
2. Compression
3. Transport
4. Storage

Capture of CO_2 is feasible at large point sources, such as fossil fuel or biomass-fed power facilities, or large industrial emitters such as refineries, steel manufacturing plants, and oil and gas processing facilities.

10.3.6.1 CO_2 Capture and Compression

There are three available capture technologies, which are currently used in various industrial processes:

- Gasification/Pre-combustion.
- Post-combustion.
- Oxy-combustion.

None of these have as yet been used in large-scale applications. The high demand for water and steam for CO_2 capture and compression decreases the power-generating capacity of a plant by approximately 30%.

During the **gasification/pre-combustion process**, coal, biomass or a coal/biomass mixture reacts with oxygen and/or steam at high temperatures and under pressure to produce carbon monoxide (CO) and hydrogen. This is commonly referred to as syngas. A simplified flowchart is shown in Figure 10.19.

To enable pre-combustion capture, the syngas is further processed in a water-gas-shift reactor, which converts CO into CO_2 while producing additional hydrogen. An acid gas or a physical solvent-based system can then be used to separate CO_2 from hydrogen.

During **post-combustion**, fuel is burned with air in a boiler to produce steam. The gas produced consists mostly of nitrogen and CO_2, which are separated by chemical solvent systems. This technology is mainly implemented in conventional coal-, oil- or gas-fired power plants. Figure 10.20 shows this process.

In the **oxy-combustion** process, fuel is burnt with a mixture of pure oxygen and a CO_2 recycle stream, as shown in Figure 10.21. The products of this process, steam and CO_2, are separated by steam condensation. Other gas constituents are filtered. Oxy-combustion produces a highly concentrated CO_2 stream (~60%). This technique is applicable to both new and existing coal-fired power plants.

All the capture technologies require an additional step of compressing the CO_2, so as to be able to transport and store the gas more efficiently. The captured CO_2 is subjected to temperatures greater than 31.1°C and a pressure of more than 72.9 atmospheres. In this supercritical state, CO_2 has the viscosity of a gas and the density of a liquid, and can easily be compressed to a smaller volume.

10.3.6.2 Transport

CO_2 can be moved around in any of the three phases: gas, liquid or solid. In the gaseous phase, the volume to be moved would be enormous for the amount of CO_2 generated. Therefore, for practical purposes, the gas is compressed into a liquid or solid form. Commercial-scale transport of the gaseous and liquid CO_2 makes use of tanks, pipelines, and ships.

Due to the large quantities of CO_2 generated by power stations, transportation from these will likely be via pipeline rather than ground transportation. CO_2 pipelines are not new, and in the USA there are more than 2,500 km of pipe carrying CO_2 to be used in enhanced oil recovery in Texas and elsewhere. However, the necessary infrastructure to carry large

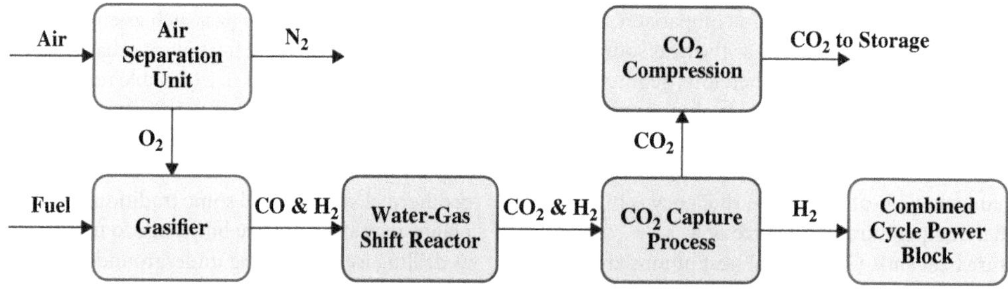

Figure 10.19 Power plant with pre-combustion CO_2 capture. *Source:* US Department of Energy, 2010.

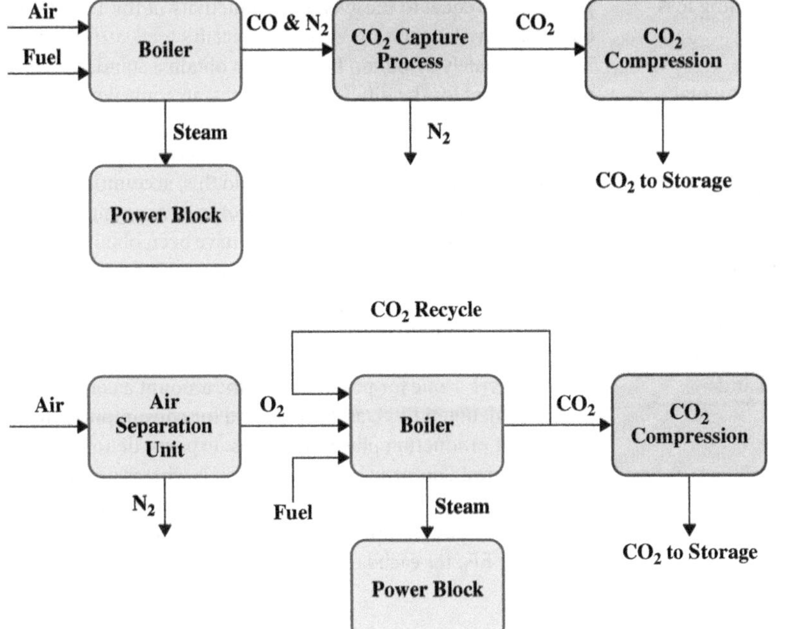

Figure 10.20 Power plant with post-combustion CO_2 capture. *Source:* US Department of Energy, 2010.

Figure 10.21 Power plant with oxy-combustion CO_2 capture. *Source:* US Department of Energy, 2010.

enough quantities to meet the targets of CO_2 reduction will be enormous, and the cost of building these new pipelines will be significant, particularly since suitable storage sites are not necessarily near generation sites.

If the storage site is offshore, tankers will be used for transportation. Since liquefied petroleum gas (LPG) and liquefied natural gas (LNG) are already frequently moved around in tanks and by ship, there are no problems other than cost associated with this mode of transport.

10.3.6.3 CO_2 Storage

There are three options for storage: geologic, ocean, and mineral carbonation.

Geologic sequestration is the process during which captured CO_2 is piped 1–4 kilometers below the Earth's surface, where it is trapped among the various geologic formations. Current use of CO_2 is for enhanced oil recovery from the reservoirs. CO_2 is injected into the oil field and pushes the oil up to the ground. Another option is to inject CO_2 into layers of porous rocks, like sandstone. The semi-liquid CO_2 is trapped, because of the existence of non-porous cap rocks, which are impermeable. The pressure put on the trapped CO_2 from the many rock layers ensure that the CO_2 will stay in this form. Once injected, it disperses slowly and eventually dissolves in the salty water or brine. Some of the injected CO_2 may be trapped in the form of carbonate minerals, because of its chemical interaction with the surrounding rock.

Another alternative for carbon storage is **oceanic** sequestration. Oceanic sequestration occurs naturally, through chemical reactions between seawater and the CO_2 in the atmosphere. Oceans absorb around one-third of the carbon emitted by human activity. Ocean sequestration can be enhanced by the addition of iron, which is a necessary micronutrient, but deficient in the oceans. On the other hand, measures should be taken to avoid ocean acidification. Deliberate injection of CO_2 into the ocean at great depth is another storage option. However, there is very limited experience in this area. Problems that would have to be addressed include determining what the environmental consequences would be, whether the public would accept the process, what the legal framework is, and what safeguards would need to be put in place.

Mineral carbonation is the fixing of CO_2 in the form of inorganic carbonates. CO_2 is reacted with metal-oxide bearing materials to form corresponding carbonates and a solid byproduct such as silica. The products of mineral carbonation are naturally occurring stable solids and would provide storage capacity on a geological time scale. However, the technology for doing this is in the early development stage. In a test case using wet carbonation of natural silicate olivine, the cost of converting 1 tonne of CO_2 was between \$50 and \$100. This translates into a 30–50% penalty on the original power plant, in addition to the 10–40% energy penalty for capturing and compressing the CO_2 (IPCC, 2005).

10.4 Addressing resource depletion and environmental degradation

In Section 10.2, we showed that anthropogenic effects on the environment are contributing to climate change, to resource depletion, and to environmental degradation. The primary causes of all these effects are population growth, accompanied by increasing resource utilization, and inadequate constraints on waste management. The Ecological Footprint shows that current trends are not sustainable, and that without major changes to how we power the planet, and how we manage resources, we are headed for perhaps catastrophic climate change, a reduction in resources such as forests and fisheries, and ecosystems that will break down, with a corresponding loss of biodiversity.

In this section, we look in more detail at some of the measures that have been developed to evaluate environmental impacts, including the Ecological Footprint and Life Cycle Assessment (LCA), and discuss how these can be used to enhance sustainable development. The environmental issues cannot be viewed in isolation. The way that societies are structured, with closely integrated economic and environmental systems, has a significant bearing on whether and how the world can move toward a sustainable future. In this context, we also look at alternative indicators to gross domestic product (GDP) and gross national product (GNP) that try to account simultaneously for environmental and societal sustainability and economic growth.

10.4.1 Environment

10.4.1.1 Ecological footprint

The Ecological Footprint is a simple and useful accounting tool to estimate human demand on the biosphere, and on the biosphere's regenerative capacity and ability to absorb waste. Representing the impact of human activity on the Earth's ecosystems, it is especially useful for revealing historical trends. If we think of **biocapacity** as the measure of the amount of biologically productive (**bioproductive**) land and sea available for human use, then the Ecological Footprint is a measure of how much of the biocapacity is used. If the Ecological Footprint is greater than the biocapacity of an area, then the population of that area is living unsustainably.

In its most basic form, the Ecological Footprint (EF) can be expressed as:

$$\mathrm{EF} = \mathrm{Demand}/\mathrm{Yield} \qquad (10.7)$$

where:
Demand = the annual demand for a product
Yield = the annual yield of that product.

Studies that are compliant with current Ecological Footprint standards use the global hectare as the unit of measure. A global hectare corresponds to one hectare of biologically productive space with world-average productivity, and this makes Ecological Footprint areas comparable throughout the world.

To calculate the Ecological Footprint, the first step is to divide the Earth into distinct biologically productive or bioproductive land types. These land types are cropland, fishing grounds, forest, grazing land, and built up land (which includes buildings and land taken up by hydropower installations) – making up approximately 12 billion hectares. In addition, there is one category of indirect demand for biocapacity, in the form of absorptive capacity for CO_2.

Using a range of global datasets developed by organizations including UN FAOSTAT, UN Comtrade, and the OECD International Energy Agency (Ewing, 2010), it is possible to determine the extent of each land type, as well as its productivity. To obtain a standard measure of productivity, one Global Hectare (gha) is defined as one hectare with a productivity equal to the average productivity of the 12 billion bioproductive hectares on Earth. The Earth's total surface area is approximately 51 billion hectares. To obtain a standardized global hectare for the different land types, an equivalence factor for each land type is used.

The next step in the calculation is to determine how much of this bioproductive land is used. To do this, accounting is done at the national level (National Footprint Accounts) using the datasets mentioned above. Results have been obtained for 240 countries, territories, and regions from 1961 to 2007, consisting of more than 800,000 data points calculated from more than 50 million source data points.

There are, in reality, two distinct Ecological Footprints at the national level – one for productivity (the amount a country produces, denoted EF_P), and the second for consumption (the measure of production plus imports less exports, denoted EF_C). The Ecological Footprint of consumption is what is generally understood to be the Ecological Footprint of an entity.

The Ecological Footprint of production for a country is the sum of the EF_P for each land use type, or:

$$\mathrm{EF}_{\mathrm{PTotal}} = \mathrm{EF}_{\mathrm{PCrop}} + \mathrm{EF}_{\mathrm{PForest}} + \mathrm{EF}_{\mathrm{PGrazing}} + \mathrm{EF}_{\mathrm{PWater}} + \mathrm{EF}_{\mathrm{PBuilt}} + \mathrm{EF}_{\mathrm{PCarbon}} \qquad (10.8)$$

The EF_P for each land use type is given as:

$$\mathrm{EF}_{\mathrm{p}} = (P/Y_{\mathrm{N}}) \times (\text{Yield Factor}) \times (\text{Equivalence Factor}) \qquad (10.9)$$

where:
P = the amount of product produced
Y_N = the national average yield for P expressed in global hectares.

The Yield and Equivalence factors will be explained in more detail below but, briefly, the yield factor is the ratio of national average yield to world average yield, calculated annually for each land use type. The equivalence factor is a productivity-based scaling factor that converts a specific land type into units of world average biologically productive area. The equivalence factor is also calculated annually for each land use type.

The Ecological Footprint of consumption (EF_C) is calculated as:

$$\mathrm{EF}_{\mathrm{C}} = \mathrm{EF}_{\mathrm{P}} + \mathrm{EF}_{\mathrm{I}} - \mathrm{EF}_{\mathrm{E}} \qquad (10.10)$$

where EF_I and EF_E are the footprints for imported and exported commodity flows, respectively.

Considering this overview, we now look in more detail at how National Ecological Footprints and Biocapacity are derived, and some of the assumptions that are made.

10.4.1.2 Calculating National Ecological Footprint and Biocapacity

The Global Footprint Network (Global Footprint Network, 2010) maintains annual National Footprint Accounts that enable nations to monitor demand for, and supply of their natural capital, and to see when demand exceeds supply. The National Footprint Accounts aim to:

- Provide a scientifically robust and transparent calculation of the demands placed by different nations on the regenerative capacity of the biosphere;
- Build a reliable and consistent method that allows for international comparisons of a nation's demand on global regenerative capacity;
- Produce information in a format that is useful for developing policies and strategies for living within the limits of the available natural capital; and
- Generate a core dataset that can be used as the basis of sub-national Ecological Footprint analyses, such as those for provinces, states, businesses, or products.

These national accounts document how much of the annual regenerative capacity of the biosphere is required to renew the resource for a given population for that year. They also show the change over time, and what portion of biocapacity demand is satisfied domestically, versus how much needs to be imported.

Accounting Method Six assumptions are used in Ecological Footprint accounting to determine how much regenerative capacity is required to maintain a given resource flow (Ewing, 2010):

- The majority of the resources people consume and the wastes they generate can be quantified and tracked.
- A substantial fraction of these resource and waste flows can be measured in terms of the biologically productive area necessary to maintain flows. Resource and waste flows that cannot be measured are excluded from the assessment, leading to a systematic underestimate of the true Ecological Footprint.
- By weighting each area in proportion to its bioproductivity, different types of areas can be converted into the common unit of global hectares, with world average bioproductivity.
- Because a single global hectare represents a single use, and each global hectare in any given year represents the same amount of bioproductivity, they can be summed to obtain an aggregate indicator of Ecological Footprint or biocapacity.
- Human demand, expressed as the Ecological Footprint, can be directly compared to nature's supply, expressed as biocapacity, when both are in global hectares.
- Areal demand can exceed areal supply if the demand on an ecosystem exceeds that ecosystem's regenerative capacity.

The National Footprint Accounts track human demand for ecological services in terms of six land use types: cropland, grazing land, forest land, fishing grounds, built-up land, and carbon footprint. We will describe these land use types in more detail later, but for now we just need to understand that the Earth's bioproductive areas are represented in a uniform way. To get to that point, we need to define the terms: yield factor; equivalence factor; and bioproductive area.

The **yield factor** takes into account the fact that a given type of land has a productive potential that is based, for example, on the climate, how it is managed, and its topography. This factor allows different areas of the same land type to be compared, based on the common denominator of yield. The national yield factor for a particular land type is calculated as the ratio of the national average yield of that land type to the world average yield of that land type. Yield factors are calculated every year for each land type in each nation. These yield factors convert one hectare of a specific land type, such as pasture, within a given country, into an equivalent number of world-average hectares of that same land type.

For each land use type (excluding cropland which is more complicated), the yield factor (YF_L) is simply given by:

$$YF_L = \text{(National Yield)/(World Yield)} \qquad (10.11)$$

This is done because these land types are considered to have a single primary product, for example fish from fishing grounds.

Table 10.2 gives the yield factors for a few representative countries.

The yield factor for built-up land is assumed to be the same as for cropland, based on the assumption that urban areas are built on or near productive agricultural lands. Areas used for hydroelectric reservoirs are presumed to have previously had world average productivity. The yield factor for carbon uptake land is assumed to be the same as that for forest land. Inland waters are assigned a yield factor of 1, due to a lack of data on freshwater ecosystem productivities.

The **equivalence factor** translates a specific land type (such as cropland or forest land) into the common unit of global hectares. This equivalence factor represents the world's average potential productivity of a given bioproductive area relative to the world average potential productivity of all bioproductive areas. For example, because the average productivity of cropland is higher than the average productivity of all other land types, it must be converted using its corresponding equivalence factor in order to be expressed in global hectares.

Equivalence factors are the same for all countries, but vary from year to year due to changes in the relative productivity usually caused by changes in the environment. The equivalence factors are derived from the suitability index of Global

Table 10.2 Yield factors for selected countries in 2007.

Yield factor	Cropland	Forest	Grazing land	Fishing grounds
World average	**1.0**	**1.0**	**1.0**	**1.0**
Germany	2.2	4.1	2.2	3.0
Japan	1.3	1.4	2.2	0.8
New Zealand	0.7	2.0	2.5	1.0
Zambia	0.2	0.2	1.5	0.0

Agro-Ecological Zones (GAEZ) (GAEZ, 2000), which is a spatial model of potential agricultural yields, combined with data on actual areas of cropland, forest land, and grazing land.

The GAEZ model divides all land globally according to its calculated potential productivity, assigning a quantitative suitability index from the following five categories:

- Very Suitable (VS) – 0.9.
- Suitable (S) – 0.7.
- Moderately Suitable (MS) – 0.5.
- Marginally Suitable (mS) – 0.3.
- Not Suitable (NS) – 0.1.

To get the equivalence factors, it is assumed that, within each country. the most suitable land will be planted with crops, followed by forest land, with the least suitable used for grazing land. The equivalence factors are calculated as the ratio of the world average suitability index for a given land use type to the world average suitability index for all land use types.

Table 10.3 shows the equivalence factors for the land use types in the 2010 National Footprint Accounts, which are based on data from 2007. We will use these numbers to calculate global hectares in the next section.

As done with the yield factors, the equivalence factor for built-up land is set equal to that of cropland, and the equivalence factor of carbon uptake land is set equal to the equivalence factor of forest land. Assuming hydroelectric reservoirs flood world average land, the equivalence factor for this area is set to 1. The equivalence factor for marine area is calculated such that a single global hectare of pasture will produce an amount of calories of beef equal to the amount of calories of salmon that can be produced by a single global hectare of marine area. Inland water has the same equivalence factor as marine water.

Table 10.3 Equivalence factors, 2007.

Area type	Equivalence factor
Cropland	2.51
Forest	1.26
Grazing land	0.46
Marine and inland water	0.37
Built-up land	2.51

Bioproductive areas are divided into the five distinct land use types mentioned previously – cropland, forest, grazing land, fishing grounds, and built-up land – making up approximately 12 billion hectares in 2007 (about 23% of the Earth's surface). In addition, there is one category of indirect demand for biocapacity, in the form of absorptive capacity for CO_2 emissions. One Global Hectare (gha) is defined as one hectare with a productivity equal to the average productivity of the 12 billion bioproductive hectares on Earth. This means that if each land use area is multiplied by its equivalence factor, the relative area of each land use type, expressed in global hectares, differs from the distribution in actual hectares as shown in Figure 10.22 for 2007.

We can see that there are fewer global hectares of grazing land than actual hectares, while there are more global hectares of cropland than actual hectares. This is because the productivity of grazing land is much lower than the productivity of cropland. We can also see that, globally, the number of unadjusted hectares equals the number of global hectares of bioproductive space.

Calculation of the Footprint We now return to the calculation we introduced in the first part of this section. As before, the Ecological Footprint of production for a particular land use type is given by:

$$EF_p = (P/Y_N) \times YF \times EQF \tag{10.12}$$

where:

P = the amount of a product harvested or CO_2 removed
Y_N = the national average yield for P (or carbon uptake capacity)
YF = the yield factor
EQF = the equivalence factor for the particular land use.

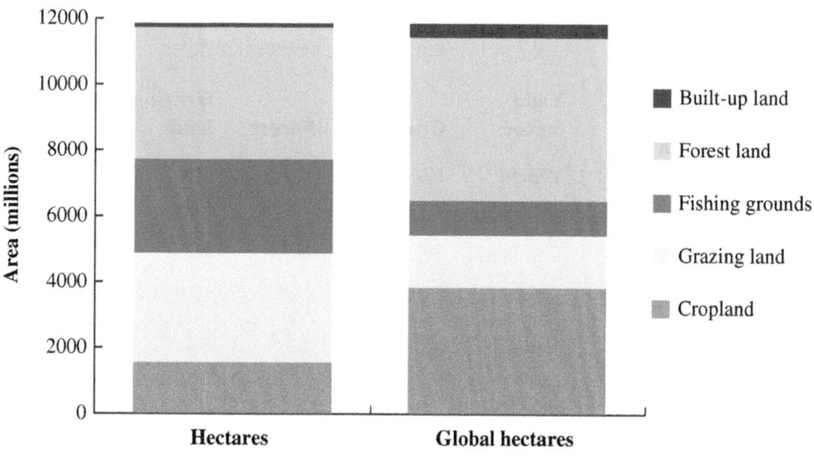

Figure 10.22 Relative area of land use types worldwide in actual hectares and in global hectares.

We now look at how P is calculated for each land use type.

Once the EF_p is calculated for each land use type, we get the EF_p for the country by summing the individual EF_ps. In other words, the primary production EF of a country is the sum of all the resources used, and all waste generated within that country's geographical borders. This includes all the areas within a country that are used to support the actual harvest of primary products (cropland, forest, grazing land , and fishing grounds), the country's infrastructure and hydropower (built-up land), and the area needed (forest) to absorb the country's CO_2 emissions (carbon footprint).

The Ecological Footprint of consumption for a given country measures the biocapacity demanded by the final consumption of all the residents of the country. All manufactured products carry with them an embodied Footprint accounting for the resources and wastes that go into their production. The accounting method takes into consideration all aspects of production and demand. For example, the national accounts distinguish between products produced within a country and products consumed by a country. The final Footprint documents actual consumption by adding imports and subtracting exports from domestic production.

As shown previously, the Ecological Footprint of consumption (EF_C) is calculated as:

$$EF_C = EF_P + EF_I - EF_E \tag{10.13}$$

where EF_I and EF_E are the footprints embodied in imported and exported commodity flows respectively.

Land Use Types in the National Accounts Each land use type is carefully delineated, so that no piece of land is counted twice in the Footprint calculation. For each type, the value of P, the amount of product harvested (or CO_2 emitted or absorbed), is calculated slightly differently. We will consider each briefly here.

Cropland is the most bioproductive of all the land use types, and comprises the area required to grow all crop products, including livestock feeds, fish-meal, oil and rubber. Globally, in 2007, there were 1.6 billion hectares designated as cropland (FAOSTAT, 2007). The EF_P of each crop type is calculated as the area of cropland that would be required to produce the harvested quantity at world average yields. This is done for 164 different crop categories.

Grazing land is defined as the area of grassland, in addition to the crop feeds used to support livestock, plus wild grasslands and prairies. 3.38 billion hectares were classified as grazing land in 2007. When calculating the Footprint of this land use type, the value of P, in this case assumed to be pasture grass, is calculated as follows:

$$P = TFR - F_{mkt} - F_{crop} - F_{res} \tag{10.14}$$

where:

TFR = the total feed requirement
F_{mkt} = the amount of feed available from general marketed crops
F_{crop} = for crops grown specifically for feed
F_{res} = crop residues.

The **Fishing grounds** Footprint is calculated using estimates of the maximum sustainable catch for a variety of fish species. These estimates are converted into an equivalent mass of primary production based on the various species' trophic levels, and this figure is then divided among the continental shelf areas of the world. Globally, there were 2.4 billion hectares of continental shelf and 433 million hectares of inland water areas in 2007. The fishing grounds Footprint is calculated based on the estimated primary production required to support the fish catch. This primary production requirement is calculated from the average trophic level of the species in question (Ewing B., 2010). The National Footprint Accounts include primary production requirement estimates for 1,439 different marine species and more than 268 freshwater species.

The **Forest land** Footprint is calculated based on the amount of lumber, pulp, timber products, and fuel wood that is consumed by a country on an annual basis. The 2007 figures show the total area of world forests at 3.9 billion hectares (FAOSTAT, 2007). It is calculated that the world average yield is 1.81 m^3 of harvestable wood product per hectare per year.

The **Carbon Footprint** is a measure of the amount of CO_2 released into the atmosphere, naturally as well as anthropogenically. It is the only land type dedicated to tracking waste products. Many different ecosystem types have the capacity to sequester CO_2. The National Footprint Accounts makes an assumption that the only land that consumes CO_2 is forest land. This use of forest land is not mutually exclusive to its use for producing wood and other forest products. The area required to absorb the CO_2 emitted by a country is calculated as:

$$A = (P_C - (1 - S_{ocean}))/Y_C \tag{10.15}$$

where:

A = the area in hectares
P_C = the annual emissions of CO_2 measured in tons
S_{ocean} = the fraction of CO_2 sequestered by the oceans in a given year
Y_C = the annual uptake of CO_2 per hectare of forest, at world average yield.

The Ecological Footprint of consumption for the Carbon Footprint is then just the adjusted area:

$$EF_C = A \times EQF \tag{10.16}$$

where A is the area and EQF is the equivalence factor defined above. CO_2 uptake land is the largest contributor to humanity's current Ecological Footprint, as shown in Figure 10.2.

The **Built-up land** Footprint is calculated based on the area of land covered by human infrastructure: transportation, housing, industrial structures, and reservoirs for hydropower. Built-up land occupied 167 million hectares of land worldwide in 2007, making up about 1.5% of the total bioproductive land.

Calculating Biocapacity As already mentioned, biocapacity is the complement to the Footprint. The Biocapacity

(*BC*) of a nation is the sum of all its bioproductive areas. Each unit of bioproductive area is transformed into global hectares (gha) by the following equation:

$$BC = A \times EQF \times YF \qquad (10.17)$$

where:
A is the area available for a given land use type
EQF and *YF* are the equivalence and yield factors, respectively.

The Biocapacity represents the maximum theoretical rate of resource supply, given prevailing technologies and management policies.

If the Ecological Footprint is greater than the Biocapacity, this is evidence of overshoot, where use exceeds natural supply. Figure 10.2 shows how the Ecological Footprint has grown since 1961. Note how Carbon Footprint is the primary cause of overshoot.

Example 10.6 Calculating total biocapacity and EF_C

Given the following data, calculate the Total Biocapacity for Germany, and the per person EF_C for Japan.

	Zambia (1)	Japan (2)	Germany (3)
Biocapacity			
Cropland	2.1	15.02	75.73
Grazing land	13.57	0.43	7.39
Forest land	11.59	43.48	53.43
Fishing grounds	0.34	9.39	6.2
Built-up land	0.21	8.01	15.73
Total biocapacity (million gha)	27.81	76.32	
EF_c			
Cropland	1.93	72.1	102.89
Grazing land	2.23	8.47	16.91
Forest land	4.29	34.99	50.08
Fishing grounds	1.03	79.54	10.8
Built-up land	0.21	8.01	222.08
Carbon footprint	1.54	399.33	15.73
Total EF_C (million gha)	11.24		418.46
Population (millions)	12.31	127.4	82.34
EF_P (per person)	0.78	3.55	4.72
EF_I	0.18	2.05	3.97
EF_E	0.05	0.87	3.6

Solution

Sum up the biocapacities of each land use type for Germany in Column (3):

$$BC_{Germany} = 75.73 + 7.39 + 53.43 + 6.2 + 15.73$$

$$BC_{Germany} = \boxed{158.48 \text{ million gha}}$$

To calculate the per person EF_C for Japan, we can either sum the EF_Cs for each land use type in Column (2), and divide by the population to get the per person value:

$$EF_{C\ Japan} = 72.1 + 8.47 + 34.99 + 79.54 + 8.01 + 399.33$$

$$EF_{C\ Japan} = 602.44 \text{ million gha}$$

$$EF_{C\ Japan}(\text{per person}) = \frac{602.44 \times 10^6}{127.4 \times 10^6} = \boxed{4.73}$$

Or, alternatively, we can use equation (10.13), since we are already given the data per person:

$$EF_C = EF_P + EF_I - EF_E$$

$$EF_C = 3.55 + 2.05 - 0.87 = \boxed{4.73}$$

To see whether Japan's EF exceeds its biocapacity, we determine the total per person biocapacity:

$$BC_{Japan}(\text{per person}) = \frac{76.32}{127.4} = \boxed{0.60}$$

Japan's overshoot is therefore:

$$4.73 - 0.60 = \boxed{4.13 \text{ gha per person}}$$

Discussion Globally, the national Ecological Footprints reveal ecological overshoot on a significant scale. However, when looked at on the national level, the sources and sites of overshoot are also evident. This should assist national governments in forming environmental policy.

The Ecological Footprint says nothing about how rapidly resource depletion is occurring, nor how long this depletion can continue before the collapse of that resource. Non-renewable resources are included in the accounts, not as depletable stocks, but only to the extent that their use damages the biosphere. And this really only applies to fossil fuels.

Resource use for which insufficient data exists is excluded from the analysis. This includes such factors as loss of biocapacity due to pollution and local impacts of fresh water use. The accounts only include the impacts of the human economy on the biosphere that the biosphere can potentially regenerate. Hazardous waste, such as that from nuclear power generation, is not accounted for in any way. Measures of societal health and environmental quality are not part of the Ecological Footprint, and need to be developed as complements to this accounting measure.

With these problems, and the fact that the accounts take an optimistic approach when there is doubt in the data, it should be obvious that the Ecological Footprint, as it is currently calculated, probably understates human demand on the environment.

10.4.2 Life cycle assessment (LCA)

This section provides a short summary of environmental life cycle assessment derived from a book on economic input-output life cycle assessment (EIO-LCA) (Hendrickson, *et al.*, 2006).

10.4.2.1 What is Life Cycle Assessment?

Life cycle assessment (LCA) studies the environmental aspects and potential impacts throughout a product's life (i.e. cradle-to-grave) from raw material acquisition through production, use, and disposal (Hendrickson *et al.*, 2006).

LCA requires careful energy and materials balances for all the stages of the life cycle, as shown in Figure 10.23. The production of a car, for example, involves the following stages:

1. The mining facilities extracting ores, coal, and other energy sources.
2. The vehicles, ships, pipelines, and other infrastructure that are used to transport the raw materials, processed materials, and subcomponents along the supply chain to the manufacturers, and transport the products to the consumer, including ships and trains transporting iron ore and oil, trucks carrying engines to automobile assembly plants and cars to dealers, and trucks transporting gasoline, oil, and tires to service stations.
3. The factories that make each of the components that go into a car, including replacement parts, and that assemble the car itself.
4. The refineries and power generation facilities that provide energy to manufacture and operate the car.
5. The facilities that handle the vehicle at the end of its life including battery recycling, shredding, and landfills for shredder waste.

It is necessary to evaluate the entire life cycle of goods and services to make informed decisions, and the results are not always apparent. For example, it has been shown that paper bags are not obviously superior to plastic bags in terms of using less energy and materials and producing less waste. Paper requires cutting trees and transporting them to a paper mill, both of which use substantial energy. Paper-making results in emissions to air and discharges to water of chlorine and biological waste. After use, the paper bag goes to a landfill, where it gradually decays, releasing methane. Plastic, in contrast, is made from petroleum, with relatively low environmental discharges. In short, it is not obvious which product is better without resorting to LCA.

10.4.2.2 The SETAC-EPA Approach to Life Cycle Assessment

Many individuals and organizations have worked to develop approaches for LCA, and several commercial software products are available (e.g. GaBi, Ecobalance, Franklin Associates). In the United States, the Society of Environmental Toxicology and Chemistry (SETAC) and the US Environmental Protection Agency (EPA) have played leading roles in standard LCA. The SETAC-EPA approach can be illustrated with the life cycle of reinforced concrete pavement, as shown in Figure 10.24.

Each of the boxes and transportation links requires a separate process model, with resource requirements and environmental impacts. In this case, there are up to 25 separate process models required, including all the transport links. The process model assessment typically consists of a detailed inventory of resource inputs and environmental outputs for the analysis period and processes considered, and each of these processes requires material and energy balances. The outputs can then be assessed for environmental harm, and design changes can be evaluated.

Figure 10.25 illustrates a unit process model for a coke oven. Coke is an essential material in the production of steel. The model here only includes the different material flows associated with coke production, which, in this case is the primary product. Coal and water are material inputs. Coke oven gas, ammonia liquor, tar, evaporated water, and purge wastewater are byproduct emissions. A more detailed analysis could include energy inputs and the

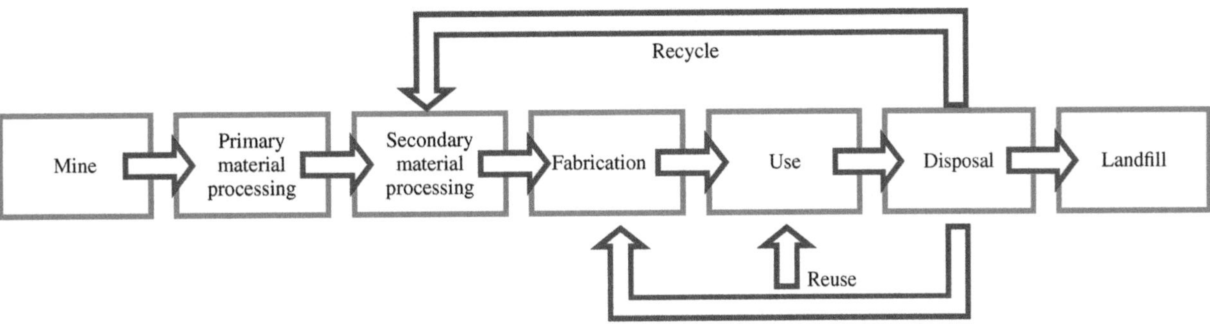

Figure 10.23 Idealized representation of some of the stages considered in life cycle assessment including raw material acquisition and processing, manufacturing, use, reuse, recycling, and disposal. *Source*: Hendrickson et *al.*, 2006. Reproduced by permission of Taylor & Francis.

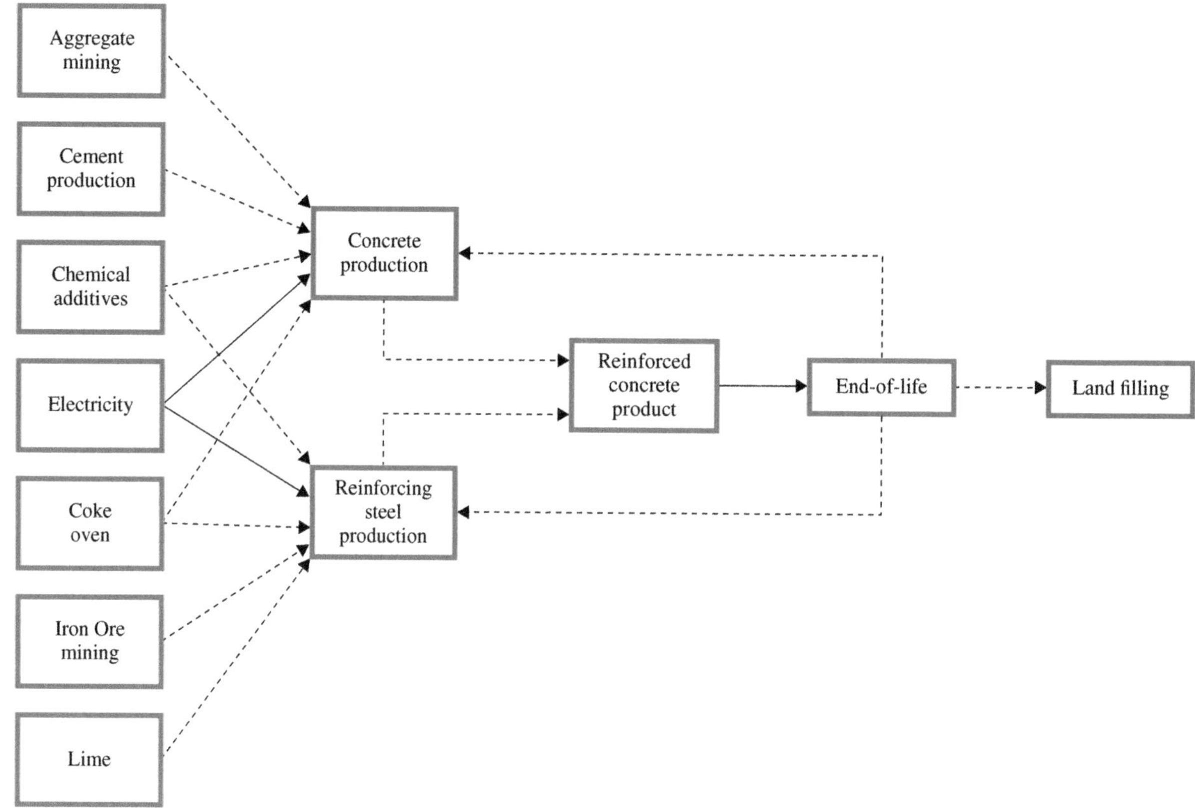

Figure 10.24 Stages in the life cycle of reinforced concrete pavement (Hendrickson *et al.*, 2006).

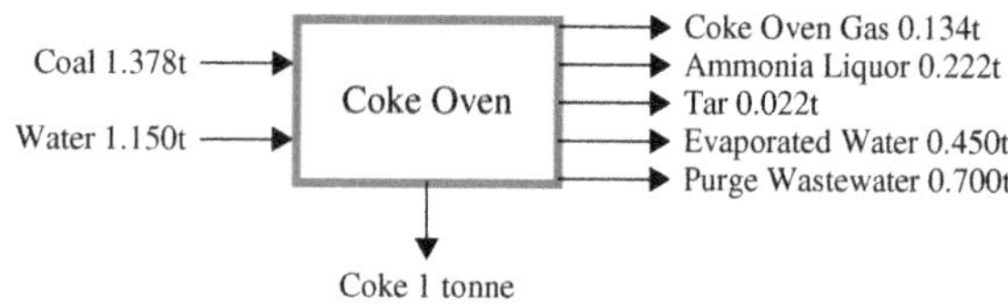

Figure 10.25 Simplified unit process model for a coke oven (Hendrickson *et al.*, 2006).

composition of gas emissions. However, to construct the full model, detailed ancillary processes for coke production should be analyzed, as shown in Figure 10.26, and the impacts from each unit process would have to been summed to get the total impact.

Thus, even for a seemingly simple unit process, such as a coke oven, considerable effort is required. Doing a complete SETAC-EPA life cycle assessment of a complicated product is therefore very time-intensive and costly. An automobile has roughly 30,000 components, involving the entire economy, and developing detailed mass and energy balances for every process is almost impossible. The SETAC-EPA approach therefore often draws a tight boundary around the processes to be investigated, analyzing only a few stages. This is the main disadvantage of the SETAC-EPA approach.

10.4.2.3 The Input-Output Approach to Life Cycle Assessment

A second approach, the economic input-output life cycle assessment (EIO-LCA) approach, takes an aggregate view of the sectors producing all of the goods and services in the economy, compared with the far more detailed SETAC-EPA process approach. EIO-LCA is based on two major assumptions:

- First, it is assumed that all production facilities that make products and provide services can be aggregated into approximately 500 sectors, such as "Sector #33610: Automobile and light truck manufacturing" and "Sector #336213: Motor home manufacturing."
- Second, the input-output flows among all these sectors are assumed to be linearly related. In other words, if 10% more output from a particular factory is needed, each of the inputs will have to increase by exactly 10%.

These two major simplifications enable the Department of Commerce to produce an "input-output" (IO) table for the US economy. Using data on required resources and environmental discharges that are appended to the IO table, researchers developed the EIO-LCA tool (available on the Internet at http://www.eiolca.net), which can be used to conduct EIO-LCA. Compared with the SETAC-EPA process approach, the EIO approach has pros and cons, as listed in Table 10.4.

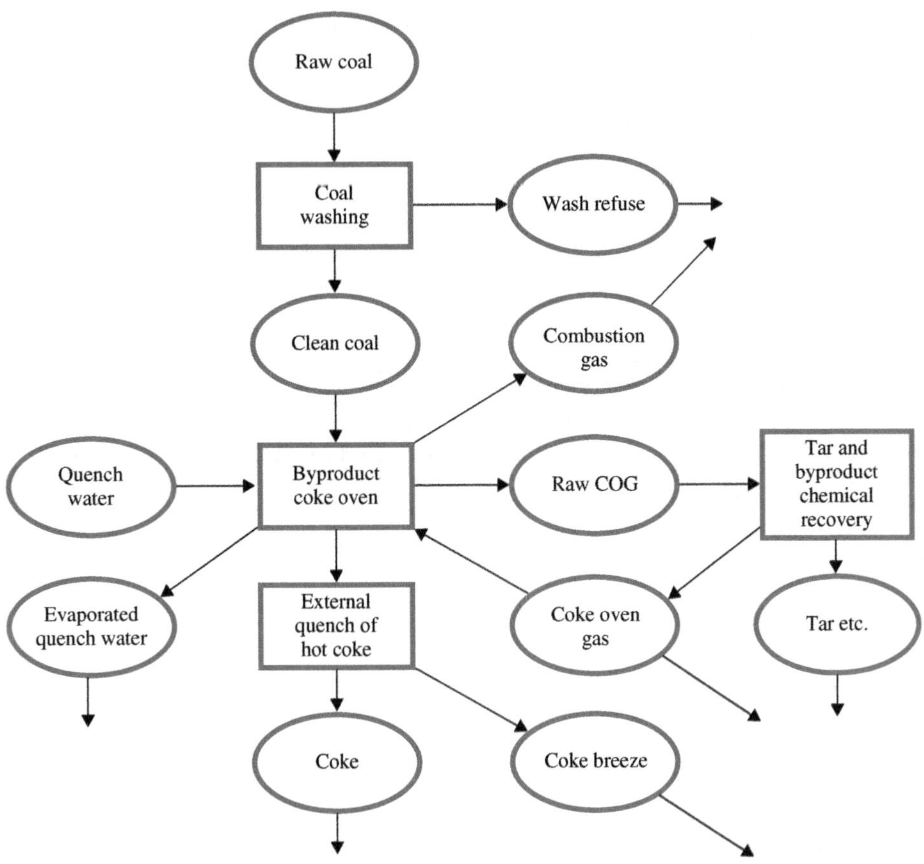

Figure 10.26 Complex unit process model for a coke oven (Hendrickson *et al.*, 2006).

Table 10.4 Comparison of the SETAC-EPA and EIO approaches to life cycle assessment.

	SETAC-EPA approach	**EIO approach**
Pros	Detailed process specific analyses. Specific product comparisons. Process improvements, weak point analyses. Future product development assessments.	Economy-wide, comprehensive assessments. System-based LCA: industries, products, services, national economy. Sensitivity analyses, scenario planning. Publicly available data, reproducible results. Future product development assessments. Information on every commodity in the economy.
Cons	System boundary setting is subjective. Tends to be time-intensive and costly. New process design difficult. Use of proprietary data. Cannot be replicated if confidential data are used Uncertainty in data.	Some product assessments contain aggregate data. Process assessments are difficult. Difficulty in linking dollar values to physical units. Economic and environmental data tend to reflect past practices. Imports treated as US products. Difficult to apply to an open economy (with substantial non-comparable imports). Non-US data availability a problem Uncertainty in data.

10.4.3 Measures of sustainability

Gross National Product (GNP[2]) and gross domestic product (GDP[3]) were created to be indicators of economic output that can be used by policy makers to monitor how the economy is performing, and for stabilizing the economy when indicators show this to be necessary. They have, over time, been co-opted by policy makers, the media, and the public to be indicators of well-being, where countries with high GDPs are thought of as rich and well-off. To truly measure well-being as well as sustainability, we need an indicator that measures more than just economic output, but one that explicitly includes measures of sustainability and well-being.

As a measure of sustainability, GDP has several shortcomings. For example, the externalities of environmental pollution and resource degradation associated with increased consumption of non-renewable resources are not considered. In these cases, GDP increases even though future growth is jeopardized. A more expensive health care system will increase GDP, even though the consequences of such a system in terms of decreased life expectancy or the elimination of disease would decrease well-being.

When trying to develop an indicator that measures both sustainability and well-being, there is another critical problem. Well-being refers to the present, while sustainability necessarily refers to the future. An indicator could better express living standards, sustainability and welfare if it incorporated factors that capture environmental, social, and economic aspects that are not included in the standard GDP measure. A number of indicators have been proposed to integrate these aspects into an indicator that could replace GDP as a measure of progress. These proposed indicators are divided into the following three categories (Stiglitz *et al.*, 2009):

- **Adjusting:** GDP is adjusted by including monetized environmental and social factors. The difficulty here is coming up with a monetary value for these factors. Examples of these indicators include Genuine Progress Indicator (GPI) and Adjusted Net Savings (ANS).
- **Replacing:** Use indicators that try to assess well-being more directly than GDP. The Human Development Index, a product of the United Nations Development Program, is used to measure a country's development. This composite index is a simple average of three indices: health and longevity; education; and living standard.
- **Supplementing:** Complement GDP with additional social and environmental information. Dashboards or sets of indicators are widely used for measuring sustainable development. In this approach, a series of indicators are gathered together with the aim of providing a comprehensive, yet manageable, indication of sustainable socio-economic progress.

In order to achieve sustainable development, human wealth, and well-being, multi-dimensional indicators are essential. The success of GDP as a measure has been its simplicity but, as a measure of sustainability or even well-being, it is obviously inadequate.

Key Words

climate change
environmental degradation
ecological footprint
economic indicators
human development
renewable energy
solar power
wind power
geothermal power
carbon sequestration
carbon footprint
life-cycle assessment
radiative forcing
greenhouse gas
greenhouse effect
sustainability

References

al-Hassan, A.Y., Hill, D.R. (1986). *Islamic Technology: An illustrated history*. Cambridge University Press, Cambridge, UK.

Archer, C.L., Jacobson, M.Z. (2005). *Evaluation of global wind power*. Retrieved from http://www.stanford.edu/group/efmh/winds/global_winds.html

Arrhenius, S. (1926). Chemistry in modern life. *Journal of the Society of Chemical Industry* **45**(29), 494.

AWEA. (2010). *US Wind Industry Annual Market Report Year Ending 2009*. American Wind Energy Association.

Betz, A. (1966). *Introduction to the Theory of Flow Machines*. Oxford: Pergamon Press.

Brundtland, G. (1987). *Our common future: The World Commission on Environment and Development*.

Carson, R. L. (1962). *Silent Spring*. Houghton Mifflin Company New York, NY.

Danish Wind Industry Association. (2003). *Roughness and Wind Shear*. Retrieved July 24, 2011 from http://www.vindselskab.dk/en/tour/wres/shear.htm.

Dodge, D.M. (2006). *Illustrated History of Wind Power Development*. Retrieved July 24, 2011 from http://telosnet.com/wind/index.html.

EERE. (2011). *Geothermal Technologies Program*. Retrieved July 24, 2011 from http://www1.eere.energy.gov/geothermal/technologies.html.

Energy Star. (2011). *CFL Purchasing Guide*. Retrieved July 24, 2011 from Energy Star: http://www.energystar.gov/index.cfm?c=cfls.pr_tips_cfls.

EPA. (2011). *Land Research*. Retrieved 2011 from EPA: http://www.epa.gov/ord/lrp/research/landfill.htm.

Eurostat. (2007). From http://epp.eurostat.ec.europa.eu/cache/ITY_OFFPUB/KS-77-07-115/EN/KS-77-07-115-EN.PDF.

EWEA. (2004). *Wind Energy – the Facts: Part I Technology*. The European Wind Energy Association.

Ewing B., D.M. (2010). *The Ecological Footprint Atlas 2010*. Global Footprint Network, Oakland: www.footprintnetwork.org.

Ewing, B.A. (2010). *Calculation Methodology for the 2010 National Footprint Accounts*. From Global Footprint Network: http://www.footprintnetwork.org/en/index.php/GFN/page/methodology/.

[2] GNP is the value of all the goods and services produced in a country, plus the value of the goods and services imported, less the value of goods and services exported.

[3] GDP is the total market value of all goods and services produced in a country equal to the total consumer, investment, and government spending, plus the value of exports, minus the value of imports.

FAOSTAT. (2007). *Food and Agriculture Organization of the United Nations (FAO) Statistical Databases*. From FAOSTAT: http://faostat.fao.org/site/291/default.aspx.

Fröhlich, C., Brusa, R.W. (1981). Solar radiation and its variation in time. *Solar Physics* **74**(1), 209–215.

GAEZ. (2000). *Global Agro-Ecological Zones*. Retrieved from: http://www.iiasa.ac.at/Research/LUC/GAEZ/index.htm

Global Footprint Network. (n.d.). *Global Footprint Network*. Retrieved October 2010, from http://www.footprintnetwork.org/.

Global Footprint Network. (2010). *Global Footprint Network*. Retrieved October, 2010 from http://www.footprintnetwork.org/.

Hendrickson, C.T., Lave, L.B., Matthews, H.S. (2006). *Environmental Life Cycle Assessment of Goods and Services: An Input-Output Approach*. Resources for the Future, Washington, DC.

Hill, D.R. (1991). Mechanical Engineering in the Medieval Near East. *Scientific American* May 1991, 64–69.

IEA. (2010). *Technology Roadmap: Solar Photovoltaic Energy*. International Energy Agency.

IPCC. (2005). *IPCC Special Report on Carbon Dioxide Capture and Storage*. Prepared by Working Group III of the Intergovernmental Panel on Climate Change Metz, B., Davidson, O. de Coninck, H.C., Loos, M., Meyer L.A. (eds.). IPCC. Cambridge: Cambridge University Press.

IPCC. (2007). *IPCC Fourth Assessment Report: Climate Change 2007*. IPCC, IPCC Geneva.

IPCC. (2007). *Mitigation of Climate Change*. United Nations.

Lund, J.W. (2007). Characteristics, development and utilization of geothermal resources. *Geo-Heat Centre Quarterly Bulletin* **28**(2), 1–9.

McKinley, G.A., Fay, A.R., Takahashi, T., Metzl, N. (2011). Convergence of atmospheric and North Atlantic carbon dioxide trends on multidecadal timescales. *Nature Geosci*, Oct 2011.

Meadows, D.H., Meadows, D.L., Randers, J., Behrens III,, W.W. (1972). *The Limits to Growth*. Universe Books, New York, NY.

Meadows, D. H., Randers, J., Meadows, D. L. (2004). *Limits to Growth: The 30-Year*. White River Junction, VT: Chelsea Green.

MIT. (2006). *The Future of Geothermal Energy, Impact of Enhanced Geothermal Systems (EGS) on the United States in the 21st Century*. Renewable Energy and Power Department, Idaho National Laboratory, Idaho Falls, ID.

Munday, P.L., Dixson, D.L., McCormick, M.I., Meekan, M., Ferrari, M.C., Chivers, D.P. (2010). Replenishment of fish populations is threatened by ocean acidification. *PNAS* **107**(29), 12930–12934.

Musgrove, P.J. (1987). Wind energy conversion: Recent progress and future prospects. *Solar & Wind Technology* **4**(1), 37–49.

NASA. (2007). *Sun: Facts & Figures*. Retrieved July 24, 2011 from http://solarsystem.nasa.gov/planets/profile.cfm?Object=Sun&Display=Facts&System=Metric

NREL. (1986a, October). Wind Energy Resource Atlas of the United States. Retrieved from http://rredc.nrel.gov/wind/pubs/atlas/.

NREL. (2010a). *New Wind Resource Maps and Wind Potential Estimates for the United States*. Retrieved July 24, 2011 from http://www.windpoweringamerica.gov/filter_detail.asp?itemid=2542

O'Grady, E. (2009). *E.ON completes world's largest wind farm in Texas*. Retrieved July 24, 2011 from http://www.reuters.com/article/2009/10/01/wind-texas-idUSN3023624320091001

Pelegrini, A.V., Harrison, D., Shackleton, J. (2007). Splitting and managing the solar spectrum for energy efficiency and daylighting. In Braganca, L., Pinheiro, M.D., Jalali, S., Mateus, R., Amoeda, R., Guedes, M.C., *Portugal SB07: Sustainable Construction, Materials and Practices – Challenge of the Industry for the New Millennium (Part I)*, p. 452. IOS Press, Amsterdam, Netherlands.

Pernick, R., Wilder, C. (2008). *Utility Solar Assessment (USA) Study*. Clean Edge Inc. and Co-op America Foundation.

Pollack, H.N., Hurter, S.J., Johnson, J.R. (1993). Heat flow from the earth's interior: Analysis of the global data set. *Review of Geophysics* **31**(3), 267–280.

Price, T.J. (2005). James Blyth – Britain's first modern wind power engineer. *Wind Engineering* **29**(3), 191–200.

Rees, W.E. (1992). Ecological footprints and appropriated carrying capacity: what urban economics leaves out. *Environment and Urbanization* **4**(2), 121–130.

REN21. (2011). *Renewables 2011 Global Status Report*. Renewable Energy Policy Network for the 21st Century.

Reuters. (2010). *S. Korea to farm Tanzania site in early 2011*. Retrieved from http://af.reuters.com/article/topNews/idAFJOE6AA08I20101111

Turcotte, D.L., Schubert, G. (2002). *Geodynamics (Second Edition)*. Cambridge University Press, New York, NY.

US DOE (2010). *DOE's Carbon Capture and Sequestration R&D Program*. Retrieved from http://www.netl.doe.gov/technologies/carbon_seq/refshelf/CCSRoadmap.pdf.

UNEP. (2009). *Towards Sustainable Production and Use of Resources: Assessing Biofuels*. United Nations Environment Programme.

US DOE. (2010). Retrieved February 2, 2011, from Carbon Capture and Sequestration R&D Program: http://www.netl.doe.gov/technologies/carbon_seq/refshelf/CCSRoadmap.pdf.

USGS. (2008). *Carbon Sequestration to mitigate Climate Change*. Retrieved from US Geological Survey: http://pubs.usgs.gov/fs/2008/3097/pdf/CarbonFS.pdf.

Wackernagel, M., Monfreda, C., Moran, D., Wermer, P., Goldfinger, S., Deumling, D., *et al.* (2005). *National Footprint and Biocapacity Accounts 2005: The underlying calculation method*. Retrieved from www.footprintnetwork.org/download.php?id=5

Wagner, H.-J., Mathur, J. (2009). *Introduction to Wind Energy Systems: Basics, Technology and Operation*. Springer.

Weart, S. (2011). *Past Climate Cycles: Ice Age Speculations*. (A.I. Physics, Producer) Retrieved June 2011, from The Discovery of Global Warming : http://www.aip.org/history/climate/cycles.htm.

Problems

1 The world average ecological footprint is about 2.7 global hectares *per capita*, while biocapacity is 2.1 global hectares per capita. Explain what this means in simple terms.

2 Identify the ultimate energy source for the following types of renewable energy: wind power, geothermal power and biofuel.

3 The Philippines, Indonesia, Japan, New Zealand, and Iceland together supply a large portion of the total geothermal electricity in the world, although they are not big with respect to land areas. Why do these countries have such large geothermal electricity generation capacity?

4 Explain why offshore wind farms are often preferable for a littoral city with a high population density.

5 Compare the pros and cons of a dry steam geothermal power plant and a dry hot rock geothermal power plant.

6 Coal and biomass both remove CO_2 from the air when they are being formed, but biomass is considered to be a carbon neutral form of energy, while coal is not. Explain why.

7 How can landfills be used to provide energy, and how does this contribute to sustainability?

8 A horizontal axis wind turbine is installed on a flat rural area with average wind speed of 8 m/s (measured at 80 m height). The tower is 30 m high, with the nacelle on the top of the tower. The airfoils are 15 m long. The electricity generator has an efficiency of 30%. Ignoring energy loss due to friction, estimate the power of this wind turbine (density of air is 1.237 kg/m^3).

9 An electricity company is planning to use the wind turbines in problem **8** to build a wind farm in a rural area. When constructing a wind farm, wind turbines cannot be packed too densely because the up-wind turbines will cast wind-shadows on the downwind ones. Experts recommend that wind turbines should not be spaced closer than five times their diameter (see figure below). At this spacing, calculate the power that the wind turbines can generate per unit land area. If wind turbines with different length of airfoils are used, how does the result change? What is your conclusion?

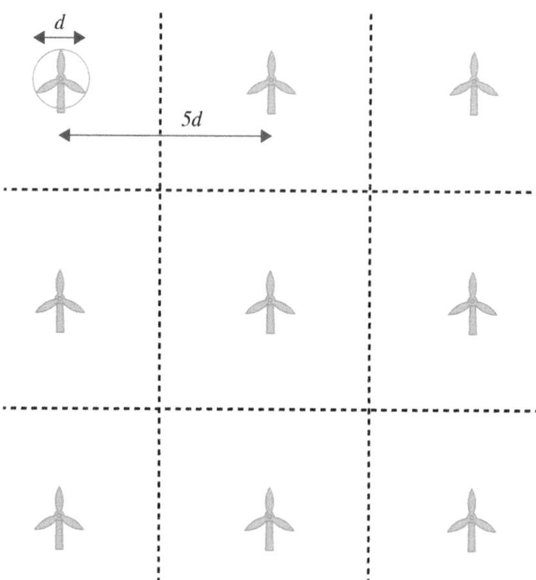

10 It is known that the solar energy received by the Earth is about 1.75×10^{17} W at the outer surface of the atmosphere. How much solar energy is received by Pluto? (diameter of the Sun is 1.39×10^9 m; effective blackbody temperature of the sun is 5,800 K; Pluto's diameter is 2,300 km; and the average distance from Pluto to the Sun is 5.866×10^{12} m). How does the solar energy received by Pluto compare to the solar energy received by the Earth?

11 A junior student drives three miles to campus every day (six miles/day considering round-trip). The fuel economy for his car is about 20 mpg (local). After learning about the importance of environmental sustainability, he decides to take the bus to campus instead of driving. Assuming this student goes to campus 160 days/year and the bus he would like to take follows the same route as that when he drives, estimate the Ecological Footprint reduction due to this change in square meters. Here are some equations and constants you may need.

a. *Fossil Energy Land Required for Bus = (10000/71000) × Distance Traveled per Year × Equivalence Factor for Different Land Types × Correction Factor for the USA.*

The unit of each term is:

Fossil Energy Land Required for Bus: square meters.
Distance Traveled per year: kilometers.

b. *Built-up Land Required for Bus = 0.02 × Distance Traveled per Year × Equivalence Factor for Different Land Types × Correction Factor for the USA.*

The unit of each term is:

Built-up Land Required for Bus: square meters
Distance Traveled per year: kilometers

c. *Fossil Energy Land Required for Gasoline = (10,000/71,000) × 35 × 1.5 × Gasoline Use per Year × Equivalence Factor for Different Land Types × Correction Factor for the USA.*

The unit of each term is:

Fossil Energy Land Required for Gasoline: square meters
Gasoline Use per Year: Liters

d. *Built-up Land Required for Gasoline = 0.6633 × Gasoline Use per Year × Equivalence Factor for Different Land Types × Correction Factor for the USA.*

The unit of each term is:

Built-up Land Required for Gasoline: square meters
Gasoline Use per Year: Liters

1 gallon = 3.785 Liters 1 mile = 1.6 kilometers

Correction factor for the USA	I) Fossil	II) Arable	III) Pasture	IV) Forest	V) Built-up	VI) Sea
Transportation	0.99	–	–	–	0.95	–

Equivalence factor for different land types					
I) Fossil	II) Arable	III) Pasture	IV) Forest	V) Built-up	VI) Sea
1.2	2.8	0.4	1.2	2.8	0.1

Chapter 11

Environmental public health

Peter Vikesland

Learning Objectives

After reading this chapter, you should be able to:

- describe toxicological pathways relevant to chemical exposures;
- define disease, epidemiology, and toxicology;
- compare and contrast acute and chronic exposures;
- describe the pertinent organ and cellular systems involved in ingestion, inhalation, dermal absorption, and placental transfer of toxic substances;
- develop dose-response relationships that can be used to define the relative toxicities of different chemicals;
- utilize potency factors and chronic daily intake values to determine the lifetime incremental risk associated with a particular exposure;
- determine LOAEL, NOAEL, and RfD values, based upon a dose-response curve for non-carcinogens;
- differentiate between disease incidence and prevalence;
- define epidemic, outbreak, and pandemic;
- utilize the epidemiology triangle to describe the dissemination of an infectious disease;
- calculate the specificity and sensitivity of a test;
- describe the differences between direct and indirect transmission of disease;
- explain vehicle-borne disease dissemination and discuss typical vehicles;
- compare and contrast common source and propagated disease epidemics;
- calculate crude mortality ratios;
- differentiate between cross-sectional, case-control, and cohort studies;
- calculate relative risks, attack rates, and odds ratios, based upon epidemiological study findings;
- compare and contrast prions, viruses, bacteria, fungi, and protozoa in terms of their sizes and morphologies;
- differentiate endotoxins from exotoxins;
- explain how the different public health measures for control of infectious disease are employed;
- understand how the 13 different factors responsible for disease dissemination collectively affect disease emergence and reemergence.

11.1 Introduction

Environmental public health is concerned with the human and ecological consequences of exposure to potentially harmful physical, thermal, chemical, or biological constituents of manmade and natural environments (Table 11.1). An **exposure** happens when proximity to or contact with one of these constituents is of sufficient duration that harmful effects may occur. Although physical and thermal dangers to humans and the environment are important, the focus of this chapter is on chemical and biological constituents, since these are of greater relevance to environmental engineering practice.

Chemical and biological constituents may be naturally occurring or anthropogenic, and may result in acute or chronic

Environmental Engineering: Principles and Practice, First Edition. Richard O. Mines, Jr.

Table 11.1 Physical, thermal, chemical, and biological constituents of natural and anthropogenic environments that may be potentially harmful.

Physical	Thermal	Chemical	Biological
Explosives	Excess heat	Pesticides	Infectious bacteria
Corrosives	Excess cold	Toxins produced by biological organisms	Infectious virus
Oxidants		Pharmaceuticals	Infectious protozoa
		Disinfection by-products	Infectious fungi
			Infectious helminths

exposures. Depending upon the identity of the contaminant and the mechanism by which an organism is exposed to it, the exposure may elicit a biological response characteristic of a particular disease. As shown in Figure 11.1, chemical and biological exposures can occur via ingestion (eating and/or drinking), inhalation (breathing), dermal absorption (through the skin), as well as placental transfer *in utero*. As discussed in section 11.2, the route a particular chemical takes is highly important, since it can alter the severity of the exposure.

For the purposes of this chapter, we define **disease** as any condition that impairs the normal function of an organism. To comprehend diseases and the mechanisms by which they are spread, it is necessary to understand two complementary fields:

1. **Toxicology** – the study of the adverse effects of biological and chemical agents on living organisms.
2. **Epidemiology** – the investigative methodology used to study the nature, cause, control, and determinants of disease transmission and distribution.

These two fields work together, with toxicology focused on developing dose-response relationships and studying mechanisms of action following exposure to toxic substances, while epidemiology is concerned with the occurrence and etiology of disease.

11.2 Toxicology

Paracelsus, a Swiss chemist and physician in the late 1400s and early 1500s, is widely attributed to have stated: "All things are poison and nothing is without poison; only the dose makes a thing not a poison." This relatively straightforward statement suggests that every substance, whether it be table salt (NaCl) or sodium cyanide (NaCN), can be considered a poison if it is administered in a dose sufficient to produce a toxic effect. For something relatively innocuous, such as table salt, the lethal dose necessary to kill 50% of the adult humans exposed to it (defined as the $\mathbf{LD_{50}}$) is approximately 2,400 mg/kg, while for

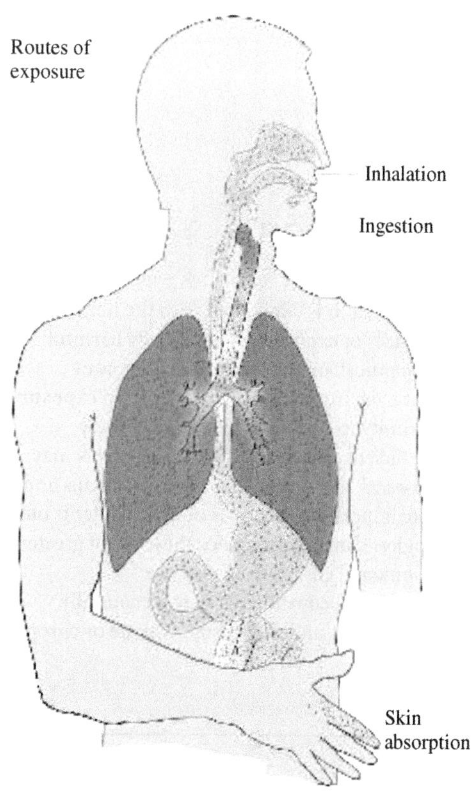

Figure 11.1 Potential exposure routes for biological and chemical contaminants.

Table 11.2 Historic acute and chronic poisoning events.

Acute poisonings	Chronic poisoning
Hiroshima and Nagasaki, Japan. Acute radiation poisoning following two atomic bombings killed > 150,000 within one day.	Long-term exposure to disinfection by-products produced by reactions between disinfectants and organic or inorganic water constituents.
Bhopal, India. Methyl isocyanate leak at a chemical plant resulted in as many as 18,000 deaths within a two-week period.	Radon gas exposures due to natural sources in the ground
	Fukushima, Japan. Radiation exposure following tsunami.
	Love Canal, Niagara Falls, New York, USA. Long term exposure to numerous chemicals dumped in the canal; first case concerning hazardous waste disposal/long term effects.

more highly toxic sodium cyanide, the LD_{50} is much lower at 10 mg/kg. For chemicals such as table salt and sodium cyanide, comparisons of LD_{50} values or other toxicological endpoints provide a straightforward mechanism to relate the potential hazards associated with a particular material.

Toxicology is dedicated to the quantitative evaluation of the lethal and sub-lethal effects of exposure to biological and chemical agents. In environmental systems, we generally deal with **chronic** low-dose exposures over extended periods of time. However, **acute** high-dose exposures can occur as a result of industrial accidents and work-place activities. Examples of historic chronic and acute environmental exposures are provided in Table 11.2. Acute responses occur within a short period of time after a single exposure to a biological or chemical agent, while chronic responses occur as a result of prolonged exposure over months or years.

A **poison** is any biological or chemical agent that can elicit a deleterious response in a biological system. The degree to which something is considered poisonous is defined in terms of its **toxicity**. This is the quantitative determination of how much of a substance is required to produce a particular deleterious response. Comparison of the LD_{50} values for table salt and sodium cyanide, as discussed previously, illustrates that sodium cyanide is more deadly than table salt, but that ingestion of sufficient quantities of table salt can nonetheless result in death. As discussed in the paragraphs that follow, endpoints other than death are often considered when evaluating the relative toxicities of substances.

A **toxic substance** has toxic properties and can be a pure chemical (e.g., lead) or a mixture of chemicals (e.g., gasoline). A **toxicant** is a toxic substance produced by anthropogenic activities, while a **toxin** is a toxic substance produced by living organisms (e.g., reptiles, insects, plants, or microorganisms).

The relative toxicity of a toxic substance is generally related to its chemical and physical properties, which, in turn, dictate how the substance interacts with an organism and its cells.

Toxicologists classify toxic substances using a number of hierarchies that are based upon chemical class, the exposure route, or the organ systems affected by a particular substance. Classifications based upon chemical identity are the simplest to produce, since they categorize chemicals with similar chemical functionalities into single groups. This type of classification scheme, however, provides limited information about the mechanism by which a particular chemical elicits a toxic effect.

An alternative approach is to classify toxic substances based upon their exposure route. In this approach, chemicals whose route of exposure is primarily via inhalation are grouped together, while those whose routes of exposure are via ingestion or dermal uptake are grouped appropriately. Unfortunately, classifications based upon the exposure route can group highly dissimilar chemicals together and, thus, can obscure differences in their mechanisms of toxicity.

The third classification system, and the one generally preferred by toxicologists, categorizes toxic substances based upon their target organs. Using this system, a chemical that, say, targets the liver is defined as a **hepatotoxin** and, as listed in Table 11.3, substances that target other organs can be defined similarly. Whichever classification system is used, the intent is to group toxic substances together in a manner that helps to explain observed biological responses.

11.2.1 Exposure

Exposures to toxic substances occur via ingestion (eating and/or drinking), inhalation (breathing), dermal absorption (through the skin), as well as placental transfer *in utero*. The route by which exposure occurs is important, since it can affect the toxicity of the substance. Once a toxic substance enters the body, it is subject to absorption into the bloodstream and transport to the body's different organ systems. Depending upon the chemical nature of the substance, it can be stored internally in lipids, eliminated from the body via excretion, or transformed into a different material. Metabolic breakdown of toxic substances generally produces metabolites with lower toxicity than the starting material; however, in some cases, more toxic byproducts are produced.

11.2.1.1 Ingestion

The human gastrointestinal tract extends from the mouth to the anus and incorporates the mouth, pharynx, esophagus, stomach, small intestine, large intestine, and anus (Figure 11.2). The digestive system is specifically designed to extract nutrients from food and liquids while moving them along the digestive tract. Because of its role in the extraction of nutrients, the digestive system is a major route for the absorption of toxic substances found in food and water.

The majority of nutrient and toxic substance absorption occurs in the stomach and upper portions of the small intestine. Within the small intestine, 0.5–1.0 mm long finger-like **villi** extend from the wall into the central cavity. As a result of their high surface densities (>40 per mm^2), these villi have tremendous absorptive capacity. Each villus contains a capillary network, an arteriole, and a lymphatic vessel (Figure 11.2).

Table 11.3 Organ systems affected by toxic substances.

Classification	Target organ		Example toxic substances
hepatotoxins	liver	Transport of toxic substances by the bloodstream enables the liver to be directly damaged. As the function of the liver is to metabolize materials, it is potentially subject to attack not only by the substance itself, but also by toxic metabolites produced by the liver.	Carbon tetrachloride, chloroform, trichloroethylene, DDT, heavy metals
nephrotoxins	kidney	The primary function of the kidneys is to filter blood, such that wastes are removed and ultimately excreted in urine. Toxic substances that attack the kidneys can alter the flow of urine and cause severe poisoning, due to the buildup of the body's waste products.	Heavy metals, chlorinated hydrocarbons
neurotoxins	nervous system	Neurotoxins act on nerve cells and typically do so by interacting with and affecting membrane proteins.	Animal venoms, saxitoxin, botulinum toxin
hematotoxins	blood system	Some hematotoxins alter the ability of the body to form the platelets necessary for blood clotting. Other hematotoxins affect the ability of blood to transport oxygen.	Carbon monoxide, nitrate

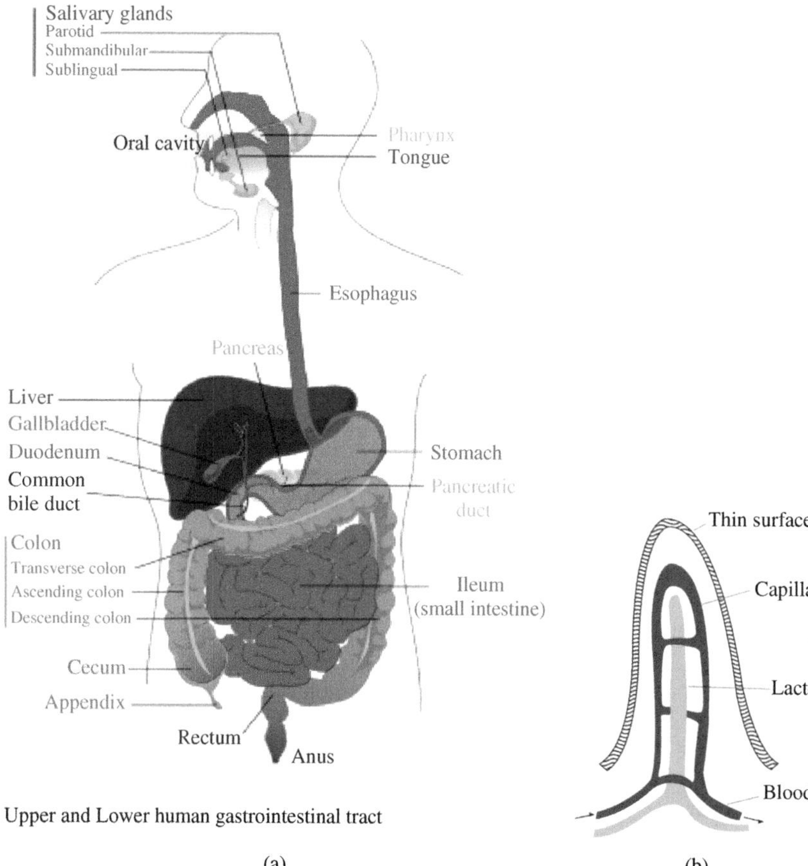

Figure 11.2 (a) Human gastrointestinal tract. *Source*: Reproduced by kind permission of LadyofHats (Mariana Ruiz Villarreal), edited by Joaquim Alves Gaspar. (b) Microvillus. *Source*: http://en.wikipedia.org/wiki/Intestinal_villus Accessed November 2013.

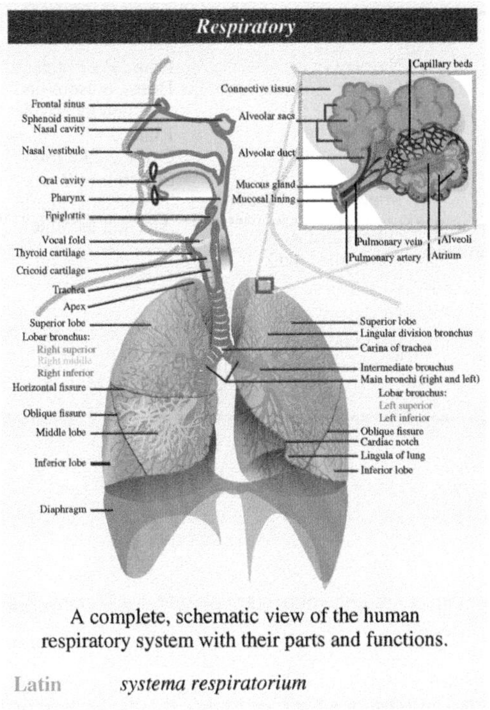

Figure 11.3 Human Respiratory tract. *Source*: Reproduced by kind permission of LadyofHats (Mariana Ruiz Villarreal).

Both nutrients and toxic substances penetrate through the epithelial cells of the villus and enter the blood and lymph vessels.

11.2.1.2 Inhalation

As shown in Figure 11.3, the human respiratory system consists of three regions: **nasopharyngeal** (nose, pharynx, larynx), **tracheobronchial** (trachea, bronchi), and **pulmonary** (alveolar sacs). The respiratory system supplies oxygen to the body, while simultaneously expelling carbon dioxide.

Air transported through the respiratory system is subjected to processing, such that any gaseous and particulate pollutants present are generally removed before the air reaches the alveoli, where air exchange occurs. Within the nasopharyngeal and tracheobronchial regions, the respiratory system walls are covered with a constantly moving mucous membrane. This constant movement is due to the collective action of cilia that beat at a rate of over 1,300 times per minute. Any materials that deposit within the mucous membrane are trapped, and are eventually either coughed up or swallowed, whereupon they are subject to digestion. In contrast, materials that reach the alveolar sacs are often retained. These retained materials can potentially be absorbed into the bloodstream.

11.2.1.3 Dermal Absorption

Dermal absorption is the transport of chemicals from the outer surface of the skin, into the skin, and ultimately into the bloodstream. The skin is the largest organ in the human body, with a total area of ~3000 in^2 that is differentiated into a number of interconnected tissues. Skin is a permeable barrier that regulates body temperature, prevents water loss, facilitates defense against microbial and viral invasion, and enables excretion of salts, water, and organic compounds from the body.

Skin consists of three layers: **epidermis, dermis**, and **subcutaneous** (Figure 11.4). The thinner, outer epidermis is attached to the thicker, underlying dermis. Below these two layers is a subcutaneous layer of glandular and adipose tissues. Because skin contains significant quantities of lipids, it is highly effective at excluding water and water-soluble materials. However, some of these materials can be absorbed through hair follicles and sweat glands. In contrast, hydrophobic (fat-soluble) materials such as petroleum products and chlorinated organic solvents are readily absorbed by the skin and can be transferred to blood vessels present in the dermis, or to the underlying subcutaneous layer.

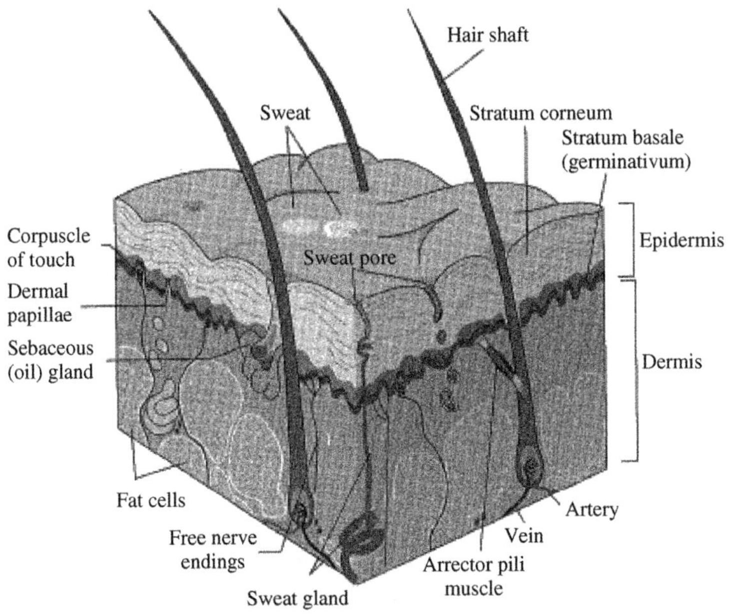

Hair shaft
Sweat
Stratum corneum
Stratum basale (germinativum)
Corpuscle of touch
Sweat pore
Epidermis
Dermal papillae
Sebaceous (oil) gland
Dermis
Fat cells
Artery
Free nerve endings
Vein
Arrector pili muscle
Sweat gland

Figure 11.4 Cross-sectional diagram of human skin. *Source*: G.S. Moore, Living with the Earth, 3rd ed. Reproduced by permission of Taylor & Francis.

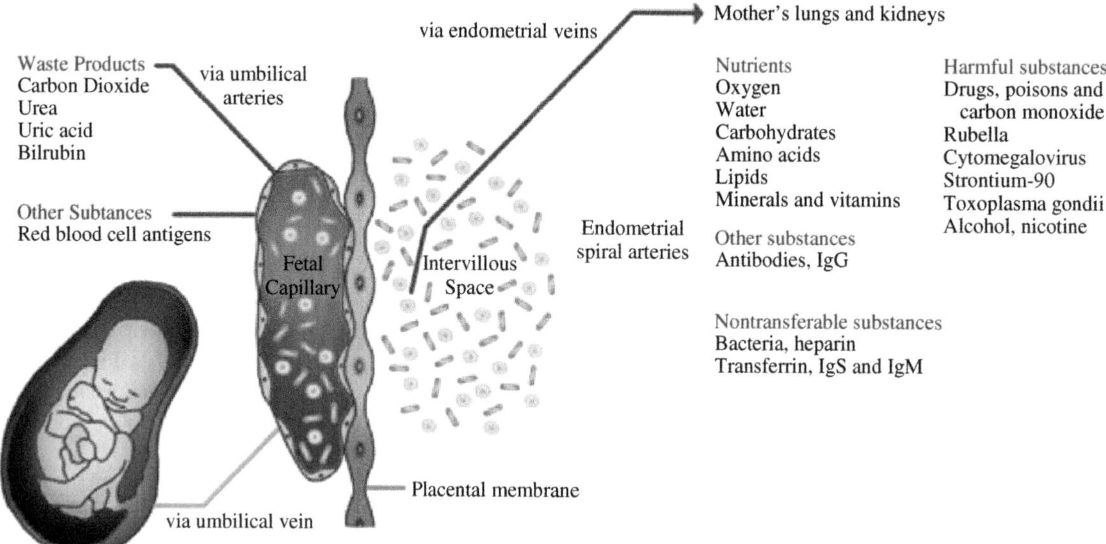

Figure 11.5 Placental transfer. *Source*: Mahan, L. Kathleen; Escott-Stump, Sylvia, editors. Krause's Food, nutrition, and diet therapy. 10th ed. Philadelphia: W.B. Saunders Company, 2000. p. 181. Reproduced by permission of Elsevier.

11.2.1.4 Placental Transfer

The potential *in utero* exposure of fetuses to environmental chemical pollutants is a particularly challenging and politically sensitive area of research. Although it is well established that chemicals present within a pregnant mother's bloodstream can cross the placenta (a complex tissue attached to the wall of the mother's womb and to the baby through the umbilical cord) and enter the bloodstream of the fetus (Figure 11.5), it is often challenging to assess the ultimate consequences of this *in utero* exposure.

However, two historic cases, one involving the tranquilizer thalidomide and the other involving the anthropogenic estrogen diethylstilbestrol (or DES), clearly indicate that *in utero* exposures can result in detrimental effects. In the late 1950s, thalidomide was often prescribed as a tranquilizer for pregnant women suffering from nausea. Unfortunately, many of the women given this drug gave birth to babies with birth defects, ranging from heart and organ malformations, to brain damage, to limb deformations. *In utero* exposure to thalidomide resulted in acute health effects, with obvious repercussions. In contrast, *in utero* exposure to DES did not typically result in immediate birth defects, but instead resulted in enhanced cancers and sterility in the DES-exposed children. These longer-term effects were much more difficult to detect, but similarly showed that *in utero* chemical exposures can have negative consequences.

11.2.2 Dose and Response

To quantify the effects of a toxic substance, a toxicologist develops a **dose-response curve** that correlates a measured biological response to the chemical dose required to elicit that response. **Response** is the biological consequence of exposure to a toxic substance. For acute exposures, the response may be organ injury, coma, or death, while chronic exposures typically result in mutagenic or carcinogenic responses. **Dose** refers to the actual amount of a toxic substance that is received by a target organism. In the laboratory, dose is often well defined. However, in environmental systems, it can be challenging to determine the actual received dose and thus, as shown in Example 11.1, environmental concentrations are generally used to define the dose.

Example 11.1 Calculating dose

If the industrial solvent trichloroethylene (TCE) is present in drinking water at a concentration of 2 ppb (μg/L), then what is the dose that a typical male adult ingests during a normal day?

Solution

When calculating doses it is assumed that an adult consumes 2 liters of water per day. Given this assumption the calculated dose is:

$$\text{Dose} = 2\ \mu\text{g/L} \times 2\ \text{L/day} = 4\ \mu\text{g/day}$$

A second assumption often made when calculating weight normalized doses is that an adult male weighs 70 kg. We use this assumed weight to normalize the dose on a weight basis:

$$\text{Weight Normalized Dose} = \frac{4\ \mu\text{g/day}}{70\ \text{kg}}$$
$$= 0.057\ \mu\text{g/day/kg}$$

Normalization of the dose by weight is done to account for weight variations between subjects. For example, if a weight of 50 kg were assumed, the calculated weight normalized dose would be higher and would be 0.080 μg/day/kg. This final result illustrates an important point that the smaller a person's body weight the greater the dose.

11.2.2.1 Measurable Responses

The interaction between an organism and a toxic substance results in a biological consequence. As described in Table 11.4, there are four principal mechanisms by which a biological consequence occurs:

1. Disruption or destruction of cellular structures.
2. Direct chemical combination with a cellular constituent.
3. Effects on enzyme activity.
4. Secondary actions that occur as a result of the presence of a pollutant.

Depending on which type of adverse effect is observed, toxic substances are placed into two general categories: non-carcinogens and carcinogens.

Mutagens are toxic substances that elicit alterations to DNA. As introduced in Chapter 4, **DNA** (deoxyribonucleic acid) is an essential component of living things, and it provides the genetic code that determines the overall characteristics of an organism. DNA is a unique biomolecule, in that it has the capacity to replicate itself precisely, such that the genetic code of an organism is readily transmitted to new cells. Mutagens, unfortunately, alter DNA such that the genetic code is no longer precisely retained or transmitted. These changes can cause cell death, cancer, reproductive failures, or birth defects in offspring.

Carcinogens are a specific class of mutagen that causes cancer. Carcinogens promote the growth of tumors by attacking or altering the DNA retained within a cell. Tumors, which consist of cells growing in an uncontrolled or abnormal manner, can be classified as either **benign** or **malignant**, depending upon whether or not they are contained within their own boundaries. If a tumor is not self-contained, it has undergone **metastasis** and is considered malignant. Malignant tumors are considerably harder to treat or remove, since they often have spread throughout the body.

Classification of a toxic substance as either a mutagen or as a carcinogen is complicated by the fact that some chemicals can elicit both mutagenic and carcinogenic activity. For example, arsenic is classified by the US EPA as both a carcinogen and as a mutagen. This dual classification is a result of arsenic's capacity to cause lung and internal (liver, kidney, lung, bladder) cancers when inhaled or ingested, while simultaneously causing a variety of non-carcinogenic skin lesions. The classification of a particular substance as either a mutagen or as a carcinogen is further complicated by the need to conduct dose-response assessments using animal models. For obvious ethical reasons, prospective toxicity testing is generally conducted using animals other than humans (Table 11.5), and these studies rely upon the underlying assumption that the effect of a particular

Table 11.5 Animal models used for toxicity testing.

Animal Model	Approximate number of animals tested per year
Rats	10–20 million
Mice	10–20 million
Guinea pigs	221,000
Rabbits	246,000
Gerbils (hamsters)	177,000
Amphibians	(no info)
Fish	(no info)
Cats	23,000
Dogs	67,000
Non-human primates (chimpanzees, monkeys, etc.)	58,000

Source: http://www.thefullwiki.org/Animal_testing and http://www.aphis.usda.gov/animal_welfare/downloads/awreports/awreport2005.

Table 11.4 Potential biological consequences of exposure to a toxic substance.

	Description	Examples
1) Disruption/destruction of cellular structure.	A toxic substance directly interacts with a cell and causes structural damage.	Acids, bases, and phenols directly combine with cellular materials and kill tissues (necrosis).
2) Direct chemical combination with a cellular constituent.	A toxic substance combines with cell constituents and forms a complex that can lead to impaired function.	Carbon monoxide in blood binds to hemoglobin and inhibits its proper function.
3) Effect on enzymes.	Toxic substances interact with enzymes and prevent their proper function.	Organophosphate inhibition of acetylcholinesterase.
4) Secondary actions as a result of the presence of a pollutant.	The presence of a "harmless" pollutant causes the release of other chemicals that are injurious to cells.	Immune system responses result in production of reactive oxygen species (hydroxyl radicals, superoxide) following inhalational dust exposures.

chemical on humans can be inferred from its effects on test animals. In some cases, however, the existing animal models do not mimic human responses adequately. Such is the case for inorganic arsenic, and it is one of many reasons why efforts to establish the current limit of 10 μg/L for arsenic in drinking water were highly controversial.

The US EPA has defined five categories (A through E) to describe how likely a given chemical is a carcinogen. These categories are as follows:

A. *Human carcinogen.* There is sufficient evidence from epidemiological studies to establish a causal association between exposure to the toxic substance and cancer.

B. *Probable human carcinogen.* Category B1: there is some limited evidence of carcinogenicity to humans. Category B2: there is a combination of inadequate human data but ample evidence of carcinogenicity in animals.

C. *Possible human carcinogen.* There is limited evidence of carcinogenicity in animals and an absence of information regarding humans

D. *Not classified.* There is inadequate human and animal evidence of carcinogenicity.

E. *Evidence of noncarcinogenicity.* No evidence of carcinogenicity for humans, based upon a lack of evidence in at least two animal tests on different species or through adequate epidemiologic and animal studies.

The classification of a toxic substance as either a mutagen or as a carcinogen is important for the development of regulations that set the maximum allowed concentrations of these substances in air, water, food, and other media.

11.2.2.2 Dose-Response Testing

The central reason to conduct dose-response testing is to identify the nature of health damage caused by a substance and the range of doses over which such damage occurs. An example dose-response curve is illustrated in Figure 11.6. In this figure, the dose is represented on the *x*-axis in terms of the mass of contaminant dose per mass of test organism (body weight).

Generally, units of mg/kg are used, since normalization by body weight allows calculation of doses for individuals of varying weight and also allows for extrapolation between doses used in animal tests and those humans are expected to receive. The measured response at each dose is then plotted on the *y*-axis and is typically presented in terms of cumulative percent response. When the dose is represented on a log-scale, such curves often exhibit a sigmoid shape with a flat portion at low dosages. This flat (or **subthreshold**) region indicates that, at low dosages, an increase in dosage elicits no measurable effect. Once the threshold is exceeded, however, there is a sharp increase in the measured response until, at some point, the maximal response is reached and the curve flattens out. The **threshold** is defined as the lowest dose at which a particular response is measurable. As discussed below, carcinogens are typically assumed to exhibit no threshold. In other words, any exposure to a carcinogen is expected to increase the risk of getting cancer.

The simplest response to measure in toxicology tests is mortality and, as explained previously, the LD_{50} is the median lethal dose that results in death in 50% of the test organism population. Other lethal doses are sometimes also used (e.g.,

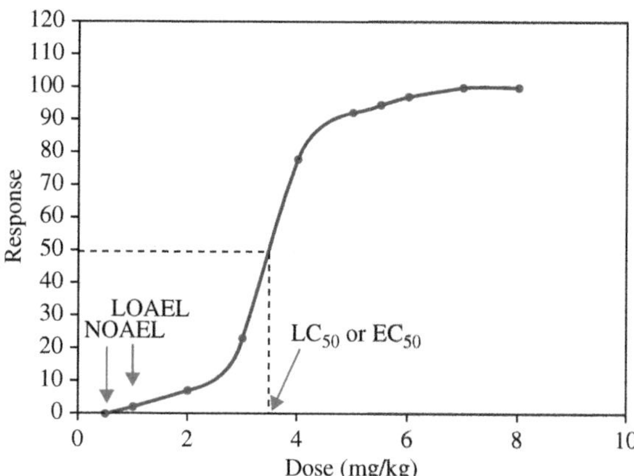

Figure 11.6 Example dose-response curve that correlates a biological response to the dose required to achieve that response. Dose is plotted on the *x*-axis in units of mass of chemical dosed divided by the organism mass. Response is presented on a percentage basis. The NOAEL and the LOAEL, as defined for a non-carcinogen, are illustrated. *Source:* P. Vikesland.

LD_{10} and LD_{90} representing 10% and 90% mortality, respectively). Determination of lethal doses is of value when considering acute, high-concentration exposures. Many environmental situations, however, involve chronic prolonged exposures to lower concentrations. For these chronic exposures, toxicological endpoints other than death are of interest, and they are represented in terms of $\mathbf{EC_{50}}$ values (the **effective concentration** that affects 50% of the population) or other effective concentrations (i.e., EC_{10}, EC_{90}).

Determination of LD and EC values reveals that not all individuals exposed in the same way to the same dose of a substance respond similarly. Some individuals are severely affected by the toxic substance at a particular dosage, while others are completely unaffected. This observation highlights how important it is to consider the potential effects of chemical exposures on susceptible populations. Often, there are fractions of a given population that are highly susceptible to the effects of chemical exposure. Potentially susceptible populations include the elderly, small children, fetuses, and immunocompromised individuals. Full evaluation of the toxicity of a given chemical requires consideration of its toxicity to susceptible populations.

To develop dose-response curves quantitatively, it is necessary to conduct a series of experiments over a range of doses, using a large number of test organisms. As noted previously, for ethical reasons toxicity testing is generally not performed on humans, and animal models are employed instead. Rats and mice are the most commonly used animal models, due to their relatively low cost and short 2–3 year lifespan. However, as indicated in Table 11.5, a variety of other animal models are also used.

One problem faced by the toxicology field is that each animal species may respond differently to a given potential toxicant. Thus, the EC_{50} value determined in a study with mice may differ considerably from that determined using a rat model, and each of these values may not be truly representative of the EC_{50} for humans. This latter point illustrates one of the major challenges of toxicology, in that no single animal species exactly duplicates the responses observed in humans. For this reason, toxicology studies are often conducted using multiple

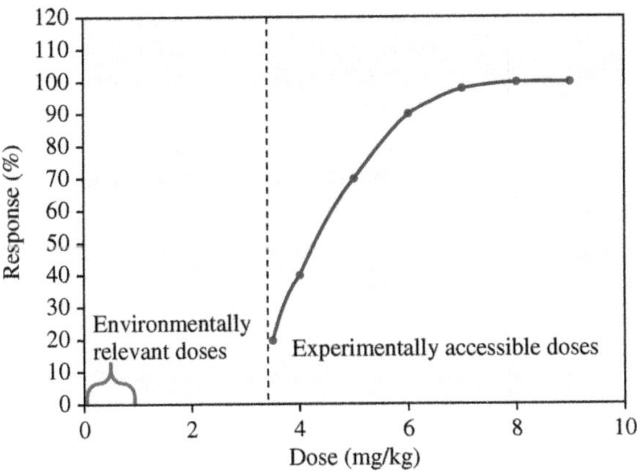

Figure 11.7 Example dose-response curve, illustrating the data gap that often exists between the experimentally accessible doses used for dose-response testing and the environmentally relevant doses. For carcinogens, a linear extrapolation is made between the lowest experimental dose and the origin. For non-carcinogens, it is assumed that a threshold dose exists below, which there is no response. *Source:* P. Vikesland.

types of organisms, and general trends associated with toxic substance exposure are determined and interpreted with respect to their implications to humans.

An additional challenge associated with the development of dose-response tests is that the doses employed in animal studies are generally much higher than the actual doses that would be encountered in normal environmental settings. For this reason, extrapolation from the high-dose region to the low-dose region is often required (Figure 11.7). Such extrapolation is not without its challenges, however, since a number of assumptions must be made as to the shape of the dose-response curve in the low-dose region.

Dose-response testing is often undertaken in order to rank chemicals according to their relative toxicities. Such rankings can be obtained simply by comparing the relative LD_{50} values (or EC_{50} values if we are dealing with a non-lethal endpoint). However, as illustrated in Figure 11.8, such simplistic

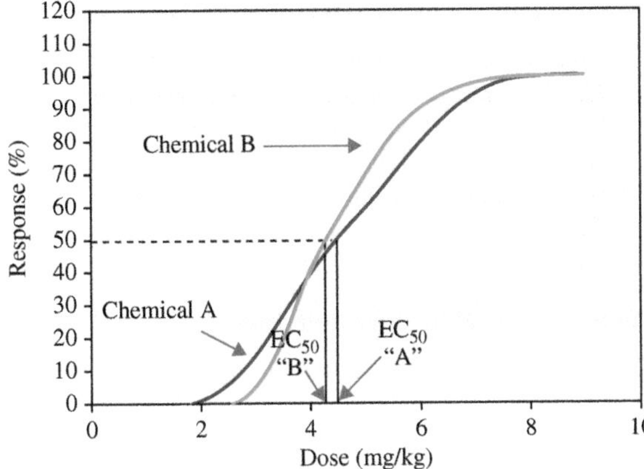

Figure 11.8 Example illustrating how simple comparisons of LC_{50} or EC_{50} values can hide low-dose or high-dose effects. Chemical B has a higher EC_{50}, but elicits a decreased response at low dosages relative to Chemical A. *Source:* P. Vikesland.

comparisons often hide low or high-dose effects, and therefore detailed collection of dose-response data is required to truly determine the relative toxicities of compounds. Example 11.2 describes the steps required to develop a dose-response curve.

Example 11.2 Dose-response assessment

Atrazine, a systemic herbicide that blocks photosynthesis is one of the most widely used agricultural pesticides in the USA. Approximately 64–75 million pounds of active ingredient are applied per year. Because of its widespread use, atrazine is one of the most commonly detected pesticides in streams, rivers, groundwater, and reservoirs, and it is the subject of multiple monitoring programs. Atrazine's frequent detection results from its volume of usage and its tendency to persist in soils and move with water. Its persistence provides the opportunity for oral exposures to atrazine via drinking water.

Based on toxicological studies, US EPA has set the maximum contaminant level (MCL) for atrazine in drinking water at 0.003 mg/L. These values are, however, currently being re-examined, and new regulatory limits may be established in the near future. As part of the re-evaluation of the standards, a series of dose-response assessments have recently been conducted. You are to evaluate the data collected from these assessments.

Due to concerns about the potential for atrazine to act as an endocrine disruptor, a study was conducted to assess the effect of atrazine on the onset of puberty in Wistar rats. The results from this study are tabulated below:

Rat #	Atrazine dose (mg/kg/day)					
	0	25	50	100	150	200
	Days until onset of puberty					
1	40	40	44	47	50	55
2	42	41	45	48	51	53
3	40	43	43	49	49	57
4	40	41	41	46	50	49
5	39	40	45	47	52	56
6	40	37	44	48	54	55
7	42	40	42	50	48	54
8	40	41	41	46	47	57
9	40	40	43	47	51	58
10	40	42	43	47	52	55
11	41	39	47	49	50	54
12	40	40	42	48	49	56
13	40	41	46	41	48	55
14	42	41	45	49	50	59
15	40	40	43	48	51	54

Using this data, develop a dose-response curve illustrating the effect of atrazine on the average onset of puberty. Determine the NOAEL, LOAEL, and RfD.

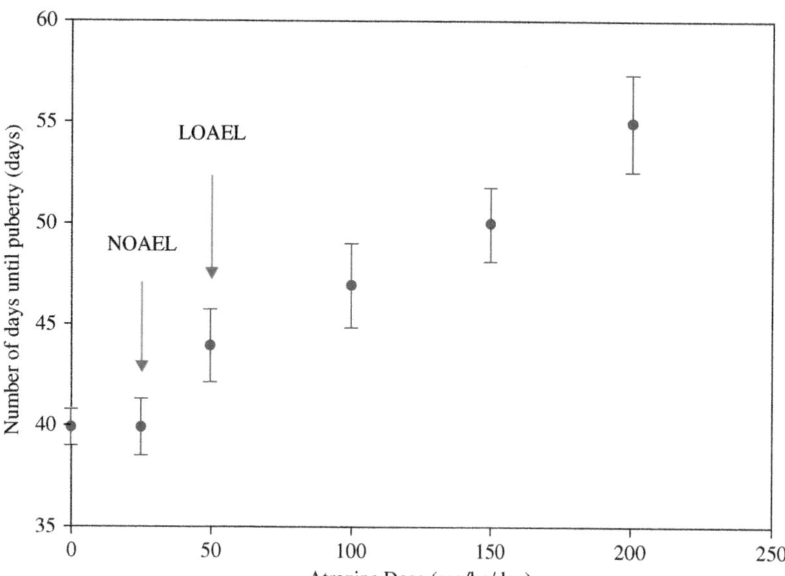

Figure E11.1 Dose-Response curve comparing the measured number of days to reach puberty relative to the atrazine dose. The NOAEL and LOAEL (see Example 11.5) are illustrated. The error bars reflect the standard deviation of each measurement.

Solution

To develop the dose-response curve for this dataset, we will first determine the average number of days until the onset of puberty for each atrazine dose. To do this, average the results for the fifteen rats used for each dose:

	Atrazine dose (mg/kg/day)					
	0	25	50	100	150	200
	Average days until onset of puberty					
Average	40	40	44	47	50	55
Standard deviation	0.9	1.4	1.8	2.1	1.8	2.4

The final step is to plot these averages versus the dose on an X-Y plot.

11.2.2.3 Dose Response Testing – Carcinogens

For carcinogens, a linear-extrapolation in the low-dose region is used since it is assumed that exposure to any amount of a carcinogen will increase an individual's likelihood of cancer. This linear extrapolation results from fits to the collected dose-response data using what is known as a linearized multistage model. This conservative model overemphasizes risk and thus is highly protective of human health. The slope of the dose-response curve in the low-dose region is defined as the **potency factor** (PF) or **slope factor** (SF) for a particular carcinogen.

$$PF = \frac{\text{Incremental lifetime cancer risk}}{\text{Chronic daily intake}} \quad (11.1)$$

The numerator defines the lifetime cancer risk and the denominator is the lifetime weight averaged dose or **chronic daily intake** (CDI). The CDI typically has units of mg/kg-day and is defined as follows:

$$CDI = \frac{\text{Average daily dose (mg/day)}}{\text{Body weight (kg)}} \quad (11.2)$$

In Equation (11.2), the average daily dose is determined based upon daily intakes of the contaminant in air, water, food, soil, or other matrices, frequency of exposure to these matrices, and estimated body weights. In addition, the exposure is then averaged over an assumed 75-year lifetime. US EPA-recommended values for the daily intake parameters, an individual frequency of exposure to a given matrix, and body weights are found in Table 11.6.

Calculations of CDIs for ingestion and inhalation of trichloroethylene from contaminated water are provided in Example 11.3.

Example 11.3 Estimating the CDI

The industrial solvent trichloroethylene (TCE) is present in drinking water at a concentration of 2 μg/L (ppb). What is the CDI for ingestion of this water by an adult (>21 years old)? Assuming that the average concentration of TCE in the air in the shower is 10 μg/m^3, what is the CDI for an adult (>21 years old) who showers in this water for a period of 15 minutes daily?

Solution – CDI for ingested dose

To calculate the CDI for the ingested dose, it is necessary to use the information detailed in Table 11.6. Upon reviewing that table, it is apparent that the following should be used:

- TCE concentration: 2 μg/L
- Water intake: 1.04 L/day

Table 11.6 US EPA recommended exposure factors.

Exposure pathway	Daily Intake	Exposure Frequency (days/year)	Exposure Duration (years)	Body Weight (kg)
Ingestion of potable water	1.04 L (adult > 21 years) 0.33 L (child 3–6 years)	350	30	80 (adult > 21 years) 18.6 (child 3–6 years)
Ingestion of soil and dust	50 mg (adult > 21 years) 100 mg (child 1–6 years)	350	24 5	80 (adult > 21 years) 18.6 (child 3–6 years)
Inhalation of contaminants	21.2 m^3 (adult >21 years, <61) 18.1 m^3 (adult 61–70 years) 10.1 m^3 (child 3–6 years)	350	30	80 (adult > 21 years) 18.6 (child 3–6 years)

Source: United States Environmental Protection Agency, 2011.

- Exposure frequency: 350 days/year
- Exposure duration: 30 years
- Assumed lifetime: 75 years
- Body weight 80 kg

$$\mathrm{CDI} = \frac{\dfrac{\left(2\frac{\mu g}{L}\right)\left(1.04\frac{L}{d}\right)\left(350\frac{d}{y}\right)(30\ \mathrm{yr})}{(75\ \text{year lifetime})}}{(80\ \mathrm{kg})} \times \left(\frac{\mathrm{mg}}{1000\ \mu\mathrm{g}}\right) = 0.0036\ \mathrm{mg/(kg \cdot d)}$$

In the calculation of the CDI, the 30-year exposure duration is used to approximate the length of time an individual resides at a residence (90th percentile).

Solution – CDI for inhaled dose

Similar to the solution for the ingested dose, the following parameters from Table 11.6 are used:

- TCE concentration in the air: 10 μg/m^3
- Daily air intake: 21.2 m^3/day
- Percentage of a day in shower: 1.04% (= 15 minutes/1440 minutes)
- Exposure frequency: 350 days/year
- Exposure duration: 30 years
- Assumed lifetime: 75 years
- Body weight 80 kg

$$\mathrm{CDI} = \frac{\dfrac{\left(10\frac{\mu g}{m^3}\right)\left(21.2\frac{m^3}{d}\right)(0.0104)\left(350\frac{d}{yr}\right)(30\ \mathrm{yr})}{(75\ \text{year lifetime})}}{(80\ \mathrm{kg})} \times \left(\frac{\mathrm{mg}}{1000\ \mu\mathrm{g}}\right) = 0.0039\ \mathrm{mg/(kg \cdot d)}$$

Based upon these calculations, it is apparent that the exposures via ingestion and via inhalation are similar with the inhalational exposure slightly larger. CDI values for additional exposure pathways can also be calculated as appropriate.

Extensive dose-response testing has enabled determination of potency factors for a number of different chemicals. Table 11.7 provides a summary list of the potency factors for some common environmental contaminants. One important inference to make upon examination of this table is that the higher the PF, the more carcinogenic a compound is. A second inference to make from this table is that potency factors can vary significantly for oral versus inhalational exposure to the same chemical. This latter effect often reflects divergent toxicological endpoints for different exposure routes.

It is possible to use tabulated potency factors along with calculated values for the CDI to estimate the incremental

Table 11.7 Potency factors for carcinogens.

Chemical	Oral potency factor[a] (kg-day/mg)	Inhalational potency factor[b] (kg-day/mg)
Arsenic	1.5	15.1
Benzene	0.015	0.029
Cadmium	–	6.3
Chloroform	0.0061	0.08
DDT	0.34	0.34
Methylene chloride	0.0075	0.00164
2,3,7,8 TCDD (a dioxin)	1.6×10^5	1.16×10^5
Trichloroethylene	0.046	0.006

[a] US EPA, IRIS database, www.epa.gov/iris.
[b] Davis and Masten, *Principles of Environmental Engineering and Science*, 2nd Edition. US EPA no longer reports Inhalational Potency Factors on IRIS.

lifetime cancer risk associated with toxic substance exposure via any route simply by rearranging Equation (11.1):

$$\text{Incremental Lifetime Cancer Risk} = \text{PF} \times \text{CDI} \qquad (11.3)$$

Comparison of the calculated incremental lifetime cancer risk with societally acceptable levels of risk is done to establish regulations for chemical contaminants. The US EPA sets regulations for water, air, and food quality based upon an incremental cancer risk that is between 10^{-6} and 10^{-4} (between 1 in a million and 1 in ten thousand). An example calculation illustrating how incremental lifetime cancer risks are calculated and used is given in Example 11.4.

Example 11.4 Estimating incremental lifetime cancer risks

What is the incremental lifetime cancer risk for the adult (>21 years old) in Example 11.3?

Solution

As listed in Table 11.7, the oral potency factor for TCE is 0.046 kg-day/mg. Using the oral CDI value calculated in Example 11.3, we can determine the overall risk associated with ingestion of this water:

$$\textit{Incremental}\ \text{Lifetime Cancer Risk}$$
$$= \text{PF} \times \text{CDI} = (0.046\ \text{kg} \cdot \text{d/mg}) \times (0.0036\ \text{mg/(kg} \cdot \text{d))}$$
$$= 1.67 \times 10^{-4}$$

As listed in Table 11.7, the inhalational potency factor for TCE is 0.006 kg-day/mg. Using the inhalational CDI value calculated in Example 11.3, we can determine the overall risk associated with ingestion of this water:

$$\textit{Incremental}\ \text{Lifetime Cancer Risk}$$
$$= \text{PF} \times \text{CDI} = (0.006\ \text{kg} \cdot \text{d/mg}) \times (0.0039\ \text{mg/(kg} \cdot \text{d))}$$
$$= 2.32 \times 10^{-5}$$

In each case, the incremental lifetime cancer risk (which is unitless) is greater than 10^{-6}, which is the risk level often used by the US EPA when setting regulations. To determine the overall risk associated with exposure to this TCE contaminated water, we can simply sum the two numbers calculated above. This is valid, since the total exposure risk is simply the additive sum of the risks associated with different exposure pathways. Accordingly,

$$\textit{Total}\ \text{Exposure Risk} = \sum risk_j$$
$$= (1.67 \times 10^{-4}) + (2.32 \times 10^{-5})$$
$$= 1.91 \times 10^{-4}$$

where the subscript j reflects each of the different exposure pathways.

11.2.2.4 Dose-Response Testing – Non-carcinogens

For non-carcinogens, a threshold dose, below which no toxic effects are observed, is typically assumed. Unfortunately, because of the high costs associated with dose-response assessment, the actual threshold dose is often challenging to accurately determine. However, as shown in Figure 11.6, the toxicologist usually can determine the following parameters when describing dose-response curves for non-carcinogens:

- **NOAEL** – No observed adverse effect level. In other words, the highest administered dose at which no adverse effects were measured. This dose often approximates the threshold dose below which no effects are observed.
- **LOAEL** – Lowest observed adverse effect level. In other words, the lowest administered dose at which adverse effects were measured.

The NOAEL can then be used to determine what is known as the **reference dose (RfD)**. The RfD is simply calculated by dividing the NOAEL (or sometimes the LOAEL) by an appropriate uncertainty factor. The magnitude of the uncertainty factor varies from 10–1000×, with a 10× uncertainty factor accounting for differences in sensitivity between individuals in a population (e.g., pregnant women, children, ill, elderly versus "normal" people), an additional factor of 10× to account for the extrapolation from animal data to humans, and a final factor of 10× when no human data is available at all and the animal data is limited in scope. Specific guidelines for uncertainty factors are listed in Table 11.8.

Determination of the NOAEL, LOAEL, and RfD is illustrated in Example 11.5.

Table 11.8 Guidelines for uncertainty factors.

Uncertainty factor	Guideline
1–10	When a NOAEL from a human study is used (account for interspecies diversity).
100	When a LOAEL from a human study is used, incorporating a factor of 10 to account for lack of NOAEL and a factor of 10 for intraspecies diversity; or when a NOAEL from an animal study is used, incorporating a factor of 10 to account for interspecies diversity and a factor of 10 for intraspecies diversity.
1000	When a LOAEL from an animal study is used; incorporates factors of 10 each for lack of NOEAL, interspecies diversity, intraspecies diversity.
1–10	Additional uncertainty factors, ranging from 1 to 10, may be incorporated on a case-by-case basis to account for database deficiencies.

Example 11.5 Determination of NOAEL, LOAEL, and RfD

Solution

As shown in Figure E11.1, the data indicates that, at a dose of zero mg/(kg · d) atrazine, the average days until the onset of puberty is 40. At the lowest administered dose of 25 mg/(kg · d), the average days until the onset of puberty remains at 40. Only at a dose of 50 mg/(kg · d) is there a statistically significant difference in the average days until the onset of puberty, relative to the no-dose situation. The value of **50 mg/(kg · d)** can thus be defined as the lowest observed adverse effect level (LOAEL), since it is the lowest dose at which an adverse effect is measured. The no observed adverse effect level (NOAEL) can subsequently be defined as **25 mg/(kg · d)**, since this is the highest dose at which no effect above the baseline was observed.

To determine the RfD, we simply divide the value determined for the NOAEL by an uncertainty factor. For this situation, an uncertainty factor of 100 is appropriate, since this value considers intraspecies diversity (i.e., diversity in the human population) as well as the need to extrapolate the results from an animal study to humans. Accordingly, the RfD will be **0.25 mg/(kg · d)**.

Similar to what was done with carcinogens, it is possible to evaluate the hazard associated with exposure to non-carcinogens. This process is done through the calculation of **Hazard Quotients** (HQ) that relate the average daily dose of a contaminant to the RfD:

$$\text{HQ} = \frac{\text{Average daily dose over period of exposure}}{\text{RfD}} \tag{11.4}$$

An important distinction must be made between Equation (11.4) and Equation (11.2). In Equation (11.4) the average daily dose is averaged *only over the exposure period*, while in Equation (11.2), the average daily dose is the total dose averaged *over an assumed 75-year lifetime*. This distinction is made due to the different mechanisms by which carcinogens and non-carcinogens cause disease. As defined by Equation (11.4), if the hazard quotient for any particular exposure is less than 1.0, there is little risk of toxicity. In contrast, hazard quotient values that exceed 1.0 suggest there is a potential risk.

11.3 Epidemiology

Epidemiology is the methodology used to detect the cause or source of diseases that produce pain, injury, illness, disability, or death in human populations or groups. Epidemiology characterizes the factors determining the frequency and distribution of disease and death in human populations, and is one of the foundations of public health and preventative medicine.

An epidemiologist characterizes the time, place, and person aspects of disease by examining the risk factors that impact, influence, provoke, and affect disease distribution in a population. By studying these factors, an epidemiologist is able to determine the needs of disease control programs, develop preventative programs and health services planning activities, and establish patterns of endemic diseases, epidemics, and pandemics. Full definitions for each of these terms are found throughout section 11.3.

11.3.1 Incidence, prevalence, and epidemics

Incidence is the extent that people within a population who do not have a disease will develop the disease during a specific time period. In other words, incidence defines the number of new cases of a disease in a population over a specific period of time. A related term is **prevalence**, which is the number of people within a population who have a certain disease at a given point in time. Comparing these two terms, it is apparent that incidence tells us about the occurrence of *new* cases, while prevalence defines the *ongoing level* of disease in a population at any point in time.

The typical prevalence of a disease within a given population or geographic area is described as the **endemic** level of disease. Depending on environmental and social conditions, some regions exhibit higher levels of endemic disease. As an example, although it is endemic in both regions, malaria generally exhibits a higher endemicity in Sub-Saharan Africa than it does in South America.

Within any particular region, some portions of the population may exhibit **hyperendemic** levels of disease. These higher levels of prevalence are typically associated with distinct populations, such as might be found in a hospital, nursing home, or other institution (e.g., bacterial meningitis amongst college students, *Staphylococcus aureus* infections amongst hospital patients).

A **holoendemic** disease is one that is highly prevalent and is commonly acquired early in life. Prior to the advent of widespread immunization in the USA, chickenpox was considered a holoendemic disease.

Disease **epidemics** occur when there is a sharp increase in the number of new cases of a disease. As shown in Figure 11.9, high levels of incidence indicate that an epidemic is at hand. Epidemics, or **outbreaks**, can arise within local populations either due to contamination by a single source or via person-to-person propagation through a community. Widespread epidemics that traverse national borders and affect multiple countries, continents, and the globe are known as **pandemics**. AIDS, cholera, and tuberculosis pandemics afflict the world in the early part of the 21st century.

11.3.2 Epidemiology triangle

Four interrelated factors contribute to outbreaks of infectious disease:

1 the host;
2 an agent or disease-causing organism;

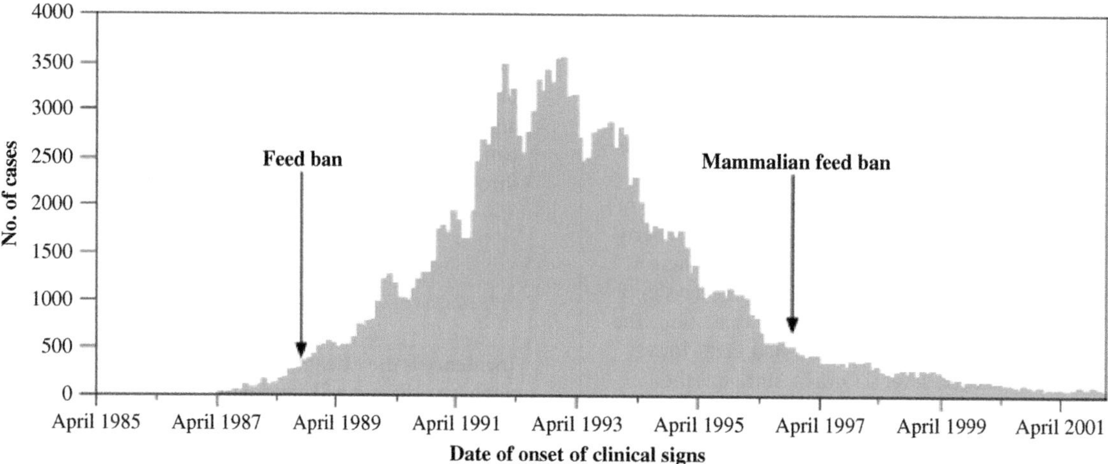

Figure 11.9 Example illustrating the epidemic curve for bovine spongiform encephalopathy (BSE) in Great Britain. High numbers of cases indicate an epidemic is occurring. *Source*: Fig. 1 from: http://www.scielosp.org/scielo.php?pid=S0042-96862003000200009&script=sci_arttext.

3 the environmental circumstances needed for a disease to thrive, survive, and spread; and
4 time.

The inter-relationship between these factors is often described in terms of the **epidemiology triangle** (Figure 11.10).

A host is an organism (typically a human or an animal) that harbors a disease agent but may not be afflicted by it. The host offers sustenance and lodging for the disease agent that is the cause of a particular disease. Bacteria, viruses, parasites, fungus, and molds are agents of infectious disease (an infection is the growth of microorganisms within a host), while natural/anthropogenic chemicals or radiation cause other types of diseases.

The interaction between the host and the agent is affected by the external environment and, in the case of infectious disease, the internal environment of the host, the reservoir, and the vector that harbor the disease. Favorable environments enable infectious agents to thrive, multiply, and facilitate disease transmission. Although internal and external biological environments are often focused upon, other environmental variables, such as the social, cultural, and physical environments can also be important.

Time is the overarching variable that accounts for incubation periods, the life expectancy of the host or the agent,

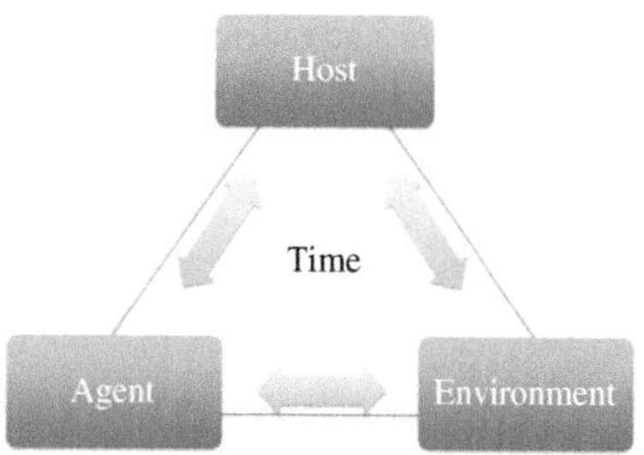

Figure 11.10 Epidemiological triangle. *Source*: P. Vikesland.

and the duration of the course of the illness. In this latter regard, time accounts for progression of the disease, which is often divided into five stages (Figure 11.11):

1 Infection.
2 Incubation period.
3 Acute period.
4 Decline period.
5 Convalescent period.

The exact time when a disease agent interacts with a susceptible host can dictate whether or not a disease is spread.

The epidemiology triangle is a construct that enables an epidemiologist to analyze the interrelatedness of each of the four factors in the epidemiology of disease. In particular, the triangle is used to determine the *influence, reactivity*, and *effect* of each factor on the other three. One of the primary goals of public health is to control and prevent disease and, to do so, the epidemiologist seeks to break one of the legs of the triangle in such a manner that disease transmission is prevented. One role of the environmental engineer is to develop public health interventions that enable this to occur.

11.3.3 Infectious disease transmission concepts

A disease **reservoir** is an animate or inanimate medium in which infectious organisms live and multiply. Potential reservoirs include humans, animals, plants, soils, and inanimate organic matter (feces or food). Infectious organisms reproduce in the reservoir in a manner that facilitates disease transmission to a susceptible host. A living reservoir is generally referred to as a **host**.

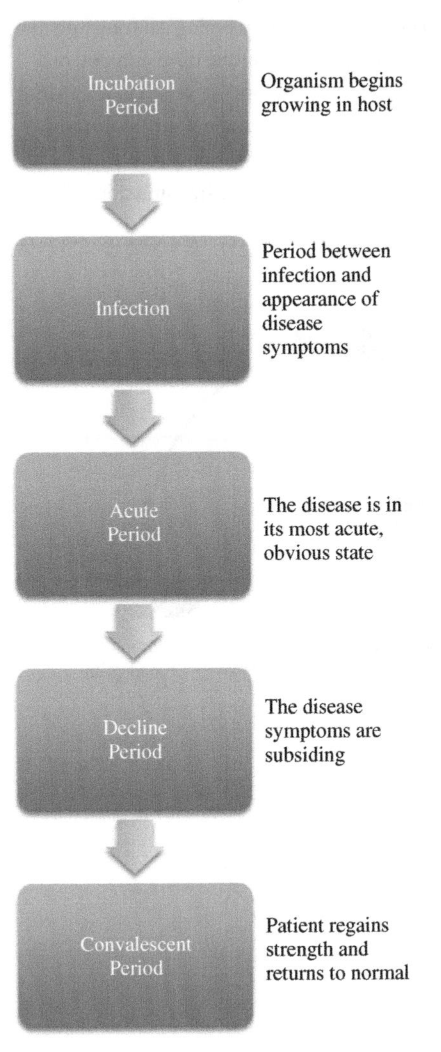

Figure 11.11 The five stages of disease progression. *Source:* P. Vikesland.

A **carrier** contains, spreads, or harbors infectious organisms, but does not show obvious signs of clinical disease. Carriers serve as a potential source of infection and disease transmission to others (both humans and animals). **Active** carriers are individuals who have been exposed to and harbor a disease-causing organism even though they have recovered from the disease. **Convalescent** carriers are individuals who have been exposed to and harbor a disease-causing organism and are in the recovery phase of the disease, but are still infectious. A **healthy** (or **passive**) carrier is an individual who has been exposed to and harbors a disease-causing organism but who has not become ill, nor has shown any of the symptoms of the disease. An **incubatory** carrier is an individual who has been exposed to and harbors a disease-causing organism, is in the beginning stages of the disease, is showing symptoms, and has the ability to transmit the disease. An **intermittent** carrier is an individual who has been exposed to and harbors a disease-causing organism and who can intermittently spread disease.

Infectious diseases can be transmitted either directly (organism-to-organism) or indirectly (organism to intermediate to organism). **Direct transmission** is the immediate transfer of the agent from a host/reservoir to a susceptible host via physical contact (skin-to-skin contact, kissing, sexual intercourse). In contrast, **indirect transmission** involves an intermediate vector or vehicle that facilitates disease agent transfer from an infected organism to a susceptible host, resulting in disease. Some infectious diseases are solely the provenance of humankind and are referred to as **anthroponotic** diseases; diseases that afflict both humans and animals are described as **zoonotic** diseases. The four different disease transmission cycles are schematically illustrated in Figure 11.12.

A **vector** is any living nonhuman carrier of disease that transports and serves the process of disease transmission. Typically vectors are insects (fly, flea, mosquito) or small animals (mouse, rat, other rodents, birds). A vector spreads an infectious agent from an infected human or animal to other susceptible humans or animals through its waste products, bite, or bodily fluids, or indirectly through food contamination.

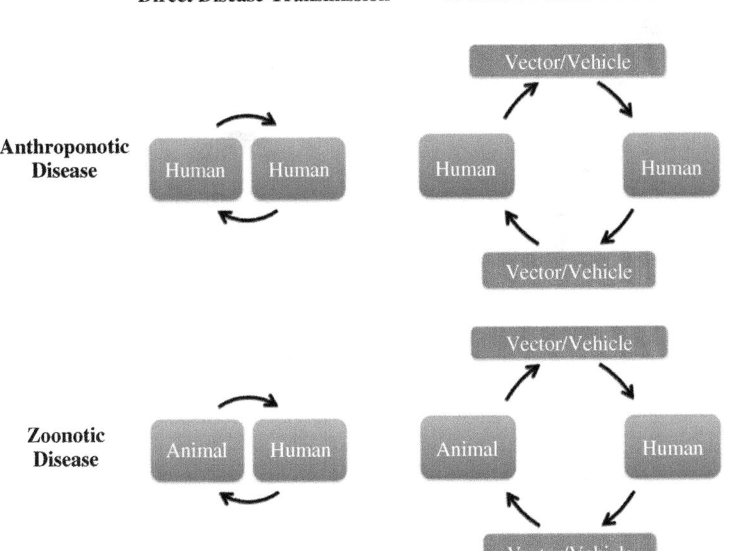

Figure 11.12 Four different cycles of infectious disease transmission. *Source:* P. Vikesland.

Vector-borne diseases can be subclassified as either occurring via **mechanical transmission** or via **biological transmission**. In the former case, the vector simply serves as a mechanism for the physical transfer of a disease agent that soils its feet or proboscis. Mechanical transmission does not require multiplication of the agent or biological development of the agent within the vector. In contrast, in biological transmission, the agent undergoes multiplication, cyclic development, or a combination of these processes, before the vector can transmit the infective form of the agent to a human. One important aspect of biological transmission is that it involves an incubation period, during which disease transmission

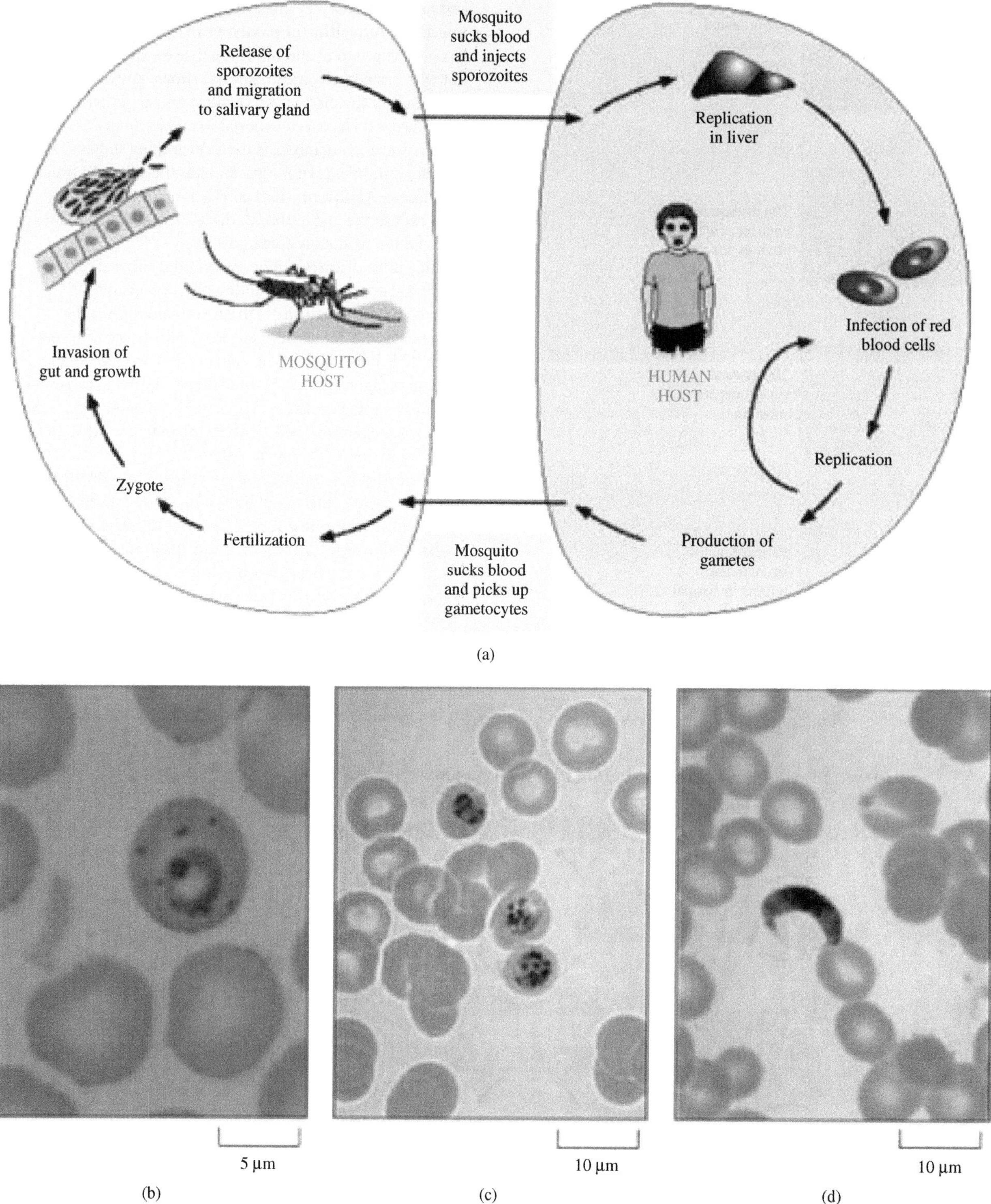

Figure 11.13 Complex life cycle of the malaria parasite. *Source*: Centers for Disease Control Division of Parasitic Diseases DPdx- from Introduction to Pathogens.

cannot occur. Malaria is a biologically transmitted disease in which the spread of the protozoan pathogen requires that part of its lifecycle occur within the female *Anopheles* mosquito (Figure 11.13).

A **vehicle** is a non-living carrier of disease. Air, water, and food are common vehicles for disease. **Airborne** diseases are spread when aerosol droplets or particles of dust carry the pathogen to the host and enable its transmission. Airborne diseases can be spread by sneezing, coughing, or via the simple act of talking. An alternative mechanism for airborne disease transmission is when droplets are carried through a building's heating or air conditioning ducts.

An infamous example of the latter was the outbreak of *Legionellosis* that occurred during an American Legion convention in Philadelphia, PA in 1976. In this outbreak, the gram negative bacterium *Legionella pneumophila* had colonized the air conditioning system within a high-rise hotel. Approximately 221 people attending the conference became ill with pneumonia-like symptoms, and 34 ultimately died as a result of this exposure.

Waterborne diseases are spread when a pathogen is present in drinking water, swimming pools, streams, or lakes used for swimming or other types of body contact. Globally, contaminated water is one of the most important sources of infectious disease. Waterborne diseases are often spread due to the contamination of water with fecal material (fecal-oral transmission). Many of the microorganisms responsible for waterborne disease grow within the intestine and leave the body in feces. If the receiving water is insufficiently treated prior to use, disease transmission can readily occur.

Foodborne diseases are spread when a pathogen or a toxic compound produced by a pathogen is present in food that is consumed. Foodborne diseases are classified in one of two categories: **food poisonings** and **food infections**. Food poisoning (sometimes referred to as food intoxication) can occur following ingestion of foods containing preformed microbial toxins. In many cases, the microorganisms that produced these toxins do not grow in the host and may not even be viable when the contaminated food is ingested. In contrast, food infection is the consumption of food that contains a sufficient dose of viable pathogens to result in infection and disease in a susceptible host. **Fomites** are inanimate objects that serve as vehicles. Fomites can be pencils, pens, drinking glasses, doorknobs, faucets, clothing, or any inanimate object that conveys infection by being contaminated with disease-causing organisms and then comes in contact with another person.

In concluding this section, it is important to highlight that many **vehicleborne** diseases are not vehicle-specific. In other words, pathogenic *Escherichia coli* can be spread potentially by food, water, aerosols, or fomites. Determination of the exact vehicle responsible for disease agent dissemination is important from an epidemiologic perspective to establish the factors leading to disease outbreaks, but the organisms responsible for the disease are generally not specific to the mechanism by which they are spread.

11.3.4 Disease epidemics

Epidemics are often classified either as **common source** or **propagated** (host-to-host). In a common source epidemic, exposure of a disease to a group of persons arises from a single source that all persons in the group had a chance to encounter. Examples of common source epidemics are food and water contamination events that occur when there is a breakdown in the sanitation processes used to protect public health. In many cases, common source diseases are those that are spread by fecal-oral transmission, wherein fecal material contaminates a food or water supply. The disease spreads when a susceptible individual ingests or otherwise uses the contaminated food or water.

Propagated epidemics occur via the direct or indirect transmission of a communicable disease from one individual to another within a susceptible population. In a propagated epidemic, there is a slow, progressive rise in the number of cases of disease, followed by a slow, progressive decline. A propagated epidemic can occur following introduction of a single infected individual into a population of people susceptible to that disease. Individuals in a communicable stage transmit disease to one or more people, and the disease then replicates in the susceptible populations. Many common infectious diseases are readily spread by host-to-host contact. Table 11.9 lists a number of diseases spread via common sources or propagated by host-to-host transmission.

As shown in Figure 11.14, common source and propagated epidemics can be differentiated from one another by the relative shapes of their epidemic curves. A common source epidemic exhibits a rapid rise in the epidemic curve, followed by a rapid fall that is generally attributable to discovery of the common source and its sanitation or removal. In contrast, propagated epidemics are generally slower to rise and slower to fall, since they are passed between members of the community.

11.3.5 Mortality and morbidity

The incidence and prevalence of disease within a community, country, or region is determined by evaluating the statistics of illness and death. By evaluating these statistics, it is possible to develop a picture of the health of a given population.

Mortality is the incidence of death in a population and can be quantitatively defined as the ratio of the number of deaths

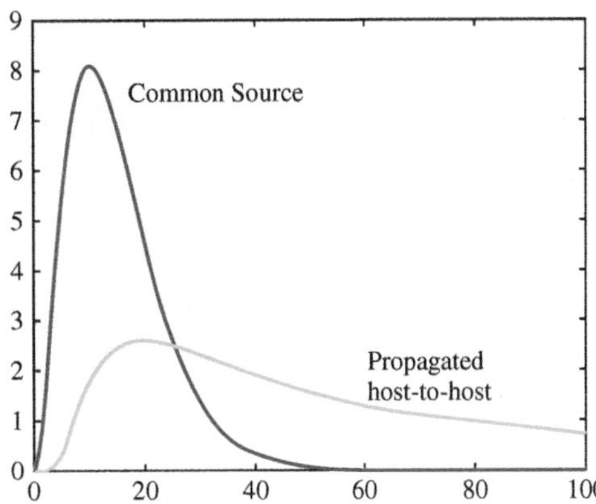

Figure 11.14 Comparison of common source and propagated outbreaks of infectious disease. Y-axis: # of cases, X-axis: time. *Source*: http://uhavax.harkord.edu/bugl/histepi.htm. Reproduced by permission of Paul Bugl.

Table 11.9 Common source and propagated diseases.

Name	Disease agent type	Causative disease agent	Gram stain response
Common source diseases			
Cholera	bacteria	*Vibrio cholerae*	negative
Legionella		*Legionella pneumophila*	negative
Typhoid fever		*Salmonella enterica* serovar Typhi	negative
Pathogenic E. coli		Shiga toxin producing *Escherichia coli* (STEC). Example: *E. coli 0157:57*. Enterotoxigenic *E. coli* (ETEC)	negative
Staphylococcus		*Staphylococcus aureus*	positive
Clostridia		*Clostridium perfringens*	positive
Botulism		*Clostridium botulinum*	positive
Salmonellosis		*Salmonella enterica*	negative
Campylobacter		*Campylobacter jejuni; C. coli; C. fetus*	all negative
Listeriosis		*Listeria monocytogenes*	positive
Giardia	protozoa	*Giardia intestinalis (lamblia)*	Not applicable
Cryptosporidosis		*Cryptosporidium parvum*	Not applicable
Toxoplasmosis		*Toxoplasma gondii*	Not applicable
Enteroviruses	virus	Poliovirus, norovirus, hepatitis A virus	Not applicable
Amebiasis	amoebae	*Entamoeba histolytica; Naegleria fowleri*	Not applicable
Propagated diseases			
Streptococcal diseases	bacteria	*Streptococcus pyogenes; S. pneumoniae*	both positive
Diphtheria		*Corynebacterium diphtheriae*	positive
Pertussis		*Bordetella pertussis*	negative
Tuberculosis		*Mycobacterium tuberculosis*	Due to waxy coatings; acid fast stains used
Leprosy		*Mycobacterium leprae*	Due to waxy coatings; acid fast stains used
Meningitis		*Neisseria meningitidis*	negative
Staphylococcus		*Staphylococcus aureus*	positive
Ulcers		*Helicobacter pylori*	negative
Hepatitis	virus	*Hepatitis A, B, C, D, E, and G*	Not applicable
Measles		*paramyxovirus*	Not applicable
Mumps		*paramyxovirus*	Not applicable
Chicken pox		*Varicella-zoster virus*	Not applicable
Rubella		*togavirus*	Not applicable

Table 11.9 *(continued)*

Name	Disease agent type	Causative disease agent	Gram stain response
Colds		*Rhinoviruses; coronaviruses*	Not applicable
Influenza		Influenza A, influenza B, influenza C	Not applicable
Sexually transmitted diseases (STDs)	virus, bacteria	Bacteria: *Neisseria gonorrhoeae* (gonorrhea); *Treponema pallidum* (syphilis); *Chlamydia trachomatis* (chlamydia). Virus: *Herpes simplex 1 virus* (HSV-1); *Herpes simplex 2 virus* (HSV-2); human papillomavirus (HPV); *Human immunodeficiency virus* (HIV)	*N. gonorrhoeae* - negative; *Treponema pallidum* - can not be detected using Gram Stain; *Chlamydia trachomatis* - negative; Gram staining not applicable for viruses

during a specific period of time relative to the number of people at risk of dying:

$$Mortality = \frac{\#\text{ deaths during a specific period of time}}{\text{Average population over a specific period of time}} \times \text{Multiplier} \qquad (11.5)$$

where 'multiplier' is typically either 100, which then makes the ratio a percentage, or is 1,000 to reflect the number of deaths on a 1,000 person basis.

Equation (11.5) defines the **crude mortality rate (CMR)** or **crude death ratio** and, as such, the denominator considers only the total population in an area and nothing about the demographics of the population. Direct comparisons of mortality defined in this manner can often be misleading, due to this lack of consideration. Accordingly, mortality rates are most readily compared when calculated for specific age (e.g., infants below the age of five, elderly above 65) or demographic groups (e.g., different ethnicities, males vs. females). The **infant mortality rate** (IMR) is defined as the number of children who die prior to their first birthday in a given year per thousand live births in that year.

Example 11.6 Calculation and comparison of mortality rates

Iraq has a population of approximately 31.5 million people and, as reported by the US CIA Factbook (https://www.cia.gov/library/publications/the-world-factbook/), there were approximately 152,000 deaths in Iraq during 2010. In comparison, the United States has a population of approximately 307 million people and there were 2.6 million deaths in the United States during 2010. What is the CMR for Iraq? What is the CMR for the United States? Use a 1,000 person basis as the multiplier.

Solution

For Iraq:

$$CMR = \frac{152{,}000 \text{ deaths in 2010}}{31{,}500{,}000 \text{ people in 2010}} \times 1{,}000$$
$$= 4.83 \text{ deaths per 1,000 people}$$

For the United States:

$$CMR = \frac{2{,}600{,}000 \text{ deaths in 2010}}{307{,}000{,}000 \text{ people in 2010}} \times 1{,}000$$
$$= 8.47 \text{ deaths per 1,000 people}$$

Simple comparison of these two numbers might suggest that the health of people in the United States is worse than that in Iraq. However, this simple comparison does not consider the fact that the median age of the people of Iraq is lower (= 20.9) than the median age of people in the United States (= 36.9). People within the US tend to live longer than those in Iraq and, as such, the higher CMR reflects its older population. A more appropriate comparison of the health characteristics of the two countries can be achieved by examining the infant mortality rates (IMR) for each country. The IMR for Iraq is 41.7 deaths/1000 live births while the IMR for the United States is 6.06 deaths/1000 live births.

Morbidity is the magnitude of illness, injury, or disability in a defined population and is quantitatively described using incidence and prevalence rates (sections 11.3.7 and 11.3.8). The overall health of a population is generally better defined in terms of morbidity statistics rather than mortality statistics, since many diseases that affect health often have low mortality. As shown in Table 11.10, the major causes of illness in high income countries are often quite different from the major causes of death, and there are also major differences between high income countries and low income countries.

To determine accurately the incidence and prevalence of a disease within a population, it is necessary to define the disease in terms of its **case definition**. A case definition clearly defines what the characteristics of a disease are, and it is structured in a manner to most effectively exclude those with a disease from those who do not have the disease. Such distinctions are important to monitor accurately trends in reported diseases, to detect unusual occurrences of disease, and ultimately to evaluate how effective a particular intervention is at controlling the disease. It is important to note that the epidemiological definition of a case is not necessarily the same as the clinical definition, since a wide variety of criteria are used to define epidemiological cases (e.g., physician diagnoses, patient

Table 11.10 Major causes of illness and death in high income countries compared to major causes of illness and death in low income countries.

LOW INCOME COUNTRIES

Causes of death[a]	Number of deaths per 100,000	Causes of infectious disease[b]	Number of reported cases
Lower respiratory infections	98	Malaria	59,600,000
HIV-AIDS	70	Tuberculosis	1,250,000
Diarrhoeal diseases	69	Cholera	177,000
Stroke	56	Measles	76,500
Ischaemic heart disease	47	Leprosy	35,200
Prematurity	43	Rubella	18,200
Malaria	38	Pertussis	7,310
Tuberculosis	32	Tetanus	5,230
Malnutrition	32		
Birth asphyxia and birth trauma	30		

HIGH INCOME COUNTRIES

Causes of death[a]	Number of deaths per 100,000	Causes of infectious disease[b]	Number of reported cases
Ischaemic heart disease	119	Mumps	136,000
Stroke	69	Tuberculosis	129,000
Trachea, bronchus, lung cancers	51	Pertussis	61,200
Alzheimers and other dementias	48	Rubella	8,540
COPD	32	Measles	6,540
Lower respiratory infections	32	Leprosy	297
Colon rectum cancers	27	Tetanus	293
Diabetes mellitus	21		
Hypertensive heart disease	20		
Breast cancer	16		

[a] *Source*: World Health Organization, Fact Sheet 2011. http://www.who.int/mediacentre/factsheets/fs310/en/index1.html
[b] *Source*: World Health Organization, World Health Statistics 2011 http://www.who.int/gho/publications/world_health_statistics/2011/en/index.html

registries, population level surveys, etc.), while clinical definitions are very tightly defined.

The prescriptive value of an epidemiological case definition to differentiate those who have a disease from those who do not can be defined in terms of its **sensitivity** and **specificity**. These two parameters can be calculated by setting up a simple 2×2 matrix. In this matrix, the rows delineate the population in terms of those who meet the case definition from those who do not. The columns are based on the number of people clinically determined to have the disease, versus those who do not have the disease.

	Clinically determined to *have* disease	Clinically determined *not to have* disease	
Meets case definition	a	b	$a+b$
Does not meet case definition	c	d	$c+d$
	$a+c$	$b+d$	

Where a = true positive, b = false positive, c = false negative, d = true negative.

The **sensitivity** of the case definition is its ability to identify truly those who have a condition. It is defined as:

$$Sensitivity = \frac{a}{a+c} \tag{11.6}$$

The **specificity** of the case definition is its capacity to exclude those who do not have a condition. It is defined as:

$$Specificity = \frac{d}{b+d} \tag{11.7}$$

A good case definition will have sensitivity and specificity that each approach 1.0. Example 11.7 illustrates how sensitivity and specificity are calculated.

Example 11.7 Calculation of sensitivity and specificity

As the laboratory director at a major biotech company, you have been put in charge of the development of a new procedure to quantify tularemia infections. As part of the testing procedure, you compared the results of the new procedure to those obtained using the existing standard procedure, which is both expensive and time-consuming. Of the 85 patients (those who have been infected) in the trial of the new procedure, 45 tested positive while 40 tested negative. Of the 74 controls (people who do not meet the clinical definition of tularemia), 8 tested positive while the remainder tested negative. What are the specificity and sensitivity of the test?

Solution

To solve this problem, we first set up a 2 × 2 matrix:

	Clinically determined to *have* disease	Clinically determined *not to have* disease	
Meets case definition	45 (= *a*)	8 (= *b*)	53
Does not meet case definition	40 (= *c*)	66 (= *d*)	106
	85	74	

Sensitivity is defined as:

$$Sensitivity = \frac{a}{a+c} = \frac{45}{85} = 0.53 \quad \text{(Equation (11.6))}$$

Specificity is defined as

$$Specificity = \frac{d}{b+d} = \frac{66}{74} = 0.89 \quad \text{(Equation (11.7))}$$

It is often desirable that the test has a minimal number of false negatives (i.e., we want to make sure that the test accurately determines those who have been exposed to tularemia), but the number of false positives is not as great a concern (i.e., positive test results for those without disease are tolerated). For this example, the sensitivity is fairly low and, thus, the new test does not meet the desired criteria to minimize the number of false negatives.

11.3.6 Epidemiological study types

There are three classic epidemiological study types: cross-sectional, case-control, and cohort studies. These three types differ with respect to their temporal nature, with cross-sectional studies occurring at a single point in time, case-control studies being retrospective (i.e., looking backwards in time), and cohort studies being prospective (i.e., looking forwards in time).

A **cross-sectional** study determines the prevalence and distribution of disease within a population at a particular point in time. A challenge associated with cross-sectional studies is that because all of the data are collected at the same time, it is impossible to establish causality between disease prevalence and possible risk factors.

In a **case-control** study, people who have an illness (cases) are compared with others who do not have an illness (controls) with respect to past exposures and risk factors. Such studies are useful for diseases that are relatively rare, since they can establish exactly what risk factors are most likely responsible for the development of disease within a particular population. Unfortunately, as was the case with cross-sectional studies, a case-control study has no causality; cases and controls often differ in terms of their characteristics.

In a **cohort study**, two groups – or cohorts – with different exposures/risk factors are monitored over time for certain health outcomes. In general, cohort studies are prospective, but they can be retrospective if necessary criteria to truly define the cohorts can be applied. The exposures can be observed (observational cohort study) or controlled (experimental).

11.3.7 Incidence and attack rates

Incidence was previously defined as the number of *new* cases of a disease that came into existence within *a certain period of time* for a *specified unit of population*. This definition can be mathematically defined as:

$$\text{Incidence Rate} = \frac{\text{\# of new cases within a population in a given time period}}{\text{\# of persons exposed to risk of developing the disease in the same time period}} \times \text{Multiplier} \quad (11.8)$$

The incidence rate provides an estimate of the risk or the probability of getting a disease within a certain time period. Ideally, the denominator of Equation (11.8) considers only the population truly at risk. However, in practice, it can be challenging to define this number accurately, because the population size changes as people move into and out of an area and the susceptible population can change as people acquire the disease or are no longer susceptible, due to immunizations or other public health interventions. In these cases, the epidemiologist develops a best estimate of the population at risk. As was the case for Equation (11.5), the multiplier in Equation (11.8) can take a number of different forms. If the incidence rate is represented on a percentage basis, it is 100, but if the incidence rate is represented on a population basis, it is often defined as 1,000.

Incidence rates are calculated to estimate the probability of, or risk of, developing a disease. Higher incidence rates indicate an increased risk of getting a disease following an exposure. Comparison of incidence rates enables evaluation of the time, person, and place aspects of disease transmission.

- *Time* – if the incidence rate for a particular disease is reliably higher during a specific time of year, the risk of developing the disease goes up at that time. This phenomenon is illustrated by the higher incidence rate of influenza in winter months relative to the summer.
- *Person* – if incidence rates are consistently higher among individuals with a specific lifestyle factor, the risk of getting the disease goes up amongst that group. For example, the incidence rate of lung cancer is higher amongst smokers versus non-smokers.

- *Place* – if the incidence rate is consistently higher among people who live in a certain place, the risk goes up for developing that disease if one lives in the area. For example, the incidence rate of malaria is higher in sub-Saharan Africa than in South America, and one thus has a greater risk of malaria infection in Africa.

Comparison of incidence rates is often done to examine the factors responsible for disease propagation within a community. One way to do so is to calculate the **relative risk**. This is the ratio of the disease incidence rate amongst those exposed to a disease, relative to the incidence rate among those not exposed.

$$\text{Relative Risk} = \frac{\text{Incidence rate of those exposed within a group}}{\text{Incidence rate of those NOT exposed within a group}} \quad (11.9)$$

Relative risk can be determined using a 2 × 2 matrix. This matrix is set up in a manner that separates those who have a particular disease from those who do not, in terms of their exposures to a particular risk factor. In this context, the rows differentiate those who have been, or who have not been, exposed to the risk factor. The columns differentiate those who have a particular disease from those who do not.

	With disease	Without disease
Exposed	*a*	*b*
Not exposed	*c*	*d*

Using this set-up, it is possible to define the relative risk (RR) numerically:

$$\text{Relative Risk} = \text{RR} = \frac{\frac{a}{a+b}}{\frac{c}{c+d}} \quad (11.10)$$

where the numerator reflects the fraction of those exposed to the risk factor who actually have the disease, while the denominator is the fraction exposed who do not have the disease.

If those ratios are the same, then exposure to the risk factor does not increase the odds of having the disease. RR values above 1.0 suggest an association between the exposure and the risk of having the disease. The larger the value for RR, the more the data suggests there is an association.

The duration of common source disease epidemics is often very short and, under these conditions, a special type of incidence rate, known as an **attack rate**, is defined. As defined mathematically, attack rates are the same as incidence rates, but in practice they are only used to analyze epidemics in which a small select population is exposed to a disease or an injury-causing event, such as food poisoning or chemical exposures.

Three different types of attack rates are often calculated. The **crude attack rate** provides a very broad overview:

$$\text{Crude attack rate} = \frac{\text{\# of persons ill with disease}}{\text{\# of persons attending event}} \times 100\% \quad (11.11)$$

While the **attack rate** is more narrowly defined:

$$\text{Attack rate} = \frac{\text{\# of new cases within a time period}}{\text{\# of persons at risk within a time period}} \times 100\% \quad (11.12)$$

And the **food-specific attack rate** is highly specific:

$$\text{Food-specific attack rate} = \frac{\text{\# of persons who ate a specific food and became ill}}{\text{\# of persons who ate a specific food}} \times 100\% \quad (11.13)$$

Example 11.8 highlights a real-world event where these three types of attack rates were of use.

Example 11.8 Calculation of attack rates and relative risk

115 people attended an evening church potluck. At the potluck, two different potato salads were served – one served in the church courtyard, the other served in the church lobby. The potato salad served in the courtyard was prepared immediately prior to the potluck, while the potato salad served in the lobby was prepared at lunchtime and left unrefrigerated on the counter until the potluck had started.

Within the first eight hours following the potluck, 21 churchgoers became ill suffering from headache, abdominal pain, severe diarrhea, fever, and nausea. Within 72 hours, a total of 47 churchgoers had become ill with similar symptoms and it was apparent that a foodborne disease outbreak had occurred. Clinical testing was done and the outbreak was ultimately attributed to consumption of *Salmonella enteritidis*-contaminated potato salad. Using a post-outbreak telephone survey of all of the potluck attendees, the following information was determined:

- 98 people remembered eating potato salad.
- 60 people exclusively got potato salad from the lobby. Of these 60 people, 43 suffered from *Salmonella*.
- 30 people exclusively got potato salad from the courtyard. Of these 30 people, 2 suffered from *Salmonella*.
- 8 people got potato salad from both the courtyard and the lobby. Of these 8 people, 2 suffered from *Salmonella*.

Using the information presented above, what was:

1. the crude attack rate for those attending the potluck?
2. the attack rate within the first 8 hours?

3 the food-specific attack rates for both potato salads?

4 the relative risk associated with consumption of the lobby potato salad compared to consumption of the courtyard potato salad?

Solution

Crude attack rate: The crude attack rate simply provides information about the total number of people who became ill after attending the potluck relative to the total number of people attending.

$$\text{Crude attack rate} = \frac{\#\text{ of persons ill with disease}}{\#\text{ of persons attending event}} \times 100\% = \frac{47}{115} \times 100\% = 40.9\%$$

This result indicates that 40.9% of the people attending the potluck became sick due to *Salmonella*.

8-hour attack rate: The attack rate for the first eight hours is time-specific and is calculated as follows:

$$\text{Attack rate} = \frac{\#\text{ of new cases within a time period}}{\#\text{ of persons at risk within a time period}} \times 100\% = \frac{21}{115} \times 100\% = 18.3\%$$

Food-specific attack rates: The attack rates for the two different potato salads are determined as follows:

$$\text{Food-specific attack rate} = \frac{\#\text{ of persons who ate a specific food and became ill}}{\#\text{ of persons who ate a specific food}} \times 100\%$$

$$\text{Lobby potato salad attack rate} = \frac{43}{60} \times 100\% = 71.6\%$$

$$\text{Courtyard potato salad attack rate} = \frac{2}{30} \times 100\% = 6.67\%$$

When calculating these two attack rates, we do not include any of the people who got potato salad from both locations. It is not possible to include them in these calculations, because their exposure history is more complicated than the other attendees and, thus, they do not fall into either of the two cohorts.

Relative Risk: The relative risk can be calculated simply by dividing the two food-specific attack rates to find that the relative risk is 10.8. This value indicates that potluck attendees eating potato salad from the lobby were 10.8 times more likely to become ill with *Salmonella* than those eating potato salad from the courtyard. An alternative way to determine the relative risk is to populate a 2 × 2 matrix as follows

	With disease	Without disease
Potato salad from lobby	43 (= a)	17 (= b)
Potato salad from courtyard	2 (= c)	28 (= d)

Accordingly the relative risk is defined as:

$$\text{Relative Risk} = \text{RR} = \frac{\frac{a}{a+b}}{\frac{c}{c+d}} = \frac{\left(\frac{43}{60}\right)}{\left(\frac{2}{30}\right)} = 10.8$$

Using the 2 × 2 matrix defined above, it is also possible to calculate two other epidemiologic statistics: **attributable risk** and **odds ratio**.

$$\text{Attributable Risk} = \text{AR} = \left(\frac{a}{a+b}\right) - \left(\frac{c}{c+d}\right) \qquad (11.14)$$

The attributable risk is the difference between the odds of having the disease following an exposure relative to the odds of having the disease without the exposure. An AR value of 0.0 suggests that there is no relationship between a particular exposure and risk. The odds ratio (Equation 11.15) provides information about the chance of exposure and the ultimate risk of getting a disease.

$$\text{Odds Ratio} = \text{OR} = \left(\frac{ad}{bc}\right) \qquad (11.15)$$

Similar to the relative risk, numbers above 1.0 suggest a relationship between exposure and risk. Odds ratios are calculated for case-control studies where it is not appropriate to calculate relative risks because the exposed group may not be directly comparable to the non-exposed group.

Example 11.9 Calculation of attributable risk and odds ratio

An outbreak of listeriosis afflicted the US in summer 2011. At the outset of this outbreak, a case-control study was conducted to determine the risk factors for this illness. From this study, it was determined that cantaloupes from Colorado had been eaten by 45 of the 60 patients (cases) and only by three of the 60 controls. Given these statistics, calculate the Attributable Risk and Odds Ratio for these exposures.

Solution

To solve this problem, we will first populate a 2 × 2 matrix.

	With disease (cases)	Without disease (controls)
Ate cantaloupe	45 (= *a*)	3 (= *b*)
Did not eat cantaloupe	15 (= *c*)	57 (= *d*)

The attributable risk is then defined as:

$$\text{Attributable Risk} = \text{AR} = \left(\frac{a}{a+b}\right) - \left(\frac{c}{c+d}\right)$$
$$= \left(\frac{45}{45+3}\right) - \left(\frac{15}{15+57}\right) = 0.70$$

The odds ratio is defined as:

$$\text{Odds Ratio} = \text{OR} = \left(\frac{ad}{bc}\right) = \left(\frac{45 \times 57}{3 \times 15}\right) = 57$$

11.3.8 Prevalence Rates

Prevalence is the number of cases of disease at a particular time in relation to the size of the population from which this number is drawn. We can define the prevalence rate as the incidence rate multiplied by the average duration of disease. Accordingly, the prevalence of a disease is directly proportional to its incidence. Furthermore, as the duration of a disease increases, so does its prevalence. Improved treatment generally decreases prevalence, with immunization preventing new cases of disease.

Similar to incidence, it is possible to explicitly calculate prevalence rates. Two types of prevalence rates are generally reported: **period prevalence** and **point prevalence**. Period prevalence is the number of individuals who have had a given disease at any time during a specified time period (typically a year). This number includes all persons with the disease who are carried over from the previous time period, as well as all of those who become ill at the end of the time period. Disease recurrences are typically included as additional cases.

$$\text{Period prevalence rate} = \frac{\text{\# of existing cases of disease within a time period}}{\text{Avg. study population within a time period}} \times 100\% \qquad (11.16)$$

Point prevalence is the number of individuals who have a disease at a single specific point in time.

$$\text{Point prevalence rate} = \frac{\text{\# of existing cases of disease}}{\text{Total study population}} \times 100\% \qquad (11.17)$$

11.3.9 Disease virulence

Virulence is the relative capacity of a disease agent to harm the host. The virulence of a particular disease agent is often defined using dose-response testing and the determination of relative LD_{50} values. Some highly virulent bacterial strains require only a few cells to establish fatal infections, while other, less virulent, strains require proportionally more. A variety of factors affect the virulence of a pathogen, including:

1. its capacity to adhere to and enter a host;
2. its ability to colonize and infect the host;
3. its aptitude to spread and/or produce exotoxins within the host.

In addition, a given host can be more or less susceptible to disease, dependent upon its age and nutritional status.

11.4 Agents of infectious disease

To this point, this chapter has generally described toxicology and epidemiology, but has yet to define the relation of these fields to environmental engineering practice. To understand how environmental engineers minimize disease transmission, it is necessary to have some understanding of the different types of agents responsible for infectious disease. Infectious diseases occur when an infectious agent invades the body. The infectious microbial agents of greatest concern are prions, viruses, bacteria, protozoa, and fungi.

11.4.1 Prions

Prions are infectious extracellular proteins that are known to cause a variety of diseases in animals. Among the different prion-associated diseases are scrapie in sheep, bovine spongiform encephalophathy (BSE of 'mad cow disease') in cattle, and kuru and Creutzfeldt-Jakob disease in humans. The mechanism(s) by which prions cause disease are a subject of significant debate; however, current thought suggests that prions cause disease by modifying the normal proteins in a host cell, such that they become misfolded. This misfolding causes the proteins to lose their normal function, to become more resistant to protease attack, and to become insoluble. Importantly, there is recent evidence suggesting that the protein responsible for BSE in cattle can infect humans, thus causing a variant form of Creutzfeldt-Jakob disease. For this reason, extensive efforts to cull BSE afflicted cattle are often warranted.

11.4.2 Viruses

Viruses are common disease agents but, because they lack internal structure and cannot replicate by themselves, they are not considered to be living cells. Only after a virus has infected a cell and hijacked its metabolic systems can it replicate. A classic definition of a virus is a "genetic element that subverts the normal cellular process for its own replication and has an extracellular form".

Viruses exhibit a wide variety of sizes and structures. The smallest viruses are ~10 nm in size, while the largest approach 1000 nm. Viruses are structurally quite diverse, as discussed below, but consistent across the structural types is the fact that the nucleic acid of the virion (an individual virus particle) is protected by a protein shell called the **capsid**, which consists of a highly symmetric arrangement of proteins that envelop the nucleic acid. Either DNA or RNA (but not both) can serve as the nucleic acid, and viruses are often described as being either a DNA virus or a RNA virus.

The symmetrical arrangement of the proteins in the capsid takes two primary forms: rod and spherical. In a spherical virus, the proteins exhibit *icosahedral* symmetry while, in a rod-shaped virus, the proteins exhibit *helical* symmetry. An icosahedron is a polyhedral structure composed of 20 identical triangular faces, 30 edges, and 12 vertices, that exhibits a nearly spherical shape. For many spherical viruses, each triangular face contains 180 repeating protein units, but some larger spherical viruses have 240 or 420 repeating protein units per triangular face. In a helical virus, the repeating protein units are arranged as a helix and the virus exhibits a more rod-like morphology.

In addition to these two basic viral types, there are more complex types, such as enveloped viruses which contain a lipid membrane that surrounds the capsid, or bacterial viruses that consist of a combination of an icosahedral head and a helical tail. Figure 11.15 provides diagrams of the different viral morphologies.

11.4.3 Bacteria

Bacteria, introduced in Chapter 4, are single-celled prokaryotic organisms. In a prokaryotic cell there are no cellular compartments, and the DNA of the cell resides directly within the cytoplasm. Even without internal structures, bacteria are able to carry on all of the essential functions of life. Current estimates suggest that there are millions, and possibly billions, of different types of bacteria. Most of these types are harmless, and only a select few are human pathogens.

The general structure of a bacterial cell is depicted in Figure 11.16. The most striking feature of this structure is the lack of any cellular compartments – all of the cellular material is contained only within the cell membrane.

The DNA of a bacterial cell is localized to a portion of the cell and generally takes the form of a single circular chromosome. Because they possess only a single chromosome, bacteria reproduce asexually. There is significant potential for

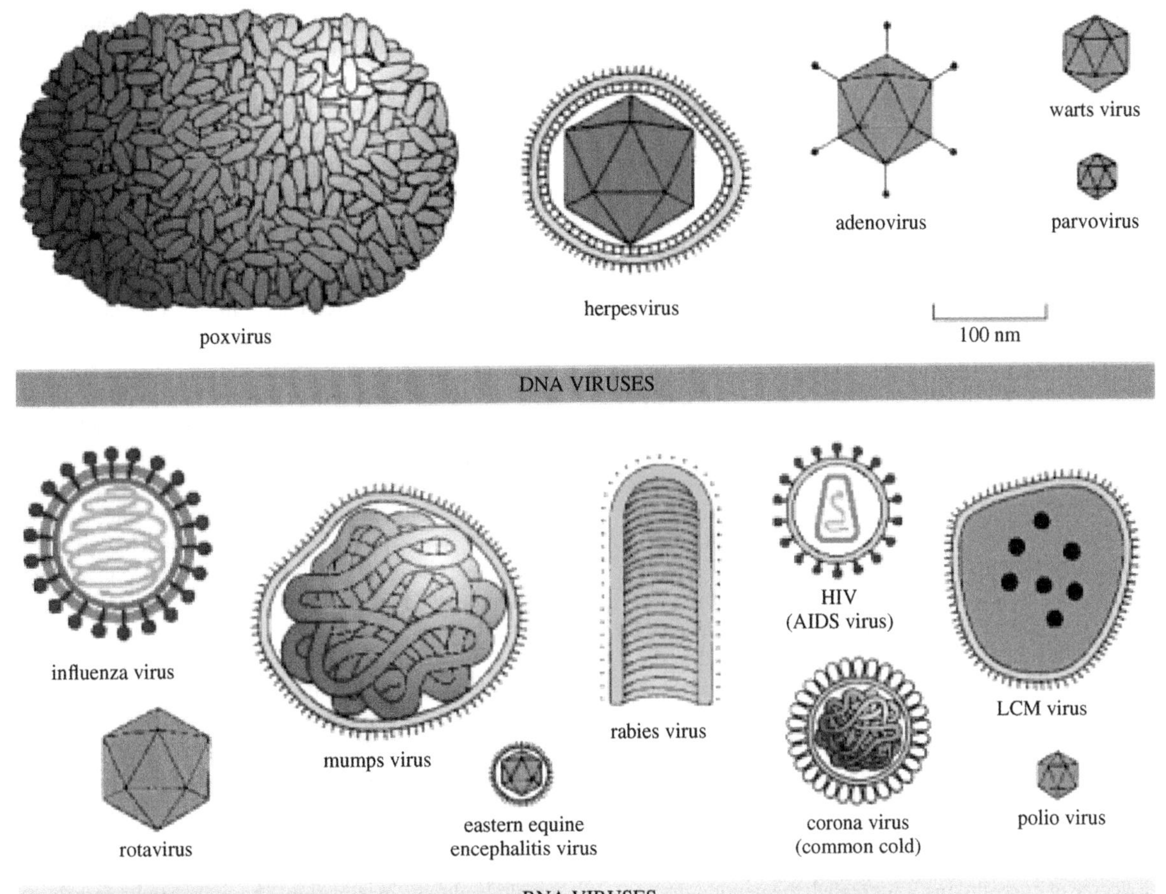

Figure 11.15 Example viral morphologies. *Source*: http://www.ncbi.nlm.nih.gov/books/NBK26917/. Reproduced by permission of Taylor & Francis.

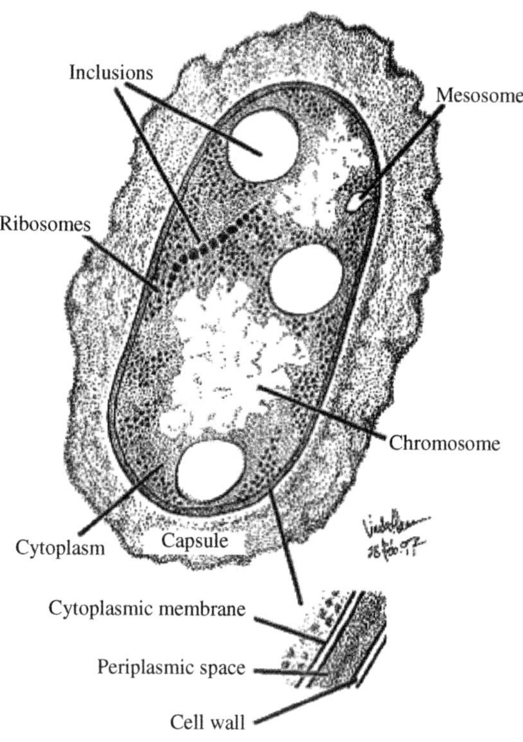

Figure 11.16 Diagram of a typical bacterial cell. *Source:* Reproduced by kind permission of LadyofHats (Mariana Ruiz Villarreal).

transfer of genetic material between bacterial cells, however, via the transfer of plasmids. **Plasmids** are separate extrachromosomal segments of DNA that are non-essential to the cell's metabolism; however, these segments often code for antimicrobial resistance elements. Other components of the bacterial cytoplasm include storage granules and **ribosomes** (consisting of RNA and protein), which are the sites for protein synthesis within the cell.

The cell membrane surrounding the cytoplasm is composed primarily of proteins (40–75%) and lipids/phospholipids (16–30%). The general structure of the cell membrane is one in which globular proteins are spread throughout a bed of lipids, with some proteins extending above the lipid bed, some extending below, and some going all the way through the lipid membrane. The lipids are oriented in a manner such that their polar (i.e., water-loving) groups are oriented either towards the exterior or interior of the cell, while their apolar (i.e., water-hating) fatty acid groups are oriented towards the interior of the cell membrane. This membrane structure enables active protein-mediated uptake of solutes required within the cell, active transport of intracellular components outwards, as well as passive uptake through the lipid membrane.

The cell wall is a rigid structure that dictates bacterial cell shape. An important component of the cell wall is a peptidoglycan layer that provides structural support to the cell. Peptidoglycan (also known as murein or mucopeptide) consists of sugar chains cross-linked together by amino acids. As the overall degree of cross-linking is increased, the peptidoglycan layer becomes increasingly rigid.

The thickness of the peptidoglycan layer in the cell wall dictates how a cell responds to a stain known as the Gram stain. Bacteria with cell walls that react positively towards the Gram stain turn purple and are described as gram-positive bacteria. The bacteria that react negatively towards the stain are gram-negative and are red-colored. The difference between gram-positive and gram-negative cells is that gram-positive cells have cell walls that are up to 90% peptidoglycan; in contrast, gram-negative cell walls contain only a thin layer of peptidoglycan, as well as proteins, lipolysaccharides, and phospholipids. The presence of these proteins and lipids enables the alcohol used during Gram-staining to penetrate the cell wall and extract the dye responsible for the purple stain. The differentiation of bacterial cells as either gram-positive or gram-negative is one important tool used in bacterial categorization. As indicated in Table 11.9, there are many gram-positive and many gram-negative bacterial pathogens.

An additional important distinction between gram-positive and gram-negative bacterial cells is the presence of an external lipopolysaccharide (or LPS) layer that covers the cell wall of gram-negative bacteria. This LPS layer is important, because it often frequently elicits a toxic response in humans and other mammals, due to reactions between a lipid in the LPS known as lipid A and cells in the host. Because of these reactions, lipid A is referred to as an **endotoxin**. Cell-bound toxins such as lipid A are released in significant amounts only when a cell lyses.

A universal symptom of endotoxin poisoning is development of fever. This response occurs because endotoxin stimulates release of pyrogen proteins that affect the temperature-control mechanisms of the brain. Additional symptoms of endotoxin poisoning include diarrhea, inflammation, and possibly death. It is important to note, however, that endotoxins are generally much less toxic than **exotoxin** proteins actively secreted by bacteria. A full comparison of the differences between endotoxins and exotoxins can be found in section 11.4.6.

Many prokaryotic bacterial species produce a non-living layer of **extracellular polymeric substance** (EPS) that coats their cellular surface. If this layer is thin (<200 μm) yet tightly bound, it is described as a microcapsule; if the layer is thicker (>200 μm) and tightly bound, it is described as a capsule; and if the layer is only loosely adhered, it is known as a slime layer. EPS is generally a byproduct of bacterial metabolism and is non-essential for bacterial survival. The EPS layer, however, does provide protection against bacteriophage viruses (viruses that attack bacteria) and against phagocytosis (engulfment) by white blood cells. Finally, there are some instances where the EPS layer plays an important role in the attachment of pathogenic bacteria to host cells. In this case, constituents within the EPS bind to specific surface receptors on the host cells.

Some types of bacteria can form **endospores**. These are a resting form of what is normally a vegetative (i.e., living) cell. The mechanism of spore formation is complex, but involves the invagination of the cytoplasmic membrane and the enclosure of cellular DNA and cytoplasm within a multilayer spore coat. This coat is resistant to heat, radiation, and desiccation (drying) and, thus, enables these cells to survive long periods under conditions that would be problematic for the vegetative cell and would lead to its inactivation. The spore can be germinated by heating, to ultimately grow into a vegetative cell. Both *Bacillus* and *Clostridium* species of gram-negative bacteria are spore-formers.

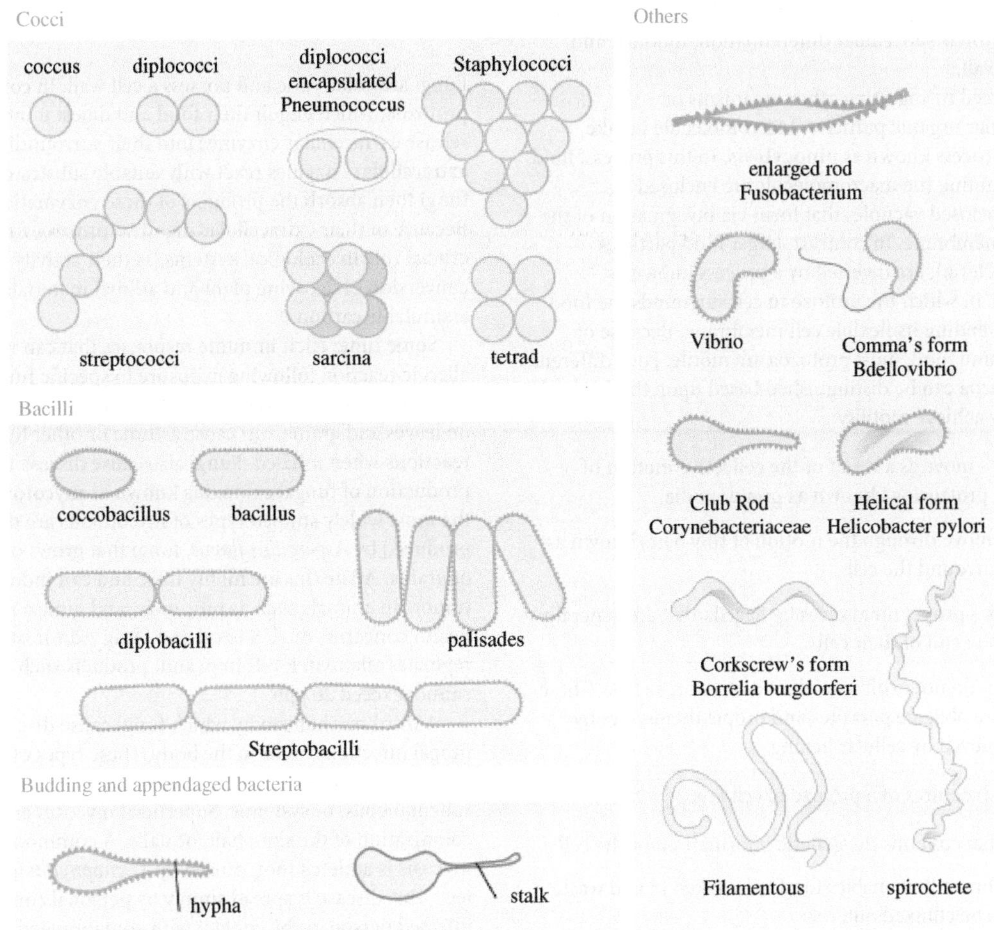

Figure 11.17 Diagrams of a typical bacterial cell morphologies. *Source*: Reproduced by kind permission of LadyofHats (Mariana Ruiz Villarreal).

Bacterial diversity arises from differences in their metabolic characteristics and their structural morphologies. Bacterial metabolism, the food sources on which bacteria thrive and the methods by which they produce energy, was discussed in Chapter 4.

Bacterial cells exhibit a wide variety of morphologies: cylindrical bacteria are referred to as rod-shaped or *bacillus* (plural form: *bacilli*); spherical or ovoid bacteria are referred to as *coccus* (plural form: *cocci*), spiral shaped bacteria are *spirilla*, while comma-shaped bacteria are *vibrio* (Figure 11.17).

Bacteria are generally larger than viruses, but smaller than protozoa or fungi. Diameters in the range of 0.2 – 1.5 μm are typical for spherical bacteria, while rod-shaped bacteria exhibit diameters between 0.4 – 4 μm and lengths of 15 – 400 μm. *Escherichia coli*, a rod-shaped bacterium, ranges from 0.4 – 0.7 μm in diameter and 1.0 – 1.3 μm in length. In contrast, the larger *Bacillus anthracis* is 1 – 1.3 μm in diameter and 3 – 10 μm long.

Many bacteria exist as single sessile cells, while others associate with one another in characteristic patterns: *Streptococcus pneumoniae* and *Pediococcus cerevisiea* form clusters of 2 – 4 cells; *Staphylococcus aureus* cluster together in structures that resemble grape clusters; *Streptococcus pyrogenes* forms long chains of cocci; and *Bacillus cereus* forms chains. Diagrams of each of these different clusters may be found in Figure 11.17.

In addition to the basic bacterial forms depicted in Figure 11.17, some types of bacteria have external appendages such as flagella and pili that facilitate motion, adhesion to surfaces, and the transfer of genetic material between bacterial cells. **Flagella**, unbranched helical filaments of uniform diameter, occur either as single structures at one end of a bacterial cell (monotrichate arrangement), as structures at each end of a cell (amphitrichate arrangement), or as tufts of filaments distributed either at one end or over the entire cell surface (peritrichate arrangement). The purpose of the flagella is to enable bacterial motility, which is accomplished by helical twisting of the flagellum. This twisting motion enables some bacteria to reach velocities as high as 200 μm/s! **Pili** are short surface appendages that often, but not always, cover the entire surface of a bacterium. In many instances, pili serve to facilitate adhesion of cells to a given surface. Recent reports have also indicated that bacterial pili can play a role in the transfer of DNA between bacterial cells, and can enable electron transfer to solid substrates.

11.4.4 Protozoa

Protozoa are unicellular, eukaryotic organisms that do not possess a cell well but do possess a nucleus. Protozoa can be differentiated from algae due to their lack of chlorophyll, and from fungi by their subcellular differentiation, motility, and lack of a cell wall.

Protozoa feed by ingesting other organisms or macromolecular organic particles. Macromolecule uptake occurs via a process known as **pinocytosis**. In this process, fluid droplets containing the macromolecule are enclosed in membrane-enclosed vacuoles that form via invagination of the exterior cell membrane. In contrast, larger food particles (including bacteria), are ingested by a process known as **phagocytosis**, in which the protozoan cell surrounds the food particle by extending its flexible cell membrane. Because of their need to find food, most protozoa are motile. Four different types of protozoa can be distinguished based upon the method by which they achieve motility:

- **Amoebae** – move as a result of the collective motion of finger-like protrusions known as pseudopodia.
- **Ciliates** – move through the motion of tiny hairs known as cilia that surround the cell.
- **Flagellates** – propel themselves by flagella that are generally located at the end of their cell.
- **Sporozoa** – do not exhibit any locomotive organelles. These protozoa are obligate parasites and propel themselves by gliding achieved by cellular flexing.

The basic structures of a protozoan cell are:

1. a nucleus that contains the genetic information for the cell;
2. a cell membrane that enables food to be taken in and waste products to be effluxed out;
3. the cytoplasm;
4. mitochondria that serve as the energy-producing structures of the cell;
5. ribosomes that enable protein assembly;
6. an endoplasmic reticulum that is the site where the ribosomes are concentrated;
7. a Golgi apparatus that serves to modify proteins;
8. lysosomes that contain the enzymes needed to digest food and cellular debris;
9. flagella, pseudopods, or cilia that enable locomotion and/or attachment to surfaces.

Protozoan shape and size vary considerably. Some protozoa are as small as 1 μm in diameter, while others are up to 2 mm in size. A variety of shapes are exhibited, with some protozoans exhibiting spherical shapes, others elongated. Importantly, some protozoa are polymorphic and exhibit different structural forms, dependent upon their life cycle. As shown in Figure 11.13, malaria is an important disease caused by a protozoan plasmodium parasite that exhibits many different cellular forms across its lifecycle. A significant challenge associated with the control of these diseases is that the protozoan cell wall often resembles that of human cells. As such, treatment of protozoan diseases can be challenging, because therapeutic agents that destroy the protozoan cell also often attack human cells.

11.4.5 Fungi

Fungi are eukaryotic and possess a cell wall. In contrast to protozoa, which engulf their food and digest it internally, fungi release extracellular enzymes into their surroundings. These extracellular enzymes react with suitable substrates, and the fungi then absorb the products of these enzymatic reactions. Because of their extracellular digestive process, fungi play a crucial role in ecological systems, as their activity facilitates the conversion of decaying plant and animal materials into assimilable carbon.

Some fungi elicit immune responses that can result in an allergic reaction following exposure to specific fungal antigens. As an example, the mold *Aspergillus*, which commonly grows on leaves and grains, can cause asthma or other hypersensitivity reactions when inhaled. Fungi also cause disease through the production of fungal exotoxins known as **mycotoxins**. Among the most widely studied types of mycotoxins are the **aflatoxins** produced by *Aspergillus flavus*, fungi that grows on many types of grains. Aflatoxins are highly toxic and can induce malignant tumors in animals such as rats, mice, and guinea pigs. Due to health concerns, the US Food and Drug Administration (FDA) regulates aflatoxin levels in peanut products such that they cannot exceed 20 ppb.

A third mechanism by which fungi cause disease is via fungal infections on or in the body. These types of infections are known as **mycoses** (singular: mycosis) and can be superficial, subcutaneous, or systemic. Superficial mycoses entail fungal colonization of the skin, hair, or nails. A common superficial mycosis is athlete's foot, caused by *Trichophyton* infection of the feet. This disease is spread simply by personal contact with an infected person, or by contact with contaminated surfaces such as bathroom floors, shower stalls, or bed linens. In contrast, subcutaneous mycoses affect the deeper layers of skin (Figure 11.4). These infections often occur when the causative organism infects a small surficial wound or abrasion. Systemic mycoses involve fungal growth within the internal organs of the body. Primary systemic infections affect normal, healthy individuals, while secondary systemic infections affect people who are immunocompromised or have predisposing medical conditions. The most common primary fungal infections in the US are *histoplasmosis*, caused by *Histoplasma capsulatum*, and *coccidioidomycosis*, caused by *Coccidioides immitis*. Each of these organisms is a normal constituent of the soil, and it is only in very rare circumstances that they cause disease. In the case of *histoplasmosis*, the disease occurs when fungal fragments are inhaled. In most cases, the result is a flu-like illness, but in more severe instances the disease can be fatal.

11.4.6 Exotoxins versus endotoxins

Exotoxins are heat-labile proteins that are actively excreted by some types of gram-positive and gram-negative bacteria. These compounds generally exhibit highly specific modes of action, due to their association with specific cell receptors. There are three general types of exotoxins: **A-B toxins**, **cytolytic toxins**, and **superantigen toxins**.

A-B toxins consist of two covalently associated subunits, A and B. The B subunit enables specificity by binding to specific cell-surface receptors. Once the B subunit has bound to a cell, the A subunit is transferred to the target cell so that it can damage the cell.

Cytolytic toxins attack cells enzymatically in a manner that causes cells to lyse. The action of cytolytic toxins is often highly nonspecific and can even result in the lysis of the bacterial cells, producing these toxins as well as the target eukaryotic cells.

Superantigen toxins, produced by staphylococci and streptococci bacteria, stimulate production of large numbers of immune response cells (e.g., T-cells) within the host. The massive production of these response cells can result in systemic inflammatory effects and, possibly, in generalized shock. **Enterotoxins** are a subclass of exotoxins that specifically act on the small intestine. Many enterotoxins cause massive secretion of fluids into the intestinal lumen and life-threatening diarrhea. The cholera A-B toxin is a classic example of an enterotoxin.

Endotoxins, introduced in section 11.4.2, are generally much less toxic than exotoxins. To illustrate this point, the LD_{50} for mice treated with endotoxin was 200–400 μg, while that for botulinum toxin produced by *Clostridium botulinum* was only 25×10^{-6} μg! A primary reason for this wide variation in toxicity is the different modes by which these toxins act. As discussed previously, the mode of action of an endotoxin is usually quite general, resulting in fever, diarrhea, and vomiting. In contrast, exotoxins are often quite specific, since they bind to specific cell receptors to elicit a toxic response. It is this specificity that enables much smaller doses of exotoxins to elicit a toxic effect.

11.5 Public health and engineering measures for control of disease

The public health engineering profession relies upon a number of tools to minimize the spread of infectious disease. The choice of the appropriate tool to employ in a given situation depends upon the reservoir the disease inhabits, whether it is disseminated via vectors or vehicles, or whether it can be controlled by immunization, quarantine, or other mechanisms of control.

11.5.1 Controls directed against disease reservoirs

Many infectious diseases are harbored by animal reservoirs. The proper controls to use to control the spread of these diseases are dependent upon the nature of the animal. If the reservoir is a *domestic animal* (e.g., livestock, pets), it is generally possible to prevent disease propagation by eliminating the disease from the infected animal population. Elimination can be achieved via immunization for diseases treatable using available vaccines. Alternately, for diseases of health concern for which no vaccines are available, elimination can be achieved via the destruction of infected animals. This latter approach has been used in recent years to minimize the spread of highly infectious diseases such as foot-and-mouth (an infectious viral disease that primarily afflicts cloven-hoofed animals and, potentially, humans) and BSE. For each of these diseases, infected cattle and sheep populations were culled throughout Europe and North America.

For diseases in which the reservoir is a *wild animal*, it is much more challenging, if not impossible, to eradicate the disease completely. An example is the anthroponotic disease *Q Fever*, caused by the bacterium *Coxiella burnetii*. This disease has both wild and domestic animal reservoirs, but is transmitted to domestic animals primarily through their contact with wild animals. For this particular disease, it is possible to control its spread by immunizing the domestic animal populations, but disease eradication is unattainable, since it would require the capture and destruction of all of the wild animal reservoirs.

For diseases in which the reservoir is an insect, it is possible to achieve effective control by eliminating the reservoir using chemical or other means of control (Table 11.11). When utilizing chemical controls it is important to consider the potential environmental consequences of their usage. As an example, DDT (dichlorodiphenyltrichloroethane) was highly effective when used against mosquito vectors in temperate climates and, as such, it has eliminated diseases such as yellow fever and malaria from much of North America and Europe. Unfortunately, as documented in the classic book *Silent Spring* by Rachel Carson, indiscriminate use of pesticides such as DDT results in devastating environmental consequences and, as such, DDT is banned from use in the USA and a majority of the developed world. Nonetheless, DDT, when used judiciously, can be used to control mosquito-borne diseases. It is estimated that approximately 30,000 tons/year of DDT is still used in developing countries

For diseases where humans are the primary reservoir, control and eradication are generally quite challenging. This fact is especially true for diseases spread via asymptomatic carriers who harbor the disease, but who do not currently suffer from its consequences. For diseases that do not have an asymptomatic phase, control can be achieved by immunization, quarantine, and surveillance. These three control strategies are discussed in the following sections.

11.5.2 Immunization

Immunization is the purposeful stimulation of immunity to infectious disease in a given individual. This stimulation of immunity is achieved either by purposely exposing an individual to a controlled dose of non-infective antigen that induces production of antibodies against the disease, or by injection of antibodies derived from an individual who already has immunity. The former approach is known as **vaccination**, while the latter results in passive immunity. In general, vaccination is used to prevent the spread of infectious disease, while antibody injection is used as a therapeutic approach to control active disease. A number of diseases that used to be quite common have all been brought under control through the use of widespread vaccination programs. Diseases such as smallpox, diphtheria, tetanus, pertussis (whooping cough), chicken pox, and poliomyelitis have been controlled in the developed world via extensive immunization programs.

Table 11.11 Control of vector-borne disease.

Type of Control	Description	Examples
Permanent control	Measures that alter the physical environment enough so that reproduction and survival of the vector is kept low or disappears.	1 Land drainage to reduce stagnant water for mosquitoes. 2 Rat-proofing and garbage control to reduce rat populations.
Temporary control	Periodic and seasonal killing of the vectors, hosts, and pests. Abatement for relief, the intensity is not planned to kill each and every one.	1 Spraying of mosquitoes in the summertime. 2 Poisoning and/or trapping of rats
Species eradication	An effort intensive enough to eradicate either the vector or the disease agent.	1 Eradication of *Anopheles* species, vectors of malaria, in the US. 2 Eradication of *Aedes egypti*, vector of yellow fever, in the US.
Naturalistic control	Measures in which natural predators or infection is introduced, or their presence is intensified among the target population.	1 Putting top-feeding minnows in mosquito breeding waters to eat larvae. 2 Bringing cats, terriers and ferrets into rat-infested places to hunt and kill the rats.

An important aspect of immunization programs is that it is generally not necessary to achieve 100% immunization to control disease. The reason for this has to do with the concept of **herd immunity**. For many diseases, the resistance of a group to infection and spread of a disease results from the immunity of a high proportion of the members of the group. If a significant proportion is protected, then the population as a whole is protected. Required percentages for herd immunity depend on the virulence of a particular disease. The more infectious a disease agent, the greater the required proportion of immune individuals in a population to prevent disease epidemics. Current estimates suggest that to effectively control against polio, 70% immunization is required; while for highly virulent diseases such as influenza, the required immunization level is much higher, at 90–95%.

11.5.3 Quarantine

Quarantine and **isolation** are used to prevent the spread of highly infectious diseases. Quarantine strictly applies to those who have been potentially exposed to an infectious disease, while isolation applies to those who are known to be suffering from the disease. In either case, the individual has his or her movements severely restricted to prevent further spread of disease. Quarantine and isolation are done in a manner such that infected individuals are not allowed to come into contact with unexposed individuals.

Because of the limits that quarantine and isolation place on individual liberties, the quarantine period is generally only as long as necessary to protect the public. This period must be sufficient to:

1 provide therapeutic health care; and

2 ensure that quarantined individuals cannot infect others.

By international convention, there are six diseases that require quarantine: smallpox, cholera, plague, yellow fever, typhoid fever, and relapsing fever. In addition, highly infectious diseases such as viral hemorrhagic fevers are also often quarantined.

11.5.4 Surveillance

Surveillance is a passive monitoring system used to observe, recognize, and report diseases as they occur. The way that this system works in the US is that doctors are instructed to look for the particular symptoms of a disease and report their observations to local and state health departments. The local and state health departments then forward this data to the US Centers for Disease Control and Prevention (CDC). This step-by-step process is illustrated in Figure 11.18.

One challenge associated with disease surveillance programs is their reliance on initial reporting by doctors. For this reason,

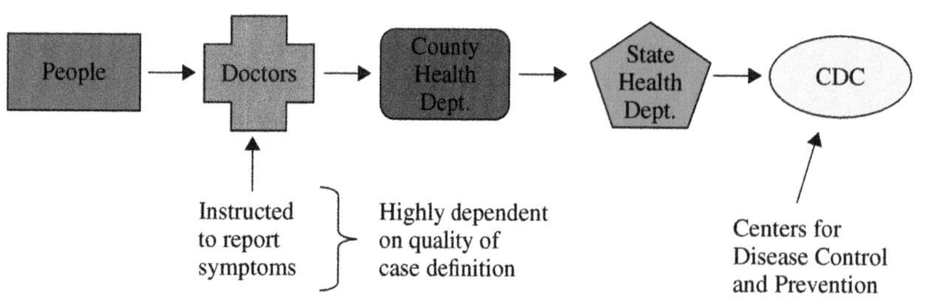

Figure 11.18 Steps in disease surveillance in the USA. *Source*: P. Vikesland.

many diseases are critically under-reported, and many episodes of accidental misreporting can occur. Successful surveillance programs are highly dependent upon the quality of the case-definition employed to describe a particular disease.

11.5.5 Controls directed against pathogen transmission

Engineering and public health controls have been developed to minimize pathogen transmission. In the case of foodborne and waterborne diseases, significant progress was made over the course of the 20th century through the institution of public health procedures that either prevent contamination of these vehicles or that destroy the pathogen in the vehicle. Examples of controls include the use of drinking water purification and the widespread implementation of wastewater treatment throughout the developed world. These have collectively led to dramatic reductions in the prevalence of infectious diseases such as typhoid fever and cholera. Similarly, milk pasteurization has helped to control bovine tuberculosis – a form of tuberculosis that only affects cows. Food protection laws greatly decrease the probability of transmission of a number of enteric pathogens to humans.

11.5.6 Pathogen eradication

The complete eradication of a pathogenic organism is theoretically possible, yet extremely challenging to implement in practice. Only through the development and implementation of extremely strict control measures can such an outcome be achieved. To date, only two diseases – smallpox and rinderpest (a viral disease that afflicts cattle and other hoofed animals exclusively) – have been eradicated. The latter disease could be eradicated because it was an anthroponotic disease requiring direct host-to-host transfer. In the smallpox eradication program, ring immunization schemes were developed in which all of the contacts and all of the contacts of the contacts were vaccinated with a non-pathogenic vaccina virus that confers immunity towards the variola virus responsible for smallpox. To prevent further transmission of the pathogen, all of the patients afflicted with smallpox were isolated until the disease had run its course.

As a result of these intensive efforts, the last naturally occurring smallpox infection occurred in 1975. At the present time, there are only two known repositories for smallpox: the Centers for Disease Control and Prevention (CDC) in the United States and the State Research Center of Virology and Biotechnology VECTOR in Koltsovo, Russia. Although there are ongoing calls for the destruction of these viral stocks by the World Health Organization and by the scientific community, there is resistance, due to persistent fears about unknown repositories for this highly virulent disease.

As of 2011, there are two diseases currently targeted for global eradication: dracunculiasis (Guinea worm disease) and poliomyelitis (polio). Dracunculiasis is a disease caused by the nematode *Dracunculus medinensis* (guinea worm) that is spread by consumption of guinea worm larvae-infested water. Preventing the spread of dracunculiasis is achieved simply by provision of clean water and the treatment of infected water with larvicides. There are no animal or environmental reservoirs of dracunculiasis and, thus, eradication can be achieved via widespread and comprehensive sanitation efforts. Poliomyelitis, or polio, is a viral disease primarily spread via the fecal-oral route. Polio has been targeted for eradication since the late 1980s, and presently is endemic in only Nigeria, Pakistan, and Afghanistan. In 1988, there were approximately 350,000 polio cases reported, while in the 2000s the cases have persistently remained below 2,000 per year.

11.6 Emergent and reemergent infectious diseases

Infectious diseases are of *global* health concern because their worldwide distribution can change dramatically and rapidly, due to alterations to the pathogen, the environment, or the host population that collectively contribute to the rapid spread of newly emergent diseases, as well as reemergent diseases that had previously been controlled. The convergence of many factors collectively contributes to emergence and reemergence of infectious disease. As described in detail in the book *Microbial Threats to Health: Emergence, Detection, and Response* published by the Institute of Medicine, the factors that affect disease emergence are:

1. Microbial adaptation and change.
2. Human susceptibility to infection.
3. Climate and weather.
4. Changing ecosystems.
5. Human demographics and behavior.
6. Economic development and land use.
7. International travel and commerce.
8. Technology and industry.
9. Breakdown of public health measures.
10. Poverty and social inequality.
11. War and famine.
12. Lack of political will.
13. Intent to harm.

Each of these factors is described in the following sections.

11.6.1 Microbial adaptation and change

Microbes have developed mechanisms that enable them to exchange and/or incorporate new genetic material into their genomes. It is well known that horizontal (lateral) transfer of

DNA commonly occurs between bacterial cells. This transfer can result in exchange of virulence genes or the genes needed to adapt to a particular host or environment. Similarly, RNA viruses (e.g., influenza, HIV, hemorrhagic fever viruses) can mutate very rapidly and very unpredictably, and this enables them to adapt quickly to changes in their external environments. Because microorganism reproduction rates are high, fairly rare mutations and adaptations can build up rapidly within viral and bacterial populations.

Horizontal gene transfer and viral mutations collectively result in antimicrobial drug resistance, which hinders the use of many once viable pharmaceutical treatments. For example, methicillin was once highly effective for the treatment of staph infections, but the *Staphylococcus aureus* pathogen has increasingly developed resistance to this antibacterial drug. In 1990, approximately 15% of *S. aureus* isolates were resistant to methicillin but, as of 2011 the resistance rate is well above 50%. Methicillin-resistant *S. aureus* (MRSA) infected over 270,000 hospital patients in 2005 and resulted in approximately 17,000 hospital patient deaths.

Viruses also have the capacity to develop resistance to medicines due to changes in their protein capsids. For example, the HIV virus rapidly develops resistance to azidothymidine and, thus, this compound is not a viable medicine unless it is used in combination with other drugs.

11.6.2 Human susceptibility to infection

The human body has many defense mechanisms that restrict the capacity for a pathogen to cause disease. The skin (Figure 11.4), mucous membranes (Figure 11.3), and intestinal epithelium (Figure 11.2) provide a protective physical and chemical barrier against pathogen entry into the body. If the skin is breached due to wounds, or if the bacterial flora that normally reside in the mucous membranes and intestinal epithelium are reduced in number (which can result from application of broad-spectrum antibiotics or other medications), there is often an increased susceptibility towards infection.

Should the skin, mucous membranes, or intestinal epithelium be breached, the human body has additional **non-specific** (innate) and **specific** (adaptive) defense mechanisms. Non-specific defense mechanisms include inflammation, fever, and phagocyte generation. **Inflammation** (redness, swelling, pain, and heat) is localized at the site of an infection and is mediated by cytokine proteins produced by white blood cells. In contrast, **fever** is an abnormal increase in the overall body temperature. Slight increases in body temperature can accelerate specific defense mechanisms, while strong fevers (>40°C) are beneficial to the pathogen, since they can damage host tissues.

Phagocytes are motile cells found throughout the body. The purpose of these cells is to engulf and destroy pathogens. During the process of phagocytosis, antigens are produced and presented to cells known as **lymphocytes**. These lymphocytes, categorized as either **B lymphocytes** or **T lymphocytes**, are involved in the specific defense mechanism. B lymphocytes (B cells) are responsible for interacting with the antigen, producing antibodies, and stimulating the memory of the immune system. T lymphocytes (T cells), in contrast, interact directly with the antigen.

The capacity for the specific and non-specific defense mechanisms to protect against disease can be altered by a number of different factors. Amongst these are genetic polymorphisms and malnutrition. Some genetic polymorphisms, such as sickle cell anemia, which increases a person's resistance to survive malaria infection but results in dramatically lower life expectancy, are clearly detrimental. **Malnutrition**, the consumption of a diet insufficient to properly provide needed vitamins and nutrients, has been known for centuries to result in increased susceptibility to infectious disease. Essentially, all of the bodily processes and physical barriers that protect against microbial threats are adversely affected by malnutrition. It is for this reason that malnutrition is associated with greater than 50% of all deaths amongst children below the age of five worldwide.

11.6.3 Climate and weather

As described by the epidemiological triangle, the physical environment influences the host, the disease agent, and the transmission of agents between hosts. Accordingly, changes in the physical environment have potentially important effects on the dissemination of infectious disease. As an example, higher levels of rainfall often increase the number of potential habitats for mosquitos and other insect vectors and this, in turn, leads to increased mosquito and insect densities and, thus, an increased opportunity for disease propagation.

It is reasonably well accepted that many infectious diseases are affected by short-term changes in weather conditions and longer-term changes in climate. A recent outbreak of disease caused by *Vibrio parahaemolyticus* was attributed to warmer Pacific Northwest ocean temperatures resulting from an *El Niño* event. Given that weather and climate influence disease propagation, there is growing concern about how projected changes in the global climate will impact infectious disease emergence. Some studies suggest that the *Anopheles* mosquito vector responsible for transmission of malaria will expand its global range as a result of climate change and, thus, the disease may become endemic in locations currently too cold for this vector to survive.

11.6.4 Changing ecosystems

Changes in ecosystems alter the capacity for pathogens to interact with humans. Transmission of many vehicle-borne and vector-borne diseases is influenced by ecosystem changes. In particular, ecological conditions are often key determinants of how vector-borne diseases are transmitted and how persistent they are. Changes in ecological conditions may alter human exposures to vectors, may increase the vector distribution and density, or may alter the longevity, activity, and habitat of the vector. Each of these alterations has the potential to affect the capacity of the vector population to transmit disease.

Estimates suggest that most emerging infections are of a zoonotic nature. These diseases, which often have significant wild animal reservoirs, are particularly challenging to control. In particular, rodent-borne viral diseases have proven to be especially challenging to control or eradicate, due to the persistence of rodents in cities and towns throughout the world. In many cases, the rodent vector transmits disease to humans through contact of a susceptible human with rodent urine, feces, or tissues.

11.6.5 Economic development and land use

As the world's human population grows and people transition from rural to urban lifestyles, there is often the concomitant consumption of natural resources, deforestation, and dam building. Each of these activities has both intended and unintended impacts on the environment that can, in turn, alter the number of interactions between human and animal reservoirs.

The conversion of forests into croplands may enhance or diminish the number of interactions with people. As an example, Venezuelan hemorrhagic fever – a disease previously unheard of – emerged as an important infectious disease following the conversion of forest into agricultural lands. In this case, agricultural fields are a highly favorable environment for the mouse reservoir that harbors the Guanarito virus responsible for this disease, and individuals who harvested crops increasingly came into contact with this reservoir.

Even conversion of former croplands back into forests can have unintended consequences. The reforestation of many areas in the eastern United States has been partially attributed to the emergence of Lyme disease over the past 20 years. In this case, the whitetail deer population, which is the natural host for the tick vectors that spread this disease, has surged, and this has led to larger numbers of ticks infected by the agent *Borrelia burgdorferi* responsible for this disease.

A final example is the mosquito-borne disease known as Rift Valley Fever. The prevalence of this viral disease increased significantly in Africa's Rift Valley following the construction of the Aswan High Dam in Egypt. After the dam was constructed, the flooded areal area was dramatically increased, and there was a resultant increase in breeding of the *Aedes* mosquito responsible for disease transmission. The first epidemic of Rift Valley fever that occurred, in 1977, affected over 200,000 people and resulted in almost 600 deaths. The disease has subsequently spread throughout the African continent, due to the capacity of the *Aedes* vector to lay eggs in shallow depressions created by the feet of cattle, sheep, and goats.

11.6.6 Human demographics and behavior

As of 2011, approximately 7 billion people inhabit the earth, and it has been projected that its population will exceed 9.3 billion by 2050. Concomitant with this growth has been a transfer of people from rural areas to urban settings, such that over 50% of the world's people now live in cities. Explosive growth in the human population and the greater number of interpersonal contacts that occur as a result of urban living have collectively enhanced the opportunity for pathogens to be transferred from one human to another.

Much of the growth in the world's population has occurred in developing regions of the world. Six countries (India, China, Pakistan, Nigeria, Bangladesh, and Indonesia) are alone responsible for almost 50 percent of the annual population growth. In these and other developing countries, the transition from a rural to an urban lifestyle is complicated by the lack of adequate water and sanitation infrastructure. Because of the insufficient availability of potable water and proper sanitation in cities and the peri-urban areas that surround them, disease propagation is often rapid and unchecked amongst their crowded denizens. Coupled with increases in population and urban living are increases in the numbers of immunocompromised peoples and those that engage in high-risk behaviors.

The immunocompromised population has steadily increased across the world, due to advances in medicine, science, and technology. For example, cancer patients are surviving longer, and patients with diseases such as AIDS and tuberculosis are living for extended periods of time relative to historic trends. These immunocompromised patients survive, but they are increasingly susceptible to fungal diseases. Increased population densities unfortunately often additionally result in increased illicit drug use and unprotected sex, which can also enhance disease dissemination.

11.6.7 Technology and industry

Numerous technological and industrial advances such as antibiotics, organ transplants, and food pasteurization have significantly improved people's lives, added years to life expectancy, and led to the control of many historically prevalent diseases. These technological advances, however, come with a price, since they have enabled some novel previously unknown disease agents to emerge.

Food production is one industry that has changed significantly over the past 100 years as a result of technological advances. It is now possible to raise thousands of poultry or beef cattle on a single farm, whereas historically it would have taken tens or hundreds of farms to handle this animal volume. To manage farms of this size, antimicrobial agents are often included in the food supply for both the treatment and prevention of infections. This often indiscriminate use of antimicrobial agents can result in the development of antimicrobial resistance, which is then disseminated either in the voluminous wastes produced by these large farms or in the meat and poultry products that are intended for human consumption. There are some who argue that the recent emergence of *enteropathogenic E. coli* as a foodborne disease is primarily a result of animal crowding at large-scale confined animal feeding operations (CAFOs), and the resultant need to feed cattle on corn and other grains that they are not adapted to.

Advances in health care have significantly improved the survival of previously vulnerable populations, but have simultaneously led to increases in the total number of hospital-based (nosocomial) infections and the further development of antimicrobial drug resistance. Many hospitals across the world continually deal with diseases that have developed resistance to antimicrobial agents. Vancomycin-resistant enterococci and methicillin-resistant staphylococci such as MRSA are endemic in hospitals throughout the world.

11.6.8 International travel and commerce

Increased international travel and the expansion of global trade have enhanced the potential for rapid disease agent and vector dissemination. Anthroponotic diseases, in particular, can rapidly circumnavigate the globe due to international travel. An infected individual can board an airplane in central India and fly to London, whereupon they can readily spread diseases, such as tuberculosis or cholera, that were not previously a problem. Cruise ships, in particular, are well known for their potential to spread disease, both amongst the passengers on the ship as well as amongst people at their ports of call. Cruise ships bring together people from extremely diverse geographic origins and then keep them together for periods of at least a few days. During these periods, the people interact quite closely, and this often results in significant disease transmission. Numerous outbreaks of infectious disease have beset the cruise industry over the past 20 years, with disease agents ranging from *S. aureus* to *Shigella* and norovirus responsible for outbreaks.

Similar to international travel, international commerce has made it possible for diseases to circumnavigate the globe. One simply has to go to the produce section at the grocery store to see the scope of the problem. There are grapes from Chile, bananas from Ecuador, green onions from Mexico, and apples from New Zealand. As these fruits and vegetables travel from their countries of origin, the cross-border transmission of infectious agents becomes increasingly possible. The foods themselves often harbor insects, plant pathogens, slugs, snails, or other vectors, and the crates and containers in which things are shipped harbor any imaginable agent or vehicle. The recent spread of West Nile Virus throughout North America has been attributed to the shipment of infected mosquitos or birds from the Middle East, where the disease has been endemic for decades, to New York City in 1999. Following its arrival in 1999, West Nile Virus has subsequently spread throughout the continental USA, with over 1,000 people infected in 2010.

11.6.9 Breakdown of public health measures

Modern society relies upon many public health measures to keep infectious diseases under control. Adequate sanitation, immunization, and vector-borne disease control are required to provide adequate protection for people. In situations where these public health measures have failed, there have been catastrophic effects in terms of disease emergence and reemergence.

Across the world, far too many infectious diseases are spread due to a lack of proper sanitation. As noted previously, many infectious diseases are spread via fecal contamination of waters used for drinking or other daily uses. Simply improving sanitation practices would significantly decrease the prevalence of infectious disease worldwide. For example, cholera is an acute gastrointestinal infection caused by the bacterium *Vibrio cholera*. This disease, which is often spread via contaminated food and water, is typically quite mild but, under some circumstances, it can cause severe, life-threatening diarrhea. Because the bacterium can survive and multiply outside the human body, it can be spread efficiently under crowded conditions. Simple sanitation practices (i.e., separating human wastes, hand-washing, water chlorination) are highly effective at controlling the spread of cholera, even in areas where it is endemic. Unfortunately, however, breakdowns in sanitation have led to recent major cholera outbreaks in Peru, Haiti, and Bangladesh. The recent cholera pandemic that started in Peru involved at least 400,000 people and led to approximately 4,000 deaths.

In developed countries, the reliance on public health measures provides great protection, except when these measures fail. Such an event occurred in Milwaukee, WI, when inadequate water treatment led to an outbreak of *Cryptosporidosis* that afflicted 370,000 people and led to 4,000 hospitalizations and 104 confirmed deaths. In this case, one of the water treatment plants in Milwaukee was in the process of changing the coagulants used in their water treatment process and, unfortunately, water contaminated with *C. parvum* oocysts was provided to the utility's customers.

11.6.10 Poverty and social inequality

Poverty and social inequality are major factors in disease emergence. Many diseases predominantly afflict the poor, due to chronic malnutrition, lack of access to clean water and sanitation, poor housing conditions, and limited funds to pay for out-of-pocket expenditures. It has been shown that the hepatitis B virus infects those with low social status, low educational attainment, and crowded urban lifestyles to a greater extent than those of higher socioeconomic status. Importantly, hepatitis B is implicated as a causative agent for liver cancer, thus suggesting a link between one's risk of developing liver cancer and one's socioeconomic status.

Developing countries chronically underfund health care relative to the developed world and, thus, they suffer from inadequate and erratic supplies of drugs and high transportation costs. Each of these factors results in poor patient compliance to doctor-initiated care, and the resultant emergence of antimicrobial resistance. Such an outcome has lead to the global emergence of multidrug-resistant tuberculosis.

11.6.11 War and famine

Between 1990 and 1998, there were 108 armed conflicts that resulted in the loss of at least 5.5 million people. Although many of these losses were on the battlefield, many more were the direct result of the breakdown in domestic stability, loss of food security, and the destruction of medical infrastructure that resulted from the armed conflict. In many cases, these "complex humanitarian emergencies" displace people and force them to live in squalid refugee camps with little to no access to medical care, protection from vectors, or adequate sanitation.

As an example, following the outbreak of war in Rwanda, there were more than 1 million Rwandan refugees who took shelter in Goma, in the Democratic Republic of the Congo. Many of the refugees in this camp obtained water from Lake Kivu that was contaminated by *Vibrio cholera*, and the resultant outbreaks of cholera and dysentery in the camp killed over 12,000 people within a three-week period.

11.6.12 Lack of political will

Diseases move rapidly around the world as a result of technology and economic independence. As globalization continues to lead to denationalization of markets, laws, and politics, there is a need for governments to cooperate with one another to prevent international dissemination of disease. Unfortunately, however, due to political, economic, and cultural differences, such cooperation is often difficult to achieve.

11.6.13 Intent to harm

The use of infectious agents to disseminate disease and cause widespread panic is a continual concern for peoples across the world. Biological weapons have been used for hundreds of years, with evidence suggesting that the Spartans used biological weapons during the Peloponnesian War (431–404 BC) that killed thousands of Athenian citizens. Smallpox was used as a biological weapon during the French and Indian Wars (1754–1767) in colonial North America. In this latter case, blankets and other inanimate fomites were purposefully infected with smallpox and then given to Native Americans fighting against the British colonists.

In the 20th century, Germany and Japan tested biological weapons during World War I and World War II, respectively and, during the Cold War, both the Soviet Union and the USA developed and tested biological weapons. In more recent years, *Bacillus anthracis* has been weaponized and sent through the US mail to disseminate anthrax, and ricin (a highly toxic (LD_{50} = 22 μg/kg) extract from the castor oil plant) was sent to the US Senate office buildings. Based on these and many other historic examples, there is a need to consider bioterror as a mechanism for disease emergence and reemergence. Recognizing this potential threat, the US has developed a classification system that places potential bioterror agents into three different categories: A, B, and C (Table 11.12).

Summary

- Disease is any condition that impairs the normal function of an organism.
- Toxicology is the study of the adverse effects of biological and chemical agents on living organisms. Epidemiology is the investigative methodology used to study the nature, cause, control, and determinants of disease transmission and distribution.
- Toxicology quantitatively evaluates the lethal and sub-lethal effects of exposure to biological and chemical agents.
  - Chronic low-dose exposures occur over extended periods of time.
  - Acute high-dose exposures can occur as a result of industrial accidents and work-place activities.
- Exposures to toxic substances occur via ingestion (eating and/or drinking), inhalation (breathing), dermal absorption (through the skin), as well as placental transfer *in utero*. The route by which an exposure occurs is important to consider, since it can affect the toxicity of the substance.
- Toxicologists develop dose-response curves correlating measured biological response to the chemical dose required to elicit that response.
  - Response is the biological consequence of exposure to a toxic substance. For acute exposures, the response may be organ injury, coma, or death, while chronic exposures typically result in mutagenic or carcinogenic responses.
  - Dose refers to the actual amount of a toxic substance that is received by a target organism.
  - Depending upon whether a toxic substance is considered a carcinogen or a non-carcinogen affects the manner in which the dose-response curve is evaluated.
    - Potency factors and chronic daily intakes are defined for carcinogens.
    - Hazard quotients are defined for non-carcinogens.
- Incidence is the extent to which people within a population who do not have a disease may develop the disease during a specific time period.
- Prevalence is the number of people within a population who have a certain disease at a given point in time.
- The epidemiology triangle depicts the inter-relatedness between hosts, disease agents, their environment, and time in terms of disease propagation.
  - A host is an organism that harbors a disease agent.
  - A disease agent is the cause of a particular disease.
  - The environment affects the interactions between the host and the disease agent.
  - Time is the overarching variable that affects the interactions between the host, the agent, and the environment.
- Infectious diseases can be transmitted either directly (organism-to-organism) or indirectly (organism to intermediate to organism).
  - Direct transmission is the transfer of the agent from a host/reservoir to a susceptible host via physical contact (skin-to-skin contact, kissing, sexual intercourse).
  - Indirect transmission involves an intermediate vector or vehicle that facilitates disease agent transfer from an infected organism to a susceptible host, resulting in disease. Common vehicles are water, air, and food. Insects and small mammals are common vectors.
- Mortality is the incidence of death in a population and can be quantitatively defined as the ratio of the number of deaths during a specific period of time relative to the number of people at risk of dying.

Table 11.12 US CDC classification of potential bioterror threats.

Category	
A	**Diseases that are easily disseminated or transmitted from person to person; result in high mortality rates; might cause public panic and social disruption; require special planning to prepare for**
	Anthrax
	Botulism
	Plague
	Smallpox
	Tularemia
	Viral hemorrhagic fevers (filoviruses – Ebola, Marburg; arenaviruses – Lass, Machupo)
B	**Moderately easy to disseminate; result in moderate morbidity and low mortality; require enhancements to current diagnostic and surveillance capabilities**
	Brucellosis
	Epsilon toxin produced by *Clostridium perfringens*
	Food safety threats (*Salmonella* species; *E. coli* O157:57, *Shigella*)
	Glanders
	Melioidosis
	Psittacosis
	Q fever
	Ricin toxin
	Staphylococcal enterotoxin B
	Typhus fever
	Viral encephalitis (alphaviruses – Venezuelan equine encephalitis, eastern and western equine encephalitis)
	Water safety threats (*Vibrio cholerae, Cryptosporidium parvum, Giardia lamblia*)
C	**Readily available emerging pathogens that could be engineered for dissemination due to their ease of production; potential for high morbidity and mortality rates and major health impacts**
	Nipah virus
	hantavirus

Source: http://www.bt.cdc.gov/agent/agentlist-category.asp. Centers for Disease Control and Prevention.

- Morbidity is the magnitude of illness, injury, or disability in a defined population and is quantitatively described using incidence and prevalence rates.
- There are three classic epidemiological study types: cross-sectional, case-control, and cohort studies.
  - A cross-sectional study determines the prevalence and distribution of disease within a population at a particular point in time.
  - In a case-control study, people who have an illness (cases) are compared with others who do not have an illness (controls) with respect to past exposures and risk factors.
  - In a cohort study, two groups, or cohorts, with different exposures/risk factors, are monitored over time for certain health outcomes.
- Comparison of incidence rates is often done to examine the factors responsible for disease propagation within a

community. One way to do so is to calculate the relative risk, which is the ratio of the disease incidence rate amongst those exposed to a disease relative to the incidence rate among those not exposed.

- The infectious microbial agents of greatest concern with respect to disease transmission are prions, viruses, bacteria, protozoa, and fungi.
  - Prions are infectious extracellular proteins that are known to cause a variety of diseases in animals.
  - Viruses are common disease agents but, because they lack internal structure and cannot replicate by themselves, they are not considered living cells. Only after a virus has infected a cell and hijacked its metabolic systems can it replicate.
  - Bacteria are single-celled prokaryotic organisms. In a prokaryotic cell, there are no cellular compartments, and the DNA of the cell resides directly within the cytoplasm.
  - Protozoa are unicellular eukaryotic organisms that do not possess a cell well, but do possess a nucleus. Protozoa can be differentiated from algae due to their lack of chlorophyll, and from fungi by their subcellular differentiation, motility, and lack of a cell wall.
  - Fungi are eukaryotic (multi-cellular) and possess a cell wall. In contrast to protozoa, which engulf their food and digest it internally, fungi release extracellular enzymes into their surroundings. These extracellular enzymes react with suitable substrates, and the fungi then absorb the products of these enzymatic reactions.
- The public health engineering profession relies upon a number of tools to minimize the spread of infectious disease. The choice of the appropriate tool to employ in a given situation is dependent upon the reservoir the disease inhabits, whether it is disseminated via vectors or vehicles, or whether it can be controlled by immunization, quarantine, or other mechanisms of control.
  - Immunization is the purposeful stimulation of immunity to infectious disease in a given individual.
  - Quarantine and isolation are used to prevent the spread of highly infectious diseases. Quarantine strictly applies to those who have been potentially exposed to an infectious disease, while isolation applies to those who are known to be suffering from the disease.
  - Surveillance is a passive monitoring system used to observe, recognize, and report diseases as they occur.
- Infectious diseases are of global health concern, because their worldwide distribution can change dramatically and rapidly due to alterations to the pathogen, the environment, or the host population that collectively contribute to the rapid spread of newly emergent diseases as well as reemergent diseases that had previously been controlled. The factors that affect disease emergence are:
  - Microbial adaptation and change;
  - Human susceptibility to infection;
  - Climate and weather;
  - Changing ecosystems;
  - Human demographics and behavior;
  - Economic development and land use;
  - International travel and commerce;
  - Technology and industry;
  - Breakdown of public health measures;
  - Poverty and social inequality;
  - War and famine;
  - Lack of political will;
  - Intent to harm.

Key Words

A-B toxins
active carriers
Acute
Aflatoxins
agent
airborne
amoebae
anthroponotic
attack rate
attributable risk
B lymphocytes
bacteria
benign
biological transmission
capsid
carcinogens
carrier
case definition
case-control
chronic
chronic daily intake (CDI)
ciliates
cohort study
common source
convalescent carriers
cross-sectional
crude attack rate
crude death ratio
crude mortality rate (CMR)
cytolytic toxins
dermis
direct transmission
disease
DNA
dose
dose-response curve
EC50
effective concentration
endemic
endospores
endotoxin
enterotoxins
environment
epidemics
epidemiology
epidemiology triangle
epidermis
exotoxin
exposure
extracellular polymeric substance (EPS)
fever
flagella
flagellates
fomites
food infections
food poisonings
food-specific attack rate
foodborne
Hazard Quotients (HQ)
healthy (passive) carrier
hepatotoxin
herd immunity
holoendemic
host
hyperendemic
incidence
incubatory
indirect transmission
infant mortality rate (IMR)
infection
inflammation
intermittent carrier
isolation
LD50
LOAEL
lymphocytes
malignant
malnutrition
mechanical transmission
metastasis
morbidity
mortality
mutagens
mycoses
mycotoxins
nasopharyngeal
NOAEL
non-specific (innate)
odds ratio
outbreaks
pandemics
period prevalence
phagocytes
phagocytosis
pili
pinocytosis
plasmids
point prevalence
poison
potency factor (PF)
prevalence
prions
propagated
pulmonary
quarantine
reference dose (RfD)
relative risk
reservoir
response
ribosomes
sensitivity
slope factor (SF)
specific (adaptive)
specificity
sporozoa
subcutaneous
subthreshold
superantigen toxins
surveillance
T lymphocytes
threshold
time
toxic substance
toxicant

toxicity	vaccination	viruses
toxicology	vector	waterborne
toxin	vehicle	zoonotic
tracheobron-chial	villi	
	virulence	

References

Buck, A. A., Aron, J.L. (2001). Epidemiological Study Designs. In Aron, J.L., Patz, J.A. *Ecosystem Change and Public Health*. 17–59. The Johns Hopkins University Press, Baltimore, MD.

Burlage, R.B. (2012). *Principles of Public Health Microbiology*. Jones & Bartlett Learning, Sudbury, MA.

Colborn, T., Dumanoski, D. *et al.* (1997). *Our Stolen Future*. Penguin Group, New York, NY.

Davis, M.L., Masten, S.J. (2009). *Principles of Environmental Engineering and Science*. McGraw-Hill, Boston, MA.

EPA (2011). *Exposure Factors Handbook: 2011 Edition*. EPA/600/R-090/052F. Environmental Protection Agency Office of Research and Development, Washington, DC.

Friis, R.H. (2007). *Essentials of Environmental Health*. Jones and Bartlett, Sudbury, MA.

Frumkin, H. (Ed.) (2010). *Environmental Health: From Global to Local*. Jossey-Bass, San Francisco.

Landis, W.G., Yu, M.-H. (2004). *Introduction to Environmental Toxicology*. CRC Press, Boca Raton, FL.

Madigan, M.T., Martinko, J.M. *et al.* (2000). *Brock: Biology of Microorganisms*. Pearson-Prentice Hall, Upper Saddle River, NJ.

Masters, G.M., Ela, W.P. (2007). *Introduction to Environmental Engineering and Science*. Pearson-Prentice Hall, Upper Saddle River, NJ.

Moore, G.S. (2007). *Living with the Earth: Concepts in Environmental Health Science*. CRC Press, Boca Raton, FL.

Pollan, M. (2006). *The Omnivore's Dilemma: A Natural History of Four Meals*. Penguin Press, New York, NY.

Rubin, E.S., Davidson, C.I. (2000). *Introduction to Engineering and the Environment*. McGraw-Hill, Boston, MA.

Smolinski, M.S., Hamburg, M.A. *et al.* (Eds.) (2003). *Microbial Threats to Health: Emergence, Detection, and Response*. The National Academies Press, Washington, DC.

Sterritt, R.M., Lester, J.N. (1988). *Microbiology for Environmental and Public Health Engineers*. E. & F. N. Spon, London, UK.

Thucydides. *History of the Peloponnesian War*.

Timmreck, T.C. (1998). *An Introduction to Epidemiology*. Jones and Bartlett, Boston, MA.

Problems

1. In 1994, a nationwide outbreak of Salmonellosis was attributed to the consumption of ice cream. At the outset of the outbreak, a case-control study was conducted to determine the risk factors for illness. From this study, it was determined that ice cream had been eaten by 12 of the 15 patients (cases) and only by 3 of the 15 controls. Given these statistics, calculate the Odds Ratio.

2. Based on the results of the case-control study in Problem 1, a nationwide recall of all of the ice cream produced from September 1 to September 30 at the ice cream plant was initiated. Following this recall, a cross-sectional study of 159 households that had purchased ice cream was conducted. This study indicated that, on average, each household contained 2.8 people who consumed ice cream and 0.4 who did not. Among those who ate ice cream, 30 people experienced symptoms of Salmonellosis (diarrhea and fever or chills). Amongst those who did not eat ice cream, only two people experienced these symptoms. What was the attack rate (answer as a percentage) for those who ate ice cream? What was the attack rate (answer as a percentage) for those who did not eat ice cream? What was the relative risk between the two groups?

3. Assuming the ice cream manufacturer in Problems 1 and 2 produced 1.06×10^6 gallons of ice cream during the month of September and each gallon was consumed by 2.8 people, estimate the total number of potential cases of gastroenteritis caused by *Salmonella enteritidis* food infection that could have resulted from this incident.

4. A new ELISA test was developed to diagnose HIV infections. Blood samples from 10,000 patients that have the disease were tested, and 9,950 were found to be positive by the new ELISA. The manufacturers then used the ELISA to test blood from 10,000 nuns who were not infected by HIV infection. 9,990 gave negative ELISA tests and ten positive results were measured. What was the specificity of the test? What was the sensitivity of the test?

5. A screening test for a newly discovered disease is being evaluated for its effectiveness and sensitivity as a rapid screening test in industry. In order to determine the effectiveness of the new test, it was administered to 765 workers; 115 of the individuals diagnosed with the disease tested positive for it. Results from the test showed a negative test finding for 65 people with the disease. A total of 30 persons not diseased tested positive for it. Calculate:
   a. the prevalence of the disease;
   b. the sensitivity of the test;
   c. the specificity of the test.

6. Given the apparent success of the ELISA test in Problem 4, it was used to test a population of 1 million people. Assuming that the sensitivity and specificity of the test remain the same as calculated above, how many people who do not have HIV will nonetheless test positive via the ELISA test? (in other words, how many false positives will there be in this population?)

7. Ozone concentrations in the air are regulated under the *National Ambient Air Quality Standards* at a level of 0.08 ppmv. Assuming a 75 year lifetime exposure to this air for an adult male, what is the estimated lifetime CDI for this concentration?

8 A 70 kg adult male is exposed to cadmium-containing vapors at his worksite. If he inhales 20 m^3 per work day and he works 8 hours a day, 5 days a week, 48 weeks a year, for 30 years, what is the maximum allowable cadmium concentration in the air if the cancer risk is to be maintained below 10^{-6}? How would the acceptable concentration change if the calculation were repeated for a 50 kg woman?

9 A laboratory trial is conducted to evaluate the risks associated with consumption of pesticide-contaminated water. 45 out of 500 mice exposed to the pesticide-contaminated water develop stomach cancer. 11 out of 500 mice that were not exposed to the pesticide contaminated water also developed cancer. What are:

a. the relative risk?
b. the attributable risk?
c. the odds ratio?

Do these epidemiological statistics suggest a potential link between consumption of this water and stomach cancer?

10 During the summer of 1996, approximately 10,000 cases of *Escherichia coli* O157:H7 infection were reported in Japan. Most cases occurred in school-age children; however, one outbreak occurred at a factory near Kyoto. In mid-July, three workers suffering from diarrhea went to the local clinic, and their stool cultures yielded *E. coli* O157:H7. The only common risk factor amongst the three was that they had all recently eaten meals at the factory cafeteria. Based on this evidence, it was suspected that a foodborne vehicle was responsible for the *E. coli* O157:H7 outbreak. To identify the vehicle, an epidemiological investigation was conducted.
After the first three cases were culture-confirmed, the management at the factory requested that workers report any symptoms of illness that they might have experienced during the month of July. Of the 3,155 employees, 74 reported gastrointestinal symptoms (i.e., diarrhea) in July, and stool samples were obtained from these workers. Of the 74 workers who reported that they were ill, 47 were culture-confirmed. Based on retrospective interviews, the onset dates for diarrhea for these 47 workers were as follows:

Date of onset	Number of cases
July 15	3
July 16	6
July 17	12
July 18	7
July 19	3
July 20	6
July 21	4
July 22	6

What was the incidence rate for this eight-day period?
Plot the epidemic curve.
What are plausible reasons why only 47 of the 74 workers who said that they had experienced gastrointestinal problems could be culture-confirmed as being infected by *E. coli* O157:H7?
Once the dates of highest probability for infection had been determined, it was subsequently found that 34 of the 1,134 people who ate lunch on the day of highest probability had diarrhea. Of the 2,021 people who did not eat lunch that day, 13 had diarrhea.

a. What was the rate of diarrhea for those who ate lunch? For those who did not eat lunch?
b. What was the Odds Ratio? Is this value significant?

To determine which food was responsible for the outbreak, the 47 sick workers (cases) and 164 non-sick workers (controls) were asked which foods they had eaten for lunch on the day of highest probability. The results of this survey were as follows (Note: Not everyone could remember exactly what they ate):

Sick workers (cases)

Food	Ate the food item	Did not eat the food item
Radish sprout salad	17	12
Boiled beef with soy sauce	8	20
Scrambled eggs	8	20

Non-sick workers (cases)

Food	Ate the food item	Did not eat the food item
Radish sprout salad	64	100
Boiled beef with soy sauce	45	100
Scrambled eggs	31	119

a. Calculate the Odds Ratio for each food.
b. Which of the three foods is most likely the cause of the E. coli O157:H7 outbreak? Which is the least likely?
c. In this example, the infected workers were all adults (Median age = 30), however, most of the E. coli O157:H7 cases in Japan that year were observed in children. Why was there such a disparity?

11 Graphically illustrate the difference between a common source epidemic and a propagated epidemic. Using online sources, find examples of recent common source and propagated epidemics.

12 Compare and contrast viruses, bacteria, and protozoa. At a minimum, consider their relative sizes as well as their internal/external organization.

13 Food and water are both well-known vehicles for common source diseases. Explain how soil could act in a similar manner.

14 A number of infectious diseases can readily afflict human populations. Using the newspaper, magazines, or the WWW find a description of a disease outbreak that occurred sometime in the past decade. Once you have found an outbreak (make sure you document your source(s)) analyze it using the following criteria: a. What factors led to the outbreak? b. What public health control measures were taken to stem the outbreak? c. Does the outbreak have the possibility of spreading beyond the region/area/locale in which it is currently located? If so, explain what steps were taken (or should be taken) to prevent the disease from being further disseminated.

15 A long-term study was conducted to evaluate the relative toxicity of a number of different organophosphate insecticides towards long-haired rabbits. One of the chemicals studied was the chemical parathion. The following data was obtained in this study:

Dose (ug/kg):	0.5	1	2	3	4	5	6	7	8
% Deaths	0	2	7	23	78	92	97	100	100

a. Based upon this data set what is the approximate LD50 for parathion? **Illustrate your answer graphically**. Report your answer using at least two significant digits.
b. Malathion, a different organophosphate insecticide, has an LD50 of ~2.0 ug/kg. Relative to parathion, is malathion more or less toxic at *environmentally relevant doses*? Choose your answer from the following:

More toxic Less toxic Unknown

Provide an explanation for your answer.

Chapter 12

Hazardous waste management

John T. Novak and Paige J. Novak

Learning Objectives

After reading this chapter, you should be able to:

- know what constitutes a hazardous compound or hazardous waste;
- recognize the important regulations governing hazardous waste management in the United States and understand the context of their promulgation;
- understand the terms "life cycle assessment" and "pollution prevention";
- recognize common groups of hazardous wastes;
- know the important physical and chemical characteristics of a contaminant that determine its fate, the treatment options available, and the likely risk and pathways for exposure;
- know Henry's Law, the systems in which it is relevant, and what it means in terms of hazardous waste management;
- know the octanol-water partition coefficient and how it can be used to evaluate potentially hazardous compounds;
- know Darcy's Law and how it can be used along with the retardation factor to determine the movement of contaminants in the subsurface;
- know what a NAPL is;
- know how redox affects remediation and treatment;
- recognize different remediation options and what considerations are relevant for deciding on an appropriate treatment option;
- understand the advantages and disadvantages of *in situ* and *ex situ* treatment.

12.1 Introduction

12.1.1 A brief history of the hazardous waste problem

There is a long history of manufacturing and chemical use in the United States, particularly in the second half of the 1940s and early 1950s, when chemical production rapidly expanded. This has led to a long history of hazardous waste problems. Several events (highlighted briefly below), however, stand out with respect to how the public thinks about hazardous waste and how the US government has responded regarding the promulgation of regulations.

- In 1962 Rachel Carson published *Silent Spring*, which described the damaging effects of DDT (dichlorodiphenyhltrichloroethane) on the terrestrial community. This spurred the environmental movement and prompted the United States to reassess its pesticide policies. In 1972, DDT was banned in the United States under The Federal Insecticide, Fungicide, and Rodenticide Act (FIFRA).
- In the 1970s, attention focused on the Stringfellow California hazardous waste disposal site, which accepted over 34 million gallons of industrial waste. Indeed, the Stringfellow Acid Pits in Riverside County, California, was considered one of the most polluted sites in California. This site contained over 200 hazardous chemicals, primarily wastes from metal finishing, electroplating, and pesticide disposal, which were disposed of during the site's 16-year operation.
- The Hudson River survey, conducted in 1974 by the Environmental Protection Agency (EPA), showed high levels of polychlorinated biphenyls (PCBs) in fish from the

Environmental Engineering: Principles and Practice, First Edition. Richard O. Mines, Jr.

river. It was estimated that approximately 1.3 million pounds of PCBs had been discharged to the river from two General Electric manufacturing plants. After a long and controversial period of studying the site, cleanup began in 2009 with the dredging of 283,000 yd^3 of contaminated sediment. Phase II dredging began in 2011. The removal of approximately 2.4 million yd^3 of sediment is planned, at an estimated cost of over $750 million.

- Love Canal was not being used as a canal, so the ends were sealed and, during the 1940s and 1950s, the site was filled with hazardous chemicals by several entities, including the Hooker Chemical Company. The site was covered and sold to the Niagara Falls School District for $1.00. The chemical company informed the school district about the prior use of the site and cautioned against excavating the site. The area was developed with many new homes, and a school was built directly on the site. By the late 1970s, a chemical odor was evident in many of the basements of nearby homes, and people began experiencing chemically induced health problems. A Pulitzer Prize-winning article exposing the problems at the site led to widespread public awareness of the problems at Love Canal, and this resulted in closure and demolition of the school and many homes in the area.
- The "Valley of the Drums" in Bullitt County, Kentucky, caught the attention of regulators when it caught fire in 1966 and burned for more than a week. In 1979, the site conditions because so bad that the EPA initiated emergency cleanup for some of the drums. This site was used by some members of Congress to justify the superfund law. The site was remediated between 1983 and 1990. Nevertheless, problems at the site, such as the discovery in 2003 of PCBs in the sediments surrounding the area, suggest that the cleanup was incomplete. As of early 2012, additional remediation was being considered.
- In 1984, in Bhopal India, 45 tons of methyl isocyanate (MIC) was released from a Union Carbide plant, resulting in nearly 300 deaths and over 350,000 injured, with an estimated 100,000 to 200,000 people with permanent disabilities. A facility using a technology similar to that used at Bhopal was located near Institute, West Virginia, and concern about the possibility of such an incident happening in the United States prompted the Congress to pass legislation allowing citizens to become aware of the hazards in their local community.

12.1.2 Important regulations

Hazardous waste management involves both the control of newly generated hazardous material and the cleanup (remediation) and disposal of existing contamination and contaminated sites. In fact, there are numerous environmental laws in the USA that deal with hazardous waste in some way. The two most important US laws are the **Resource Conservation and Recovery Act** (RCRA, passed in 1976) and the **Comprehensive Environmental Response, Compensation and Liability Act** (CERCLA or Superfund, passed in 1980).

Both solid and hazardous wastes are defined for the first time in RCRA. In addition, regulations to control the transport and disposal of hazardous material are contained in RCRA. This legislation was later significantly amended by the **Hazardous and Solid Waste Amendments** (HSWA) in 1984. CERCLA focuses on existing contamination and contaminated sites, including describing clearly the processes to be followed for their identification and remediation. CERCLA was significantly amended in 1986 by the **Superfund Amendments and Reauthorization Act** (SARA). This set of regulations (RCRA, HSWA, CERCLA, and SARA) governs the handling and disposal of hazardous material and contaminated sites in the US.

12.1.2.1 Resource Conservation and Recovery Act

RCRA was enacted in 1976 due to concerns over the handling of hazardous materials at industrial sites. RCRA implemented a hazardous material tracking system referred to as "cradle to grave", that uses the tracking program for radioactive wastes as its model. In addition, a definition of what constitutes a hazardous material was provided. The definition included two features: either the waste material is on a specific list or it has the "characteristics" of a hazardous waste. The lists outlined by RCRA contain over 500 separate chemicals or chemical types (See Watts, 1998). Characteristic hazardous wastes are materials that fit RCRA-defined criteria of toxicity, flammability, corrosiveness, or reactivity. Of these, the toxicity test procedure, called the toxicity characteristic leaching procedure (TCLP) is most used.

RCRA also restricted the use of underground storage tanks, provided fines and prison sentences for intentional violators, and provided for exemptions for some waste categories, such as household wastes and wastes from coal-burning utilities.

RCRA was modified in 1984 by the Hazardous and Solid Waste Amendments, and these regulations generally placed additional restrictions on the methods of disposal, increased fines and prison terms and put into place the so-called "hammer provisions" that forced the EPA to meet specific deadlines for promulgating industry-specific hazardous waste regulations. The hammer provisions were put into place because of concern that the EPA was "dragging its feet" in meeting the regulations specified in RCRA.

12.1.2.2 Comprehensive Environmental Response, Compensation and Liability Act

CERCLA was created in 1980 in response to several of the events described briefly above, namely Love Canal and the "Valley of the Drums." CERCLA was initially established as a five-year program to force responsible parties to clean up their sites. CERCLA, also known as "Superfund", was designed to deal with old and abandoned hazardous waste sites, while RCRA addressed newly created hazardous materials. An allotment of $1.6 billion, the superfund, was provided to clean up sites where no responsible party could be found or the responsible party(s) were financially unable to pay for the cost of the cleanup. CERCLA established a procedure for forcing cleanup action by placing sites on the **National Priorities List** (NPL), based on the **Hazardous waste site Ranking System** (HRS), which ranks the site based on risks to human health and the environment.

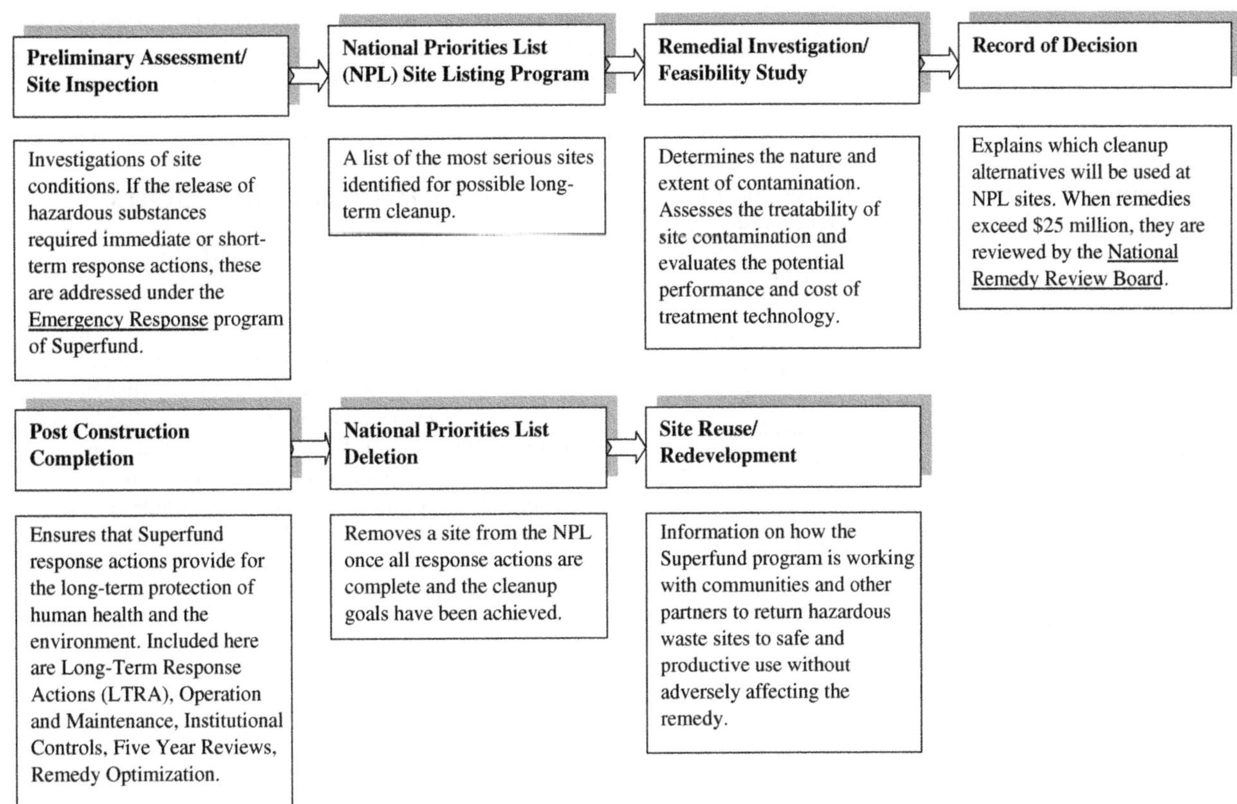

Preliminary Assessment/ Site Inspection
National Priorities List (NPL) Site Listing Program
Remedial Investigation/ Feasibility Study
Record of Decision
Investigations of site conditions. If the release of hazardous substances required immediate or short-term response actions, these are addressed under the Emergency Response program of Superfund.
A list of the most serious sites identified for possible long-term cleanup.
Determines the nature and extent of contamination. Assesses the treatability of site contamination and evaluates the potential performance and cost of treatment technology.
Explains which cleanup alternatives will be used at NPL sites. When remedies exceed $25 million, they are reviewed by the National Remedy Review Board.
Post Construction Completion
National Priorities List Deletion
Site Reuse/ Redevelopment
Ensures that Superfund response actions provide for the long-term protection of human health and the environment. Included here are Long-Term Response Actions (LTRA), Operation and Maintenance, Institutional Controls, Five Year Reviews, Remedy Optimization.
Removes a site from the NPL once all response actions are complete and the cleanup goals have been achieved.
Information on how the Superfund program is working with communities and other partners to return hazardous waste sites to safe and productive use without adversely affecting the remedy.

Figure 12.1 CERCLA flow pathway from discovery to cleanup.

Under CERCLA, once a site is listed on the National Priorities List, a series of activities takes place that ends with the site being remediated. The flow pathway from discovery to cleanup is shown in Figure 12.1.

Once the site is designated as a hazardous site, a remedial investigation (RI) is carried out to determine the sources, levels, and types of contaminants in the soil, groundwater and surface water and the need for remediation. A feasibility study (FS) is then carried out to consider alternatives for cleanup. The FS considers cost, effectiveness, community acceptance, and the protection of human health and the environment. Once the FS is completed, a record of decision (ROD) is issued and this serves as the remedy that is authorized by the EPA. The remedy is then designed and carried out. Following remediation, long term monitoring is required.

In 1986, the Superfund Amendments and Reauthorization Act (SARA) was enacted to extend superfund for five more years and add $8.5 billion to the fund. The money for superfund came from a tax on the chemical and petrochemical industries – a tax that they successfully lobbied to discontinue once the second five-year reauthorization ended. In addition to the extension of CERCLA, the **Leaky Underground Storage Tank** (LUST) trust fund, the **Emergency Planning and Community Right-to-Know Act** (EPCRA) and the **Toxics Release Inventory** (TRI) were established. EPCRA and TRI were created in response to the 1984 Bhopal incident and concern that such an incident could happen in the United States. Companies that process more than 25,000 lb/yr (11,340 kg/yr) or use 10,000 lb/yr (4536 kg/yr) of any of the 654 EPA listed hazardous chemicals must report all waste generated, in addition to releases and transfers of hazardous chemicals. Companies must also develop emergency response plans that provide for the protection for the surrounding community.

12.1.2.3 Europe and European Regulations

The European Commission (EC) has some environmental laws that are similar to those in the US regarding landfilling, incineration, and the transport of waste. In contrast, however, the EC has adopted a broad precautionary stance regarding the production and use of chemicals that could be considered hazardous (http://ec.europa.eu/environment/index_en.htm). This "precautionary principle," as it is known, is exemplified by the law REACH, which was promulgated in 2007. According to the EC, the goal of REACH is, "to ensure a high level of protection of human health and the environment," primarily by:

1. replacing chemicals that are considered particularly hazardous; and
2. placing the burden *on industries* to prove that chemicals are safe for use (http://ec.europa.eu/environment/chemicals/index.htm).

Indeed, many feel that this movement toward preventing the use of chemicals that may be hazardous, rather than treating wastes after they are produced is the future of hazardous waste management. This concept of **pollution prevention**, which is reflected in REACH, is discussed below.

12.1.3 The future of hazardous waste management

Where are we going in terms of hazardous waste management? In some ways, the future of waste management looks much like it did in the past. Nevertheless, as mentioned above, there is a growing emphasis on the prevention of hazardous waste and on the approach of determining, and then minimizing, the overall environmental impact of a product, from its manufacture through its use and final disposal.

In 1984, HSWA stated that it was to be "a national policy of the US that, where feasible, the generation of hazardous waste is to be reduced or eliminated as expeditiously as possible," (HSWA, 1984) and, on each RCRA-mandated manifest, hazardous waste generators must certify that they have a program to "reduce the volume or quantity and toxicity of such waste to the degree determined by the generator to be economically practicable" (HSWA, 1984). Therefore, there is an existing regulatory basis for pollution prevention.

The idea of pollution prevention began in the 1970s, so it is not new. Nevertheless, not all companies have embraced pollution prevention and the economic and liability benefits that it brings. Pollution prevention can be as simple as covering tanks to prevent chemical volatilization and adding automatic flow controllers to prevent tank overflows, or as complex as reformulating products, including changing input chemicals or manufacturing processes. In the end, however, the focus of pollution prevention remains on reducing environmental impact during the manufacture phase of product development – the generation of a new automobile, for example – and not on the product's use or final disposal phase.

The integration of an entire product lifecycle – mining and processing raw materials, manufacture, use, and disposal – into a complete picture of an "environmental footprint" is the domain of **lifecycle assessment** (LCA). This concept goes beyond simple recycling and pollution prevention, incorporating the environment into the entire design of a product. This is a move away from looking at single risks (e.g., liver disease in humans) and looking at global sustainability. For LCA, all environmental implications are investigated and, if possible, quantified: energy and material use, manufacture and packaging, transportation, consumer use, reuse, recycling, and disposal. In addition, traditional design considerations (cost, quality, manufacturing process, and efficiency) need to be considered as well (Graedel & Allenby, 1998).

Although LCA is embraced by more companies every year, two major problems remain:

1. a lack of quantitative detail regarding the impact of various materials, processes, and chemicals on the environment; and
2. the difficulty in assessing and addressing trade-offs with respect to environmental impacts.

For example, how do you weigh the health impacts of lead in solder with the scarcity and environmental impacts of using a replacement such as indium? How do you weigh local groundwater impacts of increased natural gas extraction to global impacts of climate change? Nevertheless, the approach of looking at waste, particularly hazardous waste, from a holistic and precautionary perspective (much like the Europeans are trying to do), is gaining traction in the US and, as of now, appears to be the future of hazardous waste management.

12.2 Common hazardous compounds and wastes

Over 700 compounds are categorized by RCRA as hazardous. Nevertheless, several classes of contaminants, including both organic and inorganic chemicals, are repeatedly encountered at manufacturing facilities and contaminated sites. This helps to limit the number of hazardous chemicals of which most engineers will need a working knowledge. Commonly encountered organic hazardous chemicals include petroleum products like gasoline and fuel oils, and chlorinated solvents such as trichloroethylene. Lead, arsenic, and chromium are among the hazardous inorganic chemicals frequently encountered.

Once knowledge is gained regarding these common chemicals, and how their chemical and physical properties dictate environmental fate and treatment, the same approach can be used to analyze other hazardous chemicals or chemical mixtures. The chemicals shown in Table 12.1 are the 20 most commonly encountered at contaminated sites. The full list can be found at www.atsdr.cdc.gov/SPL/index.html.

12.2.1 Petroleum products

Petroleum products consist of a mixture of aliphatic and aromatic organic chemicals. The number of these individual chemicals and the specific concentrations present in a mixture depend on the composition of the original crude oil and the processing method used to generate specific petroleum products. Aliphatic hydrocarbons are generally straight or branched carbon chains containing hydrogen atoms and sometimes oxygen, nitrogen, or sulfur. Alkanes contain only singly-bonded carbon atoms, while alkenes contain double bonds and alkynes contain triple bonds. Representative compounds for each category are shown in Figure 12.2. While some of these compounds resist biodegradation, most are of low toxicity and are, therefore, considered to be of less concern than the aromatic hydrocarbons (LaGrega, *et al.*, 2001).

Aromatic compounds range from single ring compounds to six- and even seven-ring compounds. The single ring compounds, shown in Table 12.2 along with their chemical characteristics, are commonly found as contaminants from gasoline spills and leaking underground storage tanks. These compounds, **b**enzene, **t**oluene, the **x**ylenes and **e**thyl benzene, are collectively called **BETX**.

The number of leaking fuel storage tanks, estimated to have exceeded 400,000 nationwide (USEPA, 2000), accounts for the widespread presence of BTEX at many locations. Of these, benzene is of most concern because it is relatively soluble and mobile in groundwater and a known and potent carcinogen (LaGrega, *et al.*, 2001). All of the BTEX compounds are regulated, however, because of health concerns. Much progress

Table 12.1 The 20 most common chemicals encountered at contaminated sites.

2011 Rank	Substance	Common current and historical uses and sources
1	Arsenic	Wood preservative, pesticide
2	Lead	Lead based paint, solder, batteries, ammunition
3	Mercury	Production of chlorine gas and caustic soda, coal combustion
4	Vinyl chloride	Manufacturing processes
5	Polychlorinated biphenyls	Insulator, coolant and lubricant in transformers and capacitors, carbonless copy paper
6	Benzene	Gasoline, plastics, resins, synthetic fibers, rubber, dyes and lubricants
7	Cadmium	Batteries, metal plating, pigments
8	Benzo(a)pyrene	Formed during combustion of organic matter, creosote
9	Polycyclic aromatic hydrocarbons	Coal combustion, petroleum refining, incomplete combustion of organic matter, creosote, coal tar and asphalt
10	Benzo(b)fluoranthene	Leaches from coal tar and asphalt, combustion of organic matter
11	Chloroform	Chemical manufacturing
12	Aroclor 1260	Lubricant, hydraulic fluid, insulator, plasticizer
13	DDT	Pesticide
14	Aroclor 1254	Lubricant, hydraulic fluid, insulator, plasticizer
15	Dibenzo(a,h)anthracene	Found in vehicle exhaust, cigarette smoke, soot, coal tar
16	trichloroethylene	Degreaser, paint remover
17	Hexavalent chromium	Chrome plating, dyes, pigments, leather tanning, wood preserving
18	Dieldren	Insecticide
19	White phosphorus	Ammunition, fertilizers, food additives
20	Hexachlorobutadiene	Produced as a by-product in the production of carbon tetrachloride and tetrachloroethylene

has been made in cleaning up contamination from underground storage tanks; nevertheless, these compounds are still widely used for industrial purposes and therefore continue to be a concern.

Polycyclic **a**romatic **h**ydrocarbons (**PAHs**) contain two or more fused benzene rings, with these fused rings sharing two carbons. These compounds are associated with the heavier fractions of petroleum products and with industrial processes and the incomplete combustion of fossil fuels. Much of the residual PAH contamination in the US is associated with byproducts from coal, oil, and chemical processing and refining. Many of these plants no longer operate, but they have left behind legacy sites that are difficult to remediate because of the limited solubility and the slow biodegradation of PAHs. Some, such as benzo(a)pyrene, a five-ring compound, are known to be highly toxic. Other PAH-contaminated sites are those near railroads that used creosote, a petroleum mixture high in PAHs, to preserve railroad ties. A listing of some of these compounds, along with their chemical properties, is shown in Table 12.3.

12.2.2 Chlorinated compounds

Chlorinated aliphatics are widely used for degreasing, as dry cleaning solvents, and as manufacturing chemicals. These compounds have contaminated many groundwater resources as a result of improper disposal practices. Some of these compounds are suspected, or known, carcinogens, and many cause kidney or liver damage. Other chlorinated compounds, often aromatic in nature, are considered **p**ersistent and **b**ioaccumulative **t**oxins (**PBTs**) and build up in the fatty tissues of biota.

Among the more widely used chlorinated compounds are tetrachloroethylene (also known as PCE or PERC), which is still widely used as a dry cleaning solvent, trichloroethylene (TCE), which is a widely used and excellent solvent and degreaser, and vinyl chloride (VC), which is used in the manufacture of polyvinylchloride (PVC) products. TCE and VC, along with dichloroethylene (DCE), are also found as microbiological degradation intermediates and/or products of PCE, so these compounds are often encountered in

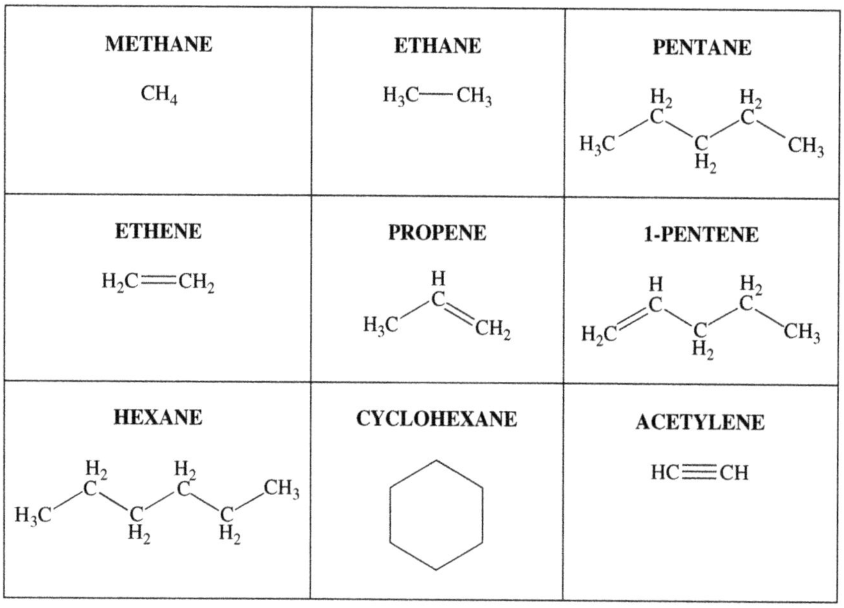

METHANE	ETHANE	PENTANE
CH_4	$H_3C—CH_3$	$H_3C-CH_2-CH_2-CH_2-CH_3$
ETHENE	**PROPENE**	**1-PENTENE**
$H_2C{=}CH_2$	$H_3C-CH{=}CH_2$	$H_2C{=}CH-CH_2-CH_2-CH_3$
HEXANE	**CYCLOHEXANE**	**ACETYLENE**
$H_3C-CH_2-CH_2-CH_2-CH_2-CH_3$		$HC{\equiv}CH$

Figure 12.2 Some representative aliphatic hydrocarbons.

Table 12.2 Some BTEX compounds and their properties.

COMPOUND	STRUCTURE	SOLUBILITY (mg/L)	LOG K_{OW}	SPECIFIC GRAVITY	HENRY'S LAW CONSTANT (atm · m³/mole)
BENZENE	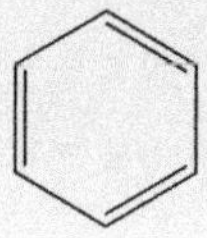	1770	2.05	0.877	0.0055
TOLUENE	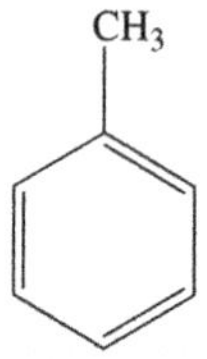 CH_3	546	2.58	0.867	0.0067
m-XYLENE	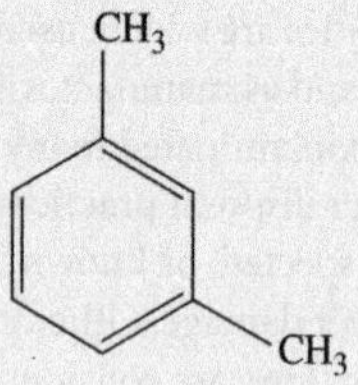 CH_3, CH_3	163	3.20	0.864	0.0063
ETHYLBENZENE	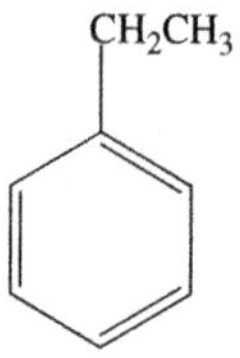 CH_2CH_3	181	3.11	0.867	0.0087

Table 12.3 Some polycyclic aromatic hydrocarbons and their properties.

Compound	Structure	Solubility (mg/L)	Log K_{OW}	Specific gravity	Henry's Law constant (atm · m³/mole)
Naphthalene	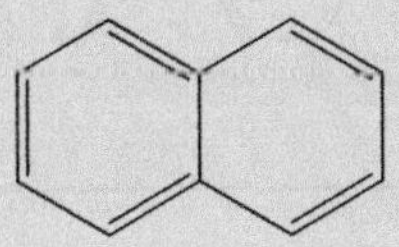	31.9	3.51	1.145	4.60×10^{-4}
Phenanthrene	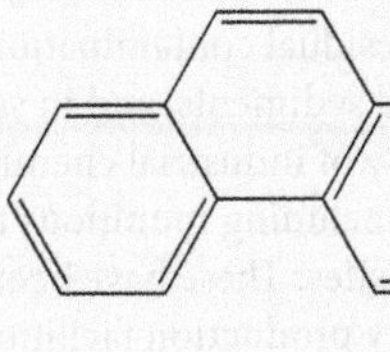	1.09	4.52	0.980	2.56×10^{-4}
Fluorene	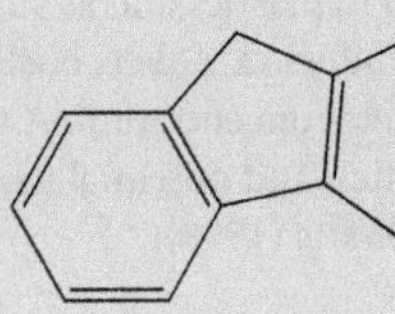	1.84	4.38	1.203	2.1×10^{-4}
Pyrene	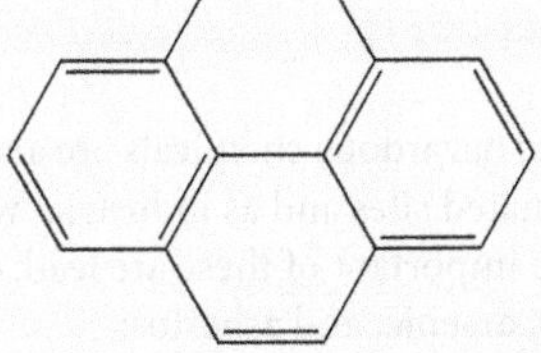	0.134	5.32	1.271	1.9×10^{-5}
Acenaphthene	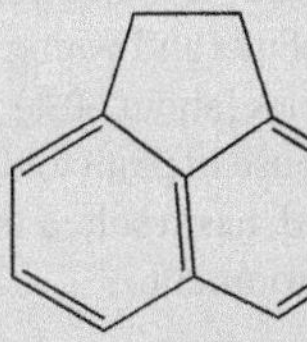	3.93	3.92	1.024	7.9×10^{-5}
Benzo(a)pyrene	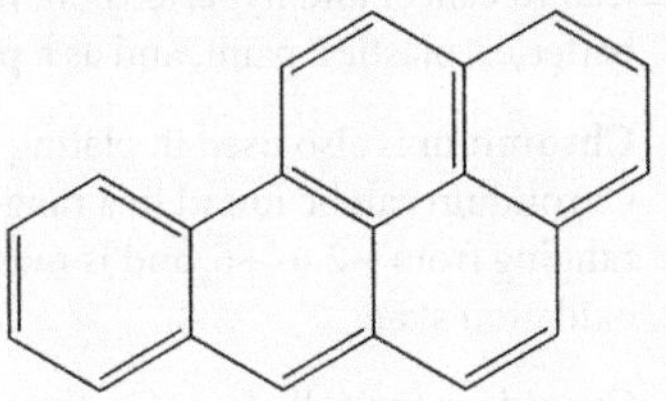	0.0038	6.06	1.351	2×10^{-6}
Benzo(a)anthracene	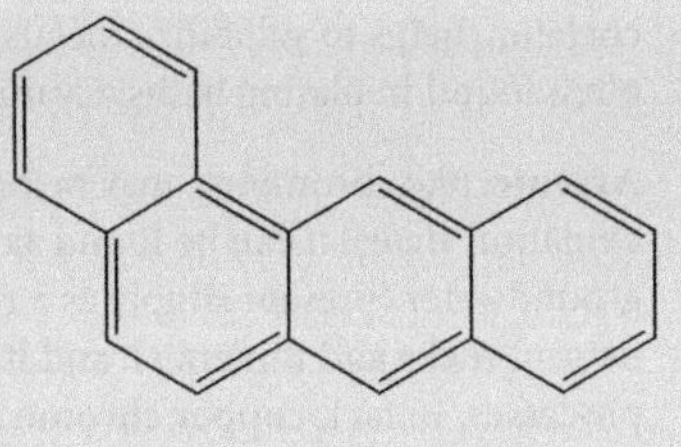	0.012	5.91	1.274	6.6×10^{-7}

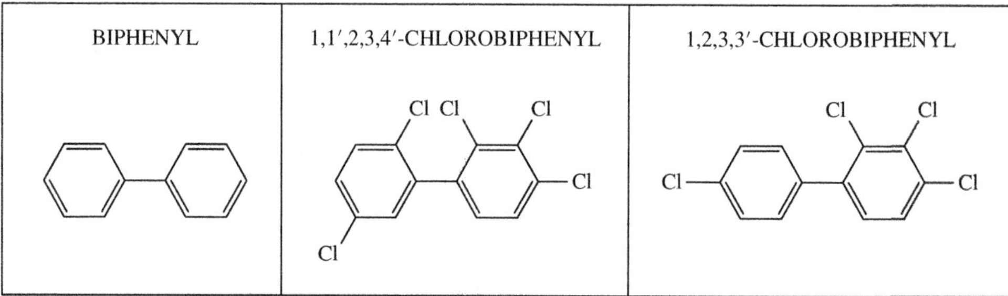

Figure 12.3 Biphenyl and two example PCB congeners.

groundwater contaminated with PCE. The NPL lists VC and TCE as among the most frequently encountered chemicals at CERCLA sites (Table 12.1).

As mentioned above, some chlorinated compounds are considered PBTs, meaning that they will accumulate in biological tissues (fish, human fat, and blood) and they tend to persist and do not degrade appreciably (microbiologically or chemically) over time. PBTs that are included among the top 20 compounds in the CERCLA Priority List include polychlorinated biphenyls (PCBs – including two specific PCB mixtures, Aroclor 1254 and 1260), the pesticide DDT and its degradate DDE, and the insecticide dieldrin, all of which are now banned in the US.

Not only do these chemicals accumulate in biological tissue over time, but they tend to magnify as one moves up the food chain as well, as organisms at higher trophic levels (large carnivorous fish, for example) eat a lot of organisms at lower trophic levels (small vegetarian or omnivorous fish). Some of these PBTs have been linked to reproductive and developmental problems, particularly in organisms at high trophic levels (e.g., Jacobson & Jacobson, 1996). More recent research has suggested that exposure to PBTs can cause even broader health effects, such as PCB exposure being linked to diabetes, and dieldrin exposure linked to Parkinson's disease (e.g., Everett *et al.*, 2010; Warner & Schapira, 2003). Some of the PBTs, such as PCBs, are actually mixtures of chemically distinct **congeners.** Congeners are chemicals that have the same general structure but differ with regard to the number and position of specific atoms; for example, PCBs are compounds that contain one or more chlorine atoms bound to a carbon on a biphenyl ring (Figure 12.3). There are 209 different possible combinations of chlorines on the biphenyl ring; therefore, these 209 distinct chemicals are all PCB congeners.

12.2.3 Other organic compounds of concern

A large number of chemicals that have been used in manufacturing processes or have been produced as pesticides, munitions, and even products designed to help with daily life (e.g., as detergents), can also be considered to be hazardous. Pesticides, herbicides, and fungicides have been used over the past 50 years in agriculture and in urban environments. Even though most of these were applied according to procedures that were considered appropriate at the time, we now know that some of these have resulted in serious environmental contamination. A number have been banned, and still more have produced residual contamination at chemical loading sites, in lakes and sediments, and in soils.

A huge variety of industrial chemicals from munitions manufacturing, including munitions themselves, can be found at contaminated sites. These have been particularly problematic at former military production facilities. Finally, some compounds used in plastics and as surfactants in household and industrial products are also considered hazardous, depending on their concentration. A summary of the structure, chemical properties, and potential hazards of these chemicals can be found in Watts (1998).

12.2.4 Inorganic compounds of concern

Inorganic hazardous chemicals are also commonly found at contaminated sites and as industrial waste products. Some of the more important of these are lead, cadmium, chromium, cyanides, arsenic, and asbestos:

- **Lead**, a chemical known to interfere with precognitive development in infants and young children, is primarily used in lead batteries (about 50%) and for a variety of other industrial uses. Its use in paint and lead pipe joints, although no longer practiced, has resulted in the ubiquitous presence of lead in the environment.
- **Cadmium** is a cumulative contaminant that is thought to lead to cancer and hypertension. It is widely used in batteries, plastics, paint, and as a plating metal.
- **Chromium** is also used in plating and as a paint pigment. Chromium can be found in a range of oxidation states ranging from −2 to +6, and is most toxic in the +6 oxidation state.
- **Cyanide** is typically found in the ionic form CN^-, or at pH values less than about 9, as HCN. Cyanide is an excellent chelator (helps to solubilize metals) and, as a result, it is often found in plating baths. Cyanide is an acute poison.
- **Arsenic**, like chromium, may be present in a range of oxidation states. It can be found as a natural contaminant in groundwater (present simply as a result of the presence of natural rocks and minerals), and it is also used in industrial processes. In fact, **copper chrome arsenate** (CCA) has been used as a wood preservative, leading to the presence of these three inorganic contaminants in landfills, some landfill leachate, and in wood preserving facilities.

- **Asbestos** is a mixture of inorganic fibers that contain silica oxides, magnesium, and iron. This material has been widely used in insulation and in asbestos-cement pipe. Inhalation of asbestos fibers can cause serious illnesses, including cancer, so, if asbestos-containing materials are removed from public buildings, it must be done carefully and by specialists.

12.3 Physical and chemical characteristics

The physical and chemical characteristics of a contaminant will determine its fate, the treatment options available, and the likely risk and pathways for exposure. Important characteristics include:

- solubility
- volatility
- density
- octanol-water partition coefficient (K_{OW})
- oxidation state

12.3.1 Solubility

The solubility of a compound in water will determine its potential for movement in the environment and the concentration that it can reach when added to water. For compounds that are sparingly soluble, the risk that they will move through the environment is reduced, but their biodegradability may also be reduced because microorganisms typically degrade compounds that are in the aqueous, or dissolved, phase. In addition, sparingly soluble compounds are more likely to be found in another phase when in contact with water. For example, they may form a solid precipitate, transfer into the gas phase, attach onto (adsorb) or move into (absorb) another phase such as soil organic matter; or they may form their own phase, as discussed below. A knowledge of the solubility of a hazardous chemical will help predict where it might end up in the environment, and can be used to design a treatment system that takes advantage of the solubility or insolubility of the chemical of interest. The chemical structure of a given compound controls its solubility. Tables of solubilities for many hazardous chemicals may be found in Watts (1998).

With respect to hazardous organic chemicals, the solubility of aromatic compounds decreases as the number of rings increases. For chlorinated aliphatics, the solubility generally decreases as the number of chlorine atoms increases (Table 12.4). This means that, although highly toxic, benzo(a)pyrene has such limited solubility that the risk of it contaminating groundwater is low. Among the least soluble hazardous organic chemicals are the dioxins and furans.

Solubility is also very important for inorganic hazardous chemicals. The solubility of metals such as lead and cadmium depend very strongly on pH, with solubility increasing at the extremes of the pH scale and minimum solubilities reached at pH values around 8–10. For chromium, the solubility depends on both the oxidation state of chromium and pH. Trivalent chromium is much less soluble than hexavalent chromium. Cyanide can exist as HCN or as CN^-, with a pK_a of about 9.1 (Watts, 1998). Both forms are soluble, but HCN is also volatile. Arsenic can range from slightly soluble to very soluble, depending on the form and oxidation state. Information about the specific arsenic species present in a given environment is therefore needed to develop appropriate treatment and handling approaches.

12.3.2 Volatility

The volatility of a compound in water is described by its Henry's Law constant and, like solubility, is a function of its chemical structure. Henry's Law describes the partitioning of a compound between the soluble and gaseous phase according to the following:

$$P = HX \tag{12.1}$$

where:
P = partial pressure, atm
X = concentration of the compound in water, moles/m^3
H = Henry's Law Constant, atm·m^3/mole.

Henry's Law constant is highly dependent on temperature and atmospheric pressure. Tables of Henry's Law constants for many hazardous chemicals can be found in Watts (1998). Compounds with Henry's Law constant values greater than 10^{-3} can be considered to be volatile, while those with a Henry's Law constant less than 10^{-7} may be considered non-volatile. Over the range of 10^{-5} to 10^{-2}, volatilization can be considered to affect the concentration of these species in water significantly over time.

For the aromatics, the BETX compounds have Henry's Law constants between 10^{-2} and 10^{-3}, meaning that they will volatilize readily. While this means that these compounds may expose a population via the gas phase, this also means that BETX-contaminated groundwater can be remediated by volatilizing it out of the water under highly controlled conditions. Many chlorinated aliphatics are also volatile and can be removed from groundwater by taking advantage of this chemical property.

PAHs and larger ring compounds tend to be less volatile, but there are a number of exceptions to this. The metals lead, cadmium, and chromium are not volatile. Cyanide in the form of HCN is volatile; therefore, pH will control cyanide volatility. Arsenic is volatile in its methylated form and can therefore be found as a contaminant in landfill gases, where CCA-treated wood has been disposed.

12.3.3 Octanol-water partition coefficient

The octanol-water partition coefficient, K_{OW}, is a general measure of the hydrophobicity of a chemical, and specifically

Table 12.4 Some chlorinated solvents and their properties.

Compound	Structure	Solubility (mg/L)	Log K_{OW}	Specific gravity	Henry's Law constant (atm · m³/mole)
Methylene chloride	CH_2Cl_2 (H–C(Cl)(H)–Cl)	18,400	1.28	1.327	2.69×10^{-3}
Chloroform	$CHCl_3$ (H–C(Cl)(Cl)–Cl)	7,870	1.94	1.483	3.2×10^{-3}
Carbon tetrachloride	CCl_4 (Cl–C(Cl)(Cl)–Cl)	911	2.73	1.594	0.03
Vinyl chloride	$ClHC{=}CH_2$	1,900	0.60	0.911	0.056
1,1-Dichloroethylene	$Cl_2C{=}CH_2$	3,920	2.13	1.218	0.021
Trichloroethylene	$Cl_2C{=}CClH$	1,310	2.33	1.464	9.1×10^{-3}
Perchloroethylene	$Cl_2C{=}CCl_2$	275	2.79	1.623	0.015

describes the tendency for a chemical to move into a separate organic phase rather than remain in water. The organic layer used in this test is *n*-octanol. A compound that has a strong affinity for octanol (a hydrophobic organic) is likely to also have a strong affinity for other organic phases, including activated carbon or the organic matter associated with soils. In this regard, the K_{OW} can be thought of as an index of hydrophobicity and, therefore, its likelihood to adsorb onto or absorb into another organic phase. Because the affinity of commonly-encountered chemicals for octanol over water can vary by over seven orders of magnitude, we frequently use the $\log_{10}$ value of K_{OW}, or log K_{OW}. The K_{OW} also is related to the solubility of a compound in water and to its volatility. Therefore, a compound with a high K_{OW} (strong tendency to partition into octanol) will be sparingly soluble and is more likely to have limited volatility.

As mentioned above, the log K_{OW} has become a valuable tool that is used for predicting the partitioning of various chemicals onto soil surfaces. For soil partitioning (K_d, or the soil distribution coefficient), the amount of a contaminant that binds to soil is typically related to its K_{OW} value and to the fraction of organic carbon on the soil.

The sorption of nonpolar hydrophobic compounds to a particular soil matrix can be described by a relationship called an isotherm, where the mass of contaminant associated with the soil surface is at equilibrium with the concentration of the contaminant in solution. This relationship is often described using a linear isotherm that follows the form:

$$K_d = s/C_L \quad (12.2)$$

where:
K_d = soil distribution coefficient, L/g
K_d = mass of contaminant sorbed (mg/g soil)/mass of contaminant dissolved (mg/L water)
s = mass of solute (contaminant) sorbed per mass of dry soil, unitless, and
C_L = liquid phase equilibrium concentration of the contaminant, mg/L or g/m^3.

The soil distribution coefficient (K_d) is related to the soil adsorption coefficient (K_{oc}) by the following equation:

$$K_d = K_{OC} \cdot f_{OC} \tag{12.3}$$

where:
K_{OC} = soil adsorption coefficient, L/g
K_{OC} = mass of contaminant sorbed (mg/g soil organic carbon)/mass of contaminant dissolved (mg/L water)
f_{OC} = the fraction of organic carbon in the soil.

The relationship between K_{OC} and K_{OW} has been described by the general equation:

$$\log K_{OC} = A \log K_{OW} + B \tag{12.4}$$

Where A and B are unitless and vary, depending on the soil type and the nature or family of contaminants.

One such relationship, provided by Kenaga and Goring (1980) is:

$$K_{OC} = 0.544 \log K_{OW} + 1.377 \tag{12.5}$$

and is appropriate for use with PAHs, PCBs, organochlorine insecticides, benzene, and a range of pesticides, chlorinated compounds, and other organic contaminants.

12.3.4 Soil and groundwater interactions

The improper disposal of hazardous wastes often impacts groundwater systems. The soil-water matrix governs the movement of contaminants through the soil, either in the vapor phase, the dissolved phase, or as a separate phase. The soil system represented in Figure 12.4 can be used to describe the soil matrix.

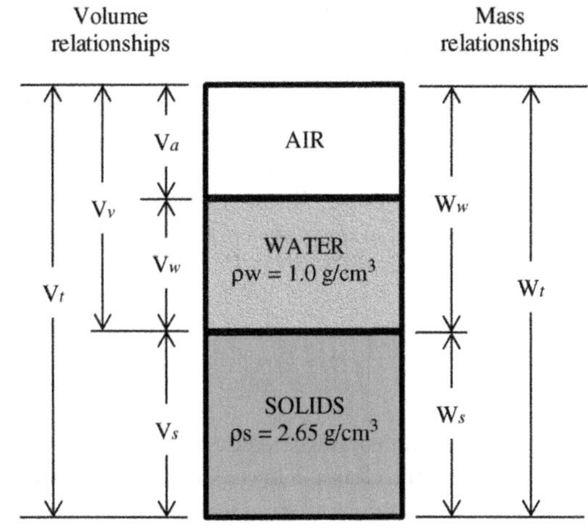

Figure 12.4 Phase diagram for partially saturated soil. *Source:* Adapted from Mines and Lackey (2009) pg. 319. Reprinted by permission of Pearson Education, Inc., Upper Saddle River, NJ.

Important relationships and properties for soil and soil constituents are shown in Table 12.5, along with a mathematical description. The total weight of soil includes the weight of solids plus the weight of water. The weight of air is neglected. The total volume includes the volume of air, water and solids, and the sum of the air and water volumes is considered to be the void volume.

12.3.4.1 Darcy's Law

Darcy's Law is a generalized relationship that is used to describe the flux (discharge per unit area) of fluid in porous media.

$$q = K_C(dh/dL) \tag{12.6}$$

where:
q = Darcy's flux along the flow path (often called the superficial velocity, discharge velocity of Darcy's velocity), m/s

Table 12.5 Volume-mass relationships of soil constituents.

Property	Symbol	Definition	Mathematical expression
Porosity	η	$\eta = \frac{\text{volume of voids}}{\text{total volume}}$	$\eta = V_v/V_t$
Volumetric water content	Θ	$\Theta = \frac{\text{volume of water}}{\text{total volume}}$	$\Theta = V_w/V_t$
Bulk density	ρ_w	$\rho_w = \frac{\text{total weight}}{\text{total volume}}$	$\rho_w = W_t/V_t$
Dry bulk density	ρ_b	$\rho_b = \frac{\text{mass of solids}}{\text{total volume}}$	$\rho_b = M_s/V_t$

Source: Mines and Lackey (2009) pg. 319 Reprinted by permission of Pearson Education, Inc., Upper Saddle River, NJ.

K_C = hydraulic conductivity, m/s
dh/dL = hydraulic gradient, dimensionless.

The actual velocity of fluid in the pores is larger than the Darcy's flux because the flow through the subsurface is limited by the available pore space in the matrix. The pore velocity, v, (also referred to as the average linear velocity and the seepage velocity, m/s) is related to the porosity (η) and the Darcy's flux by:

$$v = q/\eta \tag{12.7}$$

Sorption of contaminants carried in groundwater results in the contaminant(s) traveling at a lower speed than the fluid flow. This characterization is referred to as retardation, and it is fundamental to understanding contaminant transport. The **retardation factor**, R (unitless), is related to the seepage velocity and the velocity of the contaminant (v_p, m/s):

$$R = v/v_p \tag{12.8}$$

A conservative, non-interactive material that travels with the speed of the seepage velocity has an R value of 1. A contaminant with an R value of 2 will travel at half the groundwater seepage velocity. If the flow is at equilibrium and sorption can be characterized according to Equation (12.9), the retardation coefficient can be estimated as:

$$R = 1 + \frac{\rho_b}{\eta}(K_d) \tag{12.9}$$

Example 12.1 Calculating mass of contaminant sorbed to aquifer solids

A groundwater is contaminated with 240 μg/L of phenanthrene. Assume the aquifer solids contain 1.5% carbon and that sorption is adequately described by a linear model. Estimate the mass of phenanthrene sorbed to the solid material.

Solution

From Table 12.3, the log K_{OW} for phenanthrene is 4.52. Use Equations (12.3) and (12.5) to estimate the soil distribution coefficient, K_d.

$$K_d = K_{OC} \cdot f_{OC} = (0.544 \cdot \log K_{OC} + 1.377) \cdot f_{OC}$$

$$K_d = [(0.544 \cdot 4.52) + 1.377] \cdot 0.015 = 5.75 \times 10^{-2} \frac{L}{g}$$

$$K_d = 5.75 \times 10^{-2} \frac{L}{g} \times \left(10^3 \frac{g}{kg}\right) = 57.5 \frac{L}{kg}$$

Using Equation (12.2) and the K_d value obtained above,

$$K_d = s/C_L \tag{12.2}$$

$$s = \frac{K_d}{C_L} = \frac{57.5 \frac{L}{kg}}{240 \frac{\mu g}{L} \times \frac{1\ mg}{1000\ \mu g}} = \boxed{240 \frac{mg\ phenanthrene}{kg\ of\ soil}}$$

12.3.5 Density

The density of a hazardous chemical becomes particularly important for hydrophobic compounds that tend to form their own phase in water once they have exceeded solubility, much like olive oil does when poured into (aqueous) vinegar. Many hazardous organic chemicals, individually and as mixtures, will form a separate liquid phase when in contact with water, as a result of their sparingly soluble nature and high K_{OW}. This separate phase is often called "free product", and it can be less dense than water (like olive oil), or denser than water. These separate phases are commonly encountered in hazardous waste treatment and remediation and are called non-aqueous phase liquids (**NAPLs**).

Those NAPLs that are light and float on top of the water phase are called light non-aqueous phase liquids (**LNAPLs**), while those that are dense and settle to the bottom of an aquifer or water-filled area are called dense non-aqueous phase liquids (**DNAPLs**). Gasoline and the individual BTEX compounds are LNAPLs (See Figure 12.5). Chlorinated aliphatics and most of the PAHs are DNAPLs (see Figure 12.6). Again, because the likelihood of forming a NAPL is related to solubility and the K_{OW} value of a chemical, some contaminants will not form a separate phase. These, such as simple alcohols and short chain fatty acids, are completely miscible with water.

An interesting phenomenon with implications for risk and remediation occurs when a NAPL containing a mixture of chemicals is present. The presence of two phases – a NAPL and water phase – can impact the solubility of a chemical, because the NAPL will "compete" with the water for the chemical of interest. For example, the solubility of benzene is reported to be in the range of 1,700 – 1,800 mg/L at 20°C. Because benzene will also partition into gasoline or other light fuels, its "effective solubility" in the presence of a NAPL will seldom exceed 100 mg/L. The combination of this effective solubility and the density of a NAPL will determine the risk associated with spills and discharges, and this will influence the remediation approach.

12.3.6 Oxidation state

A common definition of oxidation is when a molecule loses electrons, and a common definition of reduction is when electrons have been added to a molecule. For hazardous chemicals, and organic chemicals, in particular, the best way to

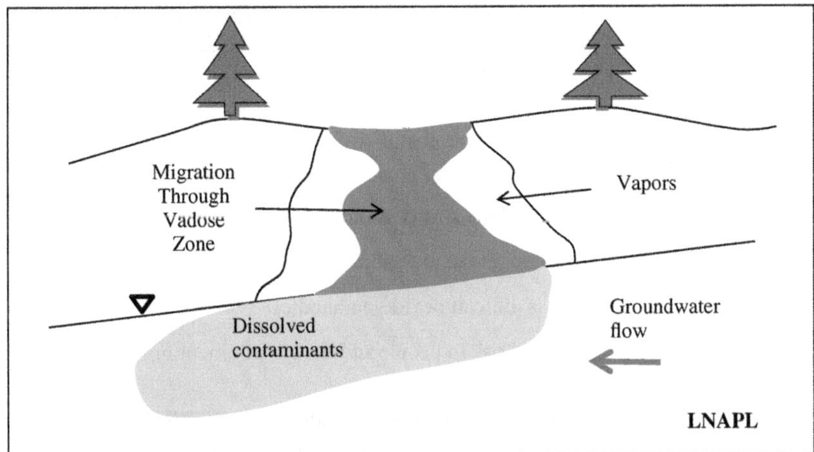

Figure 12.5 Schematic of light non-aqueous phase liquid contaminant. *Source*: Adapted from USEPA, EPA/540/S-95/500.

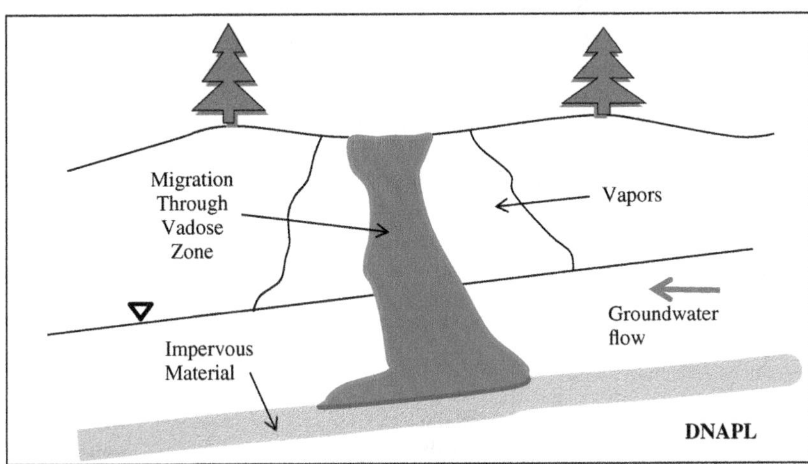

Figure 12.6 Schematic of dense non-aqueous phase liquid contaminant. *Source*: Adapted from USEPA, EPA/540/4-91-002.

deal with a hazardous chemical may be simply to change the chemistry of that compound, or react it, to make it non-toxic or less toxic. One commonly used option to accomplish this is to oxidize or reduce that hazardous chemical, chemically or microbiologically. Chemicals that are oxidized are typically easier to reduce, and those that are reduced are typically easier to oxidize.

One can determine the oxidation state of atoms in a molecule, discussed in Chapter 2, to determine whether oxidation or reduction is a better choice, or one can look for common molecular clues, described below, that help to determine whether one should attempt to oxidize or reduce a hazardous chemical.

Common molecular clues that suggest that the oxidation of a chemical is appropriate:

- The presence of electron donating groups on a carbon (OH, NH_2, CH_3).
- The chemical is lightly (1 – 2) chlorinated.
- Unsubstituted benzene rings are present.

Common molecular clues that suggest that the reduction of a chemical is appropriate:

- The presence of electron-withdrawing groups on a carbon (Cl, NO_2).
- The chemical is heavily (3+) chlorinated.

In order to oxidize or reduce a chemical, an oxidant or reductant is often added. These are discussed in detail in Section 12.4 below.

Example 12.2 Determining whether oxidation or reduction is the appropriate treatment choice

Identify whether oxidation or reduction would be more appropriate for the following chemicals: nitrate, tribromomethane, and toluene.

Solution

a) First, look up the structure of the chemicals. We find the following:

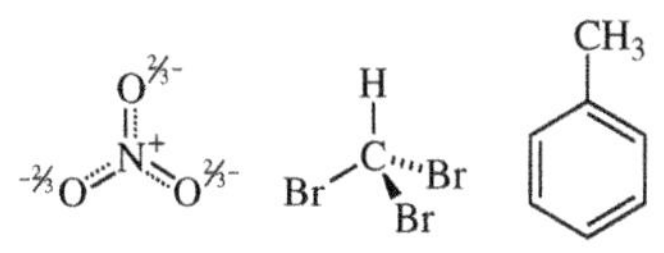

b) Next, determine what common molecular clues exist that suggest whether oxidation or reduction is likely.

In nitrate, the N has an oxidation state of +5 (O has an oxidation state of −2 or zero when bound to itself). N oxidation states range from +5 to −3; therefore, in NO_3^-, N is as oxidized as it can be. This means that NO_3^- can only be reduced.

For tribromomethane, the molecule has three electron-withdrawing groups and no electron-donating groups. This suggests that the reduction of tribromomethane is appropriate.

Finally, for toluene, the molecule contains one electron-donating group, no electron-withdrawing groups, and consists of a lightly substituted benzene ring. This suggests that oxidation is more appropriate.

12.4 Remediation

12.4.1 Considerations for determining treatment options

Many options exist for treatment of contaminated material and contaminated sites. The treatment will depend on the nature of the material, the physical and chemical characteristics, as discussed in the previous section of this chapter, and its physical location. Some additional considerations in selecting a treatment approach and treatment technologies are listed below.

- Is the contamination in the groundwater, the soil, or above ground?
- Is it a liquid, a solid, or a sludge?
- Is it organic, inorganic, or a mixture?
- It is difficult or risky to handle?
- Is the risk to people and the environment immediate?

The most important consideration above is the immediacy of the risk. This will require efforts to isolate the contamination. Because of the successful remediation of the worst sites in the USA under CERCLA, however, few sites remain where immediate treatment is warranted, unless the contamination is a newly discovered site or a fresh spill.

Based on the considerations listed above, the location of the contaminants, and the physical and chemical characteristics of the material, the appropriate treatment approach can be determined (see Table 12.6). As mentioned above, many options exist for treatment of contaminated material and contaminated sites, and there are often several options that may work equally well.

Table 12.6 Proposed treatment approach for remediation of various contaminants.

Treatment Process	BTEX, light hydro-carbons	PAHs	Halogenated solvents	Halogenated PBTs	Aqueous metals	Metal solids and sludges	Cyanide	Chromium	Aqueous pesticides/ herbicides/ fungicides
Carbon adsorption				X					X
Air or steam stripping	X		X						
Vacuum extraction	X								
Ion exchange					X		X	X	
Chemical precipitation					X			X*	
Oxidation/reduction				X			X	X*	
Incineration		X		X					
In situ bio-degradation	X		X						
Pump and treat bio-degradation	X		X						
Solids slurry reactor biotreatment		X		X					
Bioventing	X								
Solidification		X		X		X			
Landfilling		X		X		X			

*Combined chemical reduction followed by chemical precipitation.

Example 12.3 Determining best treatment option for various contaminants

How would you choose to treat a waste stream containing PAHs, Ni^{+2}, and Cd^{+2}?

Solution

a) First, we must recognize that there are a number of appropriate options for treatment.
b) Next, look at each waste component and determine if it is best volatilized, adsorbed, absorbed, oxidized, or reduced.

Ni^{+2} and Cd^{+2} are metals and cannot be changed into a non-toxic form. The most appropriate way to treat these two compounds is to remove them from solution, making them as insoluble as possible. This is best accomplished by altering the pH of solution to 8–10.

PAHs (shown in Table 12.3) tend to be reduced compounds, making them easier to oxidize. We can therefore chemically or biologically oxidize the PAHs. We can also see in Table 12.3 that PAHs generally have high K_{OW} values and varying, but generally low, Henry's Law constants. This suggests that adsorption onto a hydrophobic solid, such as activated carbon, is also an option for treatment. Further processing of the carbon will be required after adsorption, such as incineration.

12.4.2 Overriding principles

The number of possible treatment options for hazardous chemicals is large, and new treatment technologies are being developed all of the time. The basic principles important for different types of treatment are discussed briefly below, as are several of the most common treatment technologies. This discussion is by no means exhaustive, however.

Treatment of contaminated sites can be performed in place (*in situ*) or the material can be removed and treatment can be performed on site or at another location (*ex situ*). Both approaches are common.

Treating *in situ* requires that the reactants used are able to reach the contamination effectively, and it also requires a very clear understanding of the geology and hydrogeology of the site. Removal of contaminants requires care to avoid spreading the contamination; nevertheless, it also facilitates the use of a variety of treatment options, with a great deal of control.

For *ex situ* treatment, it is common to excavate contaminated soil and treat the entire mass of soil above ground. Similarly, for contaminated groundwater, the water is typically pumped to the ground surface, where it is then treated by a number of methods. This "pump and treat" approach may be successful for contaminants that have a low affinity for soil surfaces (low K_{OW}). For contaminants that adhere to soils, however (high K_{OW}), the slow release of material from the soil into the water may require pumping times of decades to centuries.

To determine how best to treat a chemical or mixture of chemicals, one must first evaluate the chemical properties described in Section 12.3. Is the material volatile? Is it soluble? Is it oxidized or reduced? Another critical criteria is whether it is biodegradable or chemically degradable. After thinking about the overriding properties of the contaminant or contaminant mixture, remediation can be considered.

12.4.3 Degradability and biodegradation

Many hazardous chemicals, particularly organic compounds, can be degraded. As mentioned in Section 12.3.6, if a compound is reduced, it is often easier to oxidize; and, if it is oxidized, it is often easier to reduce. Compounds can also be hydrolyzed or may undergo other chemical reactions, such as substitution. The goal with degrading an organic compound is to form a non-toxic or less toxic product, such as CO_2 or C_2H_4.

To chemically degrade a compound, a reactant is often added. In the case of oxidation or reduction, this would be a chemical that removes or adds electrons to the species. These reactants are often strong chemicals that are hazardous in themselves and must, therefore, be handled with care. Excess quantities of the reactants are often added to ensure that the target hazardous chemical(s) are degraded efficiently. The kinetics of reaction (Chapter 5) are often critical in such cases.

12.4.3.1 Introduction to Biodegradation

Biodegradation or microbiological treatment is a frequently used remediation technology because it is usually inexpensive and, if designed properly, it destroys the compounds of interest and changes them into innocuous (or non-harmful) products. To degrade a compound microbiologically, microorganisms (often bacteria, but sometimes fungi) serve as catalysts to enhance the rate at which hazardous chemicals are degraded. Reactants are still typically added, to ensure the efficient degradation of the target chemical(s), but these are often much more benign than those added to achieve chemical degradation. To design a system to take advantage of microbial processes, one needs to understand the basics of microbial metabolism.

12.4.3.2 Metabolism and Metabolic Needs

Because microorganisms are alive, they have the same basic metabolic needs that other living organisms, such as humans, have. These are:

1. a **carbon source** for cell synthesis;
2. an **energy source or electron donor** for cell maintenance and growth (this can often be the same as the carbon source, if the carbon source is an organic compound);
3. an **electron acceptor** (this is essentially what the organism "breathes" – in humans, oxygen is our electron acceptor); and
4. **nutrients.**

Because microorganisms are alive, they also need water, the correct temperature range (typically 20–50°C for engineered systems), the correct pH range (typically 6.5–7.5 for engineered systems), the absence of toxic compounds, adequate contact, and sufficient time for the reaction to occur. Often, hazardous chemicals serve as the electron donor/carbon source or the electron acceptor for microorganisms. To serve as an electron donor, the compound must be sufficiently reduced so that electrons can be taken from it to "feed" the microorganism. To serve as the electron acceptor, the compound must be sufficiently oxidized such that it can accept electrons. Some organisms are able to hyper-accumulate certain hazardous compounds, such as metals.

Example 12.4 Determining whether a contaminant will serve as an electron donor or acceptor

Identify whether HMX ($C_4H_8N_8O_8$), a powerful explosive, will be more likely to serve as an electron donor or electron acceptor during biological degradation.

Solution

a) First, look up the structure of HMX. We find the following:

b) Next, determine what common molecular clues exist that suggest whether oxidation or reduction is likely. We see that four NO_2 groups exist, which are electron-withdrawing. No electron-donating groups exist. This suggests that the reduction of HMX is appropriate.

c) Finally, we consider that, when reduction is appropriate, the chemical is more likely to serve as an electron acceptor. This assumes that the right enzymes and biological capability is present, as well as water, the correct environmental conditions (temperature and pH), the absence of toxic compounds, adequate contact, and sufficient time for the HMX to be utilized.

12.4.4 Physical and chemical treatment

12.4.4.1 In Situ Processes

NAPLs are often targeted for removal first, so that the continuous release of hazardous chemicals from this separate phase is stopped. LNAPLs are typically easier to locate, and they can be removed from soil and groundwater by a variety of techniques, including pumping the concentrated "free product" from the surface of the groundwater table. In this manner, the attempt is to collect material that can be cleaned and returned to productive use. DNAPLs can be particularly difficult to locate in the subsurface, because they can sink and flow into cracks and fissures at the bottom of aquitards. If located, DNAPL free product is also removed for reprocessing and reuse.

Pump and treat processes are widely used, especially for gasoline spills containing relatively low K_{OW} chemicals (Figure 12.7). Wells are placed such that they intercept the flow of the groundwater and the pumped water is treated. The choice of treatment methods will depend on the specific contaminants present and their concentration. Chemically enhanced pump and treat can be used for hazardous chemicals that have a higher K_{OW} value. This is accomplished by the use of surfactants or cosolvents to promote the release of sorbed contaminants. These chemicals can be added to water, upgradient of the contamination, and this will increase the contaminant yield as the surfactant and cosolvents help to increase the solubility of hazardous chemicals. In this manner,

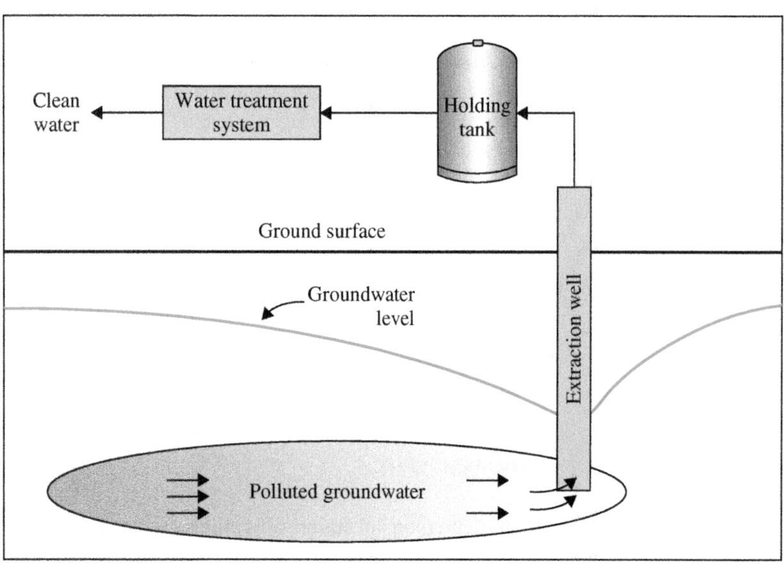

Figure 12.7 A typical pump and treat system. *Source*: US EPA (2000), *A Citizen's Guide to Pump and Treat*. United States Environmental Protection Agency.

pump and treat can proceed much faster and result in a more complete removal of contamination.

Because light petroleum products, such as gasoline, diesel oil, and fuel oils, are volatile, they can be volatilized by either drawing a vacuum directly over the contaminant plume (vacuum extraction) or by adding air directly into the plume to promote movement of the contamination to the water-air interface (air sparging). These two techniques can be combined, or combined with biological processes (**bioventing**), where the gaseous organics are drawn through the soil more slowly so that microbes have an opportunity to degrade the contaminants.

Some of the treatment technologies associated with pump and treat are described below, along with other physical-chemical processes. Bioremediation is described below in Section 12.4.5.

12.4.4.2 Ex Situ Processes

For contaminated groundwater that is pumped to the surface for treatment, the treatment will depend on the nature of the contamination. Fuels and BTEX can be treated by air stripping (if they have a high Henry's Law Constant), adsorption onto activated carbon (if they have a high K_{OW} value), or biological treatment (see Section 12.4.5). Halogenated solvents can also be stripped, adsorbed, or biologically treated. Stripping is accomplished as described for *in situ* treatment, except that contaminated water often flows downward through a packed column, while air is pumped or pulled upward. In this way, the contaminants move from the aqueous phase to the gaseous phase, where they can then be treated (often by incineration or carbon adsorption) prior to venting into the atmosphere.

Activated carbon is a hydrophobic solid with an extremely high surface area. It is often used to remove dilute hydrophobic hazardous chemicals from an aqueous or gas stream. This occurs because chemicals with a higher K_{OW} value prefer to stick to a hydrophobic surface, rather than reside in the water or gaseous phase. The high surface area of the carbon provides an ample surface for this adsorption.

Cyanides may be oxidized using chlorine gas, hypochlorite (most common), ozone, or hydrogen peroxide. Cyanide can also be oxidized using oxygen, but this requires activated carbon and a copper catalyst (Bernardin, 1973). Chromium (VI) can be reduced to chromium (III) by the addition of a strong chemical reductant – sulfur dioxide or ferrous sulfate are common. The chromium (III) is then precipitated by increasing the pH (see below). Chromium (present as negatively charged chromate ions) can also be removed by ion exchange (see below). Halogenated PBTs and solvents can be dehalogenated and detoxified by chemical or biological reduction, through the addition of a strong chemical, or in a manner similar to *in situ* bioremediation (see below).

Ion exchange can be used to remove dilute ions from solution, such as cyanide, radionuclides, or heavy metals. In ion exchange, the solid packing material, such as an engineered polymer bead, contains fixed ionic groups (for example, COO^-). These fixed ionic groups are electrostatically balanced by mobile counterions of opposite charge, which are free to move within the pore spaces of the packing material (e.g., H^+). During ion exchange, the mobile counterions exchange with the hazardous ion in solution (e.g., Cd^{+2}). Electroneutrality must be maintained so, in the example above, two H^+ would exchange for one Cd^{+2}. Ion exchange resins can be very general or very selective. Clays and some soils can also function as ion exchangers, releasing, for example, Fe^{2+} and exchanging it for a hazardous metal ion (again, Cd^{2+} for example) after a spill or accidental release.

For groundwater that contains heavy metals, chemical precipitation is frequently used. Because heavy metals are more soluble at the extremes of the pH scale, this usually requires pH adjustment. A flocculating chemical may also be added, much like in drinking water treatment, to improve precipitation. Precipitation results in the production of a metal-laden sludge that requires further treatment or disposal. These metals can be re-smelted from the sludge and reused; or, alternatively, they can be disposed of at a hazardous waste landfill. If this is the case, they may be solidified prior to placement in the landfill.

The addition of specialty chemicals or pozzolanic materials for fixation or solidification can greatly reduce the potential for leaching of the metals once the sludge is placed in the landfill or disposal site. Oily sludges can also be solidified, but this treatment has had limited success, and few application of this technology are now being approved. As of early 2012, 21 hazardous waste landfills holding current USEPA identification numbers existed. Eight of the 21 operating commercial hazardous waste landfills held a Toxic Substances Control Act (TSCA) permit for disposal of PCB-contaminated materials.

For solid organic wastes, such as PAH sludges and sediment contaminated with halogenated PBTs, incineration can be used to destroy the hazardous chemicals. In incineration, the materials are essentially oxidized completely to CO_2 and HCl, or partially to a carbonized char. It is difficult to get permits for hazardous waste incinerators, and they can be very expensive and energy-intensive to operate, as air pollution control is a major consideration. The number of hazardous waste incinerators is therefore limited. As of early 2012, there were approximately 20 facilities in the US that were permitted to incinerate hazardous waste. Three of the facilities also held a Toxic Substances Control Act (TSCA) permit to incinerate PCB-contaminated materials.

12.4.5 Bioremediation

Bioremediation can take place *in situ* or *ex situ* and is often used for the oxidation or reduction of hazardous waste. Typically, oxidation and reduction reactions take place under very specific conditions. Oxidation reactions occur most rapidly and efficiently under aerobic conditions (with oxygen present) and, in this case, the hazardous chemical(s) serves as the electron donor. Reduction reactions occur most rapidly and efficiently under anaerobic conditions (with no oxygen present and other oxidized compounds, such as nitrate, absent) and, in this case, the hazardous chemical(s) serves as the electron acceptor. It is helpful to remember that organisms thrive where they have a competitive advantage over other organisms. Part of good engineering is providing this competitive advantage for microorganisms capable of degrading hazardous chemicals, through the targeted addition of a carbon source, electron donor, electron acceptor, and/or nutrients.

In situ treatment often consists of the addition of nutrients and sometimes an electron acceptor, electron donor, or microorganisms to a site through the use of wells. This is called **biostimulation**, and it has the goal of stimulating microorganisms already present, such that they degrade the hazardous chemical(s) of interest. If the target chemical is reduced, such as BTEX, nutrients and an electron acceptor such as oxygen are often added so that the BTEX compounds can serve as the electron donor (or "food"). If the target chemical is oxidized, such as PERC, an electron donor (such as molasses or vegetable oil) is often added to "feed" organisms, such that they use the PERC as an electron acceptor. Microorganisms from off-site can also be added to degrade hazardous chemicals. This is called **bioaugmentation**, as the site is augmented by microorganisms. This practice is less common for *in situ* treatment because, as pointed out above, organisms evolve to thrive in a particular niche, where they have a competitive advantage; exogenous microorganisms are, therefore, less likely to thrive when introduced into a foreign habitat, where the various niches are already occupied.

Ex situ treatment is similar, except that much greater control can be exerted, allowing for much better contact between the microorganisms, the hazardous chemical, and any added substances, much tighter control of pH, temperature, and residence time, and the ability to add exogenous organisms more effectively.

Summary

- Hazardous wastes are defined as wastes that are likely to cause a threat to the human or environmental health as a result of their toxicity or flammable, explosive, or corrosive nature.
- RCRA defines a hazardous waste as, "a solid waste, or combination of solid wastes, which because of their quantity, concentration, or physical, chemical, or infectious characteristics, may:
  - cause, or significantly contribute to an increase in mortality or an increase in serious irreversible, or incapacitating reversible illness, or
  - pose a substantial present or potential hazard to human health or the environment when improperly treated, stored, transported, or disposed of, or otherwise managed." (Watts, 1998).
- CERCLA broadens this definition to include, "any chemical regulated under the Clean Water Act, the Clean Air Act, the Toxic Substances Control Act, or the Resource Conservation and Recovery Act; also, any other chemical or agent that will or may reasonably be anticipated to cause harmful effects to human or ecological health." (Watts, 1998).
- RCRA and CERLA are the two of the most important laws regarding the management of hazardous waste.
  - RCRA is generally considered to describe the management of hazardous compounds as they are generated.
  - CERCLA describes the management of old or abandoned hazardous waste sites.
- A large variety of treatment options exist for hazardous wastes, dictated by the chemical and physical properties of the waste itself.
- Some compounds will be chemically reduced and, hence, amenable to oxidation.
- Others will be volatile and amenable to treatment via air sparging or stripping.
- Knowledgeable treatment of hazardous waste thus requires a strong grasp of the fundamentals of chemistry and biology.

Key Words

BETX
bioaugmentation
biostimulation
bioventing
carbon source
CERCLA
congener
electron acceptor
electron donor
energy source
ex situ
HSWA
hydrophobicity
in situ
lifecycle assessment
NAPL
nutrients
PAHs
PBTs
pollution prevention
RCRA
SARA
solubility
volatility

References

Bernardin, F. (1973) Cyanide Detoxification Using Adsorption and Catalytic Oxidation on Granular Activated Carbon. *Journal of the Water Pollution Control Federation* **45**(2), 221–231.

Everett C.J., Frithsen I, Player M. (2011). Relationship of polychlorinated biphenyls with type 2 diabetes and hypertension. *Journal of Environmental Monitoring* **13**, 241–251.

Eweis, J.B., Ergas, S.J., Chang, D.P.Y., Schroeder, E. D. (1998). *Bioremediation Principles*, WCB/McGraw-Hill, New York, NY.

Graedel, T.E, Allenby, B.R. (1998). *Industrial ecology and the automobile*. Prentice Hall, Upper Saddle River, NJ.

Hazardous and Solid Waste Amendments of 1984, PL 98–616, November 9, 1984.

Jacobson J.L., Jacobson S.W. (1996). Sources and Implications of Interstudy and Interindividual Variability in the Developmental Neurotoxicity of PCBs. *Neurotoxicology and Teratology* **18**(3), 257–264.

LaGrega, M.D., Buckingham, P.L., Evans, J. C. (2001). *Hazardous Waste Management*, 2nd Edition. McGraw-Hill, New York, NY.

Mines, R.O., Lackey, L.W. (2009). *Introduction to Environmental Engineering*. Prentice Hall, New York, NY.

Warner T.T., Schapira A.H.V. (2003). Genetic and Environmental Factors in the Cause of Parkinson's Disease. *Annals of Neurology* **53**(suppl 3), S16–S25.

Watts, R.J. (1998). *Hazardous Wastes: Sources, Pathways, Receptors*. John Wiley and Sons, New York, NY.

US EPA, http://www.epa.gov/wastes/hazard/wastemin/minimize/faqs.htm rd/wastemin/minimize/faqs.htm.

US EPA. (2000). *A Citizen's Guide to Pump and Treat.* Solid Waste and Emergency Response, EPA 542-F-01-025.

US EPA. (2001). *Report to Congress on a Compliance Plan for the Underground Storage Tank Program*, Solid Waste and Emergency Response, EPA 510-R-00-001.

Problems

1 In Table 12.1, the top 20 environmental contaminants are listed. Using the source from the book, www.atsdr.cdc.gov/SPL/index.html, list the next five contaminants on the list (numbers 21–25).

2 In Table 12.2, the structures of some gasoline constituents, including *m*-xylene are listed. List the structures of *o*-xylene, *p*-xylene, and *o*-monochlorophenol.

3 In Table 12.1, the structures and chemical characteristics of some polycyclic aromatic compounds are listed.
a. Using the data in the table, plot the log of the Henry's constant (H) versus the log K_{OW}.
b. Using Table 12.1, plot the log of the solubility (s) versus the log K_{OW}.
c. Plot the log K_{OW} versus the number of rings in the aromatic compound.
d. What is the relationship between the number of rings in an aromatic compound and the solubility and Henry's Law constant?

4 If a compound in a beaker containing water and octanol has a concentration of 376 mg/L in the octanol phase and 0.0014 in the water phase, what is the K_{OW}?

5 A groundwater is contaminated with 160 μg/L of trichloroethylene. Assume the aquifer solids contain 2% carbon and that sorption is adequately described by a linear model. Estimate the mass of TCE sorbed to the solid material.

6 A soil sample contains carbon tetrachloride at a concentration of 860 ppbv in the gas phase. Calculate the expected concentration of carbon tetrachloride in the adjacent pore water.

7 A 1 L sample of moist soil weighs 1,400 g. After drying, the weight is 1,050 g. What is the soil dry bulk density, porosity, and volumetric water content? Assume that the soil density is 2.65 g/cm^3.

8 A 500-gallon underground storage tank contains water and trichloroethylene. How far will the TCE travel in one year, assuming no degradation or loss? Assume the following hydrogeologic characteristics: soil porosity of 0.4, hydraulic gradient of 0.1 ft/ft, soil organic content of 1.6%, soil specific gravity of 2.5, aquifer conductivity of 0.0005 cm/s.

9 Using the Virginia DEQ web site, look up the superfund program information.
a. How many active superfund sites are there in the state?
b. What percent of the sites have lead as a contaminant?
c. How many are wood-preserving sites?

10 a. Using the EPA website, find the list of superfund sites for your state.
b. How many sites in total have been placed on the National Priorities List?
c. How many sites have been removed from the NPL?
d. How many were found to be non-hazardous?

11 Write a paragraph describing the potential health effects, uses and contamination sources for pentachlorophenol.

12 Write a paragraph describing the potential health effects, uses and contamination sources for nickel.

13 Describe when biodegradation would be a good treatment option for hazardous waste and list three issues that must be considered if this option is to be taken.

14 What type of treatment methods would be appropriate for the following wastes? Draw a flow chart showing all influent and effluent streams, including influent streams added to encourage treatment.
a. An industrial stream consisting of nickel (30 mg/L), chloroform (10 mg/L), and grease (100 mg/L) with a pH of 4.
b. A contaminated groundwater consisting of nitrate (100 mg/L), 2,4-D (herbicide) (10 mg/L), and benzene (100 mg/L) with a pH of 8.

See chart below for additional information about 2,4-D, chloroform, and benzene

Parameter	2,4-D	Benzene	Chloroform
Mean water solubility (mg/L)	697	1770	7870
Specific gravity	1.416	0.877	1.483
Vapor pressure (mm Hg at 20°C)	0.0047	76	160
Henry's Law constant (atm-m^3/mole)	0.0195 (at 20°C)	0.00548 (at 25°C)	0.0032 (at 25°C)

15 A metal-containing hazardous sludge contains 9% solids by weight and has a density of 1.0471 kg/L. The sludge is fed to a centrifuge that treats 20,000 gal/day. The centrifuge removes 93% of the solids and produces a solid cake containing 25% solids by weight. Determine:
a. The density of the dry solids
b. The cake production rate
c. The centrate production rate
d. The centrate solids content

16 Identify whether each of the following hazardous compounds will be more likely to serve as an electron donor or electron acceptor during biological degradation.

a. TNT (a common explosive; suspected carcinogen, causes liver damage).
b. Trichloroethylene (used as a common degreaser and solvent; suspected carcinogen, causes liver damage).
c. Pyrene (a component of coal tar, crude oil, and soot; suspected carcinogen).
d. Triclosan (a common antibacterial for personal use now found in most rivers and streams – not sure of harmful effects but it does react to form a dioxin when exposed to light).
e. Carbon tetrachloride (used as a fumigant, industrial solvent; suspected carcinogen, causes liver damage).
f. Benzene (a component of gasoline; carcinogen).

17 Identify whether each of the following hazardous compounds will be more likely to be chemically oxidized or reduced.

a. Carbon tetrachloride
b. *m*-Cresol
c. Pentachlorophenol
d. 1,2,3,4-Nitrobenzene

The Periodic Table

Key	
0.98	Pauling electronegativity
3	Atomic number
Li	Element
6.941	Atomic mass (^{12}C)

1 2.20 **H** 1.008	2 **He** 4.003

d transition elements (Groups 3–12)

Group 1	Group 2	Group : 3	4	5	6	7	8	9	10	11	12	Group 13	Group 14	Group 15	Group 16	Group 17	Group 18
3 0.98 **Li** 6.941	4 1.57 **Be** 9.012											5 2.04 **B** 10.811	6 2.55 **C** 12.011	7 3.04 **N** 14.007	8 3.44 **O** 15.999	9 3.98 **F** 18.998	10 **Ne** 20.179
11 0.93 **Na** 22.990	12 1.31 **Mg** 24.305											13 1.61 **Al** 26.98	14 1.90 **Si** 28.086	15 2.19 **P** 30.974	16 2.58 **S** 32.064	17 3.16 **Cl** 35.453	18 **Ar** 39.948
19 0.82 **K** 39.102	20 1.00 **Ca** 40.08	21 **Sc** 44.956	22 **Ti** 47.90	23 **V** 50.941	24 **Cr** 51.996	25 **Mn** 54.938	26 **Fe** 55.847	27 **Co** 58.933	28 **Ni** 58.71	29 **Cu** 63.546	30 **Zn** 65.37	31 1.81 **Ga** 69.72	32 2.01 **Ge** 72.59	33 2.18 **As** 74.922	34 2.55 **Se** 78.96	35 2.96 **Br** 79.909	36 **Kr** 83.80
37 0.82 **Rb** 85.47	38 0.95 **Sr** 87.62	39 **Y** 88.906	40 **Zr** 91.22	41 **Nb** 92.906	42 **Mo** 95.94	43 **Tc** (99)	44 **Ru** 101.07	45 **Rh** 102.91	46 **Pd** 106.4	47 **Ag** 107.87	48 **Cd** 112.40	49 1.78 **In** 114.82	50 1.96 **Sn** 118.69	51 2.05 **Sb** 121.75	52 2.10 **Te** 127.60	53 2.66 **I** 126.90	54 **Xe** 131.30
55 0.79 **Cs** 132.91	56 0.89 **Ba** 137.34	57 **La** 138.91	72 **Hf** 178.49	73 **Ta** 180.95	74 **W** 183.85	75 **Re** 186.2	76 **Os** 190.2	77 **Ir** 192.22	78 **Pt** 195.09	79 **Au** 196.97	80 **Hg** 200.59	81 2.04 **Ti** 204.37	82 2.32 **Pb** 207.19	83 2.02 **Bi** 208.98	84 **Po** (210)	85 **At** (210)	86 **Rn** (222)
87 **Fr** (223)	88 **Ra** 226.025	89 **Ac** 227.0	104 **Rf** (261)	105 **Db** (262)	106 **Sg** (263)	107 **Bh**	108 **Hs**	109 **Mt**	110 **Uun**	111 **Uuu**	112 **Unb**						

58 **Ce** 140.12	59 **Pr** 140.91	60 **Nd** 144.24	61 **Pm** (147)	62 **Sm** 150.35	63 **Eu** 151.96	64 **Gd** 157.25	65 **Tb** 158.92	66 **Dy** 162.50	67 **Ho** 164.93	68 **Er** 167.26	69 **Tm** 168.93	70 **Yb** 173.04	71 **Lu** 174.97
90 **Th** 232.04	91 **Pa** (231)	92 **U** 238.03	93 **Np** (237)	94 **Pu** (242)	95 **Am** (243)	96 **Cm** (247)	97 **Bk** (247)	98 **Cf** (249)	99 **Es** (254)	100 **Fm** (253)	101 **Md** (253)	102 **No** (256)	103 **Lw** (260)

Source: Prichard and Barwick (2007), Quality Assurance in Analytical Chemistry, pp. 287. Reproduced by permission of John Wiley & Sons Ltd.

Index

Figures are indicated by italic page numbers, Examples and Tables by bold numbers.

Environmental Engineering: Principles and Practice, First Edition. Richard O. Mines, Jr.

Printed and bound by CPI Group (UK) Ltd, Croydon, CR0 4YY